BASIC LABORATORY METHODS FOR BIOTECHNOLOGY:

Textbook and Laboratory Reference

Lisa A. Seidman, Ph.D.

Program Director, Biotechnology Laboratory Technician Program,
Madison Area Technical College

Cynthia J. Moore, Ph.D.

Assistant Professor of Biological Sciences,
Illinois State University

Prentice Hall, Upper Saddle River, New Jersey 07458

Library of Congress Cataloging-in-Publication Data

Seidman, Lisa A.
    Basic laboratory methods for biotechnology : textbook and laboratory reference / Lisa
A. Seidman, Cynthia J. Moore.
        p.   cm.
    Includes bibliographical references and index.
    ISBN 0-13-795535-9
        1. Biotechnology—Laboratory manuals.   I. Moore, Cynthia J.   II. Title.

TP248.24 .S45   1999
660.6′078—dc21   99-048785

Acquisitions Editor: Halee Dinsey
Editor in Chief: Paul Corey
Assistant Vice President of Production and Manufacturing: David Riccardi
Special Projects Manager: Barbara A. Murray
Production Editor: V&M Graphics
Manufacturing Manager: Trudy Pisciotti

© 2000 by Prentice-Hall, Inc.
Upper Saddle River, New Jersey 07458

This book supported in part by grant number DUE-9454555,
the National Science Foundation, Advanced Technology Education Initiative.

Printed in the United States of America

10 9 8 7 6 5 4 3 2 1

ISBN 0-13-795535-9

Prentice-Hall International (UK) Limited, *London*
Prentice-Hall of Australia Pty. Limited, *Sydney*
Prentice-Hall Canada Inc., *Toronto*
Prentice-Hall Hispanoamericana, S.A., *Mexico*
Prentice-Hall of India Private Limited, *New Delhi*
Prentice-Hall of Japan, Inc., *Tokyo*
Prentice-Hall (Singapore) Pte. Ltd.
Editora Prentice Hall do Brasil, Ltda., *Rio de Janeiro*

This book is dedicated to Mom, Percy, and Ilana
and to Steve

TABLE OF CONTENTS

PREFACE

This is an exciting time to work in biotechnology. The Human Genome Project is generating fundamental genetic information at a breathtaking rate; basic research findings are being applied in medicine, agriculture, and the environment; and a variety of new biotechnology products are moving into production. Behind each of these accomplishments are teams of scientists and technicians whose everyday work makes such achievements possible.

For the past twelve years, we have been working with students who are beginning their careers as technicians and bench scientists in biotechnology laboratories. In order to best assist our students, we, and our colleagues elsewhere in the United States, have explored what entry level biotechnologists do at work and what abilities they need to perform this work. We have been impressed with the complexity and diversity of technical roles and responsibilities, and the importance of the skills that bench workers bring to their jobs. This book emerges partly from our experiences working with students and our explorations into the nature of the laboratory workplace*.

This book also results from our personal experiences in the laboratory. As graduate students we struggled to master the "laboratory lore" that was passed among "post-docs" and graduate students in a not always coherent chain. Some of what is in this book is the systematic introduction to laboratory lore that we wish we had received.

The result of our efforts is not a laboratory manual; this text contains few step-by-step procedures. Nor is it a book about molecular genetics, immunology, or cell culture—there are already many excellent specialized texts and manuals on these topics. This book rather is a textbook/reference manual on basic laboratory methods and the principles that underlie those methods. These basics are important to every biotechnologist, regardless of whether one is cloning DNA or purifying proteins, whether one is working in an academic setting or is employed in a company.

We intend this book to assist students preparing to become biotechnology laboratory professionals, those who already work in the laboratory, and biology students who are learning to operate effectively in the laboratory. Others who may also find this book helpful include high school teachers and their advanced students, and industry trainers. We have endeavored to make this text accessible to beginning college students with a limited science and math background. Some sections, such as the math review in Unit III, could be skipped or skimmed by more experienced readers. At the same time as we tried to make this book practical and accessible, we also endeavored to provide enough background theory so that readers will understand the methods they use and will be prepared to solve the unavoidable problems that arise in any laboratory.

Although we focus on the biotechnology laboratory, the majority of topics we cover are of importance to individuals working in any biology laboratory. A few topics, such as quality regulations and standards, are included because they are important for those working in biotechnology companies. As biotechnology companies mature, their focus shifts from research into commercial production. As this maturation occurs, scientists and technicians often find that they must add terms like "GMP", "ISO 9000", and "quality systems" to their technical vocabulary. This book therefore weaves a conversation about regulations and standards into many chapters.

We are aware that the basic methods in this book (such as how to mix a solution or weigh a sample) are less glamorous than learning how to manipulate DNA, or how to clone a sheep. However, we also know that, in practice, the most sophisticated and remarkable accomplishments of biotechnology are possible only when the most basic laboratory work is done properly.

*The results of some of these discussions about the biotechnology workplace are summarized in the National Voluntary Skill Standards Documents in Agricultural Biotechnology and the Biosciences. (FFA, "National Voluntary Occupational Skill Standards: Agricultural Biotechnology Technician," National FFA Foundation, Madison, WI, 1994 and "Gateway to the Future, Skill Standards for the Bioscience Industry," Education Development Center, Newton, MA, Inc., 1995.)

ABOUT THE AUTHORS

Lisa Seidman received her Ph.D. from the University of Wisconsin and has taught for over twelve years in the Biotechnology Laboratory Technician Program at Madison Area Technical College, Madison, Wisconsin. Cynthia Moore received her doctoral degree from Temple University School of Medicine and has a career both in research and as a science educator. She is currently a faculty member in the Department of Biological Sciences at Illinois State University, Normal, Illinois.

The authors welcome comments and feedback from readers. You can e-mail Lisa Seidman at lseidman@madison.tec.wi.us and Cynthia Moore at cjmoor1@ilstu.edu.

ACKNOWLEDGMENTS

It is always the case that many people contribute to a book such as this and we appreciate all their help. We thank Dr. Joy McMillan for her continuous, crucial support. We thank Noreen Warrren for her thorough research, her ability to find passages that make no sense, and her many contributions that have made this a much better book. Thank you to Susan Haugen for technical expertise and support. Thank you to our colleagues Jeanette Mowery, Joseph Lowndes, Diana Brandner, Judy DaWalt, and Rebecca Pearlman, and to all our students who contributed in many ways to this book. We appreciate the support and the skills of our editors, Linda Schreiber, Barbara Murray, and Dawn Murrin. Thank you also to the expert reviewers who helped make this a much better book: Michael Barazia (Scientific Protein Laboratories, Inc.), James M. Barry, (Section of Occupational Health, Madison, WI), Gail Baughman (Contra Costa College), Gerald Bergtrom (University of Wisconsin-Milwaukee), Jane Breun (Madison Area Technical College), Kendra Cawley (Portland Community College), Thomas Cole (San Diego State University), Jane Cramer (University of Wisconsin-Madison), Sandra DeMars (Promega Corporation), Jennifer Gray (Madera Community College Center), Irene Grohar (Loma Linda University), Kathy Hafer (Washington University at St. Louis), Judi Heitz (San Diego High School), David Heyse (Washington University at St. Louis), Kim Karcher, Richard Killen (College of Lake County), Russ Kumai (Madison Area Technical College), Kathleen Kennedy Norris (Baltimore City Community College), Ann P. McMahon (Science Education Services), Ron Niece (University of Wisconsin-Madison), William Thieman (Ventura College), David T. Redden (University of Alabama-Birmingham), Charles Rohde (Johns Hopkins University), David B. Shaw (Madison Area Technical College), David Singer and students (San Diego City College), Michael H. Simonian (Beckman Instruments), Ellen Schwartz (Abbott Laboratories). Many thanks also to Sandra Bayna, a graduate of the Biotechnology Program at Madison Area Technical College. Sandy drew a number of sketches while she was a student and we included these sketches in the early versions of this manuscript. Some of Sandy's drawings appear in this book in Unit V. Thank you also to Dana Benedicktus for her patience and technical assistance and to Vicki May for her moral support and encouragement.

While we do not wish to in any way endorse the products of any company over those of any other, we do want to acknowledge and thank the many companies that so graciously provided us with illustrations, technical reviews, and information.

This material is based in part upon work supported by the National Science Foundation, Advanced Technology Education Initiative, under grant number DUE-9454555.

Any opinions, findings, and conclusions or recommendations expressed in this material are those of the authors and do not necessarily reflect the views of the National Science Foundation.

UNIT 1:
Introduction to the Biotechnology Workplace

Chapters in This Unit

Biotechnology is a rapidly growing enterprise from which many worthwhile products emerge. This makes biotechnology laboratories intriguing, challenging places in which to work. This first unit, together with Unit II, introduces the biotechnology industry and the laboratory workplace.

This unit is organized as follows:

Chapter 1 briefly surveys the science, methods, and applications of biotechnology and talks about the work environments of biotechnologists.

Chapter 2 introduces important principles related to creating a safe workplace.

Chapter 3 is an overview of the relationship between the government and the biotechnology industry.

TERMINOLOGY FOR THE INTRODUCTORY UNIT

Biologic (according to FDA). A regulatory term that refers to any virus, therapeutic serum, toxin, antitoxin, vaccine, blood, blood derivative, blood component, monoclonal antibody, somatic cell gene therapy product, or allergenic product used to prevent, treat, or cure injuries or diseases in humans.

Biopharmaceutical. A term commonly used to refer to a drug product, such as insulin, that is manufactured using genetically modified organisms as a production system.

Cell Culture. The *in vitro* growth of cells isolated from multicellular organisms.

Cell Line. Cells in culture that have acquired the ability to multiply indefinitely.

Center for Biologics Evaluation and Research, CBER. FDA (Food and Drug Administration) division that regulates biologics for human use.

Center for Devices and Radiological Health, CDRH. FDA division that regulates medical devices and many, but not all, *in vitro* diagnostic kits.

Center for Drug Evaluation and Research, CDER. FDA division that regulates drug products such as insulin, heparin, aspirin, and erythropoietin.

Center for Food Safety and Applied Nutrition, CFSAN. FDA division that promotes and protects the public's health and economic interest by ensuring that food is safe, nutritious, and honestly labeled.

Chemical Hygiene Plan, CHP. A written manual that outlines specific information and procedures necessary to protect workers from hazardous chemicals.

Clearance Studies. Studies of a process in which a contaminant is intentionally added to the system and the ability of the process to remove it is tested.

Code. A set of regulations centered on a specific topic.

Code of Federal Regulations, CFR. Published annually, this document contains U.S. federal government regulations, including those related to medical products and foods.

Documentation. Written records that guide activities and record what has been done.

Downstream Processing. The stages of processing, including isolation, purification, and packaging, of a desired product that takes place after the product is made by a fermentation or cell culture process.

Drug Product. A dosage form, such as a tablet or solution, intended for use in the diagnosis, treatment, or prevention of disease in humans or animals.

Environmental Protection Agency, EPA. A U.S. government agency that, among other responsibilities, is involved in the regulation of environmental releases of genetically modified organisms.

Expression (in genetics). The process in which the product coded for by a specific gene is made.

Federal Register. A daily publication from the U.S. Government Printing Office that contains major revisions to regulations and announcements of proposed new regulations. It also contains announcements of meetings (e.g., of FDA or EPA), seminars, and guidance documents.

Fermentation (in biotechnology). A process in which a product is produced by the mass culture of microorganisms under controlled conditions.

Fermenter. A vessel in which microorganisms or cells are grown under controlled conditions of temperature, nutrient levels, aeration, pH, and mixing.

Food and Drug Administration, FDA. The federal agency empowered by Congress to regulate the production of cosmetics, electronic products, such as X-ray machines, food, medical devices, diagnostic devices, and prescription and over-the-counter pharmaceuticals for human and veterinary use.

Food, Drug and Cosmetics Act, FDCA. An act passed by the U.S. Congress that requires that food and drugs not be "adulterated" and empowers the FDA to enforce food and drug laws.

Gene Therapy. Replacing a gene that is missing, or correcting the function of a faulty gene, in order to treat or cure an illness.

Genetic Engineering. A term commonly referring to methods that allow a gene from one organism to be transferred to another.

Genetically Modified Organism. A cell or organism into which a gene from another organism has been transferred.

Good Clinical Practices, GCP. The FDA's minimum standards, regulations, and guidelines for conducting clinical trials (studies in humans) of potential pharmaceutical products.

Good Laboratory Practices, GLP. 1. Of FDA: The procedures and practices that must be followed when investigating the safety and toxicological effects of new drugs in animals. GLP requirements are specified in the Code of Federal Regulations. 2. Of EPA: The procedures and practices that must be followed when investigating the effects of pesticides and other agrochemicals. 3. With small letters (i.e., good laboratory practices, glp): A general term used to refer to all quality practices in any laboratory.

Good Manufacturing Practices, GMPs. The accepted minimum requirements followed by industry for the manufacture, testing, packaging, and distribution of pharmaceutical products.

Guidelines (from FDA). Documents produced by the FDA that establish principles and practices. Guidelines are not legal requirements; rather they are practices that are acceptable to and recommended by the FDA.

"Guidelines for Research Involving Recombinant DNA Molecules" (NIH Guidelines). An influential document written by The Recombinant DNA Advisory Committee that covers topics such as: methods of assessing the risk of a recombinant DNA experiment, methods of classifying organisms based on risk, and methods of containing recombinant organisms both on an experimental scale and in production settings.

Hazard. The equipment, chemicals, and conditions that have a potential to cause harm.

Hazard Analysis Critical Control Points, HACCP. A quality program being implemented in the meat and poultry industry that emphasizes reducing hazards throughout the production, slaughter, processing, and distribution of meat and poultry.

HCS, Federal Hazard Communication Standard. A federal standard that focuses on the availability of information concerning employee hazard exposure and applicable safety measures.

Hybridoma. A cell that results from the fusion of an antibody-producing cell and a cultured cell and which produces monoclonal antibodies.

Infectivity Assay. A method of testing cells for pathogens in which susceptible cells are grown alongside the cells being tested to see if they become infected.

International Conference on Harmonisation, ICH. An organization that brings together the regulatory authorities of Europe, Japan, and the United States and experts from the pharmaceutical industry to discuss scientific and technical aspects of pharmaceutical product regulation.

Job Safety Analysis, JSA. A detailed analysis of each step in a procedure, which identifies potential hazards and outlines accident prevention strategies.

Master Cell Bank, MCB. Stored cells that are the source of cells used in a process. The stored cells are usually derived from a single cell and are aliquoted and stored cryogenically to assure genetic stability. The MCB is the source for the working cell bank.

Material Safety Data Sheet, MSDS. A legally required technical document provided by chemical suppliers, describing the specific properties of a chemical.

Medical Device. Items that are not drugs or biologics, but which are used in the diagnosis or treatment of disease or injury. Examples include sterilizers, test kits, cell counters, and pacemakers.

Monoclonal Antibodies. Exceptionally homogeneous populations of antibodies directed against a specific target and produced by hybridoma cells.

Occupational Exposure to Hazardous Chemicals in Laboratories Standards (29 CFR Part 1910). A set of federal standards that adapt and expand the HCS to apply to academic, industrial, and clinical laboratories.

Occupational Safety and Health Administration, OSHA. The main federal agency responsible for monitoring workplace safety.

Phase I Clinical Trials. Clinical studies of potential pharmaceuticals performed on healthy human volunteers to determine the safety of potential products and their metabolic, pharmacological, and toxicological properties.

Phase II Clinical Trials. Studies of potential pharmaceuticals performed on a small number of diseased patients to determine the product's effectiveness and potential side effects.

Phase III Clinical Trials. Studies of potential pharmaceuticals performed on a population of several hundred to several thousand patients to establish therapeutic effects, side effects, and recommended dose levels.

Pilot Plant. A moderate size production facility where development of efficient production methods occurs.

Plasmid. A circular molecule of DNA found in many types of bacteria, which exists separately from the bacterial chromosome, and which can replicate independently. Plasmids are often used as vectors.

Points to Consider (from FDA). Guidance documents issued by FDA that contain information about new technologies. Points to Consider establish general concerns that should be addressed by companies, but which are not enforceable regulations.

Product. Results of activities or processes. A product may be a service, a processed material, or knowledge.

Pyrogen. Material derived from lipopolysaccharides that are released during the breakdown of gram negative bacteria. Pyrogens trigger a dangerous immune response and induce fever, hence their name *pyrogen* ("heat producing").

Quality Assurance, QA. Can be defined as all the activities and people who work to ensure the final quality of products.

Quality Control, QC. Can be defined as a subsection of QA chiefly responsible for monitoring processes and performing laboratory testing.

Recombinant DNA. DNA that contains sequences of DNA from different sources that were brought together by techniques of molecular biology and not by traditional breeding methods.

Recombinant DNA Advisory Committee, RAC. An expert committee assembled by The National Institutes of Health that has reviewed and interpreted what is known about the risks associated with genetic manipulation methods and has published guidelines for safely working with recombinant DNA.

Regulations. Requirements imposed by the government and having the force of law.

Regulatory Agency. A government body that is responsible for enforcing and interpreting legislative acts. For example, FDA is a regulatory agency responsible for ensuring the safety, effectiveness, and reliability of medical products, foods, and cosmetics as set forth in federal legislation.

Research and Development, R & D. The organizational unit in a company that finds ideas for products, performs research and testing to see if the ideas are feasible, and develops promising ideas into actual products.

Research Laboratory. A space set aside to study the complexities of nature in a controlled manner.

Resource Conservation and Recovery Act of 1976, RCRA. Legislation that provides a system for tracking hazardous waste, including poisonous or reactive chemicals, from generation to disposal.

Risk. The probability that a hazard will cause harm.

Risk Assessment. Estimation of the potential for human injury or property damage from an activity.

Safety. The elimination of hazards and risk, to the extent possible.

Safety Guideline. A procedure designed to prevent accidents, by controlling the risk of hazards in situations where the hazards cannot be eliminated entirely.

Scale-up. The transition between producing small amounts of a product to production of large quantities for commercial sale.

Standard. An operating principle or requirement.

Testing Laboratory. A place where analysts test samples.

Toxic Substance Control Act, TSCA. Legislation to regulate chemicals that pose health or environmental risks.

Transgenic. A plant or animal that is genetically modified by the introduction of foreign DNA using the techniques of biotechnology.

Underwriters Laboratories, UL. An organization that has developed codes for safe electrical devices.

United States Department of Agriculture, USDA. A U.S. government agency that enforces requirements for purity and quality of meat, poultry, and eggs, and which is involved in nutrition research and education. The USDA is one of the government agencies that plays a role in the regulation of genetically engineered food plants.

Vector. An entity, such as a plasmid or a modified virus, into which a DNA fragment of interest is integrated, and which carries the DNA of interest into a host cell.

Workers' Compensation. A no-fault state insurance system designed to pay for the medical expenses of workers who are injured on the job, or develop work-related medical problems.

Working Cell Bank, WCB. Cells derived from the master cell bank that are expanded and used for production of a product.

REFERENCES FOR THE INTRODUCTORY UNIT

There are a number of excellent books on biotechnology. A few examples of general books on this subject include:

Recombinant DNA, James D. Watson, Michael Gilman, Jan Witkowski, and Mark Zoller. 2nd edition. W.H. Freeman and Co., New York, 1992. (Readable, more technical introduction.)

Biotechnology: A Guide to Genetic Engineering, Pamela Peters. Wm. C. Brown, Dubuque, IA, 1993.

DNA Technology: The Awesome Skill, I. Edward Alcamo. Wm. C. Brown, Dubuque, IA, 1996.

Biotechnology Unzipped: Promises and Realities, Eric S. Grace. Joseph Henry Press, Washington D.C., 1997. (A nontechnical introduction to biotechnology and some of the issues it raises.)

Recombinant DNA and Biotechnology: A Guide For Teachers, Helen Kreuzer and Adrianne Massey. ASM Press, Washington, D.C., 1996.

References on safety in the biotechnology laboratory will be provided in Unit VIII.

There is a vast, evolving literature in the area of regulation of pharmaceuticals and foods. The regulations that affect biotechnology are changing so quickly that it is essential to use the Internet to find current information. A few examples of relevant readings include:

Biotechnology. Volume 12, Legal, Economic and Ethical Dimensions, edited by H.-J. Rehm, G. Reed, A. Pühler, and P. Stadler. 2nd edition. VCH Publisher, New York, 1991. (Contains a number of relevant readings by different authors.)

Regulatory Practice for Biopharmaceutical Production, edited by Anthony S. Lubiniecki and Susan A. Vargo. Wiley-Liss, New York, 1994. (Excellent, broad resource with good case studies and basic principles)

Guidelines for Research Involving Recombinant DNA Molecules (NIH Guidelines), National Institutes of Health. Effective June 24, 1994. Federal Register, July 5, 1994 (59FR34496).

Biotechnology Inspection Guide Reference Materials and Training Aids, FDA, November 1991.

International Conference on Harmonisation; Final Guideline on Quality of Biotechnological Products: Analysis of the Expression Construct in Cells Used for Production of r-DNA Derived Protein Products, Food and Drug Administration [Federal Register: February 23, 1996 (Volume 61, Number 37) Pages 7006-7008].

Points to Consider in the Manufacture and Testing of Monoclonal Antibody Products for Human Use, Food and Drug Administration Center for Biologics Evaluation and Research, February 28, 1997.

Exercise of Federal Oversight Within Scope of Statutory Authority: Planned Introductions of Biotechnology Products Into the Environment, Office of Science and Technology Policy, Executive Office of the President, 1992.

"How FDA Approved Chymosin: A Case History," Eric L. Flamm. *Bio/technology*, Volume 9:349-351. 1991.

HACCP: A Practical Approach, Sara Mortimore and Carol Wallace. Chapman and Hall, London, 1994. (Excellent overview of principles and application of HACCP.)

RECOMMENDED INTERNET RESOURCES

The Code of Federal Regulations is available at: http://www.access.gpo.gov/nara/cfr

FDA Homepage: http://www.fda.gov/fdahomepage.html

USDA Homepage: http://www.usda.gov

OSHA Homepage: http://www.osha.gov

EPA Homepage: http://www.epa.gov

ICH Homepage: http://www.ich.org

CHAPTER 1

Biotechnology and the Workplace

I. **A VERY BRIEF INTRODUCTION TO THE SCIENCE, METHODS, AND APPLICATIONS OF BIOTECHNOLOGY**

   A. **Introduction**

   B. **The Technologies of Modern Biotechnology**

   C. **Applications of Biotechnology**

       i. Medical/Veterinary Applications
      ii. Agricultural/Food Related Applications
     iii. Other Applications

II. **THE ORGANIZATION OF A BIOTECHNOLOGY COMPANY**

   A. **Research and Development**

   B. **Production**

   C. **Quality Control/Quality Assurance**

   D. **Other Functions in a Company**

   E. **The "Cultures" Within a Company**

III. **BIOTECHNOLOGY WORKPLACES**

I. A *VERY BRIEF* INTRODUCTION TO THE SCIENCE, METHODS, AND APPLICATIONS OF BIOTECHNOLOGY

A. Introduction

It seems appropriate to begin this book with a definition of *biotechnology*. As we will see, however, there is no single, definitive meaning for this word. Biotechnology sometimes refers to the use of organisms, or materials derived from organisms, to make useful products. In this sense, biotechnology is not a recent phenomenon: Humans have used organisms to make products such as wine, cheese, and bread for thousands of years. When people use the term *biotechnology*, however, they are usually referring to the more dazzling products of "modern biotechnology", such as cloned sheep, gene therapies, and DNA fingerprints, and not to bread and cheese production.

Modern biotechnology is deeply rooted in basic research from the biology laboratory; research that broadens and deepens our understanding of how living beings work. Over the years, researchers in many laboratories have explored the intricacies of cellular function, the mechanisms by which genetic information is passed from generation to generation, how the complex immune system is coordinated, and other fundamental issues in biology. The application of scientific knowledge to make products useful to humans is *technology*. In an analogous way, biotechnology can be broadly understood as the application of knowledge about living systems in ways useful to humans. Thus, in recent decades, knowledge that has emerged from biology research laboratories has been applied to create modern biotechnology.

This chapter begins with a very brief overview of the methods that comprise modern biotechnology, and their many useful applications. We will then discuss biotechnology companies, the sites where science is transformed into commercial, manufactured products. The final section of this chapter will discuss biotechnology workplaces.

It is beyond the scope of this book to explore all the science, methods, applications, and implications of biotechnology. If you want to learn more about these topics, see the references to this unit for some introductory readings.

B. The Technologies of Modern Biotechnology

Modern biotechnology both provides and makes use of powerful tools to manipulate DNA. In fact, the term *biotechnology* is often used to mean the manipulation of DNA. These tools include enzymes that cut DNA at specific sites, enzymes that splice DNA strands together, techniques to visualize DNA, techniques to separate DNA fragments from one another, techniques to identify fragments of DNA with specific sequences, techniques that can amplify DNA (which generate tens of thousands of copies of a specific gene fragment), and techniques to sequence and synthesize DNA.

The availability of tools to manipulate DNA allows biotechnologists to introduce genetic information (DNA) from one organism into another. *When a biologist causes a cell or organism to take up a gene from another organism, we say the cell or organism is* **genetically modified**, *or* **genetically engineered**, Figure 1.1. The term **recombinant DNA** *refers to DNA that contains sequences of DNA from different sources that were brought together using the tools of biotechnology.* This ability to produce genetically modified organisms is so powerful that the term *revolutionary* is often applied to

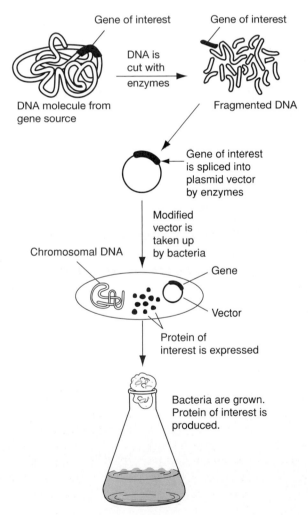

Figure 1.1.   The Genetic Modification of Bacteria. A gene of interest is isolated and inserted into a vector. The vector is taken up by the bacteria, which are now said to be "transformed" or genetically modified. The transformed bacteria can be grown up in flasks or larger containers. Under the proper conditions, the bacteria express the protein coded for by the foreign DNA.

it. It is therefore important to discuss the genetic modification of organisms in more detail.

DNA contains coded information that can direct a cell in making a specific protein. *When a protein is produced by a cell, we say that the DNA coding for that protein was* **expressed**. Under the proper conditions, a genetically modified cell can express (i.e., produce) a protein coded for by an introduced gene.

DNA can be transferred to a recipient cell in a number of ways. The gene of interest is sometimes enzymatically spliced into a plasmid. **Plasmids** *are small, circular molecules of DNA found in many types of bacteria, which exist separately from the bacterial chromosome.* Under proper conditions plasmids are readily taken up by recipient bacterial cells. Plasmids are called **vectors** *when they carry a desired gene into recipient cells.* Other methods of transferring DNA to a host cell include transfer via viral vectors and direct injection.

Human insulin was among the first commercial products made by a genetically modified organism. Prior to the 1980s, all insulin to treat diabetics was purified from the pancreas of animals slaughtered for human consumption. Animal insulin is similar, but not identical, to human insulin; therefore, some diabetics developed allergies to the drug. In addition, insulin derived from animals requires a large supply of material from slaughterhouses, which is not consistently available. In the late 1970s, researchers developed methods to transfer the gene coding for insulin into bacteria. The researchers were also able to induce the bacteria to express the human insulin gene.* This is a remarkable accomplishment; bacteria have absolutely no use for insulin and would never produce it without human intervention. Bacteria that contain the insulin gene can be grown in large quantities. *In biotechnology, this large-scale cultivation of microorganisms is called* **fermentation**. The resulting insulin can then be isolated and purified using protein separation techniques, thus providing an efficient, reliable production system for this medically important protein.

Bacteria were the first genetically engineered organisms used to make a commercial product; however, researchers soon learned to genetically modify other types of cells as well. It is often possible to introduce a gene of interest into cultured mammalian cells (or cells from other multicellular organisms) that can produce a desired protein. **Cultured cells** *are those grown in flasks, dishes, vats, or other containers outside a living organism.* The techniques that are used to grow mass amounts of cultured cells for production purposes are analogous

*The actual production of insulin by bacteria is more complicated than indicated here because the insulin protein that is made in human cells requires processing before it is active. For more information on how insulin was first produced using genetically modified bacteria, see for example, *Recombinant DNA,* James D. Watson, Michael Gilman, Jan Witkowski, and Mark Zoller. 2nd edition. W.H. Freeman and Co., New York, 1992.

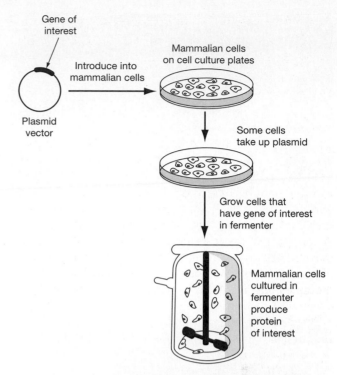

Figure 1.2.   The Use of Genetically Modified Cultured Mammalian Cells to Produce a Protein of Interest. Once the cells have been transformed with the gene of interest, they can be grown in large quantities in a fermenter, also called a *bioreactor,* and the protein product can be isolated.

to those used in bacterial fermentation. Mammalian cells, however, grow more slowly than bacteria, are more fragile, and have more complex nutrient requirements. Figure 1.2 illustrates the use of genetically modified mammalian cells to produce a protein product.

We have so far talked about introducing foreign DNA into cells in culture, either microorganisms or cells from multicellular organisms. Although it is more complex, it is also possible for biologists to introduce a gene of interest into plants and animals. *A plant or animal whose cells are genetically modified using the techniques of biotechnology is called* **transgenic**. Figure 1.3 shows the first transgenic animal. The two animals in this photo are litter mates. The mouse on the left, however, was genetically modified by the introduction of the gene for rat growth hormone. This was accomplished by microinjecting the growth hormone gene into a fertilized mouse egg; thus, all the cells in the resulting mouse contained the foreign gene. This gene was expressed, causing the transgenic animal in the photograph to be unusually large.

At this point, the question naturally arises as to whether it is possible to introduce a gene of interest into humans. This possibility does raise a number of critically important societal and ethical issues, especially if genes are introduced in such a way that they are passed on to the children of that individual. As we will see later, there are experiments underway to treat life-threaten-

Figure 1.3. The First Transgenic Animal. The gene for rat growth hormone was microinjected into a fertilized mouse egg, which was then implanted into a surrogate mother. The resulting transgenic mouse expressed the gene, resulting in an exceptionally large animal.

ing illness in humans using gene therapy. It should be noted, however, that these genetic modifications are intended to cure the individual and would not affect the children of the treated person. In addition, not every cell of a treated individual is genetically modified.

Thus, we have seen so far that biologists can produce genetically modified microorganisms and cultured cells, produce transgenic plants and animals, and potentially modify human cells as well. This technology has had a tremendous impact on biological research. Such modified organisms are used to explore the role of genes, the mechanisms by which gene expression is controlled, and other important questions. There are also many commercial applications for genetically modified organisms. These organisms can be used to produce proteins of value, such as insulin. Transgenic plants and animals may be produced that have desirable characteristics; for example, transgenic plants may have enhanced nutritional value. Figure 1.4 summarizes some of these points regarding the genetic modification of organisms.

The production of monoclonal antibodies is another technology that is often associated with modern biotechnology. Antibodies are proteins, made by the immune system, that recognize and bind to substances invading the body. A particular antibody binds only a particular target, somewhat like a specific key fits only

into a particular lock. **Monoclonal antibodies** *are exceptionally homogeneous populations of antibodies directed against a specific target.* Monoclonal antibodies are produced by **hybridoma** *cells that result from the fusion of an antibody-producing cell and a cultured cell.* Because of their specificity, monoclonal antibodies have found many applications in research and in medical diagnosis. For example, home pregnancy test kits are based on monoclonal antibodies that specifically bind to a hormone that increases during pregnancy.

C. Applications of Biotechnology

i. MEDICAL/VETERINARY APPLICATIONS

Some of the most commercially important products of biotechnology are those used in medicine. For example, as was previously discussed, genetically modified bacterial cells are used to produce insulin. **Biopharmaceuticals** *are pharmaceutical products, like insulin, that are manufactured using genetically modified organisms as production systems.* Other examples of biopharmaceutical products are shown in Table 1.1.

It is also possible to produce biopharmaceuticals in transgenic farm animals and plants. Under the proper circumstances, a transgenic animal can produce the desired protein, which is typically secreted into her milk. The protein in the milk is then isolated and purified. Genes of interest can similarly be introduced into crop plants, which are cultivated and harvested to obtain the product of interest. These products of transgenic plants and animals are not yet of great commercial significance, but they may be in the future.

An intriguing medical application of biotechnology is the production of vaccines in foods. Proteins found on the surface of pathogens can often be used as vaccines because they induce an immune response (are antigenic) when administered to humans or other mammals. If scientists can isolate the gene coding for an antigenic protein, then they can often insert the gene into a plant. The transgenic plant sometimes expresses the protein in an edible fashion, as in a banana or other fruit. Individuals who eat the fruit containing the antigenic protein are thereby immunized. Experiments on such food vaccines are in progress. These vaccines may be especially valuable in countries where vaccines that require refrigeration and/or injection are impractical.

Biopharmaceuticals, such as insulin, are very valuable, but the products themselves are often not novel. Insulin and Factor VIII were purified from animal or human sources before modern methods of biotechnology existed. Biotechnology provides a new production system, which is often more reliable and safer than traditional sources. In contrast, gene therapy was not possible prior to the introduction of modern methods of biotechnology. **Gene therapy** *involves replacing a gene that is missing, or correcting the function of a faulty gene, in order to treat or cure an illness.* One strategy for gene

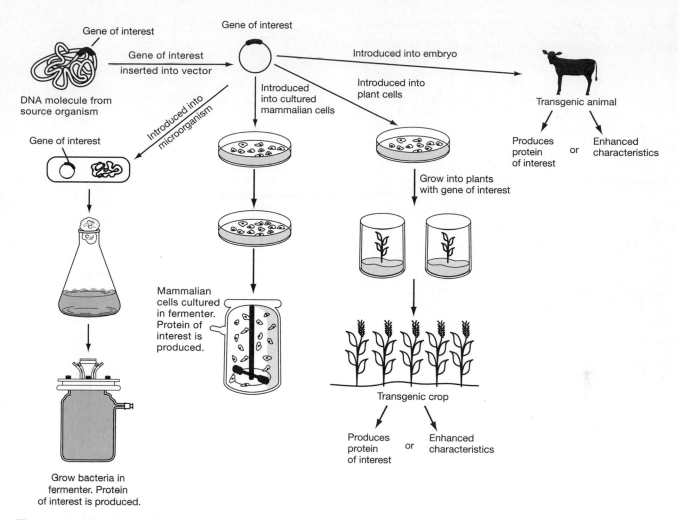

Figure 1.4.   The Genetic Modification of Organisms Using Biotechnology.

therapy involves removing cells from the patient, such as blood cells, genetically modifying them, and then returning them to the patient. The first successful gene therapy trial was begun in 1990 when an immune system defect, ADA deficiency, was apparently ameliorated in two children. Blood cells with normal ADA genes were injected into the children who lacked the ADA gene. After treatment, the children appeared to have experienced improvements in the function of their immune systems, Figure 1.5.

A second approach to gene therapy that is under investigation is the targeting of cells within the patient for genetic modification. In this approach, genes are carried into the patients by vectors, which are often infectious viruses whose pathogenic effects have been attenuated. Gene therapy trials are currently underway to treat various diseases, such as cystic fibrosis and cancer.

Another medical application of biotechnology is in identifying individuals who carry a gene that makes them susceptible to a certain illness. For example, there are tests to identify people carrying genes that will make them more susceptible than average to certain types of cancers. The use of these tests raises a number

of complex, unresolved societal issues, such as how this information may be used by insurance companies.

One of the most ambitious biotechnology projects in the world is the Human Genome Project, which will ultimately provide the entire nucleotide sequence of human DNA. This information is expected to have major consequences in medicine. Researchers hope that knowing the human genetic code will lead to better diagnosis and treatment of disease, and will provide other, as yet, unforeseen benefits.

ii. AGRICULTURAL/FOOD RELATED APPLICATIONS

Traditional methods of plant and animal breeding have long been used to enhance the characteristics of crops and livestock. It is possible, however, to use the methods of biotechnology to genetically modify plants and animals more quickly, and with better control, than is possible using traditional methods. For one thing, traditional breeding does not allow genes from different plant or animal families to combine because the plants or animals do not breed with one another. In addition, in traditional breeding, there is an uncontrolled mixing of all the genes from both parents. Recombinant DNA

Table 1.1 EXAMPLES OF BIOPHARMACEUTICAL PRODUCTS

RECOMBINANT DNA PRODUCTS PRODUCED IN BACTERIA (E. COLI)

Human insulin   Approved 1982

Used to treat diabetes

Comments: Before insulin from genetically modified bacteria was available, the only source of insulin to treat diabetics was that purified from the pancreas glands of animals intended for human consumption. Insulin derived from genetically modified bacteria provides a more reliable source than animals and is less likely to cause allergic reactions in patients.

Human Growth Hormone   Approved 1985

Used to treat dwarfism

Comments: Prior to the introduction of recombinant DNA–derived human growth hormone, dwarfism was treated with hormone purified from pituitary glands collected from human cadavers. Some children who received this hormone isolated from human sources eventually died of the neurodegenerative disease, Creutzfeldt-Jakob disease.

Interferon α-2b   Approved 1986 for treatment of leukemia

Used to treat a variety of viral diseases and cancers, including hairy cell leukemia, AIDS-related Kaposi's sarcoma, renal cell carcinoma, and chronic hepatitis B

Comments: Interferon is not available from natural sources in the amounts and purity required for experimentation and clinical trials. Therefore, although the potential of interferons in the treatment of disease was recognized in 1957, it was not available for large scale use in patients until it was produced by recombinant methods.

RECOMBINANT DNA PRODUCT PRODUCED IN YEAST CELLS

Hepatitis B vaccine   Approved 1986

Used as a vaccine to prevent infection with hepatitis B

Comments: Hepatitis B is a common and serious viral illness for which there is no treatment currently available. Before the recombinant vaccine was developed, a vaccine was prepared from the plasma of hepatitis-infected humans. This source of vaccine was limited and there were concerns about its purity. The substance required to make the hepatitis B vaccine is now produced by genetically modified yeast which makes the vaccine more available and reduces the possibility of contamination.

RECOMBINANT DNA PRODUCTS PRODUCED IN MAMMALIAN CELLS IN CULTURE

Erythropoietin   Approved 1989

Used to treat anemia due to renal failure or AZT treatment (used for AIDS patients)

Comments: Erythropoietin (EPO) is a glycoprotein, produced in the kidney, which stimulates the production and maturation of red blood cells. EPO occurs naturally in very small quantities in human urine and therefore, prior to its production by recombinant DNA methods, it had never been available in sufficient quantity for clinical testing. The conventional treatment for anemia due to renal failure was blood transfusion. When recombinant EPO was introduced, it provided a new approach to the treatment of anemia.

Factor VIII   Approved 1992

Used to treat hemophilia

Comments: Before the introduction of recombinant Factor VIII many hemophiliac patients contracted AIDS from Factor VIII derived from human plasma.

MONOCLONAL ANTIBODY PRODUCT

Murine Monoclonal antibody to CD3   Approved 1986

Used to suppress organ rejection by patients receiving kidney transplants

Comments: This product is a highly purified antibody which attacks the T cells of patients. The T cells are involved in transplant rejection. This antibody has been effective in treating patients who do not respond to conventional antirejection treatments.

methods allow the introduction of only a specific gene of interest into the offspring.

The methods of biotechnology are being used, for example, to insert genes into plants that enhance their resistance to insect, bacterial, and fungal disease. Plants can also be genetically modified to enable them to survive environmental stresses, such as drought and frost, to retard spoilage, and to produce plants with enhanced nutritional value. Similarly, methods of biotechnology are being used with livestock to produce animals that are faster growing, produce more meat, milk, eggs, or wool, or have altered nutritional value.

The use of recombinant DNA methods in food production is among the most controversial of its applications. There is concern that food produced using these methods might not be as safe as conventional foods, that animals may be stressed by the introduction of foreign genes, and that the introduction of genetically modified organisms into the environment may have unforeseen, adverse effects. On the other hand, in many

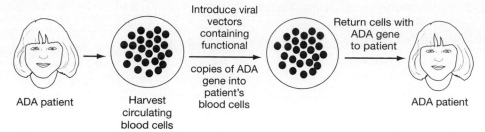

Figure 1.5. Gene Therapy to Treat ADA Deficiency. Blood cells were removed from the patient. Copies of the ADA gene were inserted into the cells, which were then returned to the patient. The genetically modified cells produced the protein that is missing in the affected children, thereby improving the function of their immune system.

countries there is a severe shortage of food, so enhancing the productivity and nutritiousness of plants and animals is potentially a great benefit to society. It is also hoped that the use of biotechnology will reduce our reliance on chemical pesticides.

iii. OTHER APPLICATIONS

There are many other applications of biotechnology. For example, microorganisms have remarkable abilities to degrade hazardous compounds into harmless by-products; therefore, they are sometimes used in the clean-up of contaminated water and soil. These degradative abilities can be enhanced by the transfer of genes from one microorganism to another. In the future, it is hoped that such genetically modified organisms will play a major role in cleaning contaminated soil and water.

Another application of DNA technology is in identification of individuals. The patterns of DNA sequences from each human are distinct (except for identical twins). The tools of modern biotechnology allow analysts to prepare a DNA "fingerprint" that is characteristic of a certain individual. This technology can be used, for example, to link a suspect's DNA to DNA found in blood, hair, or semen from a crime scene. Similar fingerprinting techniques can be used to determine the father of a child in paternity disputes.

II. THE ORGANIZATION OF A BIOTECHNOLOGY COMPANY

A. Research and Development

In biotechnology companies, the transformation of knowledge into a commercial product begins in the research and development (R & D) unit. **R & D** *is the organizational unit that finds ideas for products, performs research and testing to see if the ideas are feasible, and develops promising ideas into actual products.*

Once an idea for a product is conceived, an often long and arduous period of laboratory testing begins to see whether the concept will work, under what conditions it will work, and whether commercial production

is economically feasible. If successful, the research phase might result in a prototype device or a description of the product and how it will be used.

The development phase includes determining the properties of the product, specifying the properties the product must have to be effective and reliable, and describing how to use the product. For a pharmaceutical product, development also involves rigorous testing in animals and then in human volunteers. The development phase is a time of evolution and evaluation. As development progresses, the characteristics of the product become established, methods of production become increasingly consistent and systematized, and the specifications for raw materials are codified. It is during this evolutionary period that quality is designed and built into the product.

A product is initially made in small quantities in the laboratory. A major responsibility of the development unit is to **scale-up** *the processes used by the research team so that larger amounts of the product can be made in a consistent, reproducible, and economically effective manner.* Depending on the quantities of the product to be sold, scale-up may be a fairly simple process of moving from one size flask to another. In other situations scale-up involves initially making the product in a medium-size **pilot plant** *where further development occurs*, and then eventually in full-scale production facilities. For products that involve whole plants, such as cotton plants, scale-up may mean moving from the greenhouse, to a small experimental plot outdoors, and finally to large fields.

Large companies may have one team assigned to perform basic research and another to develop potential products. Smaller companies may combine R&D into one functional unit. Small start-up biotechnology companies may be almost entirely devoted to R & D. The responsibilities of R&D are summarized in Table 1.2.

B. Production

The systems used for production in biotechnology companies are diverse. Manufacturing a product may involve growing bacteria in laboratory flasks and isolat-

Table 1.2 *The Responsibilities of the Research and Development Unit in a Biotechnology Company*

Finding a potential product with commercial value
performing research relating to the potential product

Characterizing the properties of the product, such as:
composition, physical, and chemical properties
strength, potency, or effect of the product
purity of the product and steps required to avoid contamination
stability and shelf-life of the product
applications of the product
safety concerns in the use of the product

Establishing product specifications (descriptions of properties that every batch of the final product must have to be released for sale)

Developing methods to test the product to ensure that it meets its specifications

Developing processes to make the product

Describing any cells or microorganisms required to make the product

Determining what raw materials are required to make the product and establishing specifications to characterize these materials

Describing equipment required to make the product

Developing a plan for production of the product

Table 1.3 *The Responsibilities of the Production Unit*

Making the product

Working with large-scale equipment and/or large-volume reactions (not applicable to all biotechnology products or companies)

Routine monitoring and control of the environment as required for the product (e.g., maintaining the proper temperature or sterility requirements)

Routine cleaning, calibration, and maintenance of equipment

Following written procedures and performing tasks associated with producing the product

Monitoring processes associated with making the product

Initiating corrective actions if problems arise

Completing forms, labeling, filling in logbooks, and maintaining other required documents

---

ing products from the cultures. In other situations, manufacturing involves growing bacteria, or other types of cells, in fermenters that are several stories tall and using industrial-scale equipment to purify products from the cultures. Production may involve growing plant cells in petri dishes, cultivating crops in a field, maintaining laboratory animals, or even keeping farm animals. Thus, the details of production vary greatly from company to company. Certain functions, however, are generally the responsibility of the production team, regardless of the nature of the product and company. These functions are summarized in Table 1.3.

C. Quality Control/Quality Assurance

The quality control/assurance unit (QC/QA) in a company is responsible for ensuring the quality of the product. You may see the terms *quality control* and *quality assurance* used differently in various sources; in fact, these two terms are sometimes used interchangeably. They do, however, have different meanings. Here, we define **quality control** (QC), *as a subsection of QA chiefly responsible for monitoring processes and performing laboratory testing.* **Quality assurance** (QA) *refers to all the activities and people who work to ensure the final quality of products.* Although the terminology varies, the goal of the quality team is to assure a quality product. (See, for example, "The Case of the q.c. Unit,"

Kenneth G. Chapman, Timothy J. Fields, and Barbara C. Smith. *Pharmaceutical Technology*, pp. 74–79, Jan., 1996, for a discussion of terminology.) The responsibilities of the quality unit are summarized in Table 1.4.

Although there appears to be a transfer of information and materials in one direction, from R&D to production to QC/QA, in practice there must be continuous flow of information to and from all departments. It is important that the company facilitate communication between individuals and functional areas. Significant decisions made in one area will almost always affect other areas. For example, there may be problems in scale-up and manufacturing, particularly when a product is first going into production. Identifying and solving a problem might require input from individuals in many areas of the company.

Table 1.4 *The Responsibilities of the QA/QC Unit*

Monitoring equipment, facilities, environment, personnel, and product

Reviewing all production procedures used in the company

Ensuring that all documents are accurate, complete, and available

Testing samples of the product and the materials that go into making the product to determine whether they are acceptable

Comparing data to established standards

Deciding whether or not to approve the product for release to consumers

Reviewing customer complaints

D. Other Functions in a Company

Although they are not the focus of this book, there are other important areas in a company as well. These include:

- *Engineering (or facilities management)*, which is responsible for ensuring that the systems that control the building environment are operating properly and that large equipment is properly installed and functioning.

- *Facility maintenance and housekeeping*, which performs critical day-to-day functions, such as cleaning and certain building repairs.

- *Receiving and shipping*, which assures that incoming raw materials are properly routed and that products go to the proper destination.

- *Dispensing*, which puts products that are produced in bulk into individual containers for consumer use.

- *Metrology*, which ensures that instruments (such as those used in laboratories and those used to monitor production conditions) operate properly.

- *Marketing and sales*, which is responsible for interacting with customers.

E. The "Cultures" Within a Company

It is sometimes said that there are three cultures within a biotechnology company—R&D, production, and QC/QA. The R&D people work in an atmosphere of change and unpredictability. They often do not know which ideas will result in a product, which methods will work best, and how much success they will have in their endeavors. Creativity and a willingness to work with uncertainty characterize most investigators in R&D and in academic research laboratories. In production situations, attention to detail is critical and deviations from established procedures must be carefully considered and approved. The production team, therefore, works in a more predictable environment. The quality team is intent on assuring product integrity. The quality team interacts closely with outside agents, such as auditors, inspectors, and customers, and also works closely with the company's R&D and production personnel. The quality assurance team must effectively manage problems and be conscious of the ramifications of particular courses of action. Thus, everyone needs to be working toward a quality product, but the particular focus of the individual may vary depending on his or her role in the company.

In small, start-up biotechnology companies, which are in the early stages of developing marketable products, the situation may be different than described above. In a start-up company the emphasis is on the science involved in researching and developing products. In these companies, a single individual may play multiple roles. For example, the research scientists who

develop a product may also initially produce and package it for sale and may also deal with quality assurance and regulatory issues. The same facilities may be used to perform research and to make products for sale. The same types of equipment may also be used in research, development, and production. Thus, the distinction between R&D, production, and QC/QA may be blurred in a small company. As the company grows, more individuals are hired, with some of them specializing only in production or quality assurance activities. As the company begins to sell more and more product, the equipment and facilities used for production become larger and distinct from those used for product development. As this maturation of the company occurs, government regulations and standards may play an increasingly important role and the quality assurance unit grows to keep up with these changes. (Regulations and standards are discussed in Chapter 3 and Unit II.) These transitions are likely to affect the culture and the "feeling" of the company.

III. BIOTECHNOLOGY WORKPLACES

There is no single biotechnology workplace. Some people work in academic or government research laboratories. Some people work in R & D laboratories, quality control laboratories, or production facilities. Because this is a book about laboratory methods, let us consider the laboratory workplace.

Research laboratories have existed for hundreds of years. The **research laboratory** *is a space set aside to study the complexities of nature in a controlled manner.* In the laboratory, observations can be made, and experiments can be performed, in which the researcher controls the factors of interest. For example, if researchers are interested in the effect of light on plant growth, they can carefully control the light that plants receive in the laboratory. Outside the laboratory, the researcher has little control over light exposure or many other important factors—such as rain, temperature, and insects—that may affect the plants' growth. Basic biological research continues to be an essential part of biotechnology, so the basic research laboratory is an important part of the biotechnology enterprise.

Some biotechnologists work in research laboratories associated with biotechnology companies. The output of such a laboratory relates to a commercial product. Many of the tasks performed by people in basic research laboratories and R & D laboratories are similar, although the purpose of the work is somewhat different.

There is another category of laboratory that can be distinguished from a research laboratory. This is the **testing laboratory**, *a place where analysts test samples.* A quality control laboratory is a type of testing laboratory where samples of products and raw materials are tested. There are other types of testing laboratories, such as clinical laboratories, where samples from patients are tested; forensics laboratories, where samples from crime scenes are tested; and environmental laboratories, where samples from the environment are tested. The product of a testing laboratory is a test result, such as a measurement of enzyme activity, a measurement of the blood glucose level in a sample, or a DNA "fingerprint."

Because there are different sorts of laboratories, it is as difficult to definitively describe "The Laboratory" as it was to define biotechnology. One definition of a laboratory is that it is a workplace whose product is knowledge, data, or information. This is a reasonable statement with a couple of caveats. First, people in laboratories also produce many tangible products such as photographs, antibodies, purified proteins, and printouts. These materials, however, are produced with the purpose of learning more about a system or a sample, answering a research question, or documenting what has been discovered. The materials that emerge from a laboratory are not intended for commercial sale. Another caveat is that small biotechnology companies often produce products for sale in facilities that are also used for research or that are similar to research laboratories. However, we call any facility used for producing a commercial product a production facility, rather than a laboratory.

As we have seen, biotechnology production settings also vary from one another. In a small biotechnology company that produces limited amounts of product, production workers might use flasks, beakers, small incubators, and other materials much like those used by researchers. A large production facility is quite a different workplace, sometimes with organisms growing in vats several stories tall, valves controlled by computers, and other large scale equipment, Figure 1.6. In a company where products are crop plants, production is very different than it is in a company manufacturing drug tablets. Ultimately, the purpose of a production facility, whether it is a small lab or a major pharmaceutical complex, is to make a tangible item intended for sale or distribution.

Regardless of the work environment, biotechnology scientists and technicians are involved in the physical aspects of acquiring data or making a product. They work with test tubes, solutions, and instruments. They grow bacteria, cells, plants, and animals, and work with materials derived from organisms. They make measurements, prepare reagents, gather and record data, and assemble data. They may also interpret data. These are individuals who know how to operate instruments and control equipment. Scientists and technicians play key roles in any biotechnology enterprise. The quality of the product, whether it is knowledge, a test result, or a tangible product for sale, will depend on the quality of work performed by these individuals.

Table 1.5 summarizes job titles of individuals who work in technical positions in biotechnology. Table 1.6 summarizes different types of biotechnology workplaces.

Figure 1.6. A Large-Scale Fermenter.

Table 1.5 EXAMPLES OF JOB TITLES

In a laboratory where research* is performed, job titles include:

Scientist
Research Scientist
Research Assistant
Research Associate
Research Technician
Laboratory Technician
Laboratory Scientist
Media Preparation Technician

Individuals who work in production may be called:

Manufacturing Operator
Manufacturing Technician
Manufacturing Supervisor
Production Technician
Pilot Plant Operator
Pilot Plant Technician

Individuals who work in quality control laboratories may have job titles like:

Quality Assurance Technician
Quality Control Technician
Assay Analyst
Documentation Specialist
Inspector
Compliance Specialist

*In academic research laboratories, undergraduate and graduate students and researchers with various titles play key technical roles in the laboratory.

Table 1.6   *DIFFERENT TYPES OF BIOTECHNOLOGY WORK ENVIRONMENTS*

Basic Biological Research Laboratories in Academic or Government Settings

In basic research laboratories scientists investigate fundamental problems in biology. The product of basic research is knowledge or information, although research may at a later time result in a commercial product.

Research and Development (R&D) Laboratories Associated with an Industry

In R & D laboratories researchers investigate questions in biology that are intended to result in commercial products. When promising ideas are identified, those in R & D develop methods that turn the concept into a marketable product.

Production Facilities

People who work in production facilities must make products which consistently have certain specified properties. Production workers often—but not always—use larger scale equipment than do people in laboratories. For example, a laboratory technician might grow bacteria that produce a product in a 1 L flask. A production technician might grow the same bacteria in a 10 L flask— or in a 100,000 L fermenter. In some cases the methods used by production technicians are similar to those performed by laboratory workers. In other cases, as the scale becomes larger and the manufacturing complexity increases, the two work environments become increasingly different.

Testing Laboratories (Such as Analytical, Quality Control, Forensic, Microbiology, Metrology, and Clinical Testing Laboratories)

Testing laboratories conduct tests and assays on samples or devices. For example, analytical laboratories might evaluate the lead levels in paint from a house or perform tests to determine the solubility of a drug. Forensic laboratories test samples from crime scenes. Quality control laboratories monitor the properties of products or of the materials that go into the production of products. Clinical laboratories test samples from patients. Metrology laboratories test instruments. Testing laboratories are sometimes independent and are sometimes part of a larger organization.

PRACTICE PROBLEM

A new, effective anticancer compound is discovered. The compound is unfortunately found in the stems of a rare, slow-growing, endangered plant; therefore, very little of the compound is available. A small, start-up biotechnology company is formed by a team of research scientists who plan to isolate the gene for this anticancer agent. After several years of dedicated and difficult work, they obtain the gene that codes for the anticancer agent. They are then able to insert the gene into bacteria and are elated to find that the bacteria make the anticancer agent. They develop methods to grow large quantities of these bacteria. They learn how to harvest the bacteria and purify the product from them. The company can now make large amounts of this anticancer compound in bacteria with no harm to the rare plants. Over several years, the following tasks were performed by workers in the biotechnology company. Label each task as being primarily the job of R&D, Production, or QC/QA. Note that the tasks are not necessarily listed in the order in which they would actually be done.

a. Identify the gene that codes for the anticancer agent.
b. Develop an assay to check whether the anticancer compound is present in a solution at a certain level.
c. Devise methods to grow large amounts of the bacteria in such a way that the compound is consistently produced.
d. Write procedures for how to sterilize glassware and prepare reagents.
e. Submit forms to the government to request permission to test the compound in human volunteers.
f. Take samples from throughout the production facility to test for microbial contaminants.
g. Devise a method to insert the gene into bacterial cells so that the cells make the desired anticancer compound.
h. Process the anticancer compound into tablets.
i. Use the assay for the anticancer compound to test whether it is present in different lots of product.
j. Design purification methods to purify the compound from the bacteria.
k. Once the new product goes into production, sterilize the equipment used in production, monitor the pH levels and temperatures.
l. Purify the anticancer product (that will be sold) from the bacteria.
m. Determine the chemical composition of the anticancer compound.

ANSWERS TO PRACTICE PROBLEM

R & D: a-d, g, j, m
Production: f, h, k, l
QC/QA: e, i

## CHAPTER 2

Introduction to a Safe Workplace

I. INTRODUCTION TO LABORATORY SAFETY

A. Safety in the Workplace

"I think good lab practice, consideration for other people, and safety are three totally related issues"
—David H. Beach, Ph.D.,
Cold Spring Harbor Laboratory

Before we continue our discussion of biotechnology and its products, we will introduce basic principles of laboratory safety. Worker safety is a crucial consideration in the workplace and is discussed in detail in Unit VIII.

Safety *is defined as the elimination of potential threats to human health and well-being.* While this is an essential goal in every profession, complete safety can never be achieved. All workplaces have the potential for **accidents,** *which are unexpected and usually sudden events that cause harm.* Even people who work at a desk in an office all day may encounter an **emergency,** *which is a situation requiring immediate action to prevent an accumulation of harm or damage to people or property.* Consider the following Case Study, which occurred in a laboratory where one of the authors worked.

CASE STUDY I

Fire in the Workplace

One evening, three laboratory co-workers were finishing experiments when one commented that he smelled smoke. Within one minute, the laboratory filled with black, acrid smoke so thick that the workers could not see their hands 12 inches from their faces. Two workers were together and quickly located the third, who was trying to gather her research notebooks and experimental materials. At this point, the smoke was so thick that conversation was impossible. The workers bent over and immediately moved to the nearest stairwell, evacuating the building. The fire department quickly responded and extinguished the fire, which was traced to electrical wires in an interstitial space. Luckily there were no injuries. Even though no flames reached the laboratory, smoke damage to equipment and materials was extensive. Nothing in this situation was specific to the laboratory. The workers could not have prevented the fire, which started in an inaccessible area; however, these workers responded quickly and appropriately to the circumstances, which could have led to serious injuries from smoke inhalation.

Although laboratories may present special safety challenges, those of us who have worked in laboratories for many years can attest to the fact that most work-related accidents are mundane in nature. They include:

- tripping on unexpected items left on the floor
- falls on slippery floors (especially around sinks and ice machines)
- slamming fingers in cabinet doors
- hallway collisions with co-workers
- minor cuts while picking up pieces of broken glass

It is estimated that 30% of all workplace accidents involve trips, slips, and falls, mostly on flat surfaces. These incidents can usually be prevented by using care and common sense, which are the best approaches to avoiding all accidents. This chapter will address general strategies to workplace safety. Laboratory work, however, does require special precautions for situations that are not found in other workplaces. These issues will be covered in Unit VIII.

B. Hazards and Risk Assessment

All laboratories, by their nature are filled with hazards. **Hazards** *are the equipment, chemicals, and conditions that have a potential to cause harm.* Heavy equipment, chemicals, electricity, radioisotopes, animals, and infectious agents are examples of hazards that are frequently present in biotechnology laboratories. For this reason, the laboratory setting should always be approached with caution and respect. Hazards can be controlled, however, which is the goal of safety programs.

Because it is impossible to remove all hazards from the biotechnology workplace, the most useful measure of safety in a laboratory is risk. **Risk** *is the probability that a hazard will cause harm.* The risk associated with a hazard can be reduced, thereby reducing the probability of an accident. Although laboratory hazards may seem intimidating, you can significantly reduce the associated risk by being knowledgeable and careful.

Risk assessment *is an attempt to estimate the potential for human injury or property damage from an activity.* Risk assessment is really a measurement of safety. It is difficult to estimate the true risk of specific activities because there are too many variables involved in human actions. Accidents result from both unexpected and predictable errors. Analyzing activities, identifying hazards, and developing strategies for reducing risk lead to the development of safety guidelines. **Safety guidelines** and standards are *procedures that are designed to prevent accidents by reducing the risk of hazards in situations where the hazards cannot be eliminated entirely.*

Given the number of hazards present, it is comforting to know that major accidents are rare in the laboratory. Most injuries are relatively minor because of the natural tendency of workers to be more cautious around major hazards than they are around small hazards or

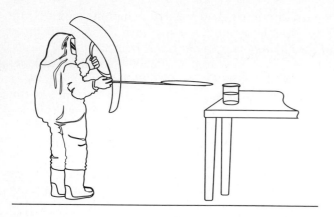

Figure 2.1.   An Overzealous Use of Caution for Most Laboratories.

routine tasks, Figure 2.1. The result is that most accidents happen under mundane circumstances involving routine tasks. A basic premise of risk management is that accidents can be avoided with caution and knowledge, which are the key factors in laboratory safety.

Every laboratory is different enough to warrant a separate approach to safety. Although it is impossible to address the safety concerns of each individual workplace here, this chapter will address general responsibilities for workplace safety. Detailed information concerning physical, chemical, and biological safety are provided in Unit VIII.

C. Who Is Responsible for Workplace Safety?

Safety is everyone's business. Federal agencies and other organizations are responsible for creating regulations and codes for safe workplaces, Table 2.1. The employer has the responsibility to provide a safe work environment and a general institutional attitude of "safety first," to train employees to work responsibly, and to develop an emergency response plan. Employees have the right to work in a safe environment, and to be well trained and informed about workplace hazards. It is the responsibility of the employee to apply this training and to implement the safety plan of the institution. This hierarchy of responsibility is illustrated in Figure 2.2.

The key to protecting yourself and others from laboratory hazards is to be continuously aware of the many types of hazards present, and to know how to handle them properly. As an employee, you have both a right and a responsibility to know the hazards in the workplace, and to learn all you can about the work being done at your place of employment.

II. LABORATORY SAFETY MANAGEMENT

A. Regulatory Agencies

There are a vast number of state, federal, and local regulations, as well as industry standards, that affect biotechnology companies. **Standards** *are operating principles or requirements related to many areas in addition to safety.* Many safety standards are voluntary. **Regulations** *are*

Table 2.1   *FEDERAL AGENCIES THAT REGULATE SAFETY AND ENVIRONMENTAL PROTECTION IN BIOTECHNOLOGY ORGANIZATIONS*

The Occupational Health and Safety Administration (OSHA)

OSHA is the federal agency charged with ensuring worker safety. OSHA promulgated the Laboratory Standard in 1990 that deals specifically with laboratory safety. The central requirement of the Laboratory Standard is that the employer develop, document, and implement a plan that protects workers from hazards.

The Environmental Protection Agency (EPA)

EPA is responsible for protecting the environment. EPA regulations affect how laboratories and companies handle and dispose of waste, what substances can be emitted into the air and water, the movement, storage, and disposal of hazardous substances, and records relating to chemicals. EPA also regulates certain types of biotechnology field work that involve releasing genetically modified organisms into the environment.

The Department of Transportation (DOT)

DOT regulates the transportation of hazardous materials, such as chemicals, compressed gas cylinders, and hazardous wastes. The regulations cover packaging, labeling, transport, and reporting procedures.

The Nuclear Regulatory Commission (NRC)

NRC is responsible for the safe use of radioactivity. Facilities that use radioactive substances for research purposes, medical applications, or in products must comply with NRC regulations, including those related to worker safety, waste disposal, and record keeping.

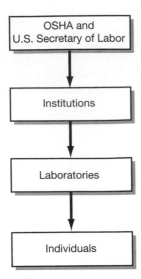

Figure 2.2. Who Is Responsible for a Safe Workplace? (Adapted from *Occupational Safety and Health in the Age of High Technology,* David L. Goetsch. Prentice Hall, Englewood Cliffs, NJ, 1996.)

operating principles which are required by law. The many regulations and standards applied to biotechnology laboratory workers can be arranged into categories:

- **Worker Safety.** For example, there are regulations that require laboratory chemicals to be labeled and which require workers to be informed about hazards (discussed in this chapter).

- **Environmental Protection.** For example, the disposal of hazardous laboratory chemicals is regulated in order to minimize impact to the environment (discussed in this chapter).

- **The Use and Handling of Animals.** For example, there are regulations regarding the cages used to house laboratory animals and regulations aimed at preventing the spread of contagious disease. These regulations both protect animals from inhumane treatment and prevent faulty experimental results due to sick animals or inconsistent treatment of animal subjects. These regulations will be discussed in Chapter 30.

- **Regulation of Radioisotopes.** For example, these regulations cover such issues as how radioisotopes should be handled and stored, who has access to such compounds, and what documentation is required when radioisotopes are used.

Table 2.1 summarizes the roles of U.S. government agencies whose regulations directly affect biotechnology companies and research laboratories. Even though this book cannot discuss all of the statutes related to laboratory safety, we will discuss some of the most frequently encountered agencies and their regula-

tions. For a more detailed and comprehensive discussion, we recommend starting with *Prudent Practices in the Laboratory. Handling and Disposal of Chemicals,* the National Research Council, National Academy Press, Washington, D.C., 1995. Significant amounts of information are also available on the Internet at the home pages for the Occupational Safety and Health Administration and the Environmental Protection Agency (http://www.osha.gov and http://www.epa.gov).

Although occupational safety is regulated primarily by federal agencies, there are many other organizations concerned with safety in the workplace. Many of these organizations establish standards or develop **codes,** *which are sets of standards centered on a specific topic.* A well-known example is the **Underwriters Laboratories (UL),** *which have developed codes for safe electrical devices.* As with many of these organizations, UL has no enforcement powers. Figure 2.3 provides a list of some of the organizations that have developed standards or codes related to worker safety. Even though the standards or codes developed by these organizations are recommendations, they are frequently the basis for federal regulations. For example, OSHA requires that protective eyewear meet American National Standards Institute (ANSI) standards for impact and penetration resistance.

Alliance for American Insurers

American Board of Industrial Hygiene

American Council of Government Industrial Hygienists

American Industrial Hygiene Association

American Insurance Association

American National Standards Institute

American Occupational Medical Association

American Society of Mechanical Engineers

American Society of Safety Engineers

American Society for Testing and Materials

Chemical Transportation Emergency Center

Human Factors Society

National Fire Protection Association

National Safety Council

National Safety Management Society

Society of Automotive Engineers

System Safety Society

Underwriters Laboratories

Figure 2.3. Professional Organizations Concerned with Workplace Safety. (From *Occupational Safety and Health in the Age of High Technology,* David L. Goetsch. Prentice Hall, Englewood Cliffs, NJ, 1996.)

Keep in mind that federal agencies and other organizations cannot provide standards to deal with every possible situation. Comprehensive rules would be prohibitively complex and lengthy. This is one reason why responsibility for safety and the use of good judgment is built into every level of the laboratory, from the organization to the individual.

B. Institutional Responsibility

i. OSHA WORKER SAFETY REGULATIONS

As part of the U.S. Department of Labor, the **Occupational Safety and Health Administration (OSHA)** *is the main federal agency responsible for monitoring workplace safety.*

> The mission of the Occupational Safety and Health Administration (OSHA) is to save lives, prevent injuries and protect the health of America's workers. To accomplish this, federal and state governments must work in partnership with the more than 100 million working men and women and their six and a half million employers who are covered by the Occupational Safety and Health Act of 1970.
>
> (From the OSHA web site—http://www.osha.gov/oshinfo/mission.html)

Since the passage of the Occupational Safety and Health Act of 1970, OSHA has both developed and enforced safety regulations that encourage employers to reduce hazards in the workplace. In 1983 OSHA created the **Federal Hazard Communication Standard (HCS)**, *which regulates the use of hazardous materials in industrial workplaces. It focuses on the availability of information concerning employee hazard exposure and applicable safety measures.* The HCS is sometimes called the "Right-To-Know" law, because it mandates that employers fulfill specific requirements for worker safety and knowledge. Employers must provide:

- workplace hazard identification
- a written hazard communication plan
- files of Material Safety Data Sheets for all hazardous chemicals
- clear labeling of all chemicals
- worker training for the safe use of all chemicals

Material Safety Data Sheets will be discussed in detail later. The original HCS applied mainly to manufacturing employers until 1987, when it was more broadly applied to all industries where workers are exposed to hazardous chemicals.

After years of development, OSHA provided a set of general safety regulations specifically aimed at labora-

Table 2.2 *REQUIRED ELEMENTS OF A CHEMICAL HYGIENE PLAN*

A CHP must provide institutional polices or procedures to address each of the following issues:

- general chemical safety rules and procedures
- purchase, distribution, and storage of chemicals
- environmental monitoring
- availability of medical programs
- maintenance, housekeeping, and inspection procedures
- availability of protective devices and clothing
- record-keeping policies
- training and employee information programs
- chemical labeling requirements
- accident and spill policies
- waste disposal programs

A CHP also generally provides information about emergency response plans, as well as the designation of safety officers.

tories. The 1990 **Occupational Exposure to Hazardous Chemicals in Laboratories Standards (29 CFR Part 1910)** *adapts and expands the HCS to apply to academic, industrial, and clinical laboratories.* The main requirement of these standards is the Chemical Hygiene Plan that each institution must develop for every laboratory. The **Chemical Hygiene Plan (CHP)** *is a written manual that outlines the specific information and procedures necessary to protect workers from hazardous chemicals.* Although institutions have considerable latitude in developing their CHP, certain issues must be addressed, as outlined in Table 2.2. Another important provision for laboratories is that all work-related injuries and health problems must be reported to OSHA.

These relatively recent regulations reflect an important and positive shift in attitudes about laboratory safety. OSHA regulations demonstrate that laboratory safety is of sufficient concern to warrant the involvement of the federal government. The regulations require that institutions provide resources that help individuals to understand hazards and work more safely. They encourage workers and their supervisors to take extra time if necessary to perform safety-related tasks. They require institutions (which may or may not have actively promoted safety in the past) to invest in safety equipment, training, and protective clothing. Of course, before the existence of these laws, laboratory workers were aware of hazards and wanted to protect themselves and others from injury. Today, however, there are more resources available, people are more knowledgeable about safety issues, and there is more willingness to institute safety practices than there was even 10 years ago.

ii. ENVIRONMENTAL PROTECTION

In addition to regulations governing worker safety, laboratories must also consider environmental issues. Generation and disposal of toxic biological and chemical wastes is a significant factor in laboratory management. The **Environmental Protection Agency (EPA)** *has primary responsibility for enforcement of laws to prevent environmental contamination with hazardous materials.* Some of the legislative regulations enforced by EPA are the Clean Water Act, the Safe Drinking Water Act, the Clean Air Act, and the Toxic Substance Control Act. The **Toxic Substance Control Act (TSCA)** *was designed to regulate chemicals that pose health or environmental risks.* It has a major impact on the chemical industry and associated laboratories. TSCA establishes chemical inventory and record-keeping requirements, allows the EPA to control or ban hazardous chemicals in commerce, and requires companies to notify the EPA of their intentions to manufacture new chemicals.

Laboratories are also affected by the requirements of the **Resource Conservation and Recovery Act (RCRA)** of 1976, *which provides a system for tracking hazardous waste, including poisonous or reactive chemicals, from creation to disposal.* RCRA provides EPA with authority for regulating transport, storage, emergency procedures, and waste management plans for toxic materials.

C. Laboratory Responsibility

i. RISK REDUCTION IN THE INDIVIDUAL LABORATORY

Although institutions have policies to implement worker safety, these policies are carried out at the level of the individual laboratory. For example, even though there is usually a CHP for the institution in general, each laboratory is required to have its own additions to the CHP to describe hazards and safety measures unique to that laboratory. Commitment to risk reduction should be a clear and constant goal for all members of the laboratory group. The supervisor or mentor is responsible for setting the tone of the daily operations, as well as for modeling safety and good laboratory practices. Every laboratory with more than three people should have a designated safety officer, who is responsible for monitoring safety practices. In addition, this individual (or a committee, in larger groups) will also:

- serve as safety advisor to the laboratory
- ensure that safety procedures are documented and understood
- act as a liaison with the institution's safety officers
- communicate policy changes to co-workers
- coordinate internal safety inspections
- ensure that equipment is properly maintained

- keep records of hazards and problems within the laboratory

ii. LABELING AND DOCUMENTATION

Labeling of hazardous chemicals is required under the HCS, as well as by common sense. This means that all containers of potentially hazardous chemicals must be labeled to an extent that makes them readily identifiable to new workers or to outsiders in case of a spill or emergency. Lack of proper labeling is one of the most common OSHA citations against laboratories. In addition, the Material Safety Data Sheet (MSDS) for every chemical used in the laboratory must be available to all workers. The **Material Safety Data Sheet (MSDS)** *is a legally required technical document provided by chemical suppliers that describes the specific properties of a chemical.* At present there is no required format for the presentation of information in an MSDS, but Table 2.3 provides a list of the information typically present.

A portion of an MSDS (some are several pages long) for the chemical acrylamide is shown in Figure 2.4. Acrylamide is commonly used in biotechnology laboratories to make polyacrylamide gels. It is a powerful neurotoxin (see p. 619), but it can be used safely with proper precautions, which are indicated in the MSDS.

Table 2.3 CONTENTS OF A MATERIAL SAFETY DATA SHEET

An MSDS generally contains the following information, although the format of presentation varies among chemical suppliers.

- Chemical name
- Chemical supplier
- Composition and ingredients information
- Potential health effects
- Exposure levels, with specific concentrations and times
- First aid procedures
- Fire fighting procedures
- Accidental release procedures
- Handling and storage procedures
- Recommended personal protection (clothing and equipment)
- Physical and chemical properties
- Stability and reactivity
- Toxicological information
- Environmental impact
- Disposal recommendations
- Transportation information
- Regulatory information

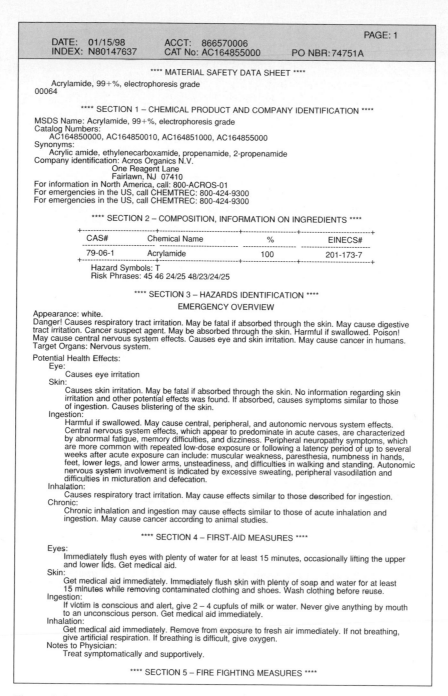

Figure 2.4. Part of a Material Safety Data Sheet for Acrylamide.

In addition to labeling chemicals, laboratory rooms and work areas must also be labeled with signs that indicate hazards. These must provide enough information to alert visitors to take appropriate cautions. Any area that is unsafe for visitors without training or specific precautions should be labeled with a "Do Not Enter" sign.

Documentation of laboratory procedures is an essential part of good laboratory practice (see Chapter 5). Every procedure that is repeated should be available in written form to allow consistency among laboratory personnel. One task that is extensively used in industry to provide both safety guidelines for personnel as well as compliance with OSHA regulations is the preparation of a Job Safety Analysis. A **Job Safety Analysis (JSA)** *is a detailed analysis of each step in a procedure, identifying hazards and outlining accident prevention strategies.* An example is provided in Figure 2.5. An effective JSA is usually prepared jointly by safety officers and individuals who perform the procedures, and can be used for both training and documentation of laboratory safety measures.

Gene Technology, Inc.
Industrial Park
Normal, IL 55555

Job Safety Analysis

Job title: _____

Analysis by: _____

Approved by: _____

Date: _____

Step-by-step sequence	Hazard	Accident prevention procedure

Figure 2.5.   Sample Job Safety Analysis Form. (Adapted from *Occupational Safety and Health in the Age of High Technology,* David L. Goetsch. Prentice Hall, Englewood Cliffs, NJ, 1996.)

iii. HOUSEKEEPING

Many hazards can be eliminated or reduced by the simple policy of good housekeeping. The majority of routine maintenance and cleaning in laboratories must be performed by the personnel who are familiar with the hazards present. In most institutions, outside staff do not clean bench tops or equipment. This prevents accidental exposure of staff to hazards that are unknown to them. It protects experimental materials from inadvertent contamination or disposal. It also means, however, that the laboratory workers themselves are responsible for maintaining a clean, orderly work space, Figure 2.6.

Safety, as well as good laboratory practice, requires that clutter on bench tops and shelves be kept to a minimum. This lessens the risk of reagent mix-ups and potential degradation of old chemicals. Fewer objects in

a work area provide fewer opportunities to accidentally contaminate equipment or containers that are not part of the current experiment. Developing a habit of regularly cleaning your work area will avoid situations where a massive effort must be undertaken to clear working space.

Laboratory bench tops should be routinely cleaned both before and after work sessions. This will guard against any residual chemical or biological contamination that may be present from a previous user, as well as prepare the space for the next user. Cleaning techniques should be based on a worst-case scenario of the hazardous contaminants that may be present. Specific cleaning methods are discussed in Unit VIII. Cleanup should always include decontaminating and rinsing glassware in preparation for dishwashing procedures. In most institutions, dish-

Figure 2.6.   I Know I Left My Experiment Here Somewhere.

Table 2.4   *BASIC EMERGENCY PRECAUTIONS FOR THE LABORATORY*

The following list provides some of the fundamental preparations that every laboratory should have in place for emergency situations.

- Everyone in the laboratory should be aware of basic emergency procedures

- There should be at least one person trained in first aid and CPR present at all times

- The first aid kit must be readily accessible and fully stocked

- All required protective devices, such as fire extinguishers and eyewash stations, must be well marked and easily accessible

- Emergency telephone numbers and instructions should be prominently posted by every telephone

- Evacuation routes should be kept clear of boxes or clutter

washing facilities will not accept laboratory items that still contain obvious residues of experimental materials.

Even though institutional regulations will require periodic laboratory inspections by external safety officers, a laboratory should not wait for these inspections to identify problems. Regular internal inspections noting housekeeping problems and potential risks can provide timely information about unsuspected hazards as well as the current state of compliance with safety regulations. It is particularly important to perform regular inspections of specific types of hazards, such as gas cylinders and chemical storage and labeling. In this way, small problems can be remedied before accidents occur.

Every laboratory needs to have a system of waste collection and disposal for specific hazards. These may include broken glass and other sharp objects, solid and liquid radioactive waste, and biologically contaminated materials. Institutions must comply with a variety of regulations concerning waste disposal, including detailed labeling of hazardous contents. Proper waste disposal is difficult and extremely expensive for unidentified materials; therefore, every laboratory should have a system for labeling waste at its source. Environmental concerns as well as common sense dictate that laboratory waste should be minimized when possible. It is also important that hazardous waste not be mixed with regular trash to avoid serious injuries for housekeeping staff, who may be unknowingly exposed to dangerous chemicals or biological contamination.

iv. EMERGENCY RESPONSE

Although all institutions have emergency plans, individual laboratories also need to prepare for potential acci-

dents and emergencies. Some of the basic preparations needed are shown in Table 2.4. Everyone in the laboratory ideally should be trained in basic first aid and CPR. The laboratory or a nearby area must be equipped with basic safety items such as fire extinguishers and fire alarms, first aid kits, chemical spill kits, a safety shower, and an eye wash station. Each worker needs to know the evacuation plan for the laboratory and the location of emergency telephone numbers and procedures. Remember that most accidents happen very quickly; there is usually no time for carefully planning a response after the accident occurs. This is the reason that emergency procedures need to be understood and practiced in advance.

Laboratory accidents, no matter how minor, must be reported as soon as possible after the occurrence. There is a natural tendency for many people to cover up minor accidents, such as cuts or spills, from either embarrassment, a dread of paperwork, or concerns about being blamed for the incident; however, these are not adequate excuses for secrecy. Accident reporting is a legal requirement and is also essential in order to prevent repeated problems. There are numerous studies that show that people learn to avoid accidents by being informed about other accidents, as well as near misses. If you are injured on the job and require medical treatment, you need to fill out a workers' compensation form. **Workers' compensation** *is a no-fault state insurance system designed to pay for the medical expenses of workers who are injured on the job, or develop work-related medical problems.* It provides a mechanism for your employer to pay for your medical treatments without the necessity of a lawsuit.

D. Personal Responsibility

i. LABORATORY SAFETY AND COMMON COURTESY

Safety is a matter of personal responsibility for every individual who works in a laboratory. These environments are hazardous by nature, but the actual risk is determined to a large extent by the actions of individuals. It is essential to accept responsibility for yourself and those who work around you. We all have a tendency to assume that accidents happen to "other people," who are probably not as smart or careful as ourselves. The fact is, accidents are by definition unexpected and sudden, and they are frequently the result of multiple mistakes or oversights. Be aware at all times of situations that could be modified to provide a safer working environment, and report any problems to a supervisor immediately. When in doubt about a situation, always ask someone with experience.

Good laboratory technique and consideration for co-workers are essential parts of laboratory safety. Proper labeling of all materials notifies co-workers of hazardous chemicals. Prompt clean up of spills, reduction of clutter, and the return to storage of unused chemicals and equipment frees working space and reduces accidental spills and mistakes. Although you may deliberately choose to expose yourself to a known risk under some circumstances, it is inexcusable to expose others who have not made that choice.

It is important that you and each of your co-workers know how to respond to an emergency. Be familiar with the emergency procedures established for the laboratory and know what to do if you or a co-worker needs medical assistance. Acquaint yourself with basic first aid and the location of the first aid materials. Know where to find and how (and when) to use the fire extinguishers, and know the evacuation plan for the building in case of fire or other physical emergency (see Case Study 1).

One of the primary guidelines for behavior in the laboratory is often referred to as the "rule of reason," where an individual asks themselves how reasonable an action is in the current situation, Figure 2.7. Considering whether an action makes sense under the circumstances frequently suggests an appropriate course of action. When tempted to cut corners, think about what your opinion would be of a co-worker who did the same thing.

ii. PERSONAL HYGIENE

This does not refer to bathing; rather, it refers to personal habits that may increase your risk of hazard exposure. Never eat, drink, smoke, chew gum, or apply cosmetics in a laboratory. Even though you are "not working with anything dangerous," you can never be sure about other hazards in the immediate area. Never use the laboratory or your pockets to store food, beverages, or anything that will be consumed. If you do smoke (outside the laboratory, of course), be aware that cigarettes in open packs can absorb vapors from volatile chemicals. Develop the habit of washing your hands every time you move away from your lab bench, and at regular intervals while you are working.

Drink only from hall fountains. Water from laboratory faucets may not be **potable** *(suitable for human consumption)*, due either to water quality or to contamination of the faucet with hazardous materials. For the

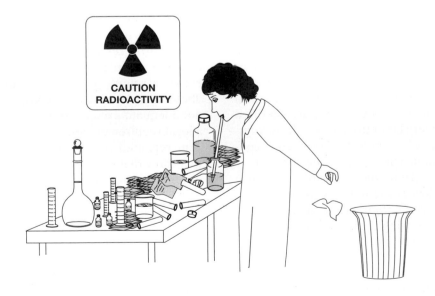

Figure 2.7. Inappropriate Laboratory Practices.

same reasons, do not consume ice taken from laboratory machines. It is impossible to know exactly what types of containers may have come into contact with the ice. Avoid storing personal items such as coats in the laboratory, where they may become contaminated with hazardous materials.

Long hair must be tied back in the laboratory to prevent a variety of problems. Hair has an unfortunate tendency to fall forward, which can lead to chemical or biological contamination of the hair, or, even worse, ignition due to contact with an open flame. A greater danger that is not often considered is that most people with long hair frequently push the hair back from their faces without conscious thought, leading to potential skin and hair contamination with hazardous substances from a gloved hand. A related issue is the wearing of beards in laboratories where biological agents are used. Many microbiological standards warn against beards because of the possibilities of experimental and personal contamination, and laboratory situations that require the use of fitted respirators may preclude beards entirely. This is an issue that is best judged in the individual situation.

Another somewhat controversial issue is the wearing of contact lenses in the laboratory. Many safety guidelines state that contact lenses should never be worn in the laboratory because of the danger of chemical splashes to the eye and an increased potential for eye damage in accidents. The American Chemical Society Committee on Chemical Safety, however, has concluded that contact lenses are suitable for most laboratory situations, as long as standard eye protection is used (*Safety in Academic Chemistry Laboratories*, p. 3. The American Chemical Society Committee on Chemical Safety. American Chemical Society, Washington, D.C., 1995). Remember that contact lenses offer no protection against chemicals or foreign objects, and always wear eye protection gear, such as safety glasses or goggles. Never insert or remove contact lenses from your eyes in or near the laboratory. Do not adjust an uncomfortable lens with a possibly contaminated finger. Be sure that your co-workers are aware that you wear contact lenses, in case of an emergency. In case of eye contamination, lenses should be left in place and emergency workers notified of their presence.

iii. WORK HABITS

Many laboratory accidents are the result of simple human carelessness, coupled with the fatigue and distractions that everyone experiences. The obvious prevention measure is to avoid working when tired or distracted, but this of course is not always practical advice. You can address this problem, however, by personally acknowledging your temporarily diminished state of attention, slowing down, and taking extra pre-

cautions. Take a few additional moments to plan what you are doing and anticipate any potential problems before they occur.

Do not work alone in the laboratory. Let your co-workers know if you will be in an isolated part of the building, such as a coldroom, for extended periods of time. It is a frequent temptation to finish experiments in the evening or on weekends when the lab is quiet, but this can be a dangerous strategy in the event of a serious accident (see Case Study 2).

CASE STUDY 2
The Dangers of Working Alone

In 1994, a senior researcher at a major university nearly died while working alone at night. While performing a familiar procedure involving small solvent volumes, a distillation flask exploded, starting a fire. Because he was not wearing a lab coat, his shirt ignited. Even worse, a flying piece of glass severed a major artery in his arm. He collapsed from loss of blood and shock before reaching the emergency phone. His life was saved because a colleague was in a nearby office, heard the fire alarm, called 911, and then performed first aid until the ambulance arrived. This university officially prohibited working alone in laboratories, but the prohibition was widely ignored. Following this incident, an enforced policy of using a "buddy system" for work at odd hours was developed. Each worker must have at least one co-worker in the immediate vicinity, in case of emergencies.

Another time for concern about hazards is when passing from the laboratory into public areas and back. Do not wear lab apparel from work, especially lab coat and gloves, into public areas. You may know that your hands and coat are uncontaminated by hazardous materials, but the people you encounter will not. If you do feel a need to wear a lab coat in public, keep a clean coat for this purpose (although this will not reassure the people who see you in the lab coat). Never wear safety gloves outside the laboratory and never handle common use items such as telephones, radios, or light switches while wearing gloves. This is an easy way to spread chemical or biological contamination. In labs where radioactive chemicals are used, radiation inspections specifically include checking door knobs, equipment controls, and laboratory desks for radioactive contamination.

If you must transport hazardous materials through public areas, handle them with one gloved hand, leaving the other hand ungloved for opening doors. Samples should be carried in sealed, double containers, to prevent spills. Try not to hurry through halls and around corners, to avoid collisions with co-workers. Many experiments have ended up on the floor (or worse, on a person) because of sudden hallway encounters.

Table 2.5 PERSONAL LABORATORY SAFETY PRACTICES

• Be sure that you are informed about the hazards that you encounter in the laboratory

• Be aware of emergency protocols

• When in doubt about a hazardous material or a procedure, ask

• Use personal protective wear such as lab coats and safety glasses at all times

• Do not eat, drink, chew gum, or smoke in the laboratory

• Avoid practical jokes or horseplay, which can unintentionally create a hazard

• Use gloves whenever in doubt (see Chapter 28 for guidelines on proper use of gloves)

• Wash your hands regularly, regardless of whether your work requires gloves

• Always wash your hands thoroughly before leaving the laboratory

• Read the labels of chemicals and reagents carefully

• Read procedures before performing them and visualize hazardous steps

• Minimize use of sharp objects and be sure that you properly dispose of them

• Clean up spills and pick up any dropped items promptly

• Label everything clearly

• Use a fume hood for any chemical or solvent that you can smell, that has known toxic properties, or that is unfamiliar to you

• Record everything in your lab notebook

• Always report accidents, however minor, immediately

Table 2.5 provides a set of general guidelines for safe laboratory practices for the biotechnology worker.

III. SAFETY TRAINING

A. Importance of a Safety Training Program

It is well documented that most injuries occur to workers with less than 2 years experience on the job. This points to the need for safety training programs aimed at all new employees. It is essential that all laboratory personnel participate in these programs because it is not possible to supervise lab work at every step.

Training program requirements originate with government agencies, and are then applied throughout institutions. The execution of these programs requires the cooperation of both individual laboratories, and the workers within them. The key aims of any safety training program are to allow workers to:

• understand the risks inherent in their jobs

• recognize their personal susceptibility to accidents

• learn about preventive measures that reduce the risk of accidents

• accept personal responsibility for accident prevention

Safety training is not just for new employees. Refresher courses should be available for more-experienced lab workers, allowing them learn about new policies and resources, and practice safety skills that have not yet been needed.

B. Contents of a Training Program

It is important to remember that even though hazards are always present in the laboratory, the risks to human health can be controlled to some extent. A basic safety training program should answer two questions for each worker:

• What are the specific hazards of this job?

• How can I minimize risks to myself and others?

Minimizing safety risks requires the development and practice of rules that can be applied to the situations that are expected in the laboratory. Many workers complain about "ten pound" institutional safety manuals that attempt to regulate virtually every job process, including breathing rates. These manuals do not tend to be effective in motivating worker compliance. In this situation, individual laboratories may need to develop a more focused and specific safety manual which directly addresses the needs of the lab workers. Figure 2.8 provides some guidelines for the development of effective safety manuals.

It is the responsibility of every individual to be certain that they have received adequate training to perform their jobs safely. Ask for a demonstration of equipment

■ Minimize the number of rules to the extent possible. Too many rules can result in rule *overload.*

■ Write rules in clear and simple language. Be brief and to the point, avoiding ambiguous or overly technical language.

■ Write only the rules that are necessary to ensure a safe and healthy workplace. Do not *nitpick.*

■ Involve employees in the development of rules that apply to their specific areas of operation.

■ Develop only rules that can and will be enforced.

■ Use common sense in developing rules.

Figure 2.8. Guidelines for Developing Safety Rules. (From *Occupational Safety Management and Engineering*, p. 406. Willie Hammer. 4th edition, Prentice Hall, Englewood Cliffs, NJ, 1989.)

Table 2.6 *ELEMENTS OF A SAFETY TRAINING PROGRAM*

Safety training programs vary widely among institutions. Employees are required by federal law to attend these programs and most institutions require employees to sign a statement indicating that they have received training. Every program should include the following topics:

- Institutional policies—hazard information, inspections, reporting systems, waste disposal
- Safety rules—practices, manual, signs, and labels
- Location and use of protective equipment—fire extinguishers, hoods, clothing, safety gear
- Emergency procedures—alarms, injuries, medical assistance
- Chemical hazard awareness, including:

  location of MSDS reference materials

  symptoms of chemical exposure

  detection methods for chemical exposure

  protective mechanisms

  emergency procedures

In addition, there are many laboratory-specific safety issues that may need to be addressed, such as radiation and biological safety.

use or procedures that are unclear. Be sure you know the location of all exits, safety equipment, and emergency contact numbers. Table 2.6 lists some of the elements that should be covered in a safety training program.

DISCUSSION PROBLEMS

1. Consider a laboratory situation where large amounts of a toxic and flammable solvent are required for experimental work. How would you approach risk reduction for this hazard?

2. Analyze Case Study 1 and list the emergency procedures that would apply to this situation.

3. Analyze Case Study 2. What basic safety precautions were ignored by the researcher involved?

4. Think of an accident that you were involved in or witnessed. What safety standards might have prevented this accident?

## CHAPTER 3

Biotechnology and the Regulation of Medical and Food Products

I. INTRODUCTION

A. A Brief History

As discussed in Chapter 1, the most commercially successful biotechnology products to date include biopharmaceuticals and diagnostic materials for medical use. There are also many agricultural and food-related products emerging from biotechnology companies. Medical products and foods are regulated by the government, and these regulations have a profound effect on the biotechnology workplace (as is further discussed in Unit II). This chapter introduces issues relating to government regulation of medical and food-related products, the relationship between the government and the biotechnology industry, and the history of that relationship. For simplicity, this chapter examines regulation primarily in the United States.

The regulated characteristics of products relate to safety, honesty (as in labeling), and, particularly for medical products, effectiveness and reliability. For example, a drug product must be correctly labeled and a food or drug product should not be contaminated with harmful microorganisms. The use of a mislabeled drug, or consumption of a contaminated food, can cause death. The individual consumer cannot determine whether a drug is properly labeled or whether meat purchased in the supermarket has a suitably low level of microbial contamination. The public, therefore, looks to the government to enact and enforce laws that provide protection from unsafe or ineffective food and drugs.

Although we take for granted that the government regulates the production of drugs and foods, this was not always the case, as is illustrated dramatically in the history of drug and food regulation in the United States. Much of this history involves tragedies and abuses that led to increasingly stringent regulation of food and drug production, labeling, and testing. A few important incidents and enactments are:

- Ten children died in 1901 from a diphtheria vaccine that was contaminated by tetanus from horse serum. This resulted in the **Virus, Serum and Toxin Act of 1901.**

- Upton Sinclair recorded filthy conditions and unacceptable practices in the food industry in his novel, *The Jungle.* As a result, the original **Food Drug and Cosmetic Act (FDCA) was passed in 1906 to prevent the commerce of unacceptable food and drugs.** The 1906 FDCA authorized regulations to ensure that pharmaceutical manufacturers did not adulterate or mislabel their products, but it did not deal with the safety or effectiveness of drugs.

- In 1927 a separate law enforcement agency was formed, first known as the Food, Drug and Insecticide Administration, and then, in 1930, as the **Food and Drug Administration (FDA),** to enforce legislation relating to food and drugs.

- In 1937, a batch of sulfanilamide was accidentally dissolved in the industrial solvent, diethylene glycol. There were 358 poisonings and 107 deaths, mostly children. As a result of this incident, the **revised Food, Drug and Cosmetic Act was passed in 1938, which required drugs to be tested for safety before release.**

- In 1955 some children vaccinated with polio vaccine contracted paralytic polio. Fifty-one people were paralyzed and ten died. The problem was traced to one manufacturer who apparently did not properly inactivate the virus used to make the vaccine. This incident, and others like it, led to increased factory inspections and testing of the safety of products before their release to the public.

- One of the great successes of the FDA occurred in the early 1960s. At that time the drug thalidomide was commonly prescribed for insomnia and nausea in pregnant women in Europe. This drug, unfortunately, led to the birth of thousands of children without arms or legs. Thalidomide was not used commercially in the United States because Dr. Frances Kelsy, who was then chief of the FDA, refused to accept it for use in the United States until it was proven safe. News of the thalidomide tragedy influenced the U.S. Congress in 1962 to pass the **Kefauver-Harris Amendments, which required that drugs be proven to be <u>both</u> safe and effective before release.**

- **In 1963 the first set of Good Manufacturing Practices (GMP) regulations** were published in the Federal Register. These regulations were intended to guide companies in the production of safe and effective drugs. (Good Manufacturing Practices are discussed in Unit II.)

- A number of injuries and deaths in the 1960s and 1970s caused by contaminated products led to the **revised GMPs in 1976.** These regulations included requirements for standard operating procedures, validated systems (validation in pharmaceutical production is discussed in Chapter 7), and extensive documentation.

- Inspections of pharmaceutical animal testing laboratories revealed that toxicology studies supporting new drug applications were poorly conceived and improperly conducted. These inadequacies led to the promulgation of the **Good Laboratory Practices Regulations in 1976,** whose goal is to assure the quality of data submitted to FDA in support of the safety of new products.

- In 1982 seven people died from taking Tylenol® that had been maliciously laced with cyanide after its distribution. This resulted in the introduction of **tamper evident packaging.**

- In 1982 insulin produced by recombinant DNA methods was approved for market and sale.

- In 1986 the first monoclonal antibody product was approved for market and sale.

- In 1987 the first recombinant DNA product produced in a mammalian cell culture line was approved for market and sale. This product is tissue plasminogen activator (TPA), used in the treatment of heart attack victims.

- In 1989 a scandal relating to the manufacture of generic drugs rocked the industry. FDA investigators discovered that a few generic drug manufacturers were guilty of illegal practices, fraud, and noncompliance with GMP, leading to defective drug products. These deficiencies led to increased inspections and FDA oversight of the industry.

This brief historical summary illustrates that the regulation of food and medical products is generally driven by a tragedy, a problem, or an advance in science and technology. The public and their elected officials respond by enacting a law(s) that is intended to reduce risks while maximizing benefits. A governmental regulatory agency

REGULATION OF PHARMACEUTICAL PRODUCTS

Congress enacted the Food, Drug, and Cosmetics Act (FDCA) which states that food and drugs should not be "adulterated" and empowers the FDA to enforce food and drug laws.
One of the key clauses in the FDCA is "A drug is adulterated . . . if the methods used in . . . its manufacture . . . do not conform to . . . current good manufacturing practice . . ."

↓

The FDA devises regulations that outline in more detail the meaning of "Good Manufacturing Practices" (GMP). These are published in **The Code of Federal Regulations (CFR)**, a document published annually, which contains all federal regulations.

↓

The FDA periodically publishes **Guidelines** to more fully interpret technical details relevant to GMP. The Guidelines do not have the force of law, but a company should follow them or be prepared to justify any deviation. "**Points to Consider**" are FDA documents that discuss issues relating to innovative technologies.

Figure 3.1. An Example of Regulation. Congress passed an act and authorized a federal agency to interpret and enforce it. The agency devises regulations and enforcement strategies to bring about compliance with the intent of Congress.

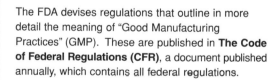

Federal Food, Drug, and Cosmetic Act
Adulterated Drugs 501 [351] *A drug or device shall be deemed adulterated—(a) (1) if it consists in whole or in part of any filthy . . . substance; or (2) (A) if it has been prepared, packed, or held under insanitary conditions whereby it may have been contaminated with filth . . . or (B) if it is a drug and the methods used in, or the facilities or controls used for, its manufacture, processing, packing, or holding do not conform to or are not operated . . . in conformity with current good manufacturing practice to assure that such drug meets the requirements of the Act as to safety and has the identity and strength, and meets the quality and purity characteristics, which it purports or is represented to possess . . .*

Figure 3.2. A Brief, but Important, Excerpt from the FDCA of 1938. This act forbids adulteration of food and drugs and authorizes the regulation of food and drug quality in the United States.

is empowered to interpret and enforce the law through a system of regulations, Figure 3.1. For example, the **Food and Drug Administration** *is a regulatory agency responsible for ensuring the safety, effectiveness, and reliability of medical products, foods, and cosmetics as set forth in various federal legislative acts, most notably, the Food, Drug, and Cosmetic Act of 1938.* A brief quote from that act is shown in Figure 3.2. Regulations change as the legislature, regulatory agencies, the public and industry respond to events and to one another.

B. Introduction to Regulatory Issues

The regulation of foods and medical products is complex and is often controversial. There is general agreement that these products should be safe and effective (or nutritious); however, the requirements of safety and efficacy must be balanced against the cost of attaining these goals. This cost may be financial; for example, it is expensive to test drugs and to inspect foods. There are other costs as well. It takes years to test a new drug to ensure its safety and effectiveness; in the meantime, lives might be lost that the drug could have saved. There is a cost if marginal food must be discarded. There is a cost if agricultural productivity is diminished because potentially toxic herbicides are not used. Thus, there is a conflict between ensuring safety and reducing costs. Another way to frame this conflict is to speak of "risk versus benefit." The field of regulations requires a continuing effort to find strategies that balance safety against cost and maximize benefits while reducing risk. Different groups often disagree as to how to achieve this balance.

New and innovative products, such as those made by the methods of biotechnology, promise some benefits and pose some risks. Modern biotechnology involves an assortment of methods used to make a variety of products. In many cases, it is the methods that are innovative; biotechnology products are often the same as, or are similar to, products made using traditional methods. For example, prior to 1982 insulin used to treat diabetics was extracted from animal organs. In 1982, human insulin, produced by bacteria containing the human insulin gene, was approved for commercial sale. In both cases the product is insulin. Regulators have therefore had to decide whether biotechnology products should be regulated differently than other products (because their method of production is innovative) or the same as traditional products (because they are often similar). This is sometimes called the "product versus process" controversy. At the heart of this debate is disagreement as to whether the very nature of biotechnology processes makes products derived by these methods inherently different from, and potentially more risky than, traditional products.

One of the events in the "product versus process" discussion was a policy statement issued in 1990 by President George Bush, "Four Principles of Regulatory Review for Biotechnology" (found in *Exercise of Federal Oversight Within Scope of Statutory Authority: Planned Introductions of Biotechnology Products Into the Environment,* Office of Science and Technology Policy, Executive Office of the President, 1992). This statement set forth four broad principles of regulation for biotechnology. The first of these principles is:

> Federal government regulatory oversight should focus on the characteristics and risks of the biotechnology product—not the process by which it is created. Products developed through biotechnology processes do not *per se* pose risks to human health and the environment; risk depends instead on the characteristics and use of individual products. Biotechnology products that pose little or no risk should not be subject to unnecessary regulatory review during testing and commercialization. This allows agencies to concentrate resources in areas that may pose substantial risks and leaves relatively unfettered the development of biotechnology products posing little or no risk.

As we will see later in this chapter, this principle has affected U.S. policy toward biotechnology, particularly in the pharmaceutical area. It states that products are to be regulated on the basis of their risk, not according to how they are produced. Thus, although there has been disagreement as to the risks associated with biotechnology products, in many cases their regulation has not involved a radical departure from established regulatory procedures. Biotechnology, particularly in the pharmaceutical area, has generally been regarded as an

innovation; one of many innovations that can be regulated within existing structures.

II. THE REGULATION OF PHARMACEUTICAL AND MEDICAL PRODUCTS IN THE UNITED STATES

A. The Role and Organization of the Food and Drug Administration

Much of the story of medical product regulation in the United States revolves around the relationship between the FDA and industry. For historical reasons, there are two related, but separate, regulatory systems within the FDA for pharmaceuticals. The first system is for products that are formally termed *drugs*. Drugs include antibiotics and other products of microbial metabolism produced by fermentation, compounds produced by chemical synthesis, and hormones extracted from animal tissues, such as insulin and human growth hormone. The second system is for products, termed *biologics,* including products extracted from human or animal plasma, vaccines produced from microbes and viruses, and allergens extracted from natural sources. Products classified as drugs are regulated by the FDA **Center for Drug Evaluation and Research (CDER)**, whereas biologics are regulated by **The Center for Biologics Evaluation and Research (CBER)**. Biotechnology products often are regulated as biologics. Note also that some biotechnology products, particularly those used for diagnostics, are regulated as medical devices by **The Center for Devices and Radiological Health (CDRH)**.

Drugs and biologics must be produced using GMP and are subject to scrutiny for safety and efficacy. There are, however, some differences in the details of regulation of drugs and biologics. For example, the regulatory submission forms for drugs and biologics differ. Table 3.1 lists products that are regulated by FDA and the division of FDA involved in their regulation.

B. The Life Cycle of a Pharmaceutical Product

Drugs and biologics have a long life cycle involving extensive development, testing, an approval process, marketing, and post-marketing surveillance. Only about 1 in 10 potential drugs performs successfully during testing and is actually marketed.

The idea for a product is researched in the laboratory during the early stages of research and development. Until recently, FDA was not involved in the early R&D phases of a product. FDA, however, is beginning to inquire about the results, documentation, and experiments that were performed during R&D because the ultimate quality of a product depends on the foundation built during development.

Table 3.1 *The Organization of FDA*

CDER, Center for Drug Evaluation and Research

Reviews applications for new drugs (such as antibiotics and hormones extracted from animals), and regulates their manufacture, labeling, and advertising.

CBER, Center for Biologics Evaluation and Research

Reviews applications for new biologics (such as vaccines produced from microbes and viruses, and allergens extracted from natural sources), and regulates their manufacture, labeling, and advertising.

CDRH, Center for Devices and Radiological Health

Regulates medical devices (such as ultrasonic cleaners for medical instruments, diagnostic kits used in medical laboratories, and ventricular by-pass devices) and many, but not all, *in vitro* diagnostic kits.

CFSAN, Center for Food Safety and Applied Nutrition

Promotes and protects the public health and economic interest by ensuring that foods are safe, nutritious, and honestly labeled.

Note: The Food and Drug Administration has an excellent world wide web site at: http://www.fda.gov.

If a potential product shows promise in the laboratory, toxicity tests are performed in animals to determine whether the substance is safe for human testing. Animal tests are also used to evaluate how much of the substance is absorbed by the blood, how the substance is metabolically altered in the body, the toxicity of metabolic by-products, and how quickly the substance and its by-products are excreted. Before the mid-1970s, the conduct of animal testing was not scrutinized by the FDA. In 1975, however, FDA inspections of several pharmaceutical testing laboratories revealed poorly conceived and carelessly executed experiments, inaccurate record-keeping, poorly maintained animal facilities, and a variety of other problems. These deficiencies led the FDA to institute the **Good Laboratory Practice* regulations (GLPs)** *to govern animal studies of pharmaceutical products.* GLPs require that testing laboratories follow written protocols and standard operating procedures (SOPs), have adequate facilities and equipment, provide proper animal care, properly record data, have well-trained and competent personnel, and conduct high-quality, valid toxicity tests.

If a potential new product appears to be safe in animal studies, then scientists prepare a plan to investigate the product in human volunteers. The company submits their plan to the FDA, including a description of the product, the results of animal tests, and the plans for further testing. The FDA then decides whether the company's materials are sufficiently complete that the company can begin testing the product in humans.

Good Clinical Practices (GCPs) *relate to the performance of clinical trials of drug safety and efficacy in human subjects.* GCPs aim to protect the rights and safety of human subjects and to ensure the scientific quality of the studies. Clinical trials are conducted in stages, each of which must be successful before continuing to the next phase.

Phase I clinical trials *are the first introduction of the proposed drug into humans.* During Phase I trials, the safety of the drug is carefully evaluated and its metabolic and pharmacologic properties in healthy humans are determined. If a drug meets the safety requirements at this phase and appears to have the desired properties, then it enters Phase II clinical trials. **Phase II trials** *are performed on a small number of diseased patients to determine the drug's efficacy.* If the drug continues to meet safety requirements and demonstrates efficacy at Phase II, then it progresses to a broader **Phase III trial** *involving more patients.* At this point, the safety and efficacy of the drug continue to be evaluated, dosages are determined, a risk versus benefit analysis is performed, drug interactions are explored, and other data are collected.

If a drug passes all three phases of testing, then the company may submit an application to the FDA that provides convincing evidence that the new drug or biologic is safe, reliable, and effective. Note that FDA itself does not actually test each drug product; rather, FDA expert reviewers examine test results and information submitted by the company to determine whether a product is acceptable. If the review team decides the evidence is sufficient, then the new product is approved (if it is a drug) or is licensed (if it is a biologic) by FDA and can be manufactured for commercial sale. This is a major accomplishment and is the beginning of the period where the company can market and sell its product. There may continue to be clinical testing of the product even after it has entered the market to provide ongoing evaluation of the product's safety and effectiveness and to allow the drug to be labeled for additional uses.

As a therapeutic agent progresses through these various phases, the requirements for its manufacture become increasingly stringent. Once in commercial production, the manufacture, labeling, packaging, shipping, storing, quality control, and marketing of the product must meet all GMP and other relevant requirements.

An important part of the enforcement of GMP is unannounced inspections of pharmaceutical facilities

---

*"Good Laboratory Practices" (GLP), when used in reference to pharmaceuticals, is a very narrow term and only refers to practices involving animal testing. These GLPs are enforced by the FDA. The Environmental Protection Agency, EPA, also mandates "Good Laboratory Practices." EPA's Good Laboratory Practices guide investigators who are studying the health and environmental effects of agrochemicals. The GLPs from FDA and EPA are not identical, but they do serve the same overall purpose of ensuring trustworthy test results. The term *good laboratory practices (glp)* is sometimes used much more broadly to refer to good practices and procedures in any type of laboratory.

by FDA inspectors. Inspectors note practices that violate GMP requirements on a form called a "483." In general, companies are able to correct deficiencies noted by inspectors. The FDA occasionally seizes and destroys products that it believes were not properly manufactured, or levies fines against a company. In more extreme cases, where individuals or companies act deceitfully or refuse to comply with regulations, individuals may be charged as criminals and may be imprisoned if convicted.

FDA thus plays many roles in the development of a pharmaceutical product. FDA has a "gate-keeping" function in approving or disapproving various applications. The agency controls the practices for testing, manufacturing, labeling, and marketing products. Thus, FDA's regulatory functions include:

- Establishing rules that ensure consumer protection
- Communicating the rules to industry and educating the public
- Monitoring industry compliance with the rules
- Performing scientific review and providing information and support to ensure that the rules are up-to-date
- Enforcing applicable laws and regulations

C. International Harmonization Efforts

An important trend in the 1990s has been the increasing internationalization of commerce. When companies begin to expand to market places outside their native country, however, they often encounter difficulties meeting the complex and varied regulatory requirements of different nations. This is particularly the case for medical products. As you can see from our discussion of FDA and its roles, the regulation of medical products in the United States is complex and stringent. Medical products are also stringently regulated in many other countries. It is difficult for a company to discover that, although their product has gone through an expensive, extensive regulatory process, it cannot be sold in another country because that country's regulatory process is different. Medical products, therefore, have been a major focus of international harmonization efforts. One facet of this effort has been in the field of medical devices, such as pacemakers, medical instruments, and diagnostic kits. In 1996, FDA announced modifications of their requirements to make them more consistent with European requirements (Federal Register: October 7, 1996, Vol. 61, Number 195 pp. 52601-52662).

Another important component of the global harmonization effort is the work of **The International Conference on Harmonisation of Technical Requirements for Registration of Pharmaceuticals for Human Use (ICH)**. ICH is an organization that brings together the regulatory authorities of Europe, Japan, and the United States, as well as experts from the pharmaceutical industry in the three regions to discuss scientific and technical aspects of pharmaceutical product regulation. ICH makes recommendations on ways to achieve greater harmonization through the establishment of technical guidelines, a number of which are relevant to biotechnology. ICH guidelines cover such areas as validation of analytical procedures, viral safety evaluation of products, and genetic stability of biotechnology products. The biotechnology industry is therefore increasingly looking to ICH for regulatory guidance.

III. BIOTECHNOLOGY PRODUCTS FOR MEDICAL USE

A. Recombinant DNA Production Systems for Pharmaceuticals

As we saw in Chapter 1, organisms that have been genetically modified by modern biotechnology methods can be used in production systems for making biopharmaceutical products. An example is the production of human insulin by bacterial host cells that have been genetically modified to contain the human insulin gene. Biotechnology production systems begin with cultured bacteria, yeast, insect, hybridoma, or mammalian **host cells** *that are modified so that they contain a* **genetic construct**. The **genetic construct** *includes the gene that codes for a protein of interest and the surrounding DNA sequences that are necessary for the gene to function.* Under appropriate conditions, the host cells express the gene and produce the protein of interest.

Before the advent of modern methods of genetic manipulation, the FDA had been regulating "traditional" biotechnology products, such as hormones and antibiotics derived from microbial fermentation. At that time, the use of cultured mammalian cell lines in pharmaceutical manufacturing was forbidden. This was because these cell lines are similar to cancer cells in that they repeatedly divide and are not subject to normal controls on growth. It was feared that the use of such continuous cells lines might introduce a "factor" into the products that would be carcinogenic in patients. Furthermore, it was feared that mammalian cells used in production systems might harbor viruses pathogenic to humans.

In the mid-1970s it became clear that recombinant DNA methods would impact the production of pharmaceutical products. In addition, it was determined that mammalian cell lines were necessary for producing some of these biotechnology products that could not be made by bacteria. It was therefore imperative that the FDA determine whether products made by new biotechnology methods were inherently more risky than traditional products and whether they would require additional or different regulation.

A number of scientific advances in the 1970s led to the FDA's acceptance of products made using modern biotechnology methods. Experiments were performed that demonstrated that genetically manipulated organisms could be handled safely. Effective methods of inactivating viruses were developed. The use of continuous cell lines was first tested in other countries without adverse effects, and studies of cancer did not reveal a carcinogenic "factor" in cell lines.

In 1974, the **National Institutes of Health (NIH)**, *a federal agency responsible for funding and overseeing research,* assembled an expert group to explore the safety of recombinant DNA technology. This committee, **The Recombinant DNA Advisory Committee (RAC)**, reviewed and interpreted what was then known about the safety and risk of genetic manipulation methods. The result of their work was a set of guidelines for conducting experiments involving recombinant DNA. These guidelines, called **Guidelines for Research Involving Recombinant DNA Molecules (NIH Guidelines)** were first published in 1976, and they have been revised several times since. The NIH Guidelines cover such topics as methods of assessing the risk of a recombinant DNA experiment, methods of classifying organisms based on their risk, and methods of containing organisms so that they cannot escape and harm workers or reach the environment. Recent revisions of the Guidelines also discuss methods of working on a large scale with recombinant organisms (e.g., in a production facility), of performing human gene therapy, and of working with plants and animals. These revised Guidelines outline basic practices that are applied to work with recombinant DNA in research and in the pharmaceutical industry, in the United States and sometimes in other countries. FDA enforces the practices established in the NIH Guidelines when appropriate. (Safety issues related to recombinant DNA research are also discussed in Chapter 30.)

As a result of the various developments throughout the 1970s, FDA determined that its existing regulatory framework for traditional biotechnology products could be extended to include products made by recombinant DNA methods. This means that (1) no additional laws or regulations were created to regulate biotechnology-derived products, (2) biotechnology companies making products for medical use must abide by GMP, GLP, and GCP, as applicable, and (3) biotechnology products must undergo the same approval processes as traditional products.

B. Technical Issues Relating to Biotechnology Production Systems

i. STABILITY

Although the same regulations apply to traditional pharmaceuticals and to those isolated from modified organisms, there are technical concerns that are specific to biotechnology products. Some of these technical issues are addressed in the NIH Guidelines. Other concerns are addressed in Points to Consider and Guideline Documents from the FDA. More recently, the ICH has become actively involved in issuing guidance documents related to technical issues in biotechnology.

One of the concerns specific to modern biotechnology products relates to stability. Biotechnology production systems depend on host cells that have been genetically modified to produce a desired product. If the host cell mutates or changes, then the product might be altered, perhaps in ways that are deleterious. Because the products of biotechnology are often complex proteins or protein derivatives, a change in the final product might be subtle and difficult to recognize. Methods to help preserve the stability of the host cells are therefore required. Sophisticated analytical methods are also needed to characterize both the product and the genetic construct so that changes in either can be detected.

A key component in maintaining genetic stability is creation of a **master cell bank (MCB)** *that is the source of the cells used for production.* For bacterial production systems, the master cell bank is derived from the original colony of bacteria that was transformed with the genetic construct. The original colony is allowed to divide a number of times and is then distributed into small aliquots (portions), each of which is placed in a vial and stored cryogenically to assure stability. If the production system uses mammalian or insect cells, then the master cell bank is similarly derived from a single cell that contains the genetic construct and is allowed to divide. The MCB is protected by the use of security measures, documentation, and alarm systems in the event of a freezer malfunction. It is also common practice to store aliquots in more than one freezer. Each time a production run is begun, *an ampule from the master cell bank is thawed and is allowed to multiply to form the* **working cell bank (WCB)**.

The history of the master cell bank and the host cells must be known in great detail. Extensive documentation is therefore required, including the DNA sequence for the genetic construct, the origin of the genetic construct, how the construct was introduced into the host cells, the sites where specific DNA-cutting enzymes can cleave the construct, and the characteristics of the host cells. This extensive characterization of the production system is an important function of the research and development laboratory.

It is necessary to regularly perform tests that are capable of detecting alterations in the production system or in the product. Testing for genetic stability typically requires sophisticated, validated methods of characterization and genetic analysis. These tests include determining the actual DNA sequence of the genetic construct, DNA fingerprinting that characterizes the DNA of the host cell, analysis of the host cell nutrient requirements, analysis of the growth and morphological characteristics of the host cells, and assaying the host cells for contaminating viruses and other agents. The protein product similarly needs to be characterized using various assays that determine the structure of the protein and its function.

ii. CONTAMINATION

Another concern relating to the use of biotechnology production systems is contamination. It is important to note that the problem of contamination is not limited to biotechnology systems; it is a serious problem throughout the medical products field. For example, many patients with hemophilia were infected by the virus causing AIDS after administration of therapeutic products isolated from human blood. Children were similarly infected with the fatal neurological disease, Creutzfeldt-Jakob disease, after administration of growth hormone isolated from pituitaries of human cadavers. In fact, obtaining drugs from continuous cell lines (rather than tissue from animals or humans) has the advantage that the lines can be extensively studied, can be evaluated for various contaminants, and can be stored until they have been thoroughly investigated. Nonetheless, there are certain types of contaminants that are of special concern when modern biotechnology methods are used. These contaminants are summarized in Table 3.2.

Table 3.2   CONTAMINANTS OF SPECIAL CONCERN IN PRODUCTS MADE USING MODERN BIOTECHNOLOGY METHODS

Viruses and Bacteria. Viral and bacterial contaminants are a concern in any pharmaceutical process because they may infect and destroy the host cells and they may contaminate the final product. Viruses can incorporate themselves into a host genome and lay dormant for long periods of time. With mammalian cell production systems there is the additional concern that these cells will harbor agents that are specifically pathogenic to humans.

Pyrogens. Pyrogens are contaminants that are by-products of bacterial degradation. They may remain in a preparation even when all bacteria have been killed. Pyrogens cause fever and elicit a dangerous inflammatory response in mammals.

Nucleic Acids from the Host Cells. Pieces of naked DNA are known to be able to enter cells where they can integrate into the cells' chromosomes. There is concern that DNA from the host cells might, for example, have oncogenic effects (induce cancer) in patients.

Proteins from the Host Cells. The concern is that such proteins might cause allergic reactions in patients, might have unwanted enzymatic effects, or might affect the regulation of genes in patients.

Yeast, Fungi, Parasites, Mycoplasma, and Other Harmful Agents Harbored in the Host Cells or in the Media. There is concern that the host cells could harbor these agents, or that they might be introduced from the growth media required by the cells.

Residuals from Processing (such as chromatography solvents). Some of these additives are specific to biotechnology production systems, but others are common in traditional processing as well.

Biotechnology products are usually complex protein molecules that cannot be sterilized by heat; therefore, they must be processed in ways that remove contaminants present in the host cells and in the production system. In addition, as with all pharmaceutical production, routine steps to prevent contaminating the product during processing are required. These routine safeguards include, for example, the use of special clean rooms and sterilization of processing equipment.

The host cells, raw materials, in-process samples, and end product must be tested for contaminants. One method of testing cells for pathogens is **infectivity assays** *where susceptible cells are grown alongside the cells being tested to see if they become infected.* Transmission electron microscopy (TEM) is a method used to actually look for viruses in cells. Assays for host cell proteins, host cell DNA, and additives are often performed. DNA content in the final product must be very low, as determined by a highly sensitive method. In addition, contaminants introduced by the recovery and purification processes should be below detectable levels using a highly sensitive analytical method.

In addition to testing for contamination, it is also possible to perform **clearance studies** *in which a contaminant is intentionally added to the production system and the ability of the process to remove it is measured.* It is common to perform viral clearance studies in which

EXAMPLE

A company develops a recombinant DNA production system in *E. coli* to produce a therapeutic protein. They establish final product specifications and assay methods to ensure that the product meets the specifications. Before a protein drug lot can be released for sale, it must be tested to ensure that it meets all these specifications:

RELEASE SPECIFICATIONS FOR PROTEIN XYZ

Characteristic	Specification	Assay Method
Appearance	Clear, colorless solution	Visual inspection
E. coli DNA	$< 0.01 \ \mu g/\mu g$ product	Southern Blot
E. coli Protein	Undetectable	Specific Immunoassays
E. coli RNA	Undetectable	Agarose Gel Electrophoresis
Residual ethanol	< 250 ppm	Gas Chromatography
Endotoxin	$< 0.1 \ EU/\mu g$	LAL assay
Sterility	No growth in 14 days	Assay in USP*
Retrovirus	Undetectable	Infectivity assay

*U.S. Pharmacopeia, a compendium of accepted methods for the pharmaceutical industry.

viruses are intentionally added and their removal by subsequent purification steps is documented. It is recommended that a series of clearance studies be performed using different types and sizes of viruses, DNA, and any additives that might be detrimental and whose removal must be documented.

Many of the issues relevant to recombinant DNA production systems also apply to materials used for gene therapy. Characterization of genetic material administered to patients is essential. Contaminants need to be removed and their absence needs to be demonstrated using appropriate test methods. Other technical issues are being explored, such as whether infectious viruses can be safely used to deliver genes to patients. Human gene therapy trials presently require approval by both FDA and NIH.

IV. THE REGULATION OF FOOD AND AGRICULTURE IN THE UNITED STATES

A. Introduction to Regulatory Agencies and Relevant Legislation

Food and its production are subject to government regulations pertaining to safety, purity, labeling, and "wholesomeness." The FDA is involved in food safety, as it is in regulating medical products. FDA is responsible for the safety of all food except meat, poultry, and frozen and dried eggs. FDA's regulatory authority for food and cosmetics comes from the Pure Food and Drugs Acts of 1906 and 1938, as well as from other acts that relate to such issues as nutritional labeling and infant formula. Like the centers that regulate medical products, FDA's **Center for Food Safety and Applied Nutrition (CFSAN)** includes chemists, nutritionists, microbiologists, and toxicologists so that its work has a scientific basis.

In addition to the FDA, **The United States Department of Agriculture (USDA)** and **The Environmental Protection Agency (EPA)** are also empowered to play major roles in the regulation of food and agriculture, Tables 3.3 and 3.4. The USDA is responsible for the safety of meat, poultry and eggs. The USDA also publishes voluntary standards for grading foods according to their quality. EPA regulates pesticides used in agriculture and substances released to the environment.

B. Food Safety

i. CONTAMINANTS

Major issues relating to food safety include avoiding contamination by pathogens, such as bacteria and parasites, minimizing adverse effects of intentional food additives, such as coloring agents and sweeteners, and minimizing adverse effects of unintentional additives, such as pesticide residues. Microbial contamination of foods is probably the most significant food-related

Table 3.3   PRIMARY FEDERAL AGENCIES INVOLVED IN THE REGULATION OF FOOD AND AGRICULTURE IN THE UNITED STATES

The United States Department of Agriculture (USDA)

The USDA enforces requirements for purity and quality of meat, poultry, and eggs, and it is involved in nutrition research and public education. The USDA regulates genetically engineered food plants through its division of Animal and Plant Health Inspection Service (APHIS).

Food and Drug Administration (FDA)

The FDA's Center for Food Safety and Applied Nutrition (CFSAN) is responsible for ensuring the safety and purity of all foods sold in interstate commerce except for meat, poultry, and eggs. (Foods that are sold only within a state are regulated by that state.) FDA also regulates the composition, quality, and safety of food and color additives.

Environmental Protection Agency (EPA)

The EPA regulates pesticides, sets tolerance limits for pesticide residues in foods (which FDA enforces), publishes directives for the safe use of pesticides, and establishes quality standards for drinking water. The EPA is also involved in the regulation of environmental releases of genetically modified organisms.

problem in the United States. It is not known how many people are affected by food-borne illness, but estimates range as high as 80 million cases of illness and 9000 deaths per year. Many of the problems with foods are traceable to the way they are handled and prepared by the consumer; therefore, the USDA and the FDA have active consumer education programs.

Until recently, prevention of meat and poultry contamination was accomplished primarily through a system of inspectors who inspected all animals slaughtered in an effort to detect diseased animals and the presence of chemical residues. This testing, however, does not prevent contamination from occurring, nor can it detect all contamination which has occurred. The federal government, therefore, recently mandated that meat and poultry producers implement quality systems based on the same principles as ISO 9000 and GMP (discussed in Unit II), that is, that quality cannot be inspected into the final product. Quality must instead be "built into" the product throughout its processing. The quality program being implemented for the meat and poultry industry is called **Hazard Analysis Critical Control Points (HACCP)**. HACCP *emphasizes reducing hazards throughout the production, slaughter, processing, and distribution of meat and poultry.* For example, hazards include the presence of pathogens, antibiotics, pesticide and hormone residues, additives, and foreign matter. The implementation of HACCP requires identifying critical points where hazardous substances might be introduced or must be eliminated. There are critical

Table 3.4 MAJOR FEDERAL LEGISLATIVE ACTS RELATING TO THE REGULATION OF FOOD AND AGRICULTURE

Food Drug and Cosmetics Act (FDCA)

The FDCA gives the FDA the authority to regulate the safety and wholesomeness of foods and food additives.

Federal Meat Inspection Act and the Poultry Products Inspection Act

This Act authorizes USDA to ensure that meat and poultry products are safe, wholesome, and accurately labeled.

Toxic Substances Control Act (TSCA)

TSCA is intended to control chemicals that may pose a threat to human health or to the environment. The act requires reporting and record-keeping by chemical manufacturers and sets requirements for notifying EPA when new chemicals are introduced. The provisions of this act primarily affect chemical manufacturers and their R&D laboratories; however, the act also authorizes the EPA to review new chemicals before they are introduced. Under this provision, EPA has established a program to review microbial products that result from the introduction of genes from one type of bacterium into another type. EPA's rationale for this program is that microorganisms engineered to contain genes from another organism are new "chemicals." This is relevant, for example, if genetically engineered microorganisms are introduced into fields to control pests, or if they are introduced into the environment for purposes of cleaning contaminated soil or water.

The Federal Insecticide, Fungicide, Rodenticide Act (FIFRA)

FIFRA makes EPA responsible for regulating the distribution, sale, use, and testing of pesticides, including those that are the result of genetic modifications to organisms.

The Federal Plant Pest Act (FPPA)

FPPA regulates the introduction or release to the environment of "plant pests"; that is, any organisms, materials, or infectious substances that can cause damage to any plants or to products of plants. As such, FPPA relates to genetically engineered organisms if they might be plant pests. This act gives APHIS (part of USDA) the authority to regulate such introductions.

---

points at the farms where the animals are raised, at the time of slaughter, during the processing and distribution of the product, and during its preparation in homes, restaurants, and institutions. As in pharmaceutical companies, the implementation of HACCP is intended to complement, but not replace, product inspections. It is the food industry's way of building quality into a food product.

ii. FOOD ADDITIVES

People want food to be safe and uncontaminated, and they also want food to be abundant, nutritious, diverse, and economical. These latter objectives are often the result of innovations, such as pesticides that make food more abundant and less expensive, or food additives, such as saccharin, which make it more diverse. Biotechnology methods can also lead to innovations in food products. Innovations sometimes lead to concerns about the safety of food. This has been the case for pesticides, nonnutritive sweeteners, and biotechnology-derived products.

FDA has the primary responsibility for regulating food additives and new food products, except meats and poultry. As is the case for drugs, new food additives must be tested for safety and then be approved by the FDA before they are marketed; however, there are exceptions for materials that are generally recognized to be safe. FDA is also responsible for ensuring that food products are produced properly and are not adulterated.

C. Recombinant DNA Methods and the Food Industry

i. SAFETY

The first recombinant DNA food ingredient was approved by the FDA in 1990. It is an enzyme, chymosin (also called renin), that is produced in *E. coli* containing the bovine chymosin gene. Renin is the main enzyme used to make milk clot during the production of cheese. The traditional source of renin is rennet, which is isolated from the fourth stomach of unweaned calves, which are killed for veal. The availability and price of rennet fluctuates depending on the availability of veal.

FDA's concerns regarding the approval of renin made by modified bacteria are basically the same as when a drug product made by recombinant DNA methods is introduced to supplement or replace a drug isolated from "natural" sources. The first concern is to establish that the recombinant product is identical to the natural product, or that any differences have no effect on its safety or effectiveness. To demonstrate this for chymosin, the company performed a variety of identity tests on both the gene itself and on the resulting protein. They also performed activity assays that showed that the recombinant chymosin clotted milk. A second aspect of approving the recombinant DNA-derived chymosin was ensuring that the preparation was safe. As with recombinant drug products, this involved demonstrating that the processing steps removed contaminants, such as DNA and protein derived from the host bacterial cells. Investigators also demonstrated that the bacterial cells used in production were themselves non-pathogenic. Once the company producing the chymosin had demonstrated its equivalence to traditional renin and its safety, the new version was approved.

FDA's general policy is currently that if chemical, microbiological, and molecular biology studies show that a food or additive is the same as, or substantially equivalent to, an accepted food-use product, then minimal extra testing or regulatory surveillance is required. A special review of a genetically engineered food product is required only when special safety issues exist. For example, special review is required if a biotechnology product can cause allergic reactions, or has different nutritional value than a traditional product.

ii. ENVIRONMENTAL RELEASE OF GENETICALLY MODIFIED ORGANISMS

A number of transgenic plants have entered the market or have been approved for field testing. There are transgenic plants that can tolerate herbicides, resist insects or viruses, or produce modified fruits and flowers. For example, insect-resistant corn plants have been created by inserting into them a bacterial gene that codes for an insecticidal protein. There are also genetically modified bacteria that have applications in agriculture (e.g., to provide nitrogen to plants). Genetically modified microorganisms are similarly being investigated to clean polluted water and soil.

There are concerns that there will be adverse effects if organisms genetically modified by the methods of biotechnology are released into the environment (as occurs when a modified crop is grown in a field or when modified microorganisms are applied to the soil). For example, it is postulated that transgenic plants could cause the development of new, hardy weed species or contribute to the loss of species diversity. It is feared that insects and other pests will become increasingly resistant to control agents because of the use of genetically modified crops. The regulation of field testing and planting transgenic crops is the responsibility of APHIS and the EPA. Their task has been difficult and particularly controversial because the effects of genetically modified organisms in the environment remain uncertain. The oversight and approval mechanisms that these agencies use have therefore been changing over the years. Regulatory agencies will presumably be able to establish consistent policies as they gain more experience with these genetically modified organisms in the environment.

UNIT II:

Product Quality and Biotechnology

Chapters in This Unit

Chapter 3 surveyed the history and the rationale behind the regulation of biotechnology products. As is clear from the history of pharmaceutical regulation, producing quality products does not happen by accident. The consistent production of quality products requires resources, planning, and commitment. This unit explores how a company, organization, or laboratory translates a commitment to quality into practices:

Chapter 4 is a broad overview of the issues and vocabulary relating to product quality systems.

Chapter 5 discusses one of the most important aspects of any quality system, that is, documentation.

Chapter 6 discusses how the principles of quality systems are applied in laboratory settings.

Chapter 7 discusses how the principles of quality systems are applied in biotechnology production facilities.

TERMINOLOGY RELATING TO PRODUCT QUALITY AND BIOTECHNOLOGY

Abbreviations used in definitions

FDA: Food and Drug Administration

FDCA: Food, Drug and Cosmetics Act

GMP: Good Manufacturing Practices

EPA: Environmental Protection Agency

ANSI: American National Standards Institute

ISO: International Organization for Standardization

*Definition from American National Standard ANSI/ISO/ASQC A8402-1994 "Quality Management and Quality Assurance Vocabulary" ASQC, Milwaukee, WI

Acceptance Criteria (according to GMP). The minimum specifications for a product, and the criteria for accepting or rejecting that product, together with a plan for sampling it.

Accuracy. The closeness of a test result to the true or accepted value.

Adulterated (according to the FDCA). A food or drug that is produced by methods that do not conform to current Good Manufacturing Practice, or which is made under unsanitary conditions, or which contains unacceptable contaminants.

American National Standards Institute, ANSI. Administrator and coordinator of the U.S. private sector, voluntary standardization system. The U.S. representative to ISO.

Batch (according to GMP). "[A] specific quantity of a drug or other material that is intended to have uniform character and quality, within specified limits, and is produced according to a single manufacturing order during the same cycle of manufacture. Note, a batch is usually smaller than a lot."

Batch Record. An exact copy of a Master Batch Record, but with an assigned lot number. The batch record is used to direct the production of a product and is the document in which formulation and manufacturing activities are recorded.

Biological Activity. The level of activity or potency of a product as determined by tests in animals or in cells in culture.

Biopharmaceutical. Refers to a drug product produced using recombinant DNA or monoclonal antibody methods.

Calibrate. To establish the relationship between values indicated by a measuring instrument and values of a certified standard.

Chain of Custody. Refers to the controlled and documented sequence of handling, storage, and disposition of a sample.

Clinical Laboratory Improvement Amendments of 1988, CLIA. A system of regulations intended to ensure the quality of results generated by laboratories that perform tests on human specimens.

Code of Federal Regulations, CFR. A document published annually that contains U.S. federal government regulations.

Compendium (of standard methods). A published collection of accepted methods in a particular field.

Current Good Manufacturing Practices, cGMP. The currently accepted minimum requirements followed by industry for the manufacture, testing, packaging, and distribution of pharmaceutical products. Note, a practice may be enforced, even if it is not in the published GMP regulations, if the practice has become accepted by industry, hence the term "cGMP."

Cytotoxic. Damaging to cells.

Deviation (in a quality environment). An unexpected occurrence.

Documentation. Written records that guide activities and record what has been done.

Downstream Processing. The isolation, purification, and packaging of a desired product after bacterial fermentation, cell culture, or other initial production process.

Electrophoresis. A technique in which molecules are separated from one another based on differences in their mobility when placed in a gel matrix and subjected to an electrical field.

Food and Drug Administration, FDA. The federal agency empowered by Congress to regulate the production of cosmetics, electronic products such as x-ray machines, food, nutritional supplements, medical devices, diagnostic devices, prescription and over-the-counter pharmaceuticals for human and veterinary use.

Form. A document, usually associated with an SOP or other document, on which an individual fills in blanks as a task is performed.

Good Clinical Practices, GCP. The FDA's minimum standards, regulations, and guidelines for conducting clinical trials (studies in humans) of potential pharmaceutical products.

Good Laboratory Practices, GLP. 1. FDA: The procedures and practices that must be followed when performing laboratory studies in animals in order to investigate the safety and toxicological effects of new drugs. 2. EPA: The procedures and practices that must be followed when investigating the effects of pesticides and other agrochemicals. 3. With small letters, that is, good laboratory practices (glp): A general term used to refer to all quality practices in any laboratory.

Good Manufacturing Practices, GMPs. See "Current Good Manufacturing Practices."

Guideline (from FDA). Documents produced by the FDA that establish principles and practices. Guidelines are not legal requirements; rather, they are practices that are acceptable to and recommended by the FDA.

Immunoassay. A test method based on the interaction between antibodies and the targets to which they bind. Immunoassays can be used to detect the presence of proteins and to quantitate how much is present.

Installation Qualification. See "Validation."

International Organization for Standardization, ISO. A worldwide federation of national standards groups from 91 countries that addresses quality management and promotes international trade and cooperation.

ISO 9000. A set of internationally accepted standards, published by ISO, that outline a system for quality management aimed at ensuring the quality of a product or service.

Laboratory Notebook. A chronological record of an individual's work; the primary document in a research laboratory.

Large-Scale Production. Production of large amounts of product, generally for commercial sale.

Limit of Detection. The lowest concentration of the material of interest that a method can detect.

Limit of Quantitation. The lowest concentration of the material of interest that a method can quantitate with acceptable accuracy and precision.

Limulus Amoebocyte Test, LAL. A sensitive test for the presence of pyrogens based on a coagulation reaction between pyrogens and the blood of horseshoe crabs.

Linearity. The ability of a method to give test results that are directly proportional to the concentration of the material of interest (within a given concentration range).

Logbook. A chronological record of the usage, status, calibration, and maintenance of a piece of equipment or an instrument.

Lot (from GMP). "[A] batch, or a specific identified portion of a batch, having uniform character and quality within specified limits . . ."

Lot Number (also, control number or batch number). An identifying number that is used to distinguish one lot from another.

Master Batch Record. Detailed step-by-step instructions for how to formulate or produce a product, including raw materials required, processing steps, controls, and required testing.

Master Cell Bank, MCB. Stored, living cells which are the source of cells used in a process. The stored cells usually are derived from a single cell and are stored cryogenically to assure genetic stability. The MCB is the source for the working cell bank.

Medical Device (according to FDA). Items that are not drugs or biologics; rather, they are used in the diagnosis or treatment of disease or injury. Examples include sterilizers, test kits, cell counters, and pacemakers.

Method. The way in which a laboratory test or assay is performed.

Method Validation. See "Validation of an Analytical Method."

Nonconformance. When a product or raw material does not meet its specifications.

Operational Qualification. See "Validation."

Points to Consider (from FDA). Guidance documents issued by FDA that contain information about new technologies. Points to Consider establish general concerns that should be addressed by companies.

Precision. The consistency of a series of measurements or tests. (See also "Reproducibility" and "Repeatability.")

Procedure.*"[S]pecified way to perform an activity."

Process.* Set of interrelated resources (such as personnel, facilities, equipment, techniques, and methods) that transforms inputs into outputs.

Process Validation. See "Validation."

Product.* Results of activities or processes. A product may be a service, a processed material, or data.

Protocol. A step-by-step outline that tells an operator how to perform a task or perform an experiment that is intended to answer a question.

Pyrogen. Materials derived from lipopolysaccharides that are released from the breakdown of gram negative bacteria. They trigger a dangerous immune response and induce fever, hence their name pyrogen ("heat producing").

Qualification Process.* "[P]rocess of demonstrating whether an entity is capable of fulfilling specified requirements."

Quality.* "[T]he totality of characteristics of an entity that bear on its ability to satisfy stated and implied needs."

Quality Assurance.* Quality assurance provides confidence that quality requirements are fulfilled, both within the organization and externally to customers.

Quality Control. The means by which it is ensured that products maintain a consistent, desired level of acceptability or quality.

Quality System.* The organization, structure, responsibilities, procedures, processes, and resources for ensuring the quality of a product or service.

Range (of a method). The range of concentrations, from the lowest to the highest, that a method can measure with acceptable results.

Raw Data. The first record of an original observation. Depending on the situation, raw data may be recorded with pen by the operator, may be a paper output from an instrument, or may be recorded directly into a computer medium.

Reference Material (ISO Guide 25). "A material or substance one or more properties of which are sufficiently well established to be used for the calibration of an apparatus, the assessment of a measuring method, or for assigning values to materials."

Regulations. Requirements imposed by the government and having the force of law.

Regulatory Agency. A government body that is responsible for enforcing and interpreting legislative acts. For example, FDA is a regulatory agency responsible for ensuring the safety, effectiveness, and reliability of medical products, foods, and cosmetics, as set forth in various federal legislation.

Regulatory Submissions. Documents that are completed to meet the requirements of a government regulatory agency.

Repeatability. The closeness of agreement between test results obtained when a test is repeated in the same laboratory under consistent conditions. Repeatability is one of the categories of the precision of a method.

Report. A document that describes the results of an executed protocol.

Representative Sample (from GMP). "[A] sample that consists of a number of units that are drawn based on rational criteria such as random sampling and intended to assure that the sample accurately portrays the material being sampled."

Reproducibility. The closeness of agreement between test results obtained when the same test is repeated in

different laboratories, or at different times or by different operators. Reproducibility is one of the categories of the precision of a test method.

Robustness. The capacity of a method to give acceptable results when there are deliberate variations in method parameters.

Sample. A subset of the whole that represents the whole (e.g., a blood sample represents all the blood in an individual).

Scale-up. The transition between producing a product in small amounts in the laboratory and producing large amounts in a production facility.

Selectivity. A measure of the extent to which a method can determine the presence of a particular compound in a sample without interference from other materials present.

Specifications. Descriptions that define and characterize the properties that a material must possess based on its intended use.

Standard. 1. Any concept, method, or way of doing things that is established by an authority, custom, or general agreement (Based on *Merriam-Webster Collegiate Dictionary,* G. & S. Merriam Co., MA, 1977.) 2. In an assay: A well-defined material that demonstrates how an assay or an instrument responds to an analyte.

Standard Operating Procedure, SOP. A set of instructions for performing a routine method, manufacturing operation, administrative process, or maintenance operation.

Statistical Process Control. The desired situation in which a manufacturing or measurement process behaves as expected. A certain amount of variability in the process results is expected, but when the process is in control the variability does not exceed previously determined limits.

Traceability.* The "ability to trace the history, application, or location of an entity . . . by means of recorded identifications. Notes:

The term *Traceability* may have one of three main meanings:

  a. in a product . . . sense it may relate to

    —the origin of materials and parts

    —the product processing history

    —the distribution and location of the product after delivery;

  b. in a calibration sense, it relates measuring equipment to national or international standards, basic physical constants or properties, or reference materials;

  c. in a data-collection sense, it relates calculations and data generated throughout the quality loop . . .

sometimes back to the requirements for quality ... for an entity."

United States Pharmacopeia, USP. (1) An organization that promotes public health by establishing and disseminating officially recognized standards for the use of medicines and other health care technologies. (2) The compendium containing drug descriptions, specifications, and standard test methods for such parameters as drug identity, strength, quality, and purity. This compendium is recognized as a legal authority by FDA.

Validation. A process or a set of activities that ensure that an individual piece of equipment, or a process, or an analytical method performs the function for which it is intended reliably and effectively.

Process validation. Activities which prove that a manufacturing process meets its requirements so that the final product will do what it is supposed to do consistently, safely, and effectively.

Equipment validation:

Installation qualification, IQ. A set of activities designed to determine if a piece of equipment is installed correctly.

Operational qualification, OQ. A set of activities designed to establish that a piece of equipment performs within acceptable limits.

Performance qualification, PQ. A set of activities that evaluates the performance of a piece of equipment under the conditions of actual use.

Validation of an analytical procedure. A process used by the scientific community to evaluate a test method or assay. This involves determining the ability of a method to reliably obtain a desired result, the conditions under which such results can be obtained, and the limitations of the method.

Validation Protocol. A document that details how a validation study will be performed.

Verification.* "[C]onfirmation by examination and provision of objective evidence that specified requirements have been fulfilled."

REFERENCES FOR PRODUCT QUALITY AND BIOTECHNOLOGY

There is a vast, evolving literature in the area of product quality, particularly as relates to pharmaceuticals. A few references to get started in this area are listed below. See also the references in the Introduction to Unit I.

Since the technology, regulations, and standards which affect biotechnology are changing quickly we recommend using the Internet to find current information. The Internet home pages for regulatory agencies (listed later) are very helpful.

Drug, Device and Diagnostic Manufacturing: The Ultimate Resource Handbook, Carol DeSain. 2nd edition. Interpharm Publishers, Illinois, 1993.

General Requirements for Accreditation, American Association for Laboratory Accreditation, Gaithersburg, MD, May 1993. (Laboratories can be certified through the American Association for Laboratory Accreditation. This document summarizes the quality requirements for accreditation and also includes the text of ISO/IEC Guide 25.)

General Requirements for the Competence of Calibration and Testing Laboratories, ISO/IEC Guide 25-1990. (This Guide is presently available at the ISO Guide 25 Home Page, http://www.microserve.net/~iso25.)

"Guidelines for a Quality Assurance Program for DNA Analysis." Prepared by the Technical Working Group on DNA Analysis Methods. Bruce Budowle, Group Chair, Federal Bureau of Investigation. *Crime Laboratory Digest* 22(2): 21–38. 1995. (This guide outlines a laboratory quality system.)

Documentation Practices, Carol DeSain and Charmaine Vercimak Sutton. Advantstar Publishers, Ohio. 1996. (A comprehensive look at documentation with many useful examples.)

"The Validation Story: Perspectives on the Systematic GMP Inspection Approach and Validation Development," Ronald F. Tetzlaff, Richard E. Shepherd, and Armand J. LeBlanc. *Pharmaceutical Technology,* pp. 100–116, March 1993. (Historical presentation of issues leading to validation.)

"A History of Validation in the United States: Part 1," Kenneth G. Chapman. *Pharmaceutical Technology.* pp. 82–96. October, 1991. (Historical presentation; includes discussion of some issues and controversies relating to validation.)

Guideline on General Principles of Process Validation, May 1987 FDA document, prepared by Dr. Arthur Shaw. Reprinted 1993. CDER, CBER, CDRH. (An FDA guideline that is an important source of information on validation.)

Guideline on Validation of Analytical Procedures, International Conference on Harmonisation, Fed Register 1996; 61(46):9315–9319. (This guideline provides the basis for method validation in the U.S. pharmaceutical and biopharmaceutical industries.)

The Good Manufacturing Practices for pharmaceuticals are found in the Code of Federal Regulations, Title 21, Parts 210 and 211.

ISO 9000: In the United States, the ISO 9000 standards are prepared by the American Society for Quality Control (ASQC) Standards

Committee for the American National Standards (ANSI) Committee on Quality Assurance. The standards are published as:

American National Standard ANSI/ISO/ASQC Q9001-1994
"Quality Systems—Model for Quality Assurance in Design, Development, Production, Installation, and Servicing," ASQC, Milwaukee, WI

American National Standard ANSI/ISO/ASQC Q9002-1994
"Quality Systems—Model for Quality Assurance in Production, Installation, and Servicing," ASQC, Milwaukee, WI

American National Standard ANSI/ISO/ASQC Q9003-1994
"Quality Systems—Model for Quality Assurance in Final Inspection and Test," ASQC, Milwaukee, WI

American National Standard ANSI/ISO/ASQC Q9000-1-1994
"Quality Management and Quality Assurance Standards—Guidelines for Selection and Use," ASQC, Milwaukee, WI

American National Standard ANSI/ISO/ASQC Q9004-1-1994
"Quality Management and Quality System Elements—Guidelines," ASQC, Milwaukee, WI

American National Standard ANSI/ISO/ASQC A8402-1994
"Quality Management and Quality Assurance Vocabulary," ASQC, Milwaukee, WI

INTERNET RESOURCES

The Code of Federal Regulations is available at: http://www.access.gpo.gov/nara/cfr

Human Drug cGMP Notes. Question and Answer format from FDA. Available at: http://www.fda.gov/cder/dmpq/cgmpnotes.htm

FDA Homepage: http://www.fda.gov

USDA Homepage: http://www.usda.gov

EPA Homepage: http://www.epa.gov

ICH Homepage: http://www.ich.org

USP Homepage: www.usp.org

CHAPTER 4

Introduction to Product Quality Systems

I. OVERVIEW

As consumers of products and services we are all familiar with the concept of product quality. At the supermarket we may find certain brands to be superior. Before purchasing a major appliance, we might consult a consumer guide to learn about the performance of various models. In this unit (and in this entire book) we are concerned with product quality, but not from the perspective of being a consumer, rather from the perspective of being the producer of a biotechnology product.

As we saw in Chapter 1, the products of biotechnology are varied. There are tangible products such as drugs, transgenic livestock, modified plants, and enzymes for research. There are laboratories whose products are test results, such as measurements of a product's activity or the performance of an instrument. The academic research laboratory is a workplace whose product is knowledge. The definition of *quality* varies for these different sorts of products. As we saw in Chapter 3, government regulations define quality for drugs, medical products, and foods. These regulated products must be safe, effective, reliable, nutritious (for foods), unadulterated, and so on. A quality test result is one that can be relied on when making decisions. A quality research study enhances our understanding of the natural world.

How does a company or a laboratory ensure that its products are of good quality? Consider two events. In 1959 the Pillsbury Company was beginning to manufacture food products for the NASA space program. It was essential that the food be completely pathogen-free to ensure the safety of the astronauts. Scientists considered testing the food for pathogens but analysis convinced them that simply testing the final food products would not ensure their purity. This is because not every food item can be tested; most testing procedures destroy the food or its packaging. Even if nondestructive methods exist, it is too time-consuming to test every item. When final products are tested, therefore, a sample is selected to represent the whole lot. The Pillsbury scientists did an analysis in which they supposed that there was one pathogenic Salmonella organism per 1000 units of food. If 20 of the 1000 units were to be tested, then there would be a 98% chance of missing the single pathogen and accepting the defective lot. The Pillsbury staff concluded that testing samples of the final food products was not an adequate method of ensuring their quality and safety. They decided instead, to make sure that the production of the foods was so tightly controlled that no contaminants could enter them.

In the 1960s and 1970s there were a number of septicemia cases in patients caused by intravenous (IV) fluids that were contaminated by bacteria. Many people became ill and some died. FDA inspection teams discovered serious problems in companies manufacturing products for IV use. They found, for example, contaminated cooling water and sterilization equipment that failed to reach sterilizing temperature. These sorts of problems in production led to contaminated products, which in turn, caused illness and deaths in patients.

The pharmaceutical companies responsible for the septicemia outbreaks did have testing programs in place to monitor samples of their final products. If the samples passed inspection, then the quality of the entire batch was assumed to be acceptable for distribution. Unfortunately, as was realized by the Pillsbury scientists, final product testing can miss contaminated items. In the IV fluids case, the acceptance of contaminated lots led to patient deaths.

The Pillsbury analysis and the septicemia deaths illustrate that final product testing, although necessary and important, does not ensure that products are of acceptable quality. As FDA emphatically states:

> Quality, safety and effectiveness are designed and built into a product. The quality of a product does not result from inspecting that product; that is, quality cannot be inspected or tested into the finished product.

How is quality designed and built into a product? There is no single answer to this question, nor is there any simple answer. Producing a quality product requires many coordinated elements (e.g., skilled and knowledgeable personnel, a well-designed and managed facility, and ready access to raw materials). All these coordinated elements together are called a **quality system**. A **quality system** *is the organizational structure, responsibilities, procedures, processes, and resources that together ensure the quality of a product or service.* All product quality systems have the goal of ensuring a quality product.

Every product is different, companies differ, and the problems that arise during production vary; therefore, there is more than one formal quality system. The general features of the quality systems that are important in biotechnology workplaces are introduced in this chapter and then explored in more detail in Chapters 5–7.

II. QUALITY SYSTEMS IN DIFFERENT WORKPLACES

A. Academic Research Laboratories

Academic scientists performing basic research are unlikely to think of their work using the term *quality system*. Research scientists do, however, have a long tradition of adhering informally to such a system, called "doing good science." "Doing good science" means using consistent, thoughtful, and effective methods; keeping honest and thorough records; verifying all results; and rigorously employing "good laboratory practices." Although the output of an academic research laboratory is not scrutinized by government inspectors, there are methods of monitoring its quality. When research scientists complete a set of work they submit it to a journal to be published. The work is then available to other scientists who consider it skeptically, evaluating its merits, flaws, and value. In addition, the work of scientists is judged by peers when scientists apply for research funding to support their studies.

B. Quality Systems in Companies that Are Regulated

The consequences of a poor product are very different in a research laboratory than in a drug or food processing company. Poorly manufactured drugs and contaminated foods have led directly to human injuries and deaths. The consequences of a poorly performed research study tend to be less severe; therefore, in contrast to the informal quality practices of research scientists, the product quality systems used in companies that make drugs, or process foods, are formal and strictly enforced.

The quality system used in companies that make drug and related medical products is called current Good Manufacturing Practices (cGMPs). The cGMPs have evolved within an ongoing relationship between the government (which seeks to protect the consumer) and the pharmaceutical industry. The cGMPs constitute a quality system that has a sweeping effect in every company that produces regulated medical products. The requirements of GMP also radiate throughout the medical products industry to impact those who package, distribute, market, sell, and use these items.

The Good Manufacturing Practices are quality principles formalized into **regulations. Regulations** *are requirements that government-sanctioned agencies, such as the Food and Drug Administration or the Environmental Protection Agency, impose on an industry and on companies within that industry.* Compliance with regulations is required by law.

Regulations are objective and generally focus on safety (or reducing risk), efficacy, and honesty. Regulations tend not to cover subjective areas of product quality, such as the flavor of cookies. Thus, local government agencies are interested in how a chocolate chip cookie baker cleans his bowls—a safety consideration—but do not care about the flavor of his cookies. A pharmaceutical company might similarly add grape flavoring to cough syrup to enhance its appeal. The individual cold sufferer is capable of evaluating whether the flavoring improves the quality of the product, and therefore, this aspect of product quality is not subject to government regulation. In contrast, the safety of that cough syrup is subject to regulation because the consumer cannot perceive whether a medication is safe or not and depends on regulations for protection.

There are other regulation-based quality systems in addition to GMP which are imposed by the government acting on behalf of the consumer; several of these quality systems were mentioned in Chapter 3. For example, The Good Laboratory Practices (GLPs) that govern laboratories using animals to test the safety of drug products is another quality system. The Environmental Protection Agency has a quality system, also called Good Laboratory Practices, that guides the testing of agrochemical products. **CLIA, "The Clinical Laboratory Improvement Amendments of 1988,"** *is a system of regulations intended to ensure the quality of results generated by laboratories that perform tests on human specimens.* The CLIA regulations play an important role in ensuring that medical test results can be trusted to make a decision about how a patient should be treated.

C. Quality Systems in Companies that Comply with Voluntary Standards

Not all biotechnology companies make products whose quality is regulated by the government. For example, many biotechnology companies produce enzymes and other molecular biology products that are used in research laboratories. These products are not food or medical products and are not used in humans; therefore, the quality of these materials is not regulated by a government agency. There are, however, **quality standards** that may apply to the production of such products. A **standard**, *broadly defined, is any concept, method, or way of doing things that is established by some authority, by custom, or by general agreement (Merriam-Webster Collegiate Dictionary,* G. & C. Merriam Co., MA. 1977). Quality standards are established by various organizations, agencies, and other entities. Like regulations, quality standards are written down and are followed consistently in different workplaces. Unlike regulations, standards are not imposed by the government and compliance with standards is usually voluntary.

There are many types of standards and many organizations that develop and disseminate standards. Standards may be objective, as, for example, standards that relate to practices in making measurements and performing chemical analyses. Standards may also be subjective, as, for example, those relating to the flavor and other characteristics of foods. Standards can be very useful because they assemble the thinking of many experts into a single, concise, available document. Standards help to ensure consistency in practices among individuals and nations.

Companies and organizations voluntarily comply with standards to improve the quality of their products and to be more competitive in the international marketplace. For example, a chocolate chip cookie baker might voluntarily join an association, such as the (fictitious) "Association of Chocolate Chip Cookie Bakers." The association might have standards with which all members comply, such as a requirement that members use only real chocolate chips (as contrasted with artificially flavored ones).

Some standards are narrow in focus (e.g., there are standards that detail how to correctly place the markings on a laboratory flask). There are also broad systems of standards that are intended to ensure the quality of a final product. These broad quality systems are much like GMP or CLIA in their scope and philosophy. **ISO 9000** *is a series of quality standards published by the* **International Organization for Standardization (ISO)**. Companies, including many biotechnology companies, comply with the requirements of ISO 9000 to improve the quality of their products, to make their processes more cost-effective, to demonstrate to potential customers that their products are well-made, and to increase the profitability of their company. ISO 9000 standards are general and so are applicable to any company that makes a product. ISO 9000 standards can also be applied to companies that provide a service, for example, disposing of hazardous waste or repairing equipment.

There is no government agency that enforces and monitors compliance with ISO 9000. If a company or organization decides to voluntarily comply with ISO 9000, then they develop and implement their own quality plan. This plan is formalized and documented in the form of a quality manual. The company then hires a certified auditor to evaluate whether they are meeting their commitments based on their own plan. If the auditor determines that the company is in compliance with its plan and if the plan includes all the required parts of a quality program, then the company achieves ISO 9000 certification. The company must periodically hire an auditor to conduct inspections to assure that the company remains in compliance with its plan. Although being ISO 9000 certified does not actually guarantee that a company is producing a high-quality product, it does show that the company has systems in place that support quality.

Although ISO 9000 and GMP both have sections that are applicable to obtaining quality results in a laboratory, they focus primarily on manufacturing facilities. **ISO Guide 25** *is a quality system that is similar to ISO 9000, but focuses on issues specific to laboratories.*

Further information about ISO 9000 is summarized in Table 4.1 and ISO 9000 is contrasted with GMP in Table 4.2.

EXAMPLE PROBLEM

Which of the following is likely to be subject to regulatory oversight; which is likely to be the subject of standards; which is likely to be neither?

a. The smoothness of peanut butter
b. The microbial content of peanut butter
c. The numbers of peanut chunks in peanut butter
d. The levels of aflatoxins in peanut butter (aflatoxins are potent carcinogens formed by certain molds)
e. The electrical wiring of a food blender
f. The color of a food blender
g. The activity of a DNA cutting enzyme
h. The activity of a substance used to treat asthma

ANSWER

a. and **c.** The smoothness and chunkiness of peanut butter are subjective features and are addressed in a standard (from USDA).

b. and **d.** Microbial and aflatoxin contamination relate to food safety and therefore are subject to regulatory oversight by FDA.

e. The wiring of devices is covered by electrical codes (regulations).

f. Color is not subject to regulations or standards.

g. Enzymes used in research or teaching laboratories are not subject to regulations or standards. (A company that makes enzymes, however, is likely to voluntarily conform to a quality system like ISO 9000 or to its own internal quality system.)

h. Substances used in the treatment of illness are subject to FDA regulation.

Table 4.1 INFORMATION ABOUT THE *ISO 9000* SERIES OF STANDARDS

1. *The ISO 9000 quality standards were first issued in 1987 by the International Organization for Standardization, based in Geneva, Switzerland.* ISO is a worldwide federation of national standards groups from 91 countries.

2. *ISO 9000 standards address quality management and promote international trade and cooperation.* The standards aim to:
   Enhance the quality of goods and services
   Promote standardization of goods and services internationally
   Promote safety practices and environmental protection
   Assure compatibility between goods and services from various nations
   Increase efficiency and decrease costs

3. *ISO 9000 is actually a series of standards with five documents:*
   ISO 9001 *covers all aspects of production ranging from research and design, development, production, installation, and servicing*
   ISO 9002 *is a subset of ISO 9001 and does not deal with design or servicing. It does deal with quality assurance in production and installation*
   ISO 9003 *addresses only quality assurance in final inspection and testing*
   ISO 9000 *provides guidance on how to select ISO 9001, 9002, or 9003*
   ISO 9004 *is a supporting document to help a company implement either 9001, 9002 or 9003*

Table 4.2   *DIFFERENCES BETWEEN THE ISO 9000 SERIES OF STANDARDS AND GOOD MANUFACTURING PRACTICES REGULATIONS*

ISO 9000	GMP
Compliance is voluntary	Compliance is required by law for companies making regulated products
Compliance is monitored by auditors who are paid by the company. The company complies voluntarily with the auditors' suggestions to improve their product quality.	GMP regulations are enforced by FDA inspectors who have enforcement authority
Standards are generic and can be applied to any manufacturing or service industry	GMP regulations are specific to the pharmaceutical/medical products industry
Requires that companies write a quality manual which provides an overview of their entire quality system	Quality manual is not required
Originated in Europe	Originated in the United States

D. Product Quality Systems and the Individual

Depending on where an individual is employed, one or another of these quality systems will impact their every-day work. For example, the work of a production operator in a biopharmaceutical laboratory is stringently monitored and controlled in order to assure the safety and efficacy of the products she helps to produce. This operator will have to follow written procedures meticulously and will have to record each activity in a specific way. An FDA inspector can come in at any time and review her records. A technician working in a testing laboratory may devote considerable attention to recording and tracking incoming samples. He will verify and document that analytical instruments are functioning properly. His activities are directed at ensuring consistent, reliable test results. A research assistant working in an academic research laboratory is not bound by the GMP regulations, nor by ISO 9000, and in fact, may not even know they exist. The researcher, however, must diligently record all her work in a laboratory notebook, and must be prepared to explain it and justify it to colleagues. Thus, whether we are talking about a production operator working to make safe products in a biopharmaceutical company or a research scientist striving to produce "good science," there are quality systems to help achieve the best possible product.

Table 4.3 summarizes quality systems in various workplaces.

E. How Quality Documents Are Written

i. LANGUAGE

Excerpts from the GMP regulations and from ISO documents are sprinkled through this unit. As you read them, observe that these documents do not provide much detail on implementation. Because every company, labo-

Table 4.3   *PRODUCT QUALITY SYSTEMS IN DIFFERENT TYPES OF WORK ENVIRONMENTS*

Basic Biological Research Laboratories in Academic or Government Settings

In these laboratories scientists investigate fundamental problems in biology. Quality is discussed in terms of "doing good science." The quality of basic research is primarily monitored through review by peers, for example, when a paper summarizing research work is submitted for publication or a grant application is submitted to a granting agency.

Research and Development (R&D) Laboratories Associated with an Industry

In R&D laboratories, researchers investigate questions that are intended to result in commercial products. The regulations and standards that affect individuals in such a laboratory will depend on the nature of their industry. In companies that make drugs and other medical products, the FDA may scrutinize laboratory notebooks and other R&D records. Some companies voluntarily comply with ISO 9000 and/or with ISO Guide 25, both of which have provisions relating to laboratory work.

Production Facilities

Production facilities must maintain tight control over their processes and systems to ensure that their products are consistent and properly made. In regulated companies and in those certified by ISO or other certification bodies, external government inspectors or auditors will conduct periodic inspections. Individuals who work in production generally follow specific written procedures and are required to record in writing that they did so properly.

Testing Laboratories (Such as Analytical, Quality Control, Forensic, Microbiology, Metrology, and Clinical Testing Laboratories)

The regulations and standards that apply to testing laboratories vary. Quality control laboratories in companies that conform to GMP regulations or ISO 9000 standards have strict requirements that they must meet and are accountable to external inspectors or auditors. Clinical laboratory personnel must similarly adhere to quality regulations. Some laboratories voluntarily adhere to quality standards in order to demonstrate to their clients that their test results are trustworthy. Still other testing laboratories may institute their own quality systems to ensure that their work is performed correctly.

ratory, and organization is different, quality regulations and standards must be written in a general fashion. For example, the GMP regulations contain the statement:

> **21CFR211.63** *Equipment used in the manufacture, processing, packing, or holding of a drug product shall be of appropriate design, adequate size, and suitably located to facilitate operations for its intended use . . .*

This statement does not specify the particular equipment of concern. The terms *appropriate, adequate,* and *suitably* are not clarified. This statement provides no concrete guidance as to how a company might achieve its demands. In the case of GMP, such generic regulatory statements are interpreted and enforced by the Food and Drug Administration (FDA). FDA publishes their interpretation of the requirements of GMPs in the form of guidance documents called "Guidelines" and "Points to Consider." Guidance documents are not laws, and are intended to help companies apply the general principles of GMP to their own situation. Each company is different and each GMP-regulated company must ultimately develop its own practices and procedures to implement the quality requirements of GMP. The ISO 9000 standards are also written in a generic fashion in order to apply to any product or service. Organizations implementing ISO 9000 standards, therefore, rely on auditors or consultants to provide guidance on how to devise a quality system for their own situation.

ii. COMMON ELEMENTS

Although various quality systems differ from one another (as was shown, for example, in Table 4.2) there are certain elements they tend to have in common. Of these common elements, documentation is probably the most important. **Documentation** *consists of written records that guide activities and substantiate and prove what occurred.* For example, written procedures guide the work of individuals, thus ensuring consistency throughout a facility. Written records show what was done, by whom, and when, thus providing accountability. There is a common saying about the importance of documentation:

> ***"If it isn't written down, it wasn't done."***

This saying emphasizes the importance of keeping good records and is applicable in any biotechnology work place, whether it is a laboratory or production facility. Another common saying is:

> ***"Do what you say and say what you do."***

Documentation is part of the job of every bench scientist, technician, and operator in every biotechnology workplace. Because this element of quality is so important, we will devote all of Chapter 5 to it.

Another common element relates to resources. Every company, laboratory, and facility needs resources to produce quality products. If any of these resources is missing or is unsuitable for its purpose, then the final product is likely to be inadequate.

Skilled personnel are one of the most important resources in any company or organization. The employer is responsible for ensuring that all employees have the education and training needed to perform consistently under ideal and unusual circumstances, employees are familiar with all quality requirements pertinent to their work, they are well-supervised, and they receive the information, equipment, and tools to do their jobs properly. In a quality environment, there are usually records for each individual to show their qualifications and to keep track of their ongoing training, education, and acquisition of skills. In almost all companies, employees participate in some sort of safety training so that they know how to deal effectively with hazards.

Employees are responsible for the accuracy and completeness of their work. Employees are required to follow instructions, document their work, observe problems and report them as appropriate, understand the impact and consequences of their actions, and undergo continuous training. The language with which GMP regulations describe personnel is:

Personnel Qualifications

> **21CFR211.25** *Each person engaged in the manufacture, processing, packing, or holding of drug product shall have education, training, and experience, or any combination thereof, to enable that person to perform the assigned functions. Training shall be in the particular operations that the employee performs and in current good manufacturing practice . . . as they relate to the employee's functions.*

Facilities, equipment, instruments, and raw materials are other resources that are necessary to make a product. There are various types of biotechnology facilities, including laboratories where research and development occur, laboratories where quality control testing is performed, greenhouses, animal facilities, fermentation plants, and purification facilities. To produce a quality product, each area must efficiently accommodate the activities that occur there, be suitably heated, cooled, and otherwise have an appropriate environment, be of a suitable size, and so on. Within the facility there must be properly maintained equipment, instruments, and materials.

III. CHANGE MANAGEMENT

Change is an important topic when discussing product quality systems. In a research laboratory, progress depends on change. Ideas are tested, the results are examined, and the system is changed and tested again.

Even though each change should be recorded and its effects evaluated, change is part of the normal, daily routine in a research environment.

In a manufacturing environment, once a product is proven safe, effective, reliable, or otherwise acceptable, the most important feature of its production is consistency. Change, which is the essence of research, can be the bane of manufacturing a quality product. Although change is seldom welcomed in a manufacturing facility, it is inevitable. Raw materials may become unavailable, equipment becomes outdated, computer methods become more powerful, regulatory requirements change, and so on. There is conflict between the need to respond to changing circumstances and the need for control over processes, materials, and documentation. Change management is therefore one of the most important and most difficult aspects of quality systems in a manufacturing environment, particularly where regulated products are made.

The basis of effectively controlling change is that changes are reviewed, evaluated, and approved before and after they are made. Changes require detailed assessment and evaluation. This is because there is the potential that a simple, seemingly unimportant change might alter the way a process runs or the characteristics of a finished product. In regulated companies some changes require approval by the proper regulatory agency.

The need for control of change requires a way of thinking that is alien to most researchers who are continuously looking for new and better ways to accomplish a task. The tension between the need for flexibility during research and the need to control change is very evident when a biotechnology company is making the transition between being involved primarily in research and development to production. During this transition, companies must institute efficient, thorough procedures to manage change, and they must educate personnel in how to use the procedures.

Some points regarding the control of planned changes are shown in Table 4.4.

EXAMPLE PROBLEM

A new quality control technician is hired in a biopharmaceutical company. She is assigned the task of routinely checking samples of the company's product for the presence of the deadly pathogen, *Badbeastea miserabilis*. The test consists of applying one *mL* of each sample to a petri dish containing special nutrient medium, allowing the plate to incubate at a specific temperature for two days, and then checking the plate for the characteristic colonies of *B. miserabilis*. The technician (who is eager to make a positive contribution in her new job) reads about a new nutrient medium that supports the growth of *B. miserabilis* better than the old medium, is less costly, and allows the assay to be performed in only one day. Elated, the technician orders the new nutrient medium

and begins checking samples of product with the new medium as soon as it arrives. She then tells her supervisor how well this new medium has worked. What do you think about this (imaginary) scenario?

DISCUSSION

Even though the enthusiasm of this new technician is laudable, her new job has not begun auspiciously. The issue in this scenario is change and how change is controlled. The assay for the *B. miserabilis* would have been developed and tested by the R&D unit. During that testing period, R&D scientists would have optimized the assay using the original nutrient medium, checked for potential problems, and written documentation for the assay. Switching to a new nutrient medium could result in an unforeseen difference in the results of the test. For example, certain strains of the pathogen might not grow on the new medium. Before switching to a new nutrient medium, there would need to be testing to see that results with the original and new media are comparable. These tests might require testing samples on both media side by side to see if the results were the same. Changing the nutrient medium would also require that proper forms be completed, that approval from a supervisor(s) and from the Quality Assurance Unit be obtained, and that new directions for performing the test be written. The new technician failed to complete the proper steps.

Note that the new technician made a serious mistake that suggests that her introduction to her job and the way in which she was supervised were inadequate. The company should therefore look seriously at its methods of training and supervising its employees.

Table 4.4   MANAGING CHANGE

1. ***Each laboratory, facility, or organization should have a procedure for making changes.*** In a nonresearch environment, this usually involves having an SOP that outlines the process for making a change.

2. ***Change should always be justified.***

3. ***In a manufacturing facility, proposed changes should be evaluated and preapproved by R&D.***

4. ***A technical review of the proposed change should be performed to assess its value, and to address risks associated with the change.***

5. ***In a regulated company, the QC/QA unit must review the change to see if it requires approval by regulatory agencies.***

6. ***If necessary, the change should be evaluated for financial implications.***

7. ***After the change is made its effects should be investigated and documented.***

EXAMPLE PROBLEM

Suppose exactly the same events transpire as in the previous example problem. This time, however, the setting is a research laboratory in a university. What do you think about the scenario now?

DISCUSSION

The technical issues are the same regardless of the type of laboratory. It is important to ascertain that the new nutrient medium provides results that are comparable to the original medium. Testing the new nutrient medium would be well-advised. The difference in a basic research laboratory, as compared with the QC laboratory, is the process for making changes. In a research laboratory there is no Quality Assurance Unit to approve this change. The technician may be able to rely on his own judgment or may need to only consult his direct supervisor. He will need to document the testing he performs of the new nutrient medium in his laboratory notebook. If he changes to the new medium, then that change must be recorded in the notebook and in the written procedure for the assay. He will not, however, need to complete change control forms.

CASE STUDY EXAMPLE RELATING TO UNCONTROLLED CHANGE

A biotechnology company that makes products for research use was producing an enzyme for use in recombinant DNA procedures. Several customers observed lot-to-lot variation in the performance of the enzyme. The company was concerned about these customer comments and began an investigation of the product. The production records showed that there had been a large increase in demand for the enzyme in recent months and so the company had combined scale-up with production. During this period, production scientists had made substitutions and changes without review. This made it very difficult to identify specific sources of the lot to lot variability. Scientists performed a number of analyses on the enzyme. They eventually observed that the enzyme's storage solution, which contained Mg^{++}, was losing the Mg^{++} over time. Mg^{++} is a cofactor required by the enzyme for activity. The loss of magnesium was eventually traced to a change in the plastic tubes in which the enzyme was stored. The tube manufacturer had begun adding an antioxidant to the plastic and the antioxidant reduced the magnesium level in the enzyme solution. In this case, a seemingly trivial change in a raw material caused a problem. Finding the cause of the problem required a costly investigation. Careful control of change aims at preventing this type of predicament, or making it easier to trace the cause of difficulties should they occur.

## CHAPTER 5

Documentation: The Foundation of Quality

I. INTRODUCTION: THE IMPORTANCE OF DOCUMENTATION

Everyone who has taken a science class is familiar with laboratory notebooks. Not every student, however, realizes their tremendous importance in the workplace, nor that laboratory notebooks are one of the basic components of a broader system of documentation. **Documentation**, *which is defined most simply as a system of records,* is essential to any quality system. The following quotes from ISO 9000 show how documentation is woven into the structure of a quality system:

> <u>Quality System</u>, ISO 4.2.1 <u>General</u>
> *The . . . [company] shall establish, <u>document</u>, and maintain a quality system as a means of ensuring that product conforms to specified requirements. <u>Quality Planning</u> 4.2.3. The . . . [company] shall define and <u>document</u> how the requirements for quality will be met . . . [Emphasis added]*

FDA's policy regarding documentation is, "if it isn't written down, it wasn't done." If the documentation relating to a particular batch of a regulated product (such as a drug) is lost, accidentally destroyed, or is badly prepared, then that batch of product cannot be sold. Such an error could cost the company millions of dollars. Regulated companies, therefore, have extensive systems in place to ensure that work is recorded, that the documents associated with every product are completed, that all documents are securely stored and can be retrieved from storage, and that documents are protected, just as is the product itself.

Documentation is equally important in a research setting. If a researcher cannot show written evidence of their results, then those results are not credible. In 1986 a scientist in an academic laboratory was accused by a colleague of publishing erroneous data. This accusation led to a series of investigations, beginning at the researcher's institution and eventually winding up before a U.S. Congressional Committee. (The project the researcher was working on was funded by federal grant money.) The Secret Service was called to scrutinize the scientist's laboratory notebooks and raw data. Secret Service analysts found evidence that the researcher did not record her observations at the time they were obtained, that she did not keep all her original observations in her laboratory notebook, but rather used a pad of lined paper, and that dates in her notebooks and on instrument recordings were incorrect. Investigators charged the scientist with intentional fraud. The researcher claimed that even though her notes were not orderly, they were not fraudulent. The investigations culminated in a 1996 hearing that involved 6 weeks of testimony and thousands of pages of written statements. The investigative panel in 1996 concluded that, although she did not follow accepted rules of documentation and had made errors, the researcher was not guilty of fraud. Over 10 years, this incident adversely affected many individuals and caused public concern about the honesty and integrity of scientists. This was an unfortunate situation that likely could have been averted if the scientist had followed accepted documentation practices.

Documents have many important functions. They provide a record of what was done, by whom, when, how, and why. They provide objective evidence that a product was made properly, that all personnel followed proper procedures, and that all equipment was operating correctly. Some of the many functions of documentation are summarized in Table 5.1.

II. TYPES OF DOCUMENTS

A. Overview

Although documentation is essential in all biotechnology work environments, the specific types of documents and the systems for documentation vary in different workplaces. In an academic research laboratory, the major documentation requirements are that investigators can reconstruct their work based on their records, solve problems and detect mistakes, prove to the scien-

Table 5.1 THE FUNCTIONS OF DOCUMENTATION

1. *Record what an individual has done and observed.*
2. *Establish ownership for patent purposes.*
3. *Tell workers how to perform particular tasks.*
4. *Establish the specifications by which to evaluate a process or product.*
5. *Demonstrate that a procedure was performed correctly.*
6. *Record operating parameters of a laboratory instrument or a manufacturing vessel.*
7. *Demonstrate by an evidence "trail" that a product meets its requirements.*
8. *Ensure traceability.* Here we define **traceability,** as does ISO, *as the ability to trace the history, applications, and location of a product and to trace the components of a product.* Traceability leads to knowledge of the origin of parts and materials that go into making a product, the production history of the product, and the distribution and location of the product after sale. Traceability helps to ensure that if problems arise in a product, then the origin of the problem can be traced to its components, and the product itself can be found and recalled if necessary. Traceability depends on an organized, well-designed system of documentation.
9. *Establish a contract between a company and consumers.* The written specifications, labels, and other documents associated with a product establish that contract.
10. *Establish a contract between a company and regulatory agencies.*

tific community that their results were properly obtained and were accurately reported, and provide a trustworthy chronological record of their work. The laboratory notebook is the primary document in a research laboratory and will become a matter of public record in patent applications, disputes over who first had an idea, and if there ever should be question about the correctness or authenticity of reported findings.

Individuals in research and development laboratories likewise rely heavily on laboratory notebooks to document findings, especially when applying for patents. R&D workers also prepare documents that describe how to manufacture the product and the properties of a product.

People in production facilities use documents other than laboratory notebooks. For example, standard operating procedures (SOPs), that describe how to perform a specific task, are essential in production facilities.

Table 5.2 summarizes various types of documents. The first part of the table focuses on documents that are commonly found in laboratory environments. These include academic research laboratories, R&D laboratories, testing laboratories, and quality control laboratories associated with production facilities. With the exception of laboratory notebooks, many of the types of documents in the first part of Table 5.2 also are used in production environments. For example, labels, SOPs, and recordings from instruments are found in both laboratories and production facilities. The second part of the table gives examples of documents that are specific to production facilities and are not used in laboratories.

EXAMPLE

Consider the documentation that might be required when performing a routine laboratory task in a regulated industry, such as mixing a 1 M solution of NaCl:

1. The technician will follow an **approved standard operating procedure** for mixing the solution.

2. The raw materials—clean glassware, NaCl, and purified water—will all have **documents associated with them that show they were tested and found satisfactory prior to being released for use by a technician**.

3. The instruments used to weigh the NaCl and measure the water will have **logbooks** or other documents showing they were properly maintained.

4. The technician will need to record information about the solution on a **form**. The form identifies the person making the solution, the date, the procedure followed, quantities and source of raw materials used,

---

Table 5.2 *TYPES OF DOCUMENTATION*

Examples of Documents that Are Common in Laboratories

1. *Laboratory notebooks* are a chronological log of everything that an individual does in a laboratory.

2. *Standard Operating Procedures (SOPs)* detail what is to be done to complete a specific task and how to document that the task was done correctly.

3. *Forms* contain blanks that are filled out by an operator to record information. Forms are typically associated with SOPs or other documents.

4. *Protocols* are similar to SOPs in that they explain how to do a task. The term *protocol*, however, is often reserved for situations where a question or hypothesis is to be investigated or the procedure is going to be performed only once.

5. *Reports* are documents generated during the execution of a protocol.

6. *Equipment/Instrument logbooks* keep track of maintenance, calibration, and problems for a given instrument or piece of equipment.

7. *Recordings from instruments.*

8. *Electronic documents.*

9. *Analytical laboratory documents* record information regarding the testing of a sample.

10. *Numbering systems* such as lot number assignments, are used to keep track of materials, equipment, and products.

11. *Labels* are attached to solutions, products, or items to identify them.

12. *Chain of custody forms* are used to trace the movement of a sample throughout a facility and to keep samples and sample test results from being confused with one another.

13. *Training reports* document that individual employees were properly trained to perform particular tasks.

Examples of Documents that Are Specific to Production Facilities

1. *Batch records* are collections of documents associated with a particular batch of a product.

2. *Regulatory submissions* are forms filled out and sent to regulatory agencies to inform them of what a company is planning and/or to ask permission to test or sell a product.

3. *Release of final product record* is filled out when a product has been approved for sale.

the storage location and conditions of the solution, the amount made, and the number of the batch.

5. The resulting solution will need an **identifying number** and there will be documentation associated with assigning this number.

6. The solution will need a **distinguishing label**.

7. If the solution is split into more than one container, **documented ID numbers** will need to be assigned to each container referring back to the original solution. Each container will need a **distinguishing label**.

B. Laboratory Notebooks

Laboratory notebooks *are assigned to individuals and are a chronological log of everything that individual does and observes in the laboratory.* A laboratory notebook should be complete enough that you or another person could exactly repeat your work based on information in the notebook. Of all the documents described in this chapter, the laboratory notebook is the most important in a research laboratory and is the most widely used form of documentation in laboratories.

Raw data *are the first records of an original observation.* Depending on the situation, raw data may be recorded into the laboratory notebook with pen by the operator, may be a paper output from an instrument, or, increasingly, may be recorded directly into a computer medium. It is essential that a scientist or operator save raw data, that it never be altered or edited, and that it be retrievable.

A laboratory notebook is an important legal document. It can be used to establish a patent claim, to assign credit for discovery of a phenomenon, to document the honesty and integrity of data for publication, and to trouble-shoot problems. It can be subpoenaed in litigations. It can be examined by the FDA, EPA, or another regulatory agency. It is critical, therefore, that the laboratory notebook be used properly and consistently. Because it is so important, there are expectations or guidelines for properly keeping a laboratory notebook. Table 5.3 outlines the components that are generally expected in a laboratory notebook. General "rules" to remember when using a laboratory notebook are summarized in Table 5.4.

C. Documents that Are Common in Laboratories

i. STANDARD OPERATING PROCEDURES

Most production facilities and many laboratories use procedures to instruct personnel in how to perform particular tasks. A **procedure** *is a written document that provides a step-by-step outline of how a task is to be performed.* Such documents are often called **standard operating procedures (SOPs),** although other names are used as well. For example, a research laboratory might simply call

these "procedures" (or even "protocols," although this term has another meaning as explained in Table 5.2). Everyone follows the same procedures to assure that tasks are performed consistently and correctly. Standard operating procedures describe what is required to perform a task, who is qualified or responsible for the work, what problems may arise and how to deal with them, and how to document that the task was performed properly. SOPs must be written so they are clear, easy to follow, and can accommodate minor changes in instrumentation.

Table 5.5 outlines the components which are generally expected in an SOP.

EXAMPLE

Discuss the considerations in writing an SOP to make chocolate chip cookies.

A. There is preliminary work that must be completed before an SOP can be written. In this example, a first step might be to list the characteristics in the product that are important, such as:

Flavor, Texture, Softness, Size, Shape, Number of chips per cookie

Each of the preceding properties must then be described (i.e., specified in such a way that it can be measured or evaluated). For example, flavor might be evaluated by individuals trained as tasters.

B. The raw materials and equipment required must be listed, such as:

Flour, Sugar, Vanilla, Chocolate chips, Mixer, Bowls, Measuring cups, Measuring spoons, Oven

There are decisions to make regarding these materials. For example, will a particular brand of flour and chocolate chips be specified? What sorts of measuring devices will be adequate? All of these decisions must be made before the SOP can be written.

C. A standard operating procedure to make cookies with the desired properties can be written. The SOP will include a title, the author, a statement of purpose, ID number and date, statement of scope, who is qualified to make the cookies, raw materials required, and the steps in making the cookies. This is, of course, a formalized "recipe." A form may also be associated with the SOP in which the baker can record information, such as the date, the lot number of the materials used, the time, and the temperature of the oven.

Consider briefly the language used in the SOP. Compare the following alternatives to describe how to mix the batter:

a. Mix the batter for 10 minutes at room temperature.

b. Mix the batter with an electric mixer until the batter is well-mixed.

c. Mix the batter with an electric mixer for 10 minutes on setting 1.

Table 5.3 *TYPICAL COMPONENTS OF A LABORATORY NOTEBOOK*

1. *In front of the notebook: an identification number for the notebook, the person to whom the book is assigned, the date of assignment, the project, the company, and any other identifying information.*

2. *A table of contents on the first pages.* The table of contents should include page numbers and descriptions with sufficient detail to allow easy searching of the notebook's contents.

3. *A page number on every page in consecutive order.*

4. *For each project, a thorough listing of the results of any literature search and any experimental information collected from colleagues.*

5. *Dates, titles, and descriptions.* Begin the record of each day's work with the date, a title, and description of the objectives for the day. It is common to begin each day's work on a new page with diagonal lines drawn across the unused part of the previous day's page.

6. *Dates and signatures on each page.* In many laboratories a corroborating witness, as well as the scientist/technician performing the work, must read, sign, and date each page. The scientist and the witness should verify that there are no blank spaces, that all tables are complete, and that the page is complete. If corrections are later made to an entry, the corrections should be signed and dated by both the scientist and witness.

7. *The rationale for each activity performed.*

8. *Relevant equations or calculations.*

9. *Complete descriptions of all instrumentation (including models and serial numbers), chemicals used (including manufacturers, catalog and lot numbers, and expiration dates), reagents used (including "recipes" or SOPs and quantities), supplies used (with a complete description), samples assayed (including complete descriptions and sample identification numbers), standards or reference materials used, and so on.*

10. *Procedural details.* If an SOP or protocol is followed, it should be referenced in a unique fashion (e.g., by title and revision date, and/or by ID number). It is usually not necessary to copy an SOP or protocol into the notebook, but any deviations and their justification should be noted. If a procedure comes from a compendium, journal, book, or manual, the complete reference for the procedure should be cited and it may be necessary to record procedural details in the laboratory notebook.

11. *Data.* Data take many forms, for example, values read from instruments, color changes observed, and instrument print-outs. Instrument print-outs may be securely affixed in the notebook with a permanent adhesive. Print-outs should be titled and dated so they can be properly replaced in the notebook, if necessary. If data (such as print-outs) are not affixed in the notebook, they may be checked, titled, signed, dated, filed, and referenced clearly in the laboratory notebook.

12. *Observations.* Observations may include, for example, changes in pH or temperature, humidity readings, and instrument operational parameters.

13. *A brief summary of the work completed.*

14. *A conclusion and brief interpretation of data collected may be appropriate.* For example, if a particular line of investigation is pursued on the basis of preliminary results, note this in the laboratory notebook. Such decisions may be relevant in patent disputes at a later date. Avoid negative or extraneous comments.

---

d. Mix the batter with an electric mixer until no unmixed flour or large chunks are visible. For example, 10 minutes on an intermediate setting on a brand Z mixer is usually sufficient.

Statement **a** is vague and does not direct the baker in how to accomplish the task or its endpoint. Statement **b** gives an endpoint, but the description is too vague to be useful. Statement **c** is useful only for a particular brand of mixer that has a setting called "1." Statement **d** has the clearest description of an endpoint and is flexible in terms of which mixer is used. The statement "intermediate setting" might be too vague, but it is probably the most applicable to most mixers. In any case, statement d explains the outcome that is the important point in this step. Note that if overmixing detracts from the quality of the cookies, then this should be noted in the procedure.

EXAMPLE

Carol DeSain (in *Documentation Practices,* Carol DeSain and Charmaine Vercimak Sutton. Advantstar Publishers, Ohio, 1996) cites an example in which the language of a procedure is clear, but the intent is not to a new employee. The statement in the SOP is: "Wash the filter press in mild detergent and rinse with WFI [purified water]. Make sure that the filter support grid is completely dry before placing a new filter on the grid." The writer expects the grid to be air or oven dried. The technician, on the other hand, focuses on the need for the grid to be <u>dry</u>. She therefore retrieves a box of laboratory wipes and wipes the grid dry. This action is not in violation of the SOP, but it results in the contamination of several batches of product with fibers from the laboratory wipes.

This example highlights the importance of knowing the audience. The degree of detail in an SOP and the particular points that are noted will, to some extent, depend on the training and background of the people who will be performing the procedure. It is important when writing an SOP to be conscious of the needs of the reader and when following an SOP to be certain that you understand its intent.

Some of the potential problems with SOPs that must be avoided are summarized in Table 5.6.

In a company, every SOP must be reviewed and accepted before it is used. The person who writes the SOP has responsibility for it and must sign it. A second individual who is knowledgeable about the work also normally shows approval with a signature and, in a company, an individual from the quality assurance unit also signs it.

SOPs periodically require changes. It is important that the old SOPs are destroyed or made unavailable when changes are made, except for a copy(ies) kept in a historical file. Only the latest revision of the SOP should be available to processing or laboratory technicians. In a regulated setting, depending on the magnitude of the change, revisions may have to be checked and approved by several levels of responsible individuals and by the quality assurance unit before the change can be implemented. Each revision of an SOP needs a date and revision number so that it can be uniquely identified. An example of an SOP is shown in Figure 5.1.

Table 5.5 Typical Components of an SOP

1. *Title.*
2. *ID number, revision number, date of revision.* Most SOPs are revised periodically. It is essential that each worker uses only the most up to date revision.
3. *Statement of Purpose* (may restate the title with a little more detail).
4. *Scope* describes when the procedure is relevant. For example, if the procedure is for verifying the performance of a balance, then it might be only for a particular brand, or a particular model.
5. *A statement of responsibility,* who does this task.
6. *Materials required, including manufacturers and identifying information.*
7. *Calculations required, preferably with an example.*
8. *The procedure itself written as a series of steps.* The actions required, how they are performed, and the endpoints of the steps should be included.
9. *References to other documents, as required.*
10. *How to document that the procedure was performed; references to any associated forms.*

Table 5.6 Problems to Avoid Relating to SOPs

1. *The SOP says what to do, but not how to do it.*
2. *The procedure was written by someone who does not have experience doing the work.*
3. *The SOP has too much detail or too little detail.*
4. *The procedure is not written in the order in which the tasks are actually performed.*
5. *The SOP is not updated as needed.*
6. *Employees cannot find the right SOP or use an older version.*

Table 5.4 Guidelines for Keeping a Laboratory Notebook

1. *Use only a bound notebook.*
2. *Make sure every page is numbered consecutively before using the notebook.*
3. *Never rip out a page.*
4. *Make all entries in ink.*
5. *Be legible, clear, and complete in your writing.* Remember that you, supervisors, colleagues, patent attorneys, and regulatory agency inspectors may review your work, perhaps years after you record it.
6. *It is essential that you enter all observations and data immediately and directly into the notebook—not onto a paper towel or the back of your hand.*
7. *Cross out errors with a single line so the underlying text is still clearly legible.* Date when the cross out was made and, in a few words, note why the entry was crossed out. Never erase mistakes. Initial crossed out items. In some settings, cross-outs must be initialed and dated also by a supervisor.
8. *Note all problems; never try to obscure, erase, or ignore a mistake; be honest.*
9. *Blank lines or unused portions of the page should be crossed out with a diagonal line so nothing may be added to the page at a later date.*
10. *Be objective in stating what is observed.*
11. *Always include __detailed__ information about instruments, reagents, samples, materials and equipment used; make sure it is possible to account for and trace all materials used.*
12. *Be certain that the laboratory notebook is stored in a secure location.*

Clean Gene, Inc.
STANDARD OPERATING PROCEDURE

SUBJECT: Identification Method for Calcium Sulfate

Effective Date: 7/1/99

SOP # 2648

Revision 01

Page 1 of 3

Prepared By *N. Warren*

Approved By *J. McMillan*_____ Date *6/24/99*

Approved By *J. Lownd*_____ Date *6/26/99*

---

1. Scope

    Provides the method for assessment of the raw material calcium sulfate.

2. Definitions—NA

3. References:

    3.1 USP, current—General Identification Tests

    3.2 Incoming Raw Material/Component specification: $CaSO_4$, Grade I

4. Reagents

    4.1   3N Hydrochloric Acid, HCl

    4.2   6N Acetic Acid

    4.3   Methyl Red

    4.4   Ammonium Oxalate

    4.5   Ammonia

    4.6   Barium Chloride

5. Responsibility

    5.1 This test is to be performed by an appropriately trained analyst. Analyst training and documentation of training will be conducted per SOP 5688. Approval of analyst data is the responsibility of the Quality Control Manager.

    5.2 The results of the test will be verified by a second qualified analyst.

6. Hazard Communication

    6.1   *3N Hydrochloric Acid, HCl*

    DANGER: Corrosive. Avoid contact with skin and eyes. Avoid inhalation of fumes and mist. Do not mix with caustics or other reactives.

    6.2   *6N Acetic Acid*

    DANGER: Corrosive. Avoid contact with skin and eyes. Avoid inhalation of fumes and mist. Do not mix with caustics or other reactives.

    6.3   *Methyl Red*   **CAUTION: Irritant. Avoid contact with skin and eyes.**

    6.4   *Ammonium Oxalate*

    CAUTION: Irritant. Avoid contact with skin and eyes.

    6.5   *Ammonia*   **CAUTION: Irritant. Avoid contact with skin and eyes.**

    6.6   *Barium Chloride*

    CAUTION: Irritant. Avoid contact with skin and eyes.

7. ATTACHMENTS

    7.1 Attachment I—Form No. 687

8. PROCEDURES

    8.1 Sample Preparation

    8.1.1   In a clean beaker or test tube, add approximately 200 *mg* of the incoming Calcium Sulfate to be tested to 4 *mL* of 3N Hydrochloric Acid and 16 *mL* reagent water.

    8.1.2   Gently warm with low stirring on a hot plate to aid dissolution.

    8.1.3   Portions of the solution should respond to the tests for calcium and sulfate.

    8.2 Calcium Identification

    8.2.1   Place about 10 *mL* of the solution from 8.1 into a clean test tube.

    8.2.2   Place 2 drops of methyl red into this solution followed by enough ammonia to turn the solution YELLOW.

    and so on . . . the entire SOP is not shown here

Figure 5.1.   A Portion of an SOP. This SOP is used to test an incoming raw material to ensure it is calcium sulfate.

CLEAN GENE, INC.   Page 1 of 1

Agarose Gel Preparation Form, Revision 01

1/31/98   Form # 992

Technician Name _____   Date _____

1. *Add between 1.8 and 2.0 grams of agarose to 100 mL of purified water.*
   Record the ID/lot number for the agarose _____
   Weight of agarose added _____
   Source of water used _____
2. *Place the mixture in a microwave oven on its highest setting and heat until the mixture just begins to boil.*
   Time to boiling _____
3. *Let the mixture cool until it is between 58 and 62°C and then pour gel.*
   Temperature when pouring gel _____

Figure 5.2.   A Portion of a Form Associated with Preparing an Agarose Solution.

ii. Forms

As mentioned in the chocolate chip cookie example, an SOP is often associated with a **form** that is filled in as the procedure is being performed. Filling in the blanks requires the individual performing the task to monitor the process as they go along, thus ensuring that everything is going smoothly. In addition, the form will remind the technician to record information about lot numbers, raw materials, times, temperatures, and other relevant information that is easy to forget to record. In some production laboratories a witness must sign key steps. Figure 5.2 shows an example of a form.

iii. Protocols

The term **protocol** *is used in some industries to refer to a procedure that tells an operator how to perform a task or an experiment that is intended to answer a question or test a hypothesis.* (The term *protocol* may also be used for a procedure that will only be performed one time.) In contrast, SOPs do not lead to the answer to a question. For example, one follows a "procedure" to clean a laminar flow hood, but one follows a "protocol" to investigate the effectiveness of cleaning a laminar flow hood with different cleaning agents.

A protocol must include information on what data are to be collected, how the data are to be gathered, what outcome proves or disproves the hypothesis, and any statistical methods that need to be used. A protocol may reference SOPs. Both research and production facilities use protocols. In research laboratories people obviously investigate questions all the time. In production facilities people investigate whether a product performs as expected, the qualities of the product under certain conditions (such as long storage), the effects of the product in a test population, and so on. Example statements or hypotheses that might be seen in an industrial protocol are:

"This study is designed to demonstrate that the cleaning process for laminar flow hoods does not leave detectable detergent residues on the surface of the hood."

"This study is designed to demonstrate that the antitumor drug XYZ has no adverse medical effects on the livers of test subjects."

The components of a protocol are somewhat different from a procedure and are summarized in Table 5.7. An example of a portion of a protocol is shown in Figure 5.3.

iv. Reports

A **report** *is a document that describes the results of an executed protocol.* The report summarizes what was done, by whom, why, the data, and the conclusions. A report is written in narrative format. Reports from basic scientific research are published in scientific journals. Reports from investigations performed in a company may or may not be published but must be available for inspection.

Table 5.7   Typical Components of a Protocol

1. *The hypothesis or question the study is designed to answer.*
2. *A description of the study.*
3. *A plan for how the study is to be conducted.*
4. *Information about how the sample(s) is to be collected, processed, and identified.*
5. *The methods that will be used to test the hypothesis.*
6. *The schedule of testing.*
7. *The way the study results and conclusions will be reported.*
8. *The criteria that will be used to reach conclusions.*

Clean Gene, Inc.

Page 1 of 3

Cleaning Protocol CL 898; Revision 01

Prepared by *Tania Seid*   Date *3/6/98*

Approved by *J. Mauey*   Date *5/15/98*   Approved by *Jon Reidman*   Date *5/16/98*

PROTOCOL

A TEST OF THE EFFICACY OF ALCOHOL FOR DISINFECTING THE SURFACE OF BIOLOGICAL HOODS

1.0   Study Question

This study is designed to determine the efficacy of alcohol for disinfecting the surface of biological hoods, BH655.

2.0   Objectives

—To determine whether surface cleaning of a contaminated biological hood with ethanol is an effective method of eliminating bacteria.

—If ethanol is effective, to establish the optimal ethanol concentration.

3.0   Plan

—Contaminate the surface of the biological hood with 10^9 *cfu* of *E. coli* strain 56298.

—Treat the surface with ethanol of different concentrations

—Determine the success of disinfection by using the Standard Wipe Method of detection of bacteria, SOP 87-68.

4.0   Equipment required

—Biological cabinet # 4

—Spectrophotometer # 7

—Overnight culture of *E. coli* strain 56298

—Ethanol at a concentration of: 60, 70, 80, 95%.

—Sterile bacteria wipes (ID No. 932-8)

—Microbial test agar (ID No. 56-84)

—Sterile forceps (ID No. 6-1-97)

—Nutrient agar plates (prepared according to SOP 87-75)

5.0   Procedure

5.1.   Grow overnight culture of *E. coli* (according to SOP 95-17).

5.2   Measure the absorbance of culture at wavelength 520 *nm*

5.3.   Calculate the number of *cfu/mL*

   1 AU = 10^9 *cfu/mL*

5.4.   Dilute a portion of the culture to get a concentration of 10^9 *cfu/mL*

5.5.   Spread 1 *mL* of diluted culture over a 2 square inch section of the hood.

5.6.   Wait 5 minutes.

5.7.   Spread 10 *mL* of ethanol across the surface of the contaminated hood.

5.8.   Wait 5 minutes.

5.9.   Using a clean, sterile absorbent wiper, wipe up the ethanol.

5.10.   Allow the area to dry for 5 minutes.

5.11.   As directed in SOP 87-68, using the sterile forceps and wipes, wipe the entire 2 square inch section of the "spill."

5.12.   Place the bacterial test swipe on nutrient medium.

5.13.   Repeat steps 5.5 through 5.12, three times for each concentration of ethanol, for a total of 12 plates.

5.14.   After 36 hours record the results, as directed in SOP 87-68.

   . . . and so on

Figure 5.3.   A Portion of a Protocol. The protocol directs the activities of the technician in conducting a study.

v. Logbooks

Logbooks *are used to record information chronologically about the status and maintenance of equipment or instruments.* Logbooks are usually bound and labeled notebooks that are associated with a specific instrument, area, or piece of equipment. When an item is used, calibrated, maintained, or repaired, this is indicated in the logbook.

vi. Recordings from Instruments

Many analytical instruments generate automatic printouts of results. Printouts are generally considered to be raw data. If the results from the instrument belong in the laboratory notebook, then it is usually acceptable to tape them there. In such cases the laboratory notebook page, signature, and date should be noted on the printout so they can be replaced in the proper place if they become detached. It is common to record on a page how many documents are taped to that page so they can be found if necessary. Alternatively, instrument printouts can also be filed and referred to in the notebook or on a form.

Some instruments monitor themselves and record their operating parameters as they operate. For example, modern autoclaves continuously record the date, time, temperature, and pressure throughout their cycle. This information is important in demonstrating that the instrument performed properly. Files may be kept of such recordings, or they may be associated with the production records for a particular batch of product (batch records will be discussed later).

Printouts from any instrument must be thoroughly identified (including, for example, the date, product name, batch number, and equipment number), and be signed and dated by the technician.

vii. Electronic Documentation

Computers are often used in the laboratory to control instruments, to monitor their performance, to record data, to analyze data, to store protocols and procedures, and for a multitude of other tasks. Computers can generate documents in the form of printouts, disks, and other media. For security reasons, computers have been slow to replace paper in regulated industries. Mechanisms to ensure the security and validity of computer documents have been and continue to be developed. The use of computers for documentation is a rapidly evolving field (see Chapter 32).

viii. Analytical Laboratory Documents

Analytical tests are those that measure a property(ies) of a sample. For example, in an environmental laboratory, analysts might test a sample of lake water to determine the level of cadmium present. In a clinical laboratory, analysts might perform a drug screen of a patient's blood. The product is the test result(s). Documentation is required that provides information both about the method used and the sample being tested. Specific types of information that must be documented are summarized in Table 5.8.

ix. Identification Numbers

Identification numbers uniquely identify items. There are many types of items that require identification, including raw materials, documents, equipment, parts, batches of product, chemicals, solutions, and laboratory samples. Identification numbers convey two pieces of information: what the item is, and which one it is. For example, the first part of an ID number might logically tell whether the item is a particular type of instrument or a particular type of solution. The next part of the ID number might tell which particular one of those instruments it is, or which batch of solution it is.

x. Labels

Labels identify equipment, raw materials, products, and other items. Information that may be found on labels is summarized in Table 5.9. An example of a label is shown in Figure 5.4.

Table 5.8   Essential Information to Document in Analytical Laboratories

1. ***Information regarding each assay method used:***
   The purpose of the test
   The limits of the test (e.g., what is the lowest level of the material of interest that the test can detect)
   The origin of the test method (e.g., whether it came from a compendium of commonly accepted methods)
   Validation information (validation is discussed in Chapter 6)
   The suitability of the test for a given purpose

2. ***Information regarding each sample:***
   The sample ID number
   How the sample was collected, by whom, and on what date
   Where the sample is stored and the conditions of storage
   How and when the sample is to be discarded

3. ***Information regarding each assay:***
   The sample tested
   The date of the test
   Who performed the test
   Reagents and materials used
   The method used to test the sample
   The raw data collected during testing
   Calculations for the sample results
   Reported conclusions based on the test

Table 5.9 *TYPICAL COMPONENTS OF A LABEL*

1. *Date the item was prepared.*
2. *The person responsible for the item.*
3. *The ID number of the item.*
4. *The lot number.*
5. *The identity, name, or composition of the item.*
6. *Safety information.*
7. *The name of the company or institution.*
8. *Storage and stability information.*

xi. CHAIN OF CUSTODY DOCUMENTATION

In laboratories that handle many samples that come from diverse subjects or sites, it is critical that samples not be confused with one another, and that information and results are always associated with the correct sample. **Chain of custody documentation** *provides a chronological history, or "paper trail," for samples.* For example, in a clinical testing laboratory patient samples must be kept in order. An environmental testing laboratory tests samples from many sites and must keep track of them. Forensics laboratories must scrupulously keep evidence from various cases in order; otherwise, their test results will be invalid in court.

Chain of custody documents are a method of organizing information about samples. Each sample must be assigned a unique ID number. Records for each sample must show the source of the sample, who collected the sample, who transported the sample, its condition upon receipt, the date of receipt, how the sample was processed and tested in the laboratory and by whom, how it was stored, and how it was disposed of, if relevant. The sample is logged in and out as it is moved and processed. The format and the exact nature of these records is variable as each organization has its own requirements.

xii. TRAINING REPORTS

Training reports are associated with individuals working in a facility. The training report shows the training the person has completed, the dates, the purpose, and so on. Training reports help to show that individuals are competent to perform their work. Training records are used in research laboratories, for example, to document that employees have been trained in the use of radio-isotopes.

An example of a training record is shown in Figure 5.5.

D. Documents that Are Specific to Production Facilities

i. BATCH RECORDS

There are many types of documents found in production facilities; a few of these will be explained here. **Batch records** accompany a particular batch of product. The **batch record** *directs the process by which a product is to be made, the raw materials required, and the SOPs to be followed.* The batch record stays with the product as it is made. The batch record usually includes blanks to be filled in by individual operators as procedures are performed.

The batch record is officially issued to the production crew by the quality department. It is essential that the batch record that is issued is complete, readable, and correct.

Table 5.10 shows some of the major components of a batch record.

Table 5.10 *ESSENTIAL COMPONENTS OF A BATCH RECORD*

1. *Product identification.*
2. *Document identification.*
3. *Company name.*
4. *Dates of manufacturing.*
5. *A step by step account of the processing and testing to be done.*
6. *The monitoring specifications—how will the operators know if the process is proceeding properly.*
7. *Raw data that must be collected and blanks to fill in to record it.*
8. *Materials and equipment that will be used.*
9. *Required signatures.*

CLEAN GENES, INC.

Solution Name:

Concentration:

Date Prepared: By: ID No.

Expiration Date:

Reagent Source: Lot No:

Storage Conditions: Special Precautions:

Figure 5.4.   An Example of a Label for a Reagent.

Clean Gene, Inc.

EMPLOYEE SOP TRAINING RECORD

Form # 875
Revision 01, 6/94

EMPLOYEE'S NAME

Rachel Allen

SOP TITLE	REVISION DATE	EMPLOYEE'S INITIALS	TRAINER'S SIGNATURE
Cleaning Hoods	6/20/97	RA	JES
Formulation 10	3/18/98	RA	LW
$CaSO_4$ Analysis	2/3/99	RA	JMM

Figure 5.5. Portion of an Employee SOP Training Record.

ii. REGULATORY SUBMISSIONS

Regulatory submissions *are documents completed to meet the requirements of an outside regulatory agency.* For example, before testing an experimental drug in humans, a company must submit an application to FDA showing its preliminary research on the drug, its plan for human studies, and other relevant information. (This document is known as an IND, or Investigational New Drug application.)

iii. RELEASE OF FINAL PRODUCT RECORDS

Companies must complete product release documents when a product has been manufactured and tested. The release document certifies the product, shows its specifications, establishes that the product documentation has been reviewed and approved, and states that the product is ready to be sold.

CASE STUDY EXAMPLES RELATING TO DOCUMENTATION

The Federal Food and Drug Administration has inspectors who periodically inspect pharmaceutical facilities and biotechnology companies that make regulated products. If the inspectors observe violations of Good Manufacturing Practices, they note the violations on forms, called "483s," and in official Warning Letters sent to the company. If companies fail to correct their deficiencies, then FDA can cause products to be seized and destroyed, fines to be levied, and, in the most extreme cases, individuals in the company may be charged as criminals and may be imprisoned if convicted. The following are excerpts from three actual warning letters sent by FDA to various companies. Improper documentation is frequently cited in warning letters. Observe in these warning letters how carefully inspectors checked for proper documentation and the details of their findings.

Warning Letter, Example 1

Dear Sir or Madam:
   During an inspection of your drug manufacturing facility . . . conducted on August 29 through September 25, 1996, our investigators documented serious deviations from the Current Good Manufacturing Practices Regulations . . . Deviations from the GMPs documented during this inspection included:
   [A list of 24 violations follows. Only some of those violations relating to documentation are excerpted here.]
   (6) Failure to maintain master production records containing: (a) signature and date of the person preparing the record, and date and signature of the person independently checking the

record; (b) strength and description of the dosage form; complete manufacturing and control instructions, sampling and testing procedures and specifications for the finished product; ... and (d) a description of the drug product containers, closures and packaging materials, including a specimen or copy of each label or other labeling signed and dated by the person or persons responsible for approval of such labeling ...

(15) Failure to establish written procedures to prevent microbiological contamination of manufactured products.

(21) Failure to document the person performing the analysis, when the analysis was conducted, and/or the name of the active ingredient in ... spectra used as an identity test of active ingredients.

The above identification of violations is not intended to be an all-inclusive list of deficiencies at your facility ... You should take prompt action to correct these deviations.

Failure to promptly correct these deviations may result in regulatory action without further notice ...

Sincerely,

[Signed by the District Director]

Warning Letter, Example 2

Dear ...

During an inspection of your firm ... on February 17–20, 1997, our investigators determined that your firm manufactures in vitro diagnostics, such as auto-immune disease test kits ... The above stated inspection revealed that ... the methods used in ... their manufacture ... are not in conformity with the Good Manufacturing Practice (GMP) for Medical Devices Regulation ... as follows:

[A number of violations follow. Only some of those violations relating to documentation are excerpted here.]

2a. The Operating Procedure for the ... Filling Machine does not reflect the current auto-immune kit microtiter plate fill volume ...

b. The Operating Procedure for the ... [device] used for sealing the packaging of the microtiter plates for the auto-immune kits does not reflect current vacuum level setting ...

This letter is not intended to be an all-inclusive list of deficiencies at your facility ...

Sincerely ...

[Signed by the district director]

Warning Letter, Example 3

Dear ...

Inspection of your unlicensed hospital blood bank on April 2–3, 1997 ... revealed serious violations ...

Inspection revealed that blood product disposition records prior to November of 1996 are not available. According to your blood bank supervisor, the missing disposition records were transferred to a computer system and were subsequently "lost" by that system. Your supervisor stated that the computer system has not been validated and is being used only for "practice".

Your supervisor also stated the blood bank had written back-up records for the data in the computer system. However, written disposition records could not be produced during the inspection for review ...

Sincerely ...

[Signed by the district director]

PRACTICE PROBLEM

Figure 5.6 is a brief portion of a batch record that covers the formulation of a particular product. Note that it consists of a series of steps to be performed by the operator and a series of blanks that the operator fills out as s/he performs each step. The operator initials each step as it is performed and a supervisor verifies that the procedure appears to have been properly executed. There are a number of errors in how this batch record was completed. Circle the errors. (Note that it is not necessary to understand the actual procedure in order to detect the errors in how the form is completed.) The answer is shown on page 73.

QUESTIONS FOR DISCUSSION

1. Consider the Warning Letters from FDA on pp. 70–71 relating to documentation inadequacies.

   a. Why is there an emphasis on signatures and dates?

   b. What components are necessary in a production record?

   c. Why is one company cited for failures to put volumes and vacuum level settings into their Standard Operating Procedures?

   d. Discuss the possible ramifications of the lost records at the blood bank.

2. The following question is modified from Scenario 5 in the "Skill Standard for the Biosciences Industries" document (Educational Development Corporation, Newton, MA, 1995).

   a. You are responsible for specimen receipt and processing. A sample is received in the laboratory for analysis. Demonstrate the procedure for processing this request.

   b. The label on the sample does not match the information on the requisition. The results are needed quickly. How should you proceed?

FORMULATION OF XYZ COMPOUND

Clean Gene, Inc.

3550 Anderson St.
MADISON, WI 54909
Revision 01
Master Batch Record # 133
Approved by _____Aaron Reid_____   _____Anna Gold_____   _____Sam Rothstein_____

           Date 2/14/98   Date 2/14/98   Date 2/16/98

Issued by: _Erin Jane_   **Date** 12/3/98   **Lot #** **15.987**

Product Name: **Very Good Product**

Strength: **10 Units/mL**   **Vial Size** **10 mL**   **Batch quantity:** **350 L**

Reference: Refer to separate Formulation SOP Q75 for quantities of each component.
           Refer to separate instrument/equipment SOP 76 for ID information

---

NOTE: PRODUCT IS TO BE STIRRED CONTINUOUSLY DURING COMPOUNDING AND FILLING

A.   COLLECTION OF WATER FOR INJECTION (WFI), USP.

A1. Collect approximately 370 L of WFI, USP, in a clean, calibrated vessel and cool to 24°C–28°C.
Vessel # ___7___ Amount collected ___365ℓ___ Initial Temperature ___23°C___ Time ___8:00___ (am)/pm
Final Temperature ___26°C___ Time ___08:15___ am/pm

                                           EJ / JM
                                         12/10/98 12/10/98

A2. Close the water for injection valves.   EJ / JM
                                         12/10/98 12/10/98

A3. Remove about 20L of the cooled WFI from step 1 and place in a clean, calibrated vessel.
     (This water will be used to bring the final formulation to the proper volume.)

Vessel # ___2___   EJ / JM
                                         12/10/98 12/10/98

A4. Remove about 5L of the cooled WFI from step 1 into a clean, calibrated vessel. (This water will be used to
     prepare the solutions used to adjust the pH.)

B.   PREPARATION OF pH ADJUSTING SOLUTIONS

B1. Collect 750 mL cool WFI from the vessel in step A4 and place into a 1000 mL volumetric flask and add
     100 g of NaOH (Sodium Hydroxide # 875) USP and dissolve.

Amount of WFI collected ___750mℓ___ Amount NaOH added ___115g___ Lot # _____

Time step completed ___09:15___ am/pm   ____ / _JM_
                                                      12/10/98

B2. Using a water bath containing cold WFI, cool the solution prepared in B1 to 25°C ± 5°C.

Final Temperature ___31°C___   EJ / JM
                                         12/10/98 12/10/98

B3. Bring the solution to 1000 mL with cool WFI from the vessel in step A4.

Approximate volume of WFI added ___300mℓ___ Time completed ___09:00___ (am)/pm

                                           EJ / JM
                                         12/10/98 12/10/98

     and so on

Figure 5.6.

ANSWER TO PROBLEM

FORMULATION OF XYZ COMPOUND

Clean Gene, Inc.

3550 Anderson St.
MADISON, WI 54909
Revision 01
Master Batch Record # 133
Approved by _____*Aaron Reid*_____   _____*Anna Gold*_____   _____*Sam Rothstein*_____

Date 2/14/98   Date 2/14/98   Date 2/16/98

Issued by: *Erin Jane*   **Date** 12/3/98   **Lot #** **15.987**

Product Name: **Very Good Product**

Strength: **10 Units/_mL_**   **Vial Size**   **10 _mL_**   **Batch quantity:**   **350 _L_**

Reference: Refer to separate Formulation SOP Q75 for quantities of each component.
Refer to separate instrument/equipment SOP 76 for ID information

---

NOTE: PRODUCT IS TO BE STIRRED CONTINUOUSLY DURING COMPOUNDING AND FILLING

A.   COLLECTION OF WATER FOR INJECTION (WFI), USP.   **wrong temperature**

A1. Collect approximately 370 *L* of WFI, USP, in a clean, calibrated vessel and cool to 24°C–28°C.
Vessel # ___7___   Amount collected ___365*L*___   Initial Temperature (23°C)   Time ___8:00___ (am)/pm
Final Temperature ___26°C___   Time ___08:15___ am/pm

↑ **neither is**   *LS* / *JM*
circled   12/10/98   12/10/98

A2. Close the water for injection valves.   *LS* / *JM*
                    12/10/98   12/10/98

A3. Remove about 20*L* of the cooled WFI from step 1 and place in a clean, calibrated vessel.
(This water will be used to bring the final formulation to the proper volume.)

Vessel # ___2___   *LS* / *JM*
                    12/10/98   12/10/98

A4. Remove about 5*L* of the cooled WFI from step 1 into a clean, calibrated vessel. (This water will be used to prepare
the solutions used to adjust the pH.)

B.   PREPARATION OF pH ADJUSTING SOLUTIONS

B1. Collect 750 *mL* cool WFI from the vessel in step A4 and place into a 1000 *mL* volumetric flask and add
100 *g* of NaOH (Sodium Hydroxide # 875) USP and dissolve.

                    wrong amount
Amount of WFI collected ___750*mL*___   Amount NaOH added (115*g*) Lot # _____ ← **missing**

Time step completed ___09:15___ am/pm   *JM*   **no initials—why did**
↑ **not circled**   12/10/98   **supervisor initial this?**

B2. Using a water bath containing cold WFI, cool the solution prepared in B1 to 25°C ± 5°C.

Final Temperature (31°C) ← **wrong temperature**   *LS* / *JM*
                    12/10/98   12/10/98

B3. Bring the solution to 1000 *mL* with cool WFI from the vessel in step A4.

Approximate volume of WFI added (300*mL*)   Time completed (09:00) (am)/pm

            ↑   ↑   *LS* / *JM*
        wrong   **time does not**   12/10/98   12/10/98
and so on   **volume**   **make sense**

CHAPTER 6

Quality Systems in the Laboratory

I. GENERAL PRINCIPLES AND LABORATORY MANAGEMENT

A. Introduction

In 1989 a woman was accused of murdering her baby by feeding him antifreeze in his bottle. The child's blood and his bottle were sent for analysis to a commercial testing laboratory and the hospital laboratory. Workers at both laboratories confirmed that there was ethylene glycol (antifreeze) in the blood and bottle. The woman was convicted of first-degree murder and sentenced to life in prison. Two scientists fortunately became interested in the case after hearing about it on a television broadcast. The scientists proved that the child had not died of poisoning, but rather because he had a rare metabolic disorder. The mother was exonerated (after serving time in prison). When the scientists investigating the case obtained the original laboratory reports, what they saw was, in their words, "scary." One laboratory said that the child's blood contained ethylene glycol even though the sample did not match the profile of a known ethylene glycol standard. The second laboratory found an abnormal component in the child's blood and "just assumed it was ethylene glycol." In fact, samples from the bottle had not showed evidence of anything unusual, yet the laboratory report claimed it contained ethylene glycol. In this case, an innocent person was convicted of murder based on the erroneous statements of laboratory workers (*Science, p. 931, Nov. 15, 1991*). These laboratories were apparently not committed to the principles of quality.

There are various laboratory quality systems, all of which are intended to ensure that data from the laboratory are trustworthy and meaningful. ISO 9000 and GMP include references to quality control laboratories which test products, raw materials, and in-process samples. An excerpt relating to these laboratory functions in the GMP regulations is as follows:

Laboratory Controls 21CFR211.160
General requirements
Laboratory controls shall include the establishment of scientifically sound and appropriate specifications, standards, sampling plans, and test procedures designed to assure that components . . . in-process materials . . . labeling, and drug products conform to appropriate standards of identity, strength, quality, and purity. Laboratory controls shall include:
(1) Determination of conformance to appropriate written specifications for the acceptance of each lot . . .
(2) Determination of conformance to written specifications and a description of sampling and testing procedures for in-process materials . . . (4) The calibration of instruments, apparatus, gauges, and recording devices at suitable intervals in accordance with an established written program.

We have previously mentioned other laboratory quality systems including: FDA's GLP (Good Labora-tory Practices), which apply to animal studies; the GLP of the Environmental Protection Agency, which apply to studies submitted in support of pesticide approval; and CLIA, which is a quality system for clinical laboratories. ISO Guide 25 is a quality system that is similar to ISO 9000, but focuses explicitly on laboratories (as contrasted with production facilities).

Every quality system relies on the commitment of individuals to produce quality results. This commitment must begin with the upper level managers and supervisors, who have the authority to purchase good equipment, hire skilled staff, and make other investments in quality. The technical tasks in a laboratory are performed by technicians, scientists, media preparation technicians and others, each of whom must be committed to obtaining quality results. The concept of "commitment to good laboratory practice" is formally stated in this brief excerpt from ISO Guide 25:

Quality System, audit and review ISO Guide 25-1990
5.1 The laboratory shall establish and maintain a quality system appropriate to the type, range and volume of calibration and testing activities it undertakes. The elements of this system shall be documented . . . The laboratory shall define and document its policies and objectives for, and its commitment to good laboratory practice.

This chapter discusses quality systems in the laboratory. Chapters 4 and 5 already surveyed many of the basic issues (such as documentation and resources), so just a few points relating to general laboratory management are reviewed here. The majority of this chapter is devoted to technical issues relating to laboratory methods.

B. Laboratory Management and Quality

i. RESOURCES

Technically skilled personnel are one of the most important resources in the laboratory. The laboratory should be managed so as to ensure that there are personnel with the required skills, that each individual's responsibilities are established, and that there is adequate supervision. Each laboratory must establish procedures to train new analysts, and that training should be documented. Experienced analysts require ongoing training to refresh their skills and to learn new technologies as they are developed. An important management issue in testing laboratories is that laboratory technicians and managers should never be placed in a situation where they feel obligated or pressured to report a particular result. Whatever test result is obtained should be exactly recorded and reported (after proper verification).

The laboratory facility must have adequate equipment to perform all required work. In addition, a suitable laboratory environment must be maintained. This includes controlling the temperature, humidity, lighting, dust levels, vibration, and other factors at suitable levels. In some situations, special facilities are required, as when pathogens are studied or when sterility of work areas is required.

Equipment and instruments required to perform tests must be available, be regularly maintained and calibrated, and their maintenance and calibration must be documented. Calibration of instruments needs to be traceable to national or international standards that ensure that the measurements made in one laboratory are consistent with measurements made in other laboratories around the world. (Calibration is addressed in more detail in Unit V.)

ii. MONITORING QUALITY

Laboratories regularly need to review their practices, progress, and results. In a research laboratory, this function might be performed through weekly laboratory meetings and discussions. Outside review occurs when papers are submitted for publication in the scientific literature and when grant proposals are submitted to granting agencies. In a testing laboratory, there are likely to be formal internal and external review processes. Internal review would involve an audit of records, data, and practices by someone within the laboratory or within the company. External review might involve audits by clients or, for laboratories in regulated companies, by inspectors. It is also possible for laboratories to voluntarily become accredited by various accrediting organizations. In this case, external reviewers will audit the laboratory to see that it operates in compliance with quality practices.

In addition to periodic audits of documents and practices, the laboratory should have checks that regularly monitor the accuracy of tests and procedures. For example, analysts might periodically verify their assays using reference standards, also called "check samples," which are known to give a particular value in the assay. If the reference standard does not give the proper result, then workers know there is a problem in the method or in its execution. This problem may have many sources, such as an out-of-calibration instrument or an environmental factor (such as temperature or humidity).

iii. A NOTE ABOUT QUALITY SYSTEMS AND RESEARCH LABORATORIES

People sometimes think that the quality issues discussed in this unit do not apply to research laboratories. As we have emphasized previously, however, research scientists have always pursued "good science." Adhering to good laboratory practices helps researchers avoid errors, ensures that experimental results can be consistently repeated, and makes the work credible to others.

Suppose a situation arises in a research laboratory where an experimental method that has been used to separate proteins from one another suddenly no longer works. A number of factors may be responsible for the problem. A raw material may have been made improperly by the manufacturer. Reagents may have been mixed improperly or may have degraded. An instrument may be malfunctioning; a technician may have made a mistake. Trouble-shooting this problem will be facilitated if lot numbers for raw materials were recorded, if there are good records of the sources and preparation of samples, if all buffers and reagents were dated and prepared using documented procedures, if the equipment used was consistently maintained and calibrated, and if the analyst kept careful records. Trouble-shooting a problem can be extremely difficult, time-consuming, and expensive if records are not kept and if these good (quality) laboratory practices were not followed. If good records are not kept throughout a research program, then there may be gaps that cause problems later on, such as difficulty reproducing experimental results or differences between prototype and production batches. Even a small change in a raw material, a procedure, an instrument, or a process, therefore, should be justified and recorded.

To a great extent, regulations and standards relating to laboratory practice simply formalize and make consistent the traditions of researchers doing "good science." Some of the specific requirements of regulatory agencies, however, are neither mandated nor practical in a research laboratory. For example, it would be unusual to require a supervisor's signature when a routine procedure is completed in a research setting, whereas such a signature may be necessary in a pharmaceutical quality control environment. The process for changing a procedure or raw material is similarly likely to be informal in a research laboratory and tightly controlled in a testing laboratory.

II. ENSURING THE QUALITY OF LABORATORY TEST RESULTS: TECHNICAL CONSIDERATIONS

A. Laboratory Methods

A wide variety of laboratory tests are performed in laboratories in biotechnology companies. For example, testing a product to see if it meets its specifications might include assays for bacterial, viral, DNA, and protein contaminants, tests that look for degradation products, and tests that measure the concentration of an active substance. Similar types of tests are often used in biological research laboratories as well.

The various tests of products, intermediates, and raw materials typically fall into one of the following categories:

1. **Tests of General Characteristics.** Examples of such tests include evaluating appearance, color, and clarity; and measuring pH, particulate, and moisture content.

2. **Tests of Identity.** Identification assays are used to ensure that a particular substance (e.g., the active

ingredient in a drug) is present. For pharmaceutical products, a single identity test is usually not sufficient; confirmation by two or more assays is usually required.

3. **Purity/impurity Testing. Purity** *is the relative absence of undesired, extraneous matter in a product.* **Impurities** *include, but are not limited to, residual moisture, pyrogenic substances, protein impurities, viruses, DNA impurities, and microorganisms.*

   There are various methods to test for purity or impurity. The concentration of the desired product is sometimes measured. Assays for individual contaminants (such as bacteria) are sometimes performed. Tests for impurities may be qualitative **limit tests**, *which simply test whether or not the impurity is present.* Impurity tests may also be quantitative, in which case the amount or concentration of the impurity is measured.

4. **Quantitation/concentration.** These assays test the amount or concentration of a substance. For example, many biotechnology products are proteins. The amount of protein in a product, or at an inter-

mediate stage in processing, is frequently determined.

5. **Potency/activity. Potency or activity** *is the specific ability of the product to produce a desired result.* For a drug, potency refers to the drug's effect in a human or animal. For other products, activity refers to the product's ability to perform as desired. For example, a restriction enzyme is intended to cleave DNA at specific sites. The activity of a restriction enzyme therefore relates to how much DNA it can specifically cleave in a set amount of time.

6. **Stability. Stability** *is the capacity of a product to remain within its specifications over time.* Stability may be evaluated, for example, by testing the product's potency, activity, pH, clarity, color, and particulate content over time. It may also be helpful to subject the product to stresses (e.g., high temperature, freezing, or exposure to air) and check the product for degradation.

Analysts use a wide variety of methods to evaluate properties of samples. Table 6.1 lists some of the types of methods that are common in biotechnology laboratories.

EXAMPLE

Purity Testing in Food

It is illegal to add commercial sweeteners (such as corn syrup) to foods without proper labeling; such sweetened foods are considered to be adulterated. Because commercial sweeteners are far less expensive than fruit juice, maple syrup, or honey, producers sometimes illegally add these inexpensive sweeteners to their products. For example, two brothers who owned a decades-old Mississippi honey and syrup business were sent to prison after

years of adulterating their products, selling them cheaply, and thereby undercutting legitimate honey and syrup producers (FDA Consumer, April 1977). Unfortunately, adulteration with sweeteners is difficult to detect. Standard chemical analyses do not reveal adulteration because the chemical composition of the sweeteners is very similar to that of honey, syrup, and fruit juice. Investigators therefore developed a gas chromatography method that is capable of distinguishing sweetened from pure products. As you can see in the figure below, adulterated foods have a distinctive "fingerprint."

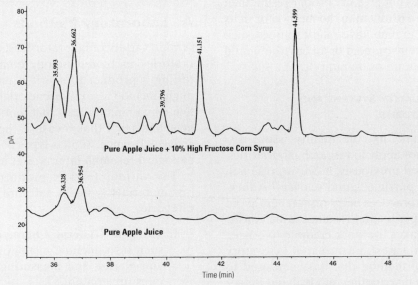

(Chromatograms courtesy of Nicholas H. Low, University of Saskatchewan, Canada; from Hewlett-Packard "Peak" Magazine, Number 2, p. 2, 1997.)

Table 6.1 *Examples of Analysis Methods Commonly Used in Biotechnology Laboratories*

BIOLOGICAL TEST METHODS

Bioassays. *Tests performed on whole animals or on cells in culture.* USED TO TEST POTENCY, ACTIVITY, OR TOXICITY OF A BIOLOGICALLY ACTIVE SUBSTANCE.

Microbiological Contaminants Testing. Various methods are available to test for the presence of microbial contaminants. These methods include the U.S. Pharmacopeia sterility test, plate counts, and specific mycoplasma tests. USED TO TEST FOR BACTERIAL CONTAMINANTS.

Viral Contaminants Testing. Various methods are used to test for viral contaminants. For example, infectivity assays involve placing the item to be tested alongside susceptible cells. If virus is present, then the susceptible cells become infected. Transmission electron microscopy can be used to visualize viral particles if cells are infected. USED TO TEST FOR VIRAL CONTAMINANTS.

Pyrogen/Endotoxin Assays. Pyrogens/endotoxins are bacterial by-products that elicit a dangerous immune response in mammals when present at very low levels. The **Limulus Amoebocyte Lysate Test (LAL)** *is a sensitive test for the presence of endotoxin contaminants that is based on the ability of endotoxin to cause a coagulation reaction in a reagent purified from the blood of the horseshoe crab.* The **USP rabbit pyrogen test** *involves injecting the sample into rabbits and monitoring their temperature to see if the test mixture elicits a fever response.* USED TO TEST FOR THE PRESENCE OF PYROGENS.

Polymerase Chain Reaction (PCR). *A technique for greatly amplifying the amount of DNA with a specific, desired sequence.* USED IN IDENTITY AND DNA IMPURITY TESTING.

Enzyme Assays. Enzyme assays test the activity of an enzyme. USED TO TEST ENZYME ACTIVITY.

ANALYTICAL TEST METHODS

Amino Acid Sequencing. *These methods are used to determine the amino acid sequence of a protein.* USED FOR IDENTITY TESTING.

Chromatographic Methods. *Chromatographic methods are used to separate and detect the components of a mixture.* USED TO DETERMINE IDENTITY, QUANTITY, AND PURITY OF MATERIAL. DURING STABILITY TESTING, USED TO LOOK FOR DEGRADATION PRODUCTS.

Electrophoresis. *Electrophoretic methods separate molecules from one another based on their relative ability to migrate when placed in an electrical field. Electrophoretic methods are widely used to separate mixtures of proteins and mixtures of DNA fragments. The separated components are visualized with dyes.* USED TO ASSAY FOR QUANTITY, IDENTITY, AND PURITY.

Immunological Methods. *Immunological methods depend on the specific interaction of an antigen and its antibody.* USED TO ASSAY FOR QUANTITY, IDENTITY, AND PURITY.

Mass Spectrometry. *Mass spectrometry is an instrumental technique that can be used to determine the primary structure of a protein.* USED FOR IDENTITY TESTING.

Peptide Mapping. *Peptide mapping is a technique used to profile a protein's structure in which the protein is broken into peptides using highly specific enzymes.* The resulting peptides are separated by chromatography or electrophoresis producing a map or profile. The peptide map for a sample can be compared to a map done on a reference standard to confirm the identity of a product. USED TO ASSAY FOR IDENTITY AND FOR PROTEIN IMPURITIES.

Spectrophotometric Methods. *These widely used instrumental methods rely on the ability of samples to absorb visible or ultraviolet light of specific wavelengths.* USED TO MEASURE QUANTITY, ACTIVITY, IDENTITY, AND PURITY.

B. The Development of Methods

i. WHERE DO METHODS COME FROM?

In a biotechnology company, the development of test methods is an important part of product development. As the product and the processes by which it is made are refined, so too are the methods that support it. In many cases there are established test methods that can be adapted for a particular application. These may be collected and published in a compendium. For example, the **U.S. Pharmacopeia (USP)** *is a compendium that contains hundreds of accepted methods for tests commonly performed in the pharmaceutical industry.* The USP methods have been validated and are accepted for use by the FDA. Another compendium is the **"Official Methods of Analysis of AOAC International,"** which contains microbiology and chemical analysis methods. Table 6.2 lists a few of the major compendia of importance in biotechnology laboratories. For novel products, compendium methods may not be available and it may be necessary to develop new test methods. In this case, the method will require intensive testing and must be demonstrated to give desired results consistently under specified conditions.

ii. SAMPLES

Laboratory tests are performed on samples. A **sample** *is a part of the whole that represents the whole.* It is important that samples be properly collected, labeled, prepared for testing, and stored. Proper sampling involves answering such questions as: How many items should be taken to represent the whole? How should the items for

Table 6.2 *A Selected List of Methods Compendia*

"Handbook of Methods for Environmental Pollutants"
U.S. EPA, Environmental Protection Agency
(Contains methods of analysis for environmental samples.)

"Official Methods of Analysis of AOAC International"
AOAC International
(Contains a wide variety of microbiological and chemical analysis methods.)

"Bacteriological Analytical Manual/Food and Drug Administration"
AOAC International, Gaithersburg, MD, 8th edition, 1995
(Contains methods for the analysis of microbiological analysis of foods.)

"USP, The United States Pharmacopeia/National Formulary"
(Contains validated standard methods of analysis for pharmaceutical and health-related products; accepted by FDA for products sold in the United States.)

"BP, British Pharmacopeia"
(Contains validated standard methods of analysis for pharmaceutical and health-related products; accepted in the United Kingdom, Australia, the British Commonwealth, and other countries.)

"EP, European Pharmacopeia"
(Contains validated standard methods of analysis for pharmaceutical and health-related products; accepted in 17 countries throughout Europe.)

the sample be selected? How should the sample be prepared for testing? Answering these questions requires an understanding of the system being evaluated and often requires a background in statistics. Once the sample is selected, it is necessary to prepare it properly, store and handle it so it is not degraded, and use the proper analytical method to evaluate it. Devising a sampling procedure is an important part of method optimization. (See also Chapter 12.)

iii. Reference Standards

Reference materials or **standards** *are well-characterized substances whose composition is known and to which samples can be compared.* For example, when a sample is tested using a chromatographic or electrophoretic method, the result is a profile or "fingerprint" that varies depending on the composition of the sample. Different materials result in different profiles. The profiles of samples can be matched to the profiles of standards of known composition. If the standard and sample fingerprints match, then they may be the same material. (Note, however, that two or more different materials may sometimes have very similar profiles using electrophoresis or chromatography.) Many analytical methods require a standard(s) for comparison.

Obtaining and characterizing reference materials is critical in the development of a method. Reference materials can sometimes be purchased from commercial suppliers, government agencies, or the USP. Standards sometimes have to be prepared in the laboratory and their composition, stability, and other qualities need to be carefully evaluated. Regardless of the source of reference materials, they must be carefully and securely stored to avoid loss, degradation, or damage. Well-characterized reference materials are a valuable resource in a laboratory.

C. Validation of Methods

Method validation *is the process used to assess the ability of a procedure to obtain a desired result reliably, to determine the conditions under which such results can be obtained, and the limitations of the procedure.* Formal validation of test methods is required in laboratories that meet GMP regulations. The various aspects of validating a method, however, are relevant in any research or testing laboratory.

The validation process may be truncated if a method comes from an accepted compendium, although each laboratory must still demonstrate that the compendium method is acceptable for their application. Methods developed in-house must be fully validated in companies that make regulated products.

There are different characteristics of test methods that can be evaluated during validation; the purpose of the test procedure will govern which characteristics need to be validated. For example, when testing for impurities, **limit of detection** is important. **Limit of detection** *is the lowest concentration of the material of interest that can be detected.* The characteristics of a method that can be evaluated include its (1) **accuracy**, (2) **precision,** (3) **limit of detection,** (4) **limit of quantitation**, (5) **selectivity**, (6) **linearity and range**, and (7) **robustness**. General descriptions of each of these assay characteristics are given in the following sections.

i. Accuracy

Accuracy *is the closeness of a test result to the true or accepted value.* Accuracy is sometimes evaluated by testing a reference standard whose true or expected value for the test is known. Accuracy is also sometimes evaluated by comparing the results of the method being validated with another method that was previously validated and is considered to be correct. Accuracy can be expressed as the difference between the result obtained for the test and the known or accepted value expected.

ii. Precision and Reproducibility

Precision *is the degree of agreement between individual test results when the procedure is applied over and over again to portions of the same sample.* There are different types of precision. Measurements repeated in succession on the same day tend to be relatively consistent. In contrast, measurements performed on different days, by different people, and using different materials and equipment, will tend to be far more variable. **Repeatability** *is the precision of measurements made under uniform conditions.* **Reproducibility** *(sometimes called **ruggedness**) is the precision of measurements made under nonuniform conditions, such as in two different laboratories.* Repeatability and reproducibility are therefore two practical extremes of precision.

iii. LIMIT OF DETECTION

Limit of detection *is the lowest concentration of the material of interest that can be detected by the method.* (The term **sensitivity** sometimes is used synonymously with limit of detection. The distinction between the two terms is discussed in Chapter 14.) Every method has a limit below which it cannot detect the analyte of interest. Analytical methods, therefore, can never prove that a particular substance is not present in a sample; rather, they show that it is not present at a concentration above a certain detection limit.

iv. LIMIT OF QUANTITATION

Limit of quantitation *is the lowest concentration of the material of interest that the method can quantitate with acceptable accuracy and precision.* The limit of quantitation is usually a higher concentration than the limit of detection.

v. SELECTIVITY

Selectivity *is a measure of the extent to which a method can determine the presence of a particular compound in a sample without interference from other materials present.** A very selective test will only give a positive result to the compound of interest. For example, the polymerase chain reaction can be used to detect a particular sequence of DNA with high selectivity, regardless of whether other DNA is present in the sample. A less selective test might erroneously give a positive result when a substance similar to the compound of interest is present. For example, an immunological test for a mouse protein that also shows up positive in the presence of a similar human protein is not highly selective.

vi. LINEARITY AND RANGE

Linearity *is the ability of a method to give test results that are directly proportional to the concentration of the material of interest (within a given concentration range).* **Range** *is defined by the limits of concentrations, from the lowest to the highest, that a method can measure with acceptable results.* Tests and assays have a particular, limited range in which they are linear. When an assay or a test is developed, it is customary to determine the linear range.

*Note that the terms *specificity* and *selectivity* are often used interchangeably. Some authors reserve the term *specific* for a test that can measure a compound of interest with absolutely no interference from other materials.

EXAMPLE

Based on the following graph, this assay is approximately linear between 0 $\mu g/mL$ of protein and about 25 $\mu g/mL$ of protein. At higher protein concentrations, the assay does not give a linear response.

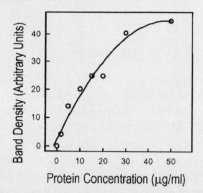

(Reprinted with permission from *SDS-Polyacrylamide Gel Electrophoretic Analysis of Proteins in the Presence of Guanidinium Hydrochloride*, James L. Funderburgh and Sujatha Prakash, *BioTechniques* 20: 376–378, March 1996.)

vii. ROBUSTNESS

Robustness *is a measure of the capacity of a method to remain unaffected when there are small, deliberate variations in method parameters.* Robustness provides an indication of the method's reliability during normal use. For example, chromatographic analytical methods usually require a particular buffer pH. A chromatographic method that has good selectivity in a range from pH 6.5 to 6.9 is more robust (relative to pH) than a method that requires a pH between 7.2 and 7.3. Robustness is important because there are often subtle and not so subtle differences between laboratories and individuals performing a particular method.

EXAMPLE

A chromatographic method of analyzing tricyclic antidepressants was evaluated for its robustness relative to pH. The amount of time it took a particular drug to run through the column was checked at pH 6.9, 7.0, and 7.1. The results were:

pH	Running Time
6.9	19.28
7.0	21.34
7.1	23.48

The percentage of difference between the running time for this drug at pH 6.9 and 7.0 is 9.65% and between pH 7.0 and 7.1 is 10.03%. This test would likely be considered to have poor robustness relative to pH. This is because it is difficult in the laboratory to absolutely ensure that a solution has a pH within 0.1 of a pH unit; it is not unusual for a solution to vary 0.1 pH units under normal laboratory conditions. In chromatography, 10% differences in running time could adversely affect the results of the test by causing sample components to shift positions and interfere with one another.

(This example is derived from *Contemporary Issues in Regulatory Compliance*, Michael E. Swartz. Presented in a workshop by Waters Corporation, October 10, 1996.)

EXAMPLE

Crime laboratories are now using DNA "fingerprinting" to identify or exclude suspects. An organization called the "Technical Working Group on DNA Analysis Methods" (TWGDAM) has developed "Guidelines for a Quality Assurance Program for DNA Analysis" for use in forensic laboratories (*Crime Laboratory Digest*, 22(2): 21–38. 1995). The guidelines contain an outline of how a crime laboratory can validate its DNA fingerprinting method to ensure that it gives trustworthy results. Some of the requirements for validation according to TWGDAM are the following studies:

a. *The DNA fingerprinting procedure must be evaluated using fresh tissues and fluids. DNA isolated from different tissues of the same individual must yield the same fingerprint.* This evaluates the robustness of the method.

b. *"Using specimens obtained from donors of known type, evaluate the reproducibility of the technique both within the laboratory and among different laboratories."* This evaluates the reproducibility of the method.

c. *"Prepare dried stains using body fluids from donors of known types and analyze to ensure that the stain specimens exhibit accurate, interpretable, and reproducible DNA types or profiles that match those obtained on liquid specimens."* This evaluates the accuracy, reproducibility, and robustness of the technique.

d. *Mixed Specimen Studies— "Investigate the ability of the system to detect the components of mixed specimens . . ."* This is a test of selectivity.

e. *Matrix Studies—"Examine prepared body fluids mixed with a variety of commonly encountered substances (e.g., dyes, soil) and deposited on commonly encountered substrates (e.g., leather, denim)."* Matrix studies are sometimes considered tests of the accuracy of the method. They also test its robustness and selectivity.

f. *Minimum Sample—"Where appropriate, establish quantity of DNA needed to obtain a reliable typing result."* This is an evaluation of the limit of detection of the method.

CASE STUDY EXAMPLES RELATING TO LABORATORY CONTROLS

The Federal Food and Drug Administration has inspectors who periodically inspect pharmaceutical facilities and biotechnology companies that make certain products. If the inspectors observe violations of Good Manufacturing Practices, they note the violations on forms, called "483s," and in official Warning Letters sent to the company. The following are excerpts from two actual warning letters that address issues relating to laboratories.

Warning Letter, Example 1

Dear Sir or Madam:
During an inspection of your drug manufacturing facility . . . conducted on August 29 through September 25, 1996, our investigators documented serious deviations from the Current Good Manufacturing Practices Regulations . . . Deviations from the GMPs documented during this inspection included:

[A list of 24 violations follows. Only some of those violations relating to laboratory functions are excerpted here.]

(16) Failure to document the standard injection used to determine the validity of the standard curve in the assay of finished products by liquid chromatography.

(17) Failure to establish a procedure describing the test for identity of ingredients by infrared spectroscopy.

(20) Failure to use standard traceable to a national standard in the assay of finished products. [The use of traceable standards is discussed in Chapter 13.]

(21) Failure to document the person performing the analysis, when the analysis was conducted, and/or the name of the active ingredient in . . . spectra used as an identity test of active ingredients.

The above identification of violations is not intended to be an all-inclusive list of deficiencies at your facility . . . You should take prompt action to correct these deviations.

Failure to promptly correct these deviations may result in regulatory action without further notice . . .
Sincerely,
[Signed by the District Director]

Warning Letter, Example 2

Dear . . .
FDA has completed its review of the report on the October 24–31, 1994, inspection of . . . manufacturing sites . . . That . . . inspection revealed a number of significant deviations from FDA's Current Good Manufacturing Practice (cGMP) regulations . . . These observations included . . . the following:

Your dissolution testing equipment was not being calibrated using the USP standard calibration tablets . . .
Sincerely . . .

PRACTICE PROBLEMS

Problems 1–3 are based on information from the FDA publication "Human Drug cGMP Notes." The notes consist of questions FDA staff are asked, and their responses. Here, the questions are posed as problems and the answers are FDA's responses. See what answers you can think of and compare yours with those of FDA. (http://www.fda.gov/cder/dmpq/cgmpnotes.htm)

1. The supplier of a raw material will always send a Certificate of Analysis with the material. The certificate includes an analysis of the material, its impurities, identity, and so on. Is it acceptable for a quality control laboratory to transcribe the information from the certificate of analysis for the raw material into a laboratory notebook? (December 1995)

2. Do the pressure-indicating gauges on high pressure gas cylinders require calibration? (Calibration establishes the relationship between the values indicated by a measuring instrument and values of a certified standard.) (September 1995)

3. Is it acceptable for a company to replace a faulty instrument with a spare or backup instrument that was previously calibrated but was stored at a remote location? (June 1997)

4. A research laboratory develops a chromatographic method to check samples for the presence of a particular hormone. The scientists take a sample and add to it some of the hormone of interest and a second hormone that is structurally similar to see if the method can distinguish between the two. Which of the following characteristics of the assay are they checking? (1) accuracy, (2) precision, (3) limit of detection, (4) limit of quantitation, (5) selectivity, (6) linearity, (7) range, or (8) robustness.

5. A research team develops a method of DNA "fingerprinting." The method involves cleaving DNA samples with restriction enzymes that recognize and cut at specific base sequences. The resulting fragments of DNA are separated from one another using electrophoresis and are made visible with dye. DNA from different sources forms different patterns when treated with this method.

   a. DNA restriction enzymes are sensitive to the level of salt in their buffer. The researchers therefore tested the method using three buffers with different salt concentrations. Which of the characteristics of an assay listed in Problem 4 are they checking?

   b. If too little DNA is used in this method, then the pattern is not visible. If too much DNA is used, then the pattern is diffuse and is not useful The researchers therefore evaluated the method to see the minimum and maximum DNA that could be used. Which of the characteristics of an assay listed in Problem 4 are they checking?

ANSWERS TO PRACTICE PROBLEMS

1. GMP requires that a company perform quality control tests of all incoming raw materials to ensure that they are acceptable for use. This means minimally that a test that confirms the identity of the material is required. It is not acceptable to rely solely on the claims of the manufacturer of the raw material. Transcribing information from the Certificate of Analysis into a laboratory notebook may be a serious error because it makes it look like the company did the tests when, in fact, it did not. The quality control laboratory must be able to show records or instrument printouts showing that they did, indeed, perform the required tests. If the laboratory workers did perform the required tests and if they clearly identified the data from the Certificate of Analysis as having been copied, then there is no problem.

2. Yes. All measuring devices in the laboratory generally require regular calibration to ensure that they are operating properly.

3. Yes, provided that the replacement instrument is still in calibration, that storage has not adversely affected its performance, and that it has not been stored longer than the established interval between calibrations. Some instruments are delicate and moving them from place to place causes them to go out of calibration. Such instruments would require recalibration before use.

4. Selectivity.

5. a. Robustness. b. Range.

CHAPTER 7

Quality Systems in the Production Facility

I. INTRODUCTION

This chapter provides a brief overview of the issues involved in manufacturing quality biotechnology products. Before discussing how companies make sophisticated products (like recombinant proteins or transgenic plants), however, let us look at a product with which most of us are more familiar.

Suppose that a man who bakes for a hobby decides that his chocolate chip cookies are so well-liked by his friends and family that he wants to open a business producing and selling them. There are many issues to be resolved as he plans his new business. He has completed the initial R&D phase, having experimented with recipes and ingredients, and having tested the quality of his cookies on a small sample of people. As the baker scales up for commercial production, he will need to identify the features that made his cookies successful and formalize these features into written specifications. The baker will need to acquire resources, such as a kitchen with large ovens, ample counter space, and refrigerators. He will need to be certain that his processes for making cookies, which were effective in his own kitchen at home, also work in a larger, commercial kitchen. He will need a program to ensure that his facilities and equipment remain clean and operate properly. He will need written recipes and assistant bakers to follow the recipes. He must find ways to ensure that his cookies always bake at exactly the right temperature, for exactly the right length of time. He will need methods to ensure that all his incoming raw materials are of acceptable quality, and will need to have suitable places to store ingredients so they remain fresh. He will need to consider packaging, labeling, and shipping of his completed cookies. Whereas personally sampling the cookies was probably adequate quality control in the early stages of his venture, the baker will need a more formal quality control program once he begins commercial production. He will have to comply with all regulations relating to food processing, general safety regulations, and environmental regulations, and he might also voluntarily comply with quality standards for cookie makers. Manufacturing quality chocolate chip cookies is a complex goal with many components, all of which need to be planned and coordinated.

The manufacturing issues in a biotechnology company resemble those facing the industrious baker. A biotechnology company must also ensure that adequate resources are available, including facilities, equipment, personnel, and raw materials. Just as the baker must control the temperature of the ovens and the length of baking time, so must a biotechnology company control its processes that turn raw materials into products. Products—whether they are cookies or monoclonal antibodies—must be properly labeled, must be tested for final quality, and must be shipped to their destination. Skilled personnel must be available to carry out all

the necessary tasks. Regulations and standards must be met. Although the basic issues are similar in the cookie kitchen and the biotechnology company, there are differences as well. A biotechnology company is likely to produce many different products and at a much larger scale than would the new bakery. Biotechnology products are likely to be far more technically complicated than chocolate chip cookies. The quality requirements in a biotechnology production facility are therefore even more complex and varied than the issues facing the bakery.

II. ISSUES RELATING TO RESOURCES

A. Facilities and Equipment

A production facility must be able to support the production of products. Consider, for example, a facility in which a drug is manufactured. The building layout should be organized so that processing steps flow in an orderly fashion from one place to another. Raw materials that have been tested and accepted for use should be stored in a space separate from unapproved materials to avoid confusing the two. Materials that are nonsterile must be physically separated from sterile ones. The paths of finished products ideally should not cross the paths of raw materials because of the possibility of confusing them or contaminating the finished product. The facility must have controls for environmental factors, such as humidity, temperature, dust, and particulates in the air. The facility must be sufficiently large to accommodate all equipment and personnel safely.

A manufacturing facility must be both well-designed and maintained properly. This involves a housekeeping program to keep the facility clean, a pest and rodent control program, and environmental monitoring and control.

The equipment and instruments in the facility likewise must be suitable for their purposes and must be properly maintained. An important aspect of equipment and facility monitoring is ensuring that all measuring devices operate properly. Measuring instruments are scattered throughout any facility. For example, thermometers are associated with freezers, refrigerators, sterilizing devices, incubators, and rooms. These thermometers must be functioning properly; otherwise, there is no assurance that materials are being held at the proper temperatures.

EXAMPLE

It is determined during the development of a product that the temperature at which the product is produced affects its quality. Temperature, therefore, must be monitored and controlled in the production area of the facility. The program for the control of temperature involves:

1. During the development of the product, laboratory tests are performed to determine the range of temperatures that is acceptable for production. Based on these tests, it is determined that the temperature of the processing area must remain between 65 and 67°F.

2. The temperature in the facility is controlled by the heating and air conditioning system.

3. A thermostat is installed in the production area. The thermostat is designed so that it:
   — monitors the temperature on the production floor
   — automatically prints out a continuous temperature recording
   — connects to the heating and air conditioning system and turns them on or off as needed.

4. A program of scheduled maintenance is implemented to check all parts of the system regularly including the thermostat and the heating and air conditioning equipment. Every week the thermostat is checked and its readings verified to be accurate with a thermometer known to be correct. The heating and air conditioning systems have monthly maintenance routines. The monitoring program is explained in documents. Records are maintained each time equipment is inspected, adjusted, or repaired.

5. If the temperature is about to go out of range in the production area, an alarm is triggered. There is a document that explains to workers what actions to take if the alarm sounds, how to record their activities and observations, and how to follow-up after the problem is resolved. This follow-up should describe what to do with product that was in process when the alarm sounded.

CASE STUDY

Example Relating to Facilities and Equipment Management

Water is one of the most important raw materials in biotechnology production facilities. The following case study describes a situation where a company failed to observe basic quality practices, resulting in contamination of their water. This negligence resulted in patient illness. (The company was subsequently able to correct their deficiencies.)

In 1995 an epidemiologist working at a hospital in Milwaukee noticed that 12 hospital patients, all of whom were on ventilators, developed Pseudomonas cepacia infections. The hospital laboratory was able to find the bacterium responsible for the infection in the patient's sputum and in bottles of "Fresh Moment" mouthwash. Although this bacterium is seldom harmful to healthy people, P. cepacia can be dangerous to people who are ill.

FDA traced the manufacture of the mouthwash to a particular company, which was subsequently inspected by an FDA investigator. The investigator found six problems thought to have caused the contamination in the mouthwash:

1. The company's purified water system had not been properly cleaned and tested since earlier in the year. The company's president told FDA officials that the company skipped the scheduled maintenance because it could not afford it.

2. The reverse osmosis membranes of the water system had not been changed for more than a year, although this should be done every 4 months, according to the company's written procedure, to prevent microbial buildup. (Water purification systems are discussed in Chapter 24.)

3. Employees failed to challenge the system with P. cepacia when they were testing the system. The challenge test would have involved intentionally adding the bacterium to water to see if their purification system could remove it.

4. Employees did not use the appropriate hose clamps on equipment. The clamps they used appeared to be the type of hose clamps that are used in the automotive trades, and they were rusty, dirty, and discolored.

5. Employees left doors open during production, allowing dust from outside to enter the manufacturing area.

(This account is excerpted from a report in The FDA Consumer, October 1996.)

B. Handling Raw Materials

Raw materials (e.g., the chemicals, empty vials, and other components that go into making a product) are resources. Unlike facilities and major equipment, raw materials flow into the company, are used and transformed into products, and flow out of the company in another form. Systems must be in place to receive materials, document their arrival, and ensure that the materials received are the ones ordered. Once received, the materials must be quarantined until they have been tested and approved by the Quality Unit. This testing confirms that the material is what it is supposed to be and that it is properly labeled. In a company that makes regulated products, raw materials are chemically tested in the laboratory to be certain of their identity. ISO 9000 has the following requirement relating to raw materials testing:

> **4.10.2 Receiving inspection and testing** *The . . . [company] shall ensure that incoming product is not used . . . until it has been . . . verified as conforming to specified requirements . . .*

Some of the principles that guide handling raw materials are summarized in Table 7.1.

CASE STUDY

Examples Relating to Resources

The Federal Food and Drug Administration has inspectors who periodically inspect pharmaceutical facilities and biotechnology companies that make certain products. If the inspectors observe violations of Good Manufacturing Practices, they note the violations on forms, called "483s," and in official Warning Letters sent

Table 7.1 Handling Raw Materials

1. *It is essential not to confuse or mislabel items.* Consider, for example, the serious consequences that could occur if a component of a drug product is accidentally mislabeled.

2. *Raw materials intended for production should not be released to production until they have been tested and found to conform with their specifications.* To ensure that materials are not used prematurely, they are quarantined. Materials are not removed from quarantine until they have been approved for use by personnel from the Quality Unit.

3. *Storage conditions should take into consideration hazards associated with the material, its perishability, and its temperature and humidity requirements.* A system should be in place so that materials in storage can be readily located when needed.

4. *Traceability of materials must be assured.* Ensuring traceability requires, for example, that all incoming and outgoing materials are recorded, records of raw material components that go into each product are maintained, and items are properly labeled.

---

to the company. The following are excerpts from three warning letters that address issues relating to facility design and cross-contamination, equipment maintenance, and raw materials testing. Observe in these warning letters how carefully inspectors checked the facilities and the details of their findings.

Warning Letter, Example #1:

Dear ...

FDA has completed its review of the report on the October 24–31, 1994, inspection of ... manufacturing sites ... That ... inspection revealed a number of significant deviations from FDA's Current Good Manufacturing Practice (cGMP) regulations ... These observations included ... the following:

We are concerned that your firm was operating, in an open area, a number of compression machines running different products that provided a potential for product cross-contamination.

Sincerely ...

Warning Letter, Example 2

Dear ...

During our August 26, 1996, through September 18, 1996, inspection of your facility ... our investigator documented deviations from Current Good Manufacturing Practice ... Regulations ... The following deviations pertaining to the manufacture of drug products were found:

1a. Failure to perform and document a specific identity test on a sample of each lot of drug raw material ...

1b. Failure to test drug raw materials that have been in inventory for extended periods of time to ensure that they will meet specifications for purity, strength, and quality, prior to being used for manufacture of drug products ...

6. Failure to maintain up-to-date cleaning logs and document cleaning, sanitizing, and maintenance of drug-manufacturing equipment, such as the 300 gallon stainless steel vat #1, 10 gallon stainless steel vat #3, and the bottling machine.

You should take prompt action to correct these deviations ... Sincerely, ...

Warning Letter, Example 3

Dear ...

Inspection of your unlicensed hospital blood bank on April 2–3, 1997 ... revealed serious violations ...

Inspection ... revealed that established written procedures for storage temperature controls, equipment maintenance, and record keeping are not being followed. The investigator documented numerous instances where temperature recorder charts on storage refrigerators exceeded specifications, whereas daily temperature logs show that refrigerator temperatures were within specified limits. Your supervisor stated that one temperature recorder is broken and personnel are allowed to bend the arm/pin back into place in an attempt to maintain the device. No documentation is available to show calibration and maintenance of the recording thermometers.

Sincerely ...

III. SPECIFICATIONS

Specifications *are descriptions that define and characterize properties that a product must possess based on its intended use.* There are specifications for raw materials and for products. Specifications are described in this quote from an FDA document (*Guideline on General Principles of Process Validation*, Food and Drug Administration, 1987).

> During the research and development ... phase, the desired product should be carefully defined in terms of its characteristics, such as physical, chemical, electrical and performance characteristics ... It is important to translate the product characteristics into specifications as a basis for description and control of the product ... For example ... [c]hemical characteristics would include raw material formulation ... The product's end use should be a determining factor in the development of product (and component) characteristics and specifications.

We briefly discussed possible specifications for chocolate chip cookies in Chapter 5. Figure 7.1 shows another example of specifications, in this case, for salt. The manufacturers' specifications for three salt products are shown: road salt, table salt, and salt for laboratory use. These specifications illustrate several points about specifications:

1. **Manufacturer's specifications describe properties that are important for that product based on its**

	Road Salt (NaCl, FW 58.44)	Table Salt (NaCl, FW 58.44)	Analytical Grade Salt (For Laboratory Use) (NaCl, FW 58.44)
Chemical Purity	Minimum 95% NaCl	Minimum 97% NaCl	Minimum 99% NaCl
Color	Clear to white, yellow, red, or black	Clear to white	Clear to white
Maximum Allowed Contaminants	Not specified	As 0.5 ppm Cu 2.0 ppm Pb 2.0 ppm Cd 0.5 ppm Hg 0.10 ppm	Al < 0.0005% As < 0.0001% Ba < 0.0005% Ca < 0.002% Cd < 0.0005% Co < 0.0005% Cr < 0.0005% Cu < 0.0005% Fe < 0.0001% and so on
Physical Requirements	90% of crystals between 2.36 mm and 12.5 mm	90% of crystals between 0.3 mm and 1.4 mm	95% of crystals between 0.18 mm and 0.3 mm
Allowed Additives	Anticaking agents of 5–100 ppm Sodium Ferrocyanide Ferric Ferrocyanide	Coating agents, hydrophobic agents	Not allowed
Moisture	2–3%	≤ 3%	Not specified

Figure 7.1. Manufacturer's Specifications for Road Salt, Table Salt, and Salt Intended for Laboratory Use

intended use. Important specifications for all three salt products include their identity (NaCl, FW 58.44) and their purity. The specifications for road salt, however, mention anticaking agents, table salt specifications mention moisture content, and salt for laboratory use specifies maximum allowable levels for a number of trace contaminants. All three products are salt, all are called "sodium chloride," yet, because they have different uses, they have specifications for different properties.

All pertinent aspects of the product that impact its performance, reliability, effectiveness, safety, and stability should be considered when establishing specifications.

2. **The specifications for the same property may vary depending on the intended use.** Although all three salts are NaCl, the salt intended for use in the analytical laboratory is required to be much purer than the other two products. It would be wasteful to spread ultra purified salt on roads and it would be unacceptable to use road salt in the laboratory. None of these products is "better" than the other; they are simply intended for different purposes.

3. **Specifications always are associated with analytical methods.** In order to tell whether a particular batch of salt meets its specifications, it is necessary to have a method to measure its purity, moisture content, and so on. In many cases the evaluation of a product to see if it meets its specifications requires a sophisticated test. For example, it is fairly easy to count the number of chocolate chips in each cookie, but it requires an analytical method to analyze the percentage of cocoa content of those chips. Analytical methods used to evaluate specifications must be proven to work properly.

4. **Specifications are written as a range of permissible values because there is inevitably some variation from product to product.** Acceptable ranges or limits should be established for each specified characteristic and should be expressed in readily measurable terms. (Note that the salt specifications express contaminants in terms of limits, that is, the maximum allowable contaminant levels. Crystal size is specified as a range of acceptable values.) The validity of the specification ranges or limits must be verified through testing the product. In addition, as mentioned earlier, the acceptable range may vary for the same compound depending on its intended use.

The preceding examples are of specifications for a final product. Specifications are also required for raw materials, equipment, and processes. For example, the quality of chocolate chips (a raw material) used to make cookies is important in determining the quality of the cookies. The requirements for the chips, therefore, must be specified. Processes also have endpoints that can be described and specified.

Establishing specifications is a key element in a quality program. Specifications document the product quality that must be achieved and are a contract between the producer and the consumer. The term *establishing specifications* means:

- Defining the characteristics of the product, material, or process
- Documenting those specifications
- Ensuring that the specifications are met

Establishing specifications occurs during the development phase of a product. During development, the specifications may evolve and change. Once large-scale production begins, specifications should remain fixed.

Deciding what properties should be specified for a product, raw material, or process is not a simple task. Establishing specifications requires knowledge of how the product, material, or process will be used and the properties that will make it suitable for that use. It is also complex to define ranges for specifications. If the range for a specification is set too tightly (i.e., if the acceptable range of values is very narrow), then some adequate product or materials might be rejected. On the other hand, if the range of values for the specifications is too broad, then the quality of the product is not protected.

Because the setting of specifications is both a key component of a quality program and a difficult task, FDA scrutinizes specifications for products it regulates. FDA will not accept specifications if they are not complete, if they are unsuitable for the product, if their range is too broad, if they are unsubstantiated by testing, or if suitable analytical methods to test them are not available.

EXAMPLE

Write specifications for an imaginary enzyme, called enzymase, which is used to destroy all enzymes, other than itself, in a reaction mixture.

The intended purpose of enzymase is to destroy other enzymes. An assay for enzyme destruction, therefore, is developed and the activity of the enzyme preparation is specified using this assay.

The enzyme must be purified before sale. R&D researchers found that if the enzyme is at least 60% pure or better, then it functions well in various applications.

The specifications for enzymase are shown in Figure 7.2.

Enzymase catalyzes the cleavage of any enzyme, destroying its activity. The reaction requires the cofactor Mg^{++}.

MOLECULAR WEIGHT: 68kD

COFACTORS: Mg^{++}

INHIBITION: 50% inhibition by greater than 100 *mM* NaCl and chelators of Mg^{++}

INACTIVATION: Inactivated by heating to 70°C or by the addition of EDTA

SOURCE: Isolated from *B. subtilis* recombinant clone

PURITY: 60–80% pure with 125–130 units/*mg* dry weight

STABILITY: Supplied as a lyophilized powder and stable at least 2 years when stored at 2–8°C. Reconstitute with 10 mM phosphate buffer at pH 5.8.

ASSAY: Enzymase is assayed by the modified enzymase assay at pH 5.8. Enzymase is incubated with purified enzyme for 60 minutes at 37°C. The extent of digestion is determined by the colorimetric ninhydrin method. The amino acids released are expressed as micromoles of leucine per milligram dry weight of enzyme. One unit is equal to one micromole of leucine released in 60 minutes.

Figure 7.2. Specifications for the Imaginary Enzyme, "Enzymase."

IV. PROCESSES

Products are made by processes. ISO 9000 defines a **process** as "*a set of interrelated* resources *and* activities *which transform* inputs *into* outputs" (American National Standard ANSI/ISO/ASQC A8402-1994 *Quality Management and Quality Assurance Vocabulary,* ASQC, Milwaukee, WI, emphasis added). Inputs are raw materials. Outputs are products or intermediates that lead to products. Resources include personnel, facilities, equipment, and procedures. Activities involve one or more people who use the raw materials, instruments, equipment, and procedures to make a product or intermediate. In order to obtain a quality product, it is essential that processes be carefully designed and developed by the R&D unit. Production operators in a biotechnology company participate in setting up, monitoring, and controlling the processes.

To illustrate a biotechnology process, consider microbial fermentation. Microbial fermentation is a process in which large numbers of microorganisms are grown in nutrient medium contained in large vats, called *fermenters*. During the fermentation process the microorganisms produce a product, such as an enzyme or antibiotic, which can be isolated and purified. The raw materials in a fermentation process include bacteria and the components of the nutrient medium. The activities involved in the fermentation process include preparing the medium, adding the medium and bacteria to the fermenters, monitoring the condition of the microorganisms, and maintaining proper conditions in the fermenters. The resources involved in the fermentation process include raw materials, fermenters, instruments to monitor conditions in the fermenters, personnel to perform the required activities, and procedures for the operators to follow.

The design of a process begins during the research and development phase and is refined as the product enters production. Every process must be designed so that the inputs are efficiently converted into the desired product. An important aspect of designing a process is finding ways to monitor the process as it occurs, and finding ways to control the process so no problems arise. To effectively design, monitor, and control a process, it is necessary to understand it. For example, if there are minor contaminants in the raw materials, will this affect the process? Is the process very sensitive to a particular factor, such as temperature or pH? Do steps of the process need to be completed within a certain time limit? The factors that are found potentially to affect the process adversely are factors that must be monitored and controlled. These factors are sometimes called "critical points" or "control points."

Monitoring and controlling a process may occur in many ways. For example, temperature and pH probes may be inserted into a fermentation broth. If the temperature or pH exceeds certain limits, then operators may perform corrective actions. Sometimes samples are removed while the process is in progress and are analyzed for a critical characteristic. Such testing is called *in-process testing*. Specifications for the in-process test results are established. If samples do not meet their specifications, then operators may follow a procedure to correct the problem, or they may initiate an investigation.

EXAMPLE

a. **Discuss the development of a fermentation process to produce the hypothetical enzyme, enzymase.**

1. The output of the process is enzymase.
2. An essential early step is to develop a method to analyze the amount of enzymase present. A spectrophotometric assay is developed that is used to measure the amount of enzymase present in samples of the broth.

3. The R&D team performs laboratory experiments to see which bacterial strain makes enzymase most efficiently; the medium formulation that results in the best enzyme yield; the temperature, pH, and aeration conditions under which the bacteria are most productive; and so on. These early experiments provide the information necessary to design the process.

4. The R&D team establishes the fermenter operating parameters, the nutrient medium formulation, and the process endpoint specifications. The endpoint specifications may include, for example, measurements of the number of bacterial cells per milliliter and the amount of enzymase produced per gram of bacteria.

   (Note that the specifications for the endpoint for the fermentation process are not the same as they are for the final product. This is because the enzymase produced by the bacteria must go through a series of isolation and purification steps before it is ready for packaging and sale. These isolation and purification steps are together referred to as **downstream processing**. Thus, the fermentation process produces an intermediate product with its own specifications.)

5. The preparation of SOPs begins in R&D and continues as the process is scaled-up. Once the process is in place and production of enzymase begins, changes to the process are tightly controlled.

b. **What processing controls are required for the fermentation of enzymase?**

1. Preliminary experiments indicate that the hypothetical process is sensitive to pH, temperature, and oxygen levels in the nutrient medium. Early experiments are therefore performed to find the acceptable range for each of these three parameters and to find means to keep them in the correct range. For example, it is found that the optimal pH for enzymase production is 6.6–7.0, and that the pH tends to drop slowly as the fermentation proceeds, eventually reaching pH 5.1. A basic solution must be slowly added to the broth to maintain an acceptable pH during the fermentation.

2. Based on these experiments, the R&D team establishes control limits, expressed as a range, for temperature, pH, and oxygen level. Methods of maintaining each parameter in the desired range are devised. When the process is scaled-up, these control ranges are confirmed to be achievable.

3. Methods of monitoring the three critical parameters are devised. Each parameter is monitored by a separate sterilized probe inserted into the media. The probe output is connected to a computer that displays and stores the values continuously. If

any of the three parameters is too high or too low, then a message is displayed and the medium is adjusted: For pH, base is added; for temperature, cooling is initiated or turned off; and for oxygen, aeration is increased or decreased. These activities are documented throughout the process.

Table 7.2 summarizes some of the steps in designing a process.

V. VALIDATION OF PROCESSES AND ASSOCIATED EQUIPMENT

A. Introduction to Validation

In a GMP facility, after processes are developed, they must be **validated.** The goal of **validation** is to ensure that product quality is built into a product as it is produced. **Process validation** *is defined by FDA as "establishing documented evidence which provides a high degree of assurance that a specific process will consistently produce a product meeting its predetermined specifications and quality attributes."*

Validation in the pharmaceutical industry emerged from problems in the 1960s and 1970s. Recall that during this period there were a number of cases of septicemia in patients caused by intravenous (IV) fluids that were contaminated by bacteria. The contamination was caused by serious problems in companies manufacturing medical products for IV use. The companies had processes in place that made products, and they had programs in place to test samples of their products; however, they still had contamination problems that led to patient deaths.

FDA's present philosophy regarding validation is easily understood in light of the septicemia history. FDA states that quality, safety, and effectiveness are designed and built into the product. Each step in the production process must be controlled to ensure that the finished product meets all quality and design specifications. Process validation is the method by which companies demonstrate that their activities, procedures, and processes consistently produce a quality result. For example, process validation of a sterilization process might involve extensive testing of the effectiveness of the process under varying conditions, when different materials are sterilized, with several operators, and with different contaminants. During this testing, the effectiveness of contaminant removal would be measured, the temperature and pressure at all locations in the sterilizer would be measured and documented, and any potential difficulties would be identified. Validation thus demonstrates that the process itself is effective. Validation and final product testing are recognized as two separate, complementary, and necessary parts of ensuring quality.

Table 7.2   DESIGNING A PROCESS

1. *The purpose of the process must be defined, that is, the desired output must be determined.*

2. *An endpoint(s) that demonstrates that the process has been performed satisfactorily must be defined.* A range of accepted values for the endpoint must be established; that is, specifications for the process must be written.

3. *A method to measure the desired endpoint(s) is required.*

4. *Raw materials and their specifications must be established.*

5. *The steps in the process must be determined, usually by experimentation.*

6. *The process must be scaled-up for production.*

7. *An analysis of potential problems must be performed.* This may involve listing steps in the process where things can go wrong. These steps are sometimes called "critical points" or "control points."

8. *Experiments must be performed to determine how the process must operate at each critical point in order to make a quality product.* For example, it might be established that the temperature during incubation must be between 20 and 25°C.

9. *Methods to monitor the process must be developed.*

10. *Methods to control the process must be developed.*

11. *Effective record-keeping procedures must be developed.*

12. *All SOPs required for the process must be written and approved.*

In addition to process validation, there is also validation of major equipment and of tests and assays. Examples of what might be validated are shown in Table 7.3.

Table 7.3   ITEMS THAT ARE VALIDATED

1. *Processes such as:*
   Cleaning and sterilization
   Fermentation
   Purification, packaging, and labeling of a product
   Chromatographic methods used to purify products
   Mixing and blending

2. *Major equipment such as:*
   Heating, ventilation, and air conditioning systems
   Water purification systems
   Autoclaves and steam generators
   Computers that control manufacturing processes
   Freeze dryers

3. *Tests and assays, such as:*
   Microbiological assays
   Contaminant tests
   Raw materials analyses
   Final product tests

B. When Is Validation Performed?

The planning of validation occurs throughout the development of the product. The actual validation process is usually performed before large-scale production and marketing of a product begins. Revalidation is required whenever there are changes in raw materials, equipment, processes, or packaging that could affect the performance of the product.

Validation is important both in "traditional" pharmaceutical manufacturing and in the production of medical products using biotechnology methods. FDA's "Guideline on General Principles of Process Validation" *(May 1987)* is a general guide that is applicable to most manufacturing situations. There are also specific, detailed guidelines directly applicable to the validation of biotechnology products for medical use. For example, the "Final Guideline on Quality of Biotechnological Products: Analysis of the Expression Construct in Cells Used for Production of r-DNA Derived Protein Products" (*International Conference on Harmonisation, in the* Federal Register: February 23, 61(37): 7006–7008, 1996) discusses validation of processes involving recombinant DNA proteins. Guidelines and resources related to validation are continuously being written and revised. The FDA website is an excellent source of current information.

Validation is a major undertaking that is expensive, time-consuming, and requires extensive planning and knowledge of the system being validated. The advantage to validation is that it helps to assure consistent product quality, greater customer satisfaction, and fewer costly product recalls. Although we are discussing validation in this text primarily from the perspective of the medical products industry, the concept of validation is generally applicable. The cost of a defective batch of product or a product recall is very high in any industry. Ensuring that all production processes function properly to make a quality product is good business sense.

C. Process Validation Planning

Every process is different. Validation begins, therefore, with an understanding of the process to be validated, how the process works, and what can go wrong. Information required for validation is gathered throughout the development phase. The actual validation of a process does not occur until the process is in place, its specifications have been established, most of the factors that can adversely affect the process have been determined, and methods of monitoring and controlling the process have been identified.

Validation requires a **validation protocol**, *which is a document that details how the validation tests will be conducted.* The components of a validation protocol are summarized in Table 7.4.

EXAMPLE

Consider the fermentation process to make the hypothetical enzymase. During R&D the specifications for enzymase were established, an assay for its activity was developed, the process for its production was established, and critical parameters, such as temperature, pH, and aeration were identified. The requirements for the fermenter equipment used to make enzymase (e.g., the size fermenter required, the presence of ports for insertion of probes, how the fermenter would be cleaned, and so on) were similarly determined during development. If necessary, a suitable fermenter was built. SOPs to perform the fermentation-related activities were written. The process is ready for validation.

D. Activities Involved in Validation

At the beginning of validation, measuring instruments must be calibrated to ensure that their readings are trustworthy. For example, a pH probe used to monitor the pH of a fermentation broth must be calibrated to ensure that it registers pH correctly. Instruments may be rechecked periodically during validation to ensure that their readings remain reliable. When validation is completed, the calibration of each instrument is rechecked to ensure that it is still correct. If any instrument has drifted out of calibration, then validation may need to be repeated.

Processes almost always involve equipment. In order for a process to proceed correctly, the equipment must be of good quality, must be properly installed, regularly maintained, and correctly operated. Equipment must therefore be validated or **qualified** *to ensure that it will function reliably under all the conditions that may occur during production.* Equipment qualification may be con-

Table 7.4 *THE COMPONENTS OF A VALIDATION PROTOCOL*

---

1. *A description of the process to be validated.*
2. *An explanation of the features of the process that are to be evaluated.*
3. *A description of the equipment, raw materials, and intermediate products that are to be evaluated.*
4. *An explanation of what samples are to be collected and how they are to be selected.*
5. *Procedures for the tests and assays to be conducted on those samples.*
6. *An explanation of how the results of the assays are to be analyzed (including statistical methods).*
7. *A statement as to how many times each test is to be repeated.*
8. *SOPs describing how to run the process.*

---

sidered and performed separately from process validation, but it is also a requirement for process validation.

Equipment qualification is often broken into steps. *First the equipment item is checked to be sure that it meets its design and purchase specifications and is properly installed; this is called* **installation qualification**. Installation qualification includes, for example, checking that instruction manuals, schematic diagrams, and spare parts lists are present; checking that all parts of the device are installed; checking that the materials used in construction were those specified; and making sure that fittings, attachments, cables, plumbing, and wiring are properly connected.

After installation *the equipment can be tested to see that it performs within acceptable limits, which is called* **operational qualification**. For example, an autoclave might be tested to see that it reaches the proper temperature, plus or minus certain limits, in a set period of time; that it reaches the proper pressure, plus or minus certain limits, and so on. The penetration of steam to all parts of the chamber, the temperature of the autoclave in all areas of the chamber, the pressure achieved at various settings, and so on, would all be tested as part of the operational qualification of an autoclave.

Once all measuring instruments are calibrated and all equipment is validated, process validation can be performed. The actual validation of the process will involve assessing the process under all the conditions that can be expected to occur during production. This includes running the process with all the raw materials that will be used and performing all the involved activities according to their SOPs. Testing includes checking the process endpoint(s) under these conditions and establishing that the process consistently meets its specifications.

Process validation also involves challenging the system with unusual circumstances. FDA speaks of the "worst case" situation(s) that might be encountered during production. For example, a sterilization process might be challenged by placing large numbers of an especially heat-resistant, spore-forming bacterium in the corner of the autoclave known to be least accessible to steam. The effectiveness of bacterial killing under these "worst case" conditions must meet the specifications for the process.

After all these validation activities have been performed, the data that were collected must be analyzed, as described in the validation protocol, and a report must be prepared. Successful validation demonstrates that a process is effective and reliable.

EXAMPLE

Discuss briefly how the steps of validation—calibration, equipment qualification, process validation, challenge, and data analysis—apply to the fermentation process used to produce enzymase.

Validation Protocol Hypothesis: Does the fermentation process for producing crude enzymase reliably and repeatedly produce a crude enzyme that meets all its specifications?

1. **Calibration.** All the measuring instruments must be calibrated to ensure that their readings are correct. This includes, for example, the pH, temperature, and oxygen probes, and the instruments that monitor agitation rates and fluid flows.

2. **Fermenter Equipment Qualification.** The fermenter is a major piece of equipment that requires qualification before the fermentation process is validated. This involves:

   a. **Installation Qualification:** Establishing that the fermenter and its ancillary equipment are properly installed and meet the specifications set by the design team. For example, the following items would be inspected and their presence documented:

      1. The vessel where the bacteria and broth are contained must have features such as smooth walls, proper fittings and seals, and a pressure relief valve.

      2. The various ports from which samples are withdrawn, and into which nutrient medium can be added, must be machined so that contaminants cannot enter the system.

      3. Heating and cooling systems must be in place, their utility requirements must be provided, they must be connected to electricity and water, as required, and so on.

      4. Pumps must be present, have the proper ratings, be connected to power, and so on.

      5. All instruction manuals and schematic diagrams must be present, and spare parts lists must be available.

   b. **Operational Qualification:** Establishing that the fermenter operates as it should. For example, the following items would be tested:

      1. The heating and cooling systems function as specified.

      2. The air and water flow rates are acceptable.

      3. All switches function correctly.

      4. Alarms sound when they should.

      5. A critical feature of a fermenter is that it can be cleaned and sterilized. This requirement is typically tested by intentionally contaminating the fermenter with a test organism which is a temperature-resistant bacterium. The fermenter is then run through its decontamination cycle, and is checked for the number of bacteria remaining.

      6. Any computer associated with the fermentation process must be qualified. It is common to use

computers to monitor and control fermenters. For example, a computer might be connected to the pH probe. The computer would continuously record the pH in the fermenter, would display the pH, might control a pump that feeds base into the fermenter as necessary, and might trigger an alarm if the pH exceeds preset limits.

3. **Process Validation.** Bacteria and nutrient medium are introduced to the system and the growth of bacteria and production of enzymase are tested. The fermentation process should be tested under varying conditions as might occur during production. Features to check include:

   1. The ability of the system to maintain the proper temperature, pH, and aeration.

   2. The absence of contamination in the system.

   3. The production of crude enzymase that meets specifications.

   4. Bacterial growth.

4. **Worst-case challenge.** A challenge to the system might include, for example, allowing the fermentation process to run extra days to see whether slow growing contaminants appear in the broth.

5. **Analysis of Data and Preparation of Report.** The information gathered during the validation is analyzed, summarized, and documented in a report.

CASE STUDY

Examples Relating to Validation

The Federal Food and Drug Administration has inspectors who periodically inspect pharmaceutical facilities and biotechnology companies that make certain products. If the inspectors observe violations of Good Manufacturing Practices, they note the violations on forms, called "483"s, and in official Warning Letters sent to the company. The following are excerpts from two actual warning letters that address issues relating to validation.

Warning Letter, Example 1:

Dear Sir or Madam:

During an inspection of your drug manufacturing facility . . . conducted on August 29 through September 25, 1996, our investigators documented serious deviations from the Current Good Manufacturing Practices Regulations . . .

Deviations from the GMPs documented during this inspection included:

(8) Failure to validate the cleaning procedure for production and process controls for drug products . . .

(9) Failure to validate the cleaning procedure for the equipment used in manufacturing drug product . . .

The above identification of violations is not intended to be an all-inclusive list of deficiencies at your facility . . . You should take prompt action to correct these deviations.

Failure to promptly correct these deviations may result in regulatory action without further notice . . .

Sincerely,

[Signed by the District Director]

Warning Letter, Example 2

Dear . . .

During a comprehensive inspection of your manufacturing facility . . . conducted January 13-March 5, 1997, investigators from this office documented serious deviations from current Good Manufacturing Practice Regulations (cGMPs) . . . as follows:

1) Manufacturing process validation for several product lines were found to be inadequate, for example:

Grifluvin Tablets-four . . . batches made in 1995–1996 were rejected for failure to meet dissolution requirements. As a result of your investigation, process changes were made and not validated. The original validation, conducted in 1989, did not consider critical process parameters to be evaluated. Other process parameters were not evaluated including the effect of additional drying steps and ranges established for mixing times, granulation and temperature . . .

Sincerely . . .

E. Unplanned Occurrences

We have so far considered many ways in which problems are avoided by careful planning, design, monitoring, and validation. Even in the best managed facility, however, there are *unexpected occurrences, called* **deviations**. Every company must be prepared for deviations and have a plan to deal with them. This plan includes having forms for documenting the deviation and a review process to decide what action to take. A supervisor and members of the Quality Unit normally review the problem and decide what, if any, corrective action to take. The corrective action and any associated investigation is documented.

When a product or a raw material does not meet its specifications, it is called a **nonconformance**. Nonconformances are also documented. The material is quarantined. Problems with a product often require an investigation to determine the source of the problem and the extent of damage. Problems may have many sources including raw material deviations, human errors, equipment malfunctions, and other unforeseen occurrences. Damage may be minor or may affect an entire batch of product. After investigation, the Quality Unit personnel make a decision as to what to do with the nonconforming material. If it is a final product, then it may need to be destroyed or reprocessed. A raw material may be returned to the vendor. As always, the investigation activities and finding should be documented and follow up should occur.

PRACTICE PROBLEMS

Problems 1–4 are based on information from FDA publication "Human Drug cGMP Notes" (http://www.fda.gov/cder/dmpq/cgmpnotes.htm). This publication consists of questions asked of FDA and the agency's response. Here, the question asked of the FDA is posed as a problem and the answer is FDA's comments. See if you can figure out what FDA's response was.

1. Is it acceptable to place an expiration date on a bottle cap instead of on the bottle label? (September 1996)

2. For drug products formulated with preservatives to inhibit microbial growth, is it necessary to test for preservatives when testing the final product? (June 1996)

3. What is the level of detergent residues that can be left after cleaning on the surface of equipment and glassware? What is the basis for arriving at this level? (June 1995)

4. If the ability of a procedure to clean a piece of equipment made of a particular material, such as 316 stainless steel, is shown to be acceptable and validated, can that material specific cleaning procedure be used without extensive validation for other pieces of equipment? (June 1995)

QUESTIONS FOR DISCUSSION

1. Consider the warning letters on pp. 95 relating to validation.

   a. Why does FDA emphasize validation?

   b. What situations do you think would require a company to revalidate a process?

   c. Procedures for cleaning equipment and work areas are extremely important in a pharmaceutical company. Describe how you think the company that received a warning letter regarding cleaning validation should have validated their procedures for cleaning equipment.

2. Consider the case study on p. 87 relating to the contaminated mouthwash. Discuss how this company might restructure to improve the quality of its operations. For example, how might they prevent mistakes such as open doors and improper hose clamps?

3. The following is an excerpt from an official warning letter sent by FDA to a company:

Dear . . .
During a comprehensive inspection of your manufacturing facility . . . conducted January 13–March 5, 1997, investigators from this office documented serious deviations from current Good Manufacturing Practice Regulations (cGMPs) . . . as follows:

Retest investigations were incomplete in the determination of assignable cause and/or lacked documentation to invalidate initial out-of-range results, for example:

An initial out-of-specification assay for Grifulvin 500 *mg* Tablets . . . was invalidated without adequate documentation to support the conclusion that a pipetting error occurred.
Sincerely . . .

In this case, laboratory workers apparently tested Grifulvin tablets and the result did not conform to the specifications for the product. Two explanations for this result are: (1) The product is defective. (2) A mistake occurred in the QC laboratory resulting in an erroneous test result. Which of these two possibilities did the company decide was the explanation; to what did the company attribute the out-of-range result? What was FDA's position? What sort of investigation should the company have performed when the result was found to be out of range?

ANSWERS TO PROBLEMS

1. No. The expiration date must appear on the immediate container. If the expiration date were to be placed on the bottle cap, there would be a greater chance of mix-up at the pharmacy or by the consumer.

2. Yes. In initial studies the company must establish the effectiveness of the preservative and how much is required to protect the product. Then, each time a batch of product is tested, it is evaluated to see whether the proper amount of preservative is present.

3. This is a specification of a process. The setting of specifications is the company's responsibility. In order to determine an acceptable range for detergent residues, the company must determine how residues affect the safety, efficacy, and stability of the product. The ability of the company's cleaning process to eliminate detergent residues is demonstrated during the validation of the process.

4. No. The design of the equipment is a major component of its cleanability; therefore, companies should demonstrate that a cleaning procedure works on each piece of equipment.

UNIT III:

Math in the Biotechnology Laboratory: An Overview

Chapters in This Unit

UNIT III. INTRODUCTION MATH IN THE BIOTECHNOLOGY LABORATORY: AN OVERVIEW

Mathematical descriptions and operations permeate daily life in the laboratory workplace. Math is an important tool for scientists and technicians, just as a hammer is a tool for a carpenter and a blender is a tool for a chef. To be a skillful laboratory biologist you do not need to "like" math any more than a carpenter needs to "like" a hammer. Most carpenters are probably neither happy nor anxious when they encounter a hammer (unless they have recently smashed their finger with one). Students in the sciences, however, sometimes do have to overcome negative feelings about math.* If you are anxious about math, remember that with practice and time people learn to perform the math required in their profession.

The purpose of this unit is to provide a brief review of common math manipulations used in bioscience laboratories and to introduce applications of math relevant to everyday laboratory work. This unit is not intended to replace a math textbook. Rather, these chapters roam through various areas of mathematics, discussing topics that relate to problems commonly encountered in the laboratory. Although these chapters review various basic math concepts, it is assumed that readers can manipulate fractions and decimals, prepare simple graphs, solve an equation with one unknown, and use a scientific calculator.

Readers who are comfortable with the mathematical operations reviewed may still want to work the "Application Problems" in each section. These problems demonstrate the ways in which familiar mathematical tools are applied in the laboratory. Readers who are less comfortable with mathematical calculations will find it beneficial to complete the "Manipulation Problems." Solving the "Manipulation Problems" will increase your comfort and efficiency when performing routine laboratory calculations.

There is often more than one strategy to solve a math problem. In some instances in this unit, we demonstrate two ways to approach a problem. For example, we show how to convert between units using both proportions and a unit canceling strategy. Both strategies, when applied correctly, will provide the correct answer. Individuals may prefer one strategy over another. (Of course, in a classroom situation students are encouraged to use strategies as directed by their instructor.) More experienced scientists and technicians often move fluidly from one problem-solving strategy to another, using whatever approach is most efficient for solving a particular problem. Experienced workers also are likely to check their answers by first using one approach to solve a problem and then testing another approach to see if it gives the same answer.

A few tips for math problem-solving include:

1. *Keep track of all information.* Begin with a sheet of paper with plenty of room for each problem. Write down relevant information and record the results of each step.

2. *Use simple sketches, arrows, or other visual aids to help define problems.*

3. *Check that each answer makes sense.*

4. *Keep track of units.* It is tempting to ignore units when performing calculations, but this practice has been the cause of much grief. (We will talk more about units later.)

5. *State the answer clearly; remember the units.*

A BRIEF NOTE ABOUT SIGNIFICANT FIGURES

If I divide 47.0 by 9.0 using my calculator, the answer reads: 5.222222222. I would be unlikely to record my answer in this fashion (with all those 2s). Rather, I would round the value to, perhaps 5.2. There are conventions that are used to guide decisions about rounding. These conventions are covered under the topic of "significant figures."

We will explore significant figures and rounding later in this book when measurements are discussed in detail. This is because the principles of significant figures relate to measurements. For example, suppose a meter stick is used to measure the height of a plant and the value obtained is 64.3 cm. This means that we can be reasonably certain of the height of the plant to the nearest tenth of a centimeter. We do not know the height of the plant to the nearest hundredth place because the meter stick did not provide that much information. When we report the result of this measurement, it is important not to mislead the reader into thinking that the height of the plant is known more "exactly" than it is. For this reason we do not record the plant's height as 64.30, or 64.33 cm. It is also incorrect to "drop" information, for example, by reporting the plant to be 64 cm.

When calculations are performed that involve numbers from measurements, the significant figure conventions provide guidance in how to round the answer. The results of calculations should be rounded in such a way as to indicate the "exactness" with which the original

---

*Many people believe that they are not good at math and/or have negative feelings about math. Sheila Tobias wrote a book about "math anxiety" (*Overcoming Math Anxiety,* Sheila Tobias. W.W. Norton and Co., New York, 1993). Tobias points out that people in Japan and Taiwan believe that hard work leads to good performance in math. In contrast, people in the United States are apt to believe that math ability leads to good performance, that one is either born with this ability or not, and that no amount of hard work can make up for the lack of math ability. Because of these beliefs, Americans may give up on math and try less hard in math classes. In reality, the ability to use math is not a genetic gift, but rather is learned with practice.

measurements were made. There are many situations in analytical chemistry and in other quantitative applications where it is essential to round values properly after calculations by applying the significant figures conventions. Although significant figures are not the focus of this math unit, some readers may prefer to read the sections on significant figures, pp. 244–247, before completing the problems in this unit in order to better understand rounding conventions.

TERMINOLOGY USED IN THE MATH UNIT

Abscissa. The X-coordinate of a point; the distance of a point along the X-axis.

Amount. How much of a substance is present. For example, 2 grams or 4 cups.

Angstrom, Å. Unit of length that is 10^{-10} meters.

Antilogarithm, antilog. The number corresponding to a given logarithm. For example, $100 = 10^2$. The log of 100 is 2. The antilog of 2 is 100.

Base (for exponents). A number that is raised to a power. In the expression 10^3, the base is 10 and the exponent is 3.

Basic Properties (in the SI system). The fundamental measured properties for which the SI system defines units. These properties are: length, mass, time, electrical current, thermodynamic temperature, luminous intensity, and amount of substance.

Centi-, c. A prefix meaning 1/100.

Centimeter, cm. A unit of length that is 1/100 of a meter.

Coefficient (as relates to scientific notation). The first part of a number expressed in scientific notation. For example, the number 235 in scientific notation is expressed as 2.35×10^2 where "2.35" is the coefficient.

Common Logarithm (also log or $\log_{10}$). The common log of a number is the power to which 10 must be raised to give that number. For example, $1000 = 10^3$, so the log of 1000 is 3. The log of 5 is 0.699 which means that $10^{0.699} = 5$.

Concentration. A ratio where the numerator is the amount of a material of interest and the denominator is the volume (or sometimes mass) of the entire solution or mixture. For example, if 1 g of table salt is dissolved in water so that the total volume is 1 L, the concentration is 1 g/L. (See also "Ratio.")

Constant. Number in a particular equation that always has the same value.

Coordinates. The address for a point on a graph; values that describe the distances horizontally and vertically for the location of a point.

Cubic Centimeter, cc, cm³. A unit of volume that is equal to 1 mL.

Denominator. The number written below the line in a fraction.

Density. The ratio between the mass and volume of a material.

Dependent Variable. A variable whose value changes depending on the value of the independent variable. For example, if an experimenter measures the growth of seedlings under different light intensities, the growth of the seeds is the dependent variable and the light intensity is the independent variable. See also "Variable" and "Independent Variable."

Diluent. A substance used to dilute another. For example, when concentrated orange juice is diluted, the diluent is water.

Dilution. Addition of one substance (often but not always water) to another to reduce the concentration of the original substance.

Dilution Series. A group of solutions that have the same components but at different concentrations.

Equation. A description of a relationship between two or more entities that uses mathematical symbols.

Exponent. A number used to show that a value (the base) should be multiplied by itself a certain number of times. The expression 10^3 means: $10 \times 10 \times 10$ which equals 1000. The base is 10 and the exponent is 3.

Exponential Equation. A relationship between two or more entities whose equation includes a variable that is an exponent, for example, $y = 2^x$.

Exponential Notation. See "Scientific Notation."

Gram, g. The basic metric unit for mass.

Half-Life. The time it takes for an amount of radioactivity to decay to half its original level.

Independent Variable. A variable whose value is controlled by the experimenter. See also "Variable" and "Dependent Variable."

Kilo-, k. A prefix meaning 1000.

Kilogram, kg. A unit of mass equal to 1000 g.

Kilometer, km. A unit of length equal to 1000 m.

Lambda, λ. An older term sometimes used to mean 1 μL.

Least Squares Method. A statistical method used to calculate the equation for the line of best fit for a series of points.

Line of Best Fit. A line connecting a series of data points on a graph in such a way that the points are collectively as close as possible to the line.

Linear Relationship. A relationship between two properties such that when they are plotted on a graph the points form a straight line.

Liter, L. A metric system unit of volume. $1 \ L = 1 \ dm^3$. (See also "Volume.")

Log, Logarithm, Log₁₀. See "Common Log."

Measurement System. A related group of units, such as inches, feet, and miles. See also "The United States Customary System," "Metric System," and "SI System."

Meter, m. The basic metric unit for length.

Metric System. A measurement system used in most laboratories and in much of the world whose basic units include meters, grams, and liters. These basic units are modified by the addition of prefixes which designate powers of 10.

Micro-, μ. A prefix meaning 1/1,000,000 or 10^{-6}.

Microgram, μg. A unit of mass that is 1/1,000,000 g, or $10^{-6} \ g$.

Microliter, μL. A unit of volume that is 1/1,000,000 L, or $10^{-6} \ L$.

Micrometer, μm. A unit of length that is 1/1,000,000 m or $10^{-6} \ m$.

Micron. A micrometer.

Milli-, m. A prefix meaning 1/1000 or 10^{-3}.

Milligram, mg. A unit of mass that is 1/1000 g.

Milliliter, mL. A unit of volume that is 1/1000 L.

Millimeter, mm. A unit of length that is 1/1000 m.

Natural Log, Logₑ, Ln. Logarithms whose base is the number 2.7183, called "e."

Numerator. The number written above the line in a fraction.

Order of Magnitude. One order of magnitude is 10^1. For example, 10^2 is said to be "two orders of magnitude" less than 10^4.

Ordinate. The Y-coordinate for a point; the distance of a point along the Y axis.

Origin. The intersection of the X and Y axes on a two-dimensional graph whose coordinates are (0,0).

Percent, %. A fraction whose denominator is 100.

pH. An expression of the concentration of H^+ ions in a solution where $pH = -\log[H^+]$.

Proportion. Two ratios that have the same value but different numbers. For example: 1/2 = 5/10.

Proportionality Constant. A value which expresses the relationship between the numerators and the denominators in a proportional relationship.

Quantitative Analysis. The determination of how much of a particular material is present in a sample.

Ratio. The relationship between two quantities, for example, "25 miles per gallon" and "10 mg of NaCl per liter."

Reciprocal. A number related to another so that when multiplied together their product equals 1. To calculate the reciprocal of a number, divide 1 by that number. For example, the reciprocal of 5 is: 1/5 or 0.2.

"Rise" (in graphing). An expression that describes the amount by which the Y coordinate increases on a two-dimensional graph.

"Run" (in graphing). An expression that describes the amount by which the X coordinate increases on a two-dimensional graph.

Scientific Notation (also called Exponential Notation). The use of exponents to simplify handling numbers that are very large or very small. For example, 4500 in scientific notation is 4.5×10^3.

Semilog Paper. A type of graph paper that has a log scale on one axis and a linear scale on the other axis.

Serial Dilutions. Dilutions made in series (each one derived from the one before) and that all have the same dilution factor. For example, a series of 1/10 dilutions of an original sample would be: 1, 1/10, 1/100, 1/1000, etc.

SI System, Systeme International d'Units. A standardized system of units of measurement, adopted in 1960 by a number of international organizations, that is derived from the metric system.

Slope (of a line). Given a straight line plotted on a graph, the slope of the line is how steeply the line rises. Slope can be calculated by choosing any two points on the line and dividing the change in their Y values by the change in their X values.

Solute. A substance that is dissolved in some other material. For example, if table salt is dissolved in water, then salt is the solute and water is the solvent.

Solvent. A substance that dissolves another. For example, if salt is dissolved in water, then water is the solvent.

Standard. A physical embodiment of a unit of measure.

Standard Curve. A graph that shows the relationship between a response (e.g., of an instrument) and a property that the experimenter controls (e.g., the concentration of a substance).

Stock Solution. A concentrated solution which is diluted to a working concentration.

Threshold (for a graph). A change in a relationship plotted on a graph.

Unit (of measure). A precisely defined amount of a property.

United States Customary System, U.S.C.S. The measurement system common in the United States which includes miles, pounds, gallons, inches and feet.

Value (in math). An assigned or calculated numerical quantity.

Variable. A property or a characteristic that can have various values (in contrast to a constant which always has the same value). (See also "Dependent Variable" and "Independent Variable.")

Volume. The amount of space a substance occupies, defined in the SI system as length × length × length = meters^3. More commonly, the liter (dm^3) is used as the basic unit of volume. (See also "Liter.")

X **Axis.** The main horizontal line on a graph.

Y **Axis.** The main vertical line on a graph.

Y **Intercept.** The point at which a line passes through the Y axis, where $X = 0$, on a two-dimensional graph.

SUGGESTIONS FOR FURTHER READING

Laboratory Mathematics: Medical and Biological Applications, June and Joe Campbell. 5th edition. Mosby Co., St. Louis, 1997.

Basic Mathematics for Beginning Chemistry, Dorothy M. Goldish. 4th edition. Macmillan Publishers, New York, NY, 1990.

CHAPTER 8

Basic Math Techniques

I. EXPONENTS AND SCIENTIFIC NOTATION

EXAMPLE APPLICATION: EXPONENTS AND SCIENTIFIC NOTATION

a. The human genome consists of approximately 3×10^9 base pairs of DNA. If the sequence of the human genome base pairs was written down, then it would occupy 200 telephone books of 10^3 pages each. How many base pairs would be on each page?

b. If it costs \$0.50 to determine the location (sequence) of a single base pair, how much would it cost to determine the location of all 3×10^9 base pairs?

Answers on page 109

A. Exponents

i. THE MEANING OF EXPONENTS

We begin this chapter by discussing exponents and scientific notation. Exponents and scientific notation are routinely used in the laboratory and so it is important to be able to manipulate them easily and quickly.

An **exponent** *is used to show that a number is to be multiplied by itself a certain number of times.* For example, 2^4 is read "two raised to the fourth power" and means that 2 is multiplied by itself 4 times:

$$2 \times 2 \times 2 \times 2 = 16$$

Similarly:

$$10^2 \text{ means: } 10 \times 10, \text{ or } 100$$
$$4^5 \text{ means: } 4 \times 4 \times 4 \times 4 \times 4 \text{ or } 1024$$

The number that is multiplied is called the **base** *and the power to which the base is raised is the* **exponent.** In the expression 10^3, the base is 10 and the exponent is 3.

A negative exponent indicates that the reciprocal of the number should be multiplied times itself. For example:

$$10^{-3} = \frac{1}{10} \times \frac{1}{10} \times \frac{1}{10} = \frac{1}{1000} = 0.001$$

Rules that govern the manipulation of exponents in calculations are summarized in Box 1.

Box 1 *CALCULATIONS INVOLVING EXPONENTS*

1. To multiply two numbers with exponents where the numbers have the same base, add the exponents:

$$a^m \times a^n = a^{m+n}$$

Two examples:
$$5^3 \times 5^6 = 5^9$$
$$10^{-3} \times 10^4 = 10^1$$

To convince yourself that this rule makes sense; consider the following example:

$$2^3 \times 2^2 = 2^5 = (2 \times 2 \times 2) \times (2 \times 2) =$$
$$2 \text{ multiplied 5 times} = 2^5 = 32$$

2. To divide two numbers with exponents where the numbers have the same base, subtract the exponents:

$$\frac{a^m}{a^n} = a^{m-n}$$

Two examples:
$$5^3/5^6 = 5^{3-6} = 5^{-3}$$
$$2^{-3}/2^{-4} = 2^{(-3)-(-4)} = 2^1 = 2$$

You can convince yourself that this rule makes sense by rewriting an example this way:

$$5^3/5^6 = \frac{\cancel{5} \times \cancel{5} \times \cancel{5}}{\cancel{5} \times \cancel{5} \times \cancel{5} \times 5 \times 5 \times 5} = \frac{1}{5 \times 5 \times 5} = \frac{1}{125} = 5^{-3}$$

3. To raise an exponential number to a higher power, multiply the two exponents.

$$(a^m)^n = a^{m \times n}$$

Two examples:
$$(2^3)^2 = 2^6$$
$$(10^3)^{-4} = 10^{-12}$$

To convince yourself that this rule makes sense, examine this example:

$$(2^3)^2 = 2^3 \times 2^3 = (2 \times 2 \times 2) \times (2 \times 2 \times 2) = 2^6$$

4. To multiply or divide numbers with exponents that have different bases, convert the numbers with exponents to their corresponding values without exponents. Then, multiply or divide.

Two examples: Multiply: $3^2 \times 2^4$

$$3^2 = 9 \text{ and } 2^4 = 16,$$
$$\text{so } 9 \times 16 = 144$$

Divide: $4^{-3}/2^3$

$$4^{-3} = \frac{1}{4} \times \frac{1}{4} \times \frac{1}{4} = \frac{1}{64} = 0.0156$$

$$\text{and } 2^3 = 8$$
$$\text{so } \frac{0.0156}{8} = 0.00195$$

5. To add or subtract numbers with exponents (whether their bases are the same or not), convert the numbers with exponents to their corresponding values without exponents.

For example: $4^3 + 2^3 = 64 + 8 = 72$

6. By definition, any number raised to the 0 power is 1.

For example: $85^0 = 1$

EXAMPLE PROBLEM

Perform the operations indicated:

$$\frac{10}{43^2 + 13^3}$$

ANSWER

The denominator involves addition of numbers with exponents. Convert the numbers with exponents to standard notation. Then perform the calculations.

$$\frac{10}{(43)^2 + (13)^3} = \frac{10}{1849 + 2197} = \frac{10}{4046} = 0.0025$$

ii. Exponents Where the Base is 10

In preparation for discussing scientific notation, consider the particular case of exponents where the base is 10. Observe the following rules as illustrated in Figure 8.1:

1. *For numbers greater than 1:*
   - *the exponent represents the number of places after the number (and before the decimal point)*
   - *the exponent is positive*
   - *the larger the positive exponent, the larger the number*

2. *For numbers less than 1:*
   - *the exponent represents the number of places to the right of the decimal point* including *the first nonzero digit*
   - *the exponent is negative*
   - *the larger the negative exponent, the smaller the number*

People commonly use the phrase *orders of magnitude* where **one order of magnitude** *is 10^1*. Using this terminology, 10^2 is said to be two orders of magnitude less than 10^4. Similarly, 10^8 is three orders of magnitude greater than 10^5.

$1,000,000 = 10^6$ = one million	(6 places before decimal point)
$100,000 = 10^5$ = one hundred thousand	(5 places before decimal point)
$10,000 = 10^4$ = ten thousand	(4 places before decimal point)
$1,000 = 10^3$ = one thousand	(3 places before decimal point)
$100 = 10^2$ = one hundred	(2 places before decimal point)
$10 = 10^1$ = ten	(1 place before decimal point)
$1 = 10^0$ = one	(no places before decimal point)
$0.1 = 10^{-1}$ = one tenth	(1 place right of the decimal point)
$0.01 = 10^{-2}$ = one hundredth	(2 places right of the decimal point)
$0.001 = 10^{-3}$ = one thousandth	(3 places right of the decimal point)
$0.0001 = 10^{-4}$ = one ten thousandth	(4 places right of the decimal point)
$0.00001 = 10^{-5}$ = one hundred thousandth	(5 places right of the decimal point)

Figure 8.1.   Using exponents where the base is 10. (When the decimal point is not written it is assumed to be to the right of the final digit in the number.)

B. Scientific Notation

i. Expressing Numbers in Scientific Notation

Scientific notation *is a tool that uses exponents to simplify handling numbers that are very big or very small.* Consider the number:

$$0.00000000000000000000000602$$

In scientific notation this lengthy number is compactly expressed as:

$$6.02 \times 10^{-23}$$

A value in scientific notation is customarily written as a number between 1 and 10 multiplied by 10 raised to a power. For example:

$$100 \text{ (\textit{Standard Notation})} = 10^2 = 1 \times 10^2 \text{ (\textit{Scientific Notation})}$$
$$1 \times 10^2 = 1 \times 10 \times 10 = 100$$

$$300 \text{ (\textit{Standard Notation})} = 3 \times 10^2 \text{ (\textit{Scientific Notation})}$$
$$3 \times 10^2 = 3 \times 10 \times 10 = 300$$

The number of bacterial cells in 1 liter of culture might be
$$100,000,000,000 \text{ (\textit{Standard Notation})} =$$
$$1 \times 10^{11} \text{ (\textit{Scientific Notation})}$$

A number in scientific notation has two parts. *The first part is called the* **coefficient.** The second part is 10 raised to some power. For example:

	First Part (Coefficient)		*Second Part (Exponential Term)*
1000 =	1	×	10^3
235 =	2.35	×	10^2

As shown in Figure 8.1, a negative exponent is used for a number less than 1. Three examples:

$$1 \times 10^{-5} = \frac{1}{10} \times \frac{1}{10} \times \frac{1}{10} \times \frac{1}{10} \times \frac{1}{10}$$

$$= \frac{1}{100,000} = 0.00001$$

$$0.000135 = 1.35 \times 10^{-4}$$

A bacterial cell wall is about 0.00000001 *m*
$$= 1 \times 10^{-8} \ m \text{ thick}$$

A procedure to convert a number from standard notation to scientific notation is shown in Box 2.

EXAMPLE PROBLEM

Convert the number **0.000348** to scientific notation.

Box 2   *A Procedure to Convert a Number from Standard Notation to Scientific Notation*

Step 1:

   a. If the number in standard notation is greater than 10, then move the decimal point to the left so that there is one nonzero digit to the left of the decimal point. This gives the first part of the notation.

   b. If the number in standard notation is less than 1, then move the decimal point to the right so that there is one nonzero digit to the left of the decimal point. This gives the first part of the notation.

   c. If the number in standard notation is between 1 and 10, then scientific notation is seldom used.

Step 2: Count how many places the decimal was moved in step 1.

Step 3:

   a. If the decimal was moved to the left, then the number of places it was moved gives the exponent in the second part of the notation.

   b. If the decimal point was moved to the right, then place a − sign in front of the value. This is the exponent for the second part of the notation.

Example

*Express the number **5467** in scientific notation.*

Step 1. This number is greater than 10. Therefore, move the decimal point to the left so that there is only one nonzero digit to the left of the decimal point: = **5467.** *move decimal point 3 places left* → **5.467**

Step 2. The decimal point was moved three places to the left.

Step 3. The exponent for the second part of the notation is therefore 3.
     This means the number in scientific notation is: **5.467 × 10^3**

ANSWER

Step 1. This number is less than 1. Move the decimal point to the right: **0.000348 → 3.48**

Step 2. The decimal point was moved four places to the right.

Step 3. The exponent is −4, so the answer in scientific notation is: **3.48 × 10⁻⁴**

So far, we have shown the customary manner of writing numbers in scientific notation, that is, with the coefficient written as a number between 1 and 10. It it is not necessary, however, always to write numbers in scientific notation in this way. For example, the value 205 may be expressed as:

$$205. = 0.205 \times 10^3$$
$$205. = 2.05 \times 10^2$$
$$205. = 20.5 \times 10^1$$
$$205. = 2050 \times 10^{-1}$$
$$205. = 20500 \times 10^{-2}$$

Similarly:

$$1.00 \times 10^4 = 10.0 \times 10^3 = 100. \times 10^2$$
$$3.45 \times 10^{23} = 0.0345 \times 10^{25} = 345 \times 10^{21}$$

There are situations where it is useful to manipulate coefficients and exponents without changing the value of the numbers. This is the case in addition and subtraction of numbers expressed in scientific notation.

EXAMPLE PROBLEM

Fill in the blank so that the numbers on both sides of the = sign are equal. (For example: $2.58 \times 10^{-2} = 25.8 \times 10^{-3}$)

$$0.0055 \times 10^4 = 5500 \times 10^{\underline{\quad}}$$

ANSWER

One way to think about this problem:

1. Convert the expression on the left to standard notation; it equals **55**.

2. The number on the right side of the expression, that is, 5500, is larger than the number on the left, that is, 55. The exponent that fills in the blank, therefore, will need to be a negative number (to make 5500 smaller).

3. 5500 times 10^{-2} equals 55. The answer is therefore **−2**.

II. CALCULATIONS WITH SCIENTIFIC NOTATION

Box 3 summarizes methods of performing calculations with scientific notation.

EXAMPLE PROBLEM

$$(3.45 \times 10^{23}) + (4.56 \times 10^{25}) = ?$$

ANSWER

The two numbers in this example would require many zeros if written in standard notation. The strategy of expressing both values in scientific notation with the same exponent, therefore, is preferred over converting both numbers to standard notation.

Step 1. Decide on a common exponent. Suppose we choose **25**.

Step 2. Express both numbers in a form with the same exponent.

$$3.45 \times 10^{23} = 0.0345 \times 10^{25}$$

Step 3. Perform the addition.

$$\begin{array}{r} 0.0345 \times 10^{25} \\ + \ 4.56 \quad\ \times 10^{25} \\ \hline 4.5945 \times 10^{25} \end{array}$$

(Note: If the coefficient is rounded according to significant figure conventions, then the answer is 4.59×10^{25})

Box 3   CALCULATIONS INVOLVING NUMBERS IN SCIENTIFIC NOTATION

1. To multiply numbers in scientific notation use two steps:

    Step 1. Multiply the coefficients together.

    Step 2. Add the exponents to which 10 is raised.

    For example:

    $$(2.34 \times 10^2)(3.50 \times 10^3) = \underset{(multiply\ the\ coefficients)}{(2.34 \times 3.50)} \times \underset{(add\ the\ exponents)}{10^{2+3}} = 8.19 \times 10^5$$

2. To divide numbers in scientific notation, use two steps:

    Step 1. Divide the coefficients.

    Step 2. Subtract the exponents to which 10 is raised.

    For example:

    $$(4.5 \times 10^5)/(2.1 \times 10^3) = \underset{(divide\ the\ coefficients)}{(4.5/2.1)} \times \underset{(subtract\ the\ exponents)}{10^{5-3}} = 2.1 \times 10^2$$

3. To add or subtract numbers in scientific notation:

a. If the numbers being added or subtracted all have 10 raised to the same exponent, then the numbers can be simply added or subtracted as shown in these examples:

$$(3.0 \times 10^4) + (2.5 \times 10^4) = ? \qquad (7.56 \times 10^{21}) - (6.53 \times 10^{21}) = ?$$

$$\begin{array}{r} 3.0 \times 10^4 \\ + \; 2.5 \times 10^4 \\ \hline 5.5 \times 10^4 \end{array} \qquad \begin{array}{r} 7.56 \times 10^{21} \\ - \; 6.53 \times 10^{21} \\ \hline 1.03 \times 10^{21} \end{array}$$

b. If the numbers being added or subtracted do not all have 10 raised to the same exponent, then there are two strategies for adding and subtracting numbers.

STRATEGY 1 Convert the numbers to standard notation and then do the addition or subtraction:

For example: $(2.05 \times 10^2) - (9.05 \times 10^{-1}) = ?$

Convert both numbers to standard notation:

$$2.05 \times 10^2 = 205$$
$$9.05 \times 10^{-1} = 0.905$$

Perform the calculation:

$$\begin{array}{r} 205 \\ - \quad 0.905 \\ \hline 204.095 \end{array}$$

(Note: If this value is rounded according to significant figure conventions, then the answer is 204.)

STRATEGY 2 Rewrite the values so they all have 10 raised to the same power:

For example: $(2.05 \times 10^2) - (9.05 \times 10^{-1}) = ?$

To convert both numbers to a form such that they both have 10 raised to the same power:

Step 1. Decide what the common exponent will be. It should be either 2 or -1. Suppose we choose 2.

Step 2. Convert 9.05×10^{-1} to a number in scientific notation with the exponent of 2:

$$9.05 \times 10^{-1} = 0.00905 \times 10^2.$$

Step 3. Perform the subtraction:

$$\begin{array}{r} 2.05 \quad \times 10^2 \\ - \; 0.00905 \times 10^2 \\ \hline 2.04095 \times 10^2 \end{array}$$

(Note: If the coefficient is rounded according to significant figure conventions, then the answer is 2.04×10^2)

A calculator will hold a limited number of places and will not accept very large or very small numbers if you try to key in all the zeros. A scientific calculator, however, works easily with large and small numbers expressed in scientific notation. On my calculator, to key in the number 1×10^3, I push the following keys:

$$1$$
$$\text{exp}$$
$$3$$

Note that I do not key in the number 10; the base in scientific notation. "exp" tells my calculator that the 10 is present. To key in the number 3×10^{-4} on my calculator, I press:

$$3$$
$$\text{exp}$$
$$4$$
$$+/-$$

The $+/-$ key tells the calculator I want a negative exponent. "EE" is sometimes used to indicate that 10 is being raised to a certain power. Consult your instruction manual to see how to key in a number in scientific notation.

EXAMPLE APPLICATION ANSWER: (From p. 104)

a. The total number of pages needed to record the genome would be

$$200 \times 10^3 = 2 \times 10^5 = 200,000$$

Then, the number of base pairs on each page would be:

$$\frac{3 \times 10^9}{2 \times 10^5} = 1.5 \times 10^4 = 15,000$$

b. $3 \times 10^9 \times \$0.50 = 1.5 \times 10^9 = 1.5$ billion dollars! (There has been much effort in reducing the cost of sequencing DNA.)

MANIPULATION PROBLEMS: EXPONENTS AND SCIENTIFIC NOTATION

1. Give the whole number or fraction that corresponds to these exponential expressions.
    a. 2^2 **b.** 3^3 **c.** 2^{-2} **d.** 3^{-3} **e.** 10^2
    f. 10^4 **g.** 10^{-2} **h.** 10^{-4} **i.** 5^0

2. Perform the operations indicated:
    a. $2^2 \times 3^3$ **b.** $(14^3)(3^6)$ **c.** $5^5 - 2^3$
    d. $5^7/8^4$ **e.** $(6^{-2})(3^2)$ **f.** $(-0.4)^3 + 9.6^2$
    g. $a^2 \times a^3$ **h.** c^3/c^{-6} **i.** $(3^4)^2$
    j. $(c^{-3})^{-5}$ **k.** $\frac{13}{43^2 + 13^3}$ **l.** $10^2/10^3$

3. Underline the larger number in each pair. Underline both if their values are equal.
    a. 5×10^{-3} cm, 500×10^{-1} cm
    b. $300 \times 10^{-3} \mu L$, $3000 \times 10^{-2} \mu L$
    c. 3.200×10^{-6} m, 3200×10^{-4} m
    d. 0.001×10^1 cm, 1×10^{-3} cm
    e. 0.008×10^{-3} L, 0.0008×10^{-4} L

4. Convert the following numbers to scientific notation.
    a. 54.0 **b.** 4567 **c.** 0.345000 **d.** 10,000,000
    e. 0.009078 **f.** 540 **g.** 0.003040 **h.** 200,567,987

5. Convert the following numbers to standard notation.
    a. 12.3×10^3 **b.** 4.56×10^4 **c.** 4.456×10^{-5}
    d. 2.300×10^{-3} **e.** 0.56×10^6 **f.** 0.45×10^{-2}

6. Perform the following calculations.
    a. $\dfrac{(4.725 \times 10^8)(0.0200)}{(3700)(0.770)}$
    b. $\dfrac{(1.93 \times 10^3)(4.22 \times 10^{-2})}{(8.8 \times 10^8)(6.0 \times 10^{-6})}$
    c. $(4.5 \times 10^3) + (2.7 \times 10^{-2})$
    d. $(35.6 \times 10^4) - (54.6 \times 10^6)$
    e. $(5.4 \times 10^{24}) + (3.4 \times 10^{26})$
    f. $(5.7 \times 10^{-3}) - (3.4 \times 10^{-6})$

7. Fill in the blanks so that the numbers on both sides of the $=$ sign are equal. For example: $2.58 \times 10^{-2} = 25.8 \times 10^{\underline{-3}}$
    a. $0.0050 \times 10^{-4} = 0.050 \times 10^{—} = 0.50 \times 10^{—} = 5.0 \times 10^{—}$
    b. $15.0 \times 10^{-3} = \underline{\quad} \times 10^{-2} = \underline{\quad} \times 10^{-1} = \underline{\quad} \times 10^1$
    c. $5.45 \times 10^{-3} = 54.5 \times 10^{—}$
    d. $100.00 \times 10^1 = 1.0000 \times 10^{—}$
    e. $6.78 \times 10^2 = 0.678 \times 10^{—}$
    f. $54.6 \times 10^2 = \underline{\quad} \times 10^6$
    g. $45.6 \times 10^8 = \underline{\quad} \times 10^6$
    h. $4.5 \times 10^{-3} = \underline{\quad} \times 10^{-5}$
    i. $356.98 \times 10^{-3} = \underline{\quad} \times 10^1$
    j. $0.0098 \times 10^{-2} = 0.98 \times 10^{—}$

II. LOGARITHMS

EXAMPLE APPLICATION: LOGARITHMS

The concentration of hydrochloric acid, HCl, secreted by the stomach after a meal is about 1.2×10^{-3} M. What is the pH of stomach acid?

Answer on Page 111

$1,000,000 = 10^6$ = one million	$\log 10^6 = 6$
$100,000 = 10^5$ = one hundred thousand	$\log 10^5 = 5$
$10,000 = 10^4$ = ten thousand	$\log 10^4 = 4$
$1,000 = 10^3$ = one thousand	$\log 10^3 = 3$
$100 = 10^2$ = one hundred	$\log 10^2 = 2$
$10 = 10^1$ = ten	$\log 10^1 = 1$
$1 = 10^0$ = one	$\log 10^0 = 0$
$0.1 = 10^{-1}$ = one tenth	$\log 10^{-1} = -1$
$0.01 = 10^{-2}$ = one hundredth	$\log 10^{-2} = -2$
$0.001 = 10^{-3}$ = one thousandth	$\log 10^{-3} = -3$
$0.0001 = 10^{-4}$ = one ten thousandth	$\log 10^{-4} = -4$
$0.00001 = 10^{-5}$ = one hundred thousandth	$\log 10^{-5} = -5$
$0.000001 = 10^{-6}$ = one millionth	$\log 10^{-6} = -6$

Figure 8.2.   Common logarithms of powers of ten.

A. Common Logarithms

Common logarithms (also called **logs** or **log$_{10}$**) are closely related to scientific notation. *The **common log** of a number is the power to which 10 must be raised to give that number,* Figure 8.2.

$$100 = 10^2$$
The log of 100 is 2 because 10 raised to the second power is 100

$$1,000,000 = 10^6$$
The log of 1,000,000 is 6 because 10 raised to the sixth power is 1,000,000

$$0.001 = 10^{-3}$$
The log of 0.001 is -3 because 10 raised to the -3 power is 0.001

$$1 = 10^0$$
The log of 1 is 0 because 10 raised to the zero power is 1 (by definition)

To what power must 10 be raised to equal the number 5? This is not obvious. We can reason that the number 5 is between 1 and 10 so the log of 5 must be greater than the log of 1 (which is 0) and less than the log of 10 (which is 1). The same is true for the numbers 2, 3, 4, 6, 7, 8, and 9; their logs are decimals between 0 and 1. There is no intuitive way to know the exact log of 5, although we know it is between 0 and 1. Rather, it is necessary to look up the log in a table of logarithms, or to find it using a scientific calculator. Usually the latter is accomplished by simply entering the number whose log you wish to know and pressing a key labeled "log." The log of 5 is **0.699** which means that $10^{0.699} = 5$. Thus:

$$5 = 10^x$$
$$\text{Log } 5 = x$$
$$x = 0.699$$
$$5 = 10^{0.699}$$

The same logic can be applied to numbers between 10 and 100. For example, what is the log of 48? The log of 10 is 1; the log of 100 is 2. The log of the number 48, therefore, must fall between 1 and 2. It is, in fact, **1.6812.** All numbers between 10 and 100 have a log between 1 and 2. We can continue and apply the same logic to all

numbers. For example, the log of 4,987,000 must fall between 6 and 7 because 4,987,000 is between 1 million (10^6) and 10 million (10^7).

What about numbers between 0 and 1? Observe in Figure 8.2 that the log of 1 is 0 and that the logs of numbers between 0 and 1 are negative numbers; therefore, all numbers between 0 and 1 have a negative log. For example:

The log of 0.130 is -0.886
The log of 0.00891 is -2.050

Note that zero and numbers less than zero do not have log values.

B. Antilogarithms

An **antilogarithm (antilog)** *is the number corresponding to a given logarithm.* For example:

The log of 5 is 0.699; therefore, 5 is the antilog of 0.699.

What is the antilog of **1.678**? We can reason that the antilog of 1.678 is a number between 10 and 100. (If this is not clear, think about the 1 in 1.678.) A scientific calculator is the simplest way to find antilogs. Consult your calculator directions to determine how to perform this function. Using a calculator I find that:

Antilog of $1.678 = 47.643$

EXAMPLE PROBLEM

What is the antilog of 3.58?

ANSWER

We can reason that the answer is between 1000 and 10,000 because of the 3.

Antilog $(3.58) = 3801.9$

EXAMPLE PROBLEM

What is the antilog of -0.780?

ANSWER

The antilog of a negative number must be a value between 0 and 1.

Antilog $(-0.780) = 0.166$

C. Natural Logarithms

Common logarithms have a base of 10. *There are also logarithms whose base is the number 2.7183, called "e." When e is the base, the log is called a **"natural"** log, abbreviated **"ln,"** or **"log$_e$."*** There are tables of natural logs and there are keys on scientific calculators to find the natural log. Note that the terms log and ln are not synonyms and should never be used interchangeably. Natural logs are less commonly used in biology laboratories than are logs with a base of 10.

D. An Application of Logarithms: pH

The proper function of aqueous biological systems depends on the medium having the correct concentration of hydrogen ions. pH is a convenient means to express the concentration of hydrogen ions in a solution.

The concentration of H^+ ions in aqueous solutions normally varies between 1.0 *M* and 0.00000000000001 *M* (where *M* is a unit of concentration). The number 0.00000000000001 can also be written as 1×10^{-14}. Søren Sørenson devised a scale of pH units in which the wide range of hydrogen ion concentrations can be written as a number between 0 and 14. If the hydrogen ion concentration in a solution is 1×10^{-14} *M*, then using Sørenson's method we say its pH is **14**. If the hydrogen ion concentration of a solution is 0.0000001 *M*, or 1×10^{-7} *M*, then we say its pH is **7**. If the hydrogen ion concentration of a solution is 1 *M*, then we say its pH is **0**. Thus, to find the pH of a solution:

Step 1. Take the log of the hydrogen ion concentration (expressed as molarity).

Step 2. Take the negative of the log.

For example:

What is the pH of a solution with a H^+ concentration of 0.000234?

Step 1. The log of 0.000234 is −3.63.

Step 2. The negative of −3.63 is 3.63. So, **3.63** is the pH of this solution.

We can express the relationship between pH and hydrogen ion concentration more formally using the following expression:

$$pH = -\log [H^+]$$

The symbol [ ] means "concentration." In words, this expression means "pH is equal to the negative log of the hydrogen ion concentration."

If we know the pH and want to find the $[H^+]$ in a solution, then we apply antilogs. To do this:

Step 1. Place a negative sign in front of the pH value.

Step 2. Find the antilog of the resulting number.

For example:

What is the concentration of hydrogen ions in a solution whose pH is 5.60?

Step 1. Place a − sign in front of the pH: −5.60.

Step 2. The antilog of −5.60 is 2.51×10^{-6}. The concentration of hydrogen ions in this solution, therefore, is $\mathbf{2.51 \times 10^{-6}}$ *M*.

The relationship between hydrogen ion concentration and pH is expressed more formally using the following expression:

$$[H^+] = \text{antilog } (-pH)$$

MANIPULATION PROBLEMS: LOGARITHMS

1. Answer the following without using a calculator:

    a. Is the log of 445 closer to 2 or 4?

    b. Is the log of 1876 closer to 3, 9, or 10?

2. Give the common log of the following numbers.

    a. 100   **b.** 10,000   **c.** 1,000,000

    d. 0.0001   **e.** 0.001

3. The log of 567 must lie between 2.0 and 3.0. For each of the following numbers, state the values the log must lie between.

    a. 7   **b.** 65.9   **c.** 89.0   **d.** 0.45   **e.** 0.0078

4. Use a scientific calculator to find the log of the following numbers.

    a. 1.50×10^4 **b.** 345   **c.** 0.0098

    d. 2.98×10^{-5} **e.** 1209   **f.** 0.345

5. Use a scientific calculator to find the antilog of the following numbers.

    a. 4.8990   **b.** 3.9900   **c.** −0.5600   **d.** −0.0089

    e. 9.8999   **f.** 1.0000   **g.** 8.9000

6. Use a scientific calculator to convert each of the following $[H^+]$ to a pH value.

    a. [0.45 *M*]   **b.** [0.045 *M*]

    c. [0.0045 *M*]   **d.** [0.00000032 *M*]

7. Use a scientific calculator to determine the hydrogen ion concentration of the solutions with the following pH values.

    a. 4.56   **b.** 5.67   **c.** 7.00   **d.** 1.09   **e.** 10.1

III. UNITS OF MEASUREMENT

Scientists and technicians often measure things such as the length of insects, the concentration of pollutants in an air sample, or the amount of air in a patient's lungs. **Measurement** *is basically a process of counting.* For example, consider an insect that is 1.55 *cm* in length. Length is the property that is measured, 1.55 is the **value** of the measurement, and centimeters (*cm*) are the **units** of measurement. Another way to write this measurement is: $1.55 \times 1 \ cm$. This latter statement tells us that the length of the insect was determined by comparison with a standard of 1 *cm* in length and was found to be 1.55 times as long as the standard. Measuring length is therefore comparable to counting how many times the standard must be placed end-to-end to equal the length of the object. Another example is the characteristic of time that can be measured by counting how many times the sun "rises" and "sets." The standard in this case is the motion of the sun and the units are days.

All measurements require the selection of a **standard** and then counting the number of times the standard is contained in the material to be measured. The terms **standard** and **unit** are related, but they can be distinguished from one another. A **unit of measure** *is a precisely defined amount of a property, such as length or mass.* A **standard** *is a physical embodiment of a unit.* Centimeters and millimeters are examples of units of measure; a ruler is an example of a standard. Because units are not physical entities they are unaffected by environmental conditions such as temperature or humidity. In contrast, standards are affected by the environment. A strip of metal used to measure meters will differ in length with temperature changes and will therefore only be correct at a particular temperature and in a particular environment.

A group of units together is a **measurement system.** *The measurement system common in the United States, which includes miles, pounds, gallons, inches, and feet, is called* **The United States Customary System (U.S.C.S.).** In most laboratories and in much of the world, the **metric system,** *with units including meters, grams, and liters, is used,* Figure 8.3.

The units in the U.S. system are not related to one another in a systematic fashion. For example, for length, 12 *in* = 1 *ft*; 3 *ft* = 1 *yd*. There is no pattern to these relationships. In contrast, in the metric system, there is only one basic unit to measure length, the meter. The unit of a meter is modified systematically by the addition of prefixes. For example, the prefix **centi** *means 1/100, so a* **centimeter** *is a meter/100.* **Kilo** *means 1000, so a* **kilometer** *is a meter $\times$ 1000.* The same prefixes can be used to modify other basic units, such as grams or liters. The prefixes that modify a basic unit always represent an exponent of 10, Table 8.1.

The most important basic metric units for biologists are **meters (*m*),** *for length;* **grams (*g*),** *for weight;* and **liters (*L*),** *for volume.* Table 8.2 shows how prefixes are used to modify the basic units. Table 8.3 shows some commonly

Table 8.1   PREFIXES IN THE METRIC SYSTEM

Decimal	Prefix	Symbol	Power of 10
1,000,000,000,000,000,000	exa-	E	10^{18}
1,000,000,000,000,000	peta-	P	10^{15}
1,000,000,000,000	tera-	T	10^{12}
1,000,000,000	giga-	G	10^{9}
1,000,000	mega-	M	10^{6}
1,000	kilo-	K	10^{3}
100	hecto-	h	10^{2}
10	deca-	da	10^{1}
1	Basic unit, no prefix		10^{0}
0.1	deci-	d	10^{-1}
0.01	centi-	c	10^{-2}
0.001	milli-	m	10^{-3}
0.000001	micro-	μ	10^{-6}
0.000000001	nano-	n	10^{-9}
0.000000000001	pico-	p	10^{-12}
0.000000000000001	femto-	f	10^{-15}
0.000000000000000001	atto-	a-	10^{-18}

(The prefixes used commonly in biology laboratories are underlined.)

used conversion factors which relate the U.S.C.S. to the metric system. In the next chapter, conversions between U.S.C.S. and the metric system are discussed.

It is always correct to convert between units that measure the same basic property, such as length. Length can be expressed in many units, including miles, meters, centimeters, and nanometers, and these units may be converted from one to the other. In general, it is not correct to convert units from one basic property to another. For example, time is a basic unit that is measured in seconds, minutes, and so on. It is unreasonable to try to convert a number of seconds to centimeters because these units measure different characteristics.

There are a few terms used by biologists that are not part of the "regular" metric system. A microliter is 1/1,000,000 liters and is normally abbreviated μL, but biologists sometimes refer to 1 μL as 1 "lambda" (λ). 10^{-10} meters may be called an "angstrom" ($\mathring{A}$). A micrometer is often called a "micron," although micrometer is the preferred term. The terms "cc" and "cm^3" (in reference to volume) are both the same as an *mL*.

The **SI measurement system** *(Systeme International d'Units) is intended to standardize measurement in all countries.* The SI system was agreed upon in 1960 by a number of international organizations at the *Conference Generale des Poids et Measures.* The International Organization for Standardization (ISO) has also endorsed the SI system.

There are seven **basic properties** in the SI system: *length, mass, time, electrical current, thermodynamic*

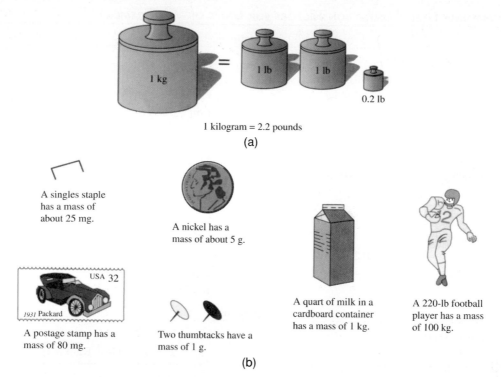

1 kilogram = 2.2 pounds

(a)

A singles staple has a mass of about 25 mg.

A nickel has a mass of about 5 g.

A postage stamp has a mass of 80 mg.

Two thumbtacks have a mass of 1 g.

A quart of milk in a cardboard container has a mass of 1 kg.

A 220-lb football player has a mass of 100 kg.

(b)

Figure 8.3. Units of Measurement. a. A comparison of the metric system kilogram and the U.S.C.S. pound. **b.** The weight of everyday objects in the metric system.

temperature, luminous intensity, and *amount of substance.* The units for these basic properties are shown in Table 8.4. To stay strictly within the SI system, the only units that can be used are those in Table 8.4 or those that are a combination of two or more units shown in this table. For example, the SI unit for **volume**, *the amount of space a substance occupies, is: length × length × length in meters, or* **meters³**. Units of liters are not part of the SI system, although liters remain the conventional metric unit for volume measurements in biology laboratories (one *liter = 1 dm³*).

EXAMPLE APPLICATION ANSWER: (From p. 111)

A $pg = 1 \times 10^{-12}$ g and
$100\ pg = 1 \times 10^{-10}$ g.

A few micrograms would be about 3×10^{-6} g; therefore, 100 picograms is thousands of times less than a few micrograms and this dose of botulinum toxin is likely to be safe.
$$(100\ pg = 0.0001\ \mu g)$$

EXAMPLE PROBLEM

Which is larger, 1 *cm* or 100 *μm*?

ANSWER

One strategy is to solve this problem is convert both values to the same units, for example, meters.
$$1\ cm = 1/100\ m = 0.01\ m$$
$$100\ \mu m = 100 \times 10^{-6}\ m = 1.00 \times 10^{-4}\ m = 0.0001\ m$$

Once both values are converted to meters, it is easy to see that the first number, 1 *cm*, is larger.

Table 8.2 COMMON METRIC UNITS IN BIOLOGY

Multiple of Ten	Mass	Abbreviation	Volume	Abbreviation	Length	Abbreviation
Basic unit	gram	*g*	liter	*L*	meter	*m*
× 10³	Kilogram	*Kg*	Kiloliter	*KL*	Kilometer	*Km*
× 10⁻¹	Decigram	*dg*	Deciliter	*dL*	Decimeter	*dm*
× 10⁻²	Centigram	*cg*	Centiliter	*cL*	Centimeter	*cm*
× 10⁻³	Milligram	*mg*	Milliliter	*mL*	Millimeter	*mm*
× 10⁻⁶	Microgram	*μg*	Microliter (formerly λ*)	*μL*	Micrometer (formerly, micron*)	*μm*
× 10⁻⁹	Nanogram	*ng*	Nanoliter	*nL*	Nanometer	*nm*
× 10⁻¹⁰					Angstrom*	*Å*
× 10⁻¹²	Picogram	*pg*	Picoliter	*pL*	Picometer	*pm*

*Older term, not part of SI system

Note: There is inconsistency in capitalization. For example, you may see "*Km*" or "*km*" and "*mL*" or "*ml*."

Table 8.3 COMMONLY USED CONVERSION FACTORS FOR UNITS OF MEASUREMENT

Length

1 *mm* = 0.001 *m* = 0.039 *in*	
1 *cm* = 0.01 *m* = 0.3937 *in* = 0.0328 *ft*	1 *in* = 2.540 *cm*
1 *m* = 39.37 *in* = 3.281 *ft* = 1.094 *yd*	1 *ft* = 12 *in* = 0.305 *m*
1 *km* = 1000 *m* = 0.6214 *mi*	1 *mi* = 5280 *ft* = 1.609 *km*

Mass

1 *g* = 0.0353 *oz* = 0.0022 *lb*	1 *oz* = 28.35 *g*
1 *kg* = 1000 *g* = 35.27 *oz* = 2.205 *lb*	1 *lb* = 16 *oz* = 453.6 *g*
	1 *tn* = 2000 *lb*

Volume

1 *mL* = 0.001 *L* = 0.034 *fl oz*	1 *fl oz* = 0.0313 *qt* = 0.0296 *L*
1 *L* = 2.113 *pt* = 1.057 *qt* = 0.2642 *gal*	1 *pt* = 0.4732 *L*
	1 *qt* = 2 *pt* = 0.9463 *L*
	1 *gal* = 8 *pt* = 3.785 *L*

Temperature

°C = (°F − 32) 0.556	°F = (1.8 × °C) + 32
K = C + 273	°R = °F + 460 (R is Rankine)

Table 8.4 THE SI SYSTEM

Basic Property Measured	Unit of Measurement	Abbreviation
Length	meter	*m*
Mass	gram	*g*
Time	second	*s*
Electrical current	ampere	*A*
Temperature	kelvin	*K*
Luminous intensity	candela	*cd*
Amount of substance	mole	*mol*

MANIPULATION PROBLEMS: MEASUREMENTS

1. For each pair, underline the larger value. If two values are the same, underline both.

   a. 1 μm, 1 *nm*   **b.** 1 *cm*, 1 *mm*

   c. 10 *cm*, 1000 *mm*   **d.** 10000 μm, 10 *m*

   e. 1000 *g*, 1 *kg*   **f.** 100 *nm*, 1000 μm

   g. 100 μg, 1 *mg*   **h.** 1 *nm*, 10 *pm*

   i. 10 *nm*, 1 Å   **j.** 100 *mg*, 1 *g*

   k. 1000 μL, 1 *mL*   **l.** 100,000 μm, 1 *m*

   m. 1 *L*, 1000 μL **n.** 1 *m*, 500 *cm*

   o. 1 *pL*, 0.001 *nL*   **p.** 10 *nm*, 0.1 μm

   q. 100 *cm*, 0.1 *m*   **r.** 0.0001 *m*, 10 μm

2. If a sample weighs 100 *g*, how much does it weigh in *kg*?

3. What is the approximate volume of a soda can in *mL*?

IV. INTRODUCTION TO THE USE OF EQUATIONS TO DESCRIBE A RELATIONSHIP

A. Equations

Much of scientific inquiry is about determining the relationship between two or more entities. For example, scientists might study how wolves affect deer populations, how size affects the rate of migration of DNA fragments in electrophoresis, or the relationship between resistance and current in an electrical circuit. Math provides various tools for describing such relationships, including equations, graphs, and statistics.

Equations *are a way to describe relationships using mathematical symbols.* An equation is often about a relationship in which two or more things are equal, although the relationship in an equation is sometimes such that one or more things are greater than (>) or less than (<) other things. Letters in a scientific equation represent the items involved in the relationship. Scientific equations may seem intimidating if they use unfamiliar symbols. However, even complicated equations with unfamiliar symbols can tell you about relationships in a system. There are many equations sprinkled throughout this book, all of which convey information about relationships.

For example, consider the equation $V = I\,R$. This equation (to be discussed in Chapter 14) describes the relationship between voltage (V), current (I), and resistance (R) in an electrical circuit. Even if you have little understanding of electricity, you can learn from this equation that:

1. *Voltage, current, and resistance in an electrical circuit are related to each other in a specific way.*

2. *If resistance increases, either the voltage must also increase, or the current must decrease. Similarly, if current increases, then either voltage must increase or resistance must decrease.*

3. *If voltage goes down, then current and/or resistance must also go down.*

For example:

Suppose that initially the voltage = 12, resistance = 4, and current = 3.

(Voltage, current and resistance have units, but we will momentarily ignore them to simplify the example.)

$$V = I R$$
$$12 = 3(4)$$

What happens if the resistance increases to 5? Perhaps voltage increases, and becomes 15. Or, perhaps current decreases and becomes 2.4:

$$V = I R \qquad\qquad V = I R$$
$$15 = 3\,(5) \qquad\qquad 12 = (2.4)(5)$$

In this case, the current did not change, so voltage must increase.   *In this case, voltage remained constant.*

You will encounter the $V = IR$ equation when performing electrophoresis, a technique that uses electricity to separate various proteins and nucleic acids from one another. During electrophoresis the resistance of the system increases as the separation of molecules occurs. The current, therefore, goes down—unless the operator increases the voltage.

Consider another equation:

$$°F = 1.8 \text{ (temperature in } °C) + 32$$

where °F *is degrees in the Fahrenheit scale*
°C *is degrees in the Celsius scale*

Celsius and Fahrenheit are scales used to measure temperature. The size of a degree is different in the Celsius and Fahrenheit scales.

1. The numbers 1.8 and 32 in this equation are called **constants** *because they are always present in the equation and always have the same value.* The constant 1.8 is needed in the equation because a Celsius degree is 1.8 times larger than a Fahrenheit degree. The number 32 is necessary because the two scales put the zero point in a different place. Celsius places zero degrees at the temperature of

frozen water. Zero degrees Fahrenheit is the lowest temperature that can be obtained by mixing salt and ice.

2. °F and °C are called **variables** *because their value can vary.* (In the previous example, V, I, and R are variables.)

3. Given either of the variables in the equation, it is possible to calculate the other variable. Thus, if we know the temperature in either Celsius or Fahrenheit, the temperature can be converted to the other scale.

B. Units and Mathematical Operations

Units can be manipulated in mathematical operations. Like numbers, units can be multiplied by themselves and can have exponents. For example, area is defined as "length times width." In equation form this is:

$$A = (L) \times (W)$$
where A *is area*
L *is length*
W *is width*

For example, if the length of a room is 15 feet and its width is 10 feet, then its area is:

$$(L) \times (W) = A$$
$$10 \text{ ft} \times 15 \text{ ft} = 150 \text{ ft}^2$$

The rules for the manipulation of exponents, as shown in Box 1, apply to units with exponents. Exponents are added for multiplication and subtracted in division.
For example:

$$\frac{40 \text{ cm}^3}{20 \text{ cm}^2} = \frac{(40) \text{ cm}^{(3-2)}}{(20)} = 2 \text{ cm}^1$$

$$(2.1 \text{ mL})(3.4 \text{ mL}) =$$
$$(2.1 \times 3.4) \times \text{ mL}^{1+1} = 7.1 \text{ mL}^2$$

When working with equations, units can cancel and can be multiplied and divided. For example:

$$2.0 \frac{cm}{mL} \times 13 \frac{mg}{mL} = 26 \frac{cm(mg)}{(mL)^2}$$

$$2.00 \frac{mg}{\cancel{mL}} \times 13.0 \cancel{mL} \times 4.00 = 104 \text{ mg}$$

Another example, which comes from a topic in statistics, is:

$$m = \left[\frac{(4716.20)(mg)(g) - \left(\dfrac{170 \text{ mg}}{7}\right)(161.03 \text{ g})}{(5750 \text{ mg}^2) - \left(\dfrac{170^2}{7}\right)(mg)^2} \right]$$

At this point, the relationship expressed by this equation is not important, but watch the units as this equation is solved algebraically:

$$m = \frac{(4716.20)(mg)(g) - 3910.7286(mg)(g)}{5750\ mg^2 - 4128.57 mg^2}$$

$$= \frac{805.4714(mg)(g)}{1621.43\ mg^2} = \frac{0.497g}{mg}$$

EXAMPLE PROBLEM

A box has a length of 3.45 *cm*, a width of 2.98 *cm* and a height of 3.00 *cm*. What is its volume?

ANSWER

$$V = (L)\ (W)\ (H)$$

$$V = 3.45\ cm \times 2.98\ cm \times 3.00\ cm =$$
$$= (3.45 \times 2.98 \times 3.00)\ (cm \times cm \times cm) = 30.8\ cm^3$$

EXAMPLE PROBLEM

Perform the following additions:

a. 23 *cm* + 56 *cm* = ?

b. 2 *cm* + 4 *s* = ?

ANSWER

a. The correct answer is **79 *cm***—not just **79**

b. The correct answer is:

  2 *cm* + 4 *s*

  This is because s and cm are different units and cannot be added together.

MANIPULATION PROBLEMS: EQUATIONS

1. Explain in words what the following relationships mean:

  a. $A = 2C$ **b.** $C = A/D$ **c.** $Y = 2\ (X) + 1$

2. An important equation in centrifugation is:

  $RCF = 11.18 \times r(RPM/1000)^2$

  It is not necessary to understand the abbreviations or to be familiar with the term "RCF" to answer the following questions:

  a. What happens to *RCF* if *RPM* increases?

  b. What happens to *RCF* if *r* is decreased?

  c. There are two constants in this equation. What are they?

3. Insert numbers into each equation in Problem 1 that will satisfy the equation. For example, if the equation is $A = 2\ C$, then $710 = 2\ (355)$ will satisfy the equation. (So will many other answers.)

4. Solve the following equations; that is, find the value of *X*. Pay attention to the units, if present.

  a. $3X = 15$ **b.** $X = 25(5-4)$

  c. $-X = 3X - 1\ mg$ **d.** $5X = 3X - 5\ mL + 34\ mL$

  e. $\dfrac{X}{2} = 25(3)cm^2$

  f. $X = 3.0\ cm\ (2.0\ mg/mL)\ (2.0)$

  g. $X = \dfrac{25\ mg}{mL}(4.0\ mL)\dfrac{3.0\ oz}{mg}$

5. Perform the following calculations.

  a. $23.4\ pounds \times 34.1\ pounds = ?$

  b. $15.2\ g/3.1\ g = ?$

  c. $\dfrac{25.2\ cm \times 34.5\ cm}{3.00}$ **d.** $\dfrac{5\ mL \times 3\ mL \times 2\ cm}{2\ cm}$

ANSWERS TO PROBLEMS

Manipulation Problems:
Exponents and Scientific Notation (from p. 109)

1. a. $2^2 = 4$

  b. $3^3 = 27$

  c. $2^{-2} = 1/4 = 0.25$

  d. $3^{-3} = 1/27 = 0.037$

  e. $10^2 = 100$

  f. $10^4 = 10,000$

  g. $10^{-2} = 1/100 = 0.01$

  h. $10^{-4} = 1/10,000 = 0.0001$

  i. $5^0 = 1$

2. a. $2^2 \times 3^3 = 4 \times 27 = 108$

  b. $(14^3)(3^6) = (2744)(729) = 2,000,376$

  c. $5^5 - 2^3 = (3125) - (8) = 3117$

  d. $\dfrac{5^7}{8^4} = \dfrac{78125}{4096} = 19.07$

  e. $(6^{-2})\ (3^2) = (1/36)(9) = 9/36 = 0.25$

  f. $(-0.4)^3 + (9.6)^2 = (-0.064) + (92.16) = 92.1$

  g. $a^2 \times a^3 = a^5$

  h. $\dfrac{c^3}{c^{-6}} = c^9$

  i. $(3^4)^2 = 3^8 = 6561$

  j. $(c^{-3})^{-5} = c^{15}$

  k. $\dfrac{13}{(43)^2 + (13)^3} = \dfrac{13}{1849 + 2197} = \dfrac{13}{4046} = 0.0032$

  l. $\dfrac{10^2}{10^3} = \dfrac{1}{10} = 10^{-1}$

3. *Larger numbers are **underlined:***

a. 5×10^{-3} cm, $\underline{500 \times 10^{-1}}$ *cm*

b. 300×10^{-3} *μL*, $\underline{3000 \times 10^{-2}}$ *μL*

c. 3.200×10^{-6} *m*, $\underline{3200 \times 10^{-4}}$ *m*

d. $\underline{0.001 \times 10^{1}}$ *cm*, 1×10^{-3} *cm*

e. $\underline{0.008 \times 10^{-3}}$ *L*, 0.0008×10^{-4} *L*

4. a. $54.0 = 5.40 \times 10^{1}$

b. $4567 = 4.567 \times 10^{3}$

c. $0.345000 = 3.45000 \times 10^{-1}$

d. $10{,}000{,}000 = 1 \times 10^{7}$

e. $0.009078 = 9.078 \times 10^{-3}$

f. $540 = 5.40 \times 10^{2}$

g. $0.003040 = 3.040 \times 10^{-3}$

h. $200{,}567{,}987 = 2.00567987 \times 10^{8}$

5. a. $12.3 \times 10^{3} = 12{,}300$

b. $4.56 \times 10^{4} = 45{,}600$

c. $4.456 \times 10^{-5} = 0.00004456$

d. $2.300 \times 10^{-3} = 0.002300$

e. $0.56 \times 10^{6} = 560{,}000$

f. $0.45 \times 10^{-2} = 0.0045$

6. a. $\dfrac{(4.725 \times 10^{8})(0.0200)}{(3700)(0.770)} =$

$\dfrac{(4.725 \times 10^{8})(2.00 \times 10^{-2})}{(3.7 \times 10^{3})(7.7 \times 10^{-1})} =$

$\dfrac{9.45 \times 10^{6}}{28.49 \times 10^{2}} = 3.32 \times 10^{3}$

b. $\dfrac{(1.93 \times 10^{3})(4.22 \times 10^{-2})}{(8.8 \times 10^{8})(6.0 \times 10^{-6})} =$

$\dfrac{8.1446 \times 10}{52.8 \times 10^{2}} = 0.154 \times 10^{-1} = 1.5 \times 10^{-2}$

c. $(4.5 \times 10^{3}) + (2.7 \times 10^{-2}) =$
$4500 + 0.027 = 4500.027 = 4.5 \times 10^{3}$

d. $(35.6 \times 10^{4}) - (54.6 \times 10^{6}) =$
$(0.356 \times 10^{6}) - (54.6 \times 10^{6}) = -54.244 \times 10^{6} =$
-5.42×10^{7}

e. $(5.4 \times 10^{24}) + (3.4 \times 10^{26}) = (0.054 \times 10^{26})$
$+ (3.4 \times 10^{26}) = 3.454 \times 10^{26} = 3.5 \times 10^{26}$

f. $(5.7 \times 10^{-3}) - (3.4 \times 10^{-6}) = (5700 \times 10^{-6})$
$- (3.4 \times 10^{-6}) = 5696.6 \times 10^{-6} = 5.7 \times 10^{-3}$

7. a. $0.0050 \times 10^{-4} = 0.050 \times 10^{-5} =$
$0.50 \times 10^{-6} = 5.0 \times 10^{-7}$

b. $15.0 \times 10^{-3} = 1.50 \times 10^{-2} = 0.150 \times 10^{-1}$
$= 0.00150 \times 10^{1}$

c. $5.45 \times 10^{-3} = 54.5 \times 10^{-4}$

d. $100.00 \times 10^{1} = 1.0000 \times 10^{3}$

e. $6.78 \times 10^{2} = 0.678 \times 10^{3}$

f. $54.6 \times 10^{2} = 0.00546 \times 10^{6}$

g. $45.6 \times 10^{8} = 4560 \times 10^{6}$

h. $4.5 \times 10^{-3} = 450 \times 10^{-5}$

i. $356.98 \times 10^{-3} = 0.035698 \times 10^{1}$

j. $0.0098 \times 10^{-2} = 0.98 \times 10^{-4}$

Manipulation Problems: Logarithms (from p. 111)

1. a. 2   **b.** 3

2. a. $\log 100 = 2$ **b.** $\log 10{,}000 = 4$

c. $\log 1{,}000{,}000 = 6$ **d.** $\log 0.0001 = -4$

e. $\log 0.001 = -3$

3. a. $\log 7$ between 0 and 1

b. $\log 65.9$ between 1 and 2

c. $\log 89.0$ between 1 and 2

d. $\log 0.45$ between 0 and -1

e. $\log 0.0078$ between -2 and -3

4. a. 4.18   **b.** 2.54   **c.** -2.0

d. -4.53 **e.** 3.082   **f.** -0.462

5. a. antilog $4.8990 = 7.925 \times 10^{4}$

b. antilog $3.9900 = 9.772 \times 10^{3}$

c. antilog $-0.5600 = 0.2754$

d. antilog $-0.0089 = 0.9797$

e. antilog $9.8999 = 7.941 \times 10^{9}$

f. antilog $1.0000 = 10.00$

g. antilog $8.9000 = 7.943 \times 10^{8}$

6. a. 0.35   **b.** 1.35   **c.** 2.35   **d.** 6.49

7. a. antilog $-4.56 = 2.75 \times 10^{-5}$ *M*

b. antilog $-5.67 = 2.14 \times 10^{-6}$ *M*

c. antilog $-7.00 = 1.00 \times 10^{-7}$ *M*

d. antilog $-1.09 = 8.13 \times 10^{-2}$ *M*

e. antilog $-10.1 = 7.94 \times 10^{-11}$ *M*

Manipulation Problems: Measurements (from p. 114)

1. Larger:

a. 1 *μm*   **b.** 1 *cm*   **c.** 1000 *mm*

d. 10 *m*   **e.** 1000 *g* = 1 *kg*   **f.** 1000 *μm*

g. 1 *mg*   **h.** 1 *nm*   **i.** 10 *nm*

j. 1 *g*   **k.** 1000 *μL* = 1 *mL*   **l.** 1 *m*

m. 1 *L*   **n.** 500 *cm*   **o.** 1 *pL* = 0.001 *nL*

p. 0.1 *μm*   **q.** 100 *cm*   **r.** 0.0001 *m*

2. $100 \ g = 0.1 \ kg$

3. $12 \ oz \sim 355 \ mL$

Manipulation Problems: Equations (from p. 116)

1. a. A is equal to the value obtained if C is doubled.

  b. C is equal to the value obtained if A divided by D.

  c. Y is equal to the value obtained if X is multiplied by 2 and then 1 is added.

2. a. RCF increases if RPM increases.

  b. RCF decreases if r decreases.

  c. 11.18, 1000

3. a. $A = 2C$ $4 = 2(2)$

  b. $C = \dfrac{A}{D}$ $12 = \dfrac{36}{3}$

  c. $Y = 2(x) + 1$ $15 = 2(7) + 1$

4. a. $3X = 15$ $X = \dfrac{15}{3} = 5$

  b. $X = 25(5-4)$ $X = 25(1) = 25$

  c. $-X = 3X - 1 \ mg$

      $1 \ mg = 4X$

      $X = \dfrac{1}{4} mg$

d. $5X = 3X - 5 \ mL + 34 \ mL$

     $2X = (34 - 5) \ mL$

     $2X = 29 \ mL$

     $X = 14.5 \ mL$

e. $\dfrac{X}{2} = 25(3) cm^2$

     $\dfrac{X}{2} = 75 \ cm^2$

     $X = 2(75) cm^2$

     $X = 150 \ cm^2$

f. $X = 3.0 \ cm \ (2.0 \ mg/mL) \ (2.0)$

     $X = 3.0 \ cm \ (4.0 \ mg/mL)$

     $X = \dfrac{12.0 \ cm \ mg}{mL}$

g. $X = \dfrac{25 \ mg}{mL} (4.0 \ mL) \dfrac{3.0 \ oz}{mg}$

     $= \dfrac{300 \ mg \ mL \ oz}{mg \ mL} = 300 \ oz$

5. a. $798 \ lb^2$

  b. $\dfrac{15.2 \ g}{3.1 \ g} = 4.9$

  c. $\dfrac{25.2 \ cm \times 34.5 \ cm}{3.00} = 290 \ cm^2$

  d. $\dfrac{5 \ mL \times 3 \ mL \times 2 \ cm}{2 \ cm} = 15 \ mL^2$

CHAPTER 9

Proportional Relationships

I. INTRODUCTION TO RATIOS AND PROPORTIONS

EXAMPLE APPLICATION: PROPORTIONAL RELATIONSHIPS

a. A transgenic animal is one that produces a protein or has a trait from another species as a result of incorporating a foreign gene(s) into its genome. In 1993, transgenic sheep were born that secrete a human protein, AAT, into their milk. AAT is valuable in the treatment of emphysema. A sheep can produce 400 liters of milk each year. If a transgenic sheep secretes 15 g of AAT into each liter of milk she produces, how many grams of AAT can this transgenic sheep produce in a year?

b. If AAT is worth $110/gram, what is the value of a year's production of AAT from this sheep?

(Information from *A Biotech Bonanza on the Hoof?*, edited by Ivan Amato. Science, 259, p. 1698, March 19, 1993.)

Answers on Page 121

A **ratio** is the relationship between two quantities. For example, we might say a car "gets 30 miles per gallon." This statement describes the relationship between gas consumption and miles traveled. The word "per" means "for every." The preparation of a cake might require 10 ounces of chocolate. The relationship between the cake and the amount of chocolate required is a ratio. Other commonplace examples of ratios are "revolutions per minute" (RPM) and "cost per pound."

A ratio can be expressed as a fraction:

$$\frac{30\ mi}{1\ gal} \text{ or, it is also correct to say } \frac{1\ gal}{30\ mi}$$

$$\frac{1\ cake}{10\ oz\ chocolate} \text{ or } \frac{10\ oz\ chocolate}{1\ cake}$$

Observe that a ratio is not a type of equation because there is no =, <, or > sign.

EXAMPLE PROBLEM

A laboratory solution contains 58.5 grams of NaCl per liter. Express this ratio as a fraction.

ANSWER

The relationship can be expressed as:

$$\frac{58.5\ g}{1\ L}$$

A **proportion** is a statement that two ratios are equal. For example, given that a single cake requires 10 ounces of chocolate, 20 ounces of chocolate are needed to bake two cakes. This can be expressed as a proportion equation showing that the two ratios are equal:

$$\frac{1\ cake}{10\ oz\ chocolate} = \frac{2\ cakes}{20\ oz\ chocolate}$$

Read as "1 cake is to 10 ounces of chocolate as 2 cakes is to 20 ounces of chocolate."

A proportion statement is an equation because there is an = sign.

Many situations in the laboratory require the same reasoning as this "chocolate cake" example. For example, if 10 g of glucose are required to make 1 L of a nutrient solution, how many grams of glucose are needed to make 3 L of nutrient solution? Of course, 30 g is the answer.

Even if you are not a great baker, you probably "knew" that if 1 cake requires 10 oz of chocolate, then 2 cakes require 20 oz, but suppose that 16 oz of chocolate are needed for 3 cakes and you want to make 5 cakes. It may not be obvious in this case how much chocolate is necessary. It is possible to use an equation as a helpful tool to solve this problem:

1. The ratio "there are 16 ounces of chocolate per 3 cakes" can be written as:

$$\frac{16\ oz\ chocolate}{3\ cakes}$$

2. The unknown, ?, is how much chocolate is required by 5 cakes.

3. If 3 cakes require 16 ounces of chocolate then how many ounces of chocolate are required for five cakes? This proportional relationship is written:

$$\frac{16\ oz\ chocolate}{3\ cakes} = \frac{?}{5\ cakes}$$

(Note that the units must be in the equation.)

4. To solve for the unknown, "cross multiply and divide":

Cross Multiply

$$\frac{16\ oz\ chocolate}{3\ cakes} \diagdown \frac{?}{5\ cakes}$$

(16 oz chocolate) (5 cakes) = (?) (3 cakes)

Divide

$$\frac{(16\ oz\ chocolate)(5\ \cancel{cakes})}{3\ \cancel{cakes}} = ?$$

The units of cake cancel:

$$\frac{(16\ oz\ chocolate)(5)}{3} = ?$$

Simplifying further:

? = 27 oz chocolate
= how much chocolate 5 cakes requires

5. Thus:

$$\frac{16 \; oz \; chocolate}{3 \; cakes} \approx \frac{27 \; oz \; chocolate}{5 \; cakes}$$

(The fractions 16/3 and 27/5 are about equal although there is a slight difference due to rounding).

Some important points about proportional relationships include:

1. *There are many problems in the laboratory that can be easily solved using the same reasoning as the "chocolate cake" example, even though the units may be less familiar and the numbers may be more complex.*

2. *It is necessary to keep track of units.* The units in the chocolate cake example are "ounces" (*oz*) (of chocolate) and "cake" (or "cakes"). In a proportion equation, the units in both denominators must be the same and the units in both numerators must be the same. For example, if one cake requires 10 *oz* of chocolate, how much chocolate is needed for two cakes? The **wrong** way to set this up is:

$$\frac{1 \; cake}{10 \; oz} = \frac{?}{2 \; cakes}$$

Cross Multiply and Divide: $? = 0.2 \; cakes^2/oz$

Clearly, these units are absurd and the answer is wrong.

It is, however, correct to either write:

$$\frac{1 \; cake}{10 \; oz} = \frac{2 \; cakes}{?} \qquad or \qquad \frac{10 \; oz}{1 \; cake} = \frac{?}{2 \; cakes}$$

EXAMPLE PROBLEM

If there are about 100 paramecia in a 20 *mL* water sample, then about how many paramecia would be found in 10^3 *mL* of this water?

ANSWER

A couple of ways to think about this problem are:

1. Strategy 1 Use a proportion equation as follows:

$$\frac{100 \; paramecia}{20 \; mL} = \frac{?}{1000 \; mL}$$

Cross multiply and divide:

$$(100 \; paramecia)(1000 \; mL) = (20 \; mL)(?)$$

$$? = \frac{(100 \; paramecia)(1000 \; mL)}{20 \; mL}$$

$$? = 5000 \; paramecia$$

2. Strategy 2 Use a proportion equation to calculate how many paramecia are in each milliliter:

$$\frac{100 \; paramecia}{20 \; mL} = \frac{?}{1 \; mL}$$

$$? = 5 \; paramecia$$

There are 5 paramecia in each milliliter; therefore, multiply 5 by 1000 to calculate how many paramecia there are altogether. The answer is 5000 paramecia.

EXAMPLE APPLICATION: PROPORTIONS ANSWERS (from p. 120)

a. This is a proportional relationship: if the sheep secretes 15 g of AAT per liter of milk, how much will she secrete into 400 *L*? Using a proportion equation:

$$\frac{15 \; g}{1.0 \; L} = \frac{?}{400 \; L}$$

$$? = 6000 \; g$$

This is the amount of AAT a single sheep might produce in a year.

b. This is also a proportional relationship:

$$\frac{\$110}{1.0 \; g} = \frac{?}{6000 \; g}$$

$$? = \$660,000$$

This is the potential value of the AAT from one sheep.

MANIPULATION PROBLEMS: PROPORTIONS

1. Solve for ?.

   a. $\dfrac{?}{5} = \dfrac{2}{10}$

   b. $\dfrac{?}{1 \; mL} = \dfrac{10 \; cm}{5 \; mL}$

   c. $\dfrac{0.5 \; mg}{10 \; mL} = \dfrac{30 \; mg}{?}$

   d. $\dfrac{50}{?} = \dfrac{100}{100}$

   e. $\dfrac{?}{30 \; in^2} = \dfrac{15 \; lb}{100 \; in^2}$

   f. ? is to 15, as 30 is to 90

   g. 100 is to 10, as 50 is to ?

2. In the following problems, first state the unknown, set up the proportion equation with the units, then solve for the unknown.

   a. If it requires 50 minutes to drive to Denver, then how long does it take to drive to Denver and back?

   b. If it requires 1 teaspoon of baking soda to make one loaf of bread, then how many teaspoons of baking soda are required for 33 loaves?

   c. If it costs $1.50 to buy one magazine, then how much do 100 magazines cost?

   d. If a recipe requires 1/4 cup of margarine for one batch, then how much margarine is required to make 5 batches?

   e. If a recipe requires 1/8 teaspoon of baking powder to make one batch, then how much baking powder is required to make a half batch?

   f. If one bag of snack chips contains 3 ounces, then how many ounces are contained in 4.5 bags of chips?

APPLICATION PROBLEMS: PROPORTIONS

The following are all proportion-type problems, like the chocolate cake problem.

1. If there are about 1×10^2 blood cells in a 1.0×10^{-2} mL sample, then about how many blood cells would be in 1.0 mL of this blood?

2. Ten milliliters of buffer are needed to fill a particular size test tube. How many mL are required to fill 37 of these test tubes?

3. If there are about 5×10^1 paramecia in a 20 mL water sample, then about how many paramecia would be in 10^5 mL of this water?

4. If there are about 315 insect larvae in 1×10^{-1} kg of river sediment, then about how many larvae would be in 5.0×10^4 grams of this sediment?

5. A fermenter is a vat in which microorganisms are grown. The microorganisms produce a material, such as an enzyme, that has commercial value. Suppose a certain fermenter holds 1000 L of broth. There are about 1×10^9 bacteria in each milliliter of the broth. About how many bacteria are present in the entire fermenter?

6. Thirty grams NaCl are required to make 1 L of a salt solution. How much NaCl is needed to make 250 mL of this solution?

7. A particular solution requires 100 mL of ethanol per liter final volume. How much ethanol is needed to make 10^5 mL of this solution?

8. A particular solution contains 50 mg/mL of enzyme. How much enzyme is needed to make 10^4 μL of solution?

9. A particular enzyme breaks down proteins by removing one amino acid at a time from the proteins. The enzyme removes 60 amino acids per minute from a protein. If a protein is initially 1000 amino acids long and the enzyme is added to it, how long will the protein chain be after 10 minutes?

10. The components required to make 100 mL of a solution are listed as follows:

Solution Q

Component	Grams
NaCl	20.0
Na azide	0.001
Mg sulfate	1.0
Tris	15.0

(Don't worry at this time about what these components are.)

Prepare a table that shows how to prepare 1.5 L of solution Q.

11. One mole of carbon has a mass of 12 grams. (Although the definition of a "mole" is not important for solving this problem, you can see p. 452 for a brief explanation.)

   a. What is the mass of 2 moles of carbon?

   b. What is the mass of 0.5 moles of carbon?

12. One mole of table salt (NaCl) has a mass of 58.5 g.

   a. What is the mass of 1.5 moles of table salt?

   b. What is the mass of 0.75 moles of table salt?

II. PERCENTS

EXAMPLE APPLICATION: PERCENTS

a. There are about 3×10^9 DNA base pairs in the human genome. Human chromosome 21 is the smallest chromosome (besides the Y chromosome) and contains about 2% of the human genome. About how many base pairs comprise chromosome 21?

b. The goal of the Human Genome Project is to find the base pair sequence of the entire human genome. If chromosome 21 is sequenced, and if the sequencing techniques used are correct 99.9% of the time, then how many of the base pairs determined will be incorrect?

c. Approximately one out of every 700 babies born has an extra copy of chromosome 21. Such children have Down syndrome, which is associated with mental retardation and heart disease. About what percent of all children are born with Down syndrome?

Answers on Page 124

A. Basic Manipulations Involving Percents

Percents are a familiar type of ratio. The % sign symbolizes a fraction and the word **percent** *means "of every hundred."* For example:

10% means 10/100 or "ten out of every hundred"
0.1% means 0.1/100 or "one tenth out of every hundred"

Suppose that 300 students are surveyed to determine their favorite computer game company. Of those, 225 students prefer Brand Z. This information can be expressed as a ratio, that is:

$$\frac{225 \ students}{300 \ students}$$

To convert this information to a percent, it is necessary to remember that percent means "out of every hundred." The ratio 225/300 can be converted to a percent using the logic of proportions:

$$\frac{225 \ students}{300 \ students} = \frac{?}{100}$$

$$? = 75$$

This means "75 out of every hundred" or 75% of students preferred Brand Z

We can generalize from this example and say that the percent of individuals with a particular characteristic is:

$$\frac{number\ with\ the\ characteristic}{total\ number} \times 100\%$$

$$= \%\ with\ characteristic$$

Thus, the percent of students who prefer Brand Z computer game is:

$$\frac{225\ students}{300\ students} \times 100\% = 75\%$$

Observe that the numerator and denominator must have the same units.

Rules for manipulating percents are summarized in Box 1.

Box 1   RULES FOR MANIPULATING PERCENTS

1. *To convert a percent to its decimal equivalent, move the decimal point two places to the left.* For example:

$$10\% = 0.10$$
$$15\% = 0.15$$
$$0.1\% = 0.001$$

2. *To change a decimal to a percent, move the decimal point two places to the right and add a % sign.* For example:

$$0.10 = 10\%$$
$$0.87 = 87\%$$
$$1 = 100\%$$

3. *To change a percent to a fraction, write the percent as a fraction with a denominator of 100.* For example:

$$30\% = 30/100 = 3/10$$
$$1.5\% = 1.5/100$$
$$100\% = 100/100 = 1$$

4. *To change a fraction to a percent:*

   Strategy 1 Use the logic of proportions:

   For example: Change 50/75 to a percent

   a. Set the problem up as a proportion equation

   $$\frac{50}{75} = \frac{?}{100}$$

   b. Cross multiply and divide

   $$? = 66.66\ldots \quad \text{So } 50/75 = 66.67\%$$

   Strategy 2 Convert the fraction to its decimal equivalent and then move the decimal two places to the right and add a % sign:

   Change 50/75 to a percent

   a. Change the fraction to a decimal: $50/75 = 0.6667$
   b. Change the decimal to a percent: $0.6667 = 66.67\%$

5. *To find a percentage of a particular number convert the percent to its decimal equivalent and multiply the decimal times the number of interest.* For example:

   To find 45% of 900

   a. Convert the percent to a decimal: $45\% = 0.45$
   b. Multiply the decimal times the number of interest: $0.45 \times 900 = 405$.

EXAMPLE PROBLEM

Show two strategies to convert the fraction 34/67 into a percent.

ANSWER

Strategy 1 Use proportions:

$$\frac{34}{67} = \frac{?}{100}$$

Cross multiply and divide

$$? = 50.7$$

Therefore, the answer is 50.7%

Strategy 2 Convert the fraction to a decimal and then move the decimal point two places to the right and add a percent sign:

$$\frac{34}{67} = 0.507$$

$$\rightarrow 50.7\%$$

B. An Application of Percents: Laboratory Solutions

Percents have many applications in the laboratory, one of which is in expressing the components of a solution. We address "percent solutions" in detail in Chapter 21 of this text. In this section, we will introduce some simple example problems relating to laboratory solutions.

EXAMPLE PROBLEM

A laboratory solution is composed of water and ethylene glycol. How could you prepare 500 *mL* of a 30% ethylene glycol solution?

ANSWER

Strategy 1 We can use the logic of proportions to calculate how much ethylene glycol is required for 500 *mL*:

$$\frac{30}{100} = \frac{?}{500 \ mL}$$

Cross multiply and divide.

$$? = 150 \ mL$$

This is how much ethylene glycol is needed.

The remainder of the solution will be water. The best way to prepare this solution is to place the 150 *mL* of ethylene glycol in a piece of glassware that is marked at 500 *mL* and then to fill it to the 500 *mL* mark with water.

Strategy 2 Since the total volume of the solution will be 500 *mL* and 30% of it will be ethylene glycol:

$$500 \ mL \times 0.30 = 150 \ mL$$

This is how much ethylene glycol is needed.

The remainder of the solution will be water. The best way to prepare this solution is to place the 150 *mL* of ethylene glycol in a piece of glassware that is marked at 500 *mL* and then to fill it to the 500 *mL* mark with water.

EXAMPLE APPLICATION ANSWERS (from p. 122)

a. 2 % = 0.02

$3 \times 10^9 \times 0.02 = 6 \times 10^7$ = the number of base pairs in chromosome 21.

b. If 99.9% of the base pairs are correct, then 0.1% will be incorrect.

0.1% = 0.001

$6 \times 10^7 \times 0.001 = 6 \times 10^4$, which is the number of base pairs that will be incorrect.

c. 1 out of 700 equals 1/700 = 0.0014 = 0.14% = percent of children born with Down syndrome.

MANIPULATION PROBLEMS: PERCENTS

1. Without using a calculator, give the approximate percent of each of the following. For example, 35 out of 354 is about 10% because it is close to 35 out of 350 which is close to 10/100 which is 10%.

   a. 98 out of 100   **b.** 110 out of 10,004

   c. 3 out of 15   **d.** 45 out of 45002

2. Express the following percents as fractions and as decimals.

   a. 34%   **b.** 89.5%   **c.** 100%

   d. 250%   **e.** 0.45%   **f.** 0.001%

3. Calculate the following.

   a. 15% of 450   **b.** 25% of 700

   c. 0.01% of 1000   **d.** 10% of 100

   e. 12% of 500   **f.** 150% of 1000

4. Express the following fractions or decimals as percents.

   a. 15/45   **b.** 2/2   **c.** 10/100

   d. 1/100   **e.** 1/1000   **f.** 6/40

   g. 0.1/0.5   **h.** 0.003/89   **i.** 5/10

   j. 0.05   **k.** 0.0034   **l.** 0.25

   m. 0.01   **n.** 0.10   **o.** 0.0001

   p. 0.0078   **q.** 0.50

5. Calculate the following without using a calculator.

   a. 20% of 100   **b.** 20% of 1000   **c.** 20% of 10,000

   d. 10% of 567   **e.** 15% of 1000   **f.** 50% of 950

   g. 5% of 100   **h.** 1% of 876   **i.** 30% of 900

6. State the following as percents:

   a. 1 part in 100 total parts

   b. 3 parts in 50 total parts

   c. 15 parts in (15 parts + 45 parts)

   d. 0.05 parts in 1 part total

   e. 1 part in 25 parts total

   f. 2.35 parts in (2.35 parts + 6.50 parts)

APPLICATION PROBLEMS: PERCENTS

1. There are 55 students in a class: 25 of them work 20 hours a week or more at jobs outside school; 14 of them work 10–19 hours per week at jobs. The rest work 0–10 hours per week at outside jobs. What percent of the students work 20 hours or more per week at jobs? What percent work 19 hours or less at outside jobs?

2. How much ethanol is present in 100 mL of a 50% solution of ethanol?

3. A laboratory solution is made of water and ethylene glycol. How could you prepare 250 mL of a 30% ethylene glycol solution?

4. A laboratory solution is required that is 10% acetonitrile, 25% methanol, and the rest is water. How could you prepare this solution? (Note that the total volume is unspecified, so you can prepare any volume you want. Also, these components come as liquids.)

5. A particular laboratory solution contains 5 mL of propanol per 100 mL solution. Express this as a percent propanol solution.

6. A solution contains 15 mL of ethanol per 700 mL total volume. Express this as a percent ethanol solution.

7. A solution contains 10 μL of methanol in 1 mL of water. Express this as a percent methanol solution.

8. A solution contains 15 mL of acetone per liter. Express this as a percent solution.

9. Suppose there is a population of insect in which 95% die over the winter, but the rest survive. Each surviving insect produces 100 offspring in the spring, and then dies. Assuming there is no mortality during the summer, and the population has 1000 insects at the beginning of the first winter, how many insects will there be after two winters?

10. Osteoporosis is a disease that leads to brittle bones and is partially caused by a lack of calcium. Each year, 250,000 people suffer osteoporosis-related hip fractures, about 15% of whom die after sustaining such fractures. How many people die each year after fracturing their hip?

11. Double-stranded DNA consists of two long strands. An adenine on one strand is always paired with a thymine on the opposite strand and a guanine on one strand is always paired with a cytosine on the opposite strand. If a purified sample of DNA contains 24% thymine, what are the percentages of the other bases?

12. Olestra is a fat substitute for food products that has been associated with gastrointestinal problems in some preliminary studies. A study was performed in which 1000 subjects were given a free soda and a movie pass. Half the subjects received an unlabeled bag of potato chips made with Olestra; half received chips without Olestra. The incidence of intestinal problems was monitored for 10 days. Of the group fed regular potato chips 17.6% experienced intestinal problems versus 15.8 % of the group given Olestra potato chips. How many individuals in each group experienced gastrointestinal problems?

(Data from "Olestra at the Movies," *The Scientist,* February 16, 1998, p. 31.)

III. DENSITY

Density, *d,* *is the ratio between the mass and volume of a material.* Thus:

$$density = \frac{mass}{volume}$$

For example, the density of benzene is

$$\frac{0.880 \ g}{mL}$$

Various materials have different densities. For example, balsa wood is less dense than lead, Figure 9.1. A lead brick, therefore, weighs more than a piece of balsa wood that occupies the same volume.

Balsa wood (0.160 kg)

Lead brick (11.3 kg)

Figure 9.1.   Density. A piece of balsa wood is lighter than a lead brick of the same volume.

Density can be expressed in various units so it is important to record the units. In addition, the density of a material changes with temperature: Most materials expand when heated and contract when cooled. It is therefore conventional to report the density of a material at a particular temperature. For example:

benzene $d^{20°} = 0.880$ g/mL

This means that the density of benzene is 0.880 g/mL at 20°C.

The densities of solids and liquids are often compared with the density of water. If the density of a material is less than water, then that material will float on water. If the material is more dense than water, then it will sink. Similarly, the densities of gases are often compared with that of air. Materials that are less dense than air (such as balloons filled with helium) rise; materials that are more dense than air do not rise.

Consider different liquids that do not dissolve in one another. If these liquids are mixed in a test tube, then they will separate according to their densities, with the most dense liquid on the bottom, Figure 9.2.

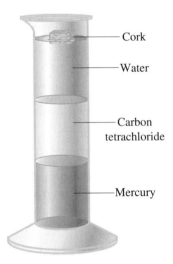
— Cork
— Water
— Carbon tetrachloride
— Mercury

Figure 9.2.   Materials of different density. The density of mercury is 13.5 g/mL, of carbon tetrachloride is 1.59 g/mL, and of water is 0.997 g/mL. Because mercury is the most dense, it is found on the bottom of the cylinder. Cork is least dense and so floats on water.

EXAMPLE PROBLEM

The density for water is:

$$d^{4°} = 1.000 \text{ g/mL.}$$
$$d^{0°} = 0.917 \text{ g/mL}$$

a. What is the volume occupied by 5.6 grams of water at 4°C?

b. Why does ice float?

ANSWER

a. This can easily be solved intuitively. The answer is simply **5.6 mL**. This can also be solved using a proportion equation:

$$\frac{1 \text{ } g}{1 \text{ } mL} = \frac{5.6 \text{ } g}{?}$$

Cross multiply and divide:

$$? = 5.6 \text{ } mL$$

b. Water is less dense at 0°C than it is at 4°C; therefore, ice (which is 0°C) floats on more dense, unfrozen, water.

EXAMPLE PROBLEM

What is the mass of 25.0 mL of benzene at 20°C? (Density = 0.880 g/mL)

ANSWER

Strategy 1 Use a proportion equation:

$$\frac{0.880 \text{ } g}{1 \text{ } mL} = \frac{?}{25.0 \text{ } mL}$$
$$? = 22.0 \text{ } g$$

Strategy 2 Since we know that 1 mL has a mass of 0.880 g, we can multiply by 25.0 to calculate the mass of 25.0 mL of benzene:

$$\frac{0.880 \text{ } g}{1 \text{ } mL} \times 25.0 \text{ } mL = 22.0 \text{ } g$$

MANIPULATION PROBLEMS: DENSITY

1. The density of olive oil is 0.92 g/mL and of water is 0.997 g/mL at 25°C. Vinegar has a density similar to that of water. Which is the top layer in vinegar and oil dressing?

2. The mass of a gold bar that is 3.00 cm^3 is 57.9 g. What is the density of gold?

3. The density of glycerol at 20°C is 1.26 g/mL. What is the volume of 20.0 g of glycerol?

IV. UNIT CONVERSIONS

EXAMPLE APPLICATION: UNIT CONVERSIONS

a. Phthalates are compounds that make plastics flexible and are used in many products, including food wrap. Studies suggest that phthalates may activate receptors for estrogen, the primary female sex hormone. There is speculation that exposure to such estrogenic compounds may increase breast cancer incidence in women, reduce fertility in men, and adversely affect wildlife. One study suggested that margarine may pick up as much as 45 *mg/kg* of phthalates from plastic wrap. If you use a half pound of margarine that has absorbed 45 *mg/kg* of phthalates to bake 3 dozen cookies, how much phthalate will three cookies contain?

b. The pesticide DDT also has estrogenic effects. Although DDT is no longer used in the United States, it is still widely used worldwide. Humans have been shown to accumulate as much as 4 $\mu g/g$ of body weight of DDT in their tissues. If a woman weighs 148 pounds, how much DDT might her body contain? (Let us simplify the situation by assuming that all tissues in her body accumulate the same maximum amount of DDT. This is probably not correct.)

*Information from: "Newest Estrogen Mimics the Commonest?," *Science News*, Vol. 148, p. 47, July 15, 1995. "Beyond Estrogens: Why Unmasking Hormone-Mimicking Pollutants Proves So Challenging," Janet Raloff. *Science News*, Vol. 148, pp. 44–46, July 15, 1995. "Endocrine-Disrupting Environmental Contaminant and Reproduction: Lessons from the Study of Wildlife," L.J. Guillette, Jr. *Women's Health Today: Perspectives on Current Research and Clinical Practice*, Parthenon Publishing Group, New York, pp. 201–207, 1994.

Answers on Page 129

A. Overview

A given quantity, amount, or length can be measured in different units. For example, a candy bar that costs $1.00 also costs 100 cents. A dollar is a larger unit of currency than a cent; therefore, it takes many cents (100) to equal the value of a single dollar. A snake that is 1 foot long is also 12 inches or 30.48 *cm* long. A foot is a larger unit of length than a centimeter. It takes more than 30 *cm* to equal the length of a single foot.

It is often necessary to convert numbers that have units in the U.S.C.S. system to numbers with metric units (e.g., from pounds to kilograms). It is also often necessary to convert from one U.S.C.S. unit to another U.S.C.S. unit (e.g., from feet to inches) or from one metric unit to another metric unit (e.g., from centimeters to meters). This section discusses such conversions.

There are two common strategies for performing unit conversions. The first approach is to think of conversion problems in terms of proportions and to use a proportion equation to solve them. The second approach is the use of conversion factors (also called the "unit canceling method" or "dimensional analysis"). Some people prefer one strategy to perform conversions; other people prefer the other. Any strategy is correct as long as it consistently yields the right answers.

B. Proportion Method of Unit Conversion

Let us illustrate the proportion approach with an example of a conversion from the U.S.C.S. system to metric units:

If a student weighs 150 pounds, how much does he weigh in kilograms?

$$(1 \ kg = 2.2 \ lb)$$

Using proportions, if 1 *kg* = 2.2 *lb*, then how many kilograms is 150 *lb*?

$$\frac{2.2 \ lb}{1 \ kg} = \frac{150 \ lb}{?}$$

$? = 68.2 \ kg$ = the student's weight in kilograms

It is a good idea to examine the answer to see if it makes sense. A kilogram is a larger unit than a pound (just as a dollar is a larger unit than a cent). When the student's weight is converted from pounds to kilograms, therefore, the number should be lower—this answer makes sense.

Another example:

A student is 6.00 *ft* tall. How tall is she in centimeters? (1 *in* = 2.54 *cm*, 1 *ft* = 12 *in*)

This is a proportion problem that, with the information given, needs to be solved in two steps:

Step 1. Convert the height in feet to inches.

$$\frac{1 \ ft}{12 \ in} = \frac{6.00 \ ft}{?}$$

$? = 72.0 \ in$ = her height in inches

Step 2. Convert the height in inches to height in centimeters.

$$\frac{1 \ in}{2.54 \ cm} = \frac{72.0 \ in}{?}$$

$? = 183 \ cm$ = her height in centimeters

Again, it is a good idea to examine the answer to see if it makes sense. A centimeter is a smaller unit than a foot; therefore, when the student's height is converted from feet to centimeters, the number will have to be larger.

Proportions can also be used to convert from one metric unit to another:

How many meters is 345 *cm*?

There are 100 *cm* in a meter. So:

$$\frac{1 \ m}{100 \ cm} = \frac{?}{345 \ cm}$$

$? = 3.45 \ m$

This answer makes sense. A meter is a larger unit than a centimeter; therefore, when centimeters are converted to meters, the value is a smaller number.

Another example:

Convert 105 *cm* to nanometers.

(1 meter = 10^2 *cm*, and 1 *m* = 10^9 *nm*)

This problem can be solved using two steps:

Step 1. $\dfrac{1\ m}{10^2\ cm} = \dfrac{?}{105\ cm}$

$? = 1.05\ m$

Step 2. $\dfrac{1\ m}{10^9\ nm} = \dfrac{1.05\ m}{?}$

$? = 1.05 \times 10^9\ nm$

This answer is reasonable. A nanometer is a very small unit of length; therefore, it will take many nanometers to equal the length of 105 *cm*. The large answer obtained, $1.05 \times 10^9\ nm$, makes sense.

C. Conversion Factor Method of Unit Conversion

A second strategy for doing conversion problems is to multiply the number to be converted times the proper conversion factor. For example:

Convert 2.80 *kg* to pounds.

A pound is 0.454 *kg*. The conversion factor is expressed as a ratio:

$$\dfrac{1\ lb}{0.454\ kg}$$

Observe that this ratio equals 1. All conversion factors equal 1.

Multiply 2.80 *kg* by the conversion factor:

$$2.80\ \cancel{kg} \times \dfrac{1\ lb}{0.454\ \cancel{kg}} = 6.17\ lb$$

The units of kilograms cancel and the answer comes out with the correct units, pounds.

As we demonstrated earlier with proportions, it is good practice to examine the answer to see that it makes sense.

Suppose you had another reference that told you not that 1 *lb* is 0.454 *kg*, but rather that 1 *kg* is 2.205 pounds. If you use this factor directly, you will get the wrong answer:

$$2.80\ kg \times \dfrac{1\ kg}{2.205\ lb} = 1.27\ kg^2/lb$$

You can tell that this is the wrong answer because the units are wrong. If you "flip over" the conversion factor, however, the kilogram units cancel, resulting in the correct units at the end:

$$2.80\ \cancel{kg} \times \dfrac{2.205\ lb}{1\ \cancel{kg}} = 6.17\ lb$$

When using conversion factors, the units guide you in setting up equations. The units must cancel so that the result has the correct units. The term *unit canceling method* is thus a good description for this strategy.

For example:

A student is 6.00 *ft* tall. How tall is she in centimeters? The conversion factors are:

$$\dfrac{1\ in}{2.54\ cm} \quad and \quad \dfrac{1\ ft}{12\ in}$$

The inches will have to cancel, the feet will have to cancel, and centimeters must remain. Let us try a couple of ways of setting this up:

$$6.00\ ft \times \dfrac{1\ ft}{12\ \cancel{in}} \times \dfrac{1\ \cancel{in}}{2.54\ cm} = ?$$

The units of feet do not cancel if the equation is set up this way and *cm* is in the denominator instead of the numerator; however, if the equation is set up:

$$6.00\ \cancel{ft} \times \dfrac{12\ \cancel{in}}{1\ \cancel{ft}} \times \dfrac{2.54\ cm}{1\ \cancel{in}} = ?$$

The units of inches and feet cancel leaving the answer with units of centimeters. The answer is 183 *cm*, which is the same as the answer obtained previously using proportions. Remember to examine the answer to see that it makes sense.

With the unit cancellation method you can string together as many conversion factors as you want into one long equation. If the equation is correct, then the units will cancel, leaving the answer in the desired units.

Consider the conversion of metric units from one another. For example:

How many meters is 345 *cm*?
The conversion factor is:

$$\dfrac{1\ m}{100\ cm}$$

Multiply $345\ \cancel{cm} \times \dfrac{1\ m}{100\ \cancel{cm}} = 3.45\ m$

The units of *cm* cancel, leaving the answer in the correct units

The answer makes sense.
Another example:

Convert 105 *cm* to nanometers.

There are $10^9\ nm$ in one meter and there are $10^2\ cm$ in 1 *m*. These are the conversion factors. Then:

$$105\ \cancel{cm} \times \dfrac{1\ \cancel{m}}{10^2\ \cancel{cm}} \times \dfrac{10^9\ nm}{1\ \cancel{m}} = ?$$

The centimeters and meters cancel, leaving the answer in nanometers:

$$105\ cm = 1.05 \times 10^9\ nm.$$

This answer makes sense.

Both the proportion and the conversion factor strategies are effective ways to convert numbers from one unit to another. Both strategies require paying attention to the units. Observe that when using proportions to do conversion problems, there is always an equals (=) sign between two ratios. In addition, when using proportions, the units must be the same in both denominators and the same in both numerators. In contrast, with the conversion factor (unit canceling) method, there is not an equals sign between two ratios. There is a multiplication (×) sign between the number to be converted and the conversion factor(s).

EXAMPLE APPLICATION: UNIT CONVERSIONS ANSWERS (FROM PP. 126–127)

a. Let us solve the first part of the question using the proportion method. First, convert 0.50 *lb* of margarine to *kg*:

$$\frac{1\ lb}{0.454\ kg} = \frac{0.50\ lb}{?}$$

? = 0.227 *kg* = weight of margarine in kilogram units.

Margarine can absorb 45 *mg* of phthalates per kilogram so 0.227 *kg* of margarine can take up:

$$\frac{45\ mg}{1\ kg} = \frac{?}{0.227\ kg}$$

? = 10.22 *mg* = milligrams of phthalates that 0.227 *kg* of margarine can absorb.

There are therefore potentially 10.22 *mg* of phthalates in 36 cookies. To calculate the milligrams of phthalates in three cookies:

$$\frac{10.22\ mg}{36\ cookies} = \frac{?}{3\ cookies}$$

? = 0.85 milligrams = *mg* of phthalates in three cookies.

b. Let us use the conversion factor method to solve part b of the problem. Conversion factors can be strung together into one long equation. First, it is necessary to convert the woman's weight from pounds to grams. A factor must then be included to account for the fact that the woman accumulated 4 *μg/g* of DDT in all her tissues. Finally, it would be helpful to covert the answer from micrograms to grams. The resulting single equation is:

$$148\ \cancel{lb} \times \frac{454\ \cancel{g}}{1\ \cancel{lb}} \times \frac{4\ \cancel{\mu g}}{\cancel{g}} \times \frac{1\ g}{10^6\ \cancel{\mu g}} =$$

= 0.269 g = accumulated DDT

MANIPULATION PROBLEMS: UNIT CONVERSIONS

(Use either the proportion method or the conversion factor method to solve these problems. Conversion factors are shown in Table 8.3)

1. Convert:

   a. 3.00 feet to *cm*   **b.** 100 *mg* to *g*

   c. 12.0 inches to miles (use scientific notation for your answer)

   d. 100 inches to *km*   **e.** 10.0555 pounds to ounces

   f. 18.989 pounds to *g*   **g.** 13 miles to *km*

   h. 150 *mL* to liters   **i.** 56.7009 *cm* to *nm*

   j. 500 *nm* to *μm*   **k.** 10.0 *nm* to inches

2. How far is a 10 *km* race in miles?

3. A marathon is 26.2 miles. Express this in kilometers.

4. How tall is a person who is 5 *ft* 4 *in* in meters?

5. In kilometers, how far is a town that is 45 miles away?

6. A car is going 55 *mph*. How fast is it moving in kilometers per hour?

7. How much does a 3.0 ton elephant weigh in the metric system?

8. Which is the least expensive jar of hot fudge?

   a. $2.50 for 12 *oz*   **b.** $3.67 for 250 *g*

   c. $4.50 for 0.300 *kg*   **d.** $2.35 for 0.75 *lbs*

Relationships

52 weeks = 1 year
1 week = 7 days
1 day = 24 hours
1 hour = 60 minutes
1 minute = 60 seconds

Fill in the blanks

9. 1 week = ____ days = ____ hours = ____minutes = ____ seconds

10. ____ year = 1 week = ____hours = ____ minutes = ____ seconds

11. 1 year = ____ weeks = ____ days = ____ minutes = ____ seconds

Relationships

1760 *yds* = 1 *mi*
1 *yd* = 3 *ft*
1 *ft* = 12 *in*

12. ____ *mi* = ____ *yds* = 1 *ft* = ____ *in*

Relationships
1 gal = 4 qts
1 qt = 2 pts
1 pt = 16 oz

13. ____ gal = ____ pt = 2 oz

Relationships
1 km = 1000 m = 10^3 m
1 m = 100 cm = 10^2 cm
1 cm = 10 mm
1 mm = 1000 μm = 10^3 μm
1 μm = 1000 nm = 10^3 nm
2.5 cm = 1 in

14. 1 km = ____ m = ____ cm = ____ mm = ____ μm = ____ nm

15. ____ km = ____ m = ____ cm = ____ mm = ____ μm = 1 nm

16. ____ km = ____ m = 2.5 cm = ____ mm = ____ μm = ____ nm

17. a. 6.25 mm = ____ μm

b. 0.00896 m = ____ mm

c. 9876000 nm = ____ mm

18. ____ km = 3 m = ____ cm = ____ in

Relationships
1 g = 10^{-3} kg
1 mg = 10^{-3} g
1 μg = 10^{-3} mg

19. 5 kg = ____ g = ____ mg = ____ μg

20. ____ kg = 0.0089 g = ____ mg = ____ μg

21. ____ kg = ____ g = ____ mg = 2×10^{-8} μg

22. a. 0.8657 g = ____ mg

b. 526 kg = ____ mg

c. 63 g = ____ μg

d. 2.63×10^{-6} μg = ____ kg

Relationships
(These units relate to the decay of radioactivity.)
(Ci is a unit called a "Curie")
1 Ci = 3.7×10^{10} dps
(disintegrations per second)
1 Ci = 1000 mCi = 10^3 mCi
1 mCi = 1000 μCi = 10^3 μCi
1 Bq (Becquerel) = 1 dps

23. 1 Ci = ____ dps = ____ dpm (disintegrations per minute)

24. 1 Ci = ____ mCi = ____ μCi = ____ dps

25. ____ Ci = ____ mCi = ____ μCi = 10^5 dps

26. ____ Ci = ____ mCi = 100 μCi = ____ dps = ____ dpm

27. 1 Ci = ____ mCi = ____ μCi = ____ dps = ____ Bq

28. ____ Ci = ____ mCi = 250 μCi = ____ dps = ____ Bq

APPLICATION PROBLEMS: CONVERSIONS

(Use either the proportion method or the conversion factor method to solve these problems. Conversion factors are shown in Table 8.3)

1. Suppose bacteria are growing in a flask. The growth medium for the bacteria requires 5 g of glucose per liter. A technician has prepared some medium and added 0.24 lb of glucose to 25 L. Did the technician make the broth correctly?

2. A recipe to make 1 L of a laboratory solution is shown as follows:

Solution X

Component	Grams
NaCl	20.00
Na azide	0.001
Mg sulfate	1.000
Tris	15.00

Prepare a table that shows how to prepare 1 mL of the same solution. Express the amounts of each component needed in mg.

3. Suppose a particular enzyme must be added to a nutrient solution used to grow bacteria. The enzyme comes as a freeze-dried powder. The manufacturer of the enzyme states that every gram of enzyme powder actually contains only 680 mg of enzyme, the rest is an inert filler that has no effect. If a recipe calls for 10.0 oz of this enzyme for every 100.0 L of broth, and if you prepare 500.0 L of broth, how much of the enzyme powder will you need to add? Remember to compensate for the inert filler.

V. CONCENTRATION AND DILUTION

A. Concentration

Concentration *is the amount of a particular substance in a stated volume (or sometimes mass) of a solution or mixture.* Concentration is a ratio where the numerator is the amount of the material of interest and the denominator is usually the volume (or sometimes mass) of the entire mixture. For example:

$$\frac{2\ g\ NaCl}{1\ L}$$

means that 2 g of NaCl is dissolved in enough liquid so that the total volume of the solution is 1 L.

The substance that is dissolved is called the **solute**. The liquid in which the solute is dissolved is called the **solvent**. In this example, NaCl is the solute and water is the solvent.

Note that the words "concentration" and "amount" are not synonyms. **Amount** *is how much of a substance is present* (e.g., 2 *g*, 4 cups, or one teaspoon). In contrast, concentration is a ratio with a numerator (amount) and a denominator (usually volume).

Because concentration is a ratio, problems involving concentrations use the same reasoning as other proportion problems. For example, a concentration of 1 *mg* NaCl in 10 *mL* of solution is the same as a concentration of 10 *mg* NaCl in 100 *mL* of solution. Similarly:

How could you make 300 *mL* of a solution that has a concentration of 10 *g* of NaCl in 100 *mL* total solution?

This is a proportion problem:

$$\frac{10 \ g}{100 \ mL \ total} = \frac{?}{300 \ mL \ total}$$

$$? = 30 \text{ grams}$$

30 g *of NaCl in 300* mL *is the same concentration as 10* g *of NaCl in 100* mL

EXAMPLE PROBLEM

Which is more pure: a chemical that contains 0.025 *g* of contaminating material in 10^4 *kg* or a chemical that contains 10^2 *mg* of contaminant in 10^4 *kg*?

ANSWER

There is more than one way to solve this problem. One approach, shown here, is to convert the concentrations of both chemicals into the same units so they can be compared more easily.

Chemical 1: $\dfrac{0.025 \ g}{10^4 \ kg}$ of contaminant

Chemical 2: Convert 10^2 *mg* to grams.
$$10^2 \ mg = 10^{-1} \ g$$

Thus, chemical 2 has a contaminant concentration of
$$\frac{10^{-1} g}{10^4 \ kg} = \frac{0.1 \ g}{10^4 \ kg}$$

Chemical 1 is more pure because it has less contaminant per 10^4 *kg* than chemical 2.

MANIPULATION/APPLICATION PROBLEMS: CONCENTRATION

1. If a solution requires a concentration of 3 *g* of NaCl in 250 *mL* total volume, how much NaCl is required to make 1000 *mL*?

2. If the concentration of magnesium sulfate in a solution is 25 *g/L*, how much magnesium sulfate is present in 100 *mL* of this solution?

3. If the concentration of magnesium chloride in a solution needs to be 1 *mg/mL*, how much magnesium chloride is required to make 15 *L* of this solution?

4. If a solution requires 0.005 *g* of Tris base per liter, how much Tris base is required to make 10^{-3} liters of this solution?

5. If there are 300 *ng* of dioxin in 100 *g* of baby diapers, how much dioxin is there in 1 *kg* of diapers?

6. A *mole* is an expression of amount (which is discussed in the Solutions Unit). If the concentration of a solute in a solution is 5.00×10^{-3} *moles/L*, how many *moles* are there in 5 *mL* of this solution?

7. If a solute has a concentration of 0.1 *moles/L*, how much solute is present in 1 *μL* of solution?

8. If a solute has a concentration of 10^{-2} *g/L*, how much solute is present in 78 *mL* of solution?

9. Enzymes are proteins that catalyze chemical reactions in biological systems. Enzymes are rated on the basis of how active they are or how quickly they can catalyze reactions. Every preparation of enzyme has a certain activity expressed in Units of Activity. Suppose you buy a vial of an enzyme called beta-galactosidase. The vial is labeled: 3 *mg* solid, 500 *Units/mg*. How many *Units* are present in the vial?

10. A vial of the enzyme horseradish peroxidase is labeled: 5 *mg* solid, 2500 *Units/mg*. How many *units* are present altogether in this vial?

11. If a solution contains 3 *g/mL* of compound A, how much compound A is present in 1 *L* of this solution?

12. A solution has 5 *μg/L* of enzyme Q. How much enzyme Q is present in:

    a. 50 *mL* of solution

    b. 500 *mL* of solution

    c. 100 *mL* of solution

    d. 100 *μL* of solution

13. A solution has 0.5 *mg/mL* of the enzyme lysozyme. How much lysozyme is present in:

    a. 5 *mL* of solution

    b. 0.5 *mL* of solution

    c. 100 *μL* of solution

    d. 1000 *μL* of solution

14. There are analytical instruments in the laboratory that are capable of detecting extremely small amounts of specific chemicals. Suppose a particular instrument can detect as little as a single molecule of benzo(a)pyrene (a carcinogen) out of 10^6 molecules of various

compounds. Is the instrument sensitive enough to detect 100 molecules of benzo(a)pyrene out of 10^9 molecules?

15. Which is more pure, a chemical that contains $1\ g$ of contaminating material in $10^6\ kg$, or a chemical that contains $10^{-2}\ mg$ of contaminant in $10^{-3}\ kg$?

B. Introduction to Dilutions: Terminology

EXAMPLE APPLICATION: DILUTIONS

Plasmids are circular DNA molecules that can transport a gene from one bacterium into another bacterium. In nature, plasmids may carry genes that make a bacterium resistant to an antibiotic. As bacteria exchange plasmids carrying resistance genes, resistance to antibiotics spreads among bacterial populations. Plasmids can also be used in the laboratory to transport useful genes into bacteria. Suppose that in a particular experiment, it is necessary to add 0.01 μg of plasmid to a tube. Suppose further that I mL of plasmid in solution at a concentration of I mg plasmid/mL is available. How might the addition of only 0.01 μg of plasmid be accomplished? Assume that it is not possible to accurately measure a volume less than I μL.

Answer on Page 138

There are many situations in the laboratory that require dilutions. A **dilution** *is when one substance (often but not always water) is added to another to reduce the concentration of the first substance. The original substance being diluted may be called the* **stock solution.**

There are various ways to speak about dilutions; unfortunately, this variation in terminology can lead to confusion. Let us illustrate dilution terminology with an example. A baker buys a bottle of food coloring and dilutes it to decorate cookies. The baker takes 1 mL of the concentrated food coloring and adds 9 mL of water so that the total volume of the diluted food coloring solution is 10 mL, Figure 9.3a. Various people might refer to this same dilution using different terminology, Figure 9.3b.

Observe that the word **to** and the symbol **:** are used inconsistently. The word **to**, or the symbol **:** is sometimes used before the volume of the *diluting substance*, the **diluent**. (In this example, the diluent is 9 mL of water.) Other times, the word **to** or the symbol **:** is used before the total volume of the final mixture. (In this example, the total volume of the final mixture is 10 mL.) The key to avoiding confusion is to keep track of whether you are talking about the *total volume of the final mixture or about the amount of diluting substance.* When reading what other people have written, try to determine what they mean.

In this book, the dilution terminology conforms to that suggested by the American Society for Microbiology (ASM) Style Manual (*ASM Style Manual for Journals*

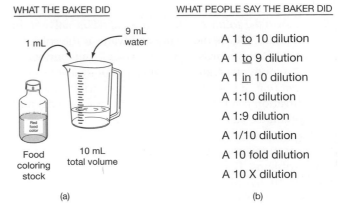

WHAT THE BAKER DID	WHAT PEOPLE SAY THE BAKER DID
	A 1 to 10 dilution
	A 1 to 9 dilution
	A 1 in 10 dilution
	A 1:10 dilution
	A 1:9 dilution
	A 1/10 dilution
	A 10 fold dilution
	A 10 X dilution
(a)	(b)

Figure 9.3. Dilutions and Terminology. a. What the baker did. **b.** Varied terminology to describe the same dilution.

and Books, American Society for Microbiology, Washington, D.C., 1991). This terminology is summarized in Box 2.

C. Dilutions and Proportional Relationships

The concepts of dilution and proportion are related. For example:

1 mL of food coloring mixed with 9 mL of water is the same dilution as 10 mL of food coloring mixed with 90 mL of water.

$$\frac{1\ mL}{10\ mL} = \frac{10\ mL}{100\ mL} = \frac{1}{10}$$

1 mL in 10 mL total = 10 mL in 100 mL total
= 1/10 dilution

Note that the units in the numerator and the denominator are the same and cancel.

All of the following are the same dilution, **1 in 10 total**:

$$
\begin{array}{r}
2\ mL\ food\ coloring \\
+\ \underline{18\ mL\ water} \\
20\ mL\ total\ volume
\end{array}
$$

$$\frac{2\ mL}{20\ mL} = \frac{1}{10}$$

$$
\begin{array}{r}
100\ \mu L\ enzyme\ solution \\
+\ \underline{900\ \mu L\ buffer\ solution} \\
1000\ \mu L\ total\ volume
\end{array}
$$

$$\frac{100\ \mu L}{1000\ \mu L} = \frac{1}{10}$$

$$
\begin{array}{r}
17\ fluid\ oz\ juice \\
+\ \underline{153\ fluid\ oz\ water} \\
170\ fluid\ oz\ total
\end{array}
$$

$$\frac{17\ oz}{170\ oz} = \frac{1}{10}$$

Box 2   DILUTION TERMINOLOGY BASED ON ASM RECOMMENDATIONS

1. ***1 part food coloring combined with 9 parts water means the food coloring is 1 part in 10* mL *total volume or 1/10 food coloring.*** The denominator in an expression with a slash (/) is the <u>total volume</u> of the solution, never the amount of the diluting substance.

2. ***An undiluted substance, by definition, is called 1/1.***

3. ***When talking about dilutions, the symbol : means parts.*** If 1 *mL* of food coloring is combined with 9 *mL* of water, that is 1 part food coloring plus 9 parts water or 1:9 food coloring **to** water. In this text, a dilution of 1 part plus 9 parts diluent is <u>not</u> referred to as a 1:10 dilution.

   For example:
   A **1:2 dilution** *means there are three parts total volume.*
   A $\frac{1}{2}$ **dilution** *means there are two parts total volume.*

   Therefore:
   1/2 is the same as 1:1
   1:2 is the same as 1/3
   1:3:5 A:B:C means that 1 part A, 3 parts B and 5 parts C are combined for a total of 9 parts.
   (The parts can be any unit. For example, this might mean 1* mL *of A, 3* mL *of B, and 5* mL *of C. Or, it might mean 1* g *of A, 3* g *of B, and 5* g *of C.)

MANIPULATION PROBLEMS: DILUTIONS A

(Follow ASM recommendations whenever applicable.)

1. Suppose you dilute 1 *oz* of orange juice concentrate with 3 *oz* of water.

   a. Express this dilution using the word *in.*

   b. Express this dilution using the word *to.*

   c. Express this dilution with a : and then with a /.

2. Express each of the following as a dilution (using a /):

   a. 1 *mL* of original sample + 9 *mL* of water.

   b. 1 *mL* sample + 10 *mL* water.

   c. 3 *mL* sample in a total volume of 30 *mL*.

   d. 3 *mL* sample + 27 *mL* water.

   e. 0.5 *mL* sample + 11.0 *mL* water.

3. Express each of the following ratios as a dilution (using a /).

   a. 1 part sample: 9 part diluent

   b. 1 part sample: 10 parts diluent

   c. 1:3   **d.** 1:1   **e.** 1:4

4. Express each of the mixtures in problem 2 as a 1:___ ratio.

5. If you take a 0.5 *mL* sample of blood and add 1.0 *mL* of water and 3.0 *mL* of reagents, what is the final dilution of the blood?

D. Calculations for Preparing One Dilution

How could you prepare 10 *mL* of a 1/10 dilution of food coloring? This is a simple dilution and is easily accomplished as illustrated:

Combine 1 *mL* of the food coloring with 9 *mL* of water. The total volume will be 10 *mL*. Thus, the dilution will be 1/10.

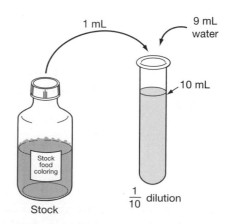

How could you make 10 *mL* of a 1/5 dilution of food coloring?

Combine 2 *mL* of food coloring with 8 *mL* of water. The total volume will be 10 *mL*. Thus, the dilution will be 2/10 = 1/5.

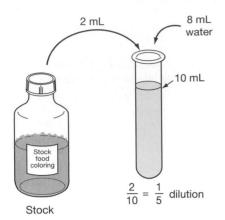

How could you make 100 *mL* of a 1/10 dilution of food coloring?

Combine 10 *mL* of food coloring with 90 *mL* of water. The total volume will be 100 *mL*. Thus, the dilution will be 10/100.

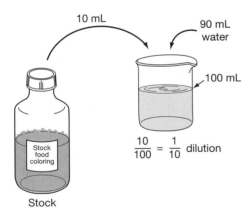

How could you make 250 *mL* of a 1/4 dilution of food coloring? This may be less obvious than the three previous examples, but it can be solved easily using the logic of proportions:

$$\frac{1}{4} = \frac{?}{250 \ mL}$$
$$? = 62.5 \ mL$$

Take 62.5 *mL* of food coloring and add enough water to get 250 *mL* total volume (187.5 *mL*). The food coloring dilution will be:

$$\frac{62.5 \ mL}{250 \ mL} = \frac{1}{4}$$

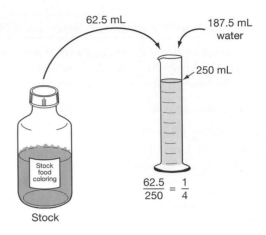

$$\frac{62.5}{250} = \frac{1}{4}$$

Proportions can thus be used to calculate how to make a particular amount of a specific dilution.

MANIPULATION PROBLEMS: DILUTIONS B

1. How would you prepare 10 *mL* of a 1/10 dilution of blood?

2. How would you prepare 250 *mL* of a 1/300 dilution of blood?

3. How would you prepare 1 *mL* of a 1/50 dilution of blood?

4. How would you prepare 1000 μL of a 1/100 dilution of food coloring?

5. How would you prepare 23 *mL* of a 3/5 dilution of solution Q?

6. Suppose you have a stock of a buffer solution that is used in experiments. The stock is 10 times more concentrated than it is used (like frozen orange juice that is sold in a concentrated form). How would you prepare 10 *mL* of the buffer solution at the right concentration?

7. Suppose you have a stock of buffer that is five times more concentrated than it is used. How would you prepare 15 *mL* at the correct concentration?

8. Suppose you need 10^3 μL of a solution. The solution is stored at a concentration that is 100 times the concentration at which it is normally used. How would you dilute the solution?

9. How would you prepare 50 *mL* of a 0.01 dilution of buffer?

E. Dilution and Concentration

When dilutions are prepared it is important to keep track of the concentration of the solute in the diluted tubes. Let's look at some examples of the relationship between concentration and dilution.

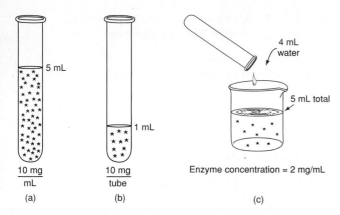

Figure 9.4.   Dilution of an Enzyme Solution. a. The original solution has a concentration of 10 *mg* enzyme/milliliter. Each ★ represents a *mg* of enzyme. **b.** 1 milliliter of solution is removed containing 10 *mg* of enzyme. **c.** The 1 *mL* of solution is diluted with 4 *mL* of water. The resulting diluted solution contains 10 *mg* of enzyme in 5 *mL*. The concentration of enzyme is therefore 10 *mg*/5 *mL* = 2 *mg/mL*.

This example is illustrated in Figure 9.4:

A stock solution contains 10 *mg/mL* of a particular enzyme. This means that every *mL* of stock contains 10 *mg* of enzyme.

The stock solution is diluted by taking 1 *mL* of the solution (which contains 10 *mg* of enzyme) and mixing it with 4 *mL* of water. What is the concentration of enzyme in the diluted solution?

Answer:

The resulting diluted solution has a total volume of 5 *mL* and contains 10 *mg* of enzyme. The concentration of enzyme is:

$$\frac{10 \; mg \; enzyme}{5 \; mL} = \frac{2 \; mg \; enzyme}{1 \; mL}$$

Thus, the stock solution had an enzyme concentration of 10 *mg/mL*. The diluted solution has an enzyme concentration of 2 *mg/mL*.

Another example:

A stock solution initially has a concentration of 20 *mg* of solute per liter.

A diluted solution is prepared by removing 1 *mL* of stock solution and adding 14 *mL* of water.

a. What is the concentration of solute in the diluted solution?

b. How much solute is present in 1 *mL* of the diluted solution?

Answer:

a. The concentration in the stock solution is 20 *mg/L*. The solution was diluted 1/15, therefore, the concentration in the diluted solution is:

$$\frac{20 \; mg}{1 \; L} \times \frac{1}{15} = \frac{1.3 \; mg}{1 \; L}$$

b. This can be solved using the logic of proportions:

$$\frac{1.3 \; mg}{1000 \; mL} = \frac{?}{1 \; mL}$$
$$? = 0.0013 \; mg$$

This is the amount of solute present in 1 *mL* of the diluted solution.

Another way to think about concentration and dilution is using this rule:

The concentration of a diluted solution is determined by multiplying the concentration of the original solution times the dilution (expressed as a fraction).

For example:

A solution of 100% ethanol is diluted 1/10. The concentration of solute in the diluted solution is:

$$100\% \; ethanol \times \frac{1}{10} = 10\% \; ethanol$$

The diluted solution has a concentration of 10% ethanol.

(Because a dilution has no units, the result has the same units as the original solution.)

Another example:

A solution has an original concentration of 10 *mg/mL* of an enzyme.

The solution is diluted 1/5. The concentration of enzyme in the diluted solution is:

$$\frac{10 \; mg}{1 \; mL} \times \frac{1}{5} = \frac{2 \; mg}{1 \; mL}$$

Note that this is the same example as in Figure 9.4, simply described in a different way.

EXAMPLE PROBLEM

A stock solution of enzyme contains 10 *mg/mL* of enzyme. 100 *mL* of this stock is diluted with 400 *mL* of buffer.

a. What is the concentration of enzyme in the resulting diluted solution?

b. How much enzyme will be present in 300 *μL* of the resulting dilution?

ANSWER

a. When 100 *mL* of stock solution is mixed with 400 *mL* of buffer, the dilution can be expressed as:

$$\frac{100 \; \cancel{mL}}{500 \; \cancel{mL}} = \frac{1}{5}$$

The concentration of enzyme in the resulting dilution, therefore, is:

$$\frac{10\ mg}{1\ mL} \times \frac{1}{5} = \frac{2\ mg}{1\ mL}$$

b. 300 μL = 0.300 mL. Because there are 2 mg/mL enzyme in the diluted solution, in 0.300 mL there are:

$$\frac{2\ mg}{mL} \times 0.300\ mL = 0.6\ mg$$

It is also possible to use the logic of proportions:

$$\frac{2\ mg}{1\ mL} = \frac{?}{0.300\ mL}$$
$$? = 0.6\ mg$$

MANIPULATION PROBLEMS: DILUTIONS C

1. If you prepare a 1/40 dilution of a 50% solution, what is the final concentration of the solution?

2. If you prepare a 1/10 dilution of a 10 mg/mL solution, what is the final concentration of the solution?

3. If you prepare a 1:1 dilution of a 10 mg/mL solution, what is the final concentration of the solution?

4. How much 1/5 diluted solution can be made if you have 1 mL of original solution?

5. To prepare 1000 mL of food coloring at a 1/100 dilution, how much of the original food coloring stock solution is required?

F. Dilution Series

A **dilution series** *is a group of solutions that have the same components but at different concentrations.* One way to prepare a dilution series is to make each diluted solution independently of the others beginning with the initial concentrated stock solution. An example is explained here and is illustrated in Figure 9.5.

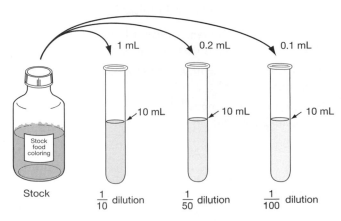

Figure 9.5.   A dilution series where each dilution is independent of the others.

An Independent Dilution Series with Three Dilutions

How could you make 1/10, 1/50, and 1/100 dilutions of food coloring from an original bottle of concentrated food coloring?

First, decide how much of each dilution to prepare; for example, 10 mL.

a. To make 10 mL of a 1/10 dilution:

Take 1 mL of the stock and bring it to a volume of 10 mL using water. This is the first diluted solution.

b. To make 10 mL of a 1/50 dilution:

$$\frac{1\ mL}{50\ mL} = \frac{?}{10\ mL}$$
$$? = 0.2\ mL$$

Thus, take 0.2 mL of the original stock and dilute to a volume of 10 mL. This is the second diluted solution.

c. To make 10 mL of a 1/100 dilution:

$$\frac{1\ mL}{100\ mL} = \frac{?}{10\ mL}$$
$$? = 0.1\ mL$$

Thus, take 0.1 mL of the original stock and dilute to a volume of 10 mL. This is the third diluted solution.

Thus, this strategy requires removing some of the original stock solution three times, once to make each dilution. Each of the three dilutions, therefore, is independent of the others. This strategy will work effectively in some situations.

Let us consider a situation where the strategy of preparing independent dilutions is not effective. Suppose you are working with bacteria and there are so many microorganisms in one mL of medium that the broth must be diluted 100,000 times to get a reasonable number of bacteria for counting. Thus, you want a 1/100,000 dilution of the original stock solution of bacteria. Assuming 10 mL final volume is needed, the proportion is:

$$\frac{1\ mL}{100,000\ mL} = \frac{?}{10\ mL}$$
$$? = 0.0001\ mL$$

To dilute the bacterial broth in one step would require taking 0.0001 mL of bacterial broth and diluting it to 10 mL. It is very difficult, however, to measure 0.0001 mL accurately. The common strategy to prepare such a dilution, therefore, is to use two or more steps. First, a dilution is prepared and then some of this first dilution is removed and used to make a second dilution. Some of the second dilution is used to make a third dilution and so on until the solution is dilute enough. The following example

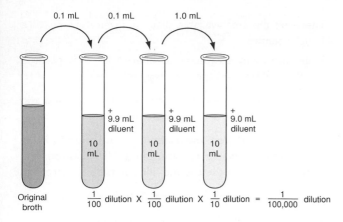

Figure 9.6. A dilution series used to dilute a bacterial stock solution 1/100,000.

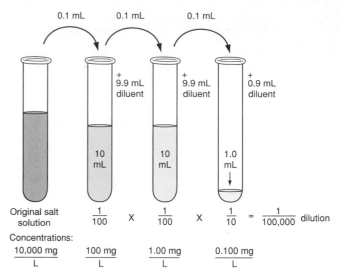

Figure 9.7. Diluting A Salt Solution.

shows how one might prepare 10 mL of a 1/100,000 dilution of bacterial cells in three steps, Figure 9.6.

A Dilution Series Where the Dilutions Are Not Independent of One Another

1. Mix 0.1 mL of the original broth with 9.9 mL of diluent. This will give a dilution of 0.1 mL/10 mL = 1/100.

⇓

2. After thorough mixing, remove 0.1 mL from the diluted bacterial broth prepared in Step 1 and add 9.9 mL of diluent; the total volume in the second dilution is 10 mL. The second dilution has a dilution factor of 1/100, as does the first dilution.

⇓

3. Remove 1 mL from the second dilution tube prepared in Step 2, and add 9 mL of diluent; the total volume in the third dilution tube is 10 mL. The third dilution is 1/10.

The bacteria were therefore diluted 1/100 in the first tube, 1/100 in the second tube, and 1/10 in the third tube. The final, total dilution is:

$$1/100 \times 1/100 \times 1/10 = 1/100,000.$$

The third tube has the desired dilution and volume.

Note in the preceding example that to determine the dilution of solute in the final dilution tube, the dilutions in each intermediate tube were multiplied by one another. Thus, the dilution of solute in the final dilution tube was 1/100 × 1/100 × 1/10 = 1/100,000. The procedure to find the concentration of solute in the final tube can be generalized into the following rule:

The concentration of a diluted solution in the final tube is determined by multiplying the concentration of the original solution times the dilution in the first tube, times the dilution in the second tube and so on until reaching the last tube.

Continuing with this bacteria example, suppose that the original broth contained 1×10^9 bacteria per milliliter. What is the concentration of bacteria in the third (final) dilution?

$$\frac{1 \times 10^9 \ bacteria}{1 \ mL} \times \frac{1}{100,000} = \frac{1 \times 10^4 \ bacteria}{1 \ mL}$$

Consider another example of a dilution series where the dilutions are not independent of one another, Figure 9.7.

A Dilution Series Where the Dilutions Are Not Independent of One Another

A solution contains 10 g salt/L. How could this solution be diluted to obtain 1 mL of solution with a salt concentration of 0.100 mg/L?

There are various strategies that will work. One example is:

1. Remove 0.1 mL of the original salt solution and add 9.9 mL of water. This is a 1/100 dilution.

The concentration of salt in this dilution is 100 mg/L because:

original concentration		dilution		concentration after first dilution
$\dfrac{10,000 \ mg}{1 \ L}$	$\times$	$\dfrac{1}{100}$	$=$	$\dfrac{100 \ mg}{1 \ L}$

⇓

2. Remove 0.1 mL from the first dilution and place it in 9.9 mL of water. This is also a 1/100 dilution.

The concentration of salt in this dilution tube is 1.00 mg/L because:

$$\underset{\substack{\text{concentration} \\ \text{after 1st dilution}}}{\frac{100\ mg}{1\ L}} \times \underset{\text{dilution}}{\frac{1}{100}} = \underset{\substack{\text{concentration after} \\ \text{second dilution}}}{\frac{1.00\ mg}{1\ L}}$$

$$\Downarrow$$

3. Remove 0.1 *mL* from the second dilution and place it in 0.9 *mL* of water. This is a 1/10 dilution. The volume in this tube is 1 *mL*, as desired.

The concentration of salt in this tube is 0.100 *mg* salt/*L* because:

$$\underset{\substack{\text{concentration after} \\ \text{2nd dilution}}}{\frac{1.00\ mg}{1\ L}} \times \underset{\text{dilution}}{\frac{1}{10}} = \underset{\substack{\text{concentration after} \\ \text{final dilution}}}{\frac{0.100\ mg}{1\ L}}$$

This is the desired final concentration and volume

A **serial dilution** *is a series of dilutions that all have the same dilution factor (for example, all are 1/10 dilutions, or all are 1/2 dilutions).* (Note that when people use the phrase *dilution series*, the dilution factor may vary.) Figure 9.8 shows an example of a 1/10 serial dilution.

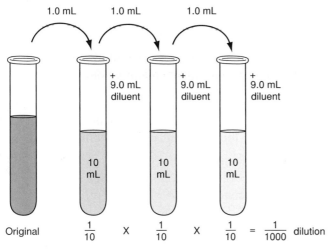

Figure 9.8. A Serial Dilution.

EXAMPLE APPLICATION: DILUTIONS ANSWERS (from p. 132)

The concentration of plasmid in the stock solution is 1 *mg/mL*, which is equal to 1 *μg/μL*. If the plasmid is drawn directly from the stock tube, only 0.01 *μL* is needed, which is a volume too small to measure accurately. Therefore, the stock must be diluted. If the stock is diluted 1000 ×, its concentration will be:

$$\frac{1\ \mu g}{1\ \mu L} \times \frac{1}{1000} = \frac{0.001\ \mu g}{1\ \mu L}$$

Then, 10 *μL* of the diluted plasmid solution will contain 0.01 *μg* of plasmid.

There are pipettes available that can accurately measure 10 *μL* volumes.

There are many strategies to dilute the stock plasmid solution 1000 times. One strategy is:

1. Remove 10 *μL* of the original plasmid stock and add 990 *μL* of buffer or water resulting in a 1/100 dilution.

2. Remove 10 *μL* from the dilution in step 1 and add 90 *μL* of buffer or water resulting in a 1/10 dilution. The total dilution in the second dilution tube is therefore:

$$1/100 \times 1/10 = 1/1000$$

The concentration of plasmid in this tube is 0.001 *μg/μL*, which is the desired concentration.

This two-step strategy involves volumes that can be measured with reasonable accuracy. This strategy also uses only a small amount of the stock solution, which might be an advantage if the stock is being saved for other experiments.

MANIPULATION PROBLEMS: DILUTIONS D

1. Dilution series: Explain how to prepare a 1/10 dilution of food coloring. Then, use the 1/10 dilution to prepare a 1/250 dilution of the original stock of food coloring. Use the 1/250 dilution to prepare a 1/1000 dilution of the stock of food coloring.

2. **a.** Explain how a dilution series could be used to prepare a $1/10^6$ dilution of bacterial cells.

   b. Explain how a 1/10 serial dilution could be used to prepare the same dilution of bacterial cells.

3. Explain how 1/5, 1/50, and 1/250 dilutions of a blood sample could be prepared.

APPLICATION PROBLEMS: DILUTIONS

1. Suppose you have 20 *μL* of an expensive enzyme and you cannot afford to purchase more. The enzyme has a concentration of 1000 *Units/mL*. You are going to do an experiment that requires tubes with a concentration of 1 *Unit/mL* of enzyme and each tube will have 5 *mL* total volume. How much enzyme does each tube require? _____ How many tubes can you prepare before you run out of enzyme? _____

2. Suppose you have 20 *μL* of an expensive enzyme that has 1000 *Units/mL*. You are going to use the enzyme in an experiment that requires tubes with 0.01 *Units/mL* of enzyme and each tube will have 5 *mL* total volume. How much enzyme will each tube require? _____ Will you be able to measure this amount of enzyme accurately? _____ Show how

you can dilute 10 μL of the original 20 μL of enzyme so that you can use it in your experiment. Use a diagram and words to demonstrate your strategy for preparing the enzyme.

3. Antibodies are frequently very concentrated relative to how much is necessary in an experiment. Show how you would use a dilution series to dilute an antibody solution 500,000 ×. Assume you have 1 mL of antibody to begin with, but you want to save at least 0.5 mL of the antibody for future experiments.

4. Counting seems like a simple mathematical process; however, counting microorganisms is not so simple. For one thing, microorganisms, such as bacteria, are not visible to the eye. Another problem is that bacteria may be present in extremely large numbers in a sample. For example, in nutrient broth, there might be 1×10^9 bacterial cells per milliliter. Therefore, microbiologists have devised various methods to count bacterial cells. One such method is called *viable cell counting*. To perform a viable cell count, a sample of bacterial cells is first diluted in series. Then, 0.1 mL of diluted cells are spread on a petri dish that contains nutrient agar. It is assumed that every living cell in the 0.1 mL placed on the agar divides to form a colony of bacterial cells. A colony contains so many individual cells that it is visible to the eye. It is also assumed that each colony originates from a single cell. It is possible to count the colonies and therefore to estimate the number of bacteria in the 0.1 mL of broth.

Assume you begin with a culture of bacteria that contains 1×10^9 *bacteria/mL*. Show how you could dilute the culture so that the final tube has a concentration of 200 *bacteria/0.1 mL*.

5. You are performing a viable cell count of bacteria. The original broth has an unknown concentration of bacteria. You dilute the culture as shown in the diagram below and plate 0.1 mL of the last three dilutions onto three petri dishes with nutrient agar. The following day you count the number of colonies on the plates with the results shown in the illustration. What was the concentration of bacteria in the original tube?

6. You are performing a viable cell count of bacteria. The original broth has an unknown concentration of bacteria. You dilute the culture as shown in the diagram on page 140 and plate 0.1 mL of the last three dilutions on three nutrient agar plates. The following day you count the number of colonies on the plates with the results shown in the illustration. What was the concentration of bacteria in the original tube?

7. Suppose you have a bacterial culture with 10^7 cells per milliliter. How would you dilute this culture so that if you plate 0.1 mL of the last dilution, you will get (in theory) 100 colonies?

8. Suppose you do an assay (test) to determine how much protein is present in a sample. To perform the

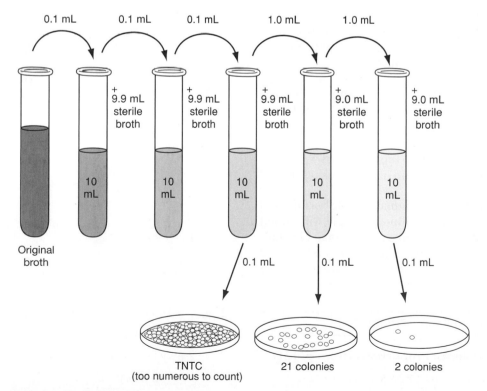

Diagram for Problem 5.

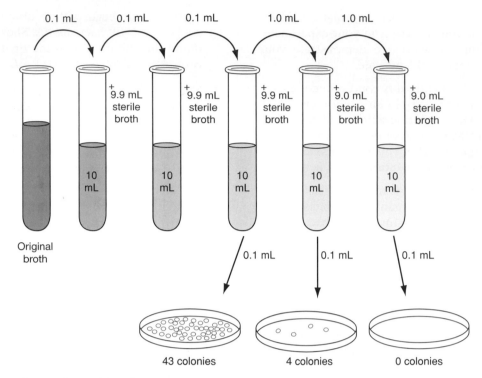

0.1 mL   0.1 mL   0.1 mL   1.0 mL   1.0 mL

+9.9 mL sterile broth   +9.9 mL sterile broth   +9.9 mL sterile broth   +9.0 mL sterile broth   +9.0 mL sterile broth

10 mL (×5)

Original broth

0.1 mL   0.1 mL   0.1 mL

43 colonies   4 colonies   0 colonies

Diagram for Problem 6.

assay, it is necessary to dilute the original sample 1/100. If the assay shows that the concentration of protein in the diluted sample was 50 *mg/mL*, what was the concentration of protein in the undiluted sample? How much protein was present in 100 *mL* of the original, undiluted sample?

9. In a protein assay, the amount of protein in 1 *mL* of diluted sample was 87 *mg*. If the original sample was diluted 1/50, what was the concentration of protein in the original sample?

10. In a protein assay, the original sample was first diluted 1/5. Then, 5 *mL* of the diluted sample was mixed with 20 *mL* of reactants to give 25 *mL* total. Five milliliters were removed from the 25 *mL* mixture. The 5 *mL* contained 3 *mg* of protein. What was the concentration of protein in the original sample?

11. In a protein assay, the original sample was diluted 1/4. Then, 10 *mL* of the diluted sample was added to 20 *mL* of reactants to give 30 *mL* total. Five milliliters were removed from the 30 *mL* mixture. The 5 *mL* contained 10 *mg* of protein. What was the concentration of protein in the original sample?

ANSWERS TO PROBLEMS

Manipulation Problems: Proportions (from p. 121)

1. a. $\dfrac{?}{5} = \dfrac{2}{10}$ $\qquad ? = \dfrac{2 \times 5}{10} = 1$

b. $\dfrac{?}{1\ mL} = \dfrac{10\ cm}{5\ mL}$

$? = \dfrac{10\ cm\,(1\ mL)}{5\ mL} = \dfrac{10\ cm}{5} = 2\ cm$

c. $\dfrac{0.5\ mg}{10\ mL} = \dfrac{30\ mg}{?}$

$? = \dfrac{(30\ mg)(10\ mL)}{0.5\ mg} = 600\ mL$

d. $\dfrac{50}{?} = \dfrac{100}{100}$ $\qquad ? = \dfrac{(50)100}{100} = 50$

e. $\dfrac{?}{30\ in^2} = \dfrac{15\ lb}{100\ in^2}$

$? = \dfrac{15\ lb(30\ in^2)}{100\ in^2} = 4.5\ lb$

f. $\dfrac{?}{15} = \dfrac{30}{90}$ $\qquad ? = \dfrac{30(15)}{90} = 5$

g. $\dfrac{100}{10} = \dfrac{50}{?}$ $\qquad ? = \dfrac{50(10)}{100} = 5$

2. a. ? = time to drive to Denver and back

$\dfrac{?}{2\ way} = \dfrac{50\ min}{1\ way}$

$? = \dfrac{(50\ min)(2\ way)}{1\ way}$ $\qquad ? = 100\ min$

b. ? = amount of baking soda for 33 loaves

$$\frac{1\,t}{1\,loaf} = \frac{?}{33\,loaf}$$

$$? = \frac{1\,t\,(33\,loaf)}{1\,loaf} \qquad ? = 33\,t$$

c. ? = cost of 100 magazines

$$\frac{\$1.50}{1\,magazine} = \frac{?}{100\,magazine}$$

$$? = \frac{\$1.50\,(100\,magazines)}{magazine} \qquad ? = \$150.00$$

d. ? = margarine for 5 batches

$$\frac{1/4\,cup}{1\,batch} = \frac{?}{5\,batch}$$

$$? = \frac{1/4\,cup\,(5\,batch)}{1\,batch} \qquad ? = 1.25\,cups$$

e. ? = baking powder for a half batch

$$\frac{1/8\,t}{1\,batch} = \frac{?}{1/2\,batch}$$

$$? = \frac{1/8\,t\,(1/2\,batch)}{1\,batch} = (1/8\,t)(1/2) \quad ? = 1/16\,t$$

f. ? = oz in 4.5 bags of chips

$$\frac{1\,bag}{3\,oz} = \frac{4.5\,bag}{?}$$

$$\frac{(3\,oz)(4.5\,bag)}{1\,bag} = ? = 13.5\,oz$$

Application Problems: Proportions (from p. 122)

1. $\dfrac{10^2\,cell}{10^{-2}\,mL} = \dfrac{?}{1\,mL}$ $\qquad$? = 10^4 cell

(10^2 cell in 10^{-2} mL = 10^4 cell/mL)

2. $\dfrac{10\,mL}{1\,tube} = \dfrac{?}{37\,tubes}$ $\qquad$? = 370 mL

3. $\dfrac{5 \times 10^1\,paramecia}{20\,mL} = \dfrac{?}{10^5\,mL}$

? = 2.5 ×10^5 paramecia

4. 5 ×10^4 g = 50 kg

$$\frac{315\,larvae}{1 \times 10^{-1}\,kg} = \frac{?}{5 \times 10^1\,kg}$$

? = 1.6 ×10^5 larvae

5. $\dfrac{1 \times 10^9\,bacteria}{1\,mL} = \dfrac{?}{1000 \times 10^3\,mL}$

? = 1 ×10^{15} bacteria

6. $\dfrac{30\,g}{1000\,mL} = \dfrac{?}{250\,mL}$ $\qquad$? = 7.5 g

7. $\dfrac{100\,mL\,ethanol}{1000\,mL} = \dfrac{?}{10^5\,mL}$

? = 10,000 mL ethanol = 10 L

8. $10^4\,\mu L$ = 10 mL

$$\frac{50\,mg}{1\,mL} = \frac{?}{10\,mL} \qquad ? = 500\,mg$$

9. $\dfrac{60\,amino\,acids}{min} = \dfrac{?}{10\,min}$ $\quad$? = 600 amino acids

1000 amino acids − 600 amino acids
= 400 amino acids left after 10 minutes

10.

	100 mL	**1500 mL**
NaCl	20.0 g	300 g
Na azide	0.001 g	0.015 g
Mg sulfate	1.0 g	15 g
Tris	15.0 g	225 g

11. a. $\dfrac{12\,g}{1\,mol} = \dfrac{?}{2\,mol}$ $\qquad$? = 24 g

$\quad$ **b.** $\dfrac{12\,g}{1\,mol} = \dfrac{?}{0.5\,mol}$ $\qquad$? = 6 g

12. a. $\dfrac{1\,mole}{58.5\,g} = \dfrac{1.5\,mole}{?}$ $\qquad$? = 87.8 g

Alternatively, 58.5 g/mole × 1.5 mole = 87.8 g

$\quad$ **b.** $\dfrac{1\,mole}{58.5\,g} = \dfrac{0.75\,mole}{?}$ $\qquad$? = 43.9 g

Manipulation Problems: Percents (from p. 124)

1. a. 98/100 = 98%

$\quad$ **b.** 110/10,004 ≈ $\dfrac{1}{100}$ = 1%

$\quad$ **c.** $\dfrac{3}{15} = \dfrac{1}{5} = \dfrac{20}{100}$ = 20%

$\quad$ **d.** $\dfrac{45}{45002} ≈ \dfrac{1}{1000}$ = 0.1%

2. a. 34% = $\dfrac{34}{100}$ = 0.34

$\quad$ **b.** 89.5% = $\dfrac{89.5}{100}$ = 0.895

$\quad$ **c.** 100% = $\dfrac{100}{100}$ = 1

$\quad$ **d.** 250% = $\dfrac{250}{100} = \dfrac{2.5}{1}$ = 2.5

$\quad$ **e.** 0.45% = $\dfrac{0.45}{100}$ = 0.0045

$\quad$ **f.** 0.001% = $\dfrac{0.001}{100}$ = 0.00001

3. **a.** 15% of 450 = 0.15 ×450 = 67.5

   b. 25% of 700 = 0.25 ×700 = 175

   c. 0.01% of 1000 = 0.0001 ×1000 = 0.1

   d. 10% of 100 = 0.1 ×100 = 10

   e. 12% of 500 = 0.12 ×500 = 60

   f. 150% of 1000 = 1.5 ×1000 = 1500

4. **a.** $\dfrac{15}{45} = 0.33 = 33\%$

   b. $\dfrac{2}{2} = 1 = 100\%$

   c. $\dfrac{10}{100} = 0.1 = 10\%$

   d. $\dfrac{1}{100} = 0.01 = 1\%$

   e. $\dfrac{1}{1000} = 0.001 = 0.1\%$

   f. $\dfrac{6}{40} = 0.15 = 15\%$

   g. $\dfrac{0.1}{0.5} = 0.2 = 20\%$

   h. $\dfrac{0.003}{89} = 0.0000337 = 0.0034\%$

   i. $\dfrac{5}{10} = 0.5 = 50\%$

   j. 0.05 = 5%   **k.** 0.0034 = 0.34%

   l. 0.25 = 25%   **m.** 0.01 = 1%

   n. 0.10 = 10%   **o.** 0.0001 = 0.01%

   p. 0.0078 = 0.78%   **q.** 0.50 = 50%

5. **a.** 20% of 100 = 20   **b.** 20% of 1000 = 200

   c. 20% of 10,000 = 2000   **d.** 10% of 567 = 56.7

   e. 15% of 1000 = 150   **f.** 50% of 950 = 475

   g. 5% of 100 = 5   **h.** 1% of 876 = 8.76

   i. 30% of 900 = 270

6. **a.** $\dfrac{1\ part}{100\ parts} = 1\%$

   b. $\dfrac{3\ parts}{50\ parts} = 6\%$

   c. $\dfrac{15\ parts}{(15 + 45)\ parts} = \dfrac{15\ parts}{60\ parts} = 25\%$

   d. $\dfrac{0.05\ parts}{1\ part} = 0.05 = 5\%$

   e. $\dfrac{1\ part}{25\ parts} = 0.04 = 4\%$

   f. $\dfrac{2.35\ parts}{(2.35 + 6.50)\ parts} = \dfrac{2.35\ parts}{8.85\ parts} = 0.266 = 26.6\%$

Application Problems: Percents (from p. 125)

1. $\dfrac{25}{55}$ work 20+ hours = 0.4545 = 45%

   $\dfrac{30}{55}$ work 19− hours = 0.545 = 55%

2. 50% = 0.5 of total; total = 100 *mL*

   100 × 0.5 = 50 *mL* ethanol

3. 30% solution of ethylene glycol

   30% of total (250 *mL*)

   0.30 × 250 = 75 *mL* ethylene glycol

   Measure out the ethylene glycol and add water until the volume is 250 *mL*

4. 10% acetonitrile for 100 *mL*:   10 *mL*

   25% MeOH (methanol):   25 *mL*

   Measure out acetonitrile and MeOH, combine, add water until the volume is 100 *mL*

5. $\dfrac{5\ mL}{100\ mL} = 0.05 = 5\%$ *propanol solution*

6. $\dfrac{15\ mL\ EtOH}{700\ mL} = 0.021 = 2.1\%$ *ethanol solution*

7. $\dfrac{10\ \mu L}{1\ mL} = \dfrac{10\ \mu L}{1000\ \mu L} = \dfrac{1}{100} = 0.01$

   = 1% *methanol solution*

8. $\dfrac{15\ mL\ acetone}{1\ liter} = \dfrac{15\ mL}{1000\ mL}$

   = 0.015 or 1.5 % *acetone solution*

9. 1000 insects, 5% survive

   1000 × 0.05 = 50 insects survive first winter

   50 insects × 100 = 5000 insects after first summer

   5000 × 0.05 = 250 insects after second winter

10. 250,000 people × 0.15 = 37,500 people die

11. Since thymine is 24%, adenine must also be 24% for a total of 48%. This leaves 52% that must be split equally between guanine and cytosine, at 26% each.

12. Of the subjects fed regular chips, 88 experienced gastrointestinal problems compared with 79 subjects fed Olestra chips.

Manipulation Problems: Density (from p. 126)

1. olive oil

2. $\dfrac{57.9\ g}{3.00\ cm^3} = \dfrac{?}{1\ cm^3}$ $? = \dfrac{19.3\ g}{cm^3}$

3. density = 1.26 g/mL

$$\frac{1.26\ g}{1\ mL} = \frac{20.0\ g}{?} \quad ? = 15.9\ mL$$

Manipulation Problems: Unit Conversions (from p. 129)

1. a. 3.00 *ft* to *cm* (proportion method):

$$\frac{?}{3.00\ ft} = \frac{12\ in}{1\ ft}$$

$$? = \frac{(12\ in)(3.00\ ft)}{1\ ft} \qquad ? = 36\ in$$

$$\frac{?}{36\ in} = \frac{2.54\ cm}{1\ in}$$

$$? = \frac{(2.54\ cm)(36\ in)}{1\ in} \quad ? = 91.4\ cm$$

3.00 *ft* to *cm* (conversion factor method)

$$3.00\ ft \times \frac{12\ in}{1\ ft} \times \frac{2.54\ cm}{1\ in} = 91.4\ cm$$

b. 100 *mg* to *g* (proportion method):

$$\frac{?}{100\ mg} = \frac{1\ g}{1000\ mg}$$

$$? = \frac{(1\ g)(100\ mg)}{1000\ mg} = 0.1\ g$$

100 *mg* to *g* (conversion factor method)

$$100\ mg \times \frac{1\ g}{1000\ mg} = 0.1\ g$$

c. 12.0 *in* to miles (proportion method):

$$\frac{?}{12.0\ in} = \frac{1\ ft}{12.0\ in}$$

$$? = \frac{(1\ ft)(12.0\ in)}{12.0\ in} = 1\ ft$$

$$\frac{?}{1\ ft} = \frac{1\ mi}{5280\ ft}$$

$$? = \frac{(1\ mi)(1\ ft)}{5280\ ft} = \frac{1\ mi}{5280} = 0.000189\ mi$$

$$? = 1.89 \times 10^{-4}\ \text{mile}$$

12.0 *in* to mile (conversion factor method)

$$12.0\ in \times \frac{1\ ft}{12.0\ in} \times \frac{1\ mi}{5280\ ft} = 1.89 \times 10^{-4}\ mi$$

d. 100 *in* to *km* (proportion method):

$$\frac{?}{100\ in} = \frac{2.54\ cm}{1\ in}$$

$$? = \frac{(2.54\ cm)(100\ in)}{1\ in} \qquad ? = 254\ cm$$

$$\frac{?}{254\ cm} = \frac{1\ m}{100\ cm}$$

$$? = \frac{(254\ cm)(1\ m)}{100\ cm} \qquad ? = 2.54\ m$$

$$\frac{?}{2.54\ m} = \frac{1\ km}{1000\ m}$$

$$? = \frac{(2.54\ m)(1\ km)}{1000\ m} \quad ? = 2.54 \times 10^{-3}\ km$$

100 *in* to *km* (conversion factor method):

$$100\ in \times \frac{2.54\ cm}{1\ in} \times \frac{1\ m}{100\ cm} \times \frac{1\ km}{1000\ m}$$

$$= 2.54 \times 10^{-3}\ km$$

e. 10.0555 *lb* to *oz* (proportion method):

$$\frac{?}{10.0555\ lb} = \frac{16\ oz}{1\ lb}$$

$$? = \frac{(16\ oz)(10.0555\ lb)}{1\ lb} \qquad ? = 160.888\ oz$$

10.0555 *lb* to *oz* (conversion factor method):

$$10.0555\ lb \times \frac{16\ oz}{1\ lb} = 160.888\ oz$$

f. 18.989 *lb* to *g* (proportion method):

$$\frac{?}{18.989\ lb} = \frac{453.6\ g}{1\ lb}$$

$$? = \frac{(18.989\ lb)(453.6\ g)}{1\ lb} \qquad ? = 8613.4\ g$$

18.989 *lb* to *g* (conversion factor method):

$$18.989\ lb \times \frac{453.6\ g}{1\ lb} = 8613.4\ g$$

g. 13 *mi* to *km* (proportion method):

$$\frac{?}{13\ mi} = \frac{1.609\ km}{1\ mi}$$

$$? = \frac{(1.609\ km)(13\ mi)}{1\ mi} \qquad ? = 21\ km$$

13 *mi* to *km* (conversion factor method):

$$13\ mi \times \frac{1.609\ km}{1\ mi} = 21\ km$$

h. 150 *mL* to *L* (proportion method):

$$\frac{?}{150\ mL} = \frac{1\ L}{1000\ mL}$$

$$? = \frac{(150\ mL)(1\ L)}{1000\ mL} \qquad ? = 0.150\ L$$

150 *mL* to *L* (conversion factor method):

$$150\ mL \times \frac{1\ L}{1000\ mL} = 0.150\ L$$

i. 56.7009 *cm* to *nm* (proportion method):

$$\frac{?}{56.7009\ cm} = \frac{1\ m}{100\ cm}$$

$$? = 0.567009\ m$$

$$\frac{?}{0.567009 \ m} = \frac{10^9 \ nm}{1 \ m}$$

$$? = 5.67009 \times 10^8 \ nm$$

56.7009 *cm* to *nm* (conversion factor method):

$$56.7009 \ cm \times \frac{1 \ m}{10^2 \ cm} \times \frac{10^9 \ nm}{1 \ m}$$

$$= 5.67009 \times 10^8 \ nm$$

j. 500 *nm* to μm (proportion method):

$$\frac{?}{500 \ nm} = \frac{1 \ \mu m}{1000 \ nm}$$

$$? = \frac{(1 \ \mu m)(500 \ nm)}{1000 \ nm} \qquad ? = 0.500 \ \mu m$$

500 *nm* to μm (conversion factor method):

$$500 \ nm \times \frac{1 \ m}{10^9 \ nm} \times \frac{10^6 \ \mu m}{1 \ m} = 0.500 \ \mu m$$

k. 10.0 *nm* to inches (proportion method):

$$\frac{?}{10 \ nm} = \frac{1 \ cm}{10^7 \ nm} \qquad ? = 1 \times 10^{-6} \ cm$$

$$\frac{?}{1 \times 10^{-6} \ cm} = \frac{1 \ in}{2.54 \ cm} \qquad ? = 3.94 \times 10^{-7} \ in$$

10.0 *nm* to inches (conversion factor method):

$$10.0 \ nm \times \frac{1 \ m}{10^9 \ nm} \times \frac{100 \ cm}{1 \ m} \times \frac{1 \ in}{2.54 \ cm}$$

$$= 3.94 \times 10^{-7} \ in$$

2. 10 *km* to miles (proportion method):

$$\frac{?}{10 \ km} = \frac{1 \ mi}{1.609 \ km}$$

$$? = \frac{(1 \ mi)(10 \ km)}{1.609 \ km} \qquad ? = 6.2 \ mi$$

10 *km* to miles (conversion factor method):

$$10 \ km \times \frac{1 \ mi}{1.609 \ km} = 6.2 \ mi$$

3. $26.2 \ mi \times \dfrac{1.609 \ km}{1 \ mi} = 42.2 \ km$

4. 5 *ft* 4 *in* to meters (proportion method):

$$\frac{?}{5 \ ft} = \frac{12 \ in}{ft} \qquad ? = \frac{(12 \ in)(5 \ ft)}{ft}$$

$$? = 60 \ in, \text{plus } 4 \ in = 64 \ in$$

$$\frac{?}{64 \ in} = \frac{1 \ m}{39.37 \ in}$$

$$? = \frac{(1 \ m)(64 \ in)}{39.37 \ in} \qquad ? = 1.6 \ m$$

5 *ft* 4 *in* to meters (conversion factor method):

$$5 \ ft \times \frac{12 \ in}{1 \ ft} = 60 \ in$$

$$60 \ in + 4 \ in = 64 \ in$$

$$64 \ in \times \frac{1 \ m}{39.37 \ in} = 1.6 \ m$$

5. 45 *mi* into kilometers (proportion method):

$$\frac{?}{45 \ mi} = \frac{1.609 \ km}{1 \ mi}$$

$$? = \frac{(1.609 \ km)(45 \ mi)}{1 \ mi} \qquad ? = 72 \ km$$

45 *mi* into *km* (conversion factor method):

$$45 \ mi \times \frac{1 \ km}{0.6214 \ mi} = 72 \ km$$

6. 55 *mph* to kilometers per hour (proportion method):

$$1 \ mph = 1.609 \ kph$$

$$\frac{1 \ mph}{1.609 \ kph} = \frac{55 \ mph}{?}$$

$$? = 88.5 \ kph$$

55 *mph* to kilometers per hour (conversion factor method):

$$55 \ mph \times \frac{1 \ kph}{0.6214 \ mph} = 88.5 \ kph$$

7. 3.0 ton to kilograms (proportion method):

$$\frac{?}{3.0 \ ton} = \frac{2000 \ lb}{ton}$$

$$? = \frac{(2000 \ lb)(3.0 \ ton)}{ton} \qquad ? = 6000 \ lb$$

$$\frac{?}{6000 \ lb} = \frac{1 \ kg}{2.205 \ lb}$$

$$? = \frac{(1 \ kg)(6000 \ lb)}{2.205 \ lb}$$

$$? = 2721.1 \ kg \quad \text{(Round to 2700 } kg\text{)}$$

3.0 ton to *kg* (conversion factor method):

$$3.0 \ ton \times \frac{2000 \ lb}{1 \ ton} \times \frac{1 \ kg}{2.205 \ lb} = 2721.1 \ kg$$

(Round to 2700 *kg*)

8. a. \$2.50 for 12 *oz* (proportion method):

$$\frac{?}{12 \ oz} = \frac{1 \ lb}{16 \ oz}$$

$$? = \frac{(1 \ lb)(12 \ oz)}{16 \ oz} \qquad ? = 0.75 \ lb$$

$$\frac{?}{0.75 \ lb} = \frac{453.6 \ g}{1 \ lb}$$

$$? = \frac{(453.6 \ g)(0.75 \ lb)}{1 \ lb} \qquad ? = 340.2 \ g$$

$$340.2 \ g/\$2.50 = 136.1 \ g/\$1.00$$

b. $3.67 for 250 g (proportion method):

$$\frac{250\ g}{\$3.67} = 68.12\ g/\$1.00$$

c. $4.50 for 0.300 kg:

$$\frac{?}{0.300\ kg} = \frac{1000\ g}{kg}$$

$$? = \frac{(1000\ g)(0.300\ kg)}{kg} \qquad ? = 300\ g$$

300 g/$4.50 = 66.7 g/$1.00

d. $2.35 for 0.75 lb:

$$\frac{?}{0.75\ lb} = \frac{453.6\ g}{lb}$$

$$? = \frac{(453.6\ g)(0.75\ lb)}{lb} \qquad ? = 340.2\ g$$

340.2 g/$2.35 = 144.8 g/$1.00 = least expensive

9. (proportion method)

$$\frac{?}{1\ week} = \frac{7\ days}{1\ week} \qquad ? = 7\ days$$

$$\frac{?}{7\ days} = \frac{24\ h}{1\ day} \qquad ? = 24 \times 7 = 168\ h$$

$$\frac{?}{168\ h} = \frac{60\ min}{1\ h}$$

$$? = 168 \times 60 = 10{,}080\ minutes$$

$$\frac{?}{10080\ min} = \frac{60\ s}{1\ min}$$

$$? = 10080 \times 60 = 604{,}800\ s$$

(conversion factor method)

$$1\ week \times \frac{7\ days}{1\ week} = 7\ days$$

$$7\ days \times \frac{24\ h}{1\ day} = 168\ h$$

$$168\ h \times \frac{60\ min}{1\ h} = 10{,}080\ min$$

$$10{,}080\ min \times \frac{60\ s}{min} = 604{,}800\ s$$

10. 1/52 year = 1 week = 168 hours = 10080 minutes = 604,800 seconds

11. 1 year = 52 weeks = 364 days = 524,160 minutes = 31,449,600 seconds

12. 1/5280 mi = 1/3 yds = 1 ft = 12 in
or 0.000189 mi = 0.333 yds = 1 ft = 12 in

13. 0.0156 gal = 0.125 pt = 2 oz

14. $1\ km = 10^3\ m = 10^5\ cm = 10^6\ mm = 10^9\ \mu m = 10^{12}\ nm$

15. $10^{-12}\ km = 10^{-9}\ m = 10^{-7}\ cm = 10^{-6}\ mm = 10^{-3}\ \mu m = 1\ nm$

16. $2.5 \times 10^{-5}\ km = 2.5 \times 10^{-2}\ m = 2.5\ cm = 25\ mm = 2.5 \times 10^4\ \mu m = 2.5 \times 10^7\ nm$

17. **a.** $6.25\ mm = 6.25 \times 10^3\ \mu m$

b. $0.00896\ m = 8.96\ mm$

c. $9876000\ nm = 9.876\ mm$

18. $3 \times 10^{-3}\ km = 3\ m = 3 \times 10^2\ cm = 1.2 \times 10^2\ in$

19. $5\ kg = 5 \times 10^3\ g = 5 \times 10^6\ mg = 5 \times 10^9\ \mu g$

20. $8.9 \times 10^{-6}\ kg = 8.9 \times 10^{-3}\ g = 8.9\ mg = 8.9 \times 10^3\ \mu g$

21. $2 \times 10^{-17}\ kg = 2 \times 10^{-14}\ g = 2 \times 10^{-11}\ mg = 2 \times 10^{-8}\ \mu g$

22. **a.** $0.8657\ g = 8.657 \times 10^{-1}\ g = 8.657 \times 10^2\ mg$

b. $526\ kg = 5.26 \times 10^2\ kg = 5.26 \times 10^8\ mg$

c. $63\ g = 6.3 \times 10^1\ g = 6.3 \times 10^7\ \mu g$

d. $2.63 \times 10^{-6}\ \mu g = 2.63 \times 10^{-15}\ kg$

23. $1\ Ci = 3.7 \times 10^{10}\ dps = 2.2 \times 10^{12}\ dpm$

24. $1\ Ci = 1000\ mCi = 10^6\ \mu Ci = 3.7 \times 10^{10}\ dps$

25. $2.7 \times 10^{-6}\ Ci = 2.7 \times 10^{-3}\ mCi = 2.7\ \mu Ci = 10^5\ dps$

26. $1 \times 10^{-4}\ Ci = 0.1\ mCi = 100\ \mu Ci = 3.7 \times 10^6\ dps = 2.2 \times 10^8\ dpm$

27. $1\ Ci = 10^3\ mCi = 10^6\ \mu Ci = 3.7 \times 10^{10}\ dps = 3.7 \times 10^{10}\ Bq$

28. $2.5 \times 10^{-4}\ Ci = 2.5 \times 10^{-1}\ mCi = 250\ \mu Ci = 9.25 \times 10^6\ dps = 9.25 \times 10^6\ Bq$

Application Problems: Conversions (from p. 130)

1. If the medium requires 5 g/1 L then 125 g of glucose are required for 25 L. The technician added:

$$\frac{?}{0.24\ lb} = \frac{453.6\ g}{1\ lb}$$

$$? = \frac{(453.6\ g)(0.24\ lb)}{lb} \qquad ? = 108.86\ g$$

The technician, therefore, added too little glucose.

2.

	For 1 L	For 1 mL
NaCl	20.00 g	20 mg
Na azide	0.001 g	0.001 mg
Mg sulfate	1.000 g	1.000 mg
Tris	15.00 g	15.00 mg

3. 680 mg = 0.680 g

so there are 0.680 g enzyme/1 g powder

need $\dfrac{10.0\ oz}{100.0\ L} = \dfrac{?}{500.0\ L}$? = 50.0 oz

$$50.0 \; oz \times \frac{1 \; lb}{16 \; oz} \times \frac{453.6 \; g}{1 \; lb} = 1417.5 \; g$$

$$\frac{0.680 \; g \; enzyme}{1 \; g \; powder} = \frac{1417.5 \; g \; enzyme}{?}$$

$$? = 2084.56 \; g \; powder \rightarrow Need \; 2.08 \; kg \; powder$$

Manipulation/Application Problems: Concentration (from p. 131)

1. $\dfrac{3 \; g}{250 \; mL} = \dfrac{?}{1000 \; mL}$

$? = \dfrac{(3 \; g)(1000 \; mL)}{250 \; mL}$ $? = 12 \; g$

2. $\dfrac{25 \; g}{1000 \; mL} = \dfrac{?}{100 \; mL}$

$? = \dfrac{25 \; g \times 100 \; mL}{1000 \; mL}$ $? = 2.5 \; g$

3. $\dfrac{1 \; mg}{1 \; mL} = \dfrac{1 \; g}{1 \; L}$ so $\dfrac{1 \; g}{1 \; L} = \dfrac{?}{15 \; L}$ $? = 15 \; g$

4. $\dfrac{0.005 \; g \; Tris \; base}{L} = \dfrac{?}{10^{-3}L}$

$? = \dfrac{(0.005 \; g)10^{-3} \; L}{L}$

$? = 0.005 \; g \times 10^{-3} \; g$

$? = 5 \times 10^{-6} \; g$

5. $\dfrac{300 \; ng \; dioxin}{100 \; g \; diapers} = \dfrac{?}{1000 \; g}$

$? = 3000 \; ng = 3 \; \mu g \; dioxin$

6. $\dfrac{?}{5 \; mL} = \dfrac{5 \times 10^{-3} \; moles}{1000 \; mL}$

$? = \dfrac{(5 \times 10^{-3} \; moles)5 \; mL}{10^3 \; mL}$

$? = 25 \times 10^{-6} \; moles = 2.5 \times 10^{-5} \; moles$

7. $(1 \; L = 10^6 \; \mu L)$

$\dfrac{?}{1 \; \mu L} = \dfrac{0.1 \; moles}{10^6 \; \mu L}$ $? = \dfrac{(0.1 \; moles) \; \mu L}{10^6 \; \mu L}$

$? = 0.1 \times 10^{-6} = 1 \times 10^{-7} \; moles$

8. $(1 \; L = 10^3 \; mL)$ $\dfrac{?}{78 \; mL} = \dfrac{10^{-2} \; g}{10^3 \; mL}$

$? = \dfrac{(10^{-2} \; g)78 \; mL}{10^3 \; mL}$

$? = 78 \times 10^{-5} \; g = 7.8 \times 10^{-4} \; g$

9. $\dfrac{500 \; U}{mg} = \dfrac{?}{3 \; mg}$

$? = \dfrac{(500 \; U)3 \; mg}{mg}$ $? = 1500 \; U$

10. $\dfrac{2500 \; U}{mg} = \dfrac{?}{5 \; mg}$

$? = \dfrac{(2500 \; U)5 \; mg}{1 \; mg}$ $? = 1.25 \times 10^4 \; U$

11. $\dfrac{?}{10^3 \; mL} = \dfrac{3 \; g}{mL}$

$? = \dfrac{(3 \; g) \times 10^3 \; mL}{mL}$ $? = 3 \times 10^3 \; g$

12. $\dfrac{5 \; \mu g}{L} = \dfrac{5 \; \mu g}{10^3 \; mL}$

a. $\dfrac{5 \; \mu g}{10^3 \; mL} = \dfrac{?}{50 \; mL}$

$? = \dfrac{(5 \; \mu g)50 \; mL}{10^3 \; mL}$

$? = \dfrac{250 \; \mu g}{10^3} = 0.25 \; \mu g$

b. $\dfrac{5 \; \mu g}{10^3 \; mL} = \dfrac{?}{500 \; mL}$

$? = \dfrac{(5 \; \mu g)500 \; mL}{10^3 \; mL}$

$? = \dfrac{2500 \; \mu g}{10^3} = 2.5 \; \mu g$

c. $\dfrac{5 \; \mu g}{10^3 \; mL} = \dfrac{?}{100 \; mL}$

$? = \dfrac{(5 \; \mu g)100 \; mL}{10^3 \; mL}$

$? = \dfrac{500 \; \mu g}{10^3} = 0.5 \; \mu g$

d. $\dfrac{5 \; \mu g}{10^6 \; \mu L} = \dfrac{?}{100 \; \mu L}$

$? = \dfrac{(5 \; \mu g)100 \; \mu L}{10^6 \; \mu L} = \dfrac{500 \; \mu g}{10^6} = 5 \times 10^{-4} \; \mu g$

13. a. $\dfrac{0.5 \; mg}{1 \; mL} = \dfrac{?}{5 \; mL}$

$? = \dfrac{(0.5 \; mg)5 \; mL}{1 \; mL}$ $? = 2.5 \; mg$

b. $\dfrac{0.5 \; mg}{1 \; mL} = \dfrac{?}{0.5 \; mL}$

$? = \dfrac{(0.5 \; mg)0.5 \; mL}{1 \; mL}$ $? = 0.25 \; mg$

c. $1 \; mL = 10^3 \; \mu L$

$\dfrac{0.5 \; mg}{10^3 \; \mu L} = \dfrac{?}{100 \; \mu L}$

$? = \dfrac{(0.5 \; mg)100 \; \mu L}{10^3 \; \mu L}$ $? = 0.05 \; mg$

d. $1000 \ \mu L = 1 \ mL$

$$\frac{0.5 \ mg}{mL} = \frac{?}{mL} \qquad ? = 0.5 \ mg$$

14. $\dfrac{100 \ molecules}{10^9 \ molecules} = \dfrac{1 \times 10^2}{10^9} = \dfrac{1}{10^7} < \dfrac{1}{10^6}$

Cannot be detected

15. $1 \ g/10^6 \ kg = \dfrac{1 \ g}{10^9 \ g}$ (more pure)

$$\frac{10^{-2} \ mg}{10^{-3} \ kg} = \frac{10^{-5} \ g}{g} = \frac{1 \ g}{10^5 \ g}$$

Manipulation Problems: Dilutions A (from p. 133)

1. a. 1 part orange juice in 4 parts total = 1/4

   b. 1 part orange juice to 3 parts water = 1: 3

   c. 1: 3 or 1/4

2. a. $1 \ mL/10 \ mL$ **b.** $1 \ mL/11 \ mL$

   c. $3 \ mL/30 \ mL$ **d.** $3 \ mL/30 \ mL$

   e. $0.5 \ mL/11.5 \ mL$

3. a. 1/10   **b.** 1/11   **c.** 1/4

   d. 1/2   **e.** 1/5

4. a. 1:9   **b.** 1:10   **c.** 1:9

   d. 1:9   **e.** 1:22

5. $0.5 \ mL$ blood

      $1.0 \ mL \ H_2O$

   + $3.0 \ mL$ reagent

      $4.5 \ mL$

   0.5/4.5 dilution = 1/9

Manipulation Problems: Dilutions B (from p. 134)

1. $\dfrac{?}{10 \ mL} = \dfrac{1}{10}$ $? = \dfrac{(1)10 \ mL}{10}$

$? = 1 \ mL$ blood

$1 \ mL + 9 \ mL$ will give 1/10 dilution

2. $\dfrac{?}{250 \ mL} = \dfrac{1}{300}$ $? = \dfrac{1(250 \ mL)}{300}$

$? = 0.833 \ mL$ blood

   $0.833 \ mL + 249.167 \ mL = 250 \ mL$

3. $\dfrac{?}{1 \ mL} = \dfrac{1}{50}$ $? = \dfrac{1(1 \ mL)}{50}$

$? = 0.02 \ mL$ blood

   $0.02 \ mL + 0.98 \ mL = 1 \ mL$

4. $\dfrac{?}{1000 \ \mu L} = \dfrac{1}{100}$ $? = \dfrac{1(1000 \ \mu L)}{100}$

$? = 10 \ \mu L$ food coloring

   $10 \ \mu L + 990 \ \mu L = 1000 \ \mu L$

5. $\dfrac{?}{23 \ mL} = \dfrac{3}{5}$ $? = \dfrac{3(23 \ mL)}{5}$

$? = 13.8 \ mL$ of Q

   $13.8 \ mL + 9.2 \ mL = 23 \ mL$

6. $\dfrac{?}{10 \ mL} = \dfrac{1}{10}$ $? = \dfrac{1(10 \ mL)}{10}$

$? = 1 \ mL$ stock

   $1 \ mL + 9 \ mL = 10 \ mL$

7. $\dfrac{?}{15 \ mL} = \dfrac{1}{5}$ $? = \dfrac{1(15 \ mL)}{5}$

$? = 3 \ mL$ stock

   $3 \ mL + 12 \ mL = 15 \ mL$

8. $\dfrac{?}{10^3 \ \mu L} = \dfrac{1}{100}$ $? = \dfrac{1(10^3 \ \mu L)}{100}$

$? = 10 \ \mu L$ stock

   $10 \ \mu L + 990 \ \mu L = 1000 \ \mu L$

9. 0.01 = 1/100

$$\frac{?}{50 \ mL} = \frac{1}{100}$$

$? = 0.5 \ mL$ buffer

   $0.5 \ mL + 49.5 \ mL$

Manipulation Problems: Dilutions C (from p. 136)

1. $\dfrac{(50\%) \ 1}{40} = 1.25\%$

2. $\dfrac{10 \ mg}{mL} \times \dfrac{1}{10} = \dfrac{10 \ mg}{10 \ mL} = 1 \ mg/mL$

3. $1:1 = \dfrac{1}{2}$ dilution

so: $\dfrac{10 \ mg}{mL} \times \dfrac{1}{2} = \dfrac{10 \ mg}{2 \ mL} = \dfrac{5 \ mg}{mL}$

4. $1 \ mL + 4 \ mL = 1/5$ dilution.

   $1 \ mL$ can make $5 \ mL$ diluted solution.

5. $1000 \ mL \times \dfrac{1}{100} = \dfrac{1000 \ mL}{100} = 10 \ mL$ original

or $\dfrac{?}{1000 \ mL} = \dfrac{1}{100}$ $? = 10 \ mL$

Manipulation Problems: Dilutions D* (from p. 138)

1. $\dfrac{1}{10}$ dilution;

   • take $0.1 \ mL$ food coloring $+ \ 0.9 \ mL$ diluent

*Note: For dilution problems, there are often several workable strategies. Only one strategy is shown in the answer key.

- $\dfrac{1}{10} \times \dfrac{1}{25} = \dfrac{1}{250}$ take 1 mL of above + 24 mL diluent

- $\dfrac{1}{250} \times \dfrac{1}{4} = \dfrac{1}{1000}$ take 1 mL of above + 3 mL diluent

2. a. $\dfrac{1}{10^6}$ dilution—use three dilutions of 1/100, 1/100, 1/100

- 0.1 mL of culture + 9.9 mL diluent = 1/100
- take 0.1 mL of above + 9.9 mL = 1/100 × 1/100 = 1/10⁴
- take 0.1 mL of above + 9.9 mL = 1/100 ×1/10⁴ = 1/10⁶

b. $\dfrac{1}{10} \times \dfrac{1}{10} \times \dfrac{1}{10} \times \dfrac{1}{10} \times \dfrac{1}{10} \times \dfrac{1}{10}$

- 0.1 mL culture + 0.9 mL diluent = 1/10
- take 0.1 mL of above + 0.9 mL = 1/10 ×1/10 = 1/10²
- take 0.1 mL of above + 0.9 mL = 1/10 ×1/10² = 1/10³
- take 0.1 mL of above + 0.9 mL = 1/10 ×1/10³ = 1/10⁴
- take 0.1 mL of above + 0.9 mL = 1/10 ×1/10⁴ = 1/10⁵
- take 0.1 mL of above + 0.9 mL = 1/10 ×1/10⁵ = 1/10⁶

3. • 1 mL blood + 4 mL diluent = 1/5
- take 1 mL of above + 9 mL = 1/10 × 1/5 = 1/50
- take 1 mL of above + 4 mL = 1/5 × 1/50 = 1/250

Application Problems: Dilutions (from page 138)

1. Each tube requires 5 U.

The enzyme concentration is:

$$\dfrac{1000\,U}{1\,mL} = \dfrac{1000\,U}{1000\,\mu L} = \dfrac{1\,U}{1\,\mu L}$$

The 20 μL contains 20 units of enzyme. You can, therefore, make four tubes with 5 U per tube.

2. Each tube requires 5 × 0.01 U = 0.05 U.

The enzyme concentration is 1 $U/\mu L$.

$$\dfrac{?}{0.05\,U} = \dfrac{1\,\mu L}{1\,U}$$

? = 0.05 μL = amount needed, but this volume is too low to measure.

Therefore, make 1/100 dilution, 10 μL enzyme + 990 μL = 1/100. After dilution, there are:

$$\dfrac{1\,U}{1\,\mu L} \times \dfrac{1}{100} = \dfrac{1\,U}{100\,\mu L} = \dfrac{10^{-2}\,U}{1\,\mu L}$$

Need 0.05 U/tube so use 5 μL of dilution for each tube to 5 mL total volume.

3. Dilute $\dfrac{1}{500,000} = \dfrac{1}{50} \times \dfrac{1}{100} \times \dfrac{1}{100}$

- 0.1 mL antibody + 4.9 mL diluent = 1/50
- take 0.1 mL of above + 9.9 mL → 1/100 × 1/50 = 1/5000
- take 0.1 mL of above + 9.9 mL → 1/100 × 1/5000 = 1/500,000

4. Have $\dfrac{1 \times 10^9\ cells}{1\ mL}$ want $\dfrac{2 \times 10^3\ cells}{1\ mL}$

$$= \dfrac{2 \times 10^2\ cells}{0.1\ mL}$$

$$\dfrac{1 \times 10^9}{2 \times 10^3} = \dfrac{1}{2} \times 10^6 \text{ or } 5 \times 10^5 \text{ dilution}$$

$$\dfrac{1}{100} \times \dfrac{1}{100} \times \dfrac{1}{50} = \dfrac{1}{5 \times 10^5}$$

- 0.1 mL culture + 9.9 mL diluent = 1/100 → (1/100) × (1 × 10⁹) = 10⁷ cells/mL
- 0.1 mL of above + 9.9 mL → 1/100 × 1/100 → (1/10⁴) × (1 × 10⁹) → 10⁵ cells/mL

- 0.1 mL of above + 4.9 mL → 1/50 × 1/10⁴
$$\rightarrow \dfrac{1}{(5 \times 10^5)} \times (1 \times 10^9) \rightarrow 2 \times 10^3 \text{ cells}/mL$$
= 200 cells/0.1 mL

5. The first plate has too many colonies to count and the third plate has fewer colonies than are desirable for a plate count. The middle plate has 21 colonies, so it is the one we will use for calculations. The concentration of cells added to this plate was about 21 cells/0.1 mL = 210 cells/mL. The dilution tube from which these cells were drawn had been diluted 1/10⁷. There were, therefore, originally about 210 × 10⁷ = 2.10 × 10⁹ bacterial cells/mL in the original broth.

6. The first plate can be used for calculations. There are about 43 colonies on the first plate so the concentration of bacteria applied to that plate was about 430 cells/mL. The dilution tube from which these cells were drawn had been diluted 1/10⁶. There were, therefore, about 430 × 10⁶ = 4.30 × 10⁸ cells/mL in the original broth.

7. 10⁷ cells/mL, dilute to $\dfrac{100\ cells}{0.1\ mL}$

= 1000 cells/mL final concentration

$$\dfrac{10^7}{10^3} = 1 \times 10^4 = \text{dilution needed}$$

- 0.1 mL culture + 9.9 diluent = 1/100 = 1/10²
→ 10⁵ colonies/mL

- 0.1 *mL* culture + 9.9 diluent = 1/100 = $1/10^2$
  → 10^3 colonies/*mL*

8. $\dfrac{1}{100}$ dilution had $\dfrac{50\ mg}{mL}$

$$\dfrac{50\ mg}{1\ mL} \times 100 = \dfrac{5000\ mg}{1\ mL} = \dfrac{5\ g}{1\ mL}$$

= concentration protein in undiluted sample.

$$\dfrac{5\ g}{1\ mL} = \dfrac{?}{100\ mL} \qquad ? = 500\ g$$

9. $\dfrac{1}{50}$ dilution had $\dfrac{87\ mg}{mL}$

$$\dfrac{87\ mg}{1\ mL} \times 50 = \dfrac{4350\ mg}{1\ mL} = \dfrac{4.35\ g}{1\ mL}$$

= concentration protein in undiluted sample.

10. The sample was diluted 1/5 and then 5/25 → (1/5)(5/25) = 5/125 total dilution. There were 3 *mg* protein in 5 *mL* = 0.6 *mg* protein/*mL*. Multiplying times the dilution: 0.6 *mg/mL* × 125/5 = 15 *mg/mL*.

11. The sample was diluted 1/4 and then 10/30 → 10/120 total dilution.
There were 10 *mg* protein/5 *mL* = 2 *mg/mL* protein in the solution. Multiplying times the dilution: 2 *mg/mL* × 120/10 = 24 *mg/mL* in original sample.

CHAPTER 10

Relationships and Graphing

I. GRAPHS AND LINEAR RELATIONSHIPS

A. Brief Review of Basic Techniques of Graphing

Graphs, like equations, are a tool for working with relationships between two (or sometimes more) variables. The general rules regarding graphing will briefly be reviewed in the first part of this chapter. The use of graphing as a tool to perform various tasks in a laboratory will be demonstrated in later sections.

A simple two-dimensional graph, Figure 10.1, consists of a vertical and a horizontal line that intersect at a point called the **origin**. *The horizontal line is the* **X axis**, *the vertical line is the* **Y axis**. The X and Y axes are each divided into evenly spaced subdivisions that are assigned numerical values. To the right of the origin on the X axis the X values are positive numbers; to the left of the origin the X values are negative. On the Y axis values above the origin are positive; below the origin, they are negative.

A basic two-dimensional graph shows the relationship between two variables, one of which is assigned to the X axis, the other to the Y axis. *For any value of a variable on the X axis, there is a corresponding value of a variable on the Y axis.* The two values are called **coordinates** because they are coordinated or associated with one another. A pair of coordinates can be plotted on the graph as shown in Figure 10.2a. *The distance of a point along the X axis is sometimes called the* **abscissa** *and the distance of a point along the Y axis is sometimes called the* **ordinate**. The first point shown on the graph has an X coordinate of 2 because it is above the 2 on the X axis, and a Y coordinate of 3, because it is across from the 3 on the Y axis. This point can be written as $X = 2$, $Y = 3$, or as (2,3). The second point on this graph has the coordinates (4,6). The point in Figure 10.2b has coordinates $X = -2$, $Y = -1$. To prepare a graph, one marks the locations of a series of points by using their coordinates.

B. Graphing Straight Lines

Two variables may be related to one another in such a way that when plotted on a graph, the points form a straight line. The two variables are then said to have a **linear relationship.** There are many applications in the laboratory that involve linear relationships.

Consider the simple equation:

$$Y = 2X$$

It is possible to find numbers that satisfy this equation. For example:

$$\text{If } X = 3, \text{ then } Y = 6$$
$$\text{If } X = 4, \text{ then } Y = 8$$
$$\text{and so on}$$

These numbers are summarized in both tabular and graphical form in Figure 10.3. Each pair of X and Y values are the coordinates of a point on the graph. The points form a straight line when connected.

In the linear equation $Y = 2X$, the value 2 is called the **slope**. In the equation $Y = 3X$ the slope is 3, and in the equation $Y = 4X$ the slope is 4. The equations $Y = 2X$, $Y = 3X$, and $Y = 4X$ are plotted on the same graph in Figure 10.4. Each equation is linear; the difference

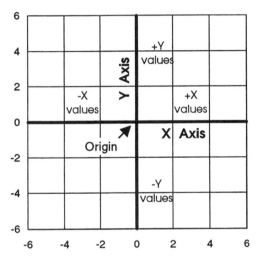

Figure 10.1. A Two-Dimensional Graph.

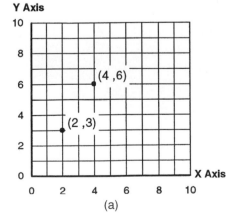
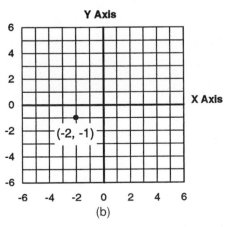

Figure 10.2. Coordinates and Graphs. a. The points (2,3) and (4,6). **b.** The point $(-2,-1)$.

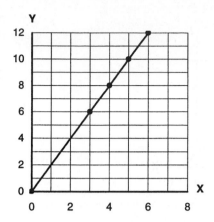

X	Y
0	0
3	6
4	8
5	10
6	12

Figure 10.3.   Graph of the Equation $Y = 2X$.

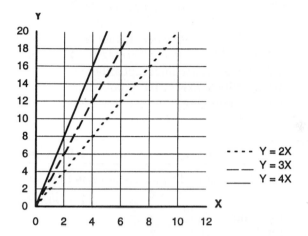

Y = 3X		Y = 4X	
X	Y	X	Y
0	0	0	0
1	3	1	4
2	6	2	8
3	9	3	12
4	12	4	16
5	15	5	20

Figure 10.4.   Slope. The greater the value for the slope of a straight line, the steeper the line.

between the three lines is their steepness. Just as a hill may be more or less steep, so a line has a slope that is more or less steep. The equation $Y = 4X$ defines a steeper line than the other two equations because its slope, 4, is the largest.

Given a straight line plotted on a graph, the slope of the line is calculated by determining how steeply the line rises. For example, consider the line for $Y = 2X$ in Figure 10.5. From point a to point b, X increases by 1 and Y increases by 2. *The amount by which the X coor-*

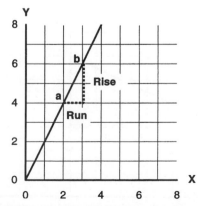

Figure 10.5.   Determining the Slope of a Line. The slope can be calculated based on the coordinates of any two points on the line. Slope = rise/run, which in this case = 2/1 = 2.

dinate increases is called the **run**; *the amount by which the Y coordinate increases is called the* **rise**. The rise divided by the run is a numerical measure of the steepness of the slope. To calculate the slope of any straight line, choose any two points on the line. The coordinates for the first point are (X_1, Y_1) and for the second point are (X_2, Y_2). Then:

$$slope = \frac{rise}{run} = \frac{change\ in\ Y}{change\ in\ X} = \frac{Y_2 - Y_1}{X_2 - X_1}$$

For the line Y = 2X:

$$slope = \frac{rise}{run} = \frac{2}{1} = 2$$

Any two points on the line $Y = 2X$ will give a slope of 2. Any two points on the line $Y = 3X$ will give a slope of 3 and any two points on the line $Y = 4X$ will yield a slope of 4.

Figure 10.6 shows the graph of a line which goes "downhill" from left to right. For this line, as the X values increase, the Y values decrease. The slope is therefore negative.

Figure 10.7 shows the plot for the equation $A = 2C + 4$. (Substitution of other symbols for X and Y does not change the basic meaning of the equation.) The difference between the plot of $Y = 2X$ and $A = 2C + 4$

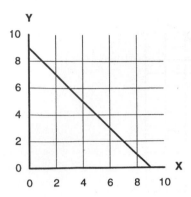

Figure 10.6. A Line with a Negative Slope. All lines that go "downhill" from left to right have a negative slope.

is that the latter line is higher and does not pass through the origin. The line $A = 2C + 4$ intercepts, or passes through, the Y axis at 4. *The point at which a line passes through the Y axis, that is, where $X = 0$, is termed the* **Y intercept**. Every straight line has a Y intercept although, if the line passes through the Y axis at the point where both X and Y are zero, the intercept is often not written. Thus, the Y intercept of the line $Y = 2X$ is 0.

If an equation gives a straight line when it is graphed, then it will have the form:

$$Y = slope(X) + Y\ intercept$$

This general equation for a straight line is sometimes written:

GENERAL EQUATION FOR A STRAIGHT LINE

$$Y = mX + a$$
where
m = the slope and
a = the Y intercept

It is possible to determine the equation for a straight line from its graph. For example, the line drawn in Figure 10.8 has a Y intercept of 3 and a slope of 0.5; therefore the equation for this line is: $Y = 0.5\ (X) + 3$. The general procedure to find the equation for any graphed straight line is shown in Box 1.

Box 1 *PROCEDURE TO FIND THE EQUATION FOR A STRAIGHT LINE ON A GRAPH*

All straight lines can be described by an equation in the form:

$$Y = mX + a$$

where m is the slope and
a is the Y intercept

1. Find the Y intercept, that is, the value of Y when $X = 0$. The intercept may be positive or negative.
2. Find the slope by picking any two points and calculating $(Y_2 - Y_1)/(X_2 - X_1)$. The slope may be positive or negative.
3. Put the slope and intercept into the proper form by filling in the blanks for slope and Y intercept:

$$Y = \underset{slope}{\underline{\qquad}}\ (X) + \underset{Y\ intercept}{\underline{\qquad\qquad}}$$

Note: The axes of a graph can have units in which case the slope and the intercept will also have units. Include the units when writing the equation for a line.

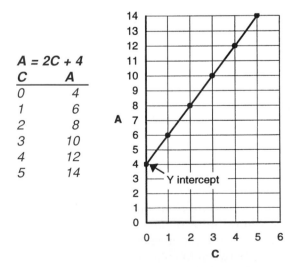

Figure 10.7. A Graph with a Nonzero Y Intercept.

$A = 2C + 4$

C	A
0	4
1	6
2	8
3	10
4	12
5	14

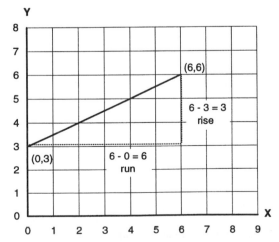

Figure 10.8. Determining the Equation for a Line.
Slope = $(6-3)/(6-0) = 0.5$. Y intercept = 3.
Equation is $Y = 0.5X + 3$.

EXAMPLE PROBLEM

Which of the following equations describe a straight line?

a. $C = 2B$
c. $Y = X^2 + 3$

b. $Q = 25.4T - 5$
d. $C = -V - 34$

ANSWER

All except c are equations for a straight line. The exponent in equation c means that it does not describe a straight line. The rest of the equations have two variables, a slope, and a Y intercept, which may or may not be zero. Equation b fits the linear equation as shown here:

$$Y = m\,X + a$$
$$Q = 25.4\,T + (-5)$$

Equation d is also the equation for a line:

$$Y = m\,X + a$$
$$C = (-1)\,V + (-34)$$

EXAMPLE PROBLEM

Heating a solution of double-stranded DNA disrupts the hydrogen bonds holding the two strands together and causes the strands to separate from one another. The midpoint of the temperature range over which the strands separate is called the *melting temperature* (T_m). There is a linear relationship between the guanine + cytosine (G + C) content of the DNA and its T_m: The higher the G + C content, the higher the melting temperature. DNA molecules from different organisms have different melting temperatures because their G + C content varies. There are many techniques performed in the laboratory that require the separation of double-stranded DNA. A graph of melting temperature versus G + C content is shown below along with questions relating to the graph.

a. What is the melting temperature for DNA which has 62% guanine + cytosine?
b. What is the melting temperature for DNA which is 15% guanine + cytosine?
c. What is the slope of the line?
d. In order to find the equation for this line, it is necessary to know the value for the Y intercept. Observe that, for convenience, the graph is drawn so that the X axis begins at 70°C. It is possible to redraw this graph so that the X axis begins at 0°C. If you were to redraw the graph with the X axis beginning at zero, you would see that the Y intercept is at −172% G + C. In reality, of course, there is no such thing as a DNA molecule with a negative G + C content, nor is there a DNA molecule whose strands separate at 0°C. Nonetheless, the Y intercept for the line on the graph is −172% G + C. What is the equation for the line?
e. Use the equation to determine the melting temperature for DNA that has 50% G + C content.

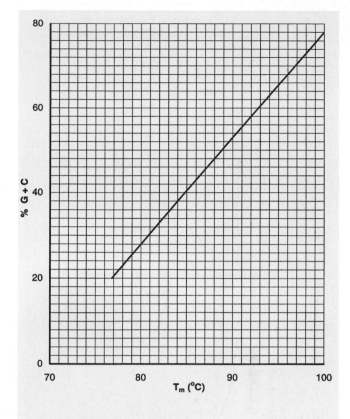

ANSWER

a. From the graph, you can see that the melting temperature for DNA with a G + C content of 62% is about 93.5°C.
b. The graph only extends as low as 20% G + C and therefore it is not possible to use it to determine the melting point for 15% guanine + cytosine.
c. The slope of the line can be calculated from any two points that are on the line. For example, 100°C on the X axis corresponds to a G+C content of 78%. 77°C on the X axis corresponds to 20% G+C. The slope is therefore:

$$\frac{(78\% - 20\%)}{(100°C - 77°C)} = \frac{2.5\%}{°C}$$

d. The equation for the line is

$$Y = \frac{(2.5\%)}{°C}(X) - 172\%$$

e. Substituting into the equation:

$$50\% = \frac{(2.5\%)}{°C}(X) - 172\%$$

$$\frac{222\%}{\left(\dfrac{2.5\%}{°C}\right)} = X$$

$$88.8°C = X$$

You can confirm that this is the right answer by looking at the graph.

MANIPULATION PROBLEMS: GRAPHING

1. For each of the following graphs, describe in words the relationship that is displayed.

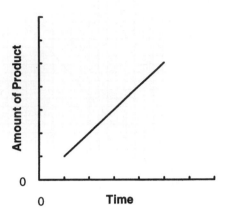

a.

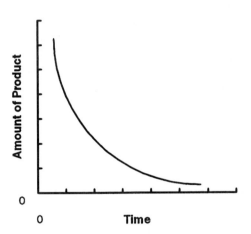

b.

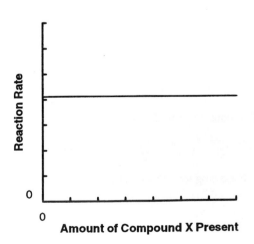

c.

2. The following figure shows a graph with four points, labeled a, b, c, and d. Write the coordinates for each of the four points.

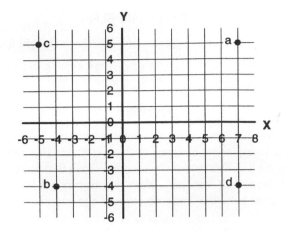

3. In the graph in Problem 2, as you move from point a to point b, by how much does the value of X change? _____ By how much does the value of Y change? _____ As you go from point b to point c, how much does the value of X change? _____ How much does the value of Y change? _____

4. Draw a graph and plot each of these points on the graph:

$$X = 4, Y = 6$$
$$X = 5, Y = -2$$
$$(-4,3)$$
$$(1,1)$$

5. A table showing three values for the equation $Y = 5X + 1$ is given. Fill in the blanks in the table, and then plot the equation.

X	Y
1	6
5	26
10	51
12	__
__	76
__	101

6. Suppose two variables are related by the equation, Variable $A = 3$(Variable Q) $- 4$. Prepare a table to show this relationship and then graph the relationship.

7. What is the slope and the Y intercept for each of these equations?
   a. $Y = 3X + 2$
   b. $C = 0.2X - 1$
   c. $Y = 0.005X$

8. What is the slope and Y intercept for lines a–d on p. 157?

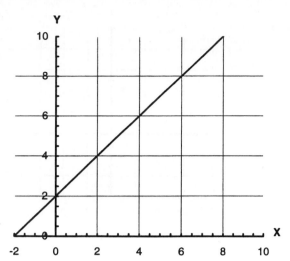

a.

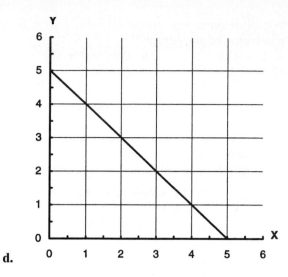

d.

9. Which of these graphs below shows a linear relationship between two variables?

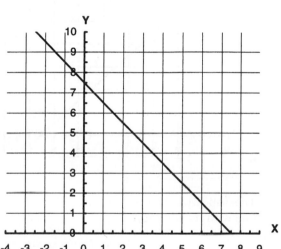

b.

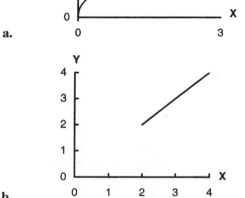

a.

b.

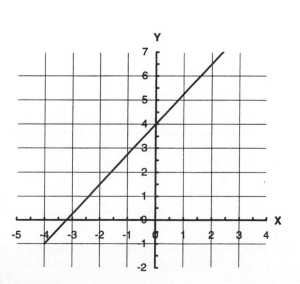

c.

c.

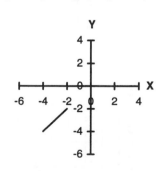

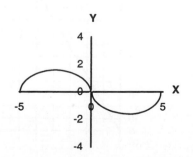

d.

10. Which of these equations will form a straight line when plotted on a graph?
  a. $Y = 45X - 1$ **b.** $C = 34D + 17$
  c. $34 + 2 - 4D = E$ **d.** $Q = R$

11. a. For each of the following equations, what is the slope?
  b. For each of the following equations, what is the Y intercept?
  c. Graph the equations.

  i $Y = (10\ cm/min)\ X + 1\ cm$
  ii $3 - X = Y$
  iii $(12\ mg) + (7\ mg/cm)\ X = Y$

12. a. Draw the line that has a Y intercept of 2, that is, $(0, 2)$, and a slope of 0.25.
  b. Draw the line that goes through the point $(-4, 3)$ and has a slope of -2.

13. Find the equations for the lines graphed in each of the following examples.

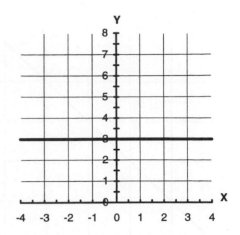
c.

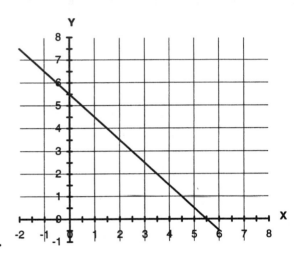
d.

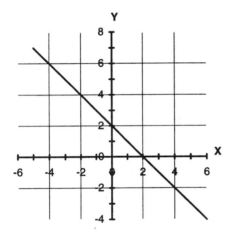
a.

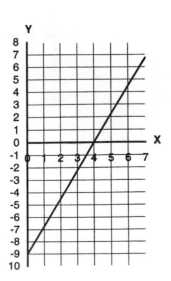
b.

14. The following figure has three lines. Match each line with its equation:

$$Y = \frac{1}{2}X + 2 \qquad Y = \frac{3}{4}X \qquad Y = 2X + 2$$

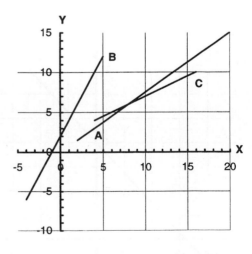

15. Plot the line that contains the following points: $(1,1)$, $(6,6)$, $(9,9)$. What is the equation for this line?

16. The relationship between the temperature in degrees Fahrenheit and degrees Celsius is given by the equation: $°F = 9/5(°C) + 32$.

  a. There are two constants in this equation. What are they?

  b. Which of the two constants is the Y intercept?

  c. Which of the two constants is the slope of the line?

  d. Graph this relationship.

C. An Application of Graphing Linear Relationships: Standard Curves and Quantitative Analysis

This section discusses an important laboratory application of graphs of linear relationships: quantitative analysis. **Quantitative analysis** *is the determination of how much of a particular material is present in a sample.* To make such determinations, a standard curve is used. A **standard curve** *is a graph of the relationship between the concentration of a material of interest and the response of a particular instrument.* Note that the term *standard curve* is used for this sort of graph; however, the desired relationship between concentration and instrument response is usually linear.

A standard curve is constructed as follows: A series of standards containing known concentrations of the material of interest are prepared. The response of an instrument to each standard is measured. A standard curve is plotted with the response of the instrument on the Y axis and the concentration of standard on the X axis. Once graphed, the standard curve is used to determine the concentration of the material of interest in the samples. This process is illustrated in the following example:

1. Standards are prepared containing 1.0 *mg/mL*, 2.0 *mg/mL*, 3.0 *mg/mL*, 5.0 *mg/mL*, 6.0 *mg/mL*, and 7.0 *mg/mL* of a compound of interest.

2. An instrument's response to each standard is measured, Table 10.1.

3. The resulting data are plotted as points on a graph with the concentration of standard on the X axis and the instrument response on the Y axis, Figure 10.9a.

4. The points are connected into a line, Figures 10.9b, c.

Table 10.1 *CONCENTRATION OF STANDARDS VERSUS INSTRUMENT RESPONSE*

Concentration of the Standard (mg/mL)	Instrument Response
1.0	0.24
2.0	0.53
3.0	0.72
5.0	1.28
6.0	1.51
7.0	1.79

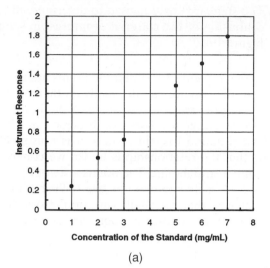
(a)

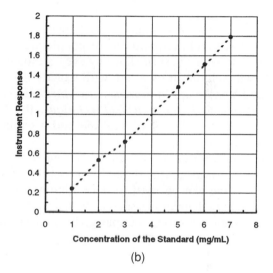
(b)

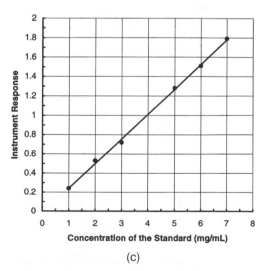
(c)

Figure 10.9.   A Standard Curve. a. Data from Table 10.1 are graphed. The concentration of compound in the standards is on the X axis, instrument response is on the Y axis. **b.** The points are not connected "dot to dot," as shown here, because this would suggest that slight variations from a straight line are meaningful. **c.** The points are connected into a "best fit line," which best averages all the points.

5. The standard curve is used to determine the concentration of the material of interest in samples, Figure 10.10.

Observe in Figure 10.9a that the points approximate a line, but they do not all fall exactly on the line. This is what we would expect due to small errors in measurement. (Measurement errors are discussed in Chapter 13.) In a situation like this, where it is reasonable to assume that the relationship is linear, we connect the points into a straight line. The points are not connected "dot to dot," as illustrated in Figure 10.9b, because this would suggest that the slight variations from the line are meaningful. Rather, the points are connected into a single line that "best fits," or best averages, all the points, as shown in Figure 10.9c.

There are two ways to get a "best fit" line for the points on a graph. One method is to place a ruler over the graph and draw a straight line that appears "by eye" to be closest to all the points. Most people are able to draw a reasonable "best fit" line "by eye," although two people will seldom draw a line with exactly the same slope and Y intercept. *A more accurate method to draw a line of best fit is the statistical technique,* **the Least Squares Method**. (This statistical method is explained in an Appendix to Chapter 20.)

Regardless of whether the points are connected into a line "by eye" or using the Method of Least Squares, the standard curve can be used to determine the level of compound in each unknown sample. In the example illustrated in Figure 10.10, a sample gives an instrument reading of 1.32. From the standard curve, one can see that a reading of 1.32 corresponds to a concentration of 5.2 *mg/mL*.

EXAMPLE PROBLEM

Biologists commonly use a protein assay to determine the concentration of protein in a solution. Various protein assays are available, many of which measure the color change in a protein solution when it reacts with various dyes. The amount of color appearing is generally proportional to the amount of protein present.

The following graph shows the relationship between the concentration of protein in a series of standards and the amount of color after the standards are reacted with dye. The amount of color is measured in terms of the amount of light absorbed by the dye.

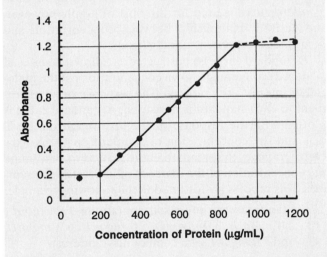

a. In what range of protein concentration does the assay give linear results? What happens at higher concentrations of protein?

b. Suppose you have a sample containing an unknown amount of protein. The sample is reacted with the dye and has an absorbance of 0.70. Based on the standard curve, what is the concentration of protein in the unknown?

ANSWER

a. The assay is linear in the middle range; above about 900 *μg/mL* and below 200 *μg/mL* the assay is not linear.

b. The concentration of the unknown is about 550 *μg/mL*.

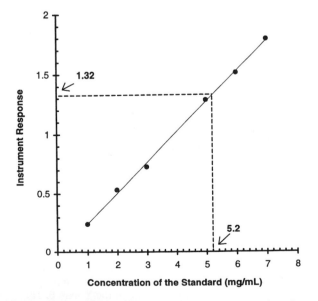

Figure 10.10. Using a Standard Curve to Determine the Concentration of a Material in a Sample. In this example a sample gives an instrument reading of 1.32 which corresponds to a concentration of 5.2 *mg/mL*.

EXAMPLE PROBLEM

The concentration of compound Z in samples needs to be determined. The response of an instrument to compound Z is related to its concentration. A series of standards are prepared and the instrument's response is measured. The resulting data are shown in the following table.

a. Plot a standard curve for compound Z based on the data in the table. Draw the line that best fits the points.

b. A sample with an unknown concentration of compound Z gives an instrument response of 1.35. Based on the standard curve, what is the concentration of compound Z in the sample?

c. Suppose that the response of the instrument used in this example is known to be inaccurate at readings above 2. If a sample has a reading of 2.67 how can you determine its concentration of compound Z?

d. If a sample is diluted 1/20 and has an instrument reading of 0.45, what was the concentration of compound Z in the original sample?

Concentration of Compound Z in Standard (mg/mL)	Instrument Response
100	0.30
150	0.44
250	0.68
400	1.13
550	1.48
650	1.82
850	2.10

ANSWER

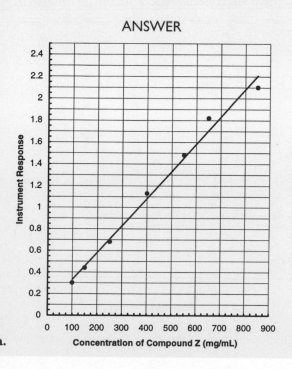

a. **Concentration of Compound Z (mg/mL)** (x-axis), **Instrument Response** (y-axis)

b. Based on the standard curve, the concentration of compound Z in the sample is 500 mg/mL.

c. It is necessary to dilute the sample before reading its absorbance.

d. A value of 0.45 corresponds to a concentration of 150 mg/mL; however, the sample was diluted and this dilution needs to be taken into consideration. We therefore multiply 150 mg/mL times the dilution factor:

$$150 \ mg/mL \times 20 = 3000 \ mg/mL$$

The concentration of compound Z in the original undiluted sample was 3000 mg/mL.

MANIPULATION AND APPLICATION PROBLEMS: QUANTITATIVE ANALYSIS

1. a. Plot a standard curve based on the data in the following table. Draw the line that best fits the values.

  b. Suppose you have a sample that gives an instrument response of 2.35. This value is higher than the reading of the highest standard; therefore, you dilute the sample 1/100. The instrument reading of the diluted sample is 0.78. What amount of the material of interest was present in the original solution?

Amount of Standard (in grams)	Instrument Response
0	0.00
10	0.14
20	0.28
30	0.41
40	0.52
50	0.71
65	0.84
75	1.04
80	1.15

2. A stock solution of copper sulfate has a concentration of 100 mg/mL. How would you dilute the stock solution to prepare each of the following standards?

  1 mg/mL

  5 mg/mL

  15 mg/mL

  25 mg/mL

  50 mg/mL

  75 mg/mL

  100 mg/mL

D. Using Graphs to Display the Results of an Experiment

Many experiments involve manipulating one variable and measuring the result of that manipulation on a second variable. For example, an investigator interested in the effect of light intensity on the rate of seedling growth could expose different groups of seedlings to different light intensities and measure their growth rates. The two experimental variables—light intensity and seedling growth rate—can be plotted on a two-dimensional graph.

When plotting data from experiments, one distinguishes between the **dependent variable** and the **independent variable**. In the preceding example, the investigator is looking at whether seedling growth rate is

dependent on the light intensity. Growth rate is therefore called the dependent variable. *A variable the investigator controls is an* **independent variable**. *A variable that changes in response to the independent variable is called a* **dependent variable**. It is conventional to plot the dependent variable on the Y axis and the independent variable on the X axis. In this example, light intensity is plotted on the X axis and seedling growth rate on the Y axis.

Let us consider how the results of a hypothetical experiment might be displayed graphically. Suppose investigators are interested in the effects of a plant hormone on the number of fruits produced by a certain plant. Investigators perform an experiment to determine the relationship, if any, between this hormone and fruit production. The investigators divide plants into 11 groups, each of which is treated identically except for the application of differing levels of the hormone. The investigators count the fruits produced by each plant. The results of this hypothetical experiment are shown in tabular form in Table 10.2 and graphically in Figure 10.11. Fruit production is the dependent variable and is plotted on the Y axis; applied hormone concentration is on the X axis. These data strongly suggest that the hormone boosts fruit production.

There are several important concepts illustrated by this graph:

1. **Thresholds.** In the central portion of the graph the points appear to form a straight line (i.e., there appears to be a linear relationship between hormone level and fruit production). Below about 5.0 *mg/L* of hormone and above about 50.0 *mg/L* of hormone the relationship changes: 5.0 *mg/L* and 50 *mg/L* are threshold values. A **threshold** *is a point on a graph where there is a change in the*

Table 10.2 *EXPERIMENTAL DATA*

Hormone Level (mg/L) (Independent Variable)	Average Fruit Production Per Plant (Dependent Variable)
0.0	3.2
3.0	3.7
5.0	3.8
10.0	6.5
15.0	12.7
20.0	15.2
25.0	17.0
30.0	24.8
40.0	32.3
50.0	36.0
55.0	36.2
60.0	36.5

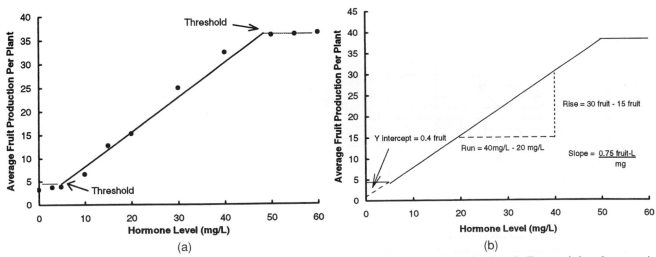

Figure 10.11. Hypothetical Experiment. a. The effect of hormone application on fruit production. **b.** Determining the equation for the line.

relationship between the variables. Thresholds at low and high values, as illustrated in Figure 10.11a, are very common when working with biological data.

2. **Best Fit Line.** The middle points on the graph are close to forming a line, and it is reasonable to conclude that the relationship between fruit production and hormone level is linear between 5.0 and 50.0 *mg/L* of applied hormone. It is acceptable, therefore, to connect these middle points into a straight line. A line may be drawn either "by eye" or using the statistical Method of Least Squares.

Once the points are connected it is possible to determine the equation for the line. The slope has units because both the X and Y axes have units. The Y axis has units of "average fruit production per plant," or, more simply, "fruit"; the X axis has units of milligrams per liter. The slope of the line, Figure 10.11b, is:

$$\frac{(0.75\ fruit)L}{mg}$$

To determine the Y intercept of the line, it is necessary to determine where the intercept would be if there were no lower threshold. By using a ruler to extend the line to the Y axis, the Y intercept can be determined to be 0.40 fruit, Figure 10.11b. The equation for this line in Figure 10.11 is therefore:

$$Y = \frac{(0.75\ fruit)\ L}{mg}X + 0.40\ fruit$$

The slope is the amount by which average production of fruit per plant increases with each milligram per liter of increased hormone. The Y intercept is the amount of fruit production expected if there were no hormone added and if the relationship was linear for all values of hormone (which is not the case).

3. **Prediction.** Equations and graphs are tools that can be used to make predictions. For example, the experiment did not involve testing the effect of 13.0 *mg/L* of hormone on fruit production. We can, however, infer from the graph that if 13.0 *mg/L* of hormone were to be applied, the average fruit production per plant would be about 10. It is also possible to use the equation to predict the amount of fruit with 13.0 *mg/L* of hormone:

$$Y = \frac{(0.75\ fruit)\ L}{mg}\frac{(13.0\ mg)}{L} + 0.4\ fruit$$
$$= 10.2\ fruit$$
= predicted average fruit production per plant

The use of equations and graphs for prediction is very powerful; however, when studying natural systems there are often thresholds. We cannot predict how much fruit production there will be with 100 *mg/L* of hormone because we do not have data at 100 *mg/L*. It is seldom possible to make predictions about biological phenomena past the range of the data.

4. **Using Graphs to Summarize Data.** The graph of the hypothetical experiment contains the same information as Table 10.2; however, it is usually easier to see a relationship between two variables when the data are displayed on a graph than when they are listed in a table.

So far, we have looked at data where two variables are indeed related to one another. What would a graph look like if the two variables studied are not related to one another? One possibility is shown in Figure 10.12a, where the points appear to be scattered without pattern on the graph. Another possibility is shown in Figure 10.12b, where the graph is "flat." The value of the Y variable is constant regardless of the value for the variable on the X axis. Observe in the hypothetical fruit and hormone experiment, Figure 10.11, the graph of the

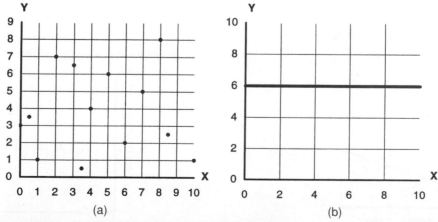

Figure 10.12. Graphs of Variables that are Not Related.

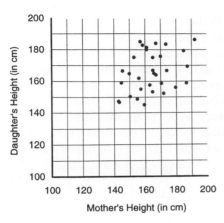

Figure 10.13. Two Variables Whose Relationship to One Another is Ambiguous.

data is "flat" at high and low hormone levels. We can, therefore, reasonably conclude that at low and high levels of hormone, the production of fruit is controlled by factors other than the level of this hormone.

Figures 10.10 and 10.11 both illustrate situations where the variable on the Y axis clearly appears to be related to the variable on the X axis. The graphs in Figure 10.12 illustrate situations where two variables clearly appear to be unrelated to one another. In practice, it is sometimes ambiguous whether or not there is a relationship between two variables. For example, in Figure 10.13 there appears to be a weak relationship between the adult height of a daughter and the height of her mother. The relationship is inconsistent, however, because maternal height is not the only factor affecting a woman's adult height. Paternal genes and nutrition also affect height. The plot of daughter versus mother's height, therefore, does not form a very "good" line. (A graph containing this type of data is called a *scatter plot* because the data points are scattered.)

APPLICATION PROBLEMS: GRAPHING

1. It is important to know whether exposure to agents such as low level radiation, pesticides, or asbestos is likely to cause cancer. Two different predictions of cancer risk due to exposure to small amounts of cancer-causing materials are shown in the graphs.

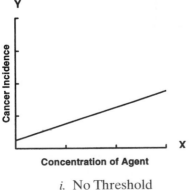

i. No Threshold

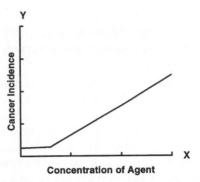

ii. Threshold at Low Exposure Levels

a. If graph *i* is correct (no threshold) and a population of humans is exposed to small amounts of potentially carcinogenic materials, will the incidence of cancer in the population increase? Explain.

b. If graph *ii* is correct, will exposure to small amounts of potentially carcinogenic materials result in an increase in the incidence of cancer?

c. Speculate as to why exposures to small amounts of carcinogenic materials may not increase the likelihood of cancer while exposure to large amounts does increase the probability of cancer.

d. Explain why neither graph *i* nor *ii* intersect the Y axis at zero.

e. Why is it important to know whether graph *i* or *ii* more accurately reflects the truth about a given material? (To learn more about this issue, see, for example, "Cancer Risk of Low-Level Exposure," Marvin Goldman. *Science* 271:1821–1822, 1996.)

2. Suppose an investigator is studying the genetics of plant productivity. The investigator determines the mass of 500 parent plants and 1000 of their offspring plants. The investigator plots parent plant mass versus the average of offspring mass.

a. Which of the following graphs (*i* or *ii*) would indicate that there is a relationship between the mass of the parent plant and the mass of the offspring? Explain.

b. Suppose the data suggest that there is no relationship between the mass of the parent plant and the mass of its offspring. Suggest a hypothesis to explain that observation and suggest an experiment to test your hypothesis.

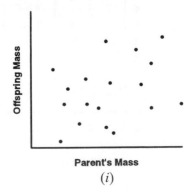

(*i*)

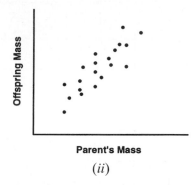

(ii)

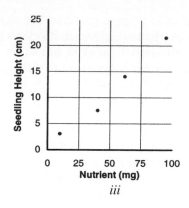

iii

3. a. Draw the best-fit line "by eye" for each of the following graphs. Be careful not to extrapolate (extend the line) past the data.

b. Calculate the slope for each of the lines—do not forget the units.

c. What is the Y intercept for each line?

d. What is the equation for each of these lines? Include the units.

e. Examine graph *i* to determine the mosquito density if there are 5 in of rain.

f. Use the equation for the line in graph *i* to predict mosquito density if there are 5 in of rain. (The answers for 3e and 3f should be the same.)

g. From graph *ii* determine the average shrub's height at 20 months of age. Confirm your determination using the equation for the line in graph *ii*.

h. From graph *iii* determine average seedling height with 50 *mg* of nutrient. Confirm your determination using the equation for the line in graph *iii*.

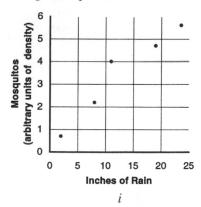

i

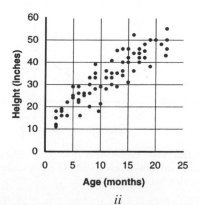

ii

4. Ten students took a midterm and final in a course. The scores for each student are plotted in the next graph.

a. What was the approximate average score for the midterm; 20, 40, 60, or 80?

b. What was the approximate average score for the final; 20, 40, 60, or 80?

c. Which exam had lower scores?

d. Was there a clear relationship between a student's midterm grade and their final grade? If the same class and the same exams were given the following year, could the teacher predict the final score of a student based on their midterm? Explain.

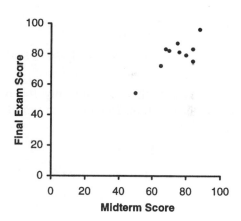

II. GRAPHS AND EXPONENTIAL RELATIONSHIPS

A. Growth of Microorganisms

i. THE NATURE OF EXPONENTIAL RELATIONSHIPS

Previous sections of this chapter focused on relationships that form a straight line when graphed. Although linear relationships are extremely important, not all relationships in the laboratory are linear. This section discusses two important examples of nonlinear relationships: (1) the relationship between the number of bacteria present and time elapsed; (2) the relationship between the amount of radioactivity present and time elapsed.

Suppose there is a single bacterial cell that divides to form two cells. The two cells each divide to form four cells, which divide into eight cells and so on. Suppose that the cells divide every hour and that the number of bacterial cells therefore doubles every hour. We say that these bacteria have a generation time of 1 hour. The relationship between time elapsed and number of bacteria in this example is summarized in Table 10.3 and graphed in Figure 10.14. Note that this relationship does not form a straight line.

How could we write an equation that describes the relationship graphed in Figure 10.14? There are two variables: the time elapsed and the number of bacteria. The equation needs to include both variables to show that the population doubles at a regular interval. The equation that describes this relationship is:

$$Y = 2^x$$

where
x = the number of generations that have elapsed
Y = the number of bacterial cells present

For example:
when x = 2, 2 generations have elapsed
$$Y = 2^2 = 4$$
the number of bacteria cells present is 4

when x = 4, 4 generations have elapsed
$$Y = 2^4 = 16$$
the number of bacteria cells present is 16

Now, suppose that there are initially 100 bacterial cells that double as before. The equation that describes these data is:

$$Y = 2^x(100)$$

This example is illustrated in tabular form in Table 10.4 and graphically in Figure 10.15. The graph where there are initially 100 cells present is much like the graph where there is initially only one bacterium, but with a Y intercept of 100.

Being able to calculate bacterial numbers is important in the laboratory in situations where we want to

Table 10.3 *BACTERIAL GROWTH*

Time Elapsed (hours)	Number of Bacterial Cells Present
0	1
1	2
2	4
3	8
4	16
5	32
6	64
7	128

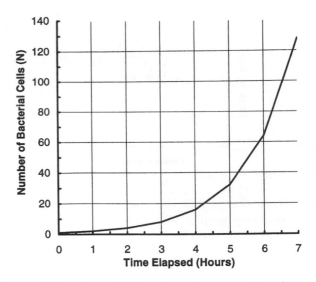

Figure 10.14.   Bacterial Growth. A plot where there is initially only one bacterial cell present and the bacteria population doubles every hour. A graph of this form is called *exponential*.

Table 10.4 *BACTERIAL GROWTH BEGINNING WITH 100 CELLS*

Time Elapsed (hours)	Number of Bacterial Cells Present
0	100
1	200
2	400
3	800
4	1600
5	3200
6	6400
7	12800

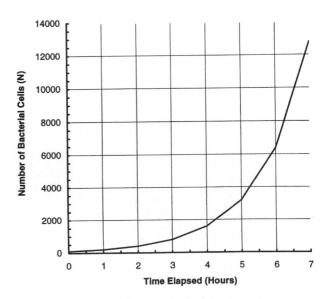

Figure 10.15.   Bacterial Growth. A plot where there are initially 100 bacterial cells present and the bacteria population doubles every hour.

predict approximately how many bacteria will be present in a culture after a certain time, or want to know how long a culture must be incubated to get a certain density of cells. It is possible to write a general growth equation that applies to any bacterial population, regardless of how many cells there are initially and regardless of how long it takes the population to double:

GENERAL EQUATION FOR BACTERIAL POPULATION GROWTH

$$Y = 2^x(N_o)$$

where

N_o = the number of bacteria initially
Y = the number of cells after x generations
x = the number of generations elapsed
(for example, if the bacteria double every 2 hours, then, after 6 hours three generations have elapsed)

EXAMPLE PROBLEM

A type of bacterium has a doubling time of 20 minutes. There are initially 10,000 bacteria present in a flask. How many bacteria will there be in one hour?

ANSWER

STRATEGY 1: USING A TABLE

Time Elapsed	Number of Generations Elapsed	Number of Bacteria Present
0	0	10,000
20 min	1	20,000
40 min	2	40,000
60 min	3	80,000

Thus, after 60 minutes we predict there will be 80,000 bacteria in the flask.

STRATEGY 2: USING THE EQUATION FOR BACTERIAL POPULATION GROWTH

Three generations have elapsed (the first at 20 minutes, the second at 40 minutes, and the third at 60 minutes). Substituting the number of generations elapsed and the number of cells originally present into the equation gives:

$$Y = 2^x (N_o)$$
$$Y = 2^3 (10,000)$$
$$Y = 8 (10,000)$$
$$Y = 80,000$$

The equation for population growth and the "intuitive" approach both give the same answer.

These relationships between the number of bacteria present and the time (or number of generations) elapsed are called **exponential** *because there is an exponent in their equation.* As you can see in Figures 10.14 and 10.15, an exponential relationship is not linear if it is plotted on a regular two-dimensional graph.

ii. SEMILOG PAPER

It is more convenient to work with graphs that are linear in form than with those that are not. It is often desirable, therefore, to convert an exponential relationship to a form that is linear when plotted. The way this is accomplished involves logarithms. If, instead of plotting the number of bacterial cells on the Y axis, we plot the log of the number of bacterial cells, then the plot is linear. This approach is illustrated in tabular form in Table 10.5 and graphically in Figure 10.16.

There is an alternative method to graph bacterial growth so that the plot is linear in form. This method involves the use of **semilog paper**, *a type of graph paper that substitutes for calculating logs.* Semilog graph paper has normal, linear subdivisions on the X axis, Figure 10.17. The divisions on the Y axis, however, are not even; rather, they are initially widely spaced and then

Table 10.5 BACTERIAL GROWTH—LOG OF NUMBER OF CELLS

Time Elapsed (hours)	Number of Cells Present	Log of Number of Cells
0	100	2.00
1	200	2.30
2	400	2.60
3	800	2.90
4	1600	3.20
5	3200	3.51
6	6400	3.81
7	12800	4.11

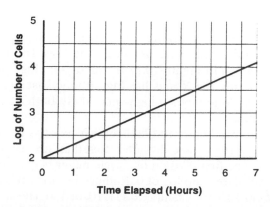

Figure 10.16. A Log Plot of Bacterial Growth. When the log of bacterial cell number is plotted versus time, a straight line is formed.

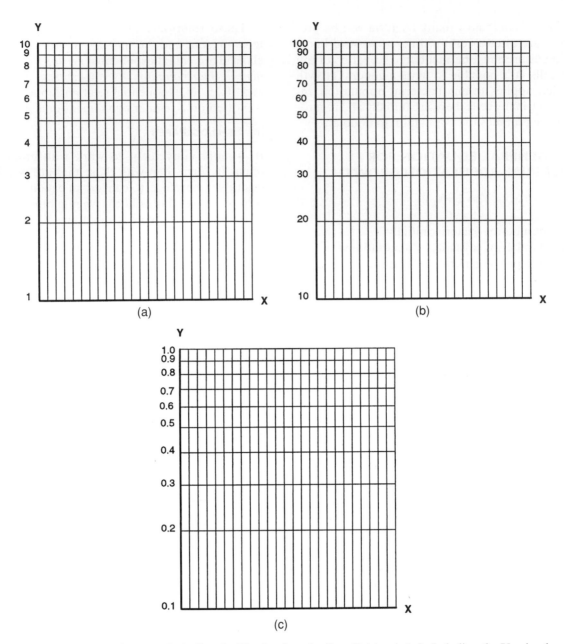

Figure 10.17. Labeling Semilog Paper. a. Labeling the Y axis when the first division is 1. **b.** Labeling the Y axis when the first division is 10. **c.** Labeling the Y axis when the first division is 0.1.

become narrower towards the top of the paper. This spacing takes the place of calculating logs. The graph paper shown in Figure 10.17 is called "semilog" paper because only the Y axis has logarithmic spacing. It is also possible to buy log-log paper where both axes are logarithmic.

Observe in Figure 10.17a that the Y axis has 10 major divisions. These ten divisions together are called a "cycle." The first division in a cycle never begins with 0 because there is no log of 0. Rather, the first division is a power of 10, for example, 0.1, 1, 10, or 100. The second division in each cycle is twice the first, the third division is three times the first, and so on. For example, the Y axis in Figure 10.17a begins at the bottom with 1. The

next division, therefore, is 2, the next 3, and so on to 10. The Y axis in Figure 10.17b begins with 10. The next division, therefore, is 20, then 30, and so on to 100. The Y axis in Figure 10.17c begin with 0.1. The next division, therefore, is 0.2, then 0.3 and so on to 1.0.

Figure 10.17 illustrates one cycle semilog paper. The graph paper in Figure 10.18a is called *two cycle semilog paper* because it repeats the same pattern twice. Three cycle semilog paper, Figure 10.18b, repeats the pattern three times. The beginning of each cycle is 10 times greater than the beginning of the previous cycle. For example, if the first cycle goes from 1 to 10, then the second cycle must range from 10 to 100, the third cycle from 100 to 1000, and so on.

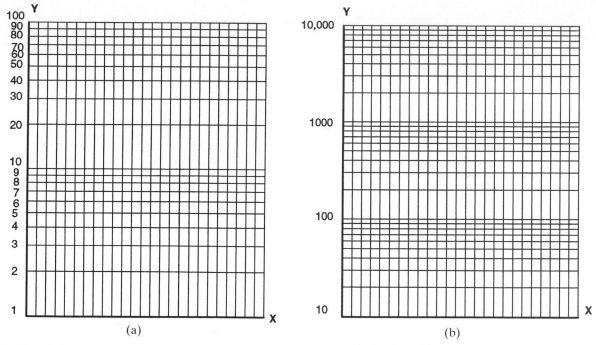

Figure 10.18. **Cycles on Semilog Paper. a.** Two cycle semilog paper. **b.** Three cycle semilog paper.

It is possible to purchase semilog paper with various numbers of cycles. To determine how many cycles you need, examine the data to be plotted. For example, in Tables 10.4 and 10.5, the Y values vary between 100 and 12,800. In this case there is no need for a cycle from 1 to 10 or from 10 to 100 so the bottom cycle can begin at 100. The first cycle required to plot these data runs from 100 to 1000. The second cycle ranges from 1000 to 10,000, and a third cycle, from 10,000 to 100,000, is needed to plot the value "12,800." Thus, three cycle semilog paper is required to plot these data. Figure 10.19 shows the semilog plot of the bacteria data from Table 10.4. Observe how the Y axis is labeled in this figure. Also, compare the plots in Figures 10.16 and 10.19. Both graphs are linear in form and either is an acceptable way to plot these data.

In principle, all biological populations have the potential to grow exponentially. If any population continued to grow exponentially for a long enough period of time, however, it would cover the earth and eventually the universe. In reality, population growth is limited by space, nutrients, waste buildup, and other factors. Bacterial growth normally looks like the plot in Figure 10.20. When a few bacteria that are not reproducing rapidly are placed in fresh media, there is initially a lag period as they utilize nutrients and prepare to divide. A period of exponential growth follows the lag period. The bacteria eventually deplete the nutrients in the broth and generate toxic waste products. Reproduction slows and stops and the population declines.

B. The Decay of Radioisotopes

Radioactive substances decay over time so that there is progressively less and less radioactivity present. The **half-life** *of a radioactive substance is the time it takes for the amount of radioactivity to decay to half its original level.* Radioactivity can be measured in various units including disintegrations/minute, Curies (*Ci*), and microCuries (μCi). For example, suppose a solution containing a radioactive substance initially undergoes 400 disintegrations/minute and that the half life of the substance is 1 hour. Then, after 1 hour, there will be 200 disintegrations/minute of radioactivity remaining in the solution. After 2 hours there will be 100 disintegrations per minute, and so on. The relationship between time elapsed and radioactivity for this example is shown in tabular form in Table 10.6 and is graphed in Figure 10.21.

An exponential equation can be used to calculate the amount of radioactivity remaining after a certain number of half lives have elapsed:

GENERAL EQUATION FOR RADIOACTIVE DECAY

$$N = (1/2)^t N_o$$

where
  N = the amount of radioactivity remaining
  t = the number of half lives elapsed
  N_o = the amount of radioactivity initially

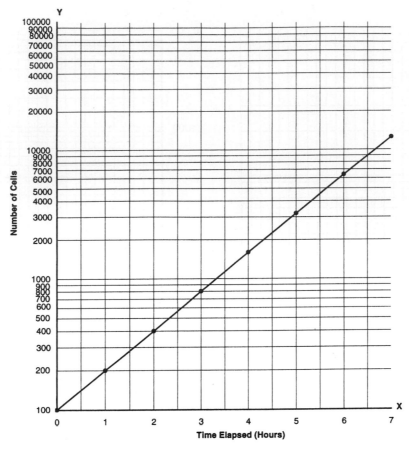

Figure 10.19.   Bacterial Growth Graphed on Semilog Paper.

For example, let us return to the example of the radioactive solution that has a half life of 1 hour and an activity of 400 disintegrations/minute initially. How much radioactivity will remain after 3 hours? We can see from Table 10.6 that there are 50 disintegrations/minute after 3 hours. It is possible to get the same answer by substituting into the General Equation for Radioactive Decay as follows (3 hours is the same as three half lives for this substance):

$$N = (\tfrac{1}{2})^3 \, (400 \; disintegrations/min)$$
$$= 50 \; disintegrations/min$$

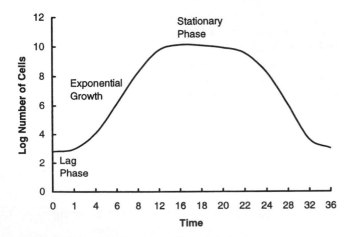

Figure 10.20.   The Limits to Bacterial Growth.

Table 10.6   RADIOACTIVE DECAY

Time Elapsed (hours)	Half Lives Elapsed	Radioactive Substance Remaining (disintegrations/minute)
0	0	400
1	1	200
2	2	100
3	3	50
4	4	25

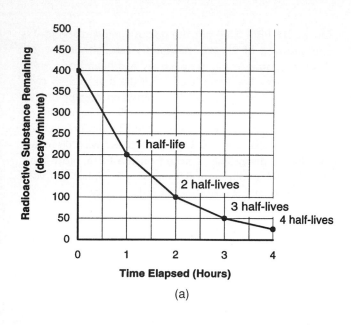

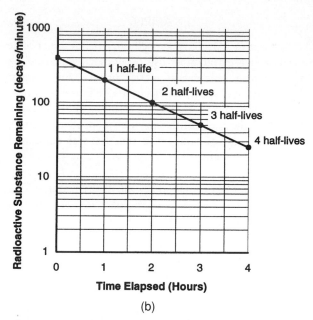

Figure 10.21. The Relationship Between Time Elapsed and Radioactivity Remaining. a. The relationship on normal graph paper. **b.** The relationship plotted on semilog paper.

EXAMPLE PROBLEM

^{131}Iodine has a half life of 8 days. If you start with 500 μCi of ^{131}I, how much will remain after 80 days?

ANSWER
STRATEGY 1: USING A TABLE

Time Elapsed (days)	Half Lives Elapsed	Radioactive Substance Remaining (μCi)
0	0	500
8	1	250
16	2	125
24	3	62.5
32	4	31.25
40	5	15.625
48	6	7.8135
56	7	3.9063
64	8	1.9531
72	9	0.9766
80	10	0.4883

The answer is 0.4883 μCi or about 0.5 μCi remain.

STRATEGY 2: USING THE GENERAL ON FOR RADIOACTIVE DECAY

1. Calculate how many half-lives have passed. The half-life of this isotope is 8 days so the number of half-lives that have passed is: 80/8 = 10.

2. Substitute into the equation:

$$N = (\tfrac{1}{2})^t (N_0)$$
$$N = (\tfrac{1}{2})^{10} (500 \ \mu Ci)$$
$$N = 0.4883 \ \mu Ci$$

The answer is that only about 0.5 μCi remain. Eight days is a relatively short half-life. Isotopes with short half-lives rapidly disappear.

EXAMPLE PROBLEM

The half-life for the radioisotope ^{32}P is 14 days. You receive 200 μCi to perform an experiment on March 3. On April 30 you must complete paperwork showing how much ^{32}P activity remains. How much is present?

ANSWER

1. Calculate how many half-lives have passed. There are 58 days between March 3 and April 30. The half-live of this isotope is 14 days so the number of half-lives that have passed is: 58/14 = 4.1.

2. Substitute into the equation: $N = (\tfrac{1}{2})^t (N_0)$

$$N = (\tfrac{1}{2})^{4.1} (200 \ \mu Ci)$$
$$N = 11.7 \ \mu Ci$$

The answer to be recorded is 11.7 μCi remain (assuming you have not used any for an experiment).

EXAMPLE

A "Radioactive Material Record" documents the receipt of radioactive material, its use (how much was used, by whom, and on what dates) and its disposal. An example of such a form is shown in Figure 10.22. In this example, the first line shows that on 12/10/98, PBM received 1.0 mCi of material containing the radioactive isotope, ^{32}P. This amount of radioactive material was contained in 1 mL total volume.

On 12/15/98, PBM removed 0.1 mL of the material. This volume contained 0.1 mCi of radioactive material. This left 0.9 mCi in the vial.

On 1/7/99, PBM filled out the disposal information form. She disposed of 0.1 mCi of liquid waste. In addition, at this time, some of the radioactivity had been lost simply due to decay. The half life of ^{32}P is 14 days and so two half lives (28 days) had elapsed between receipt of the radioisotope on 12/10/98 and 1/7/99. Therefore, of the 0.9 mCi that was in the vial, only 0.225 mCi remained. This means that $0.9 - 0.225 = 0.675$ mCi was "disposed of" due to decay. This is noted on the form. Note that there is still a volume of 0.9 mL on 1/7/99, but this volume only contains 0.225 mCi of radioactivity.

RADIOACTIVE MATERIAL RECORD

USE INFORMATION					DISPOSAL INFORMATION					
Activity (μCi or mCi)		Volume (μL or mL)		Date Initials	WASTE TYPES AND ACTIVITY (μCi or mCi)					Date & Initials
removed	remaining	removed	remaining		Solid	Liquid	Animal	Decay	Total	
0	1.0	0	1.0	12/10/98 PBM						
0.1	0.9	0.1	0.9	12/15/98 PBM						
						0.1			0.1	1/7/99 PBM
0.675	0.225	0	0.9	1/7/99 PBM				0.675	0.675	1/7/99 PBM

Figure 10.22. Radioactive Material Record. A form like this is used to document the receipt, use, and disposal of radioactive material.

There is similarity between the equations for the decay of radioactivity and the growth of microorganisms. You can see this similarity by comparing Figures 10.15 and 10.21a. Note also the similarity in their equations, which can be expressed as shown here:

A COMPARISON OF THE EQUATIONS FOR BACTERIAL GROWTH AND RADIOACTIVE DECAY

The growth of microorganisms: $N = 2^t (N_o)$

The decline of radioactivity $N = (\frac{1}{2})^t (N_o)$

where

N = amount (of bacteria or radioactivity) present after a certain period of time

t = time elapsed (number of generations or number of half lives)

N_o = initial amount (of bacteria or radioactivity)

APPLICATION PROBLEMS: EXPONENTIAL EQUATIONS

1. There is an insect population in which each adult leaves 3 offspring and then dies. The generation time is 10 months, that is, the offspring grow up and reproduce every 10 months. In this type of organism, the adults die after they reproduce. The population begins with 20 adults and reproduces for four generations. The resulting population growth is shown in this table:

Generation	Time (months)	Number
0	0	20
1	10	60
2	20	180
3	30	540
4	40	1620

a. Graph this relationship.

b. Determine the equation that describes this relationship.

2. The half-life of the radioisotope ^{32}P is 14 days. If there are initially 300 μCi of this radioisotope:

  a. How much will be left after 28 days?

  b. How much will be left after 140 days?

  c. How much will be left after 200 days?

  d. How much will be left after one year?

3. The half-life of the radioisotope ^{22}Na is 2.6 years. If there are initially 600 μCi of this radioisotope, how much will be left after:

  a. 1 year

  b. 2 years

  c. 3 years

4. Fill in the blanks in the form below. The radioisotope involved is ^{32}P with a half-life of 14 days.

5. It is difficult to kill all microorganisms in a material. A solution or material to be sterilized is typically heated under pressure. The following graph shows the effect of time of exposure to heat and bacterial death (based on information from *Principles and Methods of Sterilization in the Health Sciences,* John J. Perkins. 2nd edition, Charles C. Thomas, Springfield, 1983).

  a. About how many bacteria were there at the beginning of the experiment, before exposure to heat?

  b. About how many bacteria were present after 2 minutes of treatment?

  c. Why did the investigators show their data on a semilog plot?

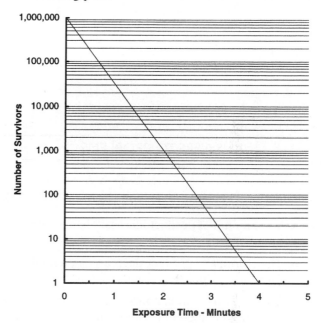

RADIOACTIVE MATERIAL RECORD

USE INFORMATION					DISPOSAL INFORMATION					
Activity (μCi or mCi)		Volume (μL or mL)		Date Initials	WASTE TYPES AND ACTIVITY (μCi or mCi)					Date & Initials
removed	remaining	removed	remaining		Solid	Liquid	Animal	Decay	Total	
0	500	0	1.0	12/3/81 NW						
100	400	0.2	0.8	12/8/81 NW						
100	—	0.2	—	12/12/81 NW						
						200			200	12/17/81 NW
—	—	0	—	12/17/81 NW				—	—	12/17/81 NW

6. Label the Y axis of this semilog graph.

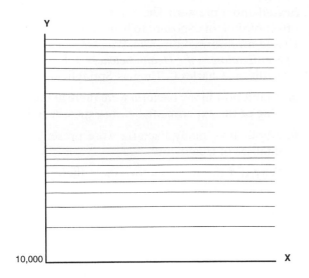

ANSWERS TO PROBLEMS

Manipulation Problems: Graphing (from pp. 156–159)

1. a. The amount of product increases in a linear fashion over time.

   b. The amount of product decreases in a nonlinear fashion over time.

   c. The reaction rate is constant, regardless of the amount of compound X present.

2. a. (7,5)

   b. (−4,−4)

   c. (−5,5)

   d. (7,−4)

3. change in $X = -11$, change in $Y = -9$, change in $X = -1$, change in $Y = 9$

4.

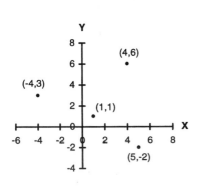

5.

12	61
15	76
20	101

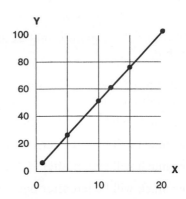

6. $A = 3Q-4$

Q	A
−1	−7
0	−4
1	−1
2	2
3	5

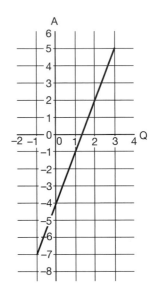

7. a. slope 3, intercept 2

   b. slope 0.2, intercept −1

   c. slope 0.005, intercept 0

8. a. slope 1, intercept 2

   b. slope −1, intercept 7.5

   c. slope 1.25, intercept 4

   d. slope −1, intercept 5

9. a. not linear

   b. linear

   c. linear

   d. not linear

10. All will form a line.

11. a., b. *i.* slope = 10 *cm/min*, Y intercept = 1 *cm*

        ii. slope = −1, Y intercept = 3

        iii. slope = 7 *mg/cm*, Y intercept = 12 *mg*

c.

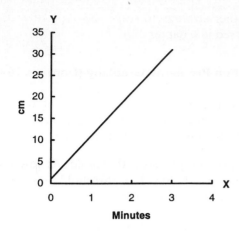

i

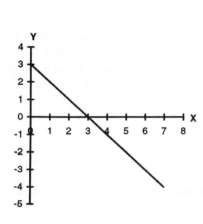

ii

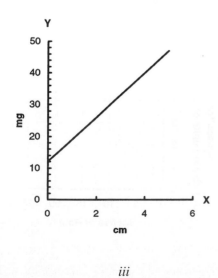

iii

12.

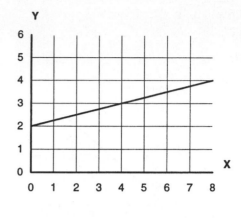

(a)

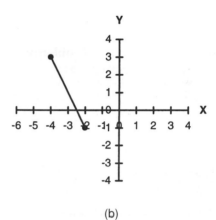

(b)

13. a. $Y = -1X + 2$ **b.** $Y = 2.25X - 9$

   c. $Y = 3$ **d.** $Y = -1X + 5.5$

14. Line A: $Y = \dfrac{3}{4}X$

   Line B: $Y = 2X + 2$

   Line C: $Y = \dfrac{1}{2}X + 2$

15. $Y = 1X + 0$

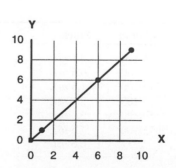

16. a. 9/5 and 32

b. 32

c. 9/5

d.

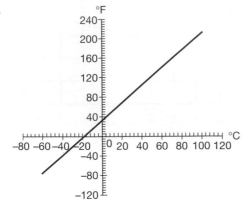

Manipulation and Application Problems: Quantitative Analysis (from p. 161)

1. a.

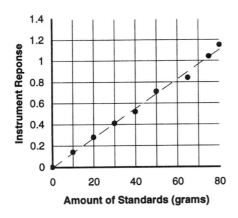

b. A 0.78 instrument response corresponds to 57 g. The sample was diluted 100× so the amount in the undiluted sample was: 57 g × 100 = 5700 g.

2. Stock solution at 100 mg/mL. First, decide what volume to make. For example, a strategy to make 10 mL of each standard:

Standard	Dilution	Stock		Diluent
1 mg/mL	1/100	0.1 mL	+	9.9 mL
5 mg/mL	5/100	0.5 mL	+	9.5 mL
15 mg/mL	15/100	1.5 mL	+	8.5 mL
25 mg/mL	25/100	2.5 mL	+	7.5 mL
50 mg/mL	50/100	5.0 mL	+	5.0 mL
75 mg/mL	75/100	7.5 mL	+	2.5 mL
100 mg/mL		10.0 mL		0 mL

(Another strategy to solve this type of problem is discussed in Chapter 22.)

Application Problems: Graphing (from pp. 164–165)

1. a. Yes. The graph shows a linear relationship between cancer incidence and exposure level, even with the lowest exposures to the agent.

b. No, the graph has a threshold. At low levels of exposure, below the threshold, no change in cancer incidence occurs.

c. Possible explanations include: (1) individuals are able to detoxify and handle low levels of the agent and (2) multiple receptors must be activated before an effect occurs.

d. A background level of cancer is found in the population even in the absence of exposure to the agent.

e. Regulatory agencies need to know whether low level exposure is safe in order to decide what levels (if any) of compound can be allowed. For example, this is important to know when deciding whether a particular pesticide can be used on crops, and if so, what levels are "safe."

2. a. Graph **b** indicates a relationship between the mass of the parent and the offspring where a higher parental mass is associated with a higher offspring mass.

b. Environmental factors may play a major role in determining mass. One experiment would be to take genetically identical clones of the plant and measure their mass under different, controlled environmental conditions. There are other experiments you might devise as well.

3. a.

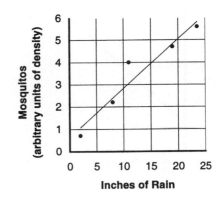

i

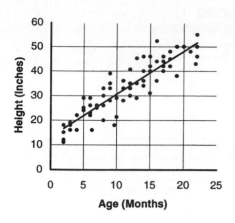

ii

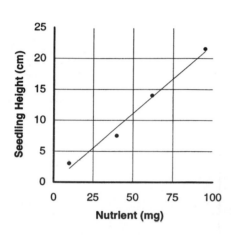

iii

b. i   slope = 0.22 mosquitos/*in* rain

   ii   slope = 1.8 *in*/month

   iii   slope = 0.24 *cm/mg*

c. i Y intercept = 0.6 mosquitos

   ii Y intercept = 13 *in*

   iii Y intercept = 0 *cm*

d. i $Y = (0.22$ mosquitos/*in* rain$) X + 0.6$ mosquitos

   ii $Y = (1.8$ *in*/month$) X + 13$ *in*

   iii $Y = (0.24$ *cm/mg*$) X + 0$ *cm*

e. 1.7 mosquitos (approximate)

f. $Y = (0.22$ mosquitos/*in*$) (5$ *in*$) + 0.6$ mosquitos
   $= 1.7$ mosquitos

g. $Y = (1.8$ *in*/month$) (20$ months$) + 13$ *in* $= 49$ *in*

h. $Y = (0.24$ *cm/mg*$)(50$ *mg*$) + 0$ *cm* $= 12$ *cm*

4. a. 80

  b. 80

  c. The slope is close to 1, so the test scores were roughly equal.

  d. The relationship is roughly linear, which means that there is a general relationship between the two scores. Students who did poorly on the midterm tended to do poorly on the final. Midterm scores appear therefore to be fairly predictive, although the next year's class could differ.

Application Problems: Exponential Equations (from pp. 172–174)

1. a.

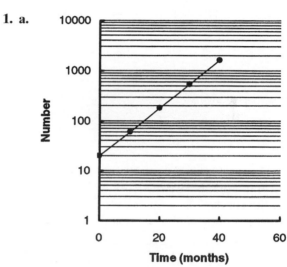

  b. $Y = 3^x (20)$ where x = number of generations

2. $N = (1/2)^t(N_o)$ $N_o = 300\ \mu Ci$ half life = 14 days

  a. t = 28 days/14 days = 2

    $N_2 = (\tfrac{1}{2})^2 (300\ \mu Ci) = 75\ \mu Ci$

  b. t = 140 days/14 days = 10

    $N_{10} = (\tfrac{1}{2})^{10} (300\ \mu Ci) = 0.293\ \mu Ci$

  c. t = 200 days/14 days = 14.3

    $N_{14.3} = (\tfrac{1}{2})^{14.3} (300\ \mu Ci) = 1.49 \times 10^{-2}\ \mu Ci$

  d. t = 365 days/14 days = 26.1

    $N_{26.1} = (\tfrac{1}{2})^{26.1} (300\ \mu Ci) = 4.17 \times 10^{-6}\ \mu Ci$

3. a. $N = (\tfrac{1}{2})^{0.38} (600\ \mu Ci) = 461\ \mu Ci$

  b. $N = (\tfrac{1}{2})^{0.77} (600\ \mu Ci) = 352\ \mu Ci$

  c. $N = (\tfrac{1}{2})^{1.2} (600\ \mu Ci) = 261\ \mu Ci$

4.

RADIOACTIVE MATERIAL RECORD

USE INFORMATION					DISPOSAL INFORMATION					
Activity (μCi or mCi)		Volume (μL or mL)		Date Initials	WASTE TYPES AND ACTIVITY (μCi or mCi)					Date & Initials
					Solid	Liquid	Animal	Decay	Total	
removed	remaining	removed	remaining							
0	500	0	1.0	12/3/81 NW						
100	400	0.2	0.8	12/8/81 NW						
100	300	0.2	0.6	12/12/81 NW						
						200			200	12/17/81 NW
150	150	0	0.6	12/17/81 NW				150	150	12/17/81 NW

5. a. 10^6

b. 10^3

c. Using semilog paper results in a straight line on the graph that is easier to work with than an exponential curve.

6.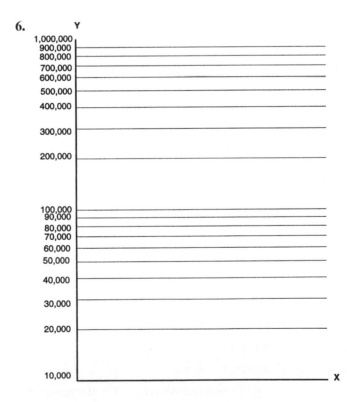

UNIT IV:

Data in the Laboratory

Chapters in This Unit

Producing data is the central function of laboratory scientists and technicians. As a broad definition, data are observations of the natural world that may take the form of numerical values, photographs, words, or other expressions. Processed, analyzed data provide the basis for the knowledge and information that emerge from the laboratory. Many sections of this book relate to the acquisition or processing of data. Unit II discussed systematic ways to ensure the quality of data (e.g., by having written standard operating procedures for all laboratory workers to follow). Unit V is about making measurements; measurements are numerical data. Chapter 32 is about processing data with the aid of computers.

This unit introduces two issues relating to laboratory data:

Chapter 11 discusses how numerical data can be organized, displayed, and described.

Chapter 12 explores some pitfalls that can arise when collecting data, as well as some ways to avoid these problems.

TERMS USED IN THIS UNIT

Acceptance (control chart usage). A decision that a process is operating in control. See also "Control Chart Limits."

Batch. A definite quantity of some product produced under uniform conditions. Note that a batch is usually smaller than a lot.

Bias (relating to measurements). Measurements are biased when there is a systematic error(s) that causes the values to be too high or too low.

Bimodal Distribution. A frequency distribution with two peaks.

Block. Originally a small plot within an experimental field within which conditions are more homogeneous than within the entire field. This now refers to any portion of experimental material, experimental environment, time, and so on, that is likely to be relatively homogeneous. The purpose of dividing experiments into blocks is to make it easier to treat the control and experimental subjects in the same way (except for the treatment of interest). Each block includes individuals from all control and experimental groups.

Coefficient of Variation, CV, (Relative Standard Deviation, RSD). A measure that expresses the standard deviation in terms of the mean.

Control (in reference to a process). A state in which the variability in a process is attributable to chance (i.e., to the normal variability inherent in the process).

Control Chart. A graphical display of test results together with limits in which the values are expected to lie if the process is in control. Control charts are useful in distinguishing between the normal variability inherent in a process and variability due to a problem or unusual occurrence.

Control Chart Limits. Boundary lines on a control chart that define the limits of acceptable values based on standard deviations as determined in preliminary studies. The **lower warning limit** (LWL) is typically the mean minus two standard deviations and the **upper warning limit** (UWL) is the mean plus two standard deviations. The warning limits define the range in which 95.5% of all values should lie. The **lower control limit** (LCL) is typically the mean minus three SD and the **upper control limit** (UCL) is the mean plus three SD. These control limits define the range in which 99.7% of all points should lie.

Control Group. In an experiment, a control group is treated in exactly the same way as the experimental group except for the treatment of interest. (See also "Positive Control" and "Negative Control.")

Data. Observations of the natural world.

Dependent Variable. A variable whose values vary in response to the independent variable.

Descriptive Statistics. Statistical methods that are used to describe and summarize data.

Deviation (of a data point). The difference between a data point and the mean.

Distribution. The pattern of variation for a given variable.

Dot Diagram. Simple graphical technique to represent a data set in which each datum is represented as a dot along an axis of values.

Experimental Group. A group that is exposed to the experimental treatment of interest.

Factor. A variable chosen to be tested in an experiment to see its effects on the process or measurement of interest.

False Negative. A test result that should be positive but erroneously appears to be negative.

False Positive. A test result that should be negative but erroneously appears to be positive.

Frequency. The number of times a particular value or range of values is observed in a data set.

Frequency Distribution. A listing of the number of times each value for a variable occurs.

Frequency Histogram. A graphical display of data in which the frequency of each value or range of values is plotted as a bar.

Frequency Polygon. A graphical display of data in which the frequency of each value or range of values is plotted as a point.

Independence (in sampling). Sampling in such a way that the choice of one member of the sample does not influence the choice of another.

Independent Variable. A variable whose value is controlled by the investigator.

Inferential Statistics. Statistical methods that are used to reach conclusions about a population based on a sample.

Lot. A definite quantity of a product obtained under conditions that are considered to be uniform.

Matched Subjects. Individuals who, before an experiment begins, are as alike as possible with regard to specific qualities. The individuals are treated differently during an experiment and their responses are compared with one another.

Mean. The average. The sum of all values divided by the number of values. Statisticians distinguish between the true mean of an entire population, represented by μ, and the mean of a sample from that population, represented by $\bar{x}$.

Measurements. Quantitative observations; numerical descriptions.

Measures of Central Tendency. Measures of the values about which a data set is centered (e.g., the mean, median, and mode).

Measures of Dispersion. Measures of the variability in a set of numerical data (e.g., range, variance, and standard deviation).

Median. A statistic that is the middle value of a data set, or the number that is half way between the highest and lowest value.

Mode. The value that is most frequently observed in a set of data.

Negative Control. When an assay is performed, a negative control is a sample that is known to respond negatively; it is used to check that the assay is working correctly.

Normal Distribution. The frequency distribution of data that have a bell shape when graphed. The peak of a perfect normal distribution is the mean, median, and mode, and values are equally spread out on either side of that central high point.

Outlier. A data point that lies far outside the range of all the other data points.

Parameter. A numerical statement about a population, comparable to a sample statistic (e.g., the true mean and standard deviation of a population are parameters).

Parametric Statistical Methods. Statistical methods that assume the variables of interest are normally distributed in the population.

Polymodal Distribution. A frequency distribution with more than two peaks.

Population. A group of events, objects, or individuals where each member of the group has some unifying characteristic(s). Examples of populations are all a person's red blood cells and all the enzyme molecules in a test tube.

Positive Control. When an assay is performed, a positive control is a sample that is known to respond positively; it is used to check that the assay is working correctly.

Probability. The likelihood of a particular outcome.

Probability Distribution. A theoretical distribution of values based on the calculated probabilities of values occurring.

Quality Control. The procedures used to ensure that products maintain a consistent, desired level of acceptability or quality.

"Random Phenomenon." A phenomenon where the outcome of a single repetition is uncertain, but the results of many repetitions have a known, predictable pattern. For example, the outcome of a single flip of a coin might be a head or a tail; the outcome of many flips is predicted to be half heads and half tails.

Random Sample. A sample drawn in such a way that every member of a population has an equal chance of being chosen.

Random Variability (Random Error). The situation where a group of observations or measurements vary by chance in such a way that they are sometimes a bit higher than expected, sometimes a bit lower, but overall average the "correct" or expected value. (See also the contrasting term, "Systematic Variability.")

Randomization. The assigning of individuals to a control or experimental group by chance alone. This concept can also be applied to the order in which samples are tested and other processes. The purpose of randomization is to prevent sample bias.

Range. (1) A range of values, from the lowest to the highest, that a method or instrument can measure with acceptable results. (2) A statistical measure that is the difference between the lowest and the highest values in a set of data.

Relative Standard Deviation. See "Coefficient of Variation."

Replication. The repetition of an experiment or a major portion of an experiment to detect variability in the experimental system.

Representative Sample. A sample that is drawn randomly and independently and therefore truly reflects the population of interest. A sample is random if all members of the population have an equal chance of being picked. A sample is independent if the choice of one member does not influence the choice of another member.

Sample. A subset of a population. For example, a person's blood sample is a subset of the population of all that person's blood.

Sample Statistic. A numerical statement about a sample (e.g., the sample mean is a statistic).

Skewed Distribution. A distribution in which the values tend to be clustered either above or below the mean.

Standard Deviation. The average amount by which each value in a set of values differs from the mean. Also, the square root of the variance.

Statistical Process Control. The desired situation in which the variability in a manufacturing or measurement process behaves as expected.

Sum of Squares. For a set of data, the sum of the squared differences between each data point and the mean.

Systematic Variability (Systematic Error). The situation where a group of observations or measurements vary in such a way that they tend to be higher or lower than the true value or the expected value; this is also called "bias." (See also the contrasting term, "Random Variability.")

Total Squared Deviation. See "Sum of Squares."

Variable. (1) A factor in a measurement or experimental system that can vary (e.g., the temperature at which an assay is performed or the pH of media) (2) A quality of a population which can be measured. For example, for the population of all 6-year-old children, one could measure heights, weights, eye color, or favorite book. Variables may be classified as follows:

1. **Discrete variables** are those that can be counted, such as litter size or number of bacterial colonies on a petri dish.
2. **Continuous variables** can be measured and can be whole numbers or any fraction of a whole number, such as weight or temperature.
3. **Qualitative variables** are attributes that are not numeric, such as color or flavor.

Variance. A measure of the variability in a set of data. The variance is the average squared deviation from the mean for all the points in a set of data.

SUGGESTIONS FOR FURTHER READING

Much of the information in this unit comes from the field of statistics. The following list includes some examples of basic statistics books.

Introductory Biological Statistics, Raymond E. Hampton. Wm. C. Brown, Dubuque, 1994.

Statistics, David Freedman, Robert Pisani, Roger Purves, and Ani Adhikari. 2nd edition. Norton, New York, 1991.

Beginning Statistics, Gene Zirkel and Robert Rosenfield. McGraw Hill Book Co., New York, 1976.

Statistics Concepts and Applications, David R. Anderson, Dennis J. Sweeney, and Thomas A. Williams. West Publishing Co., St. Paul, 1986.

Statistics for Experimenters: An Introduction to Design, Data Analysis, and Model Building, George E.P. Box, William G. Hunter, and J. Stuart Hunter. John Wiley and Sons, New York, 1978.

Experimental Design in Biotechnology, Perry Haaland. Marcel Dekker, Inc., New York, 1989.

Use of Statistics to Develop and Evaluate Analytical Methods, Grant T. Wernimont, William Spindle, ed. AOAC, Arlington, VA, 1985. (A helpful, somewhat more advanced book on the applications of statistical methods.)

Quality Control in Analytical Chemistry, G. Kateman and L. Buydens. 2nd edition. John Wiley and Sons, Inc., New York, 1993.

Statistics for Analytical Chemistry, J.C. Miller and J.N. Miller. 3rd Edition. Prentice Hall, New York, 1993. (Good explanations of how statistics are applied in the laboratory.)

"Fundamental Issues in Experimental Design," Bert Gunter. *Quality Progress*. pp. 105-113. June 1996. (A short, well-written article that addresses some basic principles of statistical thinking.)

A Handbook of Biological Investigation, Harrison W. Ambrose and Katharine Peckham Ambrose. 4th edition. Hunter Textbooks Inc., North Carolina, 1987. (This short handbook includes practical statistical information in an easy to understand format.)

## CHAPTER 11

Descriptions of Data

I. INTRODUCTION

A. Populations, Variables, and Samples

The natural world is filled with variability: organisms differ from one another; the weather changes from day to day; the earth is covered by a mosaic of different types of habitats. The natural world is also characterized by uncertainty and chance. The outcome of a card game, the weather, and the personality of a newborn child are all uncertain. Scientists strive to observe and to understand this world, which is characterized by variability and uncertainty.

Statistics is a branch of mathematics that has developed over many years to deal with variability and uncertainty. Statistics provides methods to summarize, analyze, and interpret observations of the natural world. Statistical methods are also used to reach conclusions, based on data, with a certain probability of being right—and a certain probability of being wrong. In this chapter, we will explore a small area of statistics, and its use in summarizing, displaying, and organizing numerical data. We will begin by introducing several fundamental statistical terms.

A **population** *is an entire group of events, objects, results, or individuals, all of whom share some unifying characteristic(s).* Examples of populations include all a person's red blood cells, all the enzyme molecules in a test tube, and all the college students in the United States.

Characteristics of a population often can be observed or measured, such as blood hemoglobin levels, the activity of enzymes, or the test scores of students. *Characteristics of a population that can be measured are called* **variables**. They are called "variables" because there is variation among individuals in the characteristic. A population can have numerous variables that can be studied. For example, the same population of 6-year-old children can be measured for height, shoe size, reading level, or any of a number of other characteristics. *Observations of a variable are called* **data** (singular "datum").

Observations may or may not be numerical. For example, the lengths of insects (in centimeters) are numerical data. In contrast, electron micrographs of mouse kidney cells are nonnumerical observations. The statistical methods discussed in this unit are used to interpret numerical data.

It is seldom possible to determine the value for a given variable for every member of a population. For example, it is virtually impossible to measure the level of hemoglobin in every red blood cell of a patient. Rather, a **sample** of the patient's blood is drawn and the hemoglobin level is measured and evaluated. Investigators draw conclusions about populations based on samples, Figure 11.1.

Statistical methods are generally based on the assumption that a sample is **representative of its population** that is, *that the sample truly reflects the variability in the original population.* In order for a sample to be representative of its population, it must meet two requirements: (1) *All members of the population must have an equal chance of being chosen.* If this is the case the sample is called **random**. (Note that the statistical meaning of "random" is different from the meaning in common English. In popular usage random means "haphazard," "unplanned," or "without pattern"; this is not its meaning in statistics.) (2) *The choice of one member of the sample should not influence the choice of another.* If this is the case, the sample is said to have been drawn **independently**.

Randomness and independence refer to how a sample is taken. For example, a lottery is random if every ticket has an equal chance of being drawn. If a deck of cards is properly shuffled and if a card is picked by chance from that deck, then the choice of the card is random.

To understand independence, consider a coin that is flipped twice. The second toss has a 50% probability of being heads regardless of whether or not the first toss was heads. This is because the outcomes of the two tosses are independent of one another. In contrast, suppose we begin with a full deck of cards and randomly draw and set aside one card, which turns out to be the five of hearts. A second draw from this deck is not independent because there is now no chance of drawing the five of hearts.

The individuals that comprise a population vary from one another; therefore, even when a sample is drawn randomly and independently, there is uncertainty when a sample is used to represent a whole population. The sample could truly reflect the features of the population from which it was drawn, but perhaps it does not. As you might expect, if a sample is properly drawn, then the larger the sample, the more likely it is to accurately reflect the entire population.

EXAMPLE PROBLEM

a. *In a quality-control setting,* 15 vials of product from a batch were tested. What is the sample? What is the population?

b. *In an experiment,* the effect of a carcinogenic compound was tested on 2000 laboratory rats. What is the sample? What is the population?

c. *A political poll* of 1000 voters was conducted. What is the sample? What is the population?

d. *A clinical study* of the effect of a new drug was tested on 50 patients. What is the sample? What is the population?

ANSWER

a. The sample is the 15 vials tested. The population is all the vials in the batch.

b. The sample is the 2000 rats. The population is presumably all laboratory rats.

c. The sample is the 1000 voters polled. The population is all voters.

Population—All 20 year old men in the United States

Variables—Many variables could be measured for this population such as: height, hair color, college entrance exam score, and birthplace.

Sample—A sample of the population of all 20 year old men is drawn

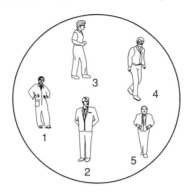

Data—Observations are made of the individuals in the sample

Individual	Height (in cm)	Hair Color	College Exam Score	Birthplace
1	172	brown	1000	WI
2	173	blond	none	IL
3	169	black	1200	NY
4	177	brown	none	WY
5	175	blond	890	MO

Figure 11.1. Populations, Samples, Variables, and Data. A sample is drawn from the population to represent that population. In this illustration, data for four variables are collected for each individual in the sample.

d. The sample is the 50 patients tested. The population is all similar patients.

EXAMPLE PROBLEM

What is meant by the statement "two out of three doctors surveyed recommend Brand X . . ."? What is the population of interest? What is the sample? Is this sample representative of the population? Does this statement ensure that the product being endorsed is better than any other?

ANSWER

Many abuses of statistics relate to poor sampling. A classic type of abuse is statements such as "two out of three doctors recommend Brand X . . ." The population of interest is all doctors. This statement suggests that two thirds of all doctors prefer Brand X to other brands; however, we have no way to know which doctors were sampled. For example, perhaps the survey only looked at

physicians who prescribe Brand X, or only physicians in a certain region. It is possible that the survey sampled doctors who were not representative of all doctors; therefore, the statement does not ensure that most doctors prefer Brand X. In addition, even if two thirds of all doctors really do recommend Brand X, it may or may not be the best brand—it may simply be the most accessible, least expensive, most advertised, and so on.

B. Describing Data Sets: Overview

When a sample is drawn from a population, and the value for a particular variable is measured for each individual in the sample, the resulting values constitute a data set. Because individuals differ from one another, the data set consists of a group of varying values.

A set of data without organization is something like letters that are not arranged into words. Like letters of the alphabet, numerical data can be arranged in ways that are meaningful. (Numerical data, like letters of the alphabet, can also be organized in ways that are confusing, deceptive, or irrelevant. Delving into the many ways that statistics are misused, however, is beyond the scope of this book.) **Descriptive statistics** *is a branch of statistics that provides methods to appropriately organize, summarize, and describe data.**

Consider, for example, the exam scores for a class of students. The variable that is measured is the exam score for each individual. A list of the scores of all the students constitutes the data set. In order to summarize and describe the performance of the class as a whole, the instructor might calculate the class's average score. The average is an example of a measure that summarizes the data. A measure that describes a sample, such as the average, is sometimes referred to as a "statistic."

The average gives information about the "center" of a data set. There are other measures, such as the median and the mode, which also describe data in terms of its center. *Such measures, which describe the center of a data set, are called* **measures of central tendency**.

Data sets A and B both have the same average:

$$\text{A: } 4\ 5\ 5\ 5\ 6\ 6$$
$$\text{B: } 1\ 2\ 4\ 7\ 8\ 9$$

Inspection of these values, however, reveals that the two data sets have a different pattern. The data in A are more clumped about the central value than the data in B. We say that the data in B are more **dispersed**, that is, *spread out*.

Measures of central tendency, such as the average, do not describe the dispersion of a set of observations; therefore, there are statistical **measures of dispersion** *which describe how much the values in a data set vary from one another.* Common measures of dispersion are: range, variance, standard deviation, and coefficient of variation.

Measures of central tendency and of dispersion are determined by calculation. The next section of this chapter discusses the calculation of these measures. It is also possible to organize and effectively display data using graphical techniques. Graphical techniques and their interpretation are discussed in Section III of this chapter. Section IV applies these ideas to the area of controlling product quality in a company.

II. DESCRIBING DATA: MEASURES OF CENTRAL TENDENCY AND DISPERSION

A. Measures of Central Tendency

Consider a hypothetical data set consisting of these values:

$$2\ 5\ 6\ 7\ 8\ 3\ 9\ 3\ 10\ 4\ 7\ 4\ 6\ 11\ 9$$

*There is another branch of statistics, called *inferential statistics,* that provides methods to make predictions about a population based on a sample. This area of statistics is beyond the scope of this book.

The simplest way to organize these values is to order them as follows:

$$2\ 3\ 3\ 4\ 4\ 5\ 6\ 6\ 7\ 7\ 8\ 9\ 9\ 10\ 11$$

Inspection of the ordered list of numbers reveals that they center somewhere around 6 or 7. The center can be calculated more exactly by determining the average or mean. The **mean** *is the sum of all the values divided by the number of values.* In this example the mean is calculated as:

$$2 + 3 + 3 + 4 + 4 + 5 + 6 + 6 + 7 + 7 + 8 + 9 + 9$$
$$+ 10 + 11 = 94$$

The average or the mean is 94/15 = 6.3

In algebraic notation, the observations are called X_1, X_2, X_3, and so on. In this example there are 15 observations, so the last value is X_{15}. The method to calculate the mean can be written compactly as:

$$mean = \frac{\sum X}{n}$$

where n = the number of values

$\sum$ (the Greek letter, sigma) is short for "add up." X means that what is to be added is all the values for the variable, X. Thus:

$$mean = \frac{\sum X}{n} = \frac{X_1 + X_2 + X_3 + \ldots + X_n}{n}$$

Statisticians distinguish between the true mean of an entire population and the mean of a sample from that population, Figure 11.2. The true mean of a population is represented by μ (the Greek letter, mu). The mean of a sample is represented by $\bar{x}$ (read as "X bar"). In practice, it is rare to know the true mean for an entire population; therefore, the sample mean, or $\bar{x}$ must be used to represent the true population mean.

The median is another measure of central tendency. The **median** *is the middle value of a data set, or the number that is half way between the highest and lowest value, when the values are arranged in ascending order.* The median for the following values is 6:

$$2\ 3\ 3\ 4\ 4\ 5\ 6\ \underline{6}\ 7\ 7\ 8\ 9\ 9\ 10\ 11$$

When there is an even number of values, the median is the value midway between the two central numbers. The median for these values is 7.5 or the average of the numbers 7 and 8:

$$3\ 5\ \underline{7\ 8}\ 9\ 11$$

Observe that in both of the following data sets the median is 5 even though their means are quite different:

2 3 4 $\underline{5}$ 6 7 8	$\bar{x} = 5$
2 3 4 $\underline{5}$ 6 100 1000	$\bar{x} = 160$

The **mode** *is the value of the variable that occurs most frequently.* For example, the mode in the following

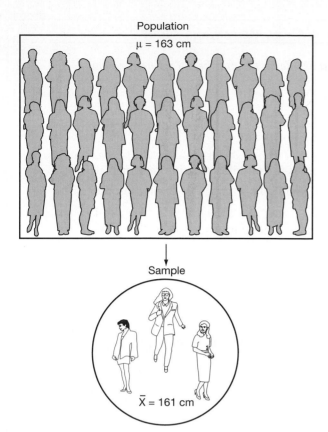

Population

$\mu = 163$ cm

Sample

$\overline{X} = 161$ cm

Figure 11.2. The True Mean of a Population is Distinguished from the Mean of a Sample. The population of interest is all women in the United States; the variable of interest is height. The mean height, μ, for all women is 163 *cm*. A sample of women is drawn and their heights are measured. The mean height in the sample, $\overline{x}$, is 161 *cm*.

data set is "3" since it appears more often than the other numbers:

$$2\ 3\ 3\ 3\ 4\ 5\ 6\ 7\ 7\ 8$$

It is possible for a set of data to have more than one mode. *If there are two modes, the data are* **bimodal**. *If there are more than two modes the data are* **polymodal**.

Box 1 summarizes calculations of these three measures of central tendency—the mean, median, and mode.

EXAMPLE PROBLEM

Suppose you are investigating the height of college students. You measure the height of every student in your classes and obtain the data shown in the following table.

a. What is the average height of the men in the class?

b. What is the average height of the women in the class?

c. What is the average height of all students in the class?

Heights of Students in the Class (in *cm*)					
Men			**Women**		
175.3	176.5	177.8	162.5	166.7	155.6
168.5	165.2	160.2	159.7	163.4	164.2
180.9	188.9	171.6	160.1	168.6	162.6
175.9	174.8	179.2	162.8	158.4	174.9
175.9	175.8	174.8	161.3	166.2	166.7

$n_{men} = 15$ (number of men) $n_{women} = 15$ (number of women)

ANSWER

a. $\overline{x} = \dfrac{2621.3\ cm}{15} = 174.75\ cm$ (men)

b. $\overline{x} = \dfrac{2453.7\ cm}{15} = 163.58\ cm$ (women)

c. $\overline{x} = \dfrac{5075.0\ cm}{30} = 169.17\ cm$ (men + women)

B. Measures of Dispersion

i. THE RANGE

The **range** is the simplest measure of dispersion. The **range** *is the difference between the lowest value and the highest value in a set of data:*

$$2\ 3\ 3\ 3\ 4\ 4\ 5\ 6\ 7\ 7\ 8\ 9\ 9\ 10\ 11$$
$$\text{The range is 2 to } 11 = 9$$

The range is not a particularly informative measure because it is based on only two values from the data set. A single extreme value will have a major effect on the range.

ii. CALCULATING THE VARIANCE AND THE STANDARD DEVIATION

The variance and the standard deviation are more informative measures of dispersion than the range because they summarize information about dispersion based on all the values in the data set. Variance and standard deviation are best explained with a simple example. Assume that the following data are lengths (in millimeters) of eight insects:

$$4\ 5\ 6\ 7\ 7\ 7\ 9\ 11$$

The mean for these data is 7 *mm*. An intuitive way to calculate how much the data is dispersed is to take each data point, one at a time, and see how far it is from the mean, Table 11.1. *The distance of a data point from the mean is its* **deviation**. There is a deviation for each data point: The deviation is sometimes positive, sometimes negative, and sometimes zero.

It is useful to summarize all the deviations with a single value. An obvious approach would be to calculate the average of the deviations, but, for any data set, the

---

BOX 1   MEASURES OF CENTRAL TENDENCY: CALCULATING THE MEAN, MEDIAN, AND MODE

1. *To calculate the mean, add all the values and divide by the number of values present.*

$$mean = \frac{\sum X}{n}$$

2. *To find the median:*

   Order the data values.

   If there is an odd number of values, the median is the middle value in the sequence.

   If there is an even number of values, the median is the average of the two middle values.

3. *Mode:* Order the data values. The mode is the most frequent observation. There may be more than one mode.

Example

Given the data set: 12, 15, 13, 12, 13, 14, 16, 19, 21, 13, 15, 14

1. Order the data 12, 12, 13, 13, 13, 14, 14, 15, 15, 16, 19, 21

2. The mean is $\dfrac{12 + 12 + 13 + 13 + 13 + 14 + 14 + 15 + 15 + 16 + 19 + 21}{12} = 14.75$

3. The median is $\dfrac{14 + 14}{2} = 14$

4. The mode is 13.

---

sum of the deviations from the mean is always zero. Mathematicians therefore devised the approach of squaring each individual deviation value to result in all positive numbers. *The squared deviations can be added together to get the* **total squared deviation**, also called the **sum of squares**. In this example, the sum of squares is 34 mm^2, Table 11.2. The sum of squares is always a positive number that indicates how much the values in a set of data deviate from the mean.

The **variance** *is the total squared deviation divided by the number of measurements:*

$$variance\ (of\ a\ population)$$

$$= \frac{total\ squared\ deviations\ from\ the\ mean}{n}$$

$$= \frac{\sum (X - mean)^2}{n}$$

where n = the number of values

In this example $n = 8$ and the variance is:

$$variance = \frac{34\ mm^2}{8} = 4.25\ mm^2$$

The standard deviation is calculated by taking the square root of the variance:

$$standard\ deviation = \sqrt{variance}$$

In this example:

$$standard\ deviation = \sqrt{4.25\ mm^2} = 2.06\ mm$$

Note that the standard deviation has the same units as the data.

The standard deviation and the variance are values that summarize the dispersion of a set of data around the mean. The larger the variance and the standard deviation, the more dispersed are the data.

iii. DISTINGUISHING BETWEEN THE VARIANCE AND STANDARD DEVIATION OF A POPULATION AND A SAMPLE

In the preceding discussion we skipped over a significant detail regarding the calculation of the variance and standard deviation. Statisticians distinguish between

Table 11.1   CALCULATION OF DEVIATION FROM THE MEAN

(value − mean)	deviation
First value − mean =	(4 − 7) = −3 mm
Second value − mean =	(5 − 7) = −2 mm
Third value − mean =	(6 − 7) = −1 mm
Fourth value − mean =	(7 − 7) = 0 mm
Fifth value − mean =	(7 − 7) = 0 mm
Sixth value − mean =	(7 − 7) = 0 mm
Seventh value − mean =	(9 − 7) = +2 mm
Eighth value − mean =	(11 − 7) = +4 mm
	0 = sum of all the deviations

Table 11.2 CALCULATION OF THE SUM OF SQUARES

(value − mean) (*mm*)	(deviation) (*mm*)	(deviation squared)
(4 − 7)	−3	9 *mm²*
(5 − 7)	−2	4 *mm²*
(6 − 7)	−1	1 *mm²*
(7 − 7)	0	0 *mm²*
(7 − 7)	0	0 *mm²*
(7 − 7)	0	0 *mm²*
(9 − 7)	+2	4 *mm²*
(11 − 7)	+4	16 *mm²*
		34 *mm²* (total squared deviation = the sum of squares)

the variance and standard deviation of a population and the variance and standard deviation of a sample, Figure 11.3. *The variance of a population is called σ^2* (read as "sigma squared"). *The variance of a sample is called S^2. The standard deviation of a population is called σ (sigma). The standard deviation of a sample is sometimes abbreviated* **S** *and sometimes* **SD**. Terminology relating to populations and samples is summarized in Table 11.3.

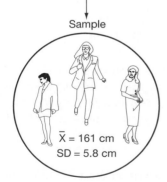

Figure 11.3. The Mean and Standard Deviation of a Population Versus a Sample. The population has a true mean and standard deviation. Every sample drawn from a population has its own mean and standard deviation which are likely to differ from the true mean and standard deviation of the population.

The formulas used to calculate the variance and standard deviation of a sample are slightly different than those for a population. In the preceding discussion we showed how to calculate the variance and standard deviation of a population. For a sample, the denominator is $n-1$ rather than n. The formulas for variance and standard deviation of a sample are thus:

$$variance\ (of\ a\ sample) =$$

$$S^2 = \frac{sum\ of\ squared\ deviations\ from\ the\ mean}{n-1}$$

$$= \frac{\sum (X - \bar{x})^2}{n-1}$$

$$standard\ deviation = SD = \sqrt{S^2}$$

$$= \sqrt{\frac{\sum (X - \bar{x})^2}{n-1}}$$

where n = the number of values

In this text we are generally looking at data from a sample and so routinely use $n-1$ in the denominator of the equation.

The standard deviation is a widely used statistic, so being able to calculate it for a given set of data is an important skill. Many scientific calculators are set up to

Table 11.3 TERMINOLOGY RELATING TO POPULATIONS AND SAMPLES

Population	Sample
An entire group of events, objects, results, or individuals where each member of the group has some unifying characteristic(s).	A subset of a population that represents the population.
Parameters:	***Statistics:***
Measures that describe a population.	Measures that describe a sample.
Mean μ	Mean $\bar{x}$
Variance σ^2	Variance S^2
Standard Deviation σ	Standard Deviation S or SD.

calculate the standard deviation easily. The operator can often enter the data and then press one key to get the standard deviation for a population and a different key to get the standard deviation for a sample. (Refer to the calculator instructions.) There is an alternative equation for standard deviation that is useful if your calculator does not have the standard deviation function built in, see Box 2. There are also statistical computer programs that are useful for analyzing large sets of data.

EXAMPLE PROBLEM

The heights for a sample of seven college women are shown in the following table. Calculate the mean ($\bar{x}$), the deviation (d) of each value from the mean, the squared deviation (d^2), the variance (S^2), and the standard deviation (SD).

Measured Value (cm)	d (cm)	d² (cm²)
162.5		
166.7		
155.6		
159.7		
163.4		
164.2		
160.1		

ANSWER

$\bar{x} = 161.74\ cm$

Measured Value (cm)	d (cm)	d² (cm²)
162.5	0.76	0.58
166.7	4.96	24.60
155.6	−6.14	37.70
159.7	−2.04	4.16
163.4	1.66	2.76
164.2	2.46	6.05
160.1	−1.64	2.69
		Summation 78.54 cm²

$n - 1 = 6$

$$S^2 = \frac{78.54\ cm^2}{6} = 13.09\ cm^2$$

$$SD = \sqrt{13.09\ cm^2} = 3.62\ cm$$

iv. THE COEFFICIENT OF VARIATION (RELATIVE STANDARD DEVIATION)

Table 11.4 shows the weights of laboratory mice and rabbits. Observe that the standard deviation for the rabbits' weights has a higher value than that for the mice, but, of course, rabbits weigh more than mice.

The dispersion of a data set can be expressed as the standard deviation divided by the mean. This is called the **coefficient of variation (CV)** or the **relative standard deviation (RSD)**. The formula for the CV is:

Table 11.4 WEIGHTS OF LABORATORY MICE AND RABBITS

Weights of Laboratory Mice (g)	Weights of Laboratory Rabbits (g)
32	3178
34	3500
24	3428
33	2908
36	2757
30	3100
36	2876
36	3369
30	3682
34	2808
$\bar{x} = 32.5\ g$	$\bar{x} = 3161\ g$
$SD = 3.75\ g$	$SD = 322.8\ g$
$CV = 11.5\%$	$CV = 10.2\%$

$$CV = \frac{standard\ deviation\ (100\%)}{mean}$$

When the dispersions of the weights of mice and rabbits are expressed in terms of the coefficient of variation, the value is slightly higher for the mouse data than it is for the rabbit data.

Note that the mean and the standard deviation have units, but the units cancel when the CV is calculated. We will come across both the standard deviation and the coefficient of variation in Unit V, where we talk about measurements.

Methods of calculating various measures of dispersion are summarized in Box 2.

EXAMPLE PROBLEM

A biotechnology company sells cultures of a particular strain of the bacterium E. coli. The bacteria are grown in batches that are freeze dried and packaged into vials. Each vial is expected to contain 200 mg of bacteria. A quality control technician tests a sample of vials from each batch and reports the mean weight and the standard deviation.

Batch Q-21 has a mean weight of 200 mg and a standard deviation of 12 mg. Batch P-34 has a mean weight of 200 mg and a standard deviation of 4 mg. Which lot appears to have been packaged in a more controlled fashion?

ANSWER

The standard deviation can be interpreted as an indication of consistency. The SD of the weights in Batch P-34 is lower than Batch Q-21. The values for Batch P-34, therefore, are less dispersed than those for Batch Q-21. The packaging of Batch P-34 appears to have been better controlled.

BOX 2   *MEASURES OF DISPERSION: THE VARIANCE, THE STANDARD DEVIATION, AND THE RELATIVE STANDARD DEVIATION*

1. To calculate the variance (S^2) for a sample, use the equation:

$$S^2 = \frac{\sum (X - \bar{x})^2}{n - 1}$$

2. To calculate the standard deviation (SD) for a sample:

$$SD = \sqrt{\frac{\sum (X - \bar{x})^2}{n - 1}}$$

There is an alternative equation that can be used to calculate SD:

$$SD = \sqrt{\frac{\sum X^2 - \frac{(\sum X)^2}{n}}{n - 1}}$$

3. To calculate the coefficient of variation (CV):

$$CV = \frac{\text{Standard deviation} (100\%)}{\text{mean}}$$

Example

Given these data from a sample:

| 1.00 *mm* | 2.00 *mm* | 2.00 *mm* | 3.00 *mm* | 3.00 *mm* | 4.00 *mm* | 4.00 *mm* | 5.00 *mm* |

$$\bar{x} = 3.00 \ mm$$

1. $S^2 = \dfrac{((1 - 3)mm)^2 + ((2 - 3)mm)^2 + ((2 - 3)mm)^2 + ((3 - 3)mm)^2 + ((3 - 3)mm)^2 + ((4 - 3)mm)^2 + ((4 - 3)mm)^2 + ((5 - 3)mm)^2}{8 - 1}$

$$= \frac{12.00 \ mm^2}{7} = 1.71 \ mm^2$$

2. The standard deviation is the square root of the variance = 1.31 *mm*

   Note that the standard deviation and the variance have units if the data have units.

   Note that using the alternate formula shown in # 2 gives the same answer:

$$\sum X^2 = 1 + 4 + 4 + 9 + 9 + 16 + 16 + 25 = 84 \ mm^2$$

$$\frac{(\sum X)^2}{n} = 1 + 2 + 2 + 3 + 3 + 4 + 4 + 5 = \frac{(24)^2}{8} = 72 \ mm^2$$

$$SD = \sqrt{\frac{84 \ mm^2 - 72 \ mm^2}{7}} = 1.31 \ mm$$

3. The coefficient of variation is:

$$CV = \frac{1.31 \ mm \times 100\%}{3.00 \ mm} = 43.67\%$$

C. Summarizing Data by the Mean and a Measure of Dispersion

It is common in a scientific paper or in other technical literature to see a set of data summarized using the mean followed by a measure of dispersion. The smaller the measure of dispersion, the closer the data points are to one another. Thus, you might encounter a mean reported in any of the following styles:

1. **A mean may be reported as the mean value ± the standard deviation**, Figure 11.4a. For example, 6.02 ± 0.23 (*SD*) means the sample mean is 6.02 and the standard deviation of the sample is 0.23.

2. **You may see a mean reported in a scientific paper as the mean value ± standard error of the mean**,

Figure 11.4b. The standard error of the mean (SEM) is not actually a measure of error; rather, it is a measure of dispersion relating to a mean. When encountered in graphs and tables, the simplest way to understand the SEM is to interpret it like the standard deviation: A larger value means there is more variability in the population than a smaller value.

3. **You may see a mean reported in a scientific paper as the mean value and the 95% (or 90% or 99%) confidence interval**, Figure 11.4c. A 95% confidence interval is a range of values that is constructed in such a manner that if 100 random samples were drawn from the population, we would expect 95 of the 100 confidence intervals to

---

EFFECT OF EXERCISE ON BLOOD PRESSURE: BLOOD PRESSURE BEFORE STARTING AN EXERCISE PROGRAM AND AFTER 16 WEEKS OF TRAINING

Variable	Baseline mean ± SD n = 43	16 weeks mean ± SD n = 43
Maximal Heart rate	153 ± 15	153 ± 11
Weight (*kg*)	97 ± 16	97 ± 17
Peak oxygen uptake (*mL/kg/min*)	21 ± 4	23 ± 4
Submaximal systolic blood pressure (*mm* Hg)	218 ± 23	187 ± 30
Submaximal diastolic pressure (*mm* Hg)	107 ± 10	94 ± 9

(a)

PHYSICAL TRAITS OF FOUR DIFFERENT SUBSPECIES OF THE WHEAT TRITICUM MONOCOCCUM

Subspecies	Number studied	Mean Value (± SEM) Seed Weight (*mg*)	Awns per spikelet
T.m. a	250	21.1 ± 0.2	2.50 ± 0.05
T.m. b	68	30.2 ± 0.7	1.00 ± 0.04
T.m. c	9	22.9 ± 0.7	1.10 ± 0.14
T.m. d	11	21.5 ± 0.9	2.33 ± 0.14

(b)

TOXICITY OF INSECTICIDE PRODUCED BY E. COLI RECOMBINANT CELLS

Insect	LC_{50} (*mg/mL*)*
Cutworm	10.9 (7.0-17.0)
Aphids	17.4 (9.7-29.2)
Beetles	147.8 (79.2-285)

*95% confidence intervals are shown in parentheses

(c)

Figure 11.4.   Examples of How the Mean is Reported in Scientific Literature. a. A study of the effect of exercise on individuals with high blood pressure. Values are reported as the mean ± the standard deviation. **b.** A study of wheat in which physical traits are reported as the mean ± the SEM. **c.** A study of insecticide toxicity on three species of insects. Results are reported as the mean and the 95% confidence interval.

include the true mean of the population. The variability of the date is considered when a confidence interval is calculated: the more variability in the data, the wider the range of the interval.

4. **You may see a mean reported graphically.** The points on the graph in Figure 11.5 represent sample means. The vertical lines are called *error bars.* The error bars may represent ± one standard deviation. Error bars may also be used to represent ± the standard error of the mean, or they may show the confidence interval. The interpretation of error bars is customarily explained in the figure caption.

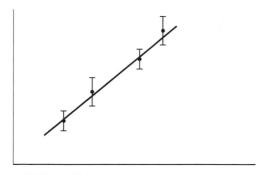

Figure 11.5.   The Representation of Sample Means Graphically; Error Bars. Each point on the line in this graph represents the mean of the observations of 20 individuals. The error bar represents ± 1 *SD.*

III.   DESCRIBING DATA: FREQUENCY DISTRIBUTIONS AND GRAPHICAL METHODS

A.   Organizing and Displaying Data

Consider a set of data consisting of the times it took 10 students to run a race (in minutes):

9.6   10.3   9.1   7.5   10.9   7.2   6.9   9.5   8.7   8.1

These values are displayed in a **dot diagram**, Figure 11.6.

Figure 11.6.   A Data Set Displayed As a Dot Diagram.

A dot diagram is a simple graphical technique used to illustrate small data sets. With a dot diagram it is easy to see the general location and center of the data. The dispersion of data is also evident in such diagrams, Figure 11.7.

Figure 11.7.   Dot Diagrams Illustrate Dispersion. a. Less dispersed data. **b.** More dispersed data.

A dot diagram is difficult to construct and interpret when a data set is large and there are other graphical methods that are more useful for organizing and displaying large sets of data. For example, suppose a researcher collects data for the weights of a number of field mice, Table 11.5. An informative way to tabulate these data is to list them by frequency where the **frequency** *is the number of times a particular value is observed.* Table 11.6 shows the values for the mouse weights and the frequency of each value.

Observe in Table 11.6 that the values for mouse weights have a pattern. Most of the mice have weights in the middle of the range, a few are heavier than average, a few are lighter, Figure 11.8. The term **distribution** *refers to the pattern of variation for a given variable.* It is very important to be aware of the patterns, or distributions, which emerge when data are organized by frequency.

The frequency distribution in Table 11.6 can be illustrated graphically as a **frequency histogram**, Figure 11.9. A frequency histogram is a graph where the X axis is the relevant units of measurement. In this example, it is weight in grams. The Y axis is the frequency of occur-

Table 11.5   THE WEIGHTS OF 175 FIELD MICE (*in grams*)

21	23	22	19	22	20	24	22	19	24	27	20	21
22	20	22	24	24	21	25	19	21	20	23	25	22
19	17	20	20	21	25	21	22	27	22	19	22	23
22	25	22	24	23	20	21	22	23	21	24	19	21
22	22	25	22	23	20	23	22	22	26	21	24	23
21	25	20	23	20	21	24	23	18	20	23	21	22
22	25	21	23	22	24	20	21	23	21	19	21	24
20	22	23	20	22	19	22	24	20	25	21	22	22
24	21	22	23	25	21	19	19	21	23	22	22	24
21	23	22	23	28	20	23	26	21	22	24	20	21
23	20	22	23	21	19	20	26	22	20	21	22	23
24	20	21	23	22	24	21	23	22	24	21	22	24
20	22	21	23	26	21	22	23	24	21	23	20	20
21	25	22	20	22	21							

Table 11.6 FREQUENCY DISTRIBUTION TABLE OF THE WEIGHTS OF FIELD MICE

Weight (g)	Frequency
17	1
18	1
19	11
20	25
21	34
22	40
23	27
24	19
25	10
26	4
27	2
28	1

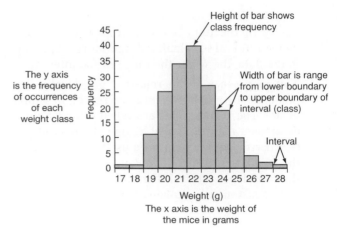

Figure 11.9. Frequency Histogram for Mouse Data.

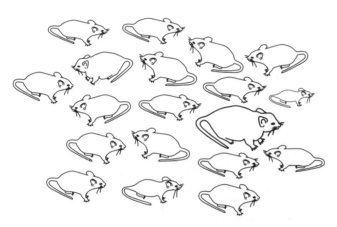

Figure 11.8. Distribution of Mouse Weights. Most mice are of about average weight: Some are a bit heavier, some a bit lighter than others. A few mice are substantially heavier, and a few mice are substantially lighter than average.

rence of a particular value. For example, 11 mice are recorded as weighing 19 g. The values for these 11 mice are illustrated as a bar. Note that when the mouse data were collected, a mouse recorded as 19 g might actually have weighed anything between 18.5 and 19.4 g. The bar in the histogram, therefore, spans an interval. In this graph the intervals are 1 g in width. You can think of the bars on the histogram as representing individual mice

piled up on the number line with each individual sitting above its score.

It is possible to construct a frequency histogram for any numerical data set consisting of the values for a single variable. The procedure for doing so is given in Box 3.

B. The Normal Frequency Distribution

The same data that are shown in Figure 11.9 are graphed in a slightly different form, as a **frequency polygon**, in Figure 11.10a. A frequency polygon does not use bars; rather, each class is represented as a single point. The point is placed halfway between the smaller and the larger limit for that class and the points are connected with lines.

Figure 11.10a shows the distribution of weights for a sample of only 175 mice. If the weights of a great many laboratory mice were measured, then it is likely that the frequency distribution would approximate a bell shape, or normal curve, Figure 11.10b. *The frequency distribution pattern that has a bell shape when graphed is called a* **normal distribution**.

There are many situations in nature and in the laboratory where the frequency distribution for a set of data approximates a normal curve, Figure 11.11. For example, the height of humans, the weight of animals, and

---

BOX 3   CONSTRUCTING A FREQUENCY HISTOGRAM

1. *Divide the range of the data into intervals, or classes.* It is simplest to make each interval the same width. There is no set rule as to how many intervals should be chosen; this will vary depending on the data. (For example, length data could be divided into intervals of: 0–9.9 *cm*, 10.0–19.9 *cm*, 20.0–29.9 *cm*, and so on.)

2. *Count the number of observations that are in each interval.* These counts are the frequencies.

3. *Prepare a frequency table showing each interval and the frequency with which a value fell into that interval.*

4. *Label the axes of a graph with the intervals on the* **X** *axis and the frequency on the* **Y** *axis.*

5. *Draw bars where the height of a bar corresponds to the frequency with which a value occurred.* Center the bars above the midpoint of the class interval. For example, if an interval is from 0 to 9 *cm*, then the bar should be centered at 4.5 *cm*.

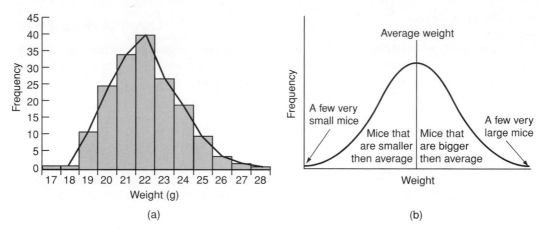

(a)   (b)

Figure 11.10.   The Frequency Distribution for Mouse Weights. a. The weights of 175 mice are graphed both as a frequency histogram and frequency polygon. **b.** If a frequency polygon were prepared for a great many mice, then the shape of the plot would approach a bell shape. The peak in the center is the average weight. Most of the mice have weights somewhere in the middle of the plot, that is, around average weight, while a few mice are substantially heavier or lighter than average. This distribution pattern is called *the normal distribution.*

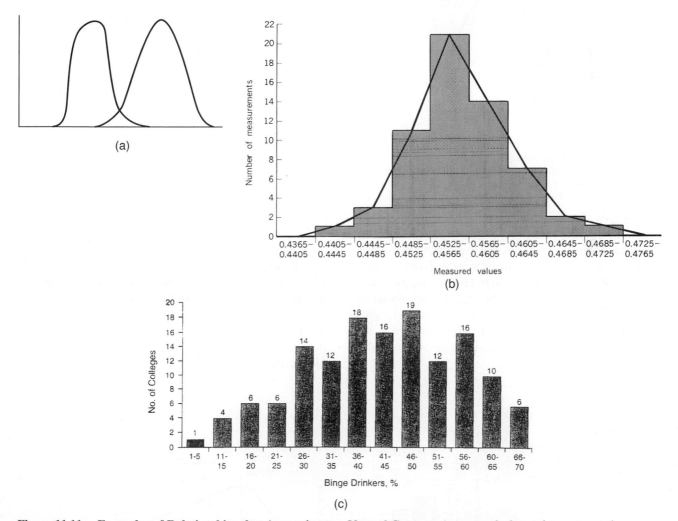

Figure 11.11.   Examples of Relationships that Approximate a Normal Curve. a. An example from chromatography, an analytical technique used to separate and detect various components in a mixture of compounds. Under ideal conditions each compound appears on the chromatographic printout as an approximately normal curve. **b.** The weights of a number of vials from the same batch. The vials vary in weight both because of variability in the manufacturing process and variability that is inherent in repeated measurements. The pattern of variation approximates a normal distribution. If more vials were weighed, then we would expect the pattern to be even closer to a normal curve. **c.** The distribution of colleges by percentage of binge drinkers. (From "Health and Behavioral Consequences of Binge Drinking in College: A National Survey at 140 Campuses." Henry Wechsler, et al. *JAMA.* 272 (21): 1672–1677. 1994; reprinted by permission.)

measurements of the same object all tend to be normally distributed.

At this point, it is possible to relate the first part of this chapter, which discussed calculated measures (such as the mean and standard deviation), to the graphical techniques in this section. The center of the peak of a perfect normal curve is the mean, median, and mode. Values are equally spread out on either side of that central high point. Moreover, the width of the normal curve is related to the standard deviation, Figure 11.12. The more dispersed the data, the higher the value for the standard deviation and the wider the normal curve.

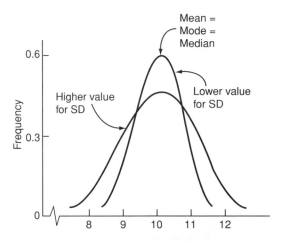

Figure 11.12.   The Normal Curve. When data are perfectly normally distributed, their mean, median, and mode all lie at the peak of the distribution plot and the standard deviation determines the width of the curve.

If we know that the pattern of variation for a variable is normally distributed, then we can make certain predictions about individuals. For example, consider a 20-year-old man, Robert Smith, who is from a large population of all 20-year-old men. Height in adults is a variable that is normally distributed. This allows us to predict that Robert is most likely about average height. He is equally likely to be either a bit taller or a bit shorter than average. He is unlikely to be substantially taller or shorter than average.

EXAMPLE PROBLEM

A student weighs himself one morning 25 times on his bathroom scale. The resulting weight values (in pounds) are shown.

a. Show these data in a frequency table and graphically.

b. Do these data appear to be approximately normally distributed?

159 159 158 161 160 158 159 157 158 160 159 158 160
159 158 161 160 159 157 159 162 161 159 158 157

ANSWER

These data can be readily summarized using six classes:

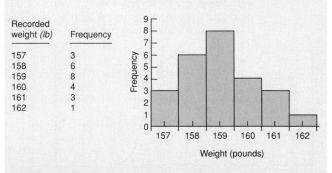

Recorded weight *(lb)*	Frequency
157	3
158	6
159	8
160	4
161	3
162	1

b. These data suggest a normal distribution. In fact, it has been found that if a great many measurements of the same item are made, and if those measurements are made properly, then there is variation in the measurements which tends to be normally distributed.

Although data are commonly normally distributed, not all data have a normal distribution. For example, the frequency distribution of a set of data may be **skewed**, *in which case the values tend to be clustered either above or below the mean*, Figure 11.13. The frequency distribution in Figure 11.14 is **bimodal**, *which means that it has two peaks*. Note that when a distribution is not normal, the mean, the median, and the mode do not coincide.

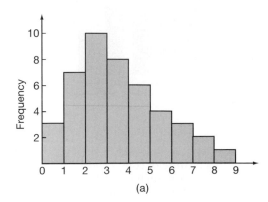

(a)

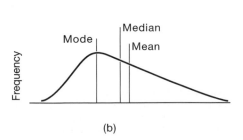

(b)

Figure 11.13.   Distributions That Are Skewed.

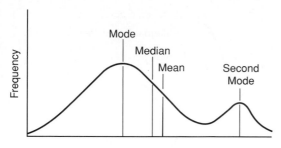

Figure 11.14. A Bimodal Distribution.

EXAMPLE PROBLEM

Multiple Choice:

What frequency distribution is illustrated in this histogram?

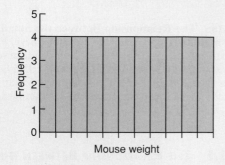

a. All the mice are of the same weight.

b. There are the same number of mice in each weight class.

c. Neither of the above.

ANSWER

b is correct. The histogram illustrates a situation where there are four mice in each weight class.

C. The Relationship Between the Normal Distribution and the Standard Deviation

The normal distribution is extremely important in interpreting and understanding data; therefore, we will now consider some of the features of this distribution in more detail.

The area under any normal curve, by definition, is equal to 100% or 1.0. The center of the peak of a normal curve is the mean. The normal curve is symmetrical; half of the area under the normal curve lies to the right of the mean, and half lies to the left of the mean, Figure 11.15.

Figure 11.16 shows four normal probability distributions. The overall pattern of the four curves is the same; each has a peak centered at the mean. Each curve is symmetrical so that half its area lies to the left of the mean and half to the right; however, the four curves are

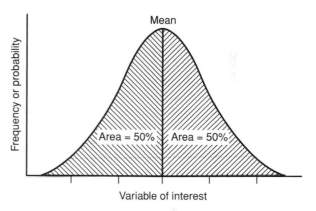

Figure 11.15. The Normal Curve. The center of the peak is the mean. The variable of interest is plotted on the X axis; either probability or frequency is on the Y axis. The normal curve is symmetrical; half its area lies to the left of the mean, half lies to the right.

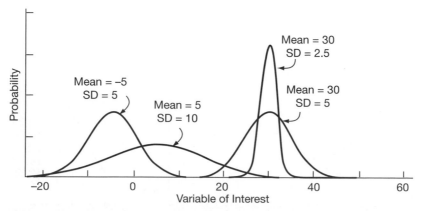

Figure 11.16. The Patterns of Four Normal Distributions Having Different Means and Different Standard Deviations. The mean locates the center of the curve along the X axis. The standard deviation determines how wide or narrow the curve is.

not identical. Three of the four distributions have different means; therefore, the curves lie at different locations along the X axis. The curves also differ in width. Those with smaller standard deviations are narrower than those with larger standard deviations. Every normal curve is thus described by two characteristics:

1. The mean, which locates the center of the curve along the X axis.

2. The standard deviation, which determines how "fat" or "thin" the curve appears.

The normal curve in Figure 11.17 shows the distribution of heights of men in the United States. Observe that this plot is "marked off" in terms of the standard deviation. The mean height for the population is about 175 *cm* and the standard deviation is about 7.6 *cm*. The mean +1 *SD* equals 175 *cm* +7.6 *cm* = 182.6 *cm*. The mean −1 *SD* is 167.4 *cm*. Observe the location of these two points on the graph. The mean height +2 *SD* is 190.2 *cm* and the mean −2 *SD* is 159.8 *cm*. Observe these values on the graph and also the location of the mean ±3 *SD*.

A graph labeled in terms of standard deviations from the mean can be constructed for any set of data that is approximately normally distributed. There is a certain pattern that is always true of such plots, Figure 11.18. The pattern is such that about 68% of the area under the normal curve lies in the section between the mean +1 *SD* and the mean −1 *SD*. About 95.5% of the total area under the curve lies in the section between the mean +2 *SD* and the mean −2 *SD*. About 99.7% of the total area under the curve lies in the section between the mean +3 *SD* and the mean −3 *SD*. The pattern shown in Figure 11.18 is consistent for any normal curve, regardless of how spread out the curve is, or the value for the mean.

The relationship between the area under a normal curve and the approximate number of standard deviations from the mean is summarized in tabular form in Table 11.7.

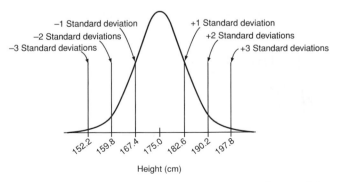

Figure 11.17.   The Normal Curve Subdivided in Terms of the Standard Deviation. The frequency distribution for the height of men in the United States in terms of the mean and the standard deviation. (The mean = 175.0 *cm*; SD = 7.6 *cm*.)

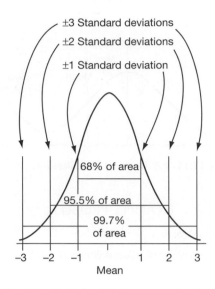

Figure 11.18.   The Relationship Between the Standard Deviation and the Area Under the Normal Curve. For any normal distribution, 68% of the area under the curve lies within ±1 standard deviation from the mean; 95.5% of the area under the curve lies within ±2 standard deviations from the mean; 99.7% of the area lies within ±3 standard deviations from the mean.

Table 11.7   THE RELATIONSHIP BETWEEN THE AREA UNDER A NORMAL CURVE AND THE APPROXIMATE NUMBER OF STANDARD DEVIATIONS FROM THE MEAN

Standard Deviations	Approximate % Area Under the Curve
±1.0	68.0
±2.0	95.5
±2.6	99.0
±3.0	99.7

EXAMPLE PROBLEM

a. What percent of the area under a perfect normal curve lies between the mean and the mean +1 standard deviation, as shown in the shaded area?

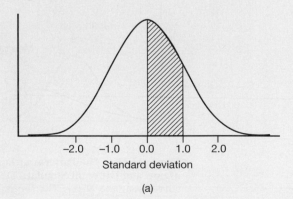

(a)

b. What percent of the area under a perfect normal curve lies in the shaded section in this graph?

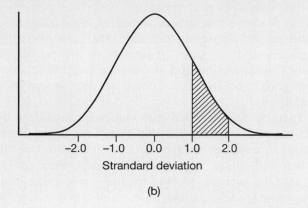

(b)

c. What percent of the area under a perfect normal curve lies in the section shown in this graph?

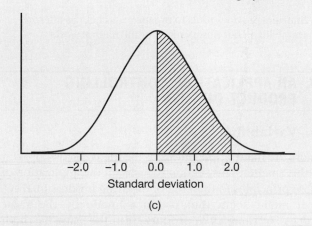

(c)

ANSWER

a. About 34% of the area lies between the mean and +1 standard deviation.

b. About 13.75% of the area lies between +1 and +2 standard deviations.

c. The area under the curve that lies between the mean and 2 standard deviation is about 47.75%.

So far, the significance of this discussion of standard deviation and the normal curve may seem obscure; however, it is actually of great practical significance.

1. *When a variable is normally distributed, the standard deviation tells us the percent of individuals whose measurements are within a certain range.* For example, height in humans is a normally distributed variable; therefore, we know that about 68% of all values measured for height will fall within 1 standard deviation on either side of the mean, Figure 11.19.

2. *When a variable is normally distributed, the standard deviation tells us the probability of obtaining a particular value if a single member of the population is randomly chosen.* For example, if a man is picked randomly there is a 68% probability that his height will be in the range of the mean ± one standard deviation.

We previously predicted that a 20-year-old man selected at random, Robert Smith, is probably about average height, might be a bit taller or shorter than average but is unlikely to be substantially taller or shorter than average. We can now be much more specific in our predictions. If Robert's name was chosen randomly, then there is about a 95.5% probability that his height is within two standard deviations on either side of the mean. The mean is 175 *cm*, so there is a 95.5% probability that he is between 159.8 *cm* and 190.2 *cm*. If we predict the height of a randomly chosen man to be between 159.8 *cm* and 190.2 *cm*, then, about 4.5% of the time, we expect to be wrong, but about 95.5% of the time we expect to be right. Thus, when data are normally distributed, knowing the standard deviation enables us to make predictions and assign them a certain probability of being correct. We can generalize to say that:

Statistics provides tools that enable us to make numerical statements and predictions in the presence of chance and variability. Statistics gives answers that are not expressed as "right" or "wrong"; rather, they are expressed in terms of probability.

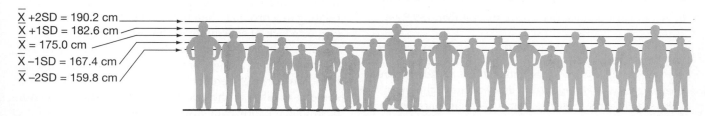

Figure 11.19. Men's Height and the Normal Distribution. About 68% of all men are of a height that falls within one standard deviation on either side of the mean. About 95.5% of men are of a height that falls within 2 *SD* on either side of the mean.

EXAMPLE PROBLEM

The average for United States women's height is about 163 cm and the standard deviation is about 6.4 cm. Assuming that women's height is normally distributed:

a. What is the probability that a woman selected at random will have a height between 156.6 cm and 169.4 cm?

b. What is the probability that a woman selected at random will have a height between 150.2 cm and 175.8 cm?

c. What percent of women have heights between 143.8 cm and 182.2 cm?

d. Show the distribution of women's heights graphically.

ANSWER:

a. This is the range of ± one standard deviation (163 cm ± 6.4 cm). We know that about 68% of all values can be expected to fall within this range, so there is a 68% chance that a woman selected at random will be in this range.

b. This is the range of ± two standard deviations (163 cm ± 12.8 cm). There is therefore a 95.5% chance that a randomly selected woman will fall in this range.

c. This is the range of ± three standard deviations (163 cm ± 19.2 cm). 99.7% of women will have heights in this range.

d.

EXAMPLE PROBLEM

The height of adult males is normally distributed with a mean of 175 cm and a standard deviation of 7.6 cm.

a. About what percent of men are taller than 190.2 cm?

b. About what percent of men are shorter than 159.8 cm?

ANSWER

We previously showed that about 95.5% of adult men have heights within two standard deviations of the mean. This means that a total of about 4.5% of men are either taller than 190.2 cm or shorter than 159.8 cm. We expect, therefore, that 2.25% of men will be taller than 190.2 cm and 2.25% of men will be shorter than 159.8 cm.

D. A Brief Summary

There is variability in nature and so there is variability in the data collected when we observe the natural world. Statistics provides methods for organizing, summarizing, and displaying such data. Measures of central tendency and measures of dispersion are calculated values that describe and summarize data. Frequency histograms and frequency polygons display variation graphically.

Patterns emerge when data values are organized by frequency. The normal distribution is a frequency distribution that is often approximated in the natural world. When values are normally distributed, they center around the average. Most individuals have values that are about average, some have values that are a bit higher or lower than average, and relatively few individuals have values that deviate greatly from the average.

Statistics provides tools to organize and help us interpret variability in our observations of the natural world.

IV. AN APPLICATION: CONTROLLING PRODUCT QUALITY

A. Variability

There are many applications of the ideas discussed so far in this chapter. This section focuses on the application of these principles to the area of ensuring product quality.

It might seem that products made by the same process, or samples of product from the same batch, or measurements of the same sample performed in the same way, ought to be identical. In fact, the goal of product quality systems [including ISO 9000 and the Good Manufacturing Practices (GMP), which govern pharmaceutical production] is to reduce variability so that products are consistently of high quality. Variability is reduced through such means as the consistent use of standard operating procedures, instrument maintenance programs, and so on. In practice, however, even when product quality systems are implemented there is always some variability in products. This is partly because there is variability inherent in any production process. In addition, measurements vary, even repeated measurements of a single characteristic of a single item. In fact, if there is no variability in a set of measurements relating to a product, then there is likely to be a malfunction or a lack of sensitivity in the system. For example, if the activity of an enzyme product is measured in 10 different vials from a single batch, or if the amount of a drug is measured in 25 tablets, then there will inevitably be some variability in the results. Product variability is not desirable; it is, however, always present.

Statistics provides important tools that can help to maintain the quality of products in the face of this inevitable variability. Statistical tools can be used to:

- determine how much variability is naturally present in a process
- determine whether there is likely a malfunction or problem in a process
- determine how certain we can be of a measurement result in the presence of variability.

Consider this example problem:

EXAMPLE PROBLEM

A technician is responsible for purifying an enzyme product made by fermentation. During the purification process she checks the amount of protein present. The results of six such measurements made during typical runs were (in grams):

23.5 31.0 24.9 26.7 25.7 21.0

One day the technician obtains a value of 15.3 g. Should she be concerned by this value? Another day the technician obtains a value of 32.1 g. Is this value a cause for concern?

ANSWER

Let us begin by looking at these values in an intuitive way. The six typical values appear to hover around about 25 g ± about 5 g. A value of 15.3 g is quite a bit lower than the other values; it is around 10 g below the approximate mean. In contrast, the value 32.1 is not as far from the mean and may therefore be less of a cause for concern.

Let us now apply statistical reasoning to this problem. The six measurements that were made during normal runs are assumed to represent values obtained when the purification process was working properly. The mean for these six values is 25.5 g and their standard deviation is 3.36 g. Based on the normal distribution, we expect 95.5% of the values to fall by chance within 2 SD of the mean. The range that includes two standard deviations is 18.8 g to 32.2 g. Thus, based on the normal distribution, the value of 15.3 g is clearly unexpected on the basis of chance alone and is therefore cause for concern. The value of 32.1 g does not immediately suggest there is a problem (although, if it signals the beginning of an upwards trend, there might in fact be a problem in the process).

Note that in this example the technician is applying statistical thinking to her everyday work. She is taking advantage of knowledge about normal distributions and about variability to identify potential problems.

B. Control Charts

There is a common graphical quality control method that extends the reasoning demonstrated in the preceding example problem. This is the use of **control charts**.

We introduce control charts with an example. A biotechnology company uses a fermentation process in which bacteria produce an enzyme that is sold for use in food processing. The pH of the bacterial broth must be controlled over time; otherwise, the pH tends to drop and production of the enzyme diminishes. A preliminary fermentation run that successfully produced enzyme was performed. The pH was monitored throughout the run and the resulting values are shown in Table 11.8.

Let us assume that this fermentation process is tested more times and that the pattern of variability illustrated in Table 11.8 is characteristic of successful runs. The mean hovers around pH 7.05 and the SD is 0.17 pH units. When the company begins to produce this enzyme for sale, the pattern of pH variability should be similar to that observed in the preliminary, successful tests. In a production setting, *a variable that has the same distribution over time is said to be* **in statistical control**, *or, is simply* **in control**. Production processes should be consistent; therefore, having "in control" processes is critical.

A control chart is a graphical representation of the variability in a process. A control chart for the data in Table 11.8 is shown in Figure 11.20. When a control chart is constructed, a line is drawn across the chart at the mean (in this case, at pH 7.05). A second line is drawn at $\bar{x} + 2\,SD$ (in this example, at 7.39). A third line is drawn on the graph at $\bar{x} - 2\,SD$ (in this example at 6.71). Two more lines are drawn at $\bar{x} + 3\,SD$ (pH 7.56 and 6.54, respectively).

The lines drawn across the control chart have names. The line at $\bar{x} - 2\,SD$ is called the **lower warning limit**, **LWL**. The line at $\bar{x} + 2\,SD$ is called the **upper warning limit**, **UWL**. The **lower control limit**, **LCL**, is $\bar{x} - 3\,SD$ and the **upper control limit**, **UCL**, is $\bar{x} + 3\,SD$.

The use of control charts is based on the important assumption that the inherent variability in production processes is approximately normally distributed. Given that this is the case, then 95.5% of all observations are expected to fall within $\pm 2\,SD$ of the mean; therefore,

Table 11.8   *The pH During Fermentation in Preliminary Studies*

Time	pH
0800	7.22
0830	7.13
0900	6.84
0930	6.93
1000	7.12
1030	7.33
1100	7.04
1130	6.79
1200	6.94
1230	7.03
1300	7.22

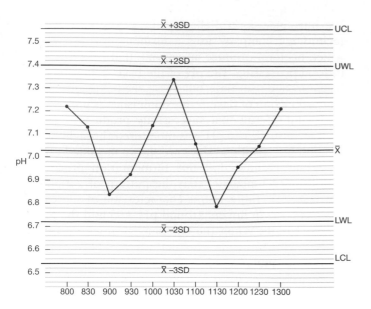

Figure 11.20.   A Control Chart for pH Values from a Fermentation Run.

the warning limits define the range in which 95.5% of all values are predicted to lie. One pH observation falling outside the warning limits is not a cause for concern because about 5% of the time values may fall outside these limits simply because of chance variability. Two or more points outside the warning limits, however, may indicate a problem in the process. The control limits define the range in which 99.7% of all points should lie. The probability that an observation, by chance, would fall outside the control lines is only 3 in 1000. It is important, therefore, to stop the process and find the problem if a value outside these action lines is obtained.

When a process is in control, its control chart has features illustrated in Figure 11.20. All the points are between the upper and lower warning limits. About half the points are above the mean and half below. It is also important that there is no trend where the points seem to be increasing in pH over time or decreasing in pH over time.

A control chart helps operators distinguish between variation that is inherent in a process and variation that is due to a problem and should be investigated. Figure 11.21 is a control chart that illustrates several problems. On days 7–10 the process is showing an upward trend. At day 10 there is a high point that is outside the con-

trol limits. At day 20 there is a point that exceeds the lower control limit. When these types of problems are observed, the process would probably be stopped and the cause(s) of the deviations would be investigated.

A point that has a value much higher or much lower than the rest of the data is called an **outlier**. In the real world, outliers may or may not signal that a process is out of control. Outliers are sometimes simply errors in how a test was performed. Errors might include, for example, improperly removing and preparing the sample, using a malfunctioning instrument, or inattentiveness on the part of the analyst. For example, an aberrant pH value in the fermentation process might indicate that the bacteria are not growing properly and that enzyme production is likely to be adversely affected. An aberrant pH reading might also be due to a malfunction in the pH meter—a testing error. In production settings the initial investigation of an outlier looks at whether it is simply due to testing error.*

Control charts are a statistical tool based on an understanding of variability and the normal distribution. Control charts allow operators to quickly distinguish between inherent, expected variability in measurement values and variation that is caused by a problem in a process. As long as the values remain within the limits of expected variability, the process is considered to be in control and variation is assumed to be due to chance alone. If a value falls outside the range predicted by chance, or if the values begin to be consistently high or low, then there may be a problem in the production process that needs to be examined. There are many types and forms of control charts other than the one illustrated

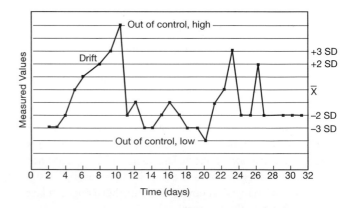

Figure 11.21.   A Control Chart Illustrating Possible Problems.

*When an aberrant or unexpected test result of a regulated product (such as a pharmaceutical product) is obtained, the company must investigate and respond. How the company must conduct its investigation and the company's response to an unexpected test result has sometimes been a cause of contention between the company and the FDA. This is because an unexpected test result may indicate that a product is defective. An unexpected test value may also indicate that an analyst made a mistake, that a piece of test equipment was malfunctioning, that a value was recorded inaccurately, or other such problems that do not relate to the quality of the product. Moreover, very high or very low measurement results occasionally occur simply by chance and are the result of normal variability in the measurement system. Companies therefore sometimes attribute unexpected test results to laboratory error or to chance when regulators are concerned that the test result might reflect a problem with a product. In at least one instance, a disagreement between a company and the FDA in how unexpected test values should be handled has been contested in the courtroom.

here. The basic purpose of a control chart is the same, however, regardless of the details of its construction.

EXAMPLE PROBLEM

The company that produces an enzyme by fermentation performs a production run. The following pH data are obtained. Prepare a control chart for these data. Is the process in control during this run? (Note that the expected pattern of variability in this process was determined in preliminary studies; therefore, use the same mean, standard deviation, UWL, LWL, UCL, and LCL as in Figure 11.20.)

Time	pH
0800	7.34
0830	7.22
0900	7.27
0930	6.88
1000	7.02
1030	7.01
1100	6.78
1130	6.88
1200	6.76
1230	6.57
1300	6.54

ANSWER

The points for this run are plotted on a control chart. Observe that the data shows a downward trend. Moreover, the final point is close to the control limit. These data strongly suggest that the process is not in control.

PRACTICE PROBLEMS

1. a. Ten vials of enzyme were chosen from a batch for quality control testing. What was the sample? What was the population?

b. The glucose level in a blood sample was measured. What was the sample? What was the population?

c. The reading ability of all first graders in Franklin Elementary School was measured. What was the sample? What was the population?

2. a. Ten measurements of a 10 g standard weight were made. Calculate the mean, standard deviation, and coefficient of variation.

10.001 g 10.000 g 10.001 g 9.999 g 9.998 g
10.000 g 10.002 g 9.999 g 10.000 g 10.001 g

b. Calculate the mean, standard deviation, and coefficient of variation for these test scores.

75% 71% 88% 99% 76% 83% 89% 91%

c. The rate of an enzyme reaction was timed in six reaction mixtures. Calculate the mean, standard deviation, and coefficient of variation.

235 sec 287 sec 198 sec
255 sec 234 sec 201 sec

d. The analyst who performed the enzyme reactions in c tried an alternative assay method and obtained the following results. Calculate the

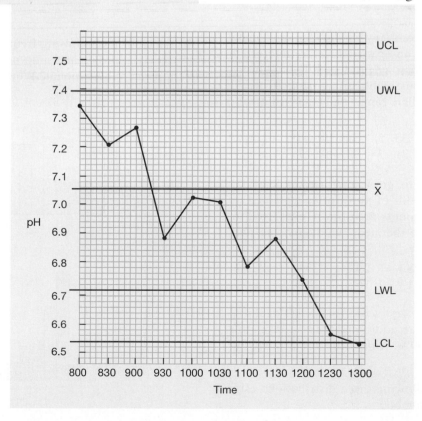

mean and the standard deviation for the alternative method. Compare the two methods.

245 sec	235 sec	256 sec	228 sec
237 sec	244 sec	234 sec	215 sec

3. (Optional: These calculations are best performed using a calculator that automatically performs statistical functions.) Calculate the mean, standard deviation, and relative standard deviation for the following data sets:

a. The weights of 33 vials from a lot were sampled. The results in milligrams are:

115 133 125 124 87 95 136 146 114 133 139
177 193 136 177 193 136 123 171 147 153 149
110 111 121 71 178 173 201 149 94 100 103

b. The weights of 25 samples were measured. The results are:

9.27 g	9.88 g	6.19 g	8.38 g	9.62 g
5.48 g	9.71 g	9.54 g	7.41 g	6.76 g
9.33 g	8.64 g	6.43 g	7.66 g	9.60 g
7.50 g	9.89 g	7.73 g	5.68 g	6.74 g
8.04 g	7.23 g	7.35 g	10.82 g	8.94 g

c. The volumes of 19 samples were measured. The results are:

1022 mL	1045 mL	1103 mL	1200 mL
1189 mL	1234 mL	1043 mL	1089 mL
1103 mL	1020 mL	1058 mL	1197 mL
1201 mL	1189 mL	1098 mL	1155 mL
1056 mL	1023 mL	1109 mL	

4. A biotechnology company is developing a new variety of fruit tree and researchers are interested in the number of blossoms produced. The following are the total number of blossoms from 20 trees. Find the mean, median, range, and standard deviation for these data:

34 36 29 27 30 35 32 31 39 30
44 30 33 43 32 21 35 33 40 27

5. A company plans to grow and market Shiitake mushrooms. The yields of mushrooms obtained in preliminary experiments are shown, expressed in kilograms. Calculate the mean, median, range, and standard deviation.

38.2	31.7	28.1	29.1	33.4	25.2
18.2	19.7	41.7	26.2	21.0	52.2

6. The special average test score for a class of thirty students is 75%. One additional student takes the exam and scores 100%. What is the new average for the class?

7. For each of the following three distributions, estimate the mean and mode.

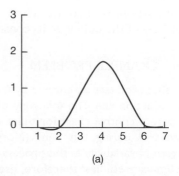

(a)

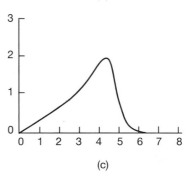

(b)

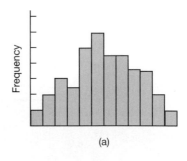

(c)

8. Which of the following frequency histograms most closely approximates a normal distribution? Which appears to be bimodal? Which appears skewed?

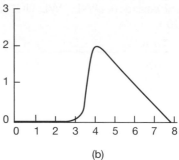

(a)

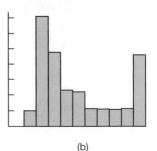

(b)

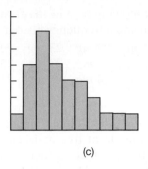

(c)

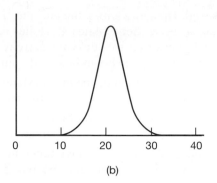

(b)

9. Which of these two distributions is less dispersed?

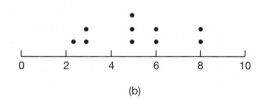

(a)

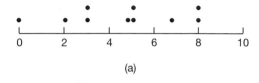

(b)

10. Which of these two distributions is less dispersed?

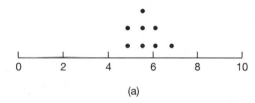

(a)

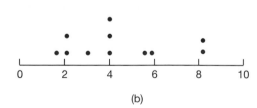

(b)

11. Which of these two distributions is less dispersed?

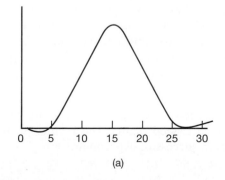

(a)

12. Which of these two distributions is less dispersed?

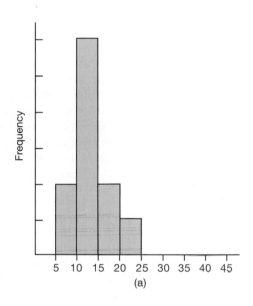

(a)

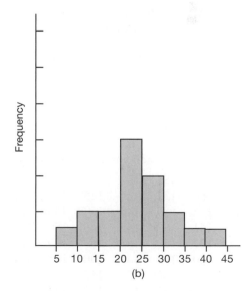

(b)

13. Cells in culture are treated in such a way that they are expected to take up a fragment of DNA containing a gene that codes for an enzyme. The activity of the enzyme in cells that take up the gene can be assayed. The more active the enzyme, the better. Suppose a researcher isolates 45 clones of treated cells and measures the enzyme activity in each clone. The results are shown in activity units.

a. Find the range, median, mean, and standard deviation for the data from these 45 treated clones.

b. Plot these data on a histogram and show on the plot where the mean and median are located.

c. Do you think these cells have taken up the gene fragment containing the enzyme, based on these data? Explain.

10.4	12.2	12.0	9.1	5.8	3.2	9.8	10.1	13.0
2.1	9.8	10.1	1.5	7.8	5.6	2.3	9.8	10.2
9.1	12.3	10.1	12.3	14.2	15.1	13.6	12.1	10.8
9.2	8.9	12.4	11.0	13.1	8.9	9.4	10.2	11.3
12.8	9.0	8.6	12.3	12.0	2.1	0.4	10.6	13.0

14. Suppose that the assay for enzyme activity (discussed in question #13) is performed on cultured cells that have not been treated with the DNA fragment and are not expected to have any enzyme activity at all. In this case, one would expect to see little or no activity when performing the assay. You test this assumption on 20 untreated cell clones that are not expected to have the enzyme and obtain the following results in activity units:

0.0	1.0	0.3	3.0	2.8	1.2	2.3	1.1	0.9
2.5	1.9	0.9	2.2	1.2	0.2	0.1	0.2	0.9
1.0	2.0							

a. Calculate the mean and the standard deviation for these results.

b. Prepare a histogram for these results.

c. Discuss the results shown in Question 13 in light of these data in Question 14.

15. The following are heights in centimeters of a common prairie aster measured from one field.

a. Determine the range and mean for these data.

b. Plot these data as a frequency histogram. (You will need to divide the data into groups. After finding the range, make 10–12 divisions of equal size and assign each value to the proper class.)

151	182	182	162	177	166	197	144
174	160	131	156	125	170	153	172
146	127	156	140	159	155	158	165
155	141	180	145	145	150	145	135
105	122	180	152	161	170	156	150
122	140	133	145	190	165	176	170
144	160	162	157	155	146	155	143
154	157	141	150	142	139	138	156
130	126	189	138	103	163	135	158
151	129	147	153	150	140	146	138
154	138	179	165	142	140	141	160
125	156	145	159	147	155	189	195
140	141	160	156				

16. Show the following data in the form of a frequency distribution and a frequency histogram and polygon.

23 *mg*	28 *mg*	22 *mg*	23 *mg*	20 *mg*	19 *mg*
22 *mg*	24 *mg*	26 *mg*	23 *mg*	23 *mg*	24 *mg*
21 *mg*	25 *mg*	24 *mg*	25 *mg*	21 *mg*	

17. As a trouser manufacturer, you are interested in the average height of adult males in your town. You take a random sample of five male customers and find the mean to be 5 feet 10 inches and the standard deviation to be 3 *in*. How certain are you of the following statements and why?

a. The average height of the population is between 0 and 50 *ft*.

b. The average height of the population is between 4 *ft* and 7.5 *ft*.

c. The next new customer will be between 5 *ft* 7 *in* and 6 *ft* 1 *in*.

d. The average height of the entire male population is 5 *ft* 10 *in*.

18. Refer to the Example Problem on p. 200 regarding women's heights.

a. What is the probability that a woman selected at random will be shorter than 150.2 *cm*?

b. What is the probability that a woman selected at random will be taller than 169.4 *cm*?

19. A pharmaceutical company finds that the average concentration of drug in each vial is 110 μg/vial with a standard deviation of 6.1 μg.

a. About what percent of all vials can be expected to have between 94.1 μg and 125.9 μg of drug?

b. What is the likelihood that a vial selected at random will have more than 122.2 μg of drug?

c. What percent of all the vials are expected to have a values between 103.9 μg and 116.1 μg?

d. Suppose 100 vials are checked. About how many of them would be predicted to have more than 125.9 μg?

20. A technician customarily performs a certain assay. The results of 8 typical assays are:

32.0 *mg*	28.9 *mg*	23.4 *mg*	30.7 *mg*
23.6 *mg*	21.5 *mg*	29.8 *mg*	27.4 *mg*

a. If the technician obtains a value of 18.1 *mg*, should he be concerned? Base your answer on estimation without performing actual calculations.

b. Perform statistical calculations to determine whether the value 18.1 *mg* is out of the range of two standard deviations.

21. A technician customarily counts the number of leaves on cloned plants. The results of nine such counts in successful experiments are:

75	54	55	61	71	67	51	77	71

a. If the technician obtains a count of 79, is this a cause for concern? Base your answer on estimation without performing actual calculations.

b. Perform statistical calculations to determine whether 79 leaves is out of the range of two standard deviations.

22. Examine this control chart. Discuss the points on March 22, April 15, May 31, and June 23.

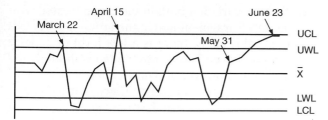

23. Suppose a biotechnology company is using bacteria to produce an antibiotic. During production of the drug, the pH of the bacterial broth must be adjusted so that it remains optimal; otherwise, production of the antibiotic diminishes. Preliminary studies were performed to determine the optimal pH. The results of a successful preliminary study are shown in the following table.

Time	pH
0800	6.12
0830	5.13
0900	5.84
0930	6.53
1000	6.12
1030	6.23
1100	6.04
1130	5.79
1200	5.94
1230	6.03
1300	6.12

a. Calculate the mean pH and the SD.

b. Construct a control chart with a central line, UWL, LWL, UCL, and LCL.

The enzyme goes into production and periodic samples of the broth are assayed for pH. The results are shown in the following table.

Time	pH
0800	6.54
0830	6.12
0900	5.87
0930	5.18
1000	4.95
1030	4.89
1100	5.03
1130	5.43
1200	5.34
1230	5.37
1300	5.38

c. By simply observing these points, what can you say about the process?

d. Plot these data on the control chart.

e. Comment on the process; did it ever reach the warning level or action required levels?

ANSWERS TO PRACTICE PROBLEMS

1. a. Sample: The ten vials. Population: All the vials in the batch.

b. Sample: The blood sample tested. Population: All the blood in that individual.

c. The sample is the first graders who were tested. The population depends on the purpose of the study. If the study was intended to test the performance of all first graders in Franklin Elementary School, then the entire population was studied. If the sample was meant to represent all first graders in a city or in a larger group, then the population is that larger group.

2. a. $\bar{x} = 10.000\,g$ $SD = 0.001\,g$ $CV = 0.012\%$

b. $\bar{x} = 84\%$ $SD = 9.49\%$ $CV = 11.29\%$

c. $\bar{x} = 235\,sec$ $SD = 33.56\,sec$ $CV = 14.28\%$

d. $\bar{x} = 237\,sec$ $SD = 12.26\,sec$ $CV = 5.18\%$

The means for the two methods are similar; however, the second method is more consistent. Based on this information, the second method seems better.

3. a. $n = 33$ $\bar{x} = 136.76\,mg$
    $SD = 32.90\,mg$ $CV = 24.06\%$

b. $n = 25$ $\bar{x} = 8.15\,g$
    $SD = 1.47\,g$ $CV = 18.09\%$

c. $n = 19$ $\bar{x} = 1112.32\,mL$
    $SD = 71.55\,mL$ $CV = 6.43\%$

4. a. $n = 20$ $\bar{x} = 33.05$ blossoms
    median = 32.5 blossoms
    range = 21–44 = 23 blossoms
    $SD = 5.57$ blossoms

5. $n = 12$ $\bar{x} = 30.39\,kg$ median = 28.6 kg
    range = 18.2 − 52.2 = 34 kg $SD = 9.90\,kg$

6. $\dfrac{\sum X}{30} = 75\%$ $\sum X = 2250\%$

2250% + 100% = new total

$\dfrac{2350\%}{31} = 75.8\%$

7. a. mean = 4, mode = 4

b. mean is approximately 5, mode is approximately 4

c. mean is approximately 3, mode is approximately 4

8. a. normal   **b.** bimodal   **c.** skewed

9. b is less dispersed

10. a is less dispersed

11. about the same

12. a is less dispersed

13. a. $n = 45$ $\bar{x} = 9.54$ activity units
    $SD = 3.60$ activity units
    range $= 0.4 - 15.1$ activity units
            $= 14.7$ activity units
    median $= 10.1$ activity units

b.

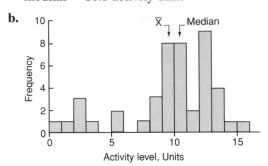

Frequency Table

Interval	Frequency
0–0.9	1
1.0–1.9	1
2.0–2.9	3
3.0–3.9	1
4.0–4.9	0
5.0–5.9	2
6.0–6.9	0
7.0–7.9	1
8.0–8.9	3
9.0–9.9	8
10.0–10.9	8
11.0–11.9	2
12.0–12.9	9
13.0–13.9	4
14.0–14.9	1
15.0–15.9	1

c. Based on just this information, we cannot tell whether the DNA fragment was taken up or not because we do not know what level of enzyme activity exists in cells that have not been treated with the fragment.

14. a. $n = 20$ $\bar{x} = 1.29$ activity units
    $SD = 0.93$ activity units

b.

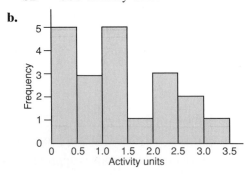

Interval	Frequency
0–0.4	5
0.5–0.9	3
1.0–1.4	5
1.5–1.9	1
2.0–2.4	3
2.5–2.9	2
3.0–3.4	1

c. There is variation both in the group treated with the DNA fragment and in the group not treated. Using this assay, there is a low level of activity exhibited even in cells not treated with the DNA fragment; however, the mean enzyme activity of the two groups is different. There is also the suggestion graphically that the clones that were treated with the DNA fell into two categories: (1) those with higher levels of activity and (2) those with lower levels, similar to clones that were not treated with the DNA. Based on these observations, it appears that some of the treated cells took up the DNA and some did not. This hypothesis could be investigated by further study.

15. $n = 100$ $\bar{x} = 152.2 \; cm$
    range $= 103–197 \; cm = 94 \; cm$

Frequency Distribution Table with data divided into 11 intervals:

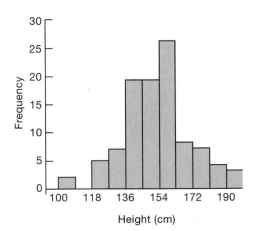

Interval *(cm)*	Frequency
100–108	2
109–117	0
118–126	5
127–135	7
136–144	19
145–153	19
154–162	26
163–171	8
172–180	7
181–189	4
190–198	3

16.

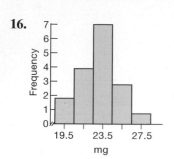 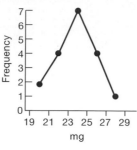

Interval (*mg*)	Frequency
19–20	2
21–22	4
23–24	7
25–26	3
27–28	1

17. a, b. Very certain. Common sense is sufficient to know this.

c. Based on the standard deviation of the sample, we would expect 68% of customers to be in this range; however, a sample of five is too small to draw firm conclusions.

d. Not very certain, a sample size of five is too small to draw firm conclusions.

18. a. 2.25%   **b.** 16%

19. a. 99% (see Table 11.7)   **b.** 2.25%

c. 68%   **d.** 0.5%

20. a. The average appears to be in the midtwenties and hovers at around ± 5; therefore, 18.1 *mg* appears a little low.

b. Mean = 27.16 *mg*, SD = 3.87 *mg*. The mean −2 SD is 19.4; therefore, 18.1 *mg* is outside the range of 2 SD and probably should be investigated.

21. a. The values appear to hover around 65 leaves give or take about 10. It is difficult to tell whether 79 is a cause for concern.

b. Mean = 64.7 leaves, SD = 9.7 leaves. The mean + 2 SD = 84.1; therefore, 79 does not appear to be unreasonable.

22. On several occasions, such as on March 22, the points lie outside the warning range; however, because later points are within the expected range, this is probably due to normal variation. The point on April 15 is out of the control range and therefore the process should be stopped and checked for problems. The points between May 31 and June 23 display an upward trend. A trend in one direction or another is also cause for concern.

23. a. Mean = 5.99 pH units, SD = 0.35 pH units

b.

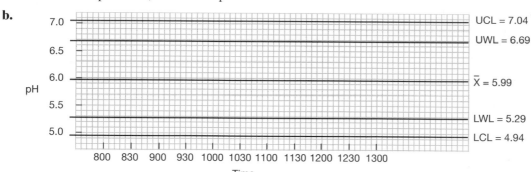

c. The pH values tend to decline with time.

d.

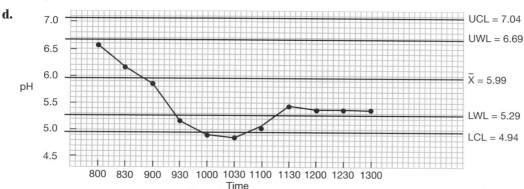

e. This process tended downward from the beginning, eventually reaching the lower control limit.

CHAPTER 12

The Collection of Meaningful Data; Avoiding Errors in the Laboratory

I. INTRODUCTION

A. The Importance of Collecting Meaningful Data

Suppose a not very savvy, alien science crew comes to earth to investigate humans. The aliens are interested in the ratio of males to females among human children. They happen to select a convenient group of families to study, all of whom are gathered in one location—for a Boy Scout camp reunion. They carefully count the number of boys and the number of girls, perform a statistical analysis, and report to their home planet that 76.23% of all human children are male. These aliens made a sampling error and so reported an erroneous conclusion based on a careful analysis of poor data.

The previous chapter in this unit explored statistical methods to summarize and display data. Regardless of how data are displayed, organized, or later interpreted, data analysis is based on the assumption that the data were properly collected. The alien scientists based their conclusion on the assumption that the sample of families they observed was representative of all human families. Unfortunately for the aliens, their sample was not representative of the population of interest. Therefore, the aliens' conclusions were incorrect.

We are unlikely to make the same mistake regarding sex ratios that was made by the aliens; however, there are many ways to make mistakes when collecting data. Like aliens from another planet, scientists investigate questions whose answers are unknown, and technologists often work with unfamiliar, complex biological systems. When our knowledge of a system is incomplete, it is easy to mistakenly collect data that are meaningless. This chapter discusses some of the pitfalls in data collection and some strategies to avoid them.

B. Random and Systematic Variability

If a coin is flipped 10 times there are various outcomes, such as seven tails and three heads, or nine tails and one head. Assuming a properly flipped fair coin, however, on the average, after many flips, heads will come up half the time. When observations vary in a random fashion, they are sometimes lower than we expect, sometimes higher, but they tend to average a predictable answer. If we were to randomly select a large number of families with children, then we would similarly find that in some families there would be more than half boys, whereas in other families there would be more than half girls, but, in general, the percent of boys and girls would each be close to 50%.

Later in this text we will see that whenever measurements are made, such as weighing an object or measuring a volume, there is some variability (even if the measurements are made correctly). Measurements are sometimes a bit higher than the correct value, and sometimes a bit lower. When we are discussing measurements, this variability is called *random error*. Although the word *error* is used, in the presence of only random variability, the average measurement result will tend to be correct.

Statistics provides powerful tools to analyze data in the presence of random variability; unfortunately, there is another source of variability in addition to random variability. This is systematic variability. **Systematic variability** *causes a series of observations to tend to be too high or too low, that is, to be* **biased**. When an experiment is being conducted, variation that causes results to erroneously be too high or too low is often called **experimental error** or **systematic error**. The term *systematic error* is used in measurements to refer to any circumstance that causes a series of measurements to tend to be too high or too low. (We will discuss random and systematic error in measurements in more detail in Unit V.)

The alien scientists investigating sex ratios in humans made a systematic error: they chose a sample of families that tended to have more boys than girls. This error caused their observations to be skewed or biased toward too many boys. Because of this error, their data were useless for answering their original question. There are many other types of mistakes that will also cause data to be biased. For example, a malfunctioning instrument, solutions that are made improperly, and errors in recording results also cause systematic error.

In the presence of systematic (experimental) error, data may be difficult or impossible to interpret. It is therefore necessary to avoid systematic errors whenever possible. An important role of the technician and bench scientist is in preventing such error by assuring that laboratory work is carefully planned and properly performed.

There are principles to guide the collection of meaningful, "good" data. In this chapter, we point out some common causes of systematic error and review basic guidelines to help avoid these errors. We cannot in this text fully explore the vast area of data collection and experimentation; however, we will highlight some common practices with which every laboratory scientist should be familiar.

C. The Human Factor

Humans are involved in the collection of data. Even if a computer is attached to an instrument which automatically makes recordings, humans programmed the computer and designed the instrument for a particular function. Humans often make observations inconsistently and/or incorrectly because of blunders, lack of knowledge, faulty expectations, and legitimate differences in judgement.

Raymond Pearl published an interesting experiment in 1911 ("The Personal Equation in Breeding Experiments Involving Certain Characters of Maize," Raymond Pearl. *Biological Bulletin*, 21:339–366. November 1911. This work is also discussed in "Solving Your Personal Equation: An Example from Genetics Studies," Lawrence C. Davis. *The American Biology Teacher*, 58(2): 90–92. February, 1996). Pearl was studying the genetics of corn. He was looking at two traits, one of which causes kernels to be yellow (as opposed to white), the other causes them to be shrunken (as opposed to "dent"). An individual kernel could be either yellow and dent, yellow and shrunken, white and dent, or white and shrunken. Pearl knew that the ratios of the four types of kernels should be consistent with Mendel's Laws of genetics.

Pearl had 15 experts sort and count corn kernels into the four classes. He was surprised to find that the 15 experts counted the kernels differently. For example, one expert counted 25 white/shrunken kernels in a particular ear of corn, but a second expert, looking at the same ear of corn, counted 79 white/shrunken kernels. In general, the counters got different results from one another. In fact, some of the counters got results that violated the principles elucidated by Mendel. This variation between different experts was puzzling.

Pearl discovered on closer analysis that the corn results did not actually contradict Mendelian genetics; however, some of the kernels were difficult to classify because their color was intermediate between yellow and white and their shape was not clearly shrunken or dent. The experts counting the kernels each decided how to classify the ambiguous kernels and each one used a different but consistent strategy; therefore, the counters got different results from one another. The differences in their strategies systematically gave each expert results that were higher or lower than the other counters. This is a good example of how human judgment can lead to systematic variability in data collection.

Pearl's study demonstrates how variability can arise from differences in judgment. Newspapers provide many examples of judgment differences between experts. For example, experts looking at the same evidence might disagree as to whether a low-salt diet is likely to reduce one's chance of developing heart disease, or they might argue about the validity of using DNA fingerprinting evidence in the courtroom.

Another important point is illustrated by the results of Pearl's corn study. Pearl suggested that some of the experts, when classifying ambiguous kernels, made decisions based on the results they expected to obtain. This suggests that observers often see what they expect to see, particularly if presented with ambiguous information. People often intend to be unbiased, yet, subconsciously make decisions that are based on predetermined expectations. There is even speculation that Mendel may have discarded results that did not correspond to his preconceived notions of what should have happened (for example, see the discussion of Mendel in *Statistics*, David Freedman, Robert Pisani, Roger Purves, and Ani Adhikari. 2nd edition. Norton, New York, 1991). Mendel came up with the right answers so most people are not likely to be alarmed if he was biased. In fact, Mendel might never have made the important contributions he did if he had no preconceptions to guide his studies. However, observing what one expects to observe, whether or not it is true, can also be disastrous, as the example on pp. 76 clearly illustrates.

The fact that human observers introduce bias into observations is well known, and this bias should be avoided whenever possible. Many of the strategies used in experimentation are designed to avoid human bias and to eliminate situations where differences in judgment between individuals might impact results.

II. ISSUES RELATING TO SAMPLES

A. Obtaining Representative Samples

Much laboratory work relates to obtaining, handling, processing, testing, and storing samples. A sample represents the population from which it is drawn. A group of patients tested with an experimental drug represents the population of all similar patients. In histology, a thin section of tissue is excised and prepared for observation under a microscope. Based on that section, conclusions may be reached about the tissue from which the section was taken. For example, a pathologist may decide whether the tissue shows signs of malignancy. In this case, the entire tissue may be considered as the population of interest, and the tissue section is the sample. (Note that statistical calculations are seldom involved in histological investigations, yet the idea of a sample representing a larger population still applies.) Consider an investigator who is interested in knowing the catalytic activity of an enzyme in a solution. Suppose the solution has a total volume of $3 \, mL$ and the analyst removes $100 \, \mu L$ to assay enzyme activity. The $100 \, \mu L$ is a sample that represents the entire $3 \, mL$ of solution.

Populations are comprised of individuals that vary from one another, so there is always uncertainty when a sample is used to represent a whole population. The field of statistics provides many tools to mathematically handle that uncertainty. However, of great importance to the laboratory scientist, the way that samples are obtained and handled will determine the quality of results that can be obtained from those samples. There is little point in analyzing data that were obtained from a poorly chosen or improperly prepared sample.

We introduced one of the common mistakes relating to sampling in the example of the hapless aliens. A representative sample should be drawn randomly (i.e., every member of the population should have an equal chance of being selected). If a sample is not drawn randomly, then the observations based on that sample are likely to be higher or lower than they should be (i.e., to be skewed in a particular direction). The population of interest to the aliens was all human children; however, their sample included only families with sons in the Boy Scouts, so their results were biased toward more boys.

It is frequently difficult to devise a strategy to obtain a representative sample. The population or the site from which the sample is drawn is often heterogeneous. For example, a classic problem in biological experimentation is how to sample an experimental agricultural plot. A field is likely to be heterogeneous with respect to its topography, soil moisture, soil type, sunlight, insects, and a host of other factors. Suppose that scientists who are performing field tests of a new strain of corn go to the field and collect a sample consisting of plants located near their parked car. These corn plants probably experienced different conditions than did plants elsewhere in the field, so they are not a representative sample of the entire corn crop.

Heterogeneity in biological systems is common and samples are typically drawn from sites or from populations characterized by variability. For example, the population of all humans is heterogeneous with respect to many variables. A fermenter is heterogenous over time. As a fermentation process occurs, the number of cells changes, the chemical composition of the broth changes, the pH changes, and so on. Multiple samples must therefore be taken over a period of time to monitor a fermentation process adequately.

Another issue in sampling is that humans usually have preferences when sampling. For example, a person casually sampling an experimental field might take only plants that are easily accessible. Even when people try not to be biased, they unknowingly have preferences. For example, when people try to list a series of numbers with no inclination for particular numbers, there still are patterns in their choices because of subconscious preferences.

Thus, obtaining a random sample is often difficult because the population or site from which the sample is taken may be heterogeneous and because people tend to have unconscious preferences. Obtaining a representative sample requires understanding the heterogeneity of the system and then devising an unbiased sampling strategy. Statisticians may be consulted to develop a sampling strategy when large and complex studies are performed. In a smaller or simpler study certain basic sampling guidelines are often employed, Box 1.

BOX 1   AVOIDING SYSTEMATIC ERRORS IN SAMPLING

1. *Pick the individuals in a sample randomly.*

   a. *Devise a numbering system to assign an identification number to all the potential members of the sample.* For example, if a sample is to be 40 students chosen at random from all the students in a college, then the students' college ID numbers can be used.

   b. *Throw a die or pick numbers from a bowl to choose the individuals who will actually be in the sample.*

   c. *As an alternative to picking numbers from a bowl or throwing a die, refer to a random number table.* A random number table is a list of numbers with no preferences for certain numbers or patterns. Close your eyes and point anywhere in the table to choose the starting point. Then, read off numbers in the table.

   d. *Computer programs that generate random numbers can be used to pick members of a sample.* Computers are particularly helpful when a sample is large.

2. *A method to randomly sample an area:*

   a. *Draw or map the area on a piece of paper.*

   b. *Place a grid with horizontal and vertical lines over the map.*

   c. *Number the boxes formed by the grid.*

   d. *Use a random number table to choose grid boxes.* Sample the areas under the chosen grid boxes.

3. *Methods to sample biological molecules and solutions in the laboratory:*

   a. *Be certain liquids are well-mixed before withdrawing a sample.*

   b. *Check liquids for precipitates before removing a sample.*

   c. *When a liquid is removed from a freezer, be sure all the ice is melted before removing a sample.* Different components of the liquid may thaw at different rates.

   d. *When taking a sample from a suspension, such as a bacterial culture, make sure the suspension is evenly distributed.*

   e. *When sampling a powder, either mix it or take a random part of the solid.*

EXAMPLE PROBLEM

Suppose a student is investigating the height of college students. She measures the height of every student in her classes. a. What is the population being studied? b. What is the sample? c. Is the sample likely to be truly representative of the population?

ANSWER

The population of interest is all college students. Her sample is convenient, but it is not random and it is probably not representative of all college students. It is possible, for example, that the student body at her college tends to have traits that are characteristic of its geographical region but are not characteristic of the entire world. Sampling the student's own college classes, therefore, does not give a random sample of all college students. Note also that the student's sample is possibly not even representative of the students in her own college. This is because the students in her classes may differ in some way from students in the college as a whole. For example, if this student takes only engineering classes, she may find that her classmates tend to differ from the students taking art classes. Box 1 provides an example of a sampling scheme to randomly sample students in a college.

B. Sample Preparation

Another key issue relating to biological samples is that they frequently require extensive preparation before they can be tested or used. These preparations may greatly affect the results of a measurement or an assay. For example, prior to electron microscopy, samples are chemically preserved, dehydrated with alcohol, embedded in plastic, thinly sliced into sections, and stained with heavy metals. The sections are then viewed with the electron microscope. The structures that are visualized depend on the way that the sections were prepared. In molecular biology, cellular extracts might be treated with harsh chemicals, centrifuged, stored, and otherwise

manipulated. There is an example on pp. 503–504 of this text of a study in which investigators were comparing cells that contained an oncogene (a cancer-causing gene) with cells that did not have the oncogene. The scientists found that their results varied depending on how the cells were prepared. The effects of sample preparation thus had the potential to obscure the effects due to the oncogene.

It is not uncommon during sample preparation that certain components of a sample will be selectively lost whereas others are retained. Structures may similarly be altered and activities, such as enzymatic activity, may be lost. All such alterations in the samples are causes of systematic error. Thus, sample preparation is a major issue in the analysis of biological samples. It is often necessary to perform preliminary investigations of the effects of sample preparation before conducting assays or experiments. A few strategies to deal with sample preparation problems are summarized in Box 2.

III. ISSUES RELATING TO EXPERIMENTAL SUBJECTS

A. Assigning Individuals to Either a Control or Experimental Group

In an experiment researchers often randomly select a sample of subjects from a population and then split the sample into groups. One or more groups are subjected to the experimental treatment. The other group is not subjected to the experimental treatment; this is called the *control group*. The experimental (treatment) group is compared with the control group. The purpose of a control group is to ensure that any effects that are observed are due to the experimental treatment of interest and not to some other factor. This section considers issues relating to the conduct of such an experiment in which a treatment group is compared with a control group.

Box 2 *AVOIDING SYSTEMATIC ERRORS IN SAMPLE PREPARATION*

1. *Prepare all the samples in a study identically.*
   a. If there are samples from experimental and control groups, be certain that they are treated in the same way.
   b. Prepare written procedures for sample handling that are followed by all analysts.
   c. Devise methods to ensure that all reagents, materials, and equipment used in sample preparation are consistent.

2. *Test for the loss of specific components.* This can often be accomplished by adding a known amount of the material of interest to a sample. (This is called "spiking" the sample.) The sample is then taken through the sample preparation steps. The amount of the material of interest remaining after sample preparation is compared with the amount initially added.

3. *Test different sample preparation procedures.* It is good practice to experiment with different sample preparation methods to determine their effects on the results.

One of the first issues is how to assign individuals to groups. The way in which this assignment is performed is critical. For example, if one is going to perform a study on the effect of a carcinogenic chemical on rats, then each rat must be equally likely to be assigned to the control or experimental group. This is a situation where random number tables can be applied. Each rat is arbitrarily assigned a number, and random numbers are then used to assign the rats to a group.

It is comparatively straightforward to assign plants or rats to groups in a random fashion. It is far more difficult to assign humans randomly. Consider an experiment that involves testing the efficacy of a new drug. The drug is to be tested by administering it to one set of patients and not to another. Much has been written on this topic because the outcomes of important, and very expensive, medical experiments may hinge on whether the assignment of individuals to groups was made properly.

Consider some of the difficulties in this assignment. A physician might suspect that a drug will be effective and might then be tempted (perhaps unconsciously) to assign to the experimental group those patients s/he thinks would benefit most from the treatment, or patients who are younger, who have children, and so on. Such nonrandom assignment will tend to make the treated experimental and untreated control groups different from one another before the experiment begins. Nonrandom assignment of subjects causes systematic error. If bias occurs in the assignment of individuals to groups, then differences in the two groups after treatment may be due to the treatment; however, they may also be due to initial differences between the groups.

Another problem in studies involving humans is that humans vary from one another. The individuals in a study are likely to vary in age, sex, education, previous health history, smoking history, and so on. This variation can again result in groups that differ from one another before an experiment begins, making it difficult to detect the effects of an experimental drug or other treatment. For this reason, some studies used **matched or paired subjects. Matched subjects** *are two individuals who are matched in as many ways as possible.* This includes pairing each individual with a corresponding individual of the same age, sex, smoking history, and so on. One of the pair is randomly assigned to receive the experimental treatment, the other is the control. Data are collected for each pair and the effects of treatment are determined.

CASE STUDY I

The Effectiveness of a Flu Vaccine

Influenza (flu) is a common cause of illness and deaths annually. Persons at high risk for complications from influenza are cur-

rently targeted to receive an annual flu vaccination. A research study was performed to see whether vaccination against flu also provides benefits for healthy, working adults who are not at high risk of complications.

There were 849 subjects recruited into the study. The subjects were assigned randomly to one of two groups using a computer-generated randomization scheme. The treatment group received a flu vaccination. The control, placebo group, received an injection as well, but it contained only the buffer in which the vaccine was diluted. The subjects did not know whether they received the real vaccine or the placebo because this knowledge might have affected their response. The health-care professionals administering the vaccine and monitoring the subjects similarly did not know who received the vaccine and who received the placebo so as to avoid bias in their reporting and/or treatment of subjects. Vaccines and placebos were available in preloaded syringes that were identical in appearance and were labeled with a code. This is called a double-blind study: Neither subjects nor health care providers knew who was in which group. Observe that the assignment of individuals to groups and the administration of the vaccine included many safeguards against human bias.

Once the assignment of subjects to either the vaccine or placebo group was made, the investigators collected basic information about each subject and used a statistical analysis to compare the two groups. The purpose of this study was to ascertain whether there were differences between the vaccine and placebo group before the experiment began. A difference between the two groups before the administration of the vaccine or placebo would be a potential source of bias. The investigators found that the treatment and control groups were not significantly different in the measured characteristics prior to the experiment.

After the vaccine or placebo was administered to the subjects, the investigators monitored upper respiratory illness in the two groups for the duration of the flu season. Individuals in both groups did contract upper respiratory illness in the winter following vaccination; however, the rate of illness was lower for the treated group than for the placebo group. In addition, the treated group missed fewer days of work due to upper respiratory illness. The investigators used statistical tools to determine whether the differences in illness observed between the vaccine and placebo groups were likely to have been due to the vaccine or might simply have been the result of normal variability in illness rates. On the basis of their data, the investigators concluded that vaccination has statistically significant health benefits.

Observe that the investigators incorporated a number of safeguards into their experiment (e.g., the use of a computer to randomly assign subjects to the vaccine or placebo group, and the use of a double-blind design) so as to ultimately arrive at meaningful, convincing conclusions.

(Based on information from "The Effectiveness of Vaccination Against Influenza in Healthy, Working Adults." Kristin L. Nichol, M.D., M.P.H., April Lind, M.S., Karen L. Margolis, M.D., M.P.H., Maureen Murdoch, M.D., M.P.H., Rodney McFadden, M.D., Meri Hauge, R.N., Sanne Magnan, M.D., PH.D., and Mari Drake, M.P.H. The New England Journal of Medicine, 333(14): 889–893, October, 1995.)

B. Treating All Individuals in an Experiment in the Same Ways

It is critical that all the subjects in the experimental group(s) and in the control group(s) be treated identically except for the variable of interest. If this basic principle is violated, then experimental error is introduced.

Although this principle of treating subjects identically (except for the experimental treatment) seems obvious, it may be difficult to accomplish in practice. For example, consider a company that is developing a formulation to promote growth in house plants. The efficacy of the product is tested by comparing the growth of plants sprayed with the formulation to the growth of plants that are not treated. If some of the plants are placed under lights whose bulbs are different, then there are two variables: the formulation and the light bulbs. This is an experimental error that is fairly obvious. More subtle factors might include different greenhouse workers handling the plants differently, an undetected pathogen that affects some of the plants, variation among bottles of the formulation, and so on. All these unidentified variables can introduce systematic error into the results.

It is easy to think mistakenly that control and experimental groups were treated identically when, in fact, they were not. Common causes of this sort of error in the laboratory include:

1. ***Using reagents and solutions for some individuals that vary from those used for others.*** This can easily occur if there is variation in raw materials, if glassware is dirty, if different technicians prepare solutions differently, and so on. Whenever a new batch of reagents is purchased or prepared, there is the possibility that the new materials will differ from the old ones. Even when a single lot of reagents is used, it may degrade or change over time.

2. ***The environmental conditions, such as temperature or humidity, may vary.*** This can easily happen if an experiment runs for more than 1 day, if it takes place outside, or if it is run in any environment that is not stable.

3. ***Data are collected by more than one person.*** It is common for different individuals to read instruments slightly differently, to work at different paces, and to vary from one another in other ways.

4. ***The response of instruments may drift over time or an instrument may malfunction.***

A strategy that is used to help ensure that all individuals in an experiment are treated consistently is the use of "blocks." To illustrate the meaning of blocks, let us return to the example of an experimental field in which crops are tested. Such a field is likely to be heterogeneous with respect to moisture, soil type, light, and so on. In this situation, plants in different parts of the field are not treated identically because of the variation in the field. To deal with this variability, researchers devised the idea of *dividing experimental fields into small plots, each of which is called a* **block**. Within a block conditions are more likely to be homogeneous than within the entire field. Investigators then plant both experimental and control plants in each block. They compare the effects of fertilizer or other substances on control and experimental groups *within* each block.

Although the concept of blocks was developed for agricultural experiments, the idea can be generalized to apply to many situations. Just as it is easier to control the conditions in a small plot than it is in an entire field, it is also easier to control the variation in a small experiment than it is in a large one, or in a small block of time than in a longer span of time. A block can refer to space, to time, or to a piece of equipment. For example, it is simpler to maintain consistency in an experiment with mice if all the mice are housed in one room than if they are housed in different facilities. In the latter case, it is necessary to consider all the things that can vary between the two facilities, such as temperature, humidity, light, animal caretakers, and food sources. A single room containing both control and experimental mice constitutes a block. Thus, blocks may differ from one another, but within each block conditions should be fairly stable.

Replication, *the repetition of an entire experiment*, is a method to identify experimental error. For example, suppose an experiment involves testing the effect of a treatment on the growth of 50 plants compared with 50 untreated controls. Replication of the experiment would involve selecting another 100 plants, assigning them to either the control or experimental groups, repeating the treatment, and obtaining a new set of growth measurements.

The reason for replicating an experiment is that there are many variables that can affect the results. The raw materials that go into making solutions might vary, environmental factors such as temperature or humidity might fluctuate, and so on. The result is experimental error. Biological systems tend to be complex, so it is difficult to think of every variable that might cause error. Replication tests whether an experiment can be duplicated: If duplication is not possible, then there is some as yet unrecognized factor affecting the results. That factor(s) needs to be identified before any conclusions can be made.

Note that replication does not mean simply to repeat a measurement on the same sample. Repeating the measurement on the same sample, or on a portion of the same sample, reveals variation in the measurement system but not in the entire experiment. For example, if you prepare a particular sample and test 1 mL of it three times using a particular instrument, you have repeated the measurement, but you have not replicated the experiment. Any differences observed between the three subsamples are likely due to the instrument used

---

Box 3   TREATING ALL INDIVIDUALS IN AN EXPERIMENT IN THE SAME WAY

1. *When possible, divide an experiment into small, homogeneous blocks. Each block should contain members of experimental and control groups.*

   a. In an agricultural study, a block is a subplot in a field.

   b. In nutritional studies, animals from the same litter may be a block.

   c. An incubator, animal room, or constant environment chamber may be a block.

   d. Environmental conditions can vary over time; therefore, a block may be a short period of time, such as a day.

2. *Use identical instruments and equipment for all individuals throughout an experiment.*

   a. Have all laboratory workers follow the same written procedures for operating instruments and equipment.

   b. Instruments and equipment, even if they are of the same model and are properly maintained, often vary subtly from one another. When possible, use a single instrument for all measurements.

   c. Make certain that all instruments and equipment are properly maintained.

3. *Ensure that reagents and solutions do not vary in an experiment.*

   a. Make certain that all solutions and reagents are consistently prepared according to written procedures. It may be helpful to have the same technician responsible for this task.

   b. Reagents and solutions can vary from batch to batch; therefore, if possible, do an entire experiment with a single lot of raw materials and reagents.

   c. Within a block, make sure both the control and experimental groups are exposed to the same lots of reagents and solutions.

4. *Avoid differences between laboratory workers in performance of tasks.*

   a. When possible, have only one person perform all the work in each part of an experiment or have one person perform the same steps for all samples.

   b. Make certain all analysts follow the same written procedures.

5. *Replication of an experiment will sometimes reveal variability due to unknown factors.*

---

to make the measurement. Replication involves preparing fresh reagents, obtaining new samples, processing those samples, and making measurements.

Box 3 summarizes some issues relating to treating all individuals in an experiment in the same way.

IV. BIOLOGICAL ASSAYS AND TESTS

There are a vast number of tests and assays that are performed on biological samples. For example, there are tests for sterility, for toxicity, and for the presence and amount of specific compounds. An assay method might show the presence of hormone receptors on cell surfaces. Another test might be used to determine whether bacteria have taken up a gene of interest and are making the protein for which it codes. This section considers certain types of problems which arise in biological assays and standard practices to avoid them.

Biological assays are often indirect (i.e., they do not evaluate the characteristic of interest directly, rather they determine an associated property). All indirect methods have potential inaccuracies and uncertainties, and they are all based on assumptions.

For example, consider counting bacteria. Bacteria are too small to see and are too numerous to count individually. It is necessary, therefore, to use an indirect method to count the number of bacteria in a broth. One such method is plate counts, which involves:

1. A small volume sample of a bacterial suspension is removed from the broth.

2. This small volume is diluted many fold.

3. The diluted sample is applied to plates that contain nutrient medium.

4. The plates are incubated to allow bacterial growth.

5. Colonies grow on the plates and are counted.

6. The number of bacteria in the original preparation is computed based on the assumptions that:

   a. Each visible colony is derived from a single bacterium that was present in the original solution.

   b. Each living bacterium in the diluted sample forms a colony on the plate.

The plate count method is indirect: The number of colonies on a plate is used as an indicator of the number of bacteria in the original broth. Although the plate count method is a widely accepted, effective way to count bacteria, there are cautions in its use, including:

1. The bacteria in the original broth must be uniformly distributed when the sample is taken.
2. The method assumes that dead or nonreproductive bacteria are unimportant because these will not be counted.
3. The method is inaccurate if adjacent bacteria merge to form a single colony.
4. The method is inaccurate if too few or too many bacteria are plated.
5. The method is inaccurate if the plates are incubated too briefly or too long.

To count bacteria properly using plate counts, it is necessary to be aware both of the assumptions of the method and of potential inaccuracies, and to take these problems into account when analyzing results. It is similarly important when using any test method to be aware of the assumptions and limitations of the method.

Let us consider a more complex class of assay methods—those that involve antibodies. Methods involving antibodies are called *immunochemical methods*. The virtue of immunochemical assays is that antibodies bind specifically to a certain target protein. Different antibodies bind to different proteins. Antibodies can be labeled with chemical tags that are in some way visible to the eye or to an instrument. For example, some tags cause an opaque or colored deposit to form under the proper conditions, whereas other tags fluoresce or are radioactive. Thus, to see whether a particular hormone receptor is present on the surface of cells, it is possible to obtain antibodies that selectively recognize and bind to this receptor. The antibodies can be complexed with a tag that forms an opaque deposit in the presence of the proper substrate. The deposit is visible under a microscope and thus labels the hormone receptor. Figure 12.1 shows tissue that has been labeled with antibodies that bind to a specific protein target.

Immunochemical methods are powerful tools to visualize biological structures that would otherwise be invisible. There are many applications of these antibody methods. They can be used in conjunction with microscopic techniques, as shown in Figure 12.1. They can be used with cells in culture and with proteins immobilized on a plate or on a plastic surface. In all these applications, however, there are problems that can arise. For example, antibodies may bind to molecules that are similar to, but not the same as, their target. This is called *nonspecific binding* and it may occur if incubation conditions are not optimized. If too little antibody is used, then the target may not be visible. If the assay conditions are wrong, then a deposit may form where it should not, or it may fail to form where it should. The sample preparation steps may affect the antibody's ability to bind to its target. All these problems can

cause systematic errors that seriously affect the results of the assay. Avoiding errors in immunochemistry requires understanding the system well, performing preliminary tests to identify as many problems as possible, optimizing the sample preparation steps, and optimizing the conditions used during the assay.

An important strategy to avoid errors in immunochemical assays is to use positive and negative controls. These types of controls are helpful in determining whether an unknown variable is affecting the results. Such controls are best explained by example.

Suppose we are using an immunochemical method to see whether a particular tissue contains a hormone receptor of interest. There are four possible outcomes when the tissue is treated:

1. *The tissue might contain the receptor and might form a visible deposit when treated by immunological methods.* This is called a **true positive result** because the result is correct and it is positive for the receptor.
2. *The tissue might contain the receptor, but because of some problem in the procedure, it might fail to form a deposit when it is treated.* This is called a **false negative result**. The result is negative, and it is incorrect.
3. *The tissue might not contain the receptor and not form a visible deposit when treated.* This is called a **true negative result**. The result is negative for the receptor and it is correct.
4. *The tissue might not contain the receptor, but it might form a visible deposit because of nonselective binding or another problem in the procedure.* This is called a **false positive result**. The result is positive and it is incorrect.

In order to help avoid false results, as in Situations 2 and 4, positive and negative controls can be used. A **negative control** *is tissue that is known not to contain the receptor of interest and which therefore should not form a visible deposit,* Figure 12.1b. The negative control tissue is treated exactly the same way as the other samples. If the negative control tissue does show a positive result, then we know there was too much antibody or there was another "glitch" in the procedure, so we would not trust the results on the samples. The routine use of negative controls helps avoid false positive results in assays.

A **positive control** *is tissue known to contain the receptor of interest and which should therefore form a visible deposit if the reagents and procedures used were good.* Thus, if the positive control fails to form a visible deposit then we know there was a problem in the procedure. The routine use of positive controls helps avoid false negative results in assays.

In addition to the preceding possibilities, the results of an assay might be ambiguous. For example, the

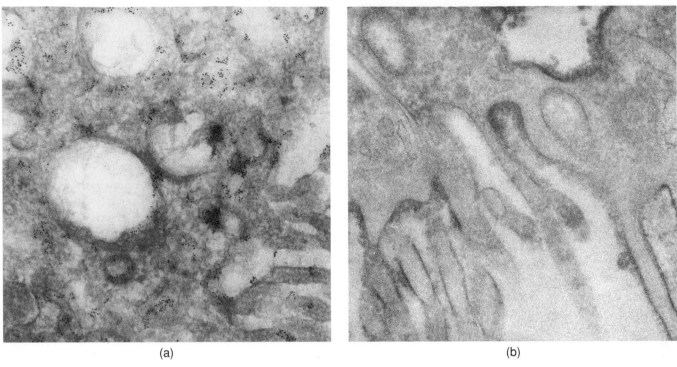

(a)   (b)

Figure 12.1.   Immunocytochemistry. a. The dark deposits in this electron micrograph represent antibodies bound to a specific target protein. **b.** A negative control. This is tissue known not to have the protein of interest. The negative control tissue was processed for electron microscopy alongside the experimental samples, and was treated exactly the same as experimental tissue.

deposit might be difficult to see, the tissue may appear to be labeled in some areas, but not in other similar areas, various tissue samples might unexpectedly differ, and so on. Ambiguous results are frustrating and typically mean that the experimental test system has not been optimized adequately. More preliminary testing may reveal conditions that can be slightly altered to give better results.

The use of positive and negative controls is applicable in many assays in addition to immunochemical ones. For example, the polymerase chain reaction (PCR) is exquisitely sensitive to DNA. Even minuscule amounts of DNA are detectable on an electrophoresis gel after PCR. Unsuspecting investigators have thought that they were investigating genes from experimental samples, when they were, in fact, investigating their own DNA that fell into the test tube, or minute amounts of DNA from other contaminants. One way to check for problems in PCR is to routinely run negative controls alongside the test samples. PCR negative controls typically lack one component of the reaction mixture and therefore should display no DNA bands after the PCR reaction, Figure 12.2.

Another situation where positive and negative controls are used is in clinical testing of patient samples. For example, a blood test might be used to see if a patient is infected with a certain pathogen. A *positive control* would be blood that is known to test positive for

the pathogen. The positive control is run alongside the patient's sample and helps detect false negative results. (A *false negative* occurs when a person who is infected shows up as uninfected.) A negative control (i.e., blood that is known not to react with the test) is used to help avoid a false positive result. (A false positive occurs when a person who is uninfected shows up as infected.)

In summary, when performing tests and assays on biological samples, it is important to be aware of the assumptions, limitations, and potential errors inherent in those tests. Most biological tests and assays are complex, so they require practice and optimization to get the best results. It is also helpful to use positive and negative controls whenever possible to detect problems in the test procedure.

CASE STUDY 2

"Review Criteria for Assessment of Human Chorionic Gonadotropin (hCG) In Vitro Diagnostic Devices (IVDs)" (The Assessment of Pregnancy Test Kits)

The hormone hCG appears around the fourth day after conception and so provides an early indication of pregnancy. Pregnancy test kits are based on an immunological assay for the hormone, hCG, in urine or serum.

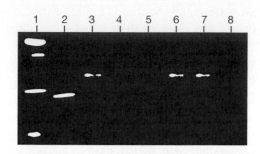

Figure 12.2.   Positive and Negative Controls in PCR. This gel is from a clinical test where PCR is used diagnostically to detect a disease. The bright bands on the gel represents DNA fragments of a particular size. The higher the band is on the gel, the larger the fragment. Lane 1 is a special type of positive control. Each of the bright bands in this lane is a DNA fragment of a different length. Mixtures of DNA fragments can be purchased from biotechnology companies where the lengths of the fragments in each band are known. Lanes 2 and 3 are also positive controls. Lane 2 is amplified DNA from a patient known to have the disease of interest. Lane 3 is amplified DNA from a healthy patient. Lanes 2 and 3 should have different, known patterns of bands. Since lanes 1–3 are positive controls, clinicians know what patterns of bands should appear in each of these three lanes. If the expected patterns had not appeared, then the analysts would know there was a problem in the procedure. Lanes 4, 5, and 8 are negative controls. Each was lacking some necessary component of the reaction mixture and therefore should have no bright bands at all after PCR. The presence of any bands in any of these negative control lanes would indicate that contamination had occurred. As expected, the negative control lanes are dark. Lanes 6 and 7 are patient samples. Both patients are considered to be healthy since their DNA pattern matches that of the healthy control.

Companies that manufacture pregnancy test kits must submit an application to the Food and Drug Administration (FDA) that contains evidence demonstrating that their kit is effective and reliable in detecting pregnancy. The FDA provides advice, in the form of a guidance document, as to how companies should acquire such evidence. This document outlines studies that should be performed, the types of data that should be collected, and the statistical analyses to be performed on those data. The procedures outlined by the FDA emphasize proper sampling and data analysis and so provide a good example of many of the principles discussed in this chapter. A few sections from this FDA guidance document are quoted and discussed in this case study.

1. Pregnancy test kits detect hCG in either urine or blood serum and so investigators must collect samples of serum and urine on which to test their kits. The FDA guidance document states:

    "Whole blood specimens should be collected into suitable tubes with/without anticoagulants, allowed to clot at room temperature and centrifuged. All specimens not tested within 48 hours of collection should be stored at −20°C. Repeated freezing and thawing should be avoided. Serum specimens showing gross hemolysis, gross lipemia, or turbidity may give false results.

    A first morning urine is recommended because hCG concentration is highest at this time. If specimens cannot be assayed immediately, they should be stored at 2–8°C for up to 48 hours.

    Note: If claims are being made to 'use any time of day,' then samples collected any time of day should be used in the study."

    Proper sample preparation is essential because the preparation and handling of a sample affects the results of assays. If a company claims that their test does not require first morning urine, this claim must be backed by collecting samples at all times of the day.

2. An important aspect of demonstrating the effectiveness of a test kit is showing that the intended users are able to get proper results. Thus, if the test kit is to be used in physician's offices, then the FDA specifies that it should be tested in physician's offices. If the kit is intended for use at home, then it should be tested in homes. The following are excerpts from the directions for evaluating a kit intended for home use:

    "At least 100 urine specimens should be used . . . FDA recommends that the home users collect the samples and perform the tests . . . For devices utilizing more than one testing procedure, i.e., urine stream or dip procedure, data must be provided to validate the equivalency of both procedures.

    Home users should be selected on a random basis as they present themselves at clinics and/or physicians offices or via advertisements. They should represent diversity of age, background, and education."

    The kit must be tested by a number of people (at least 100). It would not be acceptable for a company to pick test subjects who are well-educated, have more background or training or in some other way are not representative of the market as a whole. An improper selection of subjects might bias the results to make it appear that the kit works better (or worse) than it actually does.

3. The company must test the sensitivity of the kit to low levels of hCG and its capacity to provide true positive and negative results. "Spiking" means that known amounts of hCG are added to specimens. This is done to test the kit's ability to give true positive results.

    Luteinizing hormone (LH) is a hormone whose structure is somewhat similar to hCG, but which does not indicate pregnancy. If the kit responds to LH, then false positives will result. Assay optimization is necessary so that the kit is sensitive to low levels of hCG (such as at the very beginning of pregnancy), yet the kit does not detect LH.

    "Sensitivity should be evaluated by spiking at least 20 clinical samples from normal, nonpregnant females or males with five different concentrations of hCG below, at and above stated sensitivity. For example, use 0, 20, 25, 50, 100 *mIU/mL* for a kit with a detection limit of 25 *mIU/mL*. . . . Assay sensitivity should be such that small quantities of hCG will be detected while false-positive results due to the presence of LH will be minimized. Additionally, to support the detection limit 100 percent (%) of the samples tested must be positive and must have reacted within the specified time frame."

The guidance document specifies more studies that we will not discuss here. These other studies are intended to ensure that the kit works consistently, that it is properly labeled, that the instructions to the users are clear, and so on.

(From the Center for Devices and Radiological Health, FDA 1/14/98. http://www.fda.gov/cdrh/ode/odec1592.html)

PRACTICE PROBLEMS

1. A company produces a product that is packaged into vials, which are then loaded onto trays. A technician tests the first vial in each tray as part of a quality control program. Is this a good method of obtaining a random sample?

Questions 2–4 relate to the following issue.

Some scientists believe that there is an increasing incidence of deformity in frogs. A number of explanations have been proposed to explain this phenomenon, ranging from an epidemic illness or parasitic infection to environmental pollutants that alter frogs' hormonal functions. Various studies are being performed to evaluate the health of frog populations. One study relies on volunteers who go to specific study sites and count frogs by listening for their calls. Every possible wetland habitat cannot be monitored with existing volunteers. Rather, the frogs in a few sites are monitored and from those sites conclusions are drawn about frogs in general. One of the challenges in this study is therefore to devise a plan for sampling sites that are truly representative of all wetlands. The following questions relate to the sampling plan for this study.*

2. Some frog surveys in the past have selected marshes or areas readily accessible from backyard porches. Volunteers conveniently monitor frog calls from their homes. What is wrong with this method of selecting study sites?

3. One approach to choosing frog monitoring sites is as follows:

   i. A grid is placed on a map of the area to be surveyed.

   ii. Sites which are wetlands are identified without prior knowledge of their capacity as amphibian breeding sites.

   iii. Wetlands are randomly selected from the sites described in (ii).

   a. What technique might be used to randomly select sites, as described in Step iii?

   b. In Step ii, wetlands are identified "without prior knowledge of their capacity as breeding sites." Explain this requirement.

4. Suppose that several volunteers know that certain wetlands have lots of frogs. The volunteers reason that these sites with large frog populations are the best to monitor because they will provide the most information; therefore, the volunteers monitor these sites with which they are familiar. What will happen to the study results in this situation?

5. A college teacher conducted a study to see whether weekly review sessions are helpful to students. Two groups of students were formed: One attended weekly review sessions, the other did not. The groups were chosen voluntarily, a student could choose to be in either group. What do you think about this method of selecting groups?

6. Multiple choice: The term "sensitivity" with reference to a biological assay is related to how low a level of target the assay can detect. The more sensitive the assay, the lower the level that can be detected. Which of the following is true:

   a. There is a risk associated with optimizing an assay to increase its sensitivity. This risk is that the number of false negative results will increase.

   b. There is a risk associated with optimizing an assay to increase its sensitivity. This risk is that the number of false positive results will increase.

7. Multiple choice: In an immunological assay, nonspecific binding tends to cause which of the following:

   a. False negative results

   b. False positive results

   c. The positive control comes up negative

   d. The negative control comes up positive

   e. b and d

   f. a and c

8. The directions for a home pregnancy test kit include the following instructions: "A pinkish-purplish line will form in the control (upper) window to tell you the test is working correctly. A positive result is two lines, one in each of the two windows (the upper and lower). A negative pregnancy test is a single line in the upper (control) window and no line in the lower window."

*(These questions are based on information from *Amphibian Biomonitoring Project Protocol* 1/14/98. http://kancrn.org/amphibian/)

Explain the upper window. Why does a line form there? What is the significance of this line?

Early in this chapter we discussed Raymond Pearl's experiments with corn kernels that were difficult to classify. Practice Problems 9–13 relate to another example of data where classification is difficult. In this case the data relate to an imaginary disease, called philolaxis disease.

The serum level of the imaginary protein, protein PHX, is elevated in patients suffering from philolaxis disease. The level of this protein varies from individual to individual and is approximately normally distributed. (You may want to review the section on the normal distribution in Chapter 11.) Figure a shows the distribution in healthy patients, Figure b shows the distribution in ill patients. The distributions for healthy and ill patients are superimposed in Figure c. The level of protein PHX is used diagnostically to test whether a person is sick with philolaxis disease. You can see that there is variability in the serum levels of PHX for both affected and unaffected individuals and there is overlap in the two distributions. For this reason, there can be difficulty in classifying some patients as having the disease or not.

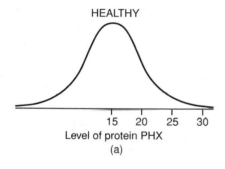

(a)

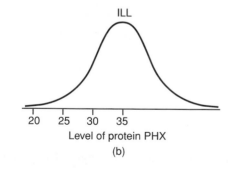

(b)

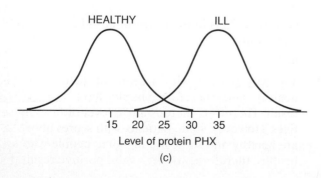

(c)

9. **a.** Does a person with a serum level of protein PHX of 10 have philolaxis disease?

   b. Does a person with a level of 20 have philolaxis disease?

   c. Does a person with a level of 25 have philolaxis disease?

   d. Does a person with a level of 40 have philolaxis disease?

10. When clinicians test for philolaxis disease based on serum levels of protein PHX, they select a cutoff score. If a patient's level of protein PHX is above the cutoff score the result is considered to be positive for philolaxis disease. If the patient's level is below the cutoff score the result is considered to be negative for the disease.

    a. Multiple choice: If the clinicians decide that the cutoff score should be 20, about what percent of the time can they expect to get a false positive result?

      (i) 0%   **(ii)** 15%   **(iii)** 50%   **(iv)** 100%

    b. If the clinicians decide that the cutoff score should be 20, about what percent of the time can they expect to get a false negative result?

      (i) 0%   **(ii)** 15%   **(iii)** 50%   **(iv)** 100%

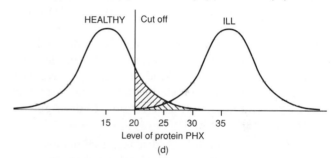

(d)

11. Multiple choice:

    a. If the clinicians decide that the cutoff score should be 25, about what percent of the time can they expect a false positive result?

      (i) 0%   **(ii)** 2.5%   **(iii)** 50%   **(iv)** 100%

    b. If the clinicians decide that the cutoff score should be 25, about what percent of the time can the clinicians expect a false negative result?

      (i) 0%   **(ii)** 2.5%   **(iii)** 50%   **(iv)** 100%

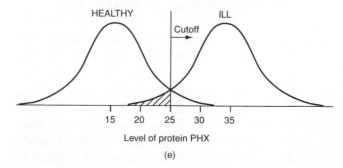

(e)

12. Multiple choice:

  a. If clinicians decide that the cutoff score should be 30, about what percent of the time can they expect to get a false positive result?

  (i) 0%   **(ii)** 15%   **(iii)** 50%   **(iv)** 100%

  b. If clinicians decide that the cutoff score should be 30, about what percent of the time can they expect to get a false negative result?

  (i) 0%   **(ii)** 15%   **(iii)** 50%   **(iv)** 100%

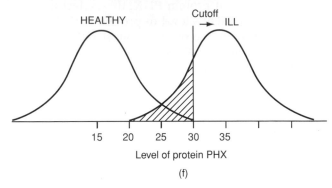

(f)

13. Consider HIV testing (for AIDS). If a person is told they do not have the virus, when in fact they do, they might unknowingly infect other people and also would not receive treatment; therefore, it is common practice to use two tests for AIDS. The first test has a high rate of false positives and a low rate of false negatives. This test is relatively inexpensive and is used to screen for all samples that might be positive. All blood that tests positive with the first method is retested with a second assay that is more expensive but is better able to distinguish a true positive result.

In the case of philolaxis disease, if clinicians want the lowest possible level of false negatives, what cutoff score should be chosen?

Note that often in the laboratory there is more than one method available to test for a certain material or phenomenon. Each test is likely to have its own strengths and weaknesses, but together, the tests provide more reliable information than any test by itself. For this reason, when possible, it is good practice to use more than one test to confirm critical results.

ANSWERS TO PRACTICE PROBLEMS

1. It depends. If the first vial in each tray tends to be different in some way from the other vials, then this selection method would introduce a systematic error.

2. "The data collected for those specific locations can only be compared among themselves and not to other habitats. For example, surveys from back porches will tell you about trends in calling amphibians heard from the porches of people inclined to participate in such surveys. The conclusions drawn from such a survey would have to be limited to the world of amphibian-loving people's backyard and associated natural habitats." (Quoted from "Amphibian Biomonitoring Project Protocol" 1/14/98 http://kancrn.org/amphibian/)

3. a. The grid squares could each be assigned a number and the sites could be selected by rolling dice or by consulting a random number table.

  b. The idea is to sample the entire region. If only sites that are suitable for breeding are selected, this will bias the frog counts so that there will be too many frogs counted for that region.

4. The study will be biased. The number of frogs reported will be too high because areas with fewer frogs were not included. In a worst case, the volunteers might contribute to a false perception that frog populations are thriving, when, in fact, they are not.

5. This is an unacceptable method of assigning experimental subjects to groups. For example, suppose that many of the students who are highly motivated in the course select to be in the review group. Conversely, many students who do not like the subject or the course choose to be in the no review group. The two groups would thus differ from one another before the review sessions ever began. Then, if differences were later found between the two groups in final exam scores, there would be no way to know whether the differences were due to the study sessions or to initial differences between the students in the two groups.

6. b

7. e

8. This is a positive control and should come up positive, whether or not the person is pregnant. If it does not, then there is likely a problem in the reagents or in the way the test was performed. The test kit instructions inform the user that if the positive control line did not appear, then the test did not work properly.

9. a. No, 10 is not within the range for ill patients.

  b. No.

  c. Maybe, 25 is in the range of both healthy and ill patients.

  d. Yes.

10. a,b. Any patient with a score of 20 or lower is healthy and therefore should have a negative result. Therefore, there is a 0% level of false negatives. However, some patients with scores above 20 are healthy and some are ill. Some people who are healthy, therefore, will get a false positive result if a cutoff score of 20 is chosen. The false positive rate will be about 15%.

11. **a,b.** Some people with a score of 25 are healthy and some are ill; therefore, there will be both false positives and false negative results. The rates of each will be equal, at about 2.5%.

12. **a,b.** Anyone with a score of 30 or higher is ill, therefore, the level of false positives will be 0; however, some people with a score of less than 30 are sick. There will therefore be about 15% false negatives.

13. 20

UNIT V:

Laboratory Measurements

Chapters in This Unit

- ✦ Chapter 13: Introduction to Quality Laboratory Measurements, Tests, and Assays

- ✦ Chapter 14: Introduction to Instrumental Methods and Electricity

- ✦ Chapter 15: The Measurement of Weight

- ✦ Chapter 16: The Measurement of Volume

- ✦ Chapter 17: The Measurement of Temperature

- ✦ Chapter 18: The Measurement of pH, Selected Ions, and Conductivity

- ✦ Chapter 19: Measurements Involving Light A: Basic Principles and Instrumentation

- ✦ Chapter 20: Measurements Involving Light B: Applications and Methods

UNIT INTRODUCTION

Measurements are *quantitative observations, or numerical descriptions*. Examples of measurements include the weight of an object, the amount of light passing through a solution, and the time required to run a race. Everyone has experience making measurements (e.g., taking the temperature of a child, weighing oneself, or measuring ingredients when cooking). **Tests** and **assays** are measurements of a property of a sample, such as the amount of cholesterol in a blood sample, lead levels in paint chips, and the activity of a DNA-cutting enzyme.

Measuring properties of samples and performing tests and assays are an integral part of everyday work in any biology laboratory. For example, solutions are required to support the activity of cells, enzymes, and other biological materials. Preparing solutions involves measuring the weights and volumes of the components. Estimating the quantity of DNA in a test tube may involve measuring how much light passes through the solution. The pH of the media in which bacteria grow during fermentation must be monitored continuously. There are countless measurements made in most laboratories, each of which must be a "good" measurement.

Although it seems obvious that laboratory measurements should be "good," it is surprisingly difficult to define a "good" measurement. One definition is that a "good" measurement is correct; however, this leads to the question of what is "correct"? Suppose a man weighs himself in the morning on a bathroom scale that reads 165 pounds. Shortly after, he weighs himself at a fitness facility where the scale reads 166 pounds. At this point the man might be somewhat uncertain as to his exact, correct weight; perhaps he weighs 165 pounds, perhaps 166. Perhaps his weight is somewhere between 166 and 165 pounds. He is likely to conclude, however, that he weighs about 165 pounds and leave it at that.

Uncertainty of a pound or so in an adult's weight is seldom of great concern. A bathroom scale that gives a weight value within 1 pound of the true weight is a reasonably "correct" instrument. In other situations, however, a measurement must be much more correct. For example, a one pound difference in the weight of an infant could mean the difference between a healthy baby and one that is severely dehydrated. In the laboratory, an error of 1 g in a measurement could mean the difference between a successful experiment and a disastrous failure. In a drug product, a 1 mg error in measurement could endanger a patient. In each of these situations a "good" measurement is one that can be trusted to make a decision. A good measurement can be trusted by a physician selecting a treatment for a patient, by a research team drawing conclusions from a study, and by a pharmaceutical company deciding whether to release a drug product to the public.

Laboratory workers play a key role in performing measurements of properties of samples. To make good, trustworthy measurements, laboratory workers must understand the principles of measurement, know how to maintain and operate instruments properly, and be aware of and avoid potential pitfalls in measurement. It takes knowledge and careful technique to produce measurements that can be trusted in a particular situation.

This unit discusses methods of making "good," trustworthy measurements in the laboratory. The first chapter in the unit introduces basic principles underlying measurements and terminology relating to metrology (the study of measurements). The second chapter introduces instrumental methods of measurement and basic principles of electricity and electronics. This second chapter serves as a transition to the next six chapters each of which discusses a specific type of measurement: weight, volume, temperature, pH, and light (spectrophotometry). Note that the specialized terminology and references for each topic follow the appropriate chapter.

Non Sequitur

CHAPTER 13

Introduction to Quality Laboratory Measurements, Tests, and Assays

I. MEASUREMENTS AND EXTERNAL AUTHORITY: STANDARDS, CALIBRATION, AND TRACEABILITY

A. Overview

This chapter discusses general terminology and concepts relating to making "good" measurements. Recall that measurements are numerical descriptions. For example, if an object is said to be "15 grams," then mass is the property, that is described, 15 is the value of that property and grams are the measurement **units**. A **unit of measure** *is an exactly defined amount of a property.*

We are accustomed to using units of measure, such as grams, pounds, inches, and centimeters. But what, exactly, is the mass of a gram? This may seem like a trivial question; obviously, the mass of a gram is a gram. However, exactly defining the meaning of a "gram," or of any other unit of measurement, is anything but trivial. A unit must be defined in some clear way, and everyone who uses the unit must agree on that definition. Establishing the meaning of a unit of measure, therefore, requires international agreement.

Metrologists, *people who work with measurements*, devote much effort toward ensuring international consistency in measurement. One result of this effort is the **SI (System Internationale) measurement system**, *which defines units of measurement* (see pp. 112–114 for more detail on the SI system). The SI definitions of units are an authority to which people in many nations refer when making measurements.

There are many organizations involved in the establishment of agreements regarding measurements, not only on definitions of units but also on measurement practices. Table 13.1 lists some of the national and international organizations that work to develop consensus measurement methods. These efforts to ensure quality and consistency in measurement have resulted in an active international dialogue.

Measurements are always made in conformance with an external authority. As a simple example, we commonly measure length using a ruler. The ruler is our external authority when measuring length. The ruler was marked by the manufacturer so that its lines are correct according to an internationally accepted definition of a "meter."

At this point, it is important to clarify the distinction between a unit of measurement and a standard, such as a ruler. A **standard** *is a physical embodiment of a unit.* Units are not physical entities. Units are unaffected by environmental conditions, but physical standards are affected by the environment. For example, units of centimeters are unaffected by corrosion, but a metal ruler, a physical embodiment of centimeters, may become corroded.

The international efforts to promote "good" measurement practices are intimately associated with the ever-increasing importance of quality systems, such as ISO 9000, ISO Guide 25, and cGMP (see Chapter 4 for explanations of these terms). Scientists and technicians have always been aware of the importance of "good" measurements. However, as people implement quality systems, they become even more concerned with establishing methods of measurement that are consistent, are widely accepted, and whose accuracy they can document.

The next sections of this chapter explore three key, interrelated words in measurement: **standard**, **calibration**, and **traceability**. Later sections explore four more interrelated terms: **error**, **precision**, **accuracy**, and **uncertainty**. As you will see in these discussions, critical terms in measurement sometimes have various meanings, depending on the context. It is paradoxical that even as the quality community has become increasingly active in establishing consistent measurement methods, the language of measurement is not "precise." Inconsistency in how words are used is probably a reflection that the field of **metrology**, *the study of measurements*, is evolving, and many of the principles on which it is based are complex.

B. Standards

i. A STANDARD AS A PHYSICAL EMBODIMENT OF A UNIT

The term *standard* was defined earlier as a physical object that embodies a unit. The most famous physical standard is probably the kilogram standard. The unit of a kilogram has been defined by international treaty to have as much mass as a special platinum-iridium bar* located at the International Bureau of Weights and Measures near Paris. All other mass standards are defined by comparison to this special metal bar. Every country that signed the treaty received a national kilogram prototype whose mass is determined by comparison with the standard in France. The United States' standard is called K_{20} and is housed at The National Institute of Standards and Technology (NIST).

The platinum-iridium bar in France and K_{20} are sometimes called **primary standards**, *standards whose values may be accepted without further verification by the user. A primary standard is used to establish the value for* **secondary standards**. Companies that manufacture **working standards**, *standards for use in individual laboratories*, typically use secondary standards as the basis for their products.

---

*There are disadvantages to using a physical object, like a platinum-iridium bar, as the ultimate definition of a unit. The bar could potentially be damaged or destroyed. Its mass changes slightly with dust and with cleaning. It requires a secure and environmentally stable storage site. Scientists are therefore considering substituting a definition for the kilogram that relies on a physical constant, not on a physical entity. This has already been done for other units. For example, the meter is defined in terms of the speed of light.

Table 13.1   *A SELECTED LIST OF AGENCIES AND ORGANIZATIONS THAT ARE INVOLVED IN THE STANDARDIZATION OF MEASUREMENTS AND ASSAYS*

ISO, International Organization for Standardization

A worldwide federation of national standards bodies from some 100 countries that promotes the development of standards and related activities in the world with a view to facilitating the international exchange of goods and services. Examples of measurement-related services provided by ISO are:

1. Publishing and updating the SI (metric) system of units.

2. Developing by consensus international standards including the ISO 9000 quality series.

ANSI, The American National Standards Institute

Administrator and coordinator of the United States private sector voluntary standardization system. ANSI does not itself develop American National Standards; rather, it facilitates their development by establishing consensus among qualified groups. ANSI is the sole U.S. representative to ISO. Examples of measurement-related services provided by ANSI are:

1. Promoting the use of U.S. standards internationally.

2. Playing an active role in the governance of ISO.

3. Promoting the use of standards, including ISO 9000, in the United States.

NIST, The National Institute for Standards and Technology (formerly known as the National Bureau of Standards, NBS)

A U.S. federal agency that works with industry and government to advance measurement science and develop standards. Examples of measurement-related services provided by NIST are:

1. Providing calibration services and primary standards for mass, temperature, humidity, fluid flow, and other physical properties.

2. Providing more than 1300 standard reference materials to be used in biological, chemical, and physical assays.

3. Performing research and development of new standards, including standards for biotechnology.

ASTM, The American Society for Testing and Materials

Coordinates efforts by manufacturers, consumers, and representatives of government and academia to develop by consensus standards for materials, products, systems, and services. Examples of measurement-related services provided by ASTM are:

1. Developing and publishing technical standards (e.g., those covering the requirements for volume-measuring glassware). More than 10,000 ASTM standards are used worldwide.

2. Providing technical publications and training courses.

NCCLS, The National Committee of Clinical Laboratory Standards

A nonprofit, educational organization that provides a communication forum for the development, promotion, and use of national and international standards for testing patient samples. A measurement-related service provided by NCCLS is:

Establishing and publishing laboratory procedures, methods, and protocols applicable to clinical laboratories.

U.S. EPA, Environmental Protection Agency

U.S. federal agency responsible for environmental quality. Examples of measurement-related services provided by EPA are:

1. Establishing standard methods for analysis of environmental samples.

2. Publishing standard methods in the compendium, *Handbook of Methods for Environmental Pollutants.*

AOAC International, formerly the Association of Official Analytical Chemists

An independent association of scientists devoted to promoting methods validation and quality measurements in the analytical sciences. Examples of measurement-related services provided by AOAC are:

1. Developing validated standard methods of analysis in microbiology and chemistry.

2. Publishing validated standard methods in the compendium, *Official Methods of Analysis of AOAC International.*

USP, The United States Pharmacopeia

An organization that promotes public health by establishing and disseminating officially recognized standards relating to medicines and other health care technologies. The methods in the USP are mandatory in the United States pharmaceutical industry and other situations where a company is cGMP compliant (see Unit II for a description of cGMP requirements). Examples of measurement-related services provided by USP are:

1. Developing validated standard methods of analysis for pharmaceutical and health-related products.

2. Publishing validated standard methods in the compendium, *The United States Pharmacopeia.*

3. Distributing reference materials to be used with the methods outlined in the Pharmacopeia.

For a standard to be useful, its properties must be known with sufficient accuracy to allow it to be used to evaluate another item. The accuracy that is required of a standard depends on the situation. For example, the value for the mass of the platinum-iridium bar in France must be known far more exactly than the value for the mass of a working standard in a teaching laboratory.

ii. REFERENCE MATERIALS FOR CHEMICAL AND BIOLOGICAL ASSAYS

Biologists measure both physical properties, such as mass or temperature, and chemical/biological properties, such as enzyme activity. To accommodate chemical and biological measurements, we must broaden our definition of a standard.

The properties of chemical or biological substances are complex and are not established with the same rigor as the mass of a mass standard. Moreover, because there are thousands of chemical and biological materials that are analyzed, there is no single reference collection of all such materials. Nonetheless, there are standards that are useful in biological and chemical assays. *Chemical or biological substances whose compositions are reasonably well-established and are used for comparison in assays are termed* **reference materials**, **reference standards**, *or simply,* **standards**.

NIST maintains more than 1300 different materials, called **standard reference materials (SRMs)**, whose compositions are determined as exactly as possible with modern methods. A wide variety of SRMs for chemistry, physics, and manufacturing are available. A biotechnology subdivision at NIST sells SRMs for biotechnology applications, such as a DNA standard for quality control in forensic DNA testing.

Chemical and biological reference materials are available from commercial companies as well as from NIST. These reference materials often have documentation associated with them that shows how their properties were determined and the certainty that the manufacturer has in their values. Figure 13.1 shows the docu-

mentation for a reference standard containing a mix of minerals. Note the care that was used to determine the levels of each mineral and the description of the variability associated with the measurements. A well-characterized standard like this mineral mixture might be used, for example, in a situation where regulatory decisions will be made based on test results.

Biologists commonly use products of biological derivation that are difficult and expensive to characterize as fully as the mineral mixture just described. Biological materials that are reasonably well-characterized are used to determine the response of a method or an instrument to the substance of interest. For example, Figure 13.2 shows examples of catalogue specifications for standards used in biological and clinical assays.

iii. STANDARD METHODS

The methods by which measurements and assays are made in many cases also must be "standardized," among individuals, laboratories, and nations. In certain fields there are organizations that make recommendations for standard measurement and assay methods (refer to Table 13.1). For example, in the United States, the Environmental Protection Agency recommends standard methods for analysis of environmental pollutants. This leads to another definition of the term **standard**: *A* **standard** *is a document established by consensus and approved by a recognized body that establishes rules or guidelines to make a procedure consistent among various people.* Figure 13.3 shows an example of a U.S. Pharmacopeia standard assay method to confirm the identity of sodium citrate in pharmaceutical products.

Table 13.2 summarizes types of measurement standards.

C. Calibration

i. CALIBRATION AS AN ADJUSTMENT TO AN INSTRUMENT OR MEASURING DEVICE

A simple definition of **calibration** *is to adjust a measuring system so that the values it gives are in accordance with an external standard(s).* For example, calibration of an instrument might involve placing a standard in or on the instrument and turning a knob or pressing keys until the instrument displays the value of the standard. After calibration the response of the instrument is in accordance with the standard. The result of calibration (according to this definition) is that the instrument or measuring device is adjusted and it gives more correct values after calibration than before.

Calibration as an adjustment to a measuring system applies to the way a manufacturer makes a measuring device. For example, glassware that is used to measure the volume of liquids is marked with lines that indicate volume. The manufacturer "calibrates" the glassware so that the lines are in the proper place.

Property	Units	Consensus Value	Std Dev
Alkalinity	*mg/L*	93.56	4.35
Calcium	*mg/L*	27.21	2.20
Chloride	*mg/L*	91.74	4.09
Conductivity	*μmho/cm*	661.00	15.60
Fluoride	*mg/L*	2.39	0.18
Magnesium	*mg/L*	12.18	0.66
Potassium	*mg/L*	9.08	0.68
Sodium	*mg/L*	83.28	4.14
Sulfate	*mg/L*	81.10	6.10
Hardness	*mg/L*	116.37	5.22

Figure 13.1. Specifications for a Mineral Mixture. The amount of each mineral in the mixture was rigorously analyzed in 230 different laboratories and the mean of the results is reported as the "Consensus Value." The standard deviation column indicates the variability in the data.

A 2153 **Fraction V Powder**
2–8°C   **Minimum 96%**
(electrophoresis)
Initial fractionation by cold
alcohol precipitation
Remainder mostly globulins.
Prepared by a modification of
Cohn, using cold ethanol, pH, and
low temperature precipitation, followed by additional
pH adjustment step prior to final drying.
The pH of a 1% (w/v) aqueous solution is
approx. 7.
Ref.: Cohn, E.J., et al., J. Am Chem. Soc., **68,** 459
(1946).

(a)

ACCUSET™ LIQUID
A 2539 **CALIBRATOR**
–0°C   **Level 1**
◆   Human serum preparation
stabilized with ethylene glycol for use in calibrating
automated analyzers. Values assigned for the
following constituents: albumin, bilirubin, calcium,
carbon dioxide, chloride, cholesterol, creatinine,
glucose, iron, magnesium, phosphorus, protein,
triglycerides, urea nitrogen and uric acid. Calibrator
values for Technicon RA-1000, Electro-Nucleonics,
Gemeni and Gemstar, BMD Hitachi 705, Gilford and
other instruments.
R: 20/21/22-36/37/38   S: 45-26-36/37/39

(b)

Figure 13.2.   Catalogue Descriptions of Materials Used as Standards in Biological and Clinical Assays. a. Description of bovine serum albumin, a protein that is often used as a standard in assays that determine the amount of protein in a preparation. **b.** A human serum preparation with known values for a number of clinically relevant substances; used to calibrate clinical instruments. (Descriptions reproduced by permission of Sigma-Aldrich Co.)

Assay for sodium citrate—
Cation-exchange column—Mix 10 *g* of styrene-divinylbenzene cation-exchange resin with 50 *mL* of water in a suitable beaker. Allow the resin to settle, and decant the supernatant liquid until a slurry of resin remains. Pour the slurry into a 15-*mm* × 30-*cm* glass chromatagraphic tube (having a sealed-in, coarse-porosity fritted disk and fitted with a stopcock), and allow to settle as a homogeneous bed. Wash the resin bed with about 100 *mL* of water, closing the stopcock when the water level is about 2 *mm* above the resin bed.
Procedure—Transfer an accurately measured volume of Oral Solution, equivalent to about 1 *g* of sodium citrate dihydrate, to a 100-*mL* volumetric flask, dilute with water to volume, and mix. Pipet 5 *mL* of this solution carefully onto the top of the resin bed in the *Cation-exchange column*. Place a 250-*mL* conical flask below the column, open the stopcock, and allow to flow until the solution has entered the resin bed. Elute the column with 60 *mL* of water at a flow rate of about 5 *mL* per minute, collecting about 65 *mL* of the eluate. Add 5 drops of phenolphthalein TS to the eluate, swirl the flask, and titrate with 0.02 *N* sodium hydroxide VS. Record the buret reading, and calculate the volume (*B*) of 0.02 *N* sodium hydroxide consumed. Calculate the quantity, in *mg*, of $C_6H_5Na_3O_7 \cdot 2H_2O$ in each *mL* of the Oral Solution taken by the formula:

$$[1.961B(20/V)] - [(294.10/210.14)C]$$

in which 1.961 is the equivalent, in *mg*, of $C_6H_5Na_3O_7 \cdot 2H_2O$, of each *mL* of 0.02 *N* sodium hydroxide. *V* is the volume, in *mL*, of Oral Solution taken, 294.10 and 210.14 are the molecular weights of sodium citrate dihydrate and citric acid monohydrate, respectively, and *C* is the concentration, in *mg* per *mL*, of citric acid monohydrate in the Oral Suspension, as obtained in the *Assay for citric acid*.

Figure 13.3.   An Example of a Standard Method from the U.S. Pharmacopeia. (Copied with permission from USP 23-NF18. All rights reserved. © 1995 The United States Pharmacopeial Convention, Inc.)

Once instruments enter the laboratory they are subject to the effects of aging and their response may be altered by changes in their environment. As a result, the response of instruments in the laboratory drifts. To correct for drift, laboratory equipment must be periodically recalibrated by the user or a service technician. For example, pH meters are very sensitive to the effects of aging and therefore require frequent calibration. Calibration involves adjusting the readings of the pH meter according to the pH of standard solutions of known pH. The calibration of pH meters and other measuring devices will be discussed in more detail in later chapters in this unit.

There is always some error when items are calibrated. The error can be reduced by using more expensive, exacting procedures, but it cannot be eliminated. **Tolerance** *is the amount of error that can be tolerated in the calibration of a particular item.* For example, the tolerance for a "500 *g*" Class 1 mass standard is ± 1.2 *mg*, which means the standard must have a true mass

Table 13.2   STANDARDS IN MEASUREMENT

There are 13 definitions for the word *standard* in the *Merriam-Webster Collegiate Dictionary* (G. & C. Merriam Co., MA. 1977). The broadest of these definitions is "something established by authority, custom, or general consent . . ." This meaning encompasses all the others described below.

1. *A standard is a physical object, the properties of which are known with sufficient accuracy to be used to evaluate another item; a physical embodiment of a unit.* For example, a metal object whose mass is accurately known can be used to determine the response of a balance.

   This definition includes the following:

   a. *Primary Standards.* Physical items whose value may be accepted without further verification by the user and which are used to determine the values for secondary standards.

   b. *Secondary Standard.* A standard whose value is determined by direct comparison with a primary standard.

   c. *Working Standard.* A physical standard that is used to make measurements in the laboratory and which is calibrated by comparison with a primary or secondary standard.

2. *In chemical or biological measurements, a standard often describes a substance or a solution that is used to establish the response of an instrument or an assay method to the analyte.* This definition includes the following:

   a. *Standard Reference Material (SRM).* A substance issued from NIST that is accompanied by documentation that shows how its composition was determined and how certain NIST is of given values.

   b. *Certified Reference Material.* A reference material from any source that is issued with documentation.

   c. *Reference Material.* Any substance for which one or more properties are established sufficiently well to allow its use in evaluating a measurement process or an assay.

   d. *Standard.* In common usage, any substance used to determine the response of an instrument or a method to the analyte of interest. The information obtained from a standard solution is often portrayed graphically in a **standard curve** (also called a **calibration curve**).

3. *A standard is a document established by consensus and approved by a recognized body that establishes rules or guidelines to make a procedure consistent among various people.* For example, ASTM specifies standard methods to calibrate volumetric glassware; the U.S. Pharmacopeia specifies methods to perform tests of pharmaceutical products.

---

between 500.0012 *g* and 499.9988 *g*. The tolerances for a standard, a measuring device, or an instrument vary depending on the purpose for which the item is being used. More examples of tolerances for specific devices are shown in later chapters.

If any instrument or piece of apparatus is not properly calibrated, its measurements will deviate unacceptably from their correct values. Improper calibration is a common cause of laboratory error; therefore, maintaining instruments "in calibration" is critical.

ii. CALIBRATION AS A FORMAL ASSESSMENT OF A MEASURING INSTRUMENT

The word *calibration* is sometimes used when an instrument, standard, or measuring device is formally assessed, but is not adjusted. During the assessment, the measuring system being calibrated is compared with a trustworthy standard. Errors in the readings of the item being calibrated are determined and documented. The result of this type of calibration is a document that certifies that the measuring item was functioning in a particular way, what corrections, if any, need to be made to its readings when it is used, and how certain the user can be of its readings. After this type of calibration the item being calibrated does not perform any better than it did before. However, the calibration document may allow the operator to use the item with better accuracy by applying correction factors to its readings. For example, it may be established by calibration that the readings of a thermometer are consistently 0.1 degree too high. Then, every time the thermometer is used, 0.1 degree must be subtracted from its readings. At present, the word *calibration* is used to sometimes mean an adjustment to a measuring device and to sometimes mean a formal, documented evaluation of an item.

iii. VERIFICATION

It is good laboratory practice to check the performance of instruments regularly to make sure they are functioning properly. **Verification** or **performance verification** *means to check the performance of an instrument or system.* Verification includes checking whether the instrument is properly calibrated. The checking done during verification is a simpler, less rigorous evaluation of the performance of a measuring item than is calibration. Verification is performed in the laboratory of the user and is typically documented in a logbook or on a form, but not with a formal document. For example, in many laboratories balance accuracy is checked regularly by weighing a standard and recording its weight. If the value for the standard's weight does not fall within a particular range, as specified by a quality control procedure, then the balance is repaired. This check of the balance's performance verifies that the balance is properly calibrated, but it should not be mistakenly called "calibrating the balance."

iv. CALIBRATION (STANDARD) CURVES

A calibration curve relates to chemical and biological assays where the response of an instrument to an ana-

Table 13.3 CALIBRATION

1. **Calibration brings a measuring system into accordance with an external standard(s).** Calibration may involve adjusting an instrument so that its readings are consistent with an external standard.

2. **Calibration is sometimes defined as a formal assessment that establishes, under specified conditions, the relationship between values indicated by a measuring instrument or measuring system and known values based on a trustworthy standard.** The result of such calibration is a document. The instrument is not adjusted according to this definition of calibration.

3. **Tolerance is the amount of error that can be tolerated in the calibration of a particular item.** National and international organizations (including ASTM and NIST) specify the tolerances allowed for particular classes of glassware, weight standards, and other measurement items.

4. **Verification is a check of the performance of an instrument or system.** Verification is a simpler, less rigorous evaluation of the performance of a measuring item than is calibration.

5. **Instrumental quantitative analysis frequently involves the construction of a standard (calibration) curve that shows graphically the relationship between the response of the instrument and the amount of reference standard present.**

---

lyte is determined. For example, it is common to evaluate the amount of protein present in a solution using a spectrophotometer. It is necessary to "calibrate" the response of the spectrophotometer to the amount of protein present. To perform this calibration, the response of the instrument to solutions containing known amounts of protein is plotted on a graph. The solutions with known amounts of protein are "standards". The resulting graph is a "standard curve" or a "calibration curve" (see pp. 159–161 for more details).

Table 13.3 summarizes terminology relating to calibration.

D. Traceability

I. THE MEANING OF TRACEABILITY

Suppose you use a balance to weigh* an object. How can you be certain, and demonstrate to others, that the balance you used was indeed properly calibrated? This is a quality assurance problem and leads to the concept of **traceability**. **Traceability** *describes the chain of calibrations that establishes the value of a standard or of a measurement.*

The concept of traceability dates at least back to the ancient Egyptians, who used the pharaoh's arm as the national standard of length. Because the pharaoh could not always be present when measurements were made, the length of his arm was reproduced using a granite rod. The rod, in turn, was duplicated to make wooden measuring sticks that were used by workers building the pyramids. This system was an early example of using an external standard for measurement and traceability to this standard. ("Chemical Measurements and the Issues of Quality Comparability and Traceability," W.P. Reed, *American Laboratory,* p. 18. December, 1994.)

Mass standards provide another example of traceability. A mass standard used in an individual laboratory might be calibrated according to a secondary standard that was in turn calibrated against K_{20}, which

was in turn calibrated relative to the international standard in France. Thus, there is a "genealogy" for the standard that is used in a particular laboratory, Figure 13.4. For traceability purposes, the genealogy for a standard must be documented in a formal certificate.

The purpose of tracing the genealogy of a standard is to ensure that measurements made with that standard, or calibrations performed using that standard, are trustworthy. A statement of traceability is a quality statement. Manufacturers use the term *traceable to NIST* in their catalogues when they have documented the genealogy of standards used in manufacturing their product, Figure 13.5.

Table 13.4 summarizes "traceability".

ii. SUMMARY: THE RELATIONSHIP BETWEEN STANDARDS, CALIBRATION, AND TRACEABILITY

To summarize the relationship between standards, calibration, and traceability, consider the example of making a measurement of weight using a balance. The balance was *calibrated* by a technician so that its readings are correct according to an internationally accepted definition of a gram. To calibrate the balance, the technician used secondary mass *standards*. The technician knows that the standards are correct embodiments of the unit called a "gram" because the comparisons of those standards to the K_{20} kilogram standard were properly performed and documented. The secondary standards are therefore *traceable* to NIST.

The ISO 9000 standards refer to this relationship between standards, traceability, and calibration in the following language:

> The supplier [any ISO 9000 compliant company or other entity] shall identify all . . . measuring . . . equipment that can affect product quality, and calibrate and adjust them at prescribed intervals, or prior to use, against certified equipment having a known valid relationship to internationally or nationally recognized standards (Section 4.11.2b).

---

*Weight and mass are not synonyms, as will be explained in a later chapter. For simplicity, however, we ignore the distinction in this chapter.

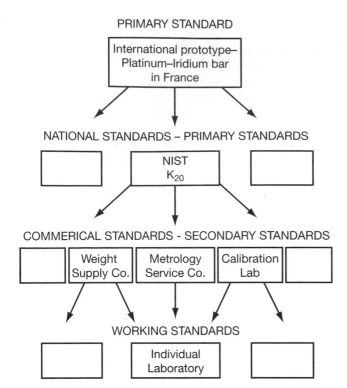

Figure 13.4.   Traceability of Mass Standards Showing Genealogy.

ASTM Class 1 calibration masses with Weight Calibration Certificate come with a statement of traceability to NIST standards, including a listing of each weight with its allowable tolerance, the actual mass value of each weight, environmental conditions during calibration, and date of calibration.

(a)

> **CERTIFICATE OF ACCURACY**
> This thermometer identified by Serial No. _____
> was compared with a standard calibrated at the National Institute of Standards and Technology (NIST), formerly the National Bureau of Standards (NBS), and was found to be accurate within ±1°C. The indications of the thermometer are traced to NIST.
> The Standard Serial Number is 55296
> The NIST Identification No. is 94415
> **ERTCO**
> **Ever Ready Thermometer Co., Inc.**

(b)

Figure 13.5.   **Catalogue Specifications Indicating Traceability to NIST. a.** A description of a mass standard. **b.** A description of a thermometer. (b. is reproduced courtesy of Ever Ready Thermometer Co., Inc.)

Table 13.4   *TRACEABILITY*

1. *Traceability describes the chain of calibrations ("genealogy") that establish the value of a standard or a measurement.*
2. *In the United States, traceability for physical and some chemical standards is often to NIST, since NIST maintains primary standards.*
3. *For the purpose of traceability, measurement values reported must include uncertainties* (discussed later in this chapter).

II. MEASUREMENT ERROR

A. Variability and Error

If you were to weigh the same standard 10 times, then would you get the same value every time? It seems reasonable to expect the 10 values to be identical if you follow the same procedure every time under uniform conditions. In reality, if you were to meticulously weigh the same object repeatedly with a high-quality balance, then the results would vary slightly each time. Twenty such weight measurements are shown in Table 13.5. Note that the first four digits in the weight values for the standard were always the same, but the last three digits varied.

It turns out that there is variability inherent in all observations of nature, including measurements. There is also uncertainty in all measurement values. What does this standard really weigh—exactly 10 g, 9.999591 g, 9.999594 g? Because of the variability in the measurements, we do not know the exact, true value for this standard. In fact, in principle, we can never be certain of the exact "true" value for any measurement, although we can approach that true value.

Error *is responsible for the difference between a measured value for a property and the "true" value for that property.* Statisticians classify measurement errors into types. One such classification scheme is:

1. **Gross errors** *are caused by blunders.* For example, if a technician were to drop a mass standard (or drop a balance), these would be gross errors. Gross errors are recognizable and, of course, should be avoided.
2. **Systematic errors** are normally more subtle and harder to detect than gross errors. There are a vast number of causes of systematic errors, such as a contaminated solution, a malfunctioning instrument, an environmental inconsistency (like a change in humidity), and so on.

An important feature of systematic error is that it results in **bias,** *that is measurements that are consistently either too high or too low.* For example, if a flask used to measure 500 mL has its 500 mL mark set slightly too low, then it will consistently deliver volumes that are a little less than they should be.

Table 13.5   *VALUES OF REPEATED WEIGHT MEASUREMENTS OF A STANDARD OBJECT*

9.999600 g	9.999601 g	9.999597 g	9.999592 g	9.999601 g
9.999595 g	9.999597 g	9.999599 g	9.999599 g	9.999598 g
9.999592 g	9.999600 g	9.999604 g	9.999600 g	9.999593 g
9.999603 g	9.999596 g	9.999595 g	9.999600 g	9.999590 g

3. Random errors *are extremely difficult or impossible to find and eliminate. Random errors cause measurement values that are sometimes too high, and sometimes too low.* Although random error is called "error," in the presence of only random error (and not gross or systematic error) measurements will average the correct, "true" value. (Note that random and systematic errors are also discussed in Chapter 12.)

What type of error is likely responsible for the variability in weight measurements shown in Table 13.5? If the technician making these measurements blundered or missed subtle factors then the variability would be due to gross or systematic error. If, however, the operator was very skilled and the method of measurement was carefully planned (as we will assume was the case), then the variability in weight measurements was due to random error, which is difficult or impossible to eliminate. Thus, there is error every time a measurement is made. Even if one uses such excellent technique that all gross and systematic errors are eliminated, there will still be random errors.

If we assume that there is no systematic error or bias in the measurements, then we must also conclude that the standard in Table 13.5 weighs slightly less than 10 *g*. If it was intended to be 10 *g*, then there was an error made at the time of its manufacture. (Note that if this standard is used to calibrate a balance, in the sense of adjusting the instrument, then all subsequent readings of this balance will be a bit too high and systematic error will be introduced.)

The word *error* has another, somewhat different, usage than we have given so far. Error is sometimes expressed as:

ERROR

Error = *True value − Measured value*

Although, in principle, we can never be certain of the exact "true" value for a measurement, in practice, this expression of error has practical application and will be explored further later in this chapter.

B. Accuracy and Precision

At this point, we introduce two important words: **accuracy** and **precision**. Measurement errors lead to a loss of accuracy and precision. These words may be defined as:

> **Precision** *is the consistency of a series of measurements or tests.*
>
> **Accuracy** *is how close a measurement value is to the true or accepted value.*

It is common to talk about the precision and accuracy of instruments, tests, assays, and methods. For example, we can speak of how consistent (precise) a balance's readings are, or the "correctness" (accuracy) of a test for blood glucose.

Accuracy and precision are not synonyms to scientists, although they are often used interchangeably in nonscientific English. It is possible for a series of measurements to be precise (consistent) but not accurate, Figure 13.6a. A series of measurements may also average the correct answer yet lack precision, Figure 13.6b. A series of measurements may be neither accurate nor precise, Figure 13.6c, or may be both accurate and precise, Figure 13.6d. One definition of a "good" measurement is that it is both accurate and precise.

Consider precision in more detail. Laboratory workers are aware that measurements repeated in succes-

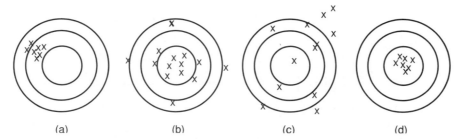

Figure 13.6.   Accuracy and Precision. Precision and accuracy are illustrated using the analogy of a target where the bull's eye is the correct value. **a.** Archer A is precise but not accurate. **b.** Archer B is inconsistent. Archer B's results average the correct value though they are not precise. **c.** Archer C is not very skilled; the results are neither accurate nor precise. **d.** Archer D is expert and is both precise and accurate.

sion on the same day tend to be relatively consistent. In contrast, measurements performed on different days, by different people, and using different materials and equipment, tend to be far more variable. There are different words for precision to make this distinction clear. **Repeatability** *is the precision of measurements made under uniform conditions.* **Reproducibility** *is the precision of measurements made under nonuniform conditions, such as in two different laboratories.* Repeatability and reproducibility are therefore two practical extremes of precision. It is challenging, but important, to work to develop procedures that give results that are as reproducible as possible when performed on different days, by different people, and so on.

EXAMPLE PROBLEM

Four students each completed a laboratory exercise in which they weighed a standard known to be 5.0000 g. The values they obtained (in grams) are shown in the following table. Comment on the accuracy and precision of each student's results.

Juan	Chris	Ilana	Mel
4.9986	5.0021	5.0001	5.1021
5.0020	5.0020	4.9998	4.9987
5.0007	5.0021	4.9999	5.0003
4.9995	5.0022	4.9998	4.9977

ANSWER

Juan's data have a mean value that is reasonably accurate; however, their precision is not as good as the data of other students.

Chris has precise but inaccurate values. It is likely that there is a systematic error in Chris's work which causes all the values to be too high (to be biased).

Ilana's values are both accurate and precise with some random fluctuations affecting only the last figure (the

fourth place to the right of the decimal point). These data are considered to be "good"; we expect *some* variability in measurements.

Mel's values are neither accurate nor precise. We might suspect that Mel was careless or had problems with the equipment.

C. The Relationship Between Error, Accuracy, and Precision

i. RANDOM ERROR AND LOSS OF PRECISION

Random error leads to a loss of precision because it leads to inconsistency in measurements; due to random error, values are sometimes too high and sometimes too low. We saw the effects of random error on the precision of weight measurements in the example in Table 13.5.

We can also consider a flask that is used to measure a volume of 500.0 *mL*. Suppose the flask is perfectly marked with a line at exactly 500.0 *mL*. Although the flask is marked correctly, every time it is used there will be a tiny variation in how much liquid is delivered due to imperceptible variations in the environment and the person using the flask. This slight variability is random error and will result in volumes being delivered that are sometimes a bit more than 500.0 *mL* and sometimes a bit less, Figure 13.7a. In this example, because the flask is perfectly marked at exactly 500.0 *mL*, if the flask is used many times, then the average volume delivered will equal the true value, 500.0 *mL*, Figure 13.7b.

It is good laboratory practice to repeat critical measurements, assays, and tests to determine the impact of random error. As measurements are repeated, you can see their variability. In addition, blunders often show up when repeating measurements. For example, one might accidentally misread a meter or remove the wrong amount of a sample but notice the error when the measurements are repeated.

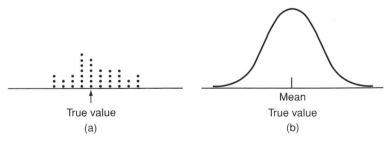

True value
(a)

Mean
True value
(b)

Figure 13.7.   Random Error. a. Hypothetical results of repeated measurements using a flask that is correctly marked at 500.0 *mL*. Each point represents one measurement. **b.** An idealized frequency distribution assuming the same volume measurement was made a great many times with a flask perfectly marked at 500.0 *mL*. The mean value and the true value are the same.

ii. ERRORS AND LOSS OF ACCURACY

Gross, systematic, and random errors all lead to a loss of accuracy. It is obvious how a gross error could cause a value to be incorrect. A systematic error will also affect the accuracy of a measurement. For example, consider a systematic error where a flask intended to measure out exactly 500.0 *mL* has its line drawn slightly too low on the flask. Every time this flask is used it will tend to deliver slightly less than 500.0 *mL*. The volume it delivers is obviously inaccurate. When there is systematic error, the system is biased, *and the mean (average) of the measurements will be too high or too low,* Figure 13.8.

Systematic error can result from many causes. For example, equipment that is improperly calibrated, is not well-maintained, or is used improperly; solutions that have degraded; and environmental fluctuations all cause systematic error. Unlike random error, the impact of systematic error is not detected by repeating measurements. If a systematic error is present, then every time the measurement is repeated it will incorporate the same error and tend to be too high or too low. To minimize systematic errors in measurement, it is necessary to be knowledgeable in the methods used, to be attentive to potential problems, and to properly maintain and calibrate equipment.

We know that random error causes a loss of precision. Does random error also cause a loss of accuracy? This question is somewhat more complex. Consider again Figure 13.6b. On the average, the accuracy of this archer is pretty good; however, if we look only at a single measurement, it may be way off center. Thus, when

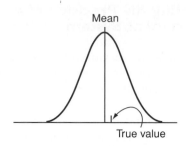

Figure 13.8.   Measurements from a Biased System. In the presence of bias, the mean is not the true value for the measurement.

considering an individual measurement, or only a few measurements, if a system is relatively imprecise, then its accuracy will also be poor. The same idea is illustrated in Figure 13.7a and 13.7b. If a single measurement from Figure 13.7a is selected, that measurement may or may not be close to the correct reading of 500 *mL* even though the average, as shown in Figure 13.7b, is accurate. Random error can thus lead to both a loss of precision and a loss of accuracy.

In practice, there are both random and systematic errors present when measurements are made in the laboratory. It is often difficult to tease apart the effects of random errors and systematic errors. Therefore, the term **inaccuracy** *is used to mean the combination of all random and systematic errors affecting the results of measurements.* **Imprecision** *is used to describe the effects of only random errors.* Terms relating to accuracy, precision, and error are summarized in Table 13.6

Table 13.6   ERROR, ACCURACY, AND PRECISION

1. *Error is sometimes defined as the difference between a measured value and the true value.*

2. *Error sometimes refers to the cause of variability and inaccuracy. Errors can be classified into three types: gross, systematic, and random.*

   a. *Gross errors. Human blunders.*

   b. *Systematic errors. Errors that cause results to generally be either too high or too low. Repeating measurements is not a useful way to detect systematic errors.*

   c. *Random errors. Errors that cause results to be sometimes too high and sometimes too low and which are difficult or impossible to eliminate completely. Repeating measurements is a useful way to determine the magnitude of random error.*

3. *Errors occur whenever laboratory measurements are performed.*

4. *Errors cause uncertainty in measurements.*

5. *Bias occurs when there is a systematic error(s) that causes measurement values to tend to be too high or too low.* In a biased system, the mean differs from the true value.

6. *Precision is the consistency of a series of measurements.* Precision is affected by random error.

   a. **Repeatability** *is the precision of measurements repeated in succession.*

   b. **Reproducibility** *is the precision of measurements performed under varying conditions.*

7. *Accuracy is sometimes defined as how close a measurement is to the true or accepted value.* In contrast, inaccuracy is how far a measurement is from the true or accepted value.

8. *Uncertainty is an estimate of the inaccuracy of a measurement that includes both the random and bias components. Uncertainty is also defined as an estimate of the range of values in which the "true" value lies.*

D. Evaluating the Precision of a Measurement System

Random error leads to variability; it causes repeated measurements to vary from one another. Variability in a series of measurements can be evaluated using statistical tools. For example, consider hypothetical values obtained by a technician analyzing lead levels in paint samples, Table 13.7. The analyst observes on February 20 that the values from two houses seem to fluctuate more than normal. This observation alerts her to the possibility that there might be a problem in the measurement system. She can analyze the variability in the data using various statistical tools including the **range,** the **standard deviation (SD)**, and the **coefficient of variation (CV)**.

The **range** *is the difference between the highest and the lowest values of a set of measurements.* Range is a simple indicator of precision: If the range is narrow, then the data are less variable than if the range is wide. Table 13.8 shows the range for the lead measurements. Note that the range of values for February 20 is greater than for the other days.

Standard deviation, SD is commonly used to evaluate the variability of a group of measurements. (See Chapter 11 for a discussion of SD and CV.) A series of measurements with a lower SD have better precision than a series of measurements from a similar system with a higher SD. The **coefficient of variation (CV)** expresses the standard deviation of a series of measurements in terms of the mean. The standard deviation and the coefficient of variation for the measurement values on February 20 were higher than on the other days, Table 13.8.

These results suggest that for some reason the data obtained on February 20 were more variable than on the other dates. There are many possible explanations for this variability. The variability might be due to a characteristic of the houses themselves. The windows from which the paint samples were taken may have had many layers of different paints applied, in which case the variability resulted from lack of homogeneity in the samples. An instrument could have been malfunctioning on February 20 and required repair. A reagent involved in the measurement procedure may have degraded. The technician at this point will want to identify the source of variability and correct any problems that may be present.

Another example of the use of standard deviation to express the precision of a measurement system is shown in Figure 13.9. Catalogue descriptions of instruments often report the precision of the instrument in

Table 13.7 *Lead Levels in Paint from Houses*

Feb. 4 House 1 Window (mg/cm^2)	Feb. 4 House 2 Window (mg/cm^2)	Feb. 8 House 3 Window (mg/cm^2)	Feb. 8 House 4 Window (mg/cm^2)	Feb. 8 House 5 Window (mg/cm^2)	Feb. 10 House 6 Window (mg/cm^2)	Feb. 20 House 7 Window (mg/cm^2)	Feb. 20 House 8 Window (mg/cm^2)
1.21	0.43	0.92	0.78	1.51	2.12	0.98	1.23
1.18	0.40	0.93	0.67	1.43	1.99	0.78	0.21
1.31	0.34	0.79	0.71	1.34	2.13	0.21	0.11
1.23	0.41	0.93	0.65	1.47	1.98	1.24	0.89

Some information for this example was taken from *Field Detection of Lead in Paint and Soil by High-Resolution XRF,* J.N. Driscoll, R. Laliberte, and C. Wood. *American Laboratory*, p. 34H, March, 1995.

Table 13.8 *Statistical Analysis of Paint Lead Levels Data*

Feb. 4 House 1 Window (mg/cm^2)	Feb. 4 House 2 Window (mg/cm^2)	Feb. 8 House 3 Window (mg/cm^2)	Feb. 8 House 4 Window (mg/cm^2)	Feb. 8 House 5 Window (mg/cm^2)	Feb. 10 House 6 Window (mg/cm^2)	Feb. 20 House 7 Window (mg/cm^2)	Feb. 20 House 8 Window (mg/cm^2)
1.21 1.18	0.43 0.40	0.92 0.93	0.78 0.67	1.51 1.43	2.12 1.99	0.98 0.78	1.23 0.21
1.31 1.23	0.34 0.41	0.79 0.93	0.71 0.65	1.34 1.47	2.13 1.98	0.21 1.24	0.11 0.89
Range = 1.31 − 1.18 = 0.13	*Range =* 0.09	*Range =* 0.14	*Range =* 0.13	*Range =* 0.17	*Range =* 0.15	*Range =* 1.03	*Range =* 1.12
Mean = 1.23	*Mean* = 0.40	*Mean* = 0.89	*Mean* = 0.70	*Mean* = 1.44	*Mean* = 2.06	*Mean* = 0.80	*Mean* = 0.61
SD = 0.06	*SD* = 0.04	*SD* = 0.07	*SD* = 0.06	*SD* = 0.07	*SD* = 0.08	*SD* = 0.44	*SD* = 0.54
CV = 4.5%	*CV* = 9.8%	*CV* = 7.7%	*CV* = 8.2%	*CV* = 5.1%	*CV* = 3.9%	*CV* = 55%	*CV* = 88%

Catalog number	pH 20
Type	pH
Range	pH 2.0 to 12.00
Resolution	0.01 pH unit
Repeatability	± 0.1 pH unit

Figure 13.9. Catalogue Descriptions of a Measurement Instrument. A pH meter description. Repeatability (± 0.1 pH unit) is the standard deviation around an undisclosed mean. Note that the term *range* is not an indicator of precision in this case, rather, it indicates the span from the highest to the lowest pH that the meter can read.

terms of the standard deviation of a series of measurements made with the instrument. The more consistent the instrument, the lower its *SD*, and, presumably, the better its quality.

EXAMPLE PROBLEM

Which result from a series of measurements is more precise: 15.0 ± 0.3 g (mean ± *SD*) or 15.00 ± 0.03 g (mean ± *SD*).

ANSWER

The *SD* "0.03" is smaller, or more precise, than 0.3 g; therefore, 15.00 ± 0.03 g is more precise.

EXAMPLE PROBLEM

A biotechnology company manufactures a particular enzyme that is used to cut DNA strands. The enzyme's activity can be assayed and is reported in terms of "units/mg". Each batch of enzyme is tested before it is sold. The results of repeated tests on four batches of enzyme are shown in the table.

a. What is the mean activity of the enzyme for each batch?

b. What is the *SD* for each batch?

c. What is the mean activity for all batches combined?

d. What is the *SD* for all batches combined?

(See p. 189 for an explanation of *SD* and p. 248 for the equation for calculating *SD*.)

ENZYME ACTIVITY (UNITS/MG)

Batch 1	Batch 2	Batch 3	Batch 4
100,900	100,800	110,000	123,000
102,000	101,000	108,000	121,000
104,000	100,100	107,000	119,000
104,100	100,800	109,100	121,000

ANSWER

Mean batch 1 = 102,750 *units/mg*   SD = 1567 *units/mg*

Mean batch 2 = 100,675 *units/mg*   SD = 395 *units/mg*

Mean batch 3 = 108,525 *units/mg*   SD = 1305 *units/mg*

Mean batch 4 = 121,000 *units/mg*   SD = 1633 *units/mg*

Mean all batches = 108,238 *units/mg*

SD all batches = 8,254 *units/mg*

Note that there is, as we might expect, more variability *between* the batches than there is *within* one batch. This is reflected in the fact that the standard deviation for all the batches is higher than the standard deviation of any one batch.

E. Evaluating the Accuracy of a Measurement System

We saw in the last section how the precision of an instrument or measuring system is evaluated by performing a series of measurements and applying statistical tests to the results. Next consider the evaluation of accuracy. Accuracy is the closeness of agreement between a measurement or test result and the true value, or the accepted reference value, for that measurement or test.

We generally do not know the "true" value for a measurement. For example, suppose an analyst is testing the level of glucose in an individual sample of blood. The analyst does not know the *true* blood glucose value for that sample—if the value was known, there would be no point in doing the test. The analyst, therefore, cannot calculate the accuracy of the measurement for that sample. The analyst can, however, evaluate the accuracy of the method itself.

The most obvious way to assess the accuracy of a method or an instrument is to use a standard. For example, the assay method for glucose can be evaluated by testing a blood sample with a known amount of glucose. This sample with a known amount of glucose is a quality control (or simply control) sample. The analyst assays the level of glucose in the control sample using the same assay method that is used for patient blood samples. The result obtained for the control is compared with the expected result to determine the accuracy of the test procedure.

Two simple ways to express accuracy (or inaccuracy) mathematically are:

Expressing Accuracy (or Inaccuracy) as "Absolute Error"

ABSOLUTE ERROR =
True Value − Average Measured Value

where "error" is an expression of accuracy (also called "inaccuracy") and the true value may be the value of an accepted reference material

Expressing Accuracy (or Inaccuracy) as "Percent Error"

$$\% \; ERROR =$$

$$\frac{True\ Value - Average\ Measured\ Value}{True\ Value} \times 100\%$$

where "% error" is an expression of accuracy (also called "inaccuracy") and the true value may be the value of an accepted reference material.

EXAMPLE PROBLEM

Suppose the glucose level in a control sample is stated to be 1000 *mg/L*. A technician performs a glucose test 10 times on this control sample and gets the following values:

996 *mg/L*, 1009 *mg/L*, 1008 *mg/L*, 998 *mg/L*, 1001 *mg/L*, 999 *mg/L*, 997 *mg/L*, 1000 *mg/L*, 1008 *mg/L*, and 1010 *mg/L*.

What are the absolute error and percent error based on these data?

ANSWER

The average value from the 10 tests is 1002.6 *mg/L*. The absolute error, based on this average is:

$$1000\ mg/L - 1002.6\ mg/L = -2.6\ mg/L$$

The percent error, based on this average is:

$$\frac{1000\ mg/L - 1002.6\ mg/L}{1000.0\ mg/L} \times 100\% = -0.26\%$$

It is good practice to check the performance of a test using one or more quality control samples on a regular basis. This practice of checking methods and instruments with controls is often a documented part of a quality control program. Based on experience, repetition of the test, and knowledge of the system, it is possible to establish a range within which values for the test should fall. If the results of a test lie outside of this range, then it is necessary to look for problems.

Figure 13.10 shows another example of the application of error, percent error, standard deviation, and relative standard deviation. A manufacturer tested their volume measuring devices at different volumes. The manufacturer specifies how accurately and precisely the volume measuring devices dispense the volumes for which they are set. (In this example, the "true" value is the volume the device should dispense.) The specifications are shown in tabular form. The first column in the table indicates the model of measuring device. The second column indicates the volume that the device should dispense (i.e., the "true volume"). The third column is the maximum percent error. The fourth column is the maximum absolute error. The fifth and sixth columns in

Model	Volume	Accuracy (mean error)		Precision (std deviation)	
	μL	%	μL	%	μL
P-2	0.2	12	0.024	6	0.012
	1.0	2.7	0.027	1.3	0.013
	2.0	1.5	0.030	0.7	0.014
P-10	0.5	5	0.025	2.8	0.014
	1	2.5	0.025	1.2	0.012
	5	1.5	0.075	0.6	0.03
	10	1	0.1	0.4	0.04

Figure 13.10.   A Catalogue Description of a Volume Measuring Device. (This illustration is copyrighted material of Rainin Instrument Co., Inc. and is used with permission.)

the table express precision in terms of the coefficient of variation and standard deviation, respectively.

EXAMPLE PROBLEM

Suppose you want to evaluate the precision of a balance. Do you need to use a standard whose mass value is traceable?

Suppose you want to check whether a balance is giving mass readings that are accurate. Do you need to use a standard whose mass value is traceable?

ANSWER

The evaluation of precision does not require a standard whose properties are known. Any object can be used. For example, you could determine the precision of the balance by checking the mass of the same pencil 20 times and calculating the standard deviation for the values.

In contrast, the evaluation of accuracy requires a standard against which the readings of the balance are compared. This is where traceability comes in: Traceability ensures that the mass of the standard used to check the balance is known.

EXAMPLE PROBLEM

An environmental testing laboratory is about to begin testing water for chromium. The company's scientists learn to perform the chromium assay and test their skills by assaying commercially prepared standards with known chromium concentrations. They use four standards and obtain the results below. The scientists want to be able to guarantee their customers that they will be able to analyze chromium in samples with less than ± 2% error. Fill in the table to show the relative percent error for each measurement. Have the scientists met their goal for accuracy based on these results?

Actual Concentration of Chromium in the Standard	Assayed Concentration of Chromium Obtained in the Laboratory	% Error
1.00 $\mu g/L$	0.91 $\mu g/L$	
5.00 $\mu g/L$	4.78 $\mu g/L$	
10.00 $\mu g/L$	9.89 $\mu g/L$	
15.00 $\mu g/L$	15.08 $\mu g/L$	

ANSWER

The percent errors are, in order: 9.00%, 4.40%, 1.10%, −0.53%

These data suggest that at low concentrations they have not achieved the desired accuracy, although they have at higher concentrations. Refinements in their technique are required.

EXAMPLE PROBLEM

An assay used to measure the amount of protein present in samples is giving erroneous results that are low by about 5 mg. What will be the percent error due to this problem if the *actual* protein present in a sample is:

i. 30 mg **ii.** 50 mg **iii.** 100 mg **iv.** 550 mg

ANSWER

i. (true − measured value)/true value
= (30 − 25 mg/30 mg) × 100 = 16.7%

ii. 10.0%   **iii.** 5.0%   **iv.** 0.91%

Note that the impact of this error is more pronounced when protein is present in low amounts than when protein is present at higher levels; the 5 mg error is a larger percent of 30 mg than it is of 550 mg.

EXAMPLE PROBLEM

A new balance must be purchased for a teaching laboratory. A teacher compares three competing brands by measuring a 1.0000 g standard five times on each balance.

a. Which balance is most accurate? Show the percent error for each balance.

b. Calculate the SD for the measurements from each balance. Which balance is most precise?

c. What is the CV for each of the balances?

d. Report the mean value for the standard from each balance ± the SD.

e. Which balance would you buy?

Brand A (grams)	Brand B (grams)	Brand C (grams)
1.0004	0.9997	1.0003
1.0005	0.9996	0.9996
1.0004	1.0003	1.0002
1.0003	1.0002	0.9995
1.0005	0.9999	1.0004

ANSWER

Balance A	Balance B	Balance C
Mean = 1.0004 g	Mean = 0.9999 g	Mean = 1.0000 g
% error = −0.04%	% error = 0.01%	% error = 0.00%
SD = 0.0001 g	SD = 0.0003 g	SD = 0.0004 g
CV = 0.01%	CV = 0.03%	CV = 0.04%
Mean ± SD = 1.0004 ± 0.0001 g	Mean ± SD = 0.9999 ± 0.0003 g	Mean ± SD = 1.0000 ± 0.0004 g

The percent error is an indication of the accuracy of the balances. Precision is shown by the SD and CV. Although the mean value of Balance C is correct, its precision is poor compared with the other balances, and any single measurement made on Balance C is likely to be inaccurate. Balance A has the best precision, but it is biased—all the readings are a bit high, which suggests a systematic problem. If Balance A can be recalibrated or otherwise be made accurate, it might be a good choice. Balance B might be an acceptable compromise, particularly if it has other qualities that are desirable, such as ruggedness and ease of use.

III. INTRODUCTION TO UNCERTAINTY ANALYSIS

A. What Is Uncertainty?

The term *uncertainty* is used often in the literature of metrology. To some extent the meaning of *uncertainty* is familiar. For example, consider a statement in the newspaper that "the distance from the earth to the sun is 156,300,000 km". Next, consider a statement that "there are 10 microscopes in our laboratory". You would know from experience that the figure given for the distance to the sun is not likely to be exactly correct; in fact, it could be off by many kilometers. In contrast, the statement that "there are 10 microscopes in the laboratory" is likely to be exactly correct. There is uncertainty in the value reported for the distance to the sun, while there is little, if any, uncertainty in a count of 10 microscopes.

As there is uncertainty in the figure reported for the distance to the sun, so there is uncertainty in the measurements we make in the laboratory. The reason there is uncertainty in laboratory measurements is that error exists. Even when people are very careful, random errors and sometimes systematic errors persist.

Metrologists try to estimate the effect of errors so they can know how much confidence to place in a measurement. Metrologists state that a "good" measurement must include not only the value for the measurement, but also a reasonable estimate of the uncertainty associated with the value. Calibration and traceability documents include statements of uncertainty. Thus, a measurement should properly be of the form:

measured value ± an estimate of uncertainty

Consider, for example, the measurements of the standard weight as shown in Table 13.5. We can be fairly certain that the standard weighs a little less than 10 *g* although we cannot be sure of its exact true weight. The best estimate of the true weight of the standard is the mean of a large number of measurements. The mean weight of the standard based on the twenty measurements in Table 13.5 is 9.999598 *g*. Therefore, to the best of our knowledge, the standard's true weight is 9.999598 *g* —"give or take" something. The "give-or-take" something is the **uncertainty in the measurement**.* **Uncertainty** *is an estimate of the inaccuracy of a measurement due to all the errors present.*

Estimating uncertainty is complex. It requires identifying to the best of one's knowledge all sources of error, estimating the magnitude of each error, and combining all the effects of error into a single value. Not surprisingly, it is difficult to state, with certainty, how much uncertainty there is in a measurement. How to best evaluate uncertainty is therefore the subject of much discussion.

B. Using Precision as an Estimate of Uncertainty

Although it is difficult to perform an in-depth uncertainty analysis, it is reasonably straightforward to evaluate the uncertainty due to random error. It is common for analysts to repeat a particular measurement and to summarize the uncertainty due to random error using the standard deviation of the repeated measurements.**

Random error is not the only type of error that may be present when a measurement is made. However, it is more difficult to estimate the uncertainty due to other types of errors. In situations where a measurement process is well understood, the analyst may make estimates of the uncertainty caused by a variety of errors. For example, a measuring system may be known to be affected by changes in barometric pressure. The analyst may therefore add in a factor for uncertainty caused by fluctuations in barometric pressure. There are statistical methods that have been developed to deal with these types of uncertainty. (For an explanation of these methods and of current practice in uncertainty measurement, consult for example, Nicholas and White, or the NIST publication "Guidelines for Evaluating and Expressing the Uncertainty of NIST Measurement Results"; both are referenced at the end of this chapter.)

C. Using Significant Figures as an Indicator of Uncertainty

i. THE MEANING OF SIGNIFICANT FIGURES

Measurements are properly displayed as a value ± an estimate of uncertainty. We previously discussed methods to estimate the figure that goes after the ± sign. In reality, biologists routinely report measurement values without explicitly adding any ± uncertainty estimate to the value. There are, however, conventional practices used to report measurement values that roughly indicate the certainty of the measurement. These practices are covered under the heading of "significant figures."

Significant figures are the most basic way that scientists show the certainty in a measurement value. A **significant figure** *is a digit within a number that is a reliable indicator of value.* It is easiest to understand significant figures by looking at practical examples from the laboratory and everyday life.

EXAMPLE 1: RULERS

Consider the length of the arrow drawn in Figure 13.11a and 13.11b.

In Figure 13.11a, the ruler's gradations divide each centimeter in half. We know the arrow is somewhat longer than 4 *cm* and it is reasonable to estimate the tenth place and report the arrow's length as "4.3 *cm*." Information will be lost if we report the length as simply "4 *cm*" because it is possible to tell that the arrow is somewhat longer than 4 *cm*. It would be unreasonable to report that the arrow is "4.35 *cm*" because the ruler gradations give no way to read the hundredths place. When measurements are recorded, it is customary to record all the digits of which

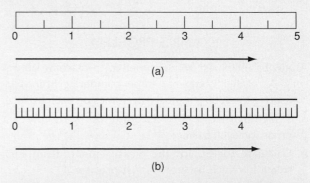

(a)

(b)

Figure 13.11. Measurements of Length with Two Rulers. a. A ruler where each centimeter is subdivided in half. **b.** A ruler where each centimeter is subdivided into tenths.

---

*In this particular example, if all the uncertainty is due to random error, then it is possible to give a value to the "give or take something". In this case, the standard error of the mean is used to estimate uncertainty. See, for example, "Statistics" by Freedman *et al.,* which is referenced at the end of this chapter.

**The standard error of the mean, SEM, is used sometimes in a similar fashion, as is a confidence interval for the mean. The calculation of both the SEM and a confidence interval begins with determining the mean and the standard deviation for a series of repeated measurements.

we are certain plus one that is estimated. In this example, the 4 is certain and the .3 is estimated, so the measurement is said to have two significant figures. Thus, by reporting the measurement to be 4.3 *cm*, we are telling the reader something about how certain we are in the length. We are sure about the 4 and not as sure about the .3.

The subdivisions in the second ruler, Figure 13.11b, are finer and divide each centimeter into tenths. With the second ruler, we can reliably say the arrow is "4.3 *cm*," and, in fact, it is reasonable to estimate that the arrow is about "4.35 *cm*." The 4.3 is certain, the .05 is estimated and so there are three significant figures in the measurement. Thus, the arrow is the same length in both Figure 13.11a and 13.11b, yet, the length *recorded* is different because the rulers are subdivided differently. The fineness of the measuring instruments, in this case, the rulers, determines our certainty of the length of the arrow. The measurement certainty is reflected by the number of significant figures used in recording the measurement.

A Note about Terminology

• Some people would say the second ruler is more accurate than the first, because it gives values with more significant figures (i.e., with more certainty).

• Some would say the second ruler is more precise than the first because it allows us to read further past the decimal point. The word *precise* thus has two meanings. Precision may refer to the fineness of increments of a measuring device; the smaller the increments, the better the precision. Precision is also used, as explained previously in this chapter, to refer to the consistency of values: the more consistent a series of measurements, the more precise.

• Some people might say the second ruler has more resolution because it is allows us to discriminate a smaller length change than the first ruler.

EXAMPLE 2: BALANCES

Suppose a particular balance can reliably weigh an object as light as 0.00001 *g*. On this balance, sample Q is found to weigh "0.12300 *g*." It would not be correct to report that sample Q weighs "0.123 *g*" because information about the certainty of the measurement is lost if the last zeros are discarded. In contrast, if sample Q is weighed on a balance that only reads three places past the decimal point, it is correct to report its weight as "0.123 *g*." It would not be correct to record the weight as "0.1230 *g*" because the balance could not read the fourth place past the decimal point.

The first balance gives more certainty about the sample weight. The difference in measurement certainty between the two balances is shown by the number of significant figures used: "0.12300 *g* " has five significant figures, whereas "0.123 *g*" has only three.

EXAMPLE 3: WEIGHTS OF A STANDARD

Suppose that the technicians who made the weight measurements of the standard in Table 13.5 used a balance that could only weigh objects to the nearest 0.1 *g* rather than six places to the right of the decimal point. Then, every time they weighed the standard, they would have recorded its weight as "10.0 *g*," a value with three significant figures. Their results would be consistent, but they would never realize that the standard truly weighs a little less than 10 *g*. Although consistent, there is therefore less certainty and fewer significant figures in the value "10.0 *g*" than in the values shown in Table 13.5.

You can see in Examples 1–3 that a basic principle in recording measurements from instruments is to report as much information as is reliable plus one last figure that is estimated and might vary if the measurement were repeated. The number of figures reported by following this principle is the number of significant figures for the measurement. The number of significant figures reported is a rough estimate of the certainty of a measurement.

Note that most modern electronic instruments show results with a digital display. There is no meter to read. With any instrument having a digital display the last place is assumed to have been estimated by the instrument.

As a final example, let us return to the value for the distance to the sun discussed at the beginning of this section.

EXAMPLE 4: ZEROS AND SIGNIFICANCE

Suppose that one source reports the distance to the sun to be 156,000,000 *km*, but another reports it as 156,155,300 *km*. Let us assume that both sources are correct, but the first rounded the number to make it easier to read. The second number, 156,155,300 *km*, is a more exact figure for the distance, which allows the reader to be more certain about the actual distance than does the first number. We say that the number 156,155,300 has more significant figures (seven) than the number 156,000,000 (which has three significant figures).

This example can be used to illustrate an important point regarding zeros. The zeros in the reported values are essential; without them the distance would be reported as a paltry 156 *km* or as 1,561,553 *km*. The zeros are "place holders" that tell us we are talking about a very large distance, but they are not correct indicators of value. Perhaps the exact distance is really 156,155,333 *km* or 156,155,389 *km*, or any of a number of other possibilities. In contrast, when I report that there are "10 microscopes in my laboratory" the zero is a reliable indicator of value. The zero shows that there are 10, not 9 or 11, microscopes. Zeros that are place holders are not called significant figures, whereas zeros that indicate value are significant.

Suppose the number given in a report is 45,000. The three zeros in this number each may be place holders or they may be indicators of value. There are various ways to tell the reader whether the zeros at the end of a number are significant. One method is to use scientific (exponential) notation. For example, there are three zeros at the end of the number 45,000. If none of the zeros are significant, then the number could be reported as 4.5×10^4; two significant figures. If there are three significant figures the number could be reported as 4.50×10^4, and so on. If all the zeros in the number 45,000 are significant, this can be shown by placing a decimal point after the number. The number 45,000. and 4.5000×10^4 both have five significant figures.

Table 13.9 summarizes rules regarding how to record measurements with the accepted number of significant figures. These rules also show how to decide when a zero is significant and when it is a place holder.

ii. CALCULATIONS AND SIGNIFICANT FIGURES

Calculations, such as addition or multiplication, bring together numbers from separate measurements. Each value in the calculation has a particular number of significant figures. The result from the calculation is limited to the certainty (number of significant figures) of the starting number that is least certain.

The result displayed on a calculator is seldom what should be recorded when a calculation is performed because calculators do not give any indication of certainty. The calculator result must be rounded to the proper number of significant figures. When numbers are brought together in calculations it is often not obvious how many significant figures there are. There are various rules to determine the correct number of significant figures (see references). Table 13.10 summarizes simple rules that are likely to be adequate for situations in biology laboratories. (These rules sometimes seem to give odd results because they simplify what is actually a complex procedure.) Remember that these rules are guides to decide where to round the result of a calculation(s).

EXAMPLE PROBLEM

Read the following meter and the digital display. How many significant figures does each measurement have?

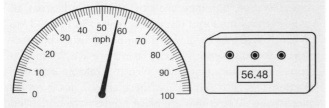

ANSWER

The meter reads **"56.6," three significant figures**; the operator estimates the last place. The digital display reads **"56.48," four significant figures**; the instrument estimates the last place.

EXAMPLE PROBLEM

Which of the following are measured values and which are counted (exact) values?

  a. The density of carbon tetrachloride at 25°C is 1.59 *g/mL*.

  b. The distance from Chicago to Milwaukee is 75 miles.

  c. Human body temperature is 37°C.

  d. There are four regional campuses in the college system.

ANSWER

Only d is counted. The other values are measured and so the rules of significant figures apply to them.

EXAMPLE PROBLEM

A catalogue advertises a particular product to be 99% pure. A competitor advertises their version of the same product to be 98.8% pure. Which one is more pure?

ANSWER

First, assume that both competitors have faithfully followed the significant figures conventions. Second, assume that the value for purity is based on a calculation(s). Then the first product might reasonably be expected to be anywhere from 98.5 to 99.4% pure because anywhere in that range the value would be rounded to 99%. The second brand would be between 98.75 and 98.84% pure. It is unclear, therefore, which brand is actually most pure. The second manufacturer, however, has presumably been able to ascertain the purity of their product with more certainty.

EXAMPLE PROBLEM

A biotechnology company specifies that the level of RNA impurities in a certain product must be less than or equal to 0.02%. If the level of RNA in a particular lot is 0.024%, does the lot meet the specification?

ANSWER

The specification is set at the hundredth decimal place; therefore, the result is also reported to that place. Rounded to the hundredth place, 0.024% is 0.02%. This lot therefore meets its specification.

Table 13.9   RULES TO RECORD MEASUREMENTS WITH THE CORRECT NUMBER OF SIGNIFICANT FIGURES

1. *The number of significant figures is related to the certainty of a measurement.* (Note that counted values, such as, there are "10 microscopes in the laboratory," or "there are 30 students in the class" are not measurements but rather are considered to be "exact." The rules of significant figures do not apply to counted values.)

2. *When reporting a measurement, record as many digits as are certain plus one digit that is estimated.* When reading a meter or ruler, estimate the last place. When reading an electronic digital display, assume the instrument estimated the last place.

3. *All non-zero digits in a number are significant.* For example, all the digits in the number 98.34 are significant; this number has four significant figures. A reader will assume that the 98.3 is certain and the 4 is estimated.

4. *All zeros between two nonzero digits are significant.* For example, in the number 100.4, the zeros are reliable indicators of value and not just place holders.

5. *Zero digits to the right of a nonzero digit but to the left of an assumed decimal point may or may not be significant.* Consider the number for the distance to the sun, 156,000,000 *km*. The decimal point is assumed to be after the last zero. In this case, the zeros are ambiguous and may or may not be reliable indicators of value. Methods of clarifying ambiguous zeros are discussed in the text.

6. *All zeros to the right of a decimal point and to the right of a nonzero digit before a decimal place are significant.* The following numbers all have five significant figures: 340.00, 0.34000 (the zero to the left of the decimal point only calls attention to the decimal point), and 3.4000.

7. *All the zeros to the left of a nonzero digit and to the right of a decimal point are not significant unless there is a significant digit to their left.* The number 0.0098 has two significant figures because the two zeros before the 98 are place holders. On the other hand, the number 0.4098 has four significant figures.

Table 13.10   RULES FOR RECORDING VALUES FROM CALCULATIONS WITH THE "CORRECT" NUMBER OF SIGNIFICANT FIGURES

1. *It is assumed that the last digit of a result from a calculation is rounded.* For example, given the result "45.6," the .6 is assumed to have been rounded; therefore, the calculated value must have been between *45.55* and *45.64*.

2. When rounding:

   a. *If the digit to be dropped is less than 5, then the preceding digit remains the same.*

   b. *If the digit to be dropped is 5 or more, the preceding digit increases by 1.*

   For example:

   54.78 is rounded to 54.8
   54.83 is rounded to 54.8
   54.65 is rounded to 54.7

   (There are different approaches to rounding when the digit to be dropped is 5. Some people, for example, round a five to the nearest *even* number. Then, 9.65 is rounded to 9.6 and 4.75 is rounded to 4.8)

3. *Round **after** performing a calculation.* If a problem requires a series of calculations, round after the *final* calculation and not after the intermediate calculations.

4. *The rule for expressing the answer after addition or subtraction is different than the rule for multiplication and division. The addition/subtraction rule focuses on the number of places to the right of the decimal point. The answer can retain no more numbers to the right of the decimal point than the number involved in the calculation having the least number of places past the decimal point.* For example, if adding the numbers 98.0008, 7.9878, and 56.2:

   98.0008
   7.9878
   56.2
   ─────────
   162.1886 Round to: 162.2

   The answer can be expressed only to the nearest tenth place because the value 56.2 has only that many places past the decimal point.

5. *In multiplication and division, keep as many significant figures as are found in the number with the least significant digits.* For example: $0.54678 \times 0.980 \times 7.899$. A calculator might display the answer as 4.232634916 but the answer should be reported as 4.23, three significant figures. Another example: $7,987 \times 12$. The answer should have only two significant figures and is therefore not 95844, but rather 96000 (the zeros are place holders). (These examples assume that all the values given are measured values.)

6. *Constants are numbers whose value is exactly known.* Constants are assumed to have an infinite number of significant figures. For example, given that "12 *in* equals a foot," 12 *in* is a constant.

PRACTICE PROBLEMS

Summary of Statistical Formulas

The variance for a sample is:

$$Variance\ for\ a\ sample:\ = \frac{\sum (X - \overline{x})^2}{n - 1}$$

The standard deviation for a sample is the square root of the variance:

$$standard\ deviation\ for\ a\ sample = \sqrt{\frac{\sum (X - \overline{x})^2}{n - 1}}$$

The relative standard deviation is:

$$RSD = \frac{Standard\ deviation\ \times (100\%)}{mean}$$

1. **a.** What is the purpose of a mass standard in a laboratory?

   b. Working mass standards need to be periodically recalibrated. What do you think is involved in recalibrating a working mass standard? Why do you think calibration needs to be repeated periodically for a working standard?

2. **a.** If a balance is calibrated with a standard which is supposed to be 100.0000 g but is actually 99.9960 g, how will this affect subsequent results from this balance?

   b. If a mercury thermometer is improperly marked so that when the temperature is 0°C the thermometer reads 0.2°C, what effect will this have on subsequent readings?

3. If a balance is improperly calibrated (in the sense of being improperly adjusted or set):

   a. Do you expect this problem to affect precision of the instrument? Explain.

   b. Will this problem affect the accuracy of the instrument? Explain.

4. If a flask that is used to measure volume is supposed to be marked at 10.00 *mL* and is actually marked at 9.98 *mL*, will this affect its accuracy? Will this affect its precision?

5. When a pH probe is placed in a sample, it requires a period of seconds to minutes to stabilize. If a technician using a pH meter sometimes allows the meter to stabilize for a minute or so and other times reads the meter immediately after placing the probe in the sample, will this affect the precision of the technician's results? Will it affect the accuracy of the results? Explain. Is this failure to allow the probe to stabilize an example of a systematic error?

6. Is *SD* a measure of precision or accuracy? Explain.

7. Suppose you prepare a solution that is intended to have 5.00 *mg/mL* of protein. You perform a protein assay on samples of this solution three times and obtain the following results:

   5.12 *mg/mL*   4.86 *mg/mL*   5.13 *mg/mL*

   If you perform a fourth assay of the solution, would you expect the result to be 5.00 *mg/mL* give or take:

   0.03 *mg/mL* or so   0.15 *mg/mL* or so
   0.06 *mg/mL* or so

8. The following graphs represent the distributions of measurements from four methods that are being compared with one another. Which method is most accurate? Which method is most precise? (Assume the values on the X axis are the same in all four cases.)

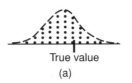

True value
(a)

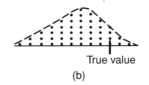

True value
(b)

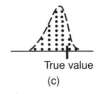

True value
(c)

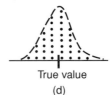

True value
(d)

9. A standard preparation of human blood serum is prepared that has 38.0 *mg/mL* of albumin. Technicians from four different laboratories analyze the standard four times in one day and obtain the results (in *mg/mL*) below. Comment on the various laboratories' accuracy and precision.

Laboratory 1	Laboratory 2	Laboratory 3	Laboratory 4
37.1	38.3	38.6	38.7
37.8	38.0	37.1	39.1
37.7	37.8	39.1	37.5
37.4	38.2	37.0	38.2

10. Technicians from Laboratory 2 repeat the analyses but this time do analyses over a period of four months. Their results are:

37.0 mg/mL 37.4 mg/mL 38.1 mg/mL 37.6 mg/mL

Comment on these results in terms of their precision. Are the results more, less, or equally precise when the tests are spread over a period of months?

11. Refer to the example problem on p. 238. Calculate the precision of each student's results using the *SD* (remember the units). Calculate the percent error for the average of each student's results.

12. Antibodies bind to antigens (such as proteins found on viruses and bacteria) and a single antibody molecule can attach to more than one antigen molecule. Monoclonal antibodies are populations of antibodies produced in such a way that they are nearly identical. A population of monoclonal antibodies are prepared that bind to the HIV virus. The number of binding sites per antibody is investigated and the results of five tests are:

3.13   3.11   3.14   3.16   3.12

What is the accuracy and precision of the test used to determine the number of binding sites?

13. The following table shows data from a study of blood calcium levels in several individuals:

Subject	Mean Calcium Level (mg/L)	# of Observations	Deviation of Results from Mean Values
1	87.5	4	0.13, 0.19, 0.05, 0.11
2	97.6	4	0.18, 0.13, 0.10, 0.02
3	104.8	4	0.09, 0.04, 0.12, 0.06

a. Calculate the *SD* for each subject's values.

b. Pool the data and calculate the mean value for blood calcium level.

14. A company manufactures buffer solutions for use in calibrating pH meters. A new lot of pH 7.00 buffer was produced. The pH of this new lot was tested on an instrument known to be properly functioning. The results of seven measurements were:

7.12   7.20   7.15   7.17   7.16   7.19   7.15

a. Calculate the mean and *SD* for these data.

b. Comment on these results if the pH of the buffer is supposed to be 7.00.

15. Samples of air in a particular factory were analyzed for lead. Each individual sample was tested three times and samples were taken on three occasions:

Sample	μg Pb/m^3 Air
1	1.4, 1.3, 1.2
2	2.3, 2.3, 2.1
3	1.6, 1.5, 1.7

a. Calculate the mean and *SD* for each sample.

b. Calculate the mean and *SD* for the pooled set of data.

c. Is the *SD* higher in a or b? Explain why.

16. A technician at LCJ Associates Environmental Laboratories evaluated a method to determine the method's repeatability and accuracy. The purpose of the method is to test levels of ammonia in water samples. To perform this assessment, the technician carefully prepared two control samples, one with 10 $\mu g/L$ ammonia in water, the other with 100 $\mu g/L$ ammonia in water. He tested each sample seven times using the method being evaluated and documented the results on the form on the next page.

a. Calculate the mean, *SD*, *RSD* (*CV*), and percent error based on these data.

b. Why did the technician prepare two standards at two different concentrations?

c. Explain how these analyses and the completion of this form are part of a quality control program.

d. How could the technician evaluate the accuracy of the method? What is the accuracy based on these data?

17. For each illustration (below), what measurement should be reported? How many significant figures does the measurement have?

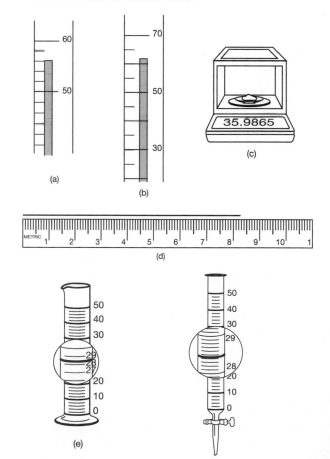

(a)

(b)

(c)

(d)

(e)

(f)

LCJ Associates Environmental Laboratories

PRECISION AND ACCURACY FORM
QUALITY CONTROL DATA

Parameter *Ammonia NH₃–N*   Date *5-6-98*

Method *Am–3*   SOP# *27931*

Reference *Water & Wastewater Manual*   Analyst *N.W.*

	Sample # *72807*	Sample # *72808*
Concentration of standard	*10 μg/ℓ*	*100 μg/ℓ*
n		
1	*09.*	*103.*
2	*10.*	*103.*
3	*08.*	*104.*
4	*11.*	*97.*
5	*10.*	*98.*
6	*13.*	*102.*
7	*09.*	*98.*
$\bar{x}$		
SD		
RSD		
% error		

18. Put a line through each of the zeros in this problem that are place holders. Place a ? above each zero that is ambiguous (i.e., may be a place holder or may convey value).
   a. 2000
   b. 1,000,000
   c. 0.00677
   d. 134,908,098

19. How many significant figures does each of the following numbers have:
   a. 45.789
   b. 0.00650

   c. 10.009
   d. 0.000878

20. Round the following numbers to the nearest tenth decimal place.
   a. 0.0345
   b. 0.98
   c. 0.55
   d. 0.2245

21. A biotechnology company specifies that a product has at least 10 *mg*/vial. Do the following lots meet the specification?

lot a.   10.2 *mg*/vial
lot b.   9.899 *mg*/vial
lot c.   7.82 *mg*/vial
lot d.   9.400 *mg*/vial

22. A biotechnology company specifies that a product has $\leq$ 0.02% of impurities. Do the following lots meet the specification?

lot a.   0.025%
lot b.   0.015%
lot c.   0.027%
lot d   0.024%

23. A spectrophotometer was used to determine the concentration of protein in a solution. The solution was analyzed six times and the absorbance values were:

0.956   0.948   0.958   0.991   0.963   0.957

According to the spectrophotometer manufacturer, the imprecision of the instrument should not exceed 1% relative standard deviation.

a. Do these results exceed 1% *RSD*?

b. If so, does this indicate that the spectrophotometer is malfunctioning or does not meet its specifications?

DISCUSSION QUESTIONS

1. If you work in a laboratory, find several examples of instruments or devices that are used to make measurements. Discuss these items in terms of calibration. How are they calibrated (in the sense of being adjusted according to a standard)? Who is responsible for their calibration? Are there standards used to calibrate the items that are traceable to national standards?

2. Examine a catalogue of scientific supplies. What devices are specified to be "NIST traceable"?

3. Obtaining the utmost accuracy and precision in a measurement or assay is generally expensive in terms of money and/or time investment. Discuss the following situations. Is it more important for the test to be as accurate and precise as possible, or is low cost and/or speed a higher priority?

a. Routine testing of the level of pollutant emissions from automobiles in a city where automobiles are required to have functioning antipollution devices.

b. Determination of the level of drug in a blood sample after an overdose to enable physicians to make a rapid decision about treatment.

c. Study of the stability of an enzyme with storage in a freezer to determine how long a product can be stored.

d. Determination of glucose levels in the urine of pregnant women to detect pregnancy-related diabetes.

TERMS RELATING TO MEASUREMENT

Some of the definitions and discussion in this glossary are either quoted from or are based on relevant quality documents and therefore use technical terminology. The text contains less technical definitions and readers may prefer to read the text before the glossary.

*Based on definitions in ASTM Standard E 456-92, "Terminology Relating to Quality and Statistics."

**Based on definitions in ISO/IEC Guide 25, "General Requirement for the Competence of Testing and Calibration Laboratories," August, 1996 edition.

***Based on "The International Vocabulary of Basic and General Terms in Metrology (VIM)," 2nd Edition, International Organization for Standardization, Geneva, 1993.

Absolute Error. The difference between the true value and the measured value. The plus (+) or minus (−) sign indicates whether the true value is above or below the measured value. *Absolute Error = true value − measured value.*

Accepted Reference Value*. "A value that serves as an agreed-upon reference for comparison, and which is derived as: (1) a theoretical or established value, based on scientific principles, (2) an assigned value, based on experimental work of some national or international organization such as the National Institute of Standards and Technology [e.g. see definition of "Standard Reference Material, SRM"], or (3) a consensus value, based on collaborative or experimental work under the auspices of a scientific or engineering group."

Accuracy. Closeness of agreement between a measurement or test result and the true value or the accepted reference value for that measurement or test. *Note—The term accuracy, when applied to a set of observed values, is affected both by random error and systematic error (bias). Because random components and bias components cannot be completely separated in routine use, the reported "accuracy" must be interpreted as a combination of these two elements. *Discussion—The use of the term *imprecision* to describe random errors and *bias* to describe systematic errors, will emphasize these distinct elements of variation. When assessing the systematic error (bias) of test methods or operations, use of the term *bias* will avoid confusion.

Analyte. A substance whose presence and/or level is evaluated in a sample using an instrument, assay, or test.

ANSI, The American National Standards Institute. An organization that promotes the use of voluntary consensus standards; the U.S. representative to ISO.

AOAC International. An independent association of scientists devoted to promoting methods validation and quality measurements in the analytical sciences.

Assay. Measurement of a property of a sample, such as the amount of cholesterol in a blood sample, lead levels in paint chips, or the activity of a DNA-cutting enzyme.

ASTM, The American Society for Testing and Materials. An organization that prepares and distributes standards to promote consistent procedures for measurement.

Bias* (relating to measurements). "The difference between the population mean of the test results and an accepted reference value. *Discussion*—Bias is a systematic error as contrasted to random error. There may be one or more systematic error components contributing to the bias. A larger systematic difference from the accepted reference value is reflected by a larger bias value."

Calibration. 1. To adjust a measuring system to bring its readings into accordance with external standards. 2. A process that establishes, under specified conditions, the relationship between values indicated by a measuring instrument or measuring system and values of a trustworthy standard. Calibration permits the estimation of the uncertainty of the measuring instrument, or measuring system. The result of a calibration is recorded in a document.

Calibration Curve. See "Standard Curve."

Compendium (of standard methods). A published collection of accepted methods in a particular field. See also "U.S. Pharmacopeia."

Confidence Interval. A range of values (calculated based on the mean and standard deviation of a sample) which is expected to include the population mean with a stated level of confidence.

Deviation (of a measurement)*. "The difference between a measurement . . . and its stated value or intended level."

DIN, Deutsche Industrial Norms. A German agency that provides engineering and measurement standards.

Error. 1. The difference between a measured value and the "true" value (see also "percent error" and "absolute error"). 2. The cause of variability in measurements. Errors are categorized as:

> **Gross error.** A human blunder.
>
> **Systematic error.** An error that causes measurement results to be biased.
>
> **Random error.** Error of unknown origin that causes a loss of precision.

Imprecision. The effect of random error on a series of measurements.

Inaccuracy. A combination of all random and systematic errors affecting the results of measurements.

ISO, The International Organization for Standardization. A worldwide federation of national standards bodies from some 100 countries that promotes the development of standardization and related activities in the world.

Mean. The average of a series of values. The sum of the values divided by the number of values.

Measurements. Quantitative observations or numerical descriptions, such as the weight of an object, the amount of light passing through a solution, or the time required to run a race.

Metrologist. A person who studies and works with measurements.

Metrology. The study of measurements.

NBS, National Bureau of Standards. See "NIST".

NCCLS, The National Committee of Clinical Laboratory Standards. An organization that promotes voluntary consensus standards for clinical laboratory testing of patient samples.

NIST, The National Institute for Standards and Technology. The national standards laboratory for the United States, formerly known as NBS, the National Bureau of Standards. NIST is a federal agency that works with industry and government to advance measurement science and to develop standards.

Percent Error. $\dfrac{True\ Value - Measured\ Value\ (100\%)}{True\ Value}$

where the true value may be the value of an accepted reference material.

Precision. 1. The consistency of a series of measurements or tests. (See also "Reproducibility" and "Repeatability.") 2. The fineness of increments of a measuring device: the smaller the increments, the better the precision.

Primary Standard. In the United States the physical items housed at NIST to which measurements are referenced (traced).

Range. The difference between the highest and the lowest values of a set of measurements.

Reference Materials (chemical or biological). Substances for which one or more properties are established sufficiently well to allow for their use in calibrating a measurement process.

> **Certified Reference Material.** Any reference material issued with documentation.
>
> **Standard Reference Material.** A certified reference material provided by NIST.

Reference Standards (for weights). Individual or combinations of weights whose mass values and uncertainties are known sufficiently well to allow them to be used to calibrate other weights, objects, or balances.

Relative Standard Deviation. A statistical measure of variability that is the standard deviation divided by the mean.

Repeatability*. "The closeness of agreement between test results obtained under repeatability conditions. Note . . . Repeatability is one of the concepts or categories of the precision of a method."

Repeatability Conditions*. "Conditions under which test results are obtained with the same test method in the same laboratory by the same operator with the same equipment in the shortest practical period of time using test units or test specimens taken at random from a singe quantity of material that is as nearly homogenous as possible . . ."

Repeatability Standard Deviation*. "The standard deviation of test results obtained under repeatability conditions."

Reproducibility*. "The closeness of agreement between test results obtained under reproducibility conditions. *Note*—Reproducibility is one of the concepts or categories of the precision of a test method."

Reproducibility Conditions*. "Conditions under which test results are obtained with the same test method on identical material in different laboratories."

Reproducibility Standard Deviation*. "The standard deviation of test results obtained under reproducibility conditions."

Resolution. The smallest detectable increment of measurement.

Secondary Standard. A standard whose value is based on comparison with a primary standard.

SI System (Systeme International d'Units). A standardized system of units of measurement, adopted in 1960 by a number of international organizations, that is derived from the metric system.

Significant Figure. A digit in a number which is a reliable indicator of value.

Standard (relating to measurement). 1. Broadly, any concept, method, or object that has been established by authority, custom, or agreement to serve as a model in the measurement of any property. 2. A physical object, the properties of which are known with sufficient accuracy to be used to evaluate another item; a physical embodiment of a unit. For example, a metal object whose mass is accurately known can be used by comparison to determine the mass of a sample. 3. In chemical or biological measurements, a standard often describes a substance or a solution that is used to establish the response of an instrument or an assay method to the analyte. 4. A document established by consensus and approved by a recognized body that establishes rules or guidelines to make a procedure consistent among various people.

Standard Curve (also called a calibration curve). A graph that shows the relationship between a response (e.g., of an instrument) and a property that the experimenter controls (e.g., the concentration of a substance).

Standard Deviation*. "The most usual measure of the dispersion of observed values or results expressed as the positive square root of the variance." A statistical measure of variability.

Standard Error of the Mean, SEM. An estimate of the standard deviation of a hypothetical distribution of sample means.

$$Standard\ Error\ of\ the\ Mean = \frac{Standard\ deviation\ of\ a\ sample}{\sqrt{N}}$$

where N = the sample size

Standard Method. A technique to perform a measurement or an assay that is specified by an external organization to ensure consistency in a field. For example, ASTM specifies standard methods to calibrate volumetric glassware; the U.S. Pharmacopeia specifies standard methods to perform tests of pharmaceutical products.

Standard Reference Material, SRM. A well-characterized material available from NIST, produced in quantity, and certified for one or more physical or chemical properties.

Test. A general term for a procedure to critically evaluate the presence, quality, or quantity of a property.

Tolerance. The amount of error allowed in the calibration of a measuring item. For example, the tolerance for a "500 *g*" Class 1 mass standard is ± 1.2 *mg* which means the standard must have a true mass between 500.0012 *g* and 499.9988 *g*.

Traceability (for measurements).*** "[The] property of the result of a measurement or the value of a standard whereby it can be related to stated references, usually national or international standards, through an unbroken chain of comparisons all having stated uncertainties."

Trueness*. "The closeness of agreement between the population mean of the measurements or test results and the accepted reference [or the true] value. Note: . . . The measure of trueness usually is expressed in terms of bias. Greater bias means less favorable trueness . . . Trueness is the systematic component of accuracy."

Uncertainty*. "An indication of the variability associated with a measured value that takes into account two major components of error: (1) bias, and (2) the random error attributed to the imprecision of the measurement process. *Discussion*—Quantitative measurements of uncertainty generally require descriptive statements of explanation because of differing traditions of usage and because of differing circumstances . . ."

Unit of Measure. An exactly defined amount of a property.

United States Environmental Protection Agency, EPA. U.S. federal agency responsible for environmental quality.

USP, United States Pharmacopeia. 1. An organization that promotes public health by establishing and disseminating officially recognized standards for the use of medicines and other health care technologies. 2. A compilation of officially recognized standards relating to medicines and other health care technologies. The methods in the USP are mandatory in the pharmaceutical industry and other situations where a company is cGMP compliant.

Variance*. "A measure of the squared dispersion of observed values or measurements expressed as a function of the sum of the squared deviations from the population mean or sample average." A statistical measure of variability.

Verification.** "Confirmation by examination and provision of objective evidence that specified requirements have been fulfilled . . . *Notes* 1. In connection with the management of measuring equipment, verification provides a means of checking that the deviations between values indicated by a measuring instrument and corresponding known values are consistently smaller than the limits of permissible error defined in a standard regulation or specification relevant to the management of the measuring equipment. 2. In connection with measurement traceability, the meaning of the terms *verification* and *verified* is that of a simplified calibration, giving evidence that specified metrological requirements including the compliance with given limits of error are met."

Working Standard. A physical standard that is used to make measurements in the laboratory and which is calibrated against a primary or secondary standard.

REFERENCES CONTAINING INFORMATION ABOUT MEASUREMENTS

General statistics books have information about standard deviation, accuracy, precision, and other important concepts in metrology. Some examples of such texts include:

Statistics, David Freedman, Robert Pisani, Roger Purves, and Ani Adhikari. 2nd edition. Norton, New York, 1991.

Statistics for Analytical Chemistry, J.C. Miller and J.N. Miller. 3rd edition. Prentice Hall, Canada, 1993.

Use of Statistics to Develop and Evaluate Analytical Methods, Grant T. Wernimont and edited by W. Spendley. AOAC, Gaithersburg, MD, 1985.

References on significant figures and uncertainty analysis include:

"Minimizing Significant Figure Fuzziness," Lawrence D. Fields and Stephen J. Hawkes. *JCST,* pp. 30–34, Sept/Oct, 1986.

"The Significance of Significant Figures," Richard C. Pinkerton and Chester E. Gleit. *Journal of Chemical Education,* 44(4):232–234, 1967.

"Guidelines for Evaluating and Expressing the Uncertainty of NIST Measurement Results," Barry N. Taylor and Chris E. Kuyatt. 1994. NIST Technical Note 1297. ("This publication . . . gives the NIST policy in expressing measurement uncertainty. Further, it succinctly summarizes, in the text of the NIST policy, the . . . ISO 100 page 'Guide to the Expression of Uncertainty in Measurement' on which the NIST policy is based . . .")

Other references include:

"Standard Reference Materials Handbook for SRM Users," John K. Taylor. NIST Special Publication 260-100. U.S. Government Printing Office, Washington D.C., 1993.

"Chemical Measurements and The Issues of Quality Comparability and Traceability," W.P. Reed. *American Laboratory,* pp. 18–21, December 1994.

Traceable Temperatures: An Introduction to Temperature Measurement and Calibration, J.V. Nicholas and D.R. White. John Wiley and Sons, Chichester, 1994. (A comprehensive introduction both to the measurement of temperature and to measurement science in general.)

ASTM Standard E 456-92. "Terminology Relating to Quality and Statistics."

ISO/IEC Guide 25, "General Requirement for the Competence of Testing and Calibration Laboratories," August, 1996 edition.

The International Vocabulary of Basic and General Terms in Metrology (VIM). 2nd edition. Geneva, International Organization for Standardization, 1993.

ANSWERS TO PRACTICE PROBLEMS

1. **a.** To calibrate and check the accuracy of a balance.

   b. Recalibration involves comparison with a primary or secondary standard.

   Recalibration is necessary because working standards may become damaged, dirty, dusty, or may otherwise be altered with time and use.

2. **a.** Readings will be too high. For example, a 100.0000 g object will be read as more than 100 g.

   b. Subsequent temperature readings will be recorded as higher than they really are.

3. We would expect all the readings made with this balance to be incorrect but their consistency is

unlikely to be affected by the improper calibration (adjustment) of the instrument. The accuracy of the instrument, therefore, would be adversely affected but not its precision.

4. See Answer # 3.

5. A pH meter needs to stabilize to give accurate results; therefore, this technician will sometimes have inaccurate results. In addition, precision requires that the operator perform a task in a consistent fashion; therefore, the precision achieved by this technician will be poor. Failure to use an instrument properly, in this case, failure to allow for stabilization, is a systematic error.

6. Precision. See text for explanation.

7. The standard deviation is 0.15 *mg/mL*; therefore, it is reasonable to expect the next measurement to be 5.00 *mg/mL* ± 0.15 *mg/mL*.

8. The most accurate measurements have the mean value at the "True Value" (d). The most precise measurements have the narrowest curve with the smallest standard deviation (c).

9.

	Lab 1	Lab 2	Lab 3	Lab 4
mean	37.5 *mg/mL*	38.1 *mg/mL*	38.0 *mg/mL*	38.4 *mg/mL*
SD	0.32 *mg/mL*	0.22 *mg/mL*	1.06 *mg/mL*	0.69 *mg/mL*

Laboratories 2 and 3 have good accuracy, but Laboratory 2 has the best precision.

10. mean 37.5 *mg/mL*
SD 0.46 *mg/mL*

The results from Laboratory 2 are more precise when the assays are completed at one time rather than over four months. (Many factors might lead to variability over time. Reagents or samples might slowly degrade, instruments may lose calibration or be replaced, personnel might change, and so on.)

11.

	Juan	Chris	Ilana	Mel
mean	5.0002 *g*	5.0021 *g*	4.9999 *g*	5.0247 *g*
SD	0.0015 *g*	0.0001 *g*	0.0001 *g*	0.0516 *g*
% error	−0.004%	−0.042%	0.002%	−0.494%

12. mean 3.13
SD 0.019

The number of binding sites for an antibody is a whole number, in this case, presumably 3. The % error (a measure of accuracy) is −4.33%. The *SD* of the assay (a measure of precision) is ± 0.019 sites.

13. a. Subject 1: $SD = 0.15$ *mg/L*
Subject 2: $SD = 0.14$ *mg/L*
Subject 3: $SD = 0.10$ *mg/L*

b. Pooled mean = 96.6 *mg/L*

14. a. $\bar{x} =$ 7.16
$SD =$ 0.03

b. Given the consistently high values and the fact that the instrument used to test the buffer was known to be properly working, we can be reasonably confident that the pH of this lot of buffer is about 7.16. This lot of buffer would be rejected.

15. a.

	Sample 1	Sample 2	Sample 3
$\bar{x}$	1.3 *μg/m³*	2.2 *μg/m³*	1.6 *μg/m³*
SD	0.10 *μg/m³*	0.12 *μg/m³*	0.10 *μg/m³*

b. Pooled sample $\bar{x}$ 1.7 *μg/m³*
SD   0.42 *μg/m³*

c. As we might expect, the standard deviation is higher for the pooled data because we would expect more variation among samples taken on different occasions than when an individual sample is tested three times.

16. a.

$\bar{x}$	10 *μg/L*	101 *μg/L*
SD	1.6 *μg/L*	2.93 *μg/L*
RSD	16.3%	2.91%
% error	0 %	−1.00%

b. The accuracy and the precision of the assay may change at two different concentrations of the analyte.

c. These analyses evaluate and document the precision and accuracy of the method. Because the assay components may change over time, this verification may be required numerous times.

d. The % error is an indication of the accuracy.

17. a. 56.0   3 significant figures
b. 62   2 significant figures
c. 35.9865   6 significant figures
d. 8.25   3 significant figures
e. 28.4 *mL*   3 significant figures
f. 28.44 *mL*   4 significant figures

18.   ???
a. 2000
    ??? ???
b. 1,000,000
c. 0.00677
d. 134,908,098

19. a. 5   **b.** 3   **c.** 5   **d.** 3

20. a. 0.0   **b.** 1.0   **c.** 0.6   **d.** 0.2

21. lot a. rounded value = 10 *mg/vial*; yes, meets specification

lot b. rounded value = 10 *mg/vial*; yes, meets specification lot c. rounded value = 8 *mg/vial*; no, does not meet specification

lot d. rounded value = 9 *mg/vial*; no, does not meet specification

22. lot a. rounded value = 0.03%, no, does not meet specification

lot b. rounded value = 0.02%, yes, meets specification

lot c. rounded value = 0.03%, no, does not meet specification

lot d. rounded value = 0.02%, yes, does meet specification

23. a, b. The *RSD* is 1.552% which, when rounded, is 2%. This is higher than the 1% specification; however, this does not necessarily indicate any problem with the spectrophotometer. Biological samples are often not homogeneous and they are often not stable. With biological materials, therefore, variability is more likely to be due to the sample than to the instrument.

CHAPTER 14

Introduction to Instrumental Methods and Electricity

I. USING INSTRUMENTS TO MAKE MEASUREMENTS

A. Introduction

The previous chapter discussed fundamental principles of measurement. The six chapters following this one each focus on a specific type of measurement, such as the measurement of volume and of weight. This chapter is a transition that introduces instrumental methods of measurement and basic vocabulary, and also concepts relating to electricity and electronics.

People made measurements before measuring instruments were invented. For example, everyone uses the senses of "feeling" to estimate the air temperature or the weight of an object. Measurements made with senses alone, however, are not very accurate or reproducible; therefore measuring instruments were invented, such as pan balances and mercury thermometers, Figure 14.1a,c. These early measuring devices did not require electricity to make a measurement and are sometimes referred to as "mechanical" instruments.

Mechanical instruments are still used in the laboratory since they are simple to understand, reliable, and often inexpensive. Electronic measuring instruments, however, are now preferred for many applications. Electronic instruments convert a physical or chemical property of a sample into an electrical signal, Figure 14.1b,d. For example, an electronic balance converts a property of an object, the force of gravity on the object, to an electrical signal. The balance then processes the electrical signal to convert it to a display of weight. Electronic instruments are generally more convenient, faster, and more consistent than mechanical ones. Electronic devices also can measure a smaller change in the property of interest than can mechanical instruments. For example, a mercury thermometer can give a temperature measurement to, at best, the closest tenth of a degree. In contrast, some electronic thermometers can detect a temperature change of a hundredth of a degree or better.

Early electronic instruments were characterized by control knobs, dials, and switches by which the analyst controlled the device. These instruments also had meters in which a needle pointed to the value of the measurement. Modern electronic instruments are usually associated with computers, so they have keypads (in place of knobs, dials, and switches) and display screens and printers (in place of meters with needles). The computer is sometimes an independent unit that is connected to the instrument via a cord or cable; other times, small computers, called **microprocessors**, are integrated into the instrument itself. Computers and microprocessors can perform complex manipulations of data, store large

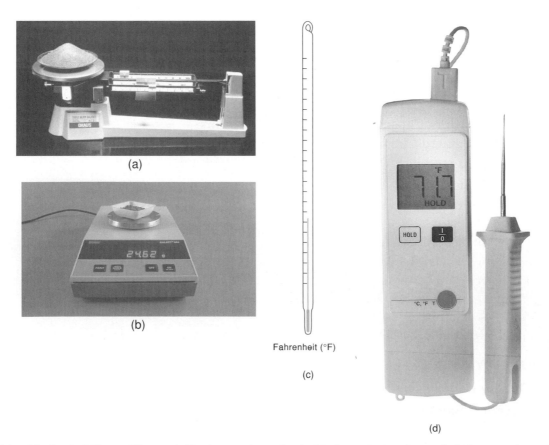

(a)

(b)

Fahrenheit (°F)

(c)

(d)

Figure 14.1.   Mechanical Versus Electronic Devices. a. A mechanical balance. **b.** An electronic balance. **c.** A mechanical thermometer. **d.** An electronic thermometer. (© Copyright 1998 Cole-Parmer Instrument Co; used by permission.)

amounts of information, control instruments so they operate consistently, and detect instrument malfunctions.

The computers associated with modern laboratory instruments are powerful, so these instruments are often simple to operate and appear "intelligent." Nonetheless, it is important to remember that the human operator, although seemingly removed from the measurement process, is still a key part of making a "good" measurement. It remains the operator's responsibility to understand the measurement system, to be aware of problems, and to be skeptical of the results. Thus, it is essential for laboratory workers to have a basic understanding of how laboratory instruments work.

B. Measurement Systems

A mercury thermometer is a relatively simple measuring device. It contains liquid mercury enclosed in a narrow glass tube or stem. If the thermometer is moved from a cooler to a warmer medium, the mercury expands and so moves up the stem. If the temperature decreases, the liquid contracts down the stem. There is a scale of "degrees" etched onto the glass stem. The operator reads the temperature of the medium by viewing the height of the mercury column relative to the scale of degrees.

In a mercury thermometer, the mercury "senses" one form of energy, the heat of the medium, and converts it to another form of energy, the mechanical expansion or contraction of mercury. There is an **interface** where the glass surface of the thermometer is in contact with the

sample. The mercury acts as a **transducer**, *a device that senses one form of energy and converts it to another form.* The degree scale etched on the stem of the thermometer is the **display** or **readout**. We can generalize from this example and say that a mechanical measurement instrument has these components:

Interface → Transducer (Sensor) → Display (Readout)

Consider a second measuring instrument, an electronic balance. The sample to be weighed is placed on the balance's weighing pan. The weighing pan is the interface. Within the balance an electronic component acts as the transducer and responds to the force of gravity on the sample. The transducer converts that force to an **electrical signal**. A **signal processor** *modifies the electrical signal so that it can be displayed as a weight value.* Thus, in an electronic instrument the components of the measuring system are:

Interface → Transducer (Sensor) →
Signal Processor → Display (Readout)

There are many different transducers. Some respond to physical energies, such as pressure, temperature, and light, while others detect chemical properties such as pH. Examples of electronic measuring instruments are illustrated in Figure 14.2.

For a measuring instrument to be useful, certain requirements must be met:

- ***The instrument's response must have a consistent and predictable relationship with the property***

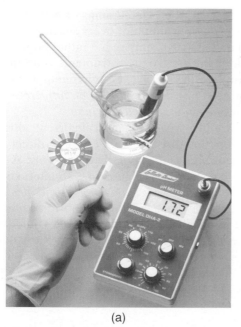

(a)

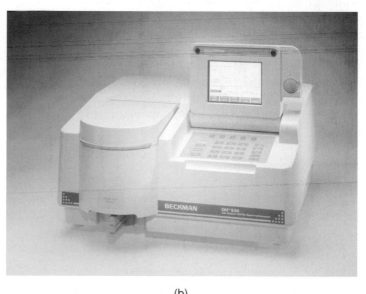

(b)

Figure 14.2.   Measurement Instruments. a. A pH Meter System. A probe that generates a signal in response to pH is placed within a liquid sample. The probe is attached via a cable to the signal processing unit and readout device. **b.** Spectrophotometer. Spectrophotometers measure the interaction between light and a sample. A photomultiplier tube located within the spectrophotometer is the transducer that senses light. A liquid sample is placed in a glass container, which is in turn placed inside a light-tight chamber inside the spectrophotometer. (Spectrophotometer image courtesy of Beckman Coulter, Inc.)

being measured. For example, a certain temperature must predictably cause the mercury to rise to a certain height. A certain weight sample must generate a consistent electrical signal.

- *The instrument's response must be related to internationally accepted units of measurement.* For example, there must be a way to relate a particular height of mercury to units of "degrees." A certain electrical signal in a balance must be related to units of "grams."

Calibration *is the process by which the response of an instrument is related to internationally accepted measurement units.* Every measuring instrument needs to be periodically calibrated and to be checked regularly to be sure it is responding predictably and consistently. Calibration and performance verification of specific instruments are discussed in subsequent chapters as these instruments are introduced.

The relationship between the property being measured and the response of an electronic instrument is most often linear. If there is a linear response, adjustment at two points (e.g., zero and full-scale) calibrates the device because two points determine a straight line. If a device is nonlinear, additional points are needed to establish the relationship between instrument response and the property being measured. For example, a pH meter is calibrated in accordance with two standards of known pH. Two standards are sufficient because there is a linear relationship between the response of the instrument and the pH of the sample, Figure 14.3. **Two-point calibration** *refers to a situation where two standards are used to calibrate an instrument whose response is linear.*

To best understand electronic laboratory instruments, it is necessary to have some knowledge of the principles and vocabulary of electricity and electronics. There are also electrical safety issues with which everyone should be familiar when operating instruments. The rest of this chapter, therefore, introduces electricity, electronics, and safety as they relate to laboratory instruments.

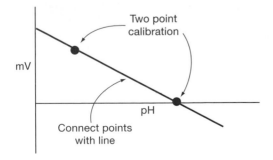

Figure 14.3. Calibration. If there is a linear relationship between instrument response and the property being measured, then two points define the relationship. This is the case for a pH meter.

II. BASIC TERMINOLOGY AND CONCEPTS OF ELECTRICITY

A. Current, Voltage, and Resistance

i. CURRENT

Electricity is understood in terms of the atomic theory of matter. In short, according to this theory, matter consists of atoms that have electrons, protons, and neutrons. Protons have positive charge, electrons have negative charge. A material is said to be **negatively charged** *if electrons are in excess* and **positively charged** *if electrons are depleted.* Objects that have the same charge repel one another, whereas objects that have opposite charges attract one another. Therefore, if two objects are separated, they will be attracted if one is positive and the other is negative. On the other hand, if both are positively or negatively charged, the objects will repel one another.

Metals have a property that is important for electricity. In metals, the outer electrons of atoms are loosely held by the atoms, so these outer electrons move easily and randomly from one atom to another. Suppose there is a metal wire with an excess of electrons at one of its ends and a deficiency of electrons at the other. In this situation, the electrons in the wire will not move randomly. Rather, they will flow away from the end with an excess of electrons and toward the end with a deficiency, Figure 14.4. It may be helpful to imagine that the flow of electrons in a wire is like water flowing in a river bed. *The flow of either water or electricity is called* **current**.

The flow of electrons, abbreviated I, is measured in units of **amperes**, abbreviated **amps** or **A**. If 6.25×10^{18} electrons pass a certain point every second, the current is said to be 1 amp. An ampere is a rather large unit. The current in many electronic instruments is on the order of **milliamperes (mA)** or even **microamperes (μA)**. A **milliampere** *is one thousandth of an ampere;* a **microampere** *is one millionth of an ampere.*

There are two types of electrical current: **alternating current (AC)** and **direct current (DC)**. In **direct current** *electrons flow in one direction through a wire.* A battery generates direct current, Figure 14.5. In **alternating current** *electrons change directions many times per second.* The current that comes from the power company to an outlet in a wall is alternating. In the United States, AC current cycles back and forth with a **frequency** of

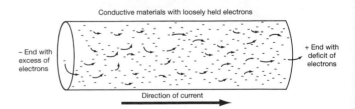

Figure 14.4. The Flow of Electricity in a Wire. Electrons move from atom to atom.

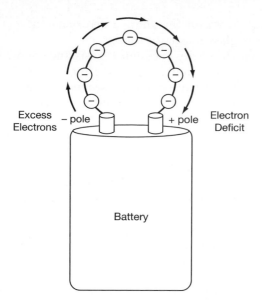

Figure 14.5.   Direct Current. In a battery, current flows directly from a pole with an excess of electrons to a pole with an electron deficit.

60 times per second. In other countries AC current has a frequency of 50 times per second. Frequency is measured in units of **Hertz (Hz)**, *where 1 Hz = 1 cycle per second.*

Most laboratory instruments require direct current. Direct current is occasionally supplied to an instrument by batteries. It is more common, however, for a **power supply** *to convert the AC current of the power company to a DC current through a process called* **rectification**.

We are accustomed to electrical current that flows through wires; however, electrical current can also flow through liquids. Electrical current in a liquid is carried by positive and negative ions derived from salts added to the liquid. This type of current flow is important in the function of pH meters, as will be discussed in Chapter 18.

ii. VOLTAGE

Energy *is defined as the ability to do work;* **potential energy** *is stored energy.* For example, water poised at the top of a waterfall has **potential energy**. As the water plummets over the waterfall, its potential energy is converted into the mechanical energy of falling. Gravity is responsible for the potential energy of the water, Figure 14.6a.

Just as water can have potential energy, there is also **electrical potential energy**. **Electrical potential** *is the potential energy of charges that are separated from one another and attract or repulse one another because of their unlike or similar charges,* Figure 14.6b. Electrical potential arises because charged objects exert attractive or repulsive forces on one another. For example, in a battery there is a negative terminal with an excess of electrons, and a positive pole with a deficit. The excess

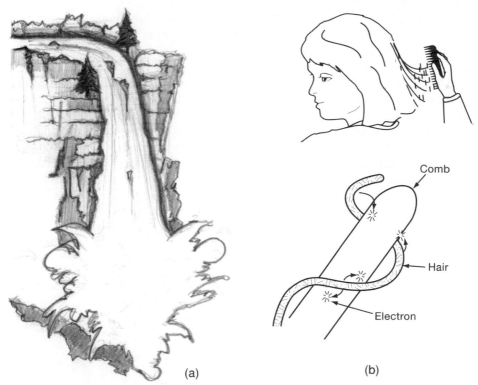

(a)   (b)

Figure 14.6.   Potential Energy. a. The potential energy of water is converted to mechanical energy in a waterfall. **b.** A familiar example of electrical potential energy. When a comb and hair are rubbed together, electrons are removed from the hair and deposited on the comb. The comb, therefore, becomes negatively charged; the hair positively charged. The positive and negative charges are thus separated from one another. Opposite charges attract, so the hair and comb attract one another. (Sketches by a Biotechnology student.)

electrons at the negative terminal can be thought of as the force that "pushes" electrons toward the positive pole. Thus, as gravity is the force that causes water to flow over a waterfall, so electrical potential is like a force that causes electrons to flow in a wire.

Electrical potential is also called **electromotive force (EMF or ε)** or **voltage**. *The units of voltage are the* **volt (V)** *and the* **millivolt (mV)**. The voltage that is supplied to a wall receptacle by the power company is either in the range from 110 to 120 *V* or is 220 *V*.

The development of voltage requires a method of separating charges from one another so that there is an excess of electrons at one site and a deficiency of electrons at another. There are many ways that voltage can be generated, all of which involve the conversion of some other form of energy into electrical potential energy. Fossil fuels, like coal, gas, and oil, can be burned to create electrical potential. It is also possible to harness the mechanical energy of a waterfall or the wind, the solar power of the sun, the energy of nuclear reactions, and the energy of chemical reactions to make electrical potential.

iii. RESISTANCE

Water in a river cannot flow if its path is blocked by a dam. Similarly, the flow of electricity is impeded if it encounters **resistance** in its path. **Resistance** *is the impedance to electron flow. The units of resistance are* **ohms**, *abbreviated,* Ω. *One* **ohm** *is the value of resistance through which 1 V maintains a current of 1 A.*

All materials (except so-called superconductors) offer some resistance to current flow. The amount of resistance depends on the material. Electricity flows most readily in **conductors**. **Conductors** *are materials, such as metals, whose outer electrons are free to flow from one atom to another.* The best conductors are those materials whose outer electrons are most loosely bound. Silver is the best conductor, followed by copper and gold. In contrast, **insulators** *are materials in which the outer electrons of atoms are not free to move and so electricity does not flow readily.* Electricity usually does not flow in the air (lightning is an exception), nor does it flow in glass, plastic, or rubber.

When electricity encounters resistance, some or all of its energy is converted into heat energy. Because even copper wire offers some resistance to current flow, as electricity flows in a wire, some energy is lost as heat. The longer a wire, the more electrical energy is dissipated as heat. This is why the use of extension cords results in a loss of power. A burner on an electrical stove is an application where we take advantage of the heat generated when electricity encounters resistance.

Semiconductors *offer intermediate resistance to electron flow and are used in the manufacture of transistors and other electronic devices.* Semiconductors are composed of crystalline silicon or germanium. Silicon and germanium do not conduct electricity well; however, if impurities are added to their crystals, their structure changes so that electrons can move more easily. Semiconductors play a critical role in electronics because engineers can adjust their resistance and thereby control the flow of current through them.

B. Circuits

i. SIMPLE CIRCUITS

Electricity only flows if there is a complete conductive pathway in which the electrons can move. At one end of the pathway there must be a source providing an excess of electrons; at the other end there must be an electron deficit. *Such a pathway for electricity is called a* **circuit**. Figure 14.5 showed a battery with a negative pole, having an excess of electrons, and a positive pole, having a deficit. The wire connecting the positive and negative poles completed the pathway for electron flow.

In practice, one would not encounter a circuit such as the one in Figure 14.5 because the flow of electrons in the wire would produce no useful work. Electricity is useful when electrical energy flowing in a circuit is transformed into another form, such as heat, light, or mechanical work. Figure 14.7a shows a diagram of a

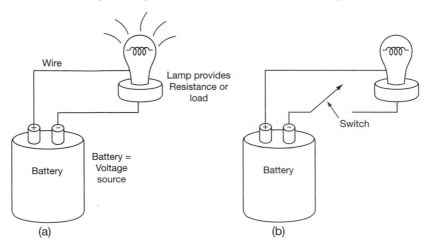

Figure 14.7.   A Simple Circuit. a. A simple circuit consisting of a voltage source, a wire, and a resistive element—in this case, a light bulb. **b.** A switch is added to the simple circuit in a.

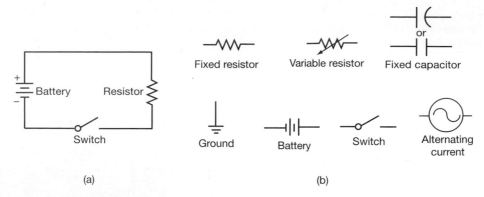

Figure 14.8.   An Electrical Diagram. a. The same circuit as shown in Figure 14.7b, but using standard electrical symbols instead of pictures. **b.** Examples of symbols used in electrical schematic diagrams.

simple circuit that includes a light bulb. As electricity flows through the bulb, a thin wire heats up and emits light. The bulb thus converts electrical energy to light. Items such as burners on a stove, light bulbs, and motors convert electrical energy into useful heat, light, and motion. Motors, bulbs, and other such elements constitute resistance to electron flow. These elements, which provide resistance, are sometimes referred to as the "load" in a circuit. As electricity flows through such devices, work is performed, and heat is generated.

Every practical circuit at a minimum consists of conductors through which electrons can travel, a voltage source, and some type of resistance. In addition, insulation is usually added, serving to confine the current to the desired paths. In Figure 14.7b a **switch** *to control electrical flow* has been added to the circuit. When the switch is closed the circuit is complete and current flows; when the switch is open the circuit is interrupted by air and current ceases to flow.

The circuit shown in Figure 14.7b has been redrawn in Figure 14.8a using schematic symbols instead of pictures. There are a number of symbols used in electrical diagrams; examples are shown in Figure 14.8b.

ii. OHM'S LAW

Ohm's Law (developed by George Simon Ohm) is an equation that relates voltage, current, and resistance in a circuit. Ohm's Law states that the current in a circuit (expressed in amps) is directly proportional to the applied voltage (expressed in volts) and is inversely proportional to the resistance (expressed in ohms). Ohm's Law is written as:

$$Voltage = (current)\ (resistance)$$
$$V = IR$$
$$or$$
$$I = \frac{V}{R}$$

For example, consider a circuit in which there is a motor that provides 5 ohms of resistance. The voltage

from the power company that supplies the circuit is 110 V. Assuming there are no other devices in the circuit, it is possible to calculate the current in the circuit:

$$V = IR$$
$$110\ V = (?)(5\ ohms)$$
$$? = 22\ amps$$

Note that the proper units must be used to apply Ohm's Law. For example, if the units in a problem are given as millivolts or microamperes, they must be converted to volts and amps.

EXAMPLE PROBLEM

According to Ohm's Law, if the resistance in a circuit suddenly becomes very low, what will happen to the current in that circuit?

ANSWER

Assuming the voltage remains constant, the current will become very high.

Electrophoresis is a technique commonly used in the biotechnology laboratory that provides a good example of how Ohm's Law is applied in practice. **Electrophoresis** *separates charged biological molecules from one another based on their rate of migration when placed in an electrical field.*

In gel electrophoresis, the sample mixture is placed in a gel matrix. The gel is positioned in a gel box and is covered with a thin layer of buffer. Positive and negative ions in the gel and the buffer are capable of conducting current. A power supply is used to provide voltage so that there is a positive pole at one end of the gel and a negative pole at the other end, Figure 14.9a. Negatively charged sample components migrate toward the positive pole; positively charged components move to the negative pole. Their speed of migration is determined by the magnitude of their charge. If two or more

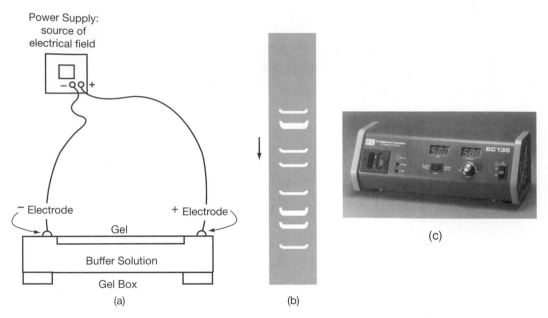

Figure 14.9.   Electrophoresis. a. An electrophoresis set-up including a gel, a gel box, and a power supply. **b.** A stained gel. The bands represent separated molecules. **c.** An electrophoresis power supply. Constant current or constant voltage can be selected. (Photo courtesy of E-C Apparatus. E-C is a registered trademark of ThermoQuest Corporation.)

components have the same charge, then the larger ones will tend to move more slowly through the gel matrix than the smaller ones. Thus, different biological molecules migrate at different rates depending on their charge and size. Once the components are separated, the gel is removed from the box, a dye is used to stain the biological molecules, and the separated components appear as bands, Figure 14.9b.

In the early stages of an electrophoresis run, the resistance of the gel increases somewhat as ions are electrophoresed out of the gel. Therefore, if the voltage supplied to the gel remains constant, then the current decreases during the run, in accordance with Ohm's Law. To maintain a constant current during electrophoresis, the voltage must be increased as the run progresses. Most power supplies designed for electrophoresis can be adjusted so that they provide either constant voltage or constant current, depending on the user's preference, Figure 14.9c. (Electrophoresis power supplies are also made that maintain constant power. Power will be discussed later.)

EXAMPLE PROBLEM

a. At the beginning of an electrophoresis run the voltage is 75 V, and the current is 30 mA. What is the resistance of the gel?

b. The voltage is held constant and the current decreases to 12 mA by the end of the separation. What is the resistance of the gel?

c. If an identical gel is run, but this time the current is maintained at 30 mA through the entire run, what will the voltage be at the end of the run?

ANSWER

According to Ohm's Law:

a. $V = IR$
$75\ V = (0.030\ A)(?)$
$? = 2.5\ k\Omega$

b. $V = IR$
$75\ V = (0.012\ A)(?)$
$? = 6.25\ k\Omega$

c. $V = IR$
$? = (0.030\ A)(6250\ \Omega)$
$? = 187.5\ V$

---

iii.   Series and Parallel Circuits

Most electrical circuits contain more than one resistive element. Circuits with multiple resistant elements (loads) can be arranged in two basic ways: in **series** or in **parallel**.

A circuit with three resistive elements is shown in Figure 14.10a. The three resistances are arranged one after the other; hence, they are said to be in series. In a **series circuit**, *current flow has only one possible pathway.* The current passing through resistance 1 is the same as the current passing through resistances 2 and 3. A familiar example of a series circuit is a string of old-fashioned Christmas tree lights. If one bulb burns out, the pathway is no longer complete and all the bulbs go out.

A **parallel circuit** with three resistive elements is shown in Figure 14.10b. The three paths split the total amperage and so the total amperage in the entire circuit is equal to the sum of the currents in the three parallel pathways. In a parallel circuit, if one pathway is broken, current can still flow in the others. Circuits in

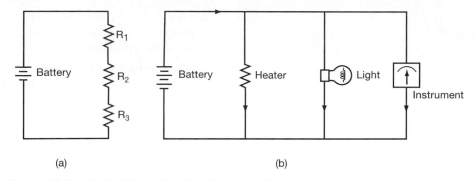

Figure 14.10.   Series Versus Parallel Circuits. a. Three resistive elements arranged in series. Current passes through all three elements. **b.** A parallel circuit. The current is split into three paths.

most modern instruments are complex and consist of both parallel and series pathways.

iv. POWER, WORK, AND CIRCUITS

Electricity is used to perform work, such as to generate light or power a pH meter. As an electric device performs work, power is consumed. *The units of power are* **watts**, abbreviated *W.* **Power** is defined as:

$$Power = (Voltage)(Current)$$
$$or$$
$$W = (V)(I)$$

Instrument manufacturers specify how much power their instruments require. The amount of power required by a device is related to how much heat that device will produce during operation. For example, a 100 *W* light bulb generates more light—and more heat—than a 25 *W* light bulb. The more power being used in a circuit, the more heat is generated in that circuit. Power ratings are useful to determine whether a circuit is being overloaded (i.e., too much heat is being generated) by devices.

EXAMPLE

A spectrophotometer requiring 300 *W*, an oven requiring 600 *W*, and a cell culture hood requiring 200 *W* are all plugged into the same laboratory circuit and are all turned on. Suppose further that the laboratory is served by a 120 *V* line with a current of 15 *A*. Will the load on the circuit exceed its capacity?

The total carrying capacity of the laboratory circuit is:

$$P = IV$$
$$= (15\ A)(120\ V) = 1800\ W$$

The total power consumed by the three instruments is:

$$300\ W + 600\ W + 200\ W = 1100\ W$$

The circuit can therefore handle these three items. If however three more spectrophotometers at 300 *W* each are plugged into the same circuit and operated at the same time, then the circuit will be overloaded and the circuit breaker or fuse protecting that circuit will blow.

---

v. GROUNDING, SHORT CIRCUITS, FUSES, AND CIRCUIT BREAKERS

Fuses or circuit breakers and grounding wires are the two types of basic electrical safety devices that are incorporated into circuits. Both types of devices are necessary to prevent electrical fires and to protect operators from serious shock.

If an unusually large amount of current flows in a wire, then that wire will become extremely hot—to the point where instruments can be damaged and/or a fire can be ignited. **Fuses** *are simple devices that protect a circuit from excess current.* A fuse contains a thin wire. Heat is generated whenever current flows through the wire; the more current, the more heat. The wire will break if too much current passes through it, thus breaking the circuit and preventing the further flow of current, Figure 14.11a,b. Fuses are rated according to volts and amps. For example, it is possible to purchase fuses for 10, 20, and 30 amps, 110 *V*. If a fuse is rated for 20 amps and more than 20 amps passes through it, it will "blow."

A circuit breaker performs the same function as a fuse; however, circuit breakers are more convenient than fuses. A blown fuse must be replaced, whereas a circuit breaker is simply switched back to its original position, once the original problem is solved.

Fuses or circuit breakers are placed in the circuits in houses and buildings. A typical house will receive from 50 to 220 amps from the power company, which is subdivided in the house into individual circuits. There is usually a main fuse or circuit breaker where the electric power line enters the building and another fuse or circuit breaker where each circuit branches from the main circuit, Figure 14.12.

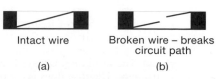

Intact wire   Broken wire – breaks
                      circuit path
(a)   (b)

Figure 14.11.   Fuses and Circuit Breakers. a. Good fuse. **b.** "Blown" fuse.

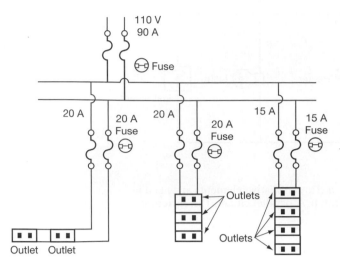

Figure 14.12.   Fuses or Circuit Breakers Protect the Circuits in Every Home and Building.

There are several common causes of excess current leading to blown fuses or tripped circuit breakers:

1. ***There may be too many devices plugged into the same circuit.*** (Anyone who has lived in an older apartment has probably experienced this problem.)

2. ***One device in the circuit may be using too much power.*** In this case, an electrician must be called to solve the problem.

3. ***There may be a short circuit in a device.*** In this case, the device should be unplugged and a service technician called.

In addition to the fuses and circuit breakers that protect the circuits in a building, most laboratory instruments have their own fuses to protect delicate components. (One of the challenges of laboratory management is keeping track of the spare fuses that go with each instrument so that they are available when needed.) These fuses blow if for any reason too much current flows through the instrument. If an instrument's fuse blows, then it can indicate a problem that requires repair by a qualified person.

Figure 14.13a illustrates a properly functioning instrument which is plugged into a wall outlet. Voltage causes current to flow from the power company, through the instrument, and then back to the wall outlet, thus completing the circuit. In this example, the electricity powers a motor and so the motor provides resistance to current flow.

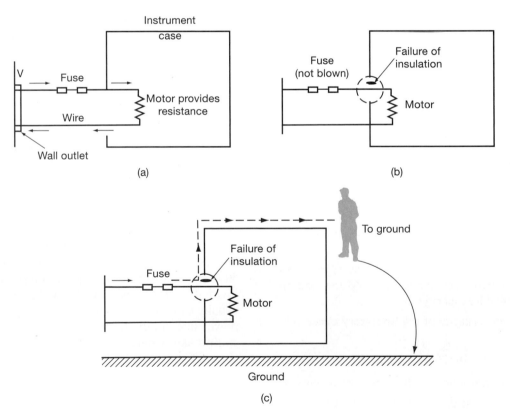

Figure 14.13.   Short Circuits and Safety Devices. a. A properly functioning instrument. Voltage causes current to flow through the instrument and then back to the wall outlet. The electricity powers a motor and so the motor provides resistance to current flow. **b.** A short circuit. The insulation is frayed so that the metal wire touches the metal case of the instrument. The fuse does not blow and the instrument may continue to be operative. **c.** If a person touches the instrument in (b), current will flow through the person to the ground causing dangerous shock.

Figure 14.13b illustrates the same instrument, but this time the insulation around the power cord is frayed so that the metal wire touches the metal case of the instrument. No current flows through the metal frame because the circuit is not complete. The fuse does not blow, and the motor may be operative. Unfortunately, in this situation, if a person touches the metal casing of the instrument that person may provide a complete path for electrons to flow to the ground, Figure 14.13c. The person will be shocked; in some cases, such shock can be fatal. This situation is called a **short circuit**, *a situation in which electrical current is able to flow without passing through the resistance, or load.*

Because the situation illustrated in Figure 14.13c is dangerous, modern electrical instruments are provided with a grounding wire, Figure 14.14. To understand this safety mechanism it is necessary to know that the earth can gain or lose electrical charges and yet remain neutral. The earth, therefore, is considered to be at zero potential and can act as a charge neutralizer. An **electrical ground** *is any conducting material connected to the earth. A* **ground wire** *is attached to the metal frame of the instrument and is connected to the earth via the third prong on the power cords of most modern instruments.* The third prong is a metal conductor that, when plugged into the wall, contacts another wire that is connected to a metal water pipe or other conductor. The water pipe in turn contacts the ground under the building. In the event of a short circuit, current flows from the instrument chassis to the third prong and is then discharged harmlessly to the ground. Note also that the fuse will blow in this situation because the current flowing to the ground will be great. The use of a three-prong cord to ground an instrument is an important safety feature and should not be by-passed by using adapters or other methods.

III. BASIC TERMINOLOGY AND CONCEPTS OF ELECTRONICS

A. Electronic Components

i. OVERVIEW

The term **electronic** *refers to electrical devices that have nonmechanical components that generate, amplify, and process electrical signals.* These components are combined and arranged in circuits. Depending on the components used and their arrangement, electronic circuits can perform a vast number of functions ranging from playing music to measuring weight in the laboratory. (The term *electrical* is typically applied to electrically operated equipment like pumps and motors.)

Although the functions that electronic devices perform are numerous, the basic components of their circuits are relatively few. Some of the most common of these components are briefly described in this section, including **resistors**, **capacitors**, **diodes**, and **transistors**. The next section discusses how these components are assembled to make functional instruments for measurement.

ii. RESISTORS, CAPACITORS, DIODES, AND TRANSISTORS

Resistors *are electronic components that impede the flow of current.* Resistors are used to control the amount of current in a circuit. There are **fixed resistors** *that have one value of resistance,* and **variable resistors** *that can be adjusted to provide different levels of resistance.* Fixed resistors are manufactured to have a specific resistance and are color coded to indicate what that resistance is.

There are two types of variable resistor: the **rheostat** and the **potentiometer**, abbreviated "pot." There are many situations where it is necessary to be able to vary

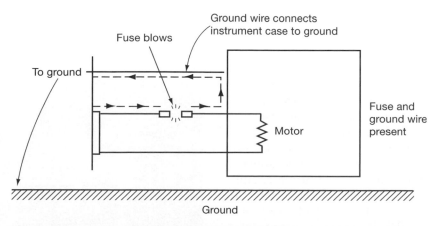

Figure 14.14.   A Properly Protected Instrument. The ground wire connects the instrument to the earth, thus harmlessly draining off the current and preventing the user from shock. A fuse is also present to prevent excess current from flowing in the circuit.

the amount of current in a circuit. For example, most centrifuges have a speed control adjustment to vary their speed of rotation. Rheostats are frequently part of such speed-adjusting mechanisms. Another example is calibration. Calibration of an electronic instrument usually involves changing the voltage or current in a particular circuit. This can be accomplished by varying the resistance in the circuit with a potentiometer.

A **capacitor** *is a device that stores electrical charge and can be used to provide current in the absence of a battery or current from an outlet.* Capacitors consist of two thin plates made of a conducting material and separated by a layer of insulating material. When current passes through a capacitor one of the plates acquires a positive charge, the other a negative charge. After current flow ceases through the capacitor, the charge can remain on the plates for days. A charged capacitor can function as if it were a DC voltage source.

A defibrillator, which is used medically to shock a patient's heart into a normal rhythm, is an example of how capacitors are used in a device. An electrical charge is stored by a capacitor in the defibrillator. When the defibrillator electrodes are connected across the patient's body, the patient forms a conducting discharge path and a large current flows for an instant.

An **electroporator** is another example of a device based on a capacitor. **Electroporation** *is used to introduce drugs, genetic material, and other molecules into living cells suspended in an aqueous medium.* This method has particular application in biotechnology because it enables scientists to introduce functional DNA into cells. In electroporation a short-duration, high-voltage electrical field is applied to a chamber containing cells. The electric field causes pores to be temporarily created in the cells' membranes through which the molecules enter. The simplest approach to create a high-voltage pulse is to charge a capacitor using a high-voltage power supply and then to discharge the capacitor into the chamber with the cells.

Diodes *are components made of a crystalline material that is capable of conducting electricity in only one direction.* The primary function of **diodes** in instruments is to *convert alternating to direct current, which is called* **rectification**.

There are also diodes, called **light emitting diodes (LEDs),** *that produce light when electrical current flows through them.* LEDs are familiar to us as the displays on digital clocks and other electronic devices.

Transistors *are devices made from semiconductor materials that are used to amplify an electrical signal.* Transistors also function as switches to turn a circuit on or off and can be used to convert direct current to alternating current. Transistors have played a key role in the development of such devices as microprocessors, pocket calculators, home computers, and video games.

iii. Integrated Circuits and Circuit Boards

Two manufacturing processes are used to make electronic circuits, conventional and integrated. Most laboratory instruments combine both circuit types. A conventional circuit contains various electronic components connected to one another by wires and attached to a base. The base is normally a **printed circuit board** *made of a thin piece of plastic or other insulating material.* Copper wires are "printed" onto the board using a special chemical process. Components can be connected to one another on the board in series, parallel, or in a combination of both, Figure 14.15.

An **integrated circuit (IC)** *is a very small electronic circuit that is assembled on a piece of semiconductor material called a* **chip**. The chip acts both as a base and as a part of the circuit. Integrated circuits contain components such as diodes, transistors, and capacitors linked together by tiny conducting wires. A single chip can contain millions of microscopic parts. The chips are packaged in protective carriers with electrical pin connectors that allow them to be plugged into a circuit board. Integrated circuits are found in almost all modern electronic laboratory instruments. ICs are often combined with one another and with other components onto a printed circuit board.

Electronic circuits are combinations of components connected to one another in various configurations to form **functional units**. Functional units perform the various tasks of electronic instruments. The next section discusses functional units as they relate to instruments that make measurements.

B. Functional Units

i. Transformers

Transformers *are used to vary the input voltage entering an instrument.* For example, a transformer can convert the 110 V AC voltage from the power company to 150 volts for an instrument. It may seem surprising that more voltage can be "created" than originally existed;

Figure 14.15. Printed Circuit Board.

however, the power supplied to the instrument remains constant. Power equals voltage multiplied by current; therefore, if the voltage supplied to an instrument is increased ("stepped up") by a transformer, then the current to that instrument is reduced. Similarly, if the voltage to the instrument is reduced ("stepped down"), then the current increases.

ii. POWER SUPPLIES

Power supplies were previously mentioned as related to electrophoresis. In electrophoresis the power supply is conspicuous because the operator attaches it to the gel box and selects how much current or voltage is to run through the gel. Most laboratory instruments contain less conspicuous internal power supplies.

The **power supply** plays a variety of roles including:

- *Converting alternating current to direct current.* The power company sends out alternating current, but most instruments require direct current. A power supply is used to convert alternating to direct current.

- *Acting as a transformer.*

- *Regulating the voltage in the event of fluctuations in voltage coming from the supplier.* It is not unusual for the voltage to surge or change unexpectedly. Instruments are protected from these variations by the power supply.

- *Distributing voltage to multiple circuits within an instrument.* Instruments are composed of multiple circuits, each of which requires a source of voltage.

iii. DETECTORS

The term *transducer* tends to be used broadly and so would include simple devices, such as the mercury in a thermometer. The term **detector** *is often used to refer to*

Table 14.1   EXAMPLES OF DETECTORS

Sample Property That Is Measured	Transducer	Electrical Output
Concentration of H^+ in a Solution	pH Electrode	Voltage
Temperature	Thermistor	Resistance
Temperature	Thermocouple	Voltage
Light intensity	Photomultiplier Tube	Current

an electronic transducer that generates an electrical signal in response to a physical or chemical property of a sample. The electrical output (signal) from a detector can be a change in voltage, current, or resistance. For example, pH meters (Chapter 18) respond to changes in the pH of a sample with a change in voltage. Thermistors (Chapter 17) respond to temperature changes with a change in resistance in a wire. Table 14.1 lists detectors that are described in later chapters in this Unit.

Because the detector generates the signal in a measuring instrument, its capabilities limit the instrument's overall performance. When evaluating detectors it is common to talk about their **detection limit**, **sensitivity**, and **range**. (These terms are used not only to refer to detectors, but also to describe methods, as is discussed in Chapter 6. Note the context when these terms are used in order to understand the author's intention.)

The **detection limit** *of a detector is the minimum level of the material or property of interest that causes a detectable signal.* The detection limit is a useful figure when comparing different techniques for measuring an analyte. In practice, the term *detection limit* is sometimes used synonymously with the term **sensitivity**, although these two terms are different. The **sensitivity** *of a detector is its response per amount of sample.* The

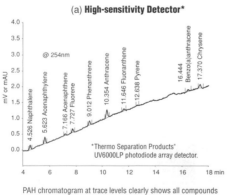

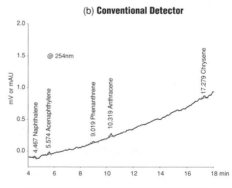

Figure 14.16.   Limit of Detection is an Important Factor in Evaluating a Detector.
a. A more sensitive detector generates signal, which appears here as "blips," in response to low levels of the material of interest. **b.** A less sensitive detector does not noticeably respond to low levels of the material of interest. (Data was collected with the Thermo Separation Products UV6000LP Photodiode Array Detector and a conventional photodiode array detector. Reprinted with permission.)

better the sensitivity, the lower the level of sample the instrument can reliably detect. Figure 14.16 is an advertisement that contrasts a high sensitivity and a low sensitivity detector. Only the high sensitivity detector can generate a detectable signal when trace concentrations of certain materials are present.

Although the sensitivity of a detector is important in determining its lower limit of detection, a second factor is also involved. This factor is **electrical noise**. Electrical noise may have two components. **Short-term electrical noise** *may be defined as random, rapid "spikes" arising in the electronics of an instrument,* Figure 14.17a. **Long-term electrical noise**, or **drift**, *is a relatively long-term increase or decrease in readings due to changes in the instrument and the electronics,* Figure 14.17b. When there is no sample in the instrument, the detector should ideally provide a steady zero response, resulting in a flat baseline on a recorder. In practice, because of short-term noise and drift, the baseline may not be absolutely stable, Figure 14.17c.

Electrical noise can arise spontaneously in an instrument's electrical circuitry. It can also be caused by other electrical devices, such as nearby power lines. In either event, noise is not related to the sample and can occur even if no sample is present. Electrical noise is a problem because it can interfere with our ability to see a small signal due to the sample, Figure 14.18.

There are various methods used to express how much electrical noise is present in an instrument. **Signal-to-noise ratio** *expresses the relationship between signal and noise in an instrumental system as the instrument response due to the sample divided by the noise present in the system,* Figure 14.19. The higher the signal to noise ratio, the better the performance of the instrument. The term **root mean square noise (RMS)** is sometimes used to indicate how much noise is present. The RMS value is based on a statistical calculation that "averages" the noise present over a period of time. The lower the value, the less noise is present in the system.

Thus, the limit of detection of a detector is affected both by its inherent sensitivity and by electrical noise present in the instrument. The **detection limit of a**

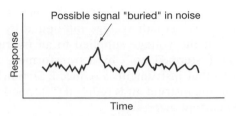

Figure 14.18. Electrical Noise Can Interfere with Our Ability to See the Signal Arising from the Detector's Response to the Sample.

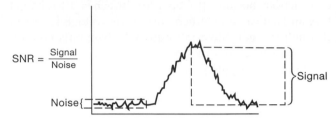

Figure 14.19. Signal-to-Noise Ratio. The relationship between signal and noise is illustrated by dividing the peak due to the signal by the average height of the spikes due to noise.

detector *is therefore often defined in practice as the minimum level of sample that generates a signal at least twice the average noise level.*

The **dynamic range** *of a detector is the range of sample concentrations that can be accurately measured by the detector.* Some detectors have a narrow dynamic range, whereas others have a wide one.

iv. SIGNAL PROCESSING UNITS

a. Overview

The electrical signal that is generated by a detector must be processed before it can be displayed by a readout device.* The processing that is required depends on the detector used and the final form of information required. For example, signal processing units may amplify the signal, count it, or convert it from one form to another, Table 14.2. A few of these processes will be discussed in more detail in this section.

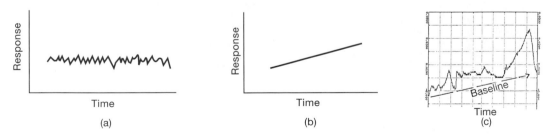

Figure 14.17. Electrical Noise. a. Electrical noise is brief, short-term spikes that occur even when no sample is present. **b.** Drift causes changes in the baseline over time, such as hours or days. **c.** Both short-term noise and drift may be present in an instrument.

*A short, helpful article on how the signal from a detector is processed is: "Choosing the Right Detector Settings," John V. Hinshaw. *LC/GC* 14(11) pp. 950–56, November, 1996.

Table 14.2 *EXAMPLES OF ELECTRICAL SIGNAL PROCESSING*

Type of Transformation	Purpose
Amplification	Boosts voltage or current in proportion to the size of the original signal
Analogue to Digital Conversion	Converts analogue signal to digital signal
Filtering	Separates and removes unwanted noise from the signal generated by a detector
Attenuation	Reduces an amplified signal to make it best fit a readout device
Log to Linear Conversion	Converts a signal which has a logarithmic relationship to the property being measured to a signal with a linear relationship*
Integrator	Calculates the area under a peak (as in chromatography)
Counting	Keeps track of and counts signals from the detector

*Important in spectrophotometry (discussed in later chapters) where transmittance values are converted to absorbance values.

b. Amplification

An **amplifier** *boosts the voltage or current from a detector in proportion to the size of the original signal.* Amplification is one of the most important types of signal processing because the current or voltage change produced in a detector is often very small. For example, suppose the output of a detector is only 3 μA. It is possible to amplify this current by a thousand times to 3 mA. In order to amplify a current there must be a power supply, so an amplifier is associated with a power supply and with other electronic components.

Gain *is the degree to which a signal can be increased or decreased.* The amplifier gain is the ratio of the output voltage from the amplifier to the input voltage arriving at the amplifier. For example:

An input voltage of 1 mV is amplified to an output voltage of 100 mV.

$$Gain = \frac{100\ mV}{1\ mV} = 100$$

In some cases, the operator will need to adjust the gain, or the amplification, of an instrument. If the signal is amplified too little, then the signal will be difficult to see and very small signals may be undetectable on the readout device. If the amplification is set too high, however, the signal may exceed the ability of the readout device to handle it.

Be aware that once a detector has generated an electrical signal, the signal-to-noise ratio cannot be changed

by simple electronic amplification because the amplification of the signal is accompanied by a corresponding boosting of the noise. There are, however, electronic filtering devices and also software programs that extract a signal from noise. These devices are used to process the signal after it leaves the amplifier.

c. Analogue to Digital Converter

There are two types of signal: **analogue** and **digital**. Signal that can change values continuously is called **analogue** *(i.e., "smoothly changing")*. In contrast, a **digital signal** *represents information by a variable that can have only a limited number of discrete values.* A meter is often used to display an analogue signal, Figure 14.20a. A meter has a needle that deflects from its zero position when a current is applied. The degree of deflection is related to the electrical signal reaching the meter. Note that the needle can point to any number, and can also point to a value anywhere in between two numbers. The speedometer of a car is a familiar example of such a meter.

The display from a digital clock radio is shown in Figure 14.20b. The clock radio can only display certain values. In this illustration the time reads 10:20. The next value it will display is 10:21. This digital clock radio cannot display any time in between 10:20 and 10:21.

The signal generated by a detector in response to a property of a sample is usually analogue. For example, a pH probe develops voltage in response to the H^+ concentration of a solution. This voltage might be 50 mV or 51 mV or any value in between.

Computers are digital instruments that represent all information using only two values: on and off. Each letter in the alphabet and every number processed by a computer is converted into a distinct pattern of on and off signals. A digital system is a collection of components that can store, process, and display information in a digital form.

Digital devices, such as computers, cannot directly manipulate an analogue signal, such as the signal produced by a detector. When the signal from a detector is to be displayed, stored or manipulated by a digital device, the analogue signal must be converted to a digital signal. An **analogue to digital converter (A/D converter)** *is a type of signal processor that very rapidly converts an analogue signal into a digital one.* There are

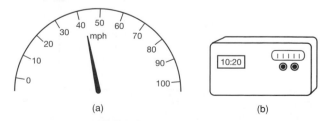

Figure 14.20. Displays of Analogue and Digital Signals.
a. A speedometer is an analogue display. **b.** A digital clock radio has a digital display.

also situations where a **digital to analogue converter** converts a digitized signal back to an analogue signal.

d. Attenuator

An amplified signal must sometimes be reduced in order to be best displayed by a readout device. This is common, for example, with gas chromatography instruments that are attached to strip chart recorders. An operator adjusts an **attenuator** *to reduce a signal.*

v. READOUT DEVICES

Various devices, such as meters, strip chart recorders, and computer screens, are used to display the information from the signal processing unit in a form that is interpretable to a human or to a computer. A display device may be either analogue or digital, Figure 14.21. Table 14.3 lists various types of readout devices, a few of which are discussed in this section.

Analogue meters that can display the result of a measurement were mentioned above. Digital devices commonly display individual measurement values with light emitting diodes (LEDs) or with a **liquid crystal display (LCD)**. LEDs produce their own light and can therefore be seen in the dark. A digital clock radio is an example of a device with an LED display. LCDs play the same role as LEDs, but they are used where lower power consumption is required. An LCD has a thin layer of a liquid crystalline material sandwiched between two plates of glass. The sandwich normally reflects light; however, if a voltage signal is passed to an area of the LCD, that area darkens. The darkened portions form the letters and numbers of the display. LCDs are used, for example, on digital wrist watches.

Strip chart recorders *are analogue display devices that record the output of a detector continuously over time.* A strip chart recorder has a pen that moves up and down over a strip of paper in response to the signal from a detector. The paper slowly moves along a track so the recorder plots the level of the electrical signal

Table 14.3 *READOUT DEVICES*

Analogue	Digital
Analogue meter	Printer
Strip chart recorder	Digital display
Oscilloscope	Disk (computer)
	Computer screen

versus time. Figure 14.22a shows an example of a strip chart recording generated as a sample flowed through a detector. The detector responded to biological compounds in the sample; the higher the signal, the higher the level of compound in the detector at that moment. In this type of recording, the area under the peak is a measure of the amount of the material being analyzed.

The signal arriving at a strip chart recorder has a range of values. The operator adjusts the recorder according to that range so that the recording fills the paper but does not "run off" the paper. For example, if the lowest value arriving at the recorder is 0 *mV* and the highest is 80 *mV*, the operator might turn a switch on the recorder to select a range from 0 to 100 *mV*. In this case, an input voltage of 100 *mV* into the strip chart recorder will move the pen all the way from the left hand edge of the paper to the right; 100 *mV* is said to be "full scale." If the recorder is set in this way, and a signal of 125 *mV* is unexpectedly produced, the pen will go "off scale," Figure 14.22b.

Meters and strip chart recorders are both useful devices and are still sometimes used in the laboratory. Most instruments are now attached to computers or incorporate microprocessors. Computers and microprocessors can take a signal and store it for later retrieval, calculate the concentration of a component in a sample, graph the signal in various ways, make corrections, produce a report, and perform many other functions. Printers are used to display the results after they

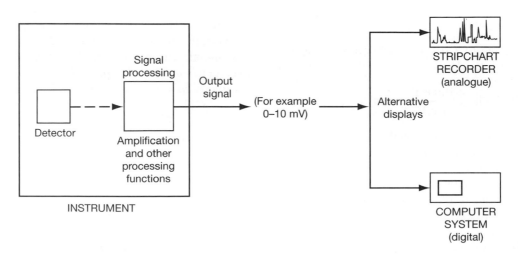

Figure 14.21. A Chromatography System.

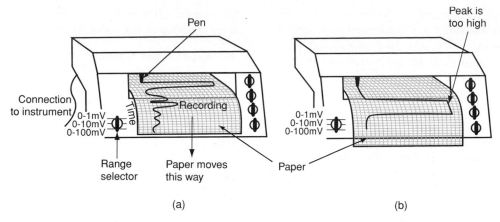

Figure 14.22. Strip Chart Recorders. a. A strip chart recorder has a pen that responds to the signal from the detector, resulting in a continuous recording of instrument response over time. In this recording, the peaks represents biological compounds. The amount of compound present is related to the area under the peak. **b.** The strip chart recorder range should be selected so that the recording fills the paper but does not flatten out, as it does in this illustration.

have been processed. Because the signal has been processed by a microprocessor or computer, the X axis need not represent time, as it does with a strip chart recorder. It can be, for example, wavelength or concentration. The Y axis similarly is not the direct signal from the detector, but it can be the result of various types of processing.

Printers associated with digital instruments do not usually have range selector knobs like those on strip chart recorders. However, the operator can usually use a keyboard or similar device to "tell" the recorder what will be the lowest and what will be the highest values on the plot.

EXAMPLE

The following figure shows the back panel of a representative instrument illustrating some of the ideas discussed so far.

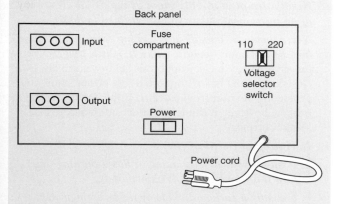

Voltage Selector. The voltage that comes to a particular receptacle from the power company may supply a voltage of either 110 or 220. An instrument can sometimes be used with either input voltage. The voltage selector is a switch that allows the user to select the

proper voltage. Note that an instrument may require a fuse to be exchanged if the selected voltage is changed.

Fuse Compartment. Instrument fuses or circuit breakers are often accessible to the user and are often located on the back panel of the instrument.

Output connectors. Each electronic measuring instrument has some sort of output signal, such as a voltage or a current. This output can be connected via cables or cords to other devices, such as a strip chart recorder, computer, or another instrument.

Input Connector. Some instruments are able to receive electrical information (signal) from other instruments. For example, chromatography instruments may be controlled by computers in which case connections are required from the computer to the instrument.

Power Connection. This is where the power cord attaches to the instrument.

IV. QUALITY AND SAFETY ISSUES

A. Quality Issues

It is important that laboratory instruments be properly maintained and repaired when malfunctions occur. **Preventive maintenance** *is a program of scheduled inspections of laboratory instruments and equipment that leads to minor adjustments or repairs and ensures that the instruments are functioning properly.* Preventive maintenance programs are intended to:

- ensure correct results
- identify components that require replacement
- ensure that instruments are safe to use
- ensure that instruments will not be shut down by a major problem
- lower the costs of repairs

Recall that **calibration** *is the adjustment of an instrument so that its readings are in accordance with the values of internationally accepted standards.* Every measuring instrument must be calibrated and this calibration may or may not be part of its routine maintenance. The frequency of calibration depends on the type of instrument and its use. The individuals who operate an instrument should ensure that it has been properly calibrated, although the operator may or may not be the one to actually perform the calibration.

Performance verification *is a process of checking that an instrument is performing properly.* One aspect of this process is to check that it is correctly calibrated. Note that we are distinguishing between "calibration" and "performance verification." We use "calibration" here to mean that adjustments to the instrument are performed so that its readings are correct. "Performance verification" is used to mean that the instrument's performance is checked, but it is not adjusted. Performance verification is usually the responsibility of the individuals who use an instrument.

Instruments can also be **validated**. **Instrument validation** *is a comprehensive set of tests done before an instrument is put in service that demonstrate that it will work and the conditions under which it will function properly.* Instrument validation is a formal process that is required in regulated laboratories and in laboratories that meet certain voluntary standards.

An important part of a quality control program in a laboratory is that all operations performed with an instrument be logged. It is common to have a logbook associated with each instrument in which performance verification checks are recorded, preventive maintenance and repairs are noted, and routine operations are logged.

Table 14.4 lists some environmental factors that may affect the performance of most types of electronic instruments. These factors should be controlled whenever possible in the laboratory.

B. Electrical Safety

The fluids in a person's body can serve as a conducting pathway for current flow. Therefore, if a person touches two objects that differ in electrical potential, current can flow through the person—this is electrical shock. Electrical shock can result in severe internal and external burns, in paralysis, in muscle contractions leading to falls, and, in extreme cases, to death. The more current that flows through a person's body, the more dangerous. As little as 1 mA is detectable, 15 mA can cause muscles to freeze, and about 75 mA is fatal.

Table 14.4 ENVIRONMENTAL FACTORS THAT COMMONLY AFFECT THE PERFORMANCE OF ELECTRONIC INSTRUMENTS

1. *Temperature.* Changes in ambient temperature often affect instruments; therefore, avoid placing instruments in areas that are subject to direct sunlight or drafts, such as near a window or an open doorway.

2. *Contaminants.* Most instruments are sensitive to smoke, acid fumes, solvent fumes, dust, and dirt. Liquid spills on the top of instruments can also cause damage. Avoid environmental contaminants.

3. *Ventilation.* Most instruments have exhaust vents to dissipate heat generated by the instrument. Do not block the ventilation holes and allow an adequate area around instruments to dissipate heat.

4. *Humidity.* The humidity in the room where instruments are used should ideally be about 40–60%.

Table 14.5 ELECTRICAL SAFETY RULES

1. *Use three-pronged power cords and receptacles to ensure proper grounding.* Do not attempt to bypass grounding provided on instruments.

2. *Avoid extension cords if possible.* When required, use only heavy duty, grounded extension cords.

3. *Use properly insulated wires and connections.* Frayed wires or connectors should be repaired by a qualified individual.

4. *Do not handle connections with wet hands or while standing on a wet floor.*

5. *Avoid operating instruments that are placed on metal surfaces, such as metal carts.*

6. *Do not operate wet electrical equipment (unless it is intended to be used in this way) or devices with chemicals spilled on them.*

7. *Be certain that electronic equipment has adequate ventilation to avoid overheating.* Many instruments have fans. Be sure there is enough space around the vents for heat to dissipate.

8. *Never attempt to trouble-shoot or repair the electronics of an instrument unless you are trained to do so.*

9. *Never touch the outside of a leaking electrophoresis gel box, or one with a puddle under it, if it is plugged into a power supply.* An electrophoresis gel box normally is closed when the box is connected to a power supply, thus protecting the user from current. If the box is cracked or broken, however, it can leak buffer and therefore also "leak" current.

10. *In case of an electrical fire:*
    a. *If an instrument is smoking or has a burning odor, turn it off and unplug it immediately.*
    b. *Use only a carbon dioxide–type fire extinguisher.* Water and foams conduct electricity and so increase the hazard.
    c. *Some electrical components (specifically selenium rectifiers) emit poisonous fumes when burning.* Avoid the fumes of burning instruments.

Table 14.6 *SIMPLE CHECKS OF AN INSTRUMENT THAT APPEARS TO BE MALFUNCTIONING*

1. *Is the instrument plugged in?*
2. *Is the power on?*
3. *If there is a power strip, is it turned on?*
4. *Is the wall outlet functioning? An outlet can be checked by plugging a lamp into it.*
5. *Is the power cord frayed? If so, have it replaced.*
6. *Is there a blown fuse or tripped circuit breaker?*
7. *Is there an accessible bulb that needs replacement?*
8. *Does the instrument have a "reset" button? If so, try pressing it.*
9. *If there is a switch to select either 110 or 220 V, is it properly set?*
10. *It sometimes helps to unplug an instrument and plug it back in.*
11. *Check the instrument manual. There may be a series of trouble-shooting steps that the user can easily and safely perform.*
12. *Call the instrument manufacturer and ask their advice.*

Human skin is an excellent insulator when it is dry. The resistance of dry skin is in the megohm range. Wet skin, unfortunately, offers far less resistance to current flow. The resistance of wet skin is about 300 ohms. If a person with wet skin is exposed to a voltage of 120 V, the current passing through that person can be as high as 0.400 A (400 mA). Take precautions, therefore, to avoid having current pass through your body. Table 14.5 lists common (and common sense) safety rules.

Most people who operate laboratory instruments are not specially trained in instrument repair and should never attempt to repair a broken instrument, other than to replace fuses, worn-out bulbs, or other simple tasks as directed by the manufacturer. Capacitors can hold a charge for weeks, so there is the risk of shock from an instrument, even when it is unplugged. Only a trained technician should bleed off the current from a capacitor. An untrained individual seeking to repair an electronic instrument can not only injure her/himself but can also damage an instrument. Table 14.6 lists simple things that can be safely checked by a user when an instrument appears to be malfunctioning.

PRACTICE PROBLEMS

1. In each of the following devices what is the voltage source? What is the resistance or load?

   a. Flashlight

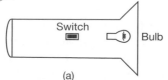

(a)

b. Washing Machine

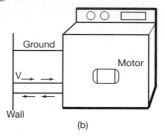

(b)

c. An electroporator

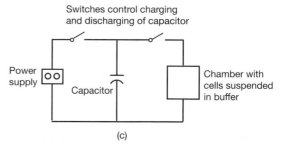

(c)

2. Most household appliances in the United States are designed so that they remain in a "standby" mode in which they continue to draw power, even when they are supposedly turned off. Manufacturers make appliances this way either because it is less expensive or because the device has a feature, such as a memory of previous settings, that requires constant power. This power consumption by devices that are not in use is responsible for billions of dollars of electricity consumption at a cost both to the consumer and to the environment.

   Suppose a home has the following devices that are constantly drawing power, 24 hours per day, 7 days a week. (For simplicity, ignore the time that the devices are actually in use and draw more power than is listed.)

   Assume that power to this house costs $0.09 per kilowatt-hour. (A kilowatt-hour is equivalent to 1000 W of power used for 1 hour.) How much would the owners of this home pay per year for standby current? How much would a city with 50,000 similar homes pay for standby current? (Information from "Must We Pull the Plug? New Programs Aim to Cut the Juice Drawn by Leaky Appliances," Janet Raloff. *Science News.* Vol. 152:266–267, October, 1997.)

Device	Power Drain While in "Standby"
Telephone Answering Machine	3.2 W
Cordless Phone	2.3 W
Portable Stereo	2.2 W
Color T.V./VCR	9.1 W
Compact Audio Unit	9.0 W
Electric Toothbrush	2.2 W
Doorbell	2.2 W
Microwave Oven	3.1 W

3. If the current in a circuit is 12 mA and the voltage is 120 mV, what is the resistance?

4. How many watts of power can be supplied by a 120 V circuit that is able to carry 20 A?

5. A flashlight is supplied with 6 V by batteries. When it is turned on, 400 mA flows through the bulb. What is the resistance of the bulb?

6. Suppose a power supply used in electrophoresis can be set at anywhere from 0 to 500 V and from 0 to 400 mA. What is the maximum amount of power that might be used by this power supply?

7. A professor wants to outfit a teaching laboratory with 15 identical electrophoresis units arranged one after the other on a laboratory bench. Each unit is powered by a power supply like the one described in Problem 6. Suppose that the circuit into which the electrophoresis units will be plugged is served by a 120 V line with a current of 20 A. If students use all 15 devices at the same time, is it possible that the load on the circuit will exceed its power rating?

8. An instrument is rated at 500 W. Assuming that the voltage is 120 V, what is the current in this instrument? What is the resistance of this instrument?

DISCUSSION QUESTION

If you work in a laboratory, list the types of electronic transducers found in the laboratory. List the types of readout devices in the laboratory.

TERMS RELATING TO INSTRUMENTAL METHODS AND ELECTRICITY

Alternating Current, AC. Electrical current that cycles between flowing first in one direction and then in the other direction.

Ammeter. A device to measure electrical current.

Ampere, A, or amp. The units by which current is expressed. When 6.25×10^{18} electrons pass a point in the path of electricity's flow every second, the path is said to be carrying a current of 1 A.

Amplification. The boosting of a weak signal.

Amplifier. A device which boosts the voltage or current from a detector in proportion to the size of the original signal.

Analogue. "Smoothly changing." Analogue measurement values are continuous, as for example, displayed by a meter with a needle that can point to any value on a scale.

Analogue to Digital Converter, A/D Converter. Device that converts an analogue signal to a digital signal.

Attenuator. An electronic component that reduces the level of the signal from an instrument.

Baseline. A line that shows an instrument's response in the absence of analyte.

Battery. A device that creates voltage by electrochemical reactions.

Calibration. In this chapter, the adjustment of an instrument so that its readings are in accordance with the values of internationally accepted standards.

Capacitor. A component of electronic circuits that stores electrical charge.

Chip. A thin piece of silicon that contains all the components of an electrical circuit.

Circuit. A complete path for current flow.

Circuit Board. A card on which electronic components are mounted and connected to one another to form a functional unit.

Circuit Breaker. A safety device that limits the current flowing in a circuit.

Conductor. A material that offers little resistance to the flow of electricity.

Current (electrical), I. The flow of charge. Current may involve the movement of electrons in a conductor or the flow of ions in a solution.

Detection Limit (of a detector). The minimum level of the material or property of interest that causes a detectable signal.

Detector. Refers to an electronic transducer that generates an electrical signal in response to a physical or chemical property of a sample.

Digital. Discontinuous measurement values (in contrast to continuous, analogue values), as for example, stored and displayed by a computer.

Digital to Analogue Converter. A signal processing device that converts a digitized signal to an analogue signal.

Diode. An electronic component that allows electricity to flow in only one direction.

Direct Current, DC. Current that always flows in the same direction.

Display (in measurement). Devices, such as meters, strip chart recorders, and computer screens, that display information in a form that is interpretable to a human or to a computer.

Dynamic Range (of a detector). The range of sample concentrations that can be accurately measured by the detector.

Electrical Ground. A conducting material that provides a pathway for current to the earth.

Electrical Potential. The potential energy of charges that are separated from one another and attract or repulse one another; measured in units of volts. Also commonly termed *voltage* and *electromotive force*.

Electromotive Force, EMF or ε. See "Electrical Potential."

Electronics. Instruments with components such as transistors, integrated circuits, and microprocessors that amplify, generate, or process electrical signals.

Electrophoresis. A method that separates charged biological molecules from one another based on their rate of migration when placed in an electrical field.

Electroporator. An instrument used to introduce drugs, genetic material, and other molecules into living cells suspended in an aqueous medium, by exposing the cells to a pulse of high voltage.

Energy. The ability to do work.

Filtering. Signal processing in which unwanted noise is removed from the signal generated by a detector.

Fixed Resistor. A resistor that provides a single, set level of resistance in a circuit.

Frequency. The rate at which alternating current cycles back and forth, measured in cycles/sec or Hertz (Hz).

Functional Units (in electronics). Combinations of electronic components connected to one another in various configurations that perform a particular function.

Fuse. A safety device that limits the current flowing in a circuit.

Gain. The degree to which an input voltage is amplified to become an output voltage.

Galvanometer. An instrument that measures small electrical currents.

Ground Wire. A separate wire attached to the metal frame of an appliance and connected to the earth through the third prong of the plug. Protects an operator in the event of a short circuit.

Hertz, *Hz*. The unit of frequency of alternating current. 1 *Hz* = 1 cycle per second.

Input Connector. A connection that enables an instrument to receive electrical signal from another device.

Instrument Response Time. The time required for an indicating device to undergo a particular displacement following a change in the property being measured.

Insulator (in electricity). Materials in which the outer electrons of atoms are not free to move and so electricity does not flow readily.

Integrated Circuit, IC. A small electronic circuit with multiple components (such as resistors, capacitors, and diodes) built on a chip.

Integrator. A signal processing mode in which the area under a peak is calculated.

Interface. Site where a measuring instrument contacts the material being measured.

Jack. A receptacle for a plug. The plug is inserted into the jack to complete a circuit.

Light Emitting Diode, LED. A diode that emits light when current flows through it.

Linearity. 1. The characteristic of a direct reading device. If a device is linear, calibration at 2 points (i.e., 0 and full-scale) calibrates the device (2 points determine a straight line). 2. A measure of how well an instrument follows an ideal, linear relationship.

Liquid Crystal Display, LCD. Digital device commonly used to display individual measurement values.

Load. The electrical demand of a device expressed as power (watts), current (amps), or resistance (ohms).

Log to Linear Conversion. Processing in which a signal that has a logarithmic relationship to the property being measured is converted to a signal with a linear relationship to the property being measured.

Microampere, μA. One millionth of an ampere.

Microchip. A computer chip.

Microelectronics. Electronics with very small components arranged into circuits.

Microprocessor. An integrated circuit that can store and process information.

Milliampere, *mA*. One thousandth of an ampere.

Millivolt, *mV*. One thousandth of a volt.

Negatively Charged. A material in which electrons are in excess.

Noise (electrical). Unwanted electrical interference which creates a signal unrelated to the property being measured. Electrical noise may have two components:

> **Short-Term.** Random, rapid spikes in the electronics of an instrument.

> **Long-Term, or Drift.** A relatively long-term increase or decrease in readings due to changes in the instrument and the electronics.

Ohm, Ω. Unit of resistance. One ohm is the value of resistance through which 1 *V* maintains a current of 1 *A*.

Ohm's Law. An equation that relates voltage, current, and resistance in a circuit:

$$Voltage = (Current)(Resistance)$$

Ohmmeter. An instrument used to measure electrical resistance in a circuit.

Output Connector. A connection that enables an instrument to send an electrical signal to another device.

Parallel Circuit. A circuit in which the components are connected so that there are two or more paths for current.

Performance Verification. A process of checking that an instrument is performing properly.

Polarity. The characteristic of having a positive or negative charge.

Polarized. An electrical device that has negative and positive identifications on its terminals. When connecting such devices to a source of voltage the negative terminal of the device is connected to the negative terminal of the voltage source.

Polarized Plug. Refers to plugs with different-sized prongs to ensure that the plug is oriented correctly in the outlet.

Positively Charged. A material in which electrons are depleted.

Potential Energy. Stored energy.

Potentiometer. A type of variable resistor that can be adjusted to control the voltage and/or current in a circuit.

Power, *P.* The rate of using energy, measured in units of Watts.

$$Power = (Volts)(Current)$$

Power Supply. A device that produces varying levels of AC and DC voltage, converts AC current to DC, and regulates the voltage level in the circuit.

Preventive Maintenance. A program of scheduled inspections of laboratory instruments and equipment that leads to minor adjustments or repairs and ensures that items are functioning properly.

Printed Circuit Board. A thin piece of insulating material onto which copper wires are printed using a chemical process. Electronic components are connected to one another on the board.

Readout. See "Display."

Receptacle. The half of a connector that is mounted on a wall, instrument panel, or other support and into which electrical devices are plugged.

Rectification. Conversion of alternating current into direct current.

Resistance, *R.* A measure of the difficulty electrons encounter when moving through a material. Measured in units of ohms.

Resistor. An electronic component that resists the flow of current.

Rheostat. A type of variable resistor that can be adjusted to control the voltage and/or current in a circuit.

Root Mean Square Noise, RMS. A statistical calculation that "averages" the noise present over a period of time; the lower the value, the less noise is present.

Semiconductor. A material that is more resistant to electron flow than a conductor but is less resistant than an insulator. Semiconductors are important in the manufacture of electronic devices.

Sensitivity (of a detector or instrument). Response per amount of sample.

Sensor. See "Transducer."

Series Circuit. A circuit configuration in which the components are connected in a "string," end to end, so that current can only flow in one pathway.

Short Circuit. An unintended path for current flow that bypasses the resistance or load.

Signal. An electrical change that conveys information. The signal may be a change in voltage, current, or resistance.

Signal-to-Noise Ratio, SNR. The instrument response due to the sample divided by the electronic noise in the system.

Signal Processing Unit (signal processor). A device that modifies an electrical signal, for example, to amplify it.

Silicon. An abundant element that is used in the manufacture of electronic components.

Solid State. An electrical circuit that uses semiconductor diodes and transistors instead of vacuum tubes. (The term *solid* is used because signal flows through a solid semi-conductor material instead of through a vacuum.)

Strip Chart Recorder. An analogue display device that records the output of a detector continuously over time.

Switch. A device that controls current flow in a circuit.

Transducer. A device that senses one form of energy and converts it to another form.

Transformer. A device used to vary the input voltage entering an instrument.

Transistor. A semiconductor component used primarily for amplification and sometimes to switch a circuit on or off.

Two-point Calibration. If the response of a device is linear, calibration at two points (for example, 0 and full-scale) calibrates the device.

VAC. An abbreviation for voltage when the current is alternating.

VDC. An abbreviation for voltage when the current is direct.

Validation (of an instrument). A comprehensive set of tests done before an instrument is put in service that demonstrates that it will work and the conditions under which it will function properly.

Variable Resistors. Device that can be adjusted to provide differing amounts of resistance to electron flow. (See also "Rheostat" and "Potentiometer.")

Voltage. An electrical potential, measured in volts.

Voltage Selector. A switch that allows the user to select the proper voltage (either 110 or 220) for an instrument.

Voltmeter. An instrument used to measure voltage.

Volts, *V*. Units of potential difference between two points in an electrical circuit.

Watts, *W*. Units of power, where Power = (Voltage) (Current).

GENERAL REFERENCES CONTAINING INFORMATION ABOUT ELECTRICITY AND ELECTRONICS

General laboratory instrumentation and physics texts often have information on electricity and on electronic instrumentation. Some examples include:

Laboratory Instruments, Larry E. Schoeff and Robert H. Williams. Mosby, St. Louis, 1993.

Clinical Laboratory Instrumentation and Automation, Kory M. Ward, Craig A. Lehmann, Alan M. Leiken. W. B. Saunders Co, Philadelphia, 1994.

Physics for the Health Sciences, Carl R. Nave, Brenda C. Nave. 3rd edition. W.B. Saunders Co, Philadelphia, 1985.

ANSWERS TO PROBLEMS

1. **a.** Voltage source—batteries
   Resistance—bulb

   b. Voltage source—power company (wall outlet)
   Resistance—motor in the machine

   c. Voltage source—power company via power supply
   Resistance—the cells and the buffer in which they are suspended

2. The total standby current is 33.3 W
   33.3 W × 24 hrs/day × 365 days = 291708 W-hours
   291708 W-hours/1000 = 291.708 kW-hours
   (291.708 kW-hrs)($0.09/$kW$-hr) = $26.25 cost for standby current per year
   Multiplied by 50,000, it comes to about $1.3 million.

3. $V = IR$
   0.120 V = 0.012 A (?)
   ? = 10 $ohms$

4. $P = VI$
   ? = (120 V) (20 A)
   ? = 2400 W

5. $V = IR$
   6 V = (0.400 A)(?)
   ? = 15 $ohms$

6. $P = VI$
   maximum power = 0.400 A × 500 V = 200 $Watts$

7. The total carrying capacity of the line in the laboratory is:

   $$P = IV$$
   $$= (20\ A)(120\ V) = 2400\ W$$

   The total power consumed by the fifteen units if they are run at their maximum power is:

   $$200\ W\ (15) = 3000\ W$$

   The circuit, therefore, cannot handle all the units if they all are using 200 W.

8. $P = VI$
   500 W = 120 V (?)
   ? = 4.2 A

   $V = IR$
   120 V = 4.2 A (?)
   ? = 28.6 $ohms$

CHAPTER 15

The Measurement of Weight

I. BASIC PRINCIPLES OF WEIGHT MEASUREMENT

Determination of the **weight** of an object is among the most basic measurements made in the laboratory. **Weight** *is the force of gravity on an object.* **Balances** *are instruments used to measure this force.*

The term *weight* is commonly used interchangeably with *mass,* although these words are not synonyms. **Mass** *is the amount of matter in an object expressed in units of grams.* An object's mass does not change when it is moved to a new location, but its weight will change if the force of gravity differs at the new site. For example, an astronaut in space is "weightless" because there is no gravity, but the astronaut's mass does not change with blast off from earth, Figure 15.1. The significance of the distinction between mass and weight is explored in Section IV of this chapter.

The process of weighing an object basically involves comparing the pull of gravity on the object with the pull of gravity on standard(s) of established mass. The weighing method illustrated in Figure 15.2a dates back to antiquity, Figure 15.2b. The object to be weighed and the standard(s) are each placed on a pan hanging from opposite ends of a **lever** or **beam**. When the beam is exactly balanced, gravity is pulling equally on the sample and the standard. They are the same weight. Thus, instruments that weigh objects in the laboratory have come to be called **"balances."*** *Balances that consist of one beam with two pans, are called* **single beam, double pan balances**.

Mechanical balances that employ the principle of balancing objects across a beam are still used in the laboratory. However, modern balances are often electronic and use a somewhat different method of comparison. With electronic balances, the comparison between the sample and the standard is made sequentially. The balance is calibrated to a mass standard at one time and the sample is weighed on the balance afterwards.

II. CHARACTERISTICS AND TYPES OF BALANCES

A. Range, Capacity, and Sensitivity

There are many types and brands of balances with various designs and features. The fundamental characteristics of a balance, however, are its **range**, **capacity**, and **sensitivity (or readability)**:

1. *Range and Capacity.* Some laboratory balances are intended to weigh heavier objects; others, lighter ones. The **range** *of a balance is the span from the lightest to the heaviest weight the balance is able to measure.* **Capacity** *is the heaviest sample that the balance can weigh.*

2. *Sensitivity and Readability.* **Balance sensitivity** *may be explained as the smallest value of weight that will cause a change in the response of the balance.* Sensitivity determines the number of places to the right of the decimal point that the balance can read accurately and reproducibly, Figure 15.3. Extremely sensitive laboratory balances can weigh samples accurately to the nearest 0.1 μg (or 0.0000001 *g*). A less sensitive laboratory balance might read weights to the nearest 0.1 *g*. In catalogue specifications manufacturers express the sensitivity of their balances by their **readability**.

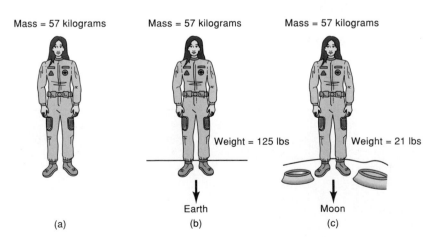

Mass = 57 kilograms   Mass = 57 kilograms   Mass = 57 kilograms

Weight = 125 lbs   Weight = 21 lbs

Earth   Moon

(a)   (b)   (c)

Figure 15.1. Mass Versus Weight. a. An astronaut is composed of matter. Her mass is a measure of that matter. **b.** The weight of the astronaut is a measure of the effect of gravity on her. **c.** On the moon, the astronaut weighs less than on earth because of the moon's weaker gravity, but her mass is unchanged.

*The word *balance* is usually reserved for laboratory instruments, whereas the common term *scale* is preferred for the less sensitive weighing instruments used in business, transportation, health care, and the home.

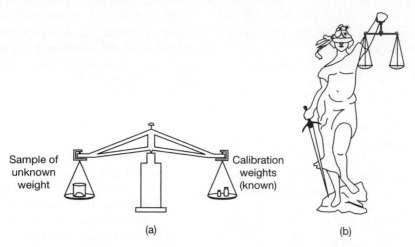

(a)   (b)

Figure 15.2.   Balances. a. A simple single-beam, double pan balance where a sample on the left pan is balanced by objects of established weight on the right pan. **b.** A double pan balance has been used to symbolize justice since ancient times. In Greek mythology Themis is the goddess of justice. She is usually represented wearing a crown of stars and holding a balance in her hand. (Double pan balance sketch by a Biotechnology student.)

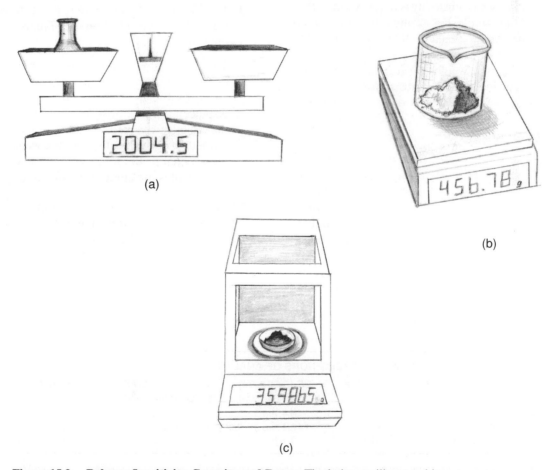

(a)

(b)

(c)

Figure 15.3.   Balance Sensitivity, Capacity, and Range. The balances illustrated in a–c are increasingly sensitive but their range becomes more limited and their capacity is lower as their sensitivity improves. **a.** A balance intended for heavier samples; readability is "0.1 *g*," range is "0–2200 *g*." **b.** Readability is "0.01 *g*" and range is "0–500 *g*." **c.** Analytical balance. Readability is "0.1 *mg*," range is "0–60 *g*." (Sketches by a Biotechnology student.)

Readability *is the value of the smallest unit of weight that can be read; it is the smallest division of the scale dial or the digital readout.*

Range, capacity, and sensitivity are interrelated. A very sensitive balance will not be able to weigh samples in the kilogram range, but a balance intended for heavier samples will not detect a weight change of a microgram.

Analytical balances *are designed to optimize sensitivity and can weigh samples to at least the nearest tenth of a milligram (0.0001 g). The most sensitive analytical balances,* **microbalances** *and* **ultramicrobalances**, *can weigh samples to the nearest microgram (0.000001 g) and the nearest tenth of a microgram (0.0000001 g), respectively.*

Figure 15.4 shows catalogue descriptions of various balances. The descriptions show the readability, range, and capacity of each balance. The descriptions also show the repeatability of the instruments and their linearity (which will be discussed later in this chapter).

Most biology laboratories have more than one type of balance. Which balance should be used depends on the weight of the sample, what sensitivity is required, and other characteristics of the sample. For example, it is not reasonable to weigh a person on a balance that has a readability of 0.0001 *g*, nor is it sensible to use an analytical balance for a laboratory sample that weighs several hundred grams. An analytical balance should be chosen, for example, to weigh milligram amounts of enzyme to use in a reaction.

B. Mechanical Balances

Balances can be broadly classified as either **mechanical** or **electronic. Laboratory mechanical balances** *typically have one or more beams; the object to be weighed is placed on a pan attached to a beam and is balanced against standards of known weight.* Mechanical balances do not generate an electronic signal when a measurement is made. Although electronic balances have replaced mechanical ones in many settings, manufacturers still produce mechanical balances that are not analytical because they are relatively inexpensive and easy to use. Examples are shown in Figure 15.5.

The analytical laboratory balances manufactured today are electronic; however older mechanical analytical balances are still found in some laboratories. These mechanical instruments give accurate and reproducible measurements if they are properly maintained and operated, so they will be discussed briefly.

Consider the mechanical analytical balance diagramed in Figure 15.6*:

1. A fine knife-shaped edge supports the beam, allowing it to swing freely. There is also a knife edge between the weighing pan and the beam. The beam is balanced on the right side by the **counterweight**, and on the left side by objects of known weight that are associated with the weighing pan. The beam is exactly balanced when the weighing pan is empty.

2. When a sample is added to the weighing pan, the beam is no longer balanced. By turning the appropriate knobs, the operator removes weights from the left side to bring the beam back to its resting, balanced position.

3. The amount of weight that must be removed to bring the beam into balance equals the weight of the sample.

4. The beam may remain slightly displaced from its original position by an amount too small to be balanced by the lightest weights. This slight residual displacement is shown on a scale and an image of the scale is projected to the user, hence, the scale is termed an **optical scale**. Thus, the weight of the sample is equal to the total weights removed from the left side plus the amount shown on the optical scale.

Mechanical analytical balances require care to operate. The knife edges and weights are easily dislocated if the balance is jostled. There is usually an "arrested" position and a "half-arrested" position in which the beam is stabilized and cannot swing freely. The balance should be in an "arrested" or "half-arrested" position when the balance is not in use, when removing or returning weights to the beam, and when moving a sample on or off the

CATALOG DESCRIPTIONS OF ANALYTICAL BALANCES				
MODEL NUMBER	CAPACITY (g)	READABILITY	REPEATABILITY (standard deviation)	LINEARITY
BA 50	50	0.1 *mg*	± 0.1 *mg*	± 0.2 *mg*
BA 300	300	0.1 *mg*	± 0.2 *mg*	± 0.5 *mg*
BA 4000	4000	0.1 *g*	± 0.1 *g*	± 0.1 *g*

Figure 15.4. Catalog Specifications for Balances. Catalogues specify the capacity, readability, repeatablity, and linearity for balances.

*The balances illustrated in Figures 15.2a and 15.6 are both mechanical balances but their design is different. "Newer" mechanical analytical balances usually have just one weighing pan and the beam does not extend equally on both sides of the support. When a sample is added to the weighing pan of the balance illustrated in Figure 15.6, standard weights are *removed* to compensate rather than being added to a second pan. Because the standard weights are removed, this type of mechanical balance is called a "substitution" balance.

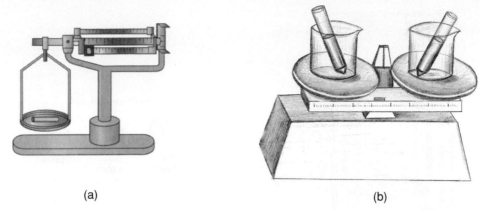

(a)   (b)

Figure 15.5.   Nonanalytical Mechanical Balances. a. Triple beam hanging pan balance. This balance has three beams. External standard weights are not needed because the weights (sometimes called "riders") slide to the right on the beams to balance the sample. This balance can be read to the nearest 0.01 *g*, and one additional place may be estimated. **b.** In centrifugation, the tubes and their contents that are opposite one another in the centrifuge must be of equal weight. A single beam, double pan balance can be used to compare the weights of centrifuge tubes easily. (Sketch (b) by a Biotechnology student.)

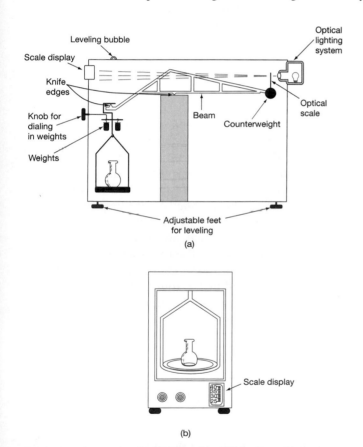

Figure 15.6.   A Mechanical Analytical Balance. a. Side view. The pan and the lever both rest on knife edges. A sample placed on the pan displaces the pan from its original balanced position. The operator turns knobs that remove weights until the beam is returned almost to its original position. The residual beam displacement after the weights are removed is shown on a scale, and an image of the scale is projected to the operator. The weight of the sample is equal to the weight removed from the left side plus a small amount shown on the optical scale. **b.** Front view showing knobs and optical scale.

weighing pan. The balance is fully released when a final measurement is being read. A mechanical balance must be perfectly level so that the beam is level. There is often a **leveling bubble** *to show whether the balance is level* and "feet" that can be adjusted up and down if necessary to even the balance, Figure 15.7.

C. Electronic Balances

Electronic balances do not have beams and use an electromagnetic force rather than internal weights to counterbalance the sample. An electronic balance produces an electrical signal when a sample is placed on the weighing pan, the magnitude of which is related to the sample's weight. The way an electronic balance works is:

1. The weighing pan is depressed by a small amount when an object is placed on it.

2. The balance has a detector that senses the depression of the pan.

3. An electromagnetic force is generated to restore the pan to its original ("null") position.* The amount of electromagnetic force required to counterbalance the object and move the pan back to its original position is proportional to the weight of the object.

4. This electromagnetic force is measured as an electrical signal that is in turn converted to a digital display of weight value. In order to convert an electrical signal to a weight value, the balance ultimately compares the electrical signal of the unknown sample to the signal of standard(s) of known weight.

Electronic balances are easy to use, automatically perform many functions, can be interfaced with com-

---

*This type of balance is sometimes called a "null position" balance because the pan is restored to its original position in order to measure the sample's weight.

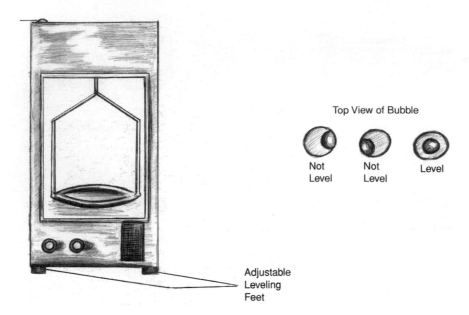

Figure 15.7. Leveling a Balance. The balance is level when the bubble is centered. For example, if the bubble is off-center to the left, then recenter it by raising the right side of the balance (or lowering the left). (Sketch by a Biotechnology student.)

puters and other instruments, permit automatic documentation, and are better able to compensate for environmental factors than mechanical balances. For these reasons, electronic balances are in widespread use.

Every model and brand of balance is operated differently; however, there are certain general processes that are required when using most balances. Box 1 outlines a generally applicable procedure for operation of an electronic balance.

III. FACTORS THAT AFFECT THE QUALITY OF WEIGHT MEASUREMENTS

A. Environmental Factors and the Operator's Technique

This section discusses factors that affect the accuracy and precision of weight measurements. These considerations are most relevant when working with analytical balances, although they may be important in other situations as well.

Recall that accuracy relates to the correctness of measurements. Precision relates to the reproducibility of a series of measurements. You might assume that the accuracy and precision achievable with a balance are dependent on the quality designed and built into that balance. It is true that a balance must be properly manufactured to give accurate, reproducible results. In normal laboratory practice, however, the accuracy and reproducibility of weight measurements are greatly affected by the conditions in that laboratory, by the user's technique, and by the routine maintenance of the balance. For this reason, manufacturers generally do not specify the accuracy of balances in their catalogue.

Mechanical balances, and to a lesser extent electronic balances, are affected by environmental conditions,

---

Box 1   GENERAL PROCEDURE FOR WEIGHING A SAMPLE WITH AN ELECTRONIC ANALYTICAL BALANCE

1. *Make sure the balance is level.*

2. *Adjust the balance to zero.* Make certain the weighing pan is clean and empty and the chamber doors (if present) are closed to avoid air currents and dust.

3. *Tare the weighing container or weigh the empty container.* Most balances can be **tared**, that is, they *can be set to subtract the container weight from the total weight automatically.* If a balance does not have the tare feature, then it is necessary for the operator to weigh the empty container and subtract its weight from the total weight of the sample plus container.

4. *Place the sample on the weighing pan and read the value for the measurement.*

5. *Remove the sample; clean the balance and area around it.*

including vibration, drafts, and temperature changes. Individuals who are accustomed to mechanical balances are often aware of the factors that interfere with obtaining accurate, reproducible measurements. In contrast, because electronic balances seem so easy to operate, users of electronic balances may be unaware of potential problems.

Temperature can have a significant effect on weight measurements. If a sample and its surroundings are at different temperatures, then air currents are generated that cause the sample to appear heavier or lighter than it really is. This problem may appear as a weight reading that slowly increases or decreases. Microbalances and ultramicrobalances have thick enclosures to protect them from thermal disturbances outside the balance.

Balances are also affected by the heat from their own electrical components. When a balance is first turned on it goes through a "warming up" period when its temperature is increasing and its readings are unstable. Many balances are therefore left on in a permanent standby mode to avoid delay while the balance warms up.

Temperature not only affects the apparent weight of a sample, it also affects the sample's actual weight. This is because there is always a thin film of moisture around objects, which is thicker when the object is colder and thinner when it is warm. Samples are therefore heavier when they are colder. (You can test this phenomenon. Weigh a flask on an analytical balance, then hold it in your hand for one minute and weigh it again. The flask should appear lighter the second time. Perspiration and oils from your hands are of minor importance; otherwise, the sample would appear heavier the second time.) Even the best electronic balance cannot compensate for this temperature effect because it actually affects the weight of the object.

Plastic and glass weighing containers and samples that are powders or granules are prone to develop a static electrical charge. Static charge occurs when different materials rub against each other. One material acquires an excess of electrons, resulting in a negative charge while the other loses electrons and acquires a positive charge. Nonconducting materials such as glass and plastic retain this static charge. If a sample or weighing vessel and its surroundings have the same charge, they repel each other; if they have opposite charges, they attract one another.

The presence of a vessel or sample that is electrostatically charged will affect the response of a balance, making its readings inaccurate and fluctuating. In extreme cases, a charged sample can literally fly out of its container and adhere to the balance chamber walls. Methods recommended to reduce the effects of static charge are listed in Table 15.1 and examples of related products are illustrated in Figure 15.8.

Temperature and static charge are only two of the factors that a careful operator must consider when making weight measurements. A number of points relating to good weighing practices are summarized in Table 15.1.

B. Calibration and Maintenance of a Balance

i. OVERVIEW

In order to give accurate readings a balance must be properly calibrated and maintained. Calibration of a

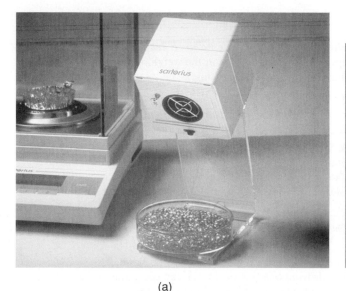

(a)

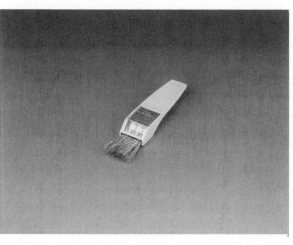

(b)

Figure 15.8. Examples of Commercial Products to Reduce the Effects of Static Charge. a. Ionizing blower, neutralizes static charges for use on samples prior to weighing. (Photo courtesy of Sartorius Corp.) **b.** Antistatic brush removes electrostatic charge when balance cases and pans are brushed. The brush contains a small amount of radioactive polonium which creates an ionized atmosphere that neutralizes ions when surfaces are brushed. (Photo courtesy of Fisher Scientific, Pittsburgh PA.).

Table 15.1   Factors that Affect Weight Measurements and Good Weighing Practices

1. *Analytical balances, especially mechanical ones, must be level.* Check that analytical balances are level before use.

2. *Vibration will affect balance readings, particularly with mechanical analytical balances.* Mechanical and microanalytical balances are usually placed on special tables designed to prevent vibration. Locate balances away from equipment that vibrates, such as refrigerators and pumps.

3. *Drafts will affect weight measurements.* Do not place balances near fans, doors, or windows. Analytical balances have shielded weighing compartments to eliminate drafts. Keep the weighing compartment doors closed when taking a reading.

4. *Balances, especially mechanical ones, should not be jostled.* Avoid leaning on the table holding the balance. Avoid moving a balance unless it is stabilized. (Consult the owner's manual for information on how to stabilize the balance.)

5. *Temperature changes will affect the actual and apparent weight of a sample.*
   a. Keep the sample, balance, and surroundings at the same temperature.
   b. Avoid touching samples and containers and placing your hands in the weighing chamber when using an analytical balance. Wearing gloves when handling samples is helpful (and is common practice), but using tongs provides better protection from temperature effects.
   c. Do not locate balances near windows, radiators, air conditioners, or equipment that produces heat.
   d. Allow a balance time to warm up before use or leave it in "stand-by" mode to avoid warm-up time.

6. *Static charge will affect measurements.* Various sources recommend the following procedures to help reduce static charge.
   a. A variety of products are available commercially to help reduce static, Figure 15.8.
   b. Charging usually occurs when the humidity is low. It is ideal to maintain the humidity between 40% and 60% in a weighing room.
   c. Metal does not tend to become electrostatically charged because it is conductive and charge readily dissipates from it. Metal weighing containers may therefore be the best choice and plastic containers the worst if electrostatic charge is a problem.
   d. Cleaning the walls of the balance chamber with alcohol or a product intended for cleaning glass may reduce charging.
   e. A slightly damp clothes softener sheet taped to the inside of the weighing chamber may be helpful.
   f. Balances should not be placed on tables with glass or plastic tops that can build up charge.
   g. Avoid wearing woolen clothing when weighing samples.

7. *Some samples gain moisture from the air and others lose volatile components to the air.* The weight of such samples will slowly increase or decrease during weighing.
   a. Work quickly but carefully to minimize such changes in the sample's composition.
   b. Maintain the humidity in a weighing room between 40 and 60% if possible because the amount of moisture loss or gain by a sample may be affected by the room's humidity.

8. *Samples that are magnetic may perturb the electromagnetic coil in an electronic balance, leading to incorrect results.* A magnetic sample will appear to have a different weight depending on its position on the weighing pan. It is helpful to distance magnetic samples from the weighing pan by placing a support under the sample.

9. *Electronic balances may give incorrect readings if the load is placed off-center.* Place the sample near the center of the pan.

10. *It is possible to damage electronic balances by placing a load that is too heavy on the weighing pan.* Newer balances have features to protect the balance from overload, but it is still good practice not to exceed the capacity of any balance.

11. *Selection of the weighing container may affect the results.* Samples can be weighed in a wide variety of containers including beakers, glassine coated **weighing paper**, and plastic or metal containers designed for weighing, called **weigh boats**. If a sample is likely to absorb or lose moisture, or contains volatile components, use a weighing vessel that is capped or has a narrow opening. Weigh samples in the smallest possible container. Be alert to sample particles that may adhere to the weighing vessel when poured out.

12. *Do not touch chemicals being weighed, magnetic stir bars, or the insides of beakers with your fingers.*

13. *Do not return unused chemicals to their storage bottles.*

14. *Balances and weighing rooms should be clean.* Sloppiness can have detrimental effects. Some chemicals are toxic, chemicals left on balances can cause corrosion and damage, and materials from one person's work can contaminate the work of someone else. Remember that balances are expensive instruments and should be well-maintained.

balance brings its readings into accordance with the values of internationally accepted standards.

Although balances are calibrated by the manufacturer when they are produced, calibration must be periodically checked in the laboratory of the user. This is because time and use (and abuse) of balances can affect their response. Calibration must also be checked when a balance is moved from one location to another. For measurements in the microgram range, the balance must be recalibrated each time the weather changes. Recalibration of the balance compensates for the effects of location and weather.

Calibration is not performed in the same way for a mechanical and an electronic analytical balance. Inside a mechanical analytical balance there are a number of standard weights, each of which must be of the proper mass

and in the proper orientation to ensure accurate weighing results. It is complex to calibrate a mechanical balance with so many internal weights and where all adjustments are made mechanically. A trained technician, therefore, is required to adjust a mechanical balance and to perform routine maintenance. (ASTM Standard E 319–85 describes in detail how to check a mechanical balance.)

In contrast to mechanical balances, it is usually straightforward to calibrate an electronic balance using a mass standard. Most manufacturers include directions for doing so in the instruction manual. Some electronic balances contain internal mass standards; others require that a standard be placed on the weighing pan during calibration. A two-point calibration procedure for electronic balances is outlined in Box 2 and illustrated in Figure 15.9.

Box 2   TWO-POINT CALIBRATION OF AN ELECTRONIC BALANCE

1. *The balance is set to zero using the appropriate knob, dial or button when the weighing pan is clean and empty.* It is simple to "zero" both electronic and mechanical balances, and they should be "zeroed" every time they are used.

2. *The second calibration point is usually at the heavier end of the balance's range.* With an electronic balance, a standard of established mass may be placed on the weighing pan and the balance set to the value of the standard using a lever, button, keypad, or other method of adjustment. Some electronic balances have a button that when pressed causes a calibration standard inside the balance to be weighed and the balance adjusts itself. In fact, there are microprocessor-controlled balances that recalibrate automatically to compensate for environmental changes without any operator intervention.

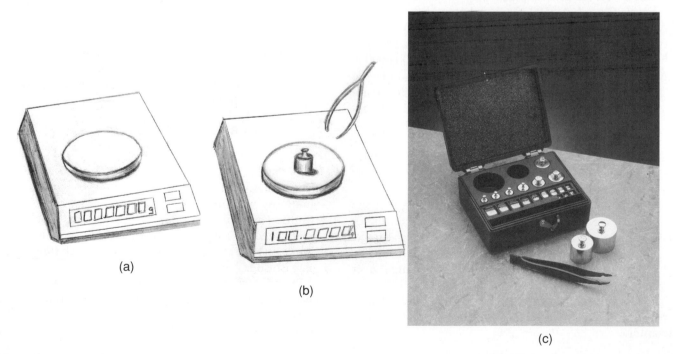

(a)

(b)

(c)

Figure 15.9.   Two-Point Calibration of a Balance. a. The balance is set to zero when the weighing pan is clean and empty. **b.** A calibration standard is placed on the pan. The balance is set to the calibration weight. (Sketches by a Biotechnology student.) **c.** Calibration standards. (Photo courtesy of Fisher Scientific, Pittsburgh, PA.)

ii. STANDARDS

The accuracy of any weight determination is limited by the accuracy of the standards used for comparison. Weight (mass) standards* are metal objects whose masses are known (within the limits of uncertainty) relative to international standards.

Standards used for calibrating balances can be purchased with a certificate showing their traceability to NIST. This certificate documents that the standard's mass went through a series of comparisons that ultimately links it to the international prototype in France. The checking of standards against one another requires stringently applied methods and specific environmental conditions.

In the United States, NIST houses official mass standards in Gaithersburg, MD. U.S. facilities that manufacture weight standards may send their standards to NIST to be checked against the U.S. standards. A standard that was tested by NIST is sometimes called "directly traceable to NIST" in the catalogues.

Standards for use in weighing can be purchased in different conformations of varying construction and with differing tolerances. The user determines the type of standards required depending on the balance being used and the requirements of the application. In the United States, the specifications for labeling and classification of standards for laboratory balances have been established by ASTM. (These specifications are reported in ASTM Standard E 617, "The Standard Specification for Laboratory Weights and Precision Mass Standards.")

The finest standards, those with the smallest tolerance, are **Class 1**. A "500 *g*" Class 1 standard, for example, must have a mass within 1.2 *mg* of 500.0000 *g* (i.e., it must be between 500.0012 *g* and 499.9988 *g*), whereas a **Class 4** standard must be within 10 *mg* of 500.0000 *g*. In routine biology laboratory work it is common to use **Class 2**, **3**, **or 4** standard weights. Class 4 is recommended for student use. The calibration standards that manufacturers provide with electronic analytical balances are usually Class 2.

Standard weights should be handled with tongs because they are damaged by skin oils and by cleaners that remove such oils. To protect their standards, a laboratory might keep one set for routine use and a second set in storage. Every 6 months or so, the standards for routine use can be checked against the set in storage. Standard weights need to be recertified periodically because even well-maintained weights change over time due to scratches, wear, and corrosion.

Information regarding the classification and features of mass standards is excerpted in the Appendix to this chapter.

C. Quality Programs and Balances

i. OVERVIEW

To ensure that balances perform properly, they must be operated correctly. They also must be regularly calibrated, cleaned, and repaired as required. The performance of a balance should be regularly checked to verify and document that it is giving accurate and precise readings. Laboratories that meet the requirements of a quality system, such as ISO 9000 or cGMP, have procedures that detail how to operate each balance and how to maintain, calibrate, and check their performance. Even if a laboratory does not conform to the requirements of a particular quality system, it is still good practice to perform these tasks on a regular basis.

The frequency of balance performance verification and calibration varies among laboratories. In some laboratories the weights of standards are checked each time the balance is used and the values are recorded. In other laboratories, this check is performed less frequently. Some laboratories have the policy that electronic analytical balances are calibrated every time they are used; in other laboratories, calibration is less frequent. Balances might similarly be checked at prescribed intervals for linearity, repeatability and other qualities of importance for that balance.

Maintenance and quality checks should be documented each time they are performed. Newer balances may have a feature that enables them to automatically record the date, time, and results of calibration for documentation purposes. Consistently checking and recording the weights of standards documents that measurements were made on a properly functioning instrument and alerts the operator to malfunctions that do occur. Whatever process is used in a laboratory to ensure the quality of balance measurements, it is good practice to have a formal, written policy and standard operating procedures that detail the process.

ii. VERIFYING BALANCE ACCURACY, PRECISION, AND LINEARITY

Performance verification of a balance involves testing its accuracy, precision, and linearity. *Accuracy* is tested simply by weighing one or more mass standards. The resulting weight must be correct within a tolerance established by a quality control procedure. This test is simple enough that it can be performed each time a balance is used.

Precision is an important aspect of the quality of a balance's performance. Precision (repeatability) is measured by weighing a sample multiple times and calculating the standard deviation. Manufacturers test the repeatability

---

*A note about terminology: Standards used for weighing are often called "weight standards"—because they are used during the process of weighing samples. Recall, however, that the weight of any object varies with its location, but its mass is constant. It is the *mass* of the standards that is established by traceability to NIST standards. This is a situation in which the terms *mass* and *weight* tend to be used interchangeably.

of their balances under ideal conditions in their testing laboratory and report the results in their balance specifications (Figure 15.4). To determine whether a balance is performing adequately, its repeatability in the laboratory can be compared with the manufacturer's specifications.

Linearity error occurs when a balance is properly calibrated at zero and full-scale (the top of its range), but the values obtained for weights in the middle of the scale are not exactly correct. Linearity can be tested by weighing individual subsets of objects and comparing the sum of the subsets to the weight of the objects all together, as shown in Box 3. If a balance is discovered to have linearity error, it is necessary to have it repaired professionally. Manufacturers report the linearity error of their balances in the balance specifications, and this figure is an indication of the quality of the instrument.

EXAMPLE PROBLEM

Suppose a technician is verifying the performance of a laboratory balance. The technician weighs a 10 g standard six times.

Weights

10.002 g	10.002 g
10.001 g	9.998 g
9.999 g	10.002 g

a. What is the standard deviation for the balance? (Remember the units.)

b. The balance specifications require that its precision (as measured by the standard deviation) be better than or equal to 0.001 g. Does it meet these specifications? If not, what should be done?

Box 3   CHECKING THE LINEARITY OF A BALANCE

Notes

a. Linearity tests do not require standard calibration weights. Any four individual objects that weigh about one fourth the capacity of the balance can be used. For example, if a balance has a capacity of 100 g, then four items with weights of about 25 g each are needed.

b. Each of the following weighings should be repeated, ideally 10 times, and the results of the 10 weighings averaged.

Test 1: Check the Balance Response at Its Midpoint

1. *Select four items whose total weight is about the capacity of the balance and label them A, B, C, D.*
2. *Weigh all four pieces together.* Record this weighing as the "full-scale value."
3. *Weigh pieces A and B together and record their weights.*
4. *Weigh and record the weight of pieces C and D together.*
5. *Add the two values from step 3 and step 4 together.* This is the "weight sum."
6. If the balance is exactly linear at its midpoint, then the weight sum will equal the full scale value. If the two are not equal, divide the difference by two because two measurements were made at this point. The result is the linearity error at the midpoint.

Test 2: Check the Balance Response at 25% of Capacity

1. *Weigh all four pieces individually.* Add their weights together and call this the "weight sum."
2. *If the balance is linear at 25% of its capacity, then the weight sum should equal the full-scale value as determined earlier.* If the two differ, divide the difference by four. This is the linearity error at 25% of capacity.

Test 3: Check the Balance Response at 75% of Capacity

1. *Divide the pieces into the following groups:*
   Group 1: pieces A, B, C
   Group 2: pieces A, B, D
   Group 3: pieces A, C, D
   Group 4: pieces B, C, D
2. *Weigh all four groups one at a time.* Add their weights together and call the result the "weight sum." Multiply the full-scale value by three and call it "75% full scale value."
3. *If the balance is linear at the 75% point, this weight sum will equal the 75% full scale value.* If the two differ, divide the result by four. This is the linearity error at 75% of capacity.

Based on information in "Assuring Balance Accuracy," Jerry Weil. Product Note. Cahn Instruments Inc., 11/91.

The standard deviation rounds to 0.002 g, which does not meet the specification. There may be a problem with the balance, although it is possible that the operator is not using good weighing practices. For example, the operator might forget to close the doors of the balance chamber or might place the standard in different locations on the weighing pan. The balance should be taken out of use and the cause of the imprecision should be investigated.

iii. OPERATING BALANCES

Modern electronic balances are easy to operate and so it may seem odd that laboratories have standard operating procedures (SOPs) to detail their operation. SOPs, however, can play a valuable role in ensuring that operators use balances properly, that the performance of each balance is checked regularly, and that these checks are documented. They also remind operators of simple problems that can lead to systematic error. For example, the SOP in Figure 15.10 specifies a warm-up period for the balance. Note also in the SOP that the balance is first calibrated and then the weight of a second standard is checked to verify that the balance is reading correctly.

IV. MASS VERSUS WEIGHT

So far in this text we have used the terms *mass* and *weight* without much discussion of the distinction between them. In this section we discuss these two words.

The value we read from a balance is the weight of an object, not its mass. This may seem surprising. After all, the object is directly compared with a standard whose mass is known. For example, consider the objects in Figure 15.11a. The standard is known to have a mass of 1 g

---

Clean Gene, Inc.
STANDARD OPERATING PROCEDURE

SUBJECT: Operation of Balance: BRP 88

Balance Location: Media prep lab

Effective Date: 7/22/98

Page 1 of 2

---

1. Scope: The routine operation and use of Balance model BRP 88
2. Resources: NIST Traceable standard mass set (2000 g)
3. References: Manufacturer's bulletin No 2
4. Responsibility: Training level of "General Lab" personnel or greater is required for use of this equipment.
5. Procedure
5.1. Pre-use calibration and verification of balance
     Frequency: Daily before use
     Form: Balance Instrument record (QF 15.3.6.3)
5.1.1   Before calibration and verification, the balance must be switched on for at least 20 minutes to allow the internal components to come to working temperature.
5.1.2   Check that the leveling bubble is centered.
5.1.2.1 If the bubble is off center, adjust the leveling feet to bring back to center.
5.1.3   Check that the balance is clean.
5.1.4   Remove any load from the pan and press "ON" briefly. When zero "0.00" is displayed, the balance is ready for operation.
5.1.5   Press and hold "CAL" until "CAL" appears in the display, release key. The required standard value flashes in the display.
5.1.6   With tongs, place the traceable standard required from 5.1.5 in the center of the pan. The balance adjusts itself.
5.1.7   When zero "0.00" flashes, remove calibration standard.
5.1.7.1 If "0.00" does not flash, the balance is not verified and must be checked by the metrology department. Refer to SOP QI 15.3.20 for procedure to remove balance from use until metrology can evaluate it.
5.1.8   Choose a NIST traceable standard whose weight is close to that of the sample to be weighed. When "0.00" is displayed the balance is ready.

(Continued)

Figure 15.10.   Example of a Standard Operating Procedure for Checking and Operating a Balance.

Clean Gene, Inc.
STANDARD OPERATING PROCEDURE

SUBJECT: Operation of Balance: BRP 88

Balance Location: Media prep lab

Effective Date: 7/22/98

Page 2 of 2

5.1.9   With tongs, place the traceable standard on the weighing pan.

5.1.10   Wait until stability detector "o" disappears.

5.1.11   Read results.

5.1.12   Record the daily verification on the form QF 15.3.6.3 with date, time, and operator's name.

5.1.13   The balance is now in the weighing mode and is ready for operation. Reverification with standards is not necessary more than once a day.

5.2. Operation

5.2.1   Remove any load from weighing pan and press "ON" briefly. When zero "0.00" is displayed the balance is ready for operation.

5.2.2   Place sample on the weighing pan.

5.2.3   Wait until the stability detector "o" disappears.

5.2.4   Read results.

5.2.5   Switch off by pressing OFF until "OFF" appears in the display. Release key. The balance may be left in this mode for the remainder of the day.

5.3. Taring weighing vessel

5.3.1   Place the empty weighing vessel on the balance, the weight of the container is displayed.

5.3.2   Press "TARE" briefly, the balance will re-zero.

5.3.3   Add sample to the container. The sample weight is displayed.

5.4. Maintenance

5.4.1   Cleaning

5.4.1.1 After working with the balance, the work surface must be checked for residual chemicals or soils. Solid materials must be removed and disposed of properly.

5.4.1.2 Each week, or more often if necessary, the unit should be wiped down with general purpose cleaning agents. Wipe down the weighing pan, baseplate, sides and lid of the balance.

5.4.1.3 In the event of a chemical spill, assess the hazard. If necessary, call hazard control team (X 3762). If nonhazardous material, remove balance pan and clean thoroughly.

5.4.2   Semi-Annual Calibration

5.4.2.1 Initial and semi-annual calibration is performed by trained service department personnel.

5.4.2.2 If the balance is out of calibration according to 5.1.7.1 it must be re-calibrated immediately by trained personnel in the QC department.

5.4.2.3 Form SOP QI 15.3.20 is to be filled out and the QC department notified for repair or recalibration.

Figure 15.10.   *(Continued)*

---

and the sample exactly balances the standard. It seems logical, therefore, that the object should also have a mass of 1 *g*; unfortunately, there is another factor that affects this system—the air.

The major force measured in weighing is the force of gravity pulling down on an object. However, there is also a slight buoyant force from air. The air around a sample or standard gives it a bit of support and makes it appear "lighter" than it really is. This means that if the same object is weighed in air and in a vacuum, the object will be slightly heavier in the vacuum. This is the **principle of buoyancy**: *Any object will experience a loss in weight equal to the weight of the medium it displaces.* Buoyancy is why ships float, balloons rise, and the weight of an object is different in a vacuum than it is in air, Figure 15.11b.

Balances are calibrated with metal mass standards. Metal has a relatively high density compared to aqueous solutions and other materials. A 1 *g* metal mass standard, therefore, displaces less air than does a 1 *g* mass of water. Because the metal mass standard displaces less air than the water, the metal standard is buoyed less by the air. This difference in buoyancy of a mass standard and a sample explains why the standard

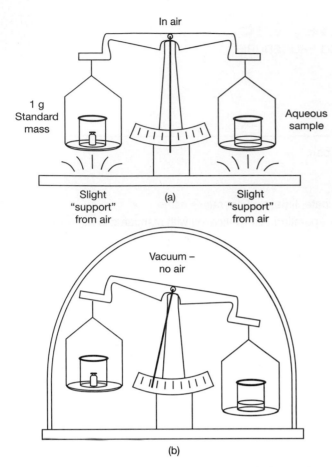

In air

1 g
Standard
mass

Aqueous
sample

Slight
"support"
from air

(a)

Slight
"support"
from air

Vacuum –
no air

(b)

Figure 15.11.   Mass Versus Weight. a. A sample is exactly
balanced against a 1 gram mass standard in air. Although the
sample and the standard have the same weight in air, they do
not have the same mass. **b.** In a vacuum, the beam will tilt to
the side with the sample.

and sample in Figure 15.11 can have the same weight
yet be of different masses. The sample is buoyed by the
air more than the standard. If a balance were calibrated
with a mass standard whose density were identical to
that of the sample, then the weight of the sample would
equal its mass. Because we use metal mass standards
and we weigh samples in air, the value we read in the
laboratory for a sample is its weight, not its mass.

This discrepancy between mass and weight is called
the **buoyancy error**. The weight readings for aqueous
solutions have a buoyancy error of roughly 1 part in
1000. This buoyancy error is small enough as to be of little concern in most applications. The distinction between
mass and weight, therefore, is generally ignored, except
when very high accuracy measurements must be made.
(If necessary, there are equations to correct for the
buoyancy effect that take into consideration the air density at particular atmospheric conditions and the density
of the object being weighed.) (See, for example, *Weighing the Right Way with Mettler: The Proper Way to Work
with Electronic Analytical and Microbalances,* Mettler-
Toledo, Switzerland, 1989.)

PRACTICE PROBLEMS

1. Suppose you need to prepare a solution with a concentration of 15 *mg/mL* of a particular enzyme. You
   try to weigh out 15 *mg* of the enzyme on the analytical
   balance, but find that it is extremely difficult to get
   exactly 15 *mg*. Suggest a strategy to get the correct
   concentration even if you cannot weigh exactly 15 *mg*.

2. How much volume of solution at a concentration of
   35 *μg/mL* can be made in each of these cases?
   a. Sample weighs 0.003560 *g*
   b. Sample weighs 0.0500 *mg*
   c. Sample weighs 1.0897 *g*
   d. Sample weighs 354.8 *μg*

3. A technician weighs a cell preparation on an analytical balance and observes that the initial weight is
   0.0067 *g*. A while later the weight is 0.0061 *g*. What
   might be happening here?

4. Which scale is more sensitive?

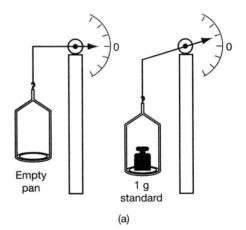

Empty
pan

1 g
standard

(a)

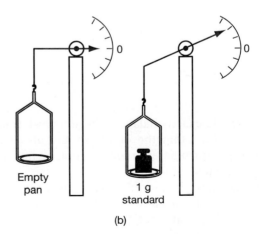

Empty
pan

1 g
standard

(b)

5. Suppose you are verifying the performance of a laboratory analytical balance. You first calibrate the balance, as directed by an SOP. You then repeatedly

weigh a Class 1 standard weight that is 10 g. The results are shown. According to the specifications for the balance, its standard deviation (precision) must be equal to or better than ±0.0001 g and its accuracy must be better than ± 0.0001 g. Does the balance meet its performance specifications?

Weight Values Obtained

10.0000 g
10.0001 g
9.9999 g
10.0000 g
9.9998 g
10.0001 g

6. A standard operating procedure to check and operate a balance is shown in Figure 15.10. Read the SOP and answer the following.

 a. How does a user determine whether the balance is operating acceptably?

 b. How does a user document that the balance is operating acceptably?

7. Suppose a 100 g standard mass is accidentally dropped on the floor. As a result, its mass is slightly less than 100 g. If the standard is used to calibrate a balance, what will happen to subsequent readings from that balance? What type of error is this?

8. The linearity of a balance is being checked. The balance has a capacity of 500 g and four weights of about 125 g each are used. The results follow. Calculate the linearity error (if any) at midpoint, 25%, and 75% capacity.

Midpoint Check
 The four pieces weighed all together weighed 500.001 g.
 A and B together weighed 250.003 g
 C and D together weighed 250.000 g

Check at 25% of Capacity
 A weighed 125.001 g
 B weighed 125.001 g
 C weighed 125.003 g
 D weighed 124.996 g

Check at 75% of Capacity
 Group 1: pieces A, B, C weighed 375.001 g
 Group 2: pieces A, B, D weighed 374.998 g
 Group 3: pieces A, C, D weighed 375.001 g
 Group 4: pieces B, C, D weighed 374.996 g

9. (Optional) **a.** The density of air is about 1.2 mg/cm^3 (depending on the temperature and atmospheric pressure). 1000 mL of water displaces 1000 cm^3 of space (from the definition of a mL). What is the weight of air displaced by 1000 mL of water?

 b. Will the measured weight of a sample change if it is weighed first in air and then in a vacuum? Will its mass change?

DISCUSSION QUESTIONS

1. The ISO 9000 standards state that companies shall "define the process employed for the calibration of inspection, measuring, and test equipment, including details of equipment type, unique identification, location, frequency of checks, check method, acceptance criteria, and the action to be taken when results are unsatisfactory" (section 4.11.2c). Explain how the SOP in Figure 15.10 accomplishes the requirements as stated by ISO 9000.

2. Consider weighing an object in the laboratory.

 a. List as many possible sources of bias as you can.

 b. What would be required to eliminate each source of bias you mentioned in part a?

3. If you work in a laboratory, write an SOP to operate a balance in your laboratory. Be sure to include steps to verify that the balance is responding properly.

TERMS RELATING TO THE MEASUREMENT OF WEIGHT

Analytical Balance. An instrument that can accurately determine the weight of a sample to at least the nearest 0.0001 g.

Balance. An instrument used to measure the weight of a sample by comparing the effect of gravity on the sample with the effect of gravity on objects of known mass.

Beam. See "Lever."

Buoyancy (principle of). Any object will experience a loss in weight equal to the weight of the medium it displaces.

Buoyancy Error. The discrepancy between mass and weight. The displacement of air by an object results in a slight buoyancy effect: Air slightly supports the object. The more air an object displaces, the more buoyant it is, and the less it appears to weigh on a balance. Standards for balance calibration are typically made of metal with a density of 8.0 g/cm^3. In contrast, water has a density of only 1.0 g/cm^3. Because water is less dense than calibration standards, a 1 kg mass of water takes up more space and displaces more air than a 1 kg mass of metal. Because balances are calibrated with metal weights, this leads to a slight buoyancy error when less dense objects, such as solutions, are weighed.

Calibration (of a balance). Bringing the readings from a balance into accordance with the values of internationally accepted standards.

Calibration Standard (for a balance). Objects whose masses are established and documented.

Capacity. Maximum load that can be weighed on a particular balance as specified by the manufacturer.

Counterbalance. A weight used in a mechanical balance that compensates for the weight of the sample.

Dampening Device. A balance component that slows the oscillations of the moving parts of a balance so that the weight value can be read more quickly.

Digital Filters. Devices used in electronic balances to stabilize their readout in the presence of drafts or vibrations.

Electronic Balance. Instrument that determines the weight of a sample by comparing the effect of a sample on a load cell with the effect of standards.

Gravimetric Volume Testing. Determination of the "true" volume of a sample based on its weight. The volume of the liquid is calculated from its weight and its density at a particular temperature. This method takes advantage of the high accuracy and precision attainable with modern balances.

International Prototype Standard. The internationally accepted unit of mass as embodied in a platinum-iridium cylinder housed in France. The mass of this cylinder is, by definition, exactly 1 kg.

Knife Edge. In mechanical balances, a support on which the beam rests, which allows it to swing freely.

Leveling. Adjusting a balance to a level, horizontal position.

Leveling Bubble. A bubble that is used to determine whether the balance is level or not. When the balance is level, the bubble is centered in the window.

Leveling Screws. The screws used to adjust the balance so that it is level.

Lever, also called "beam." A rigid bar used in mechanical balances on which the sample to be weighed is balanced against objects of known weight.

Linearity (for a balance). The ability of a balance to give readings that are directly proportional to the weight of the sample over its entire weighing range.

Load. The item(s) exerting force on the balance.

Load Cell. Device that uses an electromagnetic force to compensate for, or counterbalance, the force of an object on the weighing pan.

Magnetic Dampening. A device that helps stabilize the weighing pan in mechanical balances when samples are placed on it or weights are added to the beam, thus reducing the time required to reach a stable reading.

Mass. The amount of matter in an object, expressed in units of "grams." Mass is not affected by the location of the object or the effect of gravity.

Maximum Tare. The maximum container weight that can be automatically subtracted from the weight of a sample.

Mechanical Balance (laboratory). A balance that uses a lever and that does not generate an electrical signal in response to a sample. The object to be weighed is placed on a pan attached to the lever and is balanced against standards of known weight.

Microbalance. An extremely sensitive balance that can accurately weigh samples to the nearest 0.000001 g.

Off-center Errors. Differences in indicated weight when a sample is shifted to various positions on the weighing area of the weighing pan.

Poise. In a mechanical balance, a sliding weight mounted on a lever used to balance against a sample.

Precision (for a balance). A measure of the consistency of a series of repeated weighings of the same object.

Range. The span from the lightest to the heaviest weight the balance is able to measure.

Readability. Value of the smallest unit of weight that can be read. This may include the estimation of some fraction of a scale division or, in the case of a digital display, will represent the minimum value of the least significant digit. [According to ASTM Standard E 319-85 (reapproved 1993).]

Sensitivity (for a laboratory balance). The smallest value of weight that will cause a change in the response of the balance that can be observed by the operator.

Single Beam, Double Pan Balance. A simple balance that compares the weight of the sample with the weight of a standard that is placed on a pan across a beam.

Standard Weight. A standard whose mass is known with a given uncertainty.

Tare. A feature that allows a balance to automatically subtract the weight of the weighing vessel from the total weight of the sample plus container.

Tolerance (of a standard mass). How much absolute error is allowed in the calibration of a standard. For example, a "500 g" Class 1 standard must have a mass within 1.2 mg of 500.0000 g (i.e., must be between 500.0012 g and 499.9988 g).

Traceability (for mass standards). A documented genealogy that links the mass standard to the international prototype in Sevres, France.

Ultramicrobalance. An extremely sensitive analytical balance that can accurately weigh samples to the nearest 0.0000001 g.

U.S. National Prototype Standards. Platinum-iridium standards with mass values traceable to the International Prototype.

Weigh Boat. A plastic or metal container designed to hold a sample for weighing.

Weighing Paper. Glassine coated paper used to hold small samples for weighing.

Weight. The force of gravity on an object.

REFERENCES FOR WEIGHT MEASUREMENTS

"Laboratory Balances," Walter E. Kupper. In *Analytical Instrumentation Handbook,* Galen Wood Ewing, ed. Marcel Dekker, Inc. New York, 1990.

Fundamentals of Mass Determination, Mettler-Toledo. Mettler-Toledo, Switzerland, 1991.

Weighing the Right Way with Mettler: The Proper Way to Work with Electronic Analytical and Microbalances, Mettler-Toledo. Mettler-Toledo, Switzerland, 1989.

Mettler-Toledo Dictionary of Weighing Terms: A Practical Guide to the Terminology of Weighing. L. Bietry. Mettler-Toledo, Switzerland, 1991.

"Honest Weight and True Mass- (They are not the Same)," Walter E. Kupper, *American Laboratory* pp. 8–9. Dec. 1990.

"Air Buoyancy Correction in High-Accuracy Weighing on Analytical Balance," Schoonover, Randall M.; Jones, Frank E. *Anal. Chem.;* 53:900–902, 1981.

Troemner Mass Standards Handbook. Troemner Inc. 6825 Greenway Avenue, Philadelphia PA. 19142.

"A Look at the Electronic Analytical Balance," Randall M. Schoonover. *Anal. Chemistry*; 54(8): 973A–980A, 1982.

ASTM Standard E 617 "The Standard Specification for Laboratory Weights and Precision Mass Standards."

ANSWERS TO PROBLEMS

1. Weigh out as close to 15 *mg* as possible and then calculate the amount of buffer required to dilute the enzyme to a concentration of 15 *mg/mL*.

2. **a.** 101.7 *mL*   **b.** 1.43 *mL*
   c. 31.134 *L*   **d.** 10.14 *mL*

3. Assuming the technician has been careful to avoid temperature effects and drafts, the preparation is probably losing moisture because it was not dry to begin with.

4. Balance (b) has a greater response for a given weight, so it is more sensitive.

5. The mean value = 9.999983333 *g*. The absolute error is 0.00002 *g*.
   The *SD* = 0.000116905. Rounded, this is 0.0001.
   The balance meets its performance specifications.

6. **a.** The user first calibrates the balance and then checks the weight of a second standard.
   b. Forms are used to document the result of the performance verification.

7. All subsequent readings with that balance will be a little too high. Dropping the standard is a gross error. The fact that subsequent readings will be a bit high is a systematic error.

8. *Midpoint*
   Full scale = 500.001 *g*
   Weight sum = 500.003 *g*
   Difference = 0.002 *g*
   Linearity error = 0.001 *g*
   25%
   Full scale = 500.001 *g*
   Weight sum = 500.001 *g*
   No error
   75%
   75 % Full scale = 500.001 × 3 = 1500.003 *g*
   Weight sum = 1499.996 *g*
   Difference = 0.007 *g*
   Linearity error = 0.00175 *g*

9. **a.** Because 1000 cm^3 of air is displaced, the weight of air displaced is:
   1000 cm^3 × 1.2 mg/cm^3 = 1200 *mg* = 1.20 *g.*
   b. The weight of a sample varies depending on its location—such as whether it is in air or a vacuum. Its mass does not change.

Appendix

Laboratory Mass Standards

(Excerpted from ASTM Standard E 617)

Standards are classified by Type, Grade, and Class.

Type Refers to the Method of Construction of the Standard

Type I standards are of one piece construction and are used when the most accurate and stable standards are required.

Type II standards do not need to be constructed of a single piece of metal.

Grade Indicates the Metal of Which the Standard Is Made

Different metals and alloys have different densities and properties. The grades specify the density of the materials used to manufacture the standard, its surface area, and other physical properties. The grades are termed **S**, **O**, **P**, and **Q**.

Class Indicates the Permitted Tolerance of the Standard

The classes are **1**, **2**, **3**, **4**, **5**, **6** where the lower the class number, the smaller the tolerance. There is also a class **1.1** for calibrating low-capacity, high-sensitivity balances. The tolerances are specified in a table; a portion of that table is excerpted below.

SUGGESTED APPLICATIONS	TYPE	GRADE	CLASS
Reference standards used to calibrate other standards.	I	S	1, 2, 3, 4
High-precision standards where the utmost accuracy is required.	I or II	S or O	1 or 2
Used to calibrate microbalances.	I	S or O	1.1
Working standards appropriate for calibrating most laboratory analytical balances.	I or II	S or O	2
Built-in weights for high quality analytical balances.	I or II	S	2
Recommended for use with balances with readabilities from 0.01 to 0.1 g.	II	P	3 or 4
Recommended for semi-analytical and student use.	II	Q	4 or 5

TOLERANCES FOR STANDARD MASSES FOR CLASSES 1–6 ($\pm mg$)

Mass of Standard	Class					
	1	2	3	4	5	6
1 kg	±2.5	5.0	10	20	50	100
500 g	1.2	2.5	5.0	10	30	50
200 g	0.50	1.0	2.0	4.0	15	20
100 g	0.25	0.50	1.0	2.0	9	10
50 g	0.12	0.25	0.60	1.2	5.6	7
10 g	0.050	0.074	0.25	0.50	2.0	2
1 g	0.034	0.054	0.10	0.20	0.50	2

EXAMPLES

When a balance is being calibrated and the utmost accuracy possible is required, it is necessary to choose the correct mass standard. It is recommended that the tolerance of the standard be four times more accurate than the readability of the balance. For example:

a. A balance has a readability of 1 *mg* and is to be calibrated with a 100 *g* standard weight. What type of mass standard should be chosen? To make this decision, divide 1 *mg* by 4; the result is 0.25 *mg*. This is the tolerance that is needed in the 100 *g* standard. Refer to the tolerance chart for weight standards. For a 100 *g* standard, the tolerance of a Class 1 standard is 0.25 *mg*. A Class 1 standard, therefore, should be chosen.

b. A balance has a capacity of 60 *g* and a readability of 1 *mg*. It is to be calibrated with a 50 *g* standard. What class standard mass is required? The readability is 1 *mg*; therefore, the tolerance of the standard should be ± 0.25 *mg*. From the previous table, observe that a 50 *g* standard of class 2 has a tolerance of 0.25 *mg*.

c. A balance has a capacity of 500 *g*, a readability of 0.1 *g* and is to be calibrated with a 500 *g* standard mass. What class standard is required? The readability is 0.1 *g*; therefore, the tolerance of the standard should be 0.025 *g* or 25 *mg*. A class 4 standard is adequate.

CHAPTER 16

The Measurement of Volume

I. PRINCIPLES OF MEASURING THE VOLUME OF LIQUIDS

A. Overview

B. Basic Principles of Glassware Calibration

II. GLASS AND PLASTIC LABWARE USED TO MEASURE VOLUME

A. Beakers, Erlenmeyer Flasks, Graduated Cylinders, and Burettes

B. Volumetric Flasks

III. PIPETTES

A. Pipettes and Pipette-Aids

B. Measuring Pipettes

C. Volumetric (Transfer) Pipettes

D. Other Types of Pipettes and Related Devices

IV. MICROPIPETTING DEVICES

A. Positive Displacement and Air Displacement Micropipettors

B. Obtaining Accurate Measurements from Air Displacement Manual Micropipettors

    i. Procedure for Operation
    ii. Factors That Affect the Accuracy of Manual Micropipettors

C. Contamination and Micropipettors

D. Pipetting Methodically

E. Verifying that Micropipettors Are Performing According to Their Specifications

F. Cleaning and Maintaining Micropipettors

I. PRINCIPLES OF MEASURING THE VOLUME OF LIQUIDS

A. Overview

Volume *is the amount of space a substance occupies.* The **liter** *is the basic unit of volume.* Various devices are used to measure the volume of liquids, depending on the volume being measured and the accuracy required. For larger volumes, biologists use glass and plastic vessels, such as **graduated cylinders** and **volumetric flasks**. **Pipettes** are usually preferred for volumes in the $1–25\ mL$ range. Various **micropipetting devices** are commonly used to measure volumes in the microliter range. These various methods of measuring volume are discussed in this chapter.

B. Basic Principles of Glassware Calibration

Manufacturers of glassware and plasticware for volume measurements (such as graduated cylinders, volumetric flasks, and pipettes) are responsible for the calibration of those items. **Capacity marks** and **graduations** *are lines marked on volume measuring devices that indicate volume.* Calibration of glass and plasticware involves placing capacity marks and graduations in such a way that they correctly indicate volume. The American Society for Testing and Materials (ASTM) distributes standards that specify exacting procedures for how certain volume measuring devices are to be calibrated, checked for accuracy, and labeled. When manufacturers calibrate devices in accordance with ASTM standards, they generally state this in their catalog.

The **meniscus** must be considered when glassware is calibrated. The **meniscus** (Greek for "crescent moon") *is a curve formed by the surface of liquids confined in narrow spaces, such as in measuring devices,* Figure 16.1. Conventional practice and ASTM standards specify that the lowest point of the meniscus is used as the point of refer-

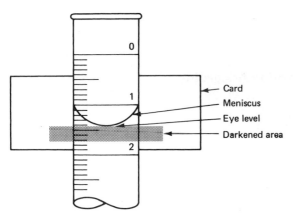

Figure 16.1. The Meniscus. A dark card may be placed behind the meniscus to make it more visible. When reading volume, the eye should be level with the bottom of the meniscus.

ence in calibrating a volume measuring device; therefore, the lowest part of the meniscus is also the reference when reading the volume of a measuring device. It may be helpful to place a dark background, such as a piece of black paper, behind the device to facilitate reading the meniscus accurately. It is important to hold your eyes level with the liquid surface when reading a meniscus.

Glassware for measuring liquid volume is calibrated either **to contain (TC)** or **to deliver (TD)**. A device that is TC *will contain the specified amount when filled to the capacity mark.* It will not deliver that amount if the liquid is poured out because some of the liquid will adhere to the sides of the container. A **TD** *device is marked slightly differently so that it does deliver the specified amount, assuming the liquid is water at 20°C and it is poured using specific techniques.* Consider the logic behind the two methods of calibration. A solution made to a specific concentration of solute can be prepared in a volumetric flask. The volume of water *in* the flask is critical in obtaining the right solute concentration, so a flask calibrated "to contain" is properly used. In contrast, if the volumetric flask is to be used to *pour* an exact volume of liquid to another container, then a flask calibrated "to deliver" should be used. (Plastics are considered to be nonwetting—that is, water does not stick to them—so there is no difference between TC and TD for plasticware.)

There is inevitably a small amount of error in the volume calibration of glassware, plasticware, and pipettes, even when they are manufactured properly. The **tolerance**, *or how much error is allowed in the calibration of a volume measuring item,* depends on the volume of the item and the purpose of the measurements to be made with it. *The most accurately calibrated glassware is termed* **volumetric**; volumetric glassware therefore has the narrowest tolerance allowed in its calibration. For example, Class A volumetric flasks are used where high accuracy is required, such as in the preparation of standard solutions. The tolerance of a $100\ mL$ Class A volumetric flask that meets ASTM standards is $\pm\ 0.08\ mL$ (i.e., its true marked volume must be between $99.92\ mL$ and $100.08\ mL$). A Class B $100\ mL$ volumetric flask, which is used where slightly less accuracy is required, must be marked within $0.16\ mL$ of $100\ mL$. The manufacturer must therefore calibrate a Class A flask more exactly than a Class B flask, and one might expect to pay more for it.

For high accuracy volume measurements, it is important to consider two effects of temperature. First, the glass or plastic forming a volume measuring device will expand and contract as the temperature changes, thus affecting the accuracy of its markings. Second, the volume of materials, including aqueous solutions, changes as the temperature changes. ASTM standards specify that manufacturers calibrate devices at 20°C using water at the same temperature. In laboratory situations where the utmost accuracy is required, volume measurements should be made at 20°C or correction factors to account for temperature should be applied (see ASTM Standard E 542).

To ensure accuracy when using volume measuring devices, it is necessary to use the device in the same manner as it was calibrated. Recalibration is required, or some accuracy may be lost, when devices calibrated with water are used to deliver other liquids. If labware is distorted by heating or is contaminated, then the calibration marks will not be accurate. Systematic error will occur if for any reason a piece of glassware does not read the correct volume. Repetition of the measurement will not reveal this error.

II. GLASS AND PLASTIC LABWARE USED TO MEASURE VOLUME

A. Beakers, Erlenmeyer Flasks, Graduated Cylinders, and Burettes

Beakers and Erlenmeyer flasks are containers intended primarily to hold liquids, not to measure volumes. These vessels are marked with graduation lines, but they are at best calibrated with a tolerance of ± 5%. This means, for example, that when the manufacturer marks a beaker at "100 *mL*," it may actually indicate anywhere between 95 and 105 *mL*.

Graduated cylinders *are cylindrical vessels calibrated with sufficient accuracy for most volume measurements in biology laboratories.* A moderately priced 100 *mL* graduated cylinder is calibrated with a tolerance of ± 0.6 *mL*. Graduated cylinders are usually calibrated "to deliver" particular volumes.

Graduated cylinders are graduated with a number of equal subdivisions. For example, a 100 *mL* graduated cylinder might be etched with 100 subdivisions, one for each *mL*, whereas a 500 *mL* cylinder might have a mark every 5 *mL*. This means that a single graduated cylinder can be used to measure various volumes; however, the correct cylinder should be chosen depending on the volume desired. For example, it is not possible to measure 83 *mL* accurately with a 500 *mL* graduated cylinder; a 100 *mL* graduated cylinder is more appropriate, Figure 16.2a–c. Note also that graduated cylinders are not designed as containers for mixing and for storing solutions because they are unstable and easily knocked over, they are expensive compared with storage bottles, and they often do not come with caps.

Burettes (also spelled *buret*) *are long graduated tubes with a stopcock at one end that are used to dispense*

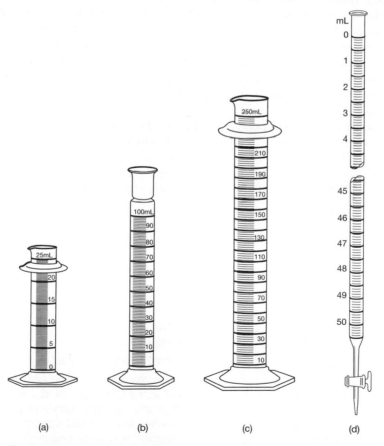

Figure 16.2.   Graduated Cylinders and Burettes. a. A 25 *mL* graduated cylinder has a graduation mark every 0.5 *mL* and is useful for volumes from 1 to 25 *mL*. **b.** A 100 *mL* cylinder has a graduation mark every 1 *mL* and is most useful for volumes from 10 to 100 *mL*. **c.** A 250 *mL* cylinder is marked every 2 *mL* and is most useful for volumes from 100 to 250 *mL*. For volumes less than or equal to 100 *mL*, a 100 *mL* graduated cylinder can be used with better accuracy. **d.** A burette that is calibrated every 0.1 *mL*.

known volumes accurately, Figure 16.2d. Burettes have a long tradition of use in chemistry laboratories for making accurate volume measurements.

B. Volumetric Flasks

Volumetric flasks *are vessels used to measure specific volumes where more accuracy is required than is attainable from a graduated cylinder.* Each **volumetric flask** *is calibrated either "to contain" or "to deliver" a single volume,* Figure 16.3. Although volumetric flasks are calibrated with high accuracy, they have the disadvantages that they are relatively expensive and that each is calibrated for only one volume (e.g., 10, 50, or 1000 *mL*). A volumetric flask cannot be used, for example, to measure 33 *mL*. Volumetric flasks, therefore, are typically used in biology laboratories only in situations where high accuracy volume measurements are required.

The ASTM tolerances for volumetric flasks of different sizes are shown in Table 16.1. Observe that there is less error tolerated in the calibration of Class A flasks than in the calibration of Class B flasks.

Manufacturers sell special, serialized (numbered) volumetric flasks, Figure 16.4. These flasks are individually calibrated using equipment whose calibration is traceable to NIST and are accompanied by a certificate of traceability. The certificate documents that a given flask is accurate within the tolerance specified for its type and class. Certified, serialized glassware may be required in facilities meeting ISO 9000 or GMP requirements.

Volumetric flasks must be used correctly to ensure accurate volume measurements. Box 1 contains points relating to the proper use of a volumetric flask.

Table 16.1 ASTM STANDARDS FOR THE TOLERANCE OF VOLUMETRIC FLASKS

Capacity (mL)	Class A Tolerances (± mL)	Class B Tolerances (± mL)
5	0.02	0.04
10	0.02	0.04
25	0.03	0.06
50	0.05	0.10
100	0.08	0.16
200	0.10	0.20
250	0.12	0.24
500	0.20	0.40
1000	0.30	0.60
2000	0.50	1.00

From ASTM Standard E 288-94, "Standard Specification for Laboratory Glass Volumetric Flasks."

III. PIPETTES

A. Pipettes and Pipette-Aids

Pipettes (also spelled "pipet") *are hollow tubes that allow liquids to be drawn into one end and are generally used to measure volumes in the 0.1–25* mL *range.* Pipettes may be made of glass or plastic and can be disposable or intended for multiple use. Presterilized disposable pipettes with cotton plugged tops are available for microbiology and tissue culture work.

Pipette-aids *are devices used to draw liquid into and expel it from pipettes,* Figure 16.5. In the past pipette-aids were less common and people would suck liquid into pipettes, like sucking a drink into a straw. This procedure, called *mouth-pipetting,* is dangerous because it is easy to accidentally inhale or swallow radioactive substances, pathogens, hazardous chemicals, or the like. Most older laboratory workers have stories of aspirating a mouthful

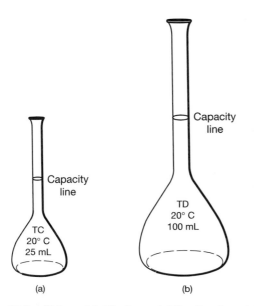

Figure 16.3. Volumetric Flasks. a. A 25 *mL* volumetric flask that is calibrated to contain 25 *mL* when filled exactly to its capacity line. **b.** A volumetric flask that is calibrated to deliver 100 *mL* when it is filled exactly to its capacity line and the liquid is then poured out.

Pyrex* Brand Serialized and Certified with $ Barrelhead Stopper

Individually serialized and supplied with a Certificate of Identification and Capacity. Individually calibrated and certified against equipment whose calibration is traceable to NIST. Calibrated "To Contain" within Class A tolerances

Figure 16.4. Catalog Description for a Serialized and Certified Volumetric Flask that is Individually Calibrated According to ASTM Standards. (Courtesy of Fisher Scientific, Pittsburgh, PA.)

Box 1   PROPER USE OF VOLUMETRIC FLASKS

1. *Choose the proper type of flask for the application: either "to contain" or "to deliver", either Class A or Class B, either serialized or not serialized.*

2. *Be sure the flask is completely clean before use.*

3. *Read the meniscus with your eyes even with the liquid surface.* The bottom of the meniscus should exactly touch the capacity line.

4. *If the flask is calibrated "to deliver," then pour as follows:*

   a. Incline the flask to pour the liquid; avoid splashing on the walls as much as possible.

   b. When the main drainage stream has ceased, the flask should be nearly vertical.

   c. Hold the flask in this vertical position for 30 seconds and touch off the drop of water adhering to the top of the flask by touching it to the receiving vessel.

5. *Never expose volumetric glassware to high temperatures because heat causes expansion and contraction that can alter its calibration.*

---

of some nauseating material. Although mouth-pipetting is now forbidden by safety regulations, its terminology persists and pipettes are often calibrated as "blow out." A pipette calibrated as "blow out" was previously literally blown into with one's mouth. Now, the last drop is ejected from a "blow out" pipette using a pipette-aid.

Pipettes come in different styles and sizes. Manufacturers place colored bands at the top of pipettes to indicate their capacity and graduation interval. These colored bands facilitate choosing the proper pipette and sorting pipettes after washing.

B. Measuring Pipettes

A measuring **pipette** *is calibrated with a series of graduation lines to allow the measurement of more than one volume.* Figure 16.6 illustrates how to read the meniscus on a measuring pipette.

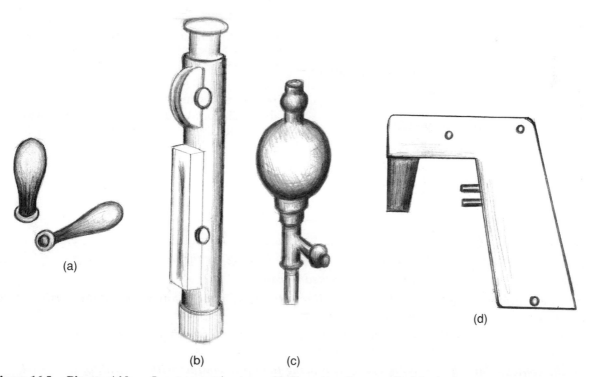

(a)

(b)   (c)

(d)

Figure 16.5.   Pipette-Aids. a. Least expensive type of bulb; not easily controlled. **b.** A pipette pump that can be used to take up and expel liquid. **c.** More expensive bulbs allow fine control of liquid. This type of pipette aid may be called a "triple valve" device because it has three buttons: The first displaces air from the bulb, the second is used to draw liquid into the pipette, and the third is used to expel the liquid. **d.** Electronic pipette aid that allows fine control and ease of use. (Sketches by a Biotechnology student.)

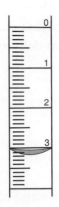

Figure 16.6. **Reading the Meniscus on a Measuring Pipette.** Liquid was drawn up to exactly the zero mark and was then dispensed. Reading the value at the bottom of the meniscus shows that 3.19 *mL* of liquid was delivered.

Measuring pipettes are sometimes classified as **serological** or **Mohr**. Both these types are calibrated "to deliver." Pipettes termed **serological** *are usually calibrated so that the last drop in the tip needs to be "blown out" to deliver the full volume of the pipette,* Figure 16.7a,b. Manufacturers place one wide or two narrow bands at the top of a pipette to indicate that it is calibrated to be "blown out." This "blow out" band(s) appears above the color coding band that indicates the capacity and graduation interval for the pipette. The proper way to use a serological pipette is summarized in Box 2.

Mohr pipettes *are calibrated "to deliver" but, unlike the serological pipettes described earlier, the liquid in the tip is not part of the measurement and the pipette is not blown out,* Figure 16.8 a,b. Serialized Mohr pipettes that have a certificate of traceability to NIST can be purchased.

C. Volumetric (Transfer) Pipettes

Volumetric (transfer) pipettes *are made of borosilicate glass and are calibrated "to deliver" a single volume when filled to their capacity line at 20°C,* Figure 16.9. Volumetric

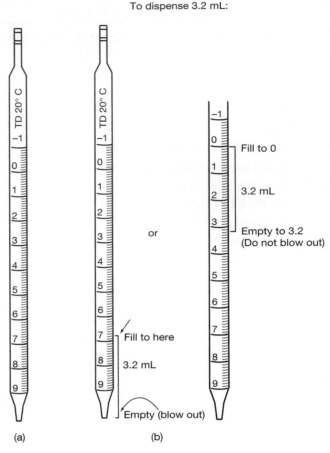

Figure 16.7. **Using Serological Pipettes. a.** A serological pipette calibrated so that the tip includes the last milliliter. The bands at the top indicate that this pipette is to be "blown-out." Note that this pipette has a scale that extends above zero to expand the calibrated capacity of the pipette. **b.** Using a serological pipette to dispense 3.2 *mL*.

pipettes are the most accurately calibrated pipette type, Table 16.2. Serialized volumetric pipettes with a certificate of traceability to NIST can be purchased.

Volumetric pipettes are calibrated so that "delivery of the contents into the receiving vessel is made with the tip in contact with the wall of the vessel and no

---

Box 2   *Using a Serological Pipette*

1. *Check that the pipette is calibrated to be "blown out" by looking for the band(s) at the top.*

2. *Examine the pipette to be sure the tip is not cracked or chipped.*

3. *Fill the pipette about 10 **mm** above the capacity line desired and remove any water on the outside of the tip by a downward wipe with lint-free tissue.*

4. *Place the tip in contact with a waste beaker and slowly lower the meniscus to the capacity line.* Do not remove any water remaining on the tip at this time.

5. *Deliver the contents into the receiving vessel by placing the tip in contact with the wall of the vessel.*

6. *When the liquid ceases to flow, "blow out" the remaining liquid in the tip with one firm "puff" with the tip in contact with the vessel wall, if possible.* Because we no longer actually "puff" on pipettes, the last drop is ejected with a pipette-aid.

To dispense 3.2 mL:

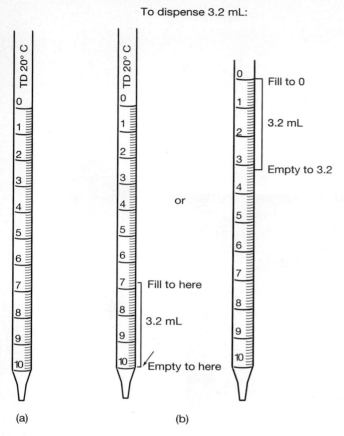

(a)   (b)

Figure 16.8.   Mohr Pipettes. a. Mohr pipettes are calibrated so that the tip does not include the last milliliter. **b.** Using a Mohr pipette to dispense 3.2 *mL*.

Capacity line

Figure 16.9.   A Volumetric Pipette.

afterdrainage period is allowed" (ASTM Standard 969). These pipettes are not "blown out." Remember to check that the tip is not cracked or chipped before using a volumetric pipette.

D. Other Types of Pipettes and Related Devices

There are various types of pipettes in addition to those described above. For example, **Pasteur pipettes** *are used to transfer liquids from one place to another.* Pasteur pipettes are not actually volume-measuring devices because they have no capacity lines, but they are convenient for transferring liquids, Figure 16.10a. There are glass pipettes for measuring volumes less than 1 *mL*. These small volume, reusable "micropipettes," however, are not widely used in biotechnology laboratories because their use has been supplanted by micropipetting instruments, as described in Section IV of this chapter. Disposable pipettes for measuring less than 1 *mL* volumes are so convenient that they are sometimes used when high accuracy is unnecessary, Figure 16.10b. Inexpensive capillary glass pipettes that deliver in the microliter range are sometimes used (e.g., in teaching laboratories), Figure 16.10c.

Table 16.2   *TOLERANCES FOR VOLUMETRIC CLASS A AND CLASS B PIPETTES COMPARED WITH GENERAL PURPOSE SEROLOGICAL PIPETTES*

Capacity (*mL*)	Volumetric Class A Tolerance	Volumetric Class B Tolerance	Glass Serological Tolerance
0.1	—	—	0.005
0.2	—	—	0.008
0.5	0.006	0.012	0.01
1.0	0.006	0.012	0.02
2.0	0.006	0.012	0.02
3.0	0.01	0.02	—
4.0	0.01	0.02	—
5.0	0.01	0.02	0.04
6.0	0.01	0.02	—
7.0	0.01	0.02	—
8.0	0.02	0.04	—
9.0	0.02	0.04	—
10.0	0.02	0.04	0.06
15.0	0.03	0.06	—
20.0	0.03	0.06	—
25.0	0.03	0.06	0.10
50.0	0.05	0.10	—
100.0	0.08	0.16	—

Tolerances are expressed as ± *mL*.
Information from ASTM Standard E 969-94 "Standard Specification for Volumetric (Transfer) Pipets" and ASTM Standard E 1044-85 (reapproved 1990) "Standard Specification for Glass Serological Pipets (General Purpose and Kahn)."

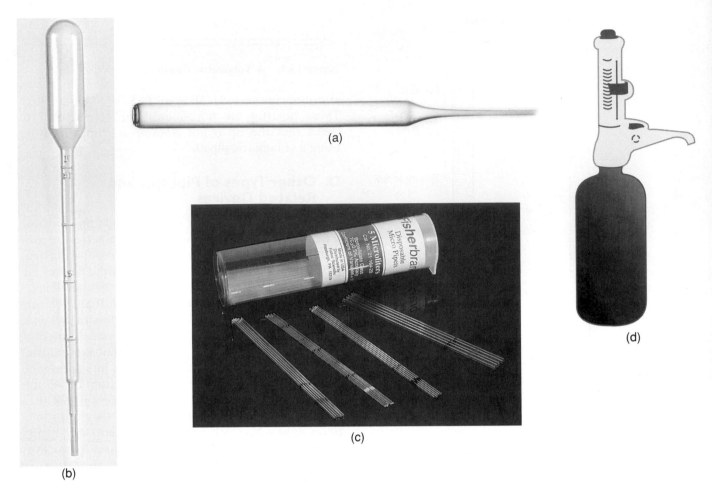

Figure 16.10. Examples of Various Types of Pipette and Liquid Dispensing Devices. a. A Pasteur pipette. **b.** Disposable plastic pipette to measure microliter volumes with an accuracy of about ± 10%. **c.** Glass capillary pipettes for measuring microliter volumes. **d.** A manual dispenser for a reagent bottle. (a–c courtesy of Fisher Scientific, Pittsburgh, PA.)

Manual dispensers for reagent bottles *are devices placed in a reagent bottle with a tube that extends to the bottom of the bottle. The dispenser has a plunger that is depressed to deliver a set volume of liquid,* Figure 16.10d. These devices effectively take the place of using a pipette or other device to remove liquids from reagent bottles.

IV. MICROPIPETTING DEVICES

A. Positive Displacement and Air Displacement Micropipettors

Glass and plastic pipettes are typically used in biology laboratories to measure volumes as small as about 1 *mL*. **Micropipettors** *are devices commonly used to measure smaller volumes, in the 1 to 1000 μL range.* These volume-delivering devices go by many names in addition to micropipettor, such as **microliter pipette**, **piston or plunger operated pipette**, and, simply, **pipettor**.

There are two distinct designs of micropipettor: **positive displacement** and **air displacement**, Figure 16.11.

Positive displacement micropipettors *include syringes and similar devices where the sample comes in contact with the plunger and the walls of the pipetting instrument.* Positive displacement devices are recommended for viscous and volatile samples. Syringes are commonly used to inject small volume samples into instruments for chromatographic analysis.

Air displacement micropipettors *are designed so that there is an air cushion between the pipette and the sample such that the sample only comes in contact with a disposable tip and does not touch the micropipettor itself.* Disposable tips reduce the chance that material from one sample will contaminate another or that the operator will be exposed to a hazardous material. Air displacement micropipettors accurately measure the volume of aqueous samples and are among the most common instruments currently used in biotechnology laboratories.

Air displacement micropipettors come in an assortment of styles, Figure 16.12. Some deliver only one volume, such as 100 μL; others, called **digital microliter pipettors**, can be adjusted to deliver different volumes over a range, such as 10–100 μL. Micropipettors can be manually or electronically driven, and some are micro-

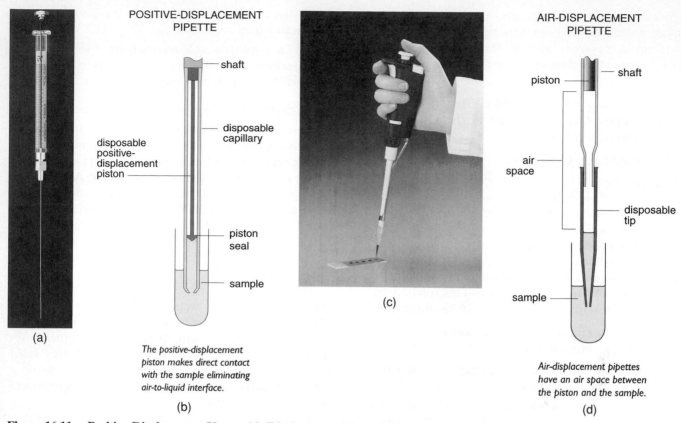

POSITIVE-DISPLACEMENT
PIPETTE

shaft

disposable
capillary

disposable
positive-
displacement
piston

piston
seal

sample

(a)

*The positive-displacement
piston makes direct contact
with the sample eliminating
air-to-liquid interface.*

(b)

(c)

AIR-DISPLACEMENT
PIPETTE

piston

shaft

air
space

disposable
tip

sample

*Air-displacement pipettes
have an air space between
the piston and the sample.*

(d)

Figure 16.11.   Positive Displacement Versus Air Displacement Micropipettors. a. Syringe used to inject small volume samples into chromatography instrument; positive displacement. (Courtesy of Fisher Scientific, Pittsburgh, PA) **b.** Diagram of a positive displacement device. **c.** Air displacement micropipettor. **d.** Diagram of an air displacement device. (Illustrations b, c and d are copyrighted material of Rainin Instrument Co. and are used with permission.)

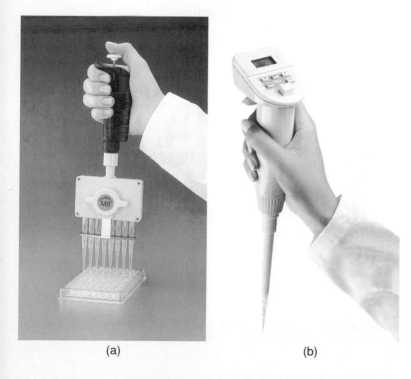

(a)

(b)

Figure 16.12.   Various Types of Air Displacement Micropipettors. a. A "multichannel" micropipettor used to simultaneously deliver the same volume to 8 wells of a 96-well plate. **b.** An electronic micropipettor that is motorized and microprocessor controlled. (These illustrations are copyrighted material of Rainin Instrument Co. and are used with permission.)

processor controlled. Repetitive pipetting can lead to injury, so, in laboratories where repetitious pipetting is performed, it is worthwhile to investigate micropipettors designed to reduce operator fatigue.

B. Obtaining Accurate Measurements from Air Displacement Manual Micropipettors

i. PROCEDURE FOR OPERATION

It is important to operate micropipettors properly or they will not deliver the correct volumes. Manual micropipettors have **plungers** *by which the operator controls the uptake and expulsion of liquids.* As the operator depresses the plunger, different "stop" levels can be felt. Although the detailed operating directions vary depending on the type and brand of micropipettor, the general operation of most manual micropipettors is similar and is summarized in Figure 16.13 and in Box 3.

ii. FACTORS THAT AFFECT THE ACCURACY OF MANUAL MICROPIPETTORS

A variety of factors affect the accuracy of volume measurements by micropipettors:

1. **The operator's technique is the most important factor affecting the performance of a manual micropipetting device.** If the operator does not operate the micropipettor consistently, smoothly, and correctly, it will not accurately deliver the specified volume. (See Box 3.)

2. **The physical and chemical properties of the liquids being measured will affect the volumes delivered.** (See Table 16.3.)

3. **Measurements are affected by the environment in which they are made.** (See Table 16.3.)

4. **The condition of a micropipettor will affect its performance.** Basic micropipettor maintenance will be discussed shortly.

C. Contamination and Micropipettors

Improper use of micropipettors can result in liquids being accidentally drawn into the micropipettor assembly. Such liquids may contaminate later experiments, damage the micropipettor, or put the operator at risk. **Aerosols** *are fine liquid droplets that remain suspended in the air.* Aerosols are readily produced during pipetting and even proper use of a micropipettor can result in aerosol contaminants being drawn into the instrument. Aerosols can be a serious problem if they contain pathogens, toxic materials, or radioactive substances.

The polymerase chain reaction (PCR) is an extremely sensitive method of amplifying and detecting DNA present in samples in tiny amounts. If DNA from one PCR sample accidentally gets into the mixture for another sample, it can have serious consequences. *DNA contamination from one sample to another sample is called* **carryover** and micropipettors are a major source of such contamination.

To avoid micropipettor contamination and carryover consider the following:

1. **Manufacturers produce special tips that have a barrier that keeps aerosols and liquids from entering the micropipettor assembly, Figure 16.14.** These tips are more expensive than normal tips, but they may be a good investment in certain circumstances.

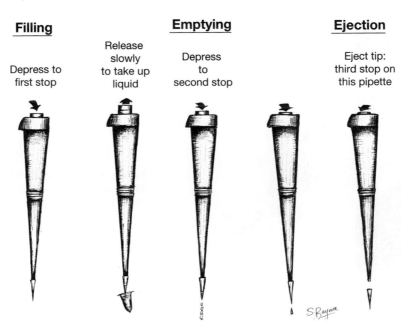

Filling		Emptying		Ejection
Depress to first stop	Release slowly to take up liquid	Depress to second stop		Eject tip: third stop on this pipette

Figure 16.13.   Using a Manual Micropipettor. Note that not all micropipettors have a third stop to eject the tip. (Sketches by a Biotechnology student.)

Box 3   *PROCEDURE TO OPERATE A MANUAL MICROPIPETTOR*   (In the Usual, "Forward Mode")

1. *Set the micropipettor (if it is adjustable to different volumes) to the desired volume.*

2. *Attach a disposable tip to the micropipettor shaft.* Press firmly to ensure an airtight seal.

3. *Optional: Prewet the tip by aspirating and dispensing the solution to be measured into a waste container or back into the original solution.*

   Note: Prewetting leaves a thin film of the liquid to be measured on the inside of the tip. Some operators prewet the tip, others do not. There is evidence that better accuracy is obtained with a prewetted tip. Since prewetting the tip will affect its performance, it is important that operators are consistent in whether they prewet the tip or not.

4. *While observing the tip and the sample, depress the plunger to the first stop, and place the tip in the liquid.*

   Note: For the utmost accuracy, it is recommended that the tip be immersed into the sample a specified distance. This distance may be specified by the manufacturer. Otherwise, general guidelines are:

   for volumes of 1–100 μL, immerse the tip 2–3 *mm*

   for volumes of 101–1000 μL, immerse the tip 2–4 *mm*

   for volumes of 1001–10,000 μL, immerse the tip 3–6 *mm*

5. *Allow the plunger to slowly return to its undepressed position as the sample is drawn into the tip.* Wait 1 second before removing the tip from the liquid.

   Note 1: *Never allow the plunger to snap up!* (If the plunger "snaps," fluid can be aspirated into the interior of the micropipettor. Also, the volume measured will be incorrect.)

   Note 2: If any liquid remains on the outside of the tip remove it carefully with a lint-free tissue, taking care not to wick liquid from the tip orifice.

6. *Place the tip so that it touches the side of the container into which the sample will be expelled; depress the plunger to the second stop.* Watch as the sample is expelled to the container. Wait about 2 seconds and be certain all the liquid is expelled. Remove the tip from the vessel carefully, *with the plunger still fully depressed.*

   Note: It is also correct to dispense the sample directly into a solution already in the tube, and to rinse the tip with the solution. Discard the tip after such use.

7. *Eject the tip using the third stop, tip ejector button, or other mechanism.*

The procedure outlined in this box is the most common method of pipetting, called "forward mode" pipetting. Forward mode pipetting is recommended for aqueous samples. There is also "reverse mode" pipetting which is less commonly used but is recommended for volatile and viscous samples. Reverse mode pipetting is described in Table 16.3. The following diagram shows the steps in forward mode pipetting.

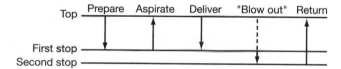

(Figure modified from one in ASTM E 1154-89 (reapproved 1993) "Standard Specification for Piston or Plunger Operated Volumetric Apparatus.")

---

2. **It is helpful to have micropipettors dedicated to a particular technique and used by only one operator.**

3. **Proper operation of the micropipettor, as described in Table 16.3, is essential to avoid contamination problems.**

Safety note: **Be certain to safely remove any radioisotopes or pathogens from an instrument before calibrating or repairing it.**

D. Pipetting Methodically

It is frequently necessary to pipette more than one liquid into a series of tubes. Each tube may receive the same ingredients or one or more components may vary from tube to tube. Although this task seems simple, it is easy to make a mistake when doing repetitious work. Moreover,

it is necessary to work quickly if unstable ingredients are being added. There are electronic, computer-controlled pipettes for repeated pipetting that help ensure consistency in volume and reduce operator fatigue. The operator, however, must still be methodical to avoid errors. Box 4 includes suggestions on how to pipette multiple ingredients into many tubes methodically.

E. Verifying that Micropipettors Are Performing According to Their Specifications

Micropipettors are calibrated by the manufacturer when they are made, but they can become less accurate as they are used. The performance of micropipettors, therefore, should be verified periodically. In some laboratories

Table 16.3 Avoiding Error in the Operation of Air Displacement Digital Micropipettors

1. *Avoid damaging the micropipettor:*
   Never drop a micropipettor.
   Never rotate the volume adjuster of an adjustable micropipettor beyond the upper or lower range of the instrument.
   Never pass a micropipettor through a flame.
   Never use a micropipettor without a tip, thus allowing liquid to contaminate the shaft assembly.
   Never lay a filled micropipettor on its side, thus allowing liquid to contaminate the shaft assembly.
   Never immerse the barrel of a micropipettor in liquid; only immerse the tip.
   Never allow the plunger to snap up when liquid is being aspirated.

2. *Use the proper disposable tip:*
   Make sure the tip matches the micropipettor. Tips come in various sizes for different size and brand pipettes. A poorly fitting tip will leak and deliver inaccurate volumes.
   Be sure the tip is firmly on the micropipettor. If the tip is not firmly attached the sample may leak and the volume dispensed will be too low.
   Use an autoclaved (sterile) tip where appropriate.
   Specialized tips are available for specific purposes, such as loading electrophoresis gels.
   Replace the tip if liquid adheres inside it and does not drain easily.

3. *Be sure the plunger moves slowly and evenly.*

4. *Wait a second or two after taking up sample to allow complete filling.*

5. *Hold the pipette as straight up and down as possible when taking up sample.*

6. *Watch the sample as it comes into and leaves the tip.* The liquid sometimes does not successfully enter the tip or is not expelled completely. If an air bubble is observed, return the sample to its original vessel and try again, immersing the tip to the proper depth and pipetting more slowly. If an air bubble returns, try a new tip. Check for drops remaining in the tip after the liquid is expelled. It may be possible to rinse the tip with the liquid into which the sample is being expelled.

7. *Avoid cross-contamination (i.e., contaminating one sample with material from another).* Change tips every time a different material is pipetted. If the same material is being added to a number of tubes, it is possible to use the same tip repeatedly. To avoid cross-contamination, expel the drop onto the side of each tube and watch to be sure the tip does not touch the liquid already in the tube.

8. *Micropipettors are calibrated with water and therefore are not accurate when used to measure liquids with densities that differ greatly from that of water.* It is possible to determine the actual volume of a liquid of any density delivered by a micropipettor at a particular setting. This determination is done empirically by setting the pipette to a particular volume, dispensing the liquid, weighing it, and determining the actual volume of that liquid dispensed at that pipettor setting. The weight of the liquid is converted to its volume based on its density. (See *The Handbook of Chemistry and Physics*, CRC Press, Inc, Boca Raton, Fl, to determine the density of a liquid.) A table can then be prepared for the liquid of interest that shows the volume dispensed at each setting of that micropipettor.

9. *Maintain the room humidity between 40 and 60%, if possible.* If the room humidity is too low, evaporation can occur quickly and may affect small volume measurements.

10. *Maintain the micropipettor and sample at room temperature, if possible.* If the micropipettor and the sample are not at room temperature (e.g., if the measurements are made in a cold box) then the volumes delivered by the micropipettor will be inaccurate. For the most accurate measurements, the micropipettor, tips, and sample must all be at the same temperature.

11. *Air displacement micropipettors are not very accurate when dispensing volatile liquids (like acetone) or viscous materials (like protein solutions). Accuracy may be improved by the following:*
    If possible, use a positive displacement micropipettor.

    Prewet the tip when pipetting volatile liquids and viscous samples. Fill the tip and empty it once or twice before the desired volume is taken up and dispensed.

    If an air displacement pipette must be used for a volatile or viscous liquid, a technique called "reverse mode" pipetting may be effective. This is accomplished by depressing the plunger to the <u>second stop</u> to take up the sample. Release the plunger after taking up sample. The tip is then touched to the side of the receiving container and the sample is expelled by depressing the plunger to the <u>first stop</u>. In this method, the liquid remaining in the tip is not "blown-out" and is discarded. Reverse mode pipetting is shown in the following diagram.

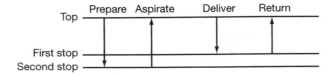

(Figure modified from one in ASTM Standard E-1154.)

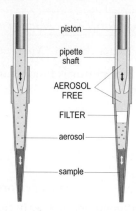

Figure 16.14. Aerosol-Resistant Tip to Help Avoid Contamination of Micropipettor and to Avoid Carryover. (This illustration is copyrighted material of Rainin Instrument Co., and is used with permission.)

users check their own micropipettors, some organizations regularly send their instruments back to the manufacturer to be inspected and repaired, and some facilities have in-house specialists and laboratory facilities for this purpose. Laboratories that are GMP or ISO 9000 compliant have written policies for micropipettor performance evaluation and routine maintenance.

Micropipettor performance is most often evaluated using a **gravimetric** *(which means "relating to weight")* procedure. Gravimetric methods take advantage of the high accuracy and precision attainable with modern balances. During gravimetric testing a desired volume of purified water is placed into or is dispensed from the device to be checked. *This desired volume is termed the* **nominal volume**. The volume measurement is then checked by weighing the liquid on a high-quality, well-maintained balance. A calculation is performed to convert the weight of the water to a volume value. If the volume measuring device was properly operating, then the nominal volume will be identical

to the volume as calculated from the weight of the water. Most laboratories have high-quality balances and can evaluate their micropipettors by a gravimetric procedure as outlined in Box 5. (It is also possible to use spectrophotometric methods to evaluate micropipettors, but those methods are not covered here.)

Performance evaluation procedures evaluate the accuracy and precision of the micropipettor. These accuracy and precision values can be compared with the manufacturer's specifications, Figure 16.15, or with

Pipetman Specifications

Model	Volume µL	Increment µL	Accuracy (mean error)		Precision (repeatability)	
			Relative %	Absolute µL	Relative %	Absolute µL
P-2	0.2	0.002	12	0.024	6	0.012
	1.0		2.7	0.027	1.3	0.013
	2.0		1.5	0.030	0.7	0.014
P-10	1	0.02	2.5	0.025	1.2	0.012
	5		1.5	0.075	0.6	0.03
	10		1	0.1	0.4	0.04
P-20	2	0.02	7.5	0.15	2	0.04
	5		3	0.15	0.9	0.045
	10		1.5	0.15	0.5	0.05
	20		1	0.2	0.3	0.06
P-100	10	0.2	3.5	0.35	1	0.1
	50		0.8	0.4	0.24	0.12
	100		0.8	0.8	0.15	0.15
P-200	50	0.2	1.0	0.5	0.4	0.2
	100		0.8	0.8	0.25	0.25
	200		0.8	1.6	0.415	0.3
P-1000	100	2	3	3	0.6	0.6
	500		0.8	4	0.2	1
	1000		0.8	8	0.13	1.3
P-6000	500	2	2.4	12	0.6	3
	2500		0.8	15	0.2	5
	5000		0.6	30	0.16	8
P-10ML	1mL	20	5	50	0.6	6
	5mL		1	50	0.2	10
	10mL		0.8	80	0.16	16

Specifications subject to change without notice.

Figure 16.15. Specifications for a Micropipettor. The specifications include values for both the accuracy and for the precision (repeatability) of the instrument. The specifications are reported for each model for more than one volume. (This illustration is copyrighted material of Rainin Instrument Co., Inc. and is used with permission. Pipetman® is a U.S.-registered trademark of Rainin Instrument Co., Inc.)

Box 4 PIPETTING METHODICALLY

1. *Plan exactly what will go into each tube before pipetting.* A table is a convenient way to display the ingredients of each tube.

2. *Label all tubes carefully before addition of liquids.*

3. *Place tubes to be filled in the front of a rack; move each one back after material is added to it.*

4. *Check off components that have been added.*

5. *Change tips each time a new material is pipetted and whenever a tip touches components previously added to a tube.* Discard a tip that does not drain completely.

6. *After adding all the components to each tube, be sure the liquids are well-mixed, are not adhering to the tube walls, and are in the bottom of the tubes.* Mixing may be accomplished by inverting the tubes, tapping them with your finger, or using a vortex mixer. Some materials, like long strands of DNA, are damaged by vortexing and rough shaking. Such samples can be mixed by gently inverting the tubes or gently passing them up and down through a wide bore pipette. Brief centrifugation will bring liquids to the bottom of the tube.

BOX 5   *A GRAVIMETRIC PROCEDURE TO DETERMINE THE ACCURACY AND PRECISION OF A MICROPIPETTOR*

(Based primarily on ASTM Method E 1154-89 "Standard Specification for Piston or Plunger Operated Volumetric Apparatus")

SUMMARY OF METHOD: This is a general procedure to verify that a micropipettor is functioning properly by checking the accuracy and precision of the volumes it delivers. The procedure is based upon the determination of the weights of water samples delivered by the instrument. The weight of the water is converted to a volume based on the density of water.

NOTES

a. Clean the micropipettor according to the manufacturer's instructions before checking its performance.

b. The analytical balance used to check the micropipettor must be well-maintained and properly calibrated. It must also be in a draft-free, vibration-free environment. It is recommended that the balance meet the minimum requirements shown in the table at the end of this Box.

c. The water used should be purified and degassed. Discard water after one use.

d. Document all relevant information including date, micropipettor serial number, temperature when the check was performed, name of person performing evaluation, and so on.

e. Allow at least 2 hours for the micropipettor, tips, vials, and water to equilibrate together to room temperature.

f. This gravimetric procedure depends on converting a weight measurement to a volume value. To make this conversion with the utmost accuracy, it is necessary to correct for the following: the evaporation of water during the test procedure, the exact temperature of the water, the barometric pressure at the time of measurement, and the buoyancy effect. These corrections are found in the notes at the end of this Box. The calculation for converting weight to volume that is shown in the body of this Box is a commonly used approximation that does not correct for all these factors. If the performance verification is being performed to meet external quality requirements, it may be necessary to apply the corrections.

g. Replicate number: For a more thorough check of the micropipettor performance, 10 replicate measurements at each volume are recommended. For a quick check of the micropipettor four or six replicate measurements are sufficient.

h. Perform the following procedure as quickly as possible, but do not compromise the consistency of volume delivery and technique. The time required to make each measurement should be as consistent as possible because evaporation will affect the results.

i. To reduce the effects of evaporation it is recommended to:
   i. Keep the humidity of the room between 50 and 60%.
   ii. Surround the weighing vessel with a reservoir of water.

j. Computer programs are available that automatically perform the required calculations. Computer-based calibration systems that interface directly with the balance used for calibration are also commercially available.

PROCEDURE

1. *Set the micropipettor to deliver a particular volume, referred to as the "nominal volume."*

2. *Optional: If corrections are to be applied, (see Note f), then measure and record the barometric pressure and the temperature of the water.*

3. *Place a small amount of water in a weighing vessel, such as a vial. The exact amount of water does not matter, but should be about 0.5 mL.*

4. *Tare the balance to the vial.* (Handle the vial with tongs.)

5. *Optional: If you customarily prewet the tip, then prewet the tip by aspirating one volume of purified water and dispensing it into a waste container.*

6. *Pipette the selected nominal volume of water into the preweighed vial using proper technique.*

7. *Weigh the vial and record the weight of the delivered water.*

Box 5   *(continued)*

8. *Repeat the measurement 4 or 10 times by repeating Steps 3–7.*

9. *Optional: If a correction for evaporation is to be applied, then perform a control blank by repeating steps 3–7 exactly as in a normal weighing, but without actually delivering any liquid to the weighing vessel.* It is suggested that this evaporation control check be performed at the beginning and end of each series of measurements and between each group of 10 samples in a larger series.

10. *For adjustable micropipettors that can be set to different volumes, it is good practice to check the micropipettor at three volumes: the maximum capacity of the micropipettor, 50% of maximum capacity, and about 10% of maximum capacity.* Perform the entire procedure at each of the three volumes.

CALCULATIONS

For each nominal volume checked:

1. *Calculate the mean weight of the water.*

2. *Convert the mean water weight to the mean volume measured.*

    Assume that the density of water is 0.9982 *g/mL* (its density at 20°C)

    1 μL of water weighs 0.9982 *mg.*

    Then:

    $$Mean\ Volume = \frac{Mean\ Weight}{Density} = \frac{Mean\ Weight\ (in\ mg)}{0.9982\ mg/\mu L}$$

    Alternatively, for the highest accuracy, apply the corrections in Step 5.

3. *Determine the % accuracy or the "inaccuracy" (also called the percent error) of the micropipettor as follows:*

    $$\%\ Accuracy = \frac{Mean\ Volume\ Measured}{Nominal\ Volume} \times (100\%)$$

    or

    $$percent\ error = \%\ Inaccuracy = \frac{Mean\ Volume\ Measured - Nominal\ Volume}{Nominal\ Volume} \times (100\%)$$

4. *Determine the precision of the micropipettor by calculating the standard deviation (SD) or the coefficient of variation (CV) as follows:*

    $$SD = \pm \sqrt{\sum (X_i - \bar{x})^2/n - 1}$$

    where n = the number of measurements, X_i = each volume measurement, $\bar{x}$ = mean volume measured

    $$CV = \pm \frac{SD \times 100\%}{\bar{x}}$$

5. **ALTERNATIVE TO STEP 2:** *Calculations to convert the mean water weight to the mean volume taking into consideration corrections for evaporation, barometric pressure, temperature, and buoyancy:*
    i. *Calculate the evaporation, e, based on the loss in weight of the blank.*
    ii. *Calculate the mean volume of the water as follows:*
    $$Mean\ volume = (mean\ weight + e)\ (z)$$

    where: *mean weight = mean weight* of the four or ten repeats at a given nominal volume; e = the evaporation; z = a conversion factor in microliters per milligram that incorporates the buoyancy correction for air at the test temperature and barometric pressure. The values for z are found in the following table.

    iii. *Continue with the calculations for accuracy or inaccuracy and precision as shown earlier.*

Box 5   *(continued)*

Z VALUES: CONVERSION FACTOR VALUES (µL/MG), AS A FUNCTION OF TEMPERATURE AND PRESSURE, FOR DISTILLED WATER

Temperature (°C)	Air pressure (kPa)						Temperature (°C)	Air pressure (kPa)					
	800	853	907	960	1013	1067		800	853	907	960	1013	1067
15	1.0016	1.0018	1.0019	1.0019	1.0020	1.0020	22.5	1.0032	1.0032	1.0033	1.0033	1.0034	1.0035
15.5	1.0018	1.0019	1.0019	1.0020	1.0020	1.0021	23	1.0033	1.0033	1.0034	1.0035	1.0035	1.0036
16	1.0019	1.0020	1.0020	1.0021	1.0021	1.0022	23.5	1.0034	1.0035	1.0035	1.0036	1.0036	1.0037
16.5	1.0020	1.0020	1.0021	1.0022	1.0022	1.0023	24	1.0035	1.0036	1.0036	1.0037	1.0038	1.0038
17	1.0021	1.0021	1.0022	1.0022	1.0023	1.0023	24.5	1.0037	1.0037	1.0038	1.0038	1.0039	1.0039
17.5	1.0022	1.0022	1.0023	1.0023	1.0024	1.0024	25	1.0038	1.0038	1.0039	1.0039	1.0040	1.0041
18	1.0022	1.0023	1.0024	1.0024	1.0025	1.0025	25.5	1.0039	1.0040	1.0040	1.0041	1.0041	1.0042
18.5	1.0023	1.0024	1.0025	1.0025	1.0026	1.0026	26	1.0040	1.0041	1.0042	1.0042	1.0043	1.0043
19	1.0024	1.0025	1.0025	1.0026	1.0027	1.0027	26.5	1.0042	1.0042	1.0043	1.0043	1.0044	1.0045
19.5	1.0025	1.0026	1.0026	1.0027	1.0028	1.0028	27	1.0043	1.0044	1.0044	1.0045	1.0045	1.0046
20	1.0026	1.0027	1.0027	1.0028	1.0029	1.0029	27.5	1.0044	1.0045	1.0046	1.0046	1.0047	1.0047
20.5	1.0027	1.0028	1.0028	1.0029	1.0030	1.0030	28	1.0046	1.0046	1.0047	1.0048	1.0048	1.0049
21	1.0028	1.0029	1.0030	1.0030	1.0031	1.0031	28.5	1.0047	1.0048	1.0048	1.0049	1.0050	1.0050
21.5	1.0030	1.0030	1.0031	1.0031	1.0032	1.0032	29	1.0048	1.0049	1.0050	1.0050	1.0051	1.0052
22	1.0031	1.0031	1.0032	1.0032	1.0033	1.0033	29.5	1.0050	1.0051	1.0051	1.0052	1.0052	1.0053
							30	1.0052	1.0052	1.0053	1.0053	1.0054	1.0055

(Copyright ASTM E 1154, reprinted with permission.)

MINIMUM BALANCE REQUIREMENTS FOR GRAVIMETRIC TEST OF A MICROPIPETTOR

Test Volume (µL)	Balance Readability (mg)	Standard Deviation (mg)
≥1	≤0.001	≤0.002
≥11	≤0.01	≤0.02
≥101	≤0.1	≤0.1
≥1000	≤0.1	≤0.2

EXAMPLE

A digital micropipettor with a range of 100–1000 µL is evaluated for performance.

The pipettor is set for 1000 µL and high purity water at 20°C is dispensed four times. The dispensed water is weighed on an analytical balance. The four aliquots of water weigh:

$$0.9931\ g \qquad 0.9948\ g \qquad 0.9928\ g \qquad 0.9953\ g$$

The micropipettor is then set for 500 µL and four aliquots of water are dispensed that weigh:

$$0.4960\ g \qquad 0.4964\ g \qquad 0.4970\ g \qquad 0.4968\ g$$

The micropipettor is then set at 100 µL and four aliquots of water are dispensed that weigh:

$$0.1010\ g \qquad 0.0901\ g \qquad 0.0959\ g \qquad 0.0947\ g$$

Determine the accuracy and the precision of this pipette (as defined in Box 5) at all three volumes using the approximation shown in Box 5. Compare them with the specifications in Figure 16.15.

For a Nominal Volume of 1000 µL:

Step 1. Calculate the mean weight of the water from the four aliquots.

$$\frac{0.9931\ g + 0.9948\ g + 0.9928\ g + 0.9953\ g}{4} = 0.9940\ g = 994.0\ mg$$

Step 2. Calculate the mean volume of the water: $Volume = \dfrac{994.0\ mg}{0.9982\ mg/\mu L} = 995.8\ \mu L$

Step 3. Calculate the accuracy, or inaccuracy. The accuracy is: $\dfrac{995.8\ \mu L}{1000\ \mu L} \times 100\% = 99.58\%$ accuracy

This may also be expressed as inaccuracy, or percent error: $\dfrac{995.8\ \mu L - 1000\ \mu L\ (100\%)}{1000\ \mu L} = -0.42\%$

Step 4. Calculate the standard deviation. The four weight measurements converted to volumes are:

$$\frac{993.1\ mg}{0.9982\ mg/\mu L} \qquad \frac{994.8\ mg}{0.9982\ mg/\mu L} \qquad \frac{992.8\ mg}{0.9982\ mg/\mu L} \qquad \frac{995.3\ mg}{0.9982\ mg/\mu L}$$

$$= 994.9\ \mu L \qquad = 996.6\ \mu L \qquad = 994.6\ \mu L \qquad = 997.1\ \mu L$$

$$SD = \sqrt{\frac{(994.9 - 995.8)^2 + (996.6 - 995.8)^2 + (994.6 - 995.8)^2 + (997.1 - 995.8)^2}{3}}$$

$$= \sqrt{\frac{0.81 + 0.64 + 1.44 + 1.69}{3}} = 1.24\ \mu L$$

$$CV = 1.24\ \mu L / 995.8\ \mu L \times 100\% = 0.12\%$$

For a Nominal Volume of 500 µL:

Mean Volume = 497.4 µL   % Error = −0.51%   SD = 0.44 µL   CV = 0.09%

For a Nominal Volume of 100 µL:

Mean Volume = 95.6 µL   % Error = −4.4%   SD = 4.49 µL   CV = 4.69%

Summary of Results for all Three Volumes

	% Mean Error (Relative)	Mean Error (Absolute*)	Precision (CV)	Precision (SD)
1000 µL	−0.42%	−4.2 µL	0.12%	1.24 µL
500 µL	−0.51%	−2.6 µL	0.09%	0.44 µL
100 µL	−4.4%	−4.4 µL	4.69%	4.49 µL

*Mean Error Absolute = (mean volume − nominal volume)

Comparing these values with Figure 16.15 shows that the results at 1000 µL and 500 µL are within the manufacturer's specifications but the results at 100 µL are not. How this problem is handled depends on the policies in the laboratory.

the requirements of the laboratory's quality control procedure. If a micropipettor is performing outside of its specifications for accuracy, then it may be possible to recalibrate it using directions from the manufacturer. In other cases, the instrument will need to be repaired by the manufacturer or an in-house specialist.

F. Cleaning and Maintaining Micropipettors

Undamaged micropipettors require periodic performance evaluation but minimal maintenance. Routine maintenance includes cleaning the exterior of the micropipettor and some manufacturers suggest replacing internal seals and o-rings periodically, Figure 16.16.

Micropipettors that are heavily used or are roughly handled may become damaged. Common symptoms of damage include:

1. **Visual inspection shows that the tip does not take up the right volume of liquid.**
2. **The micropipettor leaks or drips.**
3. **The plunger jams, sticks, or moves erratically.**

4. **The micropipettor is blocked and will not aspirate liquid.**

The first two symptoms are frequently caused simply by poorly fitting pipette tips. Replacing the tip will sometimes help; other times, switching brands of tips will resolve these problems. A "sticky" plunger may be bent and needs replacement. All four of these symptoms may also be caused by a contaminated or damaged piston. Damaged seals or a loose shaft will also cause the micropipettor to leak or take up the wrong volume of liquid. Many users return damaged micropipettors to the manufacturer for repair. If you decide to service a micropipettor yourself, determine whether the instrument is under warranty, and, if so, whether servicing it will invalidate the warranty. Consult the manual for your micropipettor if you decide to repair the device yourself. Table 16.4 has tips regarding cleaning and simple repairs that are applicable to various styles of micropipettor.

After a micropipettor is serviced, it must be checked to see that it is performing according to its specifications. The gravimetric method described in the previous section is suitable for this purpose.

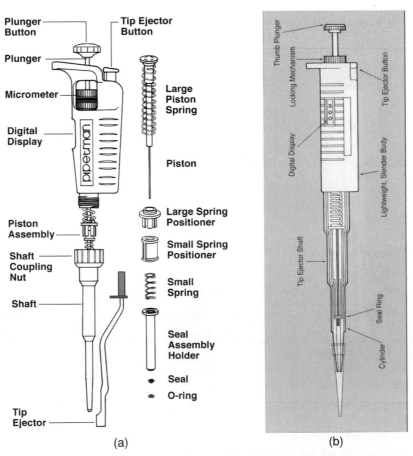

(a)   (b)

Figure 16.16. **The Parts of a Manual Micropipettor. a.** A digital micropipettor that is used for volumes from 2 to 20 μL. (From "Instructions for Pipetman Continuously Adjustable Digital Microliter Pipettes," Rainin Instrument Co., Woburn, MA. Used with permission. Pipetman® is a U.S. registered trademark of Rainin Instrument Co., Inc.) **b.** A schematic illustration of an Oxford Brand micropipettor. (Courtesy of Fisher Scientific, Pittsburgh, PA.)

Table 16.4   NOTES ON CLEANING AND SIMPLE REPAIR OF MICROPIPETTORS

1. *Consult the manufacturer's manual before attempting to clean or repair a micropipettor.* Follow the manufacturer's recommendations, as applicable.

2. *Use safety precautions when handling instruments that have been used to pipette hazardous materials.*

3. *Manufacturers recommend cleaning both the exterior of micropipettors and internal parts with distilled water alone, with soap solution followed by a rinse in distilled water, and/or with isopropanol (depending on the type of contamination).* Dry with a lint-free tissue.

4. *If you disassemble a micropipettor, keep track of the order and orientation of the parts as you remove each one.* There are springs inside the micropipettor assembly, so when the micropipettor is unscrewed, small parts may spring loose.

5. *If liquids are aspirated into the micropipettor accidentally, it is most effective to clean the device immediately and not let liquids dry inside it.* The micropipettor can usually be unscrewed and interior parts can be cleaned with distilled water. Dry the instrument after cleaning and then reassemble it.

6. *Many types of micropipettors have internal seals and o-rings that can become worn and can be replaced by the user.*

7. *Volatile organic compounds can cause the seals to swell, which in turn makes the piston move erratically.* Replace the seals.

8. *Manufacturers may recommend that the piston, seals, and o-rings be greased after cleaning.* Check the manual.

EXAMPLE PROBLEM

Suppose you are going to prepare 100 *mL* of physiological saline solution, which is 0.9% NaCl in water.

a. How much solute is necessary? What is a good way to measure the solute?

b. How much solvent is necessary? What is a good way to measure it?

c. List some common measurement errors that must be avoided to ensure that the solution is made properly.

ANSWER

a. The solute is 0.9 *g* of NaCl. This would be measured on a balance that reads at least to the nearest 0.01 *g*. (Recall that the last digit on an instrument may be estimated.)

b. Close to 100 *mL* of water will be required to bring the solution to a volume of 100 *mL*. A graduated cylinder or volumetric flask would commonly be used to make this solution.

c. There are many potential sources of error in preparing this solution, such as improper reading of the meniscus, use of dirty glassware, use of an improperly maintained balance, failure to pour all the salt from the weighing container into the mixing container, and so on. Care is required to avoid these major sources of error. There is also likely to be a small amount of error due to the fact that some error is tolerated in the calibration of a graduated cylinder or volumetric flask. This error, however, is very small compared with the potential error of the mistakes listed above. If a balance is properly calibrated, the amount of error due to the balance is insignificant compared to these other factors.

PRACTICE PROBLEMS

1. What is the volume of each of the following liquids?

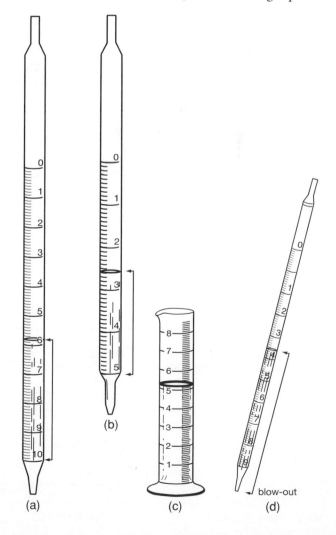

(a)   (b)   (c)   (d)

2. What type of device(s) is used commonly in biology laboratories to measure a volume of:

   a. 100 *mL*   **b.** 95 *mL*   **c.** 2 *mL*

   d. 100 μL **e.** 5 μL **f.** 500 μL

3. List common sources of error that might affect each of the measurements in Question 2.

4. Each of the devices given as an answer in Question 2 has a certain allowed error, or tolerance associated with it. Because of this tolerance there is some uncertainty in its measurements.

   Based on information given in this chapter, what is the tolerance of each device you gave as an answer to Question 2?

5. Suppose you are planning to purchase a new micropipettor to pipette volumes in the 100–200 µL range. You consult a catalog and find the following information for 3 brands of pipettor: Brand A, Brand B, and Brand C. Based on the catalog specifications, which micropipettor is most accurate? Which micropipettor is most precise? Which would you purchase?

	Volume Range	Accuracy (Expressed as % Error)	Precision (*CV*)
Brand A	40 to 200 μL	± 1%	0.5%
Brand B	100 to 200 μL	± 0.5%	0.3%
Brand C	100 to 200 μL	± 0.3%	0.4%

6. Suppose you are checking the accuracy of a micropipettor. Working at 20°C you weigh 10 empty tubes. You set the micropipettor to deliver 100 μL and carefully dispense that volume to each preweighed tube. You weigh the filled tubes on a calibrated balance and determine the weight of the water in each tube. The mean value for the weight of the water is 0.0994 *g*.

   a. The nominal volume is 100 μL. What do you expect the mean value for the weight of the water to be?

   b. What is the accuracy of your micropipettor (expressed as percent error) based on this result?

   c. The manufacturer specifies that the micropipettor should have an accuracy (expressed as % error) at least as good as ±2%. Does this pipettor meet the specification for accuracy?

7. The accuracy and precision of a micropipettor was checked using the procedure in Box 5. The micropipettor was set to deliver 500 μL. A standard form was used to record the results of the evaluation. A portion of the form and the results are shown below.

   a. Fill in the rest of the form.

   b. Based on the information in Figure 16.15, does this micropipettor meet its specifications? Place the answer in the pass/fail blank on the form.

Clean Gene, Inc.
VERIFICATION OF PERFORMANCE OF MICROPIPETTOR REPORT
FORM 232
Effective Date: 2/8/99

CALIBRATION TECHNICIAN   *J.E.S.*

CALIBRATION DATE   *9/22/99*   NEW DUE DATE FOR NEXT CALIBRATION   *3/22/2000*

MICROPIPETTOR ID NUMBER *2127*   MANUFACTURER   *Finestt Pipettes Inc.*

MODEL NUMBER *micro B*   LOCATION   *Lab #27*   TEMPERATURE   *20°C*

RANGE *100–1000µL*   PRIMARY USER   *R.E.S.*

SUMMARY:

PREVERIFICATION CLEANING AND ADJUSTMENTS   *Changed Seals & O-rings*

_____ PASS/FAIL _____

STATUS _____

Volume 1

NOMINAL VOLUME (µL)	TUBE NUMBER	INITIAL WEIGHT OF TUBE	WEIGHT AFTER H_2O DISPENSED	NET WEIGHT OF DISPENSED H_2O	H_2O VOLUME
500	1	1.9355 *g*	2.4352 *g*		
500	2	1.9877 *g*	2.4876 *g*		
500	3	1.9787 *g*	2.4789 *g*		
500	4	1.9850 *g*	2.4873 *g*		
500	5	1.9755 *g*	2.4763 *g*		
500	6	1.9387 *g*	2.4387 *g*		

Mean Water Volume ($\overline{x}$) _____ % INACCURACY _____ SD _____ CV _____

TERMS RELATING TO THE MEASUREMENT OF VOLUME

Aerosols. Fine liquid droplets that are suspended in the air.

Air Displacement Micropipettor. Device for measuring microliter volumes; designed so that there is an air cushion between the micropipettor and the sample.

"Blow Out" Pipette. A type of pipette that is calibrated so that the last drop is to be forcibly expelled from the tip of the pipette.

Borosilicate Glass. A strong, temperature-resistant glass that does not contain discernable contamination by heavy metals and is used to manufacture general purpose laboratory glassware.

Burette (also spelled "buret"). Long graduated tube with a stopcock at one end that is used to accurately dispense known volumes.

Calibration (of a volume measuring device). 1. Placement of capacity lines, graduations, or other markings on the device so that they correctly indicate volume. 2. Adjustment of dispensing devices so they dispense accurate volumes.

Capacity Line. A line marked on an item of glassware to indicate the volume if the item is filled to that mark.

"Carryover." Material from one sample that is carried to and contaminates another sample.

Density. The weight per volume of a substance under specified environmental conditions.

Digital Microliter Pipettor. An instrument used to dispense volumes in the microliter range that can be adjusted to deliver different volumes.

Graduated Cylinders. Cylindrical vessels calibrated to deliver various volumes.

Graduations. Lines marked on glassware, plasticware, and pipettes that indicate volume.

Gravimetric Method. A method that involves the use of a balance; gravimetric methods are used to calibrate volume measuring devices.

Liter. The basic unit of volume; it equals 1 dm^3.

Manual Dispenser (for reagent bottle). Device that dispenses set volumes from a bottle.

Measuring Pipette. A type of pipette calibrated with a series of graduation lines to allow the measurement of more than one volume.

Meniscus. Greek for "crescent moon." The surface of liquids in narrow spaces forms a curve known as a meniscus. The bottom of the meniscus is used as the point of reference in calibrating and using volumetric labware.

Microliter Pipette or Micropipette. A term used in this book to refer to various small volume pipettes that measure volumes in the 1–1000 μL range.

Micropipettor. A term used in this book to refer to an instrument that measures volumes typically in the 1–1000 μL range. This term is not generally used to refer to a simple hollow tube pipette, but rather to a more complex instrument.

Mohr Pipette. A type of pipette that is calibrated so that the liquid in the tip is not part of the measurement.

Nominal Volume. The desired volume for which a pipetting device is set.

O-Ring (for a micropipettor). Rubber ring used as a seal to prevent the leakage of air.

Pasteur Pipette. A type of pipette used to transfer liquids from one place to another, but not to measure volume.

Pipette (also spelled "pipet"). Hollow tube that allows liquids to be drawn in and dispensed from one end; generally used to measure volumes in the 0.1–25 mL range.

Pipette-aid. A device used to draw liquid into and expel it from pipettes.

Plunger. Part of a manual micropipettor that is depressed by the operator as liquids are taken up and expelled.

Positive Displacement Micropipettor. Volume measuring device, such as a syringe, where the sample comes in contact with the plunger and the walls of the pipetting instrument.

Serological Pipette. Term that usually refers to a calibrated glass or plastic pipette that measures in the 0.1–25 mL range. These pipettes are calibrated so that the last drop is in the tip. This drop needs to be "blown out" to deliver the full volume of the pipette.

To Contain, TC. A method of calibrating glassware so that it contains the specified amount when exactly filled to the capacity line. The device will not deliver that amount if the liquid is poured out because some of the liquid will adhere to the sides of the container.

To Deliver, TD. A method of calibrating glassware so that it delivers the specified amount when poured.

Tolerance (of a volume measuring device). How much error is allowed in the calibration of a volume measuring item. For example, the tolerance of a 100 mL Class A volumetric flask that meets ASTM standards is ± 0.08 mL (i.e., its true marked volume must be between 99.92 mL and 100.08 mL).

Volume. The amount of space a substance occupies, commonly measured in units of "liters."

Volumetric Glassware. A term used generally to refer to accurately calibrated glassware intended for applications where high accuracy volume measurements are required.

Volumetric (Transfer) Pipette. A pipette made of borosilicate glass and calibrated "to deliver" a single volume when filled to its capacity line at 20°C.

Z Factor. A conversion factor that incorporates the buoyancy correction for air at a particular test temperature and barometric pressure.

REFERENCES RELATING TO VOLUME MEASUREMENTS

Any general chemistry textbook has a discussion of volume measurements. Manufacturer's catalogs are also good sources of information. The following ASTM standards contain valuable, specific information about volume measurements:

ASTM Standard E 969-94 "Standard Specification for Volumetric (Transfer) Pipets."

ASTM Standard E 1044-85 (reapproved 1990) "Standard Specification for Glass Serological Pipets (General Purpose and Kahn)."

ASTM Standard E 288-94 "Standard Specification for Laboratory Glass Volumetric Flasks."

ASTM Standard E-694-94 "Standard Specification for Laboratory Glass Volumetric Apparatus."

ASTM Standard E 542-94 "Standard Specification for Calibration of Laboratory Volumetric Apparatus."

ASTM Standard E 1154-89 (reapproved 1993) "Standard Specification for Piston or Plunger Operated Volumetric Apparatus." (Particularly recommended for individuals responsible for checking the calibration of micropipetting devices.)

ANSWERS TO PRACTICE PROBLEMS

1. a. 4.2 *mL*

   b. 2.4 *mL*

   c. 5.2 *mL*

   d. 6.4 *mL*

2. a. a volumetric flask or a graduated cylinder.

   b. a graduated cylinder

   c. a pipette or a micropipettor

   d. a micropipettor

   e. a micropipettor

   f. a micropipettor

3. There are many possible answers. Errors in use of a graduated cylinder or volumetric flask include the use of dirty glassware, improper reading of the meniscus, and use of a TC flask where TD is appropriate and *vice versa*. For micropipettor errors, see Boxes 3 and 4 and Tables 16.3.

4. a. Graduated cylinder, ± 0.6 *mL*, volumetric flask, Class A 0.08 *mL*, Class B 0.16 *mL*.

   b. Graduated cylinder, ± 0.6 *mL*.

   c. A serological pipette ± 0.02 *mL*

   d–f From Figure 16.15, mean error in μL, as low as:

   d. ± 0.8

   e. ± 0.075

   f. ± 4

5. Brand B has the best precision and Brand C has the best accuracy, but they are basically similar and either brand is probably acceptable based on these specifications.

6. a. The expected value is ? /0.9982 $mg/\mu L$ = 100 μL. ? = 99.82 mg

   b. The mean volume for the water in μL is 99.40 mg/0.9982 $mg/\mu L$ = 99.58 μL

$$\% \text{ Error} = \frac{99.58 \ \mu L - 100.0 \ \mu L}{100 \ \mu L} \times 100\%$$

$$= -0.42\%$$

   c. Yes, it is within the specification for accuracy.

7. Form on next page.

Clean Gene, Inc.

VERIFICATION OF PERFORMANCE OF MICROPIPETTOR REPORT FORM 232

Effective Date: 2/8/99

CALIBRATION TECHNICIAN _____ *J.E.S.* _____

CALIBRATION DATE _____ *9/22/99* _____   NEW DUE DATE FOR NEXT CALIBRATION _____ *3/22/2000* _____

MICROPIPETTOR ID NUMBER _____ *2127* _____   MANUFACTURER _____ *Finestt Pipettes Inc.* _____

MODEL NUMBER _____ *2127* _____   LOCATION _____ *Lab #27* _____   TEMPERATURE _____ *20°C* _____

RANGE _____ *100–1000 µL* _____   PRIMARY USER _____ *R.E.S.* _____

SUMMARY:

PREVERIFICATION CLEANING AND ADJUSTMENTS _____ *Changed Seals & O-rings* _____

_____ PASS/FAIL _____ *Pass* _____

STATUS _____ *Returned to lab for use* _____

Volume 1

NOMINAL VOLUME (µl)	TUBE NUMBER	INITIAL WEIGHT OF TUBE	WEIGHT AFTER H_2O DISPENSED	NET WEIGHT OF DISPENSED H_2O	H_2O VOLUME
500	1	1.9355 g	2.4352 g	0.4997 g	500.60 µL
500	2	1.9877 g	2.4876 g	0.4999 g	500.80 µL
500	3	1.9787 g	2.4789 g	0.5002 g	501.10 µL
500	4	1.9850 g	2.4873 g	0.5023 g	503.20 µL
500	5	1.9755 g	2.4763 g	0.5008 g	501.70 µL
500	6	1.9387 g	2.4387 g	0.5000 g	500.90 µL

Mean Water Volume _____ *501.38 µL* _____ % INACCURACY _____ *0.28%* _____ SD _____ *0.97 µL* _____ CV _____ *0.19%* _____

CHAPTER 17

The Measurement of Temperature

I. INTRODUCTION TO TEMPERATURE MEASUREMENT

A. The Importance and Definition of Temperature

Biological systems in the laboratory are sensitive to temperature. For example, enzymes require a specific temperature for optimal activity. The polymerase chain reaction (PCR, an enzymatic process that is used to greatly amplify specific sequences of DNA) requires that a reaction mixture cycle between certain specific temperatures. Microorganisms and cells growing in culture must be incubated at a specific temperature. The calibration and operation of laboratory instruments requires temperature control. Temperature is one of the key process control variables in the fermentation industry and is closely monitored and controlled in the pharmaceutical industry. The laboratory contains many devices to control temperature, including ovens, heaters, incubators, water baths, refrigerators, freezers, freeze dryers, and PCR thermocyclers. It is thus essential that there be convenient and accurate methods to measure temperature in the laboratory.

Although temperature is familiar to everyone and has been studied for hundreds of years, its physical basis is not inherently obvious. Physicists now explain temperature as being related to the continuous, random motion of the molecules that make up substances. **Heat** *is a form of energy that is associated with this disordered molecular motion.* **Heating** *is the transfer of energy from an object with more random internal energy to an object with less.* For example, a burn results when a large amount of energy from a hot object is transferred to a person's skin.

Temperature *is a measure of the average energy of the randomly moving molecules in a substance. Temperature tells us which way heat will flow:* Objects with a higher temperature lose heat to objects with a lower temperature. *When two objects are at the same temperature, they are said to be in* **thermal equilibrium**; there is no net transfer of energy from one to the other. Temperature is measured by allowing a thermometer to come to thermal equilibrium with the material whose temperature is being determined. For example, taking a person's temperature involves placing a thermometer in contact with that person and waiting until the thermometer and person reach thermal equilibrium. The thermometer reading then indicates the person's temperature.

B. Temperature Scales

i. A BIT OF HISTORY

There are three temperature scales most commonly used in biology laboratories: Fahrenheit, Celsius (also previously called centigrade), and Kelvin. It is informative to discuss briefly how there came to be three scales in widespread use.

The Fahrenheit scale was invented in the early 1700s by Daniel Gabriel Fahrenheit. Fahrenheit made thermometers similar to the mercury thermometers used today. It is not entirely clear how Fahrenheit calibrated his thermometers, but it appears to have been more or less as follows. Fahrenheit chose two reference points. The first point was the coldest system he could produce, that is, a mixture of salt, water, and ice. He called the temperature of this mixture "zero." The second reference point was the body temperature of a person, which he called "96." (His methods were slightly in error because the temperature of a healthy person is 98.6 degrees on the Fahrenheit scale.) Fahrenheit placed mercury inside thin glass tubes. He positioned this assembly in the salt, water, and ice mixture and placed a mark on the glass tube to indicate the height of the mercury. He then placed the assembly in contact with a person and similarly marked the glass tube to indicate the height to which the mercury rose. Next, he subdivided the length of the glass tube between the "zero" and "96" marks into 96 equal divisions, each of which he called a **degree**. The result was a thermometer marked in degrees according to the **Fahrenheit temperature scale**. Note that this method is based on the assumption that the relationship between mercury expansion and temperature is linear.

Anders Celsius also made mercury thermometers in the 1700s. Celsius chose as his fixed reference points the boiling point of water and the freezing point of water. Celsius divided the interval between his two reference points into 100 divisions, each called a degree. The result is a thermometer marked according to the **Celsius temperature scale**. It is interesting to note that Celsius originally assigned the boiling point of water "0" and the freezing point "100." This was, of course, later reversed so we now think of temperature as going down when it gets colder, rather than up.

Over the years, some people preferred and used Celsius thermometers, some used Fahrenheit's, and some people invented other temperature scales based on other reference temperatures. For example, there were temperature scales based on the temperature at the first frost and the temperature of underground caverns.

The Kelvin scale was devised in the early 1800s by various scientists including William Thomson (later, Lord Kelvin). *The Kelvin scale uses a unit, called a* **kelvin**, *which is the same size as a degree in the Celsius scale.* However, the Kelvin scale sets the zero point at **absolute zero**, *the temperature at which, theoretically, molecules stop their internal random motion.* Zero kelvin is $-273.15°C$; the zero point on the Celsius scale corresponds to 273.15 K. (The degree sign, "°," is not used when temperature is expressed in kelvins.) There are 100 K between the freezing and boiling points of water, just as there are 100°C between these two points.

ii. CONVERSIONS FROM ONE TEMPERATURE SCALE TO ANOTHER

As a result of various people's work in temperature measurement, there are now three temperature scales which you are likely to encounter, each of which assigns different temperature values to materials that are the same temperature. For example, the temperature at which water freezes is variously called 32°F, 0°C, and 273.15 K, Table 17.1 and Figure 17.1.

Although Celsius is the most common scale used in biology, laboratory procedures may be written using any of the three temperature scales. It is therefore nec-essary to be able to convert temperatures from one scale to another. Simple equations can be used for such conversions, as shown in Box 1.

Table 17.1 *COMPARISON OF THE FAHRENHEIT, CELSIUS, AND KELVIN TEMPERATURE SCALES*

Absolute Zero	−460°F	−273°C	0 K
Freezing Point of Water	32°F	0°C	273 K
Average Room Temperature	68°F	20°C	293 K
Normal Human Temperature	98.6°F	37°C	310 K
Boiling Point of Water	212°F	100°C	373 K

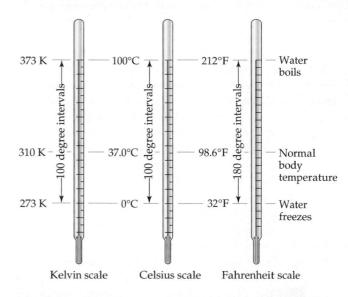

Figure 17.1. Comparison of Kelvin, Celsius, and Fahrenheit Temperature Scales.

Box 1 *CONVERSION FROM ONE TEMPERATURE SCALE TO ANOTHER*

1. *To convert from degrees Fahrenheit to degrees Celsius:*

$$°C = (°F − 32°) \, 0.556$$

   Subtract 32° from the Fahrenheit reading
   Multiply the result by 0.556 or 5/9

   Example: Convert human body temperature, 98.6°F, to degrees Celsius

$$°C = (98.6° − 32°) \, 0.556 = 37.0°C$$

2. *To convert from degrees Celsius to degrees Fahrenheit:*

$$°F = (°C × 1.8) + 32°$$

   Multiply the Celsius reading by 1.8 or 9/5
   Add 32 degrees

   Example: Convert human body temperature, 37.0°C, to degrees Fahrenheit

$$°F = (37.0° × 1.8) + 32° = 98.6°F$$

3. *To convert from degrees Celsius to kelvin and from kelvin to degrees Celsius:*

$$°C = K − 273$$
$$kelvin = °C + 273$$

   Example: Convert human body temperature, 37.0°C, to kelvin.

$$kelvin = 37.0°C + 273 = 310 \text{ K}$$

4. *To convert degrees Fahrenheit to kelvin:*
   Convert Fahrenheit to Celsius
   Convert Celsius to kelvin

   Example: Convert human body temperature, 98.6°F to kelvin.

$$°C = (98.6° − 32°) \, 0.556 = 37.0°C$$
$$kelvin = 37.0°C + 273 = 310 \text{K}$$

iii. FIXED REFERENCE POINTS AND THERMOMETER CALIBRATION

We have seen that temperature scales are based on two or more fixed reference points. **Fixed reference points** *are systems whose temperatures are determined by some physical process and hence are universal and repeatable.*

Like Celsius, metrologists today use **phase transitions** as the references for calibration of laboratory thermometers and verification of their performance. A **phase transition** *is where a liquid turns to a solid or a vapor, or a vapor turns to liquid.* **Freezing point** *is the temperature at which a substance goes from the liquid phase to the solid phase.* **Boiling point** *is the temperature at which a substance in the liquid phase transforms to the gaseous phase* (under specified conditions of pressure). People also use the **triple point** of various substances to define temperatures. The **triple point of water** *is a single temperature and pressure at which ice, water, and water-vapor coexist with one another in a closed container.* The triple point temperature for water (at atmospheric pressure) is 0.01°C. At all other temperatures and pressures, only two phases can coexist (i.e., either water–ice or water–vapor).

The boiling and freezing points of water are adequate references for thermometers used for most biological applications. Scientists who study nonliving systems use the freezing points, boiling points, and triple points not only of water but also of oxygen, hydrogen, and various metals to extend the range of temperatures at which there are fixed references.

The reason that phase transitions are used for thermometer calibration is that given the proper apparatus and conditions, phase transitions always occur at a predictable temperature and pressure. It is possible, therefore, to consistently calibrate thermometers anywhere in the world.

II. THE PRINCIPLES AND METHODS OF TEMPERATURE MEASUREMENT

A. Overview

There are various transducers (sensors) used to measure temperature, all of which respond to the effects of temperature on a physical system. For example, mercury thermometers, like those described earlier, measure the expansion or contraction of mercury in a glass tube in response to temperature changes. Types of temperature transducers include:

1. *Liquid expansion devices*
2. *Bimetallic expansion devices*
3. *Change-of-state indicators*
4. *Metallic resistance devices*
5. *Thermistors*
6. *Thermocouples*

The first two types of sensors—liquid expansion and bimetallic devices—are similar in that they are both based on the fact that most materials expand as the temperature increases and contract when it decreases. Mercury thermometers are liquid expansion devices. Mercury thermometers are simple to use and relatively inexpensive.

Mercury thermometers, bimetallic devices, and change-of-state indicators are mechanical measuring instruments (i.e., they do not require electricity to make a measurement nor do they generate an electrical signal). These devices are usually read by human eyes, so they cannot be directly interfaced with computers or electronic instrumentation. In contrast, resistance thermometers, thermistors, and thermocouples are electronic devices that generate an electrical signal and are readily interfaced with other equipment. Electronic thermometers are accurate, precise, versatile, compact, can be used at a wide range of temperatures, and can be used for remote sensing. As these electronic thermometers decrease in cost and become increasingly available, they are taking the place of mercury thermometers in many laboratories.

Because mercury thermometers are the type of thermometer traditionally most often handled in the biology laboratory, they will be discussed first. Bimetallic expansion thermometers and change-of-state indicators will then be described. The latter part of this section will outline the basic features of electronic temperature sensing devices.

B. Liquid Expansion (Liquid-in-Glass) Thermometers

i. DESCRIPTION

Liquid expansion thermometers *contain liquid that expands or contracts with temperature changes and moves up or down within a narrow glass capillary tube, or* **stem.** When the temperature of the surrounding medium increases, the liquid expands more than the glass, so the liquid moves up the stem. When the liquid and the substance whose temperature is being measured are in thermal equilibrium, the liquid stops rising. The temperature of the bulb is read by noting where the top of the liquid column coincides with a scale marked on the stem. If the temperature decreases, the liquid contracts down the stem until thermal equilibrium is reached. As was previously discussed, the scale of a liquid-in-glass thermometer is determined by marking the stem at two (or more) reference points and subdividing the interval between the marks. Liquid-in-glass thermometers are typically used in the range from −38°C to 250°C.

Mercury and alcohol are the liquids commonly used in liquid expansion thermometers. Mercury is used because it does not adhere to glass and its silvery appearance makes it easy to read. Alcohol (also called "spirit"), which is dyed red to make it easier to read the meniscus, is a safer material and poses less threat to the environment than mercury. Alcohol can be used at temperatures below −38.6°C, where mercury freezes. Mercury thermometers have a more linear response with temperature than alcohol and for this reason are more often used in the laboratory.

Safety note: **A mercury spill kit should always be available in laboratories where mercury thermometers are used. See pp. 632–633 for information on spill clean-up.**

Liquid-in-glass thermometers have certain components, Figure 17.2. These components are:

1. **Liquid**, *usually mercury or alcohol*
2. **Stem**, *a glass capillary tube through which the mercury or organic liquid moves as temperature changes*
3. **Bulb**, *a thin glass container at the bottom of the thermometer that is a reservoir for the liquid*
4. **Scale**, *graduations that indicate degrees, fractions of degrees, or multiples of degrees*

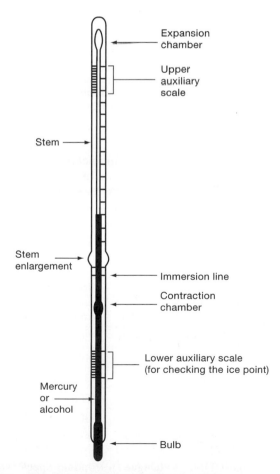

Figure 17.2.   The Parts of a Mercury Thermometer.

5. **Contraction chamber** *(not present on all thermometers), an enlargement of the capillary bore that takes up some of the volume of the liquid thereby allowing the overall length of the thermometer to be reduced*
6. **Expansion chamber**, *an enlargement of the capillary bore at the top of the thermometer that prevents buildup of excessive pressure*
7. **Stem enlargement** *(not present in all thermometers), a thickening of the stem that assists in the proper placement of the thermometer in a device (e.g., in an oven)*
8. **Immersion line** *(not present on all thermometers), a line etched onto the stem to show how far the thermometer should be immersed into the material whose temperature is to be measured*
9. **Upper and lower auxiliary scales** *(not present on all thermometers), extra scale markings at the zero degree and 100 degree areas to assist in calibration and verification of the performance of the thermometer*

Liquid-in-glass thermometers come in many styles that differ in their scale divisions, range, length, accuracy, intended applications, and other qualities. For example, a thermometer for use in a refrigerator or freezer might have a range from −40°C to 25°C. A thermometer for monitoring the temperature in an autoclave might measure temperatures from 85–135°C and may be specially designed to hold and display the maximum temperature reached during sterilization, even after the autoclave has cooled.

ii. IMMERSION

The depth of immersion of a liquid thermometer into the material whose temperature is to be measured affects the accuracy of the measurement. This is because the liquid in the part of the stem that is not immersed is not at the same temperature as the liquid in the bulb. The manufacturer takes this difference into account when the thermometer is calibrated and indicates on the thermometer how far it is to be immersed. There are two basic immersion types of the thermometer: partial immersion and total immersion. A third type, complete immersion, is less common.

A **partial immersion thermometer** *is manufactured to indicate temperature correctly when the bulb and a specified part of the stem are exposed to the temperature being measured*, Figure 17.3a. Partial immersion thermometers are easiest to read because their liquid column is not fully immersed in the medium. They are ideal for low volume solutions. A partial immersion thermometer is marked with a line to indicate how far it should be immersed.

A **total immersion thermometer** *is designed to indicate temperature correctly when that portion of the ther-*

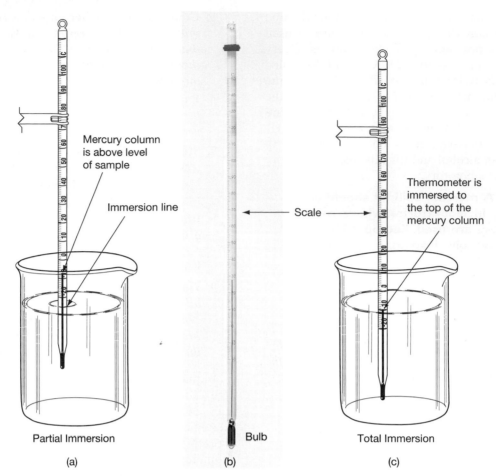

Figure 17.3. Partial and Total Immersion Thermometers. a. Partial immersion thermometer. A line on the stem indicates the required immersion depth. The thermometer is immersed only to the immersion line. **b.** Total immersion thermometer. **c.** A total immersion thermometer is positioned so that all the mercury (or alcohol) within the thermometer is immersed in the material whose temperature is being measured. (Photo courtesy of Fisher Scientific, Pittsburgh, PA.)

mometer containing the liquid mercury or alcohol is exposed to the medium whose temperature is being measured, Figure 17.3b,c. Total immersion thermometers are more accurate than partial immersion thermometers and are suggested for applications where accuracy is critical or when a thermometer is used for calibrating other thermometers or equipment.

If a total immersion thermometer is not immersed to the proper depth there will be inaccuracies in its readings. In general, at the temperatures used in a biology laboratory (e.g., between 0 and 100°C), the correction is less than 1 degree. Nonetheless, it is good practice to always immerse a thermometer properly.

There are several ways to distinguish a partial from total immersion thermometer. A partial immersion thermometer has an immersion line marked on the stem; a total immersion thermometer does not. The immersion depth, in millimeters, is sometimes written on the stem of a partial immersion thermometer. Manufacturers always specify the immersion type for each thermometer in their catalog, Figure 17.4.

A total immersion thermometer properly indicates temperature when that portion of the thermometer containing the liquid mercury or alcohol is exposed to the medium whose temperature is being measured. A total immersion thermometer is not intended to actually be placed inside an oven or to be immersed *completely* in

Temp Range (°C)	Div (°C)	Total length (mm)	Immersion depth (mm)
24 to 38	0.05	305	76
−1 to 51	0.1	460	Total
−1 to 101	0.1	610	Total
−20 to 110	1	305	76
−100 to 50	1	305	76
−1 to 201	0.2	610	Total

Figure 17.4. Catalogue Descriptions Specify How a Thermometer Should be Immersed.

a hot liquid. A total immersion thermometer will be inaccurate if it is completely immersed, and there is the possibility that the thermometer will break at high temperatures if placed inside an oven or incubator. Most laboratory ovens have a slot in which the thermometer is inserted so that the proper length of the thermometer is inside the oven and the top part protrudes visibly outside the oven. There are **complete immersion thermometers** *that are designed so that the entire thermometer, from top to bottom, is exposed to the medium whose temperature is being measured.* Figure 17.5 shows a complete immersion thermometer specifically designed to be placed inside an oven, incubator, or freezer. Observe that this thermometer is protected in case of breakage.

iii. LIMITATIONS OF LIQUID EXPANSION THERMOMETERS

Although liquid expansion thermometers are routinely used with acceptable results, they have inherent limitations. Liquid expansion thermometers have fairly wide tolerances (i.e., they may be in error by a fair amount). It is not unusual for manufacturers to specify that the tolerance for a general purpose mercury thermometer is ± 1°C or ± 5°C (± 2°F or ± 9°F). A thermometer may be in error by as much as its maximum tolerance. This means, for example, that if you are using a thermometer whose tolerance is ± 5°C, and you measure the temperature of a solution as 37°C, the actual temperature of that solution may be anywhere from 32 to 42°C.

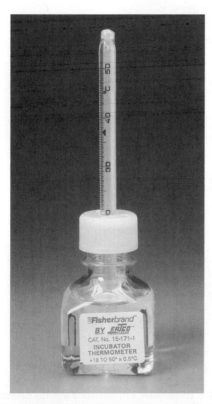

Figure 17.5.   Complete Immersion Thermometer Intended for Use in a Freezer, Oven, or Incubator. (Courtesy of Fisher Scientific, Pittsburgh PA.)

ASTM classifies thermometers and specifies how much error is tolerable for a particular class. Tolerances for partial immersion thermometers are greater than for total immersion thermometers. More expensive "precision thermometers" have narrower tolerances than other thermometers. To meet ASTM requirements, precision thermometers must have an error of less than 0.1–0.5°C, depending on their range. A precision thermometer whose range does not include 0°C will have a second, **auxiliary scale** *that shows the zero point so that the performance of the thermometer can be verified with an ice point bath* (discussed later).

Other limitations of liquid expansion thermometers are described in Table 17.2.

iv. PROPER USE OF LIQUID EXPANSION THERMOMETERS

Many of the problems that arise with liquid expansion thermometers are easy to diagnose. The most common problem is breakage. Another common problem with liquid thermometers is that the liquid column develops separations due to thermal or mechanical shock during shipping or handling. If the liquid column is discontinuous there will be substantial errors in the thermometer's readings. Thermometers should be inspected before use to be sure the liquid column is continuous. Box 2 shows several procedures that are used to reunite a liquid column.

Table 17.2   LIMITATIONS OF LIQUID EXPANSION THERMOMETERS

1. *The capillary bore within the thermometer in which the liquid moves must be very smooth so that the liquid is not impeded at any spot.* This is difficult to achieve, and less expensive thermometers may have slight imperfections leading to inaccuracy in their measurement.

2. *The scale must be very carefully etched onto the stem to ensure accuracy.* Not all thermometers are accurately subdivided into degrees.

3. *The working range of liquid thermometers is limited.* Mercury freezes at −38.6°C and alcohol at −200°C. In practice, liquid thermometers are seldom used above about 250°C. The limited range of liquid expansion thermometers is not usually a problem for biologists because life exists only in a fairly narrow range of temperatures.

4. *Glass thermometers are fragile.* The glass comprising the bulb provides a thin interface between the liquid and the substance whose temperature is to be measured. This means the glass is thin and delicate and may easily be deformed or broken.

5. *Mercury, commonly used in liquid expansion thermometers, is hazardous to human health and the environment.* Mercury fumes and solids are toxic, and mercury from a broken thermometer can contaminate both the air and a water bath, a counter, or other site.

BOX 2 REUNITING THE LIQUID COLUMN OF LIQUID EXPANSION THERMOMETERS

1. *Lightly tap the thermometer by making a fist around the bulb with one hand and gently hitting your fist into the palm of the other hand.*

2. *Clamp the thermometer vertically. Slowly immerse the bulb of the thermometer in a freezing bath of salt, ice, and water.* The liquid should all slowly contract into the bulb, causing it to reunite. Be careful to cool the bulb only, not the stem. It may be helpful to remove the bulb periodically to slow the motion of the liquid. Remove and allow the thermometer to warm while it is supported in an upright position. Repeat several times if necessary.

3. *If 1 and 2 do not work, and if the separation is in the lower part of the column, then try cooling the thermometer, as described in 3, but use a mixture of dry ice and either acetone or alcohol.* Wear cold-resistant gloves. Cool only the bulb, not the stem. Remove the bulb from the cold. Do not touch the bulb until it has warmed to room temperature; it might shatter.

4. *As a last resort, particularly if the separation is in the upper part of the mercury column, warm the thermometer.*
   a. Use this method only for thermometers with expansion chambers above the scale.
   b. Place the thermometer in a beaker of water or other nonflammable liquid and heat it slowly.
   c. As the thermometer heats, the liquid should expand into the upper expansion chamber.
   d. When the liquid has moved into the expansion chamber, gently tap the thermometer with a gloved hand to reunite the column.
   e. Watch to be sure the expansion chamber does not overfill since if it does, the thermometer will break.
   f. Allow the thermometer to cool slowly.
   g. *Never use an open flame to heat the bulb.*

5. *If the column appears to be successfully reunited, check the ice point. If the ice point is shifted from its previous position, discard the thermometer.*

6. *If the liquid column cannot be reunited, discard the thermometer.*

Thermometers must be used properly to give correct measurements. Table 17.3 summarizes points regarding the proper use of liquid expansion thermometers.

C. Bimetallic Expansion Thermometers

Materials generally expand as the temperature increases and contract when it decreases. Each material has a characteristic ***thermal expansion coefficient*** *that measures how much the material expands for a specific increase in temperature.* If two metals with different coefficients of expansion are bonded together and are then subjected to heat, the strips of metal will distort and bend. The amount of bending is related to the temperature, Figure 17.6a.

A bimetallic dial thermometer *is constructed by fusing together two metal strips, typically one of brass and one of iron, which have different coefficients of expansion.* When the temperature changes, the two metals respond unequally, resulting in bending. One end of the strips is fastened and cannot move, the other moves along a scale to display the temperature, Figure 17.6b.

Bimetallic expansion thermometers are simple and convenient, but they are generally not intended to be as accurate as liquid expansion thermometers. They do

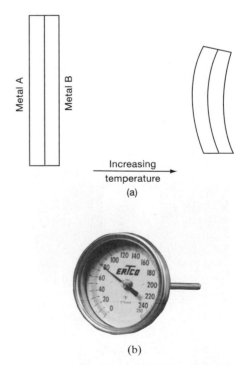

Figure 17.6. Bimetallic Expansion Thermometers. a. The unequal bending of two metals. **b.** A bimetallic thermometer. (Photo courtesy of Ever Ready Thermometer Co.)

Table 17.3 *PROPER USE OF LIQUID EXPANSION THERMOMETERS*

1. *Visually inspect a thermometer before use for:*
   breaks in the liquid column or bubbles
   foreign material in the stem
   distortions in the scale

   If the liquid is separated, reunite it as described in Box 2. If the thermometer contains foreign material or has a distorted scale, it should be discarded.

2. *Avoid tapping a glass thermometer against a hard surface to prevent small fractures.*

3. *Support the thermometer by its stem; do not allow the bulb to rest on a surface.* The bulb is fragile.

4. *Do not stir solutions with thermometers.*

5. *Never subject a glass thermometer to rapid temperature changes.* For example, do not move a thermometer from an ice bath and place in a boiling water bath. This may cause the thermometer to break or the liquid filling solution to separate.

6. *Never place a thermometer in an environment above its maximum indicated temperature.*

7. *Immerse the thermometer to the depth indicated by the manufacturer.*

8. *Read the meniscus with your eyes at an even level with the top of the liquid column.* For mercury, the top of the meniscus is used as the indicator. For organic liquids, the bottom of the meniscus is the part that is used for a reading.

9. *Store thermometers as instructed by the manufacturer.* Mercury thermometers are most often stored horizontally in a protective case to avoid damage. It is sometimes recommended that alcohol thermometers and thermometers used at temperatures below 0°C be stored vertically in a rack.

10. *Place thermometers properly in equipment whose temperature is being measured.*
    a. Do not allow thermometers to touch the sides of equipment whose temperature is being measured.
    b. Place the thermometer in a manner to avoid breakage.
    c. Safely mounted thermometers for use in refrigerators, incubators, and other equipment are available commercially. Use these when possible.
    d. For freezers and refrigerators place the thermometer so that the stem does not touch any metal surface or ice. Air must be able to flow unobstructed around the stem of the thermometer.
    e. For large freezers or refrigerators it is good practice to check the temperature at more than one location.

have many applications, however, such as for checking the temperature of chemicals in a photography darkroom and as oven thermometers in the home. Note that bimetallic laboratory thermometers should be discarded if the stem is bent or deformed.

D. Change-of-State Indicators

Change-of-state indicators *are varied products, all of which change color or form when exposed to heat.* For example, items can be painted with lacquers that change appearance when heated or marked with crayons that melt when a particular temperature is reached. Labels are available that can be attached to items and change colors at specific temperatures, Figure 17.7a. There are encapsulated liquid crystals that change to various colors when exposed to particular temperatures. Crystals change colors reversibly and so can be reused. A color guide is supplied with the crystals to indicate the temperature reached, Figure 17.7b. Crystals have a fairly narrow temperature range, but are useful at temperatures applicable to biological systems.

E. Resistance Thermometry: Metallic Resistance Thermometers and Thermistors

Metallic resistance thermometers and **thermistors** are two types of thermometers based on the principle that the electrical resistance of materials changes as their temperature changes. **Metallic resistance thermometers (RTDs)** *use metallic wires in which resistance increases as the temperature increases.* **Thermistors** *use semiconductor materials in which the resistance decreases as the temperature rises.*

Platinum wire resistance thermometers are considered to be the most accurate type of thermometer available. They are frequently used as references against which other thermometers are calibrated.

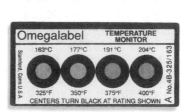

(a)   (b)

Figure 17.7. Change-of-State Indicators. a. A label that changes colors at specific temperatures. **b.** Liquid crystal display. (Photos © copyright Omega Engineering, Inc. All rights reserved. Reproduced with the permission of Omega Engineering, Inc., Stamford, CT 06907.) *(See color insert C-1.)*

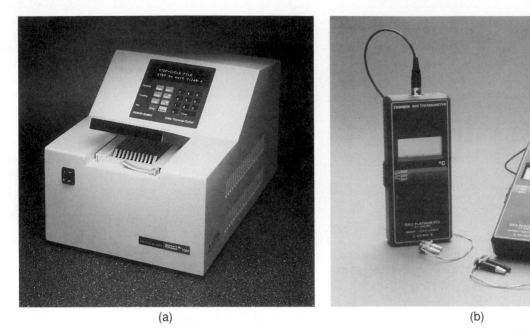

(a) (b)

Figure 17.8. Platinum Resistance Thermometer Used to Verify the Temperature in a Thermocycler.
Thermocyclers must accurately and reproducibly provide the proper temperature. A manufacturer developed a special platinum resistance probe to enable users to verify the temperatures in their instruments routinely. The probe is designed to fit into a sample well and accurately record the temperature that is applied to a reaction tube.
a. Thermocycler. **b.** Thermometers consisting of a platinum resistance probe connected to a meter by a cable. (Courtesy of Perkin Elmer.)

Figure 17.8 shows an example of how a platinum resistance thermometer is used for a biological application. A platinum resistance thermometer was designed specifically to verify the temperature in a thermocycler. Amplifying DNA by PCR requires that a sample mixture be cycled between specific temperatures for set lengths of time. **Thermocyclers** *are instruments that hold multiple tubes containing reaction mixtures and that repeatedly heat and cool the tubes to specified temperatures.* A thermocycler must reliably achieve the correct temperatures; otherwise, the PCR reaction is adversely affected. Hence, there is need for an accurate thermometer to measure temperatures in the thermocycler.

Thermistors are constructed of a hard, ceramic-like material made up of a compressed mixture of metal oxides. The material can be molded into many shapes. Thermistors respond rapidly to temperature changes, can be miniaturized, function in a wide range of temperatures, and are very accurate. Thermistors are often used in a production setting as part of a temperature control device.

F. Thermocouples

A **thermocouple** *consists of two wires made of different metals that are joined together,* Figure 17.9. It is observed that the presence of two dissimilar wires joined in this way generates a voltage. The magnitude of the voltage depends on the temperature at the junction between the wires. The voltage that develops in the wires, therefore, can be converted to a temperature measurement.

Thermocouples are rugged and versatile. They come in various types that are assigned a letter designation depending on the types of metals used in their construction. For example, a Type T thermocouple has a copper wire and a wire made from a copper-nickel alloy. This type of thermocouple is used from 0 to 350°C. Each type of thermocouple is useful in different applications and has a different temperature range.

Figure 17.10 shows a thermocouple probe that is connected to a continuous recording device, thus providing continuous temperature documentation.

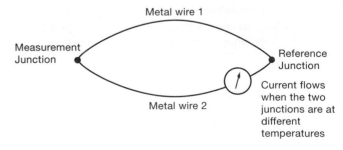

Metal wire 1

Measurement Junction

Reference Junction

Current flows when the two junctions are at different temperatures

Metal wire 2

Figure 17.9. A Simplified Thermocouple.

Figure 17.10.   A Thermocouple Connected to a Continuous Recording Device. (© Copyright 1998 Cole-Parmer Instrument Co.; used by permission.)

III.   VERIFYING THE PERFORMANCE OF LABORATORY THERMOMETERS

Thermometers have many uses in biology laboratories, including checking the temperatures of solutions and monitoring the temperatures of water baths, incubators, ovens, refrigerators, and freezers. It is good laboratory practice to verify that the thermometers used to perform these tasks are performing acceptably. Performance verification of a thermometer demonstrates that its readings are correct within its specified tolerance range.

For most thermometers, about 95% of all possible malfunctions will affect the ice-point reading (*Traceable Temperatures: An Introduction to Temperature Measurement and Calibration,* J.V. Nicholas and D.R. White. John Wiley and Sons, Chichester, England, 1994). At the ice point, a Celsius thermometer should read at the "0" mark and a Fahrenheit thermometer should read at the "32" mark; otherwise, the thermometer is not performing adequately. Box 3A describes a procedure for checking whether a thermometer is reading correctly at the ice point.

An ice point check is useful, but it only verifies the performance of a thermometer at one temperature. It is possible to check a thermometer more thoroughly by comparing its readings with those of a standard thermometer that is certified to be "NIST-traceable." A NIST-traceable thermometer is manufacturer-calibrated against a standard thermometer that, in turn, was calibrated at NIST. The readings of the thermometer that is to be checked are compared with the readings of the NIST-traceable thermometer at several temperatures spanning the thermometer's range. A procedure to check the performance of a thermometer by comparing it with a NIST-traceable thermometer is shown in Box 3B.

Box 3   VERIFYING THE PERFORMANCE OF A THERMOMETER

A. Checking the Ice Point* of the Thermometer

1. *Visually inspect the thermometer whose performance is to be tested.* Check the column for separation or bubbles. Reunite the column and remove bubbles present.

2. *Make certain the outside of the thermometer is clean.*

3. *Prepare an ice point apparatus:*

   a. *A Dewar flask (a vacuum-insulated flask for holding cold materials) is preferable to contain the ice point apparatus.*

   b. *Fill the flask one third full with distilled water. Then add ice shaved or smashed into tiny chips.* The ice should be made from distilled water.

   c. *Compress the ice water mixture into a tightly packed slush.* It is necessary to maintain tightly packed ice so that the thermometer is in close thermal contact with the ice-water mixture. Lack of good thermal contact is the major potential source of error in this procedure. Use only enough water to maintain good contact with the thermometers. The ice should not float. Remove excess water. It may be helpful to place a siphon tube that touches the bottom of the flask. The purpose of the siphon is to drain off excess water that collects at the bottom of the Dewar. The ice should appear clear and not white. White ice is a sign that the ice is at a temperature slightly below zero degrees. Avoid handling the ice so as not to contaminate it.

   d. *Wait 15–20 minutes for the mixture to reach equilibrium.*

   e. Periodically remove water from the bottom of the flask and add more ice.

Box 3   *continued*

---

4. *Immerse the thermometer in the apparatus to a depth approximately one scale division below the 0°C grada-tion.* Allow the thermometer to equilibrate to the temperature of the ice bath. The thermometer should read 0°C. It is recommended to use a 10X magnifying glass to read the scale display. Be certain your eyes are level with the display. The correct temperature is read from the top of the meniscus for a mercury thermometer.

B. Checking a Thermometer against a Certified Reference Thermometer at Temperatures other than the Ice Point

If a certified reference thermometer is available, then the thermometer can be checked against the reference at various temperatures. The certified standard thermometer should have calibrated intervals ten times more pre-cise than the working thermometer. (For example, to verify a thermometer with 0.1°C intervals, use a reference with intervals of 0.01°C.)

1. *Adjust a stable water heating bath to the temperature required for the analysis of interest.*

2. *Place the reference thermometer and the thermometer to be checked in the water.* The thermometers should be placed close to one another, but with sufficient space between them to ensure adequate circulation in the bath. Immerse the thermometers to their proper depth and allow time for them to equilibrate to the temperature of the bath.

3. *After thermal equilibrium is reached (several minutes for liquid-in-glass thermometers) determine the tem-perature reading for both thermometers.* Note that reference thermometers come with a report of calibration that may include a correction factor. If there is a correction, then add or subtract that factor to or from the reading of the reference thermometer to obtain the actual temperature. It is good practice to take several read-ings of each thermometer and average them.

4. *Using a heated liquid bath, sequentially adjust the temperature of the bath to read each temperature for which the reference thermometer is calibrated.* For example, if the reference thermometer is calibrated at 25, 30, and 37°C, then sequentially adjust the water bath to these temperatures, as shown on the reference thermometer. Place the thermometer(s) whose performance is being checked into the water. The reading of the thermometer being checked should be the temperature of the water bath within the tolerance for that thermometer.

5. *Document the results of performance verification in a logbook or on appropriate forms.*

Based on ASTM Standard E 77-92 "Standard Test Method for Inspection and Verification of Thermometers."

*The freezing (ice) point of water is essentially independent of the environment, but the boiling point of water depends on the atmospheric pressure; therefore, the freezing point is more convenient for verification than the boiling point.

A thermometer should give the correct readings at the ice point and/or when compared with a standard thermometer. (Recall, however, that a given thermome-ter may have a fairly wide tolerance range.) Liquid expansion thermometers cannot be repaired or ad-justed (except to reunite the liquid column). If the readings of a general purpose liquid expansion ther-mometer fall outside its tolerance range, it is recom-mended that the thermometer be discarded.

Note that standard mercury thermometers used to verify the performance of other thermometers may have correction factors. This means that the reading of the thermometer was tested at a specific tempera-ture and was found to be slightly "off" by the amount of the correction factor. The correction factor is added to or subtracted from the reading of the stan-dard thermometer to get the correct temperature. Thus, standard thermometers that have known cor-rections are not discarded even if they are marked slightly imperfectly.

PRACTICE PROBLEMS

1. A thermocycler is adjusted to cycle a reaction mix-ture between the temperatures shown. Convert these temperatures to °F.

94°C	(for 1 minute)
37°C	(for 1 minute)
72°C	(for 1 minute)

2. A thermophilic bacterium is found in nature at tem-peratures near 81°C. Convert this to °F.

3. A child is running a fever of 103.2°F. Convert this to °C.

4. **a.** The relationship between resistance and tempera-ture in a platinum (Pt) wire is shown in the fol-lowing graph. If a platinum resistance thermome-ter is calibrated using two reference points (e.g.,

the freezing and boiling points of water), would you expect the calibration to result in accurate results?

b. The relationship between resistance and temperature in a nickel (Ni) wire is shown in the graph. If a resistance thermometer were made using a nickel wire and if it was calibrated using two reference points (e.g., the freezing and boiling points of water), would you expect the calibration to result in accurate results?

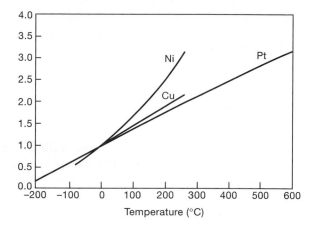

(From "Traceable Temperatures: An Introduction to Temperature Measurement and Calibration," J.V. Nicholas and P.R. White. Copyright John Wiley and Sons Ltd. Reproduced by permission)

5. A 24-hour temperature recording from a fermenter is shown in the following figure. What was the temperature at 1:00 A.M.? What was the lowest temperature in the fermenter? What was the highest temperature? Is this an analogue or a digital form of a display?

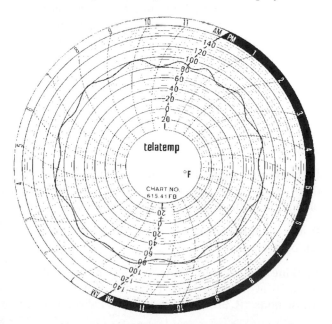

(Courtesy of Telatemp Corp., Fullerton, CA.)

DISCUSSION QUESTION

If you work in a laboratory, examine the laboratory thermometers and answer these questions. For liquid expansion thermometers, what is the filling liquid? What is the thermometer's range? Identify each thermometer as either partial or total immersion. If a thermometer is a partial immersion type, what is its specified immersion depth? Do any of your thermometers have an auxiliary scale? Determine whether the laboratory thermometers have an expansion bulb.

Are there electronic thermometers in your laboratory? If so, what is their purpose?

TERMS RELATED TO THE MEASUREMENT OF TEMPERATURE

Absolute Zero. The temperature at which thermal energy is virtually nonexistent; defined as 0 kelvin or −273.15°C.

Ambient Temperature. The average temperature of the surrounding air that contacts the instrument or system being studied; often room temperature.

Atmospheric Pressure. Pressure due to the weight of the air that comprises the atmosphere and presses down on every object on earth.

Auxiliary Scales. Extra scale markings at zero degrees and 100° to assist in calibration and verification of the performance of a thermometer.

Bimetallic Expansion Thermometers. Type of thermometer made of two different metals, each of which expands and contracts to a different extent as the temperature changes.

Boiling Point. The temperature at which a substance in the liquid phase transforms to the gaseous phase (under specified conditions of pressure).

Bulb (of a liquid-in-glass thermometer). A thin glass container at the bottom of a thermometer that is a reservoir for the liquid.

Calibration (of a thermometer). The process by which the display of a thermometer is associated with reference temperature values.

Celsius Scale (also previously called centigrade scale), C. A temperature scale defined so that 0° is the temperature at which pure water freezes and 100° is the temperature at which water boils.

Change of State Indicators. Products that change color or form when exposed to heat.

Complete Immersion Thermometer. A liquid expansion thermometer that is designed to indicate temperature correctly when the entire thermometer is exposed to the temperature being measured. Compare with "Partial Immersion Thermometer" and "Total Immersion Thermometer." (Based on definition in ASTM Standard E 344-96 "Terminology Relating to Thermometry and Hydrometry.")

Contraction Chamber (of a liquid-in-glass thermometer). An enlargement of the capillary bore that holds some of the liquid volume, thereby allowing the overall length of the thermometer to be reduced.

Degree. An incremental value or division in a temperature scale. For example, the Celsius scale is divided so that there are 100° between the freezing point and the boiling point of water, whereas the Fahrenheit scale is divided into 180° between the same two reference points.

Emergent Stem Correction. A method that corrects for the error introduced when a liquid-in-glass thermometer is immersed to a different depth than that at which it was calibrated.

Expansion Chamber. An enlargement of the capillary bore at the top of the thermometer to prevent buildup of excessive pressure.

Expansion Thermometers. Thermometers that rely on the expansion and contraction of a material in response to temperature.

Fahrenheit Scale, F. A temperature scale that sets the freezing point of water at 32° and the boiling point at 212°.

Fixed Reference points. Physical systems whose temperatures are fixed by some physical process and hence are universal and repeatable, such as phase transitions.

Freezing Point. The temperature at which a substance goes from the liquid phase to the solid phase.

Freezing Point Depression. A lowering of the freezing point of a liquid due to the addition of impurities.

Heat. The energy associated with the disordered motion of molecules in solids, liquids, and gases; thermal energy.

Heating. The transfer of energy from the object with more random internal energy to an object with less.

Ice Point Check. A method of verifying the performance of a thermometer by checking that it reads 0°C (or 32°F) when it is in an ice bath.

Immersion (of a liquid-in-glass thermometer). The depth to which a thermometer is immersed in the material whose temperature is to be measured.

Immersion Line. A line etched onto the stem of a liquid-in-glass thermometer to show how far it should be immersed into the material whose temperature is to be measured.

Infrared Thermometer. A thermometer that senses the infrared energy emitted by an object, converts the infrared energy into an electrical signal that is in turn converted to a temperature reading.

International Temperature Scale-90, ITS-90. An internationally accepted temperature scale established in 1990 by the 18th General Conference on Weights and Measures. The scale uses reference temperatures based on the freezing, boiling, and triple points of various materials including water, hydrogen, and several metals.

Kelvin Scale. A temperature scale based on the Celsius degree with 100 units between the freezing point and boiling point of water; however, the Kelvin scale sets the zero point as absolute zero, rather than the freezing point of water. 0°C = 273.15 K. (Units are in kelvin, K.)

Liquid Expansion Thermometer. See "Liquid-in-Glass Thermometer."

Liquid-in-Glass Thermometer (also called Liquid Expansion Thermometer). A thermometer in which liquid, usually mercury or alcohol, expands or contracts with temperature changes and moves up or down the bore within a glass tube.

Melting Point. The temperature at which a substance transforms from the solid phase to the liquid phase.

Mercury Thermometer. A liquid expansion thermometer containing liquid mercury.

Metallic Resistance Thermometer. Thermometers having metal wires whose resistance increases as the temperature increases.

Partial Immersion Thermometer. A liquid expansion thermometer designed to indicate temperature correctly when the bulb and a specified part of the stem are exposed to the temperature being measured. Compare with "Total Immersion Thermometer" and "Complete Immersion Thermometer." (Based on definition in ASTM Standard E 344-96 "Terminology Relating to Thermometry and Hydrometry.")

Phase Transition. Condition where a liquid turns to a solid or a vapor, or a vapor turns to liquid.

Platinum Resistance Thermometer. A metallic resistance thermometer based on the fact that the resistance in a platinum wire increases as the temperature increases.

Probe (temperature). A generic term for many types of temperature sensors.

Rankine Temperature Scale. A temperature scale which places 0 at absolute zero (like the Kelvin scale) and whose units, °R, are the same size as a Fahrenheit degree.

Reference Junction. The cold junction in a thermocouple circuit that is held at a stable known temperature.

Reference Point. A temperature at which the reading of a thermometer is calibrated or verified.

Scale (of a thermometer). Graduations that indicate degrees, fractions of degrees, or multiples of degrees.

Spirit Thermometer. A liquid expansion thermometer containing an alcohol-based liquid.

Stem (of a liquid-in-glass thermometer). A glass capillary tube through which the mercury or organic liquid moves as temperature changes.

Stem Enlargement. A thickening of the stem that assists in the proper placement of the thermometer in a device (e.g., in an oven).

Temperature. A measure of the average energy of the randomly moving molecules that make up a substance. The tendency of a substance to lose or gain heat.

Temperature Scale. A scale derived by choosing two or more fixed reference points.

Thermal Conductivity. The ability of a material to conduct heat.

Thermal Equilibrium. The condition where two objects are at the same temperature, so that there is no net transfer of internal energy from one to the other.

Thermal Expansion. An increase in the size of a material due to an increase in temperature.

Thermal Expansion Coefficient. The amount that a particular material expands with a given rise in temperature.

Thermistor. A thermometer consisting of a semiconductor material that has a large change in resistance when exposed to a small change in temperature.

Thermocouple. A temperature-sensing device that consists of two dissimilar metals joined together. The thermocouple has a voltage output proportional to the difference in temperature between the junction whose temperature is being measured and the reference junction.

Thermocycler. An instrument that holds multiple tubes and which repeatedly heats and cools the tubes to specified temperatures.

Total Immersion Thermometer. A liquid expansion thermometer designed to indicate temperature correctly when that portion of the thermometer containing the liquid is exposed to the temperature being mea-

sured. Compare with "Partial Immersion Thermometer" and "Complete Immersion Thermometer." (Based on definition in ASTM Standard E 344-96 "Terminology Relating to Thermometry and Hydrometry.")

Triple Point of Water. The temperature and pressure at which solid, liquid, and gas phases of a given substance are simultaneously present in equilibrium. The triple point for water is 0.01°C.

REFERENCES RELATING TO THE MEASUREMENT OF TEMPERATURE

Traceable Temperatures: An Introduction to Temperature Measurement and Calibration, J.V. Nicholas and D.R. White. John Wiley and Sons, Chichester, England, 1994. (A more advanced book covering temperature measurement and metrology in general.)

Physics for the Health Sciences, Carl R. Nave and Brenda C. Nave. Third edition. W.B. Saunders Company, Philadelphia. 1985. (A general physics reference with useful information for biologists.)

Relevant ASTM Standards:

ASTM Standard E 1-95 "Standard Specifications for ASTM Thermometers."

ASTM Standard E 77-92 "Standard Test Methods for Inspection and Verification of Thermometers."

ANSWERS TO PROBLEMS

1. 94°C = 201.2°F
   37°C = 98.6°F
   72°C = 161.6°F

2. 81°C = 177.8°F

3. 103.2°F = 39.6°C

4. a. The relationship between resistance and temperature appears to be linear for platinum; therefore, a two-point calibration will give accurate results.

   b. This is not the case for nickel.

5. 80°F. Lowest temperature is about 67°F. Highest temperature is about 100°F. Analogue.

CHAPTER 18

The Measurement of pH, Selected Ions, and Conductivity

I. THE IMPORTANCE AND DEFINITION OF pH

The chemistry of life is based on water. Life evolved in the presence of water and cells contain 80–90% water. **pH**, *which is a measure of the acidity of an aqueous solution*, is intrinsically related to water chemistry. Maintaining the proper pH is essential for living systems. For example, plants do not grow if the soil pH is wrong, animals die if their blood pH deviates from normal, and microorganisms require their growth medium to be a particular pH. Many industrial processes, such as food processing, sewage treatment, water purification, and pharmaceutical production, are likewise sensitive to pH. Biological systems and many industrial processes, therefore, must be monitored and maintained at the proper pH. This chapter briefly defines pH and discusses how it is measured.

Pure water naturally dissociates to a limited extent to form **hydrogen (H^+)** and **hydroxide (OH^-)** ions:

$$H_2O \leftrightarrows H^+ + OH^-$$

In pure water at 25°C, an equilibrium is established such that the concentration of H^+ ions is the same as the concentration of OH^- ions; it is 1×10^{-7} *mole/L*. (See p. 452 for an explanation of molarity.) Pure water, and other solutions with a H^+ ion concentration of 1×10^{-7} *mole/L*, are called **neutral**.

When **acids** *are added to water, they release hydrogen ions into solution.* On the other hand, when **bases** *are added to water, they cause hydrogen ions to be removed from solution.* Thus, the addition of acids to water causes the concentration of hydrogen ions to be greater than 1×10^{-7} *mole/L*. The addition of bases to water causes the concentration of hydrogen ions to be less than 1×10^{-7} *mole/L*.

Strong acids and **strong bases** completely dissociate when dissolved in water and therefore have a strong effect on the hydrogen ion concentration. For example, hydrochloric acid (HCl) dissociates completely to release H^+ ions and Cl^- ions, thus increasing the H^+ ion concentration of the solution:

$$HCl \text{ in water} \rightarrow H^+ + Cl^-$$
(all is in this form)

NaOH is an example of a strong base that dissociates completely in water as follows:

$$NaOH \text{ in water} \rightarrow Na^+ + OH^-$$
(all is in this form)

The OH^- ions react with H^+ ions to form water, thereby lowering the concentration of hydrogen ions.

Weak acids and **weak bases** do not completely dissociate in water and therefore have a smaller effect on the concentration of hydrogen ions in the solution. For example, acetic acid forms about one hydrogen ion for every 100 molecules:

$$CH_3COOH \text{ in water} \leftrightarrows H^+ + CH_3COO^-$$
(most of the acid is in this form)

The weak base, ammonia (NH_3) removes hydrogen ions from water with the formation of OH^- ions:

$$NH_3 + H_2O \leftrightarrows NH_4^+ + OH^-$$
(most of the base is in this form)

pH is a convenient way to express the hydrogen ion concentration* in a solution, or the solution's acidity. *pH is the negative log of the H^+ concentration when concentration is expressed in moles per liter.*

To illustrate the definition of pH, consider the pH of pure water:

The H^+ concentration of pure water is 1×10^{-7} *mole/L*.
The log of 1×10^{-7} is -7.
The negative log of 10^{-7} is $-(-7) = 7$.
The pH of pure water is 7.

Thus, the H^+ concentration of pure water is 10^{-7} *mole/L* and its pH is 7. By definition, the pH of any neutral solution is 7. The practical range of pH is between 0 and 14. (Very concentrated solutions of acid or base can fall outside this pH range and require special care when handling to protect both the technician and the equipment.) Solutions with a pH less than 7 are acidic while those with a pH greater than 7 are basic.

The ultimate importance of H^+ concentration is that it profoundly affects the properties of aqueous solutions. For example, a strongly acidic solution, such as concentrated sulfuric acid, can dissolve iron nails. (Strong acids and bases thus require safety precautions in the laboratory.) A weaker acid, citric acid, contributes to the taste of foods like lemon and orange juice, Table 18.1.

The pH scale is logarithmic. For example, a solution with a pH of 5.0 has 10 times more hydrogen ions than a solution with a pH of 6.0 and a solution of pH 7.0 is 100 times less acidic than a solution at pH 5.0. As hydrogen ion concentration, acidity, increases, the pH value decreases, Table 18.2.

Because maintaining the proper pH is critical to biological systems, methods are required to measure the concentration of hydrogen ions in solutions. Two common approaches to pH measurement in the laboratory

---

*The term *pH* properly should be defined as the negative log of the H^+ activity, not its concentration. The activity of hydrogen ions is primarily related to their concentration, but it is also affected by other substances in the solution, by the solvent, and by the temperature of the solution. Defining pH in terms of the apparent concentration of hydrogen ions is an approximation that is very commonly used and is adequate for most purposes in a biology laboratory.

Table 18.1 *THE pH OF COMMON COMPOUNDS*

pH Value	Compounds
0.3	Sulfuric acid (battery acid)
2	Lemon juice
3	Wines
3–4	Oranges
4–5	Beers
5–6	Measured pH of purified laboratory water
7	Pure water
7.3–7.5	Blood (human)
7–8	Egg whites
8	Sea water
8.4	Sodium bicarbonate
10	Milk of magnesia
11.6	Household ammonia
12.6	Bleach
14	1 *M* Sodium Hydroxide

Table 18.2 *THE RELATIONSHIP BETWEEN [H^+] AND pH*

Hydrogen Ion Concentration (*mole/L*)		pH
Acidic		
10^0	1.0	0
10^{-1}	0.1	1
10^{-2}	0.01	2
10^{-3}	0.001	3
10^{-4}	0.0001	4
10^{-5}	0.00001	5
10^{-6}	0.000001	6
Neutral		
10^{-7}	0.0000001	7
Basic		
10^{-8}	0.00000001	8
10^{-9}	0.000000001	9
10^{-10}	0.0000000001	10
10^{-11}	0.00000000001	11
10^{-12}	0.000000000001	12
10^{-13}	0.0000000000001	13
10^{-14}	0.00000000000001	14

are the use of **indicator dyes** and the use of **pH meter/electrode** measuring systems. We discuss first indicator dyes and then pH meter measuring systems.

EXAMPLE PROBLEM

a. What is the pH of a solution with an H^+ ion concentration of 10^{-4} *mole/L*?

b. What is the pH of solution with an H^+ ion concentration of 5.0×10^{-6} *mole/L*?

ANSWER

a. $pH = -\log [H^+] = -\log 10^{-4} = -(-4) = 4$

b. $pH = -\log [H^+] = -\log 5.0 \times 10^{-6} = -(-5.3) = 5.3$

EXAMPLE PROBLEM

What is the concentration of H^+ ions in a solution with a pH of 9.0?

ANSWER

$pH = -\log [H^+]$
$9.0 = -\log [H^+]$
$-9.0 = \log [H^+]$
antilog $(-9.0) = 1 \times 10^{-9}$ *mole/L*

II. pH INDICATORS

pH indicators *are dyes whose color is pH dependent and which change colors at certain pH values.* Indicators can be directly dissolved in a solution or can be impregnated into strips of paper which are then dipped into

the solution to be tested. **Phenolphthalein** is a well-known indicator dye (and formerly the main ingredient in some laxatives). Phenolphthalein changes from colorless to red between pH 8 and 10. **Litmus**, which is extracted from lichens, is probably the oldest pH indicator. Litmus paper is pink in acidic solutions and blue in basic ones.

Indicators do not change colors sharply at a single pH; rather, they change over a range of one or two pH units. Thus, indicators are often used to broadly categorize a solution as acidic, basic, or neutral, or to determine roughly the pH of a solution. It is also possible to purchase pH indicators or pH paper that contain mixtures of dyes. These mixtures exhibit a range of colors and can be used to determine the pH of a sample to ± 0.3 pH units.

Indicators are a useful means of monitoring pH for some applications. For example, human cells can be grown in the laboratory in nutrient medium. Proper pH of the medium is critical for cell survival but is easily altered by metabolites from the cells, contamination from microorganisms, and improper culture conditions. Therefore a nontoxic, nonreactive pH indicator dye, phenol red, is added to cell culture media. Changes in the indicator dye color provide immediate, visible information about the condition of the culture without opening the culture plates and exposing them to potential contaminants. When the medium is a certain shade of pink, the culture is healthy. A yellowish tinge indicates that the medium is becoming acidic and requires attention. Another example is the use of pH paper to measure the pH of toxic or radioactive solutions. The paper can be dipped in the hazardous solution and discarded, thus avoiding contamination of a nondisposable pH measuring device.

pH indicators are inexpensive and simple to use; however, pH indicators show, at best, the pH of a solution within a range of about 0.3 pH units. In addition, indicators cannot be used in certain types of solutions. pH meter measuring systems should be used in situations where pH indicators provide inadequate information. Table 18.3 summarizes potential sources of error when using pH indicators. Table 18.4 lists common pH indicators.

III. THE DESIGN OF pH METER/ELECTRODE MEASURING SYSTEMS

A. Overview

pH meter/electrode *systems are the most common method of measuring pH in the biology laboratory.* Although more expensive and more complicated to use than chemical indicators, pH meters provide greater accuracy, sensitivity, and flexibility. When used properly, pH meters can measure the pH of a solution to the nearest 0.1 pH unit or better, and they can be used with a variety of samples.

A pH meter/electrode measuring system consists of (1) a **voltmeter** that measures voltage, (2) two **electrodes** connected to one another through the meter, and (3) the **sample** whose pH is being measured. The term **pH meter** *is commonly used to refer both to the voltage meter and its accompanying electrodes.*

When the two electrodes are immersed in a sample, they develop an electrical potential (voltage) that is measured by the voltmeter. (The Appendix to this chapter has a brief explanation of how electrodes develop potential.) The magnitude of the measured voltage depends on the hydrogen ion concentration in

Table 18.3 *Potential Sources of Error when Using pH Indicators*

1. *Acid–base error.* Indicators are themselves acids or bases. When indicators are added to unbuffered or weakly buffered solutions, such as distilled water or very dilute solutions of strong acids or bases, the indicator causes a change in pH of the sample solution.

2. *Salt error.* Salts at concentrations above about 0.2 *M* can affect the color of pH indicators, leading to inaccuracy.

3. *Protein error.* Proteins react with indicators and can have a significant effect on their color. Indicators, therefore, should not be used to measure the pH of protein solutions.

4. *Alcohol error.* The solvent in which a sample is dissolved can affect the color of an indicator; therefore, a sample dissolved in alcohol may be a different color than if it is dissolved in buffer.

Table 18.4 Examples of Common Indicators

Indicator	Color Change
Blue Litmus	Changes from blue to red denoting a change from alkaline to acid
Brilliant Yellow	Changes from yellow at pH 6.7 to red at pH 7.9
Bromocresol Green	Changes from yellow at pH 4.0 to blue at pH 5.4
Bromocresol Purple	Changes from yellow at pH 5.2 to purple at pH 6.8
Congo Red	Changes from red at pH 3.0 to blue at pH 5.0
Cresol Red	Changes from yellow at pH 7.2 to red at pH 8.8
Methyl Red	Changes from red at pH 4.2 to yellow at pH 6.2
Neutral Litmus	Red in acid conditions and blue in alkaline conditions
Neutral Red	Changes from red at pH 6.8 to orange at pH 8.0
Phenol Red	Changes from yellow at pH 6.8 to red at pH 8–8.2
Phenolphthalein	Changes from colorless at pH 8.0 to red at pH 10.0
Red Litmus	Changes from red to blue denoting a change from acid to alkaline

The U.S. Pharmacopeia ("U.S. Pharmacopeia/National Formulary," The United States Pharmacopeial Convention, Rand McNally Publishers, 1995) is a good source of information about pH indicators, their chemical nature, solubility, and preparation, because indicators are used in testing drugs.

the solution; therefore the voltage reading can be converted by the instrument to a pH value, Figure 18.1.

B. The Basic Design and Function of Electrodes

i. Electrodes and the Measurement of pH

The electrodes are the heart of the pH measuring system and require some discussion. A pH meter is connected to two electrodes. One is a **pH measuring electrode**, the other is a **reference electrode**. The **pH measuring electrode** has a thin, fragile glass bulb at its tip and is therefore often called a **glass electrode**, Figure 18.2. The glass that composes the bulb is of a special type that is sensitive to the concentration of H^+ ions in the surrounding medium. An electrical potential develops between the inner and outer surfaces of the glass bulb when the measuring electrode is immersed in a sample solution. It is this potential whose magnitude varies depending on the concentration of hydrogen ions in the solution. The glass electrode contains a buffered chloride solution. A metal wire inside the electrode acts as a lead connecting the glass bulb to the voltmeter.

You might wonder why two electrodes are required because the glass measuring electrode by itself is sensitive to pH. In practice, it has been found that it is not pos-

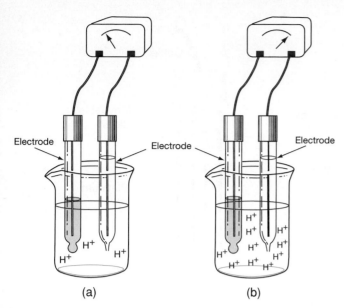

(a)   (b)

Figure 18.1.   Overview of a pH Meter Measuring System.
The system consists of two electrodes that are immersed in
the sample and which are connected to one another through
a meter. **a.** The concentration of H$^+$ ions in the solution is
relatively low. **b.** A higher concentration of H$^+$ ions leads
to a change in the voltage difference between the two
electrodes.

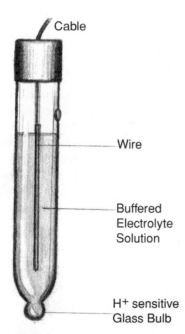

Figure 18.2.   pH Measuring Electrode. (Glass Electrode).
The bulb of this electrode is composed of glass that develops
an electrical potential when placed in solution. The
magnitude of the potential depends on the concentration of
hydrogen ions. (Sketch by a Biotechnology student.)

sible to measure the potential of a single electrode. The
pH measuring electrode must be paired with a second,
reference electrode, *which has a stable, constant voltage.*

The reference electrode has a strip of metal and an
electrolyte solution (filling solution) enclosed in a glass

or plastic tube. **Electrolytes** *are substances, such as acids,
bases, and salts, that release ions when dissolved in water.*
In the reference electrode, the metal and electrolyte
react with one another to develop a constant voltage.

The two electrodes are placed in the sample whose
pH is to be measured. The following events occur when
electrodes are in the sample solution:

1. The electrical potential (voltage) of the glass elec-
    trode varies depending on the sample's hydrogen
    ion concentration.
2. The electrical potential of the reference electrode
    is constant.
3. The two electrodes are connected to one another
    through the voltage meter.
4. The meter measures the voltage difference be-
    tween the reference electrode and the measuring
    electrode. This voltage is an electrical signal.
5. The electrical signal is amplified and converted to
    a display of pH.

ii. MORE ABOUT REFERENCE ELECTRODES

The reference electrode is the component of the pH
measuring system that is most likely to cause difficulties; therefore, it is important to be familiar with reference electrode design. Observe in Figure 18.3a that the
end of the reference electrode that is submerged in the
sample has a *porous plug*, *called a* **junction**, or **salt
bridge**. The filling solution, containing salts, flows slowly
out of the reference electrode, through the junction,
into the sample. This slow flow of ions through the junction is necessary to maintain electrical contact between
the reference electrode and the sample solution whose
pH is being measured. This flow of ions constitutes electrical current. (Recall that electrical current is a flow of
charge. When current flows in a solution, it consists
of ions. This is in contrast to the flow of electricity in a
wire, where electrical current is due to the movement of
electrons.) Because electrolyte flows out of the electrode, the sample solution cannot enter the junction
and contaminate the electrode. The flow of filling solution is normally slow enough that it does not significantly contaminate the sample whose pH is being
measured. There are different classes of junctions that
vary in their flow rates and design. Classes of junction
are described in Table 18.5.

Reference electrode filling solution is lost through
the junction, so many electrodes have a **filling hole** that
is used to replenish the electrolyte, Figure 18.3a. Other
reference electrodes are manufactured with a gel-like
filling solution that cannot be replenished. With any reference electrode, pH cannot be properly measured if
the electrolyte is depleted or if the junction is blocked
so that electrolyte does not contact the sample.

There are two different classes of reference electrodes for pH meters: silver/silver chloride (abbreviated

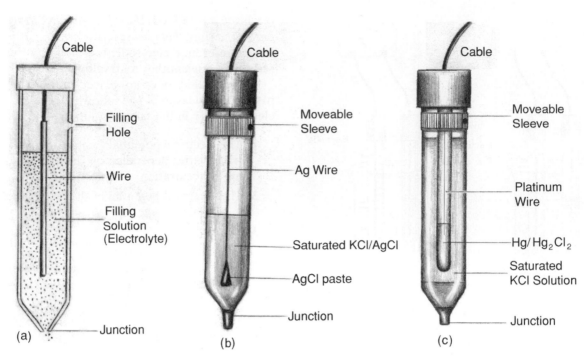

Figure 18.3. Reference Electrodes. a. The basic features of a reference electrode. **b.** A silver/silver chloride reference electrode. **c.** A calomel reference electrode. (Sketches by a Biotechnology student.)

Ag/AgCl) and calomel. An **Ag/AgCl electrode** *contains a strip of silver coated with silver chloride*. The strip is immersed in an electrolyte solution which is normally saturated in KCl and silver chloride, Figure 18.3b. A **calomel electrode** *contains a platinum wire coated with calomel, which is a paste consisting of mercury metal and mercurous chloride*. The calomel is in contact with a filling solution that is normally saturated KCl, Figure 18.3c.

Reference electrodes must maintain a constant electrical potential. Both Ag/AgCl and calomel reference electrodes are designed so that their potential is sensitive to the concentration of Cl^- in their electrolyte filling solution. The concentration of Cl^- in the solution is usually held constant by using a filling solution that contains saturated KCl. Because the filling solution is saturated it is not unusual to see KCl crystals on the end of reference electrodes.

The pH meters in biology laboratories usually appear to have only one electrode, not two. This is because they have convenient, compact, **combination electrodes. Combination electrodes** *are made by combining the reference and pH measuring electrodes into one housing*, Figure 18.4 a,b. Combination electrodes work efficiently for most purposes but are not suitable for measuring the pH of certain "difficult" samples. (Difficult samples will be discussed later in this chapter.)

pH meters can be made still more compact by combining the electrodes and the meter into a single housing, Figure 18.5. These compact pH meter systems are useful for field work and for making rapid measurements.

C. Selecting and Maintaining Electrodes

i. SELECTING ELECTRODES

There are many variations of the basic electrode designs discussed above. Electrodes come in different sizes, shapes, and brands and have features that vary from one electrode to another. Some electrodes are "general purpose"; others are intended for specific applications. It is best to consult the manufacturers and purchase electrodes that are recommended for the types of samples whose pH is measured in your laboratory. General considerations regarding selecting electrodes are summarized in Table 18.5.

ii. MAINTAINING ELECTRODES

Electrodes require proper care. Manufacturers provide specific information regarding care of their electrodes. General considerations regarding maintenance of electrodes are summarized in Table 18.6.

IV. OPERATION OF A pH METER SYSTEM

A. Calibration

Calibration, also called **standardization**, is a critical step in operating a pH meter because **calibration** *tells the meter how to translate the voltage difference between the measuring and reference electrodes into units of pH*. pH meter systems need to be calibrated every day or every

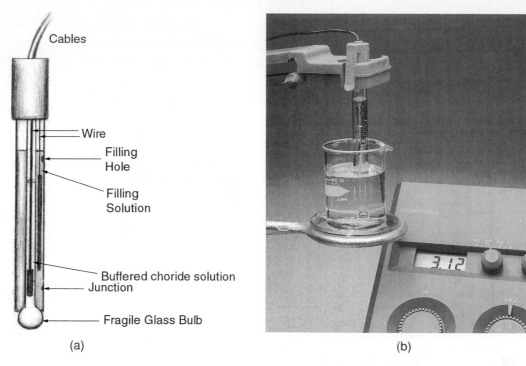

Cables

Wire

Filling Hole

Filling Solution

Buffered choride solution

Junction

Fragile Glass Bulb

(a)

(b)

Figure 18.4.   Combination Electrodes. a. Diagram of a combination electrode. **b.** A combination electrode and pH meter. (Sketch by a Biotechnology student.)

time the system is used because the response of electrodes declines over time.

There is a linear relationship between the voltage measured by the system and the pH of the sample. Because the relationship is linear, two buffers of known pH are used to calibrate a pH meter, Figure 18.8. The

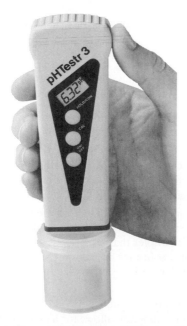

Figure 18.5.   A Compact, Portable pH Meter. (© Copyright 1998 Cole-Parmer Instrument Co. Used with permission.)

first buffer is normally pH 7.00. The second is typically pH 4.00, 10.00, or 12.00, although other standardization buffers are available. If the sample to be measured is acidic, an acidic calibration buffer is used and if the sample is basic, a basic calibration buffer is chosen. If very high accuracy is required, the second standard buffer pH should be within 1.5 pH units of the sample pH.

The general method of calibration is to first immerse the electrodes in pH 7.00 buffer. Using an appropriate knob, dial, keys, or button, the meter is set to display pH 7.00. The meter internally adjusts itself so that a pH of 7.00 corresponds to 0 millivolts (mV). The adjustment for pH 7.00 buffer may be called "set," "zero offset," or "standardize" because the meter "sets to" 0 mV. The electrodes are then placed in the second buffer and by using the appropriate knob, dial, keys, or button the meter is adjusted to display the pH of the buffer. The adjustment for the second buffer may be called "slope," "calibrate," or "gain" because internally the meter uses the second reading to establish a calibration line with a particular slope. The result of this process is that the meter constructs a calibration line that can be used to convert any voltage to its corresponding pH.

The slope of the calibration line is a measure of the response of the electrodes to pH. The steeper the slope, the more sensitive the electrodes are to hydrogen ions. The slope ideally should be close to -59.2 mV/pH unit when the temperature is 25°C, Figure 18.9a. As electrodes age or become dirty, they are less able to generate a potential and the slope of the line declines, Figure 18.9b.

Table 18.5   Considerations in Selecting Electrodes

1. *Size and shape.* It is possible to purchase electrodes in various lengths and shapes to accommodate different samples.

2. *Connections.* Electrodes are connected to pH meters with plug-in cables that must match the meter, Figure 18.6.

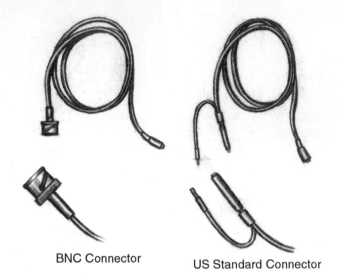

BNC Connector   US Standard Connector

Figure 18.6.   pH Meter Cables and Connectors.

3. *Measuring electrodes.* There are several types of measuring electrode that vary in the type of glass used to make the bulb. A "general purpose" measuring electrode is acceptable unless one is working with difficult samples, such as very pure water, solutions that are poorly conducting, or nonaqueous liquids.

 At pH values above about 9 or 10, measuring electrodes begin to respond to Na^+ and K^+, as well as H^+. This response is termed **alkaline error.** There are specially designed measuring electrodes available that minimize alkaline error at high pH.

4. *Reference electrodes.* "General purpose" reference electrodes are usually Ag/AgCl.

 Calomel electrodes are more resistant than Ag/AgCl to certain solutions, such as Tris buffers, which are frequently used in molecular biology. Calomel electrodes will therefore last longer and are recommended in laboratories where Tris is frequently used.

 Calomel electrodes cannot be used in solutions hotter than about 80°C and they contain mercury, which is hazardous if the electrode is broken.

5. *Junction.* The junction is a part of the reference electrode. Junctions come in various styles, sizes, and shapes to allow faster or slower flow of filling solution. (Flow rates vary from about 0.5 to 100 $\mu L/hr$). The choice of junction depends on the sample. Combination electrodes have a general purpose junction with a moderate flow rate. The classes of junctions include:

 A fibrous junction is composed of a fibrous material such as quartz or asbestos. The fibers may be packed loosely or tightly; the looser the packing, the more rapid the flow of filling solution. Quartz fiber junctions are commonly used as a general purpose junction.

 A fritted junction is a white ceramic material consisting of many small particles pressed closely together. Filling solution can leak through open spaces between the particles. The flow rate across a fritted junction is relatively slow. This type of junction is recommended when the sample should not be contaminated by filling solution or has a pH greater than 13.

 A sleeve junction is made by placing a hole in the side of the glass or plastic housing of the electrode. The hole is covered with a glass or plastic sleeve. The sleeve junction is the fastest flowing type, is the least likely to become clogged, and is easiest to clean, since the sleeve is removable. The sleeve junction is recommended for samples that are viscous or have a high concentration of solids; and for brines, colloidal suspensions, nonaqueous samples, solutions of low ionic strength, strong acids and bases, and strong oxidizing and reducing agents. Faster flowing junctions have the disadvantage that the sample is more contaminated by filling solution than with slower flowing junctions.

 A double junction is a less common and more complex reference electrode configuration, Figure 18.7. It is used when the sample is incompatible with the reference electrode filling solution. The double junction allows for a second filling solution that contacts the sample to be placed outside the normal reference electrode junction. This configuration separates the reference electrode filling solution from the sample.

Table 18.5   *(continued)*

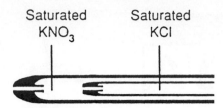

Figure 18.7.   Reference Electrode with a Double Junction. (Courtesy of Radiometer Analytical A/S)

6. *Combination versus separate electrodes.* Combination electrodes are more convenient than separate electrodes; however, there are samples that are difficult to pH and require a type of junction or a type of glass electrode that is not available in the combination configuration. Separate electrodes are necessary to pH such samples.

7. *Solid state electrodes.* New electrode/meter systems are now appearing on the market that incorporate a small silicon chip sensitive to H^+ ions into the head of a solid state electrode. These electrodes are different than conventional glass bulb electrodes. They do not require filling solution, are easily cleaned, and can measure very small volume samples.

Table 18.6   *CONSIDERATIONS IN MAINTAINING ELECTRODES*

1. *Filling solution.* Reference electrodes contain filling solution, generally saturated KCl or AgCl/KCl depending on the type of electrode and the sample.

   Gel-filled electrodes contain gelled filling solution that is never refilled, so the electrode must be discarded when the electrolyte is depleted.

   Refillable electrodes are periodically refilled through the filling hole. Use the filling solution recommended by the manufacturer for the electrode. Fill the electrode nearly to the top of the filling solution chamber to ensure proper flow out the junction.

   The filling hole must be open when a sample's pH is measured so that filling solution can flow from the junction properly. The filling hole must be closed during storage so that filling solution is not lost.

2. *Storage.* There is no single rule for how to store all types of electrodes; therefore, consult the manufacturer's instructions. Some considerations include:

   Silver/silver chloride reference electrodes can be stored in a saturated KCl solution.

   Calomel electrodes may be stored in a 1/10 dilution of their filling solution.

   Glass electrodes should not be stored in solutions that are very high in sodium or potassium chloride because the glass bulb can exchange H^+ ions for potassium or sodium, thus reducing its sensitivity to hydrogen ions.

   A combination electrode must be stored in such a way as to protect both the reference electrode and the glass electrode. A combination electrode with a Ag/AgCl reference should not be stored in a solution with low chloride. The glass bulb should not be stored in solutions with high levels of potassium or sodium chloride. Manufacturers therefore sometimes recommend that combination electrodes stored for long periods of time be stored dry.

   Do not store electrodes in distilled water.

   New electrodes, an electrode stored dry, and electrodes that have been cleaned should be conditioned before use by soaking for at least 8 hours in pH 7 buffer.

3. *Cleaning.* The measuring electrode bulb, the reference junction, and filling solution must be kept clean. Cleaning protocols are in the section on "Trouble-Shooting."

4. *Measuring the pH of strongly acidic or basic solutions.* Take a reading of such solutions quickly and then rinse the electrodes thoroughly.

5. *Nonaqueous samples.* Frequent immersion in nonaqueous samples will cause the glass measuring bulb to dry and no longer function. If measuring the pH of nonaqueous samples, periodically rehydrate the bulb by soaking it in distilled water. Consult the manufacturer for details.

6. *Fragility.* The measuring bulb of an electrode is fragile and cannot be repaired if it is broken or cracked. Keep the electrode bulb away from moving stir bars.

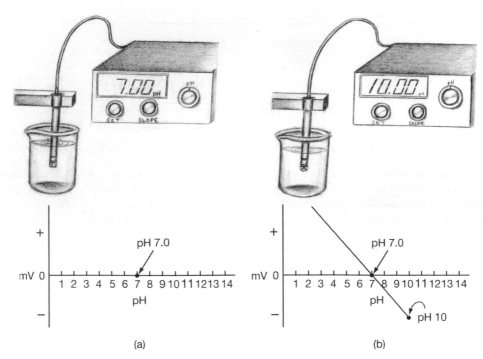

(a)   (b)

Figure 18.8.   pH Meter Calibration. The process of calibration establishes the relationship between pH and mV so that the meter can convert any mV reading to a pH value. **a.** The meter is adjusted to display pH 7 when the electrodes are immersed in pH 7 calibration buffer. **b.** The meter is adjusted to display the value of a second calibration buffer. The slope of the resulting line ideally is $-59.2\ mV$/pH unit at 25°C. (Sketches by a Biotechnology student.)

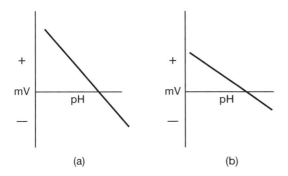

(a)   (b)

Figure 18.9.   Electrode Response and Slope. a. Normal electrode response. **b.** Effect of aging or dirt on electrode response.

As the electrode response deteriorates, the system is eventually no longer able to adjust to the value of both buffers (as discussed below in the "Trouble-Shooting Section").

Some pH meters display the slope of the calibration line as an indication of the condition of the electrodes. The slope is often expressed as a percent of the theoretical value. For example, a 97% slope is equivalent to a slope of $-57.4\ mV$/pH unit.

When operating an older pH meter, the first calibration buffer should be pH 7.00. This means, for example, that it is not recommended to calibrate an older meter using standard buffers of pH 4.00 and pH 1.68, as might be desirable when measuring the pH of acidic samples. Newer, microprocessor-controlled meters often have a feature so that the first buffer does not need to be

pH 7.00 and any two standard buffers that bracket the pH of the samples may be used for calibration. Moreover, some microprocessor-controlled meters allow the use of more than two standard buffers to determine the calibration line. These features of newer instruments can improve the accuracy of pH measurements. Consult the manufacturer's instructions to determine the features of a specific meter.

Good quality calibration buffers are essential for calibration of the pH measuring system. Calibration buffers are commonly purchased as ready to use solutions, often in various colors where each color is a different pH. Tips regarding calibration buffers are in Table 18.7.

B. Effect of Temperature on pH

Temperature is very important in the measurement of pH. Temperature has two effects:

1. The **measuring electrode's response to pH** (the potential that develops) is affected by the temperature, Figure 18.10.

2. The **pH of the solution** that is being measured may increase or decrease as its temperature changes.

It is important to consider both temperature effects when operating a pH meter.

Most pH meters can compensate for the first factor (i.e., temperature dependent changes in the electrode response). The meter needs to "know" the temperature

Table 18.7 *CALIBRATION BUFFERS*

1. ***Reaction with CO₂.*** Some buffers react with CO_2 from the air, resulting in a change in their pH. Buffer containers, therefore, should be kept closed. Remove a small amount of buffer each time an instrument is calibrated and throw it away after use. pH 10.0 buffer is particularly sensitive to CO_2. Its pH will drop to 9.6 with exposure to air.

2. ***Expiration date.*** Premixed buffers come with an expiration date. For best accuracy, do not use buffers after their expiration date.

3. ***Temperature.*** The pH of a buffer will change as its temperature changes. For example, a common standardization buffer has a pH of 7.00 at 25°C, but at 0°C, its pH is 7.12. When calibrating with this buffer at 25°C, set the meter to exactly 7.00, but if the buffer is at 0°, set the meter to 7.12. The pH listed on the container of a commercially purchased buffer is generally its pH at 25°C. Manufacturers can provide a table that lists the pH of a standardization buffer at various temperatures. Allow sufficient time for the electrode to equilibrate to the temperature of the buffer. If the change in temperature is extreme, or the pH of the buffer is very high or low, equilibration may take as long as 3 or 4 minutes.

4. ***Contamination.*** Buffers that are visibly contaminated by mold growth or contain precipitates must be discarded. Never pour buffer back into its original bottle because this will contaminate the entire stock.

of the solution to make this adjustment. It is possible to measure the temperature of a solution with a thermometer and, using the proper knob or dial, "tell" the pH meter the solution's temperature. There are also *electronic temperature probes*, **ATC (automatic temperature compensating) probes**, that can be placed in the sample alongside the pH electrodes. The probes are connected to the pH meter and automatically measure the sample temperature and report it to the meter, which compensates accordingly. Compact devices may even have a temperature probe built into the electrode housing.

The second temperature effect is that the pH of some solutions changes as their temperature changes. For example, if Tris buffer is prepared to pH 8.0 at room temperature, its pH will be close to 7.6 when used at body temperature and 8.8 when placed on ice. It is best, therefore, to measure the pH of a solution when the solution is at the temperature at which it will be used.

When measuring the pH of samples that are not at room temperature, it is common practice to calibrate the meter at room temperature using standardization buffers at room temperature. Then, when it is time to measure the pH of the sample solution, the meter is adjusted to the temperature of the sample and its pH is determined. This practice generally gives sufficiently accurate results.*

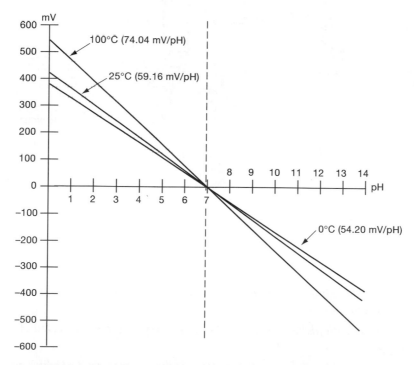

Figure 18.10. The Effect of Temperature on Electrode Response. Observe that the slope of the calibration line changes at different temperatures. (© Copyright Omega Engineering, Inc. All rights reserved. Reproduced with permission Omega Engineering, Inc., Stamford, CT 06907.)

*It is also possible to calibrate a pH meter with two calibration buffers that are equilibrated to be the same temperature as the sample. The pH meter temperature setting is adjusted to this temperature before calibration (or an ATC probe is slused). Because the pH of calibration buffers varies with temperature, the meter must be adjusted to the pH of each calibration buffer at the temperature of interest. (For example, the pH of a typical pH 7.00 calibration buffer is 7.00 at 25°C and it is 6.98 at 37°C.)

C. Summary of the Steps in Operating a pH Meter

Although each brand and style of pH meter differs, there are certain steps that are usually performed when measuring the pH of a sample. These steps are summarized in Box 1:

Box 1 CONVENTIONAL METHOD FOR MEASURING THE pH OF A SOLUTION

1. *Allow the meter to warm up as directed by the manufacturer.*

2. *Open the filling hole; be certain that the filling solution is nearly to the top.* (Refillable electrodes only.)

3. *If the meter has a "standby" mode, use it when the electrodes are not immersed in the sample. Use the "pH" mode to read the pH of a sample or standard.*

4. *Calibrate the system each day or before use:*

   a. *Adjust the meter temperature setting to room temperature with the appropriate control, or use an ATC probe.*

   b. *Obtain two standard buffers; one that has a pH of 7.00 at room temperature, the other that is acidic if the sample is acidic and basic if the sample is basic.*

   c. *Rinse the electrodes with distilled water and blot dry.* Do not wipe the electrodes as this may create a static charge leading to an erroneous reading.

   d. *Immerse the electrodes in room temperature, pH 7.00 calibration buffer.* Be certain that the junction is immersed and that the level of sample is below the level of the filling solution, Figure 18.11. Disengage the standby or follow the manufacturer's instructions. Allow the reading to stabilize.

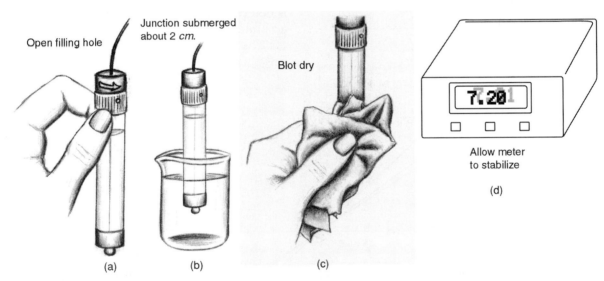

Figure 18.11 Points to Remember in Using a pH Meter. **a.** Open the filling hole before measuring pH; close it afterward. **b.** Make sure the junction is submerged in the solution whose pH is being measured. **c.** Rinse and blot the electrode. **d.** Allow time for the reading to stabilize. (Sketches a–c by a Biotechnology student.)

   e. *Adjust the meter to read 7.00 using the correct knob, dial, button, or other method of adjustment.* Reengage the standby mode, if present.

   f. *Remove the electrodes, rinse with distilled water, and blot dry.* You can also rinse the electrodes with the next solution and do not dry.

   g. *Place the electrodes in the second standardization buffer, set the meter to read pH, and allow the reading to stabilize.* Adjust the meter to the pH of the second buffer with the proper method of adjustment. Remove, rinse, and blot the electrodes.

   h. *(This step is not necessary with newer, microprocessor-controlled pH meters.) Recheck the pH 7.00 buffer as in Step d and readjust as necessary.* Then, recheck the second buffer and readjust the meter as necessary. Continue to rinse the electrodes after each solution.

Box I *(continued)*

i. *(This step is not necessary with newer, microprocessor-controlled pH meters.) Read both calibration buffers and readjust as needed up to three times.* If the readings are not within 0.05 pH units of what they should be after three adjustments, the electrode probably needs cleaning, as described later.

5. ***Optional: It is possible to perform quality control checks at this point.***

a. ***Linearity Check.*** *To check the linearity of the measuring system, take the reading of a third calibration buffer.* For example, if the meter was calibrated with pH 7.00 and pH 10.00 buffers, check the pH of a pH 4.00 calibration buffer. Immerse the electrodes in the third buffer, place the meter in pH mode, allow the reading to stabilize, and record the value. *Do not adjust the meter to this third calibration buffer; the purpose of this third buffer is to check the system's linearity.* If the reading is not within the proper range, as defined by your laboratory's quality control procedures, the electrodes should be serviced as described in the "Trouble-Shooting" section below.

b. *A second quality control check is to test the pH of a control buffer whose pH is known and which has a pH close to the pH of the sample.* It is common to set the maximum allowable error of the control buffer to be ± 0.10 pH units. Do not adjust the meter to the pH of the control buffer. The purpose of this buffer is to check the accuracy of the system. If the pH reading of the control buffer is not within the required tolerance, the electrodes require maintenance.

6. ***Set the meter to the temperature of the sample or place an automatic temperature probe in the sample.***

7. ***Place the electrodes in the sample.***

8. ***Allow the pH reading to stabilize.*** In general, the reading should stabilize within a minute. The pH reading initially changes rapidly, then it slowly stabilizes, Figure 18.12. One of the causes of error in using pH meters is not allowing the reading to stabilize.

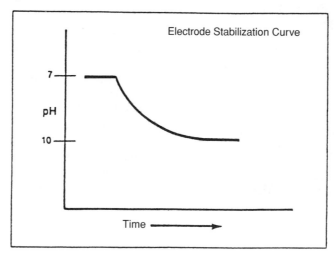

Figure 18.12. Stabilization of a pH Reading. (Courtesy of Beckman Coulter, Inc.)

Note the following points:

a. If you wait *too* long, the pH of some samples will change due to exposure to air, chemical reactions, or other factors.

b. Solutions of high or low ionic strength and nonaqueous solutions may require longer to stabilize.

c. Many new pH meters have an "autoread" feature that automatically determines when the electrode has stabilized and locks on this reading. For example, the meter may lock on the reading if it changes less than 0.004 pH units over 10 seconds. The autoread feature may help standardize readings made in a laboratory because it is consistent from reading to reading and between operators. It also ensures that impatient operators will wait long enough when taking a reading.

9. ***Record all relevant information.*** Be sure to note the temperature of the solution when its pH was measured.

10. ***Remove the electrodes from the sample, rinse them, and store them properly.*** Close the filling hole (refillable electrodes only).

D. Measuring the pH of "Difficult Samples"

The sample is a key part of the measurement system. The sample must be homogeneous to avoid variation and drift as pH is measured. The sample must also be equilibrated to a particular temperature. Some samples undergo chemical reactions that cause their pH to change over time, either among substances in the sample itself, or with CO_2 from the air.

Some samples are inherently difficult to pH and may require special techniques or specific types of electrodes. Difficult sample types include those that:

- **Are nonaqueous**
- **Are high purity water, containing little salt**
- **Contain high salt concentrations**
- **Contain high protein concentrations**
- **Contain S^{-2}, Br^-, or I^-**
- **Contain Tris buffer**
- **Are viscous**
- **Are turbid**

Table 18.8 includes tips for handling difficult samples.

E. Quality Control and Performance Verification of a pH Meter

As with any instrument, pH meters require a quality control plan. At the least this includes keeping a log book, performing regular calibration, and cleaning and replacing electrodes as needed. Although each laboratory must develop its own plan, Box 2 summarizes common, routine quality control procedures for pH meter measuring systems.

V. TROUBLE-SHOOTING pH MEASUREMENTS

A. General Considerations

i. Overview

A pH meter system must be able to respond to a range of hydrogen ion concentrations from 10^0 to 10^{-14} *moles/L*, must be responsive to very low concentrations of hydrogen ions and to very small changes in hydrogen ion concentration, and must be able to detect hydrogen ions selectively in the presence of other ionic species. Compared with many other measuring instruments, therefore, a pH meter must have a wide range of response and be very sensitive and specific. Considering these stringent requirements, it is not surprising that problems sometimes arise when using a pH meter. This section discusses sources of common problems and general strategies for identifying and solving them.

The first step in trouble-shooting is recognizing that there is a problem. Although this seems obvious, prob-

Table 18.8 Difficult Samples

1. *Nonaqueous solvents.* pH relates to aqueous solutions and it is difficult to obtain a meaningful pH reading for a nonaqueous sample. Nonaqueous solvents, like acetone, can dry the glass electrode membrane, which must be hydrated to function properly. After measuring the pH of solvents, it may be necessary to soak the measuring electrode in an aqueous solution. Consult the manufacturers for further advice.

2. *High purity water.* High purity water (e.g., acid rain, well water, distilled, high purity water, or boiler feedwater) is very difficult to pH because it does not readily conduct current and it absorbs CO_2 from the atmosphere, which affects its pH. Tips for measuring the pH of such samples include:

    Avoid the absorption of CO_2 by taking a reading soon after immersing the electrodes deep within the sample.

    Cover the sample container with a stopper that has holes for the electrodes.

    Blanket the surface of the sample with nitrogen gas.

    Use special, low ionic strength buffers for calibration. (Such buffers are available commercially.)

    Use additives to increase the ionic strength of the sample without affecting its pH. (Advertised, for example, by Orion, Boston, MA).

3. *High salt samples.* In samples with large amounts of salt, sample ions compete with the reference filling solution ions. This competition affects the voltage read by the meter. It may be possible to compensate for this effect by using calibration buffers that also have a high salt concentration. An electrode with a double junction may also be helpful.

4. *Sample-electrode compatibility.* The sample must be compatible with the reference electrode. Tris buffer, sulfides, proteins, Br^-, and I^- can precipitate or complex with the silver in Ag/AgCl electrodes, leading to a clogged junction. A calomel electrode is therefore recommended when working with these materials.

5. *Slurries, sludges, viscous, and colloidal samples.* These types of sample may require a fast flowing junction, usually the sleeve type. These samples also may coat the electrode bulb. Cleaning procedures are shown in the "Trouble Shooting" section below.

    Consult manufacturers for more information on specific electrodes to match particular types of samples.

lems with pH meter systems may be subtle and are sometimes overlooked. Symptoms of pH system problems include:

1. **The reading drifts in one direction and either will not stabilize or takes an unusually long time to stabilize.**

2. **The reading fluctuates.**

Box 2   COMMON ROUTINE QUALITY CONTROL PROCEDURES FOR pH METER SYSTEMS

Quality Control Activity	Each Use	Frequency Each Day	As Needed
Calibrate with two buffers, pH 7.00 and one other	✔		
Perform linearity check (as described in Box 1)	✔		
Read and record the pH of a control buffer whose pH is close to that of the sample	✔		
Check level of filling solution		✔	
Clean and recondition electrodes			✔

3. The meter cannot be adjusted to both calibration buffers.

4. The apparent pH value for a buffer or sample seems to be wrong.

5. There is no reading at all.

ii. THE COMPONENTS OF A pH MEASURING SYSTEM AND THEIR POTENTIAL PROBLEMS

Problems with a pH meter system can arise in any of its components: (1) the reference electrode, (2) the measuring electrode, (3) the meter, (4) the calibration buffers, or (5) the sample.

The reference electrode junction is probably the most common source of problems. If the junction is dirty and partially occluded, then sample ions can migrate into the junction and set up new electrical potentials that interfere with the proper reading of the electrode. The first indication that the junction is occluded is a long stabilization time where the reading drifts slowly toward the correct pH. Slow equilibration, however, can also be caused by changes in the temperature of the sample (such as is caused by a warm stir plate), by reactions occurring within the sample, or by incompatibility between the reference electrode and the sample. Methods of checking for junction occlusion are given in Table 18.9.

If the junction is completely plugged the reading may never stabilize. This is also caused by broken electrodes and by some problems within the meter. The best way to check which component is at fault is to substitute in a new reference or combination electrode. If the reading stabilizes with a new electrode, then the old one was faulty.

The measuring electrode bulb must be clean so it can be in close contact with the solution. Certain samples (e.g., those with high levels of protein) can leave deposits on the bulb that must be periodically cleaned, as described in Box 5.

pH measuring electrodes age, even if they are properly maintained, so they have a finite lifespan. Aqueous solutions slowly attack the glass membrane and form a gel layer on the bulb surface. As the gel layer thickens, the response of the electrode becomes sluggish and the slope of the calibration line decreases.

There is a great deal of resistance to current flow in a pH meter measuring system and the potential developed by the pH electrode bulb is very small. The voltmeters that are used to measure pH, therefore, are specially designed for these conditions. Complete lack of response (or sometimes, dramatically unexpected readings) is likely caused by problems with the meter itself (in contrast to the electrodes). The meter, however, is the least likely component of the system to cause problems.

If the buffers used to calibrate the instrument are not correct, then all readings made will be incorrect. This problem may be detected if a solution of known pH is available and its pH is checked.

If the sample is not homogenous, or if its temperature is not stable, then the pH readings will fluctuate or drift. If the sample is "difficult," the readings may be unusually slow to stabilize and/or may be incorrect.

B. Trouble-Shooting Tips

i. SIMPLE MISTAKES

Once a problem is noted, the first step in trouble-shooting a measurement instrument is always to look for and correct simple (embarrassing) mistakes. For pH meter systems these include:

- **The electrode measuring bulb and the junction are not immersed in the sample.**
- **The meter is not turned on or plugged in, or the electrode cables are not connected to the meter.**
- **The reference electrode is not filled with electrolyte.**
- **The reference electrode filling hole is closed.**
- **The sample is not well-stirred.**
- **The calibration buffers are not fresh.**
- **An electrode is cracked or broken and must be replaced.**

ii. TROUBLE-SHOOTING PROBLEMS BASED ON SYMPTOMS

If no simple mistake is discovered, then the symptoms observed provide clues for trouble-shooting. Tables 18.9–18.12 provide tips for trouble-shooting pH meter problems based on symptoms noted.

Table 18.9 TROUBLE-SHOOTING pH METERS: SYMPTOM 1—THE READING DRIFTS, WILL NOT STABILIZE, OR TAKES AN UNUSUALLY LONG TIME TO STABILIZE

Drift is often due to (1) changes in the sample, (2) problems with the reference electrode, or (3) problems with the measuring electrode.

1. *The temperature of the sample may be unstable, resulting in a fluctuating or drifting pH reading.* Wait until the temperature is stable and adjust the meter to the sample temperature.

2. *The sample may not be homogeneous, leading to a fluctuating or drifting reading.* Stir the sample.

3. *The sample may be undergoing chemical reactions or may be reacting with CO_2 from the atmosphere.* Consider the chemistry of the sample. Read the pH as soon as the reading appears to stabilize.

4. *The sample may be inherently difficult to pH and may require special techniques or specific types of electrodes.* See Table 18.8.

5. *The filling solution may be contaminated.* (Refillable electrodes only.) If so, replace the filling solution with fresh electrolyte (Box 3).

   Excessive crystallization can occur if the electrode was stored for a long time with the filling hole open or was stored in the cold.

   If the electrode was used to pH a sample when the filling solution was low, then the filling solution might have been contaminated by the entrance of sample.

6. *The reference electrode junction may be occluded by crystals or may be dirty.* This is a common cause of sluggish or drifting response. With a liquid filled reference electrode, a clogged junction can be detected by applying air pressure to the filling hole. A bead of electrolyte should appear readily on the junction. If not, the junction is likely fully or partially occluded. If the junction is occluded, clean it as described in Box 4.

7. *Long immersion in high purity water can damage the junction, resulting in reduced or no ion flow and drifting or fluctuating readings.* The reference electrode may need replacement.

8. *The silver or calomel element is damaged.* This is a less likely problem, but it can occur due to long exposure to sulfides, proteins, heavy metals, and pure water. If this is the problem, it is necessary to replace the electrode.

9. *The measuring electrode bulb may be dirty.* To clean the bulb, see Box 5.

10. *The response of the glass that constitutes the measuring bulb can decline due to prolonged use, excessive alkaline immersion, or high temperature operation.* This can be tested by performing a "linearity check" as described in Box 1. If the reading is out of range, then try rejuvenating the glass electrode as shown in Box 6.

Table 18.10 TROUBLE-SHOOTING pH METERS: SYMPTOM 2—DIFFICULTY ADJUSTING THE METER TO BOTH CALIBRATION BUFFERS

1. *The measuring electrode may be dirty.* Clean the bulb as described in Box 5.

2. *The response of the glass that constitutes the measuring bulb can decline.* Rejuvenate the electrode as shown in Box 6.

Table 18.11 TROUBLE-SHOOTING pH METERS: SYMPTOM 3—THE VALUE READ FOR A BUFFER OR SAMPLE SEEMS TO BE WRONG

1. *The calibration buffers have deteriorated.* This is the most likely problem; try fresh calibration buffers.

2. *The sample may not be homogenous.* Stir the sample.

3. *The measuring electrode may be dirty.* Clean the bulb as described in Box 5.

4. *The response of the glass that constitutes the measuring bulb has declined.* Rejuvenate the electrode as described in Box 6.

Table 18.12 TROUBLE-SHOOTING pH METERS: SYMPTOM 4—NO RESPONSE AT ALL

1. *Check that the meter is plugged in, turned on, and the electrode cables are properly connected.*

2. *Consult the pH meter manufacturer.* The operating manual may have a trouble-shooting guide.

C. Procedures for Cleaning and Maintaining Electrodes

Follow the manufacturer's instructions if available. The following procedures are general, commonly recommended suggestions.

BOX 3   REPLACING ELECTROLYTE IN A REFERENCE ELECTRODE

1. *Wash the electrode in warm water.*
2. *Shake until the crystals dissolve.*
3. *Remove the old filling solution.* It may be helpful to use a narrow plastic pipette or a syringe attached to plastic tubing to remove the solution.
4. *Insert fresh solution of the proper type.*
5. *It may be helpful to soak the electrode overnight in 1 M KCl before using.*

BOX 4   CLEANING THE JUNCTION OF A REFERENCE ELECTRODE

These methods are suitable for combination electrodes or individual reference electrodes. The first methods are least drastic and should be tried first. Continue to the next step only if the previous one has failed.

1. *Inspect the reference cavity for excessive crystallization.* If there are excessive crystals, replace the old filling solution as discussed in Box 3. (Refillable electrodes only.)
2. *For gel-filled and calomel references, soak the electrode tip in warm water (about 60°C) for 5–10 minutes.* Be careful not to heat the water much above 60°C.
3. *Dissolve crystals from the end of the electrode by soaking it in a solution of 10% saturated KCl and 90% distilled water for between 20 minutes and 3 hours.* Warm the solution to about 50°C. Immerse the electrode about 2 *in* into the solution. Refillable electrodes can be soaked overnight in this solution.
4. *Use a commercially available junction cleaner.*
5. *Remove any proteins that coat the outside of electrodes or penetrate into the junction.* Recommendations for removing protein deposits include rinsing in enzyme detergent like Terg-A-Zyme® (from Alconox), using pepsin/HCl cleaners (5% pepsin in 0.1 M HCl), or soaking the electrode in 8 M urea for about 2 hours. After removing the protein deposits, rinse the electrode and empty and replace the filling solution.
6. *Immersion in ammonium hydroxide may be helpful for refillable Ag/AgCl electrodes that are clogged with silver chloride.* Empty the filling solution and immerse the electrode in 3 to 4 M NH_4OH for only 10 minutes. Ammonia can damage the internal element of the electrode if the electrode is soaked in it too long. (NH_4OH is very caustic and the fumes are irritating. Use it in a fume hood.) Rinse the inside and outside of the electrode thoroughly with distilled water and add fresh filling solution. If the electrode is a combination electrode, soak it 10–15 minutes in pH 4 buffer.
7. *Immersion in concentrated hydrochloric acid for 5–10 minutes may be helpful for calomel electrodes.* Use adequate ventilation and safety precautions to avoid injury by the acid.
8. *It may be possible to clear the junction by forcing filling solution through the junction by applying pressure to the filling hole or vacuum to the junction tip.*
9. *As a last resort, for ceramic junctions only, sand the junction with #600 emery paper.* For combination electrodes, be careful not to contact the glass bulb.

(A "sleeve junction" has a collar or sleeve that can be loosened to unclog the junction. When the sleeve is loosened, the filling solution empties. With this type of junction, simply loosen the collar, clean the junction, and refill the electrode.)

BOX 5   CLEANING THE MEASURING ELECTRODE BULB

Use a soft toothbrush, Q-tip, or tissue to clean the bulb. Gently wipe or pat the bulb; do not rub.

1. *Remove dirt with warm soapy water.* A mild dish washing detergent without hand lotion can be used.
2. *Remove protein contaminants by washing in warm water with an enzyme detergent like Terg-A-Zyme® (from Alconox) or 5% pepsin in 0.1 N HCl followed by rinsing in water.* If your application routinely involves many pH measurements of protein solutions it is good practice to clean the bulb on a regular basis.
3. *Remove inorganic deposits by washing with EDTA, ammonia, or 0.1 N HCl.* Rinse the bulb with water and soak it in dilute KCl.
4. *Remove grease and oil with methanol or acetone.* Rinse the bulb with water and soak it in dilute KCl.
5. *To remove fingerprints, wipe the bulb gently with a 50/50 mixture of acetone and isopropyl alcohol.* Soak the electrode in a commercial soaking solution or pH 4 buffer for 1 hour.

Box 6   "REJUVENATING" A MEASURING ELECTRODE BULB

Note that solutions used for soaking and conditioning a pH measuring electrode are usually slightly acidic. This is because H^+ ions from the soaking solution replace contaminants in the glass of the bulb.

Some manufacturers recommend the following steps when the linearity of the system is poor or the system response is sluggish. The first methods are least drastic and should be tried first. Continue to the next step only if the previous one has failed.

1. *Soak the bulb in pH 4.00 calibration buffer overnight.*

2. *Soak the bulb in 1M HCl for 30 minutes.*

3. *Immerse the bulb in 0.1 M HCl for about 15 seconds, rinse with water, then immerse in 0.1 M KOH (or 0.1 M NaOH) 15 seconds.* Repeat several times and rinse. Soak the electrode in pH 4 buffer overnight.

SAFETY NOTE: Manufacturers sometimes recommend rejuvenating glass bulbs with ammonium bifluoride or with hydrofluoric acid. Ammonium bifluoride is hazardous and should be used with caution. Hydrofluoric acid is extremely hazardous! It will degrade glass containers. It can cause serious or fatal injuries even at low concentrations. We do not recommend its use except by those who are specially trained and have access to appropriate safety equipment.

VI. OTHER TYPES OF SELECTIVE ELECTRODES

An **indicator electrode** *responds selectively to a specific molecule in solution.* For example, there are indicator electrodes that respond to dissolved CO_2 and that respond to dissolved O_2. Biologists use dissolved O_2 electrodes to monitor the oxygen level in the broth during fermentation because the growth of most microorganisms is sensitive to oxygen level.

Ion-selective electrodes (ISE) *are a class of indicator electrode, each of which responds to a specific ion.* A pH measuring electrode is an example of an ion-selective electrode. Other ISEs include those that respond to Ca^{++}, Na^+, K^+, Li^+, F^-, and Cl^-. ISEs have many applications. For example, sodium levels in human blood are severely disturbed by cardiac failure and liver disease. Sodium levels in a patient's blood serum can be tested using a sodium ion selective electrode. ISEs can be used to measure fluoride levels in drinking water, nitrates in soils, plants, or foods, calcium in milk, and other applications. Figure 18.13 shows a catalog listing of different types of ISEs. There are many methods published in official compendia that utilize ISEs. For example, ASTM, EPA, and AOAC all have ISE methods for measuring levels of fluoride in water.

An ion-selective electrode measuring system is diagramed in Figure 18.14. An ISE has a membrane that, like the glass bulb of a pH electrode, develops an electrical potential in response to a particular ion. The ISE is connected to a reference electrode through a voltage meter. Although the electrodes are different, a standard pH meter can be used for these measurements, as long as it is able to display *mV* as well as pH.

Specifications and Ordering Information: Orion ISE's

Type	Concentration Range	Interferences
Gas-Sensing Models		
Ammonia	1 to 5×10^{-7}M; 17,000 to 0.01ppm.	Volatile amines.
Carbon Dioxide	10^{-2} to 10^{-4}M; 440 to 4.4ppm.	Volatile weak acids.
Nitrogen Oxide	5×10^{-3} to 4×10^{-6}M; 230 to 0.18ppm.	CO_2, volatile weak acids.
Solid-State Models		
Bromide	1 to 5×10^{-6}M; 79,900 to 0.40ppm.	S^-, I^-, CN^-, high levels of Cl^-, NH_3.
Cadmium	10^{-1} to 10^{-7}M; 11,200 to 0.01ppm.	Ag^+, Hg^{++}, Cu^{++}, high levels of Pb^{++}, Fe^{++}.
Cadmium Comb.	10^{-1} to 10^{-7}M; 11,200 to 0.01ppm.	Ag^+, Hg^{++}, Cu^{++}, high levels of Pb^{++}, Fe^{++}.
Chloride	1 to 5×10^{-5}M; 35,500 to 1.8ppm.	OH^-, S^- Br^-, I^-, CN^-.
Chloride Comb.	1 to 5×10^{-5}M; 35,500 to 1.8ppm.	OH^-, S^- Br^-, I^-, CN^-.
Chlorine Comb.	3×10^{-4} to 10^{-7}M; 20.0 to 0.01ppm.	Strong oxidizing agents.
Cupric	10^{-1} to 10^{-8}M; 6350 to 6.4×10^{-4}ppm.	Ag^+, Hg^{++}, high levels of Cl^-, Br^-, Fe^{++}.
Cyanide	10^{-2} to 8×10^{-6}M; 260 to 0.2ppm.	S^-, I^-, Br^-, Cl^-.

Figure 18.13.   A Catalogue Listing of Various Types of Ion-Selective Electrodes. (Courtesy of Fisher Scientific, Pittsburgh, PA.)

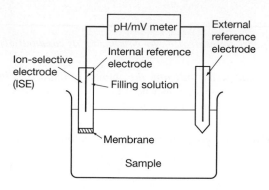

Figure 18.14.   An Ion-Selective Electrode Measuring System.

When working with indicator electrodes it is necessary to determine the relationship between electrical potential and the concentration of the ion of interest. A calibration curve must be constructed, as shown in Figure 18.15.

Each type of indicator electrode is different; therefore, it is important to read the manual that comes with the electrode. Table 18.13 lists some general points of concern regarding indicator electrodes.

VII. MEASUREMENT OF CONDUCTIVITY

A. Background

There are situations where it is useful to measure the combined concentration of all the ions in an aqueous solution, Table 18.14. The total ion concentration is determined by measuring the **conductivity** of the solution.

Conductivity *is the ability of a material to conduct electrical current.* Metals are extremely conductive; electrons carry current through metals at almost the speed of light. Electrical current can also flow through

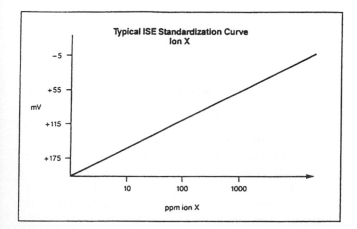

Figure 18.15.   A Calibration Curve for an Ion-Selective Electrode. (© Copyright Omega Engineering, Inc. All rights reserved. Reproduced by permission of Omega Engineering, Inc., Stamford, CT 06907.)

Table 18.13   PRACTICAL CONSIDERATIONS FOR INDICATOR ELECTRODES

1. *The sensing tips of the electrodes are fragile and should be protected from breakage and drying.*

2. *The response of the electrodes may be affected by the combined concentration of all the ions in the solution.* The standards used to prepare the calibration curve and the samples should be roughly comparable in their total ionic concentration. It is possible to purchase ionic strength adjusting buffers that are used to prepare standards and to dilute samples in order to achieve comparable ionic strength.

3. *Temperature affects the response of ISEs so the calibration curve must be prepared at the same temperature as the samples are measured.*

4. *The calibration standards should have the same pH as the samples.*

5. *Although indicator electrodes are usually specific for one substance, under certain conditions other substances may interfere with the measurement of interest.* Be aware of contaminants that might adversely affect the measurement.

6. *The sample must be well stirred.*

7. *Ion-selective electrodes are generally slower to stabilize than pH electrodes.* At least 15 minutes is recommended for stabilization.

8. *Consult the manufacturer's manual for storage directions.*

solutions by the movement of ions, although this movement is not as rapid as current flow in metals. Thus, aqueous solutions containing ions are conductive. The higher the concentration of mobile ions in a solution, the more conductive the solution; therefore, conductivity is an indicator of total ion concentration.

We are accustomed to thinking of water as a good conductor of electricity (and so would avoid being on a lake in a metal canoe during a thunderstorm). However, because it is ions that actually conduct electricity in a solution, ultrapure water has a very low conductivity.

Some substances ionize more completely than others when dissolved in water, so their solutions are more conductive. **Electrolytes** *are substances that release ions when dissolved in water.* **Strong electrolytes** *ionize completely in solution,* whereas **weak electrolytes** *partially ionize.* **Nonelectrolytes** *(e.g., sugar, alcohols, and lipids) do not ionize at all when dissolved.* The magnitude of current that will flow through an aqueous solution depends on the combined concentration of each electrolyte present and the degree to which each ionizes.

B. The Measurement of Conductivity

i. THE DESIGN OF CONDUCTIVITY INSTRUMENTS

The measurement of conductivity, like the measurement of pH, involves the use of an electrochemical system. A simple device to measure conductivity consists of two flat

Table 18.14 SOME USES OF CONDUCTIVITY MEASUREMENTS

1. *If there is only one type of salt in a solution, it is possible to determine its concentration by comparing its conductivity with standards of known salt concentration.*

2. *Pure water does not readily conduct electricity, so a measurement of conductivity indicates the amount of impurities in the water.* Very low conductivity indicates pure water. There are a vast number of applications where the purity of water is important, such as in the production of pharmaceuticals, maintaining cells in culture, and ensuring the proper operation of industrial heating and cooling systems.

3. *Detectors that measure the conductivity of a solution can be attached to chromatography instruments and used to monitor the elution of sample components, as shown in Figure 18.16.*

Organic Acid Standards

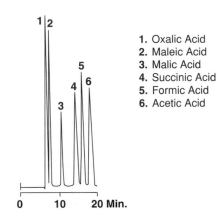

1. Oxalic Acid
2. Maleic Acid
3. Malic Acid
4. Succinic Acid
5. Formic Acid
6. Acetic Acid

Column: Wescan Anion Exclusion, 100 x 7.5mm
Mobile Phase: 1mM Sulfuric Acid
Flow Rate: 0.6mL/min
Detector: Conductivity

Figure 18.16. Display Formed by a Conductivity Detector Used to Monitor the Elution of Sample Components from a Chromatography Column. (Chromatogram courtesy of Alltech Associates, Inc.)

4. *Conductivity can be used clinically to indicate electrolyte levels in body fluids.*

5. *Conductivity can be used to approximate "total dissolved solids" in a solution.* This measurement is frequently used, for example, in the pharmaceutical industry as an indicator of purity.

6. *Some enzyme-catalyzed reactions release ions, so changes in conductivity can be used to monitor some enzyme reactions.*

7. *Electrophoresis, a method commonly used to separate charged biological macromolecules, is dependent on using buffers with the proper conductivity.*

8. *In industrial processes, conductivity may be monitored as an indication of the concentration of an important chemical.*

9. *Conductivity may be monitored in fish tanks as an indicator of water quality.*

electrodes that are dipped into the sample solution and are connected to a battery and to an **ammeter**, *a current measuring instrument*, Figure 18.17. The battery generates electric current that can flow only if the sample solution is conductive and completes the circuit path. The electrode connected to the positive pole of the battery is positive, whereas the electrode connected to the negative pole is negative. Positive ions in the solution are attracted to the negative electrode; similarly, negative ions are attracted to the positive electrode. At the positive electrode negative ions give up electrons that flow in the wire to the positive pole of the battery. At the negative electrode positive ions take electrons. These processes result in continuing electron flow toward the negative electrode and continuing

movement of ions through the solution. The more ions present, the more current is registered by the ammeter.

ii. THE CELL CONSTANT, K

The amount of current flow that is measured by a conductivity instrument will vary depending on *the distance between the two electrodes, d, and the area of the electrodes, A*, Figure 18.18. The farther apart the electrodes, the less current is registered. Conversely, the larger the electrode area, the more current that is measured. For samples with low conductivity it is best to configure the measuring device so that it is as sensitive as possible by having a small distance between the electrodes and/or a large electrode surface. If the sample is very conductive,

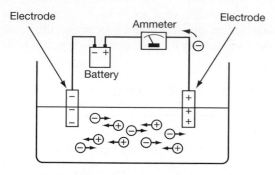

Figure 18.17.   **The Measurement of Conductivity.** The meter consists of two flat electrodes, a battery, and a meter that measures current. Positive ions flow toward the negative electrode and negative ions toward the positive electrode. Electrons flow in the wires between the electrodes and the battery and meter. This makes a complete circuit for current flow. The more ions present in the solution to carry current, the more current registered by the ammeter.

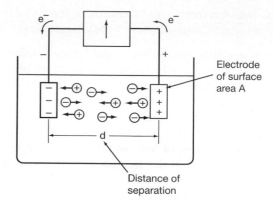

Figure 18.18.   **The Configuration of a Conductivity Measuring System.** The surface area of the electrodes and the distance between them will affect the measurement.

however, then a wider spacing between electrodes and/or smaller electrodes are more effective.

Manufacturers make conductivity instruments with sample holders, or **sample cells**, having different configurations of distances and electrode surface areas. The configuration is described by the **cell constant** for the device. The **cell constant**, *abbreviated* K, *is defined as:*

$$K = d/A$$

where d = the distance between the electrodes
A = their area

The units of K are therefore $cm/cm^2 = 1/cm$.

The smaller the value for K, the shorter the distance between electrodes and/or the larger the electrodes. A device with a lower K value, therefore, is more sensitive than a device with a higher K value. For measuring the conductivity of pure water, a sensitive device with a K value of about 0.1/cm is preferred. For samples of moderate conductivity, a cell with a K value of around 0.4/cm is useful, and for samples with very high conductivities, a cell with a K value of around 10/cm is chosen. Figure 18.19 shows how a catalog describes a conductivity instrument including the K values for three probes that are available to match different type of samples.

Model CDH-70

✓ Three Measurement Ranges
✓ Chemically Resistant Housing
✓ Conductivity Cells with Integral Temperature Compensator

Perfect for field measurements of conductivity and temperature, the CCH-70 is housed in a chemically resistant case with membrane keypad for range and function selection. The meter features direct readout of conductivity, temperature and cell constant. Three measurement ranges are provided enabling the best resolution to be obtained for the sample under test. Measurement ranges may be increased further by using optional conductivity cells with cell constants of 0.1 and 10. All results are automatically temperature compensated and referenced to 25°C. Direct temperature measurement is provided over the range −30 to +150°C using the optional PHAT-70.

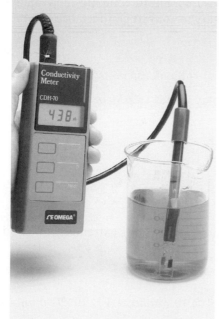

Specification
RANGE

Conductivity: 0 to 199.9 μS, 5.0 to 1999 μS, 0 to 19.99 mS.
Resolution: 0.1 μS, 1 μS, 0.01 mS.
Accuracy: ±1.5%, ±2 digits all range
Temperature: −30 to 150°C
Resolution: 0.1°C
Accuracy: ±0.5°C
Temperature Compensation: 0 to 50°
Temperature Coefficient: 2% per °C
Cell Constant: 0.75 to 1.50 digitally setible
Reference Temperature: 25°C
Power: 9 V battery
Weight: 8.8 g (0.31 *lbs*)

To Order (Specify Model Number)	
Model No.	**Description**
CDH-70	Portable conductivity meter
Accessories	
PHAT-70	Temperature compensation probe
CDE-70-10	Conductivity cell K = 10
CDE-70-01	Conductivity cell K = 0.1
CDE-70-1	Conductivity cell K = 1
CDCC-70	Carrying Case

Ordering Example: CDH-70 Conductivity meter, PHAT-70 temperature probe. Each unit is supplied with meter, conductivity cell K = 1, battery and complete operator's manual.

Figure 18.19.   **A Conductivity Meter and Its Specifications.** Different probes are available for the meter. Each probe has a different configuration and therefore is useful for samples with differing conductivities. (© Copyright Omega Engineering, Inc. All rights reserved. Reproduced by permission Omega Engineering, Inc., Stamford, CT 06907.)

iii. "Conductivity" and "Conductance"; Units Used for "Conductivity" Measurements

There is a distinction between the terms *conductivity* and *conductance*. **Conductivity** is an inherent property of a material. **Conductance** *is a measured value that depends on the configuration of the conductivity measuring device used.* Conductance is sometimes also called "measured conductivity." For example, the *conductance* of a sample of pure water will vary depending on the value for *K* in the measuring instrument used. The *conductivity* of the water sample is the same, regardless of instrument used to measure it. (The terms *resistivity* and *resistance* can similarly be distinguished from one another.)

Thus, there are four terms and four slightly different units used to describe a solution's ability to conduct current. These various units, all of which are used to describe the ability of a solution to carry current, are summarized in Box 7. Table 18.15 shows the resistivity and the conductivity of some common types of solutions.

Table 18.15 *The Conductivities and Resistivities of Common Types of Solutions*

Solution	Conductivity	Resistivity
Ultrapure Water	0.055 $\mu S/cm$	18,000,000 ohm-cm = 18M Ω-cm
Distilled Water	1 $\mu S/cm$	1,000,000 ohm-cm = 1 M Ω-cm
Deionized Water	80 $\mu S/cm$	12,500 ohm-cm
0.05% NaCl	1,000 $\mu S/cm$	1,000 ohm-cm
Seawater	50,000 $\mu S/cm$	20 ohm-cm
30% H_2SO_4	1,000,000 $\mu S/cm$	1 ohm-cm

Values shown are approximate.

Box 7 *Units of Measurement of Conductivity and Resistivity*

1. **Resistance.** The more readily a solution conducts current, the less resistance there is to current flow.

   Resistance is measured in *ohms* (abbreviated Ω)
   1 megohm (MΩ) = 1,000,000 Ω

2. **Conductance.** The conductance of a solution is the inverse of its resistance.

   Conductance = 1/resistance
   1/*ohm* = 1 *mho* = unit of conductance
   1 *mho* = 1 Siemen (abbreviated *S*) = another unit of conductance
   1 micromho (μmho) = 1 microSiemen (μS) = 1 *mho*/1,000,000

3. **Conductivity.** Conductance is the property that is measured, but to compare the results from various instruments, or from one laboratory, with another, it is necessary to convert conductance to conductivity. This is done using the following equation:

   Conductivity = conductance (*K*)
   The units of conductivity therefore are: (Siemens)(1/*cm*) = Siemens/*cm*

   In practice, a *S/cm* is too large, so most instruments display values as $\mu S/cm$ or *mS/cm*. In catalogs, $\mu S/cm$ and *mS/cm* may be abbreviated as μS and *mS*, respectively.

   1 $\mu S/cm$ = 0.001 *mS/cm* = 0.000001 *S/cm* = 1 $\mu mho/cm$

4. **Resistivity.** This term is:

   1/conductivity
   The units of resistivity, therefore, are *ohm-cm*

   Resistivity is often used in the measurement of ultrapure water due to its extremely low conductivity.

5. It is possible to convert between conductance and resistance
   For example:
   Convert a conductance of 40 μS to resistance
   Conductance = 40 μS = 40 $\times 10^{-6}$ S = 4 $\times 10^{-5}$ S = 4 $\times 10^{-5}$ *mho* = 4 $\times 10^{-5}$/1 *ohm*
   Resistance = 1/conductance
   Therefore, resistance = (4 $\times 10^{-5}$/1 *ohm*)$^{-1}$ = 1 *ohm*/4 $\times 10^{-5}$ = 25,000 *ohm*

iv. CONDUCTIVITY MEASUREMENTS

As with any instrument, conductivity meters require calibration. Calibration standards for conductivity measurements are commercially available.

Temperature has a large effect on conductivity. Conductivity standards will have a table to show their conductivity at various temperatures. Further notes about calibration and other practical comments about conductivity measurements are summarized in Table 18.16.

Table 18.16   PRACTICAL CONSIDERATIONS FOR CONDUCTIVITY MEASUREMENTS

1. **Standards.** Conductivity meters should be calibrated with a standard before use. Choose a standard whose conductivity is similar to that of the solution to be measured. Standards readily pick up contaminants from the air. Keep the standards' containers closed to avoid contamination and evaporation. A standard solution exposed to air will degrade within a day. Remember too, that evaporation will increase the conductivity of a standard.

2. **Temperature.** Raising the temperature increases the conductivity of a solution. The effect of temperature is different for every ion, but the conductivity typically increases by about 1–5% per °C. Many conductivity meters have temperature-sensing elements and are able to compensate automatically for temperature.

3. **Depth of immersion.** The depth to which the electrodes are immersed can affect the current readings. Do not immerse to the bottom of the solution if there are particulates settled there.

4. **Air bubbles.** Make sure that air bubbles are not trapped in the probe when making measurements.

5. **Electrode maintenance.** Probes should be rinsed with distilled water after each series of measurements. Probes may be allowed to dry during storage, but they may require several minutes to stabilize when placed in a solution. Do not allow dried salts or particulates to build up on the probe.

6. **Meter configuration.** Conductivity meters come with various designs that affect the size of the electrodes and the distance between the electrodes, Figure 18.18. It is necessary to choose a conductivity meter that has the proper configuration to match the sample.

7. **Acids and bases.** Most acids and bases are considerably more conductive than their salts because H^+ and OH^- ions are extremely mobile. For example, NaOH is more conductive than NaCl.

8. **Particulates.** For greatest accuracy, make sure that there are no particulates suspended in the sample. Filter or settle out particulates.

9. **Linearity.** The relationship between conductivity and ion concentration is generally, but not always, linear. In highly concentrated solutions, the relationship between conductivity and concentration may be nonlinear due to interactions between ions.

10. **Total dissolved solids (TDS) measurement.** Conductivity is related to the concentration of all dissolved electrolytes in a solution. Conductivity, therefore, is sometimes used to indicate the concentration of "total dissolved solids" (often minerals like $CaCO_3$) in a solution. In these cases, the meter should be calibrated with a dissolved solids standard that contains the same type of solids as the sample to be tested. Otherwise, there will be serious errors in the measurement because each type of dissolved substance has a different effect on conductivity. Commercially available TDS standards include calcium carbonate (often used to test boiler water), sodium chloride (often used to test brines), potassium chloride, and a mixture of 40% sodium sulfate, 40% sodium bicarbonate, and 20% sodium chloride (often used to test water from lakes, streams, wells, and boilers).

11. **Cleaning the electrodes.** Clean cells with mild liquid detergent. Dilute HCl may also be used. Rinse the cells several times with distilled water and recalibrate before use.

PRACTICE PROBLEMS

1. Calculate the hydrogen ion concentration of a solution whose pH is 9.0.

2. Calculate the hydrogen ion concentration of blood whose pH is 7.3.

3. The concentration of hydrochloric acid, HCl, secreted by the stomach after a meal is about 1.2×10^{-3} *moles/liter*. What is the pH of stomach acid?

4. Calculate the pH of 0.01 *M* HNO_3, nitric acid. Nitric acid is a strong acid that dissociates completely in dilute solutions. (Clue: because the nitric acid dissociates completely, the concentration of H^+ ions will be 0.01 *M*.)

5. (Optional) What is the pH of 0.0001 *M* NaOH? (Sodium hydroxide is a strong base.) Clue: in aqueous solutions, $[H^+][OH^-] = 10^{-14}$.

6. Consider the response of an electrode to a change in pH as shown in this graph.

   a. What is the slope of the line at 25°C?

   b. When the pH changes by one pH unit, for example, goes from 5.0 to 6.0, how much does the voltage change at 25°C?

c. What is the slope of the line at 0°C?

d. When the pH changes by one pH unit, for example, goes from 5.0 to 6.0, how much does the voltage change at 0°C?

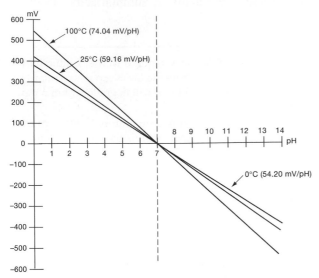

Note: We consider the slope to be negative (see Chapter 10). However, in this question, we are concerned with the magnitude of the slope and not its direction.

7. A recording from an industrial fermentation run is illustrated below. During this run, bacteria reproduced many times over and were harvested to be sold for use in commercial food production. A pH probe was present in the fermentation broth and a continuous recording of the pH was made. When the microorganisms (the seed culture) were first added to the broth, the pH was about 6.8. The pH of the culture changed during the active growth phase. Answer the following questions about this recording.

a. What happened to the pH as the bacterial population grew?

b. Speculate as to what might be the difficulties in designing pH probes for use in fermentation.

c. Why do you think the company generates a continuous recording of pH during each fermentation run?

DISCUSSION QUESTIONS

1. Compare and contrast the calibration of electrodes, pipettes, and balances. Based on this comparison, define the term *calibration* in your own words.

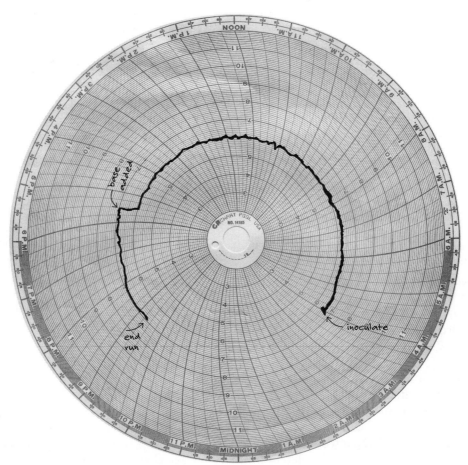

2. How does temperature affect the measurement of volume, weight, and pH?

3. Suggest some likely causes for the following pH measurement problems.

    a. The pH reading slowly drifts.

    b. The pH reading fluctuates.

    c. There is no reading at all.

4. Suggest a way to check each possibility you suggested in Question 3.

5. If you work in a laboratory write a Standard Operating Procedure to calibrate and operate a particular pH meter that is available in your laboratory.

6. a. Discuss how you would go about demonstrating that the pH meter is giving consistent, accurate results when measuring the pH of samples from a sewage treatment plant.

    b. Write an SOP to measure the pH of samples from a sewage treatment plant. Include safety precautions to protect the operator. Include precautions to avoid contamination of the meter and to avoid cross contamination (where one sample contaminates another sample).

TERMS RELATING TO ELECTROCHEMICAL MEASUREMENTS

Acid. A chemical that dissociates and releases H^+ ions when dissolved in water.

Acid–Base Error. A cause of inaccuracy when using pH indicator dyes. Indicators are themselves acids or bases and when added to unbuffered or weakly buffered solutions will cause a change in the pH of the sample.

Acidic Solution. An aqueous solution with a pH less than 7.

Activity. The effective concentration of a solute that accounts for interactions among solutes, temperature, and other effects.

Alcohol Error. A cause of inaccuracy when using pH indicator dyes. Alcohols may cause indicators to be a different color than they appear to be in aqueous solutions.

Alkaline Error. An error in the response of a pH measuring electrode in which the electrode responds to Na^+ and K^+. Alkaline error occurs when the pH is above 9 or 10.

Ammeter. An instrument that measures electrical current.

Anode. A positive electrode.

Automatic Temperature Compensating, ATC, Probe. A probe placed alongside pH electrodes that automatically reports the sample temperature to the meter.

Base. (1) A chemical that causes H^+ ions to be removed when dissolved in aqueous solutions. (2) A chemical that releases OH^- ions when dissolved in water.

Basic Solution. An aqueous solution with a pH greater than 7.

Battery. A device that harnesses the potential of electrochemical reactions to do work.

Blue Litmus. A pH indicator dye that changes from blue to red, denoting a change from alkaline to acid.

Brilliant Yellow. A pH indicator dye that is yellow at pH 6.7 and changes to red at pH 7.9.

Bromocresol Green. A pH indicator dye that changes from yellow to blue at pH 4.0–5.4.

Bromocresol Purple. A pH indicator dye that changes from yellow at pH 5.2 to purple at pH 6.8.

Buffer. A substance or combination of substances that, when in aqueous solution, resists a change in H^+ concentration even if acids or bases are added.

Calibration (for an ion-selective electrode). The use of standards with known concentrations of the ion of interest to determine the relationship between the voltage response of the electrodes and the ion concentration in the sample.

Calibration (for a pH meter). The use of standards of known pH to determine the relationship between the voltage response of the electrodes and the pH of the sample.

Calomel. A paste consisting of mercury metal and mercurous chloride, Hg/Hg_2Cl_2.

Calomel Electrode. A type of reference electrode containing calomel.

Cathode. A negative electrode.

Cell Constant (for conductivity measurements), K. The ratio of the distance between the electrodes and the area of the electrodes.

Combination Electrode. A measuring electrode and reference electrode that are combined into one housing.

Conductance. A measurement of conductivity; the reciprocal of the resistance in ohms when measured in a

$1\ cm^3$ cube of liquid at a specific temperature. Conductance = 1/resistance. The units are: $1/ohm = 1\ mho = 1$ Siemen (S). 1 micromho (μmho) = $1\ mho/1,000,000$

Conductivity. The inherent ability of a material to conduct electrical current.

Conductivity Meter. A measuring instrument that measures the conductivity of a solution.

Congo Red. pH indicator dye that changes from red at pH 3.0 to blue at pH 5.0.

Cresol Red. pH indicator dye that changes from yellow at pH 7.2 to red at pH 8.8.

Current. The flow of charge.

Double Junction. An electrode junction configuration used when the sample is incompatible with the reference electrode filling solution. A double junction separates the reference electrode filling solution from the sample.

Electrical Potential. See "Voltage."

Electrochemical Reaction. Chemical reaction that occurs when metals are in contact with electrolyte solutions.

Electrode. A metal and an electrolyte solution that participate in an electrochemical reaction.

Electrolyte. Substance, such as an acid, base, or salt, that releases ions when dissolved in water.

Electrolyte Solution. A solution containing ions that conducts electrical current.

Fibrous Junction. A type of pH reference electrode junction composed of a fibrous material such as quartz or asbestos; commonly used as a general purpose junction.

Filling Hole. Opening in a reference electrode by which filling solution is introduced into the electrode.

Filling Solution. A solution of defined composition within an electrode. The filling solution sealed inside a pH measuring electrode is called *internal filling solution*, and it normally consists of a buffered chloride solution. The filling solution that surrounds the reference electrode metal element is called *reference filling solution* and provides contact between the reference electrode metal element and the sample.

Fritted Junction. A type of relatively slow flowing pH reference electrode junction composed of a ceramic material consisting of many small particles pressed closely together.

Glass Electrode. See "pH Measuring Electrode."

Hydrogen Ion. H^+.

Hydroxide Ion. OH^-.

Indicator Electrode. An electrode that responds selectively to a specific type of ion or molecule in solution.

Ion Selective Electrode, ISE. A class of indicator electrode that responds to a specific ion in solution.

Ionic Strength. A measure of the concentration of all ions in a solution.

Junction. A part of a reference electrode that allows filling solution to flow into the solution whose pH is being measured. Also called a **salt bridge**.

Linearity Check. A pH meter system is normally calibrated at two pHs (e.g., at pH 10.00 and pH 7.00). A linearity check tests whether the meter's response is linear through its entire range (e.g., also at pH 4.00).

Millivolt, *mV*. Unit of voltage equal to 1/1000 *V*.

Methyl Red. pH indicator dye that changes from red at pH 4.2 to yellow at pH 6.2.

Neutral Litmus. pH indicator dye that is red in acid conditions and blue in alkaline conditions.

Neutral Red. pH indicator dye that changes from red at pH 6.8 to orange at pH 8.0.

Neutral Solution. An aqueous solution with an equal number of hydrogen and hydroxide ions and a pH of 7.

Nonelectrolyte. Substance that does not ionize when dissolved in solution.

pH. A measure representing the relative acidity or alkalinity of a solution; expressed as the negative log of the H^+ concentration when concentration is expressed in moles per liter.

pH Indicator Dyes. Dyes whose color is pH dependent. Indicators can be directly dissolved in a solution or can be impregnated into strips of paper that are then dipped into the solution to be tested.

pH Measuring Electrode. An electrode whose voltage depends on the concentration of H^+ ions in the solution in which it is immersed, also called a **glass electrode**.

pH Meter. A term that is commonly used to refer to a specialized voltage meter and its accompanying electrodes.

Phenol Red. pH indicator dye that changes from yellow at pH 6.8 to red at pH 8.2.

Phenolphthalein. pH indicator dye that changes from colorless at pH 8.0 to red at pH 10.0.

Potential Energy. Energy is the ability to do work; potential energy is stored energy.

Protein Error. A cause of error when using pH indicator dyes, when proteins react with indicators and affect their colors.

Red Litmus. pH indicator dye that changes from red to blue denoting a change from acid to alkaline.

Reference Electrode. An electrode that maintains a stable voltage for comparison with the pH measuring electrode or an ion-selective electrode.

Resistance (to electrical current). The tendency of a material to impede the flow of electricity, measured in units of ohms (Ω) or megohms ($M\Omega$) where 1 megohm = 1,000,000 Ω.

Resistivity. 1/conductivity. The units are *ohm-cm*.

Salt Bridge. See "Junction."

Salt Error. A cause of error when using pH indicator dyes. Salts at concentrations above about 0.2 *M* can affect the color of pH indicators.

Siemen (S). A unit of conductance.

Silver/Silver Chloride, Ag/AgCl, Electrode. A type of reference electrode that contains a strip of silver coated with silver chloride and immersed in an electrolyte solution of KCl and silver chloride.

Sleeve Junction. A type of reference electrode junction made by placing a hole in the side of the electrode housing and covering the hole with a glass or plastic sleeve. The sleeve junction is relatively fast flowing and is unlikely to become clogged.

Solid State Electrode. Newer type of electrode that relies on a small electronic chip to detect ions.

Strong Acid. A chemical that completely dissociates in water to release hydrogen ions.

Strong Base. A chemical that completely dissociates in water to release hydroxide ions.

Strong Electrolyte. A substance that ionizes completely in solution.

Total Dissolved Solids, TDS. A value representing all the solids dissolved in a solution.

Voltage, V. The potential energy of charges that are separated from one another and attract or repulse one another.

Voltmeter. An instrument used to measure voltage.

Weak Acid. A chemical that partially dissociates in water with the release of hydrogen ions.

Weak Base. A chemical that partially dissociates in water with the release of hydroxide ions.

Weak Electrolyte. A substance that partially ionizes when dissolved in water.

REFERENCES RELATING TO THE MEASUREMENT OF pH, SELECTED IONS, AND CONDUCTIVITY

General References:

Analytical Chemistry for Technicians, John Kenkel. 2nd edition. Lewis Publishers, 1994. (General chemistry reference with sections on pH and conductivity.)

"The Effect of Temperature on pH Measurements," Martin S. Frant. *American Laboratory*, pp. 18–23, July, 1995. (Short article summarizing the effects of temperature on pH measurements.)

There are a number of helpful, practical guides available from manufacturers. Examples include:

The Beckman Handbook of Applied Electrochemistry, Beckman Corp. Bulletin No. 7739. (Practical guide to pH measurements with information on different types of samples and systems.)

Calibration of pH Electrodes, Dr. W. Ingold AG. Ingold, Urdorf/Switzerland. 1982.

Orion Catalog, Boston, MA.

pH Indicator Papers, *Test Papers*, Gallard-Schlesinger Industries, Inc. U.S. Distributors for Machery-Nagel, Duren, West Germany.

The pH and Conductivity Handbook, Omega Engineering Inc. technical reference section, 1995 catalog.

ANSWERS TO PRACTICE PROBLEMS

1. $\text{pH} = -\log[\text{H}^+]$
   $9 = -\log[\text{H}^+]$
   $[\text{H}^+] = 10^{-9}$ *moles/liter*

2. $\text{pH } 7.3 = -\log[\text{H}^+]$
   $= 5.0 \times 10^{-8}$ *moles/liter*

3. $[\text{H}^+] = 1.2 \times 10^{-3}$ *moles/liter*
   $\text{pH} = -\log[1.2 \times 10^{-3}] = 2.9$

4. $\text{pH} = -\log[0.01] = 2$

5. $[1 \times 10^{-4}][\text{H}^+] = 1 \times 10^{-14}$
   $[\text{H}^+] = 1 \times 10^{-10}$
   $\text{pH} = 10$

6. **a.** 59.16 *mV*/pH unit. **b.** 59.16 *mV*
   c. 54.2 *mV*/pH unit   **d.** 54.2 *mV*

7. **a.** The pH goes down as the bacteria actively metabolize nutrients and their population grows.

b. pH probes for fermentation are difficult to produce. Difficulties include probe sterilization, which usually involves exposure to pressurized steam. Probes are immersed in a turbid solution that contains numerous materials that can contaminate the electrode surface. They must be reliable because it is difficult to substitute a new probe during a fermentation run.

c. The pH of a culture is an important indicator of its condition and therefore must be monitored. It may be necessary to adjust the pH during a fermentation run. An unexpected value may indicate a problem. It is important to document that the pH values were those expected during a "good" run. The print recording provides a record of the pH.

Appendix

A Brief Introduction to Electrochemistry

Understanding how electrodes generate electrical potential requires some knowledge of electrochemical reactions. Although the theory of electrochemistry is complex, it may be helpful to realize that this theory was devised by scientists to explain concrete, empirical observations of nature. This section briefly describes some of the early experimental observations in electrochemistry and the theory that arose to explain these observations.

Voltage (electrical potential) *is the potential energy of charges that are separated from one another.* Electrical potential arises because charged objects exert attractive or repulsive forces on one another. The development of voltage therefore requires a method of separating charges. Electrochemical reactions are one of the ways by which positive and negative charges become separated.

Some of the most important observations relating to electrochemistry were made by Alessandro Volta in the early 1800s (e.g., see *Volts to Hertz: The Rise of Electricity*, Sanford P. Bordeau. Burgess Publishing, Minneapolis, MN, 1982). Volta placed disks of two different metals adjacent to one another, but separated by a piece of cloth soaked in an electrolyte solution. Volta observed that one of the metal disks would acquire a positive charge and the other a negative charge. Volta assembled large stacks of disks separated by pieces of felt soaked in salt water. He found that if he touched the top and bottom disks in the stack at the same time, he received an electrical shock.

Scientists explain the shock Volta received as being due to **electrochemical reactions** *in which metals either release electrons or take up electrons, becoming either positively or negatively charged.* Electrochemical reactions caused separation of positive and negative charges from one another, leading to the development of voltage. When Volta touched the top and bottom disks, the electrical potential caused current to flow between the disks through his body so that he was shocked.

Although Volta did not understand electricity as it is understood today, he was able experimentally to learn a great deal about electrochemical reactions. Volta made stacks with different metals and discovered that metals could be arranged in order of how likely they are to become positively charged. For example, he found that when zinc and copper were used in the stacks, the zinc became positively charged and the copper negatively charged. Iron and copper also left the copper negatively charged, but the shock Volta experienced was milder from stacks made with iron than with zinc. If copper and silver were used, the copper became positively charged. Scientists now interpret Volta's results to mean that metals differ in their tendency to lose electrons; some metals lose electrons more readily than others. The electrochemical reactions in Volta's stacks resulted in movement of electrons from the metal that had more tendency to lose electrons to the other metal. The ordering of metals according to their tendency to lose electrons is now taught as the "activity series of metals." Scientists have investigated many metals and given numerical values to their relative tendency to lose electrons.

The idea of combining a piece of metal and an electrolyte solution in order to develop voltage has useful applications. The most familiar example is a **battery** *which harnesses the voltage due to electrochemical reactions to do useful work.* A battery consists of two **electrodes**, *metal strips in contact with an electrolyte solution, where electrochemical reactions occur.* As in Volta's stacks, the reactions lead to the development of voltage. When the two electrodes are connected via a metal wire, current flows.

Ion selective electrodes are another application that takes advantage of electrochemical reactions. Volta observed that the magnitude of the potential developed by his stacks was controlled by both the nature of the metals involved and the concentration of certain ions in the electrolyte solution. The higher the concentration of the ions, the higher the potential. This relationship between ion concentration and potential is the basis for ion-selective electrodes, such as pH measuring electrodes, which develop a greater or lesser potential depending on the concentration of a particular ion in a solution. Thus, the design of pH meter measuring systems is based on the observations of Volta and other scientists.

## CHAPTER 19

Measurements Involving Light A: Basic Principles and Instrumentation

I. LIGHT

A. Introduction

Previous chapters discussed the measurement of certain physical properties of a sample, such as its weight, volume, and temperature. This and the next chapter discuss measurement of another physical characteristic of a sample: its ability to absorb light.

Light is essential for life. Almost every living creature on earth is reliant on the energy from sunlight; for example, light enables humans to see, and light sustains plant growth. Light is able to play these roles because it interacts with molecules.

In the laboratory, we take advantage of the interaction of light with materials in samples to obtain information about those samples. **Spectrophotometers** *are instruments used to measure the effect of a sample on a beam of light.* Spectrophotometric assays are used, for example, to determine how much DNA is present in a cellular extract, the purity of protein in an enzyme preparation, the activity of an enzyme, and to confirm the identity of an ingredient in a drug formulation.

This chapter begins by exploring the nature of light. It will then discuss the design, operation, and performance of instruments that can detect and measure light. The next chapter will explain how measurements involving light are used to obtain meaningful information about biological systems.

B. Electromagnetic Radiation

i. THE ELECTROMAGNETIC SPECTRUM

Light is a type of **electromagnetic radiation**, that is, *energy that travels through space at high speeds.* There are different categories of electromagnetic radiation including gamma rays, x-rays, ultraviolet (UV) light, visible (Vis) light, infrared (IR) light, microwaves, and radio waves. *All the types of electromagnetic radiation together are termed the* **electromagnetic spectrum.** Visible, ultraviolet, and infrared radiation are all called "light" even though our eyes are only sensitive to visible light.

Each type of radiation within the electromagnetic spectrum has a characteristic way of interacting with matter. For example X-rays, which have great energy, can easily penetrate matter. We take advantage of this characteristic when producing X-ray images for medical use. Cells in our eyes are able to interact with light in the visible region of the electromagnetic spectrum, leading to vision.

Scientists describe electromagnetic radiation as having a "dual nature." This is because sometimes electromagnetic radiation is best explained as consisting of waves moving through space with crests and troughs, much like waves in a pond. Other times electromagnetic radiation is portrayed as "packages of energy" moving through space. *Each "package of energy" is called a* **photon.**

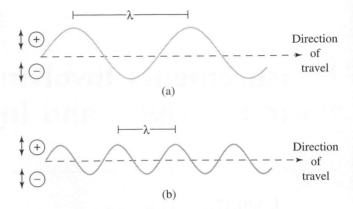

Figure 19.1. Electromagnetic Radiation Traveling as Waves. Electromagnetic radiation is often visualized with peaks and troughs, similar to waves of water. Unlike water waves, however, electromagnetic radiation is not associated with the motion of matter. Rather, electromagnetic radiation is associated with regular changes in electric and magnetic fields. The electromagnetic radiation in **a.** has a longer wavelength and carries less energy than the radiation in **b.**

Figure 19.1 illustrates the wavelike nature of light. *The distance from the crest of one wave to the crest of the next is the* **wavelength** *(abbreviated λ) of the radiation.* The wavelength of electromagnetic radiation varies greatly depending on its type. Some types of electromagnetic radiation have wavelengths so short they are measured in nanometers, but other types of electromagnetic radiation have wavelengths thousands of meters in length. For example, the wavelengths of X-rays are measured in nanometers, whereas radio waves can approach 10,000 meters, Figure 19.2.

Different types of electromagnetic radiation not only have different wavelengths, they also carry different amounts of energy. The shorter the wavelength of electromagnetic radiation, the more energy is carried by its photons. X-rays have very short wavelengths and consist of photons which carry a great deal of energy. In contrast, radio waves, which have long wavelengths, have photons with much less energy.

ii. THE VISIBLE PORTION OF THE ELECTROMAGNETIC SPECTRUM

Consider the portion of the electromagnetic spectrum to which the human eye is sensitive. Humans perceive different wavelengths of visible light as being different colors, Table 19.1. For example, if a light source emits light with wavelengths between about 505 and 555 *nm*, that light is perceived as green. "White" light emitted by the sun or an electric light bulb is **polychromatic**, *it is a mixture of many wavelengths.*

iii. THE UV PORTION OF THE ELECTROMAGNETIC SPECTRUM

Ultraviolet light is categorized into three types: UV-A (315–400 *nm*), UV-B (280–315 *nm*), and UV-C (180–280 *nm*), Table 19.2. UV-A is the type used in tanning booths and "black lights." The illumination from

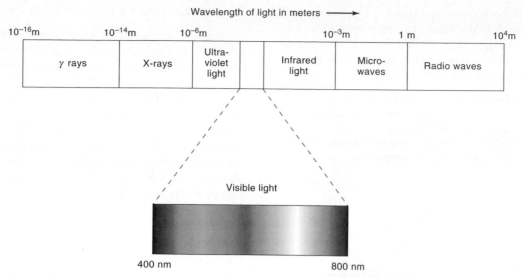

Figure 19.2.   Types of Electromagnetic Radiation and their Wavelengths. Visible light occupies a small portion of the electromagnetic spectrum. Within the visible portion of the spectrum, different wavelengths are associated with different colors. *(See color insert C-1.)*

Table 19.1   THE RELATIONSHIP BETWEEN THE WAVELENGTH OF VISIBLE LIGHT AND THE COLOR PERCEIVED

λ *of the Light (nm)*	*Color Our Eyes Perceive*
380–430	Violet
430–475	Blue
475–495	Greenish blue
495–505	Bluish green
505–555	Green
555–575	Yellowish green
575–600	Yellow
600–650	Orange
650–780	Red

Because color is subjective, these wavelengths are approximations. The particular ranges here are based on various sources.

Table 19.2   ULTRAVIOLET LIGHT

Type	*Wavelength*	*Examples of Applications*
UV-A ("long-wave")	315–400 *nm*	Used in "black lights" and tanning booths.
UV-B ("medium-wave")	280–315 *nm*	Used in electrophoresis to visualize DNA tagged with ethidium bromide.
UV-C ("short-wave")	180–280 *nm*	Used in bacteriocidal lamps and for inducing mutations in cells

a "black light" causes certain pigments to emit visible light; these materials are called **fluorescent**.

UV-B light has sufficient energy to damage biological tissues and is known to cause skin cancer. The atmosphere blocks most UV-B light; however, as the ozone layer thins, this form of radiation has become more of a health concern. UV-B light is routinely used in the biotechnology laboratory when visualizing DNA. The dye, ethidium bromide, intercalates into the grooves of DNA, forming a complex that fluoresces when exposed to UV-B radiation. The fluorescence makes the DNA visible to the eye and allows it to be photographed. There are other dyes that similarly fluoresce when exposed to UV light and can be attached to biological molecules. Fluorescence emitted by molecules allows investigators to visualize their location.

UV-C, with high energy and short wavelengths, is rarely observed in nature because it is absorbed by oxygen in the air. UV-C is used in germicidal lamps to kill bacteria. Most hoods manufactured for culturing cells are equipped with UV-C lamps to help prevent bacterial contamination of the cultures. UV-C light damages DNA in cells and therefore can be used in research to intentionally induce mutations in bacteria and cultured cells.

Safety Note: There are sources of UV light in the laboratory and it is important to avoid exposing your skin or eyes to this type of radiation. Safety glasses and goggles which specifically block UV radiation are available from scientific supply companies.

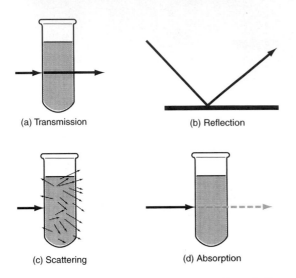

Figure 19.3. Events That Can Occur when Light Strikes Matter. a. The light may be transmitted through the object without interaction. **b.** The light may be reflected by the surface of the object. **c.** The light may be scattered by small individual particles. **d.** The light may lose energy to the matter in which case it is absorbed.

C. Interactions of Light with Matter

Various events can occur when light strikes matter, Figure 19.3. (1) The light can be **transmitted** (*i.e., pass without interaction through the material, as when light passes through glass*). (2) The light can be **reflected** (*i.e., change directions, as is light reflected by a mirror*). (3) The light can be **scattered**, *in which case the light is deflected into many different directions*. Scattering occurs when light strikes a substance composed of many individual, tiny particles. (4) The light can be **absorbed**, *in which case it gives up some or all of its energy to the material*. When light is absorbed, energy carried by photons is transferred to the electrons within a material. As this transfer occurs, light energy is converted to heat energy. Light absorption provides the basis for spectrophotometric measurements.

Because absorption is the basis of spectrophotometry, we consider it in more detail. Every material has a particular arrangement of electrons and of bonds involving electrons. For this reason, different materials vary from one another in terms of which wavelengths of light they absorb. The color of an object is determined by which wavelengths of light it absorbs. If an object absorbs light of a particular color, then that color does not reach our eyes when we look at the object. If an object absorbs all colors of light, then that object appears to be black, Figure 19.4a. If an object reflects visible light of a particular color, we see that color when we look at the object. An object appears white if it reflects all colors of light, Figure 19.4b. The object in Figure 19.4c appears to be orange because it reflects orange light and absorbs all other colors.

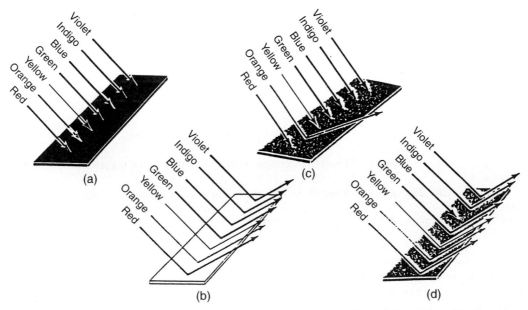

Figure 19.4. The Colors of Solid Objects. a. An object appears black if it absorbs all colors of light. **b.** An object appears white if it reflects all colors. **c.** An object appears orange if it reflects only this color and absorbs all others. **d.** An object also appears orange if it reflects all colors except blue, which is complementary to orange. *(See color insert C-1.)*

Observe that the object in Figure 19.4d reflects all colors except blue. When blue light is removed from the mixture of wavelengths, our eyes perceive the object to be orange. Orange and blue are called *complementary colors*. Figure 19.5 shows a color wheel in which complementary colors are across from one another. Looking at the color wheel, we can determine, for example, that if green is removed from a beam of light, then the light will appear to be red.

In this discussion of color, we have so far considered only solid objects that either absorb or *reflect* light. When we use a spectrophotometer, we are almost always working with samples that are in liquid form. These liquids may either absorb or *transmit* light of a particular wavelength. The spectrophotometer enables us to quantify how much light is absorbed and how much is transmitted by a solution at a specific wavelength.

Table 19.3 is a guide that shows the general relationship between the color of light absorbed by a liquid solution and the color the solution appears to the eye. Observe, for example, that if a solution absorbs blue light (light in the range from about 430 to 475 *nm*), then the solution will appear to be its complementary color, orange.

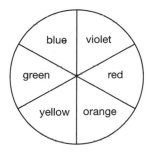

Figure 19.5.   A Color Wheel Showing Complementary Colors. *(See color insert C-2.)*

EXAMPLE PROBLEM

What color will a solution appear to be if it absorbs primarily light of that is 590 *nm*?

ANSWER

We know that 590 *nm* is yellow light. From Table 19.3 you can see that a solution that absorbs primarily yellow light will appear to be violet. Yellow and violet are complementary colors.

Consider the example illustrated in Figure 19.6. In Figure 19.6a, white light is shining on a test tube that contains a solution of red food coloring. The food coloring absorbs incoming light that is green. It also absorbs some light that is bluish-green and greenish blue. Light of other colors is transmitted through the food coloring. Because blue and greenish light is removed from the white light shining on the red food coloring, our eyes perceive the solution to be orange-red (see Table 19.3). A spectrophotometer enables us to graphically display this pattern of light absorption by the red food coloring solution, Figure 19.6b. In this display, wavelength is on the X axis. A measure of the amount of light absorbed by the food coloring is on the Y axis. Observe on the graph that there is a peak between about 490 *nm* and 530 *nm*. This peak represents the absorption of light at those wavelengths. The graph in Figure 19.6b, *that shows the extent to which a particular material absorbs different wavelengths of light, is called an* **absorbance spectrum**.

Note that red food coloring also absorbs light that has wavelengths shorter than 350 *nm*. This light, however, is in the ultraviolet range and therefore has no effect on the solution's apparent color.

Figure 19.7a shows the absorbance spectrum, as determined using a spectrophotometer, for the compound riboflavin. The absorbance spectrum of riboflavin has a pattern of "peaks" and "valleys" that is

Table 19.3   *THE ABSORPTION OF LIGHT OF PARTICULAR WAVELENGTHS AND THE COLOR OF SOLUTIONS*

Light Absorbed by the Solution		
Wavelength (nm)	Color	Color the Solution Appears To Be
380–430	Violet	yellow
430–475	Blue	orange
475–495	Greenish blue	red-orange
495–505	Bluish green	orange-red
505–555	Green	red
555–575	Yellowish green	violet-red
575–600	Yellow	violet
600–650	Orange	blue
650–780	Red	green

Note: Colors that appear alongside one another on this table are said to be "complementary" colors.

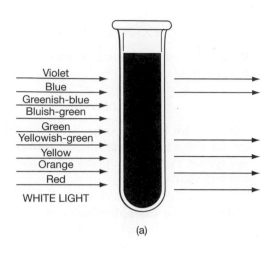

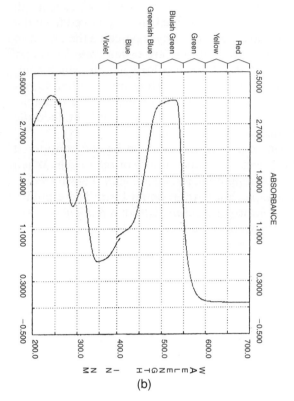

Figure 19.6. The Absorption of Light by a Solution of Red Food Coloring. a. When white light is shined on a tube containing red food coloring, blue to green colored light is absorbed. **b.** The absorbance spectrum for red food coloring, generated by a spectrophotometer. The X axis of an absorbance spectrum is wavelength, the Y axis is the amount of light absorbance. A peak of visible light absorption occurs between about 490 and 530 *nm*.

different than that of red food coloring. Every compound has its own characteristic absorbance spectrum. The absorbance spectrum of a particular compound always has the same pattern of peaks and valleys, regardless of how much of that compound is present, Figure 19.7b.

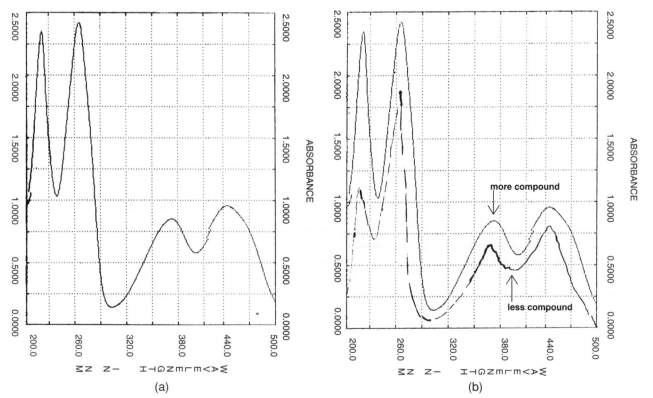

Figure 19.7. The Absorbance Spectrum for Riboflavin. a. The peaks and valleys of this absorbance spectrum are characteristic of riboflavin. **b.** Increasing amounts of riboflavin have the same spectral pattern; however, the height of the peaks is greater if more compound is present.

EXAMPLE PROBLEM

Solutions of proteins and of nucleic acids are colorless. Does this mean they do not absorb any light?

ANSWER

It means they do not absorb any light in the *visible* range of the electromagnetic spectrum. Proteins and nucleic acids do, in fact, absorb light in the *UV* range of the spectrum.

EXAMPLE PROBLEM

What color is the solution whose absorbance spectrum is shown in the following graph?

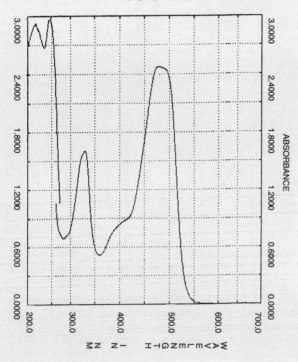

ANSWER

This compound has an absorbance peak in the greenish-blue region of the spectrum. From Table 19.3 we would expect it to be orange and it is, in fact, the dye Orange G.

EXAMPLE PROBLEM

Chlorophyll is one of the pigments responsible for the green color of leaves. Chlorophyll enables the leaves to absorb light from the sun and convert the energy of the light into food. Does chlorophyll capture all wavelengths of light from the sun?

ANSWER

No. Leaves are green because chlorophyll does not absorb much green light and those wavelengths are reflected.

EXAMPLE PROBLEM

What happens if red light is shined on a red solution?

ANSWER

The solution will not absorb the light; rather, it will transmit the light incident on it.

II. THE BASIC DESIGN OF A SPECTROPHOTOMETER

A. Transmittance and Absorbance of Light

i. TRANSMITTANCE

Figure 19.8 illustrates the basic principles underlying the operation of a spectrophotometer. In this illustration, two solutions are placed inside a spectrophotometer. The first is red food coloring dissolved in water. The analyte is red food coloring and water is the solvent. The second solution is pure water (i.e., the solvent used to dissolve the red food coloring). The water is called the **blank**. (We consider the meaning of a blank in more detail later.) A beam of green light is incident on (shines on) both solutions. As we know from Figure 19.6, the red food coloring will transmit little or no green light. In contrast, the water will readily transmit green light. A spectrophotometer electronically compares the amount of light transmitted through the sample with that transmitted through the blank. *The ratio of the amount of light transmitted through a sample to that transmitted through the blank is called the* **transmittance**, abbreviated **t**:

$$\frac{light\ transmitted\ through\ the\ sample}{light\ transmitted\ through\ the\ blank} =$$
$$transmittance = t$$

It is also common to speak of **percent transmittance**, **%T**, *which is the transmittance times 100%:*

$$\%\ T = t \times 100\%$$

In the example illustrated in Figure 19.8, the transmittance is less than 1 because the amount of light transmitted through the sample is less than the amount transmitted through the blank. (The red food coloring in the sample absorbs some of the light.) In another situation, where both the sample and the blank transmit the same amount of light, the transmittance is 1, or %T is 100%. If the sample transmits no light at all, then the transmittance is 0. Thus, transmittance can range from 0 to 1 and percent transmittance from 0% (no light passed through the sample relative to the blank) to 100% (all light was transmitted).

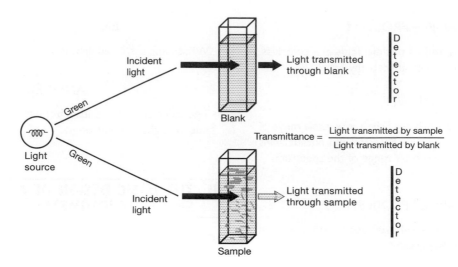

Figure 19.8.   Transmittance. Two solutions are placed inside a spectrophotometer: red food coloring and a blank consisting of pure water. Green light is shined on the two solutions and the amount of light transmitted through each is measured by a detector. The ratio of the amount of light transmitted through the sample to that transmitted through the blank is called the *transmittance*.

ii. ABSORBANCE

All spectrophotometers measure transmittance (i.e., the ratio between the amount of light transmitted through a sample and the amount of light transmitted through a blank). Analysts, however, are generally interested in the amount of light absorbed by a sample, for example as illustrated in Figures 19.6 and 19.7. It is customary, therefore, to convert transmittance to a measure of light absorption, called **absorbance**. **Absorbance** *is calculated based on the transmittance* as:*

$$A = -\log_{10}(t)$$

Absorbance is also called **optical density, OD**.

Absorbance does not have units, but absorbance values are sometimes followed by the abbreviation A or AU. For example, an absorbance value might be written as "1.6," or "1.6 A," or "1.6 AU," or "OD 1.6." Transmittance and absorbance are measured at a particular wavelength, so it is customary to record both the absorbance and the wavelength. For example, "A_{260} = 1.6," means "an absorbance of 1.6 was measured at an analytical wavelength of 260 *nm*."

Figure 19.9a shows the relationship between how much light is transmitted through a solution and the concentration of analyte in that solution. Observe that the relationship between analyte concentration and transmittance is not linear, Figure 19.9a. In contrast, there is a linear relationship between the absorbance of light by a sample and the concentration of analyte, Figure 19.9b. For this reason, absorbance values are more convenient than transmittance values.

*An alternative equation to calculate the absorbance is $A = 2 - \log_{10}(\%T)$. Both this equation and the one in the body of the text will give the same answer.

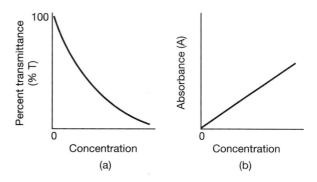

Figure 19.9.   The Relationship between % Transmittance and Absorbance of Light and the Concentration of Analyte. **a.** Concentration of analyte in a sample versus percent transmittance is not linear. **b.** In contrast, the relationship between absorbance and analyte concentration is linear.

As one would expect, when the absorbance of light increases, the transmittance decreases, and vice versa, Table 19.4. For example, if the percent transmittance is 100%, then there is no light absorbance by the sample relative to the blank. Suppose in another case that a sample absorbs so much light that the percent transmit-

Table 19.4   THE RELATIONSHIP BETWEEN ABSORBANCE AND TRANSMITTANCE, $A = -LOG_{10}(t)$

t	$\%T$	A
0.000	0.00	∞
0.001	0.10%	3
0.010	1.00%	2
0.100	10.0%	1
1.000	100%	0

tance is only 0.1%. In this example, the transmittance (*t*) is 0.001 and the absorbance is:

$$A = -\log(0.001) = -\log(10^{-3}) = -(-3) = 3$$

Most spectrophotometers can convert transmittance values to absorbances, so they can display either absorbance or transmittance. It is also possible to use a calculator to convert transmittance to absorbance, and vice versa.

EXAMPLE PROBLEM

Convert a transmittance value of 0.63 to absorbance.

ANSWER

$$A = -\log(t)$$
$$A = -\log(0.63)$$
$$A = -(-0.20) = 0.20$$

EXAMPLE PROBLEM

Convert an absorbance value of 0.380 to transmittance and % transmittance.

ANSWER

$$0.380 = -\log(t)$$
$$\text{antilog}(-0.380) = t$$
$$0.417 = t \text{ or } 41.7\% = \% T$$

EXAMPLE PROBLEM

A solution contains 2 *mg/mL* of a light absorbing substance. The solution transmits 50% of incident light relative to a blank. a. What is the absorbance in this case? b. What is the expected absorbance of a solution containing 8 *mg/mL* of the same substance? c. What is the expected transmittance of a solution containing 8 *mg/mL* of the same substance?

ANSWER

a. The absorbance of 2 *mg/mL* of this substance is equal to $-\log(0.50) = 0.30$.

b. The relationship between absorbance and concentration is linear (within a certain range, as discussed in the next chapter). Assuming linearity, therefore, if 2 *mg/mL* has an absorbance of 0.30, then 8 *mg/mL* should have an absorbance of 1.20.

c. If the absorbance is 1.20, the transmittance = antilog $(-1.20) = 0.063 = 6.3\%T$. Observe that the relationship between concentration and transmittance is not linear.

iii. THE BLANK

A spectrophotometer is used to determine the absorbance of light by the analyte in a sample. In order to make this determination, a spectrophotometer always compares the interaction of light by a sample with its interaction with a blank, as was illustrated in Figure 19.8. Let us consider in more detail the composition of the blank.

An analyte is seldom alone in a sample solution; rather, it is dissolved in a solvent and sometimes various reagents are added to the sample as well. When the sample is placed in the spectrophotometer, not only the analyte may absorb light, but the solvent and added reagents may absorb light as well. Moreover, the sample is contained in a *holder, called a* **cuvette** (or **cell**) which may absorb a small amount of light. The **blank** *should contain no analyte, but it should contain the solvent and any reagents that are intentionally added to the sample.* The blank is held in a cuvette that is identical to that used for the sample, or the blank is alternately placed in the same cuvette as the sample. In the example in Figure 19.8, the blank is pure water because the analyte, red food coloring, was dissolved in water. In another situation, the blank may have a different composition.

B. Basic Components of a Visible/UV Spectrophotometer

i. OVERVIEW

We can now consider in more detail the various components that make up a spectrophotometer. A basic spectrophotometer consists of a source of light, a wavelength selector, a sample holder, a detector, and a display, Figure 19.10.

ii. LIGHT SOURCE

The **light source**, also called a **bulb** or **lamp**, *produces the light used to illuminate the sample.* There are various types of lamps available for spectrophotometry. All lamps produce polychromatic light, but each type emits a different range of wavelengths. *The light used for visible spectrophotometry is usually produced by bulbs with a* **tungsten filament**. A **deuterium arc lamp** *is commonly used to produce light in the UV range.* Spectrophotometers that can be used in both the visible and UV ranges have two light sources.

Every type of lamp emits more light of some wavelengths than others. For example, a tungsten lamp has maximum emission at wavelengths above about 650 *nm*. The UV emission maximum for deuterium arc lamps is at about 240 *nm*.

iii. WAVELENGTH SELECTOR

Light emitted by the spectrophotometer light source is polychromatic, but the absorption of light by a sample is wavelength-specific. Spectrophotometry measures

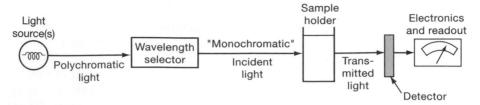

Figure 19.10.   Simplified Design of a Spectrophotometer. Polychromatic light is emitted by the light source. A narrow band of light is selected in the wavelength (radiation) selector. The light passes through the sample (or the blank), where some or all of it may be absorbed. The light that is not absorbed falls on the detector.

how much light of a particular wavelength is absorbed by a sample. Spectrophotometers, therefore, require a wavelength selecting device. The wavelength selecting component of a spectrophotometer isolates light of a particular wavelength (in practice, it isolates a narrow range of wavelengths) from the polychromatic light emitted by the source lamp.

Some instruments, called **photometers**, use filters as an inexpensive wavelength selector. A **filter** *blocks all light above or below a certain cutoff,* Figure 19.11a, b. Two filters can be combined, as shown in Figure 19.11c, to make a filter that transmits light in a certain range.

Filters cannot produce **monochromatic light** *(i.e, light of a single wavelength).* It is more effective to use a **monochromator** to produce light of a very narrow range of wavelengths. **Monochromators** *are devices that can disperse (separate) polychromatic light from the light source into its component wavelengths, and that can be used to select light of desired wavelengths.* A monochromator disperses light, much as light is dispersed into a rainbow by a prism, Figure 19.12.

The three main components of a monochromator, Figure 19.13, are:

1. **Entrance slit.** Light from the lamp is emitted in all directions. *The entrance slit shapes the light entering the monochromator into a parallel beam that strikes the dispersing element.*

2. **Dispersing element.** In earlier spectrophotometers, prisms were used to disperse light into its components. In modern instruments, **diffraction gratings** are the most common type of dispersing element. **Diffraction** *is the bending of light that occurs when light passes an obstacle.* **Diffraction gratings** *consist of a series of microscopic grooves etched onto a mirrored surface.* The grooves diffract light that strikes them in such a way that polychromatic light is dispersed into its component wavelengths.

3. **Exit slit.** *The exit slit permits selected light to exit the monochromator.*

After light is dispersed by the diffraction grating it is necessary to select light of the desired wavelength. This is accomplished by rotating the grating so that the light that reaches the exit slit is of the desired wavelength, Figure 19.14. The accuracy of wavelength selection in a spectrophotometer depends on the correct mechanical operation of the diffraction grating.

If a perfect monochromator could be produced, and if the spectrophotometer were set to select a particular wavelength, then only light of that single wavelength would illuminate the sample, Figure 19.15a. In reality, no spectrophotometer is perfect and the light that emerges from a monochromator consists of light span-

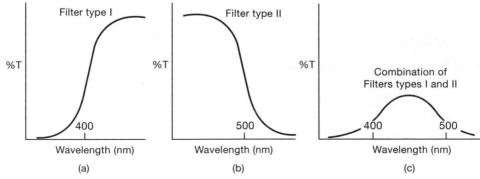

Figure 19.11.   Filters. a. A filter that blocks the transmission of light below 400 *nm.* **b.** A filter that blocks the transmission of light above 500 *nm.* **c.** Two filters can be combined so as to allow only light in a certain wavelength range to pass. In this example, light between 400 and 500 *nm* is transmitted through the filter.

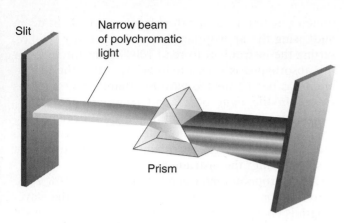

Figure 19.12. The Dispersion of Light into its Component Wavelengths by a Prism. *(See color insert C-2.)*

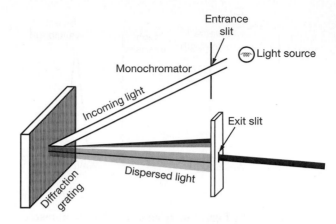

Figure 19.13. The Components of a Monochromator. The monochromator consists of an entrance slit, a dispersing element, and an exit slit. The dispersing element is a diffraction grating which consists of a series of microscopic grooves etched on a mirrored surface.

ning a narrow range of wavelengths, centered at the chosen wavelength, Figure 19.15b.

Modern spectrophotometers are well-designed so that, in most situations, the light emerging from the monochromator has a sufficiently narrow span that it can be treated as if it were truly monochromatic. There are situations, however, where the "monochromaticity" of the light incident on the sample affects an analysis. The Appendix to this chapter discusses issues relating to the "monochromaticity" of light exiting the monochromator.

iv. Sample Chamber

The sample (or the blank) is contained in a cuvette, which is placed inside a light-tight **sample chamber**. The **path length** *is the distance that the light travels through the sample or blank; it is determined by the interior dimension of the cuvette.* It is customary to use a 1 *cm* path length when working with sample volumes of 1–5 *mL*. Manufacturers

make systems, including cuvettes and cuvette holders, that permit the analysis of small volumes, down to the microliter range. Being able to work with small volumes is useful in molecular biology applications where reaction mixes and samples are often less than 50 *μL*. Note that the path length in small volume systems may not be 1 *cm*.

v. Detector

The **detector** *senses the light coming from the sample (or the blank) and converts this information into an electrical signal.* The magnitude of the electrical signal is proportional to the amount of light reaching the detector.

The most common type of detector in spectrophotometers is called a **photomultiplier tube (PMT)**. A PMT contains a series of metal plates coated with a thin layer of a **photoemissive material** *that emits electrons when it is*

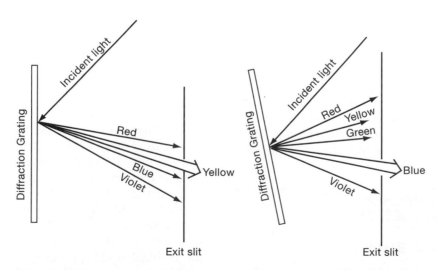

Figure 19.14. Rotating the Grating so that Light of a Selected Wavelength Falls on the Exit Slit.

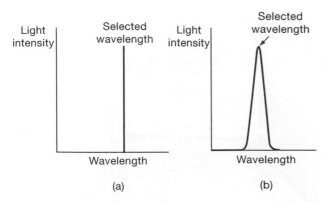

Figure 19.15. The Light Emerging from a Monochromator.
a. If the light from a monochromator were truly monochromatic, then when the instrument was set for a certain wavelength, only that wavelength would be incident on the sample. **b.** In reality, the light incident on a sample consists of a narrow band of wavelengths, centered on the selected wavelength. (Note that this figure is not an illustration of a sample spectrum, as are Figures 19.6b and 19.7. Rather, it illustrates the wavelengths of light that are incident on the sample.)

struck by photons. Electrons are held loosely by the photoemissive material so the energy of light is sufficient to "knock" electrons from the plates. Many electrons are released by each incident photon, and each electron can knock out even more electrons, resulting in a cascade that amplifies the original light signal. The stream of emitted electrons constitutes an electrical signal that is processed by the detector assembly and is converted to a reading of transmittance and/or absorbance.

All detectors have limitations. For one thing, detectors are not accurate at very high or very low levels of transmittance. It is best, therefore, to work with samples whose absorbance is in the midrange of the instrument. Second, detectors are more responsive to some signals than others. In spectrophotometry, this means that the detector will respond more strongly to some wavelengths of light than others.

vi. DISPLAY

There are many ways that the signal from the detector can be manipulated and displayed. The simplest method is to use a meter to display the resulting value. A meter has a needle that moves to indicate the value either on a scale of transmittance or absorbance. The signal can also be converted into a digital display. Modern instruments send the signal to a printer and/or to a computer for display, storage, and analysis.

III. MAKING MEASUREMENTS WITH SPECTROPHOTOMETERS

A. More About the Blank

Recall that the transmission of light through a sample is always compared with the transmission of light through a blank. This is accomplished in a simple spectropho-

tometer by first inserting the blank into the light beam and, using the appropriate knob, dial, or key strokes, setting the instrument to read 100% transmittance (or the absorbance is set to read zero). After the instrument has been adjusted with the blank, the sample is placed in the light beam and transmittance and/or absorbance is measured and displayed. In the normal operation of a spectrophotometer, it is never correct to attempt to measure the absorbance of a sample without first "blanking" the instrument.

A spectrophotometer needs to be set to 100% T (or zero absorbance) using the blank every time the wavelength is changed. This is because, as noted previously, the lamp emission intensity varies at different wavelengths. The monochromator efficiency and the detector's sensitivity to light also change as the wavelength changes. It is necessary to compensate for these instrumental variations by resetting the instrument to 100% T (or zero absorbance) using a blank every time the wavelength is changed. Moreover, even if the wavelength of an instrument were never changed, the spectrophotometer's responsiveness to a sample would change over time due to power line fluctuations, aging of the bulb, and other factors. Setting the instrument with the blank compensates for all these factors. Thus, "blanking" the instrument compensates for the effects of:

- absorption of light by materials in the sample other than the analyte, such as the solvent and reagents
- absorption of light by the cuvette
- variations in light emission as the light source ages
- fluctuations in power levels
- variations in light emission intensity with wavelength
- variations in the sensitivity of the detector to different wavelengths of light

B. The Cuvette

Glass and certain types of plastic are transparent to light in the visible range, but they block UV light. These materials are therefore used to make cuvettes for work in the visible range, Figure 19.16a. Quartz glass is transparent to visible and UV light, so it is used to make cuvettes for both visible and ultraviolet work, Figure 19.16b. There are also disposable plastic cuvettes that may be used down to about 285 *nm*, Figure 19.16c. Note that the optical quality of plastic cuvettes may not be suitable for some applications.

Even the highest quality cuvettes absorb a small amount of light. It is possible, therefore, to purchase paired cuvettes that are manufactured to be identical in their light absorbing properties. One of the pair is used for the blank and the other for the sample.

Cuvettes are carefully manufactured, precision optical devices that must be properly maintained. Small scratches and dirt will absorb or scatter light and there-

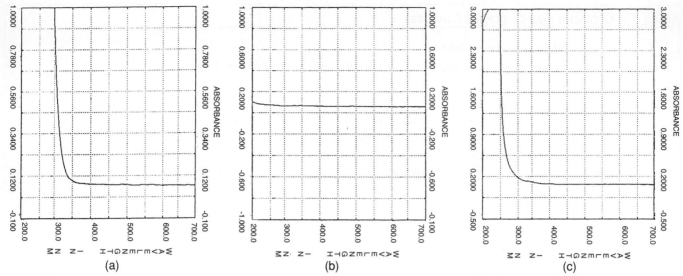

Figure 19.16.   Absorbance of Light by Cuvettes Made of Various Materials. a. The absorbance spectrum of a glass cuvette intended for work in the visible range. This cuvette is unsuitable for work below about 330 *nm*. **b.** The absorbance spectrum of a quartz cuvette that is suitable for work down at least 200 *nm*. **c.** The absorbance spectrum of a disposable polyacrylate plastic cuvette sold for work in the UV range down to about 290 *nm*. (Air was used as the blank for all three spectra.)

fore damage the cuvette. Table 19.5 summarizes considerations relating to cuvettes. Box 1 outlines methods for determining whether two cuvettes match, for checking the cleanliness of cuvettes, and for cuvette cleaning.

C. The Sample

Samples should be well-mixed and homogenous without air bubbles. Particulates should be avoided (except in certain applications where turbid samples, such as bacterial suspensions, are assayed).

The solvent is an important part of the sample. Although solvents, such as water, buffers, or alcohol, appear transparent, they absorb light at certain wavelengths in the UV range. At wavelengths below a particular cutoff value solvents absorb so much light that they interfere with the analysis of a sample. Table 19.6 is a general guide to typical solvents for UV/Vis work and the approximate cutoffs at which they begin to absorb light. The actual absorbance cutoffs for solvents vary

Table 19.6   APPROXIMATE UV CUTOFF WAVELENGTHS FOR COMMONLY USED SOLVENTS

Solvent	UV Cutoff (*nm*)
Acetone	320
Acetonitrile	190
Benzene	280
Carbon tetrachloride	260
Chloroform	240
Cyclohexane	195
95% Ethanol	205
Hexane	200
Methanol	205
Water	190
Xylene	280

depending on their grade and purity, the pH, the calibration of the spectrophotometer, and other factors. When working in the UV range, therefore, it may be

---

Table 19.5   PROPER SELECTION AND CARE OF CUVETTES

1. *Use quartz cuvettes for UV work; glass, plastic, or quartz are acceptable for work in the visible range.*

2. *Matched cuvettes are manufactured to absorb light identically so that one of the pair can be used for the sample and the other for the blank.*

3. *Make sure the cuvette is properly aligned in the spectrophotometer.*

4. *Do not touch the base of a cuvette or the sides through which light is directed.*

5. *Do not scratch cuvettes; do not store them in wire racks or clean with brushes or abrasives.*

6. *Do not allow samples to sit in a cuvette for a long period of time.*

7. *Disposable cuvettes are often recommended for colorimetric protein assays because dyes used for proteins tend to stain cuvettes and are difficult to remove.*

BOX 1   CLEANING AND CHECKING CUVETTES

---

1. *Be certain to only use clean cuvettes.*

2. *Wash cuvettes immediately after use. Recommended washing methods include:*

   Rinse with distilled water immediately after use.

   Wipe the outside of glass and quartz cuvettes with lens paper.

   After an aqueous sample, wash in warm water, rinse with dilute detergent, thoroughly rinse with distilled water. Avoid detergents with lotions. Be certain all detergent is removed after washing.

   For organic samples, rinse thoroughly with a spectroscopy grade solvent.

   If simple rinses are not sufficient to remove all traces of samples, an acid cuvette washing solution may be used. For example:

   > To make 1 liter of washing solution
   > 425 *mL* of distilled water
   > 525 *mL* of ethanol
   > 50 *mL* of concentrated acetic acid

   Do not allow acid to remain in the cuvette for more than 1 hour. Rinse thoroughly with distilled water.

   An enzyme-containing detergent, such as "Tergazyme," is sometimes recommended to remove dried proteins from cuvettes.

   Avoid blowing air into cuvettes to dry them.

3. *For high accuracy work, the cleanliness of cuvettes and matching response of two cuvettes can be checked as follows (from ASTM E-275-93, "Standard Practice for Describing and Measuring Performance of Ultraviolet, Visible, and Near-Infrared Spectrophotometers"):*

   1. Set the instrument to 100% T (or 0 A) with only air in the sample holder.

   2. Fill the cuvette with distilled water and measure its absorbance at 240 *nm* (quartz cuvette) or 650 *nm* (glass). For scanning instruments scan across the spectral region of interest. The absorbance should not be greater than 0.093 for 1 *cm* quartz cuvettes or 0.035 for glass.

   3. Rotate the cuvette 180 degrees in its holder and measure the absorbance again. Rotating the cuvette should give an absorbance difference less than 0.005.

   4. If an instrument is to be used with two cuvettes, one for the sample and one for the blank, check how well the two cuvettes are matched. Fill each with solvent and measure the absorbance of the sample cuvette. The absorbance difference between the two cuvettes should be less than 0.01.

---

necessary to prepare an absorbance spectrum of the solvent to check whether it absorbs appreciable amounts of light at the analytical wavelength(s). Solvents that are manufactured specifically for use in spectrophotometry are labeled "Spectroscopy" or "Spectro-Grade."

Figure 19.17a shows the absorbance spectrum for a commonly used biological buffer, Tris. Tris absorbs light in the UV region of the spectrum. Figure 19.17b shows the absorbance for denatured alcohol (ethanol). Note the strong absorbance peak at around 270 *nm*, a region of importance in the analysis of DNA, RNA, and proteins. Figure 19.17c is the spectrum for absolute (pure) ethanol, which has little absorbance above about 240 *nm*. Figure 19.17d is the spectrum for acetone.

EXAMPLE PROBLEM

You place a blank in a spectrophotometer but the blank has so much absorbance that the instrument cannot be set to zero absorbance. How should you proceed?

ANSWER

Think of factors that might account for this absorbance, such as:

1. The blank might be cloudy or turbid. Check its appearance. Cloudy or turbid samples and blanks generally should be avoided.

2. If the wavelength the instrument is set to read is in the ultraviolet range, make sure the cuvette is quartz or plastic suitable for UV use. If not, use a different cuvette.

3. Check that the cuvette is properly placed in the sample holder and is in the correct orientation. Many cuvettes have two sides that are transparent to light and two sides that are not transparent. The cuvette must be placed properly in the sample holder so that light can penetrate the cuvette.

4. Check the cuvette to make certain it is clean, without fingerprints, or liquid droplets on the outside.

5. Some solvents appear transparent but absorb light in the UV range. Be certain that the solvent does not absorb light at the wavelength being used, Table 19.6.

IV. DIFFERENT SPECTROPHOTOMETER DESIGNS

A. Scanning Spectrophotometers

The spectrophotometer illustrated in Figure 19.10 is a simple, **single beam** instrument. Instruments with this basic design have been used successfully for years. Basic single-beam spectrophotometers, however, have the disadvantage that the operator must reset the instru-

ment to 100% transmittance (or zero absorbance) with the blank each time the wavelength is changed. This process of manually selecting each wavelength and repeatedly "blanking" the instrument is tedious when an absorbance spectrum is being prepared. Manufacturers, therefore, have developed instruments called **scanning spectrophotometers** *that are capable of rapidly scanning through a range of wavelengths and constructing an absorbance spectrum.*

A **double-beam scanning spectrophotometer** is illustrated in Figure 19.18. Double-beam instruments have two sample holders. Both the sample and the blank are placed in the instrument at the same time so that the absorbance of the sample can be continuously and automatically compared with that of the blank. The instru-

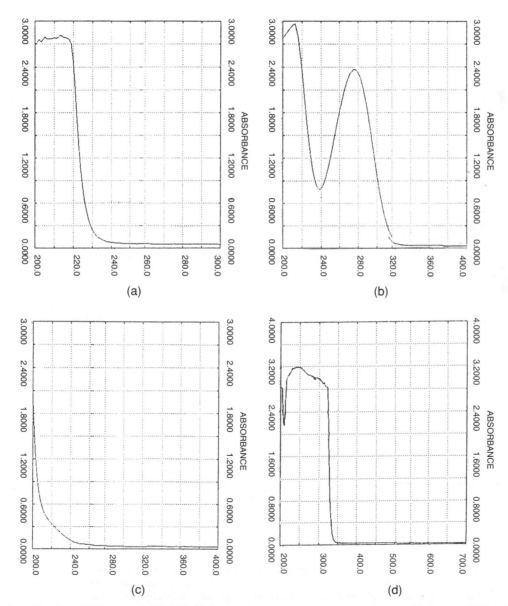

Figure 19.17.   The Absorbance Spectra of Various Solvents. a. 1 *M* Tris (Sigma) buffer absorbs strongly below about 220 *nm*. **b.** Denatured ethanol absorbs strongly in the UV region. (Fisher Scientific ethyl alcohol containing 1% ethyl acetate, 1% methyl isobutyl ketone, 1% hydrocarbon solvent, and methanol.) **c.** The absorbance spectrum of pure ethanol (Fisher Scientific). **d.** The absorbance spectrum for acetone (general use grade, Caledon Laboratories Limited).

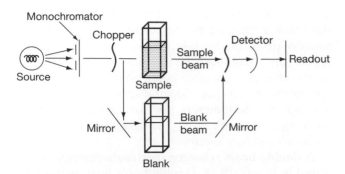

Figure 19.18.   A Double-Beam Scanning Spectrophotometer. Light emitted by the source passes through a monochromator. After the monochromator, the "chopper" rapidly alternates the beam between the sample and the blank. The output of the detector is proportional to the ratio between the blank and the sample beams and so is a measure of transmittance.

ment illustrated alternates the light beam between the blank and the sample many times per second. (Note that Figure 19.8 also illustrates the principle of a double beam instrument.)

Another type of scanning spectrophotometer takes advantage of microprocessor technology. In this type of instrument, the operator first places the blank in the sample compartment. The spectrophotometer scans the absorbance of the blank at all the wavelengths of interest as a microprocessor inside the instrument "memorizes" the values. The blank is then removed from the sample compartment and the sample is similarly scanned across the wavelength range. The microprocessor compares the absorbance of the blank to the absorbance of the sample at each wavelength and generates a spectrum automatically.

Scanning spectrophotometers often allow the operator to control the *rate at which the sample is scanned, the* **scan speed.** The scan speed is restricted to the rate at which the instrument can respond. A scan speed that is too fast creates an effect called *tracking error.* When **tracking error** *occurs, the absorbance peaks recorded for a sample are slightly shifted from their true locations.*

B. Instruments with Photodiode Array Detectors

All the single- and double-beam instruments described so far have a photomultiplier tube (PMT) detector. Instruments with PMTs measure the transmittance (and absorbance) of a sample one wavelength at a time and therefore acquire a spectrum over a period of seconds or minutes. Manufacturers also make "higher end" spectrophotometers that are equipped with a different type of detector called a **photodiode array detector (PDA).** A **PDA** *can determine the absorbance of a sample over the entire UV/Vis range virtually instantaneously.*

Photodiode array detectors are based on the electronic components, diodes. Each diode in a PDA is calibrated to respond to a specific wavelength. When struck with light of that wavelength, the voltage of the diode changes. This voltage change can be converted to an electrical signal. A PDA contains a two-dimensional array of hundreds of diodes spaced closely together.

In an instrument with a PDA, all the light from the source is directed at the sample rather than being dispersed in a monochromator. There is a monochromator located *after* the sample chamber that disperses the light that has been transmitted through the sample or the blank, Figure 19.19. The dispersed, transmitted light strikes the photodiode array. Each diode responds to light at its calibrated wavelength and sends a signal to a

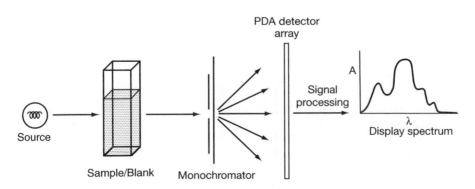

Figure 19.19.   The Optical Design of an Instrument with a PDA. The monochromator is located after the sample compartment. Each diode in the PDA array responds to light at its calibrated wavelength and sends a signal to a computer. The computer compares the amount of light transmitted by the sample and the blank and converts this information to a complete absorbance spectrum for the sample.

computer. For every wavelength in the spectrum, the computer compares the amount of light transmitted by the sample and the blank. The computer converts this information to a complete absorbance spectrum for the sample.

C. Microprocessors and Spectrophotometers

Microprocessors play a wide range of roles in modern spectrophotometers. Microprocessors control wavelength scanning, generate absorbance spectra, store and retrieve information, perform calculations and statistics on data, and plot data on graphs. A variety of software is available that simplifies routine operations. Many of the applications described in Chapter 20 are now automated.

V. QUALITY CONTROL AND PERFORMANCE VERIFICATION FOR A SPECTROPHOTOMETER

A. Performance Verification

i. OVERVIEW

The performance of a spectrophotometer needs to be periodically checked and the results documented. If necessary, the instrument must be repaired and/or calibrated to bring its readings into accordance with the values of accepted standards. To decide whether an instrument is performing satisfactorily, its performance is compared with specifications established by the manufacturer for each model. (In some cases, spectrophotometers will also need to meet specifications established by regulatory authorities). The methods that are used for verifying the performance of an instrument come from various sources including manufacturers, the U.S. Pharmacopeia, ASTM, and other standards and regulatory authorities.

The frequency with which the performance of a spectrophotometer is verified depends on the laboratory, the uses of the instrument, and whatever regulations or standards apply. Instruments that are handled roughly, or are used in environments that are dusty, have chemical vapors, or vibration will require more frequent calibration and maintenance than other instruments.

Modern spectrophotometers are sufficiently complex that the individuals who operate the instruments are usually not the ones who check their performance. Performance verification is typically performed by trained service personnel, as is most repair and adjustment. Regardless of who performs the performance checks and repairs on a spectrophotometer, however, the operator should be aware of the characteristics of the instrument that affect its capabilities, range, and its

ability to give accurate, reproducible results. Understanding these characteristics is important, for example, in the following situations:

- *When purchasing a spectrophotometer it is important to choose a model whose capabilities match the applications for which it will be used.*
- *When developing a spectrophotometry method for others to follow, the originator must describe the instrument characteristics required to duplicate the method.*
- *When following standard methods written by others, an analyst must ascertain that her/his spectrophotometer meets the requirements of the method.*
- *In situations where "difficult" samples are analyzed or high accuracy is required, it is important to have an instrument with the necessary features.* For example, if the absorbance of samples with very high levels of analyte is to be measured, some models of spectrophotometer will perform better than others.

The performance characteristics that determine the accuracy, precision, and range of operation of a spectrophotometer include (but are not limited to):

Calibration, which has two components:

  Wavelength Accuracy

  Photometric Accuracy

Linearity of Response

Stray Light

Noise

Baseline Stability

Resolution (*discussed briefly here and in more detail in the Appendix to this chapter*)

Note: For microprocessor-controlled instruments, the performance of the software should also be verified as described by the manufacturer.

This section explains performance characteristics of a spectrophotometer. More information on calibration and performance verification is provided in the articles listed in the reference section.

ii. CALIBRATION

Calibration of a spectrophotometer brings the readings of an individual instrument into accordance with nationally accepted values. Calibration is therefore part of routine quality control/maintenance for a spectrophotometer. There are two parts to calibrating a spectrophotometer: **wavelength accuracy** and **photometric accuracy**. **Wavelength accuracy** *is the agreement between the wavelength the operator selects and the actual wave-*

length that exits the monochromator and shines on the sample. **Photometric accuracy**, *or* **absorbance scale accuracy**, *is the extent to which a measured absorbance or transmittance value agrees with the value of an accepted reference standard.*

Consider wavelength accuracy. Recall that a monochromator always generates a small range of wavelengths, not a single wavelength (Figure 19.15). If an instrument is well-calibrated with respect to wavelength accuracy, then when it is set to 550 *nm*, the center of the wavelength peak incident on the sample will be 550 *nm* plus or minus a certain tolerance.

Wavelength accuracy is determined using certified standard reference materials (SRMs) available through NIST, or standards that are traceable to NIST. An absorbance spectrum for the reference standard is prepared in the instrument whose performance is being checked. The absorbance peaks for reference standards are known, so the wavelengths of the peaks generated by the instrument can be checked for accuracy. The manufacturer specifies the wavelength accuracy of a given instrument. For example, a high-performance instrument may be specified to a wavelength accuracy with a tolerance of ± 0.5 *nm*. A less-expensive instrument may be specified to have a wavelength accuracy with a tolerance of ± 3 *nm*. Figure 19.20 shows the spectrum for a commonly used reference standard, a **holmium oxide filter**.

Wavelength accuracy needs to be periodically checked. Wavelength accuracy can deteriorate if the source lamp, associated mirrors, and other parts are not properly aligned. This might occur, for example, after replacing a bulb. Problems leading to wavelength inaccuracy can also arise in the monochromator and in the display device.

Photometric accuracy ensures that if the absorbance of a given sample is measured in two different spectrophotometers at the identical wavelength and under the same conditions, the readings will be the same, and will correspond to accepted values. This is a difficult objective to achieve, even with properly calibrated instruments, because spectrophotometers differ in their optics and designs.

Absorbance scale accuracy is determined using SRMs available through NIST, or standards that are traceable to NIST. NIST establishes the transmittance of its standards using a special high performance spectrophotometer. Thus, in the United States, the meaning of a transmittance value (and therefore of an absorbance value) is defined by NIST.

For many biological applications of spectrophotometry, photometric accuracy is not critical. As long as an individual instrument is internally consistent and its readings are linear and reproducible, it will work effectively to measure the concentration of analytes, as is discussed in detail in the next chapter. There are, however, a few applications where a properly calibrated instrument is essential. This is the case, for example,

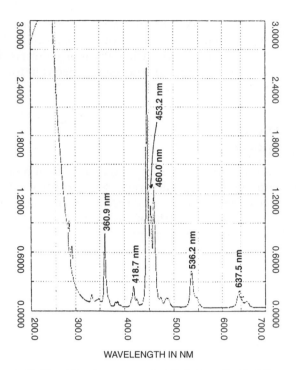

Figure 19.20. The Spectrum of a Holmium Oxide Filter Used to Test the Wavelength Accuracy of a Spectrophotometer. The filter is placed in the spectrophotometer and its spectrum is generated. If the instrument is properly calibrated with respect to wavelength, then the peaks in the spectrum will correspond to the specified wavelengths. (Based on wavelengths in ASTM E275 "Standard Practice for Describing and Measuring Performance of Ultraviolet, Visible, and Near-Infrared Spectrophotometers.")

where spectrophotometric measurements from various laboratories are being compared.

iii. STRAY LIGHT (STRAY RADIANT ENERGY)

Stray light *is radiation that reaches the detector without interacting with the sample.* This may occur if light is scattered by various optical components of the monochromator or sample chamber. This scattered light does not pass through or interact with the sample, but some of it may fall on the detector, resulting in an erroneous transmittance value. Scratches, dust, and fingerprints on the surface of cuvettes can also deflect light from its proper path. Stray radiation is reduced by keeping surfaces and cuvettes clean and by avoiding fingerprints and scratches. Another source of unwanted light is leaks of room light into the sample chamber.

When the absorbance of a sample is very high, the amount of transmitted light reaching the detector becomes vanishingly small. If there is stray light present, that stray light will be erroneously considered as transmitted light. The presence of stray light, therefore, is a factor that limits the accuracy of a spectrophotometer at high absorbance values.

Figure 19.21 shows the effect of stray radiation on the relationship between concentration and absorbance.

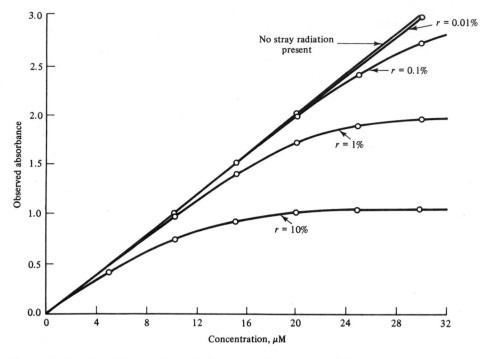

Figure 19.21. The Effect of Stray Light. In the absence of stray light, the relationship between concentration and absorbance is perfectly linear. When stray light is present, it causes the relationship to deviate from linearity at higher absorbance levels. The more stray light (r) present (expressed here as a percent), the lower the absorbance level at which this deviation occurs.

That relationship should ideally be linear at all absorbance values. In Figure 19.21, r is a value that represents the level of stray light: The higher the value for r, the more stray light present. Observe that given an instrument with high levels of stray light (i.e., where $r = 10\%$), the relationship between absorbance and concentration begins to be nonlinear at absorbances of less than 1. A moderately priced spectrophotometer might be expected to have a stray light value on the order of 1%; therefore, it can be expected to be linear in response until the absorbance approaches 2. More expensive spectrophotometers may have negligible levels of stray light and may respond in a linear fashion at absorbances of 3 or more.

EXAMPLE PROBLEM

A sample is expected to have an absorbance of 2 absorbance units. If a significant level of stray light is present (e.g., 1%), will the absorbance of the sample be greater than, less, than or equal to 2?

ANSWER

The absorbance will appear to be lower than 2 because stray light reaching the detector is interpreted as transmitted light.

iv. PHOTOMETRIC LINEARITY

Linearity of detector response *is the ability of a spectrophotometer to yield a linear relationship between the intensity of light hitting the detector and the detector's response.* It is critical that the detector's response is proportional to the amount of light incident on it. A spectrophotometer may fail to respond in a linear fashion due to stray light, problems in the detector, the amplifier, the readout device, or a monochromator exit slit that is too wide.

v. NOISE AND SPECTROPHOTOMETRY

Recall from Chapter 14 that electrical noise includes background electrical signal arising from random, short-term "spikes" in the electronics of an instrument. These electrical "spikes" occur in the absence of sample.

Signal in a spectrophotometer is electrical current arising in the photodetector that is related to the interaction of light with a sample (or with a blank). Signal is what we are interested in when using an instrument. At low concentrations of analyte, the electrical noise in a spectrophotometer can interfere with the measurement of absorbance, as illustrated in Figure 19.22. Figure 19.22a shows the absorbance spectrum for an undiluted sample of methylene blue. Methylene blue has three main peaks at about 240 nm, 280 nm, and 665 nm; the latter peak has a "shoulder" at about 620 nm. This spectrum, with its

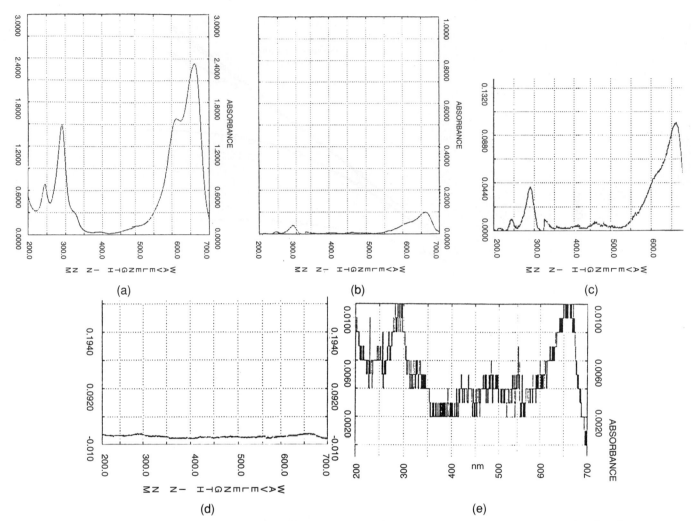

Figure 19.22. The Relationship between Signal and Electrical Noise Illustrated with Methylene Blue. a. The absorbance spectrum for methylene blue. Electrical noise is not evident. **b.** The spectrum of methylene blue diluted 1/25. **c.** The same plot as shown in (b), but with the Y axis expanded to run from zero to 0.1320. Noise becomes visible. **d.** The spectrum of the methylene blue diluted 1/150. **e.** The same plot as in (d), but the Y axis is expanded. The signal of the methylene blue peaks can be barely discerned "buried" in the electrical noise. (All spectra were plotted on a Beckman DU 64 spectrophotometer.)

peaks and valleys, is the "signal." Note that in Figure 19.22a that the absorbance scale on the Y axis runs from 0.0000 to 3.0000. In Figure 19.22a electronic noise is not evident because there is ample signal from the sample. Figure 22b is the spectrum of a diluted solution of methylene blue that was prepared by taking 1 part methylene blue and adding 24 parts of water. The pattern of peaks and valleys characteristic of methylene blue is difficult to distinguish. In order to better visualize the spectrum of this diluted sample, it is replotted in Figure 19.22c, with the Y axis expanded to run from zero to 0.1320 absorbance. In the expanded plot in Figure 19.22c the peaks and valleys characteristic of methylene blue are reasonably clear, but the "spikes" due to electrical noise also become barely visible. Figure 19.22d shows the spectrum of a very dilute sample of methylene blue that was prepared by taking 1 part of the original solution of methylene blue and adding 149 parts of water. The char-

acteristic pattern of methylene blue is not detectable. If the Y axis is expanded so that it runs from zero to 0.0100, then the signal of the methylene blue peaks can be barely discerned "buried" in the electrical noise. Thus, at low levels of sample, electrical noise limits our ability to distinguish the pattern due to the sample.

vi. RESOLUTION

Figure 19.23 shows three absorbance spectra for toluene; however, the three spectra have a different appearance. In the first spectrum eight peaks can be clearly distinguished from one another. In the middle spectrum the peaks are less clearly distinct. In the third spectrum they largely have "run together." We say that the first spectrum has better **resolution**, *the individual peaks can be better distinguished from one another.*

Resolution is a property of an instrument. A more expensive "high performance" spectrophotometer is able

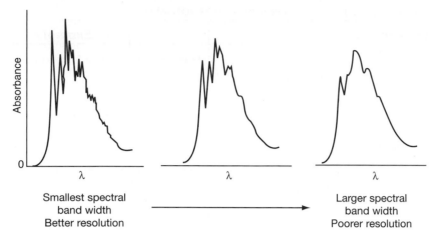

Figure 19.23.   Resolution and an Absorbance Spectrum. The better the resolution of an instrument, the better the peaks of an absorbance spectrum can be distinguished. (The first toluene spectrum used a spectral bandwidth of 0.5 *nm*, the second 1 *nm*, and the third 2 *nm*.) (Spectra reprinted by permission from *Principles of Chemical Instrumentation*, Gary T. Bender. W.B. Saunders Co., Philadelphia, 1987.)

to provide better resolution for a given sample than is a less expensive instrument. Having an instrument with "good" resolution is important in qualitative analysis where distinctive peaks are used to identify a substance. "Good" resolution is also important in the analysis of a sample containing more than one compound whose absorbance peaks are close together. The better the resolution of the instrument, the better substances with close absorbance peaks can be distinguished.

The resolution of an instrument is primarily determined by how "monochromatic" the light exiting the monochromator is. The narrower the range of wavelengths incident on the sample, the better the resolution of the instrument. High-performance spectrophotometers are designed to minimize the wavelength range incident on the sample. In an instrument's specifications, the "monochromaticity" of the instrument and therefore its resolution may be expressed in terms of the "spectral band width" or "spectral slit width." The smaller the value given, the more monochromatic the light and the better the resolution of the instrument. Resolution and spectral band width are discussed in more detail in the Appendix to this chapter.

A practical way to check the resolution of an instrument is to prepare an absorbance spectrum of a reference material that has two or more peaks whose wavelengths are close together. The instrument's ability to distinguish these peaks is an indication of its resolution. For example, a holmium oxide filter has three peaks between 440 and 470 *nm*. If a holmium oxide filter is placed in the sample compartment and scanned, the three peaks should be distinct, given a properly functioning instrument with "good" resolution (Figure 19.20).

B. Performance Specifications

The choice of a spectrophotometer depends on the applications for which it will be used. For research and for qualitative work, a more expensive instrument with better resolution is desirable. A spectrophotometer that can scan an absorbance spectrum and which has both UV and visible capabilities is often required. In laboratories where a spectrophotometer is used only for routine colorimetric assays, less expensive models may be sufficient. The samples analyzed will also determine any special features required. Molecular biology experiments frequently involve very low volume samples, so microvolume cuvette systems become valuable. Another important issue in choosing a new spectrophotometer is its software capabilities. Ease of operation and availability of software to simplify biological applications are factors to consider. Table 19.7 explains the performance specifications for a high-quality, high-resolution spectrophotometer.

Note: **The references and glossary for this chapter are included with those in Chapter 20.**

PRACTICE PROBLEMS: SPECTROPHOTOMETRY, PART A

---

1. Microwaves involve wavelengths in the range from 100 μm to 30 *cm*. Express the wavelengths of microwave radiation in terms of *nm*.

2. Which has the most energy: X-rays, green light, or ultraviolet light? _____ Which has with the least energy? _____

Table 19.7 *Performance Specifications for a Spectrophotometer*

Specification	Example
Capability (Minimally, states whether it has both UV and visible capability and scanning capability.)	**Scanning UV/Vis with Microcomputer Electronics**
Optics (Describes whether single- or double-beam. The monochromator type is also specified.)	**Double-Beam with Concave Holographic Grating with 1053** *lines/nm*
Wavelength range (Indicates the wavelengths at which the spectrophotometer can be used.)	**190-325 (UV) 325–900 (Vis)**
Sources (A UV/Vis spectrophotometer has a source to produce light in both spectral ranges.)	**Pre-aligned Deuterium and Tungsten-halogen Lamps**
Wavelength accuracy (Agreement between displayed and actual wavelength.)	**± 0.1** *nm*
Wavelength repeatability (A measure of the ability of a spectrophotometer to return to the same spectral position.)	**± 0.1** *nm*
Spectral slit width (An indication of the resolution attainable. The lower the value for spectral slit width, the better the resolution.)	**± 1** *nm*
Photometric accuracy (The ability of the detector to respond correctly to transmitted light.)	**± 0.003A at 1.00** *A* **measured with NIST 930 Filters**
Photometric stability (An indication of drift.)	**± 0.002** *A/hr* **at 0.00** *A* **at 500** *nm*
Noise (Random electrical signals. RMS is a statistical method of measuring noise. The lower the value, the less noise.)	**0.00015** *A* **RMS**
Stray light (The amount of light reaching the detector that was not transmitted through the sample.)	**0.05% at 220** *nm* **and 340** *nm*
Scanning speed (The rapidity at which the instrument changes from one wavelength to another during scanning.)	**750** *nm/min*
Microprocessor capabilities: Data handling programs (Specific programs to simplify operation or expand the data analysis capabilities of the instrument.)	**Standard Curves, Linear and Nonlinear; Kinetics; Spectral Scanning**
Accessories (Special features.)	**Special Sample Holder for Small Volumes Available Gel Scanning Accessories Available Temperature-Controlled Sample Chamber Accessories Available**

3. Are spectra A, B, and C probably from the same compound, or are they probably the spectra of different compounds? _____ Explain.

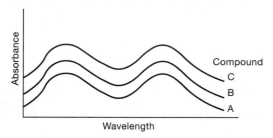

4. If blue light is shined on a solution of orange food coloring will the light be absorbed?

5. What color are the dyes whose absorbance spectra are shown here? Explain.

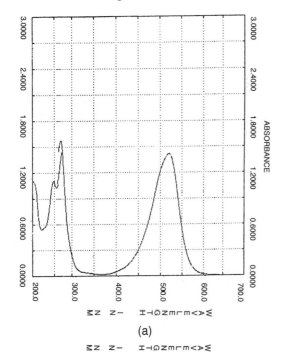

(a)

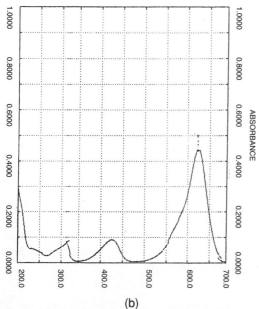

(b)

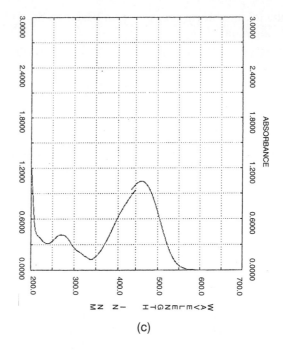

(c)

6. A spectrum for the plant pigment α-carotene, is shown in the following graph. Like chlorophyll, this pigment absorbs light, allowing the plants to make carbohydrates.

   a. What wavelengths of light does this pigment allow the plants to use?

   b. What color would a plant be if it contained only α-carotene?

   (The peaks for α-carotene are at 420, 440, and 470.)

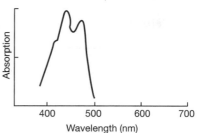

7. The absorbance spectrum for the plant pigment phycocyanin is shown in the following graph.

   a. What color light does this pigment absorb?

   b. What color would a type of algae be if it contained primarily phycocyanin?

   c. Do you think this pigment would be more common in red algae or in blue-green algae?

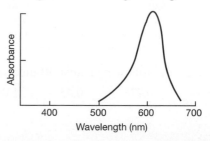

8. a. Plot an absorbance spectrum based on the following data.

b. What color is this substance based on the absorbance spectrum?

c. Optional: What is the natural band width for this substance? (See the Appendix.)

Wavelength (nm)	Absorbance
380	0.01
382	0.00
384	0.01
386	0.02
388	0.07
390	0.10
392	0.15
394	0.38
396	0.46
398	0.54
400	0.55
402	0.52
404	0.42
406	0.37
408	0.26
410	0.11
412	0.08
414	0.03

9. The values for the transmittance of a compound at different wavelengths are given.

a. Graph these data.

b. What color is this compound?

Wavelength (nm)	t
550	0.01
552	1.06
554	1.02
556	1.01
558	1.08
560	0.98
562	0.91
564	0.45
566	0.36
568	0.20
570	0.09
572	0.06
574	0.06
576	0.21
578	0.35
580	0.46
582	0.98
584	1.04

10. Convert the following transmittance values to % T.

0.876   0.776   0.45   1.00

11. Convert the following values to absorbance values.

$t = 0.876$ % $T = 25\%$

$t = 0.776$ % $T = 15\%$

$t = 0.45$ % $T = 95\%$

$t = 1.00$ % $T = 45\%$

12. Fill in the following table:

Absorbance	transmittance (t)	% Transmittance (T)
0.01	___	___
___	0.56	___
___	___	1.0%

13. Convert the following absorbance values to transmittance.

1.24   0.95   1.10   2.25

14. The specifications for two spectrophotometers are given on the following page. Answer the following questions about these spectrophotometers.

a. Which spectrophotometer should be purchased to do qualitative analysis in the ultraviolet range?

b. Which spectrophotometer would be expected to be best able to measure a sample with very high absorbances?

c. Which spectrophotometer should be used to measure the density of bands in electrophoresis gels?

d. Which would probably be better as a low-cost instrument for routine assays or for a teaching instrument.

e. Which spectrophotometer would be expected to have the best resolution?

f. Which spectrophotometer would be best able to distinguish the absorbance from compound A, which has a peak absorbance at 550 *nm* and compound B, whose peak absorbance is at 557 *nm*?

15. Two laboratories are comparing the performance of their spectrophotometers. They take the identical sample and measure its absorbance. (Assume the temperature is the same in both laboratories.) One laboratory gets an average absorbance reading of 0.72 AU; the other laboratory gets 0.87 AU. What factors might account for the difference between the two laboratories?

16. 50 $\mu g/mL$ of a certain compound is expected to have an absorbance of 1.95 on a certain spectrophotometer. If significant stray light is present, will the apparent absorbance be greater than, less than, or equal to 1.95?

Instrument A

Specification

Capability	Scanning Vis range with microcomputer electronics
Optics	Double beam with concave holographic grating with 1053 *lines/nm*
Wavelength range	325–850 (Vis)
Sources	Pre-aligned tungsten-halogen lamps
Wavelength accuracy	± 0.5 *nm*
Wavelength repeatability	± 0.5 *nm*
Spectral slit width	± 12 *nm*
Photometric accuracy	± 0.0013 *A* at 1.00 *A* measured with NIST 930 filters
Photometric stability	± 0.006 *A/hr* at 0.00 *A* at 500 *nm*
Noise	0.00030 *A, RMS*
Stray light	0.15% at 220 *nm* and 340 *nm*
Scanning speed	750 *nm/min*
Microprocessor capabilities: Data handling programs	Standard curves, linear and nonlinear; kinetics; spectral scanning
Accessories	Small volume holder available

Instrument B

Specification

Capability	Scanning UV/Vis with microcomputer electronics
Optics	Double beam with concave holographic grating with 1053 *lines/nm*
Wavelength range	170–325 (UV) 325–850 (vis)
Sources	Pre-aligned deuterium and tungsten-halogen lamps
Wavelength accuracy	± 0.1 *nm*
Wavelength repeatability	± 0.1 *nm*
Spectral slit width	± 1 *nm*
Photometric accuracy	± 0.0005 *A* at 1.00 *A* measured with NIST 930 filters
Photometric stability	± 0.002 *A/hr* at 0.00 *A* at 500 *nm*
Noise	0.00014 *A, RMS*
Stray light	0.05% at 220 *nm* and 340 *nm*
Scanning speed	550 *nm/min*
Microprocessor capabilities: Data handling programs	Standard curves, linear and nonlinear; kinetics; spectral scanning
Accessories	Small volume holder available
	Gel scanning
	Temperature control

DISCUSSION QUESTION

If you work in a laboratory with a spectrophotometer, find the answers to the following questions:

a. Does the spectrophotometer have the capability to measure absorbance in both the UV and visible range?

b. What is the smallest volume sample that can be measured using the cuvettes and sample holders available in your laboratory?

c. What is the instrument's specification for stray light?

d. Optional: What is the instrument's spectral slit width?

ANSWERS TO PRACTICE PROBLEMS: SPECTROPHOTOMETRY, PART A

1. $100 \ \mu m = 100 \times 10^3 \ nm = 100,000 \ nm$
   $30 \ cm = 30 \times 10^7 \ nm = 300,000,000 \ nm$

2. X-rays have the most energy; green light the least.

3. The spectra are probably of the same compound. The heights of the peaks varies depending on the concentration of compound present.

4. Yes.

5. From Table 19.3: a. Safranin O red absorbs light at 525 *nm*, appears red.

b. Brilliant green absorbs light at 620 *nm*, appears greenish blue.

c. Methyl orange absorbs light at 465 *nm*, appears orange.

6. a. α-carotene absorbs, and makes available to the plant, the energy of light in the violet-blue region of the spectrum, between about 420 and 480 *nm*.

b. A yellow-orange color.

7. a. Orange   **b.** greenish blue.   **c.** Blue-green algae

8.

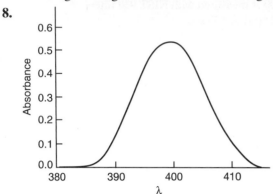

b. yellow   **c.** band width about 15 *nm*

9.

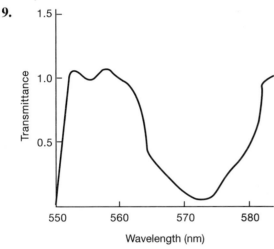

b. violet to red-violet

10. $0.876 = 87.6 \% \ T,$ $0.776 = 77.6 \% \ T,$
$0.45 = 45 \% \ T,$ $1.00 = 100 \% \ T$

11.

	A		A
$t = 0.876$	$= 0.057$	$\% \ T = 25\%$	$= 0.60$
$t = 0.776$	$= 0.110$	$\% \ T = 15\%$	$= 0.82$
$t = 0.45$	$= 0.35$	$\% \ T = 95\%$	$= 0.02$
$t = 1.00$	$= 0.00$	$\% \ T = 45\%$	$= 0.35$

12.

Absorbance	transmittance (t)	% Transmittance (T)
0.01	0.98	98%
0.25	0.56	56%
2.0	0.01	1.0%

13. $1.24 = 0.058 \ t$ $0.95 = 0.11 \ t$
$1.10 = 0.079 \ t$ $2.25 = 0.00562 \ t$

14. a. Instrument B. **b.** Instrument B (less stray light). **c.** Instrument B has gel scanning accessories. **d.** Instrument A would be expected to be less costly and is probably suitable for routine work in many laboratories. **e.** Instrument B (smaller spectral slit width). **f.** Instrument B because it has better resolution.

15. Path length, cuvettes, photometer accuracy, wavelength accuracy.

16. Less than 1.95.

Appendix
Spectral Band Width and Resolution

i. SPECTRAL BAND WIDTH AND THE ACCURACY OF A SPECTROPHOTOMETER

Spectral band width *is a measure of the range of wavelengths emerging from the monochromator when a particular wavelength is selected.* Recall that although a spectrophotometer may be set to a single, specific wavelength, in practice a narrow range of wavelengths emerges from the monochromator. Spectral band width is a value that describes the ability of the spectrophotometer to isolate a portion of the electromagnetic spectrum.*

The spectral band width is an instrument characteristic that affects the accuracy of the measurements and the resolution obtainable from a particular spectrophotometer. In general, the narrower the range of wavelengths emerging from the monochromator, the better the instrument.

Spectral band width is determined by replacing the instrument's normal light source with a special line source (i.e., a lamp that emits light only at specific, individual wavelengths). (A method for measuring spectral band width using a mercury lamp is detailed in ASTM Standard E-958-93 *"Standard Practice for Measuring Spectral Bandwidth of Ultraviolet-Visible Spectrophotometers."*) The intensity of light reaching the detector from the line source is plotted versus wavelength. The width of the emission peak at half the peak's height is measured. In the example illustrated in Figure 19.24, the wavelength at which the peak light intensity occurs is 550 *nm*. The peak height at this position is 6.0 *cm* above the baseline. Half the peak height is 3.0 *cm*. The width of the peak at half its height is 2 *nm*, so the spectral band width is 2 *nm*.

Spectral band width should not be confused with **natural band width**. **Natural band width** *describes the range of wavelengths absorbed by a particular substance.* For example, the natural band width of DNA is around 45 *nm*, as shown in Figure 19.25. The natural band width of a compound is an intrinsic property of that compound and is independent of the instrument's characteristics.

The fact that the light incident on a sample is not truly monochromatic leads to some inaccuracy in measurements of absorbance or transmittance. The amount of inaccuracy depends on both the spectral band width and the natural band width. The narrower the spectral

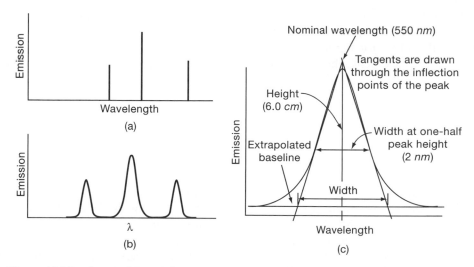

Figure 19.24. Spectral Band Width. a. The actual emission spectrum of a line source as it would appear in a "perfect" spectrophotometer. **b.** The spectrum of a line source as it appears in practice. **c.** The wavelength span is described by the spectral band width that is measured at half the height of the peak.

*There are several terms used to describe the "monochromaticity" of light. These include *spectral band width, bandwidth* or *band width, effective band width,* and *bandpass* or *band pass.* These terms are not always used in the same way by different authors. The definitions in this text are consistent with those of ASTM (ASTM Standard E-131-68 "Standard Definition of Terms and Symbols Relating to Molecular Spectroscopy").

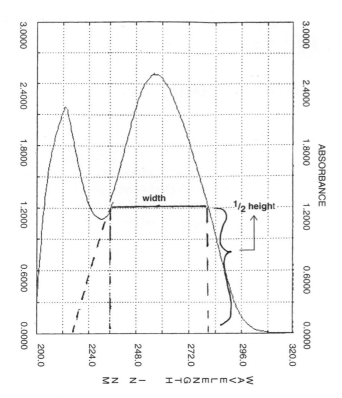

Figure 19.25.   The Natural Band Width of DNA.

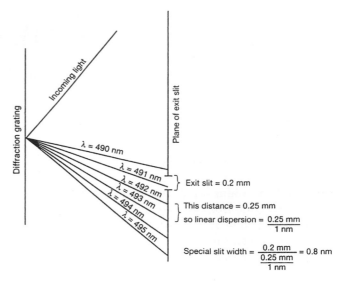

Figure 19.26.   Spectral Slit Width. The spectral slit width in this example is 0.8 *nm*.

band width (i.e., the more monochromatic the light) the more accurate the instrument. If the natural band width of a compound is relatively wide (as is, for example, the natural band width of DNA), then the inaccuracy caused even by a wide spectral band width is relatively insignificant. In contrast, if the natural band width of a compound is comparatively narrow, then a narrow spectral band width is required to get accurate measurements. A general rule is that for the most accurate measurements at an absorbance peak, the spectrophotometer must have a spectral band width equal to or less than one tenth of the natural band width of the compound to be measured. (For DNA, this means the spectral band width of the instrument should be less than or equal to 4.5 *nm*.)

Suppose the absorbance of the same sample is measured in two instruments. The first instrument has a very narrow spectral band width; the second has a broader spectral band width. It seems reasonable to suppose that the sample would have the same absorbance in both instruments. In fact, if all other factors are equal, the first instrument will give a higher absorbance reading for the sample than would the second. This is because the apparent absorptivity (absorptivity and absorptivity constants are discussed in Chapter 20) of an analyte approaches its maximum, "true," value as the spectral band width decreases. The "true" absorptivity constant for a particular compound at a particular wavelength is a theoretical concept measurable only in the presence of truly monochromatic light.

ii. THE FACTORS THAT DETERMINE THE SPECTRAL BAND WIDTH OF AN INSTRUMENT

The spectral band width is controlled by the design of the monochromator. Some monochromators can provide more monochromatic light than others. There are two features of the monochromator which affect the "monochromaticity" of selected light. The first is the **slit width**, *which is the physical width of the entrance or exit slit of the monochromator.* The narrower the slit, the narrower the range of wavelengths selected.

The second feature of the monochromator which controls the range of wavelengths selected is the **linear dispersion of the monochromator. Linear dispersion** *is a measure of the dispersion of light by the diffraction grating.* The linear dispersion of the monochromator is not controlled by the user; rather, it depends on the manufacturing process used to produce the grating. Values for linear dispersion are obtained from the manufacturer of a spectrophotometer. Linear dispersion is generally expressed in terms of how many wavelengths (in nanometers) are included per millimeter at the exit slit.

Spectral slit width combines the value for the slit width of an instrument and its linear dispersion into a single value. **Spectral slit width** *is defined as the physical width of the monochromator exit slit, divided by the linear dispersion of the diffraction grating.* For example, consider the spectrophotometer in Figure 19.26. The width of the exit slit is 0.2 *mm*. The monochromator disperses light in such a way that at the plane of the exit slit there are 0.25 *mm/nm*. Thus, the spectral slit width is:

$$\frac{0.2\ mm}{0.25\ mm/nm} = 0.8\ nm$$

The term *spectral slit width* is not synonymous with *spectral band width.* Both terms, however, relate to the "monochromaticity" of light emerging from the monochromator. Spectral band width is difficult to measure so, in practice, the spectral slit width is used as an indicator of light "monochromaticity."

We have seen that the more monochromatic the light from the monochromator, the more accurate the readings of the spectrophotometer and the better its resolution. It seems reasonable, therefore, that manufacturers would always design instruments with the narrowest possible spectral slit width. In fact, it is true that more sophisticated spectrophotometers have narrower spectral slit widths than do less sophisticated instruments; however, there are practical limitations to how narrow the spectral slit width can be.

One way to make an instrument with a narrow spectral slit width is to make the exit slit of the monochromator very narrow. Although it is possible to manufacture an extremely narrow exit slit there are two problems with this approach. First, as the slit gets narrower and narrower, less and less light passes through it. Eventually, so little light is incident on the sample that electronic noise becomes significant relative to the amount of transmitted light. Noise then limits the ability of the spectrophotometer to make accurate measurements. Second, if the exit slit is too narrow, then its edges diffract light and the slit acts as a second, unwanted diffraction grating.

iii. RESOLUTION AND SPECTRAL BAND WIDTH

Recall that resolution affects our ability to distinguish peaks from one another if their wavelengths are close together. Spectral band width affects resolution. A high-performance instrument has a narrow spectral band width; therefore, it also has optimal resolution. Less expensive instruments tend to have wider spectral band widths and less ideal resolution.

When reading instrument specifications, the spectral slit width of the monochromator is commonly used as an indication of the resolution of a spectrophotometer. The lower the value for spectral slit width, the more monochromatic the light and the better the resolution, Figure 19.27.

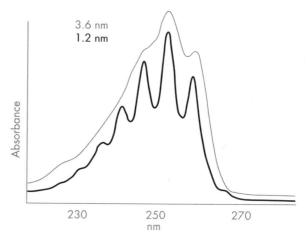

Figure 19.27.   Spectral Slit Width and Resolution. A narrower spectral slit width results in much improved resolution. *(Courtesy of Waters Corp.)*

## CHAPTER 20

Measurements Involving Light B: Applications and Methods

I. INTRODUCTION

The previous chapter described the nature of light and the design, operation, and performance verification of spectrophotometers that measure light. This chapter explains how measurements involving light are used to obtain information about biological samples.

Spectrophotometers can be used to help answer essential questions in the laboratory, including: (1) What is the identity or nature of the component(s) of a sample? (2) How much of an analyte is present in a sample? *Assays that identify the components of a sample are* **qualitative**, *whereas those that determine how much analyte is present are* **quantitative**, Figure 20.1.

There are many spectrophotometric analytical methods. For example, there are spectrophotometric methods to confirm the identity of drug products (a qualitative application), to measure levels of water pollutants (quantitative), to determine the levels of proteins in a sample (quantitative), and to study the activity of enzymes (quantitative). These spectrophotometric analytical methods must be developed, optimized, and validated. Note that ensuring that an analytical method works effectively is a separate task from verifying that the instrument involved is properly functioning.

This chapter begins by explaining the general principles of qualitative and quantitative analysis by spectrophotometry. The chapter then looks at the development, optimization, and validation of spectrophotometric methods. In addition, there are three sections that briefly introduce instrumental methods related to spectrophotometry.

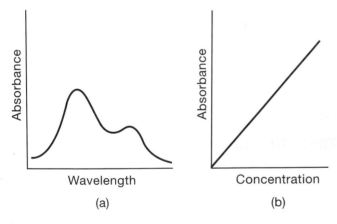

Figure 20.1.   Applications of Spectrophotometry.
a. Qualitative applications use absorbance spectra to give information about the component(s) of a sample. [Recall that an absorbance spectrum is the absorbance measured for a sample (Y axis) over a range of different wavelengths (X axis)]. **b.** Quantitative applications determine the concentration of a substance in a sample. Concentration (or amount) is on the X axis, and, as for qualitative applications, absorbance is on the Y axis. (See pp. 159–161 for an introduction to standard curves.)

II. QUALITATIVE APPLICATIONS OF SPECTROPHOTOMETRY

Qualitative applications of spectrophotometry use the spectral features of a sample to obtain information about the nature of the sample's component(s). It is sometimes possible to use an absorbance spectrum like a "fingerprint" to identify an unknown substance. The infrared (IR) spectra of organic compounds are complex and form particularly distinctive "fingerprints." Organic chemists therefore frequently use IR spectra to identify compounds.

UV/Visible spectra are less complex and distinctive than IR spectra, so they are used less commonly for identification of unknown substances. There are, however, situations where information about a sample can be obtained from its UV/Vis spectrum. For example, the U.S. Pharmacopeia includes a number of tests of drug identity in which an absorbance spectrum from 200 to 400 *nm* is used to confirm the identity of a drug. In these identity tests, the spectrum of the test sample is compared with the spectrum of a pure reference material. If the sample and reference have identical spectra, then it is likely that they are the same compound. UV/Vis identity tests are frequently used because they are simple, rapid, and the equipment to prepare an absorbance spectrum is widely available. A problem with identifying substances using UV/Vis spectrophotometry is that some compounds whose structures are similar have spectra that cannot be distinguished. UV spectra, therefore, are frequently used in conjunction with other identification methods.

UV/Vis spectra can also be used to obtain information about structural characteristics of a substance. For example, a spectrophotometer can be used to monitor hemoglobin as it binds to oxygen and other compounds. Figure 20.2a shows the difference between the spectra of hemoglobin bound to oxygen and hemoglobin bound to carbon monoxide. Similarly, the spectrum of the protein bovine serum albumin shifts when the protein becomes denatured (loses its normal structure). This shift can be monitored with spectrophotometry, Figure 20.2b.

III. INTRODUCTION TO QUANTITATION WITH SPECTROPHOTOMETRY: STANDARD CURVES AND BEER'S LAW

A. Constructing a Standard Curve

Quantitative applications of spectrophotometry have the purpose of determining the concentration (or amount) of an analyte in a sample. Quantitative analysis with a spectrophotometer typically involves a "standard curve." A **standard curve (calibration curve)** *for spectrophotometric analysis is a graph of analyte concentration (X axis) versus absorbance (Y axis).*

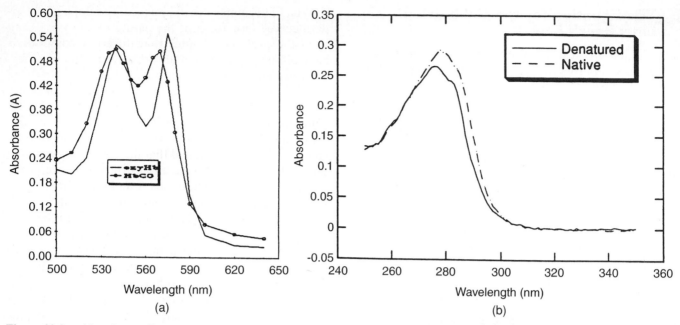

Figure 20.2.   Absorbance Spectra and Structure. a. The difference in the absorbance spectra of hemoglobin bound to oxygen and to carbon monoxide. (Used by permission from "Spectrophotometric Properties of Hemoglobin: Classroom Applications," Roger Frary. *The American Biology Teacher.* 59 (2):104–107, 1997.) **b.** The absorbance spectrum of bovine serum albumin changes when the protein loses its normal structure. (Used by permission from *Methods for Protein Analysis,* Robert A. Copeland. p. 175. Chapman and Hall, New York, 1994.)

To construct a calibration curve, standards are prepared with known concentrations of analyte. The absorbances of the standards are determined at a specified wavelength and the results are graphed. Given a standard curve, it is possible to determine the concentration of an analyte in a sample based on the sample's absorbance.

EXAMPLE

a. Construct a standard curve for red food coloring.

b. Determine the concentration of red food coloring in a sample with an absorbance of 0.50.

Step 1. Prepare a series of standards of known concentration by diluting a stock solution with distilled water.

Step 2: Prepare one tube that has no dye. This is the blank. (In this case, the blank is simply distilled water.)

Step 3. Place the blank in the spectrophotometer at the specified wavelength and adjust the spectrophotometer to 0 absorbance or 100% transmittance.

Step 4. Read the absorbance of each standard at the specified wavelength. Example results are shown in the following table.

Step 5. Plot the data on a graph with standard concentration on the X axis and absorbance on the Y axis, as shown in the following graph.

Step 6. Draw a best fit line to connect the points. (See Chapter 10 for a discussion of best fit lines and the Appendix to this chapter for a statistical method to draw a best fit line.)

Step 7. Read the absorbance of the unknown at the specified wavelength.

Step 8. Determine the concentration of the unknown based on the standard curve.

EXAMPLE DATA

Concentration of Standard	Absorbance
0.0 *ppm*	0.00
2.0 *ppm*	0.14
4.0 *ppm*	0.26
6.0 *ppm*	0.45
8.0 *ppm*	0.59
10.0 *ppm*	0.75
12.0 *ppm*	0.91
unknown	0.50

a. The standard curve is:

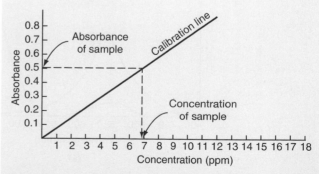

b. Based on the standard curve, the unknown has a concentration of 6.9 *ppm*.

B. The Equation for the Calibration Line; Beer's Law

Information about how an analyte interacts with light can be obtained from the line on a calibration curve. This section explores the features of that line in more detail.

The line on a calibration curve has an equation, as does every line. (See pp. 152–155 for a review of lines and linear equations.) Recall that the general equation for a line is:

$$Y = mX + a$$

where
m = the slope
a = the Y intercept

For the line on any calibration plot, the Y axis is absorbance and the X axis is concentration (or amount). Therefore, in the equation for the line on any spectrophotometry calibration curve, we can substitute A (for absorbance) and C (for concentration) as follows:

$$A = mC + a$$

For a calibration line, the Y intercept is normally zero. (A Y intercept of zero means that both the absorbance and analyte concentration are zero. A blank, containing no analyte, is used routinely to set the instrument to zero absorbance.) Thus, the equation for the line on a calibration curve is:

$$A = mC + 0$$
or simply
$$A = mC$$

in words:

$$\text{Absorbance} =$$
(slope of calibration line)(Concentration of analyte)

Consider the slope of a calibration line. If the slope is relatively steep, it means that there is a dramatic change in absorbance as the concentration increases. On the other hand, if the slope is not very steep, then as the concentration of analyte increases, the absorbance does not increase dramatically.

What determines the steepness of the slope of a standard curve? One factor is the nature of analyte. Recall from the previous chapter that different compounds have differing patterns of absorbance of light at different wavelengths. (For compounds that absorb light in the visible range, this pattern relates to the color of the compound.) For example, observe in Figure 20.3a that compound A absorbs more light at 550 *nm* than does compound B. (Assume that the concentrations of both compounds were equal and that the conditions were the same when the spectra were constructed.) If we plot a standard curve for compounds A and B at 550 *nm*, as is shown in Figure 20.3b, the line for compound A will have a steeper slope than for compound B. This is because Compound A has a greater inherent tendency to absorb light at 550 *nm* than does compound B. Thus, at a given wavelength, the slope of a calibration line will vary from one compound to another.

A second factor that affects the slope of the calibration line is the path length. If the light has to pass through a longer path, then there is more absorbing material present and more of the light will be absorbed. Thus, the two main factors that affect the slope are:

1. The tendency of the compound of interest to absorb light at the wavelength used

2. The path length

The inherent tendency of a material to absorb light at a particular wavelength is called its **absorptivity**. The greater the absorptivity of a material, the more it absorbs light at that wavelength. The absorptivity of a specific compound at a particular wavelength is constant.

The value of the absorptivity constant of a particular compound at a specific wavelength has many names, including the **absorptivity constant** (abbreviated with a Greek alpha, α) and the **absorption coefficient**. It is sometimes termed the **extinction coefficient** because it is an indication of the tendency of a compound to "extinguish" or absorb light. The absorptivity constant

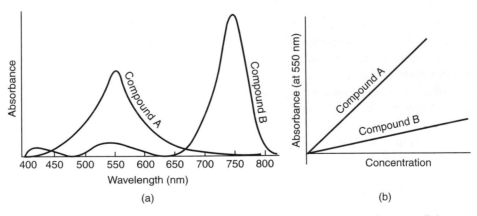

Figure 20.3.   The Slope of the Calibration Line. a. Compound A absorbs more light at 550 *nm* than does compound B. **b.** The calibration line for Compound A is steeper than for Compound B at 550 *nm*.

has units that vary depending on the units in which the concentration of the analyte is expressed. Units for expressing concentration include moles per liter, milligrams per milliliter, and parts per million. If concentration is expressed in moles per liter, then the absorptivity constant is variously termed: **the molar absorptivity constant**, ε **(the Greek letter epsilon)**, **the molar extinction coefficient**, or **the molar absorption coefficient**. Because the terminology varies, it is necessary to note the units in your own calculations and to be aware of the units in the calculations of others.

Returning to the equation for the calibration line, we can rewrite the equation to include the path length and the analyte's absorptivity, both of which contribute to the slope. The equation thus becomes:

$$A = mC$$
$$A = (\alpha b)C$$

where
A = the absorbance
α = the absorptivity for that compound at that wavelength
b = the path length
C = the concentration
$(\alpha b) = m$ = the slope

This equation, which shows the relationship between absorbance, concentration, absorptivity, and path length, is famous. It forms the basis for quantitative analysis by absorption spectrophotometry. Its discovery is variously attributed to Beer, Lambert, and Bouguer, but it is generally referred to simply as **Beer's Law**. **Beer's Law** *states that the amount of light emerging from a sample is reduced by three things:*

1. The concentration of absorbing substance in the sample (C in the equation)
2. The distance the light travels through the sample (path length or b)
3. The probability that a photon of a particular wavelength will be absorbed by the specific material in the sample (the absorptivity or α)

EXAMPLE PROBLEM

a. Using Beer's Law, what is the concentration of analyte in a sample whose molar absorptivity constant is 15,000 *L/mole-cm* and whose absorbance is 1.30 AU in a 1 *cm* cuvette?

b. Suppose that the absorbance of the preceding analyte is measured in a 1.25-*cm* cuvette. Will its absorbance be less than, more than, or equal to 1.30?

c. Will changing to a 1.25-*cm* cuvette affect the molar absorptivity constant?

ANSWER

a. Substituting into the equation for Beer's Law:

$$A = \alpha \quad b \quad C$$

$$1.30 = \frac{(15{,}000\ L)\ (1\ cm)\ C}{mole\text{–}cm}$$

The *cm* cancel:

$$1.30 = \frac{(15{,}000\ L)\ C}{mole}$$

Solving for the concentration:

$$C = \frac{1.30}{\dfrac{(15{,}000\ L)}{mole}}$$

$$= 8.67 \times 10^{-5}\,\frac{mole}{L}$$

b. This change in cuvette will affect the path length. Because the path length is longer, the absorbance will be greater than 1.30.

c. The absorptivity constant is an intrinsic property of the analyte and is unaffected by the cuvette.

EXAMPLE PROBLEM

DNA polymerase is an enzyme that participates in the assembly of DNA strands from building blocks of nucleotides. The absorptivity constant of this enzyme at 280 *nm* (under specified pH conditions) is 0.85 *mL/(mg)cm*. (*Worthington Enzyme Manual: Enzymes and Related Biochemicals*, edited by Von Worthington. Worthington, Biochemical Corp., Freehold, NJ, 1993.) If a solution of DNA polymerase has an absorbance of 0.60 at 280 *nm* (under the proper conditions), what is its concentration?

ANSWER

Substituting into the equation for Beer's Law:

$$A = \alpha \quad b \quad C$$

$$0.60 = \frac{0.85\ mL}{cm(mg)} \times 1\ cm \times C$$

$$C = \frac{0.60}{\dfrac{0.85\ mL}{mg}}$$

$$= 0.71\frac{mg}{mL}$$

C. Calculating the Absorptivity Constant from a Standard Curve

The absorptivity constant is an important property of an analyte. It is a useful skill to be able to calculate and report the absorptivity constant for a particular compound of interest based on data from your own spectrophotometer and in your own laboratory. This section shows two strategies to determine the absorptivity constant for a particular compound at a specific wavelength. The first strategy requires measuring the absorbance of a single standard and then calculating the absorptivity constant based on that single measurement. The second strategy is more reliable and involves constructing a standard curve and calculating the absorptivity constant based on its slope.

STRATEGY 1: CALCULATING AN ABSORPTIVITY CONSTANT BASED ON A SINGLE STANDARD

Beer's Equation can be rearranged as follows:

$$absorptivity\ constant = \frac{Absorbance}{(path\ length)(concentration)}$$

If the absorbance and concentration of a single standard is known, these values can be entered into the preceding equation to determine the absorptivity constant. This method is illustrated in the following example:

EXAMPLE

A standard containing 75 ppm of Compound Y is placed in a 1 cm cuvette. The absorbance of the standard is 1.20 at 450 nm. Assuming that there is a linear relationship between the concentration of Compound Y and the absorbance, and assuming that the standard was diluted properly, what is the absorptivity constant for this compound at 450 nm?

Substituting into the equation for Beer's Law:

$$absorptivity\ constant = \frac{Absorbance}{(concentration)(path\ length)}$$

$$\alpha = \frac{1.20}{75\ ppm\ (1\ cm)}$$

$$= 0.016/ppm\text{-}cm$$

The absorptivity constant at 450 nm, based on this single standard, therefore, is 0.016/ppm-cm.

This first strategy will give a value for the absorptivity constant. However, it is not the best method to use because it is based on only one standard. If that single standard is diluted incorrectly or if for some reason the absorbance value is slightly off what it should be, then the absorptivity constant calculated will also be inaccurate.

STRATEGY 2: CALCULATING THE ABSORPTIVITY CONSTANT FROM A STANDARD CURVE

A better way to calculate the absorptivity constant for a particular compound at a specified wavelength is to base it on the absorbance of a series of standards. This is readily accomplished using a standard curve. Recall that the slope of the calibration line is *the absorptivity constant (α) multiplied by the path length*. Further, recall that the equation for the slope of a line is:

$$m = \frac{Y_2 - Y_1}{X_2 - X_1}$$

where X_1 and X_2 are the X coordinates for any two points on the line

Y_1 and Y_2 are the corresponding Y coordinates for the same two points

Determining the absorptivity constant based on the slope of a standard curve is illustrated in the following example and is summarized in Box 1.

EXAMPLE

The calibration curve at 550 nm for Compound Q is shown. What is the absorptivity constant for this compound at 550 nm? (Assume the path length is 1 cm.)

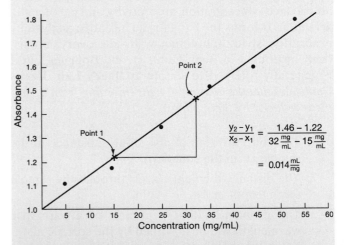

Step 1. Calculate the slope of the calibration line. Note the units. In this case, the slope is: 0.014 *mL/mg*

Step 2. The path length is 1 cm.

Step 3. From Beer's Law, the slope of the calibration line is:

$$slope = (path\ length)(absorptivity\ constant)$$

so the *absorptivity constant* $= \dfrac{slope}{path\ length}$

In this example:

$$absorptivity\ constant = \frac{0.014\ mL}{(1\ cm)\ (mg)}$$

BOX 1   DETERMINATION OF THE ABSORPTIVITY CONSTANT FROM A CALIBRATION CURVE

1. *Prepare a calibration line with concentration on the X axis and absorbance on the Y axis.*
2. *Calculate the slope of the calibration line using the equation:*

$$m = \frac{Y_2 - Y_1}{X_2 - X_1}$$

3. *Determine the path length for the instrument.* (The path length is generally 1 *cm*, assuming a standard type of cuvette and holder is used.)
4. *Solve the equation:*

$$Absorptivity\ constant = \frac{slope}{path\ length}$$

D. Variations on a Theme

i. OVERVIEW

Now that we have explored the Beer's Law equation, we will consider more examples of how it is used to determine the concentration (or amount) of analyte in a sample. In general, the best way to determine the concentration of an analyte in a sample is to construct a calibration curve based on a series of standards. There are, however, alternatives to constructing a calibration curve each time a sample is analyzed. These variations can be used successfully, but it is imperative to be aware of (and avoid) their potential pitfalls.

ii. VARIATION 1: USING A CALIBRATION CURVE FROM A PREVIOUS ANALYSIS

In laboratories where a particular quantitative analysis is done routinely, it may be determined that a calibration curve needs to be produced and checked only once in a while. The same calibration curve is then used regularly for all samples. This method is accurate only if the following assumptions are true:

- **The instrument must remain calibrated and in good working order over time.** For example, if the spectrophotometer becomes unaligned so the wavelength is not the same as it was when the curve was first constructed, then using the standard curve will lead to inaccurate results.

- **The solvents and reagents must be consistent.** Reagents may change with storage. If a batch of reagents used when making the standard curve varies from a batch used for samples at a later time, then the results may be inaccurate.

- **Assay conditions, such as incubation times and temperatures, must be consistent.**

- **A proper blank must be used when constructing the standard curve and when measuring the absorbances of the samples.**

Note that it is good practice to regularly check the absorbance of a control sample with a known analyte concentration. If the control has an unexpected reading, then the analyst is alerted that there may be a problem.

iii. VARIATION 2: DETERMINING CONCENTRATION BASED ON THE ABSORPTIVITY CONSTANT

A second alternative is to determine the absorptivity constant as shown in Box 1. Then, for subsequent samples the concentration of analyte in a sample can be determined based on Beer's Law:

$$A = \alpha bC$$
therefore
$$C = A/\alpha b$$

This method is illustrated in the example problems on p. 405. This method is accurate only if the following assumptions are true:

- **The relationship between absorbance and concentration must be linear.**

- **The instrument must remain calibrated and in good working order over time.**

- **The solvents and reagents must be consistent.**

- **Assay conditions, such as incubation times and temperatures, must be consistent.**

- **A proper blank must be used.**

- **The sample's concentration must be in the linear range of the assay (discussed in more detail later).**

iv. VARIATION 3: DETERMINING CONCENTRATION BASED ON AN ABSORPTIVITY CONSTANT FROM THE LITERATURE

A still less accurate method to determine the concentration of analyte in a sample is to use a value for the absorptivity constant published in the literature. Absorptivity constants for various compounds at various wavelengths are frequently reported in articles, catalogues, and other technical literature. In principle, spectrophotometers can be calibrated so that if the absorbance of the same sample is measured in various instruments, the same absorbance values will be obtained. In practice, two instruments often do not read identical absorbance values for a given sample. Moreover, the conditions when spectrophotometric measurements are made (such as solutions used, temperature, pH, etc.) are likely to vary from one laboratory to another. An absorptivity constant derived in one laboratory is therefore unlikely to be exactly reproducible in other circumstances. Nonetheless, there are situations where constants from the literature can be used to give estimates of concentrations. For example, if an analyst is comparing two methods of purifying a protein to see which is more efficient, it is reasonable to estimate and compare the amount of protein in the two preparations based on extinction coefficients from the literature. In contrast, if the analyst wants to know how much protein is actually in each of the preparations, then a standard curve for that protein must be prepared by the analyst.

EXAMPLE PROBLEM

Suppose a manufacturer provides an extinction coefficient for an enzyme.

a. What features of your spectrophotometer must match those of the manufacturer in order for you to get the same extinction coefficient for this enzyme?

b. If your instrument does not match the company's instrument, is the extinction coefficient listed in the catalogue useful information?

ANSWER

a. To reproduce an extinction coefficient obtained on one instrument using another instrument:

Both instruments must be calibrated for wavelength and photometric accuracy. In addition, the degree of "monochromaticity" of light exiting the monochromator must be the same. (The more monochromatic the light, the higher the absorptivity constant will be for a given compound.)

b. An extinction coefficient from the literature can be useful as an approximation. In addition, if the enzyme is measured consistently on the same instrument, then it can be compared in a relative way from batch to batch or from assay to assay.

v. Variation 4: Using a Single Standard Rather Than a Series of Standards

It is possible to use a single standard each time samples are analyzed. Then, the amount of analyte in the sample is proportional to the amount in the standard, as illustrated in the following example.

EXAMPLE

A standard is prepared with 10 mg/mL of analyte.

The absorbance of the standard = 1.6.

The sample's absorbance is 0.8.

What is the concentration of analyte in the sample?

$$\frac{10\ mg/mL}{1.6} = \frac{?}{0.8}$$
$$? = 5\ mg/mL$$

This method is accurate only if the following assumptions are true:

- **The single standard must be prepared accurately.** Manufacturers sometimes provide carefully tested standards for a particular method.

- **The relationship between analyte concentration and absorbance must be linear.**

- **The instrument must be set to zero absorbance (or 100% T) using a properly made blank containing no analyte.**

- **The sample and the standard must have absorbances in the linear range of the assay.**

EXAMPLE PROBLEM

A standard containing 20 mg/mL of Compound Z is placed in a 1 cm cuvette. The absorbance of the standard is 1.20 at 600 nm. A sample containing Compound Z has an absorbance of 0.50. Assuming that there is a linear relationship between the concentration of Compound Z and the absorbance, and assuming that the standard was diluted properly, what is the concentration of Compound Z in the sample?

There are two ways to solve this.

STRATEGY 1: USING PROPORTIONS

This is a proportional relationship; therefore,

$$\frac{20\ mg/mL}{1.2} = \frac{?}{0.5}$$
$$? = 8.33\ mg/mL$$

STRATEGY 2: BASED ON THE ABSORPTIVITY CONSTANT

Calculate the absorptivity constant based on this one standard. Substituting into the equation for Beer's Law:

$$A = \alpha\,b\,C$$
$$1.20 = \alpha\,(1\ cm)\,(20\ mg/mL)$$
$$\alpha = \frac{1.20}{20\,\frac{mg(cm)}{mL}}$$

The absorptivity constant, based on this single standard, is:

$$\alpha = 0.06 \ mL/mg(cm)$$

From Beer's Law, therefore, if the absorbance of the sample is 0.50, its concentration is:

$$A = \alpha \ b \ C$$

$$0.50 = \frac{0.06 \ mL}{cm(mg)} (1 \ cm) \ C$$

$$C = \frac{0.50}{0.06 \ \frac{mL}{mg}}$$

$$C = 8.33 \ mg/mL$$

E. Deviations from Beer's Law

We have now seen how Beer's Law, which states that the absorbance of a compound is directly proportional to its concentration, is the basis for quantitation using UV/Vis methods. In many systems, Beer's Law holds true and it is therefore a very useful equation. There are situations, however, where real systems deviate from Beer's Law. The causes of deviation from Beer's Law include:

1. **At high absorbance levels, stray light causes spectrophotometers to have a nonlinear response.**

2. **At very low levels of absorbance a spectrophotometer may be inaccurate.**

3. **There may be one or more components in the sample that interfere with the measurement of absorbance of a particular analyte.** We will discuss sample components that interfere with analysis in a later section of this chapter.

Consider what occurs when absorbance is very high or low. Observe that the calibration line drawn in Figure 20.4 has two thresholds. At very low absorbance levels, the calibration curve deviates from linearity. At high absorbance levels the instrument is not able to measure

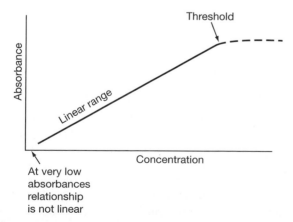

Figure 20.4.   The Relationship between Absorbance and Concentration is Linear at Intermediate Concentrations of Analyte, but Not at High or Low Concentrations.

the small amount of light passing through the sample accurately. Thus, at high and low concentrations of analyte, the relationship between absorbance and concentration is not linear; therefore, Beer's Law does not apply.

All quantitative spectrophotometric assays have a range of concentration within which the values obtained are reliable. Above this concentration range the absorbance readings are too high to be useful and below this range the absorbance values are too low. It is essential that the standards and samples all have concentrations such that their absorbance falls in the range where the values obtained are accurate. Thus, a very important cause of deviation from Beer's Law is failure to stay within the linear range of the assay.

EXAMPLE PROBLEM

Suppose you have prepared a standard curve, as shown in the following graph. What is the concentration of analyte in a sample whose absorbance is 1.95?

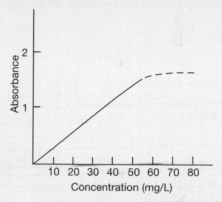

ANSWER

The absorbance of the sample is too high; it does not fall on the linear range of the standard curve. All we can tell is that the concentration of analyte in the sample is above about 50 *mg/L*. To determine the concentration of analyte, it is necessary to dilute some of the sample and repeat the assay.

EXAMPLE PROBLEM

Suppose you repeat the assay as discussed in the preceding Example Problem, this time taking 1 *mL* of the sample and diluting it with 9 *mL* of distilled water. Now, the absorbance of the sample is 1.2. What is the concentration of analyte in this sample?

ANSWER

Reading from the standard curve, the concentration of the sample is about 40 *mg/L*. However, the sample was diluted 10X. Therefore, multiply by ten, to give an answer of 400 *mg/L*.

Although there is a lower limit of concentration below which an analyte cannot be detected, spectrophotometry is generally a method used to evaluate analytes that are present at very low levels in a sample. You can easily demonstrate that this is so by taking food coloring and trying to measure its absorbance. You will find that the absorbance of undiluted food coloring is far too intense to measure in a spectrophotometer. In fact, food coloring must be diluted a great deal before its absorbance falls into the linear range of a spectrophotometer. Samples often must be diluted before they can be analyzed spectrophotometrically.

EXAMPLE PROBLEM

Assume that a spectrophotometer is able to read accurately in the range from 0.1 to 1.8 AU. The molar absorptivity constant for NADH is 15,000 L/mole-cm at 260 nm. Using Beer's Law, calculate the concentration range of NADH that can be quantitated accurately at this wavelength based on the limits of the spectrophotometer.

ANSWER

This involves calculation of the molar concentrations that will produce absorbances of 0.1 and 1.8. From Beer's Law:

$$C = \frac{A}{\alpha b}$$

Substituting 0.1 and 1.8 into the equation (and assuming a 1.0 cm path length):

$$C = \frac{0.1}{\dfrac{(15,000\ L)}{(cm)\ mole}(1.0\ cm)} = 6.7 \times 10^{-6}\ mole/L$$

$$C = \frac{1.8}{\dfrac{(15,000\ L)}{(cm)\ mole}(1.0\ cm)} = 120 \times 10^{-6}\ mole/L$$

The range of NADH concentrations that can be detected at this wavelength with this spectrophotometer is from 6.7×10^{-6} mole/L to 120×10^{-6} mole/L. These are dilute solutions of NADH.

IV. QUANTITATION WITH SPECTRO-PHOTOMETRY: COLORIMETRIC ASSAYS, TURBIDOMETRY, AND KINETIC ASSAYS

A. Colorimetric Assays

A visible spectrophotometer can only be used to analyze a material that absorbs visible light. Biological materials by themselves are usually colorless (i.e., they do not absorb visible light). **Colorimetric assays** are used to analyze materials that are naturally colorless. A **colorimetric assay** *is one in which a colorless substance of interest is exposed to another compound and/or to conditions that cause it to become colored:*

$$\text{substance without color} + \text{reagent(s)} \xrightarrow{\text{proper conditions}} \text{product with color}$$

For example, there are several commonly used colorimetric assays to measure the concentration of proteins in a sample. One such assay, the Biuret method, involves dissolving copper sulfate in an alkaline solution and adding it to the protein sample. Complexes form between the copper ions and nitrogen atoms in the proteins. These complexes produce a blue color that is measured in a spectrophotometer at 550 nm. The more protein present, the more intense the blue color.

When performing a colorimetric assay, the standards, the samples, and the blank should be handled identically. For example, if the samples are subjected to heating, cooling, the addition of various reagents, or other treatments, then the standards and the blank should also be treated in these ways. Note also that when performing colorimetric assays, if a sample is too concentrated to be in the linear range of the instrument, then it must be diluted before the reagent(s) are added. Diluting the final reaction mixture will not produce accurate results.

B. Turbidometry and the Analysis of Bacterial Suspensions

Turbid solutions *contain small suspended particles that both absorb and scatter light.* Scattered light is generally deflected away from the detector, so it appears to have been absorbed, when, in reality, it was not. Turbidity can also cause an apparent shift in absorbance peaks because shorter wavelengths are scattered more readily than longer ones. How much a beam is attenuated by scattering depends on the optics of the instrument, the orientation of the cuvette, and the uniformity of the suspension. Any inconsistencies in a turbid system, therefore, will give inconsistent absorbance readings. Turbid samples should normally be avoided in spectrophotometry. Suspended materials can be removed by centrifugation or filtration of the sample.

A useful exception to the rule of avoiding turbid samples is in evaluating the concentration of microorganisms in a sample. The greater the concentration of bacteria in a sample, the more they scatter light and the higher the apparent absorbance of the solution. Although absorbance is not actually being measured, within limits, the relationship between the apparent absorbance reading and the concentration of microorganisms is linear. Apparent absorbance, therefore, can be used as a measure of cell density in a suspension.

C. Kinetic Assays

Kinetic spectrophotometric assays *measure the changes over time in concentration of reactants or products in a chemical reaction.* Kinetic assays are useful in the analysis of enzymes. In an enzymatic reaction, one or more substrates are acted upon by the enzyme and converted to product(s):

$$\text{substrate(s)} \xrightarrow{\text{enzyme}} \text{product(s)}$$

As enzymatically catalyzed reactions proceed, the substrate(s) is consumed and product(s) appear. In an enzyme assay, therefore, either the appearance of product or the disappearance of substrate over time is monitored.

Although enzyme assays may be considered as a class of quantitative assay, the quantitation of enzymes is different from that of other proteins. Proteins other than enzymes are typically measured in terms of their concentration, for example, in terms of milligrams per milliliter. Enzymes can be measured not only in terms of their concentration but also in units of activity. Activity is a measure of the amount of substrate that is converted to product by the enzyme in a specified amount of time under certain conditions. Thus, time is an important factor in enzyme assays. Figure 20.5 shows a plot of a kinetic assay.

V. SPECTROPHOTOMETRIC METHODS

A. Developing Effective Methods for Spectrophotometry

i. OVERVIEW

Spectrophotometric methods delineate the steps for analyzing a sample using a spectrophotometer. For example, a colorimetric method might involve (1) adding a reagent to samples, standards, and the blank, (2) heating the mixtures at a certain temperature for a certain time, (3) cooling them for a set period, (4) reading their absorbances, (5) constructing a standard curve, and (6) determining the concentration of analyte in the samples based on the standard curve. A spectrophotometric identification method might involve scanning a sample over a range of wavelengths and comparing peaks in the spectrum to those of a standard. There are thousands of published spectrophotometric analytical methods available. Sources of methods include laboratory manuals, research articles, the U.S. Pharmacopeia, and the manuals of AOAC and ASTM.

Spectrophotometric methods are widely used partly because spectrophotometers are present in most laboratories and are relatively easy to operate. Another important advantage to spectrophotometric methods is that an individual substance in a mixture can be analyzed without separating the components of the sample

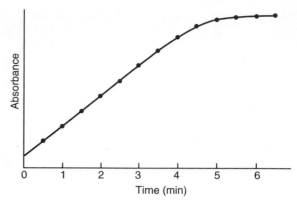

Figure 20.5.   An Example of a Kinetic Assay. Ingested alcohol is rapidly distributed throughout the bloodstream. The removal of alcohol from the body is a time-dependent, enzymatically catalyzed process that occurs in the liver. Alcohol dehydrogenase is the enzyme responsible for the initial metabolism of alcohol. The enzymatic reaction catalyzed by alcohol dehydrogenase results in the formation of NADH, which absorbs light at 340 *nm*. This graph shows the results of an assay for alcohol dehydrogenase in which the activity of alcohol dehydrogenase is monitored by following the change in absorbance at 340 *nm* over time.

from one another. For quantitative analysis, the presence of multiple substances in a sample is not a problem as long as a wavelength exists where only the material of interest absorbs light. Spectral analysis of samples with multiple components may be possible if features of their spectra can be distinguished.

Developing and optimizing a spectrophotometric method requires finding a basis for the method (e.g., a colorimetric reaction) that is specific for the material of interest, finding the optimum wavelength for analysis (for quantitative assays), determining the type of sample for which the method is appropriate, determining the concentrations of sample for which the assay is accurate, discovering characteristics of the sample that might interfere with the accuracy of the method, and determining what instrument performance features are necessary. Some of these factors are discussed in more detail in the upcoming sections. Examples of methods that are commonly used in biology laboratories are explored to illustrate some of the issues that arise when using spectrophotometry.

ii. THE SAMPLE AND INTERFERENCES

The sample is a key part of a spectrophotometric method. As was already discussed, each method is accurate only within a certain range of analyte concentrations. The analyte concentration in the sample, therefore, must match the requirements of the assay. Another important requirement is that the sample not contain substances that interfere with the assay. An **interfering substance**, *in its broadest sense, may be considered as any material in the sample that leads to an inaccurate result in the analysis.*

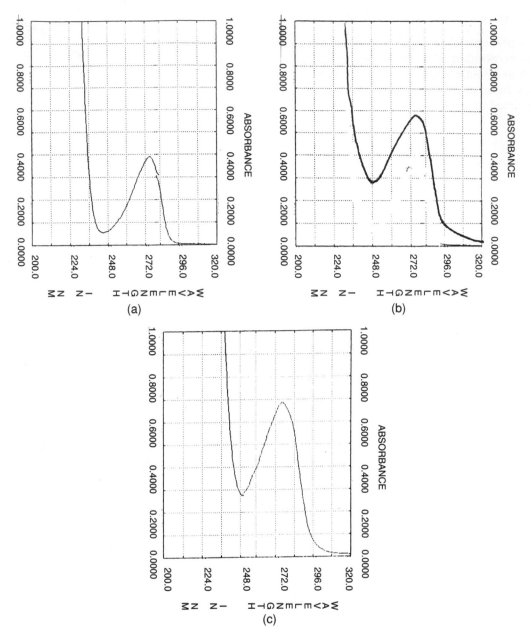

Figure 20.6. Triton X-100 as an Interference in the Analysis of Protein. a. The absorbance spectrum of Triton X. **b.** The absorbance spectrum of 1 *mg/mL* of the protein bovine serum albumin (BSA). **c.** The absorbance spectrum of the same sample of BSA as shown in **b.**, but with a small amount (0.03%) of Triton X-100 added. Although the spectrum for BSA appears similar to that in **b.**, in fact, the absorbance at 280 *nm* is raised due to the contamination of Triton X.

Interferences may cause absorbance readings to be incorrectly high or low. In absorption spectrophotometry, interferences are often compounds that absorb light at the same wavelength as the analyte and therefore result in an absorbance value that is too high. Impurities that reduce (quench) light readings are often encountered in fluorescence assays (discussed later).

Figure 20.6 illustrates the effect of an interfering substance in a sample. The interfering substance in this example is a detergent, Triton X-100, which is sometimes used in biological solutions to solubilize membrane proteins and to prevent protein aggregation. Detergents are also commonly used as cleaning agents. Both proteins and Triton X-100 absorb light at 280 *nm*. Residual detergent, therefore, must be removed from protein preparations before they are analyzed by spectrophotometry at 280 *nm*.

Colorimetric reactions are frequently used to visualize an analyte in the presence of other substances. For example, consider a sample containing a mixture of proteins, nucleic acids, and other cellular components. To analyze only the proteins, reagents can be

Table 20.1 *Symptoms That an Interfering Substance is Present in a Colorimetric Assay*

1. *Interference is sometimes detectable if the color formed by a sample mixture is not of the same hue as that of the standards.*

2. *Interference is indicated if a precipitate forms or turbidity occurs when the sample and reagent are mixed together, but not when the standards and reagent are mixed together.* (This behavior is occasionally due to a concentration of analyte in the sample that is too high. In this case, simply diluting the sample will eliminate the problem.)

3. *If color does not form at the same rate or with the same stability in a sample as in the standards, an interfering substance(s) may be present in the sample.*

added to the sample that selectively react with the proteins to form a colored product. There are situations, however, where a color-forming reagent(s) reacts with a substance in the sample in addition to the analyte. A substance that also reacts in an undesired fashion with colorimetric reagents is an interference. Table 20.1 summarizes practical points relating to interfering substances in colorimetric analyses.

It is not possible to compensate for the effect of an interfering substance by using a blank. A blank contains the solvent and reagents that are intentionally added to the sample. Interfering materials, by their nature, are variable and unpredictable substances in the sample that cannot be intentionally included in the blank.

The methods used to handle interferences vary depending on the situation. In some cases, it is possible to find an analytical wavelength where the analyte absorbs light and the interfering substance(s) does not. In other situations, where the nature of an interfering substance is known, it is possible to analyze more than one compound: the analyte, and the interference(s). Because each component individually obeys Beer's Law, and because both compounds are known, the concentration of each can be determined. A simple example of this approach will be discussed later in the section on UV methods. An alternative spectrophotometric method sometimes exists that is insensitive to the interference(s) present in a sample. In other situations, where the nature of an interfering substance is unknown and a suitable analytical wavelength cannot be found, the sample must be purified to eliminate the interference.

iii. Choosing the Proper Wavelength for Quantitative Analysis

Quantitative analyses are performed at a specific wavelength. The optimal wavelength is established based on the absorbance spectra for standards containing the analyte and based on representative samples.

The most important requirement when choosing the analytical wavelength is that the substance of interest absorb light at that wavelength. The more strongly the analyte absorbs light at the chosen wavelength, the better the assay method will be able to detect low levels of compound. A second factor to consider is the nature of the absorbance peak. It is desirable that the peak of absorbance not be too narrow (i.e., the natural band width of the peak should be fairly wide). If the peak is narrow, then any small error in the wavelength setting of the spectrophotometer will result in a large change in absorbance. For example, observe in Figure 20.7 that at λ_2 this compound strongly absorbs light, but the peak is very sharp, so a small error in wavelength setting will result in a large change in the absorbance measured. At λ_1 there is a peak with a broader natural band width that is probably the best choice of wavelength for analysis of this compound. If very low levels of analyte must be measured, then the maximum sensitivity is required and λ_2 might be used for analysis.

The presence of interfering substances in the sample is a complicating factor in choosing an analytical wavelength. If interfering substances are present, it may be possible to choose a wavelength for analysis at which the analyte absorbs less light but the interfering substances are not absorptive.

Thus, the factors that are important in choosing the optimal wavelength are:

1. **The analyte should have a peak of absorbance at the wavelength chosen.**

2. **The absorbance peak ideally should be broad.**

3. **Interfering substances should not be present that absorb light at the chosen wavelength.**

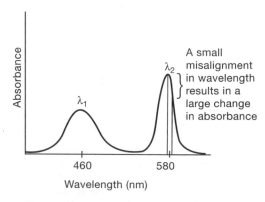

Figure 20.7. Determining the Optimal Wavelength for Quantitative Analysis. The wavelength chosen must be one at which the analyte absorbs light. There is a strong absorbance peak at 580 *nm*; however, because the peak is sharp, a small misalignment in the wavelength adjustment will cause a major change in absorbance. Unless very low levels of analyte must be measured, 460 *nm* therefore appears to be a better choice.

EXAMPLE PROBLEM

Based on the following absorbance spectrum of Compound A, what is the optimal wavelength for performing quantitative analysis of Compound A?

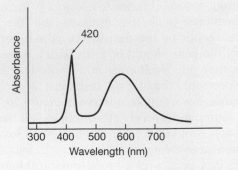

ANSWER

600 *nm* appears to be a good choice because the compound absorbs light at that wavelength and the peak is not as "sharp" as it is at 420 *nm*.

EXAMPLE PROBLEM

Based on the following absorbance spectra, what is the optimal wavelength for performing quantitative analysis of Compound A in samples?

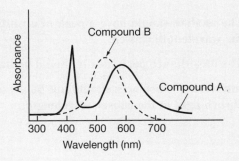

ANSWER

There is an interfering substance, Compound B, that absorbs light at 600 *nm*. 420 *nm* appears to be a suitable wavelength where compound A absorbs light and there is minimal interference.

B. Method Validation

Validation of an analytical method is intended to provide documented evidence that a method will perform as is required. The criteria used to evaluate methods include linearity and range, limit of detection and limit of quantitation, selectivity, ruggedness, accuracy, and precision. These criteria address the limitations of an assay, its ability to give accurate results, the types of samples for which it is suitable, and the conditions under which it is useful. Although formal validation of

analytical methods is required only in laboratories that are regulated or meet certain standards, evaluating an analytical method based on relevant criteria is good practice in any laboratory. (See pp. 80–82 for a discussion of method validation in the context of quality systems.) This section discusses how these criteria apply to spectrophotometric methods.

The **range of an assay** *is the interval between the upper and lower levels of analyte that can be measured accurately and with precision.* Factors that determine the range of an assay include the absorptivity of the analyte or the intensity of color produced in a colorimetric reaction, the capabilities of the spectrophotometer being used, and the choice of wavelength. In addition, the conditions of the assay may affect the range. For example, the intensity of color of some solutions is pH-dependent.

The range of a spectrophotometric assay is usually considered to be the range in which the relationship between absorbance and concentration is linear. Figure 20.8 shows a standard curve for a spectrophotometric assay. At concentrations of analyte above about 130 $\mu g/mL$ the relationship between concentration and absorbance is not linear. The linear range for this assay is from about 5 to 130 $\mu g/mL$.

There are situations where a calibration curve can be used that is not linear. Figure 20.9 illustrates a nonlinear calibration curve from an assay used to quantitate proteins. Although the relationship between concentration and absorbance is not strictly linear at higher concentrations, this assay is still useful given a standard curve with sufficient points to clearly show the relationship.* (Remember that Beer's Law does not apply if the relationship between absorbance and concentration is nonlinear.)

Sensitivity is the ratio of the change in the instrument response to a corresponding change in the stimulus. For a spectrophotometer, this is the ratio of the

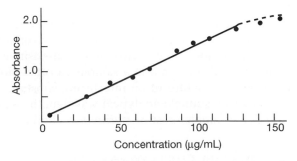

Figure 20.8. Linear Range of an Assay. This assay is useful in the linear range from about 5 to 130 $\mu g/mL$.

*Many spectrophotometers are equipped with software that automatically draws the best fit line connecting the points of a standard curve. If an assay is known to be nonlinear, these automatic line fitting programs should not be used.

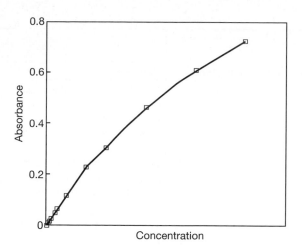

Figure 20.9.   The Standard Curve from a Lowry's Assay for Protein. The assay gave linear results at the lower concentrations, but it was not linear at higher concentrations.

change in absorbance to a corresponding change in analyte concentration:

$$Sensitivity = \frac{Change\ in\ absorbance}{Change\ in\ analyte\ concentration}$$

This concept was illustrated in Figure 20.3. Compound A had more inherent tendency to absorb light at 550 *nm* than compound B. At 550 *nm*, therefore, a spectrophotometer is more sensitive to compound A than compound B. Thus, in spectrophotometry, the sensitivity of a method is affected largely by the absorptivity of the analyte or by the intensity of color developed in a colorimetric assay.

The **limit of detection** of an assay *is the lowest concentration of analyte in a sample that can be detected, but not necessarily quantified.* **The limit of quantitation** *is the lowest concentration of analyte that can be reliably quantified.* The limits of detection and quantitation of an assay are affected by the sensitivity of a method. The more sensitive, the lower the level of analyte that can be detected. In addition, the limits are affected by noise. As the concentration of the analyte approaches zero, the signal becomes buried in the noise, and the level of analyte falls below the detection limit of the instrument. Thus, if a method needs to be able to detect trace levels of analyte, the method should be very sensitive and the noise in the spectrophotometer should be as low as possible.

Selectivity *is the ability of an assay to distinguish the compound of interest in the presence of other materials.* Selectivity in spectrophotometry, therefore, relates to the ability of a method to measure an analyte accurately in the presence of potentially interfering substances in the sample.

Ruggedness *is the degree of reproducibility of test results obtained by the analysis of the same samples under a variety of normal test conditions.* A rugged method gives accurate results over a range of conditions. Many spectrophotometric assays are affected by solution conditions, such as pH, temperature, and salt concentration. This is because the properties and structures of biological molecules change as their solution conditions change. As the molecules change form, so too, their absorptivity may change. Colorimetric reactions similarly may be affected by solution conditions; therefore, these conditions need to be controlled in spectrophotometric methods.

Accuracy is the closeness of a test result to the "true" value. One of the most important and basic ways to ensure accuracy in a quantitative spectrophotometric assay is to prepare a standard curve each time an assay is performed, using carefully prepared standards.

Precision and **long-term reproducibility** are often the most important aspects of an assay. For example, suppose spectrophotometry is being used to monitor the amount of protein present in a preparation as the protein is purified. Protein purification has multiple steps that may take place over a number of days. An assay method must give consistent results over a period of time; otherwise, it is useless to monitor the purification.

C. Colorimetric Protein Assays; An Example

Colorimetric assays of proteins are an important application of spectrophotometry in biology. Most proteins do not naturally absorb light in the visible range. Colorimetric methods, therefore, have been devised in which proteins are reacted with certain dyes with the development of an intense color. Protein assays illustrate some of the issues that arise in using colorimetric methods.

Three common colorimetric protein assay methods are summarized in Table 20.2. The methods vary in features such as their range and their tendency to be affected by interfering substances. It is necessary, therefore, to choose the assay method that best suits your samples. It is important that the range of analyte concentration that can be measured by the assay matches the range of concentrations expected in the samples. Thus, if low levels of protein are to be measured, the Lowry or the Bradford methods are preferred over the Biuret method.

Protein assays are often not linear. As shown in Figure 20.9, these assays can still be used, but a standard curve is always required to establish the relationship between absorbance and concentration. Colorimetric protein assays are all affected to some extent by interfering substances. As can be seen in Table 20.2, interferences are particularly problematic for the Lowry assay.

The Lowry and the Bradford methods both develop more intense color with some proteins than with others. To get accurate results with these methods, therefore, it is necessary to construct a standard curve for

Table 20.2 *PROTEIN QUANTITATION METHODS*

Method	Principle	Approx. Concentration Range	Linearity	Interfering Substances	Comments
Biuret	Peptides react with Cu^{2+} in alkaline solution to yield a purple complex that has an absorption maximum at 540 *nm*.	500–8000 $\mu g/mL$	Not linear at all concentrations	Some materials interfere, including Tris buffer.	The Biuret method is least susceptible to protein to protein variation because it measures the peptide bonds in a protein. The major disadvantage of the Biuret method is its relative insensitivity to low levels of proteins.
Lowry	Color results from the reaction of Folin phenol reagent with amino acids in proteins.	1–300 $\mu g/mL$	May be nonlinear above about 40 $\mu g/mL$	Many compounds interfere including phenols, glycine, ammonium sulfate, and Tris buffer.	This method is sensitive. The Lowry method is not totally specific for proteins; substances beside proteins also react with the reagent and thus interfere. The amount of color development depends on the percentage of tyrosine and tryptophan amino acids in the protein; therefore, some proteins react more intensely than others.
Bradford	Reaction under acidic conditions with Coomassie Brilliant Blue G-250 reagent. Reagent has two color forms: red and blue. The red form turns to the blue form when the dye binds to protein.	25–1400 $\mu g/mL$	May become nonlinear at higher concentrations.	Relatively few interfering substances. Some detergents interfere.	A specific and sensitive assay.

Some of the information in this table is from *Methods for Protein Analysis,* Robert A. Copeland. Chapman and Hall, New York, 1994.

each protein of interest. For example, if the protein to be measured is DNA polymerase, then the standard curve should be constructed with DNA polymerase to get the best accuracy. In practice, it is common to construct a standard curve for a protein assay using bovine serum albumin (BSA) because the protein of interest is often expensive or difficult to obtain in large quantities, whereas BSA is readily available and is relatively inexpensive. Using BSA as the standard will give results that are not accurate; however, using BSA to make the standard curve will give a reasonable estimate of protein concentration in many situations. For example, proteins are purified in a series of steps. After each step, the amount of protein present is measured. If this measurement is consistently made with reference to BSA, then the relative effectiveness of each purification step can be determined. In contrast, if an analyst wants to know the absolute amount of a specific protein in a preparation, then a standard curve for that protein must be prepared.

EXAMPLE PROBLEM

A Lowry assay was performed. A standard curve was constructed as shown using the protein bovine serum albumin. Three samples of another protein, Protein X, were then analyzed. The absorbances of the samples were:

Sample A, 0.10
Sample B, 0.58
Sample C, 1.3

a. What is the concentration of protein in the three samples based in the standard curve?

b. Is it reasonable to use BSA to prepare the standard curve when the protein of interest is Protein X?

c. Would it affect the reproducibility of the assay if a new batch of Folin reagent were used? (Folin phenol reagent is the color-producing reagent in the Lowry assay.)

d. Would it affect the reproducibility of the assay if the bulb on the spectrophotometer became less bright between assays?

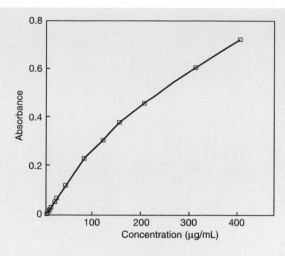

ANSWER

a. Sample A has a concentration of about 40 $\mu g/mL$. Sample B has a concentration of 300 $\mu g/mL$. The absorbance of Sample C exceeds the range of the standard curve. In this case, it is necessary to repeat the assay, diluting the sample before the color-developing reagents are added.

b. Because the Lowry method primarily detects tyrosine and tryptophan amino acids, and because the percent tyrosine and tryptophan varies from protein to protein, it is not desirable to prepare a standard curve with a protein different from the one being evaluated. This practice is necessary, however, in situations where sufficient amounts of purified protein of interest to make a standard curve are unavailable or excessively costly.

c. As long as a new standard curve is prepared each time the assay is performed, and both batches of reagent are properly made, there should be no problem. (The results may not be accurate if a standard curve is not prepared each time the assay is run.)

d. Changes in the bulb will not affect the reproducibility of the assay as long as a new standard curve is prepared each time the assay is performed.

D. Analysis of DNA, RNA, and Proteins; Examples of UV Spectrophotometry and Multicomponent Analysis

i. THE UV ABSORBANCE SPECTRA OF NUCLEIC ACIDS AND PROTEINS

Although most biological molecules do not intrinsically absorb light in the visible range, they do absorb ultraviolet light. Biologists take advantage of UV absorbance to quickly estimate the concentration and relative purity of DNA, RNA, and proteins in a sample. Points of general importance regarding spectrophotometric methods are illustrated by these applications.

Proteins have two UV absorbance peaks: one between 215 and 230 nm, where peptide bonds absorb, and another at about 280 nm due to light absorption by aromatic amino acids (tyrosine, tryptophan, and phenylalanine*). DNA and RNA have an absorbance maximum at approximately 260 nm and an absorbance minimum at about 230 nm. Certain subunits of nucleic acids (purines) have an absorbance maximum slightly below 260 nm, whereas others (pyrimidines) have a maximum slightly above 260 nm. Therefore, although it is common to say that the absorbance peak of nucleic acids is 260 nm, in reality the absorbance maxima of different fragments of DNA vary somewhat depending on their subunit composition. Table 20.3 summarizes the UV wavelengths relevant to the measurement of nucleic acids and proteins.

Figure 20.10 shows the absorbance spectra for DNA and proteins. Note that although proteins have lower absorbance at the absorbance peak of nucleic acids, 260 nm, both proteins and nucleic acids absorb light at 280 nm. Therefore, if nucleic acids and proteins are

Table 20.3 *WAVELENGTHS THAT ARE RELEVANT TO THE MEASUREMENT OF NUCLEIC ACIDS AND PROTEINS*

Wavelength	Significance	Comments
215–230 nm	Minimum absorbance for nucleic acids Peptide bonds in proteins absorb light	Measurements are generally not performed at this wavelength because commonly used buffers and solvents, such as Tris, also absorb at these wavelengths.
260 nm	Nucleic acids have maximum absorbance	Purines absorbance maximum is slightly below 260; pyrimidines maximum is slightly above 260. Purines have a higher molar absorptivity than pyrimidines. The absorbance maximum and absorptivity of a segment of DNA, therefore, depends on its base composition. Proteins have lower absorbance at this wavelength.
270 nm	Phenol absorbs strongly	Phenol may be a contaminant in nucleic acid preparations.
280 nm	Aromatic amino acids absorb light	Nucleic acids also have some absorbance at this wavelength.
320 nm	Neither proteins nor nucleic acids absorb	Used for background correction because neither nucleic acids nor proteins absorb at this wavelength.

*The structure of DNA, RNA, and proteins is briefly reviewed in Chapter 23.

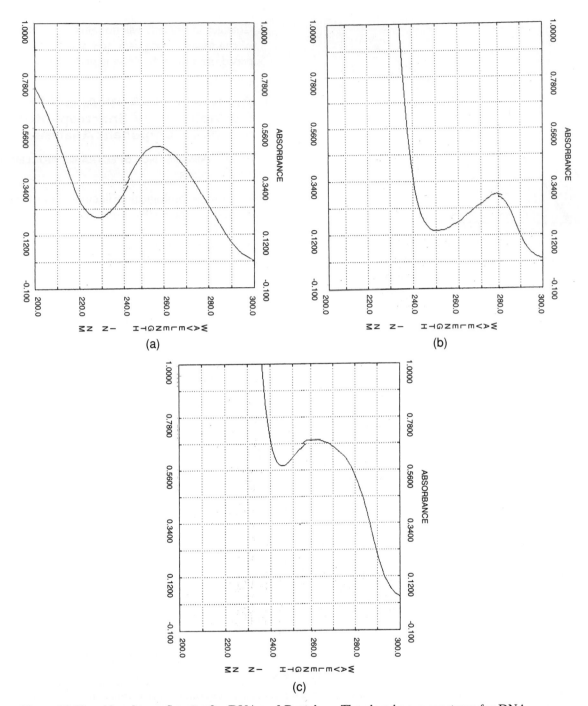

Figure 20.10. Absorbance Spectra for DNA and Protein. a. The absorbance spectrum for DNA. **b.** The absorbance spectrum for protein (BSA). **c.** The absorbance spectrum for a mixture of DNA and protein. Distinct peaks for DNA and protein cannot be resolved.

mixed in the same sample, their spectra interfere with one another.

ii. CONCENTRATION MEASUREMENTS OF NUCLEIC ACIDS AND PROTEINS

Consider two UV methods of determining concentration in a "pure" sample containing only proteins or nucleic acids. The first method involves constructing a standard curve; the second is a "short-cut" method based on absorptivity constants from the literature.

It is possible to determine the concentration of nucleic acids or proteins based on their absorbance at a wavelength of 260 *nm* or 280 *nm*, respectively. A calibration curve using standards of known concentration can be constructed. For accurate results, the standard curve should be prepared using the protein of interest or DNA that is similar to that in the sample being measured. The linear range for DNA values is reported to be from about 5 to 100 *µg/mL*. Depending on the protein, UV analysis of proteins at 280 *nm* has a linear range from about 0.1 to 5 *mg/mL*.

Biologists commonly use a "short-cut" to provide a rough estimate of the concentration of nucleic acid or protein in a sample based on the sample's absorbance at 260 or 280 *nm*. This short-cut method uses absorptivity constants. Recall that given an absorptivity constant, it is possible to by-pass the preparation of a standard curve by applying Beer's Law. (See "Variation 3", pp. 407–408.)

The absorptivity constant for a particular protein at 280 *nm* depends on its composition. Proteins that contain a higher percentage of aromatic amino acids have higher absorptivities at 280 *nm* than do those with fewer. The absorptivity constant for a nucleic acid depends on its base composition and on whether it is single-stranded or double-stranded. Despite the fact that different proteins and nucleic acid fragments vary in their absorptivity, analysts commonly use "average" absorptivity constants to estimate the concentration of nucleic acid or protein in a sample. The average absorptivity constants for proteins and nucleic acids lead to the following relationships:

- **If a sample containing pure double-stranded DNA has an absorbance of 1 at 260 *nm*, then it contains approximately 50 $\mu g/mL$ of double stranded DNA.**

- **If a sample containing pure single-stranded DNA has an absorbance of 1 at 260 *nm*, then it contains approximately 33 $\mu g/mL$ of DNA.**

- **If a sample containing pure RNA has an absorbance of 1 at 260 *nm*, then it contains approximately 40 $\mu g/mL$ of RNA.**

- **Values for proteins vary. A very rough rule is that if a sample containing pure protein has an absorbance of 1 at 280 *nm*, then it contains approximately 1 mg/mL of protein.** For example, 1 mg/mL of bovine serum albumin is reported to have an A_{280} value of 0.7. Antibodies (which are a type of protein) at a concentration of 1 mg/mL are reported to have an A_{280} between 1.35 and 1.2. (Values from *Antibodies: A Laboratory Manual*, E. Harlow, and D. Lane. p. 673. Academic Press, New York, 1988.)

EXAMPLE PROBLEM

A laboratory scientist isolated DNA from bacterial cells and was interested in estimating the amount of DNA present in the preparation. The preparation had a volume of 2 *mL*. The scientist removed 50 μL from the preparation, added 450 μL of buffer, and then read the absorbance of the dilution at 260 *nm*. The absorbance was 0.65. Assuming the sample was pure, about how much double-stranded DNA was present in the original 2000 μL preparation?

ANSWER

A proportion equation can be set up based on the relationship that a sample containing 50 $\mu g/mL$ of pure double-stranded DNA has an absorbance of 1 at 260 *nm*:

$$\frac{1}{50 \ \mu g/mL} = \frac{0.65}{?}$$

$$? = 32.5 \ \mu g/mL$$

Taking into account that the preparation was diluted, the concentration was:

$$32.5 \ \mu g/mL \times 10 = 325 \ \mu g/mL$$

To estimate the amount of DNA in the original preparation, note that there were 2 *mL* of isolated product. The concentration of DNA in that preparation was about 325 $\mu g/mL$. The 2 *mL* of the original preparation, therefore, had about:

$$2 \ mL \times 325 \ \mu g/mL \ '= 650 \ \mu g \ \text{double-stranded DNA}$$

iii. ESTIMATION OF THE PURITY OF A NUCLEIC ACID PREPARATION

It is possible to use UV spectrophotometry to estimate the purity of a solution of nucleic acids. This method involves measuring the absorbance of the solution at two wavelengths, usually 260 *nm* and 280 *nm*, and calculating the ratio of the two absorbances:

- **An A_{260}/A_{280} ratio of 2.0 is characteristic of pure RNA.**

- **An A_{260}/A_{280} of 1.8 is characteristic of pure DNA.**

- **An A_{260}/A_{280} ratio of about 0.6 is characteristic of pure protein.**

A ratio of 1.8–2.0 is therefore desired when purifying nucleic acids. (Note that this method does not actually distinguish DNA and RNA from one another.) A ratio less than 1.7 means there is probably a contaminant in the solution, typically either protein or phenol.

EXAMPLE PROBLEM

Returning to the previous Example Problem, the laboratory scientist wants to determine whether the DNA preparation is contaminated by proteins. The absorbance at 260 *nm* was 0.65 and at 280 *nm* was 0.36. Is this solution likely to contain pure nucleic acids?

ANSWER

The A_{260}/A_{280} ratio is 0.65/0.36 = 1.81. This ratio is consistent with a pure preparation of nucleic acids.

iv. MULTICOMPONENT ANALYSIS; THE WARBURG-CHRISTIAN ASSAY METHOD

(Multicomponent analysis is a more complex application of spectrophotometry and some readers may prefer to skip this section.)

It is sometimes desirable to estimate the concentrations of protein and of nucleic acids in a sample that contains both. The basic premise of an analysis such as this, a **multicomponent analysis**, is that each substance in the mixture individually obeys Beer's Law; therefore, the absorbances of two or more components in a mixture add together. For example, if nucleic acids and proteins are mixed in a sample, then the total absorbances at 280 *nm* and 260 *nm*, are:

$$A_{280} = A_{280} \text{ proteins} + A_{280} \text{ nucleic acids}$$
$$A_{260} = A_{260} \text{ proteins} + A_{260} \text{ nucleic acids}$$

To relate the absorbances at 260 *nm* and 280 *nm* to the concentrations of proteins and nucleic acids, these equations can be rewritten in terms of absorptivity constants for proteins and nucleic acids at each wavelength. Warburg and Christian performed these calculations using absorptivity constants for a yeast protein and for yeast RNA and published the results in 1941 ("Isolierung und Kristallisation des Gärungsferments Enolase," Otto Warburg and Walter Christian. *Biochem. Z.* 310, pp. 384–385, 1941/1942). They derived these two equations:

$$[\text{nucleic acid}] \approx 62.9 \, A_{260} - 36.0 \, A_{280} \text{ (in units of } \mu g/mL)$$
$$[\text{protein}] \approx 1.55 \, A_{280} - 0.757 \, A_{260} \text{ (in units of } mg/mL)$$

where the brackets indicate "concentration."

To determine the concentrations of proteins and nucleic acids in a mixture one measures the absorbances at 260 and 280 *nm* and solves the preceding equations.

EXAMPLE PROBLEM

A technician wants to get a rough estimate of the concentration of proteins and nucleic acids in a cell preparation. The absorbance of the preparation at 260 is 0.940 and at 280 is 0.820. Based on the Warburg-Christian equations, what is the approximate concentration of nucleic acids and proteins in the preparation?

ANSWER

Substituting into the Warburg-Christian equations:

$$[\text{nucleic acid}] \approx 62.9 \, (0.940) - 36.0 \, (0.820) = 29.6 \, \mu g/mL$$
$$[\text{protein}] \approx 1.55(0.820) - 0.757(0.940) = 0.559 \, mg/mL$$

Thus, based on the Warburg-Christian equations, the concentration of nucleic acids in the preparation is estimated to be 29.6 *μg/mL* and of proteins to be 0.559 *mg/mL*.

The Warburg-Christian equations are approximations based on absorptivity constants derived for the yeast protein, enolase, and yeast RNA. These equations, therefore, cannot be expected to give accurate results when applied to other proteins and nucleic acids. It is possible, however, to generalize the logic used by Warburg and Christian by substituting into the equations absorptivity constants determined for your analytes at specific wavelengths. If the correct absorptivity constants and the optimal wavelengths are used, it is possible to quantitate multiple components in a sample, even if their spectra overlap to some degree. (For more information about multicomponent analysis, see, for example, *Instrumental and Separation Analysis*, C.T. Kenner. Charles E. Merrill, Columbus, Ohio, 1973.) In addition, spectrophotometer manufacturers have developed software and methods to facilitate multicomponent analyses.

v. CONSIDERATIONS AND CAUTIONS

It is common for analysts to calculate and report purities based on A_{260}/A_{280} ratios and concentrations based on Warburg-Christian calculations. In fact, these calculations are performed so routinely that many spectrophotometers can perform them automatically and display the resulting values. It is important, however, to be aware that values obtained by these methods are only approximations and sometimes may be deceptively inaccurate. The reasons for inaccuracy include:

1. The A_{260}/A_{280} method is based on the spectral characteristics of "average" proteins and nucleic acids. In reality, proteins and nucleic acids vary from one another, so their spectra may vary from one another.

2. The Warburg-Christian method is based on absorptivity constants derived from one protein and nucleic acid from one organism. The absorptivity constants for other proteins and nucleic acids vary.

3. These UV methods assume that the spectrophotometer is accurately calibrated. If the spectrophotometer is displaced by as little as 1 *nm*, the values may be significantly affected. (See, for example, "Value of A260/A280 Ratios for Measurement of Purity of Nucleic Acids," Keith L. Manchester. *BioTechniques*, 19(2):208–210, 1995.)

4. These UV methods are affected by the pH and ionic strength (salt concentration) of the buffer,

resulting in variability. (See, for example, "Effect of pH and Ionic Strength on the Spectrophotometric Assessment of Nucleic Acid Purity," William W. Wilfinger. *BioTechniques,* 22(3):474–480, 1997.)

These cautions apply not only to UV analysis of nucleic acids and proteins, but to other spectrophotometric methods. It is always important to be aware of potential inaccuracy when using absorptivity constants reported in the literature. The type of spectrophotometer used and its calibration will affect results in almost any method. It is also important to consider matrix effects (i.e., effects due to the solvent and other components in the sample).

Box 2 summarizes the various UV methods described in this section.

E. Avoiding Errors in Spectrophotometric Analyses

The accuracy of spectrophotometric assays is affected by systematic errors. For example, in colorimetric assays, how long color is allowed to develop may affect the results. Color may be slow to develop in samples of low concentration and fast where concentration is high. Inconsistency in timing or a poor choice of development time may therefore lead to inaccurate results. Errors may occur if too little color developing reagent is added to samples so that the assay underestimates the amount of analyte present in concentrated samples. Using a blank that contains a different solvent than the samples will lead to error. Many (but not all) errors can be avoided by preparing a standard curve each time an assay is performed rather than relying on an extinction coefficient from previous assays, a single standard, or a standard curve prepared at a different time.

Table 20.4 summarizes practical considerations relating to obtaining good results with spectrophotometric methods.

EXAMPLE PROBLEM (and a true-life tale)

An analyst in our laboratory was attempting to check the absorbance spectrum of a solvent to see whether it was suitable for use in the UV range. She filled a cuvette with the solvent and prepared an absorbance spectrum, which is shown in the graph. There is a problem here—can you figure out what it is?

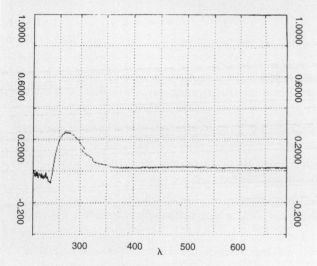

ANSWER

Based on this spectrum, she could not reach any conclusion about the solvent—the cuvette was dirty. The peak at 280 *nm* was due to protein that was left in the cuvette by another user and had dried onto the cuvette walls. The analyst soaked the cuvette for a couple of hours in distilled water and rinsed it thoroughly. She then obtained a spectrum with no peak at 280 *nm*, indicating that the solvent has little absorption in the UV range (and that soaking and rinsing the cuvette with distilled water cleaned it effectively in this case).

EXAMPLE PROBLEM

Suppose you are performing a colorimetric assay that includes a boiling step. Five minutes of boiling are necessary for complete color development. You become distracted and accidentally stop the boiling at only 3 minutes. How will this affect your results?

ANSWER

Assuming you prepare a standard curve each time you do the assay, and assuming the reagent still acts in such a way that color is proportional to the amount of analyte, then the standards should be affected by the shortened boiling time in the same way as the samples. The results, therefore, will still be accurate. You may observe, however, that the range for the assay is different than usual. Lower concentrations of analyte that are normally in the range of the assay may in this case not cause a detectable color change. (When you are familiar with an assay, any change in its performance should alert you to the fact that there is a problem.) In addition, note that if you do not prepare a standard curve each time the assay is performed, this type of mistake will lead to undetected errors in the results.

Table 20.4   CONSIDERATIONS IN OBTAINING ACCURATE SPECTROPHOTOMETRIC MEASUREMENTS

FACTORS RELATING TO THE INSTRUMENT AND ITS OPERATION

1. *Always use a well-maintained and calibrated instrument.*

2. *Use the correct wavelength.* Assays are optimized at a particular wavelength and irreproducibility or inaccuracy may occur at the improper wavelength.

3. *Allow the spectrophotometer light source to warm up, as directed by the manufacturer.* In most laboratories the lamps are turned off when the instrument is not in use because lamps have a finite life span and are relatively expensive to replace. Note that the UV source may take longer to warm up than the visible light source.

4. *Use the proper cuvettes.*

   a. For high-accuracy work use high-quality cuvettes that are clean, unscratched, and whose sides are exactly parallel.

   b. Use quartz cuvettes for work in the UV range.

   c. Use matched cuvettes for the sample and the blank, or use the same cuvette sequentially for both the sample and blank.

5. *Make certain cuvettes and glassware are clean.* In addition, drops or liquid smeared on the outside of the cuvette can cause erratic readings.

6. *Avoid "fogging."* When the air is humid and the sample or blank is cold, moisture can condense on the cuvette leading to "fogging."

FACTORS RELATING TO THE ASSAY METHOD AND THE SAMPLE

7. *Avoid samples and standards whose absorbance is very high or very low.* Depending on the instrument, the recommended range of absorbance will probably be about 0.5–1.8.

8. *Be aware of sample characteristics that affect absorbance.*

   a. The color and the absorbance of some samples varies with the pH of the solution; in such cases, make sure the pH of standards and samples is the same.

   b. Sample absorbance may vary with temperature; in such cases, water jacketed accessories are available for some spectrophotometers to hold the temperature constant during measurement. Otherwise, use a water bath or other device to control the temperature of samples and standards when they are not in the spectrophotometer.

   c. In principle, the absorptivity of an analyte should be constant regardless of its concentration. In reality, some analytes change absorptivity as their concentration increases. In these cases, it is difficult to get a linear response.

   d. The ultraviolet absorption of a compound may vary in different solvents. Be sure, therefore, to record the solvent used and, for aqueous solvents, the pH. When following established methods, use the specified solvent.

9. *Be aware of time as a factor in analysis.* The color intensity of samples reacted with color-developing reagents may increase or decrease over time, or the color may shift in absorbance peak. Time is a particularly important factor in assays involving enzymatic reactions.

10. *Avoid deteriorated reagents and standards.* Many color developing reagents deteriorate over time. It may be helpful to check reagents regularly by measuring their absorbance versus a blank of distilled water. Initial deterioration of reagents can often be recognized by changes in the reagent's absorbance. Discard reagents that become turbid or cloudy, that change colors, or that contain precipitate.

11. *Be aware that some samples are naturally fluorescent.* There are materials that fluoresce when irradiated with certain wavelengths of light. This results in erroneous transmittance values if the fluorescence falls on the detector. Fluorescent samples can be evaluated if a filter is available that blocks light of the fluorescent wavelength while allowing transmitted light to pass. The filter is placed between the sample and the detector.

12. *Avoid turbidity (except in some special cases where measurements of turbidity provide useful information).* Turbidity is a very common source of error (e.g., environmental water samples are often turbid). Particles in suspension affect the transmittance of light and may cause there to appear to be more absorbance than there really is. If the sample is not homogeneous, turbidity will result in inconsistent and incorrect results. Suspended material should generally be removed from samples prior to spectrophotometry by centrifugation or filtration. Buffers and water used for spectrophotometry may also require filtering.

13. *Use spectroscopy grade solvents.*

14. *Avoid substances that interfere with the assay.*

Box 2 APPROXIMATING THE CONCENTRATION AND PURITY OF DNA, RNA, OR PROTEIN IN A SAMPLE

1. *Concentration of double-stranded DNA $\approx$ 50 μg/mL $\times$ the absorbance at 260 nm*
2. *Concentration of single-stranded DNA $\approx$ 33 μg/mL $\times$ the absorbance at 260 nm*
3. *Concentration of RNA $\approx$ 40 μg/mL $\times$ the absorbance at 260 nm*

Turbidity causes an apparent increase in the absorbance of a sample, leading to incorrect readings. To compensate for slight turbidity, a background correction can be used. Proteins and nucleic acids do not absorb at 320 *nm*. If a sample absorbs at 320 *nm*, the absorbance is due to turbidity. The absorbance at 320 *nm* can be subtracted from the readings at 260 *nm* and 280 *nm*:

4. *Concentration of double-stranded DNA $\approx$ 50 μg/mL $(A_{260} - A_{320})$*
5. *Concentration of single-stranded DNA $\approx$ 33 μg/mL $(A_{260} - A_{320})$*
6. *Concentration of RNA $\approx$ 40 μg/mL $(A_{260} - A_{320})$*

7. *Concentration of protein $\approx$ 1 mg/mL $\times$ the absorbance at 280 nm*
8. *Concentration of protein $\approx$ 15 mg/mL $\times$ the absorbance at 215 nm*

   Notes: Values for proteins vary. The rule that 1 *mg/mL* of protein has an absorbance of 1 is approximate.

   Tris and other common solvents also absorb light at 215 *nm*. For this reason, 280 *nm* is far more commonly used for protein measurements.

9. *Purity*

   An A_{260}/A_{280} ratio of 2.0 is characteristic of pure RNA

   An A_{260}/A_{280} of 1.8 is characteristic of pure DNA

   An A_{260}/A_{280} of 0.6 is characteristic of pure protein

   Notes: A ratio of 1.8–2.0 is desired when purifying nucleic acids.

   This method does not distinguish DNA and RNA from one another.

   A ratio of less than 1.7 means there is probably a contaminant in the solution, usually either protein or phenol.

Warburg-Christian equations

For Solutions Containing a Mixture of Nucleic Acids and Proteins

10. *[nucleic acid] $\approx$ 62.9 A_{260} − 36.0 A_{280} (in units of μg/mL)*
11. *[protein] $\approx$ 1.55 A_{280} − 0.757 A_{260} (in units of mg/mL)*

VI. ASSOCIATED SPECTROPHOTOMETRIC AND FLUORESCENCE METHODS

A. Introduction

There are instruments in addition to spectrophotometers that measure interactions between light and a sample to obtain information about that sample. Some of these instruments allow samples to be studied that are not liquids in a cuvette. Other instruments are based on principles that are related to, but are not the same as absorption spectrophotometry. There are also modified spectrophotometers that find a variety of uses in the laboratory. These instruments and associated technologies expand the applications that can be performed using light measurements. This section discusses three examples of light-measuring instrumentation that are of particular importance in the biotechnology laboratory.

B. Gel Scanning/Densitometry

Electrophoresis is a widely used method in which mixtures of proteins or nucleic acids are separated from one another by allowing them to migrate through a gel matrix in the presence of an electrical field. The result of electrophoresis is that the macromolecules of different sizes are separated into individual bands in the gel. The separated bands are invisible until they are stained with some type of dye. After staining, the band pattern on the gel, or a photograph of the gel, can be examined and used to obtain information about the sample.

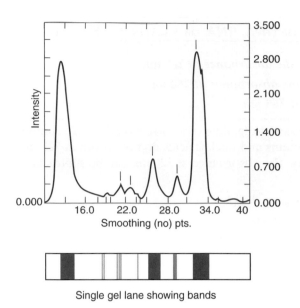

Single gel lane showing bands

Figure 20.11.   Densitometry of the Bands on an Electrophoresis Gel. The peaks represent the amount of material in each band. (The Y axis represents absorbance) (Graph courtesy of Beckman Coulter, Inc.)

Sometimes only the pattern of the bands in a gel is analyzed; other times, analysts are interested not only in the pattern, but also in quantifying the amount of nucleic acid or protein in each band. **Densitometry** *allows investigators to determine the amount of material in a band based on the intensity of its stain.* Densitometry uses light to measure how much material is present in a specific place on a gel or other matrix. When light is shined on a band in a gel, the amount of light absorbed is related to how much stain is present, which in turn depends on how much material is in the band, Figure 20.11. The absorbance of light by a band is compared with the absorbance of light by the support (gel) alone. Thus, an area of the support without stain is the blank.

The source of light in dedicated densitometers is sometimes a laser that produces a very high-intensity red light. Laser densitometers can pick up very fine details in gels, assuming the bands absorb red light, as most do.

Densitometry can be performed in various ways. The gel itself can be analyzed or a photograph or slide of the gel can be scanned. Densitometry can be performed either using conventional spectrophotometers with special adaptations or by using instruments dedicated to densitometry. When densitometry is performed with a conventional spectrophotometer, a special holder is required to place the gel into the sample chamber. The holder must be moveable so that each band can be exposed to the light beam, one at a time. Optical arrangements are needed to limit the light beam to a specific band, and a mechanism must be provided to move the gel in the sample chamber.

C. Spectrophotometers or Photometers as Detectors for Chromatography Instruments

There are many instances in biology where the components of a sample mixture are separated from one another before analysis. Chromatography is a commonly used separation method. During chromatography samples plus solvents flow through a matrix held in cylindrical columns or on flat plates. Depending on their chemical and physical nature, some components of a mixture move more quickly through the column or across the plate than others, so the components are separated from one another by their rate of motion.

As drops of sample and solvent leave a chromatography column, they can be routed to flow through a detector. There are many types of detectors available, but spectrophotometers and photometers are most common. A spectrophotometer or photometer used for chromatography does not involve cuvettes. Rather, there is a **flowcell** *through which the sample continuously flows.* As the sample moves through the flowcell, a continuous absorbance reading is produced, Figure 20.12. Because most biological materials absorb light in the UV range, these spectrophotometric detectors are frequently set to a UV wavelength (e.g., 280 *nm*).

Chromatography instruments may use a photometer that can measure light absorbance at only a single wavelength, or a few discrete wavelengths. Variable wavelength detectors have a monochromator and can

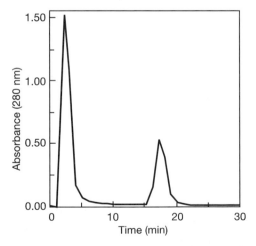

Figure 20.12.   A Spectrophotometer as a Detector in Chromatography. The absorbance of a sample eluted from a chromatography column. Time is on the X axis and absorbance on the Y axis. The peaks represent the components of the sample that absorb light at 280 *nm*.

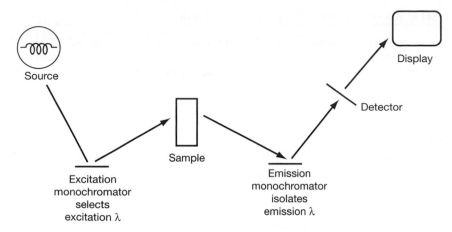

Figure 20.13.   A Fluorescence Spectrometer. The excitation beam is directed at the sample. Fluorophores in the sample are excited by the excitation beam and emit fluorescent light, which passes through a second monochromator. The second monochromator isolates light of a wavelength matching that emitted by the fluorophore. A photomultiplier tube detects the light emitted by the sample.

be adjusted to any wavelength in a certain range. In some cases, spectrophotometers with photodiode array detectors (PDAs) are used in conjunction with a chromatography instrument. The advantage to a PDA is that the absorbance spectrum of each material can be instantaneously generated as it elutes from the column. The resulting spectra are useful in identifying the separated components of the sample.

D. Fluorescence Spectroscopy

So far, we have discussed primarily the absorption of light by a sample. **Fluorescence** is another type of interaction that can give useful information about a sample. In **fluorescence**, *light interacts with matter in such a way that photons excite electrons to a higher energy state. The electrons almost immediately drop back to their original energy level with the reemission of light.* The fluorescence of clothing or posters when exposed to "black lights" are familiar examples of this phenomenon.

Fluorophores *are molecules that fluoresce when exposed to light.* Most biological molecules are not naturally fluorescent. Fluorescent compounds, therefore, are used to label or tag desired biological structures. For example, ethidium bromide is a fluorophore that is hydrophobic, planar, and positively charged. Because of these properties, ethidium bromide spontaneously inserts into DNA molecules. One of the most common molecular biology techniques is to soak an electrophoresis gel containing fragments of DNA in a solution containing ethidium bromide. The ethidium bromide intercalates

into the DNA. Ethidium bromide complexed with DNA fluoresces when exposed to light in the UV-B range, so it lights up the bands of DNA when the gel is placed on a UV light source.

Fluorophores can be used to tag antibodies that bind to their target biological structures with high affinity. Fluorescent tags have also been used to label cancer cells, genes, proteins coded by specific genes, intracellular proteins, pathogens, and a host of other materials of biological importance.

The basic design of a fluorimeter (fluorescence spectrometer) is shown in Figure 20.13. Like a spectrophotometer, the fluorimeter has a light source that emits polychromatic light. The light enters a monochromator that functions to isolate light of a narrow range of wavelengths. In fluorescence spectroscopy, *the light that shines on the sample is called the* **excitation beam**. The wavelength of the excitation beam must be selected by the operator because different fluorophores are excited by different wavelengths of light. The excitation beam shines on the sample. Fluorophores in the sample are excited by the light and emit fluorescence. At this point, a fluorescent spectrophotometer differs from a UV/Vis spectrophotometer. The emitted light passes through a second monochromator that is adjusted by the operator to isolate the light that was emitted by the fluorophore. A photomultiplier tube detects the light emitted from the sample.

Table 20.5 summarizes issues relating to fluorescence measurements.

Table 20.5 ISSUES RELATING TO FLUORESCENCE MEASUREMENTS

1. **Quenching** *occurs when an interference in the sample prevents the analyte from fluorescing.* This is a difficult problem to detect and often shows up as less sensitivity in an assay than is expected. Some fluorescent systems are quenched by oxygen from the air. Low concentrations of metal ions, as are present in some cleaning agents, may cause quenching.

2. **Fluorescence is much more sensitive to changes in the environment than is absorption.** Fluorescence increases as temperature decreases. On warm days, therefore, there may be noticeably less fluorescence than on cool days. Standards and samples should be at the same temperature. Some instruments control temperature to eliminate this variable.

3. **pH may affect the fluorescence of a sample.**

4. **There are fluorescent additives in plastics.** These additives may leach into samples that are stored in plastic containers.

5. **Most detergents used to wash glassware are strongly fluorescent.** Check that the detergent used in the laboratory does not contribute to fluorescence at the wavelength being measured.

6. **Stopcock grease and some filter papers contain fluorescent contaminants.**

7. **Microorganisms contaminating a solution may contribute fluorescence or may scatter emitted light.**

8. **The strong light sources used to excite the sample may bleach the fluorescent compounds.** Bleaching may be reduced by the choice of a different excitation wavelength, or by reducing the intensity of the excitation beam.

PRACTICE PROBLEMS

1. **a.** Title the X and Y axis for each of these two graphs.

   b. Which one is called an absorbance spectrum? Which one is called a standard curve?

   c. Which is associated with quantitative analysis? Which is associated with qualitative analysis?

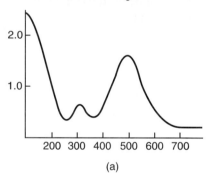

(a)

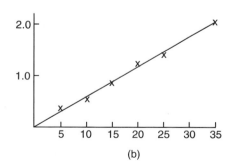

(b)

2. Label each of the following as a qualitative assay or a quantitative assay.

   a. Using an assay to determine the concentration of DNA in a sample.

   b. Using an infrared spectrum to identify an organic compound.

   c. Testing the amount of alcohol in a drivers blood.

   d. Determining the levels of lead in a child's blood.

   e. Determining whether a blood sample contains amphetamines.

   f. Determining whether or not a compound is oxygenated.

   g. Measuring the amount of product formed in an enzymatic reaction.

3. The following are absorption spectra for three compounds, A, B, and C, respectively. Assume all three compounds were present at the same concentration when the spectra were plotted and that the analysis conditions were the same.

   a. What color is each compound?

   b. Which of these compounds will have the highest absorptivity constant at a wavelength of 540 *nm*? Explain.

   c. Which of these will be detectable in the lowest concentration at 540 *nm*? Explain.

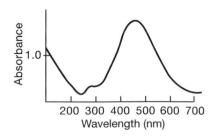

(a)

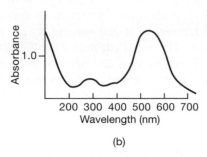

(b)

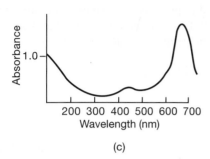

(c)

4. What is the linear range for the following assays?

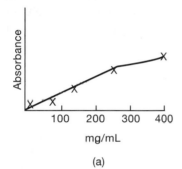

(a)

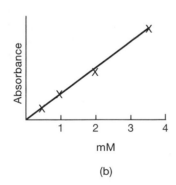

(b)

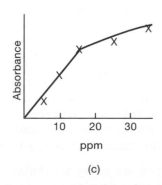

(c)

5. a. Prepare a standard curve from the data below.

b. Is this graph linear? If all or part of the graph is linear, what is the equation for the line?

c. If all or part of the graph is linear, determine the value of the absorptivity constant based on the line. Remember to include the units. (Assume the path length is 1 *cm*.)

d. Does this graph show a threshold? If so, where? If so, why is there a threshold?

Standard #	Concentration in *mg/liter*	Absorbance
1	2	0.22
2	4	0.45
3	10	1.11
4	12	1.34
5	20	2.23
6	30	2.21
7	40	2.22
8	50	2.21

6. A certain compound is detected at 520 *nm*. A series of dilutions were prepared and the data below were collected.

a. Plot the data.

b. Calculate the absorptivity constant based on the slope of the line. (Assume a path length of 1 *cm*.)

c. Determine the amount of compound in the samples.

Standard #	Concentration in *mM*	Absorbance
1	0	0.0
2	80	0.2
3	160	0.4
4	240	0.8
5	320	1.0

Sample	Absorbance
A	0.18
B	0.31
C	0.96
D	1.0

7. a. Prepare a standard curve from the data below.

b. Determine the absorptivity constant. (Assume the path length is 1 *cm*.)

c. What is the concentration of an unknown which is diluted 5 fold and has an absorbance of 0.30?

Standard #	Concentration in *mM*	Absorbance
1	1	0.22
2	2	0.46
3	5	1.08
4	6	1.34
5	10	2.18

For problems 8–18, assume Beer's Law applies.

8. A certain sample has an absorbance of 1.6 when measured in a 1 *cm* cuvette.

    a. If the same sample's absorbance is measured in a 0.5 *cm* cuvette, will its absorbance be greater than, less than, or equal to 1.6?

    b. What is the absorbance of this sample in the 0.5 *cm* cuvette?

9. A 25 ppm solution of a certain compound has an absorbance of 0.87. What do we expect will be the absorbance of a 40 ppm solution? Explain.

10. A 10 *mg/mL* solution of a compound has a transmittance of 0.680.

    a. What is the absorbance of this solution?

    b. What will be the absorbance of a 16 *mg/mL* solution of this compound?

    c. What will be the transmittance of a 16 *mg/mL* solution of this compound?

11. A solution contains 30 *mg/mL* of a light-absorbing substance. In a 1 *cm* cuvette its transmittance is 75% at a certain wavelength.

    a. What will be the transmittance of 60 *mg/mL* of the same substance?

    b. What will be the absorbance of 60 *mg/mL* of the same substance?

    c. What will be the transmittance of 90 *mg/mL* of the same substance?

    d. What will be the absorbance of 90 *mg/mL* of the same substance?

12. A standard solution of bovine serum albumin containing 1.0 *mg/mL* has an absorbance of 0.65 at 280 *nm*.

    a. Based on the BSA result, what is the protein concentration in a partially purified protein preparation if the absorbance of the preparation is 0.15 at 280 *nm*?

    b. Suggest two possible sources of error in this analysis.

13. The molar absorptivity constant of ATP is 15,400 *L/mole-cm* at 260 *nm*. (Value from *Biochemical Calculations*, Segel, 2nd edition, John Wiley and Sons, New York, 1976.) A solution of purified ATP has an absorbance of 1.6 in a 1 *cm* cuvette. What is its concentration?

14. The molar absorptivity constant for NADH is 15,000 *L/mole-cm* at 260 *nm*. (Value from *Biochemical Calculations*, Segel, 2nd edition, John Wiley and Sons, New York, 1976.) A sample of purified NADH has an absorbance of 0.98 in a 1.2 *cm* cuvette. What is its concentration?

15. The molar absorptivity constant for NADH is 15,000 *L/mole-cm* at 260 *nm* and is 6220 *L/mole-cm* at 340 *nm*. The molar extinction coefficient of ATP is 15,400 *L/mole-cm* at 260 *nm* and zero at 340 *nm*. (Values from *Biochemical Calculations*, Segel, 2nd edition, John Wiley and Sons, New York, 1976.) How could you determine the concentration of NADH in a sample that contains ATP as well as NADH?

16. Calculate the range of molar concentrations that can be used to produce absorbance readings between 0.1 and 1.8 if a 1 *cm* cuvette is used and the compound has a molar absorptivity of 10,000 *L/mole-cm*.

17. a. A sample of purified double-stranded DNA has an absorbance of 4 at 260 *nm*. The technician therefore dilutes it, mixing 1 *mL* of the DNA solution with 9.0 *mL* of buffer. This time the A_{260} reading is 1.25. What is the approximate concentration of the DNA in the original sample based on the equations in Box 2?

    b. A sample of purified RNA has an absorbance at 260 *nm* of 2.4. The technician therefore dilutes it by mixing 1 *mL* of the RNA solution with 4.0 *mL* of buffer. This time the A_{260} reading is 0.63. Approximately how much RNA was in the original sample?

18. a. A sample of reasonably pure double-stranded DNA has an A_{260} of 0.85. What is the approximate concentration of DNA in the solution?

    b. A sample of reasonably pure RNA has an A_{260} of 3.8. Because a spectrophotometer is not accurate reading absorbances that are so high, the analyst diluted the sample by adding 1 part sample to 9 parts buffer. The absorbance was then 0.69. What is the concentration of RNA in the solution?

19. Suppose a spectrophotometer has not been recently calibrated and its photometric accuracy is incorrect.

    a. Will the results of a quantitative assay be inaccurate if a standard curve is used?

    b. Will the results of a quantitative assay be inaccurate if a single standard is used?

    c. Will the results of a quantitative assay be inaccurate if the concentration of sample is calculated based on an absorptivity constant from the literature?

20. Suppose in a colorimetric assay, depending on the concentration of analyte in the sample, full color development requires anywhere from 3 to 5 minutes. You do not realize that there is a timing effect and allow the reaction to proceed for only 2 minutes. Will this cause an error in your results?

21. Suppose you are using a colorimetric assay that involves a color-developing reagent that deterio-

rates over time. You accidentally use an old bottle of the reagent that is less effective than it should be in causing a color change. What effect will this have on your results?

22. You are performing a spectrophotometer assay in which you use a buffer to prepare the standards and for the blank; however, the samples are centrifuged blood. Is this a potential source of error?

23. (Note that this problem is based on data obtained in our laboratory during a study of UV estimation methods for nucleic acids and proteins.)

An analyst was evaluating the methods summarized in Box 2 to establish whether they would be useful in her investigations. She prepared three standards: (1) a standard containing pure double-stranded DNA, (2) a standard containing pure bovine serum albumin, and (3) a standard containing both double-stranded DNA and BSA mixed together. She measured the absorbance of the three standards at 260 and 280 nm and prepared an absorbance spectrum for all three standards. She then applied the equations shown in Box 2 to her data. Because she had prepared the standards from pure DNA and protein, she knew what the results of the calculations should have been. The steps she performed and the data she obtained are shown:

Step 1. The technician dissolved 0.06 mg of DNA in 1 mL of a low salt buffer. (Low salt buffer was used as the blank also.) The DNA spectrum she obtained is shown in Figure 20.10a.
  The A_{280} for the DNA was 0.260.
  The A_{260} for the DNA was 0.497.

Step 2. She dissolved 0.5 mg of BSA in 1 mL of a low salt buffer. The protein spectrum is shown in Figure 20.10b.
  The A_{280} for the protein was 0.276.
  The A_{260} for the protein was 0.176.

Step 3. She prepared a solution containing 0.06 mg/mL DNA and 0.5 mg/mL BSA in a low salt buffer. The spectrum for the mixture is shown in Figure 20.10c.
  The A_{280} for the mixture was 0.547.
  The A_{260} for the mixture was 0.685.

a. Estimate the concentration of DNA in the DNA standard based on its UV absorbance and on Equation 1 in Box 2. The concentration of DNA in the standard was 0.06 mg/mL (based on its weight). Comment on the accuracy of the UV estimation method.

b. Estimate the concentration of protein based on its UV absorbance. Use the reported value that 1 mg/mL of BSA has an absorbance of 0.7 at

280 nm. The concentration of protein in the standard was 0.5 mg/mL (based on its weight). Comment on the accuracy of the UV estimation method.

c. Check whether the absorbances at 260 nm and at 280 nm were additive as predicted by the equations:

$$A_{260} = A_{260} \text{ proteins} + A_{260} \text{ nucleic acids}$$
$$A_{280} = A_{280} \text{ proteins} + A_{280} \text{ nucleic acids}$$

Substitute into the first equation the A_{280} reading from the pure BSA sample and the A_{280} reading from the DNA sample. Substitute into the second equation the A_{260} readings for both pure samples. Comment as to whether the absorbances were additive.

d. Calculate the A_{260}/A_{280} ratios for all three standards. Comment as to whether the ratios are as expected.

e. (Optional) Substitute the A_{260} and A_{280} values obtained for the mixture into the Warburg-Christian Equations, 9 and 10 in Box 2. Comment as to the accuracy of the Warburg-Christian equations.

f. Comment on all the preceding results.

DISCUSSION QUESTION

A mixture of phenol, chloroform, and isoamyl alcohol (25:24:1) is used in the separation of nucleic acids from cellular extracts (see page 499 for a brief discussion of phenol extractions). Phenol is normally removed from the DNA preparations by precipitating the DNA from solution and then resuspending it in buffer. The absorbance spectrum for a dilute solution of the phenol mixture (0.1%) is shown in the graph. Explain the significance of this spectrum to analysts working with nucleic acids and proteins.

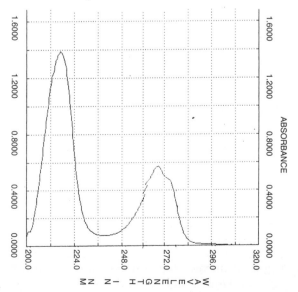

TERMS RELATING TO SPECTROPHOTOMETRY

Absorbance, A (also called "optical density"). A measure of the amount of light absorbed by a sample defined as:

$$A = -\log_{10}t$$

where t = transmittance

Note: Spectrophotometers actually detect and measure transmittance, not absorbance. *Absorbance* is a calculated value.

Absorbance Scale Accuracy. See "Photometric Accuracy."

Absorbance Spectrum. The pattern of absorbance of a particular sample when exposed to light of various wavelengths, typically plotted with wavelength on the X axis versus absorbance on the Y axis.

Absorption. The loss of light of specific wavelengths as the light passes through a material and is converted to heat energy.

Absorptivity. The inherent tendency of a material to absorb light of a certain wavelength. (Note that absorbance is a measured value that depends on the instrument used to measure it, whereas absorptivity is an intrinsic property of a material.) (See also "Beer's Law.")

Absorptivity Constant, a or α. A value that indicates how much light is absorbed by a particular substance at a particular wavelength under specific conditions (such as temperature and solvent). It is sometimes calculated as:

$$\alpha = A/bC$$

where
A = absorbance
b = path length
C = concentration
α has units that vary depending on the units of concentration and path length (See also "Beer's Law.")

Absorptivity Constant, Molar, ε. The absorptivity constant when the concentration of analyte is expressed in units of moles per liter.

Analyte. The particular compound, material, or substance being evaluated or measured in an analytical procedure.

Analytical Wavelength. The wavelength at which absorbance measurements are made in a particular assay.

Atomic Absorption Spectrophotometry. A spectrophotometric technique based on the absorption of radiant energy by atoms; used to measure the concentration of metals.

Background (in spectrophotometry). Light absorbance caused by anything other than the analyte.

Band Pass. See "Spectral Band Width."

Beer's Law. Also **Beer-Lambert** or **Beer-Bouguer Law.** A rule that states that the absorbance of a homogeneous sample is directly proportional to the concentration (C) of the absorbing substance and to the thickness of the sample in the optical path (b).

$$A = \alpha\, bC$$

where
A = the amount of light absorbed by the analyte
C = the concentration of analyte
b = the path length that the light travels through the sample
α = the absorptivity constant

Blank. A reference that contains no analyte but does contain the solvent (for a liquid sample) and any reagents that are intentionally added to the sample. The blank is held in a cuvette that is identical to that used for the sample, or the blank is alternately placed in the same cuvette as the sample. A spectrophotometer compares the interaction of light with the sample and with the blank in order to establish the absorbance due to the analyte.

Note: Air may be used as the blank for certain samples, such as glass filters and other solid materials.

Calibration (for a spectrophotometer) (1) The use of standards with known amounts of analyte to determine the relationship between light absorbance and analyte concentration. (2) Bringing the transmittance and wavelength values of a spectrophotometer into accordance with the values for externally accepted standards.

Calibration Curve or Line. See "Standard Curve."

Cell. See "Cuvette."

Chromatophore, Chromophore. An atom or group of atoms or electrons in a molecule that absorb light.

Colorimeter. An instrument used to measure the interaction of visible light with a sample.

Colorimetry. Technique for measuring color.

Cuvette. A sample "test tube" that is designed to fit a spectrophotometer and is made of an optically defined material that is transparent to light of specified wavelengths.

Dark Current. Signal that arises in the photodetector electronic circuits when no light shines on its surface.

Densitometry. A method used to quantify the amount of material on a solid medium (such as a photographic negative) by measuring its absorbance of light.

Detector (in a spectrophotometer). Device used to measure the amount of light transmitted through a sample.

Deuterium Arc Lamp. Source that produces light in the UV region, from about 185 to 375 *nm*.

Diffraction. Bending of light that occurs when light passes an obstacle or narrow aperture.

Diffraction Grating. Device consisting of a series of evenly spaced grooves on a surface that is used to separate polychromatic light into its component wavelengths.

Dispersing Element (in a spectrophotometer). A part of a monochromator that separates polychromatic light into its component wavelengths.

Dispersion. The separation of light into its component wavelengths.

Double-Beam Spectrophotometer. A type of spectrophotometer in which the sample and blank are placed simultaneously in the instrument so the absorbance of the sample can be continuously and automatically compared to that of the blank.

Electromagnetic Radiation. A form of energy that travels through space at high speeds. Electromagnetic radiation is classified into types based on wavelength, including: gamma rays, X-rays, ultraviolet (UV) light, visible (Vis) light, infrared (IR) light, microwaves, and radio waves.

Electromagnetic Spectrum. The range of all types of electromagnetic radiation from radiation with the longest wavelengths to those with the shortest.

Emission Wavelength. In fluorescence, the wavelength of light emitted by the sample as it fluoresces.

Entrance Slit Width. The size of the slit through which light enters the monochromator after being emitted by the source.

Excitation Wavelength. In fluorescence, the wavelength of light that is absorbed by the molecule of interest and causes the compound to emit fluorescence.

Exit Slit. The slit through which light exits the monochromator.

Exit Slit Width. The size of the slit through which light emerges from the monochromator.

Extinction. Absorbance.

Extinction Coefficient. See "Absorptivity constant."

Filter. Material that blocks the passage of radiant energy in a particular manner with respect to wavelength.

Filter, Neutral. A filter that attenuates radiant energy by the same factor for all wavelengths within a certain spectral region.

Fluorescence. Light emitted by an atom or molecule after it absorbs light with a shorter wavelength.

Fluorescence Spectrometer. An instrument used to analyze a sample based on its fluorescence.

Fluorophore. A molecule that fluoresces when exposed to light.

Frequency (of electromagnetic radiation). The number of waves of electromagnetic radiation that pass a given point per second, expressed in units of Hertz (Hz).

Incident Light. Light that strikes, or shines on, a substance.

Infrared Radiation. The region of the electromagnetic spectrum from 780 to 2500 *nm*.

Interference. Any substance in a sample that leads to an incorrect result in an analysis.

Kinetic Spectrophotometric Assay. An assay that measures the changes over time in concentration of reactants or products in a chemical reaction.

Light. Electromagnetic radiation with wavelengths from 2500 to 180 *nm*; includes the infrared, visible, and ultraviolet regions of the electromagnetic spectrum.

Light Scattering. An interaction of light with small particles in which the light is bent away from its initial path.

Light Source (in spectrophotometry). The bulb or lamp that emits light that shines on the sample.

Limit of Detection. Lowest amount of analyte that may be detected above the "noise" of an instrument.

Linear. Two variables that are related in such a way that they form a straight line when graphed.

Linear Dispersion of a Monochromator. A measure of the ability of the monochromator to separate light into its component wavelengths.

Monochromatic. Light that is of one wavelength. In practice, "monochromatic" light is composed of a narrow range of wavelengths.

Monochromator. Device used to separate polychromatic light into its component wavelengths and to select light of a certain wavelength (or narrow range of wavelengths).

Multicomponent Spectrophotometric Analysis. Methods that allow simultaneous quantitation of more than one analyte in a sample.

Natural Band Width. The width of the peak generated for a specific material when its absorbance is plotted versus wavelength; an intrinsic characteristic of the substance. Natural band width is measured as the width of the peak at half its height.

Noise. Electrical spikes generated by the detector and other electrical components, not related to the sample of interest.

Optical Density, OD. See "Absorbance."

Path Length. The distance light passes through the sample; typically measured as the length of the cuvette in centimeters or millimeters.

Photodetector. A device that responds to light by producing an electrical signal that is proportional to the amount of incident light.

Photodiode Array Detector, PDA. A type of detector that can simultaneously determine the absorbance of a sample at a wide range of wavelengths.

Photoemissive Surface. A surface coated with a material that gives off electrons when bombarded by light.

Photometer. An instrument that measures light absorbance, consisting of a light source, a filter to select the wavelength range, a sample holder, a detector, and a readout device.

Photometric Accuracy (also called **absorbance scale accuracy**). The extent to which a measured transmittance (or absorbance) value agrees with the nationally or internationally accepted value.

Photometric Linearity. The ability of a spectrophotometer to give a linear relationship between the intensity of light hitting the detector and the transmittance display of the instrument.

Photomultiplier Tube, PMT. A common type of detector used in spectrophotometers that consists of a series of metal plates coated with a thin layer of a photoemissive material. Light transmitted through the sample "knocks" electrons from the surface, ultimately resulting in an electrical signal.

Photon. Packet of energy of electromagnetic radiation.

Polychromatic Light. A combination of light of many wavelengths.

Protein Assay. A test of the amount or concentration of protein in a sample.

Qualitative Analysis. The analysis of the identity of substance(s) present in a sample.

Quantitative Analysis. The measurement of the concentration or amount of a substance of interest in a sample.

Radiant Energy. Energy of electromagnetic waves.

Reference (in spectrophotometry). See "Blank."

Reflection (of light). An interaction of light with matter in which the light strikes a surface, causing a change in the light's direction.

Resolution (in spectrophotometry). The separation (in nanometers) of two absorbance peaks that can just be distinguished.

Sample Chamber. The location in which the sample is placed in a spectrophotometer.

Scanning (in spectrophotometry). Process of determining the absorbance of a sample at a series of wavelengths.

Scattering (of light). Redirection of light in many directions by small particles.

Signal. The output of a detector due to its response to a sample.

Signal-to-Noise Ratio. The ratio of sample signal to the noise generated by the detector and associated electronics.

Solvent. The liquid used to dissolve the analyte.

Solvent Cutoff. The wavelength at which a particular solvent absorbs a significant amount of light. Solvent absorbance interferes with the analysis of the analyte.

Source. The lamp or bulb used to provide light in a spectrophotometer.

Spectral Band Width. Light emerging from the monochromator and incident on a sample is not truly monochromatic, but is instead a mixture of wavelengths with an upper and lower limit. Spectral band width is a measure of the range of wavelengths emerging from the monochromator when a particular wavelength is selected.

Spectral Slit Width. The physical width of the monochromator exit slit divided by the linear dispersion of the diffraction grating.

Spectrometer. Any instrument used to measure light's interactions with matter.

Spectrophotometer. An instrument that measures the effect of a sample on an incident light beam. The instrument consists of a source of light, entrance slit, monochromator, exit slit, sample holder, detector, and readout device.

Spectroscopy. The study and measurement of interactions of light with matter.

Spectrum. Electromagnetic radiation separated or distinguished according to wavelength.

Standard (in spectrophotometry). A mixture including a known concentration of the analyte of interest dissolved in solvent and used to determine the relationship between absorbance and concentration for that analyte.

Standard Curve (in spectrophotometry). A graph that shows the relationship between absorbance (on the Y axis) and the amount or concentration of a substance (on the X axis).

Stray Light. Radiation that reaches the detector without interacting with the sample.

Tracking Error. A problem that can occur when an absorbance spectrum is scanned too quickly, resulting in absorbance peaks that are slightly shifted from their true locations.

Transmittance, t. The ratio of the amount of light transmitted through the sample to that transmitted through the blank.

Transmitted Light. Light that passes through an object.

Tungsten Filament. A thin metal wire that emits light when heated and provides visible light in visible spectrophotometry.

Turbid Solution. One that contains numerous small particles that both absorb and scatter light.

Ultraviolet Radiation. The region of the electromagnetic spectrum from about 180 to 380 nm.

Visible Radiation. The region of the electromagnetic spectrum from about 380 to 780 nm. Light of different wavelengths in this range is perceived as different colors.

Wavelength, λ. The distance from the crest of one wave to the crest of the next wave.

Wavelength Accuracy. The agreement between the wavelength the operator selects and the actual wavelength that exits the monochromator and shines on the sample.

REFERENCES

General References

There are many helpful general references on spectrophotometry, some of which are found in books on laboratory instrumentation. A few examples include:

Spectrophotometry and Spectrofluorimetry: A Practical Approach, D.A. Harris and C.L. Bashford; eds. IRL Press, Oxford, 1987.

Instrumental and Separation Analysis, C.T. Kenner. Charles E. Merrill, Columbus, Ohio, 1973.

Principles of Chemical Instrumentation, Gary T. Bender. W. B. Saunders Co., Philadelphia, 1987.

Analytical Chemistry for Technicians, John Kenkel. 2nd edition. Lewis Publishers. Chelsea, Michigan, 1994.

"Getting Started With UV/Vis Spectroscopy," Emil W. Ciurczak and Jerome Workman Jr. Published as a supplement to *Spectroscopy Magazine,* 1996.

Instrument Operation

Manufacturers provide detailed references on instrument operation, performance, and performance verification. Consult each company for its most up-to-date literature. NIST also provides useful publications, one of which is referenced in the following list. Examples of publications include:

"Glass Filters as a Standard Reference Material for Spectrophotometry—Selection, Preparation, Certification and Use of SRM 930 and SRM 1930," R, Mavrodineanu, R.W. Burke, J.R. Baldwin, M.V. Smith, J.D. Messman, J.C. Travis, and J.C. Colbert. NIST Special Publication 260-116. U.S. Department of Commerce/Technology Administration. March 1994. (A good detailed source of information for those who are verifying the performance of spectrophotometers.)

"Good Laboratory Practice with UV-Visible Spectroscopy System," Application Note, Anthony Owen. Hewlett Packard, Germany. Publication number 12-5963-5615E, 1995.

"Operational Qualification and Performance Verification of UV-Visible Spectrophotometers," Technical Note, Hewlett Packard, Germany, 1997. (http:www.hp.com/go/chem)

"The Determination of Detection Limits with the DU-6 and DU-7 UV-VIS Spectrophotometers," Joseph L. Bernard. Technical Information. Beckman. T-1547-UV83-6.

"Calibration Science for UV/Visible Spectrometry, Part III in a Series on Quality Issues in Spectrophotometry", John Hammond, Doug Irish, and Steve Hartwell. *Spectroscopy* 13(2): 64–71, 1998. (Good discussion of issues relating to calibration of spectrophotometers.)

Biological Samples

Information on spectrophotometry of DNA, RNA, and proteins is available from manufacturers, articles, and books. A few examples include:

"Use of UV Methods for Measurement of Protein and Nucleic Acid Concentrations," K.L. Manchester. *BioTechniques,* 20(6): 968–970, 1996.

"Validity of Nucleic Acid Purities Monitored by 260 nm/280 nm Absorbance Ratios," J.A. Glasel. *BioTechniques,* 18(1): 62–63, 1995.

"Value of A_{260}/A_{280} Ratios for Measurement of Purity of Nucleic Acids," K.L. Manchester. *BioTechniques,* 19(2): 208–210, 1995.

"Effect of pH and Ionic Strength on the Spectrophotometric Assessment of Nucleic Acid Purity," W.W. Willfinger, K. Mackey, and P. Chomczynski. *BioTechniques,* 22(3): 474–481, 1997.

"Biochemical Applications for UV/Vis Spectroscopy: DNA, Protein and Kinetic Analysis," H. Müller and B. Schweizer. Perkin-Elmer Corp., 1996.

Methods for Protein Analysis, Robert A. Copeland. Chapman and Hall, New York, 1994.

Miscellaneous Topics

"Spectrophotometric Multicomponent Analysis (MCA): An Ideal Tool for the Pharmaceutical and Life Science Laboratory," Dr. Klaus Weber, J.L. Bernard and R.J. Jamutowski, eds. Beckman Applications Data Sheet ADS 7794. 1989. (Good summary of applications of this methodology; also has analysis of math involved in MCA.)

"The Editor's Page: A Literature Search." G.B. Levy. *American Laboratory,* p. 10, October, 1992. (A brief discussion of who discovered "Beer's" Law, for those with historical interests.)

Relevant ASTM Standards

ASTM E-131-68. "Standard Definitions of Terms and Symbols Relating to Molecular Spectroscopy."

ASTM E-169-93. "Standard Practices for General Techniques of Ultraviolet-Visible Quantitative Analysis."

ASTM E-958-93. "Standard Practice for Measuring Practical Spectral Bandwidth of Ultraviolet-Visible Spectrophotometers."

ASTM E-275-93. "Standard Practice for Describing and Measuring Performance of Ultraviolet, Visible, and Near-Infrared Spectrophotometers."

ASTM E-925-83 (Reapproved 1994). "Standard Practice for the Periodic Calibration of Narrow Band-Pass Spectrophotometers." (Includes methods to verify quantitative performance on an ongoing basis for routine work, where absorbance on a sample is compared with a standard, and for comparing one instrument's performance with another.)

ANSWERS TO PROBLEMS: SPECTROPHOTOMETRY, PART B

---

1. **a,b.** A is an absorbance spectrum. X axis, wavelength; Y axis, absorbance. B is a standard curve. X axis, concentration (or amount of analyte); Y axis, absorbance.

   c. An absorbance spectrum is associated with qualitative analysis. A standard curve is associated with quantitative analysis.

2. **a.** quantitative   **b.** qualitative   **c.** quantitative

   d. quantitative   **e.** qualitative   **f.** qualitative

   g. quantitative

3. **a.** Compound A is orange. Compound B is red. Compound C is green.

b. Compound B has the greatest absorptivity constant at 540 *nm*.

c. Compound B will be detectable at 540 *nm* at a lower concentration than the other compounds because it has the largest signal at that wavelength.

4. Graph A: The linear range is 0–250 *mg/mL*.
   Graph B: The linear range is 0–3.5 *mM*.
   Graph C: The linear range is 0–15 *ppm*.

5. **a.**

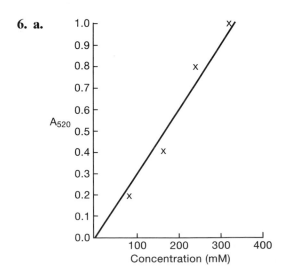

b. The graph is linear from 2 to 20 *mg/L* concentrations.

$$\text{Slope} = m = \frac{Y_2 - Y_1}{X_2 - X_1} = 1.1/10 \; mg/L = \frac{0.11 \; L}{mg}$$

$$A = \frac{0.11 \; L}{mg} C + O$$

c. *Absorptivity constant* $= \alpha = \dfrac{slope}{path \; length}$

$$= \frac{0.11 \; L}{(1 \; cm) \; mg}$$

d. This graph has an upper threshold where the graph plateaus. This is an indication of the limitation of the spectrophotometer at low light transmittances due to stray light.

6. **a.**

b. $\alpha = \dfrac{slope}{pathlength}$

$slope = 0.3/100\ mM = 0.003/mM$

$\alpha = \dfrac{0.003}{(mM)\,(1\ cm)}$

c. $A = \alpha\,b\,C$

$C = \dfrac{A\,(mM)}{0.003}$

	Absorbance	Concentration (in *mM*)
Sample A	0.18	60
Sample B	0.31	103.3
Sample C	0.96	320
Sample D	1.0	333.3

7. a.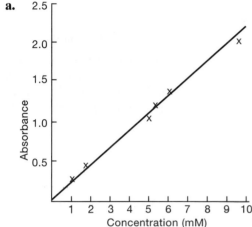

b. $\alpha = \dfrac{slope}{path\ length}$

$slope = \dfrac{(1.0 - 0.5)}{(4.5 - 2.1)\ mM}$

$\alpha = \dfrac{0.21}{mM\,(cm)}$

c. $C = \dfrac{0.30\,(cm)(mM)}{0.21\,(1\ cm)}$

$C = 1.43\ mM$ diluted

$C = 1.43\ mM \times 5$

$= 7.15\ mM$ undiluted

8. a. When the path length decreases the absorbance decreases.

b. 0.8

9. $\dfrac{25\ ppm}{0.87\ AU} = \dfrac{40\ ppm}{?}$ $? = 1.39\ AU$

The absorbance is a linear function of concentration; therefore, this is a proportion problem.

10. $t = 0.680$

$A = -\log\,(t)$

a. $A = -\log\,(0.680) = 0.167$

b. Because there is a linear relationship between absorbance and concentration:

$\dfrac{A}{16\ mg/mL} = \dfrac{0.167}{10\ mg/mL}$

$A = 0.267$

c. $A = -\log\,(t)$

$0.267 = -\log\,(t)$

$t = 0.540$

11. $A = -\log\,(0.75) = 0.12$

a,b. $\dfrac{A}{60\ mg/mL} = \dfrac{0.12}{30\ mg/mL}$

The absorbance of $60\ mg/mL = 0.24$
transmittance $= 0.58$

c,d. $\dfrac{A}{90\ mg/mL} = \dfrac{0.12}{30\ mg/mL}$

Absorbance $= 0.36$ transmittance $= 0.44$

12. $A_{280} = 0.65$ of $1\ mg/mL$

a. $\dfrac{?}{0.15} = \dfrac{1\ mg/mL}{0.65}$

$? = 0.23\ mg/mL$

b. The proteins in the preparation may not have the same aromatic amino acid composition as the BSA. There might be nucleic acids in the partially purified preparation that interfere at $280\ nm$. There are other possible answers as well.

13. $\alpha_{ATP} = 15,400\ L/mole\text{-}cm$ at $260\ nm$

$A = \alpha\,b\,C$

$1.6 = \dfrac{15400\ L\,(1\ cm)\,C}{(mole\text{-}cm)}$

$C = \dfrac{1.6\ mole}{15400\ L} = 1.04 \times 10^{-4}\ mole/L$

14. $\alpha_{NADH} = \dfrac{15000\ L}{mole\text{-}cm}$ at $260\ nm$

$A = \alpha\,b\,C$

$0.98 = \dfrac{15000\ L}{mole\text{-}cm}\,(1.2\ cm)\,C$

$C = \dfrac{0.98\ mole\text{-}cm}{15,000\,(1.2)\ cm\text{-}L}$

$C = 5.44 \times 10^{-5}\ mole/L$

15. Although the highest absorbance peak for NADH is at 260 *nm*, it may be practical to use the second peak at 340 *nm* to determine the concentration because there is no interference from ATP at this wavelength.

16. $A = \alpha\,b\,C$

For lower limit: $0.1 = \dfrac{10{,}000\,L\,(1\,cm)(C)}{mole\,(cm)}$

$C = \dfrac{0.1\,mole}{10{,}000\,L}$

$C = 1 \times 10^{-5}\,mole/L$

For upper limit:

$1.8 = \dfrac{10{,}000\,L\,(1\,cm)\,(C)}{mole-cm}$

$C = \dfrac{1.8\,mole}{10{,}000\,L}$

$C = 1.8 \times 10^{-4}\,mole/L$

The range of concentrations that can be measured for this compound, therefore, is from $1.8 \times 10^{-4}\,mole/L$ to $1 \times 10^{-5}\,mole/L$.

17. **a.** $C = 50\,\mu g/mL\,(1.25)$

   $= 62.5\,\mu g/mL$

   Because the solution was diluted, however, it is necessary to multiply the answer by 10, so $625\,\mu g/mL$ = concentration of DNA in the original sample

   b. $C = (0.63)\,(40\,\mu g/mL)$

   $= 25.2\,\mu g/mL$

   $(25.2)\,\mu g/mL \times 5$

   $= 126\,\mu g/mL$ = concentration of RNA in the original sample

18. **a.** Substituting into equation 1 from Box 2:

   Concentration = $50\,\mu g/mL\,(0.85) = 42.5\,\mu g/mL$

   b. Substituting into equation 3 from Box 2:

   $40\,\mu g/mL\,(0.69) = 27.6\,\mu g/mL$

   $27.6\,\mu g/mL \times 10 = 276\,\mu g/mL$

19. **a.** No. The results of an assay using a standard curve should be accurate because the values determined are compared with those on the standard curve.

   b. If the single standard was properly prepared and the instrument was properly adjusted with a blank, and if the standard and sample are in the linear range of the assay, then the results should be accurate.

c. Yes, the value for the absorptivity constant was presumably based on a correctly calibrated instrument and therefore will not be relevant to this spectrophotometer.

20. This problem may be evident when you examine your standard curve if it causes the results to be nonlinear or low. If you prepare a standard curve alongside your samples, therefore, it may be evident that there is a problem.

21. Assuming you prepare a standard curve each time you do the assay, and assuming the reagent still acts in such a way that color is proportional to the amount of analyte, then the standards should be affected by the reagent in the same way as the samples. The results, therefore, will still be accurate. You may observe, however, that the range for the assay is different than usual. Low concentrations of analyte that are normally in the range of the assay may not cause a detectable color change.

22. Yes. If there are substances in the blood that affect absorbance at the analytical wavelength, then the buffer blank will not compensate and the standard curve will not be correct for the samples.

23. **a.** Concentration of double-stranded

   DNA ≈ $50\,\mu g/mL\,(A_{260})$

   ≈ $50\,\mu g/mL\,(0.497) = 24.9\,\mu g/mL$

   Thus, based on what she weighed, she had a solution of $60\,\mu g/mL$ DNA; based on the absorbance, the solution contained about $25\,\mu g/mL$ DNA. This is not particularly "close," the UV method estimated the concentration of DNA at less than half of what it should have been.

   b. Concentration of protein ≈ $0.276/0.7 = 0.394\,mg/mL$

   Thus, based on weight, she had a $0.50\,mg/mL$ solution of BSA. Based on its absorbance, she had a solution with about $0.40\,mg/mL$ BSA. The UV method was thus gave a closer estimate for the protein concentration than for the DNA concentration.

   c. $A_{260} = 0.176 + 0.497 = 0.673$

   $A_{280} = 0.276 + 0.260 = 0.536$

   The A_{260} value for the mixture was 0.685 and the A_{280} value for the mixture was 0.547.

   The values for the mixture are very close to the summed values for the pure samples. This means that the absorbances at both wavelengths were additive, as predicted.

   d. For pure DNA: $A_{260}/A_{280} = 0.497/0.260 = 1.91$

   For pure BSA: $A_{260}/A_{280} = 0.176/0.276 = 0.638$

   For the mixture: $A_{260}/A_{280} = 0.685/0.547 = 1.25$

   All three ratios are consistent with the predicted ratios.

e. [nucleic acid] $\approx 62.9(0.685) - 36.0\,(0.547)$
$= 23.4\ \mu g/mL$

[protein] $\approx 1.55\,(0.547) - 0.757\,(0.685)$
$= 0.329\ mg/mL$

Thus, based on what she weighed, the DNA should have been 60 $\mu g/mL$ rather than 23.4 $\mu g/mL$. The protein should have been 0.5 mg/mL rather than 0.329 mg/mL.

f. The UV methods did not give results that exactly matched the expected results for the standards.

Many possible explanations of the differences are discussed in the text. Whether or not the estimates provided by the UV methods are "close enough" depends on the situation and the decisions that will be made based on the results. For example, if several DNA purification methods are being compared, then these UV estimates would be adequate to compare the efficacy of the purification methods. If a pharmaceutical company is trying to determine whether a recombinant protein is pure enough for drug use, however, these UV methods are not sufficiently accurate.

Appendix to Chapter 20
Calculating the Line of Best Fit

i. Overview

A **standard curve** *displays the relationship between the concentration (or amount) of an analyte present and an instrument's response to that analyte.* Figure 20.14 shows a standard curve involving a spectrophotometer. Recall that the variable that is controlled by the investigator (in this example the concentration of compound) is termed the *independent variable* and is plotted on the X axis. The variable that varies in response to the independent variable (in this example the absorbance) is the *dependent variable* and is plotted on the Y axis. When a standard curve is prepared, we expect that the response of the instrument will be entirely determined by the concentration of analyte in the standard. If this expectation is met (and if the relationship between concentration and instrument response is linear), then all the points will lie exactly on a line; however, there are sometimes slight errors or inconsistencies when a standard curve is constructed. These small errors cause the points to not lie exactly on a line, as shown in Figure 20.14.

As is shown in Figure 20.14 the points can easily be represented with a line, although the points do not all fall exactly on the line. In a situation like this, where a series of points approximate a line, we can use a ruler to draw "by eye" a line that seems to best describe the points. Drawing a line "by eye," however, is not the most accurate method, nor does it give consistent results, Figure 20.15. It is more accurate to use a statistical method to calculate the equation for the line that best describes the relationship between the two variables. *The statistical method used to determine the equation for a line is called* **the method of least squares**. The least squares method calculates the equation for a line such that the total deviation of the points from the line is minimized. *The equation for the line calculated using the method of least squares is called the* **regression equation**. *The line itself is called the* **best fit line**.

Recall that the equation for a line has the form:

$$Y = \text{slope}(X) + Y \text{ intercept}$$
or
$$Y = mX + a$$

where
m = the slope
a = the Y intercept
X = the independent variable
Y = the dependent variable

The least squares method is used to calculate: (1) the slope and (2) the Y intercept of the line that best fits the points. (The equations to perform these calculations are shown later.)

When the least squares method is applied to the data graphed in Figure 20.14, the slope for the linear regression line is calculated to be 0.029 *mL/mg*. The Y intercept is calculated to be 0.045. Therefore, the equation for the line best fitting the points in Figure 20.14 is:

$$Y = \left(0.029\,\frac{mL}{mg}\right)X + 0.045$$

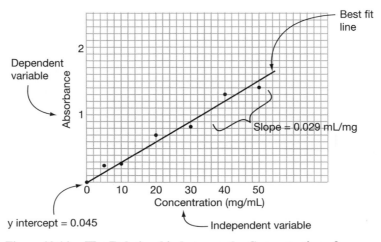

Figure 20.14. The Relationship between the Concentration of Analyte and the Response of an Instrument; a Standard Curve. Due to slight inconsistencies, the points do not all lie exactly on a line.

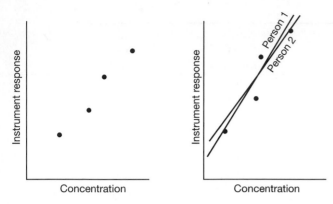

Figure 20.15.   Drawing a Best Fit Line "By Eye." Two people connect the concentration points into a best fit line by eye with slightly different results.

Note that the slope and the Y intercept can have units. (In this example, absorbance has no units; therefore, the Y intercept has no units.)

Most modern spectrophotometers can automatically calculate the equation for the line of best fit for a series of points. Computers and some calculators can also be used to facilitate the determination of the equation for a line of best fit.

ii. CALCULATING THE LINE OF BEST FIT

This section describes the method of least squares that is used to calculate the linear regression equation. It is based on the following assumptions:

- *The values of the independent variable, X, are controlled by the investigator.*

- *The dependent variable, Y, must be approximately normally distributed.* (For example, if the same sample of compound is measured many times with a spectrophotometer, the absorbance values will vary slightly from reading to reading due to random error. These various measurements are expected to be normally distributed.)

- *If there is a relationship between the two variables, it is linear.*

(1) To calculate the slope of the line of best fit, the equation is:

$$m = \frac{\sum XY - \dfrac{\sum X \sum Y}{N}}{\sum X^2 - \dfrac{(\sum X)^2}{N}}$$

(2) To calculate the Y intercept, the equation is:

$$a = \overline{Y} - m\overline{X}$$

where $\overline{X}$ and $\overline{Y}$ are the means of the X and Y values, respectively

$$\overline{X} = \frac{\sum X}{N} \text{ and } \overline{Y} = \frac{\sum Y}{N}$$

N = the sample size

These calculations are summarized in Box 3.

As an example we show how the linear regression equation is calculated for the data graphed in Figure 20.14. Table 20.6 shows the data for this example along with preliminary calculations.

BOX 3   CALCULATION OF THE LINEAR REGRESSION EQUATION

Note: Keep track of the units when calculating and expressing the equation of a line.

1. **Arrange the data into a table of X and Y pairs.**
2. **Calculate the sum of the X values, $\sum X$.**
3. **Calculate $\overline{X}$.**
4. **Calculate $\sum X^2$ by squaring each X value and adding the results together.**
5. **Calculate $(\sum X)^2$ by squaring the sum of the X values.**
6. **Calculate $\sum Y$.**
7. **Calculate $\overline{Y}$.**
8. **Calculate $\sum X \sum Y$.**
9. **Multiply each X by its corresponding Y value and add the products to obtain $\sum XY$.**
10. **Calculate the slope, m, using the preceding equation.**
11. **Calculate the Y intercept using the preceding equation.**
12. **Place the values calculated for the slope and the Y intercept into the equation:**

$$Y = mX + a$$

Table 20.6 Calculating the Line of Best Fit

Concentration of Analyte X (mg/mL)	Absorbance Y (Absorbance has no units)	XY (mg/mL)	X² (mg²/mL²)
0	0	0	0
5	0.25	1.25	25
10	0.26	2.60	100
20	0.70	14.00	400
30	0.82	24.60	900
40	1.30	52.00	1600
50	1.40	70.00	2500

$\sum X = 155\ mg/mL$ $\sum Y = 4.73$ $\sum XY = 164.45\ mg/mL$ $\sum X^2 = 5525\ mg^2/mL^2$

$\bar{x} = 22.14\ mg/mL$ $\overline{Y} = 0.676$ $N = 7$ $\sum X \sum Y = 733.15\ mg/mL$ $(\sum X)^2 = 24025\ mg^2/mL^2$

The values from Table 20.6 can be "plugged into" the equations for the best fit line. The slope of the line is:

$$m = \frac{164.45\ mg/mL - \dfrac{733.15\ mg}{7\ mL}}{5525\ \dfrac{mg^2}{mL^2} - \dfrac{24025\ mg^2}{7\ mL^2}} = 0.0285\ mL/mg$$

The Y intercept is:

$$a = 0.676 - (0.0285\ mL/mg)(22.14\ mg/mL) = 0.045$$

The regression equation for this line, therefore, is:

$$Y = \left(0.029\ \frac{mL}{mg}\right)X + 0.045$$

UNIT VI:

Laboratory Solutions

Chapters in This Unit

- ✦ Chapter 21: Preparation of Laboratory Solutions A: Concentration Expressions and Calculations

- ✦ Chapter 22: Preparation of Laboratory Solutions B: Basic Procedures and Practical Information

- ✦ Chapter 23: Laboratory Solutions to Support the Activity of Biological Macromolecules and Intact Cells

- ✦ Chapter 24: Solutions: Associated Procedures and Information

LABORATORY SOLUTIONS:
UNIT INTRODUCTION

Biological systems evolved in an aqueous environment. The recombinant DNA techniques that allow investigators to manipulate genes are based on enzymatic reactions that occur only in the proper aqueous solutions. Mammalian cells grown in culture require a carefully controlled aqueous medium. The first step in nearly every biological laboratory procedure, whether for research or production purposes, is the preparation of aqueous solutions containing the proper components to support the biological system of interest.

A **solution** *can be defined as a homogeneous mixture in which one or more substances is (are) dissolved in another.* **Solutes** *are the substances that are dissolved in a solution. The substance in which the solutes are dissolved is called the* **solvent**. Although solutes and solvents may be gases, liquids, or solids, in biological applications, the solutes are solids or liquids and the solvent is a liquid, most often water. In this unit, the term **biological solution** *refers broadly to biological materials and the liquid environments in which they are contained in the laboratory.*

This unit discusses a number of considerations relevant to the preparation and use of solutions in biology laboratories. These topics include:

- how the components of solutions are expressed in "recipes" and the calculations associated with solution preparation
- practical considerations in mixing and preparing solutions
- the components of solutions used to support the structure and function of proteins, nucleic acids, and whole cells
- the purification of water for solutions
- how glassware and plasticware are cleaned
- how solutions are sterilized and stored

TERMINOLOGY USED IN THE SOLUTIONS UNIT

Absorption. Process in which one substance is taken up by another, as when a cotton ball absorbs water.

Acid. Any substance that releases hydrogen ions when dissolved in water.

Activated Carbon. Material that adsorbs and removes organic contaminants from water. Derived from wood and other sources that are charred at a high temperature to convert them to carbon. The carbon is "activated" by oxidation from exposure to high temperature steam.

Adsorption. Process in which a material sticks to the surface of a container, membrane, bead, or other solid.

Aliquot. To divide a solution into small, equal individual containers; also refers to the individual containers.

Alkalinity. A measure of the capacity of water to accept H^+ ions; that is, its acid-neutralizing ability. Carbonate, bicarbonate, and hydroxide ions are common contributors to alkalinity.

Alpha Helix. See "Secondary Structure."

Amino Group. Part of the core structure of amino acids, $-NH_2$.

Amount. How much of a component is present (e.g., 10 *g*, 2 cups, and 30 *mL*).

Anion. A negatively charged ion.

Anion Exchange Resin. A solid matrix material that has negatively charged ions available for exchange with negatively charged ions in a solution.

Atomic Weight. See "Gram Atomic Weight."

Autoclave. A laboratory pressure cooker that sterilizes materials with pressurized steam.

β-Pleated Sheet. See "Secondary Structure."

Base. Any substance that releases hydroxide ions when dissolved in water. Also, any substance that can accept hydrogen ions.

Biological Macromolecule. A large and complex molecule that has biological function (e.g., a hemoglobin molecule or a strand of RNA).

Biological Solution. A laboratory solution that supports the structure and/or function of biological molecules, intact cells, or microorganisms in culture.

Borosilicate Glass. Type of glass used to make general purpose laboratory glassware that is strong, resistant to heat and cold shock, and does not contain discernable contamination by heavy metals.

BTV, Bring to Volume. The procedure in which solvent is added to solute (typically in a volumetric flask or graduated cylinder) until the total volume of the solution is exactly the final volume desired.

Buffer (Biological). A solution that maintains a biological system at the proper pH.

Calcium. An element often found in water, typically as dissolved calcium carbonate; causes water hardness.

Carbon Adsorption. A method of removing dissolved organic contaminants from water by using activated carbon.

Carboxyl Group. Part of the core structure of amino acids, $-COOH$.

Cation. A positively charged ion.

Cationic Ion Exchange Resin. A solid matrix material that has positively charged ions available for exchange with positively charged ions in a solution.

Chelator. An agent that binds to and removes metal ions from solution.

Chromatin. The native form of DNA that is composed of DNA bound to several types of proteins.

Clean in Place, CIP. The process of cleaning large, non-movable items (e.g., fermentation vessels) in their place. CIP often involves pumping and circulating cleaning agents through the items.

Coenzyme. A complex organic molecule required by an enzyme for activity.

Cofactor. A chemical substance required by an enzyme for activity. Cofactors include metal ions, such as Fe^{++}, Mg^{++}, and Zn^{++}.

Colony Forming Unit, CFU. A measure of the number of bacteria in a sample. Measuring CFUs requires incubating a sample and counting the resulting colonies; each colony is assumed to have been derived from one bacterium.

Compound. A substance composed of atoms of two or more elements that are bonded together.

Concentration. An amount per volume; a fraction. The numerator is the amount of the solute of interest and the denominator is usually the volume of the entire solution, the solvent and the solute(s) together. For example, "2 g of NaCl/L" means there are 2 g of NaCl dissolved in enough liquid so that the total volume of the solution is 1 L.

Condenser (in distillation). Site where cooling water lowers the temperature of water vapor, causing it to condense back to a liquid form.

Conductivity. A measure of the ability of water to conduct current. Used in water treatment to determine the level of ionized impurities present.

Corex®. A type of glass that is resistant to physical stress and is often used to make centrifuge tubes.

Covalent Bond. Strong molecular bond formed by the sharing of electrons between atoms.

Covalent Peptide Bond. A bond that connects adjacent amino acids together, thus forming chains; these bonds are relatively strong and are not readily dissociated by environmental changes, such as raised temperature.

"dpK$_a$/dt". The change of the pH of a buffer in pH units with change in its temperature in degrees Celsius. The larger this value, the more the buffer will change pH with each degree of temperature change. Negative values for dpK$_a$/dt mean that there is a decrease in pH with an increase in temperature and vice versa.

Dalton, D. A unit of mass nearly equal to the mass of a hydrogen atom, 1.0000 on the atomic mass scale. A kilodalton (kD) is a unit of mass equal to 1000 D.

Deionization. A process in which dissolved ions are removed from a solution by passing the solution through a cartridge containing ion exchange resins. The resins are composed of beads that exchange hydrogen ions for cations in the solution and hydroxyl ions for anions in the solution. The ionic impurities remain associated with the beads, whereas the hydrogen and hydroxyl ions combine to form water.

Denaturation (of DNA). Separation of the two complementary strands of DNA from one another.

Denaturation (of proteins). Unfolding of a protein so that its normal three-dimensional structure is lost but its primary structure remains intact.

Deoxyribonucleic Acids, DNA. Linear polymers of nucleotide subunits; carriers of genetic information in cells.

Depth Filter. A filter made of matted fibers or sand that retains particles through its entire depth by entrapment.

Detergent. Substances that have both a hydrophobic "tail" and a hydrophilic "head." Detergents therefore have two natures. They can be both water-soluble and lipid-soluble.

   Ionic Detergent A detergent whose hydrophilic portion is ionized in solution.

   Nonionic Detergent A detergent whose hydrophilic portion is not ionized in solution.

Disinfection. Destruction of most, but not all, microorganisms by means of heat, chemicals, or ultraviolet light.

Dissolved Inorganics. Water contaminants, not derived from plants, animals, or microorganisms, that usually dissociate in water to form ions.

Dissolved Organics. Water contaminants broadly defined to contain carbon and hydrogen. Organic materials in water may be the result of natural vegetative decay processes or may be human-made substances such as pesticides.

Distillation. A water purification process in which impurities are removed by heating the water until it vaporizes. The water vapor is then cooled to a liquid and collected, leaving nonvolatile impurities behind.

Disulfide Bond. A covalent bond that forms in proteins when sulfurs from two cysteine molecules bind to one another with the loss of two hydrogens.

Electrostatic Interactions. Attractions between positive and negative sites on macromolecules. These weak interactions bring together amino acids and stabilize protein folding.

Endotoxin. A contaminating substance released by the breakdown of bacteria. In water treatment, the term usually refers to pyrogens. See "Pyrogen."

Endotoxin Units/mL. A measure of the concentration of pyrogens in water. See "*Limulus* Amebocyte Lysate Test."

Equivalent Weight (in reference to acids and bases). For an acid, 1 equivalent is equal to the number of grams of that acid that produces 1 mole of H^+ ions. For a base, 1 equivalent is equal to the number of grams of that base that supplies 1 mole of OH^-.

Equivalent Weight (in reference to the concentration of solutes in body fluids). The number of grams of the solute that will produce 1 mole of ionic charge in solution.

Extractables. Contaminants that are leached into water from the materials used to construct filters, storage vessels, tubing, and other items.

Feedwater. The source water that enters a treatment process.

Filtration. A method of removing particles from a liquid. The liquid passes through pores or spaces of a filter leaving particles behind.

Formula Weight, FW. The weight, in grams, of 1 mole of a given compound. The FW is calculated by adding the atomic weights of the atoms that compose the compound.

Fouling. When contaminants, such as packed bacteria, form a crust on a filter, membrane, or other surface, thus blocking further flow.

Good Manufacturing Practices, GMP. Quality standards required of companies producing regulated products, such as pharmaceuticals, and administered and monitored by the Food and Drug Administration.

Gram Atomic Weight. The weight, in grams, of 6.02×10^{23} (Avogadro's number) atoms of a given element. For example, 1 mole of the element carbon weighs $12.0 \, g$.

Growth Media. Solutions that support the survival and growth of cells in culture. These solutions include nutrients, vitamins, salts, and other required substances.

Hardness (in reference to water). An indication of the level of calcium and magnesium salts in water.

Hollow Fiber. An efficient design used for reverse osmosis and ultrafiltration membranes. The membranes are rolled into thin tubes, fibers, that are supported by a thicker porous casing. The fibers are bundled into groups of 1000 or more and placed inside a cartridge. Water is forced to flow through the centers of the tubes. Purified water is collected from the outside of the tubes, whereas contaminants remain in the center.

HPLC, High Performance Liquid Chromatography. A chromatographic method used to separate and detect components of a mixture.

Hybridization. The binding of single-stranded DNA or RNA to strands of complementary DNA or RNA.

Hydrates. Compounds that contain chemically bound water. The weight of the bound water is included in the *FW* of hydrates. For example, calcium chloride can be purchased either as an anhydrous form with no bound water ($CaCl_2$, *FW* = 111.0) or as a dihydrate ($CaCl_2 \cdot 2H_2O$, *FW* = 147.0).

Hydrogen Bond. A type of weak chemical bond formed when a hydrogen atom bonded to an electronegative atom (like F, O, or N) is shared by another electronegative atom.

Hydrophilic. "Liking" water; a substance that readily dissolves in water.

Hydrophobic. "Hating" water; a substance that is relatively insoluble in water, such as an oil.

Hygroscopic Compound. A compound that absorbs moisture from the air.

Inorganic. In water treatment, matter not derived from plants, animals, or microorganisms and that usually dissociates in water to form ions.

In Vitro. "In glass"; refers to processes occurring in a test tube or other laboratory medium.

In Vivo. "In life"; refers to processes occurring in a living organism.

Ion. An atom or group of atoms with a net charge as a result of having lost or gained electrons. See "Anion" and "Cation."

Ion Exchange. The process in which ions in solution are exchanged with similarly charged ions associated with ion exchange resins.

Ionic Strength. A measure of the charges from ions in an aqueous solution.

ISO 9000. Voluntary quality standards established by international consensus that require stringent conformance to a quality plan.

Isotonic Solution. A solution that has the same solute concentration as the inside of a cell.

Leach. A process in which water dissolves substances from materials over which it flows. For example, minerals are leached as water flows over rocks.

***Limulus* Amebocyte Lysate Test, LAL.** Test used to quantify the concentration of pyrogens in a sample that requires an extract of blood from the horseshoe crab, *Limulus polyphemus*. Pyrogens initiate clotting in the presence of this blood extract.

Lipopolysaccharide (in reference to water quality). Molecules containing carbohydrate and lipid groups found in the outer wall of gram-negative bacteria. See "Pyrogen."

Lyophilization. Freeze-drying; a preservation method, commonly used for proteins, that involves rapid freezing and drying under vacuum.

Lysis Buffer. A solution whose primary function is to lyse the cell membrane and/or cell wall.

Megohm-cm. A unit used to measure the resistivity of water. The fewer dissolved ions in water, the higher its resistivity. The theoretical maximum ionic purity for water is 18.3 *megohm-cm* at 25°C.

Melting Temperature (for DNA), T_m. The temperature at which DNA denatures. The bonds between guanines and cytosines are stronger than those between adenines and thymines; therefore, a DNA molecule that is comparatively rich in G–Cs denatures at a higher temperature than one that is rich in A–Ts.

Microfiltration Membrane Filter. Plastic polymeric filter that prevents the passage of microorganisms and particles physically larger than the filter's pore size.

Micromolar, μM. A concentration expression that is 1/1,000,000 M. For example 1 M NaCl is 58.44 g of NaCl in 1 L of solution and 1 μM NaCl is 0.00005844 g of NaCl in 1 L of solution.

Micromole, $\mu mole$. An amount that is 1/1,000,000 of a *mole*. For example, 1 *mole* of NaCl is 58.44 g and 1 $\mu mole$ of NaCl is 0.00005844 g of NaCl.

Millimolar, mM. A concentration expression that is 1/1000 molar. For example 1 M NaCl is 58.44 g of NaCl in 1 L total solution and 1 mM NaCl is 0.05844 g in 1 L total volume.

Millimole, $mmole$. An amount that is 1/1000 of a *mole*. For example, 1 *mole* of NaCl is 58.44 g and 1 $mmole$ of NaCl is 0.05844 g.

Mixed Bed Ion Exchanger. A design in which both cation and anion exchange resins are housed in the same cartridge.

Molality, m. An expression of the concentration of a solute in a solution; the number of molecular weights of solute (in grams) dissolved in 1 kg of solvent. Molality has units of weight in both the numerator and denominator.

Molarity, M. An expression of the concentration of a solute in a solution that is the number of moles of a solute dissolved in 1 L of total solution.

Mole. A mole of any element contains 6.02×10^{23} (Avogadro's number) atoms. Because some atoms are heavier than others, a mole of one element weighs a different amount than a mole of another element. A mole of a compound contains 6.02×10^{23} molecules of that compound.

Molecular Weight, MW (of a compound). See "Formula Weight."

NCCLS, National Committee for Clinical Laboratory Standards. An organization that sets standards for clinical laboratories.

Nominal Pore Size (in filtration). Rated pore size; a large percent (often 99.9%) of particles above the nominal pore size should be retained by the filter.

Normality. An expression of the concentration of a solute in a solution that is the number of equivalent weights of the solute per liter of solution. See "Equivalent Weight."

Nuclease. Any member of a class of enzymes that break the covalent bonds of DNA and RNA.

Nucleotides. Subunits of RNA and DNA that consist of a phosphate group, a five carbon sugar (ribose in RNA or deoxyribose in DNA), and one of four nitrogenous bases (adenine, guanine, cytosine, or uracil in RNA; adenine, guanine, cytosine, or thymine in DNA).

Organic. A broad term that refers to materials containing carbon and hydrogen.

Osmolality. The number of osmoles of solute per kilogram of water.

Osmolarity. The number of osmoles of solute per liter of solution.

Osmole. An osmole of any substance is equal to 1 g molecular weight of that substance divided by the number of particles formed when the substance dissolves. For example, KCl ionizes when dissolved to form one K^+ and one Cl^- ion; therefore, 1 *osmole* of potassium chloride is equal to its gram molecular weight, 74.6 g, divided by 2 = 37.3 g. For a substance that does not ionize when dissolved, 1 *mole* = 1 *osmole*.

Osmosis. The diffusion of a solvent through a semipermeable membrane from a less concentrated solution to a more concentrated solution.

Osmotic Equilibrium. A condition where the rate of water flow into and out of a cell is equal.

Osmotic Support. The cells of plants, yeasts, and microbes are sensitive to changes in osmotic pressure when their cell walls are removed. Materials such as sucrose, sorbitol, mannitol, and glucose may be added to solutions containing these cells to protect the cell's integrity when the cell wall is removed.

Oxidation. The removal of electrons from a substance.

Ozone Sterilization. A method of killing bacteria by exposing them to ozone.

Particulate. A general term used to refer to a solid particle large enough to be removed by filtration.

Parts per Billion, *ppb*. The number of parts of solute per billion parts of solution.

Parts per Million, *ppm*. An expression of the concentration of a solute in a solution that is the number of parts of solute per 1 million parts of total solution. Any units may be used but must be the same for the solute and total solution.

Percent Solution. An expression of the concentration of a solute in a solution where the numerator is the amount of solute (in grams or milliliters) and the denominator is 100 units (usually milliliters) of total solution. There are three types of percent expressions which vary in their units:

> **Weight per Volume Percent, w/v.** The grams of solute per 100 mL of solution.

> **Volume Percent, v/v.** The milliliters of solute per 100 mL of solution.

> **Weight Percent, w/w.** The grams of solute per 100 g of solution.

Phosphodiester Bond. A covalent bond that connects the nucleotides that compose DNA or RNA. A phosphodiester bond is formed when the 5′ phosphate group of one nucleotide is joined to the 3′-hydroxyl group of the next nucleotide through a phosphate group bridge.

Physiological Saline. 0.9% NaCl.

Physiological Solution. An isotonic solution that is used to support intact cells and microorganisms in the laboratory.

pK$_a$. The pH at which a buffer stabilizes and is least sensitive to additions of acids or bases.

Plasmid. Small circular molecule of DNA, found in bacteria, that exists separate from the bacterial chromosome.

Polished Water. A term used to refer to high-quality water after it has gone through a final phase of treatment that removes "all" impurities.

Potable Water. Water that is suitable for drinking.

Pretreatment (of water). Initial steps in a water treatment process that remove a large portion of contaminants and prolong the life of filters and cartridges used in later, more expensive steps.

Primary Structure (of proteins). The linear sequence of amino acids that comprise a protein and are connected by covalent peptide bonds.

Product Water. Water produced by a purification process.

Protease. An enzyme that breaks peptide bonds, thus degrading proteins.

Protease Inhibitor. An agent that inhibits protease activity.

Proteolysis. The breaking apart of the primary structure of a protein by the action of enzymes (proteases) that digest the peptide bonds connecting the amino acids.

Purified Water, USP. Water that meets USP requirements and is used for pharmaceutical and cosmetic applications. "Purified water" is defined by its chemical specifications and may be produced by a variety of techniques.

Pyrogen. Materials derived from lipopolysaccharides that are released into water from the breakdown of gram-negative bacteria. They trigger a dangerous immune response and induce fever, hence their name pyrogen (heat-producing).

Q.S. Latin abbreviation, "quantum sufficit," "as much as sufficient." See "BTV, Bring to Volume."

Quaternary Structure. A protein complex formed when two or more protein chains associate with one another.

"R Group" (of proteins). The 20 amino acids are distinguished from one another by a side chain, or "R group," which has a particular structure and chemical properties.

Reagent-Grade Water. Water suitable to be a solvent for laboratory reagents or for use in analytical or biological procedures.

Recirculation (of water). A process in which purified water is continuously recirculated and repurified. Recirculation helps prevent microbial growth and leaching of materials into the water from storage containers.

Reducing Agents (in biological solutions). A chemical added to a solution to simulate the reduced intracellular environment and prevent unwanted oxidation reactions.

Regeneration (of ion exchange resins). A process in which the capacity of used ion exchange resins is restored. An acid rinse is used to restore cation resins and a sodium hydroxide rinse is used to restore anion resins.

Renaturation (of DNA). Process in which complementary base pairs reestablish hydrogen bonds, bringing together two complementary strands of DNA.

Resin (ion exchange). Beads of synthetic materials that have an affinity for certain ions.

Resistivity. The resistance of water to the flow of electricity. The fewer ions in solution, the higher the resistivity; therefore, resistivity is used to measure the ionic purity of water.

Restriction Endonucleases. Enzymes that cleave DNA at sites with specific base pair sequences.

Reverse Osmosis. A process in which water is forced through a semi-permeable membrane, leaving behind impurities.

Reverse Osmosis Membranes. Very restrictive filters that prevent the passage of materials as small as dissolved ions.

Ribonucleic Acid, RNA. Linear polymer of nucleotide subunits; functions to carry information about amino acid sequence from the nucleus to the cytoplasm (mRNA), to transfer amino acids to peptide chains (tRNA), and to form part of the structure of ribosomes (rRNA).

Salt. A compound made up of positive and negative ions.

Scale. Deposits of calcium carbonate that, for example, coat the inside surface of boilers or the surfaces of membranes.

Secondary Structure (of proteins). Regularly repeating patterns of twists or kinks of an amino acid chain, held together by hydrogen bonds. Two common types of secondary structure are the α-helix and β-pleated sheet. Regions of proteins without regularly repeating structures are often said to have a "random coil" secondary structure.

Semipermeable Membrane. A membrane that does not have measurable pores, but which allows water and some small molecules to pass through.

Softened Water. Water that has cations replaced with sodium ions.

Solute. A substance (usually a solid or liquid) that is dissolved in a solution.

Solution. A homogeneous mixture in which one or more substances is (are) dissolved in another.

Solvent. A substance in which solutes are dissolved. Although solvents may be a variety of gases, liquids, or solids, in most biological applications, the solvent is water.

Spores. Dormant microorganisms.

Spore Strips. Dried pieces of paper to which large numbers of nonpathogenic spores are adhered; used to test the success of sterilization.

Standard Operating Procedure, SOP. A written procedure that provides a standard method for performing a particular task.

Star Activity. When a restriction enzyme "mistakenly" cleaves DNA at sequences that are not its proper target.

Sterilization. The destruction of virtually all viable organisms.

Stock Solution. A concentrated solution that is diluted before use.

Stringency. The reaction conditions used when single-stranded complementary nucleic acids are allowed to hybridize. At high stringency, binding occurs only between strands with perfect complementarity, whereas at lower stringency, some mismatches of base pairs are tolerated.

Suspended Solids. Undissolved solids that can be removed by filtration.

Tertiary Structure. The three-dimensional globular structure formed by bending and twisting of a protein molecule.

Total Bacteria Count. An estimation of the total number of bacteria in a solution based on an assay that involves incubating a sample and counting the number of colony-forming units (CFU).

Total Organic Carbon, TOC. A measure of the concentration of organic contaminants in water.

Total Solids, TS. In water treatment, a measure of the concentration of both the dissolved (TDS) and suspended solids (TSS) in a solution.

Trace Analysis. Analysis of a material that is present in very low (e.g., parts per million or parts per billion) concentrations.

Traceability (with reference to solutions). The process by which it is ensured that every component of a product can be identified and documented.

Tris [Tris(hydroxymethyl)aminomethane)]. One of the most common buffers in biotechnology laboratories. Tris buffers over the normal biological range (pH 7 to pH 9), is nontoxic to cells, and is relatively inexpensive.

Tris Base. Unconjugated Tris; has a basic pH when dissolved in water.

Tris-HCl. Tris that is conjugated to HCl.

Ultrafiltration Membrane. A filter with pores small enough to remove large molecules. These membranes are rated in terms of their molecular weight cutoff.

Ultrapure Water. Type I or better water.

Ultrasonic Washing. A method of cleaning items by exposing them to high pitch sound waves that effectively penetrate and clean crevices, narrow spaces, and other difficult to reach spaces.

USP, United States Pharmacopoeia. A compendium of methods and requirements for the pharmaceutical industry.

UV Oxidation. A process in which ultraviolet light breaks down organic impurities.

Viable Organism. A living organism that can reproduce.

Water for injection, WFI. Water used to make drugs for injection. The USP specifies the quality of WFI and its method of production. In the United States at this time, WFI must be produced either by distillation or reverse osmosis and must meet stringent requirements for pyrogen levels.

Water Softener. An ion exchange device that exchanges positive ions that make water hard, primarily calcium and magnesium, with sodium ions.

Weak Molecular Interactions. Attractive interactions between atoms that are more easily disrupted than covalent bonds.

"X Solution." A stock solution where X means how many times more concentrated the stock is than normal. A 10 X solution is 10 times more concentrated than the solution is normally prepared.

GENERAL REFERENCES ON LABORATORY SOLUTIONS

For general information about molarity, molality, normality, and percent solutions, consult any basic chemistry textbook. Other useful resources include:

Chemical Technicians' Ready Reference Handbook, Gershon Shugar and Jack T. Ballinger. Fourth Edition. McGraw-Hill Book Co., New York, 1996. (This reference has useful information on solutions.)

Manufacturers' catalogues are valuable sources of information about chemicals. For example, the Sigma Company catalogue contains formula weights, compound names and formulas, and hydrated forms for many chemicals. (Sigma Company, P.O. Box 14508, St. Louis, MO 63178.)

The Practical Handbook of Biochemistry and Molecular Biology, Gerald D. Fasman, editor. CRC Press, Boca Raton, Florida, 1989. (This reference contains a great deal of information about biological materials and has information about buffers for biological systems.)

The Merck Index: An Encyclopedia of Chemicals, Drugs, and Biologicals, Susan Budavari, editor. Twelfth edition. Merck, Whitehouse Station, New Jersey, 1996. (This reference is an encyclopedia of chemicals, drugs, and biologicals. It contains information about thousands of biologically relevant chemicals, including their structures, densities, molecular weights, and solubilities.)

Laboratory Solutions for the Science Classroom, Sharon Grobe Mitchell. Flinn Scientific, Inc., Batavia, Illinois, 1991. (This book contains practical tips about solution making and a number of "recipes" for commonly used reagents.)

GENERAL REFERENCES CONTAINING INFORMATION ABOUT THE COMPONENTS OF SOLUTIONS

Guide to Protein Purification, Murray P. Deutscher. Academic Press, CA, 1990. *Methods in Enzymology*, Vol. 182.

Protein Purification: Principles and Practice, Robert K. Scopes. 3rd edition. Springer-Verlag, NY, 1994.

Current Protocols in Molecular Biology, Fred Ausubel, Roger Brent, Robert E. Kingston, David D. Moore, J.G. Seidman, John A. Smith, Kevin Struhl, eds. Wiley, NY, 1995.

Molecular Cloning: A Laboratory Manual, J. Sambrook, E.F. Fritsch, and T. Maniatis. 2nd edition. Cold Spring Harbor Laboratory Press, Cold Spring Harbor, 1989.

OTHER REFERENCES

Companies that produce water purification systems can often provide excellent literature. Examples include:

"The Water Book," Barnstead/Thermolyne Corp., Dubuque, IA.

"The Water Purification Primer," Millipore Corporation, Bedford, MA.

"Pure Water Handbook," Osmonics, Inc., Minnetonka, MN.

"Stopping Residue Interference on Labware and Equipment," Malcolm McLaughlin. *American Laboratory News Edition*, June, 1992.

"Making Water For Injection with Ceramic Membrane Ultrafiltration," William Reinholtz. *Pharmaceutical Technology*, pp. 84–96. September, 1995.

Principles and Methods of Sterilization in the Health Sciences, John J. Perkins. 2nd ed. Charles C. Thomas, 1983. (An in-depth reference on sterilization methods.)

CHAPTER 21

Preparation of Laboratory Solutions A: Concentration Expressions and Calculations

I. OVERVIEW

Solution preparation involves various basic laboratory procedures, such as weighing compounds and measuring the volumes of liquids. Solution preparation also requires interpreting the "recipe" (procedure) for the solution. The following two chapters discuss the basic procedures and the calculations that are the cornerstones of solution making.

Preparing laboratory solutions can be compared with baking. Suppose you decide to bake a batch of apple crisp. The first step is to find a recipe, Figure 21.1, that states:

1. The *components* of apple crisp
2. *How much* of each component is needed
3. *Advice for preparing* (mixing and baking) the components properly

To make apple crisp successfully you need to understand the terminology in the recipe and to measure, combine, and prepare the ingredients in the proper fashion.

Preparing a laboratory solution has much in common with baking. You need a recipe (procedure) for preparing the solution. You follow the procedure and combine the right *amount(s)* of each *component* (solute) in the right volume of solvent to get the correct *concentration* of each solute at the end. You may also need to adjust the solution to the proper pH, sterilize it, or perform other manipulations. There are differences, however, between baking and preparing laboratory solutions. One important difference is that it is appropriate to creatively modify a recipe when baking, but laboratory procedures should be strictly followed.

The recipe for apple crisp lists the *amount* of each component needed. A procedure to make a laboratory solution may similarly list the amount of each solute to use. For example, Laboratory Recipe I, Figure 21.1, lists the amount of each of the four solutes required: Na_2HPO_4, KH_2PO_4, NaCl, and NH_4Cl. Water is the solvent. The total solution has a volume of 1 L. In overview, the procedure for preparing Laboratory Recipe I is:

1. Weigh out the needed amount of each solute.
2. Dissolve the solutes in less than 1 L of water.
3. Add enough water so that the final volume is 1 L.

Laboratory Recipe I is easy to interpret. Many recipes for laboratory solutions, however, do not show the *amount* of each solute required; rather, they show the *concentration* of each solute needed. **Amount** and **concentration** are not synonyms. **Amount** *refers to how much of a component is present.* For example, 10 *g*, 2 cups, and 30 *mL* are amounts. **Concentration** *is an amount per volume; it is a fraction.* The numerator is the amount of the solute of interest. The denominator is usually the volume of the entire solution, the solvent and the solute(s) together. For example, "2 *g* of NaCl/*L*" means there are 2 *g* of NaCl dissolved in enough liquid so that the total volume of the solution is 1 *L*.

Concentration is always a fraction, although concentrations are sometimes expressed in ways that do not look like fractions. For example, 2% milk is an expression of concentration that does not at first glance appear to be a fraction. But, in fact, the % symbol represents a numerator and denominator separated by a line. "2%" milk has 2 *g* fat/100 *mL* of liquid. ("Percent" comes from Latin and means *per hundred*.) Molarity, molality, and normality, which are covered later in this chapter, are also expressions of concentration where the numerator and denominator are not explicitly shown.

There are various ways to express concentration, each of which is associated with certain mathematical calculations. The next sections discuss these concentration expressions and calculations and gives examples that biologists are likely to encounter.

Kitchen recipe I Apple crisp	Laboratory recipe I	
Preheat oven to 375°F	Na_2HPO_4	6 *g*
12 oz can frozen	KH_2PO_4	3 *g*
lemonade	NaCl	0.5 *g*
3 lbs sliced apples	NH_4Cl	1 *g*
Reconstitute juice with water. Cover apples with juice and refrigerate.	Dissolve in water Bring to a volume of 1 *L*	
Blend together: 3 c rolled oats 1/2 c honey 1 c flour 3/4 c butter 1/2 tsp cinnamon		
Place apples in bottom of 9 x 13 inch baking pan. Crumble oat mixture over apples. Pour juice over apples. Bake 1 hour.		

Figure 21.1. Comparison of a Kitchen Recipe and a Laboratory Recipe.

II. TYPES OF CONCENTRATION EXPRESSIONS AND THE CALCULATIONS ASSOCIATED WITH EACH TYPE

A. Weight per Volume

The simplest way to express the concentration of a solution is as a fraction with the amount of solute, expressed as a weight, in the numerator and the volume in the denominator. This type of concentration expression is often used for small amounts of chemicals and specialized biological reagents. For example, a solution of

2 *mg/mL* proteinase K (an enzyme) contains 2 *mg* of proteinase K for each milliliter of solution.

When concentrations are expressed as fractions, proportions can easily be used to determine how much solute is required to make the solution. For example:

How much proteinase K is needed to make 50 *mL* of proteinase K solution at a concentration of 2 *mg/mL*?

$$\frac{?}{50\ mL} = \frac{2\ mg\ \text{proteinase K}}{1\ mL}$$

? = 100 *mg* = amount proteinase K needed.

EXAMPLE PROBLEM

How many micrograms of DNA are needed to make 100 *μL* of a 100 *μg/mL* solution?

ANSWER

The amount of DNA required is found by using a proportion, but, first, 100 *μL* must be converted to milliliters, or milliliters must be converted to microliters so the units are consistent.

$$100\ \mu L = 0.1\ mL$$

$$\frac{?}{0.1\ mL\ \text{of solution}} = \frac{100\ \mu g\ \text{DNA}}{1\ mL\ \text{solution}}$$

? = 10 *μg* = amount of DNA needed.

Table 21.1 summarizes some practical considerations regarding the preparation of solutions.

Table 21.1   PRACTICAL INFORMATION

1. **SOLVENT.** When the solvent is not stated, assume it is distilled or otherwise purified water. (Water purification is discussed in Chapter 24.)

2. **"BRINGING A SOLUTION TO VOLUME."** In the proteinase K example, the enzyme will take up some room in the solution. If you add the proteinase K to 50 *mL* of water, the volume will be slightly more than 50 *mL*. The most accurate way to prepare the solution, therefore, is to dissolve the proteinase K in *less* than 50 *mL* of water and then bring the solution to a final volume of exactly 50 *mL* using a volumetric flask or graduated cylinder.

   This procedure is called **bringing the solution to the desired final volume (BTV or "bring to volume").** Figure 21.2 illustrates a method of bringing a solution to volume.

   Note that in practice 100 *mg* of proteinase K takes up so little volume that it is sensible simply to add the enzyme to exactly 50 *mL* of water. In general, however, the correct way to make a solution is to "bring it to volume."

PRACTICE PROBLEMS: WEIGHT PER VOLUME

1. How would you prepare 100 *mL* of an aqueous solution of $AgNO_3$ of strength 0.1 *g* $AgNO_3$/*mL*?
2. How many milligrams of NaCl is present in 50 *mL* of a solution that is 2 *mg/mL* NaCl?
3. How would you prepare 5 *mL* of proteinase K at a concentration of 100 *μg/mL*?

B. Molarity

Molarity *is a concentration expression that is equal to the number of moles of a solute that are dissolved per liter of solution.* Molarity is used to express concen-

(a)

(b)

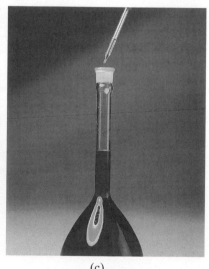
(c)

Figure 21.2.   Bringing a Solution to Volume. a. The desired amount of solute is weighed out on a balance. **b.** The solute is dissolved with mixing in less than the total solution volume. (A beaker is often used when stirring the solution at this step.) **c.** The solution is brought to its final volume by slowly filling the flask to its mark.

tration when the number of molecules in a solution is important. For example, in an enzyme-catalyzed reaction the numbers of molecules of each reactant and the enzyme are important. If there is too little enzyme relative to the number of molecules of reactants present, the reaction may be incomplete, whereas adding too much enzyme is costly and inefficient.

A brief review of terminology: A **mole** *of any element always contains* 6.02×10^{23} *(Avogadro's number) atoms.* Because some atoms are heavier than others, a mole of one element weighs a different amount than a mole of another element. *The weight of a mole of a given element is equal to its atomic weight in grams, or its* **gram atomic weight**.* Consult a periodic table of elements to find the atomic weight of an element. For example, one mole of the element carbon weighs 12.0 *g.*

Compounds *are composed of atoms of two or more elements that are bonded together.* A mole of a compound contains 6.02×10^{23} molecules of that compound. The **gram formula weight (***FW***)** or **gram molecular weight (***MW***)** *of a compound is the weight in grams of 1 mole of the compound,* Figure 21.3. The *FW* is calculated by adding the atomic weights of the atoms that make up the compound. For example, the gram molecular weight of sodium sulfate (Na_2SO_4) is 142.04 *g:*

> 2 sodium atoms $2 \times 22.99\ g =$ 45.98 *g*
> 1 sulfur atom $1 \times 32.06\ g$ $=$ 32.06 *g*
> 4 oxygen atoms $4 \times 16.00\ g =$ <u>64.00 *g*</u>
>   142.04 *g*

A 1 molar solution of a compound contains 1 *mole* **of that compound dissolved in 1** *L* **of total solution.**

For example, a 1 molar solution of sodium sulfate contains 142.04 *g* of sodium sulfate in 1 liter of total solution, Figure 21.4. We also say that a "1 molar" solution has a "molarity of 1." Note that a "mole" is an expression of *amount* and "molarity" and "molar" are words referring to the *concentration* of a solution.

The word *molar* is abbreviated with an upper case *M*. It is also common in biology to speak of "milli-molar" (*mM*) and "micromolar" (*μM*) solutions. A **millimole** *is 1/1000 of a mole;* a **micromole** *is 1/1,000,000 of a mole.* For example:

1 *M* NaCl   = 1 *mole* or 58.44 *g* of NaCl in 1 *L* of solution
1 *mM* NaCl = 1 *mmole* or 0.05844 *g* of NaCl in 1 *L* of solution
1 *μM* NaCl = 1 *μmole* or 0.00005844 *g* of NaCl in 1 *L* of solution

---

EXAMPLE PROBLEM

How much solute is required to prepare 1 L of a 1 M solution of copper sulfate ($CuSO_4$) solution?

ANSWER

Calculate the *FW* or find it on the chemical's bottle. By calculation:

> 1 copper = $1 \times 63.55\ g = 63.55\ g$
> 1 sulfur   = $1 \times 32.06\ g = 32.06\ g$
> 4 oxygens = $4 \times 16.00\ g =$ <u>64.00 g</u>
>
>   *FW* = 159.61 *g*

159.61 *g* of $CuSO_4$ is required to make 1 L of 1 M $CuSO_4$.

58.5 g table salt

18.0 g water

201 g mercury

63.5 g copper

180 g aspirin

254 g iodine

342 g table sugar

Figure 21.3. 1 Mole of Various Substances.

*Weight and mass are not synonyms and it is correct to speak of gram molecular *mass* in the context of molarity. It is common practice, however, to speak of "atomic weight," "molecular weight," and "formula weight," as we do in this book.

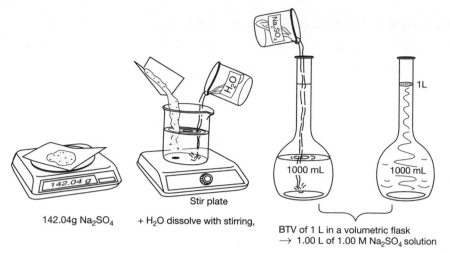

142.04g Na_2SO_4 · · · · + H_2O dissolve with stirring,

Stir plate

1000 mL · · · 1000 mL · · · 1 L

BTV of 1 L in a volumetric flask
→ 1.00 L of 1.00 M Na_2SO_4 solution

Figure 21.4. A 1 *M* Solution of Sodium Sulfate (Na₂SO₄). The *FW* of Na_2SO_4 is 142.04. This amount is dissolved in water so that the total volume is 1.00 *L*.

EXAMPLE PROBLEM

How much solute is required to prepare 1 *L* of a 1 *mM* solution of $CuSO_4$?

ANSWER

The formula weight of copper sulfate is 159.61; therefore,

$$1 \text{ mmole} = \frac{159.61 \text{ } g}{1000} = 0.15961 \text{ } g$$

1 *L* of 1 *mM* $CuSO_4$ requires 0.15961 *g*.

EXAMPLE PROBLEM

How many micromoles are there in 17.4 *mg* of NAD, *FW* 663.4? (NAD, nicotinamide adenine dinucleotide, is an organic compound that is important in metabolism.)

ANSWER

First, convert 17.4 *mg* to grams. 17.4 *mg* = 0.0174 *g*. Using a proportion:

$$\frac{1 \text{ mole}}{663.4 \text{ } g} = \frac{?}{0.0174 \text{ } g}$$

$$? = 2.62 \times 10^{-5} \text{ moles} = 26.2 \text{ micromoles}$$

Molarity is a concentration expression; concentration expressions are fractions. We can determine how to prepare solutions of different molarities or different volumes by using straightforward proportional relationships. For example:

How much solute is required to prepare 1 *L* of 0.25 *M* sodium chloride solution? Using the reasoning of proportions, if a 1 *M* solution of NaCl requires 58.44 *g* of solute, then:

$$\frac{?}{0.25 \text{ } M} = \frac{58.44 \text{ } g}{1 \text{ } M}$$

$$? = 14.61 \text{ } g = \text{amount of solute to make 1 } L \text{ of 0.25 } M$$
NaCl.

EXAMPLE PROBLEM

How much solute is required to prepare a 1 *L* solution of NaCl of molarity 2.5?

ANSWER

If a 1 *M* solution of sodium chloride requires **58.44** *g* of solute, then:

$$\frac{58.44 \text{ } g}{1 \text{ } M} = \frac{?}{2.5 \text{ } M}$$

$$? = 146.1 \text{ } g = \text{amount of solute required}$$
to make 1 *L* of 2.5 *M* NaCl.

It is sometimes necessary to make more or less than 1 L of a given solution. In these cases, proportions can again be used to determine how much solute is required. For example:

How much solute is required to make 500 *mL* of a 0.25 *M* solution of NaCl?

We know that 14.61 g *of solute is required to make 1* L *of 0.25* M *NaCl.*

$$\frac{?}{500 \text{ } mL} = \frac{14.61 \text{ } g}{1000 \text{ } mL}$$

$$? = 7.31 \text{ } g \text{ NaCl} = \text{amount of solute to}$$
make 500 *mL* of 0.25 *M* NaCl.

Thus, it is possible to use proportions to calculate how much solute is needed to make a solution with a molarity other than 1 *M* and a volume other than 1 *L*. In practice, however, most often people use a formula to calculate the amount of solute required to make a solution of a particular volume and a particular molarity. This can be written as shown in Formula 1.

454   UNIT SIX   LABORATORY SOLUTIONS

Formula 1   CALCULATION OF HOW MUCH SOLUTE IS REQUIRED FOR A SOLUTION OF A PARTICULAR MOLARITY AND VOLUME

Solute Required = (grams/1 mole)(Molarity)(Volume)

where:
Solute Required is the amount of solute needed, in grams
Grams per mole is the number of grams in 1 mole of solute
Molarity is the desired molarity of the solution expressed in moles per liter
Volume is the final volume of the solution, in liters

Note: To use this formula, volume must be expressed in units of liters. To convert a volume expressed in milliliters to liters, simply divide the volume in milliliters by 1000. For example:

To convert 100 mL to liters:

Divide by 1000

$$\frac{100}{1000} = 0.1$$

Therefore, 100 mL = 0.1 L

Similarly, 3420 mL = $\frac{3420}{1000}$ = 3.420 L

The following example problem illustrates both the proportion strategy and the use of Formula 1 to calculate how much solute is required to make a molar solution.

EXAMPLE PROBLEM

How much solute is required to prepare 300 mL of a 0.800 M solution of calcium chloride? (*FW* = 111.0)

ANSWER

Strategy 1—Proportions

First, determine how much solute is needed to make 1 L of 0.800 M. We know that 111.0 g is required to make 1 L of a 1 M solution. So:

$$\frac{?}{0.800\ M} = \frac{111.0\ g}{1\ M}$$

$? = 88.8\ g$ = amount of solute to make 1 L of 0.800 M solution.

Second, determine how much solute is needed to make 300 mL.

$$\frac{?}{300\ mL} = \frac{88.8\ g}{1000\ mL}$$

$? = 26.64\ g$ = grams of solute required to make 300 mL of 0.800 M solution.

Strategy 2—Formula

First, convert 300 mL to liters by dividing by 1000:

$$\frac{300}{1000} = 0.300 \quad \text{Therefore, 300 m}L = 0.300\ L$$

Plug into Formula 1:

Solute Required =

$$\frac{(grams/mole)}{\underset{mole}{(111.0\ g)}} \quad \frac{(Molarity)}{\underset{L}{(0.800\ mole)}} \quad \frac{(Volume)}{(0.300\ L)}$$

$= 26.64\ g$ = grams of solute required.

Observe that the units cancel, leaving the answer expressed in grams.

We have now shown how to calculate the amount of solute required to prepare a solution of a given molarity and volume using two different strategies. Either strategy when performed correctly will give the right answer. Box 1 outlines the procedure for making a solution of a particular volume and molarity.

EXAMPLE PROBLEM

How would you prepare 150 mL of a 10 mM solution of $Na_2SO_4 \cdot 10H_2O$ using the procedure outlined in Box 1?

ANSWER

Step 1. Find the *FW* of the solute. The *FW* is 322.04.

Step 2. Determine the molarity required. The molarity required is 10 mM, which is equal to 0.010 M.

Step 3. Determine the volume required. The volume required is 150 mL, which is equal to 0.150 L.

Step 4. Determine how much solute is necessary.

BOX 1   PROCEDURE TO MAKE A SOLUTION OF A PARTICULAR VOLUME AND MOLARITY

1. **Find the _FW_ of the solute.** It is possible to calculate the _FW_ of a compound by adding the atomic weights of its constituents. For compounds with complicated formulas or which come in more than one form, it is better to find the _FW_ on the label on the chemical's container or by looking in the manufacturer's catalogue.
2. **Determine the molarity required.**
3. **Determine the volume required.**
4. **Determine how much solute is necessary either by using proportions or Formula 1.**
5. **Weigh out the amount of solute required as calculated in step 4.**
6. **Dissolve the weighed out compound in less than the desired final volume of solvent.**
7. **Place the solution in a volumetric flask or graduated cylinder. Add solvent until exactly the right volume is reached (BTV).**

Notes

i **Hydrates** _are compounds that contain chemically bound water._ The bound water does not make the compounds liquid; rather, they remain powders or granules. The weight of the bound water is included in the _FW_ of hydrates. For example, calcium chloride can be purchased either as an anhydrous form with no bound water, or as a dihydrate. Anhydrous calcium chloride, $CaCl_2$, has a formula weight of 111.0. The dihydrate form, $CaCl_2 \cdot 2H_2O$, has a formula weight of 147.0 (111.0 plus the weight of two waters, 18.0 each). When hydrated compounds are dissolved, the water is released from the compound and becomes indistinguishable from the water that is added as solvent.

ii Compounds that come as liquids can be weighed out, but it may be easier to measure their volume with a pipette or graduated cylinder. Convert the required weight to a volume based on the density of the compound. (See pp. 125–126 for a discussion of density.)

Strategy 1—Using Proportions to Calculate How Much Solute is Necessary

If it requires 322.04 g to make 1 L of a 1 M solution, then:

$$\frac{322.04\ g}{1.000\ M} = \frac{?}{0.010\ M}$$

$? = 3.2204\ g =$ solute needed to make 1 L of 10 _mM_ (i.e., 0.010 _M_) solution.

To make 150 _mL_:

$$\frac{?}{150\ mL} = \frac{3.2204\ g}{1000\ mL}$$

$? = 0.4831\ g =$ amount of solute required to make 150 _mL_ of 10 _mM_ solution.

Strategy 2—Using Formula 1 to Calculate How Much Solute is Necessary

Convert 150 _mL_ to liters, 150/1000 = 0.150 L.

Substitute values into Formula 1:

Solute Required = (grams/1 mole) (Molarity) (Volume)

$$\frac{322.04\ g}{1\ mole} \times \frac{0.010\ mole}{L} \times 0.150\ L$$

$$= 0.4831\ g$$

Step 5. Weigh out the amount of solute required as calculated in step 4, that is, 0.4831 g of $Na_2SO_4 \cdot 10H_2O$.

Step 6. Dissolve the weighed out compound in less than the desired final volume (150 mL) of solvent.

Step 7. Place the solution in a volumetric flask or graduated cylinder. Add solvent until exactly 150 mL is reached (BTV).

PRACTICE PROBLEMS: MOLARITY

Assume water is the solvent for all questions.

1. If you have 3 _L_ of a solution of potassium chloride at a concentration of 2 _M_, what is the solute? _____ What is the solvent? _____ What is the volume of the solution? _____ Express 2 _M_ as a fraction. _____
2. How much solute is required to prepare 250 _mL_ of a 1 molar solution of KCl?
3. How would you prepare 10 _L_ of 0.3 _M_ KH_2PO_4?
4. How would you prepare 450 _mL_ of a 100 _mM_ solution of K_2HPO_4?
5. How much solute is required to make 600 _mL_ of a 0.4 _M_ solution of Tris buffer, (_FW_ of Tris base = 121.10)
6. Suppose you are preparing a solution that calls for 25 _g_ of $FeCl_3 \cdot 6H_2O$. You look on the shelf in your laboratory and find anhydrous ferric chloride, $FeCl_3$. What should you do?
7. How many micromoles are there in 150 _mg_ of D-ribose 5-phosphate, _FW_ = 230.11?

C. Percents

When concentration is expressed in terms of percent, the numerator is the amount of solute and the denominator is 100 units of total solution. There are three types of percent expressions which vary in their units.

TYPE I: *WEIGHT PER VOLUME PERCENT.*
GRAMS OF SOLUTE PER 100 mL OF SOLUTION.

A **weight per volume** *expression is the weight of the solute (in grams) per 100* mL *of total solution*, Figure 21.5. This is the most common way to express a percent concentration in biology manuals. If a recipe uses the term % and does not specify type, assume it is a weight per volume percent. This type of expression is abbreviated as **w/v**. For example:

20 g of NaCl in 100 *mL* of total solution
is a 20%, *w/v*, solution.

Box 2 shows the procedure for preparing a *w/v* solution.

EXAMPLE PROBLEM

How would you prepare 500 *mL* of a 5% (*w/v*) solution of NaCl?

ANSWER

1. Percent strength is 5% (*w/v*). Total volume required is 500 *mL*.

2. Expressed as a fraction, $5\% = \dfrac{5\ g}{100\ mL}$

3. $\dfrac{(5\ g)}{100\ mL}\,(500\ mL) = 25\ g$

   $= $ amount of NaCl needed

4. Weigh out 25 g of NaCl. Dissolve it in less than 500 *mL* of water.

5. In a graduated cylinder or volumetric flask, bring the solution to 500 *mL*.

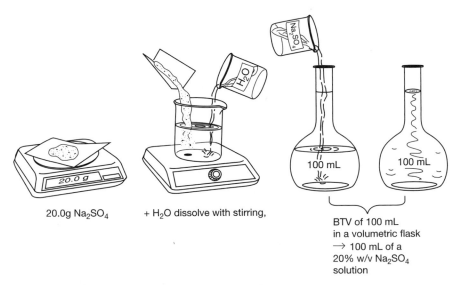

20.0g Na_2SO_4 + H_2O dissolve with stirring,

BTV of 100 mL
in a volumetric flask
$\rightarrow$ 100 mL of a
20% w/v Na_2SO_4
solution

Figure 21.5. A 20% Weight per Volume Percent Solution of Sodium Sulfate (Na_2SO_4).
Twenty grams of Na_2SO_4 is dissolved in water and brought to a volume of 100 *mL*.

Box 2 *PROCEDURE FOR PREPARING A WEIGHT PER VOLUME PERCENT SOLUTION*

1. **Determine the percent strength and volume of solution required.**

2. **Express the percent strength desired as a fraction (g/100 *mL*).**

3. **Multiply the total volume desired (Step 1) by the fraction in Step 2.**

4. **Dissolve the amount of material needed as calculated in Step 3.** (Assume water is the solvent if solvent is not specified.)

5. **Bring the solution to the desired final volume.**

TYPE II: *VOLUME PERCENT.*
MILLILITERS OF SOLUTE PER 100 mL OF SOLUTION.

In a **percent by volume expression**, abbreviated **v/v**, *both the amount of solute and the total solution are expressed in volume units.* This type of percent expression may be used when two compounds that are liquid at room temperature are being combined. For example:

100 *mL* of methanol in 1000 *mL* of total solution is a 10% by volume solution.

Box 3 shows the procedure for making a *v/v* solution.

EXAMPLE PROBLEM

How would you make 100 *mL* of a 10% by volume solution of ethanol in water (*v/v*)?

ANSWER

1. Percent strength is 10 % (*v/v*). Total volume wanted = 100 *mL*.

2. $10\% = \dfrac{10 \; mL}{100 \; mL}$

3. $\dfrac{10 \; mL}{100 \; mL} \times 100 \; mL$
= 10 *mL* of ethanol needed.

4. Place 10 *mL* of ethanol in a 100 *mL* volumetric flask. Add water to 100 *mL*.

Note: you cannot assume that 10 *mL* of ethanol + 90 *mL* of water will give 100 *mL* total volume; their combined volume may be slightly less than 100 *mL*. The most accurate way to prepare this solution, therefore, is to bring it to volume.

TYPE III: *WEIGHT PERCENT.*
GRAMS OF SOLUTE PER 100 GRAMS OF SOLUTION

Weight (mass) percent, **w/w**, *is an expression of concentration in which the weight of solute is in the numerator and the weight of the total solution is in the denominator.* This type of expression is uncommon in biology manuals, but you may encounter it if you work with thick, viscous solvents whose volumes are difficult to measure. For example:

5 *g* of NaCl plus 20 *g* of water is a 20% by mass solution.

Because:

The weight of the NaCl is 5 *g*.

The total weight of the solution = 20 *g* + 5 *g* = 25 *g*

$$\text{Weight percent} = \frac{\text{weight of solute}}{\text{total weight of solution}} \times 100$$

$$\frac{(5 \text{ g NaCl})}{25 \text{ g}} \times 100 = 20\%$$

Box 4 shows the procedure for preparing a *w/w* solution.

Box 3 PROCEDURE FOR PREPARING A PERCENT BY VOLUME SOLUTION

1. **Determine the percent strength and volume required.**

2. **Express the percent desired as a fraction (*mL*/100 *mL*).**

3. **Multiply the fraction from Step 2 by the total volume desired (Step 1) to get the volume of solute needed.**

4. **Place the volume of the material desired in a graduated cylinder or volumetric flask. Add solvent to BTV.**

Box 4 PROCEDURE FOR PREPARING A PERCENT BY WEIGHT SOLUTION

1. **Determine the percent and weight of solution desired.**

2. **Change the percent to a fraction (*g*/100 *g*).**

3. **Multiply the total weight of the solution from Step 1 by the fraction in Step 2 to get the weight of the solute needed to make the solution.**

4. **Subtract the weight of the material obtained in Step 3 from the total weight of the solution (Step 1) to get the weight of the solvent needed to make the desired solution.**

5. **Dissolve the amount of the solute found in Step 3 in the amount of solvent from Step 4.**

EXAMPLE PROBLEM

How would you make 500 g of a 5% NaCl solution by weight (w/w)?

ANSWER

1. Percent strength is 5% (w/w). Total weight of solution desired is 500 g.

2. $5\% = \dfrac{5\ g}{100\ g}$

3. $\dfrac{5\ g}{100\ g} \times 500\ g = 25\ g = $ NaCl needed.

4. 500 g − 25 g = 475 g = the amount of water needed.

5. Dissolve 25 g of NaCl in 475 g of water to get a 5% NaCl solution by weight.

PRACTICE PROBLEMS: PERCENTS

1. How would you prepare 35 mL of a 95% (v/v) solution of ethanol?

2. How would you prepare 200 g of a 75% (w/w) solution of resin in acetone?

3. How would you prepare 600 mL of a 15% (w/v) solution of NaCl?

4. Suppose you have 50 g of solute in 500 mL of solution.
   a. Express this solution as a %.
   b. Is this a w/w, w/v, or v/v solution?

5. Suppose you have 100 mg of solute in 1 L of solution.
   a. Express this as a %.
   b. Is this a w/w, w/v, or v/v solution?

6. What molarity is a 25% w/v solution of NaCl?

D. Parts

i. THE MEANING OF "PARTS"

Parts solutions tell you how many parts of each component to mix together. The parts may have any units but must be the same for all components of the mixture. For example:

A solution that is 3:2:1
ethylene:chloroform:isoamyl alcohol is:
3 parts ethylene, 2 parts chloroform,
1 part isoamyl alcohol.

There are many ways to prepare this solution. Two examples are:

Combine:		Combine:
3 L ethylene	or	3 mL ethylene
2 L chloroform		2 mL chloroform
1 L isoamyl alcohol		1 mL isoamyl alcohol

EXAMPLE PROBLEM

How could you prepare 50 mL of a solution that is 3:2:1 ethylene:chloroform:isoamyl alcohol? (Clue: The amount required of each component is calculated using a proportion equation.)

ANSWER

1. Add up all the parts required. In this case, there are 3 + 2 + 1 parts = 6 parts.

2. Use a proportion to figure out each component. If you need 3 parts of ethylene out of 6 parts total, then:

Ethylene: $\dfrac{3}{6} = \dfrac{?}{50\ mL}$? = 25 mL

You need 25 mL of ethylene.

If you need 2 parts of chloroform out of 6 parts total, then:

Chloroform: $\dfrac{2}{6} = \dfrac{?}{50\ mL}$? = 16.7 mL

You need 16.7 mL of chloroform.

If you need 1 part of isoamyl alcohol out of 6 parts total, then:

Isoamyl alcohol: $\dfrac{1}{6} = \dfrac{?}{50\ mL}$? = 8.3 mL

You need 8.3 mL of isoamyl alcohol.

Thus, this solution requires   25   mL of ethylene
                          16.7 mL of chloroform
                           8.3 mL of isoamyl alcohol
For a total of 50.0 mL

ii. PARTS PER MILLION AND PARTS PER BILLION

A variation on the theme of parts is the concentration expression "parts per million (*ppm*)." **Parts per million** *is the number of parts of solute per 1 million parts of total solution.* Any units may be used but must be the same for the solute and total solution. **Parts per billion (*ppb*)** *is the number of parts of solute per billion parts of solution.* (Percents are the same class of expression as *ppm* and *ppb*. Percents are "parts per hundred.") For example:

5 *ppm* chlorine might be:
5 g of chlorine in 1 million g of solution
5 mg chlorine in 1 million mg of solution
5 lbs of chlorine in 1 million lbs of solution
and so on

iii. CONVERSIONS

Concentration is most often expressed in terms of *ppm* (or *ppb*) in environmental applications. This expression of concentration is useful when a very small amount of something (such as a pollutant) is dissolved in a large volume of solvent. For example, an environmental scientist might speak of a pollutant in a lake as being present at a concentration of 5 *ppm*.

To prepare a 5 *ppm* solution in the laboratory, you must convert the term 5 *ppm* to a simple fraction expression such as grams per liter or milligrams per milliliter to determine how much of the solute to weigh out. Grams per liter, however, has units of weight in the numerator and volume in the denominator, but *ppm* and *ppb* expressions have the same units in the numerator and denominator.* To get around this problem, convert the weight of the water into milliliters based on the conversion factor that 1 *mL* of pure water at 20°C weighs 1 g. For example:

$$5 \text{ ppm chlorine} = \frac{5 \text{ g chlorine}}{1 \text{ million g water}}$$

$$= \frac{5 \text{ g chlorine}}{1 \text{ million } mL \text{ water}} = \frac{5 \text{ g}}{1000 \text{ L}}$$

To make the expression simpler, it is possible to divide the numerator and denominator both by 1 million:

$$5 \text{ ppm chlorine} = \frac{5 \times 10^{-6} \text{ g chlorine}}{1 \text{ mL water}}$$

5×10^{-6} g is the same thing as 5 μg.

So, 5 *ppm* chlorine in water is the same as $\frac{5 \mu g}{mL}$ of chlorine in water.

For any solute:

$$1 \text{ ppm in water} = \frac{1 \mu g}{mL}$$

Also, $1 \text{ ppb in water} = \frac{1 \text{ ng}}{mL}$

EXAMPLE

How would you prepare a 500 *ppm* solution of compound A?

Recall that:

$$1 \text{ ppm in water} = \frac{1 \mu g}{mL}$$

So 500 *ppm* must be equal to $\frac{500 \mu g}{mL}$

500 μg equals 0.500 *mg*. Therefore:

$$500 \text{ ppm} = \frac{500 \mu g}{mL} = \frac{0.500 \text{ mg}}{mL}$$

This solution might be prepared by weighing out 0.500 *mg* of compound A, dissolving it in water, and bringing to a volume of 1 *mL*; however, bringing a solution to a volume of only 1 *mL* is difficult. It is easier to prepare a solution of, say, 1 L. How much compound A is required to make 1 L of solution? If 1 *mL* of solution requires 0.500 *mg*, then 1 L of solution requires 0.500 g. Thus:

$$500 \text{ ppm} = \frac{500 \mu g}{mL} = \frac{0.500 \text{ mg}}{mL} = \frac{0.500 \text{ g}}{L}$$

Thus, an acceptable method to prepare 500 *ppm* of compound A is to weigh out 0.500 g, dissolve it in water, and BTV 1 L.

PRACTICE PROBLEMS: PARTS

1. Suppose you have a recipe that calls for 1 part salt solution to 3 parts water. How would you mix 10 *mL* of this solution?
2. Suppose you have a recipe that calls for 1 : 3.5 : 0.6 chloroform:phenol:isoamyl alcohol. How would you prepare 200 *mL* of this solution?
3. How would you mix 45 μL of a solution which is 5 : 3 : 0.1 Solution A:Solution B:Solution C?

   Assume water is the solvent in problems 4–6.
4. Convert 3 *ppm* to:
   a. milligrams per milliliter
   b. milligrams per liter
5. Convert 10 *ppb* to milligrams per liter.
6. How would you prepare a solution that has 100 *ppm* cadmium?

E. Molality and Normality

i. MOLALITY

Molality and **Normality** are two ways to express concentration that are used much more commonly in chemistry than biology manuals. We will define them briefly here, but refer to a chemistry text for more information.

Molality, abbreviated with a lower case *m*, *means the number of molecular weights of solute (in grams) per kilogram of solvent*, Figure 21.6. Molality has units of weight in both the numerator and denominator.

> **A 1 molal solution of a compound contains 1 mole of the compound dissolved in 1 *kg* of water.**

For example:

58.44 g of sodium chloride (*FW* = 58.44)
in 1 *kg* of water, is a 1 *m* solution.

In practice, the difference between a 1 molar solution and a 1 molal solution is:

- When preparing a *1 molal* (1 *m*) solution one adds the solute(s) to 1 *kg* of water.
- When preparing a *1 molar* (1 *M*) solution, one adds water to the solute(s) to bring to the final volume of 1 liter.

---

*Percent expressions are actually "parts per hundred." Thus, technically the units should be the same in the numerator and denominator. When we use % expressions with units of weight in the numerator and volume in the denominator, we are using the approximation that 1 *mL* of solvent weighs 1 g, which is reasonable if the solvent is water.

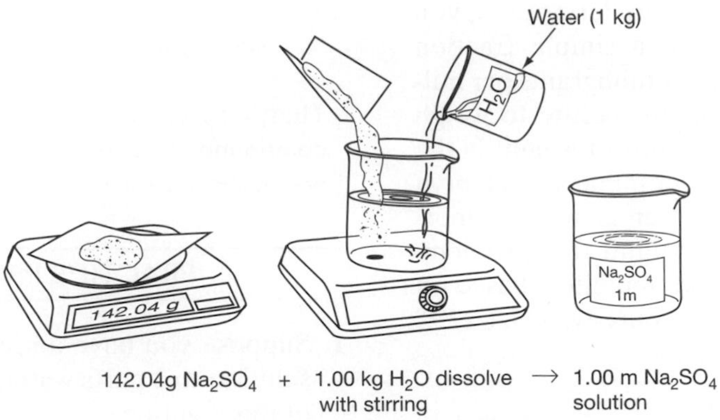

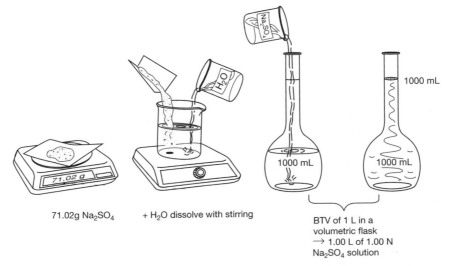

Figure 21.6.   A 1 _m_ (1 molal) Solution of Sodium Sulfate (Na_2SO_4).
142.04 _g_ of Na_2SO_4 is dissolved in 1 _kg_ of water.

ii. NORMALITY

Normality, abbreviated _N_, is another way to express concentration. **Normality** _is the number of "equivalent weights" of solute per liter of solution_, Figure 21.7. Normality is used when it is important to know how many "reactive groups" are present in a solution rather than how many molecules.

The key to understanding normality expressions is knowing the terms _reactive groups_ and _equivalent weight_. Because biologists are most likely to encounter normality in reference to acids and bases, consider these terms as they relate to acids and bases. Recall that acids dissociate in solution to release H^+ ions, and bases, like NaOH, dissociate to release OH^- ions. The H^+ and OH^- ions can participate in various reactions; hence, they are "reactive groups." _For an acid_, **1 equivalent weight** _is equal to the number of grams of that acid that reacts to yield 1 mole of H^+ ions, that is, 1 mole of reactive groups. For a base_, **1 equivalent weight** _is equal to the number of grams of that base that supplies 1 mole of OH^-, that is, 1 mole of reactive groups_. For example:

$$NaOH \rightarrow Na^+ + OH^-$$

$$H_2SO_4 \rightarrow SO_4^{-2} + 2H^+$$

In the first reaction, 1 mole of NaOH dissociates to produce 1 mole of Na^+ and 1 mole of OH^-. Because 1 equivalent weight of NaOH is the number of grams that will produce 1 mole of OH^- ions, the equivalent weight of NaOH is the same as its molecular weight. The number of grams of sodium hydroxide that will produce 1 mole of OH^- is 40.0 _g_. In contrast, 1 mole of sulfuric acid dissociates to form 2 moles of H^+ ions. It only requires 0.5 mole of H_2SO_4 to produce 1 mole of H^+ ions. Therefore, the _FW_ of H_2SO_4 is 98.1, but its equivalent weight is $98.1 \times 0.5 = 49.1$.

> A I normal solution of a compound contains
> I equivalent weight of compound dissolved in
> enough water to get I L.

Figure 21.7.   A 1 _N_ Solution of Sodium Sulfate (Na_2SO_4). When 1 mole of Na_2SO_4 dissolves, 2 moles of Na^+ ions are formed. The _FW_ of Na_2SO_4 is 142.04; the equivalent weight is 71.02. (In this case, an equivalent is the number of grams of solute that will produce 1 mole of ionic charge in solution.)

Thus:

To make a 1 N solution of NaOH dissolve 40.0 g of solute in enough water to make 1 L— the same as a 1 M solution.

To make a 1 N solution of H_2SO_4, however, dissolve 49.1 g of solute in water to make 1 L of solution—*not* the same as a 1 M solution.

Biology laboratory manuals seldom express concentration in terms of normality with the exceptions of HCl and NaOH. Because HCl dissociates to produce one H^+ ion and NaOH dissociates to produce one OH^- ion, their equivalent weight and FW are the same. For NaOH and HCl, 1 M = 1 N.

III. SUMMARY

Recipes in biology manuals sometimes list the *amounts* of each component required, other times, the *concentration* of each component required. There are various methods of expressing concentration. These methods are summarized in Tables 21.2 and 21.3.

A couple of helpful rules are:

1. ***When do you need to know*** the FW *of a compound?*

   When you have a recipe expressed in molarity, molality, or normality; or when you are converting to or from molarity, molality, or normality. With % solutions you do not need to know the FW of the compound (unless you are converting a % to molarity).

2. ***How do you convert molarity expressions to percents (w/v), and vice versa?***

   There are several strategies to do these conversions. Some people find the following equation useful:

   $$\text{Molarity} = \frac{w/v \% \times 10}{FW}$$

   (This equation works because % is $g/100 \ mL$. Multiplying the percentage by 10, therefore, gives $g/1000 \ mL$ or grams per liter.)

   For example,

   Convert 20% (*w/v*) NaCl to molarity

In this case, *w/v* % is 20

The FW is 58.44

Substituting into the equation gives:

$$\text{Molarity} = \frac{20 \times 10}{58.44} = 3.42 \text{ M}$$

Another example:

Convert 3 M NaCl to a percent

The unknown, ?, is the percent (*w/v*)

$$3 \ M = \frac{? \times 10}{58.44}$$

? = 17.53%

Table 21.2 SUMMARY OF METHODS USED TO EXPRESS CONCENTRATION

A. Weight per Volume

A fraction with the weight of the solute in the numerator and the total volume of the solution in the denominator.

B. Molarity

The number of moles of solute per liter of solution.

C. Percents

i. MASS PER VOLUME PERCENT (*w/v* %)
   Grams of solute per 100 mL of solution.

ii. VOLUME PERCENT (*v/v* %)
   Milliliters of solute per 100 mL of solution.

iii. MASS PERCENT (*w/w* %)
   Grams of solute per 100 g of solution.

D. Parts

i. Amounts of solutes are listed as "parts." The parts may have any units, but the units must be the same for all components of the mixture.

ii. PARTS PER MILLION AND PARTS PER BILLION Parts per million: the number of parts of solute per 1 million parts of total solution. Parts per billion: the number of parts of solute per 1 billion parts of solution.

E. Expressions of Concentration Not Common in Biology Manuals

i. MOLALITY Moles of solute per kilogram of solvent.

ii. NORMALITY Number of gram-equivalents of solute per liter of solution. (Normality is reaction-dependent.)

Table 21.3 A COMPARISON OF METHODS OF EXPRESSING THE CONCENTRATION OF A SOLUTE

Concentration of Solute (Na_2SO_4)	Amount of Solute	Amount of Water
1 M	142.04 g Na_2SO_4	*BTV* 1 L with water
1 m	142.04 g Na_2SO_4	Add 1.00 kg of water
1 N	71.02 g Na_2SO_4	*BTV* 1 L with water
1%	1 g Na_2SO_4	*BTV* 100 mL with water

ANSWERS TO PRACTICE PROBLEMS

Practice Problems: Weight per Volume (from p. 451)

1. $\dfrac{0.1\ g}{1\ mL} = \dfrac{?}{100\ mL}$ $? = 10\ g$

Dissolve 10 g of $AgNO_3$ in water (distilled or purified) and bring to 100 mL total solution in a graduated cylinder or volumetric flask.

2. $\dfrac{2\ mg}{1\ mL} = \dfrac{?}{50\ mL}$ $? = 100\ mg$

3. $\dfrac{100\ \mu g}{1\ mL} = \dfrac{?}{5\ mL}$ $? = 500\ \mu g = 0.0005\ g$

Dissolve 0.0005 g of proteinase K in purified water, bring to a volume of 5 mL.

Practice Problems: Molarity (from p. 455)

1. If you have a 3 L of a solution of potassium chloride at a concentration of 2 M, what is the solute? potassium chloride What is the solvent? water (purified) What is the volume of the solution? 3 L Express 2 M as a fraction.

$$\dfrac{2\ \text{moles of potassium chloride}}{\text{liter of solution}}$$

2. Atomic weight of K = 39.10, Atomic weight of Cl = 35.45. The *FW* of KCl, therefore, is: 39.10 + 35.45 = 74.55

74.55 g/mole $\times$ 1 mole/L $\times$ 0.250 L = 18.64 g of solute is required.

3. *FW* of KH_2PO_4 = 136.09.

136.09 g/mole $\times$ 0.3 mole/L $\times$ 10 L = 408.27 g = solute required.
Dissolve the solute in less than 10 L of purified water and BTV.

4. *FW* of K_2HPO_4 = 174.18.

174.18 g/mole $\times$ 0.1 M $\times$ 0.450 L = 7.84 g = solute required.
Dissolve the solute in less than 450 mL and BTV.

5. If 1 L of 1 M Tris buffer requires 121.10 g of solute, then 600 mL of 1 M solute requires:

$\dfrac{?}{600\ mL} = \dfrac{121.10\ g}{1000\ mL}$

$? = 72.66\ g$

If 600 mL of 1 M Tris buffer requires 72.66 g of solute, then 0.4 M requires:

$\dfrac{?}{0.4\ M} = \dfrac{72.66\ g}{1\ M}$

$? = 29.06\ g$ = solute required.

6. You can use the anhydrous form, but you will need to add less than 25 g. The molecular weight of the hydrated form is 270.3 and the molecular weight of the anhydrous form is 162.2. To calculate how much of the anhydrous form to use, set up a proportion:

$\dfrac{25\ g}{270.3\ g} = \dfrac{?}{162.2\ g}$

$? = 15.0\ g$

Use 15.0 g of the anhydrous form instead of 25 g of the hydrated form.

7. 150 mg = 0.150 g. Using a proportion:

$\dfrac{1\ \text{mole}}{230.11\ g} = \dfrac{?}{0.150\ g}$

$? = 6.52 \times 10^{-4}$ moles = 652 μmoles

Practice Problems: Percents (from p. 458)

1. 35 mL $\times$ 0.95 = 33.25 mL

Place 33.25 mL of 100% ethanol in a graduated cylinder and BTV 35 mL.

2. 200 g $\times$ 0.75 = 150 g
200 g − 150 g = 50 g

Combine 150 g of resin and 50 g of acetone.

3. 600 mL $\times \dfrac{15\ g}{100\ mL} = 90\ g$

Dissolve 90 g of NaCl in water and BTV of 600 mL.

4. $\dfrac{50\ g}{500\ mL} = \dfrac{?}{100\ mL}$

$? = 10\ g$, **a.** so this is a 10% solution **b.** w/v.

5. 100 mg = 0.1 g and 1 L = 1000 mL

$\dfrac{0.1\ g}{1000\ mL} = \dfrac{?}{100\ mL}$

$? = 0.01\ g$, **a.** so this is a 0.01% solution **b.** w/v.

6. 25% = 25 g/100 mL of NaCl = 250 g/1000 mL

1 M NaCl = 58.44 g/1000 mL

Using a proportion, if 58.44 μg/L is 1 M, then 250 g/L is 4.28 M.

$\dfrac{58.44\ g/L}{1\ M} = \dfrac{250\ g/L}{?}$ $? = 4.28\ M$

A second strategy:

Using the equation in the Chapter Summary:
$M = w/v\ \% \times 10/FW$, you get,

$M = 25 \times 10/58.44 = 4.28\ M$

so, 25% NaCl is the same as a 4.28 M solution.

Practice Problems: Parts (from p. 459)

1. 1 part + 3 parts = 4 parts total

$$\frac{1}{4} = \frac{?}{10 \, mL} \qquad ? = 2.5 \, mL$$

$$\frac{3}{4} = \frac{?}{10 \, mL} \qquad ? = 7.5 \, mL$$

Combine 2.5 mL of salt solution with 7.5 mL of water.

2. 1 part + 3.5 parts + 0.6 parts = 5.1 parts total

$$\frac{1}{5.1} = \frac{?}{200 \, mL} \qquad ? = 39.2 \, mL \text{ chloroform}$$

$$\frac{3.5}{5.1} = \frac{?}{200 \, mL} \qquad ? = 137.3 \, mL \text{ phenol}$$

$$\frac{0.6}{5.1} = \frac{?}{200 \, mL} \qquad ? = 23.5 \, mL \text{ isoamyl alcohol}$$

Combine the preceding volumes of each component.

3. 5 parts + 3 parts + 0.1 parts = 8.1 parts total

$$\frac{5}{8.1} = \frac{?}{45 \, \mu L} \qquad ? = 27.8 \, \mu L \text{ Solution A}$$

$$\frac{3}{8.1} = \frac{?}{45 \, \mu L} \qquad ? = 16.7 \, \mu L \text{ Solution B}$$

$$\frac{0.1}{8.1} = \frac{?}{45 \, \mu L} \qquad ? = 0.56 \, \mu L \text{ Solution C}$$

Note that it would be very difficult to accurately pipette 0.56 μL of a solution. You would probably mix a larger volume and remove 45 μL from it.

4. a. $3 \, ppm = \dfrac{3 \, g}{1 \times 10^6 \, g} = \dfrac{3 \, g}{1 \times 10^6 \, mL}$

$$= \frac{3000 \, mg}{1 \times 10^6 \, mL} = \frac{3 \times 10^{-3} \, mg}{mL}$$

Another way to think about this is to recall that

$$1 \, ppm \text{ in water} = \frac{1 \, \mu g}{mL}$$

so $3 \, ppm = \dfrac{3 \, \mu g}{mL} = \dfrac{3 \times 10^{-3} \, mg}{mL}$

b. $3 \, ppm = 3 \, mg/L$

5. $10 \, ppb = \dfrac{10 \, g}{1 \times 10^9 \, g} = \dfrac{10 \, g}{1 \times 10^9 \, mL} = \dfrac{10 \times 10^3 \, mg}{1 \times 10^9 \, mL}$

$$= \frac{10 \times 10^{-3} \, mg}{1000 \, mL} = \frac{0.01 \, mg}{L}$$

6. $100 \, ppm = \dfrac{100 \, g}{1 \times 10^6 \, mL} = \dfrac{0.1 \, g}{L}$

You could therefore dissolve 0.1 g of cadmium in water and BTV 1 L.

## CHAPTER 22

Preparation of Laboratory Solutions B: Basic Procedures and Practical Information

I. PREPARING DILUTE SOLUTIONS FROM CONCENTRATED SOLUTIONS

A. The $C_1V_1 = C_2V_2$ Equation

In the laboratory we frequently use concentrated solutions that are diluted before use. *The concentrated solutions are sometimes called* **stock solutions**. A familiar example is frozen orange juice that is purchased as a concentrate in a can. The orange juice is prepared by diluting it with a specific amount of water. In the laboratory, a 2 *M* stock solution of Tris buffer might be prepared that will be used at concentrations of 1 *M* of less. The Tris stock would then be diluted to the proper concentration whenever needed.

There is an extremely helpful equation that is used to determine how much stock solution and how much diluent (usually water) to combine to get a final solution of a desired concentration.

The $C_1V_1 = C_2V_2$ Equation:
How to Make a Less Concentrated Solution from a More-Concentrated Solution
Concentration$_{stock}$ × Volume$_{stock}$ = Concentration$_{final}$ × Volume$_{final}$
This equation can be abbreviated: $C_1V_1 = C_2V_2$

Box 1 outlines the procedure to use the $C_1V_1 = C_2V_2$ equation and the example problems illustrate its application.

EXAMPLE PROBLEM

How would you prepare 100 mL of a 1 *M* solution of Tris buffer from a 2 *M* stock of Tris buffer?

ANSWER

Consider first, is this a situation where a less concentrated solution is being made from a more concentrated solution? Yes. It is appropriate, therefore, to apply the $C_1V_1 = C_2V_2$ equation:

1. Concentrated solution = 2 *M* = C_1.

2. The volume of concentrated stock necessary is ? (what you want to calculate) = V_1.

---

3. The concentration you want to prepare is 1 *M* = C_2.

4. The volume you want to prepare is 100 *mL* = V_2.

$$C_1 \ V_1 = C_2 \quad V_2$$
$$2 \ M(?) = 1 \ M(100 \ mL)$$
$$2 \ M(?) = 100 \ M(mL)$$
$$? = \frac{100 \ M(mL)}{2 \ M}$$
$$? = 50 \ mL$$

5. Take 50 *mL* of the concentrated stock solution and BTV 100 *mL*.

A stock solution is sometimes written as an "X" solution where X means "times"; *how many times more concentrated the stock is than normal.* A 10 X solution is 10 times more concentrated than the solution is normally prepared. To work with "X" solutions, use the $C_1V_1 = C_2V_2$ equation.

EXAMPLE PROBLEM

A recipe says to mix:

10 X buffer Q	1 μL
Solution A	2 μL
Water	7 μL

What is the concentration of buffer Q in the final solution?

ANSWER

First consider whether this is a situation where a less concentrated solution is being made from a more concentrated solution. Yes. It is therefore appropriate to apply the $C_1V_1 = C_2V_2$ equation:

C_1 = 10 X $\qquad C_2$ = unknown
V_1 = 1 μL $\qquad V_2$ = 10 μL = total solution volume
$$C_1 \ V_1 = C_2 \ V_2$$
$$10 \ X \ (1 \ \mu L) = ? \ (10 \ \mu L)$$
$$? = 1 \ X$$

This means that 1 μL of a 10 X stock in a final volume of 10 μL will give a final concentration of 1 X. (Similarly, 1 mL of the 10 X buffer in a solution with a final volume of 10 mL would give a final concentration of 1 X.)

Box 1 USING THE $C_1V_1 = C_2V_2$ EQUATION

1. *Determine the initial concentration of stock solution. This is* C_1.

2. *Determine the volume of stock solution required. This is usually the unknown, ?.*

3. *Determine the final concentration required. This is* C_2.

4. *Determine the final volume required. This is* V_2.

5. *Insert the values determined in Steps 1–4 into the formula and solve for the unknown, ?.*

Note: The $C_1V_1 = C_2V_2$ equation is only used when calculating how to prepare a LESS CONCENTRATED SOLUTION FROM A MORE CONCENTRATED SOLUTION. Do not try to use this equation to calculate amounts of solute needed to prepare a solution of a given molarity or percent.

In the following example, Recipe Version A begins with buffer that is 10 times more concentrated than it is usually used. In Recipe Version B, the buffer is 5 X more concentrated than its usual concentration. Either 5 X or 10 X buffer stock can be used, but how much is needed varies. If 10 X buffer is used, then only 1 μL is required; however, if 5 X buffer is used, then 2 μL are required. The volume of the total solution must not vary, so 7 μL of water is used in Recipe Version A. In Recipe Version B, only 6 μL of water is required.

Recipe Version A		Recipe Version B	
10 X buffer Q	1 μL	5 X buffer Q	2 μL
Solution A	2 μL	Solution A	2 μL
Water	7 μL	Water	6 μL
	10 μL		10 μL

B. Dilutions Expressed as Fractions

Dilutions of concentrated stocks are sometimes expressed in terms of <u>fractions</u>. (Refer to Chapter 9 for a detailed discussion of such dilutions.) When a dilution is expressed as a fraction, most people do not use the $C_1V_1 = C_2V_2$ expression.

For example:

A 95% solution is diluted 1/10. What is its final concentration? 95% $\times$ 1/10 = 9.5% so the final concentration is 9.5%.

A 1 M solution is diluted 1/5. What is its final concentration?

1 M $\times$ 1/5 = 0.2 M so the final concentration is 0.2 M.

Suppose you want to make 10 mL of a 1/50 dilution of food coloring from an original bottle of concentrated food coloring. Then,

Set up a proportion:
$$\frac{1}{50} = \frac{?}{10 \; mL}$$

$$? = 0.2 \; mL$$

So, take 0.2 mL of the original stock and dilute to a volume of 10 mL.

II. MAINTAINING A SOLUTION AT THE PROPER pH: BIOLOGICAL BUFFERS

A. Practical Considerations Regarding Buffers for Biological Systems

i. THE USEFUL pH RANGE

Biological systems require a particular pH, and so the interior of cells, blood, and other fluids are buffered *in vivo* (in the living organism). In the laboratory, comparable buffering systems must be artificially reproduced. **Laboratory buffers** *are laboratory solutions that help maintain a biological system at the proper pH.* When a solution is buffered, it resists changes in pH, even if H^+ ions are added to or lost from the system.

There are many combinations of chemicals and some individual chemicals that have the ability to act as buffers. Each of these buffers has chemical characteristics that make it more or less useful for a particular purpose. (Recipes for a number of different buffers are given in the *CRC Handbook of Biochemistry and Molecular Biology, Physical and Chemical Data*, 3rd edition, Vol. 1, CRC Press, Boca Raton, Florida, 1975.) Some of these chemical properties that distinguish buffers will briefly be described.

The behavior of a buffer is displayed graphically in Figure 22.1. To prepare this graph, NaOH was slowly added to a buffer solution and the resulting pH was measured. Observe that the buffer tends to resist a change in pH at around pH 4.8. In other words, there is a pH at which OH^- or H^+ can be added to this solution with little change in its pH. The **pK_a** *of a buffer is the pH at which the buffer experiences little change in pH upon addition of acids or bases.* The buffer illustrated in Figure 22.1 is an acetate buffer that has a pK_a of 4.76. As shown in Table 22.1, different buffers have different pK_as. Buffers are effective in resisting a change in pH within a range of about 1 pH unit above and below their pK_a. For example, the pK_a for acetate buffer is 4.8 and acetate buffer is effective in the range from about pH 3.8 to 5.8. Thus, different buffers are effective in different pH ranges. Note also that a buffer may have more than one pK_a. For example, phosphate buffer has a pK_a at 2.15, 7.20, and 12.33.

Every biological system has an optimal pH. Stomach enzymes work at around pH 2, whereas DNA is synthesized at pH 6.9. In the laboratory, it is necessary to choose a buffer whose pK_a matches that of the system of interest. If one wants to work with a DNA solution at pH 6.9, acetate buffer should not be chosen.

ii. CHEMICAL INTERACTIONS WITH THE SYSTEM BEING STUDIED

A laboratory buffer should be inert in the system being studied. For example, Tris buffer is unsuitable for some

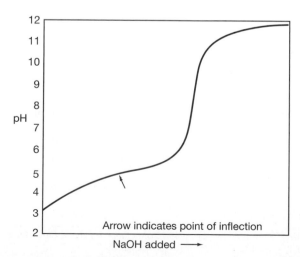

Figure 22.1. The Effect of Adding OH⁻ to Acetate Buffer. At around pH 4.8, the pK_a for acetate buffer, adding base has relatively little effect on the pH.

Table 22.1 The pK_a Values for Some Common Buffers and the Effect of Temperature Change on the pH of these Buffers*

Trivial name	Buffer name	pK_a	dK_a/dt
Phosphate (pK_1)	—	2.15	0.0044
Malate (pK_1)	—	3.40	
Formate	—	3.75	0.0
Succinate (pK_1)	—	4.21	−0.0018
Citrate (pK_2)	—	4.76	−0.0016
Acetate	—	4.76	0.0002
Malate	—	5.13	—
Pyridine	—	5.23	−0.014
Succinate (pK_2)	—	5.64	0.0
MES	2-(N-Morpholino)ethanesulfonic acid	6.10	−0.011
Cacodylate	Dimethylarsinic acid	6.27	—
Dimethylglutarate	3,3-Dimethylglutarate (pK_2)	6.34	0.0060
Carbonate (pK_1)	—	6.35	−0.0055
Citrate (pK_3)	—	6.40	0.0
Bis-Tris	[Bis(2-hydroxyethyl)imino]tris(hydroxymethyl)methane	6.46	0.0
ADA	N-2-Acetamidoiminodiacetic acid	6.59	−0.011
Pyrophosphate	—	6.60	—
EDPS (pK_1)	N,N'-Bis(3-sulfopropyl)ethylenediamine	6.65	—
Bis-Tris propane	1,3-Bis[tris(hydroxymethyl)methylamino] propane	6.80	—
PIPES	Piperazine-N,N'-bis(2-ethanesulfonic acid)	6.76	−0.0085
ACES	N-2-Acetamido-2-hydroxyethanesulfonic acid	6.78	−0.020
MOPSO	3-(N-Morpholino)-2-hydroxypropane-sulfonic acid	6.95	−0.015
Imidazole	—	6.95	−0.020
BES	N,N-Bis-(2-hydroxyethyl)-2-aminoethanesulfonic acid	7.09	−0.016
MOPS	3-(N-Morpholino)propanesulfonic acid	7.20	0.015
Phosphate (pK_2)	—	7.20	−0.0028
EMTA	3,6-Endomethylene-1,2,3,6-tetrahydrophthalic acid	7.23	—
TES	2-[Tris(hydroxymethyl)methylamino]ethanesulfonic acid	7.40	−0.020
HEPES	N-2-Hydroxyethylpiperazine-N'-2-ethanesulfonic acid	7.48	−0.014
DIPSO	3-[N-Bis(hydroxyethyl)amino]-2-hydroxypropanesulfonic acid	7.60	−0.015
TEA	Triethanolamine	7.76	−0.020
POPSO	Piperazine-N,N'-bis(2-hydroxypropanesulfonic acid)	7.85	−0.013
EPPS, HEPPS	N-2-Hydroxyethylpiperazine-N'-3-propanesulfonic acid	8.00	—
Tris	Tris(hydroxymethyl)aminomethane	8.06	−0.028
Tricine	N-[Tris(hydroxymethyl)methyl]glycine	8.05	−0.021
Glycinamide	—	8.06	−0.029
PIPPS	1,4-Bis(3-sulfopropyl)piperazine	8.10	—
Glycylglycine	—	8.25	−0.025
Bicine	N,N-Bis(2-hydroxyethyl)glycine	8.26	−0.018
TAPS	3-{[Tris(hydroxymethyl)methyl]amino}propanesulfonic acid	8.40	0.018
Morpholine	—	8.49	—
PIBS	1,4-Bis(4-sulfobutyl)piperazine	8.60	—
AES	2-Aminoethylsulfonic acid, taurine	9.06	−0.022
Borate	—	9.23	−0.008
Ammonia	—	9.25	−0.031
Ethanolamine	—	9.50	−0.029
CHES	Cyclohexylaminoethanesulfonic acid	9.55	0.029
Glycine (pK_2)	—	9.78	−0.025
EDPS	N,N'-Bis(3-sulfopropyl)ethylenediamine	9.80	—
APS	3-Aminopropanesulfonic acid	9.89	—
Carbonate (pK_2)	—	10.33	−0.009
CAPS	3-(Cyclohexylamino)propanesulfonic acid	10.40	0.032
Piperidine	—	11.12	—
Phosphate (pK_3)	—	12.33	−0.026

Source: Reprinted by permission from *Guide to Protein Purification*, Murray P. Deutscher, editor. Academic Press, 1990. *Methods in Enzymology* 182.
*Selected buffers and their pK_a values at 25°C.

protein assays because it reacts with the assay components, giving erroneous results. Phosphate buffers contribute phosphate ions to a solution, which inhibit some types of enzyme reactions. An important biotechnology example is the use of Tris-Borate-EDTA (TBE) buffer versus Tris-Acetate-EDTA (TAE) buffer when DNA fragments are separated from one another by gel electrophoresis. TBE is most commonly used for electrophoresis of DNA because it gives good resolution of DNA fragments and has good buffering capacity. It is difficult, however, to recover DNA from gels that were run with TBE buffer because the borate interacts with oligosaccharides in the gel. TAE, therefore, is the buffer of choice when DNA is to be extracted from gels after electrophoresis.

Some buffers bind ions, such as Ca^{++} or Mg^{++}, making these ions inaccessible. If the biological system of interest requires such ions, then a buffer that binds them is inappropriate; in other cases, metal ion chelation is desirable. (For more information about specific buffers, see, for example, *Biochemistry LabFax*, J.A.A. Chambers and D. Rickwood, eds., Bios Scientific, Oxford, UK, 1993.)

iii. CHANGES IN PH WITH CHANGES IN CONCENTRATION AND TEMPERATURE

It is convenient to prepare buffers as concentrated stock solutions that are diluted before use. Some buffers, however, change pH when diluted. Tris buffer is fairly stable when diluted. Its pH decreases about 0.1 pH units for every 10-fold dilution—which is acceptable in most applications. If the change with dilution of a buffer is not acceptable in an application, then the use of stock solutions should be avoided.

Temperature changes may markedly affect the pH of a buffer. For example, the pH of a Tris buffer will decrease approximately 0.028 pH units with every degree Celsius increase in temperature. Thus, Tris buffer prepared to pH 8.00 at room temperature (25°C) will be close to pH 7.64 when used at body temperature (37°C) and pH 8.70 when placed on ice. Buffers that are temperature-sensitive should be prepared at the temperature at which they will be used.

Some buffers are more sensitive than others to temperature. Table 22.1 gives values for various buffers that are used to calculate the change in pH with each degree change in temperature. The larger the numeric value for "dpK_a/dt," the more the buffer will change pH with each degree of temperature change. Negative values for dpK_a/dt mean that there is a decrease in pH with an increase in temperature, and vice versa.

Note that there are factors that affect the temperature of a solution while its pH is being measured. Magnetic stirring devices generate heat. Some chemical reactions are endothermic or exothermic (remove or generate heat); so as chemicals are combined, their temperature may change.

iv. OTHER CONSIDERATIONS

Buffer solutions should always be prepared with high-quality water. It is often good practice to sterilize buffers by autoclaving or ultrafiltration to prevent bacterial and fungal growth that can degrade organic buffers. Note, however, that some solution components are sensitive to heat and cannot be autoclaved.

Note also that in biology manuals, solutions that are called "buffers" may have functions in addition to maintaining the proper pH. For example, "buffers" may also contain salts, EDTA, β-mercaptoethanol, and so on. Thus, biologists use the term *buffer solution* in a general way, as compared with a strict chemical definition.

B. Adjusting the pH of a Buffer

i. OVERVIEW

When a buffer is prepared, the correct amount of the buffer compound(s) is weighed out and dissolved in water to get the desired concentration. For example, Tris buffer comes as a crystalline solid. (Tris that is not conjugated to any other chemical is called *Tris base*.) Tris base has a *FW* of 121.1 and a pK_a of 8.06. A 1 *M* solution of Tris can be prepared by dissolving 121.1 *g* of Tris base in 1 *L* of water. The pH of such a solution, however, will be greater than 10. This pH is far from the pK_a for Tris, so the solution will have little buffering capacity. Moreover, few biological systems function at such a high pH. Before the Tris solution is useful as a biological buffer, its pH therefore must be lowered. Most buffering compounds are not at their pK_a when they are dissolved and must be adjusted to the desired pH. There are several strategies to obtain a buffer that has both the correct pH and the correct concentration. This section describes these strategies and illustrates them using Tris and phosphate buffers as examples.

Strategy 1.

The buffer is titrated with a strong acid or base. The buffer is dissolved in water, but it is not brought to its final volume. If the pH of the solution is lower than the desired pH, then a strong base (often NaOH) is added to raise the pH. Conversely, if the pH of the solution is above the desired pH, then a strong acid (often HCl) is added to lower the pH. Once the desired pH is reached, the solution is brought to its final volume. This procedure is summarized in Box 2 and an example of this method (Tris buffer) is given.

There are situations where the sodium from NaOH, or the chloride from HCl, interferes with the system of interest. In such cases, other acids or bases are substituted.

Strategy 2.

The buffer is prepared from two reagents, one of which is more acidic, the other of which is more basic. By combining the two forms in the correct proportions, a solution with the desired pH can be obtained. This method will be illustrated for Tris buffer.

Box 2   General Procedure to Bring a Solution to the Correct pH Using a Strong Acid or a Strong Base

1. *Determine the amount of solute(s) required to make a buffer of the correct concentration.*
2. *Mix the solute(s) with most, but not all, the solvent required. Do not bring the solution to volume.* (To ensure consistency, a standard operating procedure will usually specify how much water to dissolve the solute(s) initially.)
3. *Place the solution on a magnetic stir plate, add a clean stir bar, and stir.*
4. *Check the pH. (Be careful not to crash the stir bar into the electrode.)*
5. *Add a small amount of acid or base, whichever is needed to bring the solution toward the desired pH.* The recipe will often specify which acid or base to use; if the recipe does not specify, it is usually safe to add HCl or NaOH.
6. *Stir again and then check the pH.*
7. *Repeat Steps 5 and 6 until the pH is correct.*
8. *Bring the solution to the proper volume once the pH is exactly correct, then recheck the pH.* (With many buffers, when you bring the solution to its final volume, the pH does not change significantly.)

Notes about Adjusting the pH of a Solution

i *Temperature.* pH the solution when it is at the temperature at which you plan to use it. Note also that some solutions change temperature during mixing. For example, the addition of acid to Tris buffer may cause its temperature to change. Restore the buffer to the correct temperature before making the final adjustment of its pH.

ii *Overshooting the pH.* If you accidentally overshoot the pH the correct procedure is to remake the entire solution. It is not recommended to compensate by adding some extra acid or base. This is because the additional acid or base adds extra ions to the solution; these ions will change the composition of your solution and may adversely affect it. In addition, if you add more acid or base sometimes (because of a mistake), but not other times, there will be an inconsistency that will be very difficult to diagnose if problems arise later.

iii *Overshooting the Volume.* If you accidentally overshoot the volume, begin again to avoid changing the concentration of solute(s) and to avoid introducing inconsistency. (Using more concentrated acid or base may help avoid this problem, but be careful not to overshoot the pH.)

Strategy 3.

The buffer is prepared as two stock solutions, one of which is more acidic and the other more basic. The two stocks are combined in the proper proportion to get the desired pH. An example of this method, phosphate buffer, will be discussed.

ii. Tris Buffer

Tris, (Tris(hydroxymethyl)aminomethane), is one of the most common buffers in biotechnology laboratories.

Tris is used because it buffers over the normal biological range (pH 7 to pH 9), is nontoxic to cells, and is relatively inexpensive. Table 22.2 summarizes some considerations regarding Tris buffers.

As mentioned earlier, the pK_a of Tris is 8.06 but when dissolved in water its pH is greater than 10. Two strategies to bring Tris to the correct pH are as follows:

Strategy 1.

The buffer is titrated with a strong acid until it is the desired pH. Because Tris base has a basic pH when

Table 22.2   Considerations When Using Tris Buffers

1. *Tris buffers change pH significantly with changes in temperature.* Final pH adjustments to Tris buffers, therefore, should be made when the solution is at the temperature at which it will be used.
2. *Tris buffers require a compatible electrode for accurate pH determination.* A calomel (mercury) reference electrode rather than a silver/silver chloride electrode is recommended.
3. *Tris is subject to fungal contamination.* Contamination can be avoided by autoclaving the Tris solution or filter sterilizing it before long-term storage.
4. *Tris is minimally sensitive to concentration changes.* Tris can therefore be stored as a concentrated stock solution that is diluted before use.

dissolved, it is necessary to add acid to bring the pH down to a value suitable for biological work. If the acid required to bring Tris to the proper pH is unspecified, it is most common to use HCl for titration, but other acids may be used as well. For example, boric acid, citric acid, and acetic acid can be used to prepare Tris-borate, Tris-citrate, and Tris-acetate buffers, respectively.

Strategy 2.

Combine Tris base with a conjugated form of Tris called Tris-HCl. Tris-HCl is a solid form of Tris which has been conjugated to HCl. If Tris-HCL is dissolved in water, its pH is less than 5. It is possible to combine Tris base with Tris-HCl to get a pH between 7 and 9. Depending on the pH desired, different amounts of the two forms of Tris are combined, as shown in Table 22.3. An upcoming example problem illustrates this strategy. Note also that Figure 22.4b contains a procedure to prepare Tris buffer based on this strategy.

Table 22.3 COMBINING TRIS BASE AND TRIS-HCL TO GET A BUFFER OF THE DESIRED pH*

pH at			0.05 M Grams/liter	
5°C	**25°C**	**37°C**	**Tris HCl**	**Tris Base**
7.55	7.00	6.70	7.28	0.47
7.66	7.10	6.80	7.13	0.57
7.76	7.20	6.91	7.02	0.67
7.89	7.30	7.02	6.85	0.80
7.97	7.40	7.12	6.61	0.97
8.07	7.50	7.22	6.35	1.18
8.18	7.60	7.30	6.06	1.39
8.26	7.70	7.40	5.72	1.66
8.37	7.80	7.52	5.32	1.97
8.48	7.90	7.62	4.88	2.30
8.58	8.00	7.71	4.44	2.65
8.68	8.10	7.80	4.02	2.97
8.78	8.20	7.91	3.54	3.34
8.86	8.30	8.01	3.07	3.70
8.98	8.40	8.10	2.64	4.03
9.09	8.50	8.22	2.21	4.36

To make 0.05 *M* Tris buffer: Dissolve the indicated amounts of Tris-HCl and Tris base in water to a final volume of 1 *L*. Tris-HCl and Tris base are somewhat hygroscopic at high humidity. For precise work, desiccation before weighing is recommended. Tris solution can be autoclaved.
Source: "Sigma Catalogue," Sigma Company, P.O. Box 14508, St. Louis, MO 63178. Reprinted by permission from Sigma-Aldrich Co.
*Note that Sigma company calls their Tris "Trizma."

EXAMPLE PROBLEM

Outline a procedure to make 300 *mL* of a 0.8 *M* solution of Tris buffer, pH 7.6 at 4°C. (*FW* of Tris base = 121.1) Use the strategy in Box 1.

ANSWER

1. First, it is necessary to calculate the amount of solute (Tris base) required. This may be accomplished using Formula 1, p. 454.

$$\text{Solute Required} = $$
$$(grams/mole) \quad (Molarity) \quad (Volume)$$
$$(121.1 \ g/mole)(0.80 \ mole/L)(0.300 \ L) = 29.06 \ g$$

2. Dissolve 29.06 g of Tris base in about 250 *mL* of 4°C, purified water. The pH will be above 10.

3. Add concentrated HCl (remember to use safety precautions) while stirring and reading the pH. (Use a pH meter equipped with a calomel electrode.)

4. Check the temperature of the solution. The temperature may change both during the initial dissolution of the Tris and during the addition of acid to the solution. Ensure that the temperature of the solution is 4°C before performing the final adjustment of its pH.

5. Continue adding HCl until the solution is exactly pH 7.6.

6. Bring the solution to 300 *mL* with water. (Because the pH of Tris is relatively insensitive to dilution, the pH should still be close to 7.6 after the solution is brought to volume.)

Note: There is an alternate strategy that allows this buffer to be prepared at room temperature. From Table 22.1, we know that the change in pH with temperature for Tris is -0.028 pH units per degree Celsius. Suppose room temperature is 22°C. Then the difference between room temperature and 4°C is -18°C. Multiplying -18°C $\times$ -0.028 gives a change of $+0.504$ pH units. The solution, therefore, can be prepared at room temperature but be brought to pH 7.096 (round to 7.1). When the buffer is used at 4°C, its pH should rise to be very close to pH 7.6. The advantage to this method is that it avoids the requirement of holding the temperature at 4°C as the solution is prepared.

EXAMPLE PROBLEM

Outline a procedure to make 300 *mL* of a 0.05 *M* solution of Tris buffer, pH 7.62 at 37°C. Use the information in Table 22.3.

ANSWER

1. Table 22.3 tells us that 1 L of 0.05 *M* buffer at pH 7.62 and at 37°C requires 4.88 *g* of Tris-HCl and 2.30 *g* of Tris base.

2. Because we do not want to make a liter of buffer, it is necessary to calculate how much of each chemical is required to make 300 mL. This can be easily accomplished using the logic of proportions:

Tris-HCl

$$\frac{4.88 \ g}{1000 \ mL} = \frac{?}{300 \ mL}$$

$$? = 1.464 \ g$$

Tris base

$$\frac{2.30 \ g}{1000 \ mL} = \frac{?}{300 \ mL}$$

$$? = 0.690 \ g$$

3. Combine 1.464 g of Tris-HCl and 0.690 g of Tris base in 250 mL of purified water that is at 37°C.

4. Ensure that the temperature of the solution is 37°C.

5. Check the pH of the solution with a pH meter having a calomel electrode.

6. If the pH is within 0.05 units of 7.62, then bring it to volume with water. (If the pH is not sufficiently close to 7.62, it may be acceptable to add small amounts of either acid or base to bring the pH to the exact pH.)

iii. PHOSPHATE BUFFER

Phosphate is another buffer commonly used for biological systems. Phosphate buffer is typically prepared by mixing solutions of sodium phosphate dibasic (Na_2HPO_4) and sodium phosphate monobasic (NaH_2PO_4). At 25°C, a 5% solution of dibasic sodium phosphate has a pH of 9.1, whereas a 5% solution of the monobasic form has a pH of 4.2. Mixtures of these two solutions will make a buffer with a pH between the two extremes.

The conventional way to prepare phosphate buffer is to prepare stock solutions of the monobasic and dibasic compounds that have the same molar concentration, for example, 0.2 M. The two solutions are then combined in the proportions given in Table 22.4 to achieve the proper pH. The concentration of the resulting buffer is

Table 22.4 PREPARING PHOSPHATE BUFFER

To Prepare 0.1 M Sodium Phosphate Buffer of a chosen pH

Step 1. **Prepare Solution A,** 0.2 M $NaH_2PO_4 \bullet H_2 0$ (monobasic sodium phosphate monohydrate, $FW = 138.0$): Dissolve **27.6 g $NaH_2PO_4 \bullet H_2 0$** in purified water. Bring to volume of 1000 mL.

Step 2. **Prepare Solution B,** 0.2 M Na_2HPO_4 (Dibasic Sodium Phosphate): Use either: **28.4 g Na_2HPO_4** (anhydrous dibasic sodium phosphate, $FW = 142.0$) or **53.6 g $Na_2HPO_4 \bullet 7H_2 0$** (dibasic sodium phosphate heptahydrate, $FW = 268.1$)

Dissolve in purified water and bring to volume of 1000 mL.

Step 3. Combine Solution A + Solution B in the amounts listed in the following table.

Step 4. Bring the mixture of Solution A + Solution B to a final volume of 200 mL.

Volume A (mL)	Volume B (mL)	pH
93.5	6.5	5.7
92.0	8.0	5.8
90.0	10.0	5.9
87.7	12.3	6.0
85.0	15.0	6.1
81.5	18.5	6.2
77.5	22.5	6.3
73.5	26.5	6.4
68.5	31.5	6.5
62.5	37.5	6.6
56.5	43.5	6.7
51.0	49.0	6.8
45.0	55.0	6.9
39.0	61.0	7.0
33.0	67.0	7.1
28.0	72.0	7.2
23.0	77.0	7.3
19.0	81.0	7.4
16.0	84.0	7.5
13.0	87.0	7.6
10.5	89.5	7.7
8.5	91.5	7.8
7.0	93.0	7.9
5.3	94.7	8.0

Tabular information from *Biochemistry LabFax*, J.A.A. Chambers and D. Rickwood, eds. Bios Scientific Publishers, Oxford, UK, 1993. and *Buffers: A Guide for the Preparation and Use of Buffers in Biological Systems*, Calbiochem Corporation, San Diego, CA

determined by the molarity of the phosphate ions. This means that if 0.2 M sodium phosphate monobasic and 0.2 M sodium phosphate dibasic are combined, the resulting concentration will still be 0.2 M. Phosphate buffer prepared with a mixture of the monobasic and dibasic forms has a pK_a of 7.2 and so its buffering range is about pH 6 to pH 8.

Phosphate buffers can also be prepared with the potassium salt of phosphate and occasionally with a mixture of the sodium and potassium salts. Phosphate buffers are less sensitive than are Tris buffers to the effect of temperature, but are more sensitive to dilution. Phosphate buffers, therefore, should not be prepared as concentrated stock solutions. They should instead be prepared to the concentration at which they will be used.

EXAMPLE PROBLEM

Using Table 22.4, outline a procedure to make 0.1 M sodium phosphate buffer, pH 7.0.

ANSWER

1. Prepare the two stock solutions, Solution A and Solution B, as directed in Table 22.4. These stocks have a molarity of 0.2 M.

2. Mix the amounts indicated on the chart, that is, 39.0 mL of Solution A and 61.0 mL of Solution B. This solution has a concentration of 0.2 M.

3. Bring the solution to a volume of 200 mL as directed in the chart. Dilution of the buffer with water to 200 mL means that the final concentration of phosphate in the solution is 0.1 M. Check the pH.

III. PREPARING SOLUTIONS WITH MORE THAN ONE SOLUTE

A. Introduction

Chapter 21 discussed how to prepare a solution containing a single solute in such a way that the solute is present at the correct concentration. This chapter has so far added another issue to the preparation of a laboratory solution; ensuring that the solution is at the correct pH as well as the correct concentration. We will now discuss yet another issue relating to solution preparation, the preparation of solutions with more than one solute. Figure 22.2 shows three examples of recipes for solutions with more than one solute.

Recipe I is relatively simple to interpret. It states the *amounts* of four solutes to weigh out and combine in water.

Recipe II also states how much of each solute is required; however, the solutes are prepared as individual solutions before being combined. To prepare Recipe II, first make a 1 M solution of $MgCl_2$, a 0.4% solution of thymidine, and a 20% solution of glucose. Then, mix 5 mL of the $MgCl_2$ solution, 10 mL of the

Recipe I		Recipe II		Recipe III
Na_2HPO_4	6 g	1 M $MgCl_2$	5 mL	0.1 M Tris
KH_2PO_4	3 g	0.4% thymidine	10 mL	0.01 M EDTA
NaCl	0.5 g	20% glucose	25 mL	1% SDS
NH_4Cl	1 g			

Dissolve in water.
Bring to a volume
of 1 liter.

Figure 22.2. Examples of Three Recipes for Laboratory Solutions that Contain More than one Solute.

thymidine solution, and 25 mL of the glucose solution. Additional water is not added to the mixture of the three solutes.

Recipe III is different in that it does not state the *amount* of each solute required. Rather, the recipe states the *final concentration* of each solute. If the recipe or procedure gives only the final concentration(s) needed for the solute(s), you will have to *calculate the amount(s)* you need of each. The reason authors often record concentrations instead of amounts is they do not know what volume you will need. Recipes like Recipe III are discussed in detail in this section.

B. A Recipe with Multiple Solutes: SM Buffer

Consider the following recipe:

SM Buffer

0.2 M Tris, pH 7.5
1 mM $MgSO_4$
0.1 M NaCl
0.01% gelatin

This is the entire recipe as it might appear in a manual. SM buffer contains four solutes that are combined in water. The recipe lists the *final concentration* of each solute; Tris is present in the final solution at a concentration of 0.2 M, $MgSO_4$ at a final concentration of 1 mM, and so on. The first instinct people sometimes have when looking at a recipe like this is to prepare four separate solutions of 0.2 M Tris, 1 mM magnesium sulfate, 0.1 M NaCl, and 0.01% gelatin, and then to mix them together; however, this instinct is wrong. If you combine any volume of 0.2 M Tris with any volume of 1 mM $MgSO_4$ or any of the other solutes, they will dilute one another, giving the wrong final concentrations.

Consider two correct strategies to prepare this solution.

Strategy 1: Preparing SM Buffer without Stock Solutions

In overview, first prepare 0.2 M Tris, pH 7.5 but do not BTV. Then, calculate how many grams of each of the other solutes is required. These solutes are weighed out and dissolved directly in the Tris buffer. Note that when the temperature is not specified, assume a solution is to be at room temperature. The steps are:

1. Decide how much volume to prepare, for example, 1 L.

2. One liter of 0.2 M Tris requires 24.2 g of Tris base (FW = 121.1). Dissolve the Tris in about 700 mL of water and bring the pH to 7.5. Do not bring the Tris to volume yet.

   (The Tris buffer is brought to the proper pH before the other solutes are added because the recipe shows *Tris* to be a particular pH.)

3. One liter of 0.1 M NaCl requires 5.84 g of NaCl. Add this amount to the Tris buffer.

4. One liter of 1 mM $MgSO_4$ requires 1/1000 of its *FW*. (Magnesium sulfate comes in more than one hydrated form. Read the container to determine the correct molecular weight.) Weigh out the correct amount, and add it to the Tris buffer. (For example, $MgSO_4 \bullet 7\ H_2O$ has a FW of 246.5. Add 0.246 g of this form of $MgSO_4$.)

5. 0.01% gelatin is 0.1 g in 1 L. Weigh out 0.1 g of gelatin and add to the Tris buffer.

6. Stir the mixture until all the components are dissolved, then BTV of 1 L with water.

7. It is good practice to check the pH at the end: It should be close to 7.5 because the Tris is fairly stable with dilution. Record the final pH, but do not readjust it (unless directed to do so by a procedure you are following).

Strategy 2: Preparing SM Buffer with Stock Solutions

In this strategy, the four solutes are each prepared separately as *concentrated stock solutions*. Then, when the four stocks are combined, they dilute one another to the proper final concentrations.

1. Prepare a stock solution of Tris buffer at pH 7.5. There is no set rule as to what concentration a stock solution should be. In this case, a useful concentrated stock solution might be 1 M (i.e., 5 X more concentrated than needed). To make 1 L of 1 M stock, dissolve 121.1 g of Tris base in about 900 mL of water. Bring the solution to pH 7.5 and then to a volume of 1 L.

2. Prepare a stock solution of magnesium sulfate (e.g., 1 M). To make 100 mL of this stock, dissolve 0.1 FW of $MgSO_4$ in water and bring to a volume of 100 mL. (For example, $MgSO_4 \bullet 7\ H_2O$ has a FW of 246.5. Use 24.6 g of this form of $MgSO_4$.)

3. Prepare a stock solution of NaCl, for example, 1 M. To make 100 mL of stock, dissolve 5.84 g in water and bring to a volume of 100 mL.

4. Prepare a stock solution of gelatin, for example, 1%, by dissolving 1 g in a final volume of 100 mL of water.

5. To make the final solution it is necessary to combine the right amounts of each stock. Because this is a situation where concentrated solutions are

being diluted, use the $C_1V_1 = C_2V_2$ equation four times, once for each solute. For example, to figure out how much Tris stock is required:

$$C_1 \quad V_1 \quad = \quad C_2 \quad V_2$$
$$(1\ M)\ (?) \quad = \quad (0.2\ M)\ (1000\ mL)$$
$$? = 200\ mL$$

To make 1000 mL of SM buffer, therefore, combine:

Components	Final Concentration
200 mL of the 1 M Tris stock, pH 7.5	200 mM Tris (0.2 M)
1 mL of the 1 M magnesium sulfate stock	1 mM $MgSO_4$
100 mL of the 1 M NaCl stock	0.1 M NaCl
10 mL of the 1% gelatin stock	0.01% gelatin

6. BTV 1000 mL and check the pH.

Both Strategy 1 and Strategy 2 are correct. It is generally efficient to make stock solutions of solutes that are used often because weighing out chemicals is time-consuming. For example, many solutions require NaCl; therefore, it might be useful to keep a 1 M stock solution of NaCl on hand. Chemicals that are used infrequently should not be kept as stock solutions because microorganisms can grow in them and they may slowly degrade. (Sterilizing solutions can sometimes extend their shelf life.)

C. A Recipe with Two Solutes: TE Buffer

Consider another example:

TE Buffer

Component	Final Concentration
Tris, pH 7.6	10 mM
Na_2-EDTA	1 mM

This buffer contains Tris at a pH of 7.6 and a final concentration of 10 mM and Na_2-EDTA at a final concentration of 1 mM. As for SM buffer, it is incorrect to prepare 10 mM Tris buffer and 1 mM Na_2-EDTA and then to mix them together because they will dilute one another. It is conventional to prepare TE buffer by combining stock solutions of Tris, at the desired final pH, and EDTA. (For example, this strategy is outlined in *Molecular Cloning: A Laboratory Manual,* Second ed. J. Sambrook, E.F. Fritsch, and T. Maniatis. Cold Spring Harbor Laboratory Press, Cold Spring Harbor, 1989.)

To make TE buffer:

1. Decide how much volume to prepare, for example, 100 mL.

2. Na_2-EDTA does not go into solution easily and is relatively insoluble in water if the pH is less than 8.0. Na_2-EDTA, therefore, is commonly prepared as a concentrated stock solution with a pH of 8.0. To make 100 mL of a 0.5 M stock of Na_2-EDTA,

add 18.6 g of EDTA, disodium salt, dihydrate (*FW* = 372.2) to 70 *mL* of water. Adjust the pH to 8.0 slowly with stirring by adding pellets of NaOH or concentrated NaOH solution. When the Na_2-EDTA is dissolved, bring the solution to volume.

3. The Tris buffer needs to be dissolved and brought to the proper pH before it is combined with the Na_2-EDTA. Like the Na_2-EDTA the Tris should also be prepared as a concentrated stock solution (e.g., 0.1 *M*). To make 100 *mL* of Tris stock, dissolve 1.21 g of Tris in about 70 *mL* of water, adjust it to pH to 7.6 with HCl, and then bring it to 100 *mL* final volume.

4. Use the $C_1V_1 = C_2V_2$ equation twice, once for each solute, to calculate how much of each stock will be required to make a solution of TE with the proper concentration of both solutes.

To calculate how much 0.1 *M* stock of Tris is required:

$$C_1 \quad V_1 = \quad C_2 \quad V_2$$
$$(0.1\ M)(?) = (0.010\ M)(100\ mL)$$
$$? = 10\ mL$$

(Note that the units must be consistent on both sides of the equation so 10 *mM* was converted to 0.010 *M*.)

To calculate how much 0.5 *M* Na_2-EDTA stock is required:

$$C_1 \quad V_1 = \quad C_2 \quad V_2$$
$$(0.5\ M)(?) = (0.001\ M)(100\ mL)$$
$$? = 0.2\ mL = 200\ \mu L$$

5. Combine 10 *mL* of Tris stock and 200 *μL* of Na_2-EDTA stock and bring the solution to the final volume, 100 *mL*, with water.

6. Check the pH.

IV. PREPARATION OF SOLUTIONS IN THE LABORATORY

A. General Considerations

Box 3 outlines a general procedure for making a solution in the laboratory and summarizes many of the issues discussed so far in these two chapters on solutions.

Box 3   A GENERAL PROCEDURE TO MAKE A SOLUTION IN THE LABORATORY

1. *Read and understand the procedure; identify and collect materials and equipment needed.*
2. *Determine the amount of each solute or stock solution that is required.*
3. *Weigh the amount of each solute required, or measure the volume of each stock solution.*
4. *Dissolve the weighed compounds, or mix the stock solutions, in less than the desired final volume of solvent.*
5. *Bring the solution to the correct pH, if necessary.*
6. *Bring the solution to volume. Check the pH, if necessary.*
7. *It may be necessary to sterilize the solution.*

Notes:

i **Water.** Unless instructed otherwise, use the highest quality (most purified) water available to make solutions.

ii **Grades of Chemicals.** Be sure that the purity and form of chemicals match your requirements. See Table 22.5.

iii **Contamination.** Do not return chemicals to stock containers so as not to contaminate the stocks of chemicals. In addition, avoid putting spatulas into stock containers, although it may be alright to use a clean spatula in some situations. To protect yourself and to avoid contaminating solutions, never touch any chemicals or the interior of glassware containers.

iv **Labeling.** ALWAYS LABEL THE CONTAINER OF EVERY SOLUTION. At a minimum include the following on the label: the name and/or formula of the solution, the concentration of the solution, the solvent used, the date prepared, your name or initials, and safety information such as whether the solution is toxic, radioactive, acidic, basic, and so on. Additional information, such as the lot number of the solution, may be required.

v **Concentrated and Dilute Acids and Bases.** The term *concentrated* refers to acids and bases in water at the concentration that is their maximum solubility at room temperature, Table 22.6. For example, concentrated hydrochloric acid has a molarity of 11.6 and is 36% acid. In contrast, concentrated nitric acid is 15.7 *M* and 72% acid. When working with concentrated acids, put the acid into water and not vice versa to avoid hazardous splashing and overheating.

vi **Hygroscopic Compounds.** **Hygroscopic compounds** *absorb moisture from the air.* A compound that has absorbed an unknown amount of water from the air may not be acceptable for use. Hygroscopic compounds, therefore, should be kept in tightly closed containers and should be exposed to air as little as possible. It may be advisable to purchase such compounds in small amounts so that once their container is opened they are rapidly used and are not stored. These compounds may be stored in desiccators to help protect them from moisture. In some laboratories, it is common practice to record the date a bottle of a chemical is first opened. A chemical that has been opened and stored too long may need to be discarded.

Table 22.5 GRADES OF CHEMICAL PURITY

Chemicals are manufactured with varying degrees of purity. More purified preparations are usually more expensive to purchase.

1. *Reagent (ACS)* or *Analytical Reagent (ACS)* grade chemicals are certified to contain impurities below the specifications of The American Chemical Society Committee on Analytical Reagents. If the bottle was not contaminated by previous use, these chemicals are usually acceptable for making solutions.

2. *Reagent* grade. If the American Chemical Society has not established specifications for a particular chemical, the manufacturer may do so and label the chemical Reagent grade. The manufacturer must then list the maximum allowable impurities for the chemical.

3. *U.S.P.* grade chemicals meet the standards of the U.S. Pharmacopeia, the official publication for drug product standards. Such chemicals are acceptable for drug use and are often also acceptable for general laboratory use.

4. *CP* means chemically pure and is usually similar to or slightly less pure than reagent grade. **Purified** means that the manufacturer has removed many contaminants. Read the manufacturer's specifications to determine whether these chemicals are acceptable for your purposes.

5. *Commercial* or *Technical* grade chemicals are intended for industrial use. **Practical** grade is intended for use in organic syntheses. In general, commercial, technical, and practical grade chemicals are not used for laboratory solutions.

6. *Spectroscopic* grade chemicals are intended for use in spectrophotometry because they have very little absorbance in the UV or IR range.

7. *HPLC* grade reagents are organic solvents that are very pure and are intended for use as the mobile phase in HPLC.

8. *Primary Standard* grade chemicals are usually the most highly purified chemicals and are intended to be used as standards.

9. *Other* Chemical manufacturers often have their own grades, such as "tissue culture grade." Read their catalogue to determine their specifications.

Table 22.6 ACIDS AND BASES

Acid or Base	Formula	FW	Moles per Liter	Density (g/mL)	% Solution (w/w)
Acetic Acid, glacial	CH_3COOH	60.1	17.4	1.05	99.5
Hydrochloric acid, concentrated	HCl	36.5	11.6	1.18	36
Nitric Acid, concentrated	HNO_3	63.0	16.0	1.42	71
Sulfuric acid, concentrated	H_2SO_4	98.1	18.0	1.84	96
Sodium Hydroxide, pellets	NaOH	40.0	Solid	—	—
Sodium Hydroxide, diluted		—	6.1	1.22	20

For more information on acids and bases, see also *The Merck Index: An Encyclopedia of Chemicals, Drugs, and Biologicals,* 12th edition, Susan Budavari ed. Merck, Whitehouse Station, NJ, 1996.

B. Assuring the Quality of a Solution

Quality practices relevant to laboratory solutions include the items discussed in the next sections.

i. DOCUMENTATION

All steps of solution preparation require documentation, including procedure(s) followed, the details of the preparation (weights, volumes, and measurements), problems or deviations from the procedure, components of the solution including the manufacturer's lot numbers, who prepared the solution, date of preparation, handling after preparation, maintenance and calibration of equipment used, and safety considerations. Such documentation is important, even in laboratories that do not adhere to a formal quality system.

It is always good practice to check and record the pH of a biological solution. It is also useful to check and record the solution's conductivity. The conductivity of a solution is a measure of its ability to conduct electrical current. The dissolved components of a solution, particularly salts, affect its conductivity. If a solution is made consistently, its conductivity should therefore always be the same. An unexpected conductivity reading is a warning that there is a problem.

Figure 22.3 is an example of a form used in a biotechnology company that has a centralized solution-making facility. The side of the form in Figure 22.3a is completed when a solution is requested by a laboratory scientist and the side in Figure 22.3b is filled in by the person who prepares the solution. In many companies, such information is entered directly into a computer system, but paper can also be used.

ii. TRACEABILITY

In production facilities, it is essential that every component used in producing a product be traceable. If a solution is used in production, documentation that identifies each component of that solution must be connected to the product. The solution itself is assigned an identification number that travels with the solution as it is used

MEDIA PREPARATION FORM (*SECTION A*)

ITEM NUMBER	LOT NUMBER

A-1.

1. NAME OF MEDIA/SOLUTION (*and pH @ Temp.*): _____

2. Requested By _____ Room No. _____ Dept. No. _____

3. Project # (*if applicable*)_____ 4. PIP # (*if applicable*)_____

5. Date and Time Ordered_____ Date and time Required _____

NOTE: Please allow a minimum of 24 hours for buffers and solutions, and 48 hours for agar-based plates and New Formulations.

A-2. TOTAL QUANTITY REQUIRED: _____

A-3. *Check Product Type:*

1. MEDIA PREP PRODUCT ☐   2. PRODUCT ☐   3. SHOP ORDER COMPONENT ☐

4. NEW FORMULATION ☐ *Check with Media Prep if unsure if request is a New Formulation*

    1) If a MEDIA PREP PRODUCT, write BPCS ITEM NUMBER in the box above. Media Prep will fill in LOT NUMBER.

    2) If a PRODUCT, write Production BPCS ITEM NUMBER and SHOP ORDER (LOT) NUMBER in the boxes above. Attach (staple) a copy of the CURRENT Procedure to the Media Preparation Form.

    3) If PRODUCTION SHOP ORDER COMPONENT, list PRODUCT ITEM # and SHOP ORDER # below.

        ITEM #: _____   SHOP ORDER #: _____

    4) If a NEW FORMULATION, identify the source of the new formulation. e.g., Notebook # and Page #, Journal article (Journal name, vol., article, and page), Reference Book (Title and page), etc. Attach a copy to this form.

    SOURCE: _____

    Complete the following sections of the Bill Of Materials on the reverse side (*SECTION B*) for NEW FORMULATIONS: Component, Component Item No., and Final Conc. for Media/Solution Components. If BPCS Item No.'s are unavailable, write vendor name (after the component) in Component Column, and vendor Item No. in the Item Column.

A-4. *Check items to be included in prep*

☐ Autoclave
☐ Filter Sterilize
☐ Filter (to remove extraneous debris)
    WATER SOURCE:
☐ Deionized (DI.)
☐ NANOpure
☐ Other (Describe):

A-5. DISTRIBUTION

# of Plates:	Vol. Liquid/Per Vol. Container:
# of Flasks:	Vol. Liquid/Per Vol. Container:
# of Bottles:	Vol. Liquid/Per Vol. Container:
# of Tubes:	Vol. Liquid/Per Vol. Container:
# of Slants:	Vol. Liquid/Per Vol. Container:
# of Carboys:	Vol. Liquid/Per Vol. Container:
Other:	

A-6. SPECIAL INSTRUCTIONS/COMMENTS: _____

_____

A-7. DELIVERY INSTRUCTIONS: _____

☐ Check this box if you wish to receive a copy of the completed Media Preparation Form.

(a)

Figure 22.3. Media Preparation Form. a. This part of the form is completed by the person requesting a solution. **b.** This information is entered into the computer system as the solution is made.

for various purposes. This system ensures that if a problem arises, every component of every solution that went into production of that product can be identified. Labeling the solution itself, is, of course, essential.

iii. STANDARD OPERATING PROCEDURES

SOPs tell laboratory workers in detail how to maintain and use instruments and how to prepare a particular solution. An SOP might describe, for example, exactly how a particular brand of balance is to be maintained and calibrated, and how the preparer of a solution is to document that maintenance and calibration were per-

formed. A solution SOP will usually show any calculations required (how much weight or volume of each component is needed) or example calculations, what type of purified water is to be used, whether the solution should be brought to a particular pH and, if so, how, whether the pH should be checked, whether the solution should be sterilized and, if so, how, and any other specific details. Each time a solution is made, the procedure must be followed and the particular procedure that was followed must be documented. Examples of industry procedures to make solutions are shown in Figure 22.4.

MEDIA PREPARATION FORM (*SECTION B*)   page ___ of ___
(*Complete all sections. Write a line or dash in sections that do not apply.*)

ITEM NUMBER	LOT NUMBER

NAME OF MEDIA/SOLUTION: _____

TARGET pH @ Temp. _____ (1 X pH @ Temp.) _____

CONDUCTIVITY RANGE/(Dilution): _____

SOP NUMBER FOLLOWED: _____

BILL OF MATERIALS   QUANTITY PREPARED: _____

Component (vendor)	Item Number	Lot Number	Final Conc.	Amount per liter	Total Amount	Equipment	Amount Added

WATER (*circle source and write location*)   Deionized/NANOpure/Ultrafiltered_____

EQUIPMENT AND MEASUREMENTS:

Balance(s) *ID #:* (1) _____ (2) _____ (3) _____

pH Meter *ID #:* _____ Conductivity Meter *ID #:* _____

Mixing Vessel: _____ Dispensing Equipment: _____

Filtered with: _____

Filter Sterilized with: _____

Autoclave(s) *ID #, cycle time, and run #:* _____

CALIBRATE pH ELECTRODE

Standard Buffers:_____ Slope/Efficiency @ Temp: _____

Initial pH: _____ pH adjusted with: _____

Final pH @ Temp: _____ 1 X pH @ Temp: _____

CONDUCTIVITY/Dilution: _____

DEVIATIONS FROM PROCEDURE (*list components, critical steps, etc., that differ from the procedure*)_____

_____

COMMENTS:_____

_____

_____

Location _____ Hours_____ Prepared by_____ Date _____

(b)

Figure 22.3.   *continued.*

Note that there is often some ambiguity in published solution recipes. For example, the temperature of the solution may be unstated, or it may be unclear as to whether a specified pH refers to a particular solute or whether it is the pH of the final solution. SOPs, therefore, play a key role in ensuring consistency and clarity in solution preparation. Even in laboratories that do not adhere to a particular quality system, it is good practice to develop standard procedures for preparation of solutions. For solutions that are not made routinely, be certain to record in a laboratory notebook all calculations and all details of preparation.

iv. INSTRUMENTATION

Balances, pH meters, and other instruments used in the preparation of a solution must be properly maintained and calibrated, and maintenance and calibration records must be kept for each instrument. When a solution is made, the instruments used are recorded. Note that the form in Figure 22.3b has blanks that relate to instrumentation.

v. STABILITY AND EXPIRATION DATE

Every solution should be labeled with an expiration date based on its stability during storage.

1.0 FORMULATION   DNA BUFFER   NDAB-00S

Component	Item Number	FW or Conc.	Amount/L	Final Concentration
2M Tris-HCl, pH 7.3	TRIS-73S	2M	5 mL	10 mM
5M NaCl	NACL-05S	5M	2 mL	10 mM
0.25M EDTA, pH 8.0	EDTA-25S	0.25M	4 mL	1mM
D.I. H$_2$O		—	Q.S.	—

Target pH: 7.50 ± 0.10 @ 25°C
Conductivity Range: 55–75 $\mu S/cm$ @ 25°C, 1:40 dilution (2.5 mL/97.5 mL D.I. H$_2$O)

2.0 SAFETY PRECAUTIONS

   2.1 Wear safety glasses, labcoat, and disposable laboratory gloves at all times during the preparation.

3.0 PROCEDURE

   3.1 Read referenced SOPs.

   3.2 Use only preapproved components as specified by the item number listed in the Formulation.

   3.3 Add components to the mixing vessel in the order listed above.

   3.4 Q.S. to final volume with D.I. H$_2$O.

   3.5 Allow to mix for minimum of 10 minutes on a stir plate or with a motorized lab stirrer (e.g., Lightning Mixer).

   3.6 Check final pH with a *calibrated* pH electrode.

   3.7 Filter Sterilize (if requested) or Autoclave (if requested).

   3.8 Store at Room Temperature (20–25°C).

(a)

Figure 22.4. Procedures to Make Solutions. a. Procedure to make DNA buffer. This procedure uses the strategy of preparing concentrated stocks and diluting them to make the desired solution. Q.S. is the same as "bring to volume." **b.** Procedure to make 2 M Tris-HCl, pH 7.3. This is one of the components of the DNA buffer.

PRACTICE PROBLEMS

1. Suppose you start with 5 X solution A and you want a final volume of 10 μL of 1 X solution A. How much solution A do you need?

2. How would you mix 75 mL of 95% ethanol if you have 250 mL of 100% ethanol?

3. How would you mix 25 mL of 35% acetic acid from 25% acetic acid?

4. If you have only 65 mL of 0.3 M Na$_2$SO$_4$, how much 0.1 mM solution can you make?

5. a. How would you prepare a 0.1 M stock solution of Tris buffer, pH 7.9? The FW of Tris is 121.1.

   b. How would you use this Tris stock to prepare a 10 mM solution of Tris, pH 7.9?

6. Using the information in Table 22.1, suggest a buffer to maintain a system at pH 8.5.

7. Calculate the pH of a Tris buffer solution that is pH 7.5 at 25°C and then is brought to a temperature of 65°C.

8. Write a procedure to first prepare stock solutions and then to use these stock solutions to prepare Breaking Buffer.

> ***Stock Solutions***
> 1 M Tris, pH 7.6 at 4°C
> 1 M Mg acetate
> 1 M NaCl
> ***Breaking Buffer***
> 0.2 M Tris, pH 7.6 at 4°C
> 0.2 M NaCl
> 0.01 M Mg acetate
> 0.01 M β-mercaptoethanol
> 5% (v/v) Glycerol

NaCl FW = 58.44

β-Mercaptoethanol FW = 78.13, comes as a liquid, density = 1.1 g/mL

Tris FW = 121.1

Magnesium acetate•4H$_2$O FW = 214.40

1.0 FORMULATION   2M TRIS-HCL, pH 7.3

Component	Item Number	FW or Conc.	Amount/L	Final Concentration
Trizma HCl	014426	157.6	274.0 g	1.74M
Trizma Base	014393	121.1	32.0 g	0.26M
D.I. H$_2$O		—	Q.S.	—

Target pH: 7.30–7.35 @ 25°C
Conductivity Range: 150–230 $\mu S/cm$ @ 25°C, 1:1000 dilution (100 μL/100 mL D.I. H$_2$O)

2.0 SAFETY PRECAUTIONS

2.1 Wear Safety Glasses, Labcoat, and disposable laboratory gloves at all times during the preparation.

3.0 PROCEDURE

Note: Quantities of Tris HCl and Tris Base are obtained from the Trizma Mixing Table (see Sigma Trizma Technical Bulletin No. 106B). Values given in the table are specific for a 0.05M solution. Multiply values by 40 to determine quantities of Tris HCl and Tris Base needed for a 2M solution. Increasing the total Tris concentration from 0.05M to 0.5M will increase the pH by about 0.05 pH units, and increasing the total Tris concentration to 2M generally causes an increase of about 0.05 to 0.10 pH units.

Note: For Tris buffers, *as Temperature increases, pH decreases.* As a general rule for Tris Buffers: *pH changes an average of 0.03 pH units per °C below 25°, and pH changes an average of 0.025 pH units per °C above 25°C.*

3.1 Read referenced SOPs.

3.2 Use only preapproved components as specified by the item number listeed in the Formulation.

3.3 Start with D.I. H$_2$O at approximately 60% of the total volume to be prepared. Add components in the order listed above, and allow to mix until dissolved. *Note:* Mixing of Tris components in water results in an endothermic reaction, and the temperature of the solution cools significantly.

3.4 Bring volume up to approximately 90% of the final volume, allow to mix and to warm up to near Room Temperature (20–25°C).

3.5 Check pH with a *calibrated* pH Electrode near Room Temperature (20–25°C).

3.6 If the pH is within ± 0.05 units @ 25°C, Q.S. to final volume with D.I. H$_2$O and allow to mix for a minimum of 5 minutes.

   3.6.1 If pH is not within accepatable limits, adjust carefully with concentrated HCl until pH is within acceptable limits. Q.S. to final volume with D.I. H$_2$O and allow to mix for a minimum of 5 minutes.

3.7 Check final pH with a *calibrated* pH electrode.

3.8 Check pH of a 0.05M sample (2.5 mL of Tris Buffer diluted to 100 mL).

3.9 Check Conductivity of the 2M solution @ 1:1000 dilution.

3.10 Filter to remove extraneous particles and dirt.

3.11 Store at Room Temperature (20–25°C).

3.12 *Note:* For large volume preparations (100 liters or more), heat approximately 30 liters of D.I. H$_2$O per 100 liters to approximately 80–90°C. Weigh out Tris powders and add to mix tank. Add approximately 10 liters of D.I. H$_2$O to the Tris powders. Add the 30 liters of heated D.I. H$_2$O. Bring to 90% of the total volume to be prepared. Perform steps 3.3 through 3.9 above.

(b)

Figure 22.4. *continued*

Notes:

i There are a couple of points to note regarding the Tris. The recipe specifies that the Tris buffer should be pH 7.6 at 4°C; therefore, bring it to the proper pH at this specified temperature.

ii β-Mercaptoethanol comes as a liquid. It can be weighed out in grams, but it is much easier to measure β-mercaptoethanol with a pipette. The density of β-mercaptoethanol is 1.1 g/mL. This means that 1 mL of β-mercaptoethanol weighs 1.1 g.

iii β-Mercaptoethanol is extremely volatile, so there are safety precautions involved in its use. Refer to its MSDS (Material Safety Data Sheet) to find out how to use this chemical safely. It is a good policy to check safety precautions whenever working with a new chemical (see Chapter 2).

iv Magnesium acetate comes as a tetrahydrate with four bound waters. The molecular weight given in the problem is for the tetrahydrate form.

9. Deoxynucleotide triphosphates (dNTPs) are the building blocks of DNA. There are four deoxynucleotide triphosphates: dATP, dTTP, dGTP, and dCTP. Suppose that you are going to repeatedly perform a procedure that calls for 10 μL of a solution that contains a mixture of:

1.25 mM each of dATP, dTTP, dGTP, and dCTP

You do not have any deoxynucleotide triphosphates in your laboratory, so you are going to buy some. You look in a catalogue and find the following entry:

DESCRIPTION: Deoxynucleotide triphosphates are provided at a concentration of 100 mM in water.

Product	Amount in Vial	Catalogue #
dATP	40 μmoles	XXXA
dCTP	40 μmoles	XXXC
dGTP	40 μmoles	XXXG
dTTP	40 μmoles	XXXT

a. Observe that the four nucleotides are purchased separately and will then need to be combined into one solution. Decide how many vials of each nucleotide to purchase. Assume you will make enough of the solution to perform the procedure 100 times.

b. Outline how you would prepare the nucleotide solution.

DISCUSSION QUESTIONS

1. Explain the concept of traceability as illustrated by the form in Figure 22.3.

2. Explain how the buffer in Figure 22.4b is brought to the proper pH.

3. The following question is modified from Scenario 4 in the "Skill Standard for the Biosciences Industries" document (Educational Development Corporation, 1995).

a. You have been given a procedure to prepare a solution. Discuss the considerations important in preparing a solution.

b. One of the materials that you require to make the solution must be purchased from a specific vendor. You find out, however, that the vendor is out of this particular chemical. What should you do?

ANSWERS TO PRACTICE PROBLEMS

1. $C_1 V_1 = C_2 V_2$

5 X (?) = 1 X (10 μL)

? = 2 μL

You need 2 μL of solution A.

2. $C_1 V_1 = C_2 V_2$

100% ? = 95% (75 mL)

? = 71.25 mL

Take 71.25 mL of 100% ethanol and BTV 75 mL. (The number 250 in the problem is extraneous.)

3. You cannot make a more concentrated solution from a less concentrated one.

4. $C_1 V_1 = C_2 V_2$

(0.3 M) (65 mL) = (0.0001 M) (?)

? = 195,000 mL = 195 L

You can make quite a lot—195 L!

5. a. Decide how much stock solution to make. For example, 1 L.

Using the formula given previously:

(121.1 g/mole) (0.1 mole/L) (1 L) = 12.11 g

Weigh out 12.11 g of Tris and dissolve in about 900 mL of water. Bring the solution to the proper pH using HCl. Then, bring the solution to 1 L. (Assume the solution is to be prepared and used at room temperature, since no temperature was specified.)

b. Decide how much solution to make, for example, 100 mL.

$$C_1 V_1 = C_2 V_2$$
$$0.1 M (?) = 0.01 M \ 100 \ mL$$
$$? = 10 \ mL$$

So, use 10 mL of Tris stock and BTV 100 mL. The pH may change slightly because of dilution, but should be acceptable for most applications. Record the final pH.

6. The pK_a of the buffer should be between 7.5 and 9.5. Any of the buffers in that range (e.g., HEPES or Tris) are acceptable on the basis of pK_a.

7. The pH of a Tris solution will decrease by 0.028 pH units for every degree increase in temperature. In this example the temperature increases 40°C. The pH therefore decreases 1.12 pH units, to 6.4.

8. Stocks

1 M Tris (pH 7.6 at 4°C).

To make 1 L of stock, dissolve 121.1 g of Tris base in water. After the Tris is dissolved, bring its temperature to 4°C by submerging it in an ice bath or placing it in the refrigerator. While the Tris is at this temperature, bring the pH to 7.6 with concentrated HCl. (Tris buffer will change pH as the temperature changes. As the temperature rises, the pH falls. The buffer will therefore require less HCl at 4°C than at room temperature.) Then BTV 1 liter.

1 M Mg acetate. To make 1 L of stock, dissolve 214.40 g of magnesium acetate (tetrahydrate) in water. Bring to volume.

1 M NaCl. To make 1 L of stock, dissolve 58.44 g in water. Bring to volume.

Breaking Buffer
To make 200 mL:

0.2 M TRIS	40 mL of 1 M TRIS (pH 7.6 at 4°C.)
0.2 M NaCl	40 mL 1 M NaCl
0.01 M Mg acetate	2 mL 1 M Mg acetate
0.01 M β-mercaptoethanol	142 μL (see note)
5% Glycerol	10 mL glycerol
Bring to volume of 200 mL with water.	

Note: You want 200 mL of a 0.01 M solution of β-mercaptoethanol, so you need 0.156 g. You can use a proportion to figure out how many milliliters will contain 0.156 g:

$$\frac{1.100\ g}{1\ mL} = \frac{0.156\ g}{?}$$

$? = 0.142\ mL = 142\ \mu L$

You therefore need 142 μL of β-mercaptoethanol.

9. Step 1. Each time a reaction is performed, 10 μL of the dNTP mixture is needed. You want to perform the reaction 100 times, therefore, you will need 10 μL × 100 = 1000 μL.

Step 2. Consider only a single dinucleotide, for example, how much of the dATP stock do you need? This is a $C_1 V_1 = C_2 V_2$ problem.

C_1	V_1	=	C_2	V_2
100 mM	?		1.25 mM	1000 μL

You need 12.5 μL of the dATP. You will similarly need 12.5 μL of the other three dNTPs.

Step 3. Combining 12.5 μL of each dNTP gives a volume of 50 μL. The remaining solution will be water:

```
 12.5 µL 100 mM dATP
 12.5 µL 100 mM dCTP
 12.5 µL 100 mM dGTP
 12.5 µL 100 mM dTTP
 950.0 µL water
1000   µL total solution
```

Note that it is difficult to bring a solution to a volume of 1000 μL. In this case it is correct to calculate that you need 950 μL of water.

Step 4. How many vials of each dNTP will you need to purchase? The volume of each vial is not listed in the catalogue. Rather, it tells you that the concentration of dNTP in each vial is 100 mM and there are 40 μmoles (0.040 millimoles) in each vial. So what is the volume per vial?

A concentration of 100 mM means that there are 100 millimoles of the dNTP per liter. So, what volume contains 0.040 millimoles?

$$\frac{100\ mmoles}{1000\ mL} = \frac{0.040\ mmoles}{?}$$
$$? = 0.400\ mL = 400\ \mu L$$

Each vial contains 400 μL of a given dNTP at a concentration of 100 mM. (This is a good time to review the difference between the words "amount" and "concentration.") One vial has more than enough dNTP to make the solution desired; purchase one vial of each dNTP.

## CHAPTER 23

Laboratory Solutions to Support the Activity of Biological Macromolecules and Intact Cells

I. INTRODUCTION

Chapters 21 and 22 discussed many principles, calculations, and other details regarding how to prepare laboratory solutions. For example, pp. 473–474 explained how to calculate the required amount of each component of SM buffer. We have not, however, addressed the question of *why* SM buffer has the various components that it does. This chapter explores the purposes of the components used in solutions in biology laboratories.

Biologists work in the laboratory with materials derived from organisms. These materials include various proteins, DNA, RNA, and cellular organelles. Biologists also work with intact cells and with microorganisms in culture. The activity of all these biological materials is intimately associated with their structure; in fact, the relationship between structure and function is one of the most important themes in biology. For example, the function of enzymes is to catalyze reactions. For catalysis to occur, the enzyme must recognize and bind the reactants involved. Recognition and binding occur because the enzyme's structure complements that of the reactants, Figure 23.1. Enzymes do not function if their normal structure is degraded. Thus, it is critical that laboratory solutions provide conditions that preserve the structural integrity of biological molecules.

There are many different solutions used in the biology laboratory. These solutions differ from one another because the requirements of various biological systems vary. For example, a solution to support the structure and activity of an enzyme is likely to differ from a solution for DNA. A solution that is used to store a material is likely to differ from a solution for isolating that material.

This chapter begins by briefly describing the structure and function of proteins. Then, the types of laboratory solutions that support proteins are explored. The second part of this chapter briefly discusses the structure and function of nucleic acids (DNA and RNA) followed by an exploration of how nucleic acids are supported in biological solutions. The final part of this chapter discusses the requirements for solutions that sustain the activity of intact cells and microorganisms.

II. MAINTAINING THE STRUCTURE AND FUNCTION OF PROTEINS IN LABORATORY SOLUTIONS

A. An Overview of Protein Function and Structure

i. PROTEIN FUNCTION AND AMINO ACIDS: THE SUBUNITS OF PROTEINS

The first part of this chapter discusses the solutions used when working with proteins in the laboratory. In order to understand why these solutions have the com-

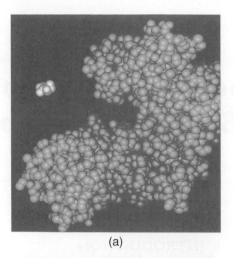

(a)

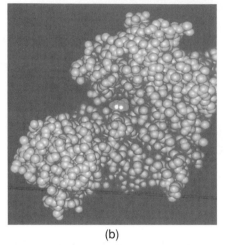

(b)

Figure 23.1.   Example of the Relationship between Protein Structure and Function. An enzyme and its substrate. The smaller substrate molecule **a.** binds within a groove in the larger enzyme that is complementary in shape and ionic properties **b.** Many proteins bind other molecules at sites whose shapes are complementary. *(See color insert C-2.)*

ponents they do, it is necessary to know something about protein function and structure.

Proteins perform a myriad of essential functions in cells, Table 23.1. This functional diversity is possible because proteins are structurally diverse. As with the English language, where an immense number of words with varied meaning are formed using only 26 letters, so diverse proteins are formed using 20 different amino acid subunits. Each of the 20 amino acids has the same core structure consisting of an **amino group**, NH_2 and a **carboxyl group**, $COOH$, attached to a central carbon atom, Figure 23.2a. What makes each amino acid distinct is the presence of a characteristic **side chain**, or **R group**, which has a particular structure and chemical properties. The diversity and versatility of proteins is due to the wide variety of characteristics conferred on proteins by the various amino acid R-groups.

A protein is formed by amino acid subunits linked together in a chain. *The bonds that connect the amino*

Table 23.1 PROTEIN FUNCTIONS

Function	Examples
Metabolism	*Enzymes* catalyze reactions
Cellular Control	*Hormones* affect metabolism of target cells
Structure	*Keratin* forms the structure of hair and nails
Storage	*Ovalbumin* (egg white) stores nutrients
Transport	*Hemoglobin* transports O_2 and CO_2
Mobility	*Actin* and *myosin* cause muscle contraction
Defense	*Antibodies* destroy invading microorganisms
Gene Regulation	*Repressor proteins* in bacteria "turn off" genes
Recognition	*Receptors* recognize specific hormones

acids to one another are **covalent peptide bonds**, Figure 23.2b. Every protein has a different sequence of amino acids that determines the final structure and function of the protein, Figure 23.2c.

ii. The Four Levels of Organization of Protein Structure

Every protein has a distinct, complex three-dimensional structure. To describe these varied protein structures, people speak of four levels of organization: primary, secondary, tertiary, and quaternary, Figure 23.3. Secondary, tertiary, and quaternary structure are sometimes collectively referred to as "higher order" structure.

1. **Primary Structure** *refers to the linear sequence of amino acids that comprise the protein.*

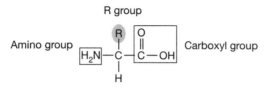

(a) Core amino acid structure

(b) Peptide bond

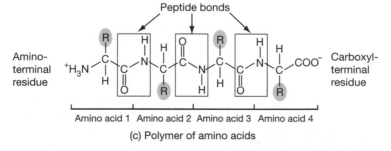

(c) Polymer of amino acids

Figure 23.2. Protein Structure. a. The core structure of an amino acid consists of an amino and a carboxyl group attached to a central carbon atom. Each of the 20 amino acids has a different R group attached. **b.** Proteins are polymers of amino acids joined by peptide bonds. A peptide bond is a covalent bond formed when the carboxyl group of one amino acid joins with the amino group of a second amino acid. **c.** Every protein has a different sequence of amino acids that determines the structure and function of the protein.

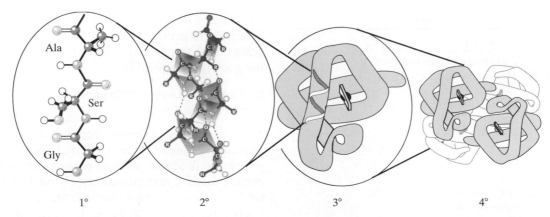

Figure 23.3.   The Four Levels of Protein Structural Organization. Four levels of protein structure: Primary (1°) structure is the sequence of amino acids, represented here by three-letter codes. Certain proteins or sections of proteins form regularly repeating patterns, called secondary (2°) structure. Many proteins require complex tertiary (3°) folding to provide the right structure and conformation for recognition, binding, and solubility. Quaternary (4°) structure occurs when folded chains associate with one another to form complexes.

2. **Secondary Structure** *refers to regularly repeating patterns of twists or kinks of the amino acid chain.* Two common types of secondary structure are called the **α-helix** and **β-pleated sheet**. *Regions of proteins that do not have regularly repeating structures are often said to have a* **random coil** *secondary structure, although their folding patterns are not truly random.* Secondary structure is held together by weak, noncovalent, molecular interactions, called *hydrogen bonds*, between nonadjacent amino acids. **Hydrogen bonds** *form when a hydrogen atom that is bonded to an electronegative atom (like F, O, or N) is also partially bonded to another atom (usually also F, O, or N).* Thus, a hydrogen bond forms because a hydrogen atom is shared by two other atoms.

3. **Tertiary Structure** *refers to the three-dimensional globular structure formed by bending and twisting of the protein.* Globular structures include regions of α-helices, β-sheets and other secondary structures that are folded together in a way that is characteristic of that protein. Tertiary structure is generally stabilized by multiple weak, noncovalent interactions. These include:

   - **Hydrogen bonds** that form when a hydrogen atom is shared by two other atoms.

   - **Electrostatic interactions** that occur between charged amino acid side chains. **Electrostatic interactions** *are attractions between positive and negative sites on macromolecules.* These weak interactions bring together amino acids and stabilize protein folding, Figure 23.4.

   - **Hydrophobic interactions** *that reflect the tendency of the aqueous environment of the cell to exclude hydrophobic amino acids.* Proteins spontaneously fold in such a way that hydro-

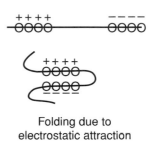

Folding due to
electrostatic attraction

Figure 23.4.   Electrostatic Interactions and Protein Folding. Interactions between positively and negatively charged amino acids are called electrostatic interactions. Proteins have various arrangements of positive and negative charges due to different amino acid side groups. Electrostatic interactions between charged side groups lead to folding of the protein.

phobic amino acids are inside, protected from the aqueous environment. In contrast, hydrophilic amino acids tend to be found on the protein's surface.

In addition to these weak, noncovalent interactions, some proteins have covalent **disulfide bonds** which stabilize their three-dimensional structure. **Disulfide bonds** *form when sulfurs from two cysteine molecules (cysteine is an amino acid) bind to one another with the loss of two hydrogens,* Figure 23.5. Disulfide bond formation is an example of oxidation. **Oxidation** *is defined as the removal of electrons from a substance.* In biological systems, oxidation often involves the removal of an electron accompanied by a proton (i.e., a hydrogen atom).

4. **Quaternary Structure** *refers to the complex formed when two or more protein chains associate with one another.* Some proteins have separately synthesized subunit chains that join together to form a functional protein complex.

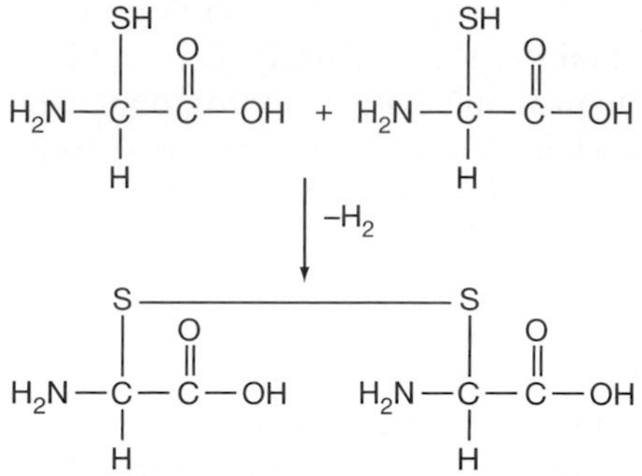

Figure 23.5.   Disulfide Bonds. Disulfide bonds contribute to tertiary folding. A disulfide bond is formed by two cysteine molecules.

The covalent bonds that link amino acids into their primary chain structure are much stronger than the noncovalent interactions that stabilize higher order structure. Relatively little energy (e.g., in the form of heat) is needed to disrupt the weaker noncovalent bonds. Higher-order structures persist because a large number of weak interactions stabilize them.

The "weakness" of the noncovalent interactions confers two important characteristics on proteins. First, proteins are flexible, and this flexibility is critical to their biological function. For example, enzymes change shape in order to bind to their substrates. Second, the normal folding and shape of proteins is readily disrupted by changes in temperature, pH, and other conditions. There are many circumstances in the laboratory, therefore, where the primary structure of the amino acid chain remains intact but the complex shape of the protein is lost.

B. How Proteins Lose Their Structure and Function

Laboratory solutions for proteins must provide suitable conditions so that the proteins' normal structure, and therefore their normal activity, is maintained. To understand how the structure of proteins is protected in laboratory solutions, it is necessary to understand how that structure can be destroyed. There are various ways by which proteins lose their normal structure and therefore their ability to function normally:

1. **Proteins can *denature* or unfold so that their three-dimensional structure is altered but their primary structure remains intact.** Because the amino acid side chain interactions that stabilize protein folding are relatively weak, environmental factors including high temperature, low or high pH, and high ionic strength can cause denaturation.

   There are many laboratory situations where denaturation occurs reversibly under one set of circumstances and proteins regain their normal conformation when suitable conditions are restored. Denaturation can also be irreversible, as when eggs are hard boiled. When boiled, the translucent egg protein, albumin, is denatured and becomes hard egg white.

2. **The primary structure of proteins can be broken apart by enzymes, called *proteases*, that digest the covalent peptide bonds between amino acids.** *This digestive process is called* **proteolysis.** Cells contain proteases sequestered in membrane-bound sacs called *lysosomes.* When cells are disrupted, the lysosomes break and release their proteases. In the laboratory, it is therefore necessary to minimize the activities of cellular proteases to protect proteins from proteolysis. Methods used to minimize proteolysis include low temperature, short working times, and addition of chemicals that inhibit proteases. Microbes also release proteases and, therefore, antimicrobial agents are often added to solutions to protect proteins.

3. **Sulfur groups on cysteines may undergo oxidation to form disulfide bonds that are not normally present.** Extra disulfide bonds can form when proteins are removed from their normal environment and are exposed to the oxidizing conditions of the air. Because disulfide bonds affect protein folding, additional disulfide bonds may produce undesirable changes in protein tertiary structure. Disulfide bond formation is an oxidation reaction; therefore, antioxidizing chemicals, called **reducing agents**, are often added to protein solutions to prevent unwanted disulfide bonding.

4. **Proteins can aggregate with one another, leading to precipitation from solution.** Each protein requires specific conditions in order to remain in solution.

5. **Proteins readily adsorb (stick to) surfaces and their activity is then lost.**

C. The Components of Laboratory Solutions for Proteins

Laboratory solutions for proteins must provide them with suitable conditions. What constitutes "suitable conditions," however, varies greatly. Because proteins differ from one another the conditions for manipulating one protein may not work for another. Even when considering a single protein, the optimal conditions differ depending on the situation. For example, the conditions under which a protein expresses maximal activity usually differ from the conditions required to isolate the protein. Yet a different set of conditions is usually required for storing a protein. Thus, the solution requirements for each protein and each task must be determined individually, often by experimentation.

Chemical agents that are used in protein solutions may perform more than one role depending on the situation. Moreover, the same agent may be desirable in one situation and problematic in another. For example, detergents play a familiar, useful role as cleaning agents. Detergents can also act to solubilize or denature proteins. Detergents are sometimes intentionally added to solutions because their effects are useful. At other times, detergents are scrupulously removed from solutions because their effects are detrimental.

Although laboratory solutions for proteins vary significantly depending on the circumstances, there are classes of agents commonly found in protein solutions. These classes of agents are summarized in Table 23.2 and general principles regarding these agents will be discussed.

i. PROTEINS AND pH: THE IMPORTANCE OF BUFFERS

Proteins that reside in the cell's cytoplasm normally have many charged amino acid side groups that interact with one another and affect the structure of the protein. The charges on these side groups are affected by the pH of their solution. The solution pH, therefore, affects the three-dimensional structure of proteins, their activity, and their solubility. There is no universal rule as to what pH is optimum, although pH extremes generally dena-

Table 23.2 CLASSES OF AGENTS USED IN LABORATORY SOLUTIONS FOR PROTEINS

Buffers
Salts
Cofactors
Detergents
Organic solvents
Denaturants and solubilizing agents
Precipitants
Reducing agents
Metal chelators
Antimicrobial agents
Protease inhibitors
Storage stabilizers

ture proteins. As a rule, the pH of a laboratory solution should be maintained as closely as possible to the pH found in the protein's native environment, which is typically somewhere between pH 6 and 8. A **buffer** is used in solutions to maintain the proper pH.

ii. SALTS

Salts *are compounds made up of positive and negative ions.* Biological systems frequently require salts, so many laboratory solutions contain salts at a specific concentration.* **Ionic strength** *is a measure of the charges from ions in an aqueous solution.* The general importance of ionic strength to biological molecules in laboratory solutions can be described as follows. Due to electrostatic interactions, two macromolecules in solution tend to attract one another if they have opposite charges, tend to repulse one another if they have the same charges, and have neither interaction if the molecules are neutral. Ions in a solution also interact with macromolecules based on charge. High concentrations of ions in a solution will tend to counteract or shield opposite macromolecular charges.

In protein solutions, therefore, salts can affect the electrostatic interactions between the side chains of amino acids. At low ionic strengths, the ions in solution have little effect on these interactions, Figure 23.6a. Increased ionic strength tends to stabilize protein structures by shielding unpaired charged groups on protein surfaces. At very high ionic strengths, many proteins are denatured, presumably because normal sites of electrostatic interaction between charged groups on amino acids are neutralized by the salt ions, Figure 23.6b.

The ionic strength of a solution affects both the three-dimensional structure of proteins and their solubility. Most cytoplasmic proteins are soluble at salt concentrations of about 0.15–0.20 M but often precipitate

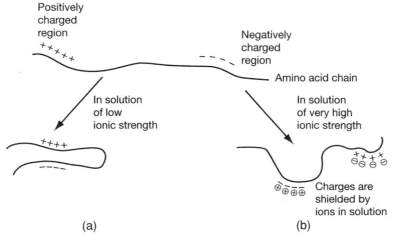

(a)   (b)

Figure 23.6.   The Effect of Ionic Strength on Macromolecular Surface Charges. a. A solution of low ionic strength. **b.** A solution of high ionic strength.

*It is not surprising that salts play many roles in biological systems; after all, life evolved in the sea. This ancestry is reflected in the fact that salts perform a number of essential roles in organisms and salt levels are rigorously controlled in living systems.

at higher or lower salt concentrations. This phenomenon is frequently exploited in the laboratory to selectively isolate proteins from other solution components. Salt is added to the solution until the proteins of interest precipitate. The precipitated proteins can then be retrieved. Salt-precipitated proteins usually regain their normal structure and solubility when the salt is removed. High salt concentrations retard microbial action; therefore, proteins are sometimes stored as salt precipitates.

The salt concentration in a solution can also be manipulated to control the binding of proteins to a solid matrix. Certain solid materials are designed so that they bind proteins when particular conditions of salt concentration and pH are met. Under these conditions, a desired protein may selectively bind to the solid matrix, leaving contaminants behind in the solution. The isolated protein is then recovered by changing the salt concentration so that the protein leaves the matrix and rinses into fresh buffer solution.

The exact effects of ionic strength on a protein depend on the protein, the type of salts involved, and the pH of the solution. Recipes for protein solutions therefore specify the pH, the type of salts, and their concentrations.

A summary of the roles that salts play in protein solutions includes:

1. **Salts affect the three-dimensional structure of proteins and may be manipulated to stabilize structure or to intentionally denature a protein.**

2. **Salts affect the solubility of proteins and may be manipulated to keep proteins in solution or to cause them to precipitate from solution.**

3. **Salts may be used to control the binding of a protein to a solid matrix.**

4. **Salts may be used to control the binding of proteins to other macromolecules** (discussed on pp. 498–499).

5. **Salts may be added as cofactors for enzymatic reactions**, as discussed in the next section.

iii. COFACTORS

Cofactors are *chemical substances that many enzymes require for activity.* Cofactors include metal ions, such as Fe^{++}, Mg^{++}, and Zn^{++}. Cofactors may also be complex organic molecules called **coenzymes**, such as **nicotinamide adenine dinucleotide (NAD)** and **coenzyme A**. Cofactors are sometimes added to or removed from protein solutions. For example, Mg^{++} is a cofactor for nucleases, enzymes that digest DNA and RNA. In some situations Mg^{++} is added to solutions while in other applications it is carefully avoided.

iv. DETERGENTS

Many types of detergents are used in the laboratory. **Detergents** *are a class of chemicals that have both a hydrophobic "tail" and a hydrophilic "head."* Detergents therefore have two natures: They can be both water-soluble and lipid-soluble. *Some detergents have hydrophilic portions that are ionized in solution* (**ionic detergents**) *whereas others have hydrophilic sections that are not ionized in solution* (**nonionic detergents**). **SDS (sodium dodecyl sulfate)** is an example of an ionic detergent; **Triton X-100** is a nonionic detergent.

Detergents interact with and modify the structure of proteins. Detergents are sometimes intentionally added to protein solutions for several reasons:

1. **Detergents may be used to solubilize proteins that are associated with lipid membranes.** Proteins that are associated with lipid membranes tend to be hydrophobic and to be difficult to bring into solution. Because detergents are both soluble in water and have hydrophobic properties, they can interact with and increase the water solubility of hydrophobic proteins. Nonionic detergents are generally preferred for solubilizing membrane-bound proteins because they tend not to be denaturing.

2. **Detergents are sometimes added to solutions to intentionally denature proteins.** Ionic detergents tend to effectively denature proteins. Denaturants are discussed below.

3. **Detergents may be added in low levels to solutions to prevent protein aggregation and prevent proteins from adsorbing to surfaces, such as glass.**

Some cautions regarding detergents are:

• *Some detergents, such as Triton X-100, absorb UV light and can interfere with spectrophotometric assays.*

• *Commercial detergents intended for cleaning may contain contaminating substances that affect proteins.* To avoid such contaminants, manufacturers purify detergents that are intended for protein work.

• *Detergents denature proteins and therefore detergent residues from cleaning glassware and equipment must be removed.*

v. ORGANIC SOLVENTS

Organic solvents, such as **ethanol** and **acetone**, are sometimes added to biological solutions. Organic solvents may be used to solubilize membrane proteins and occasionally to cause them to precipitate from a solution as part of a purification strategy. Organic solvents, however, cause most proteins to denature irreversibly and should generally be avoided when protein activity must be retained.

vi. DENATURANTS, SOLUBILIZING AGENTS, AND PRECIPITANTS

There are situations where it is desirable to denature the higher order structure of proteins. For example, polyacrylamide gel electrophoresis (PAGE) is a method used to separate proteins from one another by forcing them to migrate through a gel-like matrix in the presence of an electrical current. Before electrophoresis, the proteins are intentionally denatured with the detergent SDS so that their three-dimensional structures do not affect their movement in the gel. Ionic detergent is preferred for this task because it is a more effective denaturing agent than nonionic detergent. (In addition to denaturing proteins, SDS also confers a uniform negative charge on all the proteins so that differences in their charges do not affect their migration in the gel. When SDS is used with PAGE, proteins separate from one another based on differences in their size.) Proteins do not need to maintain their normal activity during electrophoresis, so denaturation is not a problem.

Urea at high concentrations (4–8 M) and **guanidinium salts** are other effective denaturants. Urea is frequently used when proteins are separated from one another by a method called *isoelectric focusing*. Like PAGE, isoelectric focusing requires that the proteins be completely unfolded. SDS, however, is not used in isoelectric focusing because SDS confers a charge on the proteins that interferes with the focusing method.

Urea, SDS, and guanidinium salts not only denature proteins, they also cause them to be soluble in aqueous solutions. Some proteins are not normally soluble in laboratory solutions and are therefore very difficult to manipulate in the laboratory. For example, bacteria may produce valuable recombinant proteins but sequester them in insoluble inclusion bodies. Guanidinium hydrochloride can be used in recovering the desired recombinant proteins from the bacterial inclusion bodies (see, for example, "SDS-Polyacrylamide Gel Electrophoretic Analysis of Proteins in the Presence of Guanidinium Hydrochloride." James L. Funderburgh and Sujatha Prakash. *BioTechniques*, 20:376–378, March, 1996).

We have just seen that some agents that denature proteins cause the proteins to be soluble in aqueous solutions. There are also agents that simultaneously denature and cause proteins to precipitate out of solution. **TCA (trichloroacetic acid)**, and the organic solvents ethanol, acetone, methanol, and chloroform are examples of agents that both denature and precipitate proteins. Such precipitation can be exploited in the laboratory to separate large macromolecules (including proteins, DNA, and RNA) from smaller molecules such as salts, amino acids, and nucleotides. Simultaneous denaturation and precipitation is useful, for example, when a protein is to be analyzed for its amino acid content. In this case, maintaining the protein's activity and three-dimensional structure is unnecessary.

There are situations where it is desirable to precipitate a protein, but the protein will later be required to exhibit normal activity. In such cases, precipitants such as **ammonium sulfate** and **PEG (polyethylene glycol)** are used. These agents cause proteins to precipitate without

irreversibly denaturing them. Ammonium sulfate is inexpensive and is commonly used to separate proteins from one another during protein purification procedures. As noted earlier, salt-precipitated proteins generally recover their normal activity when the salt is removed.

vii. Reducing Agents

Recall that when proteins are exposed to the oxidizing conditions of the air, sulfur groups on cysteines may undergo oxidation to form unwanted disulfide bonds. Laboratory workers sometimes add **reducing agents** to solutions to simulate the reduced intracellular environment and prevent unwanted disulfide bond formation. Common reducing agents are **β-mercaptoethanol** and **DTT (dithiothreitol)**.

viii. Metals and Chelating Agents

Metal ions, such as Ca^{++} and Mg^{++}, are frequently present when cells are disrupted and are often introduced as contaminants in reagents and water. Metals can accelerate the formation of undesired disulfide bonds and can act as cofactors for proteases. **Chelating agents**, therefore, are often added to laboratory solutions *to bind and remove metal ions from solution*. The most common chelator is **EDTA (ethylenediaminetetraacetic acid)**.

ix. Antimicrobial Agents

Low concentrations of antimicrobial agents, such as **sodium azide**, are used to kill microbes and to prevent proteolysis caused by microbes. Sodium azide is toxic and may affect the activity of some proteins.

x. Protease Inhibitors

Proteases are enzymes that are released into the cell extract when cells are disrupted. Until the proteins of interest are purified from the extract, protease inhibitors may be necessary to prevent protein degradation. EDTA is considered a protease inhibitor because it removes metal ion cofactors from solution. **PMSF (phenylmethanesulfonyl fluoride)** and **pepstatin** are examples of agents used to inhibit specific classes of proteases. Protease activity is also reduced by keeping protein solutions cold and working quickly to isolate proteins of interest from the cell extract.

D. Physical Factors Relating to Proteins in Solution

i. Temperature

Temperature has profound effects on proteins. Most proteins exhibit their maximum biological activity at physiological temperatures (30–37°C). Extremely high temperatures denature proteins, as when eggs are hardboiled. The optimal temperature for a protein depends on the situation. For example, the optimal storage temperature is seldom the same temperature at which a protein has maximal activity. Cold temperatures can be used to reduce microbial growth and slow the degradation of proteins by proteolytic enzymes. Ice water baths are used in the laboratory to maintain a biological system at 4°C. Refrigerator and freezer temperatures are used to hold biological solutions when not in use. A cold room or large refrigerator is frequently used to perform operations like enzyme purification and column separations.

Table 23.3 shows some temperatures that are significant in the laboratory.

Table 23.3   SIGNIFICANT BIOLOGY LABORATORY TEMPERATURES

121	Autoclave temperature used to sterilize equipment and solutions
100	Water boils
90	DNA separates into two strands
65–70	Activity range of certain thermophilic bacteria
37	Temperature of warm-blooded mammals
20–25	Room temperature
4–8	Refrigerator temperature
0	Ice water bath
0	Water freezes/melts
−20	Freezer temperature
−70	Ultra low temperature freezer
−196	Temperature of liquid nitrogen

Temperatures in °C.

ii. ADSORPTION

Surface charges on proteins make them "sticky," and many proteins will **adsorb**, *stick to*, almost any surface. Glassware and plasticware provide large surfaces that can adsorb proteins from solution, thus reducing protein activity. In addition, when protein solutions are filtered (as is sometimes done to sterilize the solution), proteins may adsorb to the filters and be lost. Some ways to reduce the problem of adsorption are shown in Table 23.4.

There are situations where protein adsorption can be exploited in the laboratory. For example, Western blotting is a technique used to identify proteins based on their interactions with specific antibodies. Western blotting requires immobilizing proteins on a solid surface so that they are not washed away during rinses. Protein immobilization is accomplished by adsorbing the proteins to a nylon or plastic membrane. Another example is the selective binding of a desired protein to a solid matrix.

E. Methods of Handling Protein-Containing Solutions

Some considerations in handling solutions containing proteins are summarized in Table 23.5.

F. Summary of Protein Solution Components

Table 23.6 summarizes the components of protein solutions in tabular form.

Table 23.4   METHODS TO REDUCE PROTEIN ADSORPTION

1. *Use labware coated with silanes, agents that make the labware "slippery".* (See, for example, *Molecular Cloning: A Laboratory Manual*, J. Sambrook, E.F. Fritsch, and T. Maniatis. 2nd ed. Cold Spring Harbor Laboratory Press, 1989.)

2. *Use plastic containers and filters that are made from materials that do not adsorb proteins.* Consult the manufacturers.

3. *If possible, maintain a high concentration of protein in a solution (for example, 1 mg/mL).* In such cases a small proportion of the protein may stick to the glassware, but most is left in solution.

4. *Add inert carrier proteins, such as BSA (bovine serum albumin), to otherwise dilute protein solutions.* The inert protein sticks to and covers up the sticky sites on the glass or plasticware so that the protein of interest remains in solution.

Table 23.5   GENERAL RULES FOR PROTEIN HANDLING TO REDUCE DEGRADATION

1. *Wear gloves when handling protein solutions to avoid introducing contaminating proteases from your hands.*

2. *Mix protein solutions gently since some proteins are denatured by vigorous shaking.*

3. *Use only clean glassware.* All glassware and plasticware should be well washed and rinsed. Residual detergents or metal ions may have deleterious effects on proteins. A final rinse with EDTA may be used to eliminate metal ions.

4. *Use sterile buffers, solutions, and glassware to reduce bacterial contamination when working with or storing proteins for long times.* Consider the use of antimicrobial agents such as sodium azide.

5. *In general, keep protein-containing solutions cold in the laboratory.*

6. *Use only high-purity water to prepare solutions.*

Table 23.6 SUMMARY OF COMMON COMPONENTS OF SOLUTIONS USED TO MAINTAIN PROTEIN STRUCTURE AND FUNCTION

Type of Agent	Function	Examples
Buffering agents	Maintain pH	*Tris, phosphate, acetate buffers*
Salts	Control ionic strength Cofactors	*NaCl, $MgCl_2$*
Antimicrobial agents	Prevent microbial contamination	*Na-azide*
Antioxidants	Prevent unwanted disulfide bond formation	*β-mercaptoethanol DTT*
Protease inhibitors	Prevent proteolysis	*EDTA, PMSF, pepstatin*
Chelators	Remove metals from solution	*EDTA, EGTA**
Detergents	Solubilize membrane proteins Prevent aggregation and adsorption	*Triton X-100, SDS*
Denaturants	Denature proteins	
solubilizing		*SDS, urea, guanidinium salts*
precipitating		*acetone, phenol/chloroform*
Precipitants	Precipitate proteins	*Ammonium sulfate, PEG, TCA*
Cofactors	Required for enzyme activity	*Mg^{++}, Fe^{++}, CoA*
Storage stabilizers	Protect proteins during freezing	*Glycerol*

***EGTA** is similar to EDTA, but it has a higher binding capacity for calcium than for magnesium. This chelator is commonly used when the regulation of calcium is desired.

APPLICATION PROBLEMS

1. *EcoR1* is an enzyme that is used in the laboratory to cleave DNA at specific sites. The first recipe below is for the storage buffer used by the manufacturer to store and ship *EcoR1* enzyme. The second recipe is for the buffer used when the enzyme cleaves DNA—when maximal activity is required. When the enzyme is used in the laboratory, a very small amount of the enzyme (in storage buffer) is removed and diluted in fresh activity buffer.

   a. Suggest what the purpose of each solution component might be.

   b. Explain the differences between the solution for storage and the solution for activity. Why are the two solutions not identical?

Storage Buffer (store at −20°C)	Activity Buffer (Incubate enzyme with DNA at 37°C for 30 min)
10 *mM* Tris-HCL, pH 7.4 400 *mM* NaCl 0.1 *mM* EDTA 1 *mM* DTT 0.15% Triton X-100 0.5 *mg/mL* BSA 50% glycerol	90 *mM* Tris-HCl, pH 7.5 50 *mM* NaCl 10 *mM* $MgCl_2$

2. Examine the following recipes that come from various protein manuals. Suggest a function for the various components.

   a. Sample Buffer Used When Loading Proteins into Electrophoresis Gel
      60 *mM* Tris-HCl (pH 6.8)
      2% SDS
      14.4 *mM* β-mercaptoethanol
      0.1% bromophenol blue (dye to visualize proteins)
      25% glycerol (increases the solution density causing the proteins to sink into the gel)

   b. Restriction Enzyme Buffer (Low Salt)
      10 *mM* Tris-HCl (pH 7.5)
      10 *mM* $MgCl_2$
      1 *mM* DTT

3. Based on the answers to Questions 1 and 2, what are common components of protein solutions? What is the purpose of these components?

III. MAINTAINING THE STRUCTURE AND FUNCTION OF NUCLEIC ACIDS IN LABORATORY SOLUTIONS

A. An Overview of Nucleic Acid Structure and Function

i. DNA STRUCTURE

DNA (deoxyribonucleic acid) is a biomolecule that comprises genes. **Genes** *encode the amino acid sequences for proteins.* Progress in understanding genes is leading to numerous biotechnology products and major research

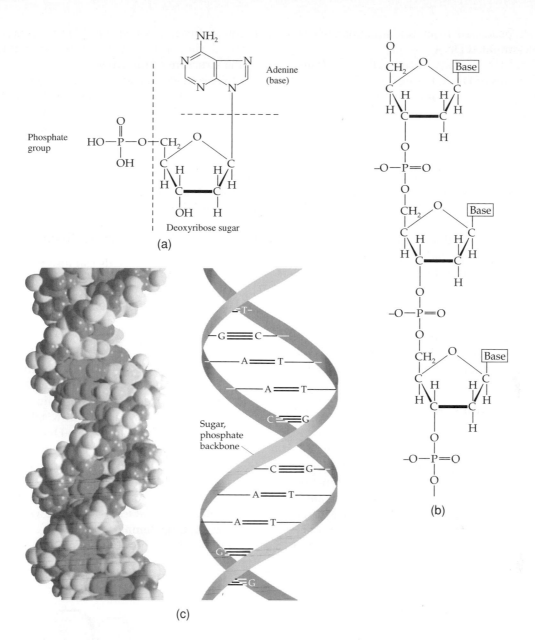

Figure 23.7.   The Structure of DNA. a. The basic subunit of DNA is the nucleotide, which consists of a sugar, a phosphate group, and a nitrogenous base, in this case, adenine. **b.** The subunits of DNA are connected by covalent phosphodiester bonds between the phosphate group on one nucleotide and the sugar on the next. The bases extend out to the side. In solution, the phosphate groups lose a H^+ and so have a negative charge. **c.** Chromosomal DNA is a double helix consisting of two strands of DNA held together by hydrogen bonds between complementary bases. Observe that three hydrogen bonds stabilize each G–C linkage, but only two hydrogen bonds stabilize each A–T linkage.

advances. At the same time, a flood of techniques have been and are being developed for working with nucleic acids in the laboratory. As with proteins, it is essential to understand the physical and chemical nature of nucleic acids when manipulating them in laboratory solutions.

DNA is a linear polymer of **deoxyribonucleotide** ("nucleotide") subunits. **Nucleotides** *are composed of a phosphate group, a five-carbon sugar (deoxyribose), and one of four nitrogenous bases (adenine, guanine, cytosine, or thymine)*, Figure 23.7a. **DNA** *consists of nucleotides connected into strands by covalent* **phosphodiester bonds**.

The bonds link the phosphate group from one nucleotide to the sugar of another. The result is a "backbone" of alternating phosphates and sugars with the nitrogenous bases sticking out to the side, Figure 23.7b.

The phosphate groups on each nucleotide readily give up H^+ ions; in fact, DNA is called an "organic acid." Thus, in solution at neutral pH, negatively charged oxygens are exposed along the backbone of DNA strands, Figure 23.7b. Nucleic acids dissolve readily in aqueous solutions because of the negatively charged backbone. The bases themselves are hydrophobic, but they face inward and are protected from the aqueous environment in double-stranded DNA.

Chromosomal DNA exists in the cell as a "double helix" of two strands of DNA, Figure 23.7c. The bases of the two strands pair with one another so that a cytosine on one strand is always across from a guanine on the opposite strand and an adenine is always across from a thymine. Cytosine and guanine are said to be "complementary"; adenine and thymine are also "complementary." Double-stranded DNA is like a ladder with the bases forming the rungs on the inside and the phosphate-sugar backbone forming the outside rails. Imagine that the ladder is taken and twisted—this twist causes DNA to be called a double "helix."

Complementary pairs of bases are held together by hydrogen bonds. Recall that hydrogen bonds are relatively weak compared to covalent bonds. Therefore, the phosphodiester bonds that link the nucleotide backbone are much stronger than the bonds holding the two strands together. This means that there are many conditions in the laboratory where the two DNA strands separate from one another, yet the strands themselves remain intact.

ii. RNA STRUCTURE AND FUNCTION

RNA (ribonucleic acid) *is a polymer of nucleotides similar to DNA except that the sugar is ribose and the nitrogenous base uracil is present in place of thymine.* There are three types of RNA with different functions and structures: messenger RNA (mRNA), transfer RNA (tRNA), and ribosomal RNA (rRNA). Messenger RNA carries information about the amino acid sequence of a protein from the nucleus to the cyto-

plasm. Transfer RNA shuttles amino acids to the ribosomes during protein synthesis. Ribosomal RNA is part of the structure of the ribosomes.

RNA is single-stranded and is shorter than chromosomal DNA. Although RNA is normally single-stranded, complementary bases *within* an RNA strand sometimes pair with one another. These bonds and other weak interactions cause RNA to fold into various complex conformations, Figure 23.8. As with proteins, the varied conformations of RNA molecules are important in controlling their functions. RNA can also pair with complementary strands of DNA.

B. Loss of Nucleic Acid Structure

Like proteins, nucleic acids can lose their normal structure in several ways:

1. **DNA can be denatured.** When speaking of DNA, **denaturation** *means that the two complementary strands separate from one another.* The relatively weak hydrogen bonds that hold together the two strands can be disrupted by changes in pH (pH greater than 10.5 will denature DNA), high temperature, or addition of organic compounds such as **urea** and **formamide.** Low salt concentrations also promote denaturation. Denaturation of DNA is reversible under suitable conditions; **renaturation** *occurs when complementary base sequences reestablish hydrogen bonds.*

The base composition of double-stranded DNA determines *the temperature at which it denatures, its* **melting temperature** *or* T_m. Three hydrogen bonds form between guanine and cytosine base

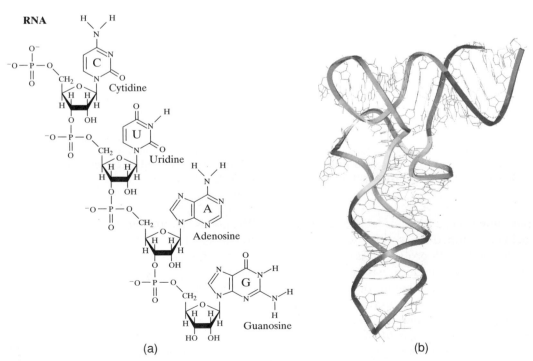

Figure 23.8. The Structure of RNA. a. Linear polymer of RNA. **b.** Example of RNA with secondary structure.

pairs, while only two form between adenine and thymine base pairs. Therefore, the total bond strength between G-C pairs is stronger than between A-Ts and a DNA molecule that is comparatively rich in G-Cs denatures at a higher temperature than one that is rich in A-Ts.

At pH 7.0 and room temperature, DNA solutions are highly viscous. When DNA is denatured, it becomes noticeably less viscous and it absorbs more UV light at a wavelength of 260 *nm*.

Local denaturation of DNA is an essential step in transcription (when RNA is copied from template DNA) and replication (when a cell divides and makes a copy of its DNA for the daughter cell). There are situations in the laboratory where denaturation is similarly desirable and other situations where denaturation should be avoided.

2. **The covalent bonds that connect nucleotides into DNA and RNA strands can be broken by enzymes** called *nucleases.* Nucleases are found on human skin, in cells, and in cellular extracts.

DNA degrading nucleases are relatively fragile, require Mg^{++} as a cofactor, and are commonly destroyed during routine DNA isolation procedures. In contrast, RNA nucleases are ubiquitous, difficult to destroy, and generally do not require metal ion cofactors to be active. The RNA nuclease, **RNase A**, can even survive periods of boiling or autoclaving. RNA nucleases frequently contaminate glassware and other laboratory items, and make RNA difficult to manipulate in the laboratory. Rules typically cited to help reduce the loss of RNA are shown in Table 23.7.

3. **Chromosomal DNA is a long, fragile molecule that is easily sheared into shorter lengths.** Vigorous vortexing, stirring, or passing a DNA solution through a small orifice, such as a hypodermic needle, will shear DNA.

Table 23.7   COMMONLY CITED GUIDELINES FOR PREVENTING THE LOSS OF RNA

1. *Ribonucleases are released into solution when cells are disrupted.* Strong protein denaturing agents are used to destroy these endogenous RNases when cells are disrupted for RNA isolation. Protein denaturing agents used in RNA solutions include **6 M urea, SDS,** and **guanidinium salts (guanidine thiocyanate** and **guanidine hydrochloride**), sometimes in conjunction with a reducing agent such as β-mercaptoethanol. Tissue may also be rapidly frozen to inactivate endogenous ribonucleases. However, freezing temperatures disrupt cells and can cause the release of ribonucleases that become active when the material is thawed.

2. *Peoples' hands are a major source of RNase contamination and gloves should be worn when working with RNA.* Once gloves have come in contact with a surface that was touched by skin (e.g., a pen, notebook, laboratory bench, pipette, etc.) the gloves should be changed.

3. *Sterile technique should always be used when working with RNA because microorganisms on airborne dust particles are a major source of ribonuclease contamination.*

4. *Disposable sterile plasticware is seldom contaminated with RNases.* Once packages of plasticware are opened, the contents must be protected from dust and contaminants.

5. *Nondisposable glassware can be baked at 200°C overnight to inactivate RNases.*

6. *The active site of RNase A contains a histidine amino acid; this active site can be destroyed by modifying the histidine.* The chemical ***DEPC (diethyl pyrocarbonate)*** in low concentrations (around 0.05–1%) can inactivate the histidine binding site. DEPC is therefore frequently used to treat equipment and solutions that will be used for RNA work. Some points regarding the use of DEPC are:

   a. DEPC is carcinogenic and should be used in a hood and handled while wearing gloves.

   b. DEPC can inhibit enzymatic reactions and can interact with nucleic acids. After DEPC inactivates RNases, it therefore needs to be completely degraded. DEPC is usually degraded by autoclaving for 30 minutes to 1 hour, which breaks the DEPC down to carbon dioxide and ethanol.

   c. Glassware can be soaked in DEPC overnight. The glassware is then autoclaved for 30 minutes to inactivate the DEPC.

   d. DEPC can be used to treat water by mixing 0.1% DEPC with the water and allowing the mixture to sit at least 6 hours at room temperature. The solution is then autoclaved to destroy the DEPC.

   e. DEPC can be used to treat various solutions *(but not Tris buffer)*. 0.1% DEPC is added to the solution and the mixture is incubated overnight. The solution is then autoclaved 1 hour.

7. *It is useful to have a separate laboratory space and separate equipment for working with RNA.* This is particularly important if DNA work is also performed in the laboratory because it is common to add RNase to DNA solutions intentionally to destroy unwanted, contaminating RNA. Once RNase is present in the laboratory, it is difficult to eliminate from pipettes, centrifuges, pH electrodes, and other devices.

8. *Products (such as buffers, enzymes, and bovine serum albumin) are all potential sources of RNase contamination.* Manufacturers produce many products that have been treated to remove ribonucleases and certify them to be RNase-free.

9. *Enzymes that are ribonuclease inhibitors, RNasins, can be purchased from molecular biology suppliers.* These inhibitors bind to and inactivate some classes of RNase.

C. Laboratory Solutions for Nucleic Acids

The structure and function of nucleic acids is not as variable as is that of proteins. For example, there are only four different nucleotides comprising DNA, but there are 20 different amino acids found in proteins. There are, however, a multitude of different tasks performed in bioscience laboratories that involve DNA, and there are numerous techniques used to manipulate nucleic acids; therefore, there are many solutions used for DNA work. A summary of the basic classes of agents found in DNA solutions are listed in Table 23.8 and then are discussed in more detail. Table 23.9 summarizes a number of solution components in terms of their functions in laboratory procedures.

When working with proteins in laboratory solutions, it is common to remove proteases carefully and to avoid conditions that will damage the proteins. In contrast, when working with nucleic acids, the first agents used are those that denature, precipitate, and destroy proteins. This is because nucleases (which are proteins) are released when cells are disrupted. These endogenous nucleases must be destroyed before the nucleic acids are degraded. Later, certain proteins may be added to nucleic acid solutions to perform a particular task. For example, *DNA may be intentionally cleaved during cloning procedures using nucleases called* **restriction endonucleases**.

i. NUCLEIC ACIDS AND pH: THE IMPORTANCE OF BUFFERS

The pH of the immediate environment can have a profound effect on the structure of nucleic acids; therefore, buffers are added to nucleic acid solutions. For example:

1. **Manipulation of pH is a technique used for controlling DNA denaturation and renaturation in certain procedures.**

2. **Treatment with mild acids, such as 5% trichloroacetic acid (TCA) at or below room temperature, will precipitate nucleic acids.**

3. **Prolonged exposure to dilute acid at or above room temperature, or briefer exposure to more concentrated acid (e.g., 15 minutes in 1 M HCl at 100°C) will cause depurination of DNA and RNA (removal of most of the purines adenine and guanine). The**

Table 23.8 *CLASSES OF AGENTS USED IN LABORATORY SOLUTIONS FOR NUCLEIC ACIDS*

Buffers
Salts
Organic solvents
Nucleases
Nuclease inhibitors
Metal chelators
Coprecipitants
Antimicrobial agents

pyrimidines (cytosine, uracil, and thymine) in these nucleic acids can also be removed by exposure to more concentrated acids or by longer periods of exposure.

4. **Very strong acid treatments will break the phosphodiester bonds that join nucleotides.**

5. **Scientists can exploit the fact that DNA and RNA react differently under mildly alkaline conditions.** For example, 0.3 M KOH at 37°C for 1 hour will cleave the phosphodiester bonds in RNA, but the phosphodiester bonds of DNA will remain intact under these conditions.

ii. SALTS

Like proteins, nucleic acids are sensitive to the ionic strength of their solution. Consider the following three important cases:

1. **Denaturation of double-stranded DNA is controlled in the laboratory by the ionic strength of the solution in conjunction with its temperature and pH.** Consider DNA dissolved in a solution of low ionic strength, Figure 23.9a. The negatively charged phosphate groups cause the two DNA strands to repel one another. In contrast, in a solution of moderate ionic strength, such as 0.4 M NaCl, Na$^+$ ions are available that are attracted to negatively charged phosphate groups. The negative charges of the phosphate groups are thus neutralized, so the two DNA strands do not repel one another, Figure 23.9b. Double-stranded DNA, therefore, is less stable and is more easily denatured in a solution with 0.01 M salt than it is in a solution of 0.4 M salt. As a result DNA denatures at a lower temperature in 0.01 M NaCl than it does in 0.4 M NaCl. In solutions with salt concentrations between 0.01 M NaCl and 0.4 M NaCl, the likelihood of denaturation is intermediate.

2. **Hybridization *(binding)* of single-stranded DNA with short strands of complementary DNA or RNA is affected by the ionic strength of the solution. Stringency** *refers to the reaction conditions used when single-stranded, complementary nucleic acids are allowed to hybridize.* At high stringency, binding occurs only between strands with perfect complementarity. (Perfect complementarity means that every guanine is base-paired with a cytosine and every adenine is base-paired with a thymine.) At lower stringency, there can be some mismatch of bases across the strands and hybridization still occurs. There are situations in the laboratory where high stringency is required and other situations where lower stringency is desirable.

Strands come together more easily when the salt concentration is high and the temperature is relatively low. Under these conditions, hybridization can occur even if there are some base pair

Table 23.9   SUMMARY OF SOLUTION COMPONENTS AND PROCEDURES FOR MANIPULATING DNA

Procedure	Components of Solutions

Isolation and Purification of Nucleic Acids from Cells
Before DNA can be studied, sequenced, or recombined, it must be isolated and purified.

Procedure	Components of Solutions
Degrade the cell wall	Enzymes: Lysozyme for bacteria
Remove membranes	Detergents: SDS, Sarkosyl
Remove proteins and RNA	Enzymes: Proteinase K, RNase A
Denature and remove proteins	Phenol/chloroform extraction
Preferentially precipitate DNA	Ethanol (65–70%) in high salt
	Isopropanol with salt
Concentrate DNA in aqueous solution	*sec*-butanol
Purify specific types of nucleic acid	CsCl density gradient
	Commercial spin columns, glass beads
	Ion exchange resins
Preferentially degrade chromosomal DNA to isolate plasmids	NaOH + SDS

Nucleic Acid Storage Buffers
Protection during storage of DNA.

Procedure	Components of Solutions
Maintain proper pH	Buffers: Tris
Chelate metals (inhibit nucleases)	EDTA
Maintain ionic strength	NaCl

Enzymatic Digestion and Modification
Enzymes are used to digest, recombine, and modify DNA. These enzymes require proper solution conditions.

Procedure	Components of Solutions
Maintain proper pH	Buffers: Tris
Prevent the loss of enzyme due to adsorption to container	Carriers: BSA
Enzyme co-factors	Divalent cations: Mg^{++}, Mn^{++}, Fe^{++}
Stabilize protein and lower freezing temperature	Glycerol
Prevent aggregation and sticking	Low levels of detergents such as Triton X
Inhibit proteases	EDTA

Electrophoresis
DNA molecules can be separated by size during migration in an electrical field.

Procedure	Components of Solutions
Gel matrix	Agarose, acrylamide
Running buffers	Tris borate EDTA, Tris acetate EDTA
Denaturing gels (lower the melting temperature of the double-stranded DNA)	Formamide, urea
Tracking dyes to visualize the migration of DNA	Xylene cyanol, bromophenol blue
Dyes to detect the presence of DNA	Ethidium bromide, methylene blue

Transformation/Transfection of purified DNA into cells
Introduction of foreign DNA into a cell. In bacteria it is necessary to prepare the cell; in eukaryotic cells it is necessary to prepare the DNA.

Procedure	Components of Solutions
Preparation of bacterial cells to be "competent"	$CaCl_2$, RbCl, DMSO
Uptake of precipitated DNA by eukaryotic cells	$CaPO_4$ precipitation of DNA
Fusion of eukaryotic cells with other membrances	Liposomes, lipid-encased DNA
Permeabilization of eukaryote membranes	DEAE dextran, DMSO, glycerol

Hybridization
Process in which denatured DNA binds (hybridizes) with complementary strands of DNA or RNA.

Procedure	Components of Solutions
Separation of DNA strands	NaOH, heat
Hybridization solution components:	
"Blocking agents" to fill nonspecific binding sites	BSA, nonfat dry milk, casein, gelatin
"Crowding agents"—inert compounds that effectively reduce the volume of the solution	Denhardt's Reagent: Ficoll, PVP, PEG
Nonspecific nucleic acid to reduce background by binding nonspecific sites	Salmon sperm DNA, tRNA

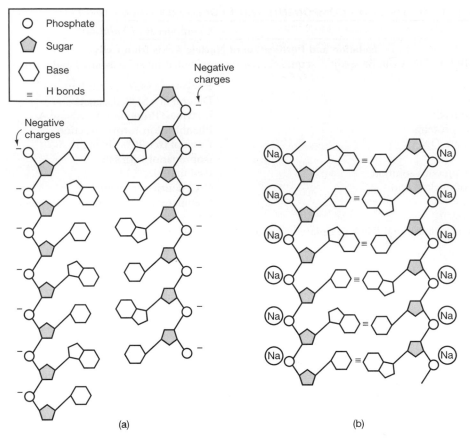

Figure 23.9.   The Effect of Ionic Strength on Nucleic Acid Denaturation and Renaturation. a. In a solution of low ionic strength the two DNA strands are negatively charged and repel one another, making denaturation more likely. **b.** In a solution of moderate ionic strength, positive ions shield negatively charged phosphates.

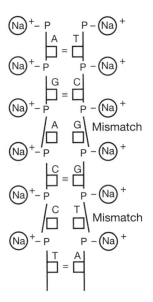

Figure 23.10.   The Effect of Ionic Strength on Stringency. At low temperatures and high ionic strength, some mismatches can be tolerated; stringency is lower than at higher temperatures and lower ionic strength.

mismatches; the conditions have lower stringency, Figure 23.10. In contrast, when the temperature is higher and the salt concentration is lowered, the conditions are more stringent.

3. **The ionic strength of the solution affects interactions between proteins and DNA.** Protein–DNA interactions are of critical importance in the cell. For example, such interactions are involved in transcription and replication. Electrostatic interactions between negatively charged phosphates on DNA and positively charged amino acids promote protein-DNA binding. However, electrostatic interactions are not specific. This means, for example, that electrostatic interactions do not cause an enzyme to find and bind to a *specific sequence* of DNA. Electrostatic interactions are affected by the concentration of salt in the solution. As salt increases, the strength of electrostatic interactions decreases. For example:

- **Chromatin** *is the native form of DNA, which is composed of DNA bound to many different proteins.* In 0.01 M NaCl chromatin is stable, but in

1 M NaCl it dissociates almost completely to yield free DNA molecules and proteins. This effect is thought to be due to the loss of electrostatic interactions that normally stabilize chromatin. When NaCl is added, the salt ions shield the various charged sites and disrupt the electrostatic bonding.

* *Restriction enzymes that cut DNA at specific sites require the correct concentration of salts to cut DNA selectively.* At lower salt concentrations, electrostatic interactions increase, allowing the restriction enzyme to nonspecifically bind to and cut the DNA. A restriction enzyme is said to exhibit **star activity** when it *"mistakenly" cleaves DNA sequences that are not its proper specific target.* Star activity is an example of nonspecific protein–DNA interaction. Reduced salt concentration is one of several factors that promote star activity.

Specific interactions between proteins and particular base sequences of DNA are not electrostatic. Unlike electrostatic interactions, specific protein–DNA interactions are frequently stable in higher salt concentrations.

iii. ORGANIC SOLVENTS

Organic solvents are commonly used to isolate DNA and RNA from other components of cellular extracts. A standard extraction procedure involves addition of the organic solvents **phenol** and **chloroform**, which simultaneously disrupt cell membranes and denature proteins (including endogenous nucleases). Phenol and chloroform are not water-soluble, so the mixture is centrifuged to separate it into an organic phenol/chloroform phase and an aqueous phase. Denatured proteins move into the organic phase or the interface between the two phases. The nucleic acids remain in the aqueous phase. A low concentration of **isoamyl alcohol** is usually added to aid in clean separation of the organic and aqueous phases. Phenol is relatively unstable in storage, so **8-hydroxyquinoline** is usually added to phenol as an antioxidant.

iv. PRECIPITANTS

Ethanol plays an important role in working with nucleic acids because it precipitates DNA and RNA, physically isolating them from a solution containing other components. Nucleic acids do not lose their structural or functional integrity when isolated with phenol/chloroform and/or ethanol.

Isopropanol is sometimes substituted for ethanol when the final solution volume must be minimized because less isopropanol is needed than ethanol to precipitate DNA.

Salt is added when nucleic acids are ethanol-precipitated because relatively high concentrations of mono-valent cations (0.1 to 0.5 M) promote aggregation and precipitation of nucleic acid molecules. Various salts are used in different situations; however, sodium acetate is most common.

v. COPRECIPITANTS

Alcohol precipitation is ineffective if only small amounts of nucleic acids are present in a solution. Inert coprecipitants, such as **yeast tRNA** or **glycogen**, are therefore sometimes added to solutions to help precipitate nucleic acids.

vi. NUCLEASES

There are many uses for nucleases in the laboratory. For example, when isolating DNA it is common to use **RNase A** to intentionally degrade contaminating RNA. Restriction endonucleases are frequently added to DNA solutions to recognize and cleave DNA phosphodiester bonds at specific sites.

vii. NUCLEASE INHIBITORS

Unwanted nucleases are released when cells are disrupted. As mentioned earlier, organic solvents used to isolate DNA and RNA denature nucleases and other proteins. **Proteinase K** is an enzyme that is also added to solutions to help degrade nucleases and histones (proteins associated with DNA in the chromosome). Proteinase K is active over a wide range of pH, salt, detergent, and temperature conditions. Chelators are also used as nuclease inhibitors because many nucleases require a metal ion cofactor for activity.

viii. ANTIMICROBIAL AGENTS

Antimicrobial agents are often added to DNA solutions to inhibit the growth of microorganisms during storage.

ix. CHELATORS AND METAL IONS

Unwanted nucleases can degrade nucleic acids, particularly during long periods of storage. There are many situations, however, when nucleases are used to intentionally break chains of nucleotides in a controlled fashion. Most nucleases require Mg^{++} as a cofactor. The levels of chelators and metal ions, therefore, are controlled in nucleic acid solutions. For example, EDTA is typically added to nucleic acid solutions during storage to reduce unwanted nuclease activity. Excess Mg^{++} is added when nuclease activity is desired.

D. Physical Factors

i. TEMPERATURE

Temperature is frequently used in the laboratory to control denaturation and renaturation of DNA. For example, DNA must be denatured whenever a procedure requires the DNA to hybridize to complementary nucleic acid fragments. Temperature is critical in the polymerase chain reaction (PCR) technique that is used to amplify the amount of a specific sequence of DNA. High temperature (around 90°C) is used to denature the DNA to be amplified. The temperature is then lowered to allow replication of the single-stranded DNA (see also Problem 4 in the problem set below).

Proteins are frequently used to manipulate nucleic acids in the laboratory. The temperature of a nucleic acid solution, therefore, may be chosen to meet the requirements of the proteins involved. For example, restriction endonucleases are most active at 37°C, so this temperature is used to digest DNA.

ii. Shearing

Because of their long strands, nucleic acids are more prone to shearing than proteins. Mixing and pipetting, therefore, must be particularly gentle to isolate high molecular weight nucleic acids.

CASE STUDY

Isolating Plasmids: The Roles of Solution Components

There are a variety of different tasks performed in bioscience laboratories that involve manipulating nucleic acids, so there are many solutions used for DNA work. The components of these solutions vary depending on the task to be accomplished. The solution components involved in several basic DNA procedures were summarized in Table 23.9.

Let us consider as an example the solutions used in a common procedure: the isolation of plasmids from bacteria. **Plasmids** are small circular molecules of DNA found in bacteria that exist separate from the bacterial chromosome. In nature, plasmids sometimes carry genes that are responsible for the spread of antibiotic resistance among bacteria.

In the laboratory, scientists can modify plasmids and use them to carry genes of interest into host bacteria. Once inside the bacteria the plasmids multiply as the bacteria divide. The plasmids can be isolated from the host bacteria using various techniques, such as the "plasmid minipreparation procedure." The minipreparation procedure takes advantage of the small, coiled nature of the plasmids to separate them from the larger, linear chromosomal DNA. The minipreparation procedure will be discussed. As we go through each step, we will explain the roles various solution components are thought to play in the separation.

"Minipreparation Procedure to Isolate Plasmids from Host Bacteria"

The Solutions

LB (Luria Broth) (1 L)
10 g tryptone
5 g yeast extract
5 g NaCl
Sterilize by autoclaving

GTE Buffer
25 mM Tris-HCl, pH 8.0
10 mM EDTA
50 mM glucose

Potassium Acetate/Acetic Acid
60 mL of 5 M potassium acetate
11.5 mL of glacial acetic acid
28.5 mL of H_2O

SDS/NaOH
1% SDS
0.2 N NaOH

TE Buffer
10 mM Tris-HCl, pH 7.5–8.0
1 mM EDTA

The Procedure

1. **Grow the plasmid-containing bacterial culture overnight in LB medium.** *LB medium contains nutrients for bacterial growth.*

2. **Remove 1.5 mL of the culture and spin down the cells in a microfuge.**

3. **Discard the supernatant and completely resuspend the cells in 100 μL of ice-cold GTE buffer.**
   The buffer is kept cold to inhibit nucleases released during the procedure. GTE Buffer contains: Tris to maintain the proper pH (pH 8.0), EDTA to bind divalent cations (Mg^{++}, Ca^{++}) in the lipid bilayer (thus weakening the cell membrane), and glucose, which may act as an osmotic support (see p. 502) or may help to prevent cells from clumping.

4. **Add 200 μL of SDS/NaOH and gently mix.**
   The bacterial cells are lysed by the SDS, which solubilizes the membrane lipids and cellular proteins. The base denatures both the chromosomal and plasmid DNA. The strands of the circular plasmid DNA remain intertwined. The suspension is mixed gently to avoid shearing the chromosomal DNA.

5. **Set on ice for 10 minutes.**
   This "resting step" allows time for the SDS/NaOH to contact all the cells and the processes in Step 4 to go to completion.

6. **Add 150 μL of cold potassium acetate/glacial acetic acid and mix gently. A white precipitate will appear.**
   The acetic acid neutralizes the sodium hydroxide added in Step 4, allowing the DNA to renature. The potassium acetate causes the SDS to precipitate, along with associated proteins and lipids. The large chromosomal DNA strands renature partially into a tangled web that precipitates along with the SDS complex. In contrast, the small, coiled renatured plasmids remain suspended. In this step, therefore, they are separated from chromosomal DNA, many proteins, and the detergent.

7. **Centrifuge the mixture; save the supernatant.**
   The discarded precipitate contains cellular material, linear DNA, protein, and detergent, whereas the supernatant contains the suspended plasmids.

8. **Remove 400 μL of the supernatant, add an equal volume of isopropanol, and mix.**
   Isopropanol, like ethanol, precipitates DNA. Isopropanol is used here because a smaller volume of iso-

propanol is required than ethanol. The alcohol quickly precipitates DNA and more slowly precipitates proteins. This step is therefore done quickly.

9. **Centrifuge and save the pellet.**

    The alcohol-precipitated plasmid DNA is in the pellet.

10. **Wash the pellet in 200 μL of 100% ethanol.**

    Although isopropanol precipitated the DNA in the preceding step, an ethanol wash helps remove extra salts and other unwanted contamination.

11. **Centrifuge and save the pellet.**

12. **Resuspend the pellet in 15 μL of Tris/EDTA buffer (TE).**

    TE is often used for short-term storage of DNA. Tris maintains the pH, whereas EDTA chelates divalent cations to inhibit nuclease activity.

Based primarily on information in DNA Science, David A. Micklos and Greg A. Freyer, Cold Spring Harbor Press, 1990.

EXAMPLE PROBLEM

Southern blotting and probe hybridization is a method in which DNA fragments separated from one another by electrophoresis are adhered to a plastic or nylon membrane and are then exposed to a single-stranded nucleic acid probe. The probe will bind to any single-stranded fragments adhered to the membrane that have a nucleotide sequence complementary to the probe. Before the DNA is adhered to the membrane, it is soaked in a solution containing NaOH. What is the function of the NaOH in this soaking solution?

ANSWER

In order for the probe to bind to DNA fragments, the DNA fragments (and the probe) must be single-stranded. The complementary bases will then be available for hybridization. The NaOH is added to the solution to denature the double-stranded DNA.

DNA : APPLICATION PROBLEMS

1. Suggest the possible purpose of the components of the following commonly used solutions.

   a. TBE buffer—electrophoresis buffer

      0.089 M Tris base (pH 8.3)
      0.089 M boric acid
      0.002 M EDTA

   b. TE buffer

      10 mM Tris-HCl (pH 8.0)
      1 mM EDTA

   c. Hybridization buffer

      (From Sambrook, et al, see references)
      40 mM PIPES Buffer (pH 6.4)
      1 mM EDTA
      0.4 M NaCl
      80% formamide

2. Examine the recipe for hybridization buffer, shown above. Comment on the salt concentration in terms of stringency.

3. (Modified from *Principles of Physical Chemistry With Applications to the Biological Sciences*, David Freifelder. 2nd edition. Jones and Bartlett, Boston, MA, 1985) The enzyme RNA polymerase binds to DNA and controls the synthesis of a complementary strand of RNA from the DNA template. The binding of RNA polymerase to DNA in cells is specific in that only a particular required sequence of DNA is copied into RNA. It has been observed that in a test tube in **0.2 M NaCl**, four RNA polymerase molecules bind to a particular DNA molecule and synthesize the same, specific RNA molecule that is found in living organisms. In contrast, in a test tube in **0.01 M NaCl**, about 50 RNA polymerase molecules bind to the DNA and a great many different RNA molecules are synthesized.

   a. What does this tell us about the interactions between DNA and RNA polymerase?

   b. What are the implications of these observations for working with RNA polymerase in the laboratory?

4. **PCR, the polymerase chain reaction**, is a method used to copy a specific region of DNA many thousands of times. PCR requires **primers** that are short, single strands of DNA. The primers are complementary to regions flanking the DNA region of interest. There are three phases to the PCR cycle:

   Phase 1: The two strands of the parent DNA molecule that includes the sequence of interest are separated by heating to about 92°C.

   Phase 2: The solution is quickly cooled to around 72°C and the single-stranded DNA hybridizes with the primers.

   Phase 3: Once the primers have annealed, the thermostable enzyme, *Taq 1* DNA polymerase, is able to copy the DNA region of interest, beginning from the location where the primers have bound.

   The PCR cycle is repeated many times, greatly amplifying the number of copies of the DNA of interest.

   a. Examine the following recipe for PCR reaction buffer. Discuss the salt concentration in terms of DNA denaturation and hybridization of the primer to the DNA.

   b. Explain the role of the $MgCl_2$ in the reaction mix.

c. *Taq 1* polymerase is functional at high temperatures. Why is this type of DNA polymerase essential for the PCR method to work?

> **PCR Reaction Buffer**
>
> 500 *mM* KCl
> 100 *mM* Tris-HCl (pH 8.0)
> 20 *mM* MgCl$_2$

5. What do you think would happen to a bacterial culture with the addition of dish washing detergent? Consider the membranes, proteins, and nucleic acids.

6. You have a sample of DNA (10 *μg* in 50 *μL*) that is stored in TE buffer (10 *mM* Tris pH 7.6 and 1 *mM* EDTA). Upon addition of 100 *μL* of ethanol, no precipitate is seen. Why not? What do you do to recover the DNA?

IV. WORKING WITH INTACT CELLS IN SOLUTION

A. Isotonic Solutions

Intact cells contain water, salts, and various biomolecules enclosed by a cell membrane. The cell membrane is **semi-permeable**, *which means that it allows water and some small molecules to flow through unimpeded, whereas other molecules are restricted.* To preserve cellular structure the amount of water that flows into a cell must equal the amount of water that leaves. If more water enters the cell than exits, the cell swells and may burst. On the other hand, if more water exits than enters, the cell shrinks.

The flow of water across the cell membrane is controlled by the concentration of particles (solutes) inside and outside the cell. The net flow of water across the cell membrane is from the side that has the lower solute concentration to the side that has the higher solute concentration. If there is a higher solute concentration outside the cell than inside, then water flows out of the cell. On the other hand, when there is a lower solute concentration outside the cell, water flows in. **Osmosis** *is the net movement of water through a semipermeable membrane from a region of lesser solute concentration to a region of greater solute concentration.* A cell is in **osmotic equilibrium** when the rate of water flow into and out of the cell is equal. To be in osmotic equilibrium a cell needs to be surrounded by an **isotonic solution** *that has the same solute concentration as the interior of a cell,* Figure 23.11. **Physiological solutions** *are isotonic solutions that support intact cells and microorganisms in the laboratory.*

Salts are used in laboratory solutions to maintain the proper osmotic equilibrium for intact cells. The most common solution used strictly to maintain osmotic equilibrium is **standard (physiological) saline**, which is 0.9% NaCl. Table 23.10 shows several solutions commonly used to maintain osmotic equilibrium.

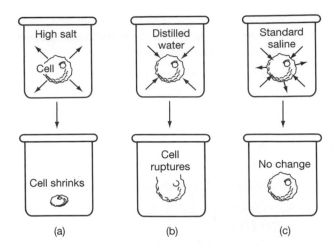

Figure 23.11. The Flow of Water Across the Cell Membrane. a. If there is a higher solute concentration outside the cell than there is inside, then water flows out of the cell and the cell shrinks. **b.** If there is a lower solute concentration outside the cell, then water flows into the cell and it may burst. **c.** An isotonic solution is in equilibrium with the interior of the cell.

Table 23.10   EXAMPLES OF SOLUTIONS USED TO MAINTAIN OSMOTIC EQUILIBRIUM

Physiological Saline (mammalian cells)	
NaCl	0.9% w/v
Phosphate Buffered Saline	
NaCl	7.2 g/L
Na$_2$HPO$_4$ (anhydrous)	1.48 g/L
KH$_2$PO$_4$	0.43 g/L
Tris Buffered Saline	
NaCl	0.9%
Tris pH 7.2	10–50 *mM*
Ringer's Solution (mammalian cells)	
CaCl$_2$	0.250 g/L
KCl	0.420 g/L
NaCl	9.000 g/L

When the cell walls of plants, yeasts, and microbes are removed, the cells become particularly sensitive to changes in osmotic pressure. **Osmotic supports** may therefore be added to their solution to protect the cells' integrity. Nonreactive sugars, including sucrose, sorbitol, mannitol, and glucose, are commonly used for this purpose.

B. Growth Media

Biologists frequently work in the laboratory with living microorganisms and cells that are cultured in special **media**. Media intended to support cell survival and growth over a period of time include other components in addition to those that maintain the solution in osmotic equilibrium. These additional components include **a buffer**, a **carbon source** to nourish the cells (such as glucose), **essential salts**, **vitamins** and various **micronutrients**. In addition, depending on the situation, other components may be added, such as **hormones** to stimulate plant cell growth, and **antibiotics** to protect mammalian cells from contamination.

CASE STUDIES

Optimizing Solutions and Avoiding Problems with Solutions

It should be clear by this point that there are many different solutions used in biology laboratories. Solutions often have multiple components, each of which must be present at the proper concentration. The early phases of research often include finding optimal solution components, sometimes by trying various combinations of agents until the best results are obtained. It is essential to understand the roles of the various components when optimizing solution recipes, as the following examples from the literature illustrate.

Example 1

An illustration of a study that looked at the effects of laboratory solutions on the results of an experiment was reported by Favre and Rudin ("Salt-Dependent Performance Variation of DNA Polymerases in Co-Amplification PCR," Nicolas Favre and Werner Rudin. BioTechniques, 21:28–30, July, 1996.) These authors were using the technique of reverse transcriptase PCR to amplify and detect specific mRNAs. The scientists were attempting to amplify two different mRNAs simultaneously in the same reaction tube. They tested reverse transcriptase enzymes from four manufacturers, in each case using the buffer solution provided by the manufacturers. They found that the enzymes/buffers from all four manufacturers were effective when the two different mRNAs were amplified individually in separate tubes. When they attempted to amplify both types of mRNA together in the same tube, however, only one manufacturer's kit was effective. Further study showed that the effective kit contained buffer with 200 mM Tris-HCl, whereas the other manufacturers provided buffer with 100 mM Tris-HCl. They concluded that the buffer concentration was critical in the amplification and detection of two mRNAs simultaneously.

Example 2

Another illustration of a study that investigated the effects of buffer components was reported by Ignatoski and Verderame ("Lysis Buffer Composition Dramatically Affects Extraction of Phosphotyrosine-Containing Proteins," Kathleen M. Woods Ignatoski and Michael F. Verderame. BioTechniques, 20:794–796, May, 1996). In order to understand this example, it is necessary to understand the role of lysis buffers:

All cells are surrounded by a cell membrane that maintains the structural integrity of the cell and also controls the interactions of the cell with its external environment. Cell membranes are composed of **phospholipids** with embedded proteins and glycoproteins. Phospholipids have a negatively charged, hydrophilic phosphate group and "tails" that are hydrophobic hydrocarbon chains. The hydrophilic "heads" face out to the cytoplasm or the exterior of the cell, whereas the hydrophobic tails face one another.

Cell membranes and cell walls must be broken (lysed) to get access to the molecules inside. Solutions whose primary function is to lyse these structures are called **lysis buffers**. Membranes are usually lysed by detergents that, like the membranes, have both hydrophilic and hydrophobic portions. Mild detergents are preferred, such as **Triton X-100** or **sarkosyl (N-lauroylsarcosine)**.

Plant and bacterial cells also have cell walls that surround their cell membranes. Cell walls have varied structures, but they are usually more difficult to break than membranes. A variety of methods are used to disrupt cell walls including physical ones, such as grinding or crushing, or the addition of enzymes, such as **lysozyme** (for lysing bacteria).

Ignatoski and Verderame were comparing proteins in cultured cells that were transformed with an oncogene (a cancer-causing gene) with proteins in cultured cells that were not transformed. The cells were broken open with lysis buffer, releasing their proteins into solution. The released solubilized proteins were analyzed. The scientists discovered that the components of the lysis buffer had a significant effect on what proteins they detected and therefore, their conclusions regarding differences between transformed and nontransformed cells.

Ignatoski and Verderame compared the effects of three different lysis buffers. All three buffers included protease inhibitors and also the following ingredients:

Lysis Buffer A	Lysis Buffer B
50 mM Tris-HCl, pH 7.5	10 mM Tris-HCl, pH 7.4
150 mM NaCl	50 mM NaCl
1% Nonidet P-40 (NP-40) (detergent)	50 mM NaF
0.25% Na$^+$ deoxycholate (detergent)	1% Triton X-100 (detergent)
10 mg/mL BSA	5 mM EDTA
	150 mM Na$_3$VO$_4$
	30 mM Na$_4$P$_2$O$_7$

Lysis Buffer C	
30 mM Tris-HCl pH 6.8	
150 mM NaCl	
1% NP40 (detergent)	
0.5% Na$^+$ deoxycholate (detergent)	
0.1% SDS (detergent)	

The authors observed that the three lysis buffers appear to "be solubilizing different subsets of cellular proteins" and they postulated that:

1. **The type and concentration of detergent could have an effect on which proteins are detected.** Ionic detergents, like Na$^+$ deoxycholate, disrupt the cell membrane more completely than do nonionic detergents, such as NP40 or Triton X-100. Thus, more proteins might have been released into solution when ionic detergents were used.

2. **The ionic strength of the buffer plays a critical role in protein solubilization.** Proteins need to be soluble in aqueous solutions to be detected. High salt con-

centrations might disrupt interactions of proteins with insoluble cellular components, thereby promoting protein solubility. The three buffers contained different salts, which may have caused the proteins solubilized in each buffer to have been different.

3. **pH can affect detergent solubility, especially of Na^+ deoxycholate, thereby changing which proteins are solubilized.** *The pH of Buffer C was different from the other two buffers; therefore, the proteins that were detected may have been affected.*

4. **The buffer effectiveness may have been affected by the presence of phosphate.** *Buffer B was the only buffer containing phosphate, and this buffer resulted in the solubilization of a different subset of proteins than the other buffers.*

The authors of the preceding study concluded that the lysis buffer components can play a role in affecting experimental results and therefore "several different lysis buffers . . . should be tested to obtain optimal experimental conditions."

As the preceding examples illustrate, the composition of solutions has a critical effect on results. Once solution recipes have been developed for a particular task, it is therefore essential to follow these recipes exactly as written (or to control and document alterations in solution components carefully).

In a production setting, optimizing a solution requires consideration not only of the effectiveness of the solution, but also the cost and availability of the components. Large quantities of raw materials will be needed when the product goes into mass production. Therefore, raw materials should be chosen that are as inexpensive as possible and yet meet the requirements of the process. Also, it is necessary to find suppliers of the solution components who can provide sufficient raw materials of consistent high quality. In a production setting, once the solution recipes have been established and suppliers of raw materials have been found, it is particularly important not to modify any aspect of the solution without careful consideration and control of the changes.

Example 3

An example of the type of problem that occurs when solution recipes are not followed was reported by McCarty (The Transforming Principle: Discovering that Genes Are Made of DNA, Maclyn McCarty, pp. 150–152. W.W. Norton Co., 1985.) in a book describing research that led to the discovery that genes are made of DNA. He reported:

I don't mean to imply that things always went smoothly. . . . There were a number of hitches along the way. . . . We even had some problems with simple fundamental operations, like growing the type III pneumococci for extraction {of DNA}. Early in the fall the central media department supplied us with a few 75-L lots of broth that either sustained the growth of the organisms very poorly or not at all. I spent some time trying to find the source of the trouble and ended up participating in the preparation of our next batches of media. The difficulty

was never pinpointed, but fortunately the difficulties disappeared with careful attention to the details of the cookbook-type recipe that had to be followed.

DISCUSSION QUESTIONS

1. Suppose you are working in a university research laboratory and are taking over a project that was initiated by another investigator. The project involves isolating DNA from a particular type of fungus. One of the problems the previous investigator had begun to explore, but had not resolved, was that the DNA isolated was inconsistent in its qualities. The DNA sometimes seemed to be fragmented, sometimes more DNA was isolated than other times, and sometimes, when the DNA was digested with restriction endonucleases, the digestion did not appear to go to completion, even when the enzymes were known to be effective. Discuss components of the various solutions involved that might be affecting the results. How could you investigate the various components of the solutions in a systematic, well-documented way?

2. Suppose you are now working in a production facility producing the DNA described in Question 1. Discuss the particular concerns you would have in this situation.

GENERAL SUGGESTIONS FOR ANSWERS TO PROBLEM SETS

Proteins. Application Problems

1. **a.** <u>Storage Buffer</u>

—Low temperature. Limits bacterial growth and degradation of product, inhibits enzyme activity

—Tris-HCl. Buffer, maintains pH

—NaCl. Salt, maintains ionic strength at moderate level

—EDTA. Chelates Mg^{++}, inhibits enzyme activity during storage

—DTT. Reducing agent, prevents unwanted disulfide bond formation

—Triton X-100. Detergent, reduces adsorption of proteins to tube walls

—BSA. Added protein, reduces loss of protein due to adsorption and protease activity

—Glycerol. Prevents freezing of protein, stabilizes enzyme at low temperature

<u>Activity Buffer</u>

—Tris. Buffer, maintains proper pH for activity

—NaCl. Salt, maintains ionic strength

—$MgCl_2$. Salt, required cofactor for enzyme

b. The two buffers have different functions. Storage requires protection of enzyme from degradation and adsorption. Activity is not required during storage. Activity buffer optimizes activity.

2. a. Sample Buffer Used When Loading Proteins into Electrophoresis Gel
 —Tris. Buffer, maintains pH
 —Glycerol. Causes proteins to sink into gel wells
 —SDS. Detergent, denatures proteins, confers consistent charge on proteins
 —β-mercaptoethanol. Reducing agent, inhibits disulfide bond formation
 —Bromophenol blue. Dye, stains proteins

 b. Restriction Enzyme Buffer (Low Salt)
 —Tris. Buffer, maintains pH
 —$MgCl_2$. Salt, maintains ionic strength, provides Mg^{++} ions as cofactors
 —DTT. Reducing agent, protects against oxidation

3. See Table 23.6.

DNA Application Problems

1. a. TBE
 —Tris-borate. Buffer, maintains pH
 —Boric acid. Bring Tris to proper pH
 —EDTA. Chelates Mg^{++}, inhibits nucleases

 b. TE Buffer
 —Tris. Buffer, maintains pH
 —EDTA. Chelates Mg^{++}, inhibits nucleases

 c. Hybridization Buffer
 —PIPES. Buffer, maintains pH
 —EDTA. Chelates Mg^{++}, inhibits nucleases
 —NaCl. Salt, maintains ionic strength
 —Formamide. Lowers the melting temperature for DNA

2. In 0.4 M salt, DNA tends to form duplexes readily; stringency is relatively low.

3. a. One explanation is that DNA and RNA polymerase can bind one another *both* by nonspecific electrostatic interactions and by another kind of interaction that is specific to a particular site on the DNA. The electrostatic interactions decrease with increasing salt concentration.

 b. This means that when working with RNA polymerase, the salt concentration will have an effect on the products made in the laboratory. One would choose a particular salt concentration based on the specificity required.

4. a. The buffer has a relatively high concentration of salt that facilitates the hybridization of DNA to primer and the synthesis of new strands at 72°C. The melting temperature is raised because of the high salt concentration.

 b. Mg^{++} is required by the *Taq 1* polymerase for its activity.

 c. If *Taq 1* were sensitive to high temperature, it would be destroyed every time the temperature is raised to denature the DNA strands. In that case, more polymerase would have to be added each cycle.

5. Most household detergents and shampoos contain SDS, which solubilizes membranes, denatures proteins, and releases DNA into solution.

6. DNA precipitation requires the presence of higher salt concentrations—bring the solution to 0.25 M Na acetate.

Appendix

Alphabetical Listing of Common Components of Biological Solutions

Acrylamide. Monomers that polymerize to form a polyacrylamide matrix used in gel electrophoresis (PAGE).

bis-Acrylamide. Cross-linker added to acrylamide monomers before polymerization. Controls the porosity of the resulting gel.

Agarose. Matrix composed of complex carbohydrates derived from seaweed, used for gel electrophoresis.

Ammonium Acetate. Salt added to DNA in solution to aid in ethanol precipitation. At a concentration of $2\,M$ it will preferentially facilitate ethanol precipitation of larger DNA and allow nucleotides to remain in solution.

Ammonium Persulfate. A free radical generator added to unpolymerized acrylamide to stimulate polymerization.

Ampicillin. Antibiotic that inhibits bacterial cell wall synthesis.

Avidin. Egg white protein that has a high affinity for biotin; may be conjugated with an enzyme or other compound as part of a detection system.

BCIP (5-Bromo-4-chloro-3-indolyl phosphate *p*-toluidine salt). A substrate for the enzyme alkaline phosphatase. In conjunction with NBT produces a blue color indicating the presence of this enzyme.

Biotin. Vitamin that has a high affinity for avidin. Biotin can be conjugated to various molecules, then bound to and detected by an indicator-linked avidin molecule.

Bromphenol Blue. Dye used to visualize sample movement in gel electrophoresis.

BSA (Bovine serum albumin). A protein often used as a carrier to help protect proteins; also used to block non-specific binding sites.

sec-Butanol. Water-insoluble alcohol used to reduce the water in aqueous DNA solutions, thus concentrating DNA into a smaller volume.

CAA (Casamino acids). Digested protein added to bacterial growth media as a source of amino acids.

CaCl$_2$ (Calcium chloride). Salt used to increase permeability of cellular membranes. Bacteria incubated in CaCl$_2$ at cold temperatures become competent to take up DNA.

Calcium Phosphate. Salt that forms a precipitate with DNA. The complex is introduced into cells in culture.

Chloramphenicol. Antibiotic used to amplify plasmids in bacteria by restricting protein synthesis but not nucleic acid synthesis. Also the substrate for CAT (chloramphenicol acetyl transferase) enzyme assay.

Coomassie (Brilliant) Blue Stain. Blue stain that binds to proteins. Used to visualize protein bands in gels after electrophoresis and in protein assays.

CsCl (Cesium chloride). Salt used to form density gradients during ultracentrifugation to separate DNA by buoyant density.

CTAB (Cetyltrimethylammonium bromide). Nonionic detergent. Precipitates DNA at salt concentrations below $0.5\,M$; at high salt concentrations, complexes with polysaccharides and protein.

DEAE Dextran (Diethylaminoethyl-dextran). Used to permeabilize mammalian cells to facilitate DNA uptake.

Denhardt's Reagent. Solution used in hybridization reactions as a buffer and as a blocking agent. Contains BSA, Ficoll, and PVP.

DEPC (Diethyl pyrocarbonate). A RNase inhibitor. Added to water, solutions, and glassware to inhibit the action of nonspecific RNases.

Detergents. Water-soluble compounds that have a hydrophobic tail. Detergents have multiple uses, including solubilization of membranes and membrane proteins, cell lysis, denaturing proteins, wetting surfaces, and emulsification of chemicals.

Dextran Sulfate. Used as a "crowding agent" to effectively increase the concentration of nucleic acids in a hybridization mixture.

DMSO (Dimethyl sulfoxide). Compound used to solubilize chemicals and to permeabilize cells.

DNase. Any member of a class of naturally occurring enzymes that digest DNA.

dNTP (Deoxyribonucleoside triphosphate). Designates any of the four deoxyribonucleotides or a mixture of the four deoxyribonucleotides that comprise DNA.

DTT (Dithiothreitol). A reducing agent added to protein solutions to inhibit the formation of unwanted disulfide bonds.

EDTA (Ethylenediaminetetraacetic acid). Chelator of divalent ions. Added to DNA solutions to reduce the activity of nucleases and modifying enzymes by binding to and removing cofactors.

EGTA [Ethylene glycol-bis(β-aminoethyl ether)-N, N, N′, N′-tetraacetic acid]. Chelator of divalent ions; binds preferentially to Ca^{++}.

Ethanol. An alcohol that dehydrates and precipitates DNA when in high salt buffer.

Ether. Volatile organic solvent; used to remove residual phenol from aqueous solutions.

Ethidium Bromide. Dye that binds to nucleic acids and fluoresces when exposed to UV light. Used to visualize and quantify nucleic acids.

Ficoll. Inert polymer that acts as a "blocking agent" and/or "crowding agent" when included in a hybridization solution. A component of Denhardt's Reagent.

Formamide. Chemical added to destabilize nucleic acid duplexes.

Glycerol. Viscous, syrupy liquid added to solutions to reduce their freezing temperature, allowing solutions to be very cold without freezing. Increases the permeability of membranes and helps to stabilize proteins.

Glycogen. Inert carrier added to help ethanol precipitate low concentrations of DNA.

Guanidine Isothiocyanate (GITC). Salt that denatures proteins; added to preparations when isolating RNA to denature and remove contaminating RNases.

HEPES. Commercially available nontoxic buffer often used in cell culture.

8-Hydroxyquinoline. Antioxidant, added to redistilled phenol to prevent unwanted oxidation products.

Isoamyl Alcohol. Added to phenol/chloroform during extraction of protein from DNA to reduce interface foaming.

Isopropanol. Alcohol that will precipitate DNA when in high salt solution.

KCl (Potassium chloride). Commonly used salt.

LiCl (Lithium chloride). Salt which at 4 M will preferentially precipitate high molecular weight RNA.

β-Mercaptoethanol (Also, 2-mercaptoethanol). Reducing agent; added to protein solutions to restrict the formation of unwanted disulfide bonds.

MgCl$_2$ (Magnesium chloride). Source of Mg^{++}, essential co-factor of many enzymes that act on nucleic acids.

NaCl (Sodium chloride). Table salt, most common salt used to increase the ionic strength of a solution.

NaOH (Sodium hydroxide). Strong base; at 0.2 M will denature DNA.

NBT (Nitro blue tetrazolium). Used with BCIP as a color indicator for detection of alkaline phosphatase activity.

Nitrocellulose. Membrane; used to immobilize nucleic acids during a Southern blot.

PBS (Phosphate buffered saline). Isotonic buffer used for intact cells.

PCA (Perchloric acid). Acid used to denature and precipitate nucleic acids and proteins.

PEG (Polyethylene glycol). Waxy polymer available in varying sizes used to precipitate virus particles and protein. Added to hybridization solutions to reduce the liquid concentration in the solution.

Phenol. Water-insoluble organic solvent used as a protein denaturant. When added to an aqueous solution of DNA and protein, the protein is denatured and separates into the phenol phase and interface, and the nucleic acids are left in the aqueous phase.

Phenol/Chloroform. Addition of chloroform to phenol increases the phenol solubility of denatured proteins.

Phenol/Chloroform/Isoamyl alcohol. Mixture of solvents used in DNA isolation to denature and remove proteins from the solution. Isoamyl alcohol is added to improve phase separation.

PMSF (Phenylmethylsulfonyl fluoride). A protease inhibitor; added to crude cell extracts to reduce degradation of protein by proteases.

Potassium Acetate. Salt commonly added with ethanol to precipitate DNA.

Protease Inhibitors. Compounds that inhibit activity of unwanted proteases.

PVP (Polyvinylpyrrolidone). Inert polymer added to Denhardt's solution as a "crowding agent".

RNase. Any one of a class of enzymes that degrade RNA.

SDS (Sodium dodecyl sulfate; also known as lauryl sulfate, sodium salt). Ionic detergent that is effective as a protein denaturant. Commonly used to denature proteins before separation by electrophoresis.

Sodium Azide (NaN$_3$). Antimicrobial agent added to some solutions to reduce contamination and degradation by microorganisms.

Sorbitol. Inert sugar sometimes used as an osmotic support to protect cells after their cell wall is removed.

Spermine. Added to stimulate T4 kinase activity.

SSC (Standard saline citrate). Buffer used in Southern blotting.

Streptomycin Sulfate. Antibiotic that inhibits protein synthesis in bacteria.

TAE (Tris acetate EDTA buffer). Buffer for gel electrophoresis of nucleic acids.

TBE (Tris borate EDTA buffer). Buffer for gel electrophoresis of nucleic acids.

TCA (Trichloroacetic acid). Protein and nucleic acid denaturant and precipitant.

TEMED (N,N,N′,N′-Tetramethylethylenediamine). Stimulates polymerization of polyacrylamide.

Tetracycline. Antibiotic that inhibits protein synthesis in bacteria.

Tris [Tris(hydroxymethyl)aminomethane]. Commercial buffer used commonly in molecular biology.

Triton X-100. Nonionic detergent that solubilizes cells and cell membranes.

Urea. RNA denaturant; destabilizes DNA by reducing the temperature at which the DNA double helix denatures; protein denaturant.

X-Gal (5-Bromo-4-chloro-3-indolyl β-D-galactopyranoside). Color indicator for the presence of the enzyme β-galactosidase.

Xylene Cyanol. Tracking dye used in nucleic acid electrophoresis.

CHAPTER 24

Solutions: Associated Procedures and Information

I. INTRODUCTION

This chapter will discuss a variety of practical topics that are related to laboratory solutions. These topics include water purification, properly choosing and cleaning glass and plasticware, and sterilization and storage of solutions and biological samples.

Water is critical for life and is the most important reagent used in biology. For example, it has been estimated that a biotechnology company might require 30,000 L of purified water to produce a single kilogram of recombinant product ("Still Waters Run Deep," John Hodgson. *Bio/Technology*, 983–987, 12 October, 1994). Obtaining acceptably purified water is an essential task in any laboratory or biotechnology facility.

Glass and plastic labware must be appropriate for its application and must be cleaned properly. The individuals in a laboratory who are responsible for clean glassware play a key role. Dirty glassware may lead to invalid chemical analyses, alter experimental results, inhibit the growth of cells and microorganisms, or add contaminants to a product. Determining how to properly clean glassware and equipment and validation of cleaning methods are critical concerns in the biotechnology/pharmaceutical industry. FDA frequently cites companies for improper cleaning.

Once prepared, solutions often must be sterilized—without affecting the activity of their components. Various methods are available for this purpose. Finally, solutions and biological samples often must be stored in a way that prevents deterioration.

There are many methods used to purify water, clean glassware, store solutions, and perform other related tasks. This chapter will introduce some of the general concerns relating to these topics.

II. WATER PURIFICATION

A. Water and Its Contaminants

The quality of water used to prepare solutions is of utmost importance. For example, *in vitro* fertilization—the joining of egg and sperm in a petri dish—will not occur if the water used to make solutions has minuscule amounts of contaminants. In electron microscopy, small particulate contaminants can be magnified hundreds of thousands of times to become major nuisances. Cells in culture will die if extremely small amounts of contaminants are present. Minor levels of impurities in water can introduce major errors into analytical procedures.

Water is an excellent solvent and it readily dissolves contaminants from a wide variety of sources, Table 24.1. In the environment, water receives contaminants from industry, municipalities, and agriculture. Water dissolves minerals from rocks and soil that make it "hard." A variety of organic materials from dead plants, animals, and microorganisms are present in water. Gases from the air dissolve in water. Suspended particles enter

Table 24.1 CONTAMINANTS OF WATER

Types of Contaminants

1. **Dissolved Inorganics.** This term refers to matter not derived from plants, animals, or microorganisms that usually dissociates in water to form ions. Calcium and magnesium, the minerals that make water "hard," are examples of dissolved inorganics, as are the heavy metals cadmium and mercury.

2. **Dissolved Organics.** Organic compounds are broadly defined to be any materials that contain carbon and hydrogen. Organic materials in water may be the result of natural vegetative decay processes, and they may also be human-made substances such as pesticides.

3. **Suspended Particles.** This category includes diverse materials, such as silt, dust, undissolved rock particles, and metal scale from pipes.

4. **Dissolved Gases.** Water can dissolve naturally occurring gases, such as carbon dioxide, as well as gases from industrial processes, such as nitric and sulfuric oxides.

5. **Microorganisms.** This category includes bacteria, viruses, and other small agents.

6. **Pyrogens/Endotoxins.** Pyrogens and endotoxins are substances, derived from bacteria, that are of special concern in industries that manufacture products for use in humans and animals. See text for an explanation of these terms.

Related Terminology

1. **Total Organic Carbon (TOC).** A measure of the concentration of organic contaminants in water.

2. **Total Solids (TS).** A measure of the concentration of both the dissolved (TDS) and suspended solids (TSS) in a solution. Total dissolved solids include both organics (such as pyrogens and tannin) and inorganics (such as ions and silica). Total suspended solids include organics (such as algae and bacteria) and inorganics (such as silt). It is common to approximate TDS by measuring the conductivity, although conductivity is actually a measure only of conductive species.

water from silt, pipe scale, dust, and other sources. In the laboratory, contaminants may leach into water from glass, plastic, and metal containers.

If water is not sterilized, unwanted microorganisms will readily grow in it and may release toxic bacterial byproducts. These by-products are termed *endotoxins* and *pyrogens*. **Pyrogens** *are derived from lipopolysaccharides that are found in the cell walls of gram-negative bacteria. Pyrogen levels are measured in* **endotoxin units (EU/mL)**. Pyrogens were thought to be toxic, so they are often called *endotoxins*. In fact, pyrogens are not directly toxic; rather, they induce a dangerous immune response in mammals. This immune response leads to fever (hence the name *pyrogen,* "heat producing"), shock, and cardiovascular failure. Pyrogens induce adverse effects when present at minute levels; therefore, their elimination is of critical importance in the manufacture of drugs for human and animal use. Even sterile water can contain pyrogens. The term *endotoxin* may be used synonymously with *pyrogen,* or it is sometimes used more broadly to refer to any contaminating by-product of bacteria.

Removing these varied contaminants from water is an important task in almost any laboratory and is a major operation and expense in bioprocessing industries. To meet these requirements for purified water, laboratories and industrial facilities use a variety of carefully engineered processes.

B. What Is Pure Water?

There actually is no such thing as "pure water." Although a variety of effective methods are used to remove contaminants from laboratory water, regardless of how stringently water is purified it is never totally free of contaminants. Professional societies therefore set standards for the relative purity of water to be used in certain applications. Societies that set standards for laboratory (reagent-grade) water include the American Society for Testing and Materials (ASTM), the American Chemical Society (ACS), the College of American Pathologists (CAP), and The National Committee for Clinical Laboratory Standards (NCCLS). Table 24.2 summarizes these standards for different types of labo-

Table 24.2 *SUMMARY OF SPECIFICATIONS FOR VARIOUS TYPES OF WATER*

Standard CAP*/NCCLS**			
	Type I	**Type II**	**Type III**
Specific Conductance (micromhos, max.)	<0.1	<0.5	<10.0
Specific Resistance (megohm, min.)	>10.0	>2.0	>1.0
Total Matter (*mg/liter*, max.)	—	—	—
Silicate (*mg/liter*, max.)	<0.5	<0.1	<1.0
Potassium Permanganate Reduction (minutes, min.)	—	—	—
Culture/Colony Count (Colony Forming Units/mL)	<10.0	10	NA
pH	NA	NA	5.0–8.0

*CAP-The College of American Pathologists
**NCCLS-The National Committee for Clinical Laboratory Standards

ASTM[1]				
	Type I	**Type II**	**Type III**	**Type IV**
Electrical Conductivity, max. $\mu S/cm$ at 298 K (25°C)	0.056	1.0	0.25	5.0
Electrical Resistivity, min. $m\Omega \bullet cm$ at 298 K (25°C)	18.0	1.0	4.0	0.2
pH at 298 K (25°C)	*	*	*	5.0–8.0
Total Organic Carbon (TOC), max., $\mu g/L$	100	50	200	no limit
Sodium, max., $\mu g/L$	1	5	10	50
Chlorides, max., $\mu g/L$	1	5	10	50
Total Silica, max., $\mu g/L$	3	3	500	no limit
	Type I[2]	**Type II[2]**	**Type III[2]**	
Maximum Heterotropic Bacterial Count	10/1000 *mL*	10/100 *mL*	100/10 *mL*	
Endotoxin, EU[3]	<0.03	0.25	NA	

*The measurement of pH in Type I, II and III reagent water has been eliminated from this specification because these grades of water do not contain constituents in sufficient quantity to significantly alter the pH.
[1]ASTM, The American Society for Testing and Materials, Volume 11.01, Section DH93-91 Dated 1992.
[2]Microbiological contamination; when bacterial levels need to be controlled, reagent grade types should be further classified.
[3]Endotoxin Units.
(Reprinted courtesy of Barnstead/Thermolyne)

Table 24.3 *WATER QUALITY AND APPLICATIONS*

Application	Biologically Pure water for applications such as: -Tissue Culture -Life Sciences -Microbiology	Organically Pure water for applications such as: -HPLC -ICP/MS -GC/MS -TOC	Ionically Pure Reagent Grade Water		
			Type I	Type II	Type III
Feedwater: Average Raw Water *(10 grains per gal. or greater)	Still Pretreatment followed by Glass Distillation or Reverse Osmosis followed by Deionization and Ultrafiltration	Reverse Osmosis followed by: Deionization w/ UV oxidation or For **BOD, ***COD only Glass Distillation	Reverse Osmosis followed by: Deionization or Still Pretreatment and Distillation followed by Deionization	Still pretreatment followed by: Distillation	Still pretreatment followed by: Distillation
High Quality Raw Water *(3–10 grains per gal.)	Glass Distillation or Deionization and Ultrafiltration	Reverse Osmosis followed by: Deionization w/UV oxidation or For **BOD, ***COD only Glass Distillation	Deionization Distillation/ Deionization	Distillation	Distillation or Single Cartridge Deionization
Central Deionization, Central Distillation or Central Reverse Osmosis	Glass Distillation or Deionization and Ultrafiltration	Deionization w/UV oxidation or For **BOD, ***COD only, Glass Distillation	Deionization Still Pretreatment/ Distillation/ Deionization	Distillation	Distillation or Single Cartridge

*These values indicate the quality of water before purification.
**Biological oxygen demand.
***Chemical oxygen demand.
   (Reprinted courtesy of Barnstead/Thermolyne.)

ratory water. Table 24.3 gives examples of situations where purified water is required, the standards applicable to those situations, and water treatment methods for each application.

The water quality standards of the various organizations are not identical, but they have general features in common. Type I is the highest class of purity. Type I water is used for most analytical procedures, such as trace metal analysis, because it has very low levels of contaminants. Type I water is also used for tissue culture, high performance liquid chromatography (HPLC), electrophoresis buffers, immunology assays, preparation of standard solutions, and reconstitution of lyophilized materials.

At this time, Type II water must be prepared by distillation to meet ASTM standards. It may have somewhat higher levels of impurities than Type I water, but it is still suitable for most routine laboratory work, such as microbiology procedures.

Type III water is acceptable for some applications, such as initial glassware rinses (final rinses should be in whatever type of water is required for the procedure to be performed) and preparing microscope slides to examine tissues.

There are applications where water that exceeds Type I standards is required. For example, analytical tests using HPLC have become extremely sensitive and can detect organic compounds in samples at levels as low as a few parts per billion. Type I water may have more impurities than this and therefore is not acceptable for extremely sensitive HPLC work. Cell culture techniques for some biotechnology applications similarly require water with lower pyrogen levels and fewer contaminants than are specified by Type I standards. There are water purification systems available that meet these stringent requirements, but they must be properly operated and maintained to deliver water of the highest quality.

The water used for producing pharmaceutical products must meet different standards than laboratory water. In the United States, these standards are found in the United States Pharmacopeia (USP). The USP standards distinguish between **water for injection** (WFI) and **purified water**. WFI is used to make drugs for injection. USP "purified water" is used more generally in a number of pharmaceutical and cosmetic applications. The USP specifies both the quality of WFI and its method of production. In the United States at this time, WFI must be produced either by distillation or reverse osmosis. WFI must also meet a stringent requirement for pyrogen levels and have no more than 0.25 endotoxin units per milliliter. "Purified water" is defined by its chemical specifications and may be produced by a variety of techniques.

C. Methods of Water Purification

i. OVERVIEW

Source or **feed water** *is the water that is to be purified.* Most laboratories and biotechnology facilities begin with municipal (tap) water that is partially purified to make it safe for drinking. Such water has reduced levels of microorganisms and other contaminants, but contains high levels of chlorine and varying levels of many other contaminants. The quality of municipal water varies greatly depending on geographic location, the season, treatment methods used, and other factors. One of the oldest biotechnology industries, the brewing industry, has a tradition of choosing manufacturing sites based on the quality of local water available.

Some water sources contain contaminants that are difficult to remove, even with the best treatment methods. Because source water is so variable, and because each application has different purity requirements, water purification systems are tailored to meet the needs of a particular laboratory or industrial facility. There are a variety of methods of purifying water that vary in their costliness and ability to remove different kinds of contaminants. These are summarized in Table 24.4. Because each purification method removes some, but not all, types of contaminants, it is common for water-purification systems to use more than one purification method.

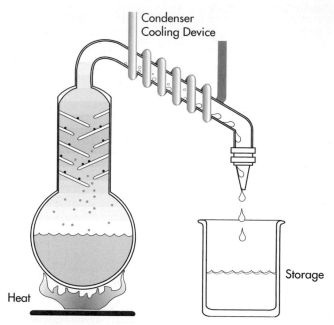

Figure 24.1.   Distillation. (Illustration courtesy of Millipore Corporation.)

ii. DISTILLATION

Distilled water is often used for making laboratory solutions. **Distilled water** *is prepared by heating water until it vaporizes,* Figure 24.1. The water vapor rises until it reaches a **condenser** *where cooling water lowers the temperature of the vapor and it condenses back to the liquid form.* The condensed liquid flows into a collection container. Most contaminants remain behind in the original vessel when the water is vaporized. Ionized solids, organic contaminants that boil at a temperature higher than water, pyrogens, and microorganisms are all removed from distilled water. A few contaminants, however, cannot be removed from water by distillation. These include dissolved gases such as carbon dioxide, chlorine, ammonia, and organic contaminants that volatilize at a low temperature.

Distillation is a well-established and frequently used method of water treatment. It is relatively expensive compared with other methods of water treatment, however, because of the substantial energy input required to heat the water.

Table 24. 4   A COMPARISON OF WATER-PURIFICATION METHODS

	Distillation	Deionization	Filtration	Reverse Osmosis	Adsorption	Ultrafiltration	UV Oxidation
Dissolved Ionized Solids	E/G	E	P	G	P	P	P
Dissolved Organics	G	P	P	G	E	G	G
Dissolved Ionized Gases	P	E	P	P	P	P	P
Particulates	E	P	E	E	P	E	P
Bacteria	E	P	E	E	P	E	G
Pyrogens	E	P	P	E	P	E	P

E = Excellent; G = Good; P = Poor (Courtesy of Barnstead/Thermolyne)

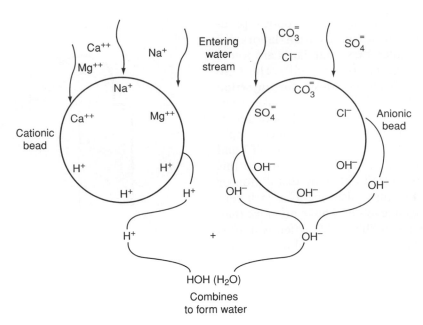

Figure 24.2.   Deionization. (Courtesy of Osmonics, Inc., Minnetonka, MN.)

iii. ION EXCHANGE

Ion-exchange *removes ionic contaminants from water.* During ion exchange water flows through cartridges packed with bead-shaped resins, called *ion exchange resins.* As the water flows past the beads, ions in the water are exchanged for ions that are bound to the beads. There are two types of resins. **Cationic resins** *bind and exchange positive ions.* **Anionic resins** *bind and exchange negative ions.*

Water softeners (such as many people have in their homes) are a type of ion exchanger. **Water softeners** *exchange positive ions that make water hard, particularly calcium and magnesium, with sodium ions.* Water softeners thus remove calcium and magnesium from water and add sodium to it. Water softeners use cationic beads that initially have Na^+ ions loosely bound to them. As water flows past these beads, positive ionic contaminants in the water exchange places with the Na^+ ions. The contaminants remain linked to the beads while the sodium ions enter the water stream. Water softening may be used as a preliminary step in water purification, but it is not adequate for laboratory water purification because it leaves sodium ions and all negative ions in the water.

Deionization *is an ion exchange process used in the laboratory to remove "all" ionic contaminants from water,* Figure 24.2. Deionization involves both cation and anion exchange beads. The anionic resin beads initially have loosely bound hydroxyl ions, whereas the cationic beads have numerous hydrogen ions loosely bound to their surfaces. As water passes by the anion exchange beads, negatively charged ions in the solution take the place of the OH^- ions on the beads. The OH^- ions enter the water stream, but the contaminants remain bound to the resin beads. Positive ionic contaminants in the water similarly exchange places with H^+ ions that are attached to the cationic beads. The hydrogen ions enter the water; the positive ionic contaminants remain bound to the beads. The H^+ ions and the OH^- ions that enter the water stream combine with one another to form more water molecules.

Deionized water is more pure than tap or softened water because the ionic contaminants have been removed, but it is generally not acceptable for making laboratory solutions because it may contain organic contaminants, pyrogens, and microorganisms.

iv. CARBON ADSORPTION

Carbon adsorption *is an effective method for removing dissolved organic compounds from water.* Carbon adsorption uses **activated carbon,** *which is traditionally made by charring wood by-products at high temperatures, then "activating" it by exposure to high temperature steam.* There is also a synthetic form of activated carbon that is produced from styrene beads. Natural activated carbon can release some ionic contaminants into the water being treated; the synthetic version is less likely to release ionic contaminants.

Activated carbon has an extremely porous, honeycomblike, structure. As water passes through the activated carbon, chlorine and organic impurities adsorb (stick) to the carbon and are thus removed, Figure 24.3. Activated carbon is sometimes used before deionization to protect the resins from contamination by organic materials and chlorine.

v. FILTRATION METHODS

Filtration removes particles as the water passes through the pores or spaces of a filter. There are four types of filters used in water treatment:

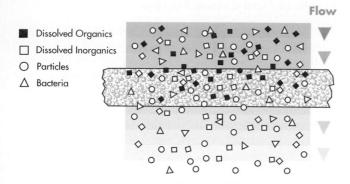

- ■ Dissolved Organics
- □ Dissolved Inorganics
- ○ Particles
- △ Bacteria

Figure 24.3. Activated Carbon. As water flows through the porous activated carbon, chlorine and dissolved organics are selectively adsorbed by the carbon. (Courtesy of Millipore Corporation.)

1. **Depth filters** *are made of sand or matted fibers and retain particles through their entire depth by entrapping them,* Figure 24.4a. Depth filters are useful for removing relatively large particles, greater than about 10 μm, from the water stream. (Pollen and some yeast cells are examples of substances in this size range.) Depth filters may be used at the beginning of a purification system as an inexpensive method to remove large particles and debris that might otherwise clog and foul the downstream elements of the system.

2. **Microfiltration membrane filters (also called screen filters or microporous membranes)** *are made by carefully polymerizing cellulose esters or other materials so that a membrane with a specific pore size is formed,* Figure 24.4b. Water treatment filters often have pore sizes on the order of 0.20 μm because bacteria are too large to penetrate such pores. Microfiltration membrane filters are useful for removing bacteria and particles above a specified pore size, but they are not useful for removing smaller, dissolved molecules.

3. **Ultrafiltration membranes** *do not have measurable pores and will prevent the passage of dissolved molecules, including most organics,* Figure 24.4c. Materials down to 1000 D in molecular weight (smaller than most proteins) can be separated from water using ultrafilters. Two important applications of ultrafilters are the removal of viruses and pyrogens from water. Small, dissolved inorganic molecules can pass through ultrafilters.

4. **Reverse osmosis (RO) membranes** *are still more restrictive than ultrafiltration membranes,* Figure 24.4d. Materials down to 300 D in molecular weight can be removed from water using RO and so this method is effective in removing viruses, bacteria, and pyrogens. Moreover, RO membranes also reject ions and very small dissolved particles, such as sugars.

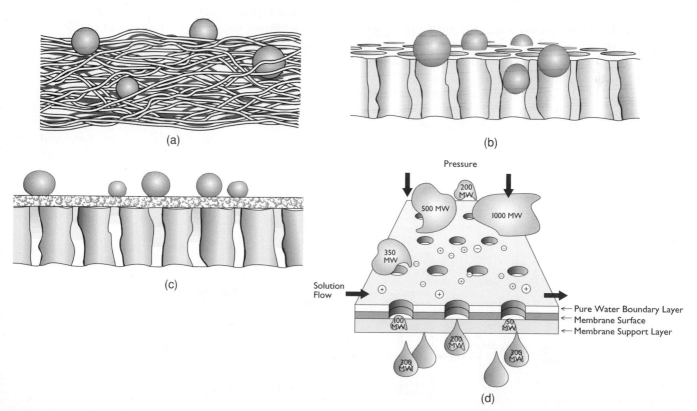

Figure 24.4. Filtration Methods. a. Depth filter. **b.** Membrane filter with pores of a particular size. **c.** Ultrafilter used to retain molecules that are above a particular molecular weight. **d.** Reverse osmosis. (**a-c.** Illustrations courtesy of Millipore Corporation. **d.** Illustration courtesy of Osmonics, Inc., Minnetonka, MN.)

In reverse osmosis, water under pressure flows over a special, thin RO membrane. The membrane allows water to pass through, but rejects a large proportion (on the order of 95–99%) of all types of impurities (particles, pyrogens, microorganisms, colloids, and dissolved organic and inorganic materials). The permeate (water that has passed through the membrane) will contain very low levels of contaminants that are able to pass through even an RO membrane, but most types of contaminants are greatly reduced. Because RO membranes are very restrictive, the flow rate through them is slow.

Reverse osmosis is similar to ultrafiltration in that both involve membranes that reject small materials. The major differences between ultrafiltration and RO include: (1) RO can retain very small solutes, including salt ions. Ultrafiltration membranes allow such very small solutes to pass through. (2) Higher pressures are used in RO to force water through the membrane. (3) An ultrafiltration membrane retains particles based almost solely on their size. An RO membrane retains materials based both on their size and on ionic charge.

vi. OTHER METHODS: ULTRAVIOLET OXIDATION AND OZONE STERILIZATION

Ultraviolet (UV) oxidation *is a method for the removal of organic contaminants from water.* Water is passed continuously (for example for 30 minutes) across the light path of an ultraviolet lamp that emits light at a wavelength of 185 *nm*. The organic compounds in the water oxidize to simple compounds such as CO_2. UV light at a wavelength of 254 *nm* will kill bacteria and is therefore sometimes used to sterilize water.

In industrial settings ozone is also sometimes used to sterilize water. Ozone kills bacteria by rupturing their membranes and acts much more quickly than chlorine. Both ozone and chlorine must be removed from the water by later purification steps.

vii. WATER PURIFICATION SYSTEMS

Many laboratories and most biotechnology and pharmaceutical facilities use water that is purified by a combination of methods, Figure 24.5. The choice of methods and system design will depend on the applications for which the water is to be used and the quality of source water available. For example, a water purification system to prepare Type I water might include the following steps:

1. *Tap (source) water is softened or distilled.*
2. *The water then passes through activated carbon to remove organics and chlorine.*
3. *Reverse osmosis is used next and the treated water is moved to a storage tank until needed.*
4. *When a user wants purified water, water from the RO tank is run through a series of ion exchange cartridges in which all ionized contaminants are removed.*
5. *The water might be filtered through an ultrafiltration membrane.* This final filtration step removes almost all microorganisms including those that the water picked up while traveling through the purification system.

viii. HANDLING OF REAGENT WATER

Highly purified water is an extremely aggressive solvent. It will readily leach contaminants from any vessel in which it is stored and will also dissolve CO_2 from the air. In addition, bacteria can multiply in purified water during storage. Type I water, therefore, cannot be stored and must be used immediately after it is produced. Type II water can be stored for short periods of time.

ix. A NOTE ABOUT TERMINOLOGY

There are a variety of methods used to purify water. Distillation, however, is the "traditional method," and it is common for people to speak of highly purified water as being "distilled" (DI) whether it is purified by RO, ultrafiltration, or actually by distillation. Deionized water is less pure and should never be referred to as "distilled" water. In this book, the term *purified water* is used to

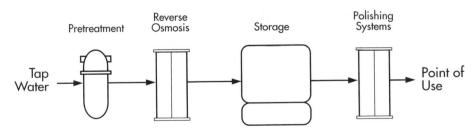

Figure 24.5. A Multistep Water Purification System. (Courtesy of Millipore Corporation.)

mean water that has been distilled or which has been run through a system as shown in Figure 24.5. Because the terminology of water purification is used inconsistently, standard operating procedures should explicitly state the source of water required.

D. Operation and Maintenance of Water Purification Systems

i. MONITORING THE SYSTEM FOR QUALITY

Purified water must be monitored periodically to be certain that it meets quality requirements. Because there are a variety of contaminants in water, no single test is adequate to measure water purity. The following are some quality parameters that can be monitored:

1. **Resistivity (resistance).** Ionic contaminants in water are easy to detect because they make water more conductive (i.e., less resistant to current flow). Resistance, therefore, is a measure of water purity: The more water resists electrical current flow, the fewer ionic contaminants it contains.

   Water purification systems typically have an attached meter to read the water resistance. A user who wants to draw water flushes the system for a minute or so, discarding the rinse water. As water is flushed through the system the resistivity increases until it reaches an acceptable level. The theoretical maximum ionic purity for water is 18.3 *megohm-cm* and 17.0 *megohm-cm* is considered an acceptable value for Type I water in many laboratories. Nonionic contaminants do not affect resistivity, so even if the resistance of water is 17.0 *megohm-cm*, it may contain contaminants.

2. **Bacterial counts.** Bacterial counts are used to monitor levels of microorganisms in water. A particular volume of the water is filtered through a membrane filter that traps bacteria on its surface. The filter is incubated on nutrient medium. *The numbers of colonies growing on the medium are counted to give the number of* **colony forming units (CFU)** *per volume of water.* The various standards for water quality specify a maximum number of colony forming units that are acceptable.

3. **Pyrogens.** Most clinical, analytical, and research laboratories do not check for pyrogen contamination but this assay is essential in the pharmaceutical industry. The *Limulus* **Amebocyte Lysate Test (LAL)** *is used to monitor pyrogens in a water sample.* The LAL test requires an extract of blood from the horseshoe crab, *Limulus polyphemus.* Different dilutions of the water to be tested are mixed with the *Limulus* blood extract and any pyrogens present cause the extract to clot. The

clotting results can be converted to **endotoxin units per milliliter (*EU/mL*).**

4. **Organic carbon.** Organic contaminants can interfere with analyses and can indicate that bacteria have contaminated the water treatment system. Organic carbon can be monitored using a chemical method (involving potassium permanganate) or instrumental methods.

5. **pH.** Ultrapure water that is exposed to air reacts with carbon dioxide, resulting in a pH of about 5.7. If it is in a covered container, pure water will have a pH of about 6.0. In some situations the pH of the water is important and is monitored.

Other water quality parameters, such as particulate and silicate levels, can be tested where necessary. Whether various tests are performed, and if so, how often, depends on the needs of the laboratory or industrial facility, Figure 24.6.

ii. MAINTENANCE

Water purification systems need to be operated and maintained properly to ensure high-quality water. The maintenance required and its frequency will vary depending on the type of laboratory, the contaminants present in the source water, and any regulations that are applicable. Some general considerations include:

1. **Distilled water apparatus.** A distilled water apparatus requires frequent, sometimes daily, cleaning; otherwise, it can become contaminated by microorganisms and their by-products. Distilled water should be prepared freshly each day because it can absorb contaminants from storage vessels and microorganisms can grow in it.

2. **Ion exchange resins.** Ion exchange resins must be regenerated once they have exchanged all their Na^+, H^+, or OH^- ions for contaminants. Regeneration is sometimes performed on-site, and sometimes cartridges are sent off-site for maintenance. Microorganisms can attach to deionization beads, thereby adding contaminants to the water. Deionization systems are therefore periodically sanitized.

3. **Filters.** Filters must be tested when installed to be certain they have no small holes and are effective. They must be periodically flushed and sanitized to remove contaminants from their surfaces. Filters must be replaced occasionally because they can develop defects over time and bacteria can actually grow through their pores.

4. **Activated Carbon.** Activated carbon may become colonized by bacteria and may become clogged by particulates. Activated carbon beds must therefore be washed to remove particulates. Activated carbon must also be periodically recharged or replaced as active sites are filled by contaminants.

Water

H$_2$O M$_r$ 18.02 [7732-18-5] EEC No 2317912

95284	*BioChemika* for molecular biology; filtered through a 0.2 μm membrane filter; DEPC-treated and auto-claved; DNases, RNases, proteases and phosphatases; none detected; d$_4^{20}$ 1.00; specific resistance at 25°C: 1.8•10^7 Ωcm, measured at production
95285	*BioChemika* for endotoxin trace analysis; ultrafiltered and autoclaved; Endotoxins: none detected (LAL-test); d$_4^{20}$ 1.00; specific resistance at 25°C: 1.8•10^7 Ωcm, measured at production
95289	*BioChemika* for cell biology; ultrafiltered and autoclaved; mycoplasma, chlamydiae and viruses: none detected (DAPI-test); d$_4^{20}$ 1.00; specific resistance at 25°C: 1.8•10^7 Ωcm, measured at production
95301	for ^{2}H-NMR-spectroscopy; deuterium-depleted; Isotopic purity: ^{2}H$_2$O < 0.0001%; d$_4^{20}$ 1.00
95311	Selectophore®; d$_4^{20}$ 1.00
95303	for UV-spectroscopy; d$_4^{20}$ 1.00

Fluka-Guarantee:

A:		$\lambda(nm)$:	200
d: 1 cm, Ref. cell: air		A$_{max}$:	0.01

95306	B&J Brand (product line of Burdick & Jackson)
95305	for inorganic trace analysis; filtered through a 0.2 μm membrane filter; d$_4^{20}$ 1.00; specific resistance at 25°C: 1.8•10^7 Ωcm, measured during production

Fluka-Guarantee:

Residue on evaporation . . . . < 0.001%   Residue on ignition . . . . < 0.0005%   KMnO$_4$-reducing matter
(as O). . . . . < 0.000003%

Bromide (Br) . . . . < 0.000001%	Ca . . . . < 0.000001%	Mn . . . . < 0.0000005%
Chloride (Cl) . . . . < 0.000001%	Cd . . . . < 0.0000005%	Mo . . . . < 0.0000005%
Iodide (I) . . . . < 0.000001%	Co . . . . < 0.0000005%	Na . . . . < 0.000001%
Nitrate (NO$_3$) . . . . < 0.000001%	Cr . . . . < 0.0000005%	NH$_4$ . . . . < 0.0000005%
Phosphate (PO$_4$) . . . . . . < 0.000001%	Cu . . . . < 0.0000005%	Ni . . . . < 0.0000005%
Sulfate (SO$_4$) . . . . < 0.000001%	Fe . . . . < 0.0000005%	Pb . . . . < 0.0000005%
Al . . . . < 0.0000005%	K . . . . < 0.000001%	Sr . . . . < 0.0000005%
Ba . . . . < 0.0000005%	Li . . . . < 0.0000005%	Zn . . . . < 0.0000005%
Bi . . . . < 0.0000005%	Mg . . . . < 0.0000005%	

95286	for organic trace analysis; trace organics < 0.000003% (GC); d$_4^{20}$ 1.00

Figure 24.6. Catalogue Specifications for Purified Water. This manufacturer produces purified water for sale. Observe that depending on the application, different characteristics of the water are featured and different purification methods are specified. (Reprinted with permission of Fluka Chemie AG.)

5. UV lamps. As UV lamps age, their light output decreases. Bulbs must therefore be replaced about once a year.

iii. DOCUMENTATION

As with any other aspect of a quality control program, the operation, maintenance, and monitoring of water purification systems requires planning and documentation. This includes developing standard operating procedures, training operators, planning routine maintenance and monitoring, establishing logbooks, developing written procedures to deal with problems, and documenting corrective actions taken.

EXAMPLE PROBLEM

a. *Ultrapure, sterile water is stored in a very clean, sterile glass bottle for 4 days. Speculate as to the quality of this water after storage.*

b. *Ultrapure, sterile water is stored in a very clean, sterile plastic bottle for 4 days. Speculate as to the quality of this water after storage.*

ANSWER

a, b. Although the bottles are clean and the conditions are sterile, ultrapure water is an extremely aggressive solvent and will leach heavy metals (which are minor contaminants of glass) from the glass container into the water. Ultrapure water will similarly leach organic carbon from the plastic and will become contaminated.

III. GLASS AND PLASTIC LABWARE

A. The Characteristics of Glass and Plastic Labware

Glass and plastic labware is used in laboratories for preparing and storing solutions, for containing cells and microorganisms, for holding experimental reaction mixtures, and so on. There is a wide variety of glass and plastic items available for laboratory use, so these items can be matched to their application.

Various grades of glass differing in strength, heat resistance, and other factors are used to make beakers, flasks, graduated cylinders, and pipettes. Glass is relatively inert when in contact with chemicals and is therefore used to make containers for chemicals. Glass, however, may adsorb metal ions and proteins from solutions and is not recommended for the storage of metal standards. Also, trace amounts of contaminants, such as sodium, barium, and aluminum can leach out of glass containers. In normal laboratory work these trace contaminants are not significant. However, in some analytical and pharmaceutical applications, they are problematic.

Borosilicate glass *is used to make general purpose laboratory glassware that is strong, resistant to heat and cold shock, and does not contain discernable contamination by heavy metals.* Borosilicate glass is sold, for example, under the trade names of Pyrex® (Corning Glass Works, Corning, N.Y.) and Kimax® (Kimble Glass Company, Vineland, NJ). Corex® (Corning) is a type of glass that is resistant to physical stress and is therefore used to make centrifuge tubes. Type I glass refers to highly resistant borosilicate glass used in the manufacture of ampules, vials, and other containers for pharmaceutical applications. Additional information about types of glass is shown in Table 24.5.

Plastics are less breakable than glass, and they come in a variety of types that vary in their properties and applications. For example, some types of plastics are more resistant than are others to heat, acids, organic solvents, and the forces in a centrifuge. Various plastics are incompatible with certain chemicals and may darken, become brittle, crack, or dissolve when exposed to them. Some plastics are porous and stored liquids may evaporate from them. Some plastics are transparent, others are not. Table 24.6 briefly summarizes the qualities of common types of plastics and their uses. Consult a chemical compatibility chart when using plastics to hold chemicals, as shown in the Appendix to this chapter.

B. Cleaning Glass and Plastic Labware

i. OVERVIEW

There are many procedures used to clean labware. Which procedure is best depends on the items to be cleaned, the soils contaminating the labware, and the application for which they will be used. Proteinaceous deposits, organic residues, metals, greases, and microorganisms are examples of contaminants that may need to be removed. Note that safety precautions are required when washing labware that was used for hazardous materials, or if the cleaning agents used are dangerous.

In general, labware washing has five steps:

1. *Prerinse.* A prerinse or soak immediately after use prevents contaminants from drying onto labware.
2. *Contaminant Removal.* Contaminants are typically removed using detergents or solvents in conjunction with a physical method such as scrubbing or exposure to jets of water.
3. *Rinse.* All traces of detergent and cleaning solvents are rinsed away.
4. *Final Rinse.* Any residues from the rinse in Step 3 are removed. Purified water is used for final rinses.
5. *Drying.*

The effectiveness of cleaning depends on four key factors: The proper choice of detergent, the effectiveness of the mechanical action, how long the items are subjected to mechanical washing, and the temperature of the wash water. Hotter water is more effective than is cooler water.

ii. DETERGENTS

There are various types of detergents used for cleaning labware that vary in their pH, additives, and other qual-

Table 24.5 COMMON TYPES OF GLASS

Type	Qualities	Purpose
Borosilicate (Pyrex®, Kimax®)	High resistance to heat and cold shock. Low in metal contaminants.	General purpose
Corex®	Aluminosilicate glass, high resistance to pressure and scratching.	Centrifugation
High Silica (quartz)	Greater than 96% silicate. Excellent optical properties.	Optical devices, mirrors, cuvettes
Low actinic*	Red-tinted to reduce light exposure of contents.	To contain light-sensitive compounds
Flint	Soda-lime glass containing oxides of sodium, silicon, and calcium. Poor resistance to temperature changes and high temperature, poor chemical resistance.	Disposable glassware, such as pipettes
Optical	Soda-lime, lead, and borosilicate.	Used in prisms, lenses, and mirrors

*Actinic light is defined as light capable of producing a photochemical reaction. This is usually near ultraviolet or blue visible light with wavelengths between 290 and 450 *nm*.
(Information in this table is primarily adapted from *Clinical Chemistry: Theory, Analysis, Correlation,* Lawrence A. Kaplan and Amadeo J. Pesce. 3rd edition, Mosby, St. Louis, 1996.)

Table 24.6   COMMON PLASTICS AND THEIR QUALITIES

Type	Qualities	Purpose
Polyethylene	Biologically and chemically inert, resistant to solvents. Strong oxidizing agents will eventually make polyethylene brittle. Not resistant to autoclaving.	Used for disposable plasticware.
Polypropylene	Translucent, can be autoclaved, resistant to solvents. More susceptible than polyethylene to strong oxidizers. Biologically inert. Thin-wall products permeable to CO_2 and O_2.	General purpose.
Polystyrene	Rigid, clear, brittle. Biologically inert. Sensitive to organic solvents. Used where transparency is an advantage. Melts in autoclave. Permeable to CO_2.	Used for disposable plasticware and tissue culture products.
Polyvinyl chloride (PVC)	Can be made soft and pliable.	Often used for laboratory tubing.
Polycarbonate	Transparent and strong. Sensitive to bases, concentrated acids, and some solvents.	Used for centrifuge-ware.
Teflon® (DuPont) (PTFE)	Highly resistant to chemicals and temperature extremes. Inert, comparatively costly.	Wide variety of uses including fittings in instruments, stir bars, stopcocks, tubing, bottle cap liners, water distribution systems.
Nylon	Strong, resistant to abrasion, resistant to organic solvents. Poor resistance to acids, oxidizing agents, and certain salts.	Widely used for membranes and filters in biology.

Based primarily on information from Nalgene Catalogue and Falcon Labware Catalogue

ities. It is necessary to choose the correct detergent and to dilute and use it properly. The choice of detergent is based on what is to be cleaned, what contaminants must be removed, and the physical method of cleaning that will be used. Table 24.7 lists some general guidelines to help determine which detergent to use.

Chromic acid washes were often routinely used in the past to remove proteinaceous and lipid contaminants from glassware. The use of chromic acid, however, has serious disadvantages. It is extremely corrosive and hazardous, so extreme caution is necessary to avoid injury when using such washes. Chromic acid contains chromium, a heavy metal, which requires costly disposal in a special hazardous waste site. Because of the hazards and costs associated with acid baths, they are now less common. Proper use of detergents, milder acids, such as 1 M HCl and 1 M HNO_3, and organic solvents usually eliminate the need for strong acid washes.

iii. PHYSICAL CLEANING METHODS

Detergents are used in conjunction with physical methods of cleaning. Depending on the situation, a variety of physical methods are used to remove contaminants from labware. These include:

Table 24.7   GENERAL GUIDELINES FOR CHOOSING A DETERGENT

1. *Alkaline Cleaners.* A broad range of organic and inorganic contaminants are readily removed from glassware by mildly alkaline cleaners.
2. *Acid Cleaners.* Metallic and inorganic residues are often solubilized and removed by acid cleaners.
3. *Protease Enzyme Cleaners.* Proteinaceous residues are effectively digested by protease enzyme cleaners.
4. *Cleaners for Radioactive Residues.* Radioactive residues are often removed by detergents with high chelating (binding) capacity. Detergents are available that are specifically designed to remove radioactive contaminants.
5. *Plasticware.* Plasticware is generally cleaned with mild, neutral pH detergents.
6. *Dishwashers.* Laboratory dishwasher detergents should be low-foaming.
7. *Manual Cleaning.* For manual, ultrasonic, and soak cleaning, a mildly alkaline (pH 8–10), foaming detergent will often work well.
8. *Corrosive Cleaners.* Some detergents are corrosive and some require expensive means of disposal. Consult the manufacturer about corrosiveness and disposal.
9. *Instructions.* Prepare and use a detergent according to the manufacturer's instructions.
10. *Phosphorus and Nitrogen Analysis.* Glassware to be used for analyzing these elements should not be washed with detergent. $NaHCO_3$ solution is recommended in the *Water and Wastewater Examination Manual* (V. Dean Adams. Lewis Publ., Michigan, 1990), followed by a soak in 6 N HCl.
11. *Tissue Culture.* Use detergents specifically formulated for cleaning items used in tissue culture.

1. **Soaking.** This simply involves submerging the items to be cleaned in a cleaning solution. The items are sometimes rinsed and used immediately, but further cleaning is usually required.

2. **Manual cleaning.** Items are washed by hand using a cloth, sponge, or brush followed by thorough rinsing. This method is not suitable for washing large numbers of items. Gloves and eye protection should be used as described by the detergent manufacturer or if labware was used for hazardous substances. Centrifuge glassware and tubes, metal equipment, and most plasticware should not be cleaned with metal brushes or other abrasives. Scratches on centrifuge tubes can lead to material fatigue and breakage when the tube experiences the force of centrifugation.

3. **Machine Washing.** Most laboratories and production facilities wash labware in commercial dishwashers. Glassware washers have a single chamber that is sequentially filled and emptied through its cycle. The cycle typically consists of a prewash with room temperature water, a wash cycle of 2–15 minutes in a detergent solution at 60–90°C, and multiple rinses. The items may be dried in the washer or in a separate drying oven. General directions include:

   - Load items so that their open ends face the spray nozzles.
   - Place difficult-to-clean articles in the center of the rack, preferably with the spray nozzles pointing directly into them.
   - Place small items in baskets.
   - Use low foaming detergents as directed by the manufacturer.
   - Use hot water, above 60°C (140°F).

4. **Automatic Siphon Pipette Washing.** Reusable glass pipettes are usually washed in long cylinders designed for this purpose. The cylinders are connected to a water supply so that the pipettes can alternately be filled with cleaning solution and drained. General directions include:

   - Presoak the pipettes by complete immersion in washing solution immediately after use.
   - Place detergent in the bottom of the washer.
   - Place the pipettes in a holder in the washer.
   - Attach the washer to the water supply and run the water so that the pipettes are filled and drained completely.
   - Continue to wash until all the detergent has washed and drained through the pipettes.
   - Use a final rinse as appropriate.

5. **Ultrasonic washing.** Ultrasonic washers expose items to high-pitched sound waves that effectively penetrate and clean crevices, narrow spaces, and other difficult-to-reach surfaces. Metal and glass are usually safely washed in these cleaners, but some types of plastic are weakened by ultrasonic cleaning. General directions include:

   - Dilute the cleaning solution and add to cleaning tank, run the machine several minutes to degas the solution, and allow the heater to come to temperature.
   - Place items in racks or baskets in the machine.
   - Align irregularly shaped items so that the long axis faces the transducer that generates sound waves.
   - Immerse the articles for 2–10 minutes.

6. **Clean in Place (CIP).** Large, nonmovable items, such as fermentation vessels or pipes, must be cleaned in place. These items are often designed so that cleaning agents can be pumped into and circulated through the device.

iv. RINSING

Rinsing is a key step in washing labware. Standard washing procedures often specify that items should be rinsed three times with water. For machine washing, this means three rinse cycles are used. For manual washing, the item is filled and emptied at least three times. Tap water may be used in certain applications for initial rinses, but it should be followed by several rinses in purified water. Deionized water heated to 60 to 70°C is an effective rinsing agent. The type of water for the final rinse depends on the application for which the item will be used.

v. DRYING

Do not dry labware by wiping with any type of towel or tissue. Towels and tissues can contaminate the item with fibrous and chemical residues. Items can be allowed to air dry on racks or can be more rapidly dried in heated drying ovens at temperatures below 140°C. Note, however, that volumetric glassware should not be heated because heat causes expansion and contraction of the glass that may alter its calibration. Some types of plastics are also sensitive to heat.

vi. STANDARD PROCEDURES FOR WASHING GLASS AND PLASTIC LABWARE

There are many appropriate procedures used to wash glassware. A common, general purpose glassware washing procedure is shown in Box 1. (This procedure is the

---

Box 1 A GENERAL METHOD FOR CLEANING LABWARE

1. *Wash containers in hot water with laboratory-grade, nonphosphate detergent.*
2. *Rinse three times with copious amounts of tap water to remove detergent.*
3. *Rinse three times with Type I water.*
4. *Oven dry.*
5. *Cool in enclosed, contaminant-free environment.*
6. *Cover to avoid exposure to dust.*

U.S. Environmental Protection Agency Procedure for Containers, Protocol B.) Table 24.8 lists some "tips" regarding glassware washing.

vii. QUALITY CONTROL OF LABWARE WASHING

It is very important that labware is properly cleaned. Each production facility and laboratory must have procedures for washing items according to the applications for which those items will be used. Laboratories that meet GMP and ISO 9000 requirements must demonstrate that their cleaning procedures are, in fact, effective. Determining the effectiveness of a cleaning procedure can be difficult and there is no standard method for doing so. Some general considerations for assuring the quality of washing operations include:

- Machine washers can be monitored to see that they maintain the proper temperature throughout their cycle, have no blocked spindles, and that the quality of the incoming water is adequate.

- The effectiveness of a cleaning method can be evaluated by intentionally soiling items with substances that are visible. After cleaning, residues can be visually detected.

- In pharmaceutical facilities and other sites where consistently and stringently clean labware is required, testing methods may be very sophisticated. Items may be periodically swabbed and the swabs tested by HPLC, spectrophotometry, or other methods to detect contaminants.

As with any other aspect of a quality control program, the operation, maintenance, and monitoring of washing systems requires planning and documentation. Standard operating procedures, operator training, plans for routine maintenance and monitoring, logbooks, written procedures to deal with problems, and documentation of corrective actions taken are components of a quality program.

Table 24.8 GUIDELINES REGARDING GLASS AND PLASTICWARE WASHING

1. *After washing, glassware should be visually inspected to be sure it is clean and not cracked.* Cracked glassware should be repaired or discarded.

2. *A simple, common test of the cleanliness of glassware is that it wets uniformly with distilled water.* Grease, oils, and other contaminants cause water to bead, or flow unevenly across the glass surface. Uneven wetting of the surface of volumetric glassware will alter the volumes measured and may distort the meniscus.

3. *Plastic containers may be preferable to glass for trace-metal analysis.* Soaking plastic for a few hours (not more than 8 hours) in 1 M HCl or 1 M HNO$_3$ followed by a rinse with high purity water will remove trace metals from plastics. It may also be helpful to clean plastics with alcohol or chloroform followed by a rinse in 1 M HCl prior to trace metal analysis. This treatment removes trace levels of organic compounds from the plastic surface. Low levels of organic compounds can contribute to the adsorption of trace elements. (Be certain the plastic is compatible with the cleaning process.)

4. *It is common to bake glassware after washing when working with RNA, in order to destroy RNA-degrading enzymes.*

5. *Labware is frequently autoclaved after washing when sterility is required.* Note that some types of plastic cannot withstand autoclaving. In addition, plastics may require longer autoclave cycle times than glass because the heat transfer properties of plastic and glass are different. Cycle lengths for plastics need to be determined empirically for each liquid and container.

6. *Solvents such as acetone, alcohols, and methylene chloride can be used with glassware to remove greases and oils.* Plasticware, however, may be dissolved by acetone and other organic solvents, so these should be used cautiously on plastic.

7. *Newly purchased glass and plastic items contain contaminants and should be washed before use.*

8. *Some plastics, including polycarbonate and polystyrene, are weakened by the heat in machine washers.* These plastics should be washed by hand in water that is less than 57°C.

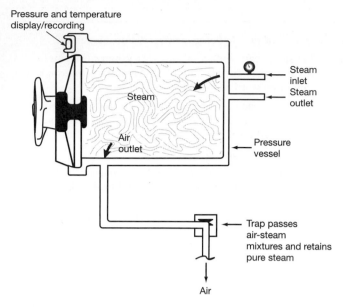

Figure 24.7.   The Basic Features of an Autoclave. Items to be sterilized are placed in the chamber and the chamber door is tightly sealed. Air is purged from the chamber at the beginning of the sterilization cycle. After the air is removed, steam under pressure is introduced into the chamber and the temperature rises to the desired level. The autoclave is maintained under pressure at a high temperature for a predetermined time. The steam is then exhausted from the chamber and the sterile materials are removed.

IV. STERILIZATION OF SOLUTIONS

There are two methods commonly used to sterilize solutions in the laboratory: **autoclaving** and **filtration**. An **autoclave** *is a pressure cooker; materials to be sterilized are heated by steam under pressure.* The pressure in an autoclave is usually held at 15–20 *lb/in²* (PSI) above normal pressure. This moist heat is very effective at destroying microorganisms; when water is present, bacteria are killed at lower temperatures than when heat alone is used. Sterilization with an autoclave is relatively fast and efficient and is the method of choice in many applications. Figure 24.7 shows the basic features of an autoclave and Table 24.9 summarizes general autoclaving guidelines.

After items have been subjected to autoclaving, it is important to know whether all microorganisms were indeed killed. The best way to test whether sterilization has occurred during autoclaving is to use spore strips. **Spores** *are dormant microorganisms.* Autoclaving should involve sufficient heat and sufficient time to kill spores. **Spore strips** *are dried pieces of paper to which large numbers of nonpathogenic, heat-resistant* Bacillus thermopholus *spores are adhered.* Spore strips are placed in the autoclave in the area that is most inaccessible to steam, such as in a flask or bottle. After auto-

claving, the strips are transferred to growth medium, being careful not to contaminate them with organisms from the air or laboratory. If sterilization in the autoclave was successful, the spores will not grow. It is important when performing spore tests to use a positive control and a negative control. The positive control is a spore strip that was not autoclaved and should grow if the spores are viable and the medium is properly prepared. The negative control is to place strips without spores in medium. No growth should occur in the negative control unless there is contamination.

Although spore strips are recommended for testing autoclave function, they require time to incubate, so they do not provide instantaneous information regarding an autoclave cycle. There are a number of convenient products used to identify materials that have been through a cycle in an autoclave. For example, autoclave tape is a material that can be wrapped on objects and changes colors when subjected to pressurized steam. Such devices are typically used in research laboratories to show that an item went through a sterilization cycle. Note that autoclave tape and other such materials may change colors before all microorganisms are killed.

Some solution components are heat sensitive and must be sterilized by methods that do not involve heat. Heat sensitive substances include proteins, vitamins, antibiotics, animal serum, and volatile chemicals. These materials are usually prepared as concentrated stock solutions and are then sterilized by filtration through a sterile filter with pores that are less than 0.2 μm in diameter. The filter-sterilized components are then added to the rest of the solution using care to maintain the sterility of the mixture. Filtration through membrane filters is effective for removing microorganisms, and is standard practice in many laboratories. Note that viruses will penetrate microfilters. In situations where viruses and pyrogens must be removed, as in the preparation of pharmaceuticals, ultrafilters can be used.

As with any other aspect of a quality program, the operation, maintenance, and monitoring of sterilization systems requires planning and documentation. Most modern autoclaves automatically record information for every cycle such as date, temperature reached, pressure, time, and so on. Other aspects of quality, such as logbooks and operator training are also essential.

V. STORAGE OF BIOLOGICAL SOLUTIONS AND MATERIALS

Biological solutions and biological materials must be properly stored to avoid degradation and contamination. If storage instructions are available from a manufacturer or previous investigator, they should usually be followed. If no instructions are available, consider the guidelines in Table 24.10.

Table 24.9 AUTOCLAVING GUIDELINES

1. *Steam sterilization requires that the steam directly contact every surface and article being sterilized.* For this reason:

   a. Do not over-crowd an autoclave. Allow spaces between items so that steam can contact all surfaces. When sterilizing instruments or fabrics, the steam must penetrate to and contact each and every surface and fiber. Anhydrous materials like oils, greases, and powders resist contact with water and therefore cannot be sterilized in an autoclave.

   b. Air can impede the contact of steam with a material. Autoclaves therefore evacuate air before introducing steam into the sterilization chamber. It is possible for air to be trapped in the material being autoclaved. For example, if a tube or needle is plugged, then air cannot escape, steam cannot enter, and that tube or needle will not be sterilized.

   c. Aqueous solutions do not require direct steam contact but they do require adequate heating.

2. *Containers should always be loosely capped.* This ensures that steam can penetrate to all surfaces. Loose caps also allow pressure to equilibrate inside and outside the container. Containers can shatter if the pressure is not equilibrated. Flasks are often covered by cotton plugs or loosely twisted caps.

3. *It is important that the autoclave be held at the proper temperature for the proper amount of time.* The minimum temperature at which sterilization occurs is 121°C (250°F). Too little time at this temperature leads to failure of sterilization; too much time can degrade media containing carbohydrates, agar, and other sensitive materials. It is standard practice to hold the autoclave at high temperature for 20 minutes. Depending on what is being autoclaved, however, 20 minutes may not be adequate. For example, plastics are poor conductors of heat. It may take much longer than 20 minutes to sterilize a 1 *L* solution in a plastic container. The minimum time required for sterilization of materials in Pyrex® vessels similarly varies depending on the container. Containers larger than 1 *L* are likely to require more than 20 minutes sterilization time. It is best to autoclave materials that are similar together, so that the optimal sterilization conditions will be the same for all items in the autoclave.

4. *The quality of the water used to make steam is important.* Volatile impurities in the water will be carried into all autoclaved items and solutions. It is common to use water that was deionized or treated by reverse osmosis.

5. *Use a "slow exhaust" setting when autoclaving solutions.* Solutions do not boil during autoclaving even though the temperature in the autoclave is above the boiling point of water. This is because the pressure of the steam prevents boiling. After sterilization, however, if the steam is allowed to exit the chamber quickly, then solutions will boil violently out of their containers. When solutions are autoclaved, the autoclave should therefore be set to exhaust the steam slowly.

6. *Note that even when steam is released slowly, solutions will lose about 3–5% of their liquid due to evaporation.* People sometimes add an extra 5% of distilled water to each flask being autoclaved to compensate for this loss.

7. *Containers with liquids should not be filled more than 75% full to allow for fluid expansion and to prevent overflow.*

8. *Some materials should be autoclaved in separate containers because they will alter one another in some way.* For example, phosphates and glucose can interact with one another when subjected to heat so that the solution is degraded.

9. *Autoclaves attain very high temperatures and pressures and must be safely operated.* Safe operation requires that the instrument be properly maintained and the operators understand how to use it safely. Note that older autoclaves sometimes lack the safety features of newer models.

PRACTICE PROBLEMS

1. Pyrogens are negatively charged molecules that range in molecular weight from about 1,000 *D* to more than 10,000 *D*. What type of filtration method can remove pyrogens?

2. Viruses are about the same size as the protein albumin and are slightly bigger than pyrogens. What type of filtration method will remove them from a solution?

3. Is table salt, NaCl, removed from a solution that is passed through an ultrafilter? Is it removed by deionization?

4. Many water purification systems recirculate already purified water back through the purification steps. Why?

5. The term *log reduction value* (*LRV*) describes mathematically the ability of an ultrafiltration membrane to remove pyrogens from water. The higher the *LRV*, the greater the rejection of pyrogens by the membrane. The equation for *LRV* is:

$$log \ reduction \ value =$$

$$log \left(\frac{feed \ water \ pyrogen \ concentration}{product \ water \ concentration} \right)$$

For example, a filter was challenged by running water through it containing pyrogens at a concentration of 10,715 *EU/mL*. The endotoxin concentration in the product water was 0.0023 *EU/mL*. What was the *LRV*?

$$LRV = log \left(\frac{10,715 \ EU/mL}{0.0023 \ EU/mL} \right) = 6.7$$

Table 24.10   *STORAGE OF BIOLOGICAL SOLUTIONS AND SAMPLES*

1. ***Buffers and salt solutions are often stored indefinitely at room temperature.*** Be aware, however, that microorganisms grow well in these solutions, and salts and other components may precipitate over time. Before use, examine solutions for visible particulates that may indicate contamination or precipitation. For some applications, solutions with some particulates are acceptable; in other applications they are not. Buffer solutions and media that have been autoclaved are usually stable at room temperature.

2. ***To prevent bacterial contamination, small amounts of bacteriocidal agents are sometimes added to solutions.*** Some solutions are best sterilized by autoclaving or filtration.

3. ***Cold tends to protect biological materials from degradation and to inhibit the growth of microorganisms.*** Some materials are best stored at 4°C, some are frozen at −20°C, others are frozen at −70°C, or in liquid nitrogen at −196°C.

4. ***Although freezing is a good way to protect many materials from degradation, potentially destructive physical events occur during freezing and thawing.*** Ice crystals, which can damage biological materials, form during freezing and melt during thawing. During freezing and thawing, different components in a solution freeze and thaw at different rates, allowing materials to be exposed to extremes of pH and salt concentration. At −20°C, concentrated solutions may not freeze at all. When working with materials extracted from cells, degradative enzymes from the cells may slowly destroy proteins and other biological materials, even in −20°C freezers; therefore, −70°C freezers are often preferable for storing cellular extracts.

5. ***Avoid multiple freeze–thaw cycles.*** Ice crystal formation during freezing can damage biomolecules, particularly proteins. To avoid multiple freeze–thaw cycles, *solutions are divided into small volumes that are frozen in individual containers, called* **aliquots.** Only as many aliquots as are needed at a given time are thawed and any unused, thawed material is discarded. This technique ensures that a protein solution is frozen and thawed only one time. (It is imperative that all aliquots are labeled carefully to avoid later confusion.) In addition, to avoid multiple freeze–thaw cycles, biological materials should be stored in freezers that are not "frost-free." Frost-free freezers go through repeated freeze–thaw cycles. If frost-free freezers must be used, it is possible to purchase insulated containers that will help protect biological substances from thawing.

6. ***Glycerol is sometimes added to protein solutions before placing in a freezer because it keeps the protein from actually freezing, prevents ice crystal formation, and, therefore, improves protein stability.*** It is common to receive enzymes that are stabilized with glycerol from suppliers. Be aware, however, that glycerol can affect subsequent experimental steps.

7. ***It is usually good practice to freeze proteins rapidly by immersing them in a dry ice bath and to thaw them rapidly by placing them in lukewarm water.*** Rapid freezing and thawing is thought to reduce exposure of the proteins to extremes of pH or salt concentration.

8. ***Living cells, such as sperm to be used for* in vitro *fertilization or cells to be used in tissue culture, are commonly stored in liquid nitrogen.***

9. ***Many biological materials, such as antibodies and enzymes, are lyophilized (freeze-dried) for long-term storage because they are sensitive to pH, temperature, and other factors when in solution.*** During lyophilization the protein solution is rapidly frozen. Water is then removed from the solution by **sublimation** under vacuum; *the water moves from a frozen state to a vapor without becoming a liquid.* Freeze-dried products are very hygroscopic and are generally stored in sealed vials to protect them from moisture and oxygen. These vials should remain refrigerated or frozen until needed. Follow the manufacturer's instructions for rehydrating dried proteins or dissolve them in an appropriate buffer. If not used immediately such materials are often aliquoted and frozen.

10. ***In the absence of nucleases, DNA is generally stable and can be stored long term either dry or in frozen solution.*** Solutions may be stored with EDTA to chelate magnesium ions that are required for endonuclease activity.

11. ***RNA is stored with EDTA in RNase-free water or Tris buffer.*** The EDTA solution should be known to be free of ribonucleases, which can be generated from contamination with microorganisms.

12. ***Low levels of nonionic detergent (0.15%) may be added to solutions to prevent protein aggregation and adsorption to surfaces.***

In another challenge, the same membrane was used to filter water containing pyrogens at a level of 4.5 *EU/mL*. After filtration, the water had less than 0.001 *EU/mL*. What was the *LRV*?

(The information in this example is from "Reducing Pyrogen Levels in Ultrapure Water, Part 2: Producing Pyrogen-Free Water Using Hollow Fiber Ultrafiltration." Millipore Technical Brief. Millipore Corp., Bedford, MA.)

DISCUSSION PROBLEMS

1. You are placed in charge of purchasing a water purification system for a laboratory. For each of the following types of laboratories, what would be important considerations to discuss with manufacturers? What standards would your purified water need to meet? What purification methods do you think would be most appropriate for your water?

a. You work in an analytical testing laboratory where you test air and water samples for pollutants.

b. You work in a biotechnology research laboratory on a variety of genetic engineering projects.

2. A biotechnology company has been successfully purifying its water for several years. Workers suddenly discover that one of their products is no longer effective and they trace the problem to the water used in production. Discuss factors that might have led to their sudden difficulties with water quality. Suggest changes they could make in their operating procedures to help avoid similar crises in the future.

3. If you have a water purification system in your laboratory, find out how it is designed and what routine maintenance it requires.

4. Write an SOP to clean glassware in your laboratory manually. If you do not work in a laboratory, write an SOP to clean glassware in a research laboratory where proteins are purified and analyzed.

5. If there is a machine washer in your laboratory, read its manual. Describe potential problems that can arise and how these problems can be recognized.

6. If you have a dishwasher at home, examine the quality of its performance. Devise a method to evaluate how effectively your dishes are cleaned.

ANSWERS TO PRACTICE PROBLEMS

1. Pyrogens are removed by ultrafiltration and reverse osmosis (also by anion exchange resins).

2. Viruses are removed by ultrafilters and reverse osmosis.

3. Table salt is smaller ($FW = 58.44$) than the cutoff for ultrafilters and so passes through them. NaCl is ionized in solution and therefore is removed by deionization.

4. Purified water is unstable. It absorbs gases from the air and leaches materials from the walls of the vessels in which it is contained. Continuous repurification is therefore used for Type I water.

5. 3.7.

Appendix
Chemical Resistance Chart for Plastics

Reproduced courtesy of Nalge Nunc International
Disclaimer: The Chemical Resistance Chart that follows contains general guidelines only. Because so many factors can affect the chemical resistance of a given product, you should test plastics under your own conditions.

Interpretation of Chemical Resistance

The Chemical Resistance Chart and Chemical Resistance Summary Chart that follow are general guidelines only. Because so many factors can affect the chemical resistance of a given product, you should test under your own conditions. If any doubt exists about specific applications of NALGENE products, please contact Technical Service, Nalge Nunc International, Box 20365, Rochester, New York 14602-0365, or call (800) 625-4327, Fax (800) 625-4363. International customers, contact our International Department at +1 (716) 264-3898, Fax +1 (716) 264-3706. In Europe, contact Nalge (U.K.) at +44 1432 26393, Fax +44 1432 351923.

Additional Chemical Resistance Information

This chemical resistance chart is to be used for all labware including containers up to 50L. For NALGENE centrifuge ware or UltraPlus centrifuge ware please refer to those charts in this catalog.

For chemical resistance of PETG (polyethylene terephthalate copolyester), see below.

For NALGENE fluorinated containers, including fluorinated high-density polyethylene (FLPE) and fluorinated polypropylene (FLPP), see inside back cover. For tanks, refer to the NALGENE Industrial Products Catalog.

Effects of Chemicals on Plastics

Chemicals can affect the strength, flexibility, surface appearance, color, dimensions or weight of plastics. The basic modes of interaction which cause these changes are: (1) chemical attack on the polymer chain, with resultant reduction in physical properties, including oxidation; reaction of functional groups in or on the chain, and depolymerization; (2) physical change, including absorption of solvents, resulting in softening and swelling of the plastic; permeation of solvent through the plastic, and dissolution in a solvent; and (3) stress-cracking from the interaction of a "stress-cracking agent" with molded-in or external stresses. Also see "Chemical Resistance Classification".

The reactive combination of compounds of two or more classes may cause a synergistic or undesirable chemical effect. Other factors affecting chemical resistance include temperature, pressure and internal or external stresses (e.g., centrifugation), length of exposure and concentration of the chemical. As temperature increases, resistance to attack decreases.

Mixing and/or dilution of certain chemicals in NALGENE labware can be potentially dangerous. The reactive combination of different chemicals or compounds of two or more classes may cause an undesirable chemical effect or result in an increased temperature which can affect chemical resistance (as temperature increases, resistance to attack decreases). Other factors affecting chemical resistance include pressure and internal or external stresses (e.g., centrifugation), length of exposure and concentration of the chemical.

Environmental Stress-Cracking

Environmental stress-cracking is the failure of a plastic material in the presence of certain types of chemicals. This failure is not a result of chemical attack. Simultaneous presence of three factors causes stress-cracking: tensile strength, a stress-cracking agent and inherent susceptability of the plastic to stress-cracking.

Common stress-cracking agents are detergents, surface active chemicals, lubricants, oils, ultra-pure water and plating additives such as brighteners and wetting agents. Relatively small concentrations of stress-cracking agent may be sufficient to cause cracking.

Mixing and/or dilution of certain chemicals may result in reactions that produce heat and can cause product failure. Pre-test your specific usage and always follow correct lab safety procedures.

ATTENTION: Please be aware that, although several polymers may have excellent resistance to various flammable organic chemicals and solvents, OSHA H CFR 29 1910.106 for flammable and combustible materials, or other local regulations, may restrict the volumes of solvents which may legally be stored in an enclosed area.

Caution

Do not store strong oxidizing agents in plastic labware except that made of FEP or PFA. Prolonged exposure causes embrittlement and failure. While prolonged storage may not be intended at time of filling, a forgotten container will fail in time and result in leakage of contents. Do not place any plastic labware in a flame.

Resin Codes:

ECTFE	Halar ECTFE** (ethylene-chlorotrifluoroethylene copolymer)
ETFE	Tefzel ETFE* (ethylene-tetrafluoroethylene)
FEP	Teflon FEP* (fluorinated ethylene propylene)
HDPE	high-density polyethylene
LDPE	low-density polyethylene
PC	polycarbonate
PETG	polyethylene terephthalate copolymer
PFA	Teflon PFA* (perfluoroalkoxy)
PMP	polymethylpentene
PP	polypropylene
PPCO*	polypropylene copolymer
PS	polystyrene
PSF	polysulfone
PVC	polyvinyl chloride
PVDF	polyvinylidene fluoride
TFE	Teflon TFE* (tetrafluoroethylene)
TMX	Thermanox
PMX	Permanox

*PPCO has replaced polyallomer (PA) in all products.

First letter of each pair applies to conditions at 20°C; the second to those at 50°C. At 20°C->EG <- at 50°C

CHEMICAL	LDPE	HDPE	PP	PPCO	PMP	PETG	FEP	TFE	PFA	ECTFE	ETFE	PC	RIGID PVC	FLEX. PVC	PSF	PS	PVDF	TMX	PMX
1,2,4-Trichlorobenzene	NN	NN	NN	NN	GF	NN	EE	EE	EE	EG	EE	NN	NN	NN	NN	NN	EE	GF	GF
1,2-Dichloroethane	NN	NN	NN	NN	NN	NN	EE	EE	EE	EG	EE	NN	FN	NN	NN	NN	EE	NN	NN
1,4-Dioxane	GF	GG	GF	GF	GF	NN	EE	EE	EE	EG	EG	GF	NN	NN	GF	NN	EE	EN	—
2,2,4-Trimethylpentane	FN	FN	FN	FN	FN	—	EE	EE	EE	EE	EE	NN	NN	NN	GF	GG	EE	FN	FN
2,4-Dichlorophenol	NN	NN	NN	NN	FN	—	EE	EE	EE	EG	EG	NN	NN	NN	EE	GG	EE	EG	EG
2-Butanol	EE	EE	EE	EE	EG	EG	EE	EE	EE	EE	EE	EG	EG	GN	EG	GG	EE	EN	EN
2-Butyl Alcohol	EG	EG	EE	EG	EG	EG	EE	EE	EE	EE	EG	EG	EG	GN	NN	GG	EE	EN	EN
2-Methoxyethanol	EE	EE	EE	EE	GN	GN	EE	EE	EE	EE	EE	NN	EG	EG	NN	EG	EE	EN	—
2-Propanol	EE	EE	EE	EE	EG	EG	EE	EE	EE	EE	EE	EG	EG	EG	EE	EG	EE	EN	EN
Acetaldehyde	GN	GF	GN	GN	GN	GN	EE	EE	EE	GF	EG	NN	NN	NN	NN	NN	EG	EG	EG
Acetamide, Sat.	EE	EE	EE	EE	EE	EE	EE	EE	EE	EE	EE	EG	EG	EG	EE	EG	EE	EE	EE
Acetic Acid, 05%	EE	EE	EE	EE	EG	EG	EE	EE	EE	EE	EE	GG	EG	GN	GG	GG	EE	EN	EN
Acetic Acid, 50%	EG	EG	EE	EG	EG	EG	EE	EE	EE	EE	EE	EG	EG	GN	EG	GG	EG	EN	EN
Acetic Acid, Glacial	EG	EG	EE	EG	GF	GF	EE	EE	EE	EE	EE	NN	EG	EG	GG	GG	EN	EN	EN
Acetic Anhydride	NN	NN	EE	NN	FN	FN	EE	EE	EE	GF	GF	NN	NN	NN	NN	NN	EG	EG	—
Acetone	EE	EE	GF	EE	FN	NN	GG	EE	EE	GF	GF	NN	NN	NN	NN	NN	NN	—	—
Acetonitrile	EE	EE	EE	EE	FN	NN	EE	EE	EE	EE	EE	NN	EF	FN	GG	EE	EE	EF	EF
Acetophenone	NN	NN	FF	FN	GN	FN	EE	EE	EE	GF	GF	NN	GN	GN	GF	GF	NN	FN	FN
Acrylonitrile	EG	EG	FF	FN	FN	—	EE	EE	EE	GF	GF	NN	GF	GN	GG	EE	EE	GN	GN
Adipic Acid	EE	EE	EE	EE	EE	EE	EE	EE	EE	EE	EE	EE	EG	EG	GG	EE	EE	EE	EE
Alanine	EE	EE	EE	EE	GF	GF	EE	EE	EE	EE	EE	GF	GF	FN	GF	GF	GF	EN	EN
Allyl Alcohol	EE	EE	EE	EE	FN	FN	EE	EE	EE	EE	EE	FN	GF	GN	GG	EE	EE	EN	—
Aluminum Chloride	EE	EE	EE	EE	EE	EE	EE	EE	EE	EE	EE	EG	EE	EE	GG	GG	EG	EE	EE
Aluminum Hydroxide	EG	EE	EE	EG	EG	EG	EE	EE	EE	EE	EE	EG	EE	EE	GG	GG	EE	GF	GF
Aluminum Salts	EE	EE	EE	EE	EE	EE	EE	EE	EE	EE	EE	EG	EE	EG	EE	GG	EE	EE	EE

E - No damage after 30 days of constant exposure.   F - Some effect after 7 days of constat exposure.   G - Little or no damage after 30 days of constant exposure.   N - Immediate damage may occur. Not recommended for continuous use.

(continued)

CHEMICAL	LDPE	HDPE	PP	PPCO	PMP	PETG	FEP	TFE	PFA	ECTFE	ETFE	PC	RIGID PVC	FLEX.PVC	PSF	PS	PVDF	TMX	PMX
Amino Acids	EE	EE	EE	EE	EE	--	EE	EE	EE	EE	EE	EE	EE	EF	EE	EE	EE	--	EE
Ammonia	EE	EE	EE	EE	EE	--	EE	EE	EE	EE	EE	NN	EE	EF	GF	GF	EE	NN	EE
Ammonia, 25%	EE	EE	EE	EE	EE	--	EE	EE	EE	EE	EE	NN	EG	GF	EG	EG	NN	NN	EE
Ammonium Acetate, Sat.	EE	EE	EE	EE	EE	--	EE	EE	EE	EE	EE	EG	EE	EG	EE	EE	EE	E-	EG
Ammonium Chloride	EE	EE	EE	EE	EE	FN	EE	EE	EE	EG	EE	EG	EE	EG	EE	EE	EE	--	EG
Ammonium Glycolate	EG	EG	EG	EG	EE	ZN	EE	EE	EE	EE	EE	EE	EE	GF	EE	EF	EE	--	EG
Ammonium Hydroxide, 0.5%	EE	EE	EG	EG	EG	--	EE	EE	EE	EF	EE	FN	EE	GF	GG	GF	EE	FN	EE
Ammonium Hydroxide, 30%	EG	EE	EG	EG	EG	--	EE	EE	EE	EE	EE	NN	EE	GF	GG	GF	EE	NN	EE
Ammonium Oxalate	EE	EE	EE	EE	EE	ZN	EE	EE	EE	EE	EE	EG	EE	GN	EE	GG	EE	GF	EE
Ammonium Salts	EE	EE	EE	EE	EE	--	EE	EE	EE	EG	EE	GF	EG	GN	EE	GF	EE	E-	EE
Amyl Alcohol	NN	FN	NN	NN	GF	--	EE	EE	EE	EG	EE	ZN	ZN	ZN	ZN	NN	EF	--	NN
Amyl Chloride	NN	EG	GF	GF	NN	--	EE	EE	EE	GN	EE	ZN	ZN	ZN	NN	NN	EE	E-	GF
Aniline	NN	NN	NN	NN	NN	--	EE	EE	EE	GF	EE	FN	EE	GG	FF	NN	EE	--	NN
Aqua Regia	NN	FN	NN	NN	GF	--	EE	EE	EE	GN	EG	FN	FN	NN	ZN	NN	EE	GN	EE
Arsenic Acid	GF	GF	GF	GF	GF	--	EE	EE	EE	EG	EG	EE	EE	GG	FF	NN	EG	EN	EG
Benzaldehyde	EG	EG	EG	EG	EG	ZN	EE	EE	EE	EG	EE	NN	EE	FN	ZN	NN	EE	--	GF
Benzenamine	EG	EG	EG	EG	EE	--	EE	EE	EE	EE	EG	FN	EE	ZN	FF	NN	EE	EF	EG
Benzene	NN	EE	NN	EE	NN	ZN	EE	EE	EE	EG	EE	NN	EE	ZN	ZN	NN	EE	--	NN
Benzoic Acid, Sat.	EE	FN	EE	EE	EE	--	EE	EE	EE	EG	EE	EE	EE	EE	ZN	EG	EE	--	EE
Benzyl Acetate	FN	FN	GF	GF	NN	--	EE	EE	EE	GF	EE	FN	EE	GN	ZN	NN	EE	--	NN
Benzyl Alcohol	NN	FN	NN	NN	NN	NN	EE	EE	EE	EG	EG	NN	GF	ZN	ZN	NN	EE	--	NN
Boric Acid	EE	EE	EE	EE	EE	--	EE	EE	EE	EG	EG	EE	EE	EG	EE	EG	EE	--	EE
Bromine	NN	FN	NN	NN	NN	--	EE	EE	EE	GF	EE	FN	GF	FN	ZN	NN	EE	--	NN
Bromobenzene	NN	NN	NN	NN	NN	ZN	EE	EE	EE	GN	EE	NN	ZN	ZN	ZN	NN	EE	--	NN
Bromoform	NN	NN	NN	NN	NN	--	EE	EE	EE	GF	EE	NN	ZN	ZN	ZN	NN	EE	--	GF
Butadiene	NN	FN	GF	GF	GF	NN	EE	EE	EE	GF	EG	FN	FN	NN	NN	NN	EE	--	FN
Butyl Acetate	GF	GF	GF	GF	GF	--	EE	EE	EE	EG	EG	NN	NN	ZN	FF	NN	GN	G-	GF
Butyl Chloride	NN	NN	NN	NN	NN	--	EE	EE	EE	GN	EG	FN	FN	NN	ZN	NN	EE	--	FN
Butyric Acid	NN	FN	NN	NN	NN	--	EE	EE	EE	EE	EG	FN	FN	NN	ZN	NN	EE	--	FN
Calcium Chloride	EE	EE	EE	EE	EE	--	EE	EE	EE	EG	EE	EE	EE	EE	EE	EE	EE	--	EE
Calcium Hydroxide, Conc.	EE	EE	EE	EE	EE	ZN	EE	EE	EE	EG	EE	NN	EG	GF	GG	GG	EE	--	EE
Calcium Hypochlorite, Sat.	EE	EE	EE	EE	EG	ZN	EE	EE	EE	EE	EG	FN	ZN	FN	EE	GF	EE	GN	EG
Carbazole	EE	EE	EE	EE	EE	--	EE	EE	EE	EE	EE	EE	EE	EF	EE	EE	EE	--	EE
Carbon Disulfide	NN	FN	NN	NN	NN	--	EE	EE	EE	EG	EE	EG	EE	GF	ZN	EG	EE	--	EE
Carbon Tetrachloride	FN	FN	GF	GF	NN	ZN	EE	EE	EE	GF	EG	GN	GF	ZN	ZN	GG	EE	--	NN
Caustic Potash, Concentrated (50%)	EE	EE	EE	EE	EE	EE	EE	EE	EE	EG	EG	NN	EE	EF	EF	GG	EE	EN	EE
Caustic Soda, 01%	EE	EE	GF	GF	EE	ZN	EE	EE	EE	EE	EE	NN	EE	FN	EF	GF	EE	--	GN
Caustic Soda, Concentrated (50%)	GG	GF	NN	NN	EE	--	EE	EE	EE	EE	EG	FN	FN	GN	FF	FF	EE	--	GN
Cedarwood Oil	NN	FN	NN	NN	EE	--	EE	EE	EE	EF	EG	GF	GF	EG	EE	EG	EE	--	GN
Cellosolve Acetate	EG	EE	EG	EG	EG	--	EE	EE	EE	EG	EE	FN	EE	GF	GG	EE	EE	--	GF
Chlorine Water	GN	GN	EG	EG	EE	ZN	EE	EE	EE	EG	EE	GF	EN	EG	GG	FF	EE	EN	GN
Chlorine, 10% (Moist)	GN	GF	FN	FN	GF	--	EE	EE	EE	EF	EG	GF	EE	FN	FN	FF	FN	--	GN
Chlorine, 10% in air	GN	EF	GN	GN	GN	--	EE	EE	EE	EG	EG	EG	EG	GN	FN	GG	FN	--	GN
Chlorine, wet gas	GN	GF	GF	GF	GF	--	EE	EE	EE	EG	EE	FN	EG	EG	GG	FF	GN	--	GN
Chloroacetic Acid	EE	EE	EE	EE	EE	E-	EE	EE	EE	EF	EE	NN	FN	NN	EE	EG	EE	EN	EG
Chlorobenzene	EE	EE	EE	EE	FN	EN	EE	EE	EE	EG	EG	NN	NN	ZN	EE	GG	EE	--	FN
Chloroform	FN	FN	GF	GF	NN	ZN	EE	EE	EE	EG	EE	NN	NN	NN	EE	EG	EE	EN	EE
Chromic Acid, 10%	EE	EE	EE	EE	EE	--	EE	EE	EE	EE	EE	GF	EE	EN	GG	EE	EE	EN	EE
Chromic Acid, 20%	EE	EE	EE	EE	EE	E-	EE	EE	EE	EG	EE	GF	EE	FN	GG	EG	EE	E-	EE
Chromic Acid, 50%	EE	EE	EE	EE	FF	ZN	EE	EE	EE	EE	EF	FN	EF	GN	FF	FF	EE	EN	GN
Chromic:Surfuric Acid Mixture, 96%	NN	NN	GF	GF	GF	--	EE	EE	EE	EF	EE	GF	GN	FN	EE	EE	EE	--	GN
Cinnamon Oil	EE	EE	EE	EE	FN	--	EE	EE	EE	EG	EG	EG	EG	EG	EE	EE	EE	EN	EE
Citric Acid, 10%	EE	EE	EE	EE	EE	E-	EE	EE	EE	EG	EE	GF	EE	GF	EE	EG	EE	E-	EE
Copper Sulfate	EE	EE	EE	EE	EE	--	EE	EE	EE	EF	EE	GN	EE	GF	EE	EE	EE	E-	EE
Cresol	NN	FN	GF	GF	GF	ZN	EE	EE	EE	EG	EE	NN	EG	EG	GG	NN	EE	--	GF
Cyclohexane	FN	FN	FN	FN	FN	--	EE	EE	EE	EF	EG	EN	EG	EG	FF	NN	EE	EN	GF
Cyclohexanone	EE	EE	EE	EE	NN	--	EE	EE	EE	EG	EG	NN	NN	NN	EE	NN	EG	--	GN
Cyclopentane	NN	FN	FF	FF	GF	ZN	EE	EE	EE	EF	EG	EG	FN	FN	GN	NN	EG	--	GN
Decahydronaphtalene	GF	GF	GF	GF	FN	--	EE	EE	EE	GF	EG	GN	EG	GF	FN	GG	EG	--	EG
Decalin	NN	FN	NN	NN	FF	--	EE	EE	EE	EG	EF	ZN	EG	GF	FN	GG	NN	--	FN
Diacetone	FN	NN	NN	NN	GG	--	EE	EE	EE	EG	EG	NN	FN	NN	ZN	NN	GN	--	FN
Diacetone Alcohol	--	--	ZN	ZN	NN	--	EE	EE	EE	EG	EE	NN	GN	NN	ZN	NN	EG	--	NN
Dibutylphthalate	NN	NN	NN	NN	GF	ZN	EE	EE	EE	EF	EG	NN	NN	NN	ZN	GG	NN	EN	EG
Diethyl Benzene	NN	NN	NN	NN	FN	--	EE	EE	EE	EG	EG	NN	NN	NN	ZN	NN	NN	--	NN
Diethyl Ether	NN	NN	NN	NN	FN	--	EE	EE	EE	EG	EG	NN	NN	NN	ZN	NN	EG	EF	FN
Diethyl Ketone	NN	NN	NN	NN	GF	--	EE	EE	EE	EG	EE	GN	EE	GF	GG	GG	EG	--	GF
Diethyl Malonate	CG	CG	GG	GG	GG	--	EE	EE	EE	EF	EG	EG	EG	GG	GG	GG	EG	EN	FF
Diethylamine	EE	EE	EE	EE	FN	--	EE	EE	EE	EG	EE	FN	FN	NN	FF	NN	NN	FN	FF
Diethylene Dioxide	NN	NN	NN	NN	FN	--	EE	EE	EE	EG	EG	NN	ZN	NN	ZN	NN	NN	--	NN
Diethylene Glycol	EE	EE	EE	EE	EE	--	EE	EE	EE	EG	EE	EE	EE	EG	EE	EE	EE	--	EE
Diethylene Glycol Ethyl Ether	EE	EE	EE	EE	EE	--	EE	EE	EE	EG	EE	FN	EE	GF	EE	EE	EE	--	EE

CHEMICAL	PMX	TMX	PVDF	PS	PSF	PVC	RIGID PVC	FLEX. PVC	PC	ETFE	ECTFE	PFA	TFE	FEP	PETG	PMP	PPCO	PP	HDPE	LDPE
Dimethyl Acetamide	FG	EN	NN	NN	NN	NN	N	N	NN	EG	EG	EE	EE	EE	–	FG	EE	EE	EE	FN
Dimethyl Formamide	EE	EN	NN	NN	NN	NN	F	Z	NN	GG	GG	EE	EE	EE	–	EE	EE	EE	EE	EE
Dimethylsulfoxide	EE	EN	NN	EG	NN	EG	N	N	NN	EG	EG	EE	EE	EE	N	EE	EE	GF	GG	EE
Dioxane	FN	FN	NN	NN	NN	FN	Z	N	FN	GG	GG	EE	EE	EE	–	FN	GF	GF	GG	GF
Dipropylene Glycol	EE	–	EE	EE	GG	GF	N	N	GF	EE	EE	EE	EE	EE	–	EG	EE	EE	EE	EE
DMSO	EE	EN	NN	EG	EE	NN	N	N	EG	EE	EE	EE	EE	EE	N	EG	EE	EE	EE	EE
Ethanol, 40%	EG	EN	EE	NN	EE	EE	E	F	NN	EG	EG	EE	EE	EE	N	FN	EE	EG	FN	EE
Ether	FN	EN	EE	NN	EG	NN	Z	N	NN	EG	EG	EE	EE	EE	N	FN	EE	EG	EE	EE
Ethyl Acetate	EG	EN	EE	NN	GG	EG	E	E	NN	EE	EE	EE	EE	EE	E	EG	EE	EG	EE	EG
Ethyl Alcohol (Absolute)	EG	EN	EE	EG	EE	EE	E	E	EG	EE	EE	EE	EE	EE	N	EN	EE	EG	EE	EE
Ethyl Alcohol, 40%	NN	EN	NN	NN	NN	EE	E	N	NN	EE	EE	EE	EE	EE	–	EN	EE	EG	EE	EG
Ethyl Alcohol, 96%	GF	–	NN	GF	EG	GF	G	F	GF	GF	EG	EE	EE	EE	–	GF	GF	GF	GF	EG
Ethyl Benzene	FN	–	NN	FN	NN	FN	N	N	FN	EG	EG	EE	EE	EE	–	FN	FN	FN	NN	GN
Ethyl Benzoate	FN	–	NN	FN	NN	FN	N	N	FN	EG	EG	EE	EE	EE	–	FN	FN	FN	GG	FF
Ethyl Butyrate	EE	–	EE	EE	FF	EE	E	E	FN	EE	EE	EE	EE	EE	–	GF	GF	GF	GF	GN
Ethyl Chloride	NN	–	NN	NN	NN	NN	N	N	NN	EE	EE	EE	EE	EE	–	FN	FN	FN	FF	FN
Ethyl Chloride, Liquid	EE	–	NN	EE	EE	EE	E	E	EE	EE	EE	EE	EE	EE	–	EE	EN	EN	FF	EE
Ethyl Cyanoacetate	FN	–	NN	NN	NN	FN	N	N	FN	EE	EE	EE	EE	EE	–	FN	FN	FN	FF	FN
Ethyl Lactate	EE	–	EE	NN	FF	NN	N	N	FN	EE	EE	EE	EE	EE	–	EE	EE	EE	EE	EE
Ethylene Chloride	GG	–	EE	NN	NN	NN	N	N	NN	EG	EG	EE	EE	EE	–	EE	FN	FN	GF	GN
Ethylene Glycol	EG	–	EE	EE	EE	EE	E	E	EE	EE	EE	EE	EE	EE	G	EG	EG	EG	EG	EG
Ethylene Glycol Monomethyl Ether	EN	EF	–	GF	GG	GN	G	G	GF	EE	EE	EE	EE	EE	G	GG	FF	FF	EE	EE
Ethylene Oxide	EG	–	GG	NN	GG	GF	G	G	EE	EG	EG	EE	EE	EE	F	GG	GG	FF	EE	GG
Ethylene Oxide Gas	EG	–	GG	GG	GG	GG	G	G	EF	EE	EE	EE	EE	EE	G	GG	GG	GG	GG	GG
Ethylene Oxide, 100%	EF	–	FF	NN	GG	FN	F	F	EE	EE	EE	EE	EE	EE	G	FF	FF	FF	FF	FF
EtO Gas	GF	–	NN	GG	GG	GN	G	G	FN	EG	EG	EE	EE	EE	F	GF	GG	GG	GG	GG
EtO	FN	–	NN	EF	EF	FN	E	N	FN	EG	EG	EE	EE	EE	–	EF	EG	EG	EG	EG
Fatty Acids	EF	EF	NN	GG	EG	EG	E	G	GF	EE	EE	EE	EE	EE	G	EF	EE	EE	GF	FN
Fluorides	FN	–	NN	NN	NN	NN	N	N	FN	EE	EE	EE	EE	EE	–	FN	EE	EE	EE	EN
Fluorine	GF	–	NN	NN	NN	GN	N	N	EE	EG	EG	EE	EE	EE	–	GN	EG	EG	EG	EG
Formaldehyde, 10%	EE	–	NN	EE	EG	GN	G	G	EG	EE	EE	EE	EE	EE	G	EE	EE	EG	EE	EE
Formaldehyde, 40%	EE	–	NN	NN	EG	GN	N	N	EE	EE	EE	EE	EE	EE	N	EE	EE	EE	EE	EE
Formalin, 10%	EE	–	NN	NN	EG	FN	N	N	EF	EE	EE	EE	EE	EE	–	EG	EG	EG	EG	EG
Formalin, 40%	EE	–	NN	NN	GN	GN	N	N	EE	EE	EE	EE	EE	EE	–	EF	EG	EG	EG	EG
Formic Acid	GG	–	EE	NN	GG	FN	G	G	EG	EE	EE	EE	EE	EE	–	EF	EG	EG	GG	GG
Formic Acid, 03%	EG	–	EE	EF	EG	EG	E	E	EF	EE	EE	EE	EE	EE	G	EG	EG	EG	EG	EG
Formic Acid, 100%	EN	–	EE	NN	GG	NN	N	N	GG	EE	EE	EE	EE	EE	F	EN	GN	GN	NN	FN
Formic Acid, 50%	EG	–	EE	EF	GG	FN	N	N	EE	EE	EE	EE	EE	EE	–	EG	EG	EG	GN	GN
Formic Acid, 85%	EG	EE	EE	NN	GG	FN	N	N	GG	EE	EE	EE	EE	EE	–	GF	EN	EN	EG	GN
Freon TF	EG	–	EE	EF	GG	GN	G	G	FN	EE	EE	EE	EE	EE	–	EF	FN	FN	EG	EG
Fuel Oil	EF	–	EE	EF	GG	EG	E	G	FF	EG	EG	EE	EE	EE	–	GF	GF	GF	GF	FN
Gasoline	GF	–	EE	GF	EF	GN	G	G	GF	EE	EE	EE	EE	EE	G	GF	GF	GF	GF	FN
Glutaraldehyde	EE	–	EE	EE	EG	GN	N	N	EE	EE	EE	EE	EE	EE	–	EE	EE	EE	EG	EG
Glutaraldehyde Disinfectant	EE	–	EE	EE	EG	GN	N	N	EE	EE	EE	EE	EE	EE	–	EE	EE	EE	EE	EE
Glycerine	EG	EF	EE	EF	GG	EG	G	G	EF	EG	EG	EE	EE	EE	–	EG	EG	EG	EG	EG
Glycerol	EG	GF	EE	EF	EG	EG	G	G	EF	EG	EG	EE	EE	EE	G	GN	EG	EG	EG	EE
Hexane	EE	–	EE	NN	EE	NN	N	N	FN	EE	EE	EE	EE	EE	–	GN	EG	EG	EE	EE
Hydrazine	NN	–	NN	NN	EE	NN	N	N	FN	EE	EE	EE	EE	EE	N	GN	FN	FN	NN	NN
Hydrobromic Acid, 69%	EE	GN	EE	EE	EE	EE	G	G	EE	EE	EE	EE	EE	EE	–	EN	NN	NN	EE	EE
Hydrochloric Acid, 05%	EG	GF	EE	EF	EG	EG	G	G	GF	EE	EE	EE	EE	EE	F	EG	EG	EG	EE	EE
Hydrochloric Acid, 20%	EE	NN	EE	EF	EE	EE	G	G	GF	EE	EE	EE	EE	EE	–	EE	EG	EG	GF	GF
Hydrochloric Acid, 35%	EG	GN	EE	EF	EE	EE	G	G	FN	EE	EE	EE	EE	EE	N	EE	EG	EG	EG	EG
Hydrofluoric Acid, 04%	EE	GF	EE	EE	EG	EE	E	E	FN	EE	EE	EE	EE	EE	–	EE	EG	EG	EE	EE
Hydrofluoric Acid, 48%	EF	EE	EE	EF	EE	EE	N	N	EE	EE	EE	EE	EE	EE	–	EE	EG	EG	EE	EE
Hydrogen Peroxide, 03%	EE	–	EE	EE	EG	EG	G	G	EE	EE	EE	EE	EE	EE	–	EG	EG	EG	EG	EG
Hydrogen Peroxide, 30%	EE	–	EE	EF	GG	EG	G	G	EF	EE	EE	EE	EE	EE	–	EG	EG	EG	EG	EG
Hydrogen Peroxide, 90%	EF	–	EE	NN	GG	FN	N	N	EF	EE	EE	EE	EE	EE	–	GN	EN	EN	NN	NN
Iodine Crystals	EF	–	EE	EF	GG	GN	G	G	FN	EE	EE	EE	EE	EE	–	EN	NN	NN	EE	EE
iso-Propanol, 100%	EF	EN	EE	EE	GG	EG	E	E	FN	EE	EE	EE	EE	EE	Z	EE	GN	GN	EE	EE
Isobutanol	GF	EN	EE	EF	EG	EG	E	E	GF	EE	EE	EE	EE	EE	–	EG	EG	EG	EG	EG
Isobutyl Alcohol	FF	EN	EE	EF	GG	EG	E	E	FF	EE	EE	EE	EE	EE	–	EG	EG	EG	EG	EG
Isopropanol	GF	EN	EE	EG	GG	EG	E	E	FN	EE	EE	EE	EE	EE	–	EG	EG	EG	EG	EG
Isopropanol, 100%	EE	–	EE	EE	GG	EG	E	E	FN	EE	EE	EE	EE	EE	Z	EE	GN	GN	EE	EE
Isopropyl Acetate	EE	EN	EE	NN	GG	EG	E	E	FN	EE	EE	EE	EE	EE	–	EE	EN	EN	EG	EG
Isopropyl Alcohol	EG	EN	EE	EE	GG	NN	G	G	EE	EE	EE	EE	EE	EE	–	EG	EN	EN	EG	EE
Isopropyl Alcohol, 100%	EG	GN	EE	EE	EE	EE	E	E	EE	EG	EG	EE	EE	EE	–	EG	EE	EE	EE	EE
Isopropyl Benzene	EE	EF	EE	EE	EE	EE	N	N	EE	EE	EE	EE	EE	EE	–	GN	GN	GN	EE	FN
Isopropyl Ether	EG	GN	EE	EE	EE	NN	N	N	EE	EE	EE	EE	EE	EE	–	GN	FN	FN	GF	NN
Jet Fuel	EG	GN	–	EG	EG	GN	G	G	EF	EE	EE	EE	EE	EE	–	GN	EG	EG	EG	EE
Kerosene	EG	EN	EE	EG	EG	GN	G	G	EG	EE	EE	EE	EE	EE	–	GN	EG	EG	EG	EG
Lacquer Thinner	NN	EN	NN	NN	GF	NN	N	N	NN	EG	EG	EE	EE	EE	F	NN	GF	GF	GG	FN
Lactic Acid, 03%	EG	EN	EG	GG	EE	EG	E	G	EG	EE	EE	EE	EE	EE	N	EG	EG	EG	EE	EG

(continued)

CHEMICAL	LDPE	HDPE	PP	PPCO	PMP	PETG	FEP	TFE	PFA	ECTFE	ETFE	PC	FLEX. PVC	RIGID PVC	PSF	PS	PVDF	TMX	PMX
Lactic Acid, 85%	EE	EE	EG	EG	EG	– –	EE	EE	EE	EG	EG	EE	EG	EE	EE	GG	GF	GN	EG
Lead Acetate	EE	EE	EE	EE	EG	– –	EE	EE	EE	EE	EE	EE	EE	EE	EE	EE	EE	EE	EE
Magnesium Chloride	NN	NN	EE	EE	NN	– –	EE	EE	EE	GF	GF	EE	GN	NN	NN	NN	NN	– –	NN
MEK	EE	EE	EE	EE	EE	GN	EE	EE	EE	EE	GF	NN	NN	NN	– –	GF	EE	EE	EE
Mercuric Chloride	EE	EE	EE	EE	EE	NN	EE	EE	EE	EE	EE	EE	EE	EE	–	EE	EE	EE	NN
Mercury=	EE	EE	EE	EE	EE	GG	EE	EE	EE	EE	EE	GF	EE	EE	GF	GF	EE	EE	EE
Methanol	EG	EG	EE	EE	EE	GG	EE	EE	EE	EE	EE	GF	EE	EE	EE	FN	EE	EN	EG
Methanol, 100%	FN	FF	GF	GF	EE	GG	EE	EE	EE	EE	EE	GF	NN	EF	NN	FN	NN	– –	EG
Methoxyethyl Oleate	EE	EE	EE	EE	EE	GG	EE	EE	EE	EE	EE	NN	NN	NN	NN	NN	EE	EN	EE
Methyl Acetate	EG	EE	GF	GF	NN	GG	EE	EE	EE	EE	EE	NN	NN	NN	NN	GN	EE	EN	EE
Methyl Alcohol	FN	NN	GF	GF	FF	– –	EE	EE	EE	GF	EE	GF	NN	NN	NN	NN	GN	– –	NN
Methyl Alcohol, 100%	EE	EE	EE	EE	EE	– –	EE	EE	EE	EE	EE	EE	NN	NN	NN	EE	EE	EE	EE
Methyl Ethyl Ketone	NN	NN	FN	FN	FN	ZZ	EE	EE	EE	GF	GF	NN	NN	NN	NN	GN	EE	EN	EF
Methyl Isobutyl Ketone	GF	EG	GF	GF	GF	NZ	EE	EE	EE	GG	GF	GF	NN	NN	NN	NN	EE	– –	FN
Methyl Propyl Ketone	FN	FN	FN	FN	FN	ZZ	EE	EE	EE	GG	EG	NN	NN	NN	NN	GN	EG	GN	FN
Methyl-t-Butyl Ether	NN	NN	NN	NN	NN	– Z	EE	EE	EE	EE	EE	GF	NN	NN	GF	NN	EN	– N	EE
Methylene Chloride	NN	NN	NN	NN	FF	– –	EE	EE	EE	GF	GF	NN	NN	NN	GN	NN	GN	– –	FN
MIBK	EE	EE	EE	EE	EE	– –	EE	EE	EE	EE	EE	EG	EE	EE	EE	EE	EE	EE	EG
Mineral Oil	FN	GN	FF	GF	FN	– –	EE	EE	EE	EE	EE	NN	NN	GZ	NN	FF	EE	EE	EE
Mineral Spirits	GF	EG	EE	EE	EG	– –	EE	EE	EE	EE	EG	NN	ZN	NN	FN	EG	EE	EE	EG
n-Amyl Acetate	GF	EG	GF	GF	EG	– –	EE	EE	EE	EE	EE	GF	NN	ZN	GF	EG	EE	– N	EG
n-Butanol	GF	GF	GF	GF	GF	GN	EE	EE	EE	EE	EE	FN	NN	NN	GF	GN	EE	EN	GF
n-Butyl Acetate	FN	EE	FN	FN	NN	ZZ	EE	EE	EE	GG	GG	NN	NN	NN	GF	GN	EE	G-	FN
n-Butyl Alcohol	FN	EG	EG	EG	EE	– –	EE	EE	EE	EF	GF	GF	EF	EF	GF	GN	EE	EN	EE
n-Decane	FN	FN	FN	FN	FN	ZZ	EE	EE	EE	EG	EE	EG	NN	NN	EE	GN	EE	GN	EE
n-Heptane	FN	GF	FN	FN	NN	ZN	EE	EE	EE	GF	EE	EG	NN	NN	EG	FN	EE	FN	FN
n-Octane	EE	EE	EE	EE	EE	Z –	EE	EE	EE	EF	EG	EG	EF	EF	EF	GN	EF	G-	EE
Nitric Acid, 10%	EE	EE	EE	EE	EG	E-	EE	EE	EE	GG	GG	GF	NN	NN	GF	GN	EE	GN	GF
Nitric Acid, 20%	EG	EG	GF	GF	GF	ZZ	EE	EE	EE	GG	GF	GF	NN	NN	FF	FN	EG	FN	FN
Nitric Acid, 50%	FN	GN	FN	FN	FN	ZZ	EE	EE	EE	EG	EG	NN	NN	NN	FF	GN	GN	EE	EE
Nitric Acid, 70%	NN	FN	NN	NN	NN	ZZ	EE	EE	EE	EG	EG	NN	EF	NN	FF	NN	NN	NN	EE
Nitrobenzene	NN	FN	NN	NN	GF	– N	EE	EE	EE	GF	GF	NN	NN	NN	FF	NN	EN	E-	EG
Nitromethane	FN	FF	FN	FN	NN	ZZ	EE	EE	EE	EF	EF	FF	NN	NN	GG	NN	EE	EN	FN
o-Dichlorobenzene	NN	NN	NN	NN	NN	ZZ	EE	EE	EE	EG	EG	NN	NN	NN	GG	NN	EE	GN	EE
Oil, Cedarwood	NN	GF	NN	NN	GF	ZZ	EE	EE	EE	GF	GF	NN	NN	NN	GG	NN	EF	G-	FN
Oil, Cinnamon	NN	NN	NN	NN	NN	ZZ	EE	EE	EE	EN	EN	NN	NN	FN	NN	NN	GN	EN	NN
Oil, Mineral	GN	GF	EE	EE	EG	N –	EE	EE	EE	EG	EG	EE	NN	NN	EE	EE	EG	EE	GF
Oil, Pine	FN	NN	EG	EG	GF	– Z	EE	EE	EE	GF	GF	FF	GF	GF	GG	EE	GF	EN	FN
Orange Oil	EE	EG	EE	EE	EE	– –	EE	EE	EE	EG	EG	GF	EG	EG	GG	EE	EE	GN	EE
Oxalic Acid, 10%	EE	EE	EE	EE	EG	– –	EE	EE	EE	EF	EF	FF	EG	EG	GG	GG	EE	GN	EE
Ozone	EE	EE	EE	EE	EE	– –	EE	EE	EE	EE	EE	EE	GF	GF	EE	EE	EE	– –	EE
p-Chloroacetophenone	FN	FN	GF	GF	GF	ZZ	EE	EE	EE	EG	EG	FN	NN	NN	FF	GN	GN	GN	GF
p-Dichlorobenzene	GN	GF	GN	GN	GN	ZZ	EE	EE	EE	EE	EF	NN	GF	GF	FF	FN	EE	FN	GN
Perchloric Acid	GN	GN	GN	GN	GN	ZZ	EE	EE	EE	EG	EG	NN	NN	NN	EG	NN	EF	NN	GN
Perchloric Acid, Concentrated (70%)	NN	NN	NN	NN	NN	ZZ	EE	EE	EE	EN	EN	NN	NN	NN	EG	NN	EE	NN	FN
Perchloroethylene	NN	NN	NN	NN	NN	ZZ	EE	EE	EE	EF	EF	FF	NN	NN	FF	NN	EE	EE	EE
Petroleum	FN	FN	FN	FN	GF	ZZ	EE	EE	EE	EG	EG	FF	FN	FN	EG	FN	EE	E-	FN
Phenol, 100%	NN	NN	NN	NN	NN	– Z	EE	EE	EE	EN	EN	NN	NN	NN	GG	NN	EE	EE	FN
Phenol, 50%	NN	NN	NN	NN	GF	ZZ	EE	EE	EE	EZ	EZ	NN	NN	NN	GG	NN	GF	GN	NN
Phenol, Crystals	GN	GF	GN	GN	NN	ZZ	EE	EE	EE	EE	EE	NN	GF	GF	GG	GG	GF	EN	EG
Phenol, Liquid	GN	NN	GN	GN	GN	ZZ	EE	EE	EE	EF	EF	EG	EG	EG	GG	GG	GF	E-	EE
Phosphoric Acid, 05%	EE	EE	EE	EE	EE	– –	EE	EE	EE	EE	EE	EE	GF	GF	GG	GF	EG	– –	EE
Phosphoric Acid, 30%	EE	EE	EE	EE	EE	– –	EE	EE	EE	EF	EF	EE	GF	GF	GG	GN	EE	EN	EE
Phosphoric Acid, 85%	EE	EE	EE	EE	EE	– –	EE	EE	EE	EF	EF	EG	GF	GF	NN	NN	EE	GN	EE
Picric Acid	EE	EE	EE	EE	EG	– –	EE	EE	EE	EE	EE	GF	GF	GF	NN	GF	EE	EE	EE
Pine Oil	FN	FN	FN	FN	FG	– –	EE	EE	EE	EG	EG	GF	GF	GF	EE	NN	GF	EN	EE
Potassium Chloride	EE	EE	EE	EE	EE	ZZ	EE	EE	EE	EE	EN	EE	FN	FN	EE	GN	EE	EN	EE
Potassium Hydroxide, 01%	EE	EE	EE	EE	EE	ZZ	EE	EE	EE	EF	EF	FN	NN	NN	EE	NN	EE	EE	EE
Potassium Hydroxide, 30%	EE	EE	EE	EE	EG	ZZ	EE	EE	EE	EF	EF	NN	GN	GN	EF	EG	GF	NN	EE
Potassium Hydroxide, Concentrated	EE	EE	EE	EE	EE	– –	EE	EE	EE	EN	EN	NN	GF	GF	EG	NN	EG	– –	EE
Potassium Permanganate	EE	EE	EE	EE	EG	– –	EE	EE	EE	EF	EF	NN	EE	EE	GG	GF	EE	EN	EE
Propane Gas	FN	FN	FN	FN	GF	ZZ	EE	EE	EE	EN	EN	EE	FN	FN	GG	NN	EE	E-	EE
Proprionic Acid	EE	EE	EE	EE	EE	ZZ	EE	EE	EE	EF	EF	FN	GF	GF	GF	FN	GN	EN	EF
Propylene Glycol	EE	EE	EE	EE	EE	– –	EE	EE	EE	EE	EE	GF	GN	GN	GG	GN	GF	EE	EE
Propylene Oxide	EG	EE	EG	EG	EG	– –	EE	EE	EE	EF	EF	GF	FN	FN	NN	EE	EE	EE	EE
Pyridine	NN	NN	NN	NN	GF	– –	EE	EE	EE	EZ	EZ	NN	NN	NN	NN	NN	FN	GN	NN
Resorcinol, 05%	EE	EE	EE	EE	EE	– –	EE	EE	EE	EF	EF	GF	GF	GF	GG	GF	EG	GN	EG
Resorcinol, Sat.	EG	EG	EG	EG	EG	– –	EE	EE	EE	EN	EN	EG	FF	FF	GG	NN	GF	EN	EE
Salicylaldehyde	EE	EE	EE	EE	EE	– –	EE	EE	EE	EE	EE	EG	GF	GF	NN	NN	EE	– –	EE
Salicylic Acid, Powder	EE	EE	EE	EE	EE	– –	EE	EE	EE	EE	EE	EE	GF	GF	EE	EG	EE	EE	EE
Salicylic Acid, Sat.	EE	EE	EE	EE	EE	– –	EE	EE	EE	EE	EE	EE	GF	GF	EE	EG	EE	EF	EE
Salt Solutions, Metallic	EE	EE	EE	, E	EE	– –	EE	EE	EE	EE	EE	EE	EG	EG	EE	GG	EE	EF	EE

E - No damage after 30 days of constant exposure. F - Some effect after 7 days of constnat exposure. G - Little or no damage after 30 days of constant exposure. N - Immediate damage may occur. Not recommended for continuous use.

CHEMICAL	LDPE	HDPE	PP	PPCO	PMP	PETG	FEP	TFE	PFA	ECTFE	ETFE	PC	RIGID PVC	FLEX. PVC	PSF	PS	PVDF	TMX	PMX
sec-Butanol	E E	E E	E E	E E	E G	- -	E E	E E	E E	E E	E E	E G	E G	G N	E G	G G	E E	E N	E G
sec-Butyl Alcohol	E G	E E	E E	E E	E G	- -	E E	E E	E E	E E	E E	E G	E G	G N	E G	G G	E E	E N	E G
Silicone Oil	E G	E E	E E	E E	E E	N N	E E	E E	E E	E E	E E	E G	E G	G N	E E	E G	E E	- -	E E
Silver Acetate	E E	E E	E E	E G	E E	N N	E E	E E	E E	E E	E E	E E	G G	G N	E E	G G	E E	- -	E E
Silver Nitrate	E E	E E	E E	E G	E G	- -	E E	E E	E E	E E	E E	N N	N Z	E G	N N	N N	E E	G N	E E
Skydrol LD4	G F	E G	E G	E G	E G	- -	E E	E E	E E	E E	E E	E G	N Z	E G	E E	G G	E E	- -	E E
Sodium Acetate, Sat.	E E	E E	E E	E E	E G	E E	E E	E E	E E	E E	E E	E G	G F	E F	E E	E G	E E	G N	E E
Sodium Carbonate	E E	E E	E E	E E	E G	- -	E E	E E	E E	E E	E E	E E	E E	E F	E E	E E	E E	G N	E E
Sodium Dichromate	E G	E G	E E	E E	E G	E E	E E	E E	E E	E E	E E	E E	E E	G N	E E	E G	E E	E N	E G
Sodium Hydroxide, 01%	G G	G G	E E	F N	E E	E N	E E	E E	E E	E E	E E	F N	E F	E F	E E	E G	E E	N N	G G
Sodium Hydroxide, 10%	G G	G G	E E	F N	E E	E N	E E	E E	E E	E E	E E	N N	N N	F N	E E	E G	E E	N N	E E
Sodium Hydroxide, Concentrated (50%)	E E	E E	E E	E E	E E	G -	E E	E E	E E	E E	E E	N N	E G	G N	G G	N N	E E	Z -	G G
Sodium Hypochlorite, 15%	E E	E E	E E	G F	E E	- -	E E	E E	E E	E G	E G	E G	E E	F N	G G	E G	E E	- -	G F
Stearic Acid	N N	F N	N N	N N	E N	- -	E E	E E	E E	E E	E E	E G	G N	G N	G G	N N	E E	- -	N N
Stearic Acid, Crystals	N N	F N	N N	N N	E N	- -	E E	E E	E E	E G	E G	E N	F Z	G N	G G	F N	E E	- -	F F
Sulfur Dioxide	F N	G F	F N	F N	F F	- -	E E	E E	E E	E E	E E	F N	F Z	G N	G G	E G	E E	- -	F N
Sulfur Dioxide, Liquid	E E	E E	E E	E E	E E	E E	E E	E E	E E	E E	E E	E E	E Z	E G	E E	E G	E E	- -	E E
Sulfur Dioxide, Wet or Dry Gas	E E	E E	E E	E E	E G	E E	E E	E E	E E	E E	E E	E E	E Z	E F	E E	G G	E E	G N	E G
Sulfur Salts	E G	E G	E G	F N	G G	E E	E E	E E	E E	E E	E E	G Z	G Z	F N	E E	N N	E E	G N	E G
Sulfuric Acid, 06%	G G	G G	F N	F N	G G	N N	E E	E E	E E	E E	E E	N Z	N Z	N N	G G	N N	E E	G N	G G
Sulfuric Acid, 20%	G G	G G	F N	F N	G G	N N	E E	E E	E E	E E	E E	N Z	N Z	N N	G G	N N	E E	G N	G G
Sulfuric Acid, 30%	F N	E E	E E	E E	E E	- -	E E	E E	E E	E E	E E	E E	E E	N N	E E	E G	E E	G N	E G
Sulfuric Acid, 60%	E G	E E	E E	E E	E E	- -	E E	E E	E E	E E	E E	E E	E E	N N	G F	N N	E E	G N	E E
Sulfuric Acid, 98%	N N	F N	F N	F N	F N	- -	E E	E E	E E	E E	E E	F N	F Z	N N	F F	N N	E F	N N	F F
Sulfuric Acid, Concentrated (96%)	F N	F N	F N	F N	F N	- -	E E	E E	E E	E E	E E	F N	F Z	N N	G F	G F	E G	G N	G F
Tartaric Acid	E G	E E	E E	E E	E E	- -	E E	E E	E E	E E	E E	E E	E E	N N	E G	N N	E E	N N	E E
TCA	E G	E E	E G	E G	E G	- -	E E	E E	E E	E F	E F	E G	E G	F N	E G	F N	E E	E N	E E
tert-Butanol	G F	G F	G F	G F	F F	- -	E E	E E	E E	E E	E E	N N	N Z	F N	N N	N N	F N	- -	F F
tert-Butyl Alcohol	N N	N N	N N	N N	N N	F N	E E	E E	E E	G F	G F	N N	N Z	N N	N Z	G F	N N	G -	N N
Tetrahydrofuran	E G	E G	E G	E G	E G	F N	E E	E E	E G	E E	E G	F N	Z -	N N	F F	N N	E F	G F	G F
THF	G F	F F	G F	G F	G F	- -	E E	E E	E E	N N	N N	F N	Z -	N N	C C	F N	E -	- -	N N
Thionyl Chloride	G F	E G	E G	G G	E E	G G	E E	E E	E E	E E	E E	E G	G G	N N	G G	G F	E G	E E	E E
Tincture of Iodine	F N	F F	G F	G F	G F	E N	E G	E E	E G	E E	E G	F N	G F	F N	F N	F N	E E	- -	G F
Toluene	N N	F N	N N	N N	N N	- -	E E	E E	E E	N N	N N	N N	N N	N N	N N	N N	E E	- -	N N
Tributyl Citrate	E E	E E	E E	E E	E E	- -	E E	E E	E E	E E	E E	E G	G F	F N	E G	E G	E E	E E	E E
Trichloroacetic Acid	E G	E G	E E	E E	E G	- -	E E	E E	E E	N N	N N	F N	G F	F N	F N	N N	E G	E N	E G
Trichloroethane	N N	F N	N N	N N	N N	- -	E G	E E	E G	E E	E E	N N	N Z	N N	N Z	F N	E E	- -	N N
Trichloroethylene	E E	E E	E E	E E	E E	- -	E E	E E	E E	E E	E E	E E	G F	F N	E G	E G	E E	E E	E E
Triethylene Glycol	E E	E E	E E	E E	E E	- -	E E	E E	E E	E E	E E	E E	G F	F N	G F	E E	E E	- -	E E
Tripropylene Glycol	E G	E G	E G	E G	E E	- -	E E	E E	E E	E G	E G	E G	G F	F N	G F	G N	E G	E -	E G
Tris Buffer, Solution	E F	E E	E E	E E	F F	- -	E E	E E	E E	E G	E G	F N	G F	F N	F F	F G	E E	E -	F F
Trisodium Phosphate	F N	F N	E E	E E	E E	- -	E E	E E	E E	E E	E E	F N	N Z	N N	N Z	G F	E E	E E	G N
Turpentine	E E	E E	E E	E E	E E	- -	E E	E E	E E	E E	E E	E G	E G	F N	E G	E G	E E	E E	E E
Undecyl Alcohol	E F	E G	E G	E G	E E	E E	E E	E E	E G	E G	E G	G F	G F	F N	G F	G G	E E	E -	E G
Urea	E E	E E	E E	E E	E E	E E	E E	E E	E E	E E	E E	E G	E G	F N	E G	E G	E E	E -	E G
Vinylidene Chloride	N N	F N	N N	N N	N N	E N	E G	E E	E G	E E	E E	N N	N Z	N N	N Z	N N	E E	G -	N N
Xylene	G N	F N	F N	F N	F N	- -	E E	E E	E E	G F	G F	N N	Z Z	N N	F F	G F	E E	G -	F N
Zinc Chloride, 10%	E E	E E	E E	E E	E E	- -	E E	E E	E E	E G	E G	E E	G G	G N	E E	G G	E E	E E	E E
Zinc Stearate	E E	E E	E E	E E	E E	- -	E E	E E	E E	E E	E E	E E	G F	G N	E E	G F	E E	G -	E E
Zinc Sulfate, 10%	E E	E E	E E	E E	E E	- -	E E	E E	E E	E E	E E	E E	E E	G N	E E	E E	E E	- -	E E

E - No damage after 30 days of constant exposure.   F - Some effect after 7 days of constnat exposure.   G - Little or no damage after 30 days of constant exposure.   N - Immediate damage may occur. Not recommended for continuous use.

(continued)

	Max. Use Temp. (°C)²	HDT¹ Temp. (°C)	Brittle-ness Temp. (°C)¹³	Trans-parency	Microwav-ability²	Sterilization⁴					Specific Gravity	Flexi-bility	Permeability (cc·mil/100in²·24 hr·atm)			Water Absorp-tion (%)	Non-Cyto-toxicity⁶	Suitability for Food and Bev. Use⁷	
						Auto-claving	Gas	Dry Heat	Radi-ation	Disin-fectants			N₂	O₂	CO₂			Rating	Reg. Part 21 CFR
ETFE	150	104	-105	Translucent	Yes	Yes	Yes	Yes	Yes	Yes	1.7	rigid	30	100	250	0.03	Yes	Yes	--
ECTFE	150	115	-105	Translucent	Yes	Yes	Yes	Yes	No	Yes	1.69	rigid	10	25	110	0.01	Yes	Yes	--
FEP	205	158	-270	Translucent	Marginal³	Yes	Yes	Yes	No	Yes	2.15	excel	320	750	2200	<0.01	Yes	Yes	177.1550
HDPE	120	65	-100	Translucent	No	No	Yes	No	Yes	Yes	0.95	rigid	42	185	580	<0.01	Yes⁹	Yes⁹	177.1520
LDPE	80	45	-100	Translucent	No	No	Yes	No	Yes	Yes	0.92	excel	180	500	2700	<0.01	Yes⁹	Yes⁹	177.1520
PC	135	138	-135	Clear	Marginal³	Yes⁵	Yes	No	Yes	Yes	1.2	rigid	50	300	1075	0.35	Yes	Yes	177.1580
PEI	170	210	--	Clear Amber	Yes	Yes	Yes	--	Yes	Yes	1.27	rigid	18.6	37	171.3	0.25	--	Yes	177.1595
PETG	70	70	-40	Clear	Yes	No	Yes	No	Yes	Yes	1.27	mod	10	25	80	0.15	Yes	Yes¹⁰	177.1315
PFA	250	166	-270	Translucent	Yes	Yes	Yes	Yes	No	Yes	2.15	excel	291	881	2260	<0.03	Yes	No	--
PK	220	218	-40	Opaque	Yes	Yes	Yes	--	Yes	Some	1.24	rigid	--	0.2	1.6	0.45	--	--	--
PMMA	50	93	20	Clear	No	Yes	No	Yes	Yes	Some	1.2	rigid	--	--	20	0.35	Yes	--	--
PMP	175	85	20	Clear	Yes	Yes	Yes	Yes	No	Yes	0.83	rigid	1100	4500	--	0.01	Yes	Yes¹¹	177.1520
PP	135	107	0	Translucent	Yes	Yes	Yes	No	No	Yes	0.9	rigid	48	240	800	<0.02	Yes	Yes	177.1520
PPCO	121	90	-40	Translucent	Marginal³	Yes	Yes	No	No	Yes	0.9	mod	45	200	650	<0.02	Yes	Yes	177.1520
PPO	100	149	-170	Opaque	--	Yes	--	No	Yes	No	1.06	rigid	--	1000	--	0.06	--	Yes	177.2460
PS	90	105	20	Clear	No	No	Yes	No	Yes	Some	1.05	rigid	55	300	1150	0.05	Yes	Yes	177.1640
PSF	165	174	-100	Clear	Yes	Yes	Yes	Yes⁵	Yes	Yes	1.24	rigid	55	300	700	0.3	Yes	Yes	177.1655
PUR	82	<23	-70	Clear	No	No	Yes	No	Yes	Yes	1.2	excel	41-119	75-327	450-1650	0.03	Yes	--	--
PVC (rigid)	70	90	-30	Clear	Yes	No	Yes	No	No	Yes	1.34	rigid	1-10	4-30	4-50	0.15-0.75	Yes	Yes¹²	--
PVC (tubing)	82	-32	-32	Clear	Yes¹⁴	No	Yes	No	No	Yes	1.34	excel	--	100-1400	20-12,000	0.15-0.75	Yes	Yes¹²	--
PVDF	150	139	-62	Translucent	--	Yes	Yes	Yes	No	Yes	1.75	rigid	9	14	505	0.05	Yes	Yes	177.2510
SAN	93	104	20	Clear	--	No	Yes	No	--	No	1.08	rigid	--	--	--	0.2	--	--	--
Silicone	200	-46	-117	Translucent	--	Yes	Yes	--	Yes	Yes	1.15	excel	43000	123000	312000	0.1	Yes	Yes	177.2600
TPE	121	<23	<-50	Opaque	Yes	Yes	Yes	No	-	Some	0.9	excel	31-145	85-646	900-8634	0.1-.042	Yes	--	--
TFE	260	200	-100	Opaque	Yes	Yes	Yes	Yes	No	Yes	2.20	rigid	--	--	--	<0.01	Yes	--	--
XLPE	65	59	-118	Translucent	No	No	Yes	No	Yes	Yes	0.93	rigid	--	--	--	<0.01	Yes	--	--
Permanox®	180	85	-10	Transparent	--	Yes	Yes	Yes	No	Yes	0.84	rigid	--	--	--	<0.01	--	--	--
Thermanox®	150	75	-60	Transparent	--	No	Yes	No	Yes	Some	1.30	mod	0.7-1.0	3-6	15-25	0.25	--	--	--

¹ Heat Deflection Temperature is the temperature at which a bar deflects 0.01" at 66 psig (ASTM D648). Materials may be used above Heat Deflection temperatures in non-stress applications; see Max. Use Temp.

² Ratings based on 5-minute tests using 600 watts of power on exposed, empty labware. CAUTION: Do not exceed Max. Use Temp., or expose labware to chemicals which heating cause to attack the plastic or be rapidly absorbed.

³ Plastic will absorb heat.

⁴ STERILIZATION:
 • Autoclaving (121°C, 15 psig for 20 minutes) -- Clean and rinse items with distilled water before autoclaving. (Always completely disengage thread before autoclaving.) Certain chemicals which have no appreciable effect on resins at room temperature may cause deterioration at autoclaving temperatures unless removed with distilled water beforehand.
 • Gas -- Ethylene Oxide, formaldehyde, hydrogen peroxide
 • Dry Heat (160°C, 120 minutes)
 • Disinfectants -- Benzalkonium chloride, formalin/formaldehyde, ethanol, etc.
 • Radiation -- gamma irradiation at 25 kGy (2.5 MRad) with unstabilized plastic.

⁵ Sterilizing reduces mechanical strength. Do not use PC vessels for vacuum applications if they have been autoclaved. Refer to Use and Care Guidelines for NALGENE Labware, for detailed information on sterilizing.

⁶ "Yes" indicates the resin has been determined to be non-cytotoxic, based on USP and ASTM biocompatibility testing standards utilizing an MEM elution technique on a WI38 human diploid lung cell line.

⁷ Resins meet requirements of CFR21 section of Food Additives Amendment of the Federal Food and Drug Act. End users are responsible for validation of compliance for specific containers used in conjunction with their particular packaging applications.

⁸ Acceptable for aqueous foods only, at temperatures up to 121°C/250°F. Not sanctioned for use with alcoholic or fatty foods at any temperature.

⁹ Acceptable for:
 • Nonacid, aqueous products; may contain salt, sugar or both (pH above 5.0)
 • Dairy products and modifications; oil-in-water emulsions, high or low fat
 • Moist bakery products with surface containing no free fat or oil
 • Dry solids with the surfaces containing no free fat or oil (no end-test required) and under all conditions as described in Table 2 of FDA Regulation 177.1520 except condition A - high temperature sterilization (e.g. over 100°C/212°F)

¹⁰ Acceptable for:
 • Alcoholic foods containing not more than 15% (by volume) alcohol; fill and storage temperature not to exceed 49°C (120°F)
 • Non-alcoholic foods of hot fill to not exceed 82°C (180°F) and 49° C (120°F) in storage.
 • Not suitable for carbonated beverages or beer or packaging food requiring thermal processing.

¹¹ Straight-sided jars, beakers and graduated cylinders only.

¹² Acceptable for aqueous, oil, dairy, acidic, and alcoholic foods up to 71°C/160°F.

¹³ The brittleness temperature is the temperature at which an item made from the resin may break or cracked if dropped. This is not the lowest use temperature if care is exercised in use and handling.

¹⁴ The tubing will become opaque from absorbed water, see the current NALGENE® Labware catalog for details.

Chemical Resistance for NALGENE Labware

Resin Codes:

ECTFE	Halar* ECTFE (ethylene-chlorotrifluoroethylene copolymer)
ETFE	Tefzel† ETFE (ethylene-tetrafluoroethylene)
FEP	Teflon† FEP (fluorinated ethylene propylene)
FLPE	fluorinated high-density polyethylene
HDPE	high-density polyethylene
LDPE	low-density polyethylene
PC	polycarbonate
PETG	polyethylene terephthalate copolyester
PFA	Teflon† PFA (perfluoroalkoxy)
PK	polyketone
PMMA	polymethyl methacrylate (acrylic)
PMP	polymethylpentene ("TPX")
PP	polypropylene
PPCO	polypropylene copolymer
PS	polystyrene
PSF	polysulfone
PUR	polyurethane
PVC	polyvinyl chloride
PVDF	polyvinylidene fluoride
TFE	Teflon† TFE (tetrafluoroethylene)
TPE	thermoplastic elastomer
XLPE	cross-linked high-density polyethylene

*Halar is a registered trademark of Allied Corporation.
† Or equivalent. Teflon and Tefzel are registered trademarks of DuPont.

Chemical Resistance Classification:

E – 30 days of constant exposure cause no damage. Plastic may even tolerate for years.

G – Little or no damage after 30 days of constant exposure to the reagent.

F – Some effect after 7 days of constant exposure to the reagent. Depending on the plastic, the effect may be crazing, cracking, loss of strength or discoloration. Solvents may cause softening, swelling and permeation losses with LDPE, HDPE, PP, PPCO and PMP. The solvent effects on these five resins are normally reversible; the part will usually return to its normal condition after evaporation.

N – Not recommended for continuous use. Immediate damage may occur. Depending on the plastic, the effect will be a more severe crazing, cracking, loss of strength, discoloration, deformation, dissolution or permeation loss.

Chemical Resistance Summary*

Classes of Substances 20°C	ECTFE/ETFE	FEP/TFE/PFA	FLPE	HDPE/XLPE	LDPE	PC	PETG	PK	PMMA	PMP	PP/PPCO	PS	PSF	PUR	PVC BOTTLES	FLEXIBLE PVC TUBING	PVDF	TPE***
Acids, dilute or weak	E	E	E	E	E	E	E	G	E	E	E	E	G	E	E	E	E	E
Acids,** strong and concentrated	G	E	E	E	N	N	G	N	E	E	F	G	F	E	F	E	F	F
Alcohols, aliphatic	E	E	E	E	E	G	E	G	N	E	E	G	F	E	G	E	E	E
Aldehydes	E	E	G	G	G	F	N	E	G	G	N	F	G	N	N	E	N	N
Bases	E	E	F	E	E	N	N	G	F	E	E	E	E	N	E	G	E	E
Esters	E	E	E	G	G	N	E	N	E	N	G	G	N	N	N	N	G	N
Hydrocarbons, aliphatic	E	E	E	G	F	F	E	E	G	F	G	N	G	E	E	F	E	N
Hydrocarbons, aromatic	E	E	E	G	F	N	E	N	F	F	N	N	N	N	N	N	E	N
Hydrocarbons, halogenated	E	E	G	F	N	N	N	N	N	F	N	N	N	N	N	N	E	N
Ketones	G	E	E	G	G	N	N	E	N	F	G	N	N	N	N	N	N	N
Oxidizing Agents, strong	F	E	F	F	F	N	N	G	N	F	F	N	G	N	G	N	F	G

*For tubing chemical resistance, other than PVC, see tubing section.
**Except for oxidizing acids: for oxidizing acids, see "Oxidizing Agents, strong."
***TPE gaskets.

UNIT VII:

Basic Separation Methods

This unit provides an introduction to the basic principles of the most common molecular separation methods used in biotechnology laboratories. The ability to separate biological materials from one another is essential in any biotechnology setting. A common technique in the biotechnology laboratory is the separation of DNA fragments of different sizes from one another using agarose gel electrophoresis. The ability to isolate specific segments of DNA is key to all recombinant DNA technology. Another example of a common separation technique is the use of centrifugation to separate cellular organelles (e.g., mitochondria and nuclei) from one another. Filtration is yet another separation technique that is frequently used in biotechnology settings. Common examples of applications of filtration are found in the removal of contaminants from biological preparations, water purification, and sterilization of solutions.

Separation techniques are used whenever specific biological products are isolated and purified for commercial sale in large-scale production settings. Separation techniques are also used whenever a biological material is purified for further study and analysis, as when gene segments are purified for later sequencing. Many separation techniques can be applied to both small- and large-scale preparations. Thus, these techniques are essential in the laboratory as well as in production facilities.

There are many different types of separation methods used for biological materials, each of which separates materials based on differences in their molecular properties. For example, filtration separates materials based on differences in their sizes—as agarose gel electrophoresis does for DNA fragments. Centrifugation separates materials based on differences in their sizes or densities. Various forms of chromatography separate materials based on properties such as size, charge, or relative solubility in solvents.

In order to isolate and purify a particular biomolecule, it is necessary to perform a series of separation techniques in succession. Consider, for example, the production of insulin to treat diabetic individuals. Insulin can be obtained either from animal pancreatic tissue or from *E. coli* bacteria that have been genetically modified to produce insulin. If pancreatic insulin is used, it must first be extracted from the tissue. A series of steps must then be performed that sequentially remove contaminants (e.g., other proteins, lipids, and cellular debris) from the insulin before it is suitable to inject into a human. Similarly, insulin isolated from bacteria must be purified of all bacterial substances. The techniques used to extract, isolate, and purify the insulin are called *bioseparation methods*. Each separation technique removes certain contaminants from the product of interest. One of the challenges to researchers is to develop a bioseparation strategy, consisting of a series of separation techniques, that most effectively isolates and purifies a biological product of interest.

The unit addresses in more detail two of the most routinely used separation methods: centrifugation and filtration. This unit then provides an overview of basic principles of bioseparation strategies.

Chapter 25 introduces the basic principles of filtration and their applications in both small- and large-scale operations.

Chapter 26 introduces the basic principles of centrifugation and their applications. It includes information about instrumentation and safety issues related to centrifugation.

Chapter 27 is an overview of the principles and strategies of product purification applied in both research and production laboratories.

There are many references available that provide detailed information about general principles of separation and about specific bioseparation techniques, such as centrifugation. These references and appropriate glossaries are provided at the end of each chapter.

CHAPTER 25

Introduction to Filtration

I. INTRODUCTION

A. The Basic Principles of Filtration

Filtration is a common separation method based on a simple principle: Particles smaller than a certain size pass through a porous filter material; particles larger than a certain size are trapped by the filter. Anyone who has strained spaghetti through a colander, made coffee with coffee filters, or played with a sieve at the beach has used filters. Gases, such as air, can be filtered as well as liquids. The air filters in cars and furnaces are commonplace examples.

The fluid and particles that pass through a filter are called the **filtrate** *or* **permeate**. *The materials trapped by the filter are sometimes called the* **retentate**. We are sometimes interested in the filtrate, as is the case with a coffee filter. We are interested other times in the material retained on the top of the filter, as is the case with spaghetti. We occasionally want to collect both the retained particles and the filtrate.

Filtration is also commonly observed in nature. Water is cleared of particulates as it passes through sandy soil to the groundwater. The kidneys are effective filtration devices that allow small, unwanted metabolites to pass from blood into the urine while retaining the relatively large blood cells.

Just as filtration plays various roles in the home and in nature, so it has a long history in the laboratory. Filtration was traditionally accomplished by folding a filter paper and placing it in a funnel for support. Liquids poured into the funnel would drip through the filter and solids would remain on the surface. To speed the process, analysts could add a vacuum pump to the flask. These simple filter paper systems are still used in laboratories to remove coarse particles from a solution, Figure 25.1.

Although the principle of filtration is simple, the study of filtration is complicated by its wide range of applications and the many types of filtration devices available. Filtration is used from the smallest scale in the laboratory, where samples of only a few microliters may be processed, to major industrial processes involving thousands of liters.

Depending on the application, the sample type, and the scale, filtration systems can vary greatly. Regardless of the simplicity or complexity of a filtration system, however, the following four components are present:

1. **The filter itself**
2. **A support for the filter (like the funnels in Figure 25.1)**
3. **A vessel to receive the filtrate**
4. **A driving force (such as gravity or vacuum) that drives the movement of fluids and particles through the filter.**

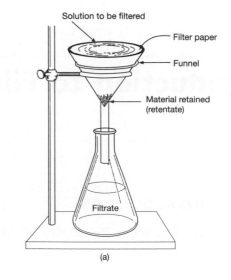

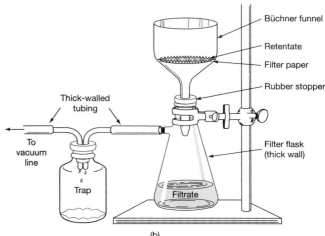

Figure 25.1. Laboratory Filtration Systems. a. The simplest laboratory filtration system consists of a folded piece of filter paper supported in a conical funnel. Liquid moves through the filter due to the force of gravity. **b.** Vacuum can be used to facilitate the movement of fluid through the filter. In this example, a Buchner style funnel is used to support the filter paper. Note that it is good practice to place a trap bottle between the vacuum flask and the source of vacuum to trap materials that are pulled into the tubing.

These four components are present both in a simple coffee maker and in a complex process filtration system in a pharmaceutical company.

B. Overview of Issues in Filtration

There are several issues that affect filtration and which must be considered in designing or selecting a filtration system. The most obvious is clogging. Filters clog if they are covered by large particles, or by aggregations of smaller particles. Oils, lipids and fats can similarly form a film on the surface of a filter that prevents filtration. In some situations clogging is "cured" simply by replacing the clogged filter with a new one. In more sophisticated systems, particularly in industrial settings, clogging is reduced by moving the liquid to be filtered across the filter surface. Such systems are discussed later in this chapter.

Some substances, like proteins, bind to certain filter materials. *When a component binds to the surface of a medium, such as a filter, it is called* **adsorption**. Adsorption tends to block the pores of a membrane, resulting in a lower rate of filtration. Adsorption also leads to loss of the adsorbing species. Some materials used in the manufacture of filters bind macromolecules readily, whereas others minimize binding.

Adsorption is different than absorption. **Absorption** *is when liquids are taken up into the entire depth of a material, as when water is absorbed by a sponge.* There is no chemical selectivity associated with absorption as there is with adsorption. Both adsorption and absorption can occur in filtration.

Extractability is another issue in filtration. **Extractables** *are compounds from the filter that leach out and enter the sample being filtered.* Fibers from the filter may similarly enter the sample being filtered. The contamination of a sample by materials from the filter can be a serious problem in some applications. For example, in cell culture, extractable substances from filters have been shown to inhibit the growth of cells. In the pharmaceutical industry, filters have many applications, such as removing bacteria and viruses from products, and purifying water. It is essential that these filtration processes not introduce impurities into the drugs. Another application of filtration is to remove particulates from samples before they are injected into instruments for analysis. Many types of instruments used in analyzing samples detect minute amounts of compounds, so impurities from filters must be meticulously avoided.

II. TYPES OF FILTRATION AND FILTERS

A. Overview

Filters are generally classified into three types according to the size of particles they retain, Figure 25.2:

1. **Macrofilters or general filters (depth filters)**, *are used for the separation of particles on the order of 10 μm or larger.* A coffee filter and a common laboratory paper filter are examples of macrofilters.

2. **Microfilters** *are typically used to separate particles whose sizes range from about 0.1 μm to 10 μm.* Bacteria and whole cells fall in this size range.

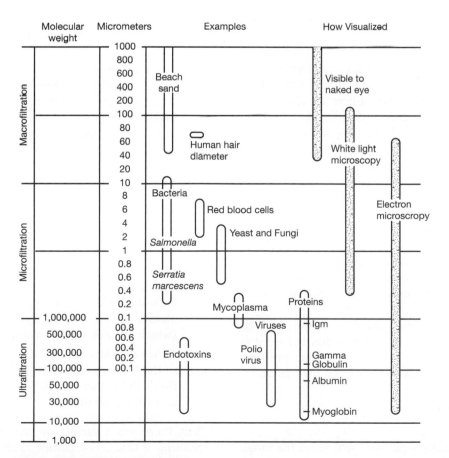

Figure 25.2. Filtration and Size.

3. Ultrafilters *are used to separate macromolecules on the basis of their molecular weight.* For example, large proteins can be separated from smaller ones using ultrafiltration.

Each of these classes of filtration is discussed in more detail in the next sections. In addition, we will briefly consider two related techniques: dialysis and reverse osmosis.

B. Macrofiltration

General filters are inexpensive devices used to remove relatively large particles, above about 10 μm, from a liquid. These filters are made of materials such as paper, glass fibers, sand, and cloth. General filters do not have pores or holes of a particular size; rather they consist of a convoluted granular or fibrous matrix. Particles are trapped both on the filter surface and within the sinuous matrix. Because these filters trap particles throughout their depth, they are also called **depth filters**.

Macrofilters for the laboratory are usually made of cellulose (paper) or of glass fibers. In practice, both types may be referred to as *filter paper.* Glass-fiber filters consist of long strands of borosilicate glass that are formed under high heat conditions. In contrast to filters made of paper, glass-fiber filters allow faster flow rates, are stable under a wide range of temperatures, and are compatible with corrosive chemicals.

Manufacturers make a variety of paper filters and glass-fiber filters. These filters vary in the density of their mesh and therefore in the sizes of particles they tend to trap. Filters that retain smaller particles tend to have slower flow rates than those that trap only larger particles. Different types of paper filters are also graded by the amount of ash residue they leave when burned. Low ash level is important in some applications in analytical chemistry where a sample solution is filtered and then the filter paper is burned to leave behind only the sample. The ash level is also a measure of the purity of the fibers used to make the filter: the lower the ash level, the higher the purity. The term **quantitative grade** *refers to filter papers that leave little ash when burned and are intended for use in certain chemical analyses.* **Qualitative grade** *papers leave significant amounts of ash.* **Hardened grade** *papers are treated to be stronger and better able to withstand vacuum filtration than unhardened papers.*

General filters are often used as "prefilters" to remove large particulates from a solution before it is filtered with a microfilter or ultrafilter. The prefilter prevents clogging of the more expensive, finer filter.

C. Microfiltration
i. MICROFILTERS

Microfiltration typically separates particles in the range of about 0.1–10 μm from a liquid or gas medium. Microfiltration filters are called **membranes** *and are manufactured to have a particular pore size.* Particles larger than the rated size are retained on the surface of the membrane and smaller particles pass through. When a manufacturer specifies the **pore size (absolute)** *they mean that 100% of particles above that size will be retained by the membrane under specified conditions.* **Pore Size (nominal)** *means that particles of that size will be retained with an efficiency below 100% (typically 90–98%).* The methods of rating the nominal pore size can vary between manufacturers. Note that, in practice, pore size ratings refer to the size of particles retained, not to the actual physical dimensions of the pores.

Microfiltration membranes can be manufactured from a variety of plastics, metals, and ceramic materials, Table 25.1. While the first factor to consider in choosing a microfilter is pore size, different types of filter membranes differ in important properties. These properties include:

1. **Resistance to organic solvents.** Some membranes are dissolved by organic solvents, others are not. When organic solvents are being filtered, it is necessary to choose a membrane made of a solvent-resistant material.

2. **Binding properties.** Some membrane materials readily bind (adsorb) single-stranded DNA, RNA, and proteins. There are applications in molecular biology where this binding is desirable because it immobilizes the macromolecules. For filtration, this binding is often undesirable because the macromolecules are "lost."

3. **Surface smoothness.** Some membranes, called *screen membranes*, have a smooth surface with regularly spaced and evenly sized pores. Other membranes are more irregular in their pore structure. Smooth membranes are useful, for example, when the particles trapped on a membrane are to be viewed with a microscope.

4. **Extractables.** Some membranes have extremely low levels of extractables, whereas with others there is extraction from the membrane.

5. **Wetting properties.** Most membranes are hydrophilic, but some are hydrophobic. Almost all biological samples are aqueous and are filtered through hydrophilic membranes (which are readily wetted by water). Gases and organic solvents are almost always filtered with hydrophobic membranes (which are not readily wetted by water).

6. **Other characteristics.** These include the strength of the filter, its resistance to heat, the rate at which fluids flow through it, whether it can be autoclaved, and so on.

Thus, there are a variety of factors to consider when choosing a filter. The choice is simplified for routine applications because manufacturers provide filtration units tailored specifically for particular purposes.

Table 25.1 *EXAMPLES OF MATERIALS COMMONLY USED TO MAKE MICROFILTERS*

Material	Features	Examples of Applications
Nitrocellulose	Hydrophilic. Frequently used. Readily binds single-stranded DNA, RNA, and proteins.	General purpose microbiology. Sterility testing.
Cellulose Acetate	Hydrophilic. Very low aqueous extractability. Very low binding. Not resistant to organic solvents. Resistant to heat.	General filtration. Sterilizing tissue culture media. General sterilization.
Nylon	Hydrophilic. Very low extractability. Readily binds proteins. Compatible with alcohols and solvents frequently used in HPLC.	Filtering organic solvents. Filtering samples before HPLC.
Polysulfone	Hydrophilic. Low extractability. Low protein binding. Wide chemical compatibility. Autoclavable.	Sterilizing tissue culture media. Manufacturing ultrafiltration membranes.
PTFE	Naturally hydrophobic. Inert to most chemically aggressive solvents, strong acids, and bases. Expensive.	Ideal for filtering gases, air. Filtering organic solvents.
PVDF	Must be rendered hydrophilic. Highly resistant to solvents. Very low binding properties.	Filtering samples before HPLC.
Polypropylene	Hydrophilic. Low fiber release. Resistant to many solvents.	Filtering samples before HPLC.

ii. APPLICATIONS OF MICROFILTRATION

One of the most important applications of microfiltration, both in the laboratory and in industry, is to remove contaminating bacteria, yeast, and fungi from solutions. Other methods of sterilization require heat or chemicals that are destructive to certain types of materials. For example, filtration is often used to sterilize cell culture media because these media contain heat-sensitive solutes, such as vitamins and antibiotics.

Various pore sizes of microfiltration membrane will retain various sized microorganisms:

- **0.10 μm**, recommended by some manufacturers to remove Mycoplasma, which is a very small type of bacterium that can contaminate cell cultures.
- **0.22 μm**, standard pore size for removing *E. coli* bacteria
- **0.65 μm**, used to remove fungi and yeast
- **0.45–0.80 μm**, used for general particle removal
- **1.0, or 2.5 or 5.0 μm**, for "coarse" particles

Microfiltration is used both in the laboratory and in industry to sterilize both liquids and also gases, such as air and carbon dioxide. In many bacterial fermentations, air is supplied continuously to the fermentation vessel to provide agitation and oxygen. Hydrophobic microfilters are placed in the air stream to remove contaminating particles and microorganisms from the air before it enters the fermentation vessel. Filters may also be attached to supply lines for carbon dioxide and air running to animal cell culture vessels. Filters are used not only to protect the contents of fermentation vessels from contamination by outside air, but also to protect the facility from the vessel contents.

Microbiologists use microfiltration membranes when they monitor the levels of coliform bacteria in drinking water. A sample of water is filtered through a membrane that retains bacteria on its surface. The membrane is then incubated on a petri dish with nutrient medium. Nutrients pass through the filter allowing the bacteria to grow into colonies. The colonies are counted to provide an estimate of the quantity of bacteria in the water supply. In an analogous way, particulates in air can be assessed by filtering large volumes of air through a membrane filter. The captured particulates can be analyzed, for example, by microscopic examination.

iii. HEPA FILTERS

High Efficiency Particulate Air (HEPA) *filters are used to remove particulates, including microorganisms, from air.* HEPA filters are manufactured to retain particles as small as 0.3 μm. Unlike microfilter membranes, however, HEPA filters are depth filters made of glass microfibers that are formed into a flat sheet. The sheets are then pleated to increase the overall surface area. The pleated sheets are separated and supported by aluminum baffles, see p. 645.

HEPA filters have many applications both in the laboratory and in industry. HEPA filters are used in laboratory biological safety hoods to protect products from contamination and/or personnel from exposure to hazardous substances. In animal care facilities, HEPA filters are used on the tops of animal cages to protect valuable laboratory animals from infection with microorganisms. In industry, HEPA filters may be used to filter the air in entire rooms to protect products from contaminants.

D. Ultrafiltration

i. ULTRAFILTERS

Ultrafilters are membranes that separate sample components on the basis of their molecular weight. Membranes used for ultrafiltration have pore diameters from 1 to 100 Å and can separate particles with molecular weights ranging from about 1000 to 1,000,000.

The term *molecular weight cutoff* (MWCO), is used to describe the sizes of particles separated by an ultrafilter. The **molecular weight cutoff** *is the lowest molecular weight solute that is generally retained by the membrane.* MWCO values are not absolute because the degree to which a particular solute is retained by an ultrafiltration membrane is not entirely dependent on its molecular weight. The shape of the solute, its association with water, and its charge also affect its permeability through an ultrafiltration membrane. For example, a membrane is less likely to retain a linear molecule than a coiled, spherical molecule of the same molecular weight. In addition, the nature of the solvent, its pH, ionic strength, and temperature, all affect the movement of solutes through membranes. By convention, if a membrane is rated to have a MWCO of 10,000, this means that the membrane will retain at least 90% of globular-shaped molecules whose molecular weight is 10,000 or greater.

ii. APPLICATIONS OF ULTRAFILTRATION

The applications of ultrafiltration can be classified as either fractionation, concentration, or desalting. **Fractionation** *is the separation of larger particles from smaller ones.* For example, proteins that are significantly different in size can be separated from one another by ultrafiltration. A protein product that is made by cells in culture can be separated from other components of the cell culture medium in this way. Intact strands of DNA can similarly be separated from nucleotides.

In **concentration**, *solvent is forced through a filter while solute is retained. The initial volume of the sample is thus reduced and the high molecular weight species are concentrated above the filter.* For example, gel electrophoresis is used to separate and visualize proteins. Before electrophoresis, the proteins must be concentrated because only a very small volume can be applied to the gel. Ultrafiltration can be used for this purpose.

In **desalting**, *low molecular weight salt ions are removed from a sample solution.* Ultrafiltration is a simple method to remove salts because they readily penetrate the membranes, leaving the solutes of interest on the membrane surface.

E. Dialysis and Reverse Osmosis

Dialysis, and **reverse osmosis** are separation processes that use membranes similar to those used for ultrafiltration. *Dialysis is based on differences in the concentrations of solutes between one side of the membrane and the other.* Solute molecules that are small enough to pass through the pores of the membrane will diffuse from the side with a higher concentration to the side with a lower concentration. The distinctive feature of dialysis is that differences in solute concentration provide the "driving force"; dialysis does not require pumps or a vacuum to force materials through the pores of the membrane.

Desalting is an example of the use of dialysis in the laboratory. During the process of purifying a protein, it is common to cause the protein to precipitate from solution by adding high concentrations of salt. Subsequent steps in the protein purification process require that the salt be removed; this is called *desalting*. The salts can be removed from the protein by placing the sample in a bag made of dialysis membrane, Figure 25.3. The dialysis bag containing the sample is sealed at both ends and is sus-

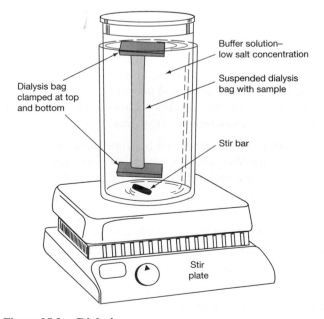

Figure 25.3. **Dialysis.**

pended in a large volume of water or buffer solution. Thus, the concentration of salts is much higher inside the bag than outside. The relatively large protein molecules cannot penetrate the pores of the dialysis membrane and so remain inside the bag, but small molecules, including salt, readily move through the membrane. Over time, low molecular weight salt molecules inside the bag diffuse out to the water or buffer solution. The concentration of salt inside the bag and outside the bag eventually equalizes and the system reaches equilibrium. The salt molecules that were originally inside the dialysis bag are now distributed throughout both the large volume of buffer (or water) and the dialysis bag. Observe that the salts are not completely removed from the sample, but their concentration is much reduced. In order to further reduce the concentration of salt in the sample, the dialysis bag can be moved into fresh water or buffer solution and the process can be repeated.

Dialysis is relatively inexpensive (compared with ultrafiltration), simple, and gentle. Because dialysis relies on passive diffusion, however, it is a relatively slow process. A number of special devices are available from manufacturers to make dialysis more efficient and convenient.

Reverse osmosis (RO) *is used to remove very low molecular weight materials, including salts, from a liquid (usually water).* Reverse osmosis is important in water purification systems (see Chapter 24). In reverse osmosis, water under pressure flows over a special, thin RO membrane. The membrane allows water to pass through, but rejects a large proportion (on the order of 95–99%) of all types of impurities (viruses, particles, pyrogens, microorganisms, colloids, and dissolved organic and inorganic materials). The permeate will contain very low levels of contaminants that are able to get by even an RO membrane, but most types of contaminants are greatly reduced. An RO membrane retains materials based both on their size and on ionic charge and it can retain smaller solutes than an ultrafiltration membrane.

III. FILTRATION SYSTEMS

A. Small-Scale Laboratory Filtration Systems

The principles of filtration are the same, whether the sample is 10 μL in the laboratory or 10,000 L in industry, but the design of filtration systems depends on the scale involved. The filter's size and shape, the support of the filter, the type of force used to move fluids through the filter, and the vessels involved can vary greatly depending on the scale.

Filtration requires a force to cause materials to flow through the filter. As was illustrated in Figure 25.1, it is conventional to use vacuum filtration in the laboratory. Figure 25.4 shows laboratory vacuum filtration systems that take advantage of membrane filters.

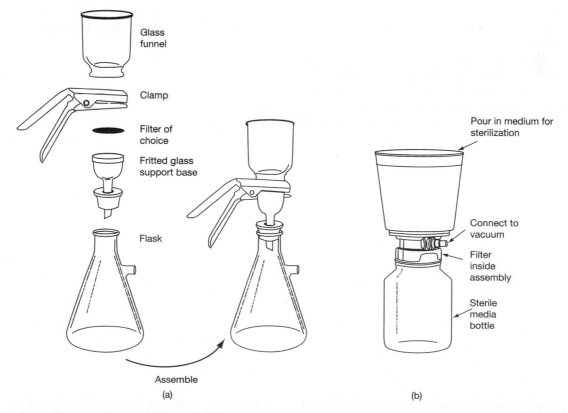

Figure 25.4.   Laboratory Filtration Systems. a. A membrane filter is mounted in a holder that clamps to the top of a flask. Vacuum is used to pull the material through the filter. **b.** A sterile filtration system used to conveniently sterilize cell culture media.

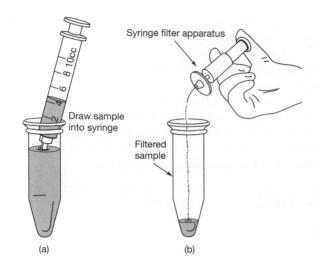

(a)   (b)

Figure 25.5.   Syringe Filter Device. a. The sample is drawn into the syringe. **b.** The filter is attached to the syringe. Sample is forced through the filter by depressing the plunger.

Another laboratory method to force samples through a filter is to use a syringe. The sample is loaded into the syringe and the filter unit is mounted on the end. The sample fluid is forced through the filter by depressing the plunger, Figure 25.5. Syringe filtration units contain microfilters of varying pore sizes. Sterile syringe filter units are available and are popular for sterilizing small volumes of sample, such as milliliter solutions of an antibiotic for cell culture. Very small syringe filter units are used for removing particulates from microliter volume samples prior to HPLC.

Spin filtration is a type of laboratory filtration system that uses centrifugal force. Spin filtration units contain ultrafilters, or sometimes microfilters, that are housed within a centrifuge tube. The sample is placed in the tube on top of the filter and the unit is spun in a centrifuge. During centrifugation, the liquid and smaller particles are forced through the filter and are captured in the bottom of the centrifuge tube. Larger molecules remain behind on the surface of the filter, Figure 25.6. Spin filters can be used to fractionate, concentrate, and

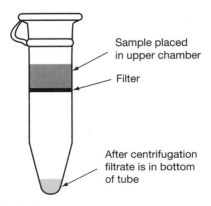

Sample placed in upper chamber

Filter

After centrifugation filtrate is in bottom of tube

Figure 25.6.   Spin Filtration Takes Advantage of Centrifugal Force. Sample is loaded into the top of the tube. Sample is forced through the filter by the force of centrifugation.

desalt samples. They are often used in the molecular biology laboratory for small volume samples of proteins, nucleic acids, antibodies, and viruses.

B. Large-Scale Filtration Systems

Process filtration *involves large volumes of liquids, as are found in biotechnology production facilities, pharmaceutical companies, and food production facilities.* For example, filtration is used in the manufacture of biopharmaceutical products, such as erythropoietin and tissue plasminogen activator. These products are made by large-scale cell culture. The desired product is often secreted by the cells into the surrounding liquid broth. The first step in processing, therefore, is to separate the cells and cell debris from the liquid containing the product. This separation step is called *clarification*. Clarification is often a costly and challenging process involving thousands of liters of material and a fragile protein product. Filtration is often preferred over centrifugation because filtration tends to be less expensive, gentler, and more convenient.

Filtration systems for industry must be designed to handle large volumes in a reasonably short time. They must be capable of being cleaned and sterilized, and their effectiveness must be validated. Industrial-scale filtration systems have sophisticated designs to meet these requirements. These systems maximize the surface area for filtration. The simplest way to do this is by using large sheets of filter membrane. More complex systems form the membranes into tubes, spirals, or pleats to maximize surface area. For example, **hollow-fiber ultrafilters** *are cylindrical cartridges packed with ultrafiltration membranes formed into hollow fibers.* The liquid to be filtered flows through the lumen of the fibers. As molecules pass through the fiber core, substances smaller than the MWCO penetrate the membrane, whereas those that are larger are concentrated in the center, Figure 25.7.

Membrane clogging is a major problem in industry. One method to reduce clogging is to use **tangential flow**, or **cross-flow filtration**, *where the fluid to be filtered flows over the surface of the filter as well as through the filter.* The sweeping motion of the fluid clears the surface of the membrane, thus reducing clogging.

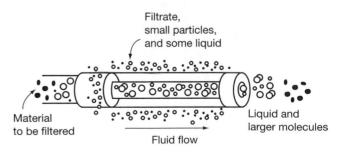

Filtrate, small particles, and some liquid

Material to be filtered

Liquid and larger molecules

Fluid flow

Figure 25.7.   Hollow Fiber Ultrafilter System. Sample flows through the lumen of the tubular filter. Small molecules pass through the filter.

PRACTICE PROBLEMS

1. Order the following items in terms of size, from the smallest to the largest.

   pollen grains (about 30 μm)   sand grains

   polio virus   red blood cells

   albumin (a protein)   NaCl ions

   Serratia marcescens (a type of bacteria)

2. For each of the following separations, state whether it will involve macrofiltration, microfiltration, or ultrafiltration.

   a. Purifying antibodies from a liquid medium.

   b. Removing viruses from a vaccine.

   c. Removing salts from a solution containing DNA.

   d. Sieving large particulates from water before it is treated in a sewage treatment plant.

   e. Sterilizing cell culture medium by removing bacteria.

   f. Removing pyrogens (fever causing agents) from a drug product.

   g. Harvesting mammalian cells from a fermenter.

   h. Removing *Mycoplasma* from bovine serum (which is often added to cell culture media).

3. The specifications for three filters are shown. Match the filters with the applications that follow:

   Ready Separation Hollow Fiber Filtration System

   The "Ready Separation" System is a compact hollow fiber filtration system capable of processing from 5 to 100 L. The system is sterilizable and can be purchased with microfiltration membranes from 0.1 to 0.80 μm or with ultrafiltration membranes from 300 to 500,000 MWCO.

   Filter Type XYZ

   The XYZ unit contains a hydrophobic, solvent-resistant, 0.2 μm membrane designed for use as a sterilizing filter for gases and liquids. The unit is a pyrogen-free, sterile, single use device.

   Filter Type ABC

   The ABC filtration unit is a low protein-binding, sterile filter for aqueous, proteinaceous substances. It is intended for applications where minimal sample loss is desired. It is a 0.22 μm filter, single use product for use with syringes.

   a. A fermentation process is being designed in which microorganisms produce an antibiotic that will be isolated from the broth. The microorganisms generate carbon dioxide which is vented from the fermenter via a plastic tube. On the end of the tube is a filtration unit that keeps organisms from the outside air from contaminating the fermenter. What type of filtration unit is used?

   b. In the laboratory, an antibiotic solution needs to be added to cell culture plates. It is heat sensitive and is therefore sterilized by filtration. Which filter would be used?

   c. An organic solvent is to be used with HPLC. It needs to be filtered to remove particulates. Which filter would be used?

   d. A biotechnology company uses a large-scale cell culture process to produce a valuable protein that is being tested as a drug. The protein is secreted by the cells into the culture medium. Which type of filter might be used in the process of isolating the protein product from the cell culture medium?

TERMS RELATING TO FILTRATION

Absorption (in filtration). The taking up of liquids into the depth of a filtering material.

Adsorption (in filtration). A phenomenon in which solutes bind to the surface of a filtration membrane.

Aerosols. Airborne particulates.

Airborne Particle Retention. A measure of the efficiency of an air filtration system. In the United States, the DOP (Dioctyl Phthalate) test is commonly used in which the filter is challenged with air containing 0.3 μm DOP particles. The efficiency of the filter system is expressed as the percent of these particles that are retained. For example, a HEPA-grade filter should retain at least 99.97% of the particles.

Anisotropic Membrane. A membrane in which the pore openings are larger on one side than on the other.

Ash Content. The percent ash residue remaining when a paper filter is burned. Ash content is important in some analytical chemistry applications and is also a measure of the purity of the filter.

Clarification (in cell culture). The removal of cells and cellular debris from a culture medium.

Clean Room. A room or enclosed area whose ventilating system incorporates filters that remove a large percent of particulates from the air.

Depth Filters. Filters that retain particles both on their surface and within their fibrous matrix.

Dialysis. A separation method based on differences in the concentrations of solutes between one side of a membrane and the other. Solute molecules that are small enough to pass through the pores of the membrane will diffuse from the side with a higher concentration to the side with a lower concentration.

Downstream (in filtration). The side of a filter facing the filtrate.

Extractables (in filtration). Compounds from a filter that are released into the material being filtered.

Filter. A device used to separate components of samples on the basis of size; particles smaller than a certain size pass through a porous filter material; particles larger than a certain size are trapped by the filter.

Filtrate. The fluid and any associated particles that have passed through a filter.

Filtration. A separation method in which particles are separated from a liquid or gas by passage through a porous material.

Flow Rate. The rate of flow of a gas or liquid through a filter at a given pressure and temperature. The flow rate determines the volume that can be filtered in a given amount of time.

High-Efficiency Particulate Air, HEPA, Filter. A filter used to remove particulates, including microorganisms, from air.

Hollow-Fiber Ultrafilter. A cylindrical cartridge packed with ultrafiltration membranes formed into hollow, tubular fibers.

Hydrophilic. "Water-loving"; in filtration, refers to a filter's ability to wet with water.

Hydrophobic. "Water-hating"; in filtration, refers to a filter's resistance to wetting with water.

Isotropic Membrane. A membrane in which the pore size is the same on both sides of the membrane.

Macrofiltration. The removal of relatively large particles, above about 10 μm, from a liquid by passage through porous materials including paper, glass fibers, and cloth.

Membrane Filter. A filter where particles are retained primarily on the surface of the membrane and in which there are pores or channels of defined size.

Microfiltration. The filtration of particles whose sizes are typically in the range from about 0.1 μm to 10 μm using membrane filters.

Molecular Weight Cutoff, MWCO. In ultrafiltration, the lowest molecular weight solute that is retained by the membrane.

Nonfiber Releasing Filter. A filter that is treated in such a way that it will not release fibers into the filtrate.

Particle Retention. A term used in reference to depth filters to indicate the smallest particle size (in micrometers) that is retained by the filter. Particle retention ratings are nominal (i.e., a small percent of particles of the rated size will penetrate the filter).

Permeate. The fluid and particles that pass through a membrane filter.

Polarization Layer. A build-up of retained particles or solutes on a membrane.

Pore Size (absolute). The size of particles that are retained with 100% efficiency by a microfilter under specified conditions.

Pore Size (nominal). The size of particles that are retained with an efficiency below 100% (typically 90–98%) by a microfilter under specified conditions.

Porosity. The percentage of a filter area that is porous.

Prefilter. A filter placed upstream from a membrane filter or from an instrument to protect the membrane or instrument from particulates that would cause clogging.

Qualitative Filter Paper. A paper filter that leaves an appreciable percent of ash when burned.

Quantitative Filter Paper. A paper filter that leaves little ash when burned.

Residue Analysis. When particles of interest are separated and retained on a filter for further analysis.

Retentate. The materials trapped by a filter.

Reverse Osmosis, RO. A separation method, commonly used to purify water, that can remove very low molecular weight materials, including salts, from a liquid.

Sterilizing Filter. A nonfiber releasing filter that produces a filtrate containing no demonstrable microorganisms when tested as specified in the United States Pharmacopeia.

Surface Filter. See "Membrane Filter."

Tangential Flow Filtration. A filtration mode used mainly in industry where the fluid to be filtered flows across the filter. The sweeping motion of the fluid clears the surface of the membrane, thus reducing clogging.

Throughput. How long a liquid can flow through a filter before the membrane clogs.

Ultrafiltration. A membrane separation technique that is used to separate macromolecules on the basis of their molecular weight.

Upstream. The side of a filter facing the incoming liquid or gas.

ANSWERS TO PROBLEMS

1. NaCl ions, albumin, polio virus, *Serratia marcescens*, red blood cells, pollen grains, sand grains

2. **a.–c.** ultrafiltration   **d.** macrofiltration
   e. microfiltration   **f.** ultrafiltration
   g. microfiltration   **h.** microfiltration

3. **a.** XYZ   **b.** ABC
   c. XYZ   **d.** "Ready Separation"

CHAPTER 26

Introduction to Centrifugation

A brief note about safety: Centrifuges pose a variety of hazards to laboratory personnel. Moreover, although centrifuges and their components appear solid and sturdy, they are in fact expensive devices that can easily be damaged by mishandling. Safe and proper handling of centrifuges is discussed throughout this chapter.

I. INTRODUCTION TO CENTRIFUGATION: PRINCIPLES AND INSTRUMENTATION

A. Basic Principles

i. SEDIMENTATION

To begin thinking about centrifugation, consider a cylinder filled with a slurry of gravel, sand, and water, Figure 26.1. We would expect that, over time, gravity would cause the particles to **sediment** (*i.e., to move through the liquid and settle to the bottom of the cylinder*). The larger gravel particles will sediment more quickly than the smaller sand particles so that there will eventually be a bottom layer of mostly gravel, a layer of mostly sand, and relatively clear liquid in the upper part of the cylinder. Thus, a separation, or fractionation process will occur in which the particles separate from

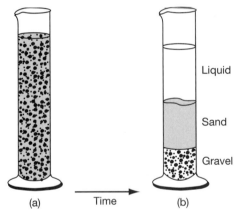

(a)   Time →   (b)

Figure 26.1.   Sedimentation Due to Gravity. a. A slurry consisting of gravel, sand, and water is placed in a cylinder. **b.** The particles sediment according to their size and density under the influence of gravity.

the liquid medium and from one another due to the force of gravity.

Biologists seldom need to separate gravel and sand from one another; rather, they are interested in such materials as cells, organelles, bacteria, and viruses. These biological materials, called *particles* in the context of centrifugation, are much smaller than sand and would take a long time to sediment due to the force of gravity alone. A **centrifuge** *is a piece of equipment that accelerates the rate of sedimentation by rapidly spinning the samples, thus creating a centrifugal force many times that of gravity.*

A simple centrifuge is illustrated in Figure 26.2. There is a central drive shaft that rotates during centrifugation. A **rotor** *sits on top of the drive shaft and holds tubes, bottles, or other sample containers.* As the drive shaft rotates, the sample containers spin rapidly, creating the force that facilitates the sedimentation of particles.

ii. CENTRIFUGAL FORCE

To appreciate the nature of the force in a centrifuge, centrifugal force, imagine that you tie a stone to the end of a string and whirl it rapidly in a circle above your head. You will feel a pull on your hand as the stone rotates; this pull is centrifugal force. *Whenever an object is forced to move in a circular path (as in a centrifuge)* **centrifugal force** is generated. Now, consider what would happen if you whirled the stone faster—you would feel more pull. Similarly, the faster a sample is spun in a centrifuge, the more force is experienced by that sample. The speed of rotation in a centrifuge is expressed as **revolutions per minute (RPM)**.

In addition to the speed of rotation, there is another (somewhat less obvious) factor that affects how much centrifugal force is experienced by a sample in a centrifuge. If you whirl a stone tied to a long string, there will be more force on the stone than if you use a shorter string. In a centrifuge, the further a particle is from the center of rotation, the more force the particle experiences. *The distance from the center of rotation to the material of interest is called* **the radius of rotation (r)**, Figure 26.3. In a centrifuge, the radius of rotation varies depending on the particular equipment being used.

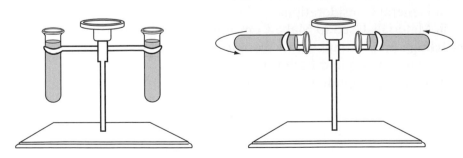

Figure 26.2.   A Simple Centrifuge. Tubes containing a sample are rapidly spun about a central shaft, creating a force many times that of gravity.

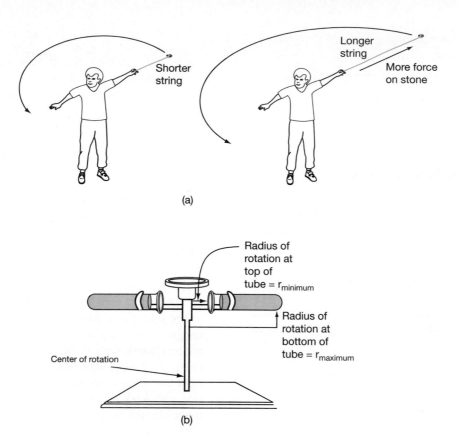

Figure 26.3. The Radius of Rotation. a. The longer a string being whirled, the greater the radius of rotation and the more force that is experienced by the stone. **b.** In a centrifuge, the further a particle is from the center of rotation, the more force it experiences.

Thus, there are two factors that determine the centrifugal force experienced by a particle in a centrifuge:

1. **The speed of rotation, expressed as *RPMs***
2. **The distance of the particle from the center of rotation, *r***

The centrifugal force acting on samples in a centrifuge is expressed as the **relative centrifugal field (RCF)**. The relative centrifugal field has units that are multiples of the earth's gravitational field (g). You may see the force in a centrifuge expressed as, for example, $10,000 \times g$ or $10,000 \; RCF$. Both these expressions mean that the force in the centrifuge is 10,000 times greater than the normal force of gravity.

It is possible to calculate the relative centrifugal field in a centrifuge if one knows the speed of rotation and the radius of rotation. The equation for the relative centrifugal field is:

$$RCF = 11.2 \times r \left(\frac{RPM}{1,000} \right)^2$$

where
RCF = the relative centrifugal field in units of $\times g$
RPM = the speed of rotation in revolutions per minute
r = the radius of rotation in centimeters

You may also see this equation rearranged as:

$$RCF = 1.12 \times 10^{-5} \, (RPM)^2 \, (r)$$

Observe that two centrifuges may be spinning at the same rate, yet they might each be subjecting the samples to a different centrifugal force. This is because the radius of rotation varies. For this reason it is more useful to report to others the relative centrifugal field when a sample is centrifuged, rather than simply the speed of rotation.

EXAMPLE PROBLEM:

Your centrifuge equipment has a maximum radius of rotation of 9.2 *cm*. If you spin a sample at a speed of 15,000 *RPMs*, what is the relative centrifugal field?

ANSWER:

Substituting into the RCF equation gives:

$$RCF = 11.2 \times 9.2 \left(\frac{15,000}{1,000} \right)^2$$

$$RCF = 11.2 \times 9.2 \, (225) = 23,184 \times g$$

The relative centrifugal field is **23,184 $\times$ g.**

Substituting into the rearranged form of the equation gives the same answer:

$$RCF = 1.12 \times 10^{-5} \, (RPM)^2 \, (r)$$
$$RCF = 1.12 \times 10^{-5} \, (15,000)^2 \, (9.2)$$
$$= 23,184 \times g$$

EXAMPLE PROBLEM

A colleague tells you that a sample was centrifuged for a certain time at a speed of 15,000 *RPM* and that the force of centrifugation was 16,000 × *g*. Your centrifuge equipment has a radius of rotation of 9.2 *cm*. What speed will you use to achieve a force of 16,000 × *g*?

ANSWER

The unknown in this case is the speed of rotation. It is possible to algebraically rearrange the *RCF* equation to solve for *RPM*s as follows:

$$RCF = 11.2 \times r \left(\frac{RPM}{1,000} \right)^2$$

$$\frac{RCF}{11.2 \, r} = \left(\frac{RPM}{1000} \right)^2$$

$$\sqrt{\frac{RCF}{11.2 \, r}} = \left(\frac{RPM}{1000} \right)$$

$$1000 \sqrt{\frac{RCF}{11.2 \, r}} = RPM$$

Then, substituting into this equation gives:

$$1000 \sqrt{\frac{16,000 \times g}{11.2 \, (9.2 \, cm)}} = RPM$$

$$12,461 = RPM$$

Thus, in order to achieve a force of 16,000 *RCF*, you will need to use a speed of 12,461 *RPM*. Observe that to obtain a force of 16,000 × *g*, the two centrifuges must be run at different speeds. **It is the force in the centrifuge that ultimately matters, not the speed of rotation.**

EXAMPLE

A colleague tells you that a sample was centrifuged at 30,000 × *g* for 15 minutes. Your rotor can achieve a maximum *RCF* of only 20,000 × *g*. What can you do?

It may be possible to compensate for the reduced force in your rotor by using a longer centrifugation run (but see the cautions at the end of this example). The time required in your rotor can be calculated using the following equation:

$$t_a = \frac{t_s \times RCF_s}{RCF_a}$$

where

t_a = time required in alternate rotor
t_s = time specified in the procedure
RCF_a = RCF of alternate rotor
RCF_s = RCF specified in the procedure

Thus:

$$t_a = \frac{15 \, min \times 30,000 \times g}{20,000 \times g} = 22.5 \, min$$

Spinning the sample for 22.5 minutes at 20,000 × *g* will duplicate the required conditions.

Note, however, that changing the time or the force of centrifugation may affect the results. Harder pelleting (more force) may damage sensitive biological materials. Longer spin times may lead to deterioration of some samples, particularly if the centrifuge is not refrigerated. Density gradient separations (discussed in a later section) are more complex than simple sedimentation and may not work if the time or *RCFs* are changed. For all these reasons, before changing centrifugation conditions, check that the changes will have no adverse effects.

iii. FACTORS THAT DETERMINE THE RATE OF SEDIMENTATION OF A PARTICLE

It is reasonable that the greater the force during centrifugation, the more quickly particles sediment. Another factor that affects the sedimentation rate is the viscosity of the liquid medium through which the particles are moving. The less viscous the medium, the faster the rate of sedimentation.

The rate of sedimentation of a particle also depends on its own characteristics. Larger particles sediment faster than smaller ones (as was also shown earlier in the example of sand and gravel sedimenting in a cylinder). It is also the case that the sedimentation rate of a particle depends on its density relative to the density of the surrounding liquid medium. Recall that density is the mass per unit volume. A stone, for example, is more dense than a beach ball. If a stone and a beach ball are thrown into a pond, the stone will sink because it is more dense than the water, but the ball will float. The density of particles similarly affects their movement in a liquid medium during centrifugation.

If a particle is more dense than the liquid medium, then the particle sediments (sinks) during centrifugation. If the densities of the particle and the liquid medium are equal, the particle does not move relative to the medium—no matter how strong the centrifugal force is. Rather, the particle remains stationary in the medium. There are also situations where the medium is more dense than a particle (as water is more dense than a beach ball). In these situations, a particle can actually "float" in a centrifuge tube, or move in the opposite direction of the force of centrifugation. Thus:

1. **If the density of a particle and the liquid medium are equal, the particle does not move in a centrifugal field.**

2. **If the density of a particle is greater than the density of the medium, the particle sediments (moves away from the center of rotation).**

3. **If the density of a particle is less than the density of the medium, the particle moves toward the center of rotation (floats "upwards" in the tube).**

If the liquid medium in centrifugation is water, or a water-based buffer, then the density of most biological materials, such as cells, organelles, and macromolecules, is greater than the density of the medium. These particles move toward the bottom of the tube. Lipids and fats, which are less dense than water, are found at the top of the tube after centrifugation.

In summary* the factors that influence the rate of sedimentation of a particular particle include:

1. **The relative centrifugal field.** The higher the relative centrifugal field, the faster the rate of sedimentation.

2. **The viscosity of the medium.** The less viscous the medium through which the particles must move, the faster the rate of sedimentation.

3. **The size of the particle.** The larger a particle, the faster its rate of sedimentation.

4. **The difference in density between the particle and the medium.**

iv. SEDIMENTATION COEFFICIENTS

You may see the term **sedimentation coefficient**, or **S value**. The sedimentation coefficient *is a value for a particular type of particle that describes its rate of sedimentation in a centrifugal field.* The larger the S value, the

Table 26.1   EXAMPLES OF SEDIMENTATION COEFFICIENTS

Biological Particle	Sedimentation Coefficient, S
Soluble Proteins	
Ribonuclease ($MW = 13,683$)	1.6
Ovalbumin ($MW = 45,000$)	3.6
Hemoglobin ($MW = 68,000$)	4.3
Urease ($MW = 480,000$)	18.6
Other Materials	
E. coli rRNA	20
Calf Liver DNA	20
Ribosomal Subunits	40
Ribosomes	80
Tobacco Mosaic Virus	200
Bacteriophage T2	1000
Mitochondria	15,000–70,000

Notes: 1. The sedimentation coefficient depends on the liquid medium and the temperature. Standard conditions for these coefficients are in water at 20°C.
2. Sedimentation coefficients are expressed in Svedberg units, S, where one Svedberg unit = 10^{-13} seconds.

faster the particle will sediment. The sedimentation coefficient is a physical constant that refers to a particular biological particle. Various proteins, different organelles, and nucleic acids each have their own sedimentation coefficient. Sedimentation coefficients have been measured for a variety of biological particles; some examples are shown in Table 26.1. Observe in this table that larger particles generally have larger sedimentation coefficients.

v. APPLICATIONS OF CENTRIFUGATION

Centrifugation has a long history in the biology research laboratory, where it is used in the separation, isolation, purification, and analysis of biological materials. In the biotechnology industry, centrifugation often plays an essential role in the separation of cells from broth after fermentation.

Centrifugation applications may be broadly classified as either preparative or analytical. **Preparative cen-**

*You may come across the following equation, which combines these various factors as follows:

$$v = \frac{d^2(\rho_p - \rho_m)\, g}{18\,\mu}$$

where
  v = velocity of sedimentation
  d = the diameter of the particle
  ρ_p = density of the particle
  ρ_m = density of the medium through which the particle is sedimenting
  g = the relative centrifugal force
  μ = the viscosity of the medium

Although this equation may seem formidable because it has several unfamiliar symbols, it is possible to obtain information from it fairly easily. Observe that the viscosity of the medium is in the denominator—the more viscous the medium, the slower the velocity of sedimentation. The g force is in the numerator—the higher the g force, the faster the sedimentation velocity. The difference in density between the particle and the medium is also in the equation. When a particle and the medium are of the same density, $(\rho_p - \rho_m) = 0$, which means the sedimentation velocity is zero—the particle does not move. If the particle is more dense than the medium the particle moves with a certain velocity toward the bottom of the tube. It is also possible for the particle to be less dense than the medium. In this case, the velocity is a negative number—that means the particle moves upwards in the tube.

Table 26.2 EXAMPLES OF APPLICATIONS OF PREPARATIVE CENTRIFUGATION

Separation of intact, single cell suspensions such as bacteria, viruses, and blood cells from a liquid medium

Separation of cellular organelles, such as mitochondria, and ribosomes from disrupted cells

Separation of biological macromolecules, including DNA, RNA, and protein from a solution

Separation of plasma (the fluid portion of blood) from blood cells

Separation of immiscible liquids from one another

Separation of cells from broth after fermentation

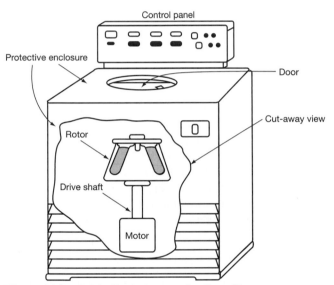

Figure 26.4. **Basic Components of a Centrifuge.**

trifugation provides separated materials for later use. Examples of applications of preparative centrifugation are in Table 26.2.

Analytical centrifugation *is used in specialized research settings to determine the molecular weight, purity, shape, and other physical characteristics of proteins and other macromolecules.* Analytical centrifugation involves specially designed, computer-controlled centrifuges that allow the operator to monitor the movement of particles. Analytical centrifugation is not nearly as common as preparative centrifugation and is not considered further in this text.

B. Design of a Centrifuge: Overview

The typical components of a centrifuge are shown in Figure 26.4. The motor, which turns the drive shaft, is located at the base of the unit. A rotor sits on top of the drive shaft and holds the sample containers. With the exception of older, low-speed instruments, centrifuges have a protective enclosure so that the rotor sits in a chamber. In modern centrifuges the chamber has a cover that is usually locked when the centrifuge is spinning. In higher speed centrifuges, the enclosure and cover are made of reinforced steel which, in the event of a serious mishap, can withstand the tremendous force of a rotor that flies off the drive shaft. Note that older high-speed centrifuges may lack the protective safety features of newer models. Most centrifuges also have a braking device to slow the rotor, a control device to set the speed, and a timer.

There are many styles and models of centrifuges, so it is convenient to classify them into types. A simple classification by speed is summarized in Table 26.3 and in the following list:

Table 26.3 CENTRIFUGES

	Type			
Characteristics	*Low Speed*	*High Speed*	*Ultracentrifuge*	*Microfuge*
Maximum Speed (*RPMs*)	10,000	30,000	120,000	15,000
Maximum Force ($\times g$)	8,000	100,000	700,000	21,000
Refrigeration	Sometimes	Yes	Yes	Sometimes
Vacuum	No	Sometimes	Yes	No
Examples of applications	Harvest intact plant and animal cells. Separate blood components. Harvest larger organelles (such as nuclei). Separate immiscible liquids.	Purify viruses. Isolate mitochondria, lysosomes, and other moderate size organelles.	Purify small organelles such as ribosomal subunits. Purify dissolved DNA, RNA, and proteins.	Purify precipitated nucleic acids. Perform small volume separations.

- **Low-speed centrifuges** typically attain rotation speeds less than 10,000 *RPM* and forces less than 8,000 × *g*.

- **High-speed centrifuges** typically attain rotation speeds up to about 30,000 *RPM* and generate forces up to about 100,000 × *g*.

- **Ultraspeed centrifuges**, or **ultracentrifuges**, typically attain rotation speeds up to about 120,000 *RPM* and generate forces up to 700,000 × *g*.

Heat is generated as a rotor turns. High-speed centrifuges and ultracentrifuges, therefore, are usually refrigerated. Refrigeration is preferred for most biological samples to protect them from adverse effects due to heat, although intact cultured cells are usually centrifuged gently at room temperature.

Ultracentrifuges have vacuum pumps that evacuate air from the chamber. This is necessary because the friction of air rubbing against the spinning rotor creates drag and heat. Vacuum is commonly achieved using a system with two pumps. The first pump is a "roughing pump" that draws air out of the system beginning at atmospheric pressure. A diffusion pump takes over when some vacuum has been established. Diffusion pumps generate heat and so require a cooling system.

Centrifuges can be classified both on the basis of speed and by their capacity or by the sizes and types of sample containers they can spin. For example, **microfuges** or **microcentrifuges** *are small centrifuges that are intended for small volume samples in the microliter to 1 or 2* mL *range*. Microfuges are useful in molecular biology laboratories for conveniently pelleting small volume samples. At the other end of the spectrum, there are industrial, large-capacity centrifuges that are used in production facilities (e.g., to separate large volumes of cells from nutrient broth).

II. MODES OF CENTRIFUGE OPERATION

A. Differential Centrifugation

Preparative centrifugation can be performed in several different modes. Of these modes, *differential centrifugation*, also called *differential pelleting*, is the simplest and probably the most common in the biology laboratory. **Differential centrifugation** *separates a sample into two phases: a* **pellet** *consisting of sedimented materials*, and a **supernatant** *consisting of liquid and unsedimented particles*. A given type of particle may move into the pellet, remain in the supernatant, or be found in both phases. Which sample components move into the pellet and which remain in the supernatant depends on the nature

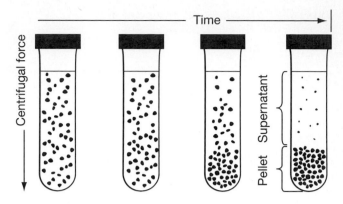

Figure 26.5. Differential Centrifugation. A suspension of particles in a liquid medium is initially uniformly distributed in the centrifuge tube. During centrifugation, particles sediment into a pellet at differing rates depending mainly on their size.

of the components, on the force, and on the duration of centrifugation.

In differential centrifugation, a uniform sample is poured into a centrifuge tube or bottle. The sample is centrifuged for a certain time with a particular centrifugal force. As the sample spins, larger particles sediment faster than smaller ones. By using a suitable combination of *g*-force and time, a pellet enriched in a particular component can be obtained, Figure 26.5.

After centrifugation, the supernatant can readily be decanted from the tube. It is possible to continue by centrifuging the supernatant again at a higher force, thus obtaining a second pellet. This process can be repeated, resulting in a series of pellets containing progressively smaller particles. For example, a common application of differential centrifugation is the isolation of cellular organelles from a cellular homogenate. (A cellular homogenate is produced by disrupting cells and releasing their organelles, membranes, and soluble macromolecules into a liquid medium.) Nuclei are the largest organelle. It is possible to obtain a pellet enriched in nuclei by centrifuging a cellular homogenate for 10 minutes at 600 × *g*. Other organelles tend not to move into the pellet given this relatively low centrifugal force and short run time. The supernatant is then removed and spun at a higher *g*-force. This process is continued until the smallest subcellular particles have been pelleted, Figure 26.6.

Although differential centrifugation is simple and widely used, it has disadvantages. When a sample mixture is initially poured into a centrifuge tube, the various sample components are dispersed throughout the tube. Thus, each pellet consists of sedimented components contaminated with whatever unsedimented particles were at the bottom of the tube initially. Moreover, the yield of smaller particles is lowered by their loss in

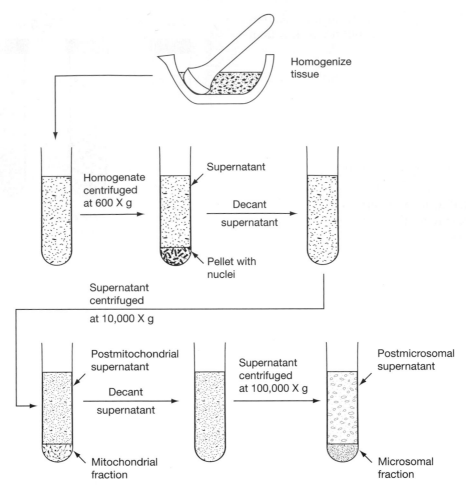

Figure 26.6.   Example of a Scheme to Isolate Cellular Organelles Using Differential Centrifugation. Various organelles are separated from one another as the sample is centrifuged at progressively higher forces.

early pellets. Differential centrifugation may also require starting and stopping the centrifuge repeatedly, which is time-consuming for the operator. Density gradient centrifugation is another mode of centrifugation that reduces or eliminates these problems.

B. Density Centrifugation

i. THE FORMATION OF DENSITY GRADIENTS

A **density gradient** *is a column of fluid that increases in density from the top of the centrifuge tube to the bottom.* **Density gradient centrifugation** *involves sedimenting a sample through a density gradient.* As different particles move through the density gradient, they separate into discrete bands, Figure 26.7.

Density gradients can be formed in three ways. Some high molecular weight solutes, such as cesium chloride (CsCl), form gradients spontaneously under the influence of a centrifugal field. With these media, the sample and the gradient material are mixed

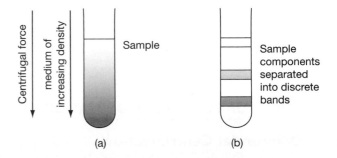

Figure 26.7.   Density Gradient Centrifugation. a. A sample is layered on top of a preformed density gradient. **b.** During centrifugation sample components separate into discrete bands as they move through the gradient.

together and centrifuged. As centrifugation proceeds, a gradient forms.

Many density gradient media do not spontaneously form gradients, so the operator prepares the gradient before the separation is begun. A preformed gradient may either be a step gradient or a continuous gradient.

A **step gradient** *is formed in layers with the densest layer on the bottom. Each layer has a different density and is clearly demarcated from the layers above and below.* A step gradient is prepared from separate solutions of the gradient medium, each of which has a different concentration and therefore a different density. The different solutions are carefully layered one on top of the other using a pipette or syringe, Figure 26.8a. The sample is loaded on top of the preformed gradient.

In **continuous gradients** *there is a smooth decrease in density from bottom to top with no sharp boundaries between layers.* Preformed continuous gradients are often produced by continuously mixing together two solutions of different density while simultaneously delivering the mixture to the centrifuge tube. This requires a special apparatus, called a **gradient maker**, Figure 26.8b. Alternatively, if a gradient maker is unavailable, a preformed step gradient will diffuse into a continuous gradient over time.

There is a wide variety of materials used to make density gradients. Sucrose and Ficoll are examples of media often used for preformed gradients. Cesium chloride is a salt that forms a gradient spontaneously when in a centrifugal field. The medium chosen must be compatible with the sample, must have a sufficient range of densities to separate all the components of the sample,

must not interfere with later steps in the procedure, must not be corrosive to the particular rotor being used, and so on. Because the requirements for the medium vary from application to application, different density gradient media have been developed and are available from manufacturers.

ii. RATE ZONAL CENTRIFUGATION

There are two methods of density centrifugation: rate zonal and buoyant density (also called isopycnic centrifugation). **Rate zonal centrifugation** *is a time-dependent method in which particles are separated as they sediment through a density gradient.* In the rate zonal method, the sample solution is layered on top of a preformed, usually continuous, density gradient. During centrifugation the particles sediment through the gradient. The rate of movement of the particles depends primarily on their size, less so on their density, so different particles move at different rates. As the particles move at different speeds through the gradient, they separate into bands, or zones (Figure 26.7).

In rate zonal centrifugation, the particles are more dense than the most dense part of the gradient. This means that if centrifugation is continued too long, the particles will all move to the bottom of the tube, mixing again into a pellet. Rate zonal centrifugation, therefore, is time dependent and is halted once the sample components have separated into discrete bands.

iii. BUOYANT DENSITY CENTRIFUGATION

Buoyant density centrifugation *is a mode in which particles are separated based on their density alone.* As a simple example, suppose that it is necessary to separate two types of particles, one of which has a density of 1.50 *g/mL*, the other a density of 1.35 *g/mL*. This can be accomplished by layering the sample on top of a liquid medium that has a density of 1.40 *g/mL*. After centrifugation, the less-dense particles will form a floating band, whereas the more dense particles will be in a pellet in the bottom of the tube.

In the simple previous example, there is not actually a density gradient; rather, there is a single density interface, called a *density barrier*. More complex buoyant density separations involve a density gradient. The gradient may be a preformed step gradient, a preformed continuous gradient, or a gradient that forms spontaneously during centrifugation. However the gradient is formed, the density of the medium at the bottom of the gradient must be greater than the density of any of the particles to be separated.

During buoyant density centrifugation, particles move away from the center of rotation if they are more dense than the medium in which they are located. Particles move toward the center of rotation if they are less dense than the medium at that point. Equilibrium is

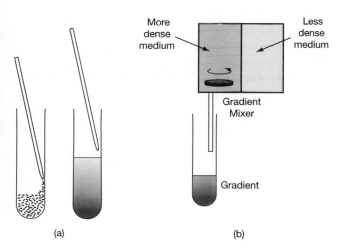

Figure 26.8. Preparing Density Gradients. a. Layering a step gradient. **b.** Preparing a continuous gradient using a gradient maker. A high-density (high concentration) solution of the gradient medium is placed in one of the chambers of the gradient maker; this chamber contains a stir bar for mixing the solution. A lower-density solution is placed in the other chamber. A narrow tube connecting the two chambers is opened. At the same time, liquid is allowed to begin flowing out of the high-density chamber into the centrifuge tube. Only dense solution initially reaches the centrifuge tube. As time goes on, more and more of the less dense solution flows into the mixing chamber, progressively diluting the medium. As a result, the medium flowing into the centrifuge tube slowly decreases in density (concentration), thus forming a continuous density gradient in the centrifuge tube.

eventually reached when all the particles are banded into the portion of the density gradient where their density equals that of the medium around them. Although the size of a particle does affect its *rate* of migration, buoyant density separations are ultimately based on density, not size. Buoyant density gradient centrifugation thus differs from rate zonal and differential centrifugation where particle size affects the final separation*.

Observe that in buoyant density centrifugation, the sample can be loaded anywhere in the tube. If the sample is placed on top of the gradient, the various particles will sediment down the tube to reach their equilibrium positions. If the sample is loaded in the bottom of the tube with the density gradient on top, the sample components will float up to reach their equilibrium positions. The sample can also be homogeneously mixed throughout a medium that forms a spontaneous gradient. During centrifugation, the gradient forms and particles move till they are in equilibrium with the surrounding medium, Figure 26.9.

After a density gradient separation is completed, there are various methods to collect the banded sample components. Figure 26.10 shows a technique in which the bottom of the tube is punctured and drops are col-

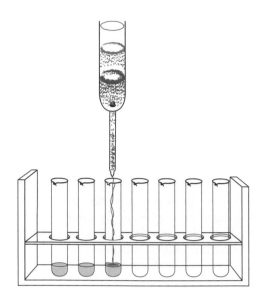

Figure 26.10.   Collecting the Fractions After Density Gradient Centrifugation. Collecting fractions from the bottom of a tube.

lected in successive test tubes. Another method of retrieving the sample is to puncture the side of the tube at the site of a specific band.

C. Continuous Centrifugation

Continuous centrifugation is a mode of centrifugation intended for large-volume samples, such as when cells must be separated from a large-volume fermentation broth. In continuous centrifugation, the sample is continuously fed into a rotor that is rotating at its operating speed. Particles sediment in the rotor while the supernatant is conveyed out of the rotor continuously. Special rotors, centrifuge equipment, and methods are required for continuous centrifugation.

The various modes of preparative centrifugation are summarized in Table 26.4.

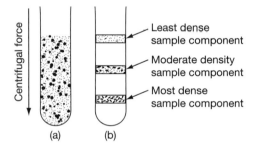

Figure 26.9.   Buoyant Density Gradient Centrifugation. a. A uniform mixture of gradient medium and sample is placed in a tube. **b.** During centrifugation, particles separate into bands such that at equilibrium their density equals that of the medium.

Table 26.4   MODES OF PREPARATIVE CENTRIFUGATION

Mode	Description
1. DIFFERENTIAL PELLETING	Results in two phases: a pellet of sedimented material and a supernatant consisting of liquid and unsedimented particles.
2. DENSITY GRADIENT CENTRIFUGATION	The liquid medium increases in density from the top to the bottom of the centrifuge tube or bottle.
a. RATE ZONAL	Sample components form bands due to differing rates of migration through the medium. Run is timed; equilibrium is not reached. Separation is based on size and density.
b. BUOYANT DENSITY	Sample components form bands where their density equals that of the surrounding medium. Equilibrium is achieved. Separation is based only on density.
3. CONTINUOUS CENTRIFUGATION	For large-volume samples. Sample enters centrifuge and supernatant exits continuously as rotor spins.

*There are many techniques that can separate materials on the basis of size, including rate zonal and differential centrifugation, electrophoresis, and chromatography. Buoyant density centrifugation is the only common method that separates materials based solely on differences in their densities.

III. INSTRUMENTATION: ROTORS, TUBES, AND ADAPTERS

A. Types of Rotors

i. HORIZONTAL ROTORS

Samples are centrifuged inside centrifuge tubes (or bottles) that are placed in compartments in rotors. Individuals who use centrifuges must be able to choose rotors, tubes, and accessories for each application. The user must also be able to properly clean, sterilize (when appropriate), and store these centrifuge components. These practical aspects of centrifuge operation are discussed in this section.

There are three commonly used styles of rotor:

- **Horizontal, also called Swinging Bucket**
- **Fixed Angle**
- **Vertical**

The horizontal (swinging bucket) rotor allows tubes to swing up into a horizontal position during centrifugation, Figure 26.11. As centrifugation proceeds, particles move toward the bottom of the tube. Particles beginning at the top of the tube move the entire length of the tube to form a pellet. This maximizes the path length through which particles move. (As we will see, the path length is shorter in other types of rotors.) The long path length results in good separations of similar particles by rate zonal centrifugation. Swinging bucket rotors are also useful for differential centrifugation because they give well-formed pellets. A disadvantage to swinging bucket rotors is that, because of their design, they must usually be run at slower speeds than fixed angle or vertical rotors. Because the path length is longer and the speed is slower in horizontal rotors than in other rotors, separations take longer.

Observe in Figure 26.11a that the radius of rotation (r) for a particle varies depending on the particle's location in the tube. A particle at the top of the tube, closest to the center of rotation, is at the point where the radius of rotation, called r_{min} (minimum radius), is shortest. In the middle of the tube the radius is called r_{ave} (average radius) and at the bottom of the tube the radius is called r_{max} (maximum radius). When calculating the g-force experienced by a particle, it is necessary to consider its position in the tube. If the position is not known, use the average radius.

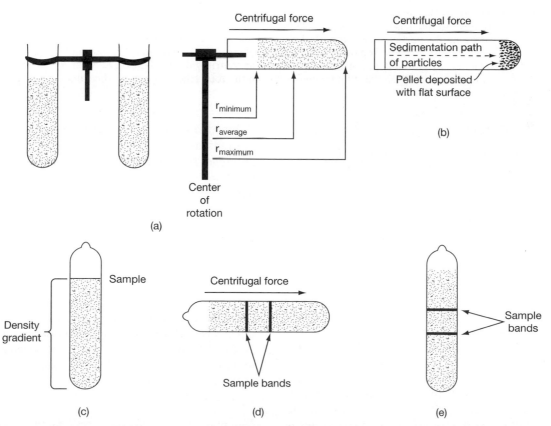

Figure 26.11. Swinging Bucket (Horizontal) Rotor. a. The rotor buckets swing outward to a horizontal position during centrifugation. Particles experience increasingly greater force as they move away from the center of rotation. **b.** Differential centrifugation in a horizontal rotor. **c.–e.** Density gradient separation. **c.** A sample is layered on top of a density gradient. **d.** Bands form during centrifugation. **e.** After centrifugation the tube returns to a vertical position; the bands remain in their same orientation.

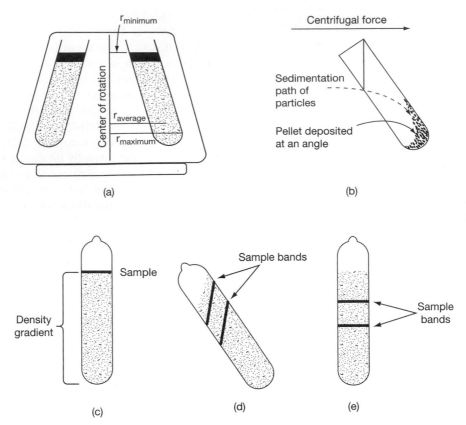

Figure 26.12. Fixed Angle Rotor. a. Fixed angle rotors hold the tubes at an angle relative to the axis of rotation. The radius of rotation varies depending on the location of a particle in the tube. **b.** Differential centrifugation in a fixed angle rotor. **c–e.** Density gradient separation in a fixed angle rotor. **c.** A sample is layered on top of a density gradient. **d.** Particles form bands as centrifugation proceeds. **e.** After centrifugation, when the tube is removed from the rotor, the contents of the tube reorient.

ii. FIXED ANGLE ROTORS

Fixed angle rotors *hold centrifuge tubes or bottles in compartments at a particular angle, usually between 15 and 40 degrees*, Figure 26.12a. The force of centrifugation causes particles to move outward toward the sides of the tubes. During differential pelleting, particles accumulate at the outer tube wall and then slide downward until they pellet along the side at the bottom of the tube, Figure 26.12b. The path length traveled by particles across the tube in a fixed angle rotor is relatively short. Fixed angle rotors, therefore, result in rapid pelleting (compared to horizontal rotors). A disadvantage to fixed angle rotors for pelleting is that fragile particles, such as intact cells, may be damaged as they are forced against the walls of the tube. Horizontal rotors are gentler.

As in a swinging bucket rotor, the radius of rotation varies for a particle depending on its location in the tube, Figure 26.12a. At the top of the tube the radius is r_{min}; at the bottom of the tube the radius is r_{max}. Particles at the bottom of a centrifuge tube in a fixed angle rotor therefore experience a higher *g*-force than particles at the top of the tube.

Fixed angle rotors are used for density gradient separations as well as for differential pelleting, Figure 26.12c–e. Note that the density gradient formed in a fixed angle rotor reorients horizontally when centrifugation ends and the tubes are removed from the rotor.

iii. NEAR VERTICAL AND VERTICAL TUBE ROTORS

Near vertical tube rotors *are a modification of fixed angle rotors that have a shallow angle of about 8–10 degrees*. These rotors are recommended for buoyant density separations where short run times are desired.

Vertical rotors *hold tubes straight up and down resulting in a very short path length and fast separations*, Figure 26.13. Their main use is for buoyant density separations and they can also be used for rate zonal separations.

iv. κ FACTORS

You may come across the term *k factor* (clearing factor) in reference to a rotor, particularly when working with ultracentrifugation. *The **k** factor provides an estimate of the time required to pellet a particle of a known sedimentation coefficient.* The *k* factor varies for different

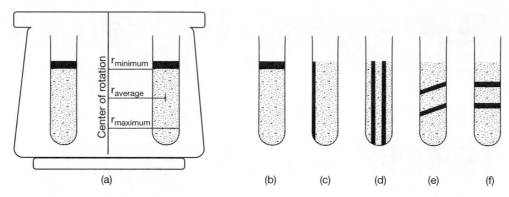

Figure 26.13.   Vertical Rotor. a. Tubes are vertical in the rotor compartments. **b–f.** Density gradient separation in a vertical rotor. **b.** A sample is layered on top of a density gradient. **c.** When centrifugation begins, the tube contents reorient under the influence of the centrifugal force. **d.** The sample sediments into vertical bands. **e–f.** When the rotor comes to a stop and the tubes are removed from the rotor, the tube contents reorient forming horizontal bands.

rotors, depending on their minimum and maximum radii and on the speed of rotation. (Manufacturers typically report the k factor at the maximum speed that the rotor can attain.) The smaller the k factor for a rotor, the faster the separations using that rotor. One use of k factors, therefore, is in comparing the efficiency of various rotors.

The k factor is calculated by the manufacturer so that it is possible to approximate a particle's sedimentation time, in hours, using the following equation*:

$$t = k/S$$

where

t = the predicted time (in hours) to move particles from the top of a tube, at r_{min}, to the bottom of a tube, at r_{max}

k = the clearing factor

S = the sedimentation coefficient in Svedberg units

(This formula assumes that the liquid medium is water at 20°C. If the medium is denser or more viscous than water, the sedimentation time will be longer.)

EXAMPLE

The manufacturer reports that the k factor for a rotor is 100. How long will it take ribosomes to sediment through water to the bottom of a tube?

Ribosomes have a sedimentation coefficient of 80 (Table 26.1). Therefore:

$$t = \frac{100}{80} = 1.25 \ hours$$

B. Balancing a Rotor

One of the most important tasks performed by the operator of a centrifuge is making sure that the rotor is used properly and that the tubes or bottles are balanced in the rotor. The purpose of balancing the load is to ensure that the rotor spins evenly on the drive shaft. This is essential for minimizing rotor stress and preventing damage to the centrifuge, rotor, and tubes. In worst-case accidents, improperly used, unbalanced, or damaged rotors have detached from the spindle and flown out of the centrifuge chamber at high velocity, wreaking havoc along their path, Figure 26.14. Fortunately, modern centrifuges are well-armored, which helps prevent rotors from exiting the centrifuge in an accident. However, if a rotor detaches from the spindle at high speed, even if it does not exit the centrifuge chamber, it will crash into the centrifuge chamber walls with great force, destroying the centrifuge chamber and smashing the rotor into small fragments. The damage to the centrifuge may be irreparable and will at best be extremely costly. A centrifuge accident will also spread the sample throughout the laboratory—a serious problem if the sample is hazardous.

A rotor is balanced by placing samples and their containers symmetrically in the rotor, Figure 26.15. Tubes or bottles across from one another should be of nearly equal weight. The manufacturer specifies how close in weight the tubes that are opposite from one another must be. For example, for a low-speed centrifuge the manufacturer might specify that opposing tubes should be balanced so that there is no more than 0.5 g difference in their weight. For an ultracentrifuge, the manufacturer might specify that opposing tubes

---

*Sorvall® Centrifuges (DuPont Company, Wilmington, DE) has a helpful application note dealing with k factors: "Using the K-Factor to Improve Centrifugation Efficiency and Reduce Run Times," Philip Sheeler. #11, June 1981. This note discusses some of the subtleties involving k factors, such as differences between full and partially filled tubes.

Figure 26.14. Scene Following an Accident in Which an Ultracentrifuge Rotor Failed During Operation. The subsequent "explosion" destroyed the centrifuge, ruined a nearby refrigerator and freezer, and made holes in the walls and ceiling. The centrifuge itself was propelled sideways and cabinets containing hundreds of chemicals were damaged. Fortunately, cabinet doors prevented chemical containers from falling to the floor and breaking. A shock wave from the explosion shattered all four windows in the room and destroyed the control system for an incubator. Luckily, no one was injured. The cause of the accident was believed to be the use of a model of rotor that was not approved by the manufacturer for the particular model of centrifuge. (Courtesy of Cornell Office of Environmental Health and Safety, Cornell University, Ithaca, NY.)

must be within 0.1 *g* of one another. Always adhere to the manufacturer's instructions for balancing tubes. Note also that for ultracentrifuges, tubes or bottles that are across from one another must contain the same density profile as well as match in weight. Guidelines regarding proper balancing of a rotor and operation of a centrifuge are in Box 1.

C. Care of Centrifuge Rotors

i. OVERVIEW

Although rotors appear to be massive and sturdy, they are subjected to tremendous forces during centrifugation and can easily be damaged. A damaged or improperly used rotor can fail during centrifugation with cata-

strophic results. Always read the instructions that come with a rotor. General guidelines for care and cleaning that are relevant to most brands of rotor are discussed in this section and are summarized in Box 2.

Recall that the force generated in centrifugation is:

$$RCF = 11.2 \times r \left(\frac{RPM}{1,000}\right)^2$$

This equation tells us that the *RCF* generated is proportional to the square of the speed. For example, if the *RPMs* are doubled, then the rotor is subjected to four times the original force. The stress on ultracentrifuge rotors, therefore, is considerably greater than that on lower speed rotors. Small imperfections in an ultracen-

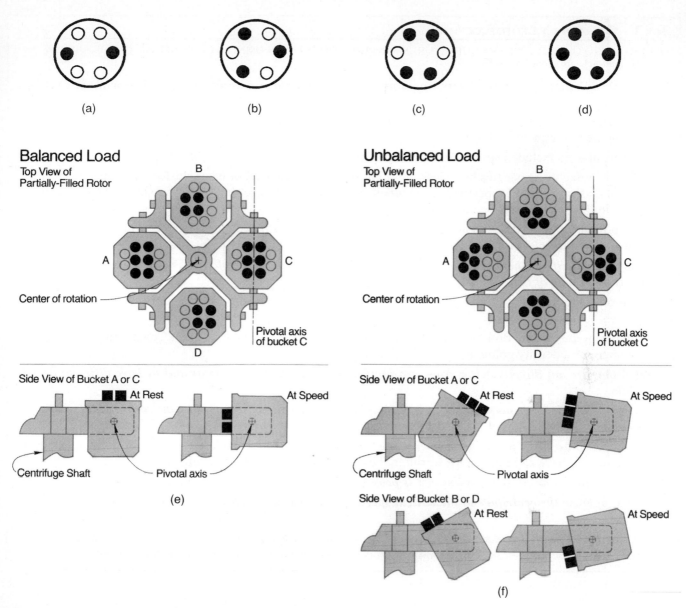

Figure 26.15.   Balancing a Rotor Load. a–d. Rotors containing two, three, four, and six tubes, respectively. Opposing tubes have the same weight (and density profile for ultracentrifuges). **e–f.** The situation is more complex in buckets holding multiple tubes. Each bucket in this rotor holds 10 tubes. **e.** The opposing bucket sets A–C and B–D are loaded with an equal number of tubes, each filled with an equal weight of sample, and are balanced across the center of rotation. Each bucket is also balanced with respect to its pivotal axis. **f.** This rotor is not balanced. None of the four buckets is balanced with respect to its pivotal axis. During centrifugation, buckets A and C will not reach a horizontal position, whereas buckets B and D will pivot past horizontal. Note also that the tubes in B and D are not arranged symmetrically across the center of rotation. (Reprinted by permission of Beckman Coulter, Inc.)

trifuge rotor can cause the rotor to fail when it is centrifuged at high speeds. For this reason, rotors, particularly those for the ultracentrifuge, are made to exacting specifications and must be correctly handled and maintained.

Each rotor is specified by the manufacturer for a maximum allowable speed above which it should never be run. Most high-speed and ultracentrifuge rotors have a protective mechanism that prevents them from being rotated at excessive speeds. This mechanism often

involves an **overspeed disk** *that has black and white alternating sectors, or lines, and is affixed to the bottom of the rotor.* The number of sectors on an overspeed disk varies depending on the maximum allowable speed for the rotor. In the centrifuge there is a photoelectric device that detects the lines on the overspeed disc during centrifugation. If the lines pass by the photoelectric device faster than an allowable limit, then the centrifuge automatically decelerates.

BOX 1 OPERATING A CENTRIFUGE

BE CERTAIN TO CONSULT THE MANUFACTURER'S INSTRUCTIONS WHEN USING A CENTRIFUGE. Guidelines that are generally applicable are shown in this box.

1. *For low- and high-speed centrifuges, opposing tubes can be balanced according to weight without considering density.*

    a. The simplest way to balance tubes or bottles is to place them on opposite sides of a two-pan balance and bring them to equal weight.

    b. Remember to include caps and seals when balancing tubes.

    c. The manufacturer specifies how close in weight the tubes must be. For example, for a low-speed centrifuge the manufacturer might specify that the tubes should be balanced so that there is no more than 0.5 *g* difference in their weight.

    d. Opposing tubes may both contain sample, or one of the tubes may be a "blank" containing only a liquid, such as water.

2. *In an ultracentrifuge, the tubes that are across from one another must be of the same type with respect to density, as well as be of the same weight.*

    a. Fill opposing tubes with materials of the same density profile and make certain that their weights also match.

    b. Opposing tubes may both contain sample, or one of the tubes may be a "blank" containing only a liquid medium or a density gradient.

3. *Tubes that are not directly across from one another can be of different weights and/or have different density profiles.*

4. *Do not exceed the maximum weight in any sample compartment, as specified by the rotor manufacturer.* Be sure to include the weight of any seals, plugs or caps.

5. *All buckets must be in place in a swinging bucket rotor, even if some contain no samples.* If the buckets are numbered and have caps, they must be run in the correct positions with their caps.

6. *Make sure the rotor cover is present and is properly tightened.*

7. *Do not operate the centrifuge at the critical speed.* At the **critical speed** *any slight imbalance in the rotor will cause vibration.*

8. *Pay attention to the brake setting.* The brake slows the rotor after a run is completed. On many centrifuges it is possible to select a brake setting. If braking is too abrupt, some types of pellets will become resuspended.

---

ii. CHEMICAL CORROSION AND METAL FATIGUE

Chemical corrosion *is a chemical reaction that causes a metal surface to become rusted or pitted.* Common salt solutions, such as NaCl, are particularly likely to cause corrosion and should be removed thoroughly if spilled on a rotor. Other corrosive agents include strong acids and bases, alkaline laboratory detergents, and salts of heavy metals like cesium, lead, silver, or mercury. Moisture that stands in the rotor compartments can also cause corrosion over time.

Even in the absence of chemical corrosion, metals fatigue, or become weakened, due to the repeated stress of centrifugation. This damage occurs inside the rotor, so it is not visible. Ultracentrifuge and high-speed rotors must therefore be derated and/or retired as time goes by and as they are used. *Derating means to run a rotor at a speed less than its originally rated*

maximum speed. **Retiring** *a rotor means to discard it or return it to the manufacturer.* The manufacturer specifies the conditions under which a rotor should be derated or retired. Because a rotor has a finite lifetime, it is essential to document its use, typically in a rotor log.

Low-speed rotors can be made of a variety of materials including aluminum, bronze, steel, and even plastic. High-speed rotors are usually made of aluminum, which is light, strong, and relatively inexpensive. Aluminum rotors, however, are especially susceptible to corrosion. For ultracentrifugation, more expensive titanium rotors are preferred. Titanium is much more resistant to corrosion than aluminum, so titanium rotors are less likely to fail when repeatedly centrifuged. There are also newer-style, lightweight rotors that are not made of metal, but of a composite material that is composed of spun or woven carbon fibers embedded in epoxy resin. Composite materials tend

---

BOX 2   SUMMARY OF GUIDELINES FOR CLEANING, MAINTAINING, AND HANDLING ROTORS, TUBES, AND ADAPTERS

1. ***Regularly clean rotor and adapters.***
   a. Use a mild, nonalkaline detergent and warm water.
   b. Thoroughly remove all disinfectants and detergents using water.
   c. With swinging bucket rotors, clean only the buckets themselves. Do not immerse the rotor portion because its pins cannot readily be dried and will rust if wetted.
   d. Air dry and store rotors upside down to allow liquids to drain. Make sure rotors remain dry during storage.
   e. Disinfect and sterilize rotors according to manufacturer's instructions.

2. ***Protect rotors from corrosive agents, including moisture, chemicals, alkaline solutions, and chloride salts.*** Aluminum is particularly sensitive to corrosion. The cover and the compartments of titanium rotors may be aluminum; therefore, they are more vulnerable than the rest of the rotor.
   a. Immediately wash the rotor if it is exposed to corrosive materials.
   b. Routinely check the rotor visually for hairline cracks, dents, white spots, and pitting. If damage is detected, do not use the rotor without consulting a service technician.

3. ***Do not scratch a rotor.***
   a. Use soft bristle brushes for cleaning.
   b. Do not place metal forceps or other implements into rotor compartments.
   c. Do not use metal implements to remove stuck tubes or to open tight lids, except as provided by the manufacturer.

4. ***Check rotor O-rings.*** Many rotors have O-rings that maintain a seal during a run. These O-rings and the surfaces they contact must be kept clean and in good condition. Some manufacturers recommend lightly greasing the O-rings with silicone vacuum grease. Check O-rings for cracks, tears, or an uneven surface; replace if necessary.

5. ***Consult the ultracentrifuge manufacturer before using a rotor made by a different manufacturer.***

6. ***Find out from the manufacturer how long, or for how many runs, a rotor is guaranteed.*** Derate or retire a rotor as directed by the manufacturer.

7. ***Never exceed the maximum speed of a rotor.***
   a. Never remove overspeed protective devices.
   b. Pay attention to the density of solutions in ultracentrifugation. It may be necessary to run the centrifuge more slowly than usual if a high-density solution is used.
   c. Uncapped tubes, partially filled tubes, and heavy tubes and/or samples may require that the rotor be used at less than its maximum rated speed.

8. ***Clean centrifuge tubes and bottles according to the manufacturer's directions.*** General guidelines include:
   a. Reusable tubes, bottles, and adapters should be cleaned by hand with a mild detergent using soft, non-scratching brushes. Commercial dishwashers are likely to be too harsh.
   b. Bottles and tubes should not be dried in an oven; rather, they should be air dried. Stainless steel tubes must be thoroughly dried.
   c. Some tubes and bottles can be autoclaved; however, note that this will reduce their life span.

9. ***<u>Never</u> use a cracked tube or bottle or one that has become yellowed or brittle with age.***

10. ***Make sure that tubes and bottles used for centrifugation are chemically compatible with the sample.***

11. ***Do not spin tubes or bottles at speeds above those for which they are rated.*** It may be necessary to reduce the speed when using plastic adapters or metal tubes.

12. ***Use adapters when necessary to ensure that tubes or bottles fit snugly in rotor compartments.*** Tubes that do not fit snugly in the rotor compartments may shift during centrifugation, resulting in breakage or a poorly formed pellet.

13. ***Maintain a rotor log and a centrifuge log.*** Document major cleaning and decontamination processes as well as all centrifuge runs.

to be resistant to chemical corrosion and to fatigue. Note that titanium and carbon composite rotors usually have some aluminum or iron components, such as tube compartments, that are susceptible to chemical damage.

iii. Situations that Require Using Rotors at Less than Their Maximum Rated Speed

A rotor is sometimes derated due to its age or the number of times it has been spun. There are other situations that also call for reducing the speed of rotation to avoid overstressing the rotor. One of these situations relates to the density of the samples being centrifuged. Every rotor is designed to centrifuge materials whose densities are below a certain maximum cutoff. This maximum density is often 1.2 g/mL. Some fixed angle and vertical tube rotors are designed to hold a density of 1.7 g/mL, which is the maximum density of CsCl solutions used for DNA separations.

It is sometimes necessary to run a sample that exceeds the density limit for the rotor. In these cases, the rotor cannot be run at its maximum rated speed. The rotor manual will usually provide information about centrifuging samples whose density exceeds the rotor design limits. If the manual does not have this information, then there is an equation to estimate the maximum safe speed when centrifuging dense solutions:

$$\text{Derated speed} =$$
$$\text{Maximum rotor speed in } RPM \sqrt{(RD/SD)}$$

where
RD = the design limit of the rotor, as specified by the manufacturer
SD = the sample density

EXAMPLE PROBLEM

A rotor is specified to have a maximum speed of 25,000 RPM and is designed for densities less than or equal to 1.2 g/mL. A density gradient is being centrifuged with a maximum density at the bottom of the tube of 1.5 g/mL. What is the maximum speed at which this gradient can be centrifuged?

ANSWER

Derated speed =

$$\text{Maximum rotor speed in } RPM \sqrt{(RD/SD)}$$
$$= 25,000 \sqrt{(1.2)/1.5}$$
$$= 22,360 \ RPM$$

This is the derated maximum rotor speed.

It is possible that the combination of rotor speed and temperature can be such that a gradient material precipitates out of solution. CsCl gradients in particular can form crystals in the bottom of the centrifuge tube at low temperatures. These crystals have a very high density (4 g/mL) that will produce stress that far exceeds the design limits of most rotors. Consult the manufacturer regarding the use of CsCl gradients in a particular type of rotor*.

There are other situations, in addition to the spinning of high-density samples, that require that a rotor be used at less than its normal maximum speed. For example, samples or tubes are sometimes used that exceed the maximum weight for which the rotor was designed. The rotor must then be used at less than its maximum speed, according to the manufacturer's directions.

iv. Rotor Cleaning

Rotors must be kept clean. Rotors that receive heavy use should be cleaned on a routine basis, and every rotor should be thoroughly washed in the event of a spill. Detergents used to clean rotors must be mild and nonalkaline. Brushes used for cleaning must be soft so as not to scratch the rotor. Manufacturers sell mild detergents and brushes for use with rotors. Detergents must be thoroughly rinsed from rotors with distilled water. Manufacturers recommend air-drying rotors and storing them upside-down in a dry environment with their lids removed to avoid having moisture collect in the compartments.

Some rotors can be safely autoclaved, although repeated autoclaving may weaken ultracentrifuge rotors. Seventy percent alcohol is compatible with all rotor materials and therefore can be used for disinfection. Some manufacturers recommend sterilizing rotors with ethylene oxide, 2% glutaraldehyde, or UV light. Rotors contaminated with pathogenic or radioactive materials should be cleaned with agents that are compatible with the rotor. Note that alkaline detergents designed to remove radioactivity are not compatible with most rotors.

Rotors generally have O-rings and gaskets to ensure tight seals. These O-rings and gaskets can typically be lubricated with a thin layer of silicone vacuum grease. Worn or cracked O-rings and gaskets should be replaced.

Aluminum rotors should frequently be inspected for signs of corrosion, including rough spots, pitting, white powder deposits (which may be aluminum oxide), and heavy discoloration. If problems are observed, consult the manufacturer before using the rotor. Titanium rotors are much more corrosion resistant and require less frequent inspection.

---

*For example, Beckman Instruments, Inc. provides useful information on CsCl gradients in its manual, "Rotors and Tubes for Preparative Ultracentrifuges: A User's Manual," Beckman Instruments, Inc., Palo Alto, CA, 1990.

D. Centrifuge Tubes, Bottles, and Adapters

i. GENERAL CONSIDERATIONS

Centrifuge tubes are containers for smaller volume samples; centrifuge bottles are containers for larger volume samples. Centrifuge tubes and bottles are available in varying styles and sizes, and made of different materials. Some tubes and bottles are used with caps or seals, others are centrifuged open. Some tubes and bottles can be centrifuged when partially filled, whereas others must be completely filled or they will collapse under the force of centrifugation. Some tubes can be washed and used repeatedly, but others must be disposed of after one use. Because there are many types of sample containers for centrifugation, it is necessary to select those that match your samples and rotors.

Capped centrifuge tubes should always be used if a sample is potentially hazardous or volatile. Although screw caps are simple to use and appear to be leak-proof when handled on the laboratory bench, they distort under the pressure of centrifugation. This distortion can allow samples to leak from the tubes. Manufacturers provide cap assemblies that in some way compress onto tubes or otherwise seal firmly.

At low and moderate *RCF*s, most tubes and bottles can be centrifuged when only partially filled. At higher *RCF*s, only thick-walled tubes can be run partially filled; thin-walled tubes collapse during centrifugation.

A single rotor can cost thousands of dollars. To maximize their versatility, manufacturers therefore design rotors so that they can flexibly accommodate different styles and sizes of tubes and bottles. This is accomplished with **adapters**, *plastic or rubber inserts that reduce the size of the sample compartment so that smaller tubes can be run in the same rotor*. There are many types of adapters, Figure 26.16. Some adapters are tube shaped and line the rotor compartment. Other adapters support the tops of tubes and still others the bottoms. It is important to purchase the correct configurations of tubes, caps, and adapters for each rotor and each type of sample that is to be used in that rotor. Be aware that it may be necessary to run a rotor at less than its maximum speed when using adapters.

ii. MATERIALS USED TO MAKE CENTRIFUGE TUBES AND BOTTLES

Materials used to make centrifuge tubes and bottles include glass, stainless steel, and various types of plastics, Table 26.5. These materials must be able to withstand the high forces of centrifugation. Depending on its composition and its design, every tube and bottle is rated for a maximum *g*-force.

Plastics are relatively light and strong, so they are the preferred material for making tubes and bottles that will experience high centrifugal forces; however, plastics have limitations. Most plastic centrifuge tubes and bottles will soften and deform if they are centrifuged at temperatures above 25°C. Completely filled, thick-wall polyallomer (a type of plastic) tubes can be used at higher temperatures, but lower speeds and shorter run times are preferred. Stainless steel can be used safely at high temperatures, but these tubes are heavy and limit the speed of centrifugation. It is good practice to pretest tubes or bottles that will be used at higher temperatures without sample to determine whether the material can withstand the conditions of centrifugation.

Another limitation is that some plastics are incompatible with various solvents (e.g., phenol, acetone, and alcohols) and with acids and bases. For example, most plastics will soften or dissolve when exposed to phenol. To determine whether a particular tube material is compatible with your sample, consult a chemical compatibility chart from the manufacturer. Even if the sample and centrifuge tube material appear to be compatible based on the chart, it may still be a good idea to perform a test run.

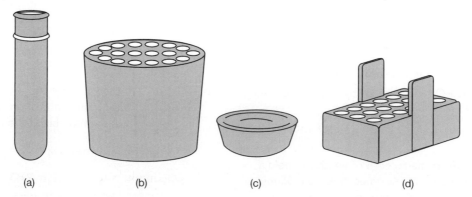

(a)   (b)   (c)   (d)

Figure 26.16.   Adapters. Adapters allow varying sizes and shapes of tubes to be used in a single rotor. **a.** An adapter that lines the rotor compartment and holds one tube. **b.** An adapter that holds eighteen tubes at once, **c.** A cushion to support a glass bottle in a horizontal low-speed rotor. **d.** An adapter to hold 18 tubes in a horizontal low-speed rotor.

EXAMPLE PROBLEM

A page from a rotor catalogue is shown. Answer the following questions about the rotor described.

a. What is the maximum speed that this rotor can withstand?

b. What is the maximum force that this rotor can withstand?

c. How many different types of tubes can be used with this rotor?

d. What is the maximum volume that can be centrifuged in this rotor?

e. Which types of tubes can withstand being centrifuged at 80,000 *RPM*?

f. What is the function of the adapters? Which tubes require adapters?

g. What happens to the maximum speed that can be used with this rotor if adapters are used?

h. What is the maximum speed at which a density gradient with a maximum density of 1.4 *g/mL* can be run in this rotor?

i. Is this rotor for a low speed, high speed, or ultracentrifuge?

FA-80 ROTOR

DESCRIPTION

25-degree fixed angle   Titanium
10 places, 3.5 *mL* maximum volume
Maximum Speed 80,000 *RPM*   *k*-Factor 25.8
Designed for samples with a density less than or equal to 1.2 *g/mL*

	RCF (× g)	RADIUS (cm)
maximum	460,000	6.55
average	350,000	5.00
minimum	250,000	3.41

TUBES

CAPACITY (mL)	DESCRIPTION	MAXIMUM SPEED	CAP ASSEMBLY	ADAPTERS
3.5	Polyallomer, Thin Walled Tube	80,000	Multipiece	None
2.7	Polycarbonate Thick Walled Tube	45,000	Screw Cap	None
3.5	Polyallomer Thin Walled	80,000	Crimp Seal	None
2.0	Polycarbonate Thick Walled Tube	55,000	None	1 per compartment required (Delrin)
2.0	Cellulose Acetate Butyrate Thick Walled Tube	40,000	None	1 per compartment required (Delrin)
0.4	Cellulose Acetate Butyrate Thick Walled Tube	40,000	None	1 per compartment required (Delrin)

ANSWER

a. The maximum speed for the rotor is 80,000 *RPM*.

b. The maximum RCF generated by this rotor is 460,000 × g at r_{max}.

c. The manufacturer specifies six types of tubes for this rotor.

d. There are 10 compartments each of which can hold a maximum of 3.5 *mL*, so the maximum volume is 35 *mL*.

e. Only the polyallomer thin walled tubes can be used at the rotor's maximum speed.

f. Adapters allow tubes smaller than the sample compartment to be used. The tubes which hold a lower volume require adapters.

g. The use of adapters reduces the maximum speed at which the rotor can be used.

h. The rotor is designed for a density ≤ 1.2 *g/mL*. The square root equation can be used to determine the maximum rate given the higher density as follows:

$$\text{Derated speed} =$$

$$\text{Maximum rotor speed in } RPM \sqrt{(RD/SD)}$$

$$= 80,000\sqrt{(1.2/1.4)}$$

$$= 74,066 = \text{maximum speed for this density}$$

i. This is an ultracentrifuge rotor.

Table 26.5 MATERIALS USED TO MAKE CENTRIFUGE TUBES AND BOTTLES

1. **Glass.** *Glass is largely inert, which makes it chemically compatible with almost all samples.* Normal soda glass is unable to withstand forces much above $3000 \times g$, whereas borosilicate glass is able to withstand at least $10,000$–$12,000 \times g$. The disadvantage to glass is the possibility of breakage. Glass therefore cannot be used at high *RCF*s. It is best to support glass tubes using a rubber pad or adapter, thus avoiding pressure between glass and metal. It is always good practice to do a test run with glass tubes or bottles before running them with potentially hazardous materials.

2. **Stainless steel.** *Stainless steel is resistant to organic solvents and to heat.* Stainless steel tubes can be reused many times as long as they are undamaged and uncorroded. These tubes are strong enough to be run partially filled. The disadvantage to metal tubes is that they are heavy and so require that a rotor be spun at less than its maximum speed.

PLASTICS

3. **Polycarbonate.** *Polycarbonate is strong, transparent, and autoclavable.* Polycarbonate can be constructed with thick walls so that tubes can be only partially filled and do not collapse when centrifuged. Polycarbonate is commonly used for high speed and ultracentrifuges. Polycarbonate, however, is not resistant to many solvents, including phenol and ethanol, both of which are commonly used in biology laboratories. Polycarbonate is also attacked by alkaline solutions, including many common laboratory detergents. Polycarbonate can become brittle and should be visually inspected before use. Discard cracked or brittle tubes.

4. **Polysulphone.** *This plastic has many of the advantages of polycarbonate, but it is also resistant to alkaline solutions.* It is, however, attacked by phenol.

5. **Polypropylene and polyallomer.** *Polyallomer is commonly used for high speed and ultracentrifuge tubes.* Both polyallomer and polypropylene can be autoclaved. Both can be made into thin-wall tubes that can be punctured readily to remove bands from gradients. Both plastics tend to be resistant to most organic solvents. Polyallomer is recommended for separating DNA in CsCl gradients because DNA does not adhere to the tube walls.

6. **Cellulose esters.** *These tubes are transparent, easy to pierce, and strong, but they cannot be autoclaved and are attacked by strong acids and bases and by some organic solvents.*

7. **Other plastics.** *Many companies make specialized plastic tubes and bottles with various properties and under their own trade names.* Check with the manufacturer before using these with acids, bases, and solvents, and before autoclaving. Note also the maximum speed at which they can be used.

E. Centrifuge Maintenance and Trouble-Shooting

i. MAINTENANCE AND PERFORMANCE VERIFICATION

The details of operation vary from centrifuge to centrifuge, so it is necessary to read the manufacturer's instructions. It is especially important to read the manual when operating an ultracentrifuge because these instruments can easily be damaged with dangerous and costly results.

Centrifuges require periodic maintenance and performance verification that are usually performed by trained service technicians. The technicians make sure that the *RPM*s that are displayed correspond to the actual speed of the rotor, that the temperature in the chamber is maintained properly, and that timers and other controls work correctly. The vacuum and refrigeration systems must also be checked. Older centrifuges may have brushes that require replacement. All service and maintenance procedures should be documented.

ii. TROUBLE-SHOOTING COMMON PROBLEMS

Some common centrifugation problems are described in Box 3.

F. Safety

Various safety issues relating to rotors and centrifuge tubes have already been addressed in this chapter. Another risk that is less dramatic than a catapulting rotor, but is equally dangerous, relates to the centrifugation of hazardous materials such as viruses, radioactivity, and carcinogenic compounds. Protection from hazardous materials is of great importance in biological centrifugation. In a worst case, a centrifuge tube that breaks during centrifugation can disperse hazardous aerosols throughout the laboratory and throughout the air-handling system in a building. There are more subtle concerns as well. A poorly sealed centrifuge tube cap can allow aerosols to escape. Screw caps on centrifuge tubes or bottles are likely to distort under the high force of centrifugation, allowing liquids and aerosols to be released. Simply opening a tube, particularly a tube with a "snap-cap," creates aerosols. Pouring and decanting fluids from centrifuge tubes and bottles creates aerosols. Spills can occur while introducing samples into tubes or removing samples from tubes.

Because centrifugation provides many opportunities for contamination, it is necessary to use special precautions when spinning pathogenic or toxic materials. Bio-

Box 3   Trouble-Shooting Tips

Symptom 1: *VIBRATION DURING OPERATION*

Vibration is due to rotor imbalance

1. Turn off the centrifuge.
2. Check that all containers are properly balanced and arranged properly.
3. Check that cushions, adapters, and caps match in all sample compartments.
4. Check that the centrifuge is on a level surface.
5. Increase speed gradually.

Symptom 2: *TUBE BREAKAGE*

1. Check that the type of tube in use is rated for the speed and force being used.
2. Check that the tubes are properly balanced.
3. Make sure that the adapters and cushions match the type of tubes.
4. Make sure the contents of the tubes are compatible with the materials used in constructing the tubes.
5. If using partially filled tubes, check that sufficient volume is being used to meet manufacturer's specifications.

Symptom 3: *FINE GRAY OR WHITE POWDER OR DUST IN CHAMBER*

This is caused by particles of glass sandblasting the inside of the centrifuge

1. Clean centrifuge thoroughly.
2. Run the centrifuge without samples several times and clean between each operation.

Box 4   Guidelines for Safe Handling of Centrifuges and Their Accessories

1. *Carefully read the manufacturer's instructions.* Pay attention to warnings provided. Note limitations of rotors. Find out how to use tubes and adapters properly. Careful attention to the manufacturer's directions is especially important when operating high speed and ultracentrifuges.
2. *Use special precautions when centrifuging pathogenic, toxic, or radioactive materials.*
   a. Make sure that tubes are tightly capped and sealed.
   b. Avoid caps that snap open.
   c. Open tubes containing dangerous materials in a suitable hood or other protected enclosure.
   d. Pour and decant hazardous samples in a hood or protected enclosure.
   e. Clean spills promptly.
   f. Use special containment instrumentation as appropriate. Sealable tubes and rotors are available to contain hazardous materials.
3. *In laboratories where toxic materials, pathogens, or radioisotopes are used, avoid touching any rotors with bare hands.* Assume that rotors and the insides of centrifuges are contaminated, but, at the same time, try to avoid contamination. Assume that gloves are contaminated after touching a rotor and remove them before touching anything else. Never place bare fingers in any rotor compartment, both because of the possibility of contaminants, and because there may be broken glass.
4. *Protect service technicians from pathogens and radioactivity by careful cleaning of equipment before any repairs are performed.* Document the decontamination process.
5. *Protect yourself from the spinning rotor.* Although many centrifuges have locking covers that prevent access to a spinning rotor, some centrifuges lack this safety feature. There are also modes of centrifuge operation where the user must have access to the rotor as it spins.
   a. Do not touch a rotor that is in motion and do not slow a rotor with your hand.
   b. If you must work with a spinning rotor, then:
      • remove jewelry, such as necklaces and bracelets
      • remove neckties and scarves
      • tie back long hair
      • roll up and secure shirt sleeves
6. *Make sure the contents of a rotor are properly balanced before centrifugation.*
7. *Do not centrifuge large volumes of explosive or flammable materials, such as ethyl alcohol, chloroform, and toluene.* (Note that in practice, small volumes of these materials are often centrifuged in low-speed centrifuges.)

logical safety cabinets can be used when filling and emptying centrifuge tubes or bottles. There are centrifuges, rotors, and centrifuge tubes that are specifically designed for hazardous materials. Some rotors, for example, are designed to safely contain spilled materials in the event of tube breakage.

Radioactive materials can readily contaminate rotors and centrifuges. The commonly used radioisotope, ^{32}P, tends to bind tightly to metal surfaces, making it difficult to remove. This problem is exacerbated by the fact that alkaline detergents intended to remove radioactive contaminants are damaging to rotors.

Thus, failure to follow proper practices when operating centrifuges can endanger laboratory staff, result in damage to the centrifuge or its accessories, and cause contamination of the centrifuge and the laboratory. Some safe handling guidelines are summarized in Box 4.

PRACTICE PROBLEMS

1. A portion of a table from a manufacturer's rotor manual is shown. Calculate the *RCF*s to fill in the blanks. Refer to the diagram to determine the values for r_{max}, r_{min}, and $r_{average}$.

SPEED (in RPMs)	RCFs GENERATED		
	r_{max}	r_{min}	$r_{average}$
20,000	40,800	17,200	29,000
30,000	_____	_____	_____
40,000	_____	_____	_____

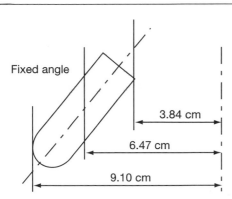

Fixed angle

3.84 cm

6.47 cm

9.10 cm

2. The following calculation is incorrect. Why?

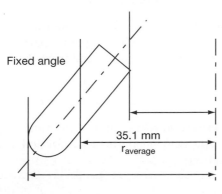

Fixed angle

35.1 mm

$r_{average}$

Speed is 45,000 *RPM*

$$RCF = 11.2 \times r \left(\frac{RPM}{1,000} \right)^2$$

$$RCF = 11.2 \times 35.1 \left(\frac{45,000}{1,000} \right)^2$$

$$RCF = 796,068$$

Incorrect answer!

3. A journal article includes the statement below. If you have a similar rotor with an r_{ave} of 2.6 *cm*, how many *RPM*s will you need to use to duplicate the force used by these authors?

"The extract was centrifuged in a fixed angle rotor at 140,000 × *g* for three hours."

4. Specifications for three rotors are shown. Answer the following questions about these rotors.

   a. Which rotor would be best for separating large volumes of bacterial cells after fermentation? Explain.

   b. Which would be used to isolate small, subcellular organelles and for buoyant density separations of macromolecules like DNA and RNA? Explain.

   c. Which rotor would be useful for centrifuging 1 *mL* volumes of precipitated DNA in a molecular biology laboratory? Explain.

	ROTOR A	ROTOR B	ROTOR C
TYPE	Fixed Angle	Fixed Angle	Fixed Angle
MAXIMUM SPEED (*RPM*)	60,000	16,000	10,000
MAXIMUM FORCE (× *g*)	400,0000	40,000	6,000
CAPACITY (# × *mL*)	6 × 25	6 × 500	12 × 2

5. A portion from a rotor catalogue is shown. Answer the following questions about the rotor described.

   a. What is the maximum speed that this rotor can withstand?

   b. What is the maximum force that can be generated with this rotor?

   c. What is the maximum volume that can be centrifuged at one time in this rotor?

   d. What is the maximum number of tubes that can be centrifuged at once in this rotor? What is the maximum volume they can contain?

   e. Is this rotor for a low speed, high speed, or ultracentrifuge?

FA13

DESCRIPTION

28° fixed angle aluminum
Six places, 315 mL maximum volume
Maximum Speed 13,000 RPM k-Factor 2026
Designed for samples with a density less than or equal to 1.2 g/mL
Maximum RCF 27,500
Maximum Radius 14.60 cm

TUBES

CAPACITY (mL)	DESCRIPTION	MAXIMUM SPEED	CAP ASSEMBLY	ADAPTERS
315	Stainless Steel Bottle	13,000	Stainless Steel	None
280	Polysulfone Bottle	13,000	Polypropylene Seal	None
150	Polycarbonate Thick Walled	13,000	None	1 per compartment required (Delrin)
30	Borosilicate Glass Tube	10,000	None	1 per compartment required (polypropylene) Each adapter holds three tubes
4	Polypropylene	13,000	Polypropylene	1 per compartment required (polypropylene) Each adapter holds 12 tubes

DISCUSSION QUESTION

High speed and ultracentrifuges are expensive pieces of laboratory equipment that must often be shared by a number of people. Moreover, centrifuge runs may be long; it is not unusual for a procedure to require that samples spin overnight or even longer. Planning, organization, and "centrifuge etiquette" are therefore required to ensure that everyone has a turn to use the centrifuge and that a centrifuge is available at critical points in a procedure. Most laboratories require individuals to schedule their centrifuge use and to sign up in advance to avoid conflicts. **Scenario 19 in the "Skill Standard for the Biosciences Industries" document (Educational Development Corporation, 1995) asks the following question:**

> One week ago you reserved time for a 12-hour spin to coincide with the completion of your assay. When you bring your samples to the centrifuge, you discover that it is currently being used. There is no indication of who may be using the centrifuge. What would you do?

TERMINOLOGY RELATING TO CENTRIFUGATION

Adapter (in centrifugation). An insert placed in a rotor compartment that permits the use of a smaller-sized tube than could otherwise be accommodated.

Analytical Ultracentrifuge. An ultracentrifuge that is designed to provide information about the sedimentation properties of particles.

Anodized Surface. Thin, protective coating of aluminum oxide deposited electrochemically on aluminum rotors to help protect them from corrosion.

Brushes. A component of a centrifuge motor that conducts electrical current. Because brushes require maintenance, brushless motors are found in newer centrifuges.

Buoyant Density Gradient Centrifugation. A method of particle separation based on the buoyant density of particles. Particles move through a density gradient until equilibrium is reached where the density of each particle matches that of the surrounding medium.

Centrifugal Force. The force that pulls a particle away from the center of rotation.

Centrifuge. An instrument that generates centrifugal force, commonly used to help separate particles in a liquid medium from one another and from the liquid.

Chemical Corrosion. A chemical reaction that causes a metal surface to become rusted or pitted.

Clearing Factor, k. A measure of the time required to sediment a particle in a particular rotor and under specified conditions.

Clearing Time, *t*. The time required to sediment a particle.

$$t = k/S$$

where
 t = time in hours
 k = the clearing factor (in hours-Svedbergs)
 S = the sedimentation coefficient (in Svedberg units, *S*).

Continuous Density Gradient. A density gradient where there is a smooth increase in density from top to bottom with no sharp boundaries between layers.

Continuous Flow Centrifugation. A mode of centrifugation (requiring special equipment) in which sample flows continuously into and out of the spinning rotor.

Critical Speed. A low speed at which any slight rotor imbalance will cause the rotor to vibrate.

Cushion (in centrifugation). Rubber or plastic pads positioned at the bottom of the compartments in a rotor, or placed underneath adapters, which support tubes and minimize breakage by distributing force over a larger area.

Density. Mass per unit volume.

Density Gradient Centrifugation. A mode of centrifugation in which the sample moves through a column of liquid of changing density. During centrifugation each particle sediments to a final position in the gradient based either on its density alone, or on its size and density. See also "Buoyant Density Centrifugation" and "Rate Zonal Centrifugation."

Derate. A situation where a rotor must be run at a speed lower than its originally specified maximum speed.

Differential Centrifugation. A mode of centrifugation in which the sample is separated into two phases: a pellet consisting of sedimented material and a supernatant.

Elutriation. A specialized centrifugation method in which the sedimentation of particles in a centrifugal field is opposed by the flow of liquid pumped toward the center of rotation. The flow rate and the centrifugation speed can be adjusted in such a way as to wash the cells out with the flowing liquid. This method minimizes the forces experienced by the particles and is useful for isolating fragile, living cells.

Fixed Angle Rotor. A rotor that holds tubes at a fixed angle.

Gradient Maker. An apparatus used to make a continuous density gradient.

Gravity, *g*. The unit of measure for the rate of acceleration of gravity. The earth's normal gravity is defined as $1 \times g$ (or 1-*g* force).

High Speed Centrifuge. A centrifuge that spins samples at rates up to about 30,000 *RPM* and generates forces up to about $100,000 \times g$.

Horizontal Rotor (also called **swinging bucket rotor**). A rotor in which tubes or sample bottles swing into a horizontal position when centrifugal force is applied.

Isopycnic Density Gradient Centrifugation. See "Buoyant Density Gradient Centrifugation."

Low-Speed Centrifuge. A centrifuge that spins samples at rates less than about 10,000 *RPM*.

Microfuge. A small centrifuge intended for small volume samples in the microliter to 1 or 2 *mL* range.

Near Vertical Tube Rotor. A modification of a fixed angle rotor having a shallow angle of about 8–10 degrees.

Overspeed Disc. A striped black disc attached to the bottom of rotors which, in conjunction with a photoelectric device in the centrifuge, prevents a rotor from being spun at rates above its maximum safe speed.

Pellet. Components of a sample that have settled to the bottom of a container after centrifugation.

Preparative Centrifugation. Centrifugation for the purpose of obtaining material for further use.

Radius of Rotation. The distance from the center of rotation to the material being centrifuged.

Rate Zonal Density Gradient Centrifugation. A method of density gradient separation in which particles move through a gradient at different rates based on their size and density.

Relative Centrifugal Field, *RCF*. The ratio of a centrifugal field to the earth's force of gravity; the units of *RCF* are in $\times g$. The relative centrifugal field developed in a centrifuge depends on the speed of rotation and the distance to the center of rotation according to the formula:

$$RCF = 11.2 \times r \, (RPM/1000)^2$$

where
 RCF = relative centrifugal field
 RPM = revolutions per minute
 r = the radius of rotation in *cm*

Retiring (a rotor). Taking a rotor out of use due to age or amount of use it has received.

Revolutions per Minute, *RPM*. A measure of the speed of rotation in a centrifuge.

Rotor. The device that rotates in the centrifuge and holds the sample tubes or other containers.

Sedimentation. The settling out of particles from a liquid suspension.

Sedimentation Coefficient, *S*. A measure of the sedimentation velocity of a particle. In practice, the sedimentation coefficient is a function of the size of a particle; larger particles have larger sedimentation coefficients.

Spontaneous Density Gradient. A density gradient made of a medium that forms a gradient spontaneously under the influence of a centrifugal field.

Step Gradient. A density gradient in which there are layers of medium with different densities, each of which is clearly demarcated from the layer above and below.

Supernatant. The liquid medium above a pellet after centrifugation.

Svedberg, T. A pioneer in centrifugation who worked on the mathematics, methods, and instrumentation for centrifugation.

Swinging Bucket Rotor. See "Horizontal Rotor."

Ultraspeed Centrifuge, or **Ultracentrifuge.** A centrifuge that rotates at speeds up to about 120,000 *RPM* and can generate forces up to 700,000 × *g*.

Vertical Rotor. A rotor in which tubes are held upright in the sample compartments.

Zonal Rotor. Rotors that accommodate large volume samples by containing the sample in a large cylindrical cavity rather than in individual tubes or bottles. With some zonal rotors, the sample is loaded and unloaded while the rotor is spinning at a slow speed; in other cases zonal rotors are loaded and unloaded when stationary. Zonal rotors are used for both rate zonal and buoyant density gradient separations, but are seldom used for pelleting.

REFERENCES

Examples of general references on centrifugation include:

An Introduction to Centrifugation, T.C. Ford and J.M. Graham. Bios, Oxford, UK, 1991. (Readable, basic primer on centrifuges.)

Centrifugation: Essential Data, D. Rickwood, T.C. Ford, and J. Steensgaard. Wiley, Chichester, UK, 1994.

Preparative Centrifugation: A Practical Approach, D. Rickwood, ed. IRL Press at Oxford University Press, Oxford, England, 1992.

There are a number of informative publications by manufacturers. Among these are:

Density Gradient Media: Centrifugation Techniques VII Gradient Formation. Nycomed Pharma AS, Division Diagnostics, P.O. Box 5012 Majorstua, N-0301 Oslo, Norway. Distributed by Mediatech and Fisher Scientific, Mediatech, 13884 Park Center Road, Herndon, VA 20171. (A good summary of density gradient methods.)

Beckman Coulter Instruments Inc., Palo Alto, CA. Beckman publishes a number of bulletins relating to centrifuge applications and care of centrifuge equipment.

Sorvall, Inc. Newtown, CT. Sorvall publishes a number of bulletins relating to centrifuge applications and care of centrifuge equipment.

ANSWERS TO PRACTICE PROBLEMS

1. $r_{max} = 9.10 \, cm$ $r_{min} = 3.84 \, cm$ $r_{average} = 6.47 \, cm$

SPEED	RCFs GENERATED		
(in RPMs)	r_{max}	r_{min}	$r_{average}$
20,000	40,800	17,200	29,000
30,000	91,700	38,700	65,200
40,000	163,100	68,800	115,900

2. The radius must be in centimeters to use this equation.

3. $1000\sqrt{\dfrac{RCF}{11.2 \, r}} = RPM$

$= 1000\sqrt{\dfrac{140,000}{11.2 \, (2.6)}}$

$= 69,338 \, RPM$

4. a. For separating large volumes of cells from fermentation, a high-capacity rotor is more important than a rotor that can spin at high *RCF*s. Rotor B is therefore preferred.

b. The separation of very small particles requires a rotor, like rotor A, which is capable of running at high *RCF*s.

c. A microfuge is used for small volume samples like this, rotor C.

5. a. The maximum speed for this rotor is 13,000 *RPM*.

b. The maximum force that can be generated with this rotor is 27,500 × *g*.

c. The maximum volume that can be centrifuged in this rotor is 6 × 315 *mL* = 1890 *mL*.

d. This rotor can hold 72 tubes each with a volume of 4 *mL* = 288 *mL*.

e. High speed.

CHAPTER 27

Introduction to Bioseparations

I. BIOSEPARATIONS IN THE BIOTECHNOLOGY LABORATORY

A. Introduction

We have so far discussed two types of bioseparation methods: filtration and centrifugation. In this chapter, we discuss how biotechnology laboratories purify specific molecules from complex mixtures. In these cases, it is necessary to devise purification strategies that involve a series of bioseparation steps. **Bioseparation methods** *are the techniques used to separate and purify specific biological products, or biomolecules (such as a particular protein).* These methods play a central role in biotechnology. **Purification**, or isolation, *is the separation of a specific material of interest from contaminants, in a manner that provides a useful end product.*

The ability to isolate specific biomolecules is essential in both research and production settings. In a research setting, a purified biological material may be used for molecular **analysis**, *which is the study of the specific chemical properties of the molecule.* Research into the chemical properties of proteins and other biomolecules can provide basic information about cellular processes and lead, for example, to the development of new drug treatments for human illness. In a production setting, safe and reliable purification processes are required to provide a previously analyzed product in sufficient quantities for large-scale testing or for commercial sale.

In order to purify a particular biological product completely, it is necessary to devise a purification strategy that involves a series of separation steps. Each separation step removes certain contaminants from the product of interest, resulting in a progressively purer preparation. This chapter will provide an overview of how purification strategies are developed, with an emphasis on general guidelines.

B. Sources of Biotechnology Products

Biomolecules *are compounds that are produced by some type of biological source, such as a plant, animal, microorganism, or cultured cell.* This source is the starting point for purification of the desired product. The optimal strategy to purify a particular material of interest will depend to some extent on the source of starting material. This is because the best strategy to purify a product depends partly on the nature of the biomolecule being purified and partly on the nature of the contaminants that must be removed. For example, the strategies used to purify insulin differ depending on whether its source is animal pancreas or recombinant *E. coli* that have been genetically engineered to produce insulin.

In an industrial setting, production methods are categorized as upstream or downstream processes. **Upstream processes**, such as fermentation or cell culture, *produce the starting material of interest.* **Downstream processes** *are the separation procedures that result in a purified*

product. Whereas downstream processes represent much of the cost of production, they are based on the choices made in the upstream system. Studies show that changes in upstream sources of biomolecules (e.g., switching from an animal source to a genetically engineered microorganism) can have a much greater effect on final production costs than changes in downstream processing.

There are a variety of biological sources that are commonly used for the production of biomolecules, Table 27.1. Natural sources such as plants, animal tissues, blood, and other biological fluids have long been used as sources of useful products. These natural sources, however, usually provide complex and relatively uncharacterized molecular mixtures, which require complex purification strategies to isolate the biomolecule of interest. Moreover, the amount of a single biomolecule of interest is often quite limited in natural sources. In some cases, as when a biomolecule is found in human tissue or an endangered plant, the natural source cannot be used for mass production purposes.

Due to the complexity of isolating pure products from natural sources and the limited amount of the desired biomolecule that may be available, biotechnologists often use as sources cultured plant or animal cells, or microorganisms grown using fermentation technology. In many cases, microorganisms or cultured cells can be genetically engineered to better produce the molecule of interest, as is the case with insulin. Microbial fermentation is a common source of commercially available biotechnology products, because of the relative ease of developing large-scale upstream operations.

The subcellular localization of a product is significant. As we will discuss shortly, it is much easier to harvest a product that is secreted from cells than one that is sequestered within the cell. Many genetically engineered sources of biomolecules have been designed to secrete proteins that normally would be found in an intracellular location.

In general, the biological source of a product should be chosen or engineered to produce the greatest amount of stable, usable product as possible, as economically as possible. Table 27.2 summarizes some of the major variables in upstream production of biomolecules.

C. Goals and Strategies for Bioseparations

Strategies to purify a specific biomolecule involve a series of fractionation steps. The first steps **extract**, or *release*, the molecule of interest from its biological

Table 27.1   *SOURCES OF BIOTECHNOLOGY PRODUCTS*

Natural Sources	Cell Cultures	Fermentation
Plant tissues	Animal cells	Bacteria
Animal tissues	Plant cells	Yeast
Microorganisms isolated from their natural environment		
Blood and other fluids		

Table 27.2 CONSIDERATIONS IN CHOOSING A BIOLOGICAL SOURCE FOR MOLECULES

There are many variables that can affect the final cost and product yield in bioseparations. These are some of the major considerations when choosing an upstream source for molecules:

- the availability of the source
- the total amount of product present
- the volume of material that will need to be processed
- the stability of the product under upstream conditions
- the amount of product relative to the amount of impurities present
- the nature of the impurities
- the potential pathogenicity of the source
- the cost of media, food, and equipment to grow the source

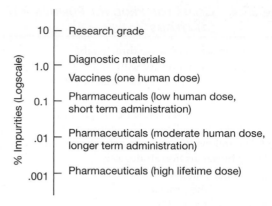

Figure 27.1. Levels of Product Purity Required for Specific Product Functions. Products that are not administered to humans may have higher levels of impurities than those intended for human use.

source. Later steps successively remove various contaminants from the product of interest.

There is no "standard" series of specific steps that can be used to purify all molecules. For example, proteins, the cellular products most likely to be purified in biotechnology facilities, differ greatly in their molecular properties. The specific series of steps used to purify Protein A, therefore, is unlikely to be effective for Protein B, unless the two proteins are closely related in structure. Many separation techniques can be performed under almost infinitely variable conditions (such as pH, temperature, and ionic strength). The conditions under which a technique is performed must therefore be individually optimized for every product. Even though there are no universal procedures applicable to every biomolecule, there are guidelines available for the design of appropriate strategies. Table 27.3 provides an outline of these guidelines, which will be discussed throughout this chapter.

The key step in developing a bioseparation strategy is to determine the goals of the process. Questions to ask include: How pure does the product need to be to work effectively? How much of the product is required? How important is low cost? The priorities of these goals vary in research and production facilities and may also vary depending on the nature of the product. For example, a drug product requires the utmost purity, even if the cost is high. In contrast, a product not intended for use in

humans may be acceptable in a less purified, cheaper form, Figure 27.1. In production facilities, the cost of materials and equipment are a major concern when designing a large-scale purification strategy. In research laboratories, minimizing required personnel time may be a primary concern. Another consideration that may be important in a biotechnology company is developing a purification strategy quickly to get a product to market as soon as possible. In an academic research laboratory, speed to publication may be a primary goal.

All purification strategies require compromises. For example, the highest product purities are usually obtained only at high cost and with reduced yield. **Yield** *is defined as the % **recovery**, or percent of the starting amount of the product of interest that can be recovered in purified form using a specific strategy.* Yield is intimately related to cost-effectiveness and maintaining product quality, and will be discussed in greater detail. Optimizing either yield or purity will usually raise the cost of a purification process.

It is important to prioritize goals before developing a strategy. Figure 27.2 shows the three primary goals of a purification strategy: optimizing purity, optimizing yield, and reducing costs. These three goals are illustrated as a triangle to emphasize that all purification strategies

Table 27.3 DESIGNING A PRODUCT PURIFICATION STRATEGY

1. Prioritize the goals of the purification process.
2. Select a source for the product.
3. Develop a specific assay for the product of interest.
4. Determine optimal methods to:
   a. release the product from its source in a soluble form.
   b. reduce the volume of the product.
   c. separate the product from impurities.
   d. purify the product of interest.
5. Verify the identity and purity of the product.

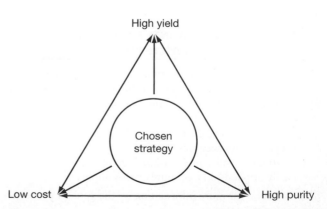

Figure 27.2. Goals for Bioseparation Strategies in the Biotechnology Laboratory.

Table 27.4 *GOALS FOR PRODUCT PURIFICATION STRATEGIES*

All purification schemes share three goals:
- maximum product purity, with
- minimum loss of product material or activity, at the
- lowest possible cost.

Each of these goals, however, varies in importance compared with the others, depending on the nature of the final product desired. The following issues must also be considered when formulating bioseparation strategies:
- safety requirements
- simplicity of the procedures
- reliability of the methods
- volumes suitable for processing
- speed to market
- requirement for product activity

require compromises. Table 27.4 summarizes some of the general concerns that should be considered when planning a product purification scheme.

II. ANALYSIS OF PRODUCT PURITY AND YIELD

A. Assaying the Product of Interest

This section will discuss how product purity and yield are measured and reported. Because biotechnology companies frequently purify proteins for commercial sale, this discussion will focus primarily on proteins.

Devising a purification strategy requires the development of a specific, quantitative assay for the biomolecule of interest. Such an assay enables analysts to detect, quantitate, and determine the degree of purity of the product. The assay is initially used as researchers develop an optimum separation strategy. Later, workers rely on the assay to routinely monitor the separation process. The desired qualities of an analytical assay are summarized in Table 27.5.

In rare cases, the product of interest is distinctive enough to detect without a complex assay. For example, the protein myoglobin can be detected by its red color,

Table 27.5 *REQUIREMENTS FOR A PRODUCT ANALYSIS METHOD*

Product analysis methods for monitoring purification processes vary widely according to the molecule of interest. The assay chosen, however, should have the following characteristics:
- sensitivity to small amounts of the material of interest
- precision and accuracy
- relatively small sample volumes required
- easily applied to large number of samples

See Chapter 6 for more information on the general properties of a biological assay.

which is derived from a heme prosthetic group. Most biological materials are not so obliging, and the detection assay can be complex and potentially destructive to the sample.

During a multistep purification process, the product of interest is successively isolated from multiple impurities. After each separation step, an aliquot of the product mixture is set aside for analysis. Assuming that the molecule to be purified is a protein, the items that are measured are:

1. the amount of the desired protein present, and
2. the amount of total protein present (this includes the protein of interest plus protein contaminants).

Based on these two measurements, it is possible to calculate the specific activity of the protein of interest, and the product yield after each step in the purification process. **Specific activity** *is the amount (or units) of the protein of interest, divided by the total amount of protein in a sample.* As protein purification proceeds, the product of interest should become more pure at each step, so the specific activity of the preparation should increase. Each purification step unfortunately results both in the removal of contaminants and in some unavoidable loss of the biomolecule of interest. As protein purification proceeds, the specific activity therefore increases, but the total amount of the desired product (calculated as the percent yield or recovery) decreases.

B. Protein Assays

In order to determine the purity of a protein product, it is necessary to know both how much of the desired protein is present as well as how much contaminating protein is present. When proteins are purified, the amount of total protein (desired protein plus contaminating protein) is measured after every step. In a purification procedure, the method used to measure total protein must be reliable, fast, sensitive to low protein concentrations, and relatively insensitive to potential interfering agents (e.g., buffers or nucleic acids). To ensure consistency, a single protein assay method should be chosen and then used throughout the purification process. For more information about protein assays, refer to page 415.

C. Specific Activity and Yield Determinations

Biologically active molecules are generally assayed by their activity. Here we will use the example of an enzyme that is assayed by its ability to catalyze a reaction. Enzymatic reactions involve the conversion of a substrate(s) into a product(s):

$$\text{SUBSTRATE(S)} \xrightarrow{\text{enzyme}} \text{PRODUCT(S)}$$

Enzyme assays generally involve the measurement of either the disappearance of substrate or the appearance of product. Enzyme activity is typically expressed as units, based on the rate of the reaction they catalyze. There are two standard expressions of enzyme activity. The **International Unit (IU) of enzyme activity** *is defined as the amount of enzyme necessary to catalyze transformation of 1.0 μmole of substrate to product per minute under optimal measurement conditions.* (Conditions, such as pH and temperature, may affect the activity of an enzyme and so must be optimized.) The **SI unit of enzyme activity** *is defined as amount of enzyme necessary to catalyze transformation of 1.0 mole of substrate to product per second under optimal measurement conditions (such as optimal pH and temperature). This is defined as a* **kat** *unit.* The ability to measure enzyme activity is dependent on knowing both the ideal reaction conditions and having a quantitative assay that measures the amount of substrate or product. Some enzymes, such as proteases, which digest substrates of variable or unknown molecular weight, cannot be quantitated in these standard units. For these proteins, units are defined according to the assay used, providing a relative measure of enzyme activity. Once the amount of the protein of interest and the amount of total protein have been measured, it is then possible to calculate the specific activity of the protein.

EXAMPLE PROBLEM

An assay for enzyme Q measures the disappearance of substrate. In 15 minutes, 12 *mmoles* of substrate are converted to product. What is the activity of the enzyme preparation in International Units?

ANSWER

12 *mmoles* = 12,000 *μmoles*. The rate of conversion of substrate to product is 12,000 *μmoles* in 15 minutes = 800 *μmoles* per minute.

The preparation therefore contains 800 *IU* of enzyme.

The following example problem shows a sample purification scheme and illustrates how specific activity is calculated and used.

EXAMPLE PROBLEM

The enzyme, β-galactosidase was purified from a culture of *E. coli*. The purification steps are:

1. Ten liters of cells were grown in a fermenter.

2. The 10 L of cells plus broth were centrifuged to separate the cells from the broth. The result was a pellet containing the *E. coli* cells.

3. The cells were suspended in buffer and were broken apart by sonication (a method that uses high-pitched sound). The resulting cell suspension was centrifuged again. The pellet and the supernatant were assayed

for β-galactosidase enzymatic activity. β-galactosidase was detected only in the supernatant. This supernatant is called a crude extract.

4. The supernatant was treated with salt to precipitate the β-galactosidase. The precipitated enzyme was collected by centrifugation and was resuspended in buffer. The resulting β-galactosidase preparation was assayed for β-galactosidase enzymatic activity and for total protein.

5. Salt was removed from the β-galactosidase solution using a method called dialysis. The resulting β-galactosidase preparation was assayed for β-galactosidase activity and for total protein.

6. The resulting solution, after dialysis, was passed through a chromatography column. This step separated the β-galactosidase from a number of impurities. The resulting β-galactosidase preparation was assayed for β-galactosidase activity and for total protein.

Observe that the β-galactosidase was extracted from *E. coli* cells and was then purified using a series of steps. After each step, the activity of the β-galactosidase and the amount of total protein were tested. It would be possible to continue with further purification steps at this point, or to stop, depending on the final purity required. The results are shown in the following table.

a. Fill in the blanks in the table.

b. What happened to yield as the purification steps proceeded?

c. What happened to specific activity as the purification proceeded?

PURIFICATION OF β-GALACTOSIDASE FROM E. COLI

Purification Step	Total Protein (mg)	Total Activity (IU)	Specific Activity (IU/mg)	Yield (%)
Crude extract	2,140	1,360	0.64	100
Salt precipitation	760	1,350	____	99
Dialysis	740	____	1.80	____
Chromatography	390	1,240	____	____

ANSWER

a. Specific activity equals the amount of activity of the protein of interest divided by the total amount of protein present. Yield is the percent of the amount of starting activity that is still present after each purification step. In this example, 1360 *IU* represents 100% yield.

PURIFICATION OF β-GALACTOSIDASE FROM E. COLI

Purification Step	Total Protein (mg)	Total Activity (IU)	Specific Activity (IU/mg)	Yield (%)
Crude extract	2,140	1,360	0.64	100
Salt precipitation	760	1,350	1.78	99
Dialysis	740	1,332	1.80	98
Chromatography	390	1,240	3.18	91

b. As we would predict, some of the β-galactosidase was lost at each step and the percent of total activity remaining decreased as the purification proceeded.

c. As the purification proceeded, the β-galactosidase became more pure; therefore, the specific activity value increased.

Information about specific activity is useful when purchasing enzymes or purified proteins. You will frequently be offered various grades of enzyme, based on either total activity or the amount of protein present. In each case, a specific activity level will be provided by the manufacturer. This will allow you to choose among purity levels depending on your application. In general, less purified enzyme grades are less expensive because less effort is required and greater product yield is possible.

EXAMPLE PROBLEM

You need to purchase enough Enzyme Z to perform 100 reactions. You determine that this will require approximately 8000 U of enzyme. The catalog from your favorite enzyme supplier offers you the following choices:

Enzyme Z

One unit of enzyme activity catalyzes the conversion of 1 $\mu mole$ of A to B in 1 minute at 25°C at pH = 7.5.

Grade 1	From aardvark liver	100 mg	$ 50
	Activity: 10,000 U/g	500 mg	$ 200
		1 g	$ 350
Grade 2	From aardvark liver	100 mg	$ 200
	Activity: 50,000 U/g	500 mg	$ 700
		1 g	$1500
Grade 3	From aardvark liver	1 g	$ 30
	Activity: 2000 U/g	5 g	$ 110
		10 g	$ 250

a. What enzyme grade would be your best choice if your main priority was to find the least expensive option?

b. What would be your best choice if you needed the purest preparation possible?

ANSWER

First you want to consider the specific activities of each enzyme grade and determine what quantity you need to purchase:

Grade 1: At a specific activity of 10,000 U/g, you will need 0.8 g to provide the 8000 activity units you need. This will require 8 × 100 mg for a total of $400, or 1 g for $350.

Grade 2: At 50,000 U/g, you will need 160 mg, or 2 × 100 mg for $400.

Grade 3: At 2000 U/g, you will need 4 g of protein, at 4 × 1 g for $120, or 5 g at $110.

a. The least expensive option is Grade 3, which is the least pure.

b. The highest specific activity, and therefore the most concentrated activity, is found in Grade 2.

D. Requirements of the Biotechnology Industry

As discussed previously, the required purity of the end product in a production setting will vary according to the intended use of the product, Figure 27.1. In all cases, undesired biological activities or effects must be eliminated. Products taken from sources containing potentially harmful contaminants will generally require higher standards of purification than products isolated from innocuous sources.

Purification strategies for regulated commercial products must meet federal and state regulations and receive approval from appropriate agencies (see Chapter 3). Once the product and its preparation procedure have been approved, the procedures for product purification must be rigorously followed and documented. In order to change these procedures, the company may again be required to prove the safety and effectiveness of the final product to regulatory agencies, which is a process that can require years of additional testing. There is considerable incentive, therefore, to develop an optimal purification strategy early in the process of product development.

Any product that is administered to humans as a drug must meet especially strict quality specifications for potency and homogeneity. It must be sterile and free of any contaminating substances, especially those such as proteins or nucleic acids that could induce adverse immune responses. Regulatory agencies require a yield analysis for all pharmaceuticals. All starting materials must be accounted for because there cannot be additional product generated by the purification process (this serves as an internal control for the methods), nor should there be significant levels of product breakdown (as opposed to product loss). Because yield obviously affects the production cost of a product, companies use yield analysis as part of their quality control tests.

When drugs are produced using recombinant DNA production systems, product consistency and thorough characterization of the DNA involved are essential. Because recombinant DNA products are being manufactured outside their usual environments (e.g., by *E. coli* rather than by pancreatic cells in the case of insulin), the possibility of incorrectly folded, processed, or aggregated proteins must be considered. Table 27.6 summarizes specific concerns for recombinant DNA products (see also Chapter 3).

Table 27.6 CONCERNS FOR RECOMBINANT DNA PRODUCTS

When manufacturing pharmaceutical products using recombinant DNA systems, biotechnology companies must demonstrate that:

- the DNA sequence that was inserted into the host organism is fully characterized
- the amino acid sequence of the produced protein is identical to the predicted product
- products are consistent among production batches, with no variation due to the recombinant DNA process
- any contaminating substances due to the genetic engineering process are removed

III. CHOOSING BIOSEPARATION METHODS

A. Parameters for Method Selection

A purification strategy consists of a series of purification steps, each of which removes certain impurities from the product of interest. Each purification step involves a different technique. Most individual purification techniques select one or sometimes two specific molecular properties as the basis for bioseparation. There are many molecular properties that can be exploited for separations. They include:

- Size or molecular weight
- Molecular charge
- Relative solubility of the molecule in water or other solvents
- Relative solubility of the molecule in the presence of salts
- Sensitivity to the effects of pH, light, oxygen, or heat
- **Affinity**, *which is the ability to bind specifically to other biomolecules*

The **selectivity** of a technique *is its ability to separate a specific component from a heterogeneous mixture based on molecular properties.* By using a sequence of steps that separates molecules based on three or more different properties, it is generally possible to purify any biomolecule, Figure 27.3.

One of the basic principles of bioseparations is never to purify the product more than necessary. As indicated earlier, the highest product yields will be obtained with the fewest steps. Four to six sequential methods are typically needed to purify and concentrate biomolecules. It is essential to work quickly between steps in order to maintain product activity (if desired), and to coordinate techniques so that the end product of one step provides a suitable substrate for the next. For example, after a centrifugation step, the pelleted product should be dissolved in a volume of a buffer that is compatible with the next technique.

Because there are many purification techniques available, it is necessary to choose the techniques to be used for a particular product, and the order in which those

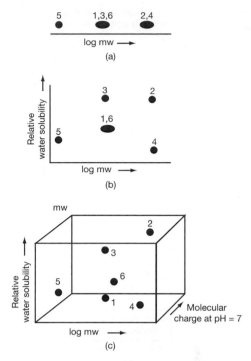

Figure 27.3. Isolation of Specific Molecules by Their Characteristics. This separation starts with a mixture of six molecules. **a.** Separation on the basis of molecular weight gives clear separation of only Molecule 5. **b.** Separation of these fractions by both molecular weight and relative water solubility purifies all of the components except Molecules 1 and 6. **c.** Adding a third separation parameter, charge, allows all components of the mixture to be purified.

techniques are performed. The decisions as to which techniques to use depend partly on the basic biochemical nature of the biomolecule of interest, partly on the probable nature of the impurities, and partly on the goals of the purification process. For example, certain purification techniques are costly and would only be used where the highest purity is required. Certain techniques similarly are most useful for low volume samples; others are suitable for higher volumes. This is a problem that requires systematic strategy development. The simplest starting point for a bioseparation strategy is to follow the procedure for a similar molecule. This technique varies widely in its success. An example of a flowchart for a purification strategy is shown in Figure 27.4.

All bioseparations start with a relatively crude mixture of materials from the biological source. One of the first priorities is to remove any live organisms from the mixture. Any known biohazardous contaminants should be removed as early in the separation process as possible. Typically the simplest techniques are used early in the separation process, when the source mixture is complex and represents a relatively large volume. Early steps will generally require the highest **capacity**, *which is the volume of sample that can be processed simultaneously by a technique.* Capacity also refers to the amount of the product of interest that can be separated by the technique. The first definition is usually more significant early

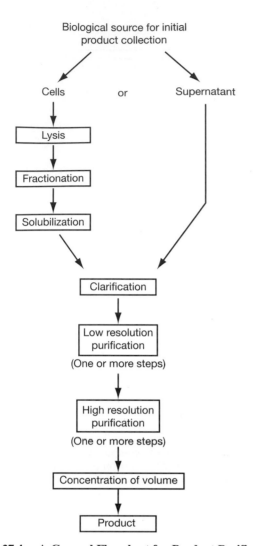

Figure 27.4.   A General Flowchart for Product Purification.

Table 27.7   THE STAGES OF A BIOSEPARATION STRATEGY

The sequential stages of a purification process should follow a logical pattern. Although there are occasional exceptions, the series of steps selected will address the following goals. Examples of appropriate techniques are provided for each goal.

1. Solubilization of the product of interest
   - disruption of cells, and/or
   - collection of supernatants containing soluble product
2. Separation of product from contaminating solid material (clarification)
   - centrifugation
   - filtration
3. Separation of product from divergent impurities
   - low resolution purification techniques, such as:
     salt fractionation
     organic or polymer extractions
4. Purification of product from similar impurities
   - high resolution purification techniques, such as:
     polyacrylamide gel electrophoresis
     high-pressure liquid chromatography
5. Product polishing, concentration, and preparation for end use
   - dialysis to reduce final volume
   - gel filtration to remove salts

B. Overview of Available Methods

i. FRACTIONATION AND CLARIFICATION METHODS

As shown in Table 27.7, bioseparation strategies begin with solubilization of the product of interest from cellular and other particulate matter. In the case of secreted molecules, liquid supernatants can simply be separated from solid matter by centrifugation or filtration. In cases where the product is found inside the cell, the starting point is the preparation of a cell homogenate. A **cell homogenate** *is a suspension of cell contents in liquid, produced by disrupting the outer cell membrane, the cell wall if present, and some of the internal structure of the cell.* Cell homogenates generally contain intact organelles, such as mitochondria.

There are many factors to be aware of when preparing a cell homogenate. For example, molecules inside the cell may be soluble in the cytoplasm or may be associated with cellular organelles. Depending on these properties, different buffers and techniques might be used for homogenization. The interior of an intact cell is buffered and protected from oxidizing agents that can destroy molecules or their biological activity. Solutions used for homogenization, therefore, usually maintain a constant pH and include antioxidizing agents. Cell homogenization should be performed as gently as possible, to avoid loss of product due to the factors indicated in Table 27.8.

Many techniques can be used to **lyse** *(break open)* cells, Table 27.9. Cells with relatively fragile membranes, such as liver cells, will require less strenuous methods than plant cells, which have a tough outer cell

in the separation process, when volumes may be high (such as a supernatant from a 1000 *L* fermentation vessel). Because it is usually more expensive to process large volume samples, volume reduction is also an early concern. These early steps in a purification strategy also usually take advantage of the major differences between the desired product and the expected impurities. For example, if the product is a large molecule contaminated by many low molecular weight materials, an early separation step based on size would be appropriate.

Later steps in a purification process usually involve techniques that remove impurities similar in nature to the product itself. Later steps are often more expensive and have a lower capacity than earlier steps. Analysts use the terms **low resolution** and **high resolution** to distinguish different types of separation techniques. **Resolution**, or **resolving power**, *is the relative ability of a technique to distinguish between the product of interest and its contaminants.* Earlier steps in the purification process are usually low resolution; later ones are of higher resolution. Table 27.7 and Figure 27.4 provide an overview of the usual stages of bioseparation strategies.

Table 27.8 DESTRUCTIVE FACTORS DURING PURIFICATION

Biomolecules removed from their intracellular location can lose biological activity and/or structural integrity due to inhospitable *in vitro* environments. It is important to consider potential destructive factors that may adversely affect your final product. These factors are present not only during homogenization procedures, but also during other purification steps.

- Mechanical stress of cell homogenization
- Dilution and removal of intracellular stabilizing factors
- Changes in buffering, pH, and temperature
- Presence of oxidizing agents
- Product destruction by cellular enzymes

Table 27.9 COMMON CELL HOMOGENIZATION METHODS

The cell homogenization method used in a specific bioseparation scheme depends on the intracellular location of the molecule of interest, the cell type involved, and other factors, as listed in Table 27.8. Some of the most commonly applied techniques are:

- Grinding
- Sonication
- Freezing and thawing
- Application of high pressure using a French press
- Osmotic lysis
- Enzymatic treatment
- Chemical treatment

wall. Most lysis techniques can be adjusted to be more or less disruptive. Roughness and length of treatment can be varied according to need. It is important to remember that mechanical disruption methods, such as grinding or sonication (the use of high-pitched sound), can generate considerable amounts of heat within the sample. These procedures should be performed under chilled conditions when possible.

After cell lysis, the resulting homogenate should be clarified. **Clarification** *is the removal of unwanted solid matter, usually by centrifugation or filtration.* If the product of interest is still located within a cellular organelle, such as mitochondria, a differential centrifugation can be performed to isolate the organelle of interest (described in Chapter 26).

ii. LOW-RESOLUTION PURIFICATION METHODS

There is a wide variety of relatively simple, low-resolution purification methods available. Earlier we defined *resolution* as the relative ability of a technique to distinguish between the product of interest and its contaminants. The resolving power of a method is directly dependent on its selectivity. **Low-resolution purification methods** *are those that generally have high capacities and can be performed quickly, but also have relatively low selectivity.* They eliminate contaminants that are highly dissimilar to the product of interest, but not those that have relatively similar

molecular properties. These methods provide an efficient starting point for a bioseparation strategy, when the molecule of interest represents only a small portion of source material.

Many of these low-resolution methods exploit the solubility characteristics of the desired product. The most common types of methods are extractions and precipitations. **Extraction methods** *are based on the premise that a specific molecule will exhibit higher solubility in one medium compared with another.* For example, if you need to isolate a **hydrophobic** *(water-hating)* compound from a cell homogenate, you would add an organic solvent such as ethyl acetate to the aqueous homogenate. By mixing the buffer and ethyl acetate well (usually by shaking), the hydrophobic compound and other hydrophobic contaminants will be extracted into the organic phase. By centrifuging the resulting mixture, you can collect the organic phase containing virtually all of the compound of interest, while removing and discarding water-soluble contaminants in the aqueous phase, Figure 27.5. Application of this method requires knowledge of the relative solubility of the molecule of interest in nonmiscible liquids.

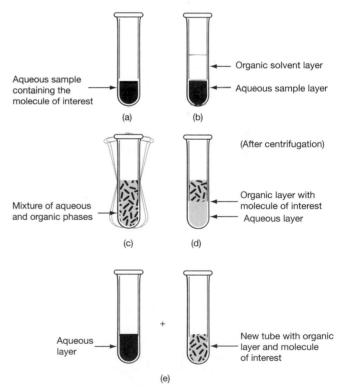

Figure 27.5. An Example of a Simple Organic Extraction. **a.** Start with an aqueous cell homogenate containing the product of interest. **b.** Add an equal volume of ethyl acetate, an organic solvent. **c.** Mix the two liquids thoroughly. **d.** Centrifuge the mixture and separate the organic and aqueous phases. **e.** Collect and assay the organic layer, which contains the product of interest. Note that in this example, the organic layer is of lower density than the aqueous layer. (Relative densities of extraction layers can be determined by consulting *The Merck Index*, Susan Budavari, editor. Twelfth edition. Merck, Whitehouse Station, New Jersey, 1996.)

Other common and simple bioseparation steps include the use of precipitation techniques. **Precipitation methods** *are based on differences between molecules in their tendency to precipitate in particular liquids.* Manipulation of specific characteristics of the liquid medium of the sample will usually result in precipitation of certain molecules, which can then be separated from the liquid. In some cases, the desired product is precipitated; in other strategies, impurities are precipitated and thereby removed. The characteristics of the liquid that are most commonly adjusted are:

- salt concentration
- pH
- temperature
- presence of alcohols, such as ethanol
- presence of organic polymers, such as polyethylene glycol (PEG)

These factors are sometimes combined, as in the precipitation of DNA with ethanol and salt. (Examples of these techniques were also discussed in Chapter 23.)

C. High-Resolution Purification Methods

i. Electrophoresis

High-resolution purification methods *are those that have relatively high selectivities, allowing separation of product from similar impurities.* These methods are generally applied after low resolution techniques have removed the bulk of impurities that are dissimilar to the product of interest, Figure 27.6. High-resolution methods frequently (but not always) have higher costs and lower capacities relative to the methods discussed in the previous section. The most commonly applied high-resolution purification methods are gel electrophoresis and chromatographic techniques. Gel electrophoresis techniques are low capacity and generally are not used for production systems. In contrast, many chromatographic techniques can be adapted to high capacity systems.

Electrophoresis *is the separation of charged molecules in an electric field.* In addition to separating specific proteins and DNA fragments of different sizes, electrophoresis techniques are frequently used to determine:

- the length of DNA fragments, in base pairs
- the molecular weight of specific proteins
- the **isoelectric point** of a protein, *the pH value at which a protein exhibits an overall neutral charge*
- the purity of an isolated protein

Many of the references at the end of this chapter describe the uses of electrophoresis in detail. Here we will briefly discuss two of the most common electrophoretic techniques: agarose gel electrophoresis, most commonly used to separate DNA fragments on the basis of base pair length, and SDS-polyacrylamide gel electrophoresis (PAGE), most commonly used to separate proteins on the basis of molecular weight. Both of these techniques use polymer gels as a separation matrix.

Digested or sheared DNA fragments are usually separated in agarose gels. **Agarose** *is a natural polysaccharide derived from agar, a substance found in some seaweeds,* Figure 27.7. When agarose powder is heated at low concentrations (0.8% is typical) in an appropriate buffer, the agarose dissolves into a viscous solution that

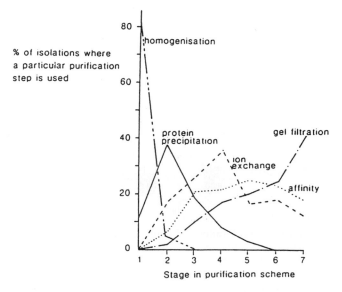

Figure 27.6. Use of Specific Separation Techniques at Progressive Stages of Purification. Methods are discussed later in the text. (Adapted from "Protein Purification: The right step at the right time," J. Bonnerjea, S. Oh, M. Hoare, and P. Dunnill. *Bio/Technology,* 4 (11): 954–958, 1986. © Nature Publishing Co.)

$$CH_2 = CHCONH_2 + CH_2(NHCOCH = CH_2)_2$$

acrylamide   N,N' methylenebisacrylamide + activator (ammonium persulfate) · catalyst (TEMED)

Figure 27.7. Subunit Structures for (a) Agarose and (b) Polyacrylamide Gels.

gels when cooled. This gelling is reversible, so that agarose solutions can be cooled and heated repeatedly. The cooled gel is chemically stable, with a relatively high gel strength for thin sheets of agarose. These sheets can be gently manipulated for staining and transfer of DNA to other media. Because agarose gels generally cannot support their own weights, electrophoresis is performed in a horizontal gel box. An agarose gel electrophoresis apparatus and example gel are shown in Figure 14.9.

Because of the regular occurrence of charged phosphate groups along DNA strands, all DNA molecules have a uniform negative charge. When DNA samples are applied to a gel and exposed to a directional electric field, all DNA fragments will therefore migrate toward the positive electrode. Because the agarose gel acts as a molecular sieve, smaller DNA fragments will move to the positively charged end of the gel faster than larger fragments, which will become entangled in the pores of the gel. As long as the electric field is applied, DNA fragments will migrate through the gel at a rate that is roughly related to the inverse log of the number of base pairs in the fragment. The bands of DNA are most commonly stained with ethidium bromide, allowing visualization of the bands under ultraviolet light. If DNA fragments of known base pair length are applied to the same gel as controls, a reasonable approximation of base pair length for unknown DNA bands can be made. Note that ethidium bromide is a known mutagen, whose handling will be discussed in Chapter 29.

Proteins are generally separated on **polyacrylamide gels**, *which are made from acrylamide* (a neurotoxin—see Chapter 29) *and bisacrylamide polymerized in the presence of an appropriate initiator (ammonium persulfate) and catalyst (N,N,N',N'-tetramethylethylenediamine, commonly called TEMED)* (see Figure 27.7b). The pore size of a gel can be adjusted by combining different amounts of acrylamide and bisacrylamide. Polyacryl-

amide gels are stronger than agarose gels of comparable thickness, so they can be used in vertical polyacrylamide gel electrophoresis (PAGE), Figure 27.8.

Unlike DNA molecules, all proteins are not uniformly charged. The presence of both positively and negatively charged amino acids at a given pH allows the PAGE technique to separate proteins on the basis of charge. A more common adaptation of PAGE, however, is SDS-PAGE. **Sodium dodecyl sulfate-PAGE (SDS-PAGE)** *is a technique that separates proteins on a polyacrylamide gel on the basis of molecular weight.* Protein samples are treated with the negatively charged detergent SDS. SDS binds to proteins at a ratio of approximately one SDS molecule for every two amino acids in the protein. The resulting negative charge conferred by the SDS overshadows any native protein charge so that all proteins will migrate toward the positive electrode in an electric field (as shown in Figure 27.8). In this case, sieving of the proteins occurs in the gel according to the molecular weight of the protein. The detergent also denatures the protein, minimizing the effect of protein shape on migration rate. Higher concentrations of acrylamide and higher levels of cross-linking provide a gel with smaller pore sizes. Adjusting pore size controls the effective range of protein sizes that are separated on an individual gel.

SDS-PAGE is frequently used both for determination of the molecular weight of proteins, and also for the determination of the purity of protein preparations. The presence of a single protein band on an SDS-PAGE gel is considered one measure of protein purity. The protein may still contain contaminants of the same molecular weight and properties, and so at least one additional method should be used to test the purity of the product.

Electrophoresis requires the same or even greater safety precautions as any other use of electricity. These

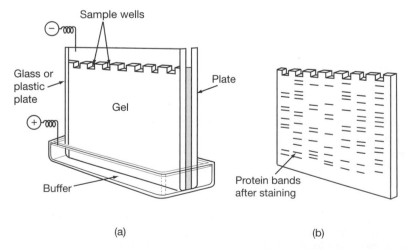

Figure 27.8.   Polyacrylamide Gel Electrophoresis (PAGE). a. The gel is poured vertically between two glass plates. Sample is applied to wells at the top of the gel. **b.** Protein bands are separated on the basis of relative molecular weight and visualized with stains. *Note:* This is an example of SDS-PAGE.

techniques require extreme care because an electric current is being deliberately applied to a liquid. Electrophoresis generates heat, which may result in the evaporation of system buffer and allow short circuits, equipment damage, or fires. Additional safety suggestions are provided in Chapter 28.

ii. CHROMATOGRAPHY

The term *chromatography* describes the most widely used set of techniques applied to the purification of biomolecules. **Chromatography** *refers to a family of separation techniques based on the differential interaction of molecules between a stationary and a mobile phase.* As suggested by their names, the **stationary phase** *is an immobile matrix* and the **mobile phase** *is a liquid or gas that moves past the stationary phase and elutes molecules.* Because molecules differ in their relative attraction to the mobile phase, they will move past the stationary phase at differing rates. Some type of chromatography is used in most laboratories, for either purification or analytical purposes (or both).

One of the advantages of chromatography is the versatility of techniques available. Separations can be based on many molecular properties, and some techniques can separate molecules based on quite subtle differences. Recovery of product is generally simple, and chromatographic techniques are usually amenable to moving from laboratory to large-scale operations.

There are many ways to classify chromatographic techniques. Table 27.10 discusses some of the molecular properties that are most commonly used as the basis for chromatographic separations. One classification scheme is shown in Figure 27.9, which divides methods based on

Table 27.10 MOLECULAR CHARACTERISTICS COMMONLY USED AS A BASIS FOR CHROMATOGRAPHY*

Molecular Property	Chromatography Method
• Polarity, or relative water-solubility	Partition chromatography
• Charge	Ion exchange chromatography
• Molecular weight	Gel permeation chromatography
• Hydrophobic properties	Hydrophobic interaction chromatography
• Specific molecular binding properties	Affinity chromatography

*Each of these methods is defined more completely in the chapter glossary. Virtually any molecular properties can be exploited with a chromatographic technique. The most commonly used properties and the corresponding techniques are shown in this table.

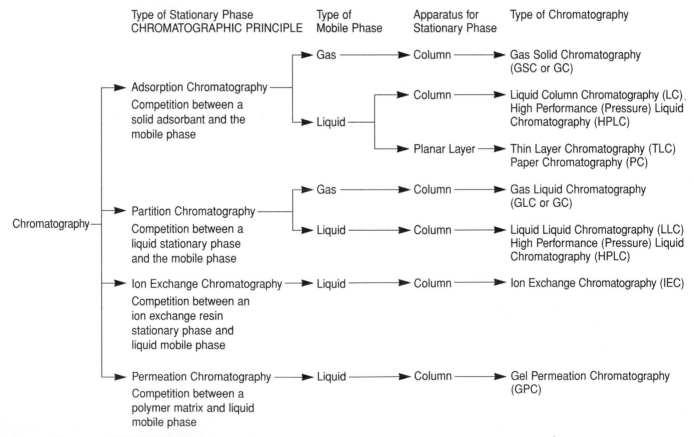

Figure 27.9. Classification Scheme for Chromatographic Methods. (For additional information about these methods, consult the chromatography references listed at the end of the chapter. From *Chromatographic Methods*, A. Braithwaite and F.J. Smith, Chapman and Hall, New York, p. 6, 1985, with kind permission from Kluwer Academic Publishers).

their underlying separation principles, and then on the nature of the mobile phase.

The first step in choosing a chromatographic method is to decide on the basis of separation (i.e., what molecular property will be exploited) and the systems available (e.g., partition chromatography on a gas–liquid chromatography system). The choice of a stationary and a mobile phase is critical in developing an appropriate chromatographic technique. Solid stationary phases in modern techniques are usually small microporous particles, which are packed into a column or sometimes spread on a plate (TLC). These particles must be nonreactive, uniform in their physical characteristics, and provide a large surface area for potential interaction with molecules in the mobile phase. The mobile phase can be either a liquid or gas. The term **LC** (not in combination with other letters) *describes any technique where the mobile phase is liquid.*

Chromatographic methods vary in their levels of required technology. Early methods used planar stationary phases such as paper. **Thin-layer chromatography (TLC)** is a technique that does not require expensive equipment or materials. In TLC, *a thin layer of stationary phase material is spread on a glass or plastic plate, and the mobile phase passes through the stationary phase by either capillary action or gravity.*

Column chromatography *is performed with the stationary phase packed into a cylindrical container that may vary in width and length.* The mobile phase, or **eluent**, passes through the column, propelled either by gravity or mechanically with a pump. **Elution** of molecules from the column *occurs as the molecules in the sample distribute themselves between the two phases according to their affinities for each phase,* Figure 27.10. Fractions of the mobile phase are collected and assayed as they leave the column.

Small columns, usually made of glass or plastic, can be filled with stationary phase by the user, although these columns are generally used only once. Prefilled reusable columns can be purchased for convenience, or to provide more delicate stationary phase materials or larger columns. With proper method development, large columns can provide very high product capacities together with high resolution. This accounts for the common use of chromatographic techniques in production settings. Capacity increases with column volume (up to a point), although this increased capacity comes at increased monetary cost as well. Large columns are usually constructed with steel walls and require pumps to elute the mobile phase.

High-performance liquid chromatography (HPLC) *is a set of chromatographic techniques that involve specially packed columns and high-pressure pumps.* HPLC columns provide stationary phases made of smaller and harder particles than traditional column chromatography. These particles allow the use of faster mobile phase flow rates. Use of HPLC speeds up sample elution 10-fold or more relative to low-pressure

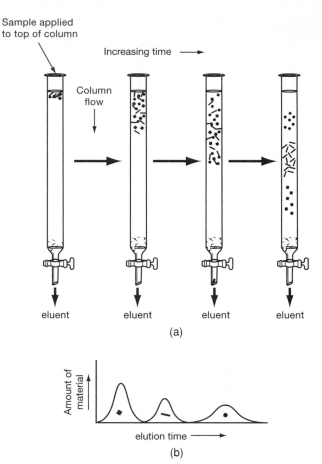

Figure 27.10. Elution of a Sample Mixture from a Chromatographic Column. a. A sample containing three different molecules is loaded onto a column. As the mobile phase carries the sample constituents through the column, movement rate of the molecules is affected by the relative tendency of the molecules to adsorb to the stationary phase, slowing their elution. **b.** The three molecules are graphed according to their elution times from the columns.

methods. This is convenient, and also allows greater recovery of sample activity. HPLC can be adapted to any of the chromatographic principles shown in Table 27.10. The disadvantage to HPLC is that it requires pumps and other pieces of equipment that are much more expensive than simple, low-pressure LC systems. A comparison of typical equipment for traditional LC and HPLC is shown in Figure 27.11.

One chromatographic method that merits separate discussion is gel filtration, or gel permeation chromatography. **Gel permeation chromatography (GPC)** *is a technique for separating molecules by relative size and molecular weight using a column filled with porous gel particles.* GPC is also known as **molecular sieving** or **size exclusion chromatography**. The gel particles are usually highly cross-linked polysaccharides (one typical material is Sephadex®, manufactured by Pharmacia) with relatively large pore sizes compared to other chromatographic media. Gel permeation is based on molecular sieving rather than on chemical interaction between the sample and stationary phase. In a gel permeation

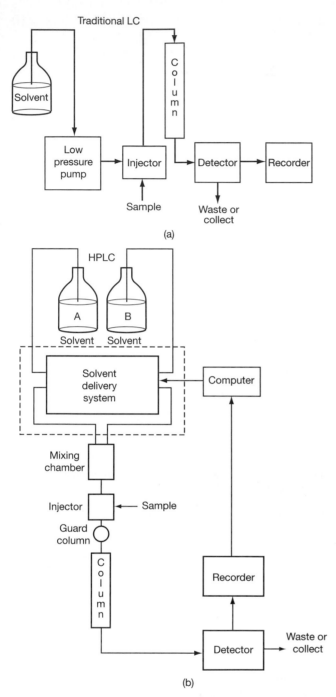

Figure 27.11. Typical Equipment Systems for Traditional Liquid Chromatography (a) and High-Performance Liquid Chromatography (b).

nique. Even though it is a low-resolution technique, it is commonly used as a final purification step to remove salts from products.

The multitude and flexibility of chromatographic techniques make the design of purification methods complex. As with other techniques, the properties of the desired product and its impurities must be considered. Because chromatographic methods (with the exception of gel permeation) frequently separate molecules based on subtle differences, small changes in stationary or mobile phase materials can make major differences in purification results. These methods frequently require extensive fine tuning to provide optimal results.

D. Common Problems with Purification Strategies

The development of an optimized purification strategy often does not proceed as smoothly or quickly as desired. Perseverance is necessary to ensure that the final strategy will be appropriate on a long-term basis. The most common problem encountered is the apparent loss of product during a specific separation step. If this occurs during initial strategy development, it may suggest that an inappropriate separation method was chosen. For example, if you are performing an affinity chromatography technique (see glossary), and your product will not elute from the column, you may have inadvertently chosen a column matrix that binds irreversibly with the product material. Always develop your separation strategy using an aliquot of a larger sample so that you can return to the results of a previous step and try another strategy.

Product is sometimes unexpectedly lost during a separation step that was previously proven to be effective. In these cases, there is usually a technical difficulty that can be identified. These same difficulties can occur during strategy development, leading to the incorrect assumption that a method is not effective for purifying your product. Whenever large amounts of product disappear during a separation step, consider whether the factors outlined in Table 27.11 may be involved.

Note that some of the problems indicated in Table 27.11 are the result of technical difficulties (e.g., using the wrong buffer) and others are the relatively unpredictable result of source contamination (e.g., the presence of enzyme inhibitors). In the latter case, either the interfering substance must be removed before the product is assayed, an alternative source of product must be used, or a different product assay technique is required.

To summarize the most basic precautions against losing product during a bioseparation strategy:

- In general, the fewer number of purification steps used, the higher the ultimate product yield.

- With many biomolecules, stability is a major concern. Maintain a constant temperature (usually

system, small molecules enter the pores of the stationary phase and are retarded on the column. Large molecules are excluded from the pores and elute from the column relatively quickly, Figure 27.12. In other words, the larger the molecule, the faster it passes through the column. This is the opposite result of the sieving in electrophoretic techniques, although each of these methods can be used to determine the approximate molecular weight of materials. Unlike other chromatography methods, gel permeation is a relatively low-resolution tech-

Table 27.11 *FACTORS THAT MAY RESULT IN PRODUCT LOSS DURING PURIFICATION PROCEDURES*

There are many factors that may contribute to a major loss of product during a bioseparation procedure. Some of these are relatively easy to identify (at least in retrospect), and others are difficult to predict. Common problems and examples include:

Problems leading to product loss

- Discarding the fraction that contains the product of interest (e.g., discarding a supernatant containing product while keeping a pellet of debris)
- Performing separations using poor quality reagents (e.g., ion exchange medium may be inadequately prepared)
- Poor product stability (may be due to enzymatic degradation by impurities in the mixture, or inherent instability of product)
- Unexpected precipitation of product (e.g., sample may be suspended in an inadequate amount of liquid)

Problems with the product assay

- Performing an inappropriate product assay (lacking adequate selectivity or specificity)
- Using an incorrect buffer for the product (e.g., certain salts may interfere with enzyme assays)
- Presence of product inhibitors (if presence of product is measured by activity)
- Loss of a required cofactor (e.g., an enzyme may require the presence of magnesium ions for activity)

chilled) and pH for the sample and proceed from one step to the next as quickly as feasible.

- Never discard any fraction during a separation procedure. Save everything until the end of the process, when the product has been recovered.

E. Large-Scale Operations

All purification strategies are developed at the laboratory scale before moving to a production setting, and so the ability to convert a strategy to a large-scale operation is essential. The purification of agricultural products, for example, can require handling tons of source and final materials. **Scaling up** *is the process of converting a small-scale laboratory procedure to one that will be appropriate for large-scale product purification.* Scale-up must be considered from the start of product development. Some laboratory purification procedures may not be economically sound or physically possible in an industrial setting. For example, using sonication to lyse cells may work well in a laboratory, but it becomes impractical when hundreds of liters of cellular material are present. Breaking cells open with grinding or high-pressure techniques is more practical on a large scale.

When testing purification strategies for commercial purposes, the research and development team selects and optimizes methods on a small scale. Development scientists then need to increase the scale of operations and test the robustness of the strategy. The working con-

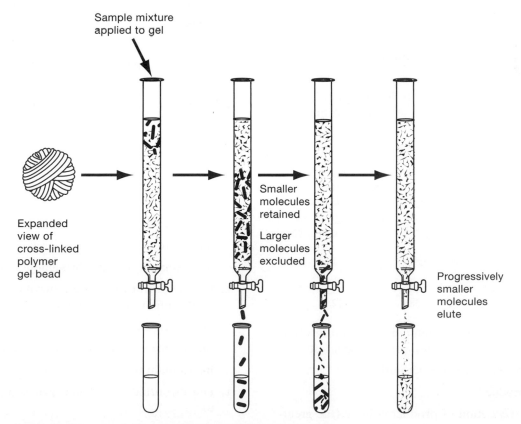

Figure 27.12. **Diagram of Gel Permeation Chromatography.**

ditions (e.g., pH, temperature, ion concentrations) must be achievable under large-scale conditions. Analysts need to consider the possibility of variations in biological source materials and must design techniques that will be relatively insensitive to the most likely variations.

Some of the key factors to be considered during scale-up are important during methods development, but assume an even greater significance under production conditions. As indicated earlier, reproducibility of the techniques in the production setting is essential. Having a simple assay method and documentation process for each purification step will contribute to cost-effectiveness and quality control. Of course, safety issues must be considered during scale-up. Strategies must be designed to eliminate any potential biohazards early in the process, and appropriate containment facilities and procedures are essential.

In reality, many potentially useful biotechnology products never reach the consumer market due to the high costs of downstream processes. The scale of required equipment and labor costs are major considerations. The ability to automate bioseparation procedures can increase cost-effectiveness to some extent, but not all procedures are easily automated. The ability to reduce product source volumes quickly, apply less expensive low-resolution methods, and achieve high product yields are key factors in reducing purification costs for biotechnology products.

PRACTICE PROBLEMS

1. You have received a vial of enzyme that is labeled as containing 4 *kat* (*SI* units) of enzyme. How many *IU* of enzyme are in the vial?

2. You are presented with the following data from a series of purification steps for the enzyme "comatase". Fill in the blanks in the table:

PURIFICATION OF COMATASE FROM E. coli

Purification Step	Volume (ml)	Total Protein (mg)	Total Activity (IU)	Specific Activity (IU/mg)	Yield (%)
I. Homogenization	100,000	12,350	9,000	___	100
II. Dialysis	5,000	10,233	8,289	0.81	92.1
III. Organic extraction	80	3,860	___	1.65	___
IV. Ion exchange chromatography	10	1,140	5,625	___	62.5
V. PAGE	2	386	4,688	12.15	___

3. Given the purification scheme data shown in Problem 2, which purification step gave the greatest:

   a. relative increase in product purity?

   b. loss of product?

   c. absolute reduction of product volume (the greatest decrease in actual volume)?

   d. relative reduction of product volume (the greatest % decrease from the previous step)?

4. Suppose you are given the assignment of isolating an enzyme, "enzylase," which is secreted into the fermentation broth of the bacterium *B. techi*. The enzyme has an approximate molecular weight of 60,000. You grow 3 *L* of *B. techi*, which you then spin down in a centrifuge. You discard the broth and weigh the cells. You break open the cells using a grinding method and centrifuge the resulting paste. You discard the cellular debris and assay the supernatant for the enzylase. You are disappointed to find no activity. You test your assay with "store-bought" enzylase and find that the assay works well. What important mistake did you make?

5. You are interested in an enzyme produced by a particular bacterium. You grow 10 *L* of the microorganism in a fermenter. You centrifuge the 10 *L* to obtain a pellet of cells and set aside the supernatant. You resuspend the cell pellet in 20 *mL* of buffer, break open the cells by sonication, and remove the cellular debris by centrifugation. The enzyme of interest is in the intracellular supernatant. (Assume there is still 20 *mL* of supernatant.) The supernatant is called the crude extract. You remove 0.5 *mL* of the crude extract and perform an enzyme assay. You find there are 500 Units of enzyme activity in the 0.5 *mL* sample of crude extract. You take another 0.5 *mL* sample of crude extract and perform a protein assay. You find there are 25 *mg* of protein present in the 0.5 *mL* of crude extract.

   a. Draw a flow chart of this procedure.

   b. What is the specific activity of enzyme in the crude extract?

   c. How many units of enzyme activity were in the original 10 *L*?

6. Choose the most probable answer to the following questions:

   a. During purification of an enzyme, the specific activity should:

      i. Increase as the purification proceeds

      ii. Decrease as the purification proceeds

      iii. Remain the same throughout the purification

   b. During purification of an enzyme, the amount of total protein present should:

      i. Increase as the purification proceeds

      ii. Decrease as the purification proceeds

      iii. Remain the same throughout the purification

   c. During purification of an enzyme, the yield is likely to:

      i. Increase as the purification proceeds

ii. Decrease as the purification proceeds

iii. Remain the same throughout the purification

7. Before developing a strategy to purify a protein, it is important to learn about the biochemical characteristics of that protein. List at least three features of the protein that are important to know when developing a strategy to purify a protein.

DISCUSSION QUESTION

Development and troubleshooting of purification strategies and procedures are an integral part of work in a biotechnology setting. Quality control and its documentation are critical aspects of product preparation. **Scenario 13 in the Skill Standards for the Bioscience Industry (Education Development Center, Inc., 1995) presents the following topic and question:**

You are responsible for following the protocol for purifying your company's product. Demonstrate the steps you take in product purification.

The 2 *L* of the crude product has a calculated maximum yield of 10 *g/L*. You expect an 80% yield. After running the column, you calculate the purified total sample yield as 22 *g/L*. Show how you would handle this result.

TERMINOLOGY RELATED TO BIOSEPARATIONS

Acrylamide. A small neurotoxic monomer that can be polymerized to form relatively nontoxic polyacrylamide gels. (See "Polyacrylamide Gels.")

Adsorption Chromatography. A chromatographic technique for separating molecules based on their relative affinities for a solid stationary phase and a gas or liquid mobile phase.

Affinity. The capacity of a material to bind specifically to other biomolecules.

Affinity Chromatography. A chromatographic technique for separating molecules by their ability to bind specifically to other molecules that are incorporated onto the surface of a stationary phase.

Agarose. A natural polysaccharide derived from agar, a substance found in some seaweeds, used in making electrophoresis gels.

Analysis. The study of the specific chemical properties of a molecule.

Biomolecules. Compounds produced by some type of biological source, such as a plant, animal, microorganism, or cultured cell.

Bioseparation Methods. Separation techniques that are used to extract, isolate, and purify specific biological products, or biomolecules (such as a particular protein).

Capacity. The relative volume of sample that can be processed simultaneously by a technique; also refers to the amount of the product of interest that can be separated by the technique.

Cell Homogenate. A suspension of cell contents in liquid, produced by disrupting the outer cell membrane and wall (if present) and some of the internal structure of the cell.

Chromatography. A group of bioseparation techniques based on the differential interaction of molecules between a stationary and a mobile phase. Because molecules differ in their relative attraction to the mobile phase, they will move past the stationary phase at differing rates.

Clarification (in bioseparations). The removal of unwanted solid matter after a bioseparation procedure, usually by centrifugation or filtration.

Column Chromatography. Chromatography performed with the stationary phase packed into a cylindrical container.

Downstream Processes. The separation procedures that result in a purified product.

Electrophoresis. The separation of charged molecules in an electrical field.

Eluent. The mobile phase in chromatography.

Elution. The passage of molecules through a chromatographic column, as the molecules in the sample distribute themselves between the two phases according to their affinities.

Ethidium Bromide. A mutagenic dye that reversibly intercalates into DNA molecules, allowing them to be visualized under ultraviolet light.

Extraction Methods. Crude separation of molecules from particulate matter in the cells; also, bioseparation techniques based on the fact that molecules differ from one another in their solubility in various liquids.

Gel Filtration. See "Gel Permeation Chromatography."

Gel Permeation Chromatography, GPC. A chromatographic technique for separating molecules by relative size and molecular weight using a column filled with porous gel particles.

High-Performance Liquid Chromatography, HPLC. Chromatographic techniques that involve specially packed columns and high pressure pumps.

High-Resolution Purification Methods. Techniques that have relatively high selectivities but relatively low capacities. (See also "Low-Resolution Purification Methods.")

Hydrophobic *(water-hating)*. The property of being relatively insoluble in aqueous liquids.

Hydrophobic Interaction Chromatography, HIC. A chromatographic technique for separating molecules based on their hydrophobic properties.

International Unit, IU, of Enzyme Activity. The amount of enzyme necessary to catalyze transformation of 1.0 $\mu mole$ of substrate to product per minute under optimal measurement conditions.

Ion Exchange Chromatography. A chromatographic technique for separating molecules by molecular charge; based on ionic stationary phases and manipulation of salt concentrations and pH in the mobile phase.

Isoelectric Point. The pH value at which a protein exhibits an overall neutral charge.

Isolation. See "Purification."

Kat. See "SI Units of Enzyme Activity."

Liquid Chromatography, LC. A general term for any chromatographic technique where the mobile phase is liquid.

Low-Resolution Purification Methods. Bioseparation techniques that generally have high capacities and can be performed quickly, but exhibit relatively low selectivity. (See also "High-Resolution Purification Methods.")

Lyse. To break open cells and release their contents.

Mobile Phase (in chromatography). The liquid or gas medium that moves past the stationary phase and elutes molecules.

Molecular Sieving. See "Gel Permeation Chromatography."

Partition Chromatography. A chromatographic technique for separating molecules between two liquid phases based on their relative solubility in each phase.

Percent Recovery. See "Yield."

Polyacrylamide Gel. A separation gel made from polymerized acrylamide and bisacrylamide in the presence of an appropriate initiator (ammonium persulfate) and catalyst (N,N,N',N'-tetramethylethylenediamine, also called TEMED).

Polyacrylamide Gel Electrophoresis, PAGE. A separation technique that involves applying an electric field to a polyacrylamide gel.

Precipitation Methods. Bioseparation techniques that are based on differences between molecules in their tendency to precipitate from various liquids.

Purification. The separation of a specific material of interest from contaminants in a manner that provides a useful end product.

Resolution (in bioseparations). The relative ability of a technique to distinguish between the product of interest and its contaminants. The resolving power of a method is directly dependent on its selectivity.

Scale-up. The process of converting a small-scale laboratory procedure to one that will be appropriate for large-scale product purification.

SDS, Sodium Dodecyl Sulfate (also known as sodium lauryl sulfate). A negatively charged detergent.

SDS-PAGE, Sodium Dodecyl Sulfate-PAGE. A technique that separates proteins on the basis of molecular size, using polyacrylamide gel electrophoresis.

Selectivity. The ability of a technique to separate a specific component from a heterogeneous mixture.

Sephadex®. A gel permeation medium manufactured by Pharmacia.

SI Unit of Enzyme Activity. The amount of enzyme necessary to catalyze transformation of 1.0 $mole$ of substrate to product per second under optimal measurement conditions; expressed in units of kat.

Size Exclusion Chromatography. See "Gel Permeation Chromatography."

Specific Activity. The amount (or units) of the protein of interest, divided by the total amount of protein in a sample.

Stationary Phase (in chromatography). The portion of a chromatographic separation system that remains immobile.

TEMED, N,N,N',N'-tetramethylethylenediamine. A chemical catalyst for the polymerization of acrylamide.

Thin Layer Chromatography, TLC. A chromatographic technique where a thin layer of stationary phase material is spread on a glass or plastic plate, and the mobile phase passes through the stationary phase by either capillary action or gravity.

Upstream Processes. Biological processes, such as fermentation or cell culture, that produce the biomolecule of interest.

Yield. The % recovery, or percent of the starting amount of the product of interest that can be recovered in purified form using a specific strategy.

REFERENCES

There are many references available for bioseparation techniques. Some of these address general strategies, whereas others cover specific types of methods in detail. The following list provides starting materials for a more in-depth study of purification methods.

General References

Although each of these references mentions proteins in their titles, many of the techniques and approaches described can be applied to other types of biomolecules as well.

Protein Methods, Daniel M. Bollag, Michael D. Rozycki, and Stuart J. Edelstein. Wiley-Liss, New York, 1996.

Guide to Protein Purification, Murray P. Deutscher, ed. *Methods in Enzymology*, Volume 182, Academic Press, London, 1990.

Protein Analysis and Purification. Benchtop Techniques, Ian M. Rosenberg. Birkhauser, Boston, 1996.

Bioseparation of Proteins, Ajit Sadana. Academic Press, London, 1998.

Protein Purification, Scott M. Wheelwright. Hanser, New York, 1991.

Specialized Techniques

Many techniques such as electrophoresis and chromatography are sufficiently specialized to be described in separate volumes. Some suggested references for detailed information beyond that provided by the general references follow. In addition, the manufacturers and suppliers of materials for these techniques are usually excellent sources of information.

Electrophoresis: Theory, Techniques and Biomedical and Clinical Applications, 2nd ed. Anthony T. Andrews. Oxford University Press, New York, 1986.

Chromatographic Methods, A. Braithwaite and F.J. Smith. Chapman and Hall, New York, 1985.

Molecular Bases of Chromatographic Separation, E. Forgacs and T. Cserhati. CRC Press, Boca Raton, 1997.

Gel Electrophoresis of Proteins: A Practical Approach, 2nd ed., B.D. Hames and D. Rickwood. IRL Press, New York, 1990.

Electrophoresis of Large DNA Molecules, Eric Lai and Bruce W. Birren, eds. Cold Spring Harbor Laboratory Press, Cold Spring Harbor, NY, 1990.

The Busy Researcher's Guide to Biomolecule Chromatography, PerSeptive Biosystems, Framingham, MA, 1996.

ANSWERS TO PROBLEMS

1. 2.4×10^8 *IU*

2. *PURIFICATION OF COMATASE FROM* **E. coli**

Purification Step	Volume (ml)	Total Protein (mg)	Total Activity (IU)	Specific Activity (IU/mg)	Yield (%)
I. Homogenization	100,000	12,350	9,000	0.73	100
II. Dialysis	5,000	10,233	8,289	0.81	92.1
III. Organic extraction	80	3,860	6,369	1.65	70.7
IV. Ion exchange chromatography	10	1,140	5,625	4.93	62.5
V. PAGE	2	386	4,688	12.15	52.1

3. In evaluating the relative merits of specific purification steps, consider the main purpose of each step. In the example given, Step II provided a major reduction in product volume, but little purification (a resulting specific activity of 0.81 vs. 0.73 for the starting material.

a. Step IV provides the greatest relative increase in product purity, with approximately a threefold increase in the specific activity of the product. Step V gave a 2.5-fold purification.

b. The loss of product material in each step can be calculated by dividing the total product activity for each step by the activity present in the preceding step. Step III, therefore, gave the greatest loss of product activity:

$(6{,}369\ IU/8{,}289\ IU) \times 100\%$
$= 76.8\%$ product recovery in this step.

c. Absolute reduction in product volume can be determined by looking at the final product volume for each step. In this example, Step II gave the greatest volume reduction, from 100,000 *mL* to 5000 *mL*. In a production setting, this volume reduction would significantly ease the cost of operations.

d. Relative reduction in product volume is calculated in a similar manner to 3b. The volume after each step is divided by the volume at the beginning of the step. Step III, therefore, gave the greatest *relative* reduction in volume size:

$(80\ mL/5{,}000\ mL) \times 100\%$
$= 1.6\%$ of the starting volume for this step vs. 5% for Step I

4. The enzyme was secreted into the broth, which was discarded after the first centrifugation step. It is good practice to keep all supernatants and pellets until you are certain you have the product of interest.

5. a. See Figure 27.4 for an example of a flow chart.

b. 20 *U/mg* protein.

c. If there are 500 units in 0.5 *mL* of the supernatant, then there are 20,000 units in the entire 20 *mL*.

6. a. i. Specific activity should increase.

b. ii. Amount of total protein should decrease.

c. ii. Yield will decrease.

7. Some possible answers:

Whether it is found inside cells or is secreted

If it is intracellular, its intracellular location

Molecular weight

Solubility characteristics in different solvents

Molecular charge

Stability at room temperature

UNIT VIII:

Safety in the Laboratory

Chapters in this Unit

✦ Chapter 28: Working Safely in the Laboratory: General
   Considerations and Physical Hazards

✦ Chapter 29: Working Safely with Chemicals

✦ Chapter 30: Working Safely with Biological Materials

It is essential in a safe workplace to recognize hazards and reduce risks to the workers. As you may recall from Chapter 2, **hazards** *are the equipment, chemicals, and conditions that have a potential to cause harm*, and **risk** *is the probability that a hazard will cause harm.* For example, even though toxic chemicals are hazardous, the risk of working with them is reduced by using smaller working volumes, proper ventilation, shorter working times, and good experimental technique. Laboratory hazards fall into several categories:

- Physical hazards
- Chemical Hazards
- Biological Hazards

This unit discusses examples of these classes of hazards and specific approaches to risk reduction.

Safety information is critical in the laboratory so that hazard exposure can be reduced or eliminated by good laboratory practices. In addition, in the event of an accident, a quick and appropriate response usually results in less harm to people and property. Knowing about the potential for laboratory injuries can help you anticipate the types of emergencies that are most likely to occur and to plan how you would react.

This unit is intended to provide practical advice that stems from a variety of general information sources, as well as the personal experience of the authors. It is not a substitute for a safety manual, which is specific for an institution or facility.

By OSHA regulation, each laboratory will have either a Chemical Hazard Communication program or a Chemical Hygiene Plan (CHP—see Chapter 2). It is a good practice for every laboratory to have at least one comprehensive reference book covering the specific type of hazards (i.e., biological, radioactive, etc.) found in that setting. There are many excellent books available that can serve as safety references in the laboratory. Each of these tends to focus on specific aspects of safety, so several may be necessary to cover all contingencies. Related references and glossaries are provided at the end of each chapter in this unit.

Unit VIII is organized as follows:

Chapter 28 surveys general risk reduction strategies, personal protective equipment, and the most common physical hazards found in laboratories.

Chapter 29 will discuss safe handling of chemicals, with special emphasis on those most likely to be found in the biotechnology laboratory.

Chapter 30 will give an overview of biosafety issues, including universal precautions, containment and sterilization strategies, animal handling, and recombinant DNA guidelines.

CHAPTER 28

Working Safely in the Laboratory: General Considerations and Physical Hazards

I. RISK REDUCTION IN THE LABORATORY

There are four general approaches to risk reduction in the laboratory that apply to all categories of hazards:

- Reduce the presence of hazards
- Reduce the risk of inevitable hazards with good laboratory design
- Establish good laboratory practices for handling hazards
- Use personal protective equipment

First, the presence of hazards should be reduced as much as possible. For example, amounts of radioactive substances and flammable solvents on the premises should be limited. Next, the risk of those hazards that cannot be eliminated should be reduced by good laboratory engineering. This means, for example, the installation of properly functioning fume hoods, protective shielding, and fire-resistant chemical storage facilities. The third approach is to establish good laboratory practices in ways that reduce risk. All personnel must take advantage of the engineered solutions such as fume hoods, be aware of proper procedures for performing hazardous operations, and exercise caution in their work behavior. Finally, the provision of personal protective equipment (commonly abbreviated PPE), such as safety goggles, is essential to create a barrier between the worker and hazards, to reduce residual hazards, and to guard against unexpected events.

Table 28.1 summarizes some general guidelines that can be applied to reduce the risk of many types of hazardous situations.

Table 28.1 *RISK REDUCTION STRATEGIES FOR THE LABORATORY*

These general strategies should be applied, preferably in the order listed, to minimize exposure of personnel to laboratory hazards.

1. Reduce the presence of hazards.
   - Eliminate the hazardous material when possible.
   - Substitute a less hazardous equivalent.
2. Design the laboratory to minimize risk to personnel.
   - Remove the employee from the hazard (e.g., using automation).
   - Contain the hazard (e.g., by using a fume hood).
   - Dilute or reduce the volume of the hazard.
3. Apply good laboratory practices at all times.
   - Provide employee training.
   - Practice good housekeeping.
4. Use personal protective equipment whenever necessary.

Adapted from *Occupational Safety and Health in the Age of High Technology*, David L. Goetsch. Prentice Hall, Englewood Cliffs, NJ, p. 383, 1996. Cited from *Tool and Manufacturing Engineers Handbook*, The Society of Manufacturing Engineers, Volume 5, pp. 12–20. Dearborn, MI, 1988,

II. PERSONAL PROTECTION IN THE LABORATORY

A. Clothing

i. GENERAL DRESS

Proper clothing is required whenever entering a laboratory. Even though a lab coat will protect you and your clothing from some hazards, what you wear under a lab coat can be just as important, Figure 28.1. It is important that clothes cover all parts of the body, including legs. For this reason, pants or long skirts are appropriate. Avoid dangling jewelry or ties and long loose hair that can fall into your experiment or get caught in moving equipment. It is also a good idea to refrain from wearing rings, bracelets, or watches in the laboratory. It is easy for chemicals to seep under these items. Any clothing worn in the laboratory should be fire-resistant and easily removable in case of chemical or biological contamination. It should also be appropriate for protection against the types of chemicals that are used in the laboratory (Figure 28.2 offers some guidelines). Many experienced laboratory workers keep a spare change of clothing handy in case of spills, or for wearing after work.

ii. LAB COATS

Lab coats should be worn at all times in the laboratory. Even when you are not using hazardous materials yourself, other people's activities in the laboratory might present unexpected hazards. Lab coats provide a barrier against harmful agents and prevent contamination of street clothes. By soaking up spills, they allow more time to recognize contamination problems and protect yourself. They also protect experiments from contaminants outside the laboratory that might be carried in on clothing.

Many types of lab coats are available and selection should be based on the hazards that are of most concern. For example, front-buttoning coats are more desirable for protection against chemical spills than for biological hazards because they can be removed quickly. All lab coats should be flame-resistant, with cotton frequently providing the best resistance to both chemicals and heat in a comfortable garment, Figure 28.2. For specialized work, such as pouring large quantities of corrosive chemicals, an impermeable apron may be most appropriate.

To be effective, lab coats must fit properly and remain buttoned at all times in the laboratory. Sleeves must be long enough to provide arm protection and should fit the arm fairly snugly to avoid flapping. Rolling up the sleeves provides a holding area for chemical and biological contaminants and is not recommended.

Lab coats should be laundered regularly at your institution, even in the absence of any known contamina-

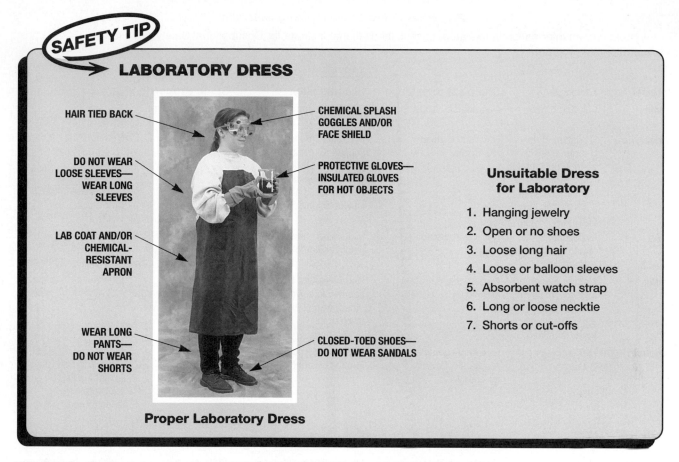

Figure 28.1. Proper Laboratory Dress. (©1998 Flinn Scientific Inc. All Rights Reserved. Reproduced with permission from Flinn Scientific Inc., Batavia, IL, USA)

tion. Never take laboratory clothing home for washing. In case of known contamination, the coat can be decontaminated in the laboratory before washing, or be discarded. Do not wear lab coats used in the laboratory into common areas such as lunch rooms or lavatories.

iii. SHOES

Proper footwear for the laboratory includes shoes with covered toes and nonslip soles, which will protect the feet from broken glass and hazardous spills. Sandals, sneakers, or woven shoes provide little protection. Low heels are generally the most comfortable while standing at the lab bench and also protect against falls.

Laboratory workers may want to consider keeping a special pair of shoes to wear only in the lab. Changing to and from street shoes prevents the tracking of hazardous materials from the lab into the outside environment, and also prevents the introduction of potential contaminants into a cleanroom. (A **cleanroom** *is a special laboratory facility where all contaminating materials and any particulate matter in the air must be limited.*) Numerous studies have shown that shoes worn in a bacterially contaminated environment may carry higher concentrations of bacteria on their soles than the floor

itself. One method to prevent contamination of shoes is the use of disposable shoe covers, which are routinely used for animal surgery and cleanroom operations, and which are removed before exiting the laboratory. Shoe covers will also prevent your shoes from carrying bacterial contamination into public areas and your home.

B. Gloves

i. CHOICE OF GLOVES

The proper use of gloves in the laboratory provides a significant measure of protection against many types of hazards. One of the most obvious benefits is the creation of a barrier between your skin and chemical or biological contamination.

If you look in a catalog, you will find a bewildering variety of gloves available, in many materials and styles. It is important to remember that although every type of glove provides a barrier, none can protect against all types of hazards. Each glove type will provide at least some protection against one or more of the following:

- corrosive or toxic chemicals
- biological contaminants

Properties of Protective Clothing Materials*

For the most current information in this area, consult the National Institute for Occupational Safety and Health (NIOSH) protective equipment program.

Properties

Material	Strength	Chemical resistance	Flammability	Static Properties	Comfort	Uses
Cotton	Fair durability.	Degraded by acids. Binds.	Special treatment for flame.	No static problems.	Comfortable, lightweight.	Lab coats.
Modacrylic	Resistant to rips & tears, but less so then polyamide fibers. Abrasion resistant but less so than nylon or polyester.	Resistant to most chemicals.	In direct flame fabric shrinks to resist flame penetration. Will not melt or drip. Self-extinguishing. Rapidly dissipates when ignition source is removed.	Has antistatic properties.	Comfortable, soft & resilient. Easy to clean & has soil release properties.	Lab coats.
Nylon	Exceptionally strong & abrasion resistant.	Not water absorbent.	Melts when heated. Requires flame retardant.	Static buildup possible. Requres antistatic agent.	Lightweight.	Lab coats.
Plastic	Usually reinforced at points of strain. Will not stick together, peel, crack, or stiffen.	Resistant to corrosive chemicals.	Can be ignited by flammable solvents & others with static discharge.	Accumulates considerable charge of static electricity.	Lightweight.	Aprons, sleeve protectors, boots.
Polyolefin	Resistant to rips & tears.	Excellent chemical resistance. Low binding for chemicals.	High melting point. Flame resistant.	Good static dissociation.	Lightweight. Good permeability. Limited moisture absorbance. May be uncomfortable if perspiring.	Bouffant caps.
Polypropylene	Strong.	Resistant to most chemicals. Oxygen & light sensitive.	Low melting point. Requires flame retardant.	Static buildup requires anti-static agent.	Lightweight.	Aprons.
Rayon	Fairly durable.			Degraded by acids. Binds some chemicals.		Lab coats.

*Based on manufacturer's claims.

Figure 28.2. Properties of Protective Clothing Materials. (Reprinted with permission from *Safety in Academic Chemistry Laboratories*, Appendix II. The American Chemical Society Committee on Chemical Safety, American Chemical Society, Washington, DC, 1995.)

- sharps
- extreme temperatures

Because no glove can provide all the necessary protective features, most laboratories have several glove types available, including:

- thin-walled gloves for dexterity
- heavy rubber gloves for dishwashing
- insulated gloves for handling hot and cold materials
- puncture-resistant gloves for handling animals

The first step in choosing the right glove for a job is deciding what protection is required. Are you trying to protect yourself or your work materials from contamination? Do you need maximum protection from a highly toxic chemical? Table 28.2 provides an overview of the most common glove materials and their advantages and disadvantages.

When choosing gloves for protection against chemicals, always consult the specific glove manufacturer's chemical resistance chart, which is usually supplied with the gloves or found in the supplier's catalog. This will provide information about the properties of specific glove materials. The information provided generally includes:

- **degradation rate**, *which indicates the tendency of a chemical to physically change the properties of a glove on contact*

- **permeation rate**, *which measures the tendency of a chemical to penetrate the glove material*

- **breakthrough rate**, *which indicates the time required for a chemical that is spilled on the outside of a glove to be detected on the inside of the glove.*

The following example indicates how to use this information.

Table 28.2 *PROTECTIVE GLOVE MATERIALS*

Type	Advantages	Disadvantages	Recommended for Protection from:
Natural rubber latex	Low cost, flexibility	Poor protection from oils, organic solvents. May be of poor quality. May trigger allergic reactions.	Bases, alcohols, bloodborne pathogens
Natural rubber blends	Low cost, flexibility, better chemical resistance than natural rubber against some chemicals	Physical properties frequently inferior to natural rubber	Same as natural rubber
Polyvinyl chloride (PVC)	Medium cost, medium chemical resistance	Plasticizers can be stripped, making gloves brittle; may be of poor quality	Strong acids and bases, salts, other aqueous solutions, alcohols
Neoprene	Medium cost, medium chemical resistance, abrasion-resistant	Not as flexible as rubber, can give poor grip	Oxidizing acids, phenol, glycol ethers
Nitrile	Low cost, abrasion-resistant, dexterity, comfortable for longer wear, hypoallergenic	Poor protection from benzene, methylene chloride, trichloroethylene, many ketones	Oils, aliphatic chemicals, xylene, bloodborne pathogens
Butyl	Specialty glove, resistant to polar organics	Expensive, poor protection from hydrocarbons, chlorinated solvents	Gases, aldehydes, glycol ethers, ketones, esters
Polyvinyl alcohol (PVA)	Specialty glove, resists a very broad range of organic solvents	Very expensive, water-sensitive, poor protection from light alcohols	Aliphatics, aromatics, chlorinated solvents, ketones (except acetone), esters, ethers
Fluoroelastomer (Viton)	Specialty glove, resistant to organic solvents	Extremely expensive, poor physical properties, poor protection from some ketones, esters, amines	Carcinogens, aromatic and chlorinated solvents
Norfoil (silver shield)	Specialty glove, excellent chemical resistance, light weight, flexible	Poor fit, easily punctured, poor grip	Use as glove liner, good for emergency use in chemical spills

EXAMPLE: CHOOSING A GLOVE FOR CHEMICAL RESISTANCE

For this example, assume that you are performing an experiment and are concerned about the possibility of acetone spills. After taking precautions to minimize the risk of skin exposure to acetone, you will still want to wear gloves. The two types of gloves you have available are made of either PVC or butyl. You then check the chemical resistance guide from the glove manufacturer and find the following information for acetone:

PVC gloves: Degradation rate >25% in 30 minutes
Permeation and Breakthrough Rate <1 minute

Butyl gloves: Degradation rate—no effect
Permeation and Breakthrough Rate >17 hours

This indicates that PVC gloves are susceptible to chemical breakdown by acetone, and that any acetone spilled on the gloves will contact your skin in less than 1 minute. Butyl, on the other hand, appears to be highly resistant to acetone; the better choice for this situation.

The best option when dealing with highly toxic agents is sometimes to double glove using two different types of gloves (e.g., using chemically resistant gloves under puncture resistant gloves). This provides the benefits of two glove types.

In addition to choosing the proper glove material, you may also have choices in the thickness of glove material. Glove thickness is usually measured in **mils**, *a unit where* 1 *mil* = 0.001 *in.* Thinner gloves generally provide more flexibility but less protection. Therefore, use the thickest glove that does not decrease the necessary dexterity for the task.

Do not allow the wearing of gloves to provide a false sense of security when working with highly toxic materials. Because all gloves are permeable to some extent, assume that your gloves may leak and do not rely on them to protect you when a spill occurs. If you are working with potentially hazardous materials that are unfamiliar to you, do not proceed with your experiments until you confirm that you have proper protection, Case Study 1.

CASE STUDY 1

Proper Gloves Could Save Your Life

The research world was horrified in June 1997 when Dartmouth Professor Karen Wetterhahn died of mercury poisoning, 10 months after what seemed at the time to be a minor incident. Dr. Wetterhahn, who was considered a careful laboratory worker by her colleagues, was following standard precautions and wearing latex gloves when she spilled one or more drops of highly toxic dimethylmercury on her gloved hand. About 3 months later, she began to develop symptoms of mercury poisoning, starting with nausea and proceeding to neurological problems. Tests revealed that she had been exposed to a single dose of mercury far above toxic levels. Later testing showed that latex disposable gloves offered virtually no protection against dimethylmercury, with a breakthrough rate of 15 seconds or less. The incident has led to improved safety information provided to the users of the chemical, along with a recommendation for double gloving with a silver laminate glove under a heavy-duty neoprene or nitrile glove. This knowledge certainly came at a very high price.

ii. Proper Use of Gloves

Glove choice is only the first step in ensuring safety. Gloves must be used properly in order to provide full protection. In fact, improper use of gloves can actually increase the risk of hazard exposure in the laboratory by spreading contamination or sealing it against the skin of the user.

To provide maximum user protection, before using a pair of gloves, check them for any holes or openings. Disposable gloves are usually mass produced, which means that a certain percent (depending on the manufacturer) will be defective. These should be discarded immediately. Any cuts or abrasions on the hands should be bandaged or covered before donning gloves because these are possible entry sites for contamination. Long or ragged fingernails can easily tear many disposable glove materials. Gloves must be long enough to provide wrist protection; if not, use arm protectors for hazardous work. It is best not to wear wristwatches that can trap contamination against the skin.

When working with hazardous materials, it is important to change gloves regularly. Always have a plentiful supply of disposable gloves nearby for quick changes. Change gloves immediately if you think they might have come into contact with hazardous material, and also change them regularly even if you think that they are clean. Remember that all gloves are permeable to some extent, and the longer you wear them, the more likely they are to develop small holes or tears. When removing fitted disposable gloves, use a removal technique that will not spread contamination from the outside of the glove to the skin. A useful technique is illustrated in Figure 28.3. Never remove thin gloves by pulling on the fingertips; these are likely to tear. Wash your hands thoroughly after glove use and in between glove changes if any contamination is suspected.

In addition to user risk, improper glove use can extend hazards to other members of the laboratory. Once the outside of a glove is contaminated with hazardous material, that contamination will be spread to any surface touched by the glove. Always remove at least one glove when opening refrigerator doors, using laboratory equipment with adjustable controls, touching doorknobs or light switches, answering telephones, or any time you leave the laboratory and pass through common areas. Be careful when writing in lab notebooks while using gloves. The pen used should be kept at the lab bench and considered contaminated. If you need to remove gloves to answer the telephone, for example, do not reuse disposable gloves; get a fresh pair to avoid the risk of a tear or skin contamination.

iii. Potential Health Risks From Disposable Gloves

Many laboratory workers report minor problems from working with disposable gloves. This frequently takes the form of general irritation from the gloves. This can be alleviated by changing gloves often and allowing your hands to dry between changes. Use larger gloves to allow more air circulation around the fingers. It may help to apply a barrier cream under your gloves. These are hand lotions (available in scientific supply catalogs) that are designed to prevent irritation from glove materials and prevent drying of hands. Cotton glove liners that absorb perspiration are also available, although these may be unsuitable for highly hazardous work.

A much more serious problem arises when laboratory workers develop allergies to rubber **latex**, a natural product that is commonly found not only in gloves, but in many pieces of laboratory equipment and household items. **Allergies** *are reactions by the body's immune system to exposure to specific chemicals*; in this case, proteins that are found in latex gloves. Although this problem is rare among the general population, latex allergies are becoming more common among health care workers, affecting up to 1 million people. The allergy apparently originates in sensitization to proteins found in natural rubber latex, and appear to be more likely to develop in users of low-quality powdered gloves ("Don't Let Latex Allergy Leave You Short-Handed," Darcy Lewis. *Safety + Health*, pp. 44–48. December, 1996). Glove powder, which is generally USP-approved cornstarch, is believed to make skin adherence or inhalation of latex proteins more likely, thereby acting as a sensitizer. A **sensitizer** *is an agent that can trigger allergies by itself, or cause an individual to develop an allergic reaction to an accompanying chemical*. The resulting reactions can range from minor skin irritation to respiratory shock and even death after exposure.

Current recommendations to avoid the development of latex allergies (assuming that your work includes the use of latex gloves) include:

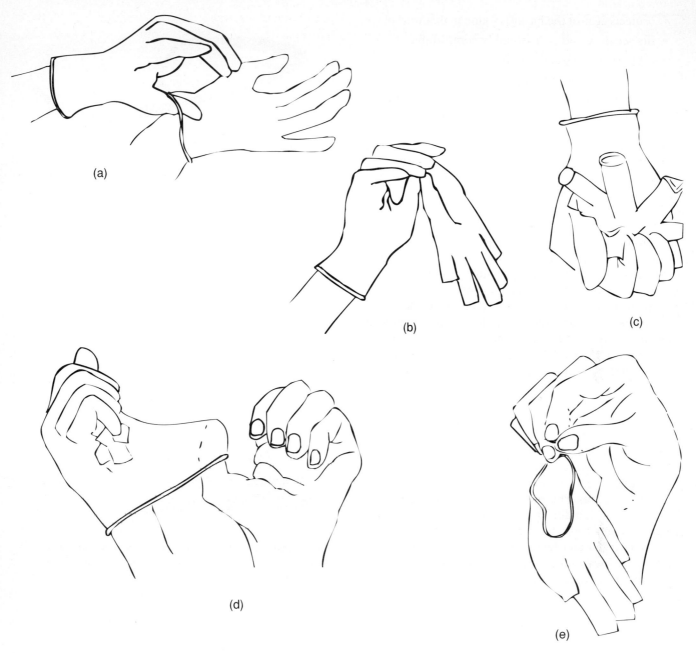

Figure 28.3. Proper Glove Removal Technique. a. Hook a finger on the cuff of one glove, being careful not to contact the skin of the wrist. **b.** Pull the glove from the hand inside out. **c.** Roll up the removed glove in the palm of the remaining gloved hand. **d.** Use an ungloved finger to hook the inside of the glove cuff. **e.** Pull off the glove inside out and discard properly. (Artist: Dana Benedicktus)

- choosing gloves marked as **hypoallergenic**, which indicates that the gloves are less likely to trigger allergic reactions than similar gloves
- choosing powder-free gloves
- checking the glove manufacturer's information sheets for data about the levels of latex protein present in the gloves, and choosing gloves with the lowest levels of latex proteins

Individuals who exhibit any symptoms of latex allergies should avoid gloves and all other items containing latex. Other glove materials should be substituted.

C. Eye Protection

According to OSHA, an estimated 1000 eye injuries per day occur in U.S. workplaces. Of the injured individuals, approximately 60% were not wearing any eye protection, and the remainder were wearing inappropriate devices. The most common type of injuries were the result of small flying particles, and about one in five injuries were caused by chemicals. Laboratory workers and visitors must be provided with adequate means of preventing eye injuries. OSHA regulations require that workplaces provide suitable eye protection gear that:

- protects against the hazards found in that workplace
- fits securely and is reasonably comfortable
- is clean and in good repair

Virtually all protective eyewear sold in scientific catalogs meet strict ANSI standards for impact resistance, but there are many other types of hazards found in biotechnology laboratories, Table 28.3.

There are three general types of eye and face protection devices that should be available in all laboratories, Figure 28.4. First, safety glasses are a minimum precaution against small splashes and minor hazards. Safety glasses must have side protection in order to provide splash protection. For this reason, regular eyeglasses are not considered adequate eye protection in the laboratory. The next step up in eye protection is goggles, which are sealed around the eyes, providing good protection against large splashes or caustic agents. As with gloves, the laboratory may stock different types of goggles. Goggle materials may provide protection against chemicals or against ultraviolet radiation, not necessarily both.

Finally, full face shields should be used when working with materials under vacuum or where there is any threat of explosion. It is essential to wear additional eye protection under a face shield, which is not sealed. Shields made of polycarbonate, coupled with appropriate eye wear, can protect against UV radiation, liquid nitrogen, or chemical splashes.

Individuals who require vision correction should wear goggles over their regular glasses, or request prescription safety glasses. The use of contact lenses in laboratories is much debated in the literature, but if they are worn, they must not be considered eye protection. If you do wear contact lenses in the lab, be sure that coworkers are aware of this, so that in case of an accident they can inform emergency personnel.

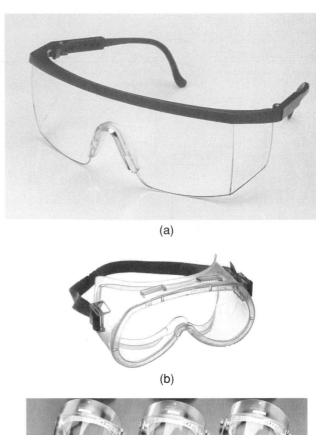

(a)

(b)

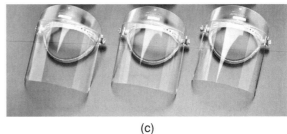

(c)

Figure 28.4. Eye And Face Protection. a. Safety glasses with side protection. **b.** Goggles. **c.** Face shields. (Photos courtesy of Fisher Scientific, Pittsburgh, PA)

Table 28.3 WHY DO YOU NEED EYE PROTECTION?

Eye protection should be worn at all times in the laboratory. This is also true for visitors, who are not aware of hazards. Chemical goggles may be uncomfortable, but modern safety glasses are often sufficient and are designed to be worn comfortably. Appropriate eye and face protection should always be used because laboratories present many hazards to vision, such as:

- danger of explosion or flying particles
- glassware under vacuum
- corrosive liquids such as acids or bases
- cryogenic materials
- liquids that may splash into the eyes
- compressed gases
- blood and other fluids containing infectious materials, that may form aerosols or splash
- radioactive materials
- ultraviolet light and other radiation

Everyone in the laboratory should be aware of the appropriate procedures to follow in case of an accident involving the eyes or face, Table 28.4. Many studies have shown that a quick emergency response can significantly reduce the possibility of permanent damage to vision. An eye wash station must be placed within 25 feet of each laboratory area. Be certain that these stations are easy to locate, that each worker knows how to use them, and that they are tested on a weekly basis, both for function and to wash out any contamination that may have accumulated in the water.

D. Ear Protection

Noise-emitting laboratory equipment, such as centrifuges, often produce sound that is uncomfortable or hazardous. Long-term exposure to high noise levels may cause loss of hearing sensitivity. Disposable or personal ear plugs should be available to reduce ambient noise when necessary. **Sonication devices**, *which are used to*

Table 28.4 *EMERGENCY EYE WASH PROCEDURES*

An important line of defense against eye injuries in the laboratory (or other workplace) is the use of the emergency eye wash station, Figure 28.5.

- Know the location of the eye wash station and how to operate it.
- If you have an accident involving the eyes or face, yell for help <u>AND</u> immediately move to the nearest eye wash station if you can.
- Be prepared to help an accident victim to the eye wash station—seconds count! Never assume that victims can take care of the problem themselves.
- Turn on the water at the station and hold the victim's eyes into the double streams of water. Do not worry about wet clothing.
- Most chemical splashes to the eye and many other injuries involve both eyes, so be sure each eye is properly flushed.
- The eyes should receive a constant stream of warm water for at least 15 minutes. (For this reason personal eye wash bottles are not adequate for an emergency.)
- Help the involved person hold their eyelids open and roll their eyes around to aid in proper flushing.
- Do not attempt to remove any particulate matter from the eye by hand (this includes contact lenses).
- In case of a chemical injury, have someone determine the chemical involved and call for medical help.
- All potentially injured eyes should be examined by a medical professional after the flushing period, even if the person does not feel pain.

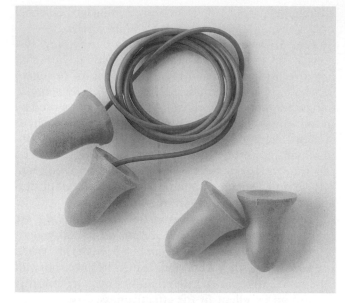

(a)

(b)

Figure 28.6. Two Styles of Ear Protection. a. Ear plugs. **b.** Ear muffs. (Photo courtesy of Fisher Scientific, Pittsburgh, PA)

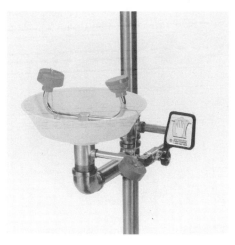

Figure 28.5. An Emergency Eye Wash Station. There are two nozzles to direct streams of water at each eye simultaneously. The hand lever should allow the water to stay on with a single push. (Photo courtesy of Fisher Scientific, Pittsburgh, PA)

disrupt cells with high-frequency sound waves, produce particularly high levels of noise and should only be used with ear protection for everyone in the area. The manufacturer may suggest proper ear protection; otherwise, standard hearing protection devices should worn. Several designs are effective at reducing noise levels, and comfort is a prime consideration when choosing ear protection, Figure 28.6.

E. Masks and Respirators

Laboratories are generally equipped with at least basic respiratory protective equipment. These items may include:

- masks, which filter dirt and large particles from the air, and provide splash protection
- air-purification or filtration respirators
- self-contained breathing systems (in specialized situations)

The majority of lab personnel never use more respiratory protection than a surgical-type mask, which mainly filters dust and larger aerosols from the air and shields the face from minor splashes. These are ideal for animal work because they can remove allergens from the air. They offer little protection against airborne infectious agents, although they may be helpful in preventing lab personnel from touching their noses or faces while gloved.

Air-purification and self-contained breathing systems are examples of respirators, which are devices that improve air quality for the user. **Respirators**, *breathing devices designed to reduce airborne hazards by manipulating the quality of the air supply*, should not be used by untrained personnel. They are required by OSHA under circumstances where toxic fumes or hazardous air contaminants cannot be removed from the environment by other means. Respirators are ineffective if not used properly, so only trained personnel should be placed in situations where respirators are required.

Air purification respirators *work by filtering the room air through canisters of various adsorbent materials that remove specific contaminants from the air*. When used properly, they can significantly reduce, but not eliminate, airborne contaminants. These respirators require an excellent fit for effectiveness, especially the half-mask models. When fit is a problem, full hood respirators should be used, Figure 28.7.

A **self-contained breathing apparatus** *contains its own air supply and is required in situations where the user is exposed to highly toxic gases*. These systems are heavy, uncomfortable, and are limited to the short period of protection provided by the gas supply. They are not routinely used and are not appropriate for use by untrained personnel.

III. PHYSICAL HAZARDS IN THE LABORATORY

A. Introduction

There are numerous physical hazards that are encountered in laboratories, which are busy places with many workers sharing the same space and equipment. Being able to work efficiently and safely in crowded spaces is essential. A few of the most common causes of physical injuries in the laboratory will be discussed here.

B. Glassware and Other Sharp Objects

One of the most common injuries in the lab is a minor cut from broken glass or some other sharp item. Even though most laboratory glassware is formulated to resist breakage, they will still develop weak spots, chips and scratches. Chipped or scratched glassware can break when under pressure, such as during centrifugation, vacuum work, or when filled with liquid. Glassware should be inspected for cracks and chips before being washed and again before laboratory use. Damaged glassware should be discarded or repaired, because it is especially fragile and can easily shatter. Cut glass plates or glass tubing and rods should be sanded or fire polished to dull the edges.

A significant number of injuries from broken glass occur during the cleaning process. Any sink used to col-

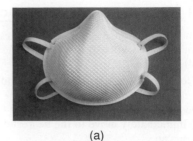

(a)

(b)

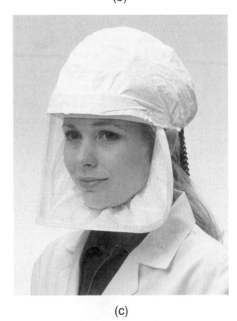

(c)

Figure 28.7.   Breathing Protection. a. Particle mask. **b.** Air-purifying half-mask. **c.** Air purifying hood. (Photos courtesy of Fisher Scientific, Pittsburgh, PA.)

lect dirty glassware should be equipped with a soft mat to prevent breakage. Discard any cracked or chipped glassware after your experiments because these items are especially dangerous for dishwashers. Do not leave broken glass in a sink, where someone else may reach in and be cut. For this and other reasons, never reach into a laboratory sink without hand protection.

Sharps *is a term that describes laboratory items, such as razor blades and needles, that can cause cuts and lacer-*

ations. Razor and scalpel blades should be handled with care, in blade holders whenever possible. Careful covering of the sharp edge with tape when not in use or before discarding can reduce accidental cuts. Never leave these items sitting on lab benches. When using a razorblade to scrape a label from a bottle, be sure to scrape away from yourself, bracing the bottle on a solid surface.

Needles should be handled with caution in the laboratory, especially if they are used with biological materials. It is safest not to reuse needles because many needle punctures are the result of attempts to recap a needle. Gloves will not provide adequate protection against a puncture by a small gauge needle. If it is absolutely necessary to reuse a needle, then recap it by placing the cap on a flat surface and then placing the needle in the cap. Do not hold the needle cap in your hand. All needles, broken glassware, and other sharps should be disposed of in a properly labeled container and not in the general trash, Figure 28.8.

C. Compressed Gases

Cylinders of compressed gas are commonly found in laboratories because certain laboratory instruments require a supply of specific gases. For example, incubators used to contain cultured mammalian cells require a steady flow of carbon dioxide gas. Gas chromatography instruments require gases such as helium and nitrogen. Gases for laboratory use are stored under high pressure in metal tanks, thus allowing a large amount of gas to be stored in a relatively small volume. Although most gases used in biotechnology laboratories are nontoxic and nonflammable, the gas cylinders themselves can be dangerous because they are under high pressure. Accidents involving the rupture of gas cylinders are rare, but quite dramatic, Figure 28.9.

Proper storage and handling of compressed gas cylinders is essential to prevent serious accidents. Cylin-

Figure 28.9.   Compressed Gas Cylinders Are Under Extreme Pressure. This cylinder exploded during storage and was thrown through the roof of the building, landing 330 feet from the explosion site. Note the rupture in the side of the tank and the bent valve stem. (Reprinted with permission from *Laboratory Safety Management and the Assessment of Risk*, Joseph R. Songer in *Laboratory Safety. Principles and Practices*, 2d ed. Diane O. Fleming, John H. Richardson, Jerry J. Tulis, and Donald Vesley, ASM Press, Washington, DC, 1995.)

ders should be handled as explosives. Wear eye protection whenever handling tanks, whether they are full or empty. Cylinders should be stored upright, attached to a wall or other solid surface with a strong canvas strap or chain. Extra or unneeded gases should not be stored in the laboratory. All cylinders should be delivered with a safety cap over the delivery valve. This cap should be in place at all times when the tank is not in use. It protects the cylinder valve from damage, and it also prevents the accidental opening of the release valve. Fuel and oxygen cylinders must be stored at least 20 feet apart, or separated by a fire-proof wall.

When selecting a gas cylinder for use, always read the label. Do not rely on color-coding, which is not uniform among manufacturers. Cylinders should be transported carefully one at a time, using a cart that allows the cylinder to be secured with a strap. Never roll a cylinder on end. These tanks are heavy, and if the cylinder falls, one of the least destructive results would be a broken foot. Cylinders can become high-energy missiles if the gas valve is accidentally knocked off on impact, so be sure that the safety cap is securely fastened when moving tanks.

Compressed gases are dispensed with gas pressure **regulators**, *which decrease and modulate the pressure of the gas leaving the cylinder*, Figure 28.10. A gas storage tank cannot be directly connected to a laboratory incubator or instrument because of the high pressure of the gas. The gas pressure regulator:

* reduces and controls the pressure of the gas flowing from the storage tank to the instrument
* displays the pressure in the storage tank and the pressure of gas flowing to the instrument

It is the responsibility of the instrument operator to adjust the flow of gas to the instrument, to check that the tank is not depleted, and to replace an empty tank when necessary.

Regulators are threaded to fit the valve outlets of cylinders designed for specific gas types. The gas sup-

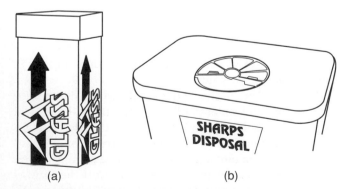

(a)   (b)

Figure 28.8.   Sharps Disposal. a. Broken glass should be collected in a rigid clearly marked container. **b.** Proper needle disposal in a marked receptacle. (From *The Medical Laboratory Assistant*, Jacquelyn R. Marshall. Prentice Hall, Inc., Englewood Cliffs, NJ, p. 110, 1990.)

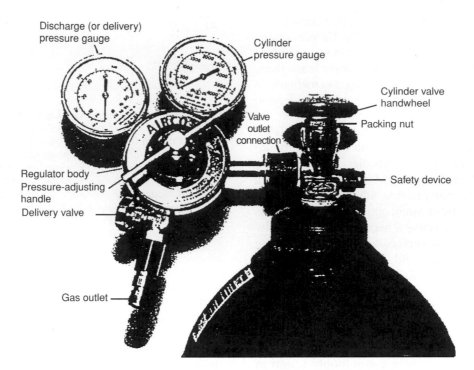

Figure 28.10.   Gas Cylinder Valve and Regulator. The cylinder valve at the top of the high-pressure tank is opened and closed by turning the handwheel. When the valve is open, gas can flow out of the cylinder. The valve outlet connection is where the regulator is attached to the tank. The cylinder pressure gauge measures the pressure in the storage tank. As gas is used, the gauge displays a gradual decrease in tank pressure. Typical values on this gauge read from 0 *lb/in²*, for an empty tank, to 3000 *lb/in²*, for a full tank. The discharge pressure gauge measures the pressure of the gas being delivered to the instrument. Typical values range from 0 to 100 *lb/in²*. The operator controls the flow of gas to the instrument using the pressure-adjusting handle. Turning the handle outward, so that it feels looser, closes the valve and reduces the flow of gas. Turning the knob inward, so that it feels tighter, opens the valve allowing more gas to flow to the instrument. The delivery valve is opened to allow flow from the regulator to the instrument; the gas outlet is where the regulator is attached to the instrument. (Used with permission from *The Foundations of Laboratory Safety. A Guide for the Biomedical Laboratory*, Stephen R. Rayburn. Springer-Verlag, New York, 1990.)

plier can help you make sure that you have the correct type of regulator. Never attempt to adapt a regulator to an outlet that does not fit. Do not use grease on the valve, washers or O-rings, or regulator fittings. Use a nonadjustable wrench to attach the regulator to the valve. Pliers or other inappropriate devices may damage the cylinder or regulator fittings.

Although there are various types of regulators, the most common type in biology laboratories is the **two-stage gas regulator**. The first stage of a two-stage regulator greatly reduces the pressure of gas leaving the cylinder. The second stage is used to "fine tune" the pressure reaching the instrument. The regulator thus controls the pressure reaching the instrument and protects it from a "blast" of pressurized gas. The parts of a regulator and storage tank are illustrated in Figure 28.10.

To use a gas cylinder with a two-stage regulator, make sure that the regulator delivery valve is closed, then slowly open the cylinder valve. Never force a valve open. If hand pressure is insufficient to open a valve, it may be damaged

and dangerous. Once the cylinder valve is open, which is indicated by a steady reading on the cylinder pressure gauge of the regulator, you can open the regulator valve or handle and adjust the gas flow to the instrument.

Before using any gas, learn about the properties and potential hazards of that gas. Because of the high pressure in gas cylinders, never direct a stream of any gas at another person or yourself.

Cylinders can be checked for leaks with a dilute solution of soap and water applied to the fittings. Leaks will appear as bubbles. Special instruments that detect leaks of certain gases are also available and are recommended by instrument manufacturers under some circumstances.

There is a proper procedure to shut off the flow of gas after use that will decrease the possibility of accidents or damage to regulators. Whenever the gas is not in use, close the cylinder valve completely. Leave the regulator open to empty the line to the gas outlet, then close the regulator. This prevents any residual pressure on the regulator that may cause leakage.

When a tank is almost empty, shut off the gas as described earlier, remove the regulator, and replace the cap over the valve. It is best not to drain gas cylinders completely, to avoid contamination of the empty cylinder with air. Mark empty cylinders with the symbol **MT** (empty) to prevent others from trying to guess whether a tank is full or not. Use tape or other large markings. Do not check the fullness of the cylinder by banging on it. Regardless of labels, treat all gas cylinders as though they were full. Table 28.5 provides a summary list of guidelines for safely handling compressed gas cylinders.

D. Heat

Laboratory burns due to hot plates, Bunsen burners, and autoclaves are not as unusual as they should be. Any heat-producing equipment can cause burns or a fire and should be treated with respect. Never leave an uncontrolled heat source, such as a Bunsen burner, unattended. Keep the gas regulated on open flames so that they are visible to any observer. Do not place a device with an open flame in a position on the lab bench or hood where you or anyone else might need to reach past it.

Another source of laboratory burns is hot plates that are left on unintentionally, or which have just been turned off. Most people would not touch a hot-plate surface that contained a boiling container, but they might contact a hot surface when it is empty. When you turn off a hot plate, it is courteous to leave a note indi-

Table 28.5 *GUIDELINES FOR HANDLING COMPRESSED GAS CYLINDERS*

* Be sure all cylinders are clearly labeled.
* Secure cylinders in an upright position at all times, using a strong strap or chain.
* Store only necessary amounts of gases.
* Store gas tanks in a well-ventilated area.
* Never store gas tanks at temperatures above 125°F.
* Do not drop or strike cylinders.
* Use a proper cart with a strap for moving cylinders—do not roll them.
* Do not move a gas tank while the regulator is attached.
* Keep the protective cap on the cylinder head when not in use.
* Use a proper gas pressure regulator.
* Open valves slowly and do not attempt to force a valve open.
* When changing tanks, close the valve between the tank and regulator, and be certain that there is no residual pressure in the gas lines.
* Keep the valves from the tanks closed when the tanks are not in use.
* Never try to repair a cylinder or valve.
* Check cylinders of toxic or flammable gas for leaks on a regular basis.
* Always use safety glasses when handling compressed gases, especially while connecting or disconnecting regulators or supply lines.

cating that the equipment is still hot (also to remove the note when the item has cooled).

Boiling or heated liquids are hazards if not handled properly. Use insulated gloves or tongs to handle hot beakers and flasks. Never heat a sealed container of liquid, because this creates a risk of an explosion and the violent splashing of hot liquid. Be cautious around liquids that may have **superheated**, *which means that they have been heated past their boiling point without the release of the gaseous phase.* Superheated liquids may boil over, sometimes violently, if jarred. This happens regularly with agarose solutions heated in microwave ovens, as well as with liquids that have been autoclaved. It is best to avoid superheating liquids, for example, by heating agarose solutions only enough to melt the gel properly. If you suspect the possibility of superheating, approach the container cautiously, and allow some cooling time before moving the solution.

Most laboratory heat burns are minor. In the case of first-degree burns, which involve reddening of skin, flood the burned area with cold water for 10–15 minutes to reduce the pain and spread of tissue damage. If any blistering occurs or any chemical contamination is involved, the burn should receive prompt medical attention.

E. Fire

Fire is a chemical chain reaction between fuel and oxygen that requires heat or other ignition source, Figure 28.11. Fuels include any flammable materials, such as paper or solvents. A **flammable** substance *is one that will ignite and burn readily in air.*

The most common source of laboratory fires is the ignition of flammable organic liquids and vapors. This hazard will be discussed more fully in the next chapter on chemical safety. The best overall fire-prevention strategy for labs is to limit sources of flammable materials. This and other general strategies for fire prevention are provided in Table 28.6.

In the event of a fire, it is critical to know what to do. Attempting to extinguish a fire (which may not be the best strategy) requires an understanding of fires and how they spread. The fire triangle is important to remember here, because all fires require fuel, heat, and oxygen. If any of these factors are removed, the fire will be extin-

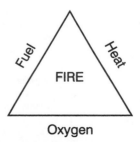

Figure 28.11. The Fire Triangle. Fuel, heat, and oxygen must each be present for fire to occur. (From *Occupational Safety and Health in the Age of High Technology*, David L. Goetsch. Prentice Hall, Englewood Cliffs, NJ, 1996.)

Table 28.6 *FIRE PREVENTION IN THE LABORATORY*

- Store only minimum amounts of flammable materials.
- Keep open solvent containers away from heat sources.
- Store flammable solvents in appropriate containers.
- Use water baths or hot plates in preference to Bunsen burners.
- Limit other ignition sources, such as sparks or static electricity.
- Never leave an open flame unattended.
- Mark heated hot plates.
- Reduce electrical hazards in the laboratory (see Table 28.7)

guished. The use of appropriate fire extinguishers can control small fires if used properly.

There are four types of fires as are summarized in Figure 28.12. Most fires in homes involve Class A combustibles such as paper, cloth, or wood. Class A fires can be extinguished by water or multipurpose dry chemical extinguishers. In the laboratory Class B fires are more likely than Class A. Class B fires usually involve organic solvents and other flammable liquids. Water is more likely to spread than to extinguish this type of fire, which should be smothered with chemical foam or carbon dioxide to remove the oxygen supply from the flames. Class C fires, which involve electrical equipment, also occur in laboratory settings, and water is a poor choice for fighting these fires. Water may increase the possibility of serious electric shock to individuals in the area. Class D fires, which involve combustible metals, are seldom a concern in biotechnology laboratories.

Dry chemical fire extinguishers are common in laboratories. These are highly effective against Class B and C (solvent and electrical) fires. Carbon dioxide extinguishers can be used on the same types of fire, although they tend to have a limited distance of effectiveness. Multipurpose chemical fire extinguishers that can be used for class A + B + C fires are convenient, but most types leave residue behind that requires significant cleanup.

Every laboratory must be equipped with fire extinguishers that are installed close to exits, are easily accessible, and are regularly checked for pressure. Do not try to use these devices, however, unless you have been trained to operate them, and you have practiced recently. Your institutional safety office should be able to provide this training. If you are certain that the fire is contained and minor, and you are qualified to extinguish it, have someone call the fire department before you start. Be sure that all people are out of danger and that they are aware of the fire. Many laboratory flammables will create toxic fumes and heavy smoke as they burn, which makes them difficult to control. Be alert for this possibility. Be sure that you have a clear route of retreat.

When a fire occurs, it is frequently best to evacuate. Everyone working in a laboratory should be familiar with evacuation routes and procedures and it is particularly important that no one stay behind after an evacu-

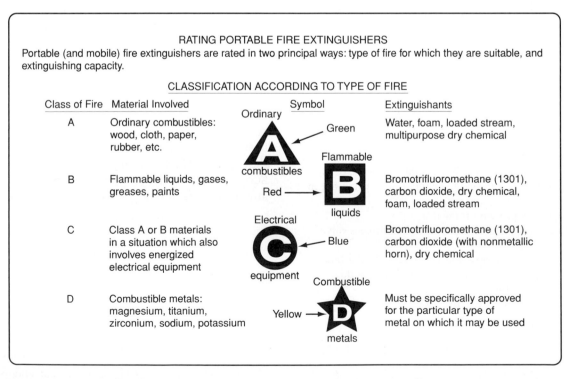

Figure 28.12.   Classification of Fires and Rating Portable Fire Extinguishers. (From *Occupational Safety Management and Engineering*, 4th ed. Willie Hammer, Prentice Hall, Englewood Cliffs, NJ, p. 426, 1989.)

ation is ordered. Close all fire doors and windows when leaving, if it is safe to do so. Evacuation procedures should include a designated meeting place away from the building. Do not attempt to reenter the building until the appropriate emergency director indicates that it is safe to do so.

F. Cold

The most common cold hazards encountered in biotechnology laboratories are low temperature freezers, dry ice baths, and liquid nitrogen. Skin contact with **cryogenic**, or *extremely cold (usually defined as temperatures below −78°C)* substances, or their cooled containers, can result in skin damage similar to heat burns. When handling cryogenic materials, proper hand and eye protection are essential. Gloves should be insulated and loose fitting for quick removal.

Cryogenic liquids are not compatible with ordinary lab glassware or, especially, plasticware. They are usually contained in **Dewar flasks**, *which are heavy multi-walled evacuated metal or glass containers.* While Dewar flasks are similar in appearance to Thermos® bottles, they are not the same. Thermos® bottles are thin-walled and will shatter if exposed to cryogenic temperatures.

Dry ice, *which is frozen carbon dioxide in solid form*, is frequently used to prepare coolant baths for samples. Most of us have handled dry ice casually at Halloween parties, where it is used to create fuming cauldrons of punch; however, dry ice, which can readily burn wet skin, should always be handled with insulated gloves. Most laboratories that use dry ice keep a supply in a special dry ice chest. If the dry ice is provided as chips, do not use a fragile Dewar flask to scoop up the chips. Use a metal scoop to place the chips in an ice bucket. If the dry ice is furnished as a large block, you will need to make chips of the appropriate size. Do not chip dry ice without goggles for eye protection. A less obvious hazard is the danger of induced **hyperventilation** (*increased respiration rate*) or even **asphyxiation** (*interruption of normal breathing*) from breathing high concentrations of carbon dioxide. Do not lean into the dry ice chest for extended periods of time.

To make a coolant bath, a Dewar flask is usually filled about one third full with an appropriate solvent, and then small pieces of dry ice are added to cool the solvent. Acetone was used historically as the solvent, but this is not recommended for safety reasons, Case Study 2. Isopropyl alcohol or various solvent mixtures that are relatively nontoxic and nonflammable are recommended (see *Prudent Practices in the Laboratory. Handling and Disposal of Chemicals*, The National Research Council, National Academy Press, Washington, DC, 1995). It is essential to wear gloves and use a face shield and eye protection while preparing the coolant bath. The dry ice chips must be added slowly, waiting until the solvent stops bubbling before adding

additional dry ice. Once the solvent is suitably cooled, the sample can be introduced to the bath. Always add the item to be cooled slowly. A sudden change in bath temperature can cause the solvent to splatter or overflow the container.

CASE STUDY 2
Acetone–Dry Ice Bath

A graduate student was making a dry ice and acetone cooling bath on a general use lab bench. Being in a hurry, he added the dry ice chips rapidly and ignored the rapid bubbling of the acetone. He did not notice that the acetone was spilling onto the smooth lab bench and spreading flammable vapors along the surface. When the acetone vapors contacted the motor of a working water bath several feet away, the solvent vapors ignited, causing a fire that completely destroyed the water bath and surrounding materials. Of more immediate concern to the worker himself was the fact that the fire, which started at the water bath, traveled along the stream of solvent vapors back to the dry ice bath. The vapors there ignited as well. The student could have been seriously injured if he had not been standing away from the flask and wearing proper eye protection.

Another cryogenic hazard frequently encountered in laboratories is **liquid nitrogen**, *a liquid form of nitrogen that is supplied in compressed gas cylinders at −198°C*. This substance is cold enough to change the physical properties of many materials, and it will cause the equivalent of third degree burns if it contacts skin. It is essential to handle liquid nitrogen with great care. When drawing it from the original cylinder, always wear a face shield and goggles, as well as protective clothing, including heavy leather gloves and leg and foot protection. Because of the high pressure, be certain that you have a firm grip on the dispensing hose, and that the container you are dispensing into is firmly secured in place.

Liquid nitrogen poses both a cryogenic hazard, and a hazard from high-pressure gas release. Liquid nitrogen is formed and stored under pressure and can convert to the gaseous form quickly, with a gas expansion ratio of about 700 volumes of nitrogen gas to 1 volume of liquid nitrogen. Special tanks for sample storage have loose fitting caps with vents. Never place liquid nitrogen into a sealed container. Always wear a face shield and eye protection when removing vials from liquid nitrogen. If liquid nitrogen has leaked into a vial during storage, the pressure buildup from nitrogen gas expansion can cause the vial to explode if warmed quickly.

G. Electricity

Laboratories are full of electrical equipment; therefore, they contain electrical hazards. These issues were discussed in Chapter 14. This is an appropriate place, however, to remind the reader that even small levels of current flowing through the human body can cause harm or even death. An **electrical shock** *is the sudden stimulation of the body by electricity, when the body becomes part of a electrical circuit.*

Make sure that all electrical equipment is in good working condition and properly grounded. When appropriate, GFI circuits should be installed. **GFI**, or **ground fault interrupt circuits**, *are safety circuits designed to shut off electrical flow into the circuit if an unintentional grounding is detected.* These are usually installed around sinks and other water sources.

High-voltage power supplies and electrophoresis equipment should be used with caution. Handle power leads one at a time. Table 28.7 provides some general suggestions for avoiding electrical hazards in the laboratory.

H. Ultraviolet Light

Ultraviolet (UV) light *is a form of nonionizing radiation that makes up the light spectrum between visible light and X-rays.* UV radiation is generally divided into three classes:

- UV-A—315–400 *nm*
- UV-B—280–315 *nm*
- UV-C—180–280 *nm*

UV-A, which is also called "black light," increases skin pigmentation and is not usually generated in laboratories. The major sources of UV in labs are transilluminators for visualizing stained DNA bands in electrophoresis gels and germicidal lamps, which operate in the UV-B and UV-C ranges. Hand-held UV lamps for various purposes usually operate around 254 *nm*. UV-B and UV-C wavelengths damage skin and eyes, and exposure to radiation shorter than 250 *nm* is considered dangerous.

UV light can cause burns and severe damage to the cornea and conjunctiva of the eyes. These tissues absorb

Table 28.7 REDUCING RISK FROM ELECTRICAL DEVICES

- Use only Underwriters Laboratory (UL)-approved electrical equipment.
- Be sure that all equipment is properly grounded.
- Use GFI circuits whenever appropriate.
- Check electrical cords regularly to be sure they are in good condition.
- Keep your hands dry when handling electrical equipment.
- Never use electrical equipment with puddles of liquid underneath.
- Avoid using extension cords.
- Do not perform repairs inside equipment unless you are qualified to do so.
- Keep equipment unplugged when not in use.
- Plug in fume hood equipment outside the hood.

Table 28.8 GUIDELINES FOR MINIMIZING EXPOSURE TO ULTRAVIOLET RADIATION IN THE LABORATORY

- Wear UV-absorbing eye protection in the presence of UV radiation.
- Protect skin surfaces from UV exposure. Remember your wrists.
- Turn on germicidal lamps in hoods or rooms only when the area is not in use.
- Be careful to avoid reflection from hand-held lamps into your eyes.
- Make sure UV light is not directed toward an unsuspecting co-worker.
- Use a UV-absorbing face shield for UV transilluminator work.

the energies of UV light, with eye irritation developing 3–9 hours after exposure. Skin can also be severely burned by UV radiation, as in a bad sunburn. UV radiation can easily be blocked with certain types of glass and plastic. Eye and skin protection is essential when UV exposure is a hazard, so check that the materials in the protection devices used will guard against UV. Table 28.8 suggests strategies for minimizing laboratory exposure to UV radiation.

I. Pressure Hazards

Many common laboratory operations, such as filtration, are carried out using a vacuum. All low-pressure work should be carried out inside a shielded enclosure, such as a fume hood, that will provide a partial safety jacket in case of an implosion. An **implosion** *is the collapsing of a vessel under low pressure compared with the outside atmosphere.* To reduce the risk of implosion, only suitably heavy-walled glassware should be used for vacuum work. All vacuum glassware should be rigorously checked for chips or cracks before use. Although the fume hood sash can be used as a partial body shield, you should still use a face shield and goggles when working with systems under strong (high) vacuum.

High vacuums require a vacuum pump, but moderate vacuums suitable for filtration and other purposes can be provided by water aspirators (see Figure 25.1). A **water aspirator** *creates a vacuum through a side arm to a faucet with flowing water.* All vacuum lines should contain a trap for any liquids or vapors that may be pulled into the vacuum source.

A common high-pressure application in biotechnology laboratories is the use of autoclaves for sterilization of glassware and solutions. All modern autoclaves are equipped with safety features to reduce risks but careful operation is still required. Table 28.9 provides some suggestions for the safe use of autoclaves.

Table 28.9 *SAFE USE OF AUTOCLAVES*

* Learn correct loading procedures to prevent broken glassware.
* Make sure glassware placed in an autoclave is not cracked or chipped.
* Be sure plastics are autoclavable (Teflon® or polypropylene are suitable for this purpose).
* Never autoclave a sealed container.
* Check that the inside is at room pressure before opening the autoclave door.
* Use protection for eyes and hands when opening an autoclave.
* Stand back while opening the autoclave door.
* Release autoclave pressure slowly for liquids to prevent boilover.
* Allow autoclaved liquids to rest for at least 10 minutes before moving the containers to prevent boilover from superheating.

PRACTICE PROBLEMS

1. You are attempting to open the valve on a cylinder of carbon dioxide gas. The valve will not turn despite your best efforts. What should you do?

2. Why are canvas shoes not recommended in the laboratory?

3. Are disposable latex gloves suitable for handling dry ice?

4. How long should an eye be flushed with water after a chemical splash?

DISCUSSION PROBLEMS

1. Think of common situations in the laboratory when gloves are worn. Make a list of possible actions you might take during a procedure and the appropriate times to remove your gloves.

2. Consider how you would handle a fire at work. What safety equipment is available? Where are the exits? Are emergency telephone numbers accessible?

TERMS USED IN THE GENERAL LABORATORY SAFETY CHAPTER

Air Purifying Respirator. A respirator that filters room air through canisters of various materials that remove specific contaminants.

Allergy. Reactions by the body's immune system to exposure to specific agents.

Asphyxiation. Interruption of normal breathing, caused by lack of oxygen or from breathing high concentrations of carbon dioxide or other gases.

Breakthrough Rate (related to gloves). A measurement of the time required for a chemical that is spilled on the outside of a glove to be detected on the inside of the glove.

Cleanroom. A special laboratory facility where contaminating materials and particulate matter in the air must be limited, usually for product protection.

Cryogenic. Refers to extremely cold substances; usually at temperatures below −78°C.

Degradation Rate (related to gloves). A measurement of the tendency of a chemical to change the physical properties of a glove on contact.

Dewar Flask. A heavy multiwalled evacuated metal or glass container, used to hold cryogenic liquids.

Dry Ice. Frozen carbon dioxide.

Electrical Shock. The sudden stimulation of the body by electricity, when the body becomes part of an electrical circuit.

Fire. A chemical chain reaction between fuel and oxygen, which requires heat or other ignition source.

Fire Triangle. The combination of heat, fuel, and oxygen required to start a fire.

Flammable. Any substance that will ignite and burn readily in air.

Ground Fault Interrupt, GFI circuits. Safety circuits designed to shut off electric flow into a circuit if an unintentional grounding is detected; these are usually installed around sinks and other water sources.

Hazard. The equipment, chemicals, and conditions that have a potential to cause harm.

Hyperventilation. An increased respiration rate that can induce dizziness.

Hypoallergenic. A product label indicating that the product is less likely to trigger allergic reactions than similar products; note that this does not mean the product is nonallergenic.

Implosion. The violent collapsing of a vessel with an internal pressure lower than the outside atmosphere.

Latex. A natural rubber product that is commonly found in gloves and many pieces of laboratory equipment and household items.

Liquid Nitrogen. The liquid form of nitrogen, supplied in compressed gas cylinders at −198°C.

Mil (related to gloves). A measurement of glove thickness, where 1 *mil* = 0.001 inch.

MT. Empty; used to mark empty gas cylinders.

Permeation Rate (related to gloves). A measurement of the tendency of a chemical to penetrate glove material.

Personal Protective Equipment, PPE. Specialized clothing or equipment worn for protection against a hazard.

Regulator (related to gas cylinders). A device that attaches to the valve of a compressed gas cylinder and decreases and modulates the pressure of the gas leaving the cylinder.

Respirator. Breathing devices designed to reduce airborne hazards by manipulating the quality of the air supply.

Risk. The probability that a hazard will cause harm.

Self-Contained Breathing Apparatus. A respirator that contains its own air supply and is appropriate for situations where the user is exposed to highly toxic gases.

Sensitizer. An agent that can trigger allergies themselves, or which cause an individual to develop an allergic reaction to an accompanying chemical.

Sharps. A term to describe laboratory items, such as razor blades and needles, that can cause cuts and lacerations.

Sonication Device or Sonicator. Laboratory equipment that disrupts cells with high frequency sound waves.

Sublimate. Change directly from a solid to a gaseous form (as in dry ice).

Superheated. Refers to liquids that have been heated past their boiling point without the release of the gaseous phase. Superheated liquids can boil over, sometimes violently, if jarred.

Two-Stage Gas Regulator. A gas regulator that provides a pair of valves that 1) greatly reduce the pressure of gas leaving the cylinder and 2) provide the operator with the ability to fine-tune the gas pressure reaching its destination.

Ultraviolet (UV) Light. A form of non-ionizing radiation which makes up the light spectrum between visible light and x-rays.

Water Aspirator. A laboratory set-up that creates a vacuum through a side arm to a faucet with flowing water.

REFERENCES

The Foundations of Laboratory Safety. A Guide for the Biomedical Laboratory, Stephen R. Rayburn. Springer-Verlag, New York, 1990. This is an excellent introduction to general laboratory safety principles.

Fisher Safety Products Reference Manual, Fisher Scientific, 1996, and *Flinn Chemical and Biological Catalog Reference Manual*, 1998. These catalogues provide sound information on a variety of safety topics related to laboratories and their hazards.

Occupational Safety and Health in the Age of High Technology, David L. Goetsch. Prentice Hall, Englewood Cliffs, NJ, 1996. This book provides information on safety in the general workplace.

See the following chapters for additional references that emphasize specific aspects of laboratory safety.

ANSWERS TO PRACTICE PROBLEMS

1. You should replace the cylinder cap, mark the cylinder as potentially damaged, and return it to the supplier. Do not attempt to use a lever to force the valve open.

2. Canvas shoes can absorb hazardous chemicals or become contaminated with biological materials. They do not provide a solid protective barrier for the feet.

3. Disposable latex gloves are too thin to provide adequate insulation from cryogenic materials. Thick gloves would be more appropriate.

4. A minimum of 15 minutes.

## CHAPTER 29

Working Safely with Chemicals

I. INTRODUCTION TO CHEMICAL SAFETY

Biotechnology laboratories contain a wide variety of chemicals with different health and environmental hazards. It would be an insurmountable task for an individual to memorize all the hazards for each compound. As discussed in Chapter 2, in 1983 OSHA created the Federal Hazard Communication Standard (HCS), which regulates the use of hazardous materials in industrial workplaces. The purpose of this law is to ensure that chemical hazards in the workplace are identified and their risks evaluated, and that this information is communicated to employees. This law requires the employer to:

- identify all chemicals used in the workplace
- have a Material Safety Data Sheet, MSDS, (see Chapter 2) available for each chemical
- label all chemicals
- provide a written program for handling the chemicals
- train employees on the proper use of all chemicals
- provide complete information to health care professionals in emergencies

Chemicals are an integral part of the working environment of the laboratory. All chemicals can be hazardous (e.g., common table salt can be related to high blood pressure in some cases; sugar can be hazardous to an individual with diabetes). When handling any chemicals, even those with no known hazards, the best strategy is to use good general laboratory practices, Table 29.1.

II. CHEMICAL HAZARDS

A. Introduction to Hazardous Chemicals

Many chemicals found in laboratories can present risks to users who do not practice good laboratory techniques. A chemical is defined as *hazardous* if:

- it has been shown to cause harmful biological effects
- it is flammable, explosive, or highly reactive
- it generates potentially harmful vapors or dust

Table 29.1 *GOOD PRACTICES FOR WORKING WITH LABORATORY CHEMICALS*

- Learn about the physical and toxic properties of a chemical before you start working (e.g., by reading the MSDS).
- Minimize the amount of chemical used.
- Handle, store, and dispose of the chemical according to recommended procedures.
- Work only in well-ventilated areas.
- Label all containers with the chemical name and hazard warnings.
- Wear appropriate personal protective clothing.

In addition, many chemicals are considered potential hazards because of their structural similarities to known hazards, even though no data are available about the chemical itself.

The preceding categories are not mutually exclusive, and many chemicals exhibit more than one of these properties. There are a number of labeling systems that chemical suppliers use to indicate hazards on chemical containers. One of the most widespread is the hazard diamond system developed by the National Fire Protection Association (NFPA). The **hazard diamond system** *rates chemicals according to their fire, reactivity, and general health hazards*, Figure 29.1. This system provides easy-to-read, color-coded information with simple numerical scales. Special hazards, such as radioactivity, are indicated by standard symbols.

It is important to have information about the hazards of laboratory chemicals on hand at all times. One excellent resource for laboratories is *Prudent Practices in the Laboratory. Handling and Disposal of Chemicals* (National Research Council, National Academy Press, Washington, DC, 1995), which provides an overview of the issues related to chemical hazards, provides general information about many laboratory chemicals, and also

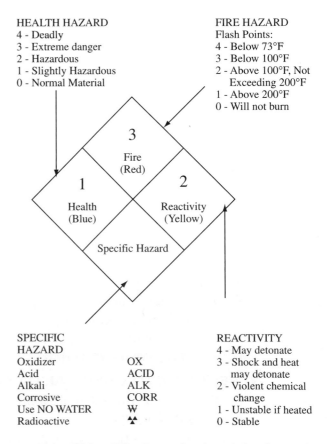

HEALTH HAZARD	FIRE HAZARD
4 - Deadly	Flash Points:
3 - Extreme danger	4 - Below 73°F
2 - Hazardous	3 - Below 100°F
1 - Slightly Hazardous	2 - Above 100°F, Not Exceeding 200°F
0 - Normal Material	1 - Above 200°F
	0 - Will not burn

SPECIFIC HAZARD		REACTIVITY
Oxidizer	OX	4 - May detonate
Acid	ACID	3 - Shock and heat may detonate
Alkali	ALK	2 - Violent chemical change
Corrosive	CORR	1 - Unstable if heated
Use NO WATER	W	0 - Stable
Radioactive	☢	

Figure 29.1. National Fire Protection Association System for Hazard Classification. This example label indicates that a chemical is slightly hazardous to human health, is flammable, and may be explosive. (Reprinted with permission from *Prudent Practices in the Laboratory. Handling and Disposal of Chemicals*, the National Research Council, National Academy Press, Washington, DC, p. 49, 1995.)

suggests appropriate reference materials for more detailed information. The Hazard Communication Standard requires that a set of MSDSs be available for all chemicals used in a particular workplace. Many laboratories keep their own set in a notebook. MSDS information is also readily available on the Internet (see the references at the end of this chapter).

B. Flammable Chemicals

As discussed in the previous chapter, laboratory fires frequently involve flammable chemicals (Class B fires). **Flammable** *refers to materials that are relatively easy to ignite and burn.* Note that the term *inflammable also refers to flammable materials.* (The term *nonflammable is sometimes applied to chemicals that do not readily ignite and burn.*) The most common flammable chemicals found in laboratories are liquid organic **solvents**, *which are chemicals that dissolve other substances.* Many of these liquids are **volatile**, *evaporating quickly at room temperature.* It is the vapor phase of flammable liquids that burns, not the liquid itself.

Flammability is a relative term because all materials can be ignited in the presence of adequate heat. **Flash point** *is the temperature where a chemical produces enough vapor to burn in the presence of an ignition source.* The lower the flash point, the more flammable the compound. Many common laboratory chemicals, such as acetone and hexane, have flash points well below 0°C, Example 1. This means that flammability is always a concern when working with these materials. Once ignition of a flammable solvent occurs, the increasing temperature increases the rate of vaporization, which in turn provides more fuel to burn. Solvent fires can spread out of control very quickly if there is not prompt corrective action. Another concern with flammable liquids is that solvent vapors can diffuse along lab benches or floors, mixing with air and eventually contacting an ignition source. The flames can then follow the invisible vapor trail back to the original container, igniting the larger source of fuel.

EXAMPLE I

Acetone

Acetone is an example of a relatively nontoxic but highly flammable solvent that is commonly found in biotechnology laboratories. It has a flash point of −18°C (NFPA Fire Hazard Rating = 3) and it is susceptible to "flash back" because of its volatility (remember Case Study 2 in Chapter 28, p. 609). The odor of acetone is detectable at concentrations in air well below toxic levels. Acetone can act as an eye and nasal irritant, but these effects are generally short-lived.

There are many systems available for rating the flammability of laboratory chemicals, usually based on

Table 29.2 *SAFE HANDLING OF FLAMMABLE CHEMICALS*

- Standard safety practices as described in Table 29.1, plus:
- Keep all flammable substances away from ignition sources.
- Never heat a flammable chemical with an open flame.
- Remember that solvent vapors can mix with air and diffuse to distant ignition sources.
- Know the appropriate fire prevention methods for the chemical.
- Keep containers tightly closed at all times when not in use.
- Never store incompatible chemicals together (keep acids separate from bases, etc.).
- Keep flammable chemicals away from reactive chemicals.
- Work only in fume hoods or other well-ventilated areas.
- Avoid static electricity discharges when working with flammable substances.
- Never pour flammable substances down a drain or into the trash.
- In case of a spill, deal with any skin contamination before beginning the laboratory decontamination process.

---

their flash points. The NFPA, for example, rates chemicals on a scale of 0–4, with 0 being nonflammable and 4 being extremely flammable. Any chemical with a rating of 3 or 4, which includes those with flash points at room temperature or below, should be treated as a fire hazard. Table 29.2 provides a set of guidelines for handling flammable chemicals to reduce the risk of fire.

Handling chemicals safely depends on knowledge of the properties of individual chemicals. Some powdered metals, for example, are **pyrophoric**, *which means that they will ignite on contact with air.* Other chemicals (e.g., elemental sodium) react violently on contact with water. These types of chemicals must be manipulated only within a controlled environment. Possibly the most dangerous flammable liquid handled regularly in biotechnology laboratories is diethyl ether, with a flash point of −45°C (NFPA Fire Hazard Rating = 4). This chemical should never be placed in proximity to electrical equipment that might produce sparks (e.g., centrifuges or regular refrigerators). It should be handled in fume hoods that will prevent a buildup of flammable vapors.

Fire codes require that 10 gallons or more of flammable liquids be stored in safety cabinets that are designed to minimize the risk of fire or explosion, Figure 29.2. These cabinets do not require venting in the absence of toxic fumes, and the lack of an outside air supply will act as a limiting agent for any fire within the cabinet. If venting is introduced, the cabinet system must be constructed with proper ducts and blowers to vent fumes from the building.

C. Reactive Chemicals

Most laboratory chemicals are reactive to some extent, but those that pose the greatest hazard to laboratory workers are those that undergo violent chemical reactions. Reactive chemicals can be categorized as:

Figure 29.2. Storage Cabinet for Flammable Chemicals. (From *Occupational Safety and Health in the Age of High Technology*, David L. Goetsch, Prentice Hall, Englewood Cliffs, NJ, p. 262, 1996.)

- those that participate in exothermic or gas-generating reactions
- unstable chemicals
- those that form peroxides and other oxidizing agents
- incompatible chemical mixtures

Reactive chemicals may produce an **explosion**, *a sudden release of large amounts of energy and gas within a confined area*. A fire may result, depending on the elements involved. General guidelines for handling reactive chemicals are provided in Table 29.3.

In a laboratory, explosions are likely to occur as a result of combining reactive chemicals in a sealed container. Sealed containers can also explode if used for any **exothermic** (*heat-producing*) or gas-forming chemical reaction, Case Study 1. This happens with some regularity in mixed chemical waste containers (which are not recommended). Glass bottles should never be used to contain potentially reactive chemical mixtures.

Table 29.3 *Safe Handling of Reactive Chemicals*

- Standard safety practices as described in Table 29.1, plus:
- Know the reactive properties of the chemical (read the MSDS).
- Never mix unknown chemicals together.
- Label containers of reactive chemicals carefully.
- Store only compatible chemicals in the same area.
- Store oxidizing chemicals away from flammable materials.

CASE STUDY 1
Mixed Waste Containers

A graduate student in one of the author's labs was working late and in a hurry. He needed to dispose of some liquid waste and poured it into a glass bottle, designated for organic wastes, inside a fume hood. Noticing that the bottle was full, he replaced the cap, pulled the hood sash 80% closed, and left the room. Within three minutes the bottle, which contained acid as well as organic waste, exploded from a gas-producing chemical reaction. The sash of the fume hood, which was constructed of shatter-proof material, contained much of the flying glass and liquid within the hood, but was sufficiently damaged to require replacement. Several large pieces of chemically-contaminated glass flew under the hood sash, through the open lab door, across the hallway, and into the facing lab—a distance of almost 50 feet. It was fortunate that no one was present in the path of the debris. Cleanup and decontamination took several hours. After the incident, only plastic bottles, with the caps removed, were used as waste containers.

Some chemicals are unstable and are susceptible to chemical breakdown with time, Example 2. This is one reason why all chemicals should be labeled with the date of receipt. In some cases, the breakdown products are shock-sensitive. The most commonly encountered examples are picric acid, dinitrophenol, and compounds that break down into organic peroxides.

A **peroxide former** *is a chemical that produces peroxides or hydroperoxides with age or air contact*. These chemicals, which include a variety of aldehydes, ethers, and ketones, are highly hazardous. They are flammable and may explode on exposure to heat or shock. It is important to refer to the MSDS for information about the hazards of specific chemicals.

EXAMPLE 2
DEPC (Diethyl Pyrocarbonate)

Diethyl pyrocarbonate (DEPC) is a toxic chemical commonly found in biotechnology laboratories. It is used to treat solutions and glassware when isolating RNA because it is an effective agent for inactivating RNA-digesting enzymes. It is also a suspected carcinogen, and should be handled only with gloves. In addition, DEPC breaks down to carbon dioxide gas and ethanol when exposed to moist air. If this decomposition takes place in a sealed bottle, pressure can build to explosive levels. DEPC should be stored under dry conditions in a refrigerator, within a desiccator. If possible, store the bottle in the original metal container to act as an explosion barrier. Always allow refrigerated DEPC to equilibrate to room temperature before opening the bottle. Because DEPC is an explosion hazard, always use goggles and a shield when handling a stock bottle.

Table 29.4 EXAMPLES OF INCOMPATIBLE CHEMICALS

Chemical	Avoid contact with:
Acetic Acid	Aldehydes, bases, carbonates, ethylene glycol, hydroxides, metals, oxidizers, perchloric acid, peroxides, permanganates, phosphates, xylene
Acetone	Acids, amines, oxidizers, plastics
Chlorates	Any finely divided combustible or organic materials
Flammable liquids	Ammonium nitrate, chromic acid, halogens, hydrogen peroxide, nitric acid, oxidizers
Hydrocarbons	Bromine, chlorine, chromic acid, fluorine, sodium peroxide
Hydrogen peroxide	Acetone, acetic acid, alcohols, carboxylic acid, combustible materials, most metals or their salts, nitric acid, phosphorus, sodium, sulfuric acid
Hydrogen sulfide	Metals, nitric acid, oxidizers, sodium
Iodine	Acetylene, ammonia (aqueous or anhydrous), hydrogen gas
Nitric acid	Acetic acid, aniline, chromic acid, flammable gases and liquids, hydrogen sulfide
Perchloric acid	Acetic anhydride, alcohol, paper, sulfuric acid, wood
Potassium permanganate	Benzaldehyde, ethylene glycol, glycerin, sulfuric acid
Sulfuric acid	Potassium chlorate, potassium permanganate (or other compounds with light metals, such as sodium or lithium)

Peroxide formation is generally limited to liquids that have evaporated and undergone autooxidation. Diethyl ether and tetrahydrofuran (THF) are the most likely peroxide formers to be found in biotechnology laboratories. These chemicals should be stored away from light and heat in carefully sealed containers. Any containers that show signs of evaporation, especially older containers of ether, should not be handled. Contact your institutional safety office for proper disposal instructions.

In addition to the preceding hazards, certain combinations of chemicals can undergo violent reactions that result in explosions or release of highly toxic gases or other products. The best protection against this phenomenon is to follow standard laboratory procedures when mixing chemicals. Never combine chemicals without an established set of instructions or without researching the reactive properties of the chemicals involved. Know the hazards associated with the chemicals with which you work. Table 29.4 provides a few examples of incompatible chemical mixtures.

D. Corrosive Chemicals

Corrosive chemicals *are those that can destroy tissue and equipment on contact.* Acids and bases are the most common corrosives found in biotechnology laboratories. Both will cause chemical burns and tissue damage on contact with skin or eyes. An even greater danger is that of inhalation of corrosive vapors, which can irritate or burn mucous membranes and potentially cause serious lung damage. Because of this inhalation hazard, strong solutions of corrosives should always be used in a fume hood. The extent of potential damage caused by a corrosive chemical will depend upon the nature of the chemical and the amount and length of exposure. Table 29.5 offers some guidelines for safe handling of corrosives.

Table 29.5 SAFE HANDLING OF CORROSIVE CHEMICALS

- Standard safety practices as described in Table 29.1, plus:
- Always work with corrosive chemicals in a fume hood to avoid respiratory irritation.
- Add acid to water, not the reverse.
- Perform neutralization reactions between acids and bases slowly to minimize gas and heat generation.
- Be sure that protective gear is appropriate to the chemical in use.
- Store acids and bases in separate areas.
- Do not work with hydrofluoric acid without specific training and precautions.

E. Toxic Chemicals

i. ACUTE VERSUS CHRONIC TOXICITY

"What thing is not a poison? All things are poison and nothing is without poison. It is the dose only that makes a thing not a poison."
Paracelsus, Sixteenth Century

Toxicity *is the term used to describe the capacity of a chemical to act as a poison, creating biological harm to an organism.* A toxic material may alter the function of essential organs of the body; the heart, lungs, liver, nervous system, or kidneys. The level of toxic hazard is dependent on the nature of the chemical itself, the concentration and length of exposure, the health of the individual, and the speed and success of corrective measures. As recognized by Paracelsus, all chemicals can be toxic at higher dose levels in some individuals, Figure 29.3. Many laboratory chemicals, however, have well-documented toxic effects at low doses. The risk of harmful effects from these substances can be reduced or eliminated by proper laboratory technique and simple precautions.

Toxicity Class	Probable Oral Lethal Dose (for a 70 *kg* Adult Human)	Examples in Toxicity Class	Rat Oral LD$_{50}$ of Example
Almost Nontoxic	More than 1 liter	Sucrose	29.7 *g/kg*
Slightly toxic	Between 0.2 and 1 liter	Ethanol	14 *g/kg*
Moderately toxic	Between 5 *ml* and 200 *ml*	Sodium Chloride	3 *g/kg*
Very toxic	Between 100 μl and 1 *ml*	Sodium Cyanide	6.4 *mg/kg*
Extremely toxic	Less than 20 μl	Strychnine	2.5 *mg/kg*

Figure 29.3. Classes of Relative Toxicity.

Toxic materials can act in a variety of ways. An **acute toxic agent** *causes damage in a short period of time, and a single exposure may be adequate to cause harmful effects.* A fast-acting poison like hydrogen cyanide is an example of an acute toxic material. **Chronic toxic agents** *have cumulative effects or may accumulate in the body with multiple small exposures.* These small doses of the toxic material may not produce an immediate effect, but instead produce injury over time. Lead poisoning in children is an example where cumulative exposure to very low levels of a toxic material can have long-term, serious health effects. Many chemicals can act as both acute and chronic toxic materials, Example 3. In some cases, exposure to mixtures of chemicals may increase their toxic effects.

EXAMPLE 3

Toxicity of Organic Solvents

Nonchlorinated flammable organic solvents, such as benzene, toluene, and xylene, can exert a wide range of toxic effects. As acutely toxic materials, they can cause headaches and dizziness after inhalation. They are respiratory, skin, and eye irritants. Because of their solvent properties, contact with skin will cause drying, leading to **dermatitis**, *a painful reddening and inflammation of the skin.* Some organic solvents, such as phenol, can act as corrosives as well. Organic solvents also may cause chronic toxicity, with long-term exposure leading to liver, lung, and kidney damage. Chronic exposure to benzene is linked to human leukemia. Virtually all of these chemicals have characteristic odors, and some lab workers mistakenly use these to indicate exposure. The vapor levels necessary for detection by smell, however, have no relationship to toxicity levels in most cases, so this is not an adequate warning system. When working with organic solvents, always use a fume hood and refer to the MSDS for hazard information.

Exposure to toxic chemicals can result in a wide variety of human health problems. The next few sections will discuss some of the types of toxic materials likely to be encountered in the laboratory. Table 29.6 provides some general suggestions for safe handling of toxic chemicals.

Table 29.6 SAFE HANDLING OF TOXIC CHEMICALS

- Standard safety practices as described in Table 29.1, plus:
- Be aware of both the acute and chronic effects of a known toxic chemical.
- Treat all chemicals as toxic unless otherwise informed.
- Be certain that the gloves you are wearing provide appropriate protection from the chemicals being used.
- Do not rely on odors to warn you of exposure to chemicals.
- Minimize exposures to any toxic substance that can accumulate in the body.
- Maximize precautions when working with any chemical known to be mutagenic or carcinogenic.
- Be alert to symptoms of toxicity or sensitization to chemicals.
- Both men and women should consider reproductive hazards in the workplace. Handle toxic chemicals carefully if you plan on starting or increasing a family, or suspect that you might be pregnant.

ii. IRRITANTS, SENSITIZERS, AND ALLERGENS

Many laboratory chemicals are **irritants**, *which produce unpleasant or painful reactions when they contact the human body.* Skin irritation frequently takes the form of redness, itching, or dermatitis. Chemical irritation can be more serious when the point of contact is the eyes or respiratory tract. It is always a good idea to minimize body contact with all chemicals, even those that seem fairly innocuous. Wear proper protective clothing, including gloves and safety glasses, whenever handling chemicals. Some chemicals with acute irritating effects can also exert chronic effects, Example 4.

EXAMPLE 4

Formaldehyde

Everyone who has taken an anatomy class has probably encountered formaldehyde. Formaldehyde gas is a toxic and highly flammable substance. Diluted formalin, a 37% solution of formaldehyde, has been used historically to preserve tissue specimens for dissection. Formalin contains 7–15% methanol to stabilize the formaldehyde. Many suppliers of preserved specimens are now using propylene glycol-based or other types of tissue preservatives because of the formidable list of the toxic properties documented for formaldehyde, including:

- respiratory and eye irritation

- slow developing burns to eyes and skin

- skin sensitization

- potential carcinogenicity

Allergies *are reactions by the body's immune system to exposure to a specific chemical,* or **allergen**. Allergies can manifest themselves as skin or respiratory reactions, depending on the route of exposure to the allergen. Just as individuals will have a wide range of sensitivities to environmental allergens, such as dust and pollens, they will also have a wide range of sensitivities and symptoms to chemical allergens and a range of symptoms associated with an allergic response. Reactions can range from a mild rash to nasal congestion to **anaphylactic shock**, *a sudden life-threatening reaction to allergen exposure.*

Before allergies occur, an individual must be sensitized to the allergen. Some chemicals can act as **sensitizers**, *and may trigger allergies themselves, or cause an individual to develop an allergic reaction to an accompanying chemical.* Dimethylsulfoxide (DMSO) is an example of a sensitizing agent that penetrates skin and carries other chemicals with it, Case Study 2. Limiting contact with potential allergens and sensitizers, working under well-ventilated conditions, and wearing proper protective clothing are wise precautions to reduce exposure to these agents.

CASE STUDY 2
Chemical Sensitization

As a beginning graduate student, one of the authors did extensive tissue culture work. This work was performed in an appropriate biological safety cabinet while wearing gloves. After about 4 months, she developed a rash on her left wrist, exactly corresponding to the shape and location of her wristwatch. Because she had worn this same watch almost continuously for more than 6 years, it seemed unlikely that she had suddenly developed an allergy to it. After examining the rash, a dermatologist questioned the author about her work. As soon as tissue culture was mentioned, the doctor asked if she worked with cells treated with DMSO (dimethylsulfoxide). When confirmed, he pointed out that DMSO is a powerful penetrating and sensitizing agent, which probably had triggered a reaction to nickel found in the watchband. Even though the author had never spilled DMSO on her wrist, enough DMSO vapors had reached above the glove line to produce the sensitization reaction. After this, the author wore longer gloves while handling DMSO, and did not wear a watch in the laboratory. The problem disappeared.

---

iii. NEUROTOXINS

Neurotoxins *are compounds that can cause damage to the central nervous system.* In many cases, the neurological effects, which may include loss of coordination and slurred speech, may develop slowly after long-term exposure to small doses of the toxic agent. Organometallic compounds, like methylmercury, act as potent neurotoxins. Acrylamide, which is routinely used to prepare gels for protein separations, is a common neurotoxin in many biotechnology laboratories, Example 5.

EXAMPLE 5
Acrylamide

Acrylamide is used in the preparation of polyacrylamide gels for protein separation. In its polymerized form, it is considered harmless, and is sold in gardening stores as a water-absorbing agent to be mixed with potting soil for plants. Acrylamide in its unpolymerized form, however, is a potent neurotoxin. It can have both acute and chronic effects. Direct contact can result in eye burns and skin rashes. Symptoms of chronic overexposure include dizziness, slurred speech, and numbness of extremities. Acrylamide is also a suspected human carcinogen. This is a chemical that should always be handled with great respect for its toxic properties. Always wear a lab coat, gloves, and dust mask when handling the solid form. Acrylamide should only be weighed in a designated balance within a fume hood. Given the documented risk from respiration of acrylamide powder, laboratories should consider the extra expense of purchasing premade acrylamide solutions. Although polymerized polyacrylamide gels are nontoxic, do not handle these gels without gloves. There may be residual unpolymerized acrylamide present that can be absorbed through the skin.

---

iv. MUTAGENS AND CARCINOGENS

Mutagens *are compounds that affect the genetic material of a cell.* They cause alterations in DNA that will be inherited by offspring cells. Mutagens are considered chronic toxic agents because the induced damage may not be apparent for years. Mutational damage is cumulative. Because mutagens can affect the genetic material of cells, they may also act as cancer-causing agents. Although the relationship between carcinogenicity and mutagenicity has not been clearly demonstrated for many chemicals, it is prudent to assume that any mutagen is potentially carcinogenic as well. Mutagenicity data are generally derived from animal studies, so the exact human risk is difficult to determine. The MSDS can provide a summary of the available data for individual chemicals. Some of the more common laboratory chemicals that are known to be animal mutagens are shown in Table 29.7.

One of the most commonly used chemicals in biotechnology laboratories is ethidium bromide. **Ethidium bromide (EtBr)** *is a fluorescent dye used to visualize*

Table 29.7   EXAMPLES OF KNOWN ANIMAL MUTAGENS

The following are examples of common laboratory chemicals that have been shown to be mutagenic in animal models and are presumed to be human mutagens as well:

- Acridine Orange
- Colchicine
- Ethidium Bromide
- Formaldehyde
- Hydroquinone
- Osmium Tetraoxide
- Potassium Permanganate
- Silver Nitrate
- Sodium Azide
- Sodium Nitrate
- Toluene

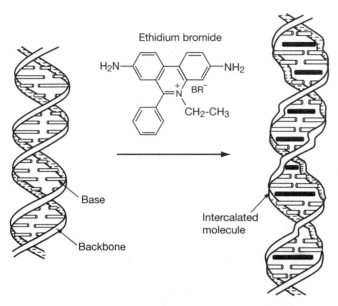

Figure 29.4.   Interaction of Ethidium Bromide with DNA.

nucleic acids in agarose gels. It acts by inserting itself into DNA molecules, Figure 29.4. This provides the mechanism for the strong mutagenicity EtBr demonstrates in animal models, Example 6.

EXAMPLE 6

Ethidium Bromide

Ethidium bromide (EtBr) is a known mutagen that is commonly used in biotechnology laboratories to visualize nucleic acids under ultraviolet light. Most laboratories treat solid EtBr and concentrated stock solutions with great respect. However, workers commonly add low concentrations of EtBr to agarose gels and electrophoresis running buffers and these dilute solutions should still be considered potential health hazards. The traditional method of decontaminating spills or solutions containing EtBr with bleach has been demonstrated to be ineffective and capable of creating breakdown products that may be more harmful than EtBr itself ("Ethidium Bromide: Destruction and Decontamination of Solutions," Lunn, G., Sansone, E. *Analytical Biochemistry*, 162:453–458, 1987). Several studies indicate that laboratories should follow more effective

procedures using special detoxification resins or a deamination procedure with sodium nitrite and hypophosphorous acid. Complete details of these procedures can currently be found on the Internet at http://www.ehs.ucdavis.edu/sftynet/chem/sn-53.html. The decontamination process can be monitored with a handheld ultraviolet light. All items that come into contact with EtBr, such as gloves, spatulas, or paper towels, should be properly decontaminated or treated as hazardous waste.

Carcinogens, or **cancer causing agents**, *can initiate and promote the development of malignant growth in tissues.* They act as chronic toxic materials, and exposure to these compounds is often unrecognized for years. Known and suspected laboratory carcinogens are identified and listed by OSHA, and other agencies including the **International Agency for Research on Cancer (IARC)**, *an agency that determines the relative cancer hazard of materials.* Carcinogenic substances need to be handled with great care. A list of types of chemicals known to have carcinogenic effects in animals is shown in Table 29.8. Information about carcinogenicity is included in the MSDS for specific chemicals.

V. EMBRYOTOXINS AND TERATOGENS

Embryotoxins *are compounds known to be especially toxic to the developing fetus.* For example, organic mercury compounds, some lead compounds, and formamide are known to be embryotoxins. **Teratogens** *are a type of embryotoxin that cause fetal malformation.* They are known to interfere with normal embryonic development, but they do not cause direct harm to the mother. The greatest susceptibility of the fetus to these compounds is

Table 29.8   EXAMPLES OF KNOWN ANIMAL CARCINOGENS

Carcinogen Type	Example
Acylating agents	β-Propiolactone
Alkylating agents	Acrolein Ethylene oxide Ethyl methanesulfonate
Aromatic amines	Benzidine
Aromatic hydrocarbons	Benzene, Benzo[a]pyrene
Hydrazines	Hydrazine
Miscellaneous inorganic compounds	Arsenic and certain arsenic compounds
Miscellaneous organic compounds	Formaldehyde (gas)
Natural products	Aflatoxins
N-Nitroso compounds	*N*-nitroso-*N*-alkylureas *N*-nitrosodimethylamine
Organohalogen compounds	Carbon tetrachloride Vinyl chloride

generally in the first 12 weeks of pregnancy, sometimes when the woman is unaware of the pregnancy.

Pregnant women and women of child-bearing age should be especially cautious when working with toxic chemicals. Substances that enter the bloodstream of the mother may be able to pass the placental barrier to the fetus. Always discuss an intended or actual pregnancy with your laboratory supervisor. It may be safest to request alternate duties during at least the first trimester of pregnancy.

Table 29.9 provides a summary of the hazards of some common biotechnology laboratory chemicals.

III. ROUTES OF CHEMICAL EXPOSURE

A. Introduction to Toxicity Measurements

Toxic chemicals can enter the body by four main routes of exposure:

- inhalation
- skin and eye contact
- ingestion
- injection

Types and levels of toxic injuries depend on the exposure route. Some chemicals are especially dangerous when inhaled; others only when ingested. Of the four routes of toxicity, inhalation and skin absorption are most likely in a laboratory workplace.

Every chemical has its own level of toxicity; some are more poisonous than others, Figure 29.3. Many chemicals are found in the foods we eat and are considered relatively nontoxic. Some substances, such as hand creams, are good for the skin, but should not be ingested. Many compounds, such as over-the-counter pain relievers, are considered safe at one level, but toxic when the dosage is increased. Other chemicals will cause quick death even in small doses.

One way to assess the toxic effects of different chemicals is to measure the amount of the compound that will cause a reaction or death in animals. Test animals receive known doses of the chemical and the results are measured. One common measure of chemical toxicity is

Table 29.9 *SUMMARY OF COMMON CHEMICAL HAZARDS IN BIOTECHNOLOGY LABORATORIES**

Chemical	Flammable	Corrosive	Reactive	Low-Dose Toxicity	Major Route of Exposure
Acetic acid		yes	yes	acute	any
Acetone	yes		yes	chronic	any
Acetonitrile	yes		yes	acute	any
Acrylamide				acute, chronic	skin, inhalation
Ammonium hydroxide		yes	yes	acute	skin, inhalation
Benzene	yes		yes	chronic, carcinogen	inhalation, skin
Chloroform		yes	yes	possible carcinogen, teratogen	inhalation
Diethyl ether	yes		yes	acute	inhalation
Dimethyl sulfoxide			yes	sensitizer	skin contact with vapors
Ethanol	yes		yes	acute, chronic	ingestion
Ethidium bromide				mutagen	skin
Ethyl acetate	yes		yes	acute, chronic	any
Formaldehyde	yes		yes	acute, sensitizer, carcinogen	any
Hexane	yes		yes	chronic, neurotoxin	inhalation
Hydrochloric acid		yes	yes	acute	any
Hydrogen peroxide			yes	irritant	skin
Mercury				acute, chronic, neurotoxin	inhalation, skin
Methanol	yes		yes	acute	ingestion
Nitric acid		yes	yes	acute	skin
Phenol		yes	yes	acute, chronic, neurotoxin, hepatotoxin	skin, inhalation
Pyridine	yes			chronic, hepatotoxin	inhalation, skin
Sodium hydroxide		yes	yes	acute	skin, inhalation
Sulfuric acid		yes	yes	acute	skin, inhalation
Toluene	yes		yes	chronic, irritant	skin, inhalation

This table was prepared from information provided in *Prudent Practices in the Laboratory. Handling and Disposal of Chemicals,* the National Research Council, National Academy Press, Washington, DC, 1995.
*This list provides a summary of the most significant hazards of some of the chemicals commonly encountered in biotechnology laboratories. Many have additional hazardous properties that are not listed here. Always consult the MSDS and other reference sources when working with an unfamiliar chemical.

the LD_{50} level. **LD_{50} (Lethal Dose—50%)** *is the amount of a chemical that will cause death in 50% of test animals.* Animals of different species are used. In order to compare these doses, the amount needed to kill 50% of the animals is recorded in amount (grams or milligrams) per kilogram of the animal's body weight. A larger absolute dose is generally necessary to kill a larger animal, although relative values may be similar (see the Example Problem). LD_{50} studies obviously cannot be performed with humans, but the information from animal studies can be used to indicate the relative toxicity of certain compounds. Animal toxicity is usually, but not always, a good predictor of human toxicity.

EXAMPLE PROBLEM

LD_{50} studies can be performed in a variety of animal species, with the results suggesting relative toxicity levels in humans. For example, consider a chemical that is tested for toxicity in rats and mice. The dose of chemical that kills 50% of the rats tested is 48 mg. The dose of chemical that kills 50% of the mice tested is 5 *mg*. How would you convert these data to LD_{50} values, and how might you extrapolate the data to humans?

ANSWER

The key additional information needed for calculating LD_{50} values is the mean body weights of the rats and mice tested. In this scenario, assume that the mean body weight of the rats is about 400 g and the mice, about 39 g. The LD_{50} values would be calculated as follows:

For rats—48 *mg* of chemical divided by 400 g per rat = 0.12 *mg* chemical per gram body weight or 120 *mg/kg* body weight = the LD_{50} in rats

For mice—5 *mg* of chemical divided by 39 g per rat = 0.13 *mg* chemical per gram body weight or 130 *mg/kg body weight* = the LD_{50} in mice

In order to reach a solid conclusion about human toxicity, you would like to have human data. In its absence, only a hypothetical calculation can be made. Because the preceding LD_{50} values are similar, it would be reasonable to predict that the LD_{50} value for humans could be comparable. Based on this assumption, for a 70 *kg* (150 *lb*) human, a lethal dose of the chemical is in the range of 8.4–9.1 grams. This would be considered a mildly toxic chemical.

LD_{50} values depend on the route of chemical exposure. Ingestion, or oral, LD_{50} values may differ significantly from skin exposure LD_{50} values. Although toxicity is a relative term, certain LD_{50} values are generally accepted as indications of high toxicity. These are:

- LD_{50} <500 *mg/kg* body weight by ingestion or injection
- LD_{50} <1000 *mg/kg* body weight by skin contact

Toxicity for inhaled chemicals is measured in a similar manner, although the results are expressed differently. With toxic chemicals that are likely to be inhaled, the **LC_{50} (Lethal Concentration—50%)** is usually provided. The **LC_{50}** *is the chemical concentration in air that will kill 50% of exposed animals.* LC_{50} is usually expressed as **ppm** *(parts per million in air)* or **mg/m^3** *(milligrams per cubic meter of air).* The approximate level indicating high toxicity is an LC_{50} <2000 *ppm* inhalation.

B. Inhalation

Toxic vapors, gases, and dusts can all enter the body through inhalation. The respiratory system is wonderfully designed with a large surface area to deliver oxygen and other airborne chemicals to the blood. This unfortunately means that hazardous chemicals in the air can easily penetrate the mucous membranes of the nose, throat, and lungs and be delivered through the bloodstream for distribution to all the tissues of the body. Reactions to inhaled chemicals can range from mild discomfort to burning sensations in the throat to asphyxiation.

For example, exposure to small amounts of the odorless gas carbon monoxide can be lethal. Exposure can cause headache, nausea, and eventually unconsciousness. Only 0.03% in air (300 *ppm*) can be fatal within 30 minutes. OSHA has set a maximum allowable limit of 50 *ppm* of carbon monoxide in air for an 8-hour period (TWA, as discussed later in this chapter).

For most volatile chemicals, inhalation exposure levels cannot be estimated by odor or other obvious characteristics. It is important, therefore, to limit exposure to the vapors of chemicals through the use of fume hoods and other methods. OSHA and other agencies have created regulations and guidelines to indicate safe exposure levels to many, although not all, hazardous chemicals. **The American Conference of Governmental Industrial Hygienists (ACGIH)** *is an organization of governmental, academic, and industrial professionals who develop and publish recommended exposure standards for chemical and physical agents.* OSHA has used these standards as the basis for many of their regulations for chemical exposure levels.

ACGIH has established several types of chemical exposure guidelines of importance to laboratory workers. The **Threshold Limit Value (TLV)** *is the airborne concentration for a chemical that most healthy workers can safely be exposed to for 8 hours per day, repeatedly, with no adverse effects.* This is considered the highest safe level for the work environment. Because chemical exposures vary during a day, recommended exposure levels are frequently designated as **TLV-TWA (time-weighted average)** values. *TWA values acknowledge that exposure levels may differ from the recommended TLV during an 8 hour day, but the TLV-TWA will represent the average exposure level.*

Table 29.10   SUMMARY OF GUIDELINES FOR THE AMOUNTS OF CHEMICALS ALLOWED IN THE AIR

Abbreviation	Name	Definition
TLV	Threshold Limit Value	Safe airborne concentration of a chemical that healthy workers can be exposed to on a daily basis.
TLV-TWA	Time Weighted Average	The average allowable airborne concentration of a chemical within an 8-hour day.
TLV-STEL	Short Term Exposure Limit	The air concentration allowed for only a short period of time (15 minutes). Only four 15-minute periods of exposure to this concentration are allowed in a day.
TLV-C	Ceiling Exposure Limit	During the day the exposure to airborne chemicals must never exceed this value.
PEL	Permissible Exposure Limit (OSHA)	Employers have the legal responsibility to keep employee exposure below this level.

For those chemicals that are extremely toxic at certain levels, ACGIH suggests upper limits to short-term exposures. The **TLV-STEL (Short-Term Exposure Limit)** *indicates the air concentration at which only 15 continuous minutes of exposure is allowed, up to four times during an 8-hour day.* Some rapidly acting toxic materials also have **TLV-C (Ceiling)** values, indicating the highest concentration allowable. During the day the airborne exposure to the chemical must never exceed this value. Exposures to ceiling levels of chemicals may have biological consequences.

The TLV values established by the ACGIH are merely advisory. OSHA used these values and additional data to set PEL values. **Permissible Exposure Limit (PEL)** *values are the legal limits set by OSHA for worker exposure to chemicals.* Employers have the legal responsibility not to allow employee exposure to exceed hazardous levels and to provide safety equipment and training to protect workers. PEL values are often (but not always) the same as or similar to TLV values. Table 29.10 summarizes the main types of exposure limit information that are available to laboratory workers.

C. Skin and Eye Contact

Direct skin or eye contact is another frequent route of laboratory injury by toxic chemicals. When working with chemicals, especially in large volumes, it is often difficult to avoid all contact with the chemical. When potentially toxic chemicals contact the skin, there are four possible outcomes:

- the skin will act as a protective barrier—no harm will occur
- the skin surface will react with the chemical—rashes or burns may result
- the chemical will penetrate the skin—allergic sensitization may occur
- the chemical will penetrate the skin and enter the blood stream—systemic toxicity may result

Skin is designed to provide a natural protective barrier for the rest of the body, guarding against both chemical and biological invasion. However, there are natural entry sites for hazards at hair follicles and sweat glands, and any cuts or abrasions of the skin will also provide entry for toxic materials.

Most laboratory workers are naturally cautious when working with known corrosive agents such as acids and bases, which can cause extensive local tissue damage. It is important to remember, however, that many other types of laboratory chemicals can cause skin irritation or burns, and virtually all chemicals will cause problems if splashed into the eyes. Organic solvents can cause direct skin irritation because of their ability to strip natural oils, diminishing the skin's natural barrier properties. This can also lead to increased skin irritation from other chemicals. It is essential to limit chemical contact by wearing proper laboratory attire, including gloves and eye protection.

In many cases, the most serious consequence of skin exposure to chemicals is the absorption of toxic substances into the bloodstream. Potent neurotoxic materials, such as mercury or acrylamide, can easily be absorbed through skin. Phenol, which is frequently found in biotechnology laboratories, can burn the skin directly and also be absorbed systemically. Exposure to phenol is extremely dangerous in large doses, Example 7.

EXAMPLE 7

Phenol

Phenol is used extensively in biotechnology laboratories to aid in the isolation of nucleic acids. It is corrosive and can cause severe chemical burns. It damages cells by denaturing and precipitating protein. Phenol burns have a characteristic white appearance when they first appear. Because phenol also has anesthetic properties, these injuries may be unnoticed at first. Phenol splashes to the eye are likely to lead to serious burns and possible blindness, so phenol must never be handled without proper eye protection and a readily available eyewash station. Phenol is a respiratory irritant and readily penetrates skin. Whether exposure is by inhalation or skin contact, phenol can enter the bloodstream in significant amounts. From there, it can damage the nervous system, liver, and kidneys. Chronic exposure to phenol vapors can result in chronic symptoms, such as headache, nausea and vomiting, diarrhea, and loss of appetite. Lethal doses of phenol can be inhaled or absorbed through skin relatively easily, so anyone who has significant contact with phenol through a spill or other mishap should seek medical attention. Some symptoms may be delayed in appearance. Phenol fortunately has a characteristic odor that is detectable well below toxic concentrations. This chemical should always be handled in a fume hood to avoid any air contamination of the laboratory.

D. Ingestion

It is common sense that laboratory chemicals taken into the mouth and swallowed can cause harm. Because no one would deliberately consume laboratory chemicals, this is not a common path for biotechnology workplace poisonings. However, chemicals can inadvertently be transferred to the mouth in toxic quantities by contaminated fingers or pencil ends. Develop the habit of keeping your hands and other objects away from your face, mouth, and eyes while in the laboratory. Never pipette by mouth. Separate refrigerators must be designated for food storage, and all food and drinks should be consumed in designated areas where laboratory chemicals are not allowed. Maintaining complete separation of work areas and food areas helps prevent accidental ingestion of chemicals.

Oral toxicity levels for chemicals are determined with animal studies that measure LD_{50} values. Another indicator of chemical toxicity in humans is the **LD_{Lo}** (*Lethal Dose—Low*), *the lowest dose of a compound that has been reported to cause a human death.*

E. Injection

Direct injection of toxic chemicals is a relatively unlikely route of poisoning in the biotechnology laboratory, although it does play a more common role in the spread of biological hazards, Chapter 30. The most likely method of chemical injection is through contact with chemically contaminated broken glass or other sharps. The injection of toxic chemicals that might not otherwise be able to pass the skin or respiratory barrier can be serious, and requires medical attention.

Table 29.11 provides information on the relative toxicities of some of the common chemicals found in biotechnology laboratories.

Table 29.11 RELATIVE TOXICITIES OF COMMON LABORATORY CHEMICALS

Chemical	LD_{50}, oral, rat (mg/kg)	TLV (ppm)	TLV (mg/m³)	PEL (ppm)	PEL (mg/m³)
Acetic acid	3310	10	25	10	25
Acetone	5800	750	1780	1000	2400
Acetonitrile	2730	40	70	40	70
Acrylamide	124	—	0.03 skin	—	0.3 skin
Ammonium hydroxide	350	25	17	35	27
Benzene	4894	10	32	1	3
Chloroform	908	10	49	50	240
Diethyl ether	1215	400	—	400	—
Ethanol	7060	1000	1880	1000	1900
Ethyl acetate	6100	400	1440	400	1200
Formaldehyde	500	0.3	0.37	0.75	1.5
Hexane	28,710	50	176	500	1800
Hydrochloric acid	—	5	7.5	5	7
Hydrogen peroxide	75	1	1.4	1	1.4
Mercury	—	—	0.025	—	0.1
Methanol	5628	200	262	200	260
Nitric acid	—	2	5.2	2	5
Phenol	384	5	19	5	19
Sodium Hydroxide	140	—	2	—	2
Sulfuric acid	2140	—	1	—	1
Toluene	2650	50	188	200	750

Data compiled from various sources; see references at the end of the chapter.

IV. STRATEGIES FOR MINIMIZING CHEMICAL HAZARDS

A. Preparing a Work Area

An important part of good laboratory practice is planning each task, anticipating hazards, and preparing an appropriate work area. For any task, you should have all supplies and small equipment nearby to minimize wasted time and movement. Always wear proper PPE. Read the MSDS and follow the prescribed safety recommendations. Prepare appropriate safety materials before work begins. For example, when working with hazardous liquids, cover work surfaces with absorbent plastic-backed paper that can absorb small spills and be disposed of after your work is finished. Another containment strategy is to perform the work within a tray that can be easily cleaned. The work area should be marked with an appropriate hazard sign while in use.

B. Using Chemical Fume Hoods

i. STRUCTURE AND FUNCTION OF FUME HOODS

Many tasks involving toxic chemicals are best performed in a chemical fume hood. A **chemical fume hood** *is a well-ventilated, enclosed chemical- and fire-resistant work area that provides user access from one side.* The hood isolates and removes toxic or noxious vapors from a working area and protects the user. A fume hood should be used whenever your work involves chemicals with the following characteristics:

- volatility
- unpleasant smells
- a TLV lower than 50 *ppm* in air

A typical laboratory fume hood is shown in Figure 29.5. The hood usually sits on top of a storage cabinet. The opening is covered by a transparent **sash**, *which is a window constructed of impact-resistant material, which can be raised and lowered by the user.* Gas and electrical outlets should be located on the outside of the hood. Inside is a smooth work surface tray, which is designed to contain spills and provide easy cleaning. Most modern fume hoods include interior **baffles**, *which direct the air flow within the hood.* **Airfoils** *located at the bottom and sides of the sash help to reduce air turbulence at the face of the hood.*

There are several types of effective fume hoods available for biotechnology laboratories. Many have a **constant air volume (CAV) design**, *where a constant air flow is pulled through the exhaust duct.* In these hoods, raising and lowering the sash changes the face velocity at the sash opening. **Face velocity** *is the rate of air flow into the entrance of the hood, measured in* **linear feet per minute (fpm)**. In a **non-bypass fume hood**, *air enters the hood only at the bottom of the sash.* When the hood sash

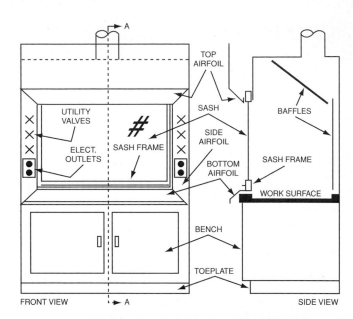

Figure 29.5.   A Standard Benchtop Fume Hood. (Reprinted with permission from *Prudent Practices in the Laboratory. Handling and Disposal of Chemicals*, The National Research Council, National Academy Press, Washington, DC, p. 186, 1995.)

is lowered, the face velocity of a CAV hood can increase dramatically, Figure 29.6. This can have the effect of blowing around paper and small items within the hood when the sash is lowered.

One partial solution to the problem of excessive face velocity is the use of a **bypass hood**, *which has an opening at the top of the hood behind the sash. This allows air*

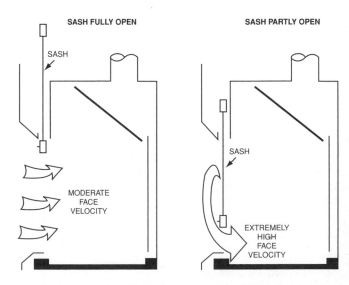

Figure 29.6.   A Non-Bypass Constant Air Volume Fume Hood. (Reprinted with permission from *Prudent Practices in the Laboratory. Handling and Disposal of Chemicals*, The National Research Council, National Academy Press, Washington, DC, p. 183, 1995.)

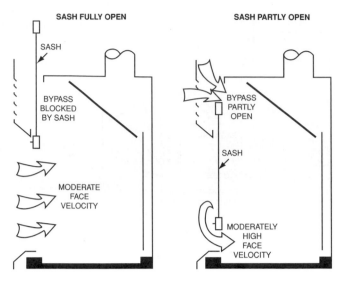

Figure 29.7. A Bypass Fume Hood. (Reprinted with permission from *Prudent Practices in the Laboratory. Handling and Disposal of Chemicals*, The National Research Council, National Academy Press, Washington, DC, p. 184, 1995.)

to enter the hood and bypass the working face, restricting face velocity when the sash is lowered, Figure 29.7. Face velocity will still increase, but not as dramatically.

There are also **variable air volume (VAV) fume hoods** available, *which maintain a relatively constant face velocity by changing the amount of air exhausted from the hood*. VAV hoods are usually designed without an air bypass.

ii. PLACEMENT OF FUME HOODS

The location of fume hoods within a laboratory can influence their effectiveness. Modern fume hood design has eliminated much, but not all, of the sensitivity of these hoods to air currents in the room. The movement of materials as well as hands and arms into and from the hood creates air drafts that can pull toxic vapors from the hood into the laboratory. The passage of other workers directly in front of the hood opening can similarly pull vapors through the face opening. Airfoils are designed to minimize this problem, but fume hoods should be installed in locations where they are isolated from traffic and drafts from doors and ventilation fans, Figure 29.8. Fume hoods should not be located in the corners of rooms, where air currents collide and create turbulence.

iii. TESTING PROCEDURES

Laboratories must test the effectiveness of their fume hoods at least annually, according to OSHA regulations. A complete performance check evaluates three parameters:

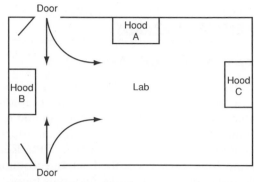

Arrows indicate air currents

Figure 29.8. Placement of Fume Hoods and the Effects of Traffic Air Currents. Hoods B and A are poorly placed to avoid air currents. Hood C is in the better location for this laboratory.

- general function and air flow patterns within the hood
- face velocity
- the uniformity of the face velocity

The third measurement is usually performed when a fume hood is first installed, and later estimated using a smoke generator.

Smoke generators *are small tubes of chemicals, frequently including titanium tetrachloride, that generate highly visible white smoke from a chemical reaction*, Figure 29.9. They are used to check the general function of the hood. The tube is ignited and slowly moved across the front of the hood and then inside at various locations, to check the direction of air flow. This will indicate whether the hood is drawing air from all parts of the work surface, and whether fumes from inside the hood are entering the laboratory. A smoke generator is also useful for checking the effects of sudden hand motions and room traffic on the effectiveness of the hood.

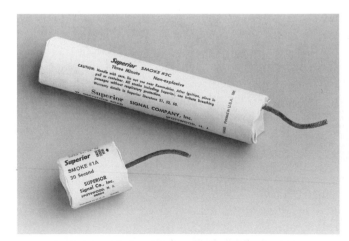

Figure 29.9. Smoke Generators for Testing Fume Hood Function. (© 1998, Flinn Scientific, Inc. All rights reserved. Reproduced by permission from Flinn Scientific, Inc., Batavia, IL, USA)

Smoke generators should be used to determine the most effective locations within the hood for drawing out toxic fumes. For example, in most hoods, volatile chemicals should be handled 5–6 *in*. back from the sash. Exhaust levels tend to increase toward the back of the hood and decrease somewhat at the edges. During the smoke test, it is helpful to mark the effective parts of the fume hood with tape, indicating to users the optimum placement for experimental materials. This is also a good time to check that large items located in the hood do not obstruct air flow. If they do, they can be relocated to optimize hood function. All equipment within the hood should be elevated at least 2 *in*. above the work surface to aid in air circulation.

A **velometer (velocity meter)** *is an instrument used to measure the face velocity of the fume hood.* The recommended face velocity for most fume hoods is 100 linear feet per minute (*fpm*). Face velocities both lower and higher than this value are less effective in containing fumes. Higher velocities create internal turbulence that may be counterproductive, allowing small papers and other items to be drawn into the exhaust vent. This decreases air flow and therefore the effectiveness of the hood. For a standard CAV fume hood, the tester raises or lowers the hood sash to achieve the proper face velocity of 100 *fpm*. This sash location should be marked with a sign or tape on the hood frame to indicate the optimum sash height.

iv. OPTIMAL USE OF A FUME HOOD

Fume hoods cannot protect the user and others in the laboratory unless they are used properly. Before working in the hood, be certain that it is functioning properly. The exhaust fan in the hood should always be running, and in many models the fan cannot be turned off. If yours is adjustable, check that it is turned on and operational. A quick test is to dangle a piece of tissue paper at the hood opening and see that it is drawn toward the hood. All supplies and equipment you will need should be loaded into the hood at the beginning of your work, so that the sash will not have to be raised again while chemicals are in use. Consider chemical compatibility when placing materials within the hood. Table 29.12 summarizes guidelines for proper hood maintenance and use.

It is essential to distinguish between fume hoods and biological safety cabinets. **Biological safety cabinets** *are enclosures designed for the containment of biological hazards.* They are equipped with special filters that remove potentially dangerous particles from the air within the cabinet, and they provide varying degrees of safety and sterility for users and cabinet contents. Fume hoods are not equipped to filter biological hazards from the exhausted air, and do not protect the environment from these agents. Biological safety cabinets can function as fume hoods if they are vented to the outside of the building. However, many of the volatile and toxic chemicals

Table 29.12 *PROPER USE OF A FUME HOOD*

Hood maintenance:
- Regularly test the face velocity and general function of the hood.
- Mark the optimum sash position to attain a face velocity of 100 linear *fpm*.
- Mark the safe interior working area with tape.
- Keep the sash closed when not in use.
- Do not use the hood work surface for chemical or other long-term storage. Note that most hoods are constructed with storage cabinets underneath.
- Keep the exhaust fan on at all times.
- Do not adjust the interior baffles unless you are trained to do so.

Hood use:
- Be sure that the hood is functioning and drawing air.
- Open the sash and load all necessary items.
- Do not overload the hood, blocking air flow.
- Be sure that all chemicals in the hood are compatible.
- Elevate all equipment at least 2 *in*. above the work surface.
- Once all items are loaded, move the sash slowly to the optimum position.
- Keep your face outside the hood and behind the sash at all times.
- Use appropriate face and eye protection.
- Move your arms slowly while using the hood.
- Never stack materials on the bottom air foil.
- Work within the marked safe interior working area.
- Avoid allowing paper or other objects to enter the exhaust ducts.
- Decontaminate the working surface properly every day after use.
- Do not use infectious materials within a fume hood.

In case of a spill or fire within the hood:
- Immediately close the sash completely if you can safely.
- Do not turn off the exhaust fan.
- Unplug all equipment within the hood (this assumes that the equipment is plugged into outlets outside the hood).
- Warn other personnel and evacuate the area.

appropriate for fume hoods will destroy the filters in the cabinets, reducing or eliminating the removal of hazardous particles from the air. Biological filters will not detoxify chemically contaminated air and may not provide adequately protect the user from toxic chemicals.

C. Limiting Skin Exposure

As discussed earlier in the chapter, it is essential to avoid skin contact with laboratory chemicals. To protect yourself, always wear a lab coat to protect the body. Wear gloves that are appropriate for the chemicals being handled. Remember that gloves differ in materials and in thickness, which can directly affect their ability to protect the hands. Table 29.13 provides data on the chemical compatibility of various types of gloves, measured by BDT in minutes. These data were compiled from several different manufacturers, as indicated by

Table 29.13 CHEMICAL COMPATIBILITY WITH GLOVE MATERIALS

Chemical	Breakthrough Detection Rates (minutes)					
	Rubber	**Neoprene**	**Nitrile**	**PVC**	**Butyl**	**Viton**
Acetic acid	31–ND	ND	240–ND	47–300	ND	ND
Acetone	0	12–35	0	0	ND	0
Acetonitrile	0–16	40–65	0	0–24	ND	0
Ammonium hydroxide	58–120	ND	240–ND	60–ND	ND	ND
Benzene	0	15–16	16–27	2–13	30–34	ND
Chloroform	0	14–23	0	0	21	ND
Diethyl ether	0	12–18	33–64	0–14	8–19	12–29
Dimethyl sulfoxide	240	0	0	60	ND	90
Ethanol	ND	ND	225–ND	20–66	ND	ND
Ethyl acetate	0–72	24–34	0–30	0	212–ND	0
Formaldehyde	ND	ND	ND	ND	ND	ND
Hexane	0–21	39–173	234–ND	0–29	0–13	ND
Hydrochloric acid	211–ND	ND	ND	ND	ND	ND
Methanol	60–82	60–226	28–118	3–39	ND	ND
Nitric acid	233–ND	ND	0–72	114–240	ND	ND
Phenol	ND	ND	ND	32	ND	ND
Sodium hydroxide	ND	ND	ND	ND	ND	ND
Sulfuric acid	ND	ND	180–ND	210–ND	ND	ND
Toluene	0	14–25	26–28	3–19	21–22	ND

ND = none detected within 5 hours; 0 = not recommended, less than 10 minutes BDT.
Note: These data are compilations from several glove manufacturers (resulting in ranges of values) and do not constitute specific recommendations. Always check the data for the specific brand and thickness of glove for intended use.

the wide range of BDT values for certain glove–chemical combinations. Always check the specific manufacturer's specifications when choosing a glove for toxic chemical work.

Another factor to keep in mind is the length of the gloves (remember Case Study 2!). The skin of the hands is generally thicker and withstands chemical penetration better than the thinner, more sensitive skin of the

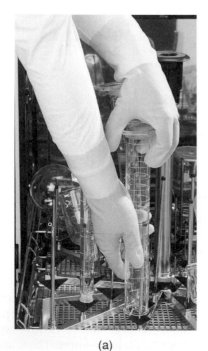

(a)

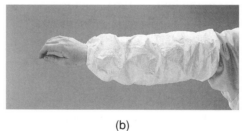

(b)

Figure 29.10. Wrist and Forearm Protection Using a. Gloves or b. Arm Protectors (Photos courtesy of Fisher Scientific, Pittsburgh, PA)

wrists or forearms. Gloves ideally should be pulled over the cuffs of your lab coat to provide complete protection for your arms, Figure 29.10a. If the available gloves are too short, disposable arm protectors are desirable when working with highly toxic chemicals, Figure 29.10b. Always wash your hands thoroughly before leaving the laboratory.

D. Storing Chemicals Properly

i. STORAGE FACILITIES

There are numerous state and federal regulations that cover the storage of laboratory chemicals. These vary by location, but certain issues are invariably regulated. Common regulations cover:

- storage of chemicals with fire and explosion risk in approved safety cabinets
- separation of stored chemicals by compatibility
- identification of all chemicals, with appropriate hazard labeling
- limiting access to radioactive materials, explosives, controlled substances, and other special hazards

Laboratory chemicals should be stored in safe locations as close as possible to the point of use. This will minimize the risk of an spill during transport, which is a frequent source of accidents. Flammable and corrosive chemicals should be transported in secondary containers, such as buckets or trays, for protection from spills or breakage. Chemical storage areas must have spill containment materials immediately available that can handle the contents of the largest container of chemical present.

Every laboratory should maintain a current inventory of chemicals, indicating their date of receipt, user, and location. Every chemical should have a specific storage location, where it can easily be found. Chemical containers removed for use should always be returned to their proper location. Only the amounts of chemical in immediate use should be kept at the lab bench or at other work areas. Chemical storage areas should be inspected regularly to be sure no leaking containers or other signs of container damage are present. Laboratories should dispose of old chemicals or those that are no longer needed regularly. This is especially important when personnel change in the laboratory, leaving behind "orphan" chemicals that are unwanted or unidentifiable. Additional guidelines are provided in Table 29.14.

ii. LABELING

All **primary chemical containers** (i.e., those supplied by the manufacturer) should immediately be labeled with a date and user name when they arrive in the laboratory. The manufacturer will provide information to determine the storage category of each chemical. Chemicals in different categories must always be stored in separate areas. Stock chemicals should not be repackaged in other con-

Table 29.14 GUIDELINES FOR CHEMICAL STORAGE

- Date all chemicals on receipt.
- Be sure all stored chemicals are labeled by contents and general hazards.
- Store only minimum amounts of chemicals.
- Keep only small quantities of chemicals at your work station for immediate use.
- Use proper storage containers and cabinets.
- Do not store flammable chemicals in a standard refrigerator; use a certified spark-free refrigerator or freezer.
- Sort chemicals by hazard and store each type separately. Some of the suggested chemical categories are:
  - acids
  - bases
  - organic oxidizers
  - inorganic oxidizers
  - flammable liquids
  - flammable solids
  - acute poisons
  - water-reactive chemicals

More information can be obtained by consulting the catalog for Flinn Scientific, Inc., Batavia, IL.

- Maintain a current inventory of chemicals for the laboratory.
- Store every chemical in a specific location.
- Never store chemicals on the floor or above eye level.
- Inspect containers weekly for signs of leakage or deterioration.
- Dispose of old chemicals properly and promptly.
- Be sure that all appropriate spill kits are easily available and fully stocked.

tainers unless the original container is damaged. Containers for immediate use should be clearly labeled with the chemical name, date, user, and appropriate hazard information. Chemical name abbreviations are not considered adequate under OSHA regulations. Handwritten labels should be legible and written in indelible ink; organic solvents are notorious for dissolving writing in nonpermanent ink. Hazard labeling should provide information about the greatest risk from the chemical. For example, strong acids should carry a label for "Corrosives," Figure 29.11. Laboratories can purchase sets of standard hazard stickers that provide effective warnings for handlers who are unfamiliar with the specific chemical. Experimental materials should also be labeled by hazard.

E. Handling Waste Materials

Waste *refers to any laboratory material that has fulfilled its original purpose and is being disposed of permanently.* Numerous OSHA and EPA regulations cover the disposal of hazardous waste, which is becoming

Figure 29.11. Common Hazardous Materials Labels.

increasingly expensive to handle. Biotechnology laboratories must maintain several types of waste containers, to sort materials by hazard. To minimize hazardous waste volume, do not dispose of nonhazardous materials in hazardous waste containers. Receptacles for nonhazardous materials, such as old notes or relatively clean paper towels, should be available away from immediate work areas. No sharps or any hazardous materials should ever be placed in these containers. In most institutions, this nonhazardous waste is handled by the janitorial staff.

Any waste materials that present a health risk should be collected in appropriately labeled containers. To the extent possible, waste should be sorted by type (e.g., separating biological waste from chemical waste). Chemical wastes should similarly be collected in a manner that keeps hazard groups separate. The EPA classifies chemical wastes by specific characteristics, such as flammability, corrosive properties, and reactive potential. Local waste disposal facilities may also designate chemical compatibility and hazard groups. The safety office at your institution will tell you how to sort waste materials and will place labels on each container for disposal. Always pay attention to these labels. Never put a chemical in a container if you are uncertain that it belongs there (see Case Study 1).

Chemical wastes are generated as the end products of procedures, or as chemical stocks that are no longer needed in the laboratory. Chemical wastes that contain mixtures of compatible chemicals should be labeled with identification of each chemical and its approximate percentage in the mixture. Solid wastes should be divided into those that can be incinerated (e.g., as lightly contaminated paper towels) and those that cannot. Dry wastes should be double bagged for safety and clearly labeled by hazard. Liquid wastes are collected in labeled glass or plastic containers, as appropriate. Never pour chemicals down a sink drain unless they have been neutralized or deactivated in accordance with local regulations for

Table 29.15 DISPOSAL OF STOCK CHEMICALS

Most laboratories have a tendency to keep chemical stocks "just in case" someone needs them in the future. Although this is a reasonable approach with some chemicals, such as sodium chloride, which is stable and used on a regular basis, it can be a hazardous practice in other cases. Dispose of chemical stocks if any of the following criteria are observed. (Consult with your institutional safety office for proper disposal procedures.)

- The chemical is more than 1 year old and not in current use.
- There is evidence of pressure buildup in a container.
- A formerly clear liquid has turned cloudy.
- Solids have clumped or show other signs of water absorption.
- Chemicals have changed color.
- Containers show signs of damage.
- Chemical identities are unclear.

These stocks should be disposed of in their original containers, with appropriate identity and hazard labeling.

Table 29.16 MINIMIZATION OF CHEMICAL WASTE

- Maintain a current chemical inventory and avoid duplication of stocks.
- Order as little of a chemical as needed for planned work.
- Date all chemicals on arrival, to eliminate doubts about age.
- Use older chemicals first.
- Be sure all chemical containers are labeled by content.
- Keep all chemicals tightly sealed to prevent deterioration from air exposure.
- Do not contaminate chemical stocks by returning materials to the container.

sewer disposal. Never use the drain for flammable solvents or potentially reactive chemicals. Even at very low concentrations, some chemicals, such as sodium azide, can form explosive mixtures within drain pipes, with predictably unpleasant results. Table 29.15 provides some guidelines for determining when stock chemicals should be discarded.

Because of the high expense of toxic waste disposal, it is important to minimize the amounts of waste generated by the laboratory. Table 29.16 contains some suggestions for reducing the amounts of chemicals that will require disposal. Unknown or unlabeled wastes are by far the most expensive type to dispose of, so all workers should be diligent about chemical labeling.

V. RESPONSE TO CHEMICAL HAZARDS

A. Chemical Emergency Response

Every laboratory is required by OSHA to have an emergency response plan for accidents involving hazardous chemicals. These plans should be explained as part of laboratory safety training, and it is the responsibility of each worker to be familiar with the appropriate actions. Emergency response needs to be an automatic reaction because accidents rarely provide time for detailed analysis and research into procedures (Case Study 3). It is important to be able to distinguish between a truly minor incident and an emergency. Table 29.17 provides some guidelines that should be considered before they are needed.

In cases where the exposure of a worker to chemicals is obvious, immediately call the Poison Control Center or a designated physician. These telephone numbers should be posted by every telephone in the laboratory area. In the case of a skin or body splash, the victim should move to the emergency shower, remove any contaminated clothing, and drench the affected area for at least 15 minutes. Eye exposure requires flushing of both eyes for at least 15 minutes as well. If chemicals have been inhaled, the person should be moved to an area with fresh air and remain quiet until medical help arrives.

Laboratory workers should be familiar with the symptoms of toxic chemical exposure, which are summarized in Table 29.18. Specific information about individual chemicals can be found in the MSDS, which should always be consulted before chemical use.

It is certainly better to prevent accidents or chemical exposures before they happen. Immediate preventive measures or notification of a supervisor is appropriate

Table 29.17 WHEN IS A "PROBLEM" INVOLVING CHEMICALS AN EMERGENCY?

Most of us are reluctant to overreact to a potentially dangerous laboratory situation. The following list includes some of the circumstances where the best course of action is to summon help immediately. You probably require assistance from a professional safety officer or emergency response team when a chemical problem:

- causes a serious injury
- involves a public area
- creates a fire hazard
- cannot be isolated or contained by those present
- creates toxic vapors that could spread through the building
- causes property damage
- requires a prolonged cleanup
- involves mercury compounds
- involves hydrofluoric acid
- involves an unknown hazard

Your laboratory should post the telephone numbers of the designated sources of local assistance near all telephones. Some of the numbers that might be useful in an emergency:

- Institutional safety office, in case of spills or questions about safety procedures
- Fire department, in case of fire
- Emergency medical assistance, in case of injury
- Poison control, in case of exposure to toxic chemicals

Table 29.18 SYMPTOMS OF CHEMICAL EXPOSURE

Any time you become aware of direct skin or eye contact with a laboratory chemical, you should take immediate action to remove any traces of the chemical and seek medical attention as needed. If you or your co-workers notice a chemical odor, especially for low TLV substances, immediate action may be required. In the absence of known contact or inhalation, the following are some of the potential signs of toxic chemical exposure. The MSDS for a chemical will provide specific details.

Acute exposure:
- headache or dizziness
- sudden nausea or vomiting
- coughing spasms
- eye, nose, or throat irritation

especially if:
- the preceding symptoms disappear with fresh air
- the symptoms reappear when work is resumed
- more than one person in the laboratory is affected

Chronic exposure:
- persistent dermatitis
- unusual body or breath odor
- a strange taste in the mouth
- discolored urine or skin
- numbness or tremors

Table 29.19 WHEN TO TAKE SPECIAL PRECAUTIONS

There are many circumstances where a worker should take steps to prevent a problem directly, or to notify an appropriate supervisor. Some examples:
- Chemical leaks or spills are noticed or anticipated.
- Possible symptoms of chemical poisoning are noticed.
- Laboratory workers notice odd smells.
- A fume hood or other safety equipment fails to operate properly.
- A procedure creates chemical exposure that may exceed toxic thresholds.
- A new or altered procedure requires chemicals with unknown properties.

under any circumstances where you have concerns about potential problems. Table 29.19 provides a few examples.

B. Chemical Spills

i. PREVENTING CHEMICAL SPILLS

In laboratory work, chemical spills occur. Most of these will be small and many can be prevented by good laboratory practices. Proper advance preparation for spills ensures that most of these incidents will remain relatively minor in terms of scope and consequences. Many spills are caused by simply knocking over open containers of chemicals. These spills can be avoided by properly arranging experimental materials so that routine arm movements will not endanger chemical containers. Containers should be closed when possible, or anchored in a rack or holder for added stability.

Only small amounts of chemicals, in unbreakable containers when possible, should be kept at lab benches. Any toxic chemicals should have secondary containment, such as spill trays. Benches or fume hood work surfaces should be covered with absorbent, plastic-backed paper for liquid chemical work. This makes clean-up quick and simple.

Many serious spills occur when transporting chemicals through the laboratory. Large reagent containers should be handled one at a time, preferably in a secondary transport container. Never carry glass reagent bottles by the cap or solely by the ring at the top. Support the bottle from underneath with one hand. Always wear gloves when transporting chemicals.

Be aware that the bottom seam of glass bottles is generally the weakest part of the container, Case Study 3. One potential cause of major spills is the transfer of glass bottles from a warm-water bath or freezer directly to a counter top. Thermal shock can weaken the bottom seam and cause it to crack. When thawing reagents in a water bath, always place the bottle in a secondary container that will contain any leaked material.

CASE STUDY 3

Emergency Response to a Major Chemical Spill

Major chemical spills require a quick response by all members of a laboratory. A biotechnology worker in one of the author's laboratories was thawing a 1 L stock bottle of frozen phenol in a water bath. Following proper procedures, she placed the bottle in a beaker for containment. When thawing was complete, she checked the bottle and beaker for any signs of leakage, and then picked up the phenol bottle at the side and began to transport it across the lab. Half way across the room, the bottom of the bottle gave way, dumping the entire liter of phenol down her leg and onto the floor. Knowing that phenol is toxic through skin contact and is also corrosive, she immediately called for help and ran for the emergency shower. She was properly attired in a lab coat, gloves, long pants, and solid shoes, which limited her contact area. In the meantime, the lab supervisor arrived on the scene and determined that this was a major spill. She ordered that the room be evacuated, and both doors be closed, locked, and labeled with a "Danger: Phenol Spill" sign. Someone else called the building engineering hotline and ordered an immediate switch of the ventilation system for the floor to total exhaust to prevent the spread of toxic fumes within the building. Another worker was assisting the technician in the shower. The phenol had soaked through her pants fairly quickly and she was convinced to remove them when someone brought a fire blanket to use as a modesty shield. Colleagues contacted the Institution Safety Office and requested an emergency spill team with respirators to clean up the phenol, and then escorted the injured person to the local emergency room. She suffered from nausea and headache for 2 days, but her chemical burns were relatively minor, and there were no apparent long-term effects. Aside from the inconvenience of being evacuated from the contaminated laboratory for several hours, no other personnel were affected. This would have been a very serious accident without the quick response and cooperation of many members of the laboratory team.

---

ii. HANDLING CHEMICAL SPILLS

When a chemical spill occurs, it is essential to assess the extent of the problem immediately. Even though a small spill is obviously easier to clean up than a larger mess, in many cases the volume of the spill is less significant than the toxicity of the spilled material. Whenever a highly toxic and volatile liquid is spilled, as in Case Study 3, it is best to evacuate personnel and inform the safety experts at your institution. Your efforts may be best spent ensuring that others are warned of the hazard and that any injuries are promptly tended.

Small spills of relatively innocuous chemicals can be cleaned up without much concern. For hazardous chemical spills, however, a well-intentioned but inappropriate clean-up effort may cause more harm than good. For example, in the case of a volatile solvent, every liter spilled can produce up to 600 L of flammable vapors. An attempt to wipe up a spill with paper towels will actually encourage vapor production by increasing the surface area of the spill. Liquid spills should be handled with appropriate chemical absorbents. Absorbent in sufficient quantities to soak up the largest volumes used in the laboratory should be readily available. Dry chemicals should never be swept up, because this will create dust that can be inhaled. Wet mopping is more appropriate in cases where absorbents are unnecessary. Table 29.20 provides a summary of proper procedures for dealing with both minor and major chemical spills.

If hazardous chemicals are spilled on clothing, the clothing should be removed immediately and then rinsed. Any skin under the clothing should be flushed for 15 minutes in an emergency shower. When removing the clothes, be careful not to spread chemical to additional areas of the body. Never pull a contaminated shirt or sweater over the face; if the chemical is toxic, cut the clothing off with scissors. Do not attempt simply to rinse the chemicals out of the clothes while wearing them. A fire blanket or spare lab coat can be used to protect modesty if required.

iii. CHEMICAL SPILL KITS

Properly assembled and conveniently located spill kits can make the cleanup of minor spills significantly easier. **Chemical spill kits** *are preassembled materials for controlling and cleaning up small to medium size laboratory spills.* Every room where chemicals are used or stored should have a kit with sufficient materials to handle a spill of the largest container in the immediate area. A list of suggested contents for a general kit are provided in Table 29.21. Having all necessary materials

Table 29.20 *CHEMICAL SPILL PROCEDURES*

It is essential that all laboratory personnel know what to do in advance in the event of a chemical spill. This topic should be discussed at periodic safety meetings, along with other routine safety procedures.

Minor Spills:

- Take care of personal contamination first.
- Notify nearby workers and evacuate them as needed.
- Turn off heat or ignition sources if flammable chemicals are involved.
- Prevent the immediate spreading of the spill by layering with towels (do not wipe).
- Avoid breathing any vapors from the spill.
- Dress properly and get the appropriate spill kit.
- First surround the spill with absorbent to contain the spill area.
- Slowly sprinkle absorbent over the spill and follow any directions that accompany the spill kit.
- Collect the contaminated absorbent using a whisk and dust pan, and avoid creating dust or aerosols.
- Place the waste in a disposable bag, seal, and label; the whisk and pan should be thoroughly cleaned or discarded.
- Finish the cleanup by washing the area several times with detergent and water.

Major Spills:

- Leave the spill site immediately.
- Warn others about the hazard.
- Tend to any injuries.
- Prevent others from approaching the spill site.
- Call for expert assistance.

Table 29.21 *CONTENTS OF A GENERAL SPILL KIT*

You can purchase general chemical spill kits or assemble your own, according to your needs. A general use spill kit should contain:

- chemical-resistant, long-sleeved gloves
- chemical-resistant goggles
- chemical-absorbent materials for any anticipated chemical type
- spill pillows for containment
- small whisk broom and dust pan
- disposable plastic bags for hazardous waste

Most laboratories will require more than one type of spill kit because different absorbents and cleanup procedures are required for acids, bases, organic solvents, and the like. There are also special spill kits that should be purchased if your laboratory contains mercury or hydrofluoric acid hazards.

assembled ahead of time, with all personnel trained to use the kits correctly, can make the difference between a minor inconvenience and a major problem.

Spill control kits may contain both loose absorbents as well as absorbent-filled pads and pillows. The absorbents should be chosen to reduce the vapor pressure from a liquid spill efficiently and to minimize any

personal contact with the chemical. The type of absorbent material most applicable in biotechnology laboratories is labeled **universal absorbent**. *These are generally polypropylene or expanded silicates and can absorb virtually any liquid, including some corrosives.* Check the manufacturer's recommendations and be certain that kits are clearly and appropriately labeled. These absorbents control liquid spills but do not reduce the toxic properties of absorbed liquids. Acids and bases, for example, may still require neutralization before disposal.

Loose absorbent works best for small spills. The absorbent should be poured around the edges of the spill to contain it and then sprinkled on the interior. There are special neutralizer absorbents available for acid spills. For larger spills, absorbent pillows can be placed around and on top of the spill. Absorbent should usually be left undisturbed for a short period of time before final cleanup.

Any laboratory that uses mercury should keep a special mercury spill kit available. These kits should only be used by trained personnel because of the extreme toxicity of mercury. Mercury is readily absorbed through the skin and by inhalation. Its vapors are odorless and tasteless, so it is difficult to judge mercury exposure levels. Always handle mercury on trays to contain any possible spills. Avoid using mercury thermometers if possible; if unavoidable, purchase thermometer shields to guard against breakage.

PRACTICE PROBLEMS

1. As described in the text, small concentrations of the odorless gas, carbon monoxide, in air can be lethal. Exposure to low levels of carbon monoxide can cause headache, nausea, and eventually unconsciousness. OSHA has set a maximum allowable limit of 50 *ppm* in air (TWA) for an 8-hour period. What is the significance of the TWA designation?

2. A scientist in your laboratory is carrying a glass bottle of organic solvent across the laboratory when he slips on a wet spot on the floor, loses his grip, and drops the solvent bottle.

   a. What are several precautions that might have prevented this accident?

   b. What immediate steps should this person take to minimize any risk to himself and other laboratory workers?

3. Is the odor of organic solvents a reasonable indicator of safe exposure levels?

4. Material Safety Data Sheets (MSDSs) are a major source of safety information and it is essential to be able to use the information they contain. Excerpts from four Material Safety Data Sheets follow. Based on the information in these MSDSs, answer the ques-

tions below. (NOTE: *These are only brief excerpts from the actual MSDSs and are intended only for purposes of study, not for use in the laboratory.*)

Material Safety Data Sheet
General Information
Item Name: ACETONE, REAGENT
Date MSDS Prepared: 01Jan95
Toxicity Information
OSHA PEL: 1000 *PPM*
ACGIH TLV: 750 *PPM*/1000 STEL
Physical/Chemical Characteristics
Appearance and Odor: Clear, Colorless, Volatile Liquid with a Characteristic Sweetish Odor
Boiling Point: 133°F, 56°C
Melting Point: −139°F, −95°C
Solubility in Water: Very Soluble
Fire and Explosion Hazard Data
Flash Point: −4°F, −20°C
Extinguishing Media: Water Spray, Dry Chemical, Carbon Dioxide, Alcohol-Resistant Foam

---

Material Safety Data Sheet
General Information
Item Name: ETHYL ALCOHOL ACS
Date MSDS Prepared: 01Jan95
Toxicity Information
OSHA PEL: 1000 *PPM*
ACGIH TLV: 1000 *PPM*
Physical/Chemical Characteristics
Appearance: Clear Colorless Liquid
Fire and Explosion Hazard Data
Flash Point: 55°F

Material Safety Data Sheet
General Information
Item Name: PHENOL, USP
Date MSDS Prepared: 01Jan95
Toxicity Information
OSHA PEL: 5 *PPM*
ACGIH TLV: 5 *PPM*
Physical/Chemical Characteristics
Appearance and Odor: Crystal
Fire and Explosion Hazard Data
Flash Point: N/A
Extinguishing Media: Water Spray, CO_2, Dry Chemical or Foam

Material Safety Data Sheet
General Information
Item Name: SODIUM CHLORIDE
Date MSDS Prepared: 01Jan95
Physical/Chemical Characteristics
Appearance and Odor: White Crystal
Boiling Point: 1413°C
Fire and Explosion Hazard Data
Flash Point: N/A
Extinguishing Media: N/A
Special Fire Fighting Proc: N/A

---

a. Rank each of the compounds in order from least toxic to most toxic based on their TLV values.

b. For acetone, the MSDS states "ACGIH TLV 750 *ppm*/1000 STEL." Explain the difference between the two numbers, 750 *ppm* and 1000 *ppm*.

c. Based on the flash points given in the MSDSs, which of these compounds poses the most significant fire hazard? How should a fire involving that compound be extinguished?

DISCUSSION QUESTION

MSDS information for many chemicals is available on the World Wide Web. They can currently be found at *http://hazard.com/msds/index.html*

Look around your house and identify three to five common household chemicals, such as cleansers or house paint. Look up the MSDS for each of those chemicals and answer the following questions about each one:

a. Is the substance hazardous?

b. If the substance is hazardous, what is the nature of the hazard(s)?

c. What safety precautions should you use when working with each of these household substances?

TERMINOLOGY RELATING TO CHEMICAL SAFETY

Absorption. The incorporation or penetration of a liquid into a solid substance; an absorbed liquid will be soaked up by the solid.

Acid. A compound that dissociates in water to release a hydrogen ion and reacts with base to form neutral salts and water; can be corrosive and reactive.

Acute. Having a rapid onset.

Acute Exposure. A short-term contact or one dose of a substance or chemical.

Acute Toxicity. The harmful effects of any single dose or short-term exposure to a chemical or other substance.

Airfoils (refers to fume hoods). Grates located at the bottom and sides of the hood sash that help to reduce air turbulence at the face opening of a hood.

Allergen. A substance that produces an allergic response in some individuals.

Allergy. A reaction by the body's immune system to exposure to a specific chemical.

American Chemical Society, ACS. Professional organization that sets standards for purity of chemical reagents.

American Conference of Governmental Industrial Hygienists, ACGIH. An organization of governmental, academic, and industrial professionals who develop and publish recommended threshold exposure limits for chemical and physical agents.

American National Standards Institute, ANSI. A national organization that sets standards related to safety and safety design.

Anaphylactic Shock. A sudden life-threatening reaction to allergen exposure.

Asphyxiant. A gaseous compound or vapor that can cause unconsciousness or death due to lack of oxygen.

Baffles (refers to fume hoods). Adjustable panels that direct the air flow within a fume hood.

Biological Safety Cabinets. Enclosures designed for the containment of biological hazards.

Bypass Fume Hood. Fume hood that has an opening at the top of the hood behind the sash, allowing air to enter the hood and bypass the working face, restricting face velocity when the sash is lowered.

Cancer. A disease that is characterized by the uncontrolled growth of cells.

Carcinogen. Compound that is capable or suspected of causing cancer in humans or animals.

Ceiling Limit. The air concentration of a chemical, not to be exceeded at any time.

Central Nervous System, CNS. The biological system that includes the brain, spinal chord, and system of neurons; the target of neurotoxins.

Chemical Abstracts Service Number, CAS Number. Unique identification number assigned to every commercial chemical compound.

Chemical Fume Hood. A well-ventilated, enclosed chemical- and fire-resistant work area that provides user access from one side.

Chemical Spill Kits. Preassembled materials for controlling and cleaning up small to medium–size laboratory spills.

Chronic Exposure. Long-term continuous or intermittent contact.

Chronic Health Hazard. An agent that can cause biological harm with long-term contact, or continuous or intermittent exposure.

Code of Federal Regulations, CFR. A numerical system for the classification and identification of all Federal Regulations; all legally established federal regulations have a CFR number.

Combustible. Substance that can vigorously and rapidly burn under most conditions.

Common Name. The nonchemical, brand name, trade name, or name in common use for a chemical compound or mixture.

Constant Air Volume, CAV, Fume Hood. A fume hood that has a constant air flow through the exhaust duct; in these hoods, raising and lowering the sash changes the face velocity at the sash opening.

Corrosive. Substance that will cause tissue damage or destruction at the site of contact.

Department of Transportation, DOT. Agency that regulates the transportation of hazardous materials.

Dermatitis. Redness, inflammation, or irritation of the skin.

Embryotoxin. A substance that is harmful to the developing fetus while showing little effect on the mother.

Entry Routes. Method of entry into the body, such as the mouth, lungs, or absorption through the skin.

Environmental Protection Agency, EPA. Federal agency that oversees the regulation and enforcement of environmental laws.

Ethidium Bromide, EtBr. A fluorescent dye used to visualize nucleic acids in agarose gels; acts by inserting itself into DNA molecules.

Exothermic. A chemical reaction that gives off heat.

Explosion. A sudden release of large amounts of energy and gas within a confined area.

Explosive. Substance that is capable of rapid combustion, causing sudden release of heat, gas, and pressure.

Face Velocity (refers to fume hoods). The rate of air flow into the entrance of a hood, measured in linear feet per minute (*fpm*).

Feet Per Minute, Linear, *FPM*, (refers to fume hoods). Measurement of face velocity.

Flammable. A substance that can ignite easily and burn quickly.

Flash Point. The minimum temperature at which a compound gives off sufficient vapors to be ignited.

Fume Hood. See "Chemical Fume Hood."

Hazard Diamond System. A system developed by the NFPA to rate chemicals according to their fire, reactivity, and general health hazards.

IDLH, Immediate Danger to Life and Health. Designation that indicates environmental conditions requiring respirator use and maximum personal protective equipment.

Incompatible (in chemical safety). Refers to combinations of chemicals that will react and cause hazardous conditions.

Inflammable. Another term for flammable materials.

Ingestion. Entering the body through the mouth and digestive tract.

Inhalation. Entering the body through the respiratory system and lungs.

International Agency for Research on Cancer, IARC. Agency that determines the relative cancer hazard of materials.

Irritant. Substance that will cause irritation to the skin, eyes, or respiratory system.

Lachrymator. Substance that is an eye irritant and stimulates tear formation.

Latency Period. Time from the first exposure to a toxic agent to the time when biological effects can be detected.

LC_{50}, Lethal Concentration 50%. Concentration of a compound in air that will kill 50% of test animals.

LC_{Lo}, Lethal Concentration Low. The lowest concentration reported to cause lethality in test animals.

LD_{50}, Lethal Dose 50%. Amount of a toxic compound given in a single dose that will cause death in 50% of test animals.

LD_{Lo}, Lethal Dose Low. The lowest single dosage known to cause lethality in a human.

Material Safety Data Sheets, MSDS. An OSHA-required technical document provided by chemical suppliers, describing the specific properties of a chemical.

Milligrams per Cubic Meter of Air, *mg/m^3*. A measure of air concentrations of chemicals.

Mutagen. A substance that can cause changes in DNA, resulting in genetic alterations.

Neurotoxins. Compounds that can cause damage to the central nervous system.

Neutralize. To react acids and bases chemically to form neutral salts and water; to bring the pH of a material or mixture to 7.0.

National Fire Protection Association, NFPA. Organization that developed the visual labeling and rating system for health, flammability, reactivity, and related hazards.

National Institute for Occupational Safety and Health, NIOSH. Public health service that tests and recommends chemical exposure limits.

Non-Bypass Fume Hood. A fume hood design where air enters the hood only at the bottom of the sash.

Nonflammable. Not easily ignited and burned.

Occupational Safety and Health Administration, OSHA. The federal agency responsible for regulating workplace safety.

Oxidizing Agent. A substance that gains electrons in a reaction or can oxidize another substance.

Parts Per Million, *PPM*. Designates air or water concentrations of chemicals.

Permissible Exposure Limit, PEL. Limit set by OSHA for the allowable concentration of a substance in air.

Peroxide Former. A chemical that produces peroxides or hydroperoxides with age or air contact.

Personal Protective Equipment, PPE. The clothing and equipment needed to protect personnel against chemical exposure.

Primary Chemical Container. Containers supplied by the manufacturer; these should immediately be labeled with a date and user name when they arrive in the laboratory.

Pyrophoric. Chemicals that will ignite on contact with air.

Reactivity. The tendency of a chemical to undergo chemical reactions.

Reducing Agent. A substance that donates electrons in a chemical reaction.

Resistant (in safety). Indicates a relative inability to react with another material.

Sensitizer. A substance that may trigger an allergy directly, or cause an individual to develop an allergic reaction to an accompanying chemical.

Smoke Generators. Small tubes of chemicals, frequently including titanium tetrachloride, that generate highly visible white smoke from a chemical reaction; used to check the general function of fume hoods.

Solvents. Chemicals that dissolve other substances. In the context of safety, this usually refers to organic liquids.

Stability (in safety). The chemical characteristic of remaining unchanged over time.

STEL, Short Term Exposure Limit. See "TLV-STEL."

Target Organ. The body part or organ most likely to be affected by exposure to a chemical or hazard.

Teratogen or Teratogenic. Compound can cause defects in a fetus when administered to the mother.

TLV, Threshold Limit Value. The air concentration of a chemical that will not pose a health threat to most normal healthy workers; determined by the ACGIH.

TLV-C, Threshold Value Limit-Ceiling. The maximum allowable concentration of a material in air; concentrations should never exceed this value.

TLV-STEL, Threshold Value Limit-Short Term Exposure Limit. The concentration of a toxic material in air to which a worker should not be exposed for more than 15 minutes; determined by the ACGIH.

TLV-TWA, Threshold Limit Value-Time Weighted Average. The acceptable air concentration of a substance averaged over an 8-hour day; determined by the ACGIH.

Toxic. Poisonous; a substance's ability to cause harm to biological organisms or tissue or to cause adverse health effects.

Toxic Materials. Substances that are poisonous.

Toxicology. The study of poisonous materials and their effects on living organisms and tissue.

TWA, Time Weighted Average. The concentration of a substance in air that is allowed when averaged over an eight hour day; during the day the actual concentrations will be higher and lower than the daily average concentration.

Universal Absorbent. Polypropylene, expanded silicates, or other materials that can absorb any liquids safely, including some corrosives; can be purchased in loose or pillow form.

Variable Air Volume, VAV, Fume Hoods. Fume hoods that maintain a relatively constant face velocity by changing the amount of air exhausted from the hood.

Velometer, Velocity Meter. An instrument used to measure the face velocity of a fume hood.

Volatile (in safety). Refers to chemicals that evaporate quickly at room temperature.

Waste. Any laboratory material that has completed its original purpose and is being disposed of permanently.

REFERENCES

The safety references provided in the previous chapter contain significant information about chemical hazards. The following references emphasize chemical safety, as well as other good laboratory practices.

Prudent Practices in the Laboratory. Handling and Disposal of Chemicals, The National Research Council, National Academy Press, Washington, DC, 1995.

Safety in Academic Chemistry Laboratories, The American Chemical Society Committee on Chemical Safety, American Chemical Society, Washington, DC, 1995.

CRC Handbook of Laboratory Safety, A. Keith Furr, ed. CRC, Cleveland, 1995.

MSDS data on Internet:
   http://www.ilpi.com/msds/index.chtml
   http://hazard.com/msds/index.html

ANSWERS TO PRACTICE PROBLEMS

1. The TWA means that the 50 *ppm* level is a time-weighted average over 8 hours. Actual air concentrations may vary from this average value.

2. **a.** Keeping the floor dry.

   Being careful walking around water sources.

   Using a secondary container to transport the solvent bottle.

   Storing the solvent in a nonglass container.

   b. Notify other personnel of the problem.

   Consider the toxic properties of the spilled liquid and order an evacuation if appropriate.

   Be certain that any personal contamination with the solvent is cleaned up first.

   If safe, proceed with proper spill cleanup techniques using a spill kit.

3. No! The air concentrations of a chemical that can be detected by the human nose are unrelated to toxic concentrations.

4. **a.** NaCl (least toxic, no TLV reported because inhalation is not a hazard); ethyl alcohol, acetone, phenol (most toxic) based on TLV values.

   b. The STEL value is the limit to which a person can be exposed for only 15 continuous minutes, up to four times during an 8-hour day. It is therefore a higher limit than the TLV value, which is the limit to which a person can be exposed 8 hours/day, 40 hours/week.

   c. Acetone. Extinguish using water spray, dry chemical, carbon dioxide, or alcohol-resistant foam.

CHAPTER 30

Working Safely with Biological Materials

I. INTRODUCTION TO BIOLOGICAL SAFETY

A. Biological Hazards

Biological safety is a major issue in biotechnology facilities. These facilities may house a wide range of life forms, ranging from viruses and bacteria to plants, and even to farm animals. For example:

- Bacteria are used for many processes in the food industry, and as tools to explore and produce new protein and enzyme products.

- **Viruses** (*particles containing either DNA or RNA surrounded by a protein coat*) can be used to introduce DNA into cells of bacteria and higher organisms.

- Molds and fungi are capable of producing complex metabolic products (such as penicillin).

- Yeasts are essential for making bread and wine. They can also be used as small eukaryotic factories for synthesis of complex proteins.

- Cells of higher organisms are grown in culture for product testing and producing recombinant proteins.

- Whole plants are studied to develop means of better food production for humans and animals, and as production systems for desirable proteins.

- Whole animals are used in research to test the products of biotechnology, and as factories for the production of antibodies or other complex proteins.

Any of these systems is potentially a **biohazard**, *which is a biological agent with the potential to produce harmful effects in humans.* Examples include microorganisms, recombinant DNA products, cultured human or animal cells, and other agents defined by laws, regulations, or guidelines. **Pathogenicity** *is a term indicating the relative capability of an organism to cause disease in humans or other living organisms.* We commonly think of bacteria and viruses when we think of pathogens, but fungi, single celled protozoa, and even multicellular organisms (such as nematodes and other worms) can be pathogenic.

In order to cause a disease, an organism must be not only pathogenic but also infectious. **Infectious** *refers to the ability of a pathogenic organism to invade a host organism. An organism that causes a specific disease in an infected host is called an* **etiological agent**. For example, the bacterium *Vibrio cholerae* is the etiological agent for cholera. Some organisms can infect a host without causing disease. In this case, the host is called a carrier. A **carrier** *is an infected individual who is capable of spreading the infecting agent to other hosts.*

In order to infect a host, an organism must be able to spread to the host and then penetrate its natural defenses. As with chemical hazards, the primary routes of exposure to biological hazards are inhalation, skin and eye contact, ingestion, and injection. Inhalation of airborne biological agents is the most likely route of laboratory exposure. Many routine laboratory procedures can produce aero-

sols. **Aerosols** *are very small particles or droplets suspended in the air.* A **bioaerosol** *is an aerosol that includes biologically active materials.* Strategies for the prevention of aerosols and aerosol exposure are discussed in detail later in the chapter.

Many of the same safety precautions that reduce risk from chemical hazards can also be applied to biological agents. The best health protection for a laboratory worker is to learn as much as possible about the biological materials present in their workplace. Table 30.1 provides criteria that can be used to evaluate the risk from a laboratory biohazard.

B. Laboratory-Acquired Infections

Biotechnology laboratories work with different types and amounts of biohazards, each with its own associated risks. There are clear incentives for laboratories to use non-pathogenic organisms because of safety issues related to workers, investment costs, and product safety validation concerns. Therefore, production work, and more than 80% of the research work conducted worldwide, involves the use of non-pathogenic organisms.

There are situations, however, where biotechnologists must work with pathogenic materials. Some laboratory pathogens have caused infections of laboratory workers, resulting in disease, illness, and even death. One study in 1978 reported on 3,921 documented cases of **laboratory-acquired infections (LAIs)**, *which are infections that can be traced directly to laboratory organisms handled by or used in the vicinity of the infected individuals.* Of these cases reported between 1924 and

---

Table 30.1 RISK ASSESSMENT FOR A BIOHAZARDOUS AGENT

It is essential for every laboratory worker to evaluate the health risks of specific biohazards in their workplace. Your evaluation should address the following issues:

- Is this a known human or primate pathogen?
- What is the history of laboratory use of this organism or agent, and what are the recognized risks?
- Has this agent been associated with laboratory-acquired infections?
- If so, what are the health consequences of these infections?
- Is there an effective treatment or preventive vaccine?
- Does this agent frequently induce sensitivities or allergies in workers?
- What is my potential susceptibility to infection with this agent as a function of age, sex, or medical condition, as documented by previous cases?
- How can I limit my exposure to this agent?
- What are the recommended safety precautions for this agent, and are they being practiced in this laboratory?
- Is the estimated risk acceptable to me?

1977, 168 were ultimately fatal ("Laboratory-associated infections: summary and analysis of 3921 cases," R.M. Pike. *Health Lab. Science*, 13: 105–114, 1978.). OSHA estimates that as many as 12,000 cases of LAIs with hepatitis B (with 200 associated fatalities) occurred annually among U.S. health care workers prior to the introduction of an effective vaccine.

It can be difficult to predict the health risks to individuals from exposure to low doses of biohazards, so the key to biological safety is prevention of human exposure to potentially harmful agents. Workers who knowingly handle biohazards are not the only individuals who contract LAIs. On average, one out of every four infections associated with a laboratory biohazard occurs among dishwashers, custodians, clerical staff, and maintenance personnel. You should therefore be familiar with the basic safety issues, not only to protect yourself, but also to ensure that others are not exposed to biological hazards. Some precautions are as simple as marking rooms containing these hazards with biohazard warning signs to indicate that only trained personnel should enter the facilities, Figure 30.1. These signs can also designate the specific biohazards present. Proper waste disposal and decontamination procedures must also be followed to protect other workers.

C. Regulations and Guidelines for Handling Biohazards

There is a wide variety of guidelines, standards, and regulations that govern the use of biological materials in the laboratory. A list of examples is provided in Table 30.2. Much of the regulatory information about biosafety is readily available on the Internet. Many large institutions place copies of their safety manuals on web pages for easy reference. The agencies whose biohazard guidelines are most often cited are the **Centers for Disease Control and Prevention (CDC)** and the **National Institutes of Health (NIH)**. The **CDC** *is an agency of the*

(a)

(b)

Figure 30.1. Standard Biohazard Warning Signals. Either the symbol or the background are solid orange or red. **a.** Sign marking entrances to laboratory and storage areas where biohazards are handled and access must be limited. **b.** Symbol for biohazardous waste containers. These containers must be autoclaved or burned before disposal.

Table 30.2 *EXAMPLES OF GUIDELINES AND REGULATIONS THAT COVER BIOLOGICAL HAZARDS IN THE LABORATORY*

Biosafety in Microbiological and Biomedical Laboratories, CDC/NIH: This set of guidelines provides recommendations for facilities, operating procedures, hazard identification, and risk assessment for laboratories working with biological hazards.

OSHA Bloodborne Pathogens Standard, 29CFR1910.1030: This OSHA standard describes the requirements for employers to develop organizational plans for handling human blood and blood-related products. These requirements are discussed later in the text.

Guidelines for Research Involving Recombinant DNA Molecules, National Institutes of Health: These NIH guidelines provide recommendations for facilities, operating procedures, hazard identification, and risk assessment for laboratories involved in recombinant DNA research.

Biological Safety Manual for Research Involving Oncogenic Viruses, National Cancer Institute: This National Cancer Institute (NCI) standard provides minimum facility requirements for NIH/NCI funded grants involving oncogenic viruses.

Guide for Care and Use of Laboratory Animals, Institute of Laboratory Animal Resources (ILAR): This set of guidelines developed through the National Research Council provides the basic procedures to be used for laboratory animal research, and is considered a primary reference on animal care and use.

Animal Welfare, USDA 9CFR Parts 1,2,3 : These regulations of the U.S. Department of Agriculture define the requirements for licensing, registration, identification, records, facilities, health, and husbandry for all animals covered by the Animal Welfare Act.

Most of the above documents are available on the Internet from the appropriate government agency. Complete references are included at the end of the chapter. In addition to these federal codes, there are also state, local, and institutional codes, guidelines, and design criteria that apply to biotechnology facilities.

federal Department of Health and Human Services whose mission is "to promote health and quality of life by preventing and controlling disease, injury, and disability" (CDC Mission Statement). The **NIH** *is a federal health agency comprised of 25 separate centers and institutions. The agency performs and funds major biomedical research initiatives and provides guidelines for laboratory safety.* The joint CDC/NIH guidelines form the basis for many of the standard biosafety precautions used in laboratories across the country.

II. STRATEGIES FOR MINIMIZING THE RISKS OF BIOHAZARDS

A. Standard Practices and Containment

The hazards of pathogenic organisms are generally reduced by strategies of good laboratory practices and containment. **Standard microbiological practices** *are the basic practices that should be used when working with all microbiological organisms.* These practices have two purposes; to separate the worker from the microorganism, and to maintain the purity of the microbial cultures. These recommended practices, which are similar to those used for chemical safety, are summarized in Table 30.3. Every biotechnologist should be familiar with the practices in this table.

Always wear appropriate PPE when working with biohazards, Figure 30.2. This minimally includes lab coat, gloves, and eye and face protection. Wash your hands as soon as possible after removal of gloves or other PPE, and immediately after any potential skin contact with biohazardous materials. All PPE must be removed before leaving the laboratory. If lab coats or other nondisposable items become contaminated, they should be disinfected and then laundered on site (not at home).

Containment *is the control of biohazards by isolation and separation of the organism from the worker.* There are many terms used in the literature to refer to con-

Table 30.3 STANDARD MICROBIOLOGICAL PRACTICES

- Access to the laboratory should be limited to trained individuals.
- Lab coats and eye protection should be worn at all times.
- Workers must wash their hands after any work with microorganisms and whenever they leave the lab.
- Eating, drinking, and smoking in the laboratory area are prohibited.
- Hand-to-mouth or hand-to-eye contact must be avoided.
- Mouth pipetting of any substance in the laboratory is prohibited.
- Steps must be taken to minimize aerosol production.
- Work should be performed on a clean hard bench top with appropriate disinfectant readily available.
- Work surfaces should be decontaminated after any spill, and at the end of every work session.
- All biological materials must be properly decontaminated before disposal.

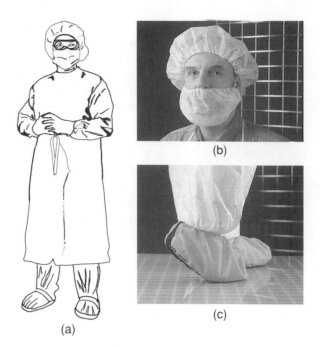

(a)   (b)   (c)

Figure 30.2. Protective Personal Apparel for Routine Work with Body Fluids. a. PPE protection for biohazardous work includes long gown, cap, eye shield, face mask, gloves, and shoe covers. Specialized PPE items are available for specific applications, such as **b.** Beard covers, and **c.** Shoe covers with anti-static strips to prevent static discharges (Photos courtesy of Fisher Scientific, Pittsburgh, PA).

tainment practices. In this text, we use the term **primary containment** *to refer to equipment and practices that protect personnel and the immediate laboratory environment from hazardous exposure.* **Secondary containment** *refers to laboratory design features, equipment, and practices that protect the general environment.* **Personal containment** *refers to standard worker practices, such as those outlined in Table 30.3, used to reduce the spread of microorganisms.* The procedures used for work with and disposal of the biohazardous material, along with the practices of proper laboratory hygiene, are examples of personal containment. **Physical containment** *includes laboratory design features and the physical barriers that workers use to isolate biohazards.* Special ventilation systems, PPE, gloves, and biological safety hoods are examples of physical containment.

B. Recommended Biosafety Levels

The CDC classifies microorganisms into four Biosafety Levels according to their risk and the containment methods required for handling them. These Biosafety Levels dictate progressively higher levels of containment for more hazardous microorganisms. **Biosafety level 1 (BSL1)** *is used for well-characterized strains of living microorganisms that are not known to cause disease in healthy adult humans.* BSL1 organisms have been used safely in many laboratories. High school and college students commonly work with BSL1 organisms. Examples of organisms in this category are non-pathogenic strains of *E. coli*, yeasts, and most plants. The fact that BSL1 organisms are generally nonpathogenic does not mean that they can be handled without caution. Standard microbiological practices must be applied to handling these organisms.

Biosafety level II (BSL2) *is designated when working with agents that may cause human disease and therefore pose a risk to personnel.* Diseases caused by BSL2 organisms are usually treatable or preventive vaccines are available. These organisms are not likely to spread to the external environment and create a general health threat. Organisms in the BSL2 category are pathogenic bacteria such as *Salmonella* and *Clostridium*, molds such as *Penicillium*, and viruses such as polio virus and rabies virus. As discussed later, human blood and tissue products are handled in a BSL2 environment.

The health risks associated with BSL2 organisms require a higher level of containment than BSL1. Written procedures and special worker training are often required. Physical containment procedures may require that all work be performed in a biological safety cabinet to minimize the release of microorganisms, especially in the form of aerosols. Waste must be decontaminated before disposal. For some infectious agents, the health of the worker may be monitored and the worker may be required to be immunized against the infectious agent. Production of blood antibodies against the microorganism may be checked to determine whether an unvaccinated worker has been exposed to a specific organism.

Biosafety Levels 3 (BSL3) and **4 (BSL4)** are generally associated with more dangerous agents that are highly infectious. These hazardous agents require significantly higher containment barriers in the form of special facilities. The airflow into and out of the room where such organisms are handled is closely monitored. Examples of organisms requiring BSL3 facilities are certain arboviruses (arthropod-borne viruses associated with several human diseases) and large cultures of *Mycobacterium tuberculosis*, the etiological agent for human tuberculosis. (Note: Handling small amounts of *M. tuberculosis* can be performed under BSL2 conditions.) Very dangerous BSL4 agents require closed glove boxes, isolated air supplies, and an enclosed breathing apparatus for the operators. BSL4 facilities are found in only a few places in the world. They are required for handling Ebola virus and related organisms that cause rapidly fatal human disease. Table 30.4 provides an overview of the containment practices and equipment required at each of the four biosafety levels. Many of the requirements shown in this table are discussed later in this chapter.

Table 30.4 *SUMMARY OF RECOMMENDED BIOSAFETY LEVELS FOR INFECTIOUS AGENTS*

Level	Agents	Special Safety Practices	Special Equipment	Facilities
1	Not associated with disease in healthy adults	Standard microbiological practices	No special equipment required	Handwashing sink required
2	Associated with human disease that is usually treatable, mild, or has an effective vaccine available	BSL1 practices plus: • Biohazard warning signs • "Sharps" precautions • Biosafety manual defining any needed waste decontamination or medical surveillance policies	Class I or II BSCs or other physical containment devices used for all manipulations of agents that may produce aerosols of infectious materials	BSL1 plus: Autoclave available
3	Associated with severe or lethal human disease, aerosol transmission	BSL2 practices plus: • Controlled access to laboratory • Decontamination of all waste • Decontamination of all lab clothing before laundering • Baseline blood serum testing for workers	BSL2 plus: • Physical containment used for all manipulations • Additional protective clothing and respiratory protection as needed	BSL2 plus: • Physical separation from access corridors • Self-closing, double-door system • Exhausted air not recirculated • Negative airflow into laboratory
4	Associated with a high risk of lethal human disease, aerosol-transmitted infections, or agents with unknown risk of transmission	BSL3 practices plus: • Clothing change before entering • Shower on exit • All material decontaminated on exit from facility	All procedures are conducted in Class III BSCs **or** in Class I or II BSCs in combination with full-body, air supplied, positive pressure personnel suit	BSL3 plus: • Separate building or isolated zone • Dedicated air supply/exhaust, vacuum and decontamination systems

BSC, Biological Safety Cabinet; PPE, Personal Protective Equipment
Adapted from *Biosafety in Microbiological and Biomedical Laboratories*, 3rd ed., U.S. Department of Health and Human Services, Public Health Service, Centers for Disease Control and Prevention and National Institutes of Health, U.S. Government Printing Office, Washington, D.C. 1993.

C. Laboratory Design

While good laboratory design can provide significant containment of biohazards, absolute containment is not achievable and generally not necessary, except in a BSL4 facility. Figure 30.3 provides examples of primary and secondary containment barriers in the laboratory.

A sampling of some of the recommended physical containment measures that are incorporated into laboratories meeting the CDC/NIH Guidelines for the four BSLs are provided in Table 30.5. The table lists only a few of the laboratory design features addressed in the guidelines.

D. Biological Safety Cabinets

i. WHAT IS A BIOLOGICAL SAFETY CABINET?

Laboratory hoods *are enclosed spaces with separate air supplies or directed air flows.* There are many types of hoods, each designed for specific safety purposes. Chem-

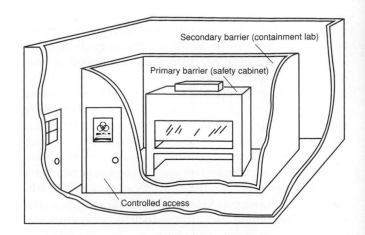

Figure 30.3. The Relationship Between Primary and Secondary Barriers for Hazard Containment. (Reprinted with permission from *The Foundations of Laboratory Safety. A Guide for the Biomedical Laboratory*, Stephen R. Rayburn, Springer-Verlag, New York, p. 105, 1990.)

Table 30.5 SAFETY-RELATED LABORATORY DESIGN FEATURES

Laboratory design feature	Biosafety Level			
	1	2	3	4
• Laboratory separated from public areas by a door	yes	yes	yes	yes
• Hand washing facilities in laboratory	yes	yes	yes	yes
• Dedicated hand washing facilities with automatic, foot, or knee controls		rec.	yes	yes
• Autoclave in building	rec.	yes	yes	yes
• Autoclave in laboratory			yes	yes
• All laboratory surfaces and furnishings are easily cleaned and disinfected	rec.	rec.	yes	yes
• Laboratory doors labeled with biohazard signs		yes	yes	yes
• Testing of biological safety cabinets meets required specifications after installation		yes	yes	yes
• Biological safety cabinets Class I		rec.	yes	no
• Biological safety cabinets Class II		rec.	yes	yes
• Biological safety cabinets Class III				yes
• Biological safety cabinets equipped with pressure monitoring gauges for all HEPA filters		rec.	yes	yes
• Doors lockable and self-closing, access limited to authorized personnel		rec.	yes	yes
• Air exhausted from laboratory at a minimum of ten room volumes per hour		rec.	yes	yes
• Standby electrical generator available for emergency support of essential equipment such as BSCs		rec.	rec.	yes
• Laboratory perimeter sealed to allow gaseous decontamination of whole room			yes	yes
• Clothing change area adjacent to containment area			yes	yes
• Body shower in containment area			rec.	yes
• Independent air supply			rec.	yes
• Ventilated airlock required for the separation of high containment area				yes

rec. = recommended

ical fume hoods, for example, are designed to remove volatile gases from the operator. This is accomplished by drawing air from the room past the working space and venting the air outside the building (see Chapter 29).

A cabinet used to handle biohazards superficially resembles a chemical fume hood, but has a different purpose and construction. **Biological safety cabinets (BSCs)** *provide containment for aerosols and separate the work material from the operator and the laboratory, while providing clean air within an enclosed area.* They are designed to filter all air that passes through the cabinet, using HEPA filters. A **HEPA (High Efficiency Particulate Air) filter** *is a specially constructed filter made of highly pleated glass and paper fibers,* Figure 30.4. HEPA filters are designed to have a 99.97% efficiency in removing particles of 0.3 μm or larger. This filter size will exclude most bacteria and microorganisms. Viruses found in aerosols are usually associated with particles larger than 0.3 μm and so HEPA filters provide protection against most airborne virus particles. It is important to remember that HEPA filters are designed to remove particulate matter from the air. They are not effective for removing chemical vapors from air.

BSCs contain fans that direct air into a non-turbulent curtain. Air is passed through the HEPA filter and directed over the work surface at a velocity that minimizes disturbances in the interior of the cabinet. This type of airflow is called laminar flow and therefore a **laminar flow cabinet** *is one that has a sterile, directed airflow.* All BSCs have laminar flow designs; however, not all laminar flow cabinets are BSCs. This is because some laminar flow hoods, such as the **clean bench**, or **horizontal laminar flow hood**, *are designed to provide a sterile work surface but not worker protection,* Figure 30.5. Air passes through the HEPA filter onto the work surface and is vented directly into the room toward the operator. This type of laminar flow cabinet is useful for media preparation or sterile work with non-harmful organisms. It is not a BSC because it does not provide protection for the operator.

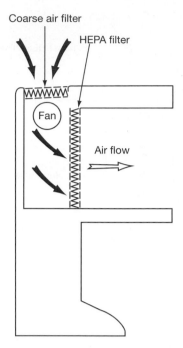

Figure 30.5. Design of a Clean Bench. Non-sterile air is shown as solid arrows, sterile air as open arrows.

ii. CLASSIFICATION OF BIOLOGICAL SAFETY CABINETS

There are three classes of BSCs that provide different levels of protection against biological hazards. A **Class I cabinet** *is designed to protect the operator from airborne material generated at the work surface. The cabinet draws air directly from the room. The air then flows over the working area and is vented back into the room after being filtered.* This cabinet will filter aerosols released from the work materials, and so can be used to reduce aerosol escape to the environment. Because the air flowing over the work surface is room air, this type of hood is not designed to maintain sterility of the work surface. A Class I cabinet is used for routine operations that might generate harmful aerosols, such as vortexing, pouring, or centrifuging solutions containing microorganisms.

A Class II BSC, which is the most common type found in biotechnology laboratories, provides protection for both the operator and the work product. It is designed to draw HEPA-filtered air across the work surface and re-filter the air before venting. Because the air is filtered before it passes over the work surface, this cabinet type will maintain a sterile work area. A Class II cabinet is suitable for microbiology and tissue culture procedures.

Class II cabinets are subdivided into several categories. The most common types are Class IIA and IIB. **Class IIA cabinets** *do not have external ducts, and release filtered air directly into the laboratory,* Figure 30.6. All airflow across the work surface is also filtered.

Class IIB cabinets *maintain a faster airflow than Class IIA cabinets, and are ducted to external exhaust systems.* This prevents any cabinet air recirculation in

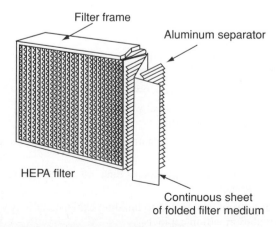

Figure 30.4. Structure of a HEPA Filter.

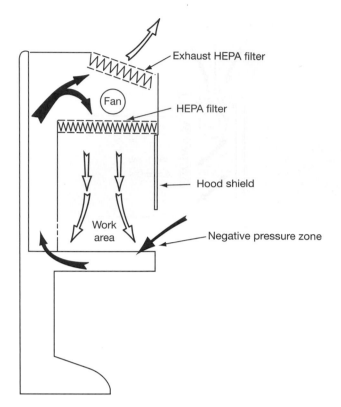

Figure 30.6. Design of a Class IIA BSC. Non-sterile air is shown as solid arrows, sterile air as open arrows.

the laboratory. Class IIB cabinets can be used with low levels of volatile chemicals, although it should be noted that this type of work shortens the life of the HEPA filter systems. There are three subtypes of Class IIB cabinets: IIB1, IIB2, and IIB3. These differ in the levels of recirculated filtered air within the cabinet, Table 30.6.

Class III cabinets, also called **glove boxes**, are gastight cabinets designed to provide total containment for extremely hazardous biological agents. They provide the highest level of protection possible for both user and environment. Class III cabinets are designed to isolate the work material completely within a closed system. A

Class III cabinet is accessed with air-tight gloves. Materials in these cabinets usually require decontamination before removal from the cabinet. Class III cabinets often have incubators, autoclaves, and air locks directly attached. A Class III cabinet would only be found when working with the most hazardous agents, and special rooms would also be required for proper use.

Table 30.6 summarizes the basic characteristics of the described biological safety cabinets.

iii. PROPER USE OF CLASS II BIOSAFETY CABINETS

Class II cabinets are appropriate for both microbiology and tissue culture work. They provide effective containment and a sterile air supply to the working surface. These cabinets need to be inspected and tested periodically to ensure the filter is in good condition and the airflow is adequate. A certified technician must decontaminate and replace the filters at regular intervals.

Many of these BSCs contain a UV germicidal lamp inside the work area. **Germicidal** *refers to methods that are capable of killing bacteria or other microorganisms.* When using a BSC, turn on the UV lamp for a minimum of 15 minutes before starting work. Because of the radiation, the lamp must be turned off when the cabinet is in use. Serious eye damage and burns to the hands can result if the light is accidentally left on during cabinet use. At least 5 minutes before using the cabinet, start the airflow to purge the hood of unfiltered air. Turn off the germicidal lamp and place all the needed materials in the hood. Only items that are needed in the workspace should be placed in the cabinet. Figure 30.7 shows a recommended layout for materials within a BSC.

All operations in a BSC should be planned to minimize disruptions in the air flow through the cabinet. Move hands and other objects slowly into and out of the cabinet interior. Strong air currents created outside the cabinet by people moving rapidly or laboratory doors opening may also disrupt the airflow pattern and

Table 30.6 COMPARISON OF TYPES OF BIOLOGICAL SAFETY CABINETS

Type	User Protection	Work Surface Protection	Face Velocity (in linear feet per minute)	Airflow Pattern
Clean bench (not a BSC)	No	Yes	Varies	Air in at top, exhausted at front toward operator
Class I	Yes	No	75	Air in at front, exhaust duct within lab
Class IIA	Yes	Yes	100	Air in at front, exhaust duct within lab, 70% air recirculation within cabinet
Class IIB1	Yes	Yes	100	Air in at front, external exhaust duct, 30% air recirculation within cabinet
Class IIB2	Yes	Yes	100	Air in at front, external exhaust duct, no air recirculation within cabinet
Class IIB3	Yes	Yes	100	Air in at front, connected to building exhaust, 70% air recirculation within cabinet
Class III	Yes	Yes	None	Filtered air ducts for supply and exhaust, gas-tight

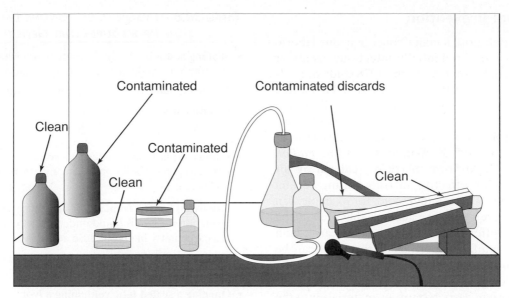

Figure 30.7.   Suggested Layout for Work Materials in a BSC. Note that all materials should be placed 4–6 inches back from the front airfoil.

reduce the effectiveness of a BSC. The use of Bunsen burners and open flames should be avoided or kept to a minimum. The heat generated produces air currents that can affect the sterility of the work area, and the added heat may damage the HEPA filter. Table 30.7 provides guidelines for the proper use of Class II BSCs.

Table 30.7   SAFE AND EFFECTIVE USE OF CLASS II BIOSAFETY CABINETS

Class II BSCs are designed to provide protection for workers, the work product, and the environment when users observe basic guidelines for working in the cabinet.

General Guidelines
• Understand how the cabinet works.
• Do not disrupt the protective airflow pattern of the BSC.
• Plan your work.
• Minimize the storage of materials in and around the BSC.

Cabinet Set-up Guidelines
• Turn on the germicidal lamp if present; wait at least 15 minutes before proceeding with any activities in the cabinet.
• Turn on the airflow in the BSC at least 5 minutes before starting to place objects in cabinet. (Note: in some laboratories, the BSC is never turned off.)
• Be sure that all vents and airfoils are clean and unblocked.
• Before starting to work, wash hands and lower arms thoroughly using a germicidal soap and wipe the work surface in the cabinet with 70% alcohol or other suitable disinfectant.
• Assemble all the materials you will need for your work and wipe each item with 70% alcohol before placing it inside the cabinet.
• Do not overload the cabinet.
• Wait another 5 minutes before beginning to work in the cabinet.

Cabinet Use Guidelines
• Keep all materials at least 4 inches inside the cabinet face, and work with biological materials as deep within the cabinet as feasible.
• Move arms slowly when removing or introducing new items into the BSC.
• Do not place any objects over the front air intake vents or blocking the rear exhaust vents.
• Segregate contaminated and clean items; work from clean areas towards contaminated areas (see Figure 30.7).
• Place a pan with disinfectant and a sharps container (if necessary) inside the BSC for pipette discard; do not use vertical pipette wash canisters on the floor outside the cabinet.
• Place any equipment that may create air turbulence (such as a microcentrifuge) at the back of the cabinet (but not blocking the rear vents).
• Clean up any spills within the cabinet immediately; wait at least 10 minutes before resuming work.
• When finished, remove all equipment and materials and wipe all interior surfaces of the cabinet with disinfectant.
• Remove PPE and wash hands thoroughly before leaving the laboratory.

E. Bioaerosol Prevention

Airborne materials pose a major concern in any laboratory or facility where potentially infectious agents or hazardous materials are used. In the 1978 study of LAIs cited on page 641, fewer than 20% could be linked to a specific spill or other accident. Researchers believe that about 70% of all LAIs result from inhalation of infectious particles (bioaerosols). (A review of this issue is: "In Pursuit of Safety" R. Williams. *J. Clin. Pathol.* 34:232–239, 1981). Airborne particles (carried by air currents) are highly effective in transmitting certain types of infections. The common cold, for example, can be spread by sneeze-generated aerosols that are inhaled by the next cold sufferer. Airborne organisms can remain suspended for significant periods of time and can be carried through a laboratory by air currents and even throughout buildings by way of the ventilation system. If inhaled, aerosols can carry hazardous materials into the lungs. Aerosols will also eventually settle on laboratory surfaces, where they can deposit hazardous substances.

One of the most important steps that you can take to minimize your risk of LAI is to reduce the presence of aerosols in your working environment. Table 30.8 lists examples of common laboratory operations that generate airborne particles. Any activity that mechanically disturbs a liquid or a dry powder has the potential to release airborne substances. Care must therefore be used whenever handling any culture containing microorganisms, cells, or infectious agents, and when manipulating solutions containing hazardous substances. Because aerosols are not visible, and because they are created during routine laboratory operations, laboratory workers are often unaware of their presence. Always be careful and deliberate when handling potentially hazardous liquid cultures, and use containment devices as appropriate.

Complete prevention of aerosol formation is not possible. By using careful technique, however, it is possible to greatly reduce their formation rate. For example, it is good practice to wait a few minutes after centrifugation before opening centrifuge tubes and rotors. This allows time for aerosols to settle. Table 30.9 includes examples of other practices that reduce aerosol formation. Even when careful practices are used, some airborne particles are still generated. Most laboratories therefore provide various containment barriers to control unavoidable aerosols. For example, BSCs can be used for operations that are likely to generate aerosols, such as pouring fluids, blending tissue samples, and handling cultures. It is also possible to purchase accessories for centrifugation that safely enclose potentially hazardous materials.

F. Use of Disinfectants and Sterilization

Bioaerosols or work-related splashes can leave contamination on laboratory work surfaces. An effective practice for preventing the spread of biohazards is therefore

Table 30.8   EXAMPLES OF ROUTINE LABORATORY PROCEDURES THAT GENERATE AEROSOLS

- **Shaking and mixing liquids**, particularly when using vigorous mixing devices like blenders and vortex mixers

- **Pouring liquids** from one container to another

- **Pipetting liquids** generates aerosols when:
  - the last drop of liquid is ejected from a pipette tip
  - a micropipettor tip is ejected into a disposal container
  - liquids splash upon entering a container
  - a used pipette is dropped into a vertical cylinder for cleaning

- **Removing the cap from a tube** can cause aerosol formation if there is liquid under the cap; "snap cap" tubes are particularly likely to cause aerosols.

- **Removing a stopper or cotton plug from a culture bottle or flask**

- **Opening a sealed tube containing a lyophilized (freeze-dried) agent**. When reconstituting lyophilized materials, add liquid to the contents slowly. Mix the contents without bubbling or excessive agitation.

- **Breaking cells open by sonication**

- **Grinding cells or tissues** with a mortar and pestle or homogenizer

- **Bubbling air into a liquid (aeration)**

- **Centrifuging samples** is particularly problematic. Note the following:
  - Centrifugation in uncapped tubes causes aerosols to be released; never centrifuge hazardous materials in uncapped tubes.
  - A centrifuge tube that breaks during spinning will spew its contents into the air.
  - Even if a tube is unbroken and capped, it is possible for the cap to deform under the force of centrifugation, allowing liquid to leak out and form aerosols during spinning.

- **Inserting a hot loop into a bacterial culture**. It is common practice to flame metal loops or needles (to kill any microorganisms) and then to place the loop or needle into a liquid culture. If the loop is hot when inserted into the culture, splattering can occur which leads to aerosol formation. Allow needles and loops to cool before placing them in a culture.

the use of proper decontamination procedures. Decontamination can be accomplished with the use of germicidal agents. Some of the most common agents are heat (autoclaving), gas exposure (using ethylene oxide, for example), and the use of liquid chemical disinfectants. **Disinfectants** *are chemicals that kill pathogenic microorganisms and other biohazardous particles*. There are many factors that determine the effectiveness of germicidal agents, such as:

- the type of contaminating organism and its susceptibility to disinfecting agents

Table 30.9 GOOD PRACTICES THAT REDUCE EXPOSURE TO AEROSOLS

- Whenever possible, use a BSC when performing any operation that may generate hazardous aerosols.
- Before opening containers whose contents have been shaken, blended, or centrifuged, wait a few minutes to allow aerosols to settle.
- Pipetting practices that reduce aerosol formation and protect the operator include:
  - Never pipette by mouth.
  - Plug the tops of pipettes with cotton.
  - Never mix a liquid that contains hazardous material by pipetting up and down.
  - Use pipettes that do not require expulsion of the last drop of liquid.
  - Do not discharge biohazardous material from a pipette in a manner that induces splashing; when possible, allow the discharge to run down the container wall.
  - Place contaminated pipettes horizontally in a pan containing enough decontaminant to allow complete immersion.
- Centrifuge practices that reduce aerosol formation and/or protect the operator include:
  - Use centrifugation rotors and tubes designed to contain hazardous materials.
  - Fill and open centrifuge rotors and tubes in a BSC.
  - Carefully and completely disinfect rotors and tubes used for hazardous materials according to the manufacturer's directions.
  - Discard chipped or cracked centrifuge tubes.
  - Routinely allow the contents of rotors and tubes to settle for a few minutes before opening.
- Avoid pouring liquids or decanting supernatants if possible; it is preferable to use a vacuum system with appropriate in-line safety reservoirs and filters to remove the liquid from a centrifuge tube or other vessel.

- the level of contamination
- the chemical composition and concentration of the decontaminating agent

- the length of exposure of the organism to the decontaminant
- the shape and texture of the surface being decontaminated

Biohazardous agents differ in their susceptibility to disinfection procedures, Table 30.10. For example, most growing bacteria are relatively easy to kill, but bacterial spores are difficult to eradicate.

There are distinctions between sanitizing, disinfecting, and sterilizing a surface or object. **Sanitization** *is the general reduction of the number of microorganisms on a surface.* Sanitization is generally the purpose of antiseptics. An **antiseptic** *is a type of mild disinfectant gentle enough to be applied to skin.* Chemical solutions that are suitable for use as antiseptics, such as 3% hydrogen peroxide or 70% isopropyl (rubbing) alcohol, may not provide enough disinfectant strength for work surfaces. **Disinfection** *is the removal of all or almost all pathogenic organisms from a surface.* Disinfectants can be classified as low, intermediate, and high level, according to their capacity to destroy microorganisms. **Sterilization** *is the highest level of disinfection; the killing of all living organisms on a surface.* High level disinfectants achieve or approach sterilization, whereas low level disinfectants are of variable effectiveness, depending on the nature of the contaminating organisms. Table 30.11 provides a summary of the sterilization and disinfection capabilities of some commonly used agents. Note that many chemical agents cannot be used effectively for sterilization purposes.

The U.S. EPA labels disinfectants as pesticides for regulatory purposes. The Federal Insecticide, Fungicide, and Rodenticide Act (FIFRA) *authorizes EPA to control the distribution, sale, and use of pesticides, and requires all manufacturers to register their pesticides and disinfectants with the EPA before marketing.* Disinfectants that are used on medical devices must be listed with the U.S. Food and Drug Administration. Com-

Table 30.10 DISINFECTION RESISTANCE OF MICROORGANISMS

Resistance Level to Disinfection	Type of Organism	Examples of Organisms
Least resistant	Hydrophobic and/or medium-sized viruses	Human immunodeficiency virus Herpes simplex virus Hepatitis B virus
↓	Bacteria	*Escherichia coli* *Staphylococcus aureus*
↓	Fungi	*Candida* sp. *Cryptococcus* sp.
↓	Hydrophilic and/or small viruses	Rhinovirus Poliovirus
↓	Mycobacteria	*Mycobacterium tuberculosis*
Most resistant	Bacterial spores	*Bacillus subtilis* spores *Clostridium sp.* spores

Table 30.11 STRENGTH OF COMMON GERMICIDAL INGREDIENTS

Ingredient	Concentration for Sterilization	Disinfection Concentration	Disinfection activity level	Example of Commercial Product
Alcohols	NA	70%	Intermediate	
Chlorine mixtures	NA	500–5000 *ppm* chlorine (1–10% bleach solution)	Intermediate	Chlorox®
Formaldehyde	6–8%	4–8%	High	
Glutaraldehyde	2% at high pH	2%	High	Microklene®
Hydrogen peroxide	6–30%	3–6%	High	
Iodophor mixtures	NA	40–50 *mg* free iodine/liter	Intermediate	Betadine®
Phenolic mixtures	NA	0.5–3%	Intermediate	Lysol®
Quaternary ammonium mixtures	NA	0.1–0.2%	Low	Roccal®

NA = not applicable
Data assembled from various sources; see references at the end of the chapter.

monly used disinfectants are also covered under the Toxic Substance Control Act (TSCA), which establishes chemical inventory and record keeping requirements.

A major consideration when choosing a disinfecting agent is the nature of the biohazardous organism that must be removed from a surface. For human safety reasons, it is best to use the least toxic chemical that can get the job done. Table 30.12 shows the relative effectiveness of a selection of chemical disinfecting agents against specific types of biohazardous organisms.

All disinfection and sterilization methods present certain safety hazards to the worker as well as to the targeted microorganisms. Table 30.13 provides exam-

ples of some of the human hazards associated with germicidal techniques.

III. SPECIFIC LABORATORY BIOHAZARDS

A. Microorganisms

Biotechnologists often work with microbiological agents such as bacteria, fungi, protozoan parasites, and viruses. Bacteria, fungi, and viruses are easily spread by aerosols or by hand-to-mouth contamination. Protozoans are more likely to pose a threat when ingested. Skin contact

Table 30.12 ANTIMICROBIAL ACTIVITY OF COMMONLY USED CHEMICAL DISINFECTANTS

Antimicrobial agent	Range of effectiveness				
	Bacteria	Mycobacterium tuberculosis	Bacterial spores	Fungal spores	Viruses
Ethylene oxide gas (500 to 800 *mg/L*)	4	4	4	4	4
Ethyl alcohol (70%)	2	2	0	2	1
Sodium hypochlorite (5% bleach)	3	2	0	2	2
Formaldehyde (37% aqueous)	2	2	2	2	2
Glutaraldehyde (2%)	2	2	3	2	2
Iodophors (1%)	2	2	1	1	2
Phenolic derivatives (1% to 3%)	2	2	0	2	1
Quaternary ammonium compounds (3%)	2	0	0	2	1

4 — Superior, 3 — Very good, 2 — Good, 1 — Fair, 0 — No activity
Adapted from *Laboratory Safety: Principles and Practices,* Diane O. Fleming, John H. Richardson, Jerry J. Tulis and Donald Vesley, 2nd edition, ASM Press, Washington, DC, 1995.

Table 30.13   HUMAN HAZARDS ASSOCIATED WITH SELECTED ANTIMICROBIAL AGENTS

Antimicrobial Agent	Potential Hazards
UV light	Eye and skin burns; mutagen; carcinogen
Steam (autoclave)	Burns; aerosols and chemical vapors; cuts from imploding glassware
Ethylene oxide	Severe respiratory and eye irritant; sensitizer; mutagen; suspected human carcinogen
Isopropyl alcohol	Flammable; toxic; severe eye irritant
Chlorine (bleach)	Gaseous form highly toxic; oxidizing agent
Formalin (37% formaldehyde)	Strong eye and skin irritant; toxic; sensitizer; mutagen; suspected human carcinogen
Glutaraldehyde	Toxic; skin irritant
Hydrogen peroxide	Oxidizing agent; corrosive; skin irritant
Iodine (iodophors)	Skin irritant; toxic vapors; sensitizer
Phenols	Chemical burns; toxic vapors; strong eye and skin irritant
Quaternary ammonium compounds	Skin irritant

with fungi and parasites also tends to induce allergic reactions, which range from annoying to life-threatening.

The pathogenic properties of many bacteria have been characterized and bacteria and other microorganisms are classified into risk groups, corresponding to the biosafety levels described previously. Most biotechnology work uses non-pathogens when possible, and any potential pathogens are usually identified at the onset of work. However, good laboratory practices are necessary when working with both pathogenic and non-pathogenic species. A worker can become infected with a non-pathogenic species and spread the organism to environments outside the laboratory. Good practices will also prevent the spread of unidentified or unknown organisms.

Molds and fungi are ubiquitous airborne organisms, with millions of fungal spores in an average cubic meter of unfiltered air. **Fungi** *are primitive eukaryotic microorganisms that have complex life cycles and can form spores.* There is an incredible diversity of fungal species. **Molds** *are filamentous forms of fungi, presenting a "fuzzy" appearance when they grow.* The genus *Penicillium* is an example of a mold that produces a useful product; in this case, the antibiotic penicillin. **Yeasts** *are single-celled forms of fungi.* The yeast we are most familiar with is *Saccharomyces cerevisiae*, which is common baker's yeast. *S. cerevisiae* is often used in laboratories as a research organism.

Fungi can sometimes be dangerous to humans through several mechanisms. A **mycosis**, or **mycotic disease**, *is a disease caused by infection with a pathogenic fungus.* Some skin mycoses, such as "athlete's foot," are relatively common and do not pose a serious threat to healthy individuals. Systemic mycoses occur occasionally, although the spread of systemic fungal disease between individuals is rare.

Many fungi produce **mycotoxins**, *which are toxic products that may render the fungus poisonous, or contaminate the environment of the fungus.* The common mold *Aspergillus flavus*, for example, produces aflatoxin, a highly potent carcinogen. Special precautions should be taken to avoid personal contamination when working with mycotoxins or any fungal species known to produce mycotoxins.

One of the most likely problems to arise from exposure to fungi in the laboratory is the development of allergies to the fungal spores. Fungal allergies are common, and are easily triggered in the presence of significant numbers of airborne spores. Because biotechnology facilities are increasingly using fungi to produce enzymes or food products, occupational exposure becomes more likely. Proper containment and reduction of airborne spores in the immediate environment are essential measures to prevent allergies and environmental contamination.

B. Viruses

Viruses *are particles that contain nucleic acid usually surrounded by a protein coat of varying complexity. They require a cellular host for replication.* Viruses can contain either DNA or RNA as their genetic material. Both RNA- and DNA-containing viruses have the ability at times to integrate their nucleic acid sequences into an infected cell's genome. In some instances this viral DNA can direct the cell to make new virus particles, which will then be released as infectious viruses.

Many types of viruses have the ability to remain dormant for long periods of time, and in some cases, to resist disinfection and sterilization procedures (as shown in Table 30.10). Because virus particles are much smaller than other microorganisms, Figure 30.8, they can aero-

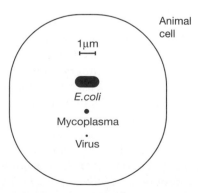

Figure 30.8.   Relative Sizes of Biological Organisms Used in Laboratories. These include animal cells, bacteria, and viruses.

solize readily, and are relatively difficult to filter from the air at high concentrations (although HEPA filters are effective at filtering virus particles associated with aerosol particles). It is essential to use standard microbiological practices when handling viruses and virally-infected cells. These practices provide the dual benefits of worker protection and protection of the work material from infection with extraneous viruses.

Viruses pose risks of acute and sometimes chronic infection. Acute infection can be transitory and harmless, or it can result in disease. Viruses are classified into the same risk levels as other microorganisms. Always know the properties of and any potential risks posed by viruses that are used in your workplace.

The greatest concern in laboratories that use human tissue or blood products is unknown contamination of materials with HBV or HIV. **HBV**, or **hepatitis B virus**, *is the etiological agent for human hepatitis B.* **HIV**, or **human immunodeficiency virus**, *is the etiological agent for acquired immunodeficiency syndrome (AIDS).* Most of us are more aware of and concerned about HIV infection, but in reality, the rate of transmission of HIV is much lower than that of HBV. For example, the apparent risk of infection with HIV after skin puncture with an HIV-contaminated needle is less than 1%, based on studies of hospital exposures. The risk of infection with HBV is at least 10-fold higher after skin puncture with an HBV-contaminated needle (in unvaccinated individuals).

Hepatitis B *is a bloodborne infection that attacks the liver and causes inflammation.* Most people who contract hepatitis B exhibit moderate to severe flu symptoms. From 1 to 5 percent of these individuals will develop chronic disease, which can eventually prove fatal. Another 5 to 10 percent of infected individuals become chronic carriers of HBV. Fortunately there is a reasonably effective (80–95%) vaccine available that should be administered to all workers who contact human blood or blood products. This vaccine requires a series of three injections. Federal guidelines mandate that this vaccine be offered free of charge to any individual who works with or may be exposed to human tissue or blood products in the workplace.

HIV is an example of a type of RNA virus called a retrovirus. A **retrovirus** *is an RNA-containing virus that also contains **reverse transcriptase**, an enzyme that transcribes RNA into DNA.* This DNA copy of the viral genetic material can then integrate into the DNA of a host cell. With some retroviruses, this alteration may be undetectable except by molecular analysis, and does not lead to further problems. In other cases, such as HIV infection, adverse effects become apparent with time.

Without treatment, HIV infection eventually leads to the development of AIDS. **AIDS**, or **acquired immunodeficiency syndrome**, *is a condition in which the T-cells of the immune system are destroyed, devastating the body's immune defenses and allowing other infectious agents to cause disease and frequently death.* Individuals with AIDS, or other conditions that weaken the immune system, such as chemotherapy, are especially susceptible to infection with microbiological agents that are considered to be relatively non-pathogenic in healthy individuals.

Although there is no vaccine currently available for HIV, there are treatments that help to delay the deterioration of the immune system and therefore the manifestation of immunodeficiency symptoms in HIV-infected individuals. These treatments can slow the development of AIDS and increase both life expectancy and quality of life for infected individuals. Fortunately, HIV is not highly transmissible under laboratory conditions, and the virus is susceptible to inactivation with desiccation or disinfection procedures.

Another potential consequence of exposure to certain types of viruses is the development of cancer. **Cancer** *includes a variety of diseases characterized by the uncontrolled growth of cells and the ability of these cells to spread to other parts of the body* (**metastasis**). **Oncogenic viruses** *are agents that can induce cancer after infecting cells.* This occurs when the viral infection either interferes with normal cellular growth control, or alters the cell's genetic material in a manner that affects growth control.

Potentially oncogenic viruses are classified as of low, moderate, or high risk to humans. A low risk virus is one that has the theoretical capability of being oncogenic, but has not yet shown any evidence of a threat. A high risk oncogenic virus would be one that has been shown to cause cancer in humans. No virus has been definitively placed in this category at present, although there are several candidates. Many DNA- and RNA-containing viruses have been classified as moderate risk (a few examples are shown in Table 30.14). It is important to note that many individuals become infected with these viruses (such as Epstein-Barr or hepatitis B and C viruses) without developing cancer as a result. The criteria for designating moderate risk oncogenic viruses are shown in Table 30.15.

Table 30.14 EXAMPLES OF VIRUSES CONSIDERED TO POSE A MODERATE RISK FOR CAUSING HUMAN CANCER

DNA-Containing Viruses	RNA-Containing Viruses
Papilloma viruses (various types)	Hepatitis B and C viruses
Herpesvirus ateles virus (HVA)	Feline leukemia virus (FeLV)
Epstein-Barr virus (EBV)	Feline sarcoma virus (FeSV)
Yaba pox virus	Wooly monkey fibrosarcoma virus (SSV-1)
Any DNA-containing virus isolated from a human tumor	Any RNA-containing virus isolated from a human tumor

Table 30.15 *CRITERIA FOR CLASSIFICATION AS A MODERATE RISK ONCOGENIC VIRUS*

- The virus has been isolated from human cancers.
- The virus causes cancer in nonhuman primates without experimental manipulation of the host.
- The virus can produce tumors in healthy nonprimate mammals.
- The virus can transform human cells *in vitro*.
- Any virus or genetic material that is a recombinant of an oncogenic animal virus and a microorganism infectious for humans is classified as a moderate oncogenic risk until its human oncogenic potential has been determined.
- Any large scale or concentrated preparation of infectious virus or viral nucleic acid is classified as a moderate oncogenic risk until its human oncogenic potential has been determined.

C. Human Blood Products

Biological materials from humans are a special problem for laboratory workers. There are no test methods available that can guarantee the absence of all infectious agents in biological materials, and infectious agents that are present are likely to be targeted to humans. OSHA therefore requires that all human blood and blood products be handled as though they were contaminated. This involves the routine use of **Universal Precautions**, *which are standard safety practices for handling human blood products*, Table 30.16. In general, Biosafety Level 2 procedures and a Class II biological safety cabinet should be used. All tissue and waste must be treated as pathogenic and be properly decontaminated.

Table 30.17 provides a list of potentially infectious materials that are considered by OSHA to constitute hazardous human products requiring special precautions. Commercially purchased human antibodies, cells derived from humans, and blood factors are sometimes used in research. The supplier will provide a description of pathogen testing with the product, but the absence of all pathogens cannot be guaranteed. These products should therefore be handled with the same precautions as other human materials.

Institutions and organizations where occupational exposure occurs must develop an Exposure Control Plan to comply with the OSHA Bloodborne Pathogens Standard 29CFR Part 1910.1030. **Occupational exposure** *is defined as reasonably anticipated skin, eye, mucous membrane, or parenteral contact with blood or other potentially infectious materials that may result from the performance of an employee's duties* (Occupational Exposure to Bloodborne Pathogens, U.S. Department of Labor, OSHA, Federal Register 56 (235): 64175–64182, 1991). In addition to specifying the use of Universal Precautions, the Exposure Control Plan must also include detailed information about:

Table 30.16 *SUMMARY OF UNIVERSAL PRECAUTIONS*

This is a summary of OSHA standards describing safe handling procedures for the materials described in Table 30.17. Notice the similarities between many of these items and those incorporated into standard microbiological practices (Table 30.3).

- Universal precautions must be followed at all times when blood, other body fluids (as described in Table 30.17), and other potentially infectious materials are handled.
- Workers must wash their hands after handling potentially infectious material, whenever changing gloves, and whenever leaving the laboratory.
- Use of needles and other sharps must be avoided whenever possible; if use is necessary, needles must not be recapped, bent, or otherwise manipulated by hand; used needles and other sharps must be disposed of in nearby puncture-resistant containers.
- Gloves and other protective clothing must be worn when there is potential for contact with blood, or other potentially infectious materials; gloves must be replaced when visibly soiled or punctured; fluid-resistant clothing, surgical caps or hoods, face masks, and shoe covers should be worn if there is a potential for splashing or spraying of blood or other potentially infectious materials.
- Blood or other potentially infectious materials must be placed in leak-proof, properly labeled containers during storage or transport.
- Mouth-pipetting is not allowed.
- Work surfaces must be clean and decontaminated with an appropriate disinfectant after any known contamination, after completion of planned work, and at the end of the day.
- All contaminated waste material must be decontaminated prior to disposal, following proper institutional procedures.
- Laboratory equipment that has been in contact with blood or other potentially infected material must be decontaminated by laboratory personnel before any servicing or shipping.
- Biohazard warning signs must be prominently placed in facilities where blood or other potentially infected material is present.

- exposure determination
- PPE and first aid materials available for employees
- provision of hepatitis B vaccination
- procedures for any post-exposure follow-up
- documentation requirements
- employee training

Employee training should include information about the OSHA standard, explanations of bloodborne pathogens and Universal Precautions, use of PPE and other equipment, availability of HBV vaccination, and emergency procedures.

It is advisable to be cautious when assisting in a medical emergency that may occur in the laboratory. In an

Table 30.17 POTENTIALLY INFECTIOUS
HUMAN PRODUCTS

The following materials are potential sources of human
bloodborne pathogens:

- Human blood and blood components
- Products made from human blood
- The following human body fluids:
  - semen
  - vaginal secretions
  - cerebrospinal fluid (fluid from the nervous system)
  - synovial fluid (fluid from the joints)
  - pleural fluid (fluid from the lungs)
  - pericardial fluid (fluid from the heart)
  - peritoneal fluid (fluid from the abdomen)
  - amniotic fluid
  - saliva in dental procedures
  - any other body fluid that is visibly contaminated with
    blood, such as saliva or vomitus
  - all body fluids in situations where it is difficult or
    impossible to differentiate between body fluids
- Any unfixed tissue or organ (other than intact skin) from
  a human (living or dead)
- HIV-containing cell or tissue cultures, organ cultures, and
  HIV- or HBV-containing culture medium or other
  solutions
- Blood, organs, or other tissues from experimental animals
  infected with HIV or HBV

From Occupational Exposure to Bloodborne Pathogens, U.S.
Department of Labor, OSHA, Federal Register 56 (235): 64175-
64182, 1991.

emergency where a loss of blood occurs, after assisting
the victim, take time to protect yourself by proper disinfection of your skin and clothing and the contaminated
areas. If appropriate, consult a health care professional.

D. Tissue Culture

Tissue culture, *the* in vitro *propagation of the cells
derived from tissue of higher organisms*, presents particular areas of biosafety concern. Cell culture generally
involves growing a monolayer of cells, covered with a
layer of appropriate growth-sustaining liquid medium,
in glass or plastic dishes. Cultured cells have a variety of
biotechnology applications, some of which are listed in
Table 30.18.

Many healthy animal cells are unable to divide and
grow *in vitro*. Cells that do propagate in culture tend to
lose some of the normal cell properties exhibited in their
natural environments. However, early cell researchers
sometimes found that tumor cells are easier to establish
in culture. Moreover, many chemical and viral agents
can transform normal animal cells so that they have
characteristics of tumor cells. In this context, **transform**
*means to cause cells to become competent to grow in
long term culture and develop some properties similar to
tumor cells.*

Table 30.18 USES FOR CULTURED CELLS IN
BIOTECHNOLOGY LABORATORIES

Cultured cells, especially from mammalian sources, have
many uses in biotechnology. These include:

- Tools for biomedical research (e.g., for the study of
  mechanisms of growth control)
- Sources of nucleic acids for the preparation of genomic
  and cDNA libraries
- Sources of species-specific proteins
- Propagation of viruses (e.g., for vaccine production)
- Production of recombinant eukaryotic proteins that
  cannot be expressed in bacteria (e.g., proteins with
  extensive post-translational modifications)

Primary cell cultures *are newly isolated cells from
tissue or blood, growing outside the body for the first
time.* After removal from the body's tissue, the cells are
potential carriers for any infectious agents that were
present in the organism. Primary cell cultures from
humans may therefore contain pathogenic agents that
can infect workers. The most likely viral hazards of primary human cell lines are HIV, HBV, and hepatitis C
virus. Non-human cells, especially from monkeys and
other primates, can also carry agents that are serious
health hazards.

Cell lines, *which are cells that are established in long
term culture*, are less likely to pose an unknown health
threat. The cell lines used in laboratories have generally
been analyzed and any problems are well documented.
Even these cells, however, may harbor hidden virus particles, which may be released upon unusual circumstances. Some established cell lines have the ability to
induce tumor development when injected into laboratory animals. In at least one reported case, a tumor developed where a laboratory worker accidentally injected a
cancer cell line into her hand ("Needlestick transmission
of human colonic adenocarcinoma," Gugel, E.A. and
Sanders, M.E. *New Eng. J. Med.*, 315 (23): 1487, 1986).
Because one of the most common injuries among tissue
culture workers is skin punctures from broken glass or
needles, it is essential to handle sharps with care.

When performing tissue culture, a sterile work surface and sterile materials are essential to avoid contaminating the cells with extraneous microorganisms. The
same practices that maintain the sterility of cell cultures
also serve to protect the worker. Use of a Class II BSC
provides a suitable environment for sterile work, although
with proper technique, tissue culture can be performed
successfully on an open laboratory bench (assuming no
pathogenic agents are involved). All pipettes, culture
dishes, dissection instruments, and other tools must be
sterilized before and after every use. When working with
biohazardous cultures, it is good practice to keep an
autoclavable pan filled with distilled water or disinfectant in the rear of the BSC. This can be used to collect all
contaminated reusable items, such as pipettes. The pan is

covered before removal from the cabinet and auto-claved before reopening. Contaminated liquid culture media for disposal should be treated with a disinfectant, such as bleach, before discarding.

E. Recombinant DNA

Many of the benefits of biotechnology are a result of recombinant DNA methods. **Recombinant DNA (rDNA)** *is a DNA molecule constructed outside of a living cell, joining DNA from more than one source and capable of replication in a host cell.* Recombinant DNA methods are being used around the world in more than 20,000 laboratories. Since 1979, rDNA technology has led to many advances, particularly in the fields of medicine and agriculture.

When the possibilities of rDNA technology were first conceived, many scientists were concerned with the safety aspects of this type of experimentation and the possible introduction of new and unpredictable life forms. In 1975 leading scientists in the field assembled in Asilomar, California, to discuss the safety implications of rDNA methods. This was one of the few times that industry assembled to discuss the safety of a new technology before any major safety incident occurred. The consensus recommendations formulated at this meeting became the basis for NIH's Guidelines for Research Involving Recombinant DNA Molecules, which were officially established in 1976, and substantially revised several times afterwards. The NIH Guidelines are currently considered the general authority on rDNA safety procedures. While the NIH Guidelines do not have the force of federal regulations, any institution receiving NIH funding must comply with their recommendations. The Guidelines for Research Involving Recombinant DNA Molecules are constantly updated as new information becomes available.

The safety aspects of rDNA have been vigorously debated in both the popular literature and the scientific press. The safety model that is currently applied worldwide assumes that the safety of rDNA work depends on the DNA fragment used, the properties of the chosen host organism, and the properties of the vector used for delivering the DNA to the host cells. Some of the criteria used for evaluation of rDNA safety are shown in Table 30.19.

The NIH Guidelines for working with rDNA are similar to the regulations for work with infectious agents. The recommended way to protect workers and the environment is with containment barriers. As risks become greater, the procedures and facilities are designed to have a greater level of containment. The NIH Guidelines classify rDNA hazards into four risk groups related to their potential pathogenicity for humans. The Guidelines also designate 4 levels of physical biosafety containment, BL1 through BL4. These levels are similar to those assigned to microorganisms. The level of contain-

Table 30.19 SOME FACTORS RELATED TO THE SAFETY OF RECOMBINANT DNA EXPERIMENTS

There are many factors that enter into a risk assessment for planned rDNA work. These factors center on the properties of the host organism and vector, as well as on the nature of the DNA sequence being transferred. Examples of these concerns include:

Host Organism and/or Vector
- Pathogenicity or infectious properties of the host organism or vector
- Potential routes of human infection (aerosols, direct or indirect contact, etc.)
- Ability of an altered host cell or vector to survive outside the biotechnology facility
- Capacity of host cells to transfer genes into other, unintended organisms

Nature of the DNA Sequence
- Level of product expression
- Biological stability of the product
- Coding for:
  - Production of toxins
  - Antibiotic resistance
  - Protein products that trigger allergies
  - Genetic elements that could increase invasive properties of the host

ment for any NIH-funded rDNA project has historically been recommended by an institutional advisory committee.

The NIH Guidelines provide an "exempt" category, to designate rDNA experiments that do not require containment measures beyond those normally used for the organisms involved. Section III-F-6 of the Guidelines states that exempt experiments are "those that do not present a significant risk to health or the environment" when usual containment procedures are followed. Approximately 80–90% of all rDNA experiments are in the exempt category. Most of this exempt work is conducted using *E. coli* strain K12, *Saccharomyces cerevisiae*, and non-spore-forming *Bacillus subtilis* host cell systems. In these cases, the host cells themselves contain the rDNA safely.

E. coli strain K12 is a particularly well-characterized bacterium. Despite the sensational news stories about *E. coli*-induced food poisonings, the vast majority of *E. coli* strains have been characterized as non-pathogenic. Strain K12 is a particularly safe variety of *E. coli* that is used in high school and college laboratories, because studies indicate that strain K12 cannot colonize humans even after deliberate inoculation.

When host cells that require stringent growth conditions and that are generally not infectious (such as *E. coli* K12) are coupled with vectors that only infect this host cell type, biological containment of rDNA is relatively complete. Of course, not every laboratory can conduct their work in these systems, so it is essential to learn about the host cell and vector system used in your laboratory, to better understand any possible biohazards involved.

F. Laboratory Animals

i. HUMANE HANDLING

The decision whether to use animals in a laboratory requires a thoughtful evaluation of health concerns for both animals and personnel. Because of concerns for animal welfare, and also because laboratory work with animals is among the most expensive investigative method available, animals are generally used only when no alternatives are available. The care and use of laboratory animals is covered under a variety of regulations and guidelines. Detailed references are provided at the end of the chapter.

Currently applied guidelines describing the appropriate use of animals in the laboratory are contained in the "Guide for Care and Use of Laboratory Animals" (NRC, 1996), as recommended by the Public Health Service (PHS) Policy on the Humane Care and Use of Laboratory Animals (1996). The PHS Policy contains animal care requirements for all PHS-supported agencies, their research, and related activities. The PHS includes NIH, FDA, CDC, and many other federal agencies. The Policy is mandated by the Health Research Extension Act of 1985 (Public Law 99-158). Included in the Policy are institutional responsibilities for Animal Welfare Assurances, Institutional Animal Care and Use Committees (IACUCs), review of programs, facility inspections, record keeping, and reporting.

All PHS-funded institutions are required to have an **Institutional Animal Care and Use Committee (IACUC),** *which among other duties, reviews the institution's programs for humane care and use of animals and inspects institutional animal facilities.* PHS recognizes two categories of institutions that perform animal research. Category 2 institutions are evaluated internally by their IACUC, which ensures that the facilities and programs involving animals meet PHS Policy. In addition to meeting all Category 2 criteria, Category 1 institutions are voluntarily accredited by the **American Association for Accreditation of Laboratory Animal Care (AAALAC).** AAALAC is an independent peer-review organization that ensures that companies, colleges, hospitals, government agencies and other research institutions surpass minimal animal care standards. Currently, more than 600 U.S. institutions are AAALAC-accredited.

One significant requirement of the agencies that oversee animal care is that all personnel who work with animals receive appropriate training. In addition to the considerations of humane treatment, animals must be handled in ways that minimize health threats to both the animals and handlers. Sick or mistreated animals will not provide useful information in any study where they are used. All personnel who perform animal surgery or any other experimental manipulations must receive training in the performance of these techniques.

ii. WORKER SAFETY

Workers who come into contact with laboratory animals can be at risk for allergies and animal bites. It is fairly common for laboratory workers who frequently contact or work in the vicinity of rodents to develop allergies to the hair and dander of these animals. These allergies can be triggered by direct contact with the animals, or by inhalation of the allergens. Workers who develop these allergies generally should request a reassignment of duties to minimize animal contact, because allergies tend to worsen with continuing exposure. If reassignment is not an option, the worker must use PPEs, including gloves and respiratory protection. Once the symptoms of an allergy occur, it is best to contact a physician with experience in animal-triggered allergies. The physician can advise you about the dangers of continuing to work with animals.

Bites are another problem that can occur when handling animals. All animal bites must be reported to the safety authorities at your institution, who are required to record all incidents. Bite areas should be washed immediately and thoroughly. If the bite is deep or extremely painful, further medical attention should be obtained. The biggest health concerns from animal bites are local infection of the wound and tetanus. The probability of infection can be minimized by proper cleansing and bandaging of the wound. Anyone bitten by a laboratory animal should receive a booster injection of tetanus vaccine unless they have a current vaccination. Always keep a record of any immunizations you have received.

Zoonotic diseases (zoonoses) *are those that can be passed between different species.* Most of the laboratory animals you are likely to encounter, such as rats or mice, carry few zoonotic disease agents that can be passed to humans, although there are some exceptions, such as hantaviruses. Any worker who comes into contact with laboratory primates must receive special training that will include information about possible health hazards from zoonotic diseases.

Special precautions are required when laboratory animals are treated with biohazardous materials. In these cases, the worker must consider both animal welfare and biohazard containment. Only personnel who have specific training in handling pathogenic agents and animals should be allowed in the areas where this work is conducted. In the case of highly pathogenic agents, special animal facilities must be used, with provisions for safe handling of animals and disposal of infectious materials. All procedures that involve direct handling of the biohazardous materials should be conducted within biological safety cabinets or other physical containment units. The formation of bioaerosols and the disposal of excrement and other wastes are major concerns. All bedding, food, and other materials that contact the infected animals must be treated as biohazardous waste. Animal cages must be disposable or decontaminated before washing. It is absolutely essential that appropriate warning signs be placed on the doors of any containment facilities used for infected animals.

IV. HANDLING BIOHAZARDOUS WASTE MATERIALS

Biohazardous waste *consists of biological and biologically contaminated materials that are no longer needed in the laboratory.* These materials include:

- Discarded cultures of bacteria and other cells
- Outdated stocks of organisms
- Human and animal waste, including blood and other body fluids
- Used culture dishes and tubes
- Biologically contaminated sharps

All biohazardous wastes should be placed in labeled, closable, leak-proof containers or bags. These waste containers should usually be placed inside secondary containers in case of accidental punctures. As with chemical wastes, every institution will have specific procedures for sorting and labeling biological wastes.

In most situations, heat sterilization by autoclaving is considered an appropriate decontamination procedure for biohazardous waste. If this method is used, the resulting materials can be discarded as regular laboratory waste (in accordance with local regulations). Table 30.20

Table 30.20 *GUIDELINES FOR SAFE DECONTAMINATION OF BIOHAZARDOUS WASTE BY AUTOCLAVING*

- Do not mix chemical and biological wastes for autoclaving.
- All biohazardous waste must be placed in orange/red biohazard bags with a heat-sensitive sterilization indicator.
- Before autoclaving, biohazard waste bags should be kept closed to prevent airborne contamination and odors; while autoclaving, however, the bag must be open to allow the steam to penetrate.
- Add at least 100 *ml* water to each biohazard bag before autoclaving, for effective steam generation.
- Autoclave biohazardous materials for at least 45 minutes at the standard 121°C and 15 PSI for a single bag, and at least 60 minutes for a run with multiple bags.
- After autoclaving, the bag should be closed and disposed of in an opaque regular waste bag.
- All decontamination autoclaves should be tested regularly for sterilization effectiveness (at least annually).

Adapted from information from the Michigan State University Office of Biological Safety (http://www.orcbs.msu.edu).

provides guidelines for effective sterilization of biological waste using an autoclave.

V. RESPONSE TO BIOHAZARD SPILLS

Despite precautions, spills of biological materials occur. All laboratories that work with biohazards must have an emergency plan available in case of accidents. This plan should be written and posted at convenient locations for easy access. It is essential that all personnel be familiar with these plans before they are needed. Many of the same precautions that apply to chemical spills are also appropriate for biological spills. Always consider the nature of the hazard before attempting to handle the situation on your own. If a highly pathogenic agent is involved, evacuate the premises and summon help.

In cases where a spill is small, involves a relatively non-pathogenic organism (BSL1 or BSL2), and a biological spill kit is available, the worker can deal with the spill directly. Spill kits are commercially available, or they can be assembled ahead of time in the laboratory, Figure 30.9. A biological spill kit should include:

- personal protective equipment (e.g., gloves, eye, and face protection)
- absorbent materials (e.g., chemical absorbent or paper towels)
- cleanup tools (e.g., a whisk and dustpan)
- waste container (e.g., biohazard bags)
- disinfectant (e.g., fresh 10% bleach solution)

The bleach solution must be made fresh on at least a weekly basis.

When cleaning up a spill, remember that there are three hazards that must be addressed. The main spill area

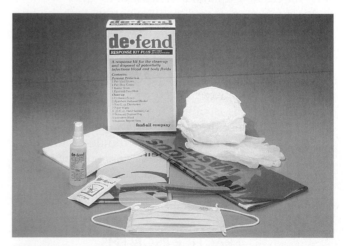

Figure 30.9. Biohazard Emergency Response Kit. These kits typically contain PPE, absorbents with disinfectant, biohazard disposal bags, and instructions for general biohazard cleanup. (Photo courtesy of Fisher Scientific, Pittsburgh, PA).

Table 30.21 GENERAL GUIDELINES FOR HANDLING MINOR BIOHAZARD SPILLS

BSL1 Organism

- Be sure you are wearing appropriate PPE, including lab coat, gloves, and eye protection.
- Soak absorbent materials such as towels in an appropriate disinfectant (usually a fresh 10% solution of bleach).
- Use these towels to soak up the spill; do not wipe the spill with dry towels.
- Dispose of the towels as biohazardous waste.
- Clean the spill area with fresh disinfectant and towels.

BSL2 Organism

- Notify all personnel in the area and evacuate for at least 10 minutes to allow aerosols to settle.
- In the meantime, remove any contaminated clothing and decontaminate them.
- Be sure you are wearing appropriate PPE, including lab coat, gloves, and eye protection.
- Soak absorbent material such as towels in an appropriate disinfectant (usually a fresh 10% solution of bleach).
- Cover the main spill area with the towels or other absorbent materials that contain disinfectant.
- Carefully (to avoid creating aerosols) flood the immediately surrounding area with disinfectant.
- Cover all items in the spill area with disinfectant and then remove them from the spill area.
- Remove any broken glassware with forceps and place in a sharps container; never pick up contaminated sharps with your hands.
- Remove the contaminated towels and any other absorbent material and place in a biohazard bag.
- Gently apply additional disinfectant to the spill area and allow at least 20 minutes for decontamination.
- Remove the disinfectant by applying paper towels or other absorbent material, which are then discarded in a biohazard bag.
- Wipe off any residual spilled material and reapply disinfectant for final clean-up.
- Collect all materials in the biohazard bag and seal for autoclaving.
- Reopen area to general use only after spill clean-up and decontamination is complete.

may be obvious, but there is also a surrounding splash area. This consists of small droplets of liquid that have splashed from the main spill site. It is essential that the cleanup effort does not spread contamination from these splash spots. Perhaps the most significant consideration at the time of the spill is the third element: the generation of aerosols. When the spill is large, or involves a pathogenic organism, it is best to leave the room, close the doors, and wait at least 10 minutes for the aerosols to settle. Keep in mind that every surface where the aerosols settle is also contaminated. Table 30.21 provides a general set of guidelines for cleaning up minor biohazardous spills.

In cases of personal contamination with biohazardous material, it is essential to take immediate steps for decontamination. Remove all contaminated clothing and soak lab coats in a bleach solution to disinfect before washing. Wash any potentially contaminated skin areas vigorously for at least 10 minutes. Use an antiseptic such as Betadine if available. In case of eye contamination, use the eye wash station and flush for at least 15 minutes. Your supervisor should be informed immediately of the accident. Depending on the biohazard involved, it may be necessary to seek medical attention and notify proper authorities as designated by your institution's biohazard plan.

PRACTICE PROBLEMS

1. A laboratory scientist drops a glass flask containing pathogenic bacteria. The glass shatters and splatters the scientist with liquid.

   a. What are several precautions the scientist should have taken ahead of time to reduce the risk of this incident?

   b. What should this person do immediately?

2. HEPA filters are designed to remove 99.97% of all particulate matter above the size of 0.3 μm from air. If a cubic meter of air containing 2,880,000 particles of about 0.5 μm passes through the filter, how many particles will remain in the air?

3. What are the similarities and differences between standard microbiological practices and Universal Precautions?

DISCUSSION QUESTIONS

Biological safety issues are of great importance in biotechnology settings. This is reflected in the Skill Standards that have been created for the industry.

1. **Scenario 1 in the Skill Standards for the Bioscience Industry (Education Development Center, Inc., 1995) presents the following topic and question:**

   One part of your laboratory responsibilities is to safely unpack and process biological samples. Demonstrate everything you would do to accomplish this.

   While unpacking samples one morning, you notice that one of the samples is leaking from the container. What should you do?

2. **Scenario 16 in the Skill Standards for the Bioscience Industry (Education Development Center, Inc., 1995) presents the following topic and question:**

Your job is to assist with cleaning, preparing, sterilizing, and inoculating a bioreactor. Show what tests and procedures you follow to perform these tasks.

After inoculation, a coworker points out that the bioreactor exit air filter cartridge is not installed. This means that there is no filter between the recombinant cells in the bioreactor and the outside environment. Demonstrate how you would handle this.

TERMINOLOGY

Note: Many of these definitions are taken from OSHA 29 CFR 1910.1030, Bloodborne Pathogens Standard (noted by "OSHA").

Aerosol. Very small particles suspended in the air.

AIDS, Acquired Immunodeficiency Syndrome. A condition in which the T-cells of the immune system are destroyed, devastating the body's immune defenses and allowing other infectious agents to cause disease and frequently death in AIDS patients.

Antiseptic. Type of mild disinfectant, gentle enough to be applied to skin.

Bioaerosol. An aerosol that includes biologically active materials.

Biohazard. A potentially dangerous biological organism or material.

Biohazardous Waste. Biological and biologically contaminated materials that are no longer needed in the laboratory.

Biological Safety Cabinet, BSC. Cabinet designed to contain aerosols and isolate the work material from the operator and the laboratory, while providing a clean workbench.

Biosafety Level 1, BSL1. The lowest biosafety level; used when working with well-characterized strains of living microorganisms that are not known to cause disease in healthy adult humans.

Biosafety Level 2, BSL2. The biosafety level used when working with agents that may cause human disease and therefore pose a hazard to personnel.

Biosafety Levels 3 and 4, BSL3 and BSL4. Biosafety levels generally associated with dangerous agents that are highly infectious.

Blood. Refers to human blood, blood components, and any products made from human blood (OSHA).

Bloodborne Pathogens. Pathogenic microorganisms that may be present in human blood and can cause disease in humans. These pathogens include hepatitis B virus (HBV) and human immunodeficiency virus (HIV).

Cancer. A variety of diseases characterized by the uncontrolled growth of cells and the ability of these cells to spread to distant parts of the body.

Carrier. An infected individual who is capable of spreading the infecting agent to other hosts; this individual may or may not show symptoms of disease.

CDC, Centers for Disease Control and Prevention. An agency of the Department of Health and Human Services, whose mission is "to promote health and quality of life by preventing and controlling disease, injury, and disability" (CDC Mission Statement).

Cell Lines. Cells that are established in long term culture.

Class I Biosafety Cabinet. A BSC designed to protect the operator from airborne material generated at the work surface.

Class IIA Biosafety Cabinet. A BSC that does not have external ducts, and releases filtered air directly into the laboratory.

Class IIB Biosafety Cabinet. A BSC that maintains a faster airflow than Class IIA cabinets, and is ducted to external exhaust systems.

Class III Biosafety Cabinet. Gas-tight cabinet designed to provide total containment for extremely hazardous biological agents; also called glove box.

Clean Bench. A type of laminar flow cabinet designed to provide a sterile work surface but not worker protection.

Clinical Laboratory. A workplace where diagnostic or other screening procedures are performed on blood or other potentially infectious materials (OSHA).

Containment. Any measure that prevents hazardous biological organisms or materials from presenting a health risk by isolating the organisms from the worker; see "Primary Containment" and "Secondary Containment."

Contaminated. Refers to the presence or the reasonably anticipated presence of blood or other potentially infectious materials on an item or surface (OSHA).

Decontamination. The use of physical or chemical means to remove, inactivate, or destroy hazardous biological materials on a surface or item to the point where the surface or item is considered safe for handling, use, or disposal (OSHA).

Disinfectants. Chemicals that kill pathogenic microorganisms and other biohazardous particles; can be classified as low, medium, and high level, according to their capacity to kill microorganisms.

Disinfection. The removal of all, or almost all, pathogenic organisms from a surface.

Ethylene Oxide. A reactive gas used for sterilization.

Etiological Agent. An organism that causes a specific disease in an infected host.

Exposure Incident. A specific eye, mouth, mucous membrane, non-intact skin or parenteral contact with blood or other potentially infectious materials that results from the performance of an employee's duties (OSHA).

Federal Insecticide, Fungicide, and Rodenticide Act, FIFRA. Legislation that controls the distribution, sale, and use of pesticides, including disinfectants, and that requires manufacturers to register their pesticides with the EPA before marketing.

Fungi. Primitive eukaryotic microorganisms that have complex life cycles and can form spores.

Germicidal. Methods that are capable of killing bacteria or other microorganisms.

Glove Box. See Class III Biosafety Cabinet.

HEPA, High Efficiency Particulate Air, Filter. A specially constructed high efficiency filter made of pleated glass and paper fibers; designed to have a 99.97% efficiency in removing particles of 0.3 μm or larger.

Hepatitis B. A bloodborne infection that attacks the liver and causes inflammation.

Hepatitis B Virus, HBV. Etiological agent for human hepatitis B.

Horizontal Laminar Flow Hood. See "Clean Bench."

Human Immunodeficiency Virus, HIV. The etiological agent for acquired immunodeficiency syndrome, or AIDS.

Impervious. Preventing passage of an organism or material.

Inactivation. Prevention of the replication of microorganisms or other biological agents.

Infection. Invasion and multiplication of microorganisms in tissue; may be associated with health effects.

Infectious. The ability of an organism to invade a host organism.

Institute of Laboratory Animal Resources, ILAR. Agency founded under the National Research Council to compile and disseminate information about laboratory animals and their care, and to promote humane care of these animals.

Institutional Animal Care and Use Committee, IACUC. An institutional committee required by the Public Health Service to review the institution's programs for humane care and use of animals and inspect institutional animal facilities, among other duties.

Laboratory Hood. Enclosed space with separate air supply or directed air flow.

Laboratory-Acquired Infection, LAI. Infection that can be traced directly to laboratory organisms handled by or used in the vicinity of the infected individual.

Laminar Flow Hood. A cabinet that has a sterile, directed airflow.

Large Scale Bioproduction. Production of biological materials in quantities greater than 10 L.

Lyophilized. Freeze-dried.

Metastasis. The ability of cancer cells to spread to distant parts of the body.

Microorganism. A microscopic organism, including bacteria, fungi, and viruses.

Mold. Filamentous form of a fungus.

Mycosis, or **Mycotic Disease.** A disease caused by fungal infection.

Mycotoxins. Toxic products produced by fungi; may render the fungus poisonous, or contaminate the environment of the fungus.

NCI, National Cancer Institute. A branch of NIH that focuses on cancer research.

NIH, National Institutes of Health. A federal health agency, comprised of 25 separate Centers and Institutions, that performs and funds major biomedical research and provides guidelines for laboratory safety.

Occupational Exposure. Reasonably anticipated skin, eye, mucous membrane, or parenteral contact with blood or other potentially infectious materials that may result from the performance of an employee's duties (OSHA).

Oncogenic. Agents, such as viruses, that can induce cancer after infecting cells.

Other Potentially Infectious Materials. Term referring to (1) Human body fluids such as semen, vaginal secretions, cerebrospinal fluid, synovial fluid, pleural fluid, pericardial fluid, peritoneal fluid, amniotic fluid, saliva in dental procedures, any body fluid that is visibly contaminated with blood, and all body fluids in situations where it is difficult or impossible to differentiate between body fluids; (2) Any unfixed tissue or organ (other than intact skin) from a human (living or dead); and (3) HIV-containing cell or tissue cultures, organ cultures, and HIV- or HBV-containing culture medium or other solutions; and blood. organs, or other tissues from experimental animals infected with HIV or HBV (OSHA).

Parenteral. Refers to piercing mucous membranes or the skin barrier through events such as needle sticks, human bites, cuts, and abrasions (OSHA).

Pathogen. Any biological agent that can cause disease.

Pathogenicity. Term indicating the relative capability of an organism to cause disease in humans or other organisms.

Personal Containment. Standard worker practices used to reduce the spread of microorganisms.

Physical Containment. Containment strategies that include laboratory design and the physical barriers used by workers.

Primary Cell Cultures. Newly isolated cells from tissue or blood, growing outside of the body for the first time.

Primary Containment. Protection of the worker and the immediate laboratory environment by the use of good microbiological technique and safety equipment.

Recombinant DNA, rDNA. A DNA molecule constructed outside of a living cell by joining DNA from one source with a segment of DNA that is able to replicate in a cell.

Regulated Waste. Liquid or semi-liquid blood or other potentially infectious materials; any contaminated items that would release blood or other potentially infectious materials in a liquid or semi-liquid state if compressed; any items that are caked with dried blood or other potentially infectious materials and are capable of releasing these materials during handling; contaminated sharps; and pathological and microbiological wastes containing blood or other potentially infectious materials (OSHA).

Retrovirus. An RNA-containing virus that also contains reverse transcriptase, an enzyme that transcribes RNA into DNA.

Sanitization. The general reduction of the number of microorganisms on a surface.

Secondary Containment. Protection of the environment outside the laboratory by the use of good laboratory design and safe practices.

Standard Microbiological Practices. The basic practices that should be used when working with all microbiological organisms.

Sterilization. Destruction of all microbial life forms on an object or surface.

Sterilize. Using a physical or chemical procedure to destroy all microbial life, including highly resistant bacterial spores (OSHA).

Tissue Culture. *In vitro* propagation of the cells derived from tissue of higher organisms; generally involves growing a monolayer of cells, covered with a layer of appropriate growth-sustaining liquid medium, in glass or plastic dishes.

Transform (in tissue culture). To cause cells to become capable of growing long term in culture and to develop some properties similar to tumor cells.

Universal Precautions. An approach to infection control, where all human blood and certain human body fluids are treated as if known to be infectious for HIV, HBV, and other bloodborne pathogens (OSHA).

Virulence. The capacity of a biological agent or material to cause harm.

Viruses. Microorganisms that contain nucleic acid surrounded by a protein coat of varying complexity.

Yeast. Single-celled form of fungi.

Zoonosis or **Zoonotic Disease.** A disease that can be passed between different species.

GENERAL REFERENCES FOR BIOLOGICAL SAFETY

Laboratory Safety: Principles and Practices, Diane O. Fleming, John H. Richardson, Jerry J. Tulis and Donald Vesley, 2nd ed. ASM Press, Washington, DC, 1995. (An excellent general reference on biological safety.)

Biosafety in Microbiological and Biomedical Laboratories, 3rd ed., U.S. Department of Health and Human Services, Public Health Service, Centers for Disease Control and Prevention and National Institutes of Health, U.S. Government Printing Office: Washington, D.C. 1993. HHS Publication # (CDC) 93-8395. (Contains tables of Biosafety Level requirements and pathogenicity levels of specific organisms.)

Biohazards Management Handbook, D.F. Lieberman, ed., Marcel Dekker, New York, 1995.

SPECIALIZED REFERENCES FOR BIOLOGICAL SAFETY

Guidelines for Research Involving Recombinant DNA Molecules, National Institutes of Health, Federal Register 49 (227):46266–46291, 1998. (Available on the Internet at http://www.nih.gov/od/orda/toc.htm.)

Biological Safety Manual for Research Involving Oncogenic Viruses, National Cancer Institute, U.S. Department of Health, Education, and Welfare, Washington, D.C.,1976. DHEW Publication # 76-1165.

Occupational Exposure to Bloodborne Pathogens (29 CFR Part 1910.1030), U.S. Department of Labor, Occupational Safety and Health Administration, Federal Register 56 (235): 64175-64182, 1991. Also reprinted in "Laboratory Safety: Principles and Practices," Diane O. Fleming et al.

Primary Containment for Biohazards: Selection, Installation and Use of Biological Safety Cabinets, CDC/National Institutes of Health. U.S. Department of Health and Human Services, Public Health Service, CDC, 1995.

Bioaerosols Handbook, C.S. Cox & C.M. Wathes, Lewis. London, 1995.

Guide for the Care and Use of Laboratory Animals, Institute of Laboratory Animal Resources Commission on Life Sciences, National Research Council, National Academy Press, Washington, D.C., 1996.

Public Health Service Policy on the Humane Care and Use of Laboratory Animals. Revised 1986, reprinted 1996. (Available at http://www.nih.gov/grants/oprr/phspol.htm.)

Education and Training in the Care and Use of Laboratory Animals, National Research Council, National Academy Press, Washington, D.C., 1991.

Occupational Health and Safety in the Care and Use of Research Animals, Committee on Occupational Safety and Health in Research Animal Facilities, National Academy Press, Washington, D.C., 1997.

WEB SITES RELATED TO BIOSAFETY

These are many current web sites that provide helpful information and links to other sites about biological safety issues. These include:

Centers for Disease Control—http://www.cdc.gov/

World Health Organization—http://www.who.int/

Howard Hughes Medical Institute— http://www.hhmi.org/science/labsafe/

ANSWERS TO PRACTICE PROBLEMS

1. **a.** Some basic precautions:
   Do not use glass containers for biohazards.
   Use a sealed plastic container.
   Use a secondary container for the flask.
   Try to avoid using liquid cultures of pathogens.
   Wear a lab coat to protect clothes.

   b. This individual's first priority is personal safety (assuming that no one else is involved in the accident). He should remove all contaminated items of clothing (within reason) and immediately leave the room where the accident occurred. The doors should be closed and no one should reenter the room for at least 10 minutes, to allow any aerosols to settle. The scientist should then wash any skin areas that have been exposed to the pathogen. He should use an antiseptic soap and vigorously scrub these areas for at least 10 minutes. Following the emergency plan for his institution, he or a colleague should notify the safety office, or set up the materials to decontaminate the laboratory. Cleanup should follow the procedures outlined in Table 30.21.

2. Approximately 864.

3. Similarities: Both refer to the importance of washing hands, decontaminating work surfaces, wearing PPE, avoiding mouth-pipetting, proper disposal of waste materials.

   Differences: Universal Precautions are OSHA regulations. They were developed for handling human blood and blood-related products. Universal Precautions are more detailed, but all of the requirements of each are compatible.

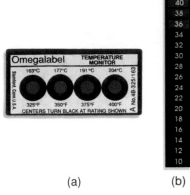

(a)

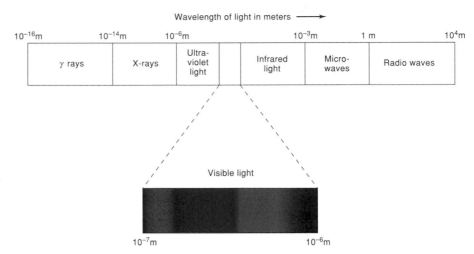

(b)

Figure 17.7.   Change-of-State Indicators. a. A label that changes colors at specific temperatures. **b.** Liquid crystal display. (Photos © copyright Omega Engineering, Inc. All rights reserved. Reproduced with the permission of Omega Engineering, Inc., Stamford, CT 06907.)

Figure 19.2.   Types of Electromagnetic Radiation and their Wavelengths. Visible light occupies a small portion of the electromagnetic spectrum. Within the visible portion of the spectrum, different wavelengths are associated with different colors.

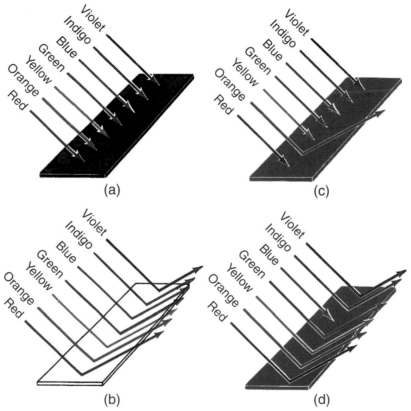

(a)

(b)

(c)

(d)

Figure 19.4.   The Colors of Solid Objects. a. An object appears black if it absorbs all colors of light. **b.** An object appears white if it reflects all colors. **c.** An object appears orange if it reflects only this color and absorbs all others. **d.** An object also appears orange if it reflects all colors except blue, which is complementary to orange.

Science & Math

featured titles

Welcome to the Prentice Hall Science and Math Home Page. ESM is a leading publisher for college and professional texts. For a comprehensive overview of titles, see our guide to books by professional interest area area below. Or if you prefer, we'll build you a custom catalog of related titles.

mail lists

Interested in writing for Prentice Hall? Please visit our on-line guide for authors. Subscribe to our electronic mailing lists and receive periodic information in the areas that interest you!

author guide

Looking for the Engineering and Computer Science pages? They've moved to http://www.prenhall.com/divisions/ecs.

course catalog

Browse our titles by course.

© Prentice-Hall, Inc.
A Simon & Schuster Company
Upper Saddle River, New Jersey 07458

custom catalog

Home

(a)

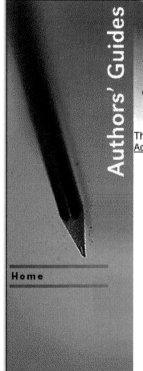

Authors' Guides

Home

The following guides are intended for authors of Prentice Hall textbooks. You can read them online, or if you have Adobe Acrobat, you may download each of them as pdf files.

General Author Guide: for higher education authors

Technical Author Guide #1: for computer science and engineering authors NOT providing us with camera ready copy (PDF)

Technical Author Guide #2: for computer science and engineering authors providing us with camera ready copy (PDF)

Non-Technical Author Guide #1: for PTR authors NOT providing us with camera ready copy (PDF)

Non-Technical Author Guide #2: for PTR authors providing us with camera ready copy (PDF)

(b)

Figure 33.4.   Hypertext in Action. a. This is the home page for the Prentice Hall Science and Math Division web site. The underlined phrases represent links to additional information on specific topics. For example, if you are interested in writing a textbook for Prentice Hall, you would select the link to *on-line guide*, which would immediately connect you to **b.** (http://www. prenhall.com/divisions/esm/)

UNIT IX:

Computers in the Laboratory

Computers have become essential laboratory equipment as well as integral parts of everyday life in this country. Each day, more than twice as many computers are sold (more than 100,000) than existed in 1975. This is rather ironic, considering that in 1943, Thomas Watson, the chairman of IBM, predicted that there would be "a world market for maybe five computers." Computers have clearly provided useful functions far beyond those originally envisioned!

One reason for the explosion of computer use is that these machines can perform certain types of functions faster and more reliably than humans. Some of these functions include:

- information storage and organization
- general data processing
- calculations
- creation of illustrations and graphics
- text editing and formatting
- local and international communication

The ability of computers to perform these tasks relies on human input. Computers systems will continue to become faster, smaller, and less expensive, but for the foreseeable future, they will require human operators. **Computer literacy** is no longer a specialized skill, but one that is required for effective job performance in many fields. Becoming computer literate *means developing a general understanding of computers and their uses.* You need to feel comfortable applying a computer to solve problems, without necessarily learning a lot about the technical workings of these machines. The essence of computer literacy is the ability to make the computer work for you.

This unit is designed to provide a general overview of how computers function and how you can use computers for data handling. We will concentrate on the familiar **microcomputers**, or personal computers, which are the self-contained units that you are most likely to encounter. These can be linked together into **networks**, providing them with tremendous power for communications. Given the pace at which new hardware models and software packages are developed, it is safe to predict that computers will play an increasingly significant role in laboratory work. The information here will provide a foundation for understanding the potential of these tools.

TERMINOLOGY

Analog. Data in the form of continuously changing values (contrast with "Digital").

Anonymous FTP sites. Internet sites that allow users without passwords to copy files from archives.

Application (related to computers). Software designed to perform specific data handling tasks.

ASCII, American Standard Code for Information Interchange. A seven-bit text coding format, used as a default for text communication between programs; an extended eight-bit code is also available.

Back Up (verb, related to computers). To copy files from one storage medium to another.

Backup (noun, related to computers). An extra copy of data files that is stored in a separate location from the original files.

Backwards Compatible. Indicates that the new version of software is compatible with old versions; files created with previous versions of the same program can be used effectively.

Bar Graph. Graphs in which data are presented as bars and that are best adapted to show proportions or comparisons among small numbers of data items.

Baud. An outdated measurement unit for modem speed. The preferred unit is bps.

BBS, Bulletin Board Service. An electronic "bulletin board" where individuals and organizations can use a modem to post messages and announcements, as well as upload and download software; these services are frequently sponsored by special interest groups.

Benchmark. A performance standard that measures the speed and performance of hardware or software.

Binary Code. A code with two options; for computers, those options are 0 or 1.

Bit, Binary Digit. The smallest information unit in a computer; a 0 or 1.

Bit Map. A graphic image stored as a set of dots or pixels, which can be individually manipulated.

BMP, Bit-Mapped Paint. Standard file format for the Paint program supplied with the Windows operating system.

Bookmark (related to computers). A URL recorded by an Internet browser, which will allow you to return to that site without retyping the address.

BPS, Bits Per Second. Transmission rate for a modem, indicating the transfer rate for data bits; sometimes inaccurately estimated in baud units.

Browse (related to computers). To move through large amounts of information quickly without changing it (usually in a database) or to move around the Internet using a browser.

Browser. Software that provides a user-friendly interface for the Internet.

Buffer (related to computers). A separate area of memory used to store information for immediate retrieval.

Bug (related to computers). Mistakes or defects in software or hardware that cause a computer system to crash or otherwise malfunction.

Bundled. Extra software or sometimes hardware items that are included in a merchandising package; for example, scanners are usually sold bundled with basic graphics software.

Bus (related to computers). The set of connecting wires that acts as a data pathway through a computer.

Byte. A sequence of binary signals that defines a character; generally 8 bits.

Cache Memory. A memory buffer that maintains a copy of data or instructions in reserve for high speed recovery when they are used repeatedly during a work session.

Carpal Tunnel Syndrome. A painful medical condition resulting from repetitive wrist and hand movements; caused by pressure on nerves due to deep tissue swelling in the wrist. An example of a repetitive stress injury.

CD-ROM, Compact Disk-Read Only Memory. An optical storage disk which can hold almost 700 MB of information.

Cell (related to spreadsheets). The location in a worksheet where a single piece of data can be entered; designated by a unique row and column grid number.

Chatroom. A cyberspace location where individuals can have interactive conversations by typing messages for real-time delivery; usually limited to specific topics.

Chip. A self-contained electronic circuit placed on a small wafer of silicon; an integrated circuit.

Clip Art. Small graphic images that are stored in files and available for pasting into other documents. Frequently sold in collections.

Clipboard (related to computers). A holding area within the computer memory for data that are being exchanged between files or applications.

Clock Speed. A measurement of how fast the CPU can carry out an instruction, reported in MHz.

Column. A vertical line of data in a spreadsheet or table.

Command-Line Interface. An interface that requires the user to type in specific formatted text commands in order to control computer operations (contrast with "GUI").

Communication Software. Programs that provide a common computer language to direct the network interface card or modem to contact other computers and instruct the computer how to send and receive data through the network.

Computer Graphics. Pictures or images, which can be bitmaps or objects (contrast with "Text").

Computer Literacy. Development of a general understanding of computers and their uses.

Copy Protection. Software mechanisms to prevent the creation of multiple copies of commercial programs.

CPU, Central Processing Unit. The microprocessor or main chip that determines the relative power and functional capabilities of a computer.

Crash (related to computers). A system failure, which could be the result of either a software or hardware problem.

Cursor. A pointer, arrow, I-bar or blinking rectangle that indicates your working location on a monitor display.

Cyber-. A prefix to indicate information or experiences on a computer network.

Cyberspace. An on-line, virtual environment or meeting area.

Cyberspeak. Computer jargon related to the Internet, including various abbreviations used to indicate emotional states.

Data. Numbers and text that are manipulated within applications; the unprocessed facts from which we derive information.

Database. A collection of organized data that can be searched and sorted by database management software.

Database Management System. Software that allows the user to sort and report data and information in a variety of formats.

Data Compression. File coding formats that allow data storage in fewer bytes than normal; useful for conserving storage space and speeding data exchange through networks. Files require decompression before reading.

Data Processing. A series of complex data manipulations, which include data recording, searching for patterns in collected data, interpreting these patterns, and presentation of organized data and conclusions.

Default. The control settings that your computer will assume if you do not change them; the best option under many circumstances.

Desktop (related to computers). The monitor display area where icons, windows, and other graphic user interface elements are displayed.

Desktop Publishing Software, DTP. Application programs to format combinations of text and graphics for publication; rapidly replacing typesetting.

Digital. Data in the form of a series of discrete values; for computers, these values are 0 or 1 (contrast with "Analog").

Digitizing Tablet. A flat touch-sensitive screen with X–Y coordinates that exactly correspond to those of the monitor display. The user moves a special pen across the tablet to enter digital data (drawings or writing) into the computer.

Disk. Hardware that stores software as magnetic signals; may be portable (floppy disks) or installed within the computer (hard disk).

Disk Drive. The hardware that reads and writes data on a disk.

DNS, Domain Name System. The address assignment system that distributes responsibility for naming web pages by assigning names to networks.

DOS, Disk Operating System. An early text-driven operating system for IBM-compatible microcomputers, with a command-line interface.

Download. To copy files or programs from a different location (e.g., a disk, the Internet, or a BBS) to a local computer (compare with "Upload").

DPI, Dots Per Inch. The measurement unit for resolution in printers and scanners.

Draw Program. A graphics program that uses vector graphics; ideal for detailed images containing clean lines with precise dimensions (e.g., blueprints and maps).

Driver (related to computers). A program that controls a peripheral device; provides compatibility between the CPU and peripherals.

E-mail, Electronic Mail. Computerized system for sending and receiving messages at high speed; can be used in a global or local network.

Emoticon, Emotion Icon. Standard combinations of typed symbols used in electronic communications to represent emotions; an example is a happy face :-).

Encryption. Translation of data into a private code to prevent unauthorized use.

Erase (related to computers). To eliminate data from a storage medium (see Chapter 31 for details).

Ergonomics. The study of the effects of environmental factors on worker health and comfort; the design of environments to increase worker health and productivity.

Ethernet. The most widespread current type of LAN.

Expansion Port. See "Port."

Expert System. A computer system designed to solve problems using knowledge and logic.

FAQ, Frequently Asked Questions. A file of basic questions and answers available to new web site visitors, or to participants in a newsgroup or chatroom; always read these before joining an electronic conversation.

FAT, File Allocation Table. The computer's "address list" that allows it to find files.

Field (related to computers). Database term that refers to the individual data items, which can be text, numbers, or logical values.

File. The basic organizational unit for data; a set of digital data that is grouped to form a single storage unit. In a database application, a set of related records.

File Compression. File formats in which the data are abbreviated according to a specified formula.

File Format. The method used by a specific program to store data in a file.

Flaming. To make offensive personal comments in a discussion group or by e-mail.

Flatfile Database. A database contained in a single computer file.

Floppy Disk. A portable and reusable magnetic storage disk.

Font. A set of characters that include all the letters, numbers, and symbols of one style and size of type.

Format (verb). See "Initialize."

FTP, File Transfer Protocol. A common protocol for transporting files over the Internet.

Function Keys. A special set of keys at the top of a keyboard that can be programmed within software to indicate specific commands.

GB, Gigabyte. ~1 billion bytes, actually 2^{30} (2 to the thirtieth power).

GIF, Graphics Interchange Format. A file format for highly compressed storage of graphic images; used for scanned (bit-mapped) graphics.

GIGO, Garbage In/Garbage Out. A common expression which means that computer output cannot be of better quality than your data input.

Gopher. A menu-driven information search and retrieval tool for the Internet.

GUI, Graphical User Interface. A user-friendly system interface that allows the user to control the computer by selecting items; based on the use of windows, icons, menus, and pointers. (Contrast with "Command-Line Interface.")

Hard Copy. A paper or printed copy of data.

Hard Disk. A set of rigid magnetic storage disks encased in a disk drive and usually installed inside the computer; portable versions are available.

Hardware. The physical components of a computer system.

High-Capacity Data Cartridge or **Disk.** Portable data storage devices with capacities of 100 MB or more; several proprietary types are available.

High-End. Refers to the most powerful and expensive hardware and software options available.

Histogram. A type of bar graph that contains numerous thin, closely spaced bars showing the distribution of a variable.

Hit (related to computers). Refers to entering a web site; site managers frequently count the number of hits per week to determine how many people view the site.

Homepage. The first screen of information that introduces visitors to a web site; generally includes a series of links, and information about the site sponsor or author.

Host Computer. The core computer in a network; usually the site of accessible information for network users.

HTML, Hypertext Markup Language. Programming language used to create and format web site content and links.

HTTP, Hypertext Transfer Protocol. The information transfer protocol used on the WWW.

Hyperlink. A link, or electronic connection, between two sites on the Internet; usually indicated by specially formatted words or icons.

Hypertext. Information presented through the use of linked pages of multimedia content.

Icon. A graphic symbol representing an object or computer function.

Illustration Program. Graphics software that contains both bit-mapped and vector graphics capabilities.

Image Editing Software. A program that provides the tools to take previously existing images, such as scanned photographs, and modify them.

Import. To copy data from one application program into another program.

Information. Processed data that have been collected, manipulated, and organized into an understandable form.

Initialize, or **Format (verb).** To erase and prepare a storage disk for use; creates writing sectors, tracks, and a file address table (FAT) on the disk.

Input Device. Any device that enters data or provides commands to the computer.

Install. To use the installation program provided with software to place a new application into your computer system; places program files at specific locations on the disk so that the application and system software can find them.

Integrated Software Packages. A basic set of programs providing word processing, spreadsheet, database management, graphics, and communications modules.

Interface. The boundary area between hardware and software where users communicate with computers. See "GUI."

Internet (noncapitalized). Two or more connected networks.

Internet, Interconnected Networks. A network of tens of thousands of loosely connected computer networks that use a common set of guidelines to send and receive information.

Internet Tools or **Services.** Tools that provide means of access and data exchange over the Internet.

Intranet. An internal, usually private network.

I/O, Input/Output. Refers to the transfer of data between the computer and peripheral devices.

IP, Internet Protocol. The standard communication protocol for the Internet.

IRC, Internet Relay Chat. An Internet service which allows interactive, real-time typed "conversations" among the participants.

ISDN, Integrated Services Digital Network. A high speed data transfer standard which can send digital multimedia signals through standard telephone lines.

ISP, Internet Service Provider. A company that provides fee-based access to a host computer that connects to the Internet.

JPEG, Joint Photographics Expert Group. A file format that compresses color bit maps for storage of graphic images; commonly used for photographic images.

Keyboard. An input device that looks like part of a typewriter, used to enter text into a computer.

Kilobyte, KB or **K.** A practical measurement unit for file size; ~1,000 bytes (actually 1024, representing 2^{10} (2 to the tenth power).

LAN, Local Area Network. A private computer network operating within a limited area such as a building, company, or school.

Laptop Computer. A portable, generally one piece computer system that can be operated on batteries.

Light Pen. A light-sensitive pointer device that is used to input graphic or command data on a monitor screen.

Line Graph. Graphs that connect data points on X–Y coordinates; show continuous data items and trends, particularly over time.

Link (related to computers). An electronic connection between two sites on the Internet; usually indicated by specially formatted words or icons. Also called a hyperlink.

Listserv. An automatic mailing list for e-mail messages; subscribers to a listserv will receive copies of all messages sent to the group.

Log In (Log On). To enter your name and password into a computer in order to gain access to a network.

LPT, Line Printer. Label found on printer ports in some microcomputers.

LQ, Letter-Quality. Refers to print that reproduces the quality provided by a good typewriter; modern printers generally provide better than LQ print.

Macintosh, Mac. The trademark name of a series of personal computers produced by Apple, Inc., since 1984.

Macro. A short program, generally written within an application, to define a commonly used sequence of commands with a single keystroke; used as a time-saver.

Mainframe. A large computer that is more powerful than a microcomputer and provides computing access for multiple users simultaneously; physical access is provided through locally available workstations.

Megabyte, MB. Approximately 1 million bytes, actually 2^{20} (2 to the twentieth power).

Memory. Temporary electronic data recording within a computer (see "RAM" and "ROM").

Menu. A graphically displayed list of command choices in a GUI or within a particular application.

Menu Bar. A horizontal bar found at the top of the desktop in GUI, displaying the titles of submenus.

MHz, Megahertz. A unit for measuring the speed of a microprocessor; one million electrical cycles per second.

Microcomputer. A personal computer; a single-user self-contained computer.

Microprocessor. The CPU of a computer; a small complex calculator that controls data input and output.

MIPS, Millions of Instructions Per Second. A measure of computer speed (see "Clock Speed").

Modem, Modulator-Demodulator. An input/output device for network computers. A modem changes (modulates) the digital signal from a computer to an analog signal that can be carried over phone lines. These signals are received by another modem, which changes the signals back to digital form.

Monitor (related to computers). The display device for computer input and output; similar in appearance to a television screen.

Mouse (related to computers). A data input device, which only vaguely resembles a live mouse, that allows you to control a pointer and select commands without using the keyboard.

Mouse Pad. A rectangular piece of rubber with a fairly smooth surface that provides more traction for manipulating a mouse than a typical desk.

Multimedia. Applications that combine text, sound, graphics, animation, and/or video output.

Multimedia Authoring. The creation of presentations that use more than one medium to convey information.

Multitasking. The ability to open and run more than one program at a time.

Navigate (verb, related to computers). To move through a series of linked sites on the Internet; usually with a purpose (contrast with "Surf").

Netiquette, Network Etiquette. A generally accepted code of manners that apply to behavior in cyberspace, particularly in newsgroups or chatrooms.

Network. A group of computers that are electronically connected for the purpose of sharing programs, files, information, and/or other resources.

Network Interface Card, NIC. Hardware that provides an appropriate interface between the computer and various networks.

Newsgroup. A bulletin board area on Usenet; messages are posted for public reading and response. Newsgroups are generally devoted to a single topic.

Number Crunching. Slang for the repetitive mathematical processing of large amounts of data; computers are ideally suited for this purpose.

Object (related to computers). A data set that is treated as a whole rather than as individual bits of information.

Object-Oriented Graphics. Programs where graphic data are described by vectors or mathematical equations; contrast with "Bit-Map."

OCR, Optical Character Recognition. The process of reading scanned text as characters rather than a single object; OCR software is necessary to prepare scanned text for editing.

OLE, Object Linking and Embedding. Technology used by Microsoft software to link objects in different files and applications.

On-Line. Refers to being currently connected to a computer network.

Open. To load a file or program from storage into RAM.

OS, Operating System. The part of the system software that operates the hardware, runs software applications, manages file storage, and monitors system operations.

Paint Program. Graphics program that provides the ability to create and modify complex images using computerized tools that are similar to those used by an artist on paper; products are stored as bit maps.

Parallel Port. A port where multiple streams of data can travel in or out simultaneously; printers are usually plugged into parallel ports.

Paste (related to computers). To move or copy text or images from one location or document to another.

PDF, Portable Document Format. A file format for text created and read by the Adobe Acrobat™ program; frequently found on the Internet for downloading.

Peripheral Device. Any separate electrical device attached to the computer.

PIC or **PICT, Picture.** A general format for graphics files.

Pie Chart. A type of graphic that shows data items as portions of a circle; not common in scientific publications.

Piracy. Illegal duplication and distribution of copyrighted application programs.

Pixel, Picture Element. An individual point of light on a monitor display; the smallest unit of an image that can be individually manipulated.

Platform (related to computers). A standard defined by the combination of microprocessor and system software found in an individual computer; application software is developed to operate on specific platforms.

Plotter. A digital graphics output device that uses one or more adjustable ink pens that write directly on a continuously moving paper surface.

Point and Click. The usual technique for selecting commands in a GUI by moving the cursor to the desired item and selecting the item; usually performed with a mouse.

Pop-Up Menu. A menu that appears after you choose an option on another menu.

Port (related to computers). A socket where peripheral devices can be plugged into the computer; microcomputers may have two general types of ports: parallel and serial.

Post (related to computers). To send an on-line message to a newsgroup or chatroom, where others can read and respond to the message.

Presentation Graphics Program. Software that provides capabilities to create and incorporate both text and graphic images into "slide" presentations, which can be presented from a computer screen, printed as handouts, or used to make actual slides.

Program. A set of computer instructions written in special computer language or code and designed to allow the computer to perform specific functions.

Programmer. A person who writes programs, the instruction sets for computers.

Protocol (related to computers). A standardized computer format for network communication. Examples include IP, FTP, and HTTP.

Public Domain. Noncopyrighted software that the author has provided for free to the public; contrast with shareware.

Query (related to computers). The process of extracting a desired subset of data from a database and then presenting it in a specific format.

RAM, Random Access Memory. Short-term electrical memory that provides the internal working area of the computer; files must be loaded into RAM before you can use them.

Raster Graphics. Graphics in which the image is created and stored as a collection of dots (or pixels or bits); bit-mapping.

Read-Only. Refers to storage media that contain readable information but cannot be used for storing new data.

Record (related to computers). The basic organizational units of information in a database.

Relational Database Management System. Database management software in which various database files are linked by common fields; updating information in one database automatically updates the other files as well.

Repetitive Stress Injury. Physical injury caused by performing constant identical motions over long periods of time.

Resolution. The sharpness of an image produced by a monitor, printer, or scanner. For printer and scanner output, resolution is measured in dpi. For monitors, the screen display is described by the number of horizontal pixels times the number of vertical lines (as in 640×480).

ROM, Read-Only Memory. Permanent computer memory that is built into a memory chip and contains instructions necessary to start the computer and perform diagnostics and other basic functions; these instructions cannot be changed by the user.

Row. A horizontal line of data in a spreadsheet or table.

RS-232, Recommended Standard-232. A popular international standard for computer ports, cables, and interfaces; used for connecting peripherals and laboratory equipment to computers or other devices.

Save (related to computers). To transfer data from temporary electrical memory to long term magnetic or optical storage.

Scanner. Hardware that changes a graphic image to digital data and sends it to a computer.

Scatter Plots. Graphs that show the general distribution of data points that are not adequately defined by a line.

Screen Saver. A utility program that blanks out or creates a moving graphic display on the monitor screen after a specified time without input; prevents damage to older monitors from etching of images.

Scroll (related to computers). To use the cursor or arrow keys to change the portions of a large document that are visible within a current window.

SCSI, Small Computer System Interface. A specific type of high transmission speed serial port that allows chains of devices to be connected.

Serial Port. A port where data travel in a single continuous stream in one direction; the most common current type in a laboratory is called an RS-232.

Server (related to computers). The combination of hardware and software that provides user access to the network and its resources.

Shareware. Software that is under copyright but easily available for trial use; users who keep the program are expected to pay a small specified fee to the software author.

SIMM, Single In-Line Memory Module. A type of memory module that can be added to computers to increase RAM.

Site License. A purchased authorization to install a program on multiple computers.

Soft Copy. Impermanent computer output such as a screen display.

Software. Instruction sets, or programs, written in computer language or code.

Spamming. Sending inappropriate and unwelcome messages to large numbers of recipients (the term derives from a Monty Python comedy sketch).

Spell Checker. Function of word processing programs that points out and usually suggests substitutions for words that are not included in the program's spelling dictionary.

Spreadsheet. A grid of numbers in rows and columns which can be mathematically manipulated by spreadsheet software.

Storage. The magnetic or optical recording of data or information onto hardware; storage media include disks, CD-ROMs, and high capacity data cartridges or disks.

Subscribe (related to computers). To add your name to a mailing list or newsgroup in order to receive automatic updates.

Suite (related to computers). See "Integrated Software Packages."

Surf (related to computers). To move through a series of linked sites on the Internet; implies relatively casual movement, usually for amusement (contrast with "Navigate").

Surge Protector. An electrical device that protects equipment from power surges and other fluctuations in electrical current by acting as a power regulator between the computer and the power outlet.

System Software. Programs that run the computer and provide the foundation, or platform, for application programs.

TCP, Transmission Control Protocol. A protocol that works with IP software to make sure that data sets arrive and are reassembled in the correct order, and that the data are intact; ensures reliable data transfer on the Internet.

Telecommunications. Electronic transmission of data in the forms of text, audio, and video over significant distances.

Telnet. A terminal emulation program that allows your computer to access remote computers.

Template. A blank standard form that can be used in an application to create many similar documents.

Text. Letters, words, and other symbols that are represented in ASCII code and individually manipulated; contrast with "Graphics."

Text Editor. The simplest type of word processing program; provides ASCII output.

TIFF, Tagged Image File Format. A widely supported and versatile graphics file format; current standard for scanned images.

Toolbar. A menu of icons that provide command options in a GUI.

Trackball. An input device similar to an upside-down stationary mouse; the user directly manipulates the rolling ball to move the screen cursor.

Trackpad. A flat touch-sensitive pad that substitutes for a mouse; the user moves a finger on the pad, causing the on-screen cursor to move in a similar direction.

Unerase Program. A utility program that can recover erased files by restoring their address in a disk directory (see Chapter 31).

UNIX. A complex multiuser, multitasking operating system that uses a command-line user interface. UNIX provided the original interface for the Internet.

Unsubscribe. To request that your name be removed from a mailing list or newsgroup.

Upgrade (related to computers). Addition of new hardware, such as memory cards or a second hard drive, to increase the performance capabilities of a computer.

Upload. To copy files or programs from a local computer to a distant location (compare with "Download").

UPS, Uninterruptable Power Supply. A power supply with a battery that will allow the user to shut down a computer properly during a central power failure.

URL, Uniform Resource Locator. An address or identifier for a site on the Internet.

Usenet. An Internet service that provides more than 22,000 discussion groups or newsgroups, each generally focused on a single topic of interest.

Utility. A "housekeeping" program that enhances computer operations; useful accessories for the user.

Vector Graphics. Graphic images that are created based on a set of lines, which can be described and stored by the software as mathematical equations.

Virtual (related to computers). Events, objects, or functions that occur on-line or within a computer, without a real world presence.

Virus (related to computers). Small programs created to enter or "infect" computers or their programs without the user's knowledge; may cause system malfunctions.

Web (related to computers). See "WWW" (World Wide Web).

Web Page. The basic units of information storage in hypertext and the components of web sites; vary in size from a sentence to multiple printed pages.

Web Site. A specific location on the WWW, designated by a URL.

Window (related to computers). A rectangular viewing and work area on the computer desktop; these create a separation of programs and data files. Note that Windows (plural, with a capital W) is a popular operating system.

Word Processing. Text handling program that allows text editing plus complex document formatting such as superscripts, spell checking, tables, and outlining.

Worksheet. A blank spreadsheet.

Workstation. A single user computer connected to a mainframe computer through a LAN.

WWW, World Wide Web. An Internet network which features multimedia sites created in hypertext; requires a web browser and a GUI.

WYSIWYG, What You See Is What You Get. In word processors, indicates that the monitor display shows an accurate representation of the final printed product.

GENERAL REFERENCES ON COMPUTERS

This short unit cannot provide comprehensive information about computers, but there are many excellent books available. Given the rapid rate of progress in this field, the most useful information will be found in very recent publications. Most computer reference books are updated frequently. For a general overview, the authors recommend:

Introduction to Computers and Software, Robert A. Szymanski, Donald P. Szymanski, and Donna M. Pulschen. Prentice Hall, Upper Saddle River, NJ, 1996. (This is an excellent overview of computer science and basic principles of computer operations.)

PCs for Dummies, Dan Gookin, 5th ed., IDG Books, San Mateo, CA, 1997, and *Macs for Dummies*, David W. Pogue, 2nd ed., IDG Books, San Mateo, CA, 1993. (Despite the annoying names, these are helpful and humorous books full of practical information and tips for beginning computer users.)

Introduction to Computers and Information Systems, 5th ed., Larry Long and Nancy Long, Prentice Hall, Upper Saddle River, NJ, 1997.

BIOLOGY AND COMPUTERS

The Internet and the New Biology: Tools for Genomic and Molecular Research, Leonard F. and Anne H. Peruski, ASM Press, Washington, D.C., 1997. (This book provides a general introduction to the tools of the Internet as well as their specific applications to biotechnology.)

CHAPTER 31

Computers: An Overview

I. INTRODUCTION

One of the major functions of laboratories is the production of data. (The word *data* is the plural form of the seldom seen *datum*, which refers to a single number, letter, or other symbol. For this reason, the word *data* requires a plural verb.) **Data** *are the unprocessed facts from which we derive information.* **Information** *is processed data that have been collected, manipulated, and organized into an understandable form.*

Computers have become essential to laboratory operations because they can convert data into information at rates that would overwhelm the abilities of humans, Figure 31.1. It is important, however, to remember that all computer functions (at least at present) are the result of human engineering and planning. These machines do not yet have significant creative capabilities nor do they "think" in terms of the vast stores of human knowledge. Although their capacities for accurately performing repetitive data manipulations are impressive, they simply do as they are told. Your goal, as a computer user, is to determine which functions can best be performed by a computer, and to provide appropriate instructions and data.

A computer system includes two basic elements: hardware and software. As the names suggest, **hardware** *includes the solid objects needed for computer functions.* **Software**, on the other hand, *includes the programs that provide the instructions that make computers useful to us.* Specific software **programs**, *which can be considered instruction sets,* allow us to "communicate" with computers, and accomplish specific data manipulation tasks.

The components of a computer system allow four general functions. These are:

- data input
- data and information processing
- information output
- data and information storage

This chapter will focus on the computer components and how they work, while the next will discuss how you can handle data using computer software, and the final unit chapter will introduce telecommunications.

II. HOW COMPUTERS WORK—HARDWARE

A. Digital Technology—Bits and Bytes

The power of computer operations stems from the digital processing system. Like other digital equipment, the computer understands only two electrical states: off and on. Not all digital technologies are new. Telegraphs provide digital data in the form of clicks, which are then reconstructed into interpretable human language. Just as you do not personally need to know Morse code in order to receive a telegram, you also do not need to understand computer code in order to use a computer. A basic understanding of how computers "see" data, however, can be helpful.

In computer language, "off" and "on" are represented as 0 or 1. *The 0 or 1 represents the smallest available information unit, also called a* **bit** (binary digit). In contrast, analog equipment measures change as a continuous pattern. As a result, analog signals are subject to distortion and noise (see Chapter 14). A computer data bit can only equal 0 or 1, eliminating ambiguity. Another advantage of digital data is that it can be reproduced indefinitely without loss of quality.

Because a single bit has only two possible values, multiple bits are required to represent the possibilities for real world data, which include letters of the alphabet, numbers, and other symbols. A computer "word" or **byte**, *is a group of continuous bits* (usually eight, providing 256—2^8—possible combinations) *that define a single understandable symbol or character.* As users, we rarely have to deal with single bytes of information, and for convenience, we measure the data in computers in **kilobytes** (***K*** or ***KB***), *which represent approximately 1,000 bytes* (actually 1024 bytes, representing 2^{10}, but the approximation is adequate for ordinary purposes), and **megabytes (*MB*)**, (2^{20} bytes), Figure 31.2. Current computers have data storage capacities measured in **gigabytes (*GB*)** (approximately 1 billion bytes, 2^{30} bytes). As a general approximation, about one half of a plain text page requires about 1 *KB* of data.

B. Central Processing Unit

The *working core of a computer* is called the **central processing unit (CPU)**. This is the main **microprocessor** that determines the relative power of the computer. A

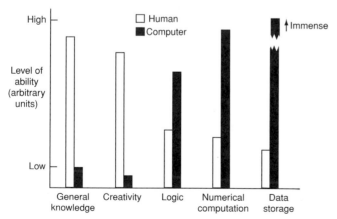

Figure 31.1. A Comparison of Relative Abilities of Humans and Computers. Bar heights represent estimated relationships. (Adapted from *Intelligent Database Tools and Applications,* Kamran Parsaye and Mark Chignell, John Wiley & Sons, New York, p. 3, 1993.) Copyright © (1993, John Wiley & Sons, Inc.). Reprinted by permission of John Wiley & Sons, Inc.

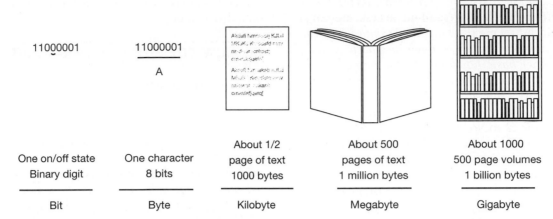

Bit	Byte	Kilobyte	Megabyte	Gigabyte
One on/off state Binary digit	One character 8 bits	About 1/2 page of text 1000 bytes	About 500 pages of text 1 million bytes	About 1000 500 page volumes 1 billion bytes

Figure 31.2.   Comparison of Computer Data Size Units. (Adapted from *Introduction to Computers & Software,* R.A. Szymanski, D.P. Szymanski, and D.M. Pulschen, Prentice Hall, p. 74, 1996.)

Figure 31.3.   A Central Processing Unit for a Personal Computer. This is an Intel Pentium III microprocessor.

microprocessor is a *small* (Figure 31.3) *but complex calculator that also controls data input and output.*

The microprocessor contains the electrical circuits that carry out the instructions found in the computer's software. Microprocessors are also found in a variety of laboratory and home appliances such as microwave ovens, but these are dedicated to specific functions. The CPU of a computer is highly versatile, because of the diversity of instructions that can be provided by other hardware as well as software.

The thinking power of a computer is defined by the type of microprocessor that it contains. There are two families of microprocessors currently available in personal computers; those manufactured by Intel for PC systems and those manufactured for Apple Macintosh computers. These two families of personal computers are discussed in the Case Study on page 679. Chip manufacturers, such as Intel, assign model numbers to specific types of microprocessors.

The main specification that a general user needs is the system **clock speed,** *the speed with which the microprocessor can carry out an instruction.* Clock speed is measured and reported in **megahertz (MHz),** *representing 1 million electrical cycles (hertz) per second.* Higher clock speeds

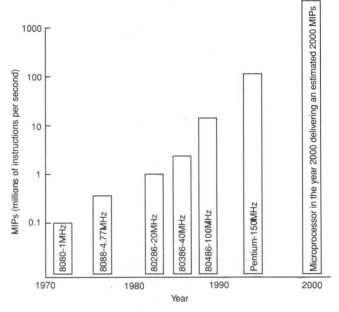

Figure 31.4.   Increasing Speed of Computer Processors. This graph shows the progress made since the 1970s in improving the speed of the Intel family of processors, which are found in approximately 90% of the personal computers in current use. The anticipated technology for the year 2000 will have about 10,000 times the speed of the Intel 8088 processor. Note the logarithmic scale on the Y-axis. (From *Introduction to Computers and Information Systems,* 5th ed., Larry Long and Nancy Long, Prentice Hall, p. C58, 1997.)

mean faster computers. Another direct measure of computer speed is **MIPs,** *or millions of instructions per second.* New microprocessor models usually provide significant increases in computer speed, Figure 31.4.

C. Memory

An important feature of a computer is its memory capacity. **Memory** *is the ability of a computer to electronically keep track of data.* The amount of memory in a typical microcomputer is measured in megabytes.

Computer memory exists in two forms, **ROM (read-only memory)** and **RAM (random access memory)**. ROM is built into a memory chip and *contains instructions necessary to start the computer and perform diagnostic and other basic functions.* As the name implies, these instructions cannot be changed. The computer operator is more interested in **RAM**, *which represents the electronic memory available to work with.* When you turn on the power to your computer, RAM is empty. With prompting from the ROM, the computer loads part of the operating system (explained later) into RAM. Once the computer is ready, you choose additional software and data to load into RAM, Figure 31.5. RAM is the working area of the computer, and items must be added to RAM before you can use them.

RAM requires electricity to hold its contents. Data found only in RAM are called volatile because they will disappear when you turn off the computer or if a power outage occurs. In order for data to become a permanent record within your computer, they must be stored. There is a critical distinction between memory and storage in a computer. **Storage** *is the magnetic recording of data or information onto hardware such as disks.* Stored information can be retrieved later. Memory is electronic and therefore temporary. One of the most important things that all computer users learn (usually the hard way!) is to **save** their data, *which means to transfer them from volatile memory to permanent storage.* If you are writing a 20-page paper on a computer and even a brief power outage occurs, all of your work in RAM

will be lost. Only the portion of the paper that is saved on a disk can be recovered.

III. HOW COMPUTERS WORK—SOFTWARE

A. System Software

Although hardware provides the tangible part of a computer, the hardware cannot accomplish anything without software. **Software** or **programs**, *are instruction sets written in computer language or code.* The **programmers** *who write these coded instructions make computer use possible for those of us who have not learned computer language.* Programs provide instructions that enable hardware to perform necessary functions.

There are two general categories of software: applications and system software. An **application** *is software that addresses specific problems, such as data organization or word processing.* These will be discussed in the next chapter. **System software** performs a series of more basic functions. *These are the programs that run the system hardware and provide the foundation for application programs.* System software does not manipulate data or create documents. In the computing world, a **platform** *is defined by the combination of CPU and system software found in an individual computer.* All application programs are written to perform on specific platforms. Examples of two current platforms are the PowerMac G3 / OS X and Pentium III / Windows 2000.

The components of system software most familiar to users are the **operating system (OS)** and the user **interface.** The OS *provides four general functions: It operates the hardware, runs software applications, manages file storage, and monitors system operation.* It also supports the user interface, which is the part of the system software that the user sees. All communication between the user and the computer hardware occurs through this interface, allowing the computer to carry out the user's commands, Figure 31.6. Some software applications provide their own user interface.

There is a limited number of OS systems in common use. The oldest system still in use in personal computers is DOS (for Disk Operating System). DOS provides a text-based interface (defined later), and is capable of running only one program at a time. DOS cannot be used in networks, and has been largely replaced by **Windows**. The Windows operating system (always spelled with a capital *W*) is more user-friendly than DOS. It is a **multitasking** OS, *which means that the user can open and run more than one program at a time.* This feature is found in all newer operating systems. A computer must have enough RAM to open multiple programs and documents in order to use multitasking effectively.

B. User Interface

Every OS must have a user interface, which is usually based on either text or graphics. Text is generally in the

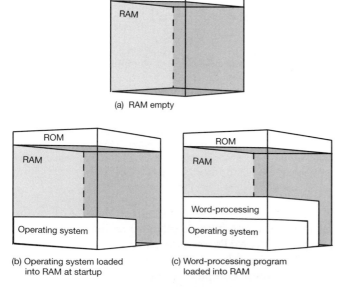

Figure 31.5. The Use of RAM and ROM within a Computer. a. Computer memory is divided into ROM and RAM sections. **b.** When operating, computers load the operating system into RAM. **c.** Additional programs are added to RAM as needed. (Adapted from *Introduction to Computers & Software,* R.A. Szymanski, D.P. Szymanski, and D.M. Pulschen, Prentice Hall, p. 78, 1996.)

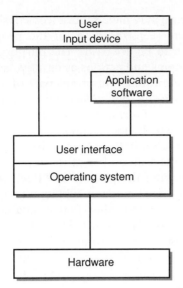

Figure 31.6.   Interactions Among Computer Hardware, Software, and User. (Adapted from *Introduction to Computers & Software,* R.A. Szymanski, D.P. Szymanski, and D.M. Pulschen, Prentice Hall, p. 115, 1996.)

form of a **command-line** or **text-based interface**, *where the user types instructions, or commands, for the computer on a designated line.* DOS and UNIX are examples of two text-based operating systems. **UNIX** *is a widely used operating system for large network computers,* which are generally operated by individuals with special training. These text interfaces require knowledge of specific commands and formatting; therefore, they are relatively difficult for beginners to use. Because a simple DOS "conversation" might look like Figure 31.7, it is no surprise that when Apple Computer introduced the Macintosh OS in 1984, its intuitive, graphics-based interface was considered a user-friendly breakthrough. Since then, the Windows OS, which is also graphics-based, has almost entirely replaced DOS. Text-based interfaces are currently used almost exclusively by computer experts.

Virtually all personal computers now provide a **graphical user interface**, or **GUI** (pronounced "gooey"). Users no longer need to remember complex commands and grammar—they can now "point and click" using a

CONVERSATION		TRANSLATION
Computer:	C:\>	I'm looking at the main directory on the hard disk.
User:	C:\>CD COMPARE	Change directories and look in the folder called "COMPARE"
Computer:	C:\COMPARE>	OK, I'm looking.
User:	C:\COMPARE>COMP.BMP	Open the Paint file called COMP.BMP.
Computer:	C:\>FILE NOT FOUND	I can't find it here.
User:	C:\COMPARE>DIR/P	Show me the directory of the folder "COMPARE," one screen at a time.
Computer:	Volume in drive C is MS-DOS_6 Directory of C:\COMPARE	OK, here it is.
	. <DIR> 04-06-98 3:37p	
	.. <DIR> 04-06-98 3:37p	
	COMWIN95 EXE 626,425 02-17-98 1:12p	
	DOC <DIR> 02-17-98 1:12p	
	CONTRAST EXE 170,496 02-17-98 1:12p	
	2 file(s) 801,901 bytes	
	3 dir(s) 398,950,400 bytes free	
User:	C:\COMPARE>C:\COMPARE\DOC\COMP.BMP	
		Now I remember that my file is in the DOC folder inside the COMPARE folder
(All this to find and open one file!)		

Figure 31.7.   A DOS Command-Line Conversation to Find and Open a File.

computer mouse rather than typing instructions with a keyboard. Instead of text commands, GUIs operate with what Steven Mandell memorably calls WIMPs—**windows**, **icons**, **menus**, and **pointers** (*Dr. Mandell's Ultimate Personal Computer Desk Reference*, Steven Mandell, Rawhide Press, Toledo, OH, 1993). This type of interface is designed to be as intuitive as possible. For example, the screen where the WIMPs appear is now called a **desktop**, representing the area where your work can be laid out.

A GUI **window** *is a rectangular viewing and work area.* Each individual document or information display appears in its own window, and many windows can be present on the desktop simultaneously. You can change the shape and size of windows according to your needs, Figure 31.8. Only one window is active at a time, but the contents of the others are visible if the windows do not overlap. *You choose the active window by moving the pointer arrow to the desired window and clicking the mouse button* (called **point and click**).

Icons *are graphic representations of specific objects and computer functions.* In some cases, such as the file folders and the trash can shown in Figure 31.8, the function may be obvious. You can select and move items into the folders for storage or into the trash can for disposal. In other cases, application programs and the files

that they create may have special icons designed by the manufacturers (e.g., the icon for the program Acrobat™ Reader in Figure 31.8). The purpose is to make selection of items and actions as intuitive as possible.

Menus are another important part of a GUI. They come in many forms, but *all have the common feature of offering a series of text options or icons for your selection.* These can be highly useful in complex applications because they eliminate the need to memorize commands. There is a menu bar at the top of Figures 31.8 and 31.9. Clicking the pointer on "File," for example, causes a drop-down menu or submenu to appear, offering choices of (among others) creating a new folder or finding a specific file, Figure 31.9. Many applications have "floating" menus (sometimes called **toolbars** if they contain icons) that can be moved (float) to any convenient position on the desktop while you are working in that program.

The final component of a GUI is the pointer, which acts as a **cursor** *to indicate your location on the desktop or in a window.* The pointer may appear as an arrow, an I-bar, or a blinking rectangle. The pointer is usually controlled with a mouse, although it can also be moved with arrow keys on a keyboard. Most graphical user interfaces make the "point and click" technique with a mouse easier than providing keyboard input.

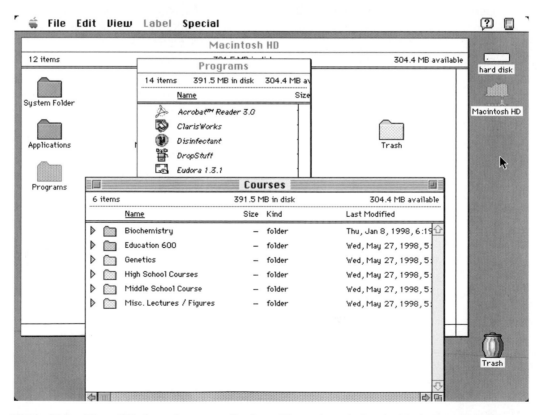

Figure 31.8.   Three Windows Open on a Desktop. The active window is titled "Courses," with "Programs" and "Macintosh HD" in the background. Note that all three windows have been specifically sized and shaped. Folder icons are visible, as are the main menu headings ("File," "Edit," etc.). The pointer arrow is visible on the right side of the desktop.

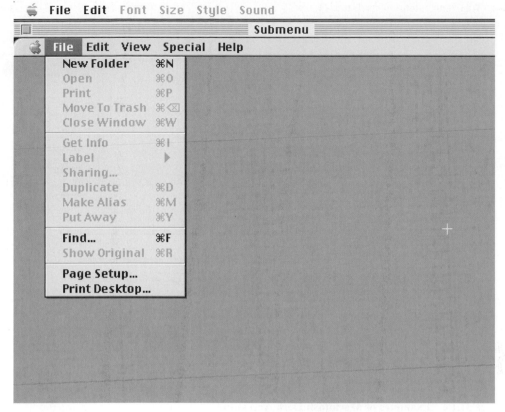

Figure 31.9.   Selection of a Submenu from a Menu Bar. The selection of the File submenu offers additional command choices. (From *Introduction to Computers & Software*, R.A. Szymanski, D.P. Szymanski, and D.M. Pulschen, Prentice Hall, p. 198, 1996.)

CASE STUDY

PC versus MAC

At the current time, there are two main types of personal computers available—the Apple Macintosh family and the PC/Intel processor family. While the PCs currently dominate the market, Macintosh users are a loyal and enthusiastic minority. The PC family of personal computers originated with the IBM and IBM-compatible microcomputer, which used DOS (for disk operating system), with a text-based interface, as an operating system. Then in 1984, Apple introduced a personal computer with a graphical user interface (GUI). The Apple computer could perform many of the same tasks as the DOS systems, but required little or no computer expertise. While these early Apple computers were slow compared to DOS machines, they developed a strong following. Microsoft eventually developed the Windows operating system to provide a GUI for DOS machines. Virtually all PC systems use the Windows operating system at present.

From the beginning, Apple chose to maintain a proprietary operating system, so that only Apple and licensed companies could produce peripherals and software for Apple computers. IBM, on the other hand, made their operating system available for other companies to exploit. Many business analysts believe that this business choice led to the current market dominance of PCs. Because it was originally intended as an ideal system for graphic design work, Macintosh still remains the standard in this

field. As of 1998, though, PCs account for approximately 90% of microcomputers sold in this country.

At the current time, most personal computer users describe themselves as either Macintosh or Windows people, based on the platform of their computers. In fact, the interfaces of these systems are similar in function, and new Macintosh and Intel microprocessors rival each other in speed. Prices for Apple systems have historically been higher, but even this difference is disappearing. Most major software packages are available in both Mac and PC formats. New Macintosh computers are equipped with standard software that enables them to read PC-formatted disks (but not programs), so that Macintosh and PC users can exchange files.

C. Application Software

Application programs are designed to provide computer solutions to human problems: A **program** *is simply a series of instructions, written in one of many computer languages, which directs the computer to perform a specific function*. The program that controls a digital wristwatch may only need a thousand instructions, while complex computer applications may require millions of instructions to work properly. This reflects the difference between the limited operations of a watch compared with the versatility of a computer running application software. For most applications, the user

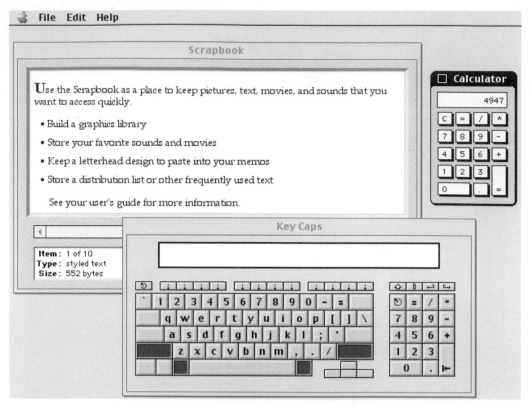

Figure 31.10.   Examples of Macintosh Desk Accessories or Utilities. Key Caps shows the characters made by the keys of the keyboard. Other accessories include a simple calculator, in the active window, and the Scrapbook, where you can paste a series of small items.

provides initial data, which is placed into a selected area in a window. The application will (with proper instructions) process and save the data in a variety of ways.

Utilities *are "housekeeping" programs that enhance computer operations.* Utilities generally do not create new documents or files. Examples include virus protection software, screensavers, and accessories such as desktop calculators, Figure 31.10. **Screen savers** *are programs that blank out or create a moving graphic display on the monitor screen after a specified time without user input.* These programs were originally designed to prevent damage to monitors from etching of images on the screen. Newer monitors are unlikely to have this problem, but most users enjoy these programs anyway.

D. Files

A computer **file** *represents a set of digital data that are grouped to form a single storage unit.* Each file stored on your computer has a name, and represents either a program or a data document. Program files have assigned names that generally should not be changed. Operating systems provide conventions for naming document files. For example, DOS and early Windows file names contain up to eight letters, then a period followed by one to three letters. Numbers may be used instead of letters, but certain punctuation marks such as commas or spaces are not allowed. In contrast, current operating systems with GUIs allow almost any file

name. File names should reflect the contents of the file so that specific information can be located quickly.

When the computer stores a document file, it uses the language and conventions of the application software that created the document. The resulting **file format** allows the appropriate application to recognize its own files, but these files may not be readable by other applications. For this reason, there are standard file formats available. The best known is probably **ASCII** (American Standard Code for Information Interchange), which is *a default format for text documents.* In ASCII, each character is defined by a seven-bit code of 0s and 1s (e.g., the code for the number *9* is "0110110"). ASCII files can be read by virtually any application. A seven-bit code can define only 128 possible characters (2^7), so there are extended versions of ASCII using more bits and providing additional combinations. Both PCs and Macs can read ASCII files from within many of their programs.

While ASCII codes are adequate for simple text, they do not provide enough information to incorporate the complex formatting you would typically find in a word-processing program. Formatting instructions are generally program-specific, Figure 31.11. One example is a **PDF** (*portable document format*) file, created and read by the Adobe Acrobat™ program. These produce complex and attractive text documents that are commonly found on the Internet and which cannot be fully interpreted by most other text programs.

CONTENTS

	FOREWORD	VI
1	**CONTENT PREPARATION**	1
	Manuscript Length	1
	Make an Outline	1
	Avoid Wordiness	2
	Write Directly to Your Readers	2
	Provide Learning Aids	3
	Don't "Date" Your Book	3
	Text Style	4
	Prentice Hall House Style Guides, 4	
	Term of Copyright	4
	Public Domain, 5	
	Government Publications, 5	
	Direct Quotations, 5	
	Permission Procedure, 6	
	Guard Against Prejudice	7
	Sexism, 7	
	Unintentional Slurs on Race, Religion, or Sexual Orientation, 7	
	Broad Accusations Against Professional Groups, 8	
	Elements of the Manuscript	8
	Submitting Sample Electronic Files, 8	
	Preparing an Electronic Manuscript, 9	
	Setting the Manuscript, 10	
	Preparing Electronic Manuscript Disks, 12	
	Front Matter, 12	
	Cross-References, 14	
	Extract Material, 14	

(a)

```
FOREWORD VI
1 CONTENT PREPARATION 1
Manuscript Length 1
Make an Outline 1
Avoid Wordiness 2
Write Directly to Your Readers 2
Provide Learning Aids 3
Don't "Date" Your Book 3
Text Style 4
Prentice Hall House Style Guides, 4
Term of Copyright 4
Public Domain, 5
Government Publications, 5
Direct Quotations, 5
Permission Procedure, 6
Guard Against Prejudice 7
Sexism, 7
Unintentional Slurs on Race, Religion,
or Sexual Orientation, 7
Broad Accusations Against Professional Groups, 8
Elements of the Manuscript 8
Submitting Sample Electronic Files, 8
Preparing an Electronic Manuscript, 9
Setting the Manuscript, 10
Preparing Electronic Manuscript Disks, 12
Front Matter, 12
Cross-References, 14
Extract Material, 14
```

(b)

Figure 31.11. Comparison of the Contents of a PDF File Read by a. Adobe Acrobat™ Reader and **b.** A simple text reading program. Note the formatting in **a.** and the loss of the header and spacing in **b.** *(Downloaded from the Prentice-Hall web site)*

When you **open** a file, *the computer transfers a copy of the contents of that file from storage into memory (RAM).* The original file still exists in storage, and any changes that you make to the file contents remain volatile in RAM until you save them. The changes can be saved under the same file name, which usually writes over the previous version, or the new version can be saved with a different name, so that both versions are stored. Opening a document file will usually start the application that created the file, although it is important to remember that closing a file does not automatically shut down the application program.

E. Storage Media (Hardware)

When comparing computer storage and memory, it may be helpful to remember that computers deal with information much the way humans do: They either keep an idea in memory or write it down for future reference. Computers write files to various forms of storage media, which are frequently disks. A **disk** *is a piece of hardware that stores software or data as magnetic signals.* All current computers contain a permanent hard disk and the ability to read and write to various types of portable storage disks or cartridges. *The hardware that reads and stores information on a specific type of disk is called a* **disk drive**. Disk drives can be either inside the computer or a plug-in item.

One of the advantages of information storage on a computer is that digital data can be copied an infinite number of times without changing the data. If you have ever copied a videotape (an analog technology), you are aware that copies lose some of the quality of the original, and that each succeeding copy is worse than the previous one. Digital recording offers the ability to make large numbers of perfect copies. The most common computer storage media are:

- floppy disks
- hard disks
- high capacity data cartridges or disks
- CD-ROMs

Current computers are equipped with a permanent hard drive and usually a floppy disk or cartridge drive, which can be used for both reading and storage of data, as well as a CD-ROM player, which can read data.

Floppy disks (or floppies) provided the main storage medium for early personal computers. The original common floppy disk was 5¼-inch and flexible (hence the name). The first versions held 360 *KB* of information and later up to 1.2 *MB*. There are still 5¼-inch disk drives in use, but all computer manufacturers have switched to the 3½-inch standard disk. These disks have the advantages of 1.4 *MB* of storage, a more portable size, and rigid construction. The term *floppy disk* is no longer literally appropriate, but it is still widely used to describe either size of portable disk.

The part of the 3½-inch disk that you handle is the plastic cover, Figure 31.12a. A sliding metal cover protects the magnetic writing surface from outside damage. One useful feature of these disks is the write-protect tab, which you can use to "lock" the disk and make it **read-only**; *this prevents anyone from erasing or writing over your stored files.* The tab can be moved back to the unlocked position if you need to make file changes.

The **hard disk** *is a set of rigid magnetic storage disks encased in a disk drive and usually installed inside the computer,* providing most of the storage capacity of the personal computer. A 1.4 *MB* floppy disk can store a text file containing approximately 1000 pages of text, but many files containing images are too large for a single disk. All personal computers now contain a hard

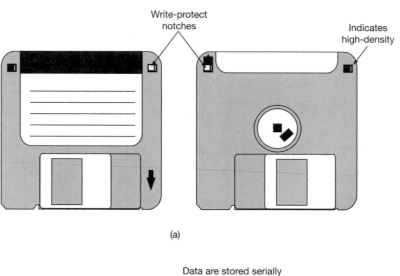

(a)

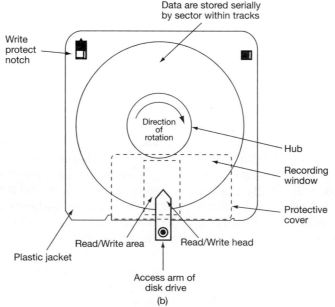

(b)

Figure 31.12.   Structure of a 3½-Inch Disk. a. The plastic disk cover holds a label and protects the magnetic medium inside. High density disks store momore information than low density disks, and are now standard. **b.** Inside the disk cover is the data recording medium. Data are read or written in circular tracks on the disk by a floating access arm in the disk drive.

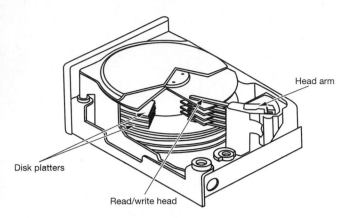

Disk platters

Head arm

Read/write head

Figure 31.13.   Internal Structure of a Hard Disk Drive. The drive casing has been removed and a section of the disks has been cut away to show the series of eight storage disks and read/write heads.

disk and drive. This is a well-protected, sealed unit containing two or more large, dense magnetic disks and a drive mechanism, Figure 31.13. A current hard disk typically has 1–10 *GB* of storage space, so it can hold the information equivalent of thousands of 3½-inch disks. Hard disks are therefore ideal for long-term storage of large files and programs. Hard disks are rigid enough to spin significantly faster than floppy disks, which means that the information stored on the disk can be accessed much faster.

High-capacity data cartridges or **disks** *are portable storage devices which can hold 100* MB *or more of data.* These systems and their drives are becoming increasingly common as storage needs increase. Creating a complete backup copy of a 2 *GB* hard disk would require significantly fewer 100 *MB* cartridges than 1.4 *MB* floppy disks. These high-capacity devices are rapidly replacing floppy disks as the storage medium of choice. They are available in a variety of proprietary formats, such as the Iomega Zip™ disk.

CD-ROMs, Compact Disk—Read Only Memory, *are optical laser disks that look exactly like music CDs, but which contain computer programs and files.* CD-ROMs are the most common form of optical disk, but like magnetic disks, they contain stored digital information that is reconstructed into the analog signals of text, sound, and video. *Programs that combine these three elements are called* **multimedia** applications. A single CD-ROM can store almost 700 *MB* of information, the storage equivalent of approximately 475 3½-inch disks. They provide an ideal medium for distribution of encyclopedias, large video-containing programs such as games, and other multimedia applications. They have the added advantage of being inexpensive to produce. The major disadvantage for consumers, the read-only nature of these disks, is rapidly disappearing with the emergence of reasonably priced CD writers. These machines can store information on blank CDs and allow users to create their own CD storage banks and multimedia applications for distribution.

IV. PERIPHERAL DEVICES

A. Introduction

A **peripheral device**, or simply peripheral, *is any hardware that you plug into your computer,* Figure 31.14. Peripherals usually act as input and/or output devices. The computing device itself is the box that contains the CPU—everything else is a peripheral. Peripherals are necessary for all user input and output. These devices require three components to function: cables, ports, and drivers.

Cables are the wiring that connects the peripherals to the computer unit. These connect through specific types of sockets, called **expansion** or **input/output (I/O) ports**, for each device. *These ports allow communication between components,* and come in two general types. **Parallel ports** *are used for printers, and allow multiple streams of data to travel simultaneously in parallel lines.* Serial ports are more versatile, connecting modems, the computer mouse, and many laboratory devices to the computer. **Serial ports** *carry a single data stream at a time* and are available in several industry standards, including the **RS-232 port**. Many Macintosh and other types of computers have special serial ports called **SCSI** (pronounced "scuzzy") **ports**, *which allow serial chains of peripheral devices to be connected through one port.* More recently, USB (Universal Serial Bus) connections have become more common.

Like all computer hardware, peripheral devices require software in order to be useful. The required program is called a **driver**, *which is the software that controls the device.* The driver translates commands from applications and the CPU into specific signals recognized by the peripheral, and vice versa. It establishes compatibility between the CPU and peripheral device.

B. Input Devices

Computers require data input into RAM before any processing can occur. All necessary programs and data must be either copied from a storage device, such as a disk, or entered into the system through a peripheral, such as a keyboard, before any actions begin. All operations take place in the RAM and the results are then transferred to one or more output devices and (usually) to a storage device, Figure 31.15. A wide variety of peripherals are available for data input and output; the most common will be described here.

i. KEYBOARD

The keyboard is the most familiar input device for the computer, Figure 31.16. A typical keyboard provides the standard typewriter keys and also:

- a numeric keypad for fast numerical data entry
- arrow keys for moving the cursor
- special editing keys

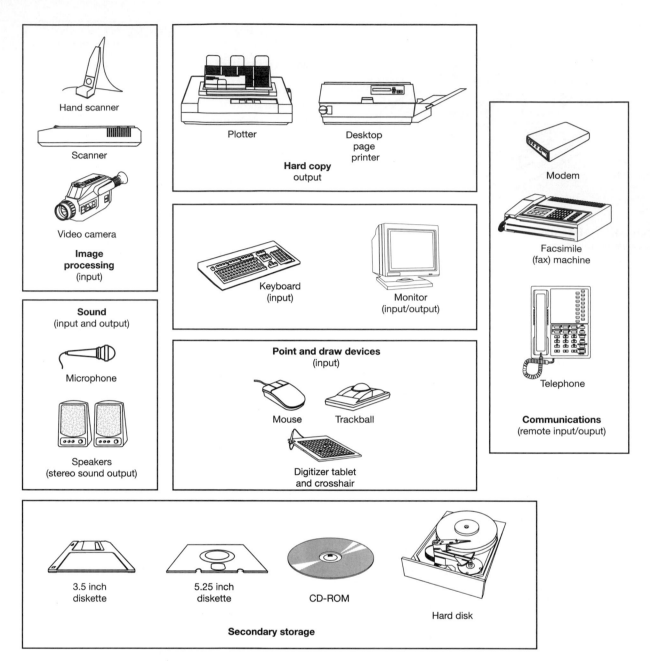

Figure 31.14. Examples of Common Peripherals for Personal Computers. (Adapted from *Introduction to Computers and Information Systems,* 5th ed., Larry Long and Nancy Long, Prentice Hall, p. C15, 1997.)

- a set of **function keys** *that may be defined within specific applications*

Both data and commands can be entered through the keyboard, although some commands may be more easily entered with a mouse in a graphical user interface.

ii. MOUSE

A **mouse** *is a pointing device that controls the on-screen cursor in a graphical user interface.* The cord, which points away from the user, supposedly resembles the tail of a mouse. In any case, a standard mouse is a smoothed rectangular device with one (on a Macintosh) or more buttons on top, Figure 31.17. Inside the mouse is a small rubber ball that protrudes slightly from the bottom sur-

face. By rolling the mouse across a smooth surface, you move the pointer on the screen in the same direction as the mouse movement. The use of a mouse is sometimes called **point and click** input *because the user rolls the mouse to point the cursor at an on-screen object, and then presses or clicks one of the mouse buttons to select or activate the object.* Because of the speed of the cursor movement, using a mouse can provide faster command entry than the keyboard in many situations.

Most people use a mouse pad as the rolling surface for their mouse. A **mouse pad** *is simply a rectangular piece of rubber with a fairly smooth surface that provides more traction for the mouse than a typical desk.* These pads can be plain or highly decorative. It is best to avoid pads with very smooth or painted surfaces,

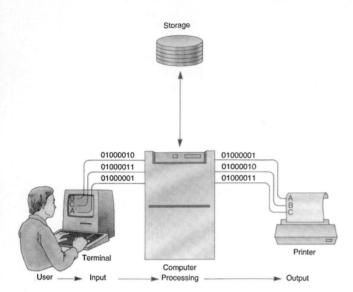

Figure 31.15. The Flow of Data Through a Computer System. (From *Introduction to Computers & Software*, R.A. Szymanski, D.P. Szymanski, and D.M. Pulschen, Prentice Hall, p. 16, 1996.)

which can provide poor traction and sometimes literally gum up the inside of the mouse.

There are several common variants on the mouse, Figure 31.17. A **trackball** *is basically a stationary, upside-down mouse.* Instead of rolling a mouse, you roll the ball directly with your fingers. These are built into many laptop computers, where a peripheral mouse would be inconvenient. **Laptop computers** *are portable computer systems that can be operated on batteries.* Some laptops have **trackpads**, which have no moving parts. A trackpad *is a flat rectangular area next to the keyboard.* The user moves a finger on the touch-sensitive pad, which causes the on-screen cursor to move in a similar direction, and then the user clicks a button next to the pad. Some users feel that these devices give finer cursor control than a standard mouse. Still others prefer a mouse pen, which is shaped and held like a ball-point pen, and rolled across the desk to move the cursor.

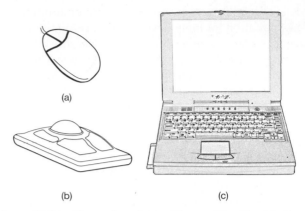

Figure 31.17. Examples of (a) a Mouse, (b) a Trackball, and (c) a Trackpad on a Laptop Computer.

iii. SCANNER

A **scanner** *is an optical device that converts graphic images to digital form and sends them to the computer for manipulation and storage.* The most common types are flatbed or hand-held scanners, Figure 31.14. Flatbed scanners are similar in appearance and performance to photocopiers, except that they do not directly produce printed copies. Hand-held scanners are manipulated by hand to scan pictures and are best suited for small images. Scanners treat blocks of text as **objects** instead of a collection of characters. An **object**, in computer language, *is a graphic image that is treated as a single item, described by a mathematical equation.* In order to recognize and edit scanned text, the computer requires **optical character recognition (OCR)** software. **OCR** technology *allows text to be scanned and converted into text in a word processing program without retyping.*

iv. VOICE RECOGNITION

For most users, the fastest and simplest way to input information and commands into a computer would be by speaking. Although the technology for **voice recogni-**

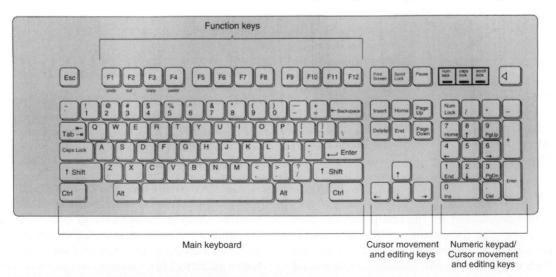

Figure 31.16. A Standard PC Keyboard. (From *Introduction to Computers & Software*, R.A. Szymanski, D.P. Szymanski, and D.M. Pulschen, Prentice Hall, p. 79, 1996.)

tion is still relatively primitive, this is undoubtedly a future direction for input devices. **Voice recognition software** *provides a computer with the capability to accept voice input by recognizing the speech of a user.* These systems are frequently used by scientific and medical professionals who need to keep their hands free for other work, as well as by physically challenged individuals.

Many voice recognition systems at present are highly dependent on the individual speaker. They generally work well for individual words, but they require "programming" by each user for continuous speech. To train the system to recognize specific words, the user pronounces words and sentences into the speaker system. The software digitizes the analog sound waves and then matches them to the words found in the software dictionary. The computer will then either obey the command or display the spoken text on the monitor. The user corrects any mistakes, which enables the computer to "learn" how each word or phrase is pronounced. Although it is currently time-consuming to program a voice recognition system, time is saved later during system use. The amount of programming necessary will decrease as voice recognition technology becomes more common and versatile.

Computers can also provide voice output, which is relatively simple compared with speech recognition. Most personal computers can be directed to "read" text from a file, using a variety of synthetic voices.

v. Other Input Devices

There are many other input devices available for specialized uses. For example, artists and other users who input very precise graphic data can use a **digitizing tablet and pen**. These tablets contain touch-sensitive screens with electronic circuitry underneath. The X–Y coordinates of the tablet screen exactly correspond to those of the monitor display screen. This allows both precise drawing and (with proper programming) handwritten input.

Many additional devices are designed for use by physically challenged individuals. There are head position trackers that substitute a headset for the mouse, positioning the cursor with head movements. These trackers have attached mouth tubes to allow "clicking" by puffing air into the tube. There are also eye trackers, which use a mounted video camera to track the position of the eye pupils of the user. Computers, particularly coupled with communications peripherals and software, provide global access to many individuals who might otherwise be isolated from the outside world.

C. Output Devices

After data are manipulated by the computer, the user generally wants to see the results of the processing. Information output is available in two forms: soft and hard copy. **Soft copy** *refers to the information displayed on the monitor.* These data can be easily manipulated, but

because they exist only in RAM, they need to be saved before the computer is turned off. **Hard copy** *is a printed paper copy of the information* (the term originated in newspaper offices for the paper translation of Morse code messages). When you do not need immediate hard copy, any information that is stored to disk is available for later printing. Virtually all computer systems have both a monitor and hard copy output devices attached.

i. Monitor

It would be theoretically possible but very difficult to use a computer without a monitor. The **monitor** *is the main soft copy display where you can see your data input and choose commands,* Figure 31.14. In a graphical user interface, the screen display is necessary for point and click interactions. It is important to remember that data shown on the monitor are in RAM, and are therefore volatile until saved.

Monitors are chosen by two parameters: screen size and resolution. A third important characteristic is the number of colors displayed, which is generally controlled by the user. Screen size is measured diagonally, and can range from the small displays of laptops to those large enough to display two full-size pages of text at printed size. These larger monitors are quite expensive at present, and most users do not require the larger display. A 14-inch screen is adequate for most applications and many new systems come with a 17-inch display.

Monitors differ from television screens in offering much higher resolution. **Resolution** *indicates the sharpness of the screen image.* Monitor screen displays are described by the number of horizontal **pixels**, *picture elements or dots,* in a line that are separately shown on the display times the number of vertical lines (e.g., described as a "640×480" display). The third factor to consider is the number of colors in the display. All monitors now provide a way to select the number of colors shown in the display, from black and white to 256 colors to thousands or millions of colors. When setting this parameter, keep in mind that choosing fewer colors will produce higher resolution in the display. While television screens offer low picture resolution (and are therefore not useful substitutes for computer monitors), they provide a nearly infinite number of colors. Software applications with a major graphics component frequently require a specific limited number of colors for proper display.

ii. Printer

Printers are by far the most common and familiar hard copy output devices. These provide paper copies of file contents, screen displays, and program information. Computer printers were originally similar in operation to typewriters, and the standard for a good printer was **letter-quality (LQ)**, *which means the production of print of typewriter quality.* Modern ink-jet and laser printers provide better than LQ print. Choosing a printer involves considerations of price, resolution, and speed. Laser printers work somewhat like photocopiers, printing whole pages at once. They are fast and offer excellent resolution, but are relatively expensive. Ink-jet printers actually squirt tiny sprays of ink onto the paper, producing surprisingly sharp print. They are much less expensive than laser printers, but also noticeably slower. Printer resolution is measured in *dots per inch* **(dpi)**, with more dpi indicating greater resolution and a sharper image. Laser printers have standard outputs of 300–600 dpi. Speed is measured in text pages printed per minute. It is best to buy a printer that uses standard photocopier paper, which is relatively inexpensive, although some color printing may require special paper for maximum resolution.

iii. PLOTTER

A computer **plotter** is a *special type of printer used specifically for graphic images.* A plotter uses one or more adjustable ink pens (or other marking devices) that write directly on a continuously moving paper surface, Figure 31.18. These are ideally suited for complex output or unusually large printouts, such as schematic diagrams or blueprints. Plotters are similar in appearance to strip chart recorders (see page 272–273), but plotters report processed digital information, whereas strip chart recorders directly record analog data. The X axis for computer plotter output frequently does not represent time as it does on a strip chart recorder.

D. Modems

A modem serves as both an input and output device for network computers. A **modem** (short for <u>mo</u>dulator / <u>dem</u>odulator) *changes (modulates) the digital signal from a computer to an analog signal that can be carried*

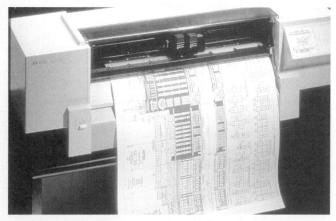

Figure 31.18. Plotters Can Provide Schematic Diagrams and Large Drawings. Although there is a physical resemblance to a strip chart recorder (see page 273), plotters report digital or processed information, and the X axis does not necessarily represent time.

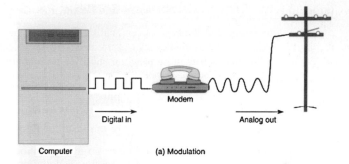

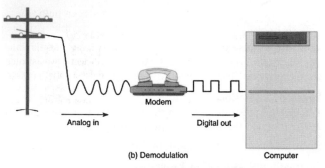

Figure 31.19. How a Modem Works. A modem "modulates" or changes a digital signal from a computer to an analog signal that can be transmitted over telephone lines. This analog signal is then reconverted to a digital signal at the end of transmission. (From *Introduction to Computers & Software,* R.A. Szymanski, D.P. Szymanski, and D.M. Pulschen, Prentice Hall, p. 258, 1996.)

over phone lines. These signals must be received by another modem, which changes the signals back to digital form, Figure 31.19. A modem can be a separate peripheral device, or be built into the computer console. Either type of modem connects directly to a phone line for signal transmission. Modem speed is measured in **bits per second (bps),** *referring to the speed of data transmission.* This is also sometimes referred to as a **baud rate,** which is a less accurate term. The current standard for modems is 56.6 *Kbps* (56,600 *bps*), although these speeds are increasing rapidly.

V. PURCHASING COMPUTER SYSTEMS

A. Software

The first consideration when buying a new computer system is deciding what you need the computer to do. This will determine the major types of software programs required. Specific software requirements will then dictate the hardware capabilities that your computer needs, especially the platform (type of microprocessor and operating system) and amount of memory. We will discuss applications types in the next chapter, but here we will describe general considerations for choosing among software packages.

Table 31.1 Tips for Choosing a Software Program

Once you have identified the type of application you need, you may find several specific programs that can provide the necessary product. You can ideally try out the program, or at least a demonstration version, before you buy. You may also want to keep the following factors in mind when choosing among your options:

Memory Requirements
- How much RAM does the program require?
- Can your computer easily provide this much RAM, with memory to spare for other operations?

Remember that most computer crashes are the result of RAM problems.

Ease of Use
- Does it come with a manual, and if so, can you read and understand it?
- Does the program include a tutorial?
- Does the program provide extensive on-screen help?
- How long will it take you to become comfortable with the program?

Many programs no longer come with a written manual, which is not necessarily a major problem. Most tutorials are useful to beginners, and the help screens can provide all the assistance you need while using the program. A well-designed program should provide an intuitive interface, which makes it easy to operate.

Speed of Operation
- Does the program perform the necessary functions quickly?
- How fast do the files load into the program?

If the program seems sluggish in the store, it will seem even worse when you know what you are doing.

Compatibility with Other Programs
- Can the program read files from other applications?
- Can the program save files in a variety of formats?

The import/export capabilities of a program can be a significant consideration. If you have already developed spreadsheets in a different program, can these data be easily transferred to your new application? What if you want to send your database file to someone who uses a different program? You are sometimes better off choosing a less than optimal application program if that program is standard among your colleagues.

For a specific type of application purchase, your major concerns are the ease of data input and the quality of program output. Be sure that the program you choose will provide the product that you need. For example, if you want software that can perform certain types of statistical analyses and then create color bar graphs, first check the program description and see if the product claims to perform these tasks. Then try out the software yourself using the same type of problem, and make certain that you can get the final product you want, and that the process is not prohibitively difficult. Additional tips are offered in Table 31.1.

All software is not equally easy to use. Computer programming requires designing, writing, and testing complex instruction sets, which may include literally millions of lines in their code. Because all programming is done by humans, this means that there are inevitably mistakes, or "**bugs**,"* in every program. In some cases, these are virtually unnoticeable, but they provide a source of constant annoyance in others. For this reason, it is an excellent policy to try out a program yourself, as well as talk to people who have used the program for a while. You can also search out a software review in one of the many computer magazines.

One way to avoid the worst bugs is to check the program version number. Most programs have a number after the title (e.g., Netscape Navigator™ 3.01). The numbers to the left of the decimal indicate new versions of a program, generally with extra features. Numbers to the right of the decimal indicate minor adjustments and improvements to this version. The example would, theoretically, be the third version of Netscape Navigator, with one minor revision to version 3, probably to fix bugs. Many computer-savvy people avoid buying programs either in their first version (1.---), or with only a zero after the decimal (---.0, called a *point-oh* version), and instead wait for later revisions to fix any original problems. For this reason, some software manufacturers skip 1.0 numbers entirely. Although it is best in many cases to buy the most recent version of an application, older versions sometimes have all the capabilities that you need. Do not rush to replace a satisfactory program simply because there is a new version on the market.

---

*Although Thomas Edison used "bug" in the 1870s to mean a mistake, the use of this term in computer science started with a moth found inside a malfunctioning Mark II computer by Grace Hopper, an early computer technology pioneer. That moth is still taped in a logbook at the Naval Museum in Dahlgren, Virginia. Since then, fixing mistakes in a computer program has been called "debugging."

B. Hardware

i. UPGRADING

Each "generation" of new computers (and a computer generation is no more than 4–6 months long) provides significantly faster and less-expensive machines than the earlier models. This does not mean, however, that you need a new system every 6 months. For many situations, a computer can perform faster than its human operator, so adding speed with a new CPU may be unnecessary except for memory-intensive applications such as graphics or video. If you find yourself outgrowing the capabilities of your computer, consider whether a system **upgrade** will solve the problem, *by adding additional power to the old system.*

Many computer problems, such as sluggish performance and full hard disks, can be addressed relatively inexpensively. New versions of complex programs frequently require large amounts of memory and disk space for storage. One of the least expensive and most useful upgrades is the addition of additional RAM. Memory upgrade chips or cards come in standard sizes of 4, 8, or 16 *MB*, which can significantly enhance computer performance, Figure 31.20. If computer performance is adequate but the hard disk is full, you can add an external hard drive to the current system for extra storage.

ii. BUYING A NEW SYSTEM

With current rates of progress in technology, it becomes more cost effective to replace a computer rather than upgrade after about three years. Choosing a new computer system, however, can be a challenging assignment. A typical mail order catalog description of a Macintosh computer might read:

Great Buy!   G3/233   32/4GB/24XCD

(An explanation of these terms is provided in Table 31.2.) It is easy to become confused about all of the specifications of a computer system, but there are three that are of greatest importance when buying a new system:

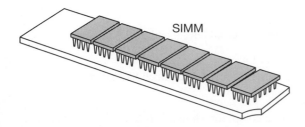

Figure 31.20. Extra RAM Can Be Added to Your Computer in the Form of Additional Memory Modules. This is an example of a SIMM (single in-line memory module), with eight memory chips on a card.

Table 31.2   *GREAT BUY!*
G3/233   32/4GB/24XCD

It is easy to figure out what the "Great Buy!" part is supposed to mean, but what about the rest of this computer description? People do not buy computers based on their appearance, but on their performance, which is indicated by the listed information. Here is an interpretation of this catalog description.

The first part—G3/233—shows the most basic specifications:
* **G3**— type of microprocessor
* **233**—clock speed, in *MHz*

The second part indicates the specific features of this system:
* **32**—RAM, in *MB*
* **4GB**—storage capacity of hard disk
* **24XCD**—speed of the CD-ROM drive

Together, these specifications provide enough information to determine the power of a particular system, and whether it can supply the needs of the user.

---

* microprocessor
* memory
* hard drive space

Before buying a system, you must determine which software applications you will need. Because software is written for specific platforms, this may specify which type of computer you purchase. Software requirements will also dictate how much memory and disk space you will minimally need.

The speed of a computer system is determined by the type of microprocessor and the clock speed in megahertz of the system. Remember that even though higher clock speeds indicate a faster computer, many users do not need the fastest system available. The performance you get from your system is likely to be most dependent on the size of the RAM, which is the working area of the computer. In general, you will want as much RAM as you can afford (and extra memory cards are not expensive). Many computers are sold with a minimum amount of RAM that may not be adequate for large applications. Even though installation of extra memory cards is fairly simple, it is easiest to have the vendor install (and guarantee) them at the time of purchase. Similarly, you will probably want as large a hard drive as practical.

If your computer is part of a network (see Chapter 33) consider whether you can save money by sharing peripheral devices through the network. Many useful peripherals, such as high-quality laser printers and flatbed scanners, may cost more than the computer system itself. You also need to consider the level of graphics and video support that you need. Some computer systems, such as the Macintosh, are designed specifically for high-quality graphics, but other systems may need added sound cards and video boards for multimedia applications. Table 31.3 offers some general buying tips.

Table 31.3   *STEPS IN BUYING A COMPUTER*

Before buying a new computer system, you should decide:

1. What realistic budget is available.

2. What types of software you need to use.

3. How much storage you could possibly need (and then double it).

4. How fast the computer needs to be.

5. Which peripheral devices you *must* have.

6. Which peripheral devices you would *like* to have if they fit your budget.

7. Which models to "test drive."

8. Which products and vendors offer the best documentation, warranty, and technical support.

VI. CARING FOR COMPUTERS IN THE LABORATORY

A. Maintenance

i. COMPUTER COMPONENTS

There are many possible causes for computer problems in the laboratory or office. Some of the most obvious are:

- environmental disasters (fire, etc.)
- vandalism or theft
- user errors
- program bugs
- hardware failures

Natural disasters are hard to avoid, but your best protection is to be sure that your computer files are backed up regularly and that the file copies are kept in a separate and safe place. Suggestions for keeping vandalism and theft to a minimum are provided in the next section. User errors are fairly common, but rarely cause permanent damage. It is impossible to make your computer explode by any actions you take at the user interface.

It is a relatively simple matter, however, to make your computer "crash." A system **crash** *means that your computer stops working.* A basic reset, which restarts the computer and reloads the system software, will usually eliminate the problem. PCs and Macs have different reset procedures and it is a good idea to keep the instructions posted near the computer. Unfortunately, after a reset, any unsaved data that was in RAM will be permanently lost. Because computer crashes are not rare events, this is excellent motivation to save files frequently while you work.

Most system crashes are related to memory problems rather than user error. Software bugs can cause these problems with some regularity (some programs are well known for this undesirable feature). Memory problems are generally the result of either too little available RAM for the operation in progress, or a conflict between different programs operating in RAM. If you have plenty of RAM and still have memory-related crashes, it may help to purchase software designed to detect application incompatibilities.

Hardware failures tend to be more serious in their consequences. Given the number of parts in a computer, hardware failures are inevitable in aging machines, but you can maximize the life of your computer with proper maintenance. Your main goal should be to keep your computer in a cool, clean, low-humidity environment. Internal computer components are easily damaged by high temperatures and dust. The fan that cools the power supply tends to pull dust into the machine through the vents, so it is important to keep the immediate area relatively clean. Table 31.4 offers some tips for regular computer maintenance.

Computers are highly sensitive to variations in electricity. Avoid any discharges of static electricity near the computer, especially the ports. Never plug any peripherals into the computer or unplug them while the system is turned on (this especially applies to the keyboard and mouse). An important accessory for every computer is a surge protector. These are relatively inexpensive and provide significant system protection in case of a power outage. A **surge protector** *acts as a power regu-*

Table 31.4   *HARDWARE MAINTENANCE GUIDELINES*

The following suggestions will maximize the life of your computer hardware:
- Keep any objects containing magnets away from the computer. This includes external fans, telephones, and noncomputer audio speakers.
- Avoid static electricity discharges in the computer area.
- Never turn the computer off and on again quickly. This can damage the hard disk and drive (much like grinding the clutch in a car).
- Keep liquids away from computer vents, the keyboard, and the mouse.
- Never block the vents in the computer console or place a dust cover over a computer while it is running. This will cause overheating of the CPU.
- Avoid any extreme temperatures and direct sunlight.
- Keep the computer area as dust-free as possible.
- Check the power cords and connecting cables on the system regularly.
- Buy and use a surge protector.

lator between the computer and the power outlet. When the power goes off, the contents of RAM are lost, but no damage occurs to the system. When power is restored, however, there is frequently an electrical spike that can damage or even destroy the CPU. A surge protector is designed to smooth out momentary spikes and fluctuations in power levels.

Even a surge protector will not guard against a direct lightning strike, which will burn out the internal components of a computer and other electrical equipment. If lightning or power outages are a frequent problem in your area, you might want to invest in an **Uninterruptable Power Supply (UPS)**, *which will deliver constant power to the computer for a brief period during a power outage.* The batteries will last 10 minutes or less, but you will have enough time to save your work and shut down your computer properly. Most UPS units also contain a circuit breaker, which will burn out in case of a lightning strike and save the computer.

On a lighter note, computers do not require much regular maintenance work. The plastic components of the console and the monitor screen can be cleaned with appropriate solutions. **Never spray cleaner directly onto the computer!** Instead, spray liquids onto a dust-free rag and then wipe down the computer surface. The only peripheral that may need regular internal servicing is the mouse, which tends to accumulate dust and grime around the rolling ball. You can remove the small plate that holds the ball in place, and carefully remove any dust or sticky substances on the rollers that contact the ball.

ii. DISK MAINTENANCE

Damaged or malfunctioning hardware is easy to replace and inexpensive compared with the consequences of data loss due to disk problems. The first lesson of computer use is to have *at least* two copies of every important data file, with the backup copy stored in a safe location. Every computer user dreads a hard drive crash, when all data stored on the hard disk becomes at least temporarily inaccessible and sometimes permanently lost. Backups should always be made onto a separate disk or other storage device. Table 31.5 offers suggestions for protecting your storage media, disk drives, and data.

It is not uncommon for a computer user to "throw away" or **erase** a file in error. This is why GUIs frequently provide a "trash can" or "recycle bin" icon for storage of discarded files—the user can retrieve the file before it is erased. Once you issue the command to empty the trash folder, these files disappear from both the folder and the disk. Because of the way disk storage works, however, these files are not truly removed. What actually happens is that the file name and address are eliminated from the **file allocation table (FAT)**, *which is the central address directory for the disk.* The disk space occupied by the file information is then made available for future storage. If you immediately realize that you

Table 31.5   DISK MAINTENANCE GUIDELINES

Disk Protection
- Keep disks away from anything magnetic. This includes telephones, speakers, fans, photocopiers, and magnetized cards.
- Keep disks away from liquids and foods (probably the most frequently violated guideline!). A disk will generally not survive a coffee spill.
- Handle only the disk case, not the magnetic medium.
- Avoid extreme temperatures, including direct sunlight.
- Keep disks stored away from dust that might enter the case.
- Store disks vertically to avoid warping.
- Do not set objects on top of disks—this can grind dust into the magnetic medium.

Drive Protection
- Never force a disk into the disk drive. Either the disk is in the wrong orientation or there is already a disk in the drive.
- Never insert or remove a disk from a disk drive if the drive is working.
- Remove portable disks from their drives before you turn off the computer.
- Use the labels that come with the disks. Any other labels may get stuck in the disk drive.

Data Protection
- Save your data files frequently while you work.
- Back up your disks on a regular basis.
- Store the backup copies away from the main computer.
- Use the write-protect tab to "lock" important disks and make them read-only.

---

have discarded an important file, it may be possible to recover it using an "**unerase**" program. *These programs* (Norton Utilities™ provides a popular example) *find the file contents and create a new directory entry.* This will work only if you have not stored other information on top of the old file, so the unerase utility must be applied as soon as possible after the file is discarded.

There is another consequence of the way computers discard files. Confidential information is not removed from the disk immediately after a file is discarded, which means that some or all of these data could be recovered by outside sources. If you need to reliably eradicate disk information, you can use true "erase" utilities, which will write over sensitive files immediately. Another way to completely erase the contents of a disk is to reformat or initialize that disk. **Initialization** *is the preparation of a disk for data storage in the appropriate computer format.* This will destroy any files and data on the disk and prepare it for reuse, making all storage space available. Never reinitialize a disk without being absolutely certain that it does not contain any useful files because this process is not reversible. In general, you will never want to initialize a hard disk without expert assistance.

B. Security and Ethics

i. Computer Security

Computer theft and vandalism unfortunately are not sufficiently rare to ignore basic precautions. The best ways to prevent these problems is to limit access to the computer area and to the computers themselves. Install any portable computers and expensive peripherals in nonpublic, locked areas when possible. Most computers can be attached to a desk or table with a strong cable and lock. Be sure that the serial numbers of all computer components are recorded.

Vandalism frequently takes the form of using the computer to damage data files or read private information. Preventing physical access is one simple precaution. Another is to set up the operating system or screen saver to require a password for CPU access. This will prevent any idle browsing of the computer contents (unless you write your password on the computer, as some users do). To prevent a more serious intruder from using the computer, do not choose an obvious password, such as your address or birthday. Experts recommend a random combination of letters and numbers, which would be almost impossible to guess, although these are also difficult to remember. Keep in mind that if you forget your password, you may be prevented from using the computer too!

If your computer contains proprietary or sensitive information, you will want to consider additional levels of data protection. There are reliable security programs available that can lock individual folders, create invisible files, and/or *store the data in coded form* (**encryption**).

ii. Viruses

Business losses due to computer viruses have been estimated at up to $2 billion each year. A computer **virus** is *a program designed to hide inside other programs and replicate or infect any computer that loads the contaminated program.* Some viruses are designed merely to cause annoyance (e.g., by displaying strange messages on your monitor screen at random intervals). Others can cause serious damage to data files, up to and including erasing the complete contents of your hard disk. To date, no one has developed a virus that will actually damage computer hardware.

If your computer is not part of a network, contains only legally purchased commercial software, and does not come into contact with portable disks used in other computers, your system is safe from virus infection. The rest of us, however, need to take precautions, Table 31.6. Viruses provide yet another reason to back up files at regular intervals, and to keep several previous versions of your hard disk contents. In this way, you can restore files that have been damaged by a virus, and also make sure that you return to a virus-free hard disk.

Table 31.6 *Preventing Computer Virus Problems*

- Use virus protection software.
- Back up your data frequently.
- Avoid using floppy disks of unknown origin.
- Avoid stolen versions of software, which may contain viruses.
- Control access to your computer.
- Do not open e-mail attachments from unknown sources.

There are many excellent antivirus programs available (some of them free!). At least one should be installed and used both to check all portable disks and to regularly scan your hard disk, especially if you download software from a public network. Virus protection software is not foolproof; it can only recognize specific viruses. For this reason, you should always use the most recent version of your protection software. Many companies offer free upgrades on the Internet.

iii. Ethical Use of Software

Programs are available with a variety of legal arrangements:

- commercial software
- public domain software, or "freeware"
- shareware

An essential part of ethical behavior when using a computer is to refrain from illegally copying commercial software from a network. Most laboratories need complex data manipulation capabilities that are only found in commercial software. These are programs that are copyrighted and licensed to users for (usually) substantial fees. When you "buy" software, you are really only purchasing a license to use the software on a single computer. If you need to install the program on multiple computers, you are legally required to purchase a **site license**, *which grants you the legal right to a specific number of program copies.* Although some software companies use **copy protection** devices on their software, *which prevent you from making more than one copy,* copy protection also prevents you from making multiple backup copies and can sometimes interfere with the use of legally installed software. Because of consumer complaints, many companies do not copy protect their software. This does not mean that you are allowed to make extra copies for distribution.

Many widely used utility programs (and games) are categorized as shareware or public domain software. **Public domain** programs are *noncopyrighted software that the author has chosen to provide for free to the public.* These programs are available through networks and computer groups, but, for obvious reasons, there are no guarantees of quality. Somewhat more reliable is **shareware**, *which is software that is under copyright but easily available for trial use.* Users who keep and use the

Table 31.7   COMPUTER ETHICS

- Do not use someone else's password without permission.
- Do not use illegal copies of commercial software.
- If you like and use a shareware program, pay for it.
- Despite the simplicity of using a scanner, consider copyright issues before reproducing artwork.
- Request the author's permission before downloading and reproducing web page contents.

program are expected to pay a small specified fee to the software author by mail. This will frequently entitle the registered user to upgrades and a manual.

Table 31.7 summarizes some basic guidelines for ethical use of computers.

C. Health Concerns

Ergonomics *is the study of the effects of environmental factors on worker health and comfort, and the design of environments to increase worker health and productivity.* Ergonomic design can prevent many of the health concerns related to computers. There are three computer-related problems that are common among individuals who use computers extensively. These are:

- eyestrain
- wrist and arm problems
- neck and back pains

These are caused by **repetitive stress**, *where the same movements and actions are repeated until physical fatigue results.* The best prevention techniques are to take frequent breaks while working at the computer and to ensure that the computer environment is both comfortable and efficient, Figure 31.21.

Eyestrain results from staring at a flickering display screen for extended time periods. It can be minimized by purchasing a high-quality monitor, reducing any glare from room lights or sun, keeping the screen clean, and working in a well-lit room. It is helpful to give your eyes a rest by frequently looking away from the screen.

Many people experience wrist and arm problems from the repetitive movements of data input. This can eventually lead to a condition called **carpal tunnel syndrome**, *which is a painful irritation of nerves in the wrist.* To minimize risk for this problem, keep your wrists straight while using the keyboard and mouse. It may be useful to purchase a soft wrist rest to keep your wrists elevated while typing, and to wear a wrist protection device, Figure 31.22. Most important, do not ignore wrist or arm pain when it starts; take immediate precautions to prevent the problem from getting worse.

Back and neck pains can usually be avoided by the use of an ergonomically designed desk chair, good posture, and frequent breaks. Figure 31.21 provides other suggestions for comfortable workplace design.

The Hardware

Monitor location (A). The monitor should be located directly in front of you at arm's length with the top at forehand level. Outside windows should be to the side of the monitor to reduce glare. *Monitor features.* The monitor should be high-resolution with anti-glare screens. *Monitor maintenance.* The monitor should be free of smudges or dust buildup. *Keyboard location (B).* The keyboard should be located such that the upper arm and forearms are at a 90-degree angle. *Keyboard features.* The keyboard should be ergonomically designed to accommodate better the movements of the fingers, hands, and arms.

The Chair
The chair should be fully adjustable to the size and contour of the body. Features should include: *Pneumatic seat height adjustment (C); Seat and back angle adjustment (D); Back-rest height adjustment (E); Recessed armrests with height adjustment (F); Lumbar support adjustment (for lower back support) (G); Five-leg pedestal on casters (H).*

The Desk
The swing space. Use wraparound work space to keep the PC, important office materials, and files within 18 inches of the chair. *Adjustable tray for keyboard and mouse (I):* The tray should have height and swivel adjustments.

The Room
Freedom of movement. The work area should permit freedom of movement and ample leg room. *Lighting.* Lighting should be positioned to minimize glare on the monitor and printed materials.

Other Equipment
Wrist rest (J). The wrist rest is used in conjunction with adjustable armrests to keep arms in a neutral straight position at the keyboard. *Footrest (K).* The adjustable footrest takes pressure off the lower back while encouraging proper posture.

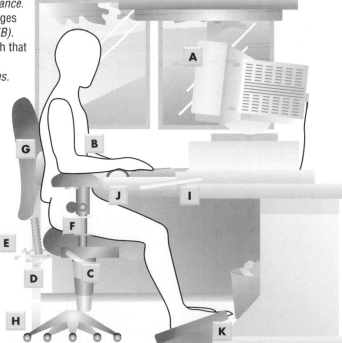

Figure 31.21.   Ergonomic Design of the Computer Workplace. (From *Computers,* 5th ed., Larry Long and Nancy Long, Prentice Hall, p. 112, 1998.)

Figure 31.22.   Prevention of Hand and Wrist Injuries from Extensive Computer Use. There are numerous wrist protection devices that help reduce or prevent repetitive injuries.

PRACTICE PROBLEMS

1. Which of the following items are hardware and which are software?

   a. keyboard

   b. GUI

   c. printer cable

   d. printer driver

   e. floppy disk

   f. unerase program

2. How many bytes are actually in a megabyte?

3. What can you infer about this program from its number: MegaWhiz 1.4?

4. If you need to make a complete backup of a 2 *GB* hard disk;

   a. How many 100 *MB* high-capacity data cartridges are required?

   b. How many 1.4 *MB* floppy disks are required?

DISCUSSION PROBLEM

Think about the computer system in your laboratory or home. How would you optimize or upgrade the available features?

ANSWERS TO PRACTICE PROBLEMS

1. **a.** hardware

   b. software

   c. hardware

   d. software

   e. hardware

   f. software

2. One megabyte = 2 to the twentieth power

   $= 2 \times 2 \times 2 \times 2 \times 2 \times 2 \times 2 \times 2 \times 2 \times 2 \times 2 \times 2 \times 2 \times 2 \times 2 \times 2 \times 2 \times 2 \times 2 \times 2 = 1,048,576$

3. This is the first major version of MegaWhiz, and there have been up to four revisions of this version, probably to remove bugs from the program.

4. **a.** 20

   b. 1,429

CHAPTER 32

Data Handling with Computers

I. INTRODUCTION TO DATA PROCESSING

A. The Significance of Data

A significant portion of the scientific process consists of gathering data. This is accomplished by performing procedures under controlled conditions and gathering and recording specific measurements or observations—data. Data take the form of numbers, text, and graphics. These data are then converted, through organization and mathematical manipulations, into usable information. When we talk about computers, this conversion is generally called *data processing*.

At one time, the term *data processing* was used to describe the simple collection and sorting of data. Scientific **data processing**, however, *is a series of complex manipulations that include data recording, storage, and retrieval, as well as searching for patterns in collected data, interpreting these patterns, and, eventually, presenting organized data and conclusions*. Scientists must be able to reanalyze and compare data sets regularly as new patterns emerge and hypotheses change.

Data must have certain properties in order to be useful. These are the same properties that were discussed in Units IV and V. Data must be:

- accurate
- well-documented, in terms of conditions and variables
- repeatable and reproducible
- complete
- concise
- understandable

The quantities of data gathered in scientific laboratories are too large for manual processing in most cases, but computers are ideally suited for large-scale data handling. It is ultimately the human user, however, who decides which data are important and how these data should be interpreted. Collection of raw (unprocessed) data is relatively simple compared with the decisions that must be made about how to process and interpret the data. The human user is also responsible for deciding how to present the data to others. The computer can help to accomplish specific goals, but those goals are always set by the end user.

B. Data Handling

One of the major advantages of computers is that a single set of data input can be processed in many different ways to provide multiple forms of output. The data input phase is the quality foundation on which all data processing depends. In computer circles, this is called the **"GIGO" principle** (Garbage In, Garbage Out), *meaning that information output can never be of higher quality than the starting data*. This is the reason that proper data

recording, documentation, and other quality assurance measures are essential. Data may be input directly into a computer through a connected instrument, or manually transferred from a primary source such as a laboratory notebook. Problems in data input into a computer system can arise from either errors in the data (e.g., lack of numerical accuracy or precision) or from the data entry process itself. Most computer programs have methods for internally checking data input based on data characteristics. For example, if you are typing a DNA sequence, most DNA analysis programs would mark any input values other than A, C, G, or T (the only expected values) as errors and bring them to your attention.

Computers excel at four data processing functions that are essential for laboratories:

- storage and retrieval of large quantities of raw data
- mathematical manipulation of numerical data
- sorting and organizing data into compact and easily accessed forms
- comparing and testing patterns among data

Data processing also includes the steps needed to analyze, organize, and prepare data for presentation. Even though a software user needs input procedures that are fast and easy, processing needs may be complex. Computer software is usually designed to make processing as simple as possible, but the user must still instruct the computer to perform the appropriate data manipulations.

Most software can provide you with many forms of data output, including forms, formal reports, graphs, tables, and other documents. Data selection and formatting are essential parts of data output. Decisions about appropriate data processing software are frequently based on the ability of programs to provide the desired end product. The software descriptions in upcoming sections provide examples of output from different types of programs.

II. COMPUTER SOFTWARE FOR DATA PROCESSING

A. Introduction

Data processing and output needs will determine the types of application software you need for a specific task. Although some users may require specialized software, virtually all laboratories use a basic set of programs providing word processing, spreadsheet, database management, graphics, and communications software. Table 32.1 describes some common functional features of applications. These programs may be purchased separately or as integrated products such as Microsoft Office or ClarisWorks. These **integrated software packages** or software **suites** *usually contain modules for all of the preceding functions*. Suites are much less expensive and usually require less disk space than separate programs. They

Table 32.1 *General Tips for Using Application Software*

Fortunately, most of us do not have to write our own software programs. However, we are expected to use application software effectively to solve problems. It is helpful to be familiar with some standard features of these programs.

Installation

- Check the manual or other documentation first to determine the installation procedure for your software.
- Many programs must be **installed**, not simply copied to the hard disk.
- This requires the installation program included with your software, which places program files at various locations on the hard disk so that the application and the system software can find them.
- In order to remove a program from your hard disk, you may need to use an "uninstall" program, which finds the scattered files and eliminates them.

Using Your Software

- Always keep your original copies of programs in a location away from your main computer.
- If your software provides an interactive tutorial, this is frequently the easiest way to learn basic commands.
- A program must be loaded into RAM in order to run.
- Generally you use the cursor to select objects or sections of the window content and then apply commands to the selection.
- Use scroll bars or arrow keys to move different parts of a document into view.
- Remember to save your documents frequently.

Customization

- Most install programs will allow a custom installation of only selected parts of the program to save disk space.
- After you become familiar with the program, consider changing some of the **default** options.
- Many applications allow you to program the function keys for your own needs.
- You can create or purchase **templates,** or blank forms, that will allow you to fill in data in a standard format. These are readily available for word processing, spreadsheet, and database programs.

Exit Routines

- Save your files!
- Remember that closing a file does not automatically quit the program.
- All applications have a standard exit routine that should be followed. This allows the computer to file information properly and prevent future problems.

Back Up Your Computer Files!

- Develop the habit of regularly backing up, or making a spare copy, of all your data files.
- Keep these copies in a safe place away from your main computer.

are often easier to use because commands are standardized among the programs, and data can be readily transferred between functions. The disadvantage of these packages is that the individual programs are not likely to be the most powerful available. For many users, however, they provide adequate capabilities for routine use.

This chapter will focus on the basic uses of word processing, spreadsheet, database management, and graphics software. Each section contains a table of suggestions on when to use that type of program, and another table that lists some basic user tips. Communications software will be discussed in Chapter 33.

B. Word Processing

Word processing is easily the most familiar and popular application of computers. Virtually all recent documents were created on some type of word processor. These programs have many significant advantages over using a typewriter. They eliminate most of the mechanical concerns of typing, such as changing sheets of paper and watching for the end of lines, and they allow the writer

to concentrate on content. Written materials can be edited without complete retyping, so multiple drafts of a manuscript are easy to create. All these factors provide speed and efficiency for handling text, Table 32.2.

There are three types of word handling (processing) programs, which are called text editors, word processors,

Table 32.2 *When to Use a Word Processor*

Use a word processing program whenever you need to create and format text. Documents may include:

- research reports and papers
- standard memos and forms that can be reused as templates
- letters that will be addressed to multiple recipients
- outlines and lists
- newsletters and product announcements
- any documents that will need multiple drafts

You do not need a word processor for:

- simple forms that will not be repeated
- single envelopes
- personal notes that should be hand-written

and desktop publishing programs, Figure 32.1. A **text editor** *allows you to enter, edit, save, and print plain (ASCII) text.* All computer systems have a basic text editor installed. **Word processing programs** *allow text editing plus complex document formatting such as superscripts, spell checking, tables, and outlining.* **Desktop publishing (DTP) programs** *provide complete page layout capabilities for manipulating graphics and text, and preparing documents for professional publication.*

For most documents, a word processing program is appropriate. Use of these programs is relatively simple: you type in text, then edit and format it, Table 32.3. The editing options include cutting and pasting blocks of text, finding and replacing words or phrases, and performing spelling and grammar checks. Spell checking functions are useful for detecting typing errors and making a first round of corrections. A **spell checker** *will point out and usually suggest substitutions for words that are not included in the program's spelling dictionary*, Figure 32.2. You can add words to the dictionary, so that scientific terms and abbreviations will not be marked as errors. Spell checkers cannot be used as a substitute for proofreading, however, because they do not know your intentions. A spell checker will not point out that you mistakenly typed *flak* when you meant *flask* because *flak* is in the program's dictionary.

General formatting functions include selection of document parameters such as page margins and tabs, as well as selection of fonts. A **font** *includes all the letters, numbers, and symbols of one style and size of type*, Figure 32.3. You can produce bold, italic, and underlined text easily. Formatting will determine how your document looks on a final printed page.

Desktop Publishing (DTP) refers to the ability to produce camera-ready copy for publication using a personal computer. These programs, which require significant computer power and user expertise, are designed to create complex page layouts combining text and graphics. Blocks of text and graphics are treated as objects, which are positioned on-screen for printing. In computer language, an **object** *is any file or portion of a file that is treated as a unified whole for purposes of movement and processing.* Text is generally created in a word processing program and then manipulated as an object in the DTP program, although some text manipulation can be performed in DTP programs.

C. Spreadsheets

One of the first data processing functions to be transferred to personal computers was spreadsheet development. **Spreadsheets** *are a template of rows and columns to*

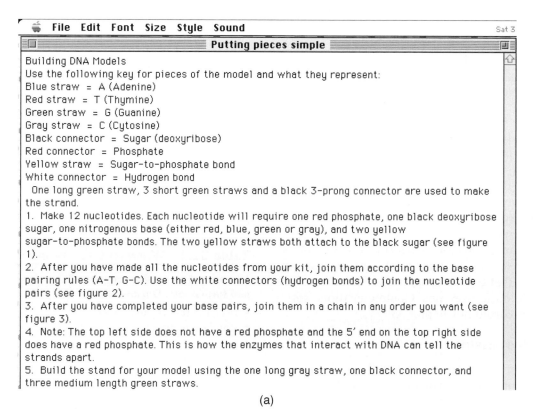

(a)

Figure 32.1. Word Handling Programs. Sample output from **a.** a text editor, **b.** a word processor, and **c.** a desktop publishing program. Note the progressively more sophisticated formatting.

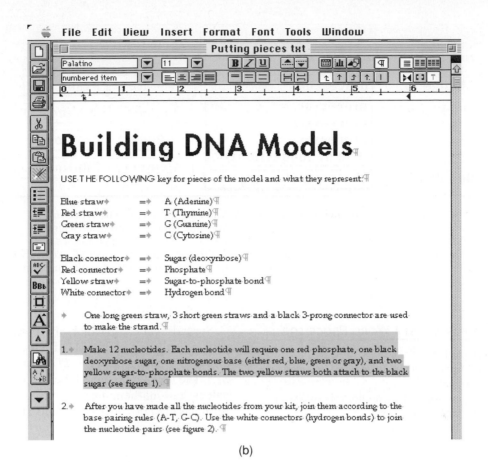

(b)

(c)

Figure 32.1. *(continued)*

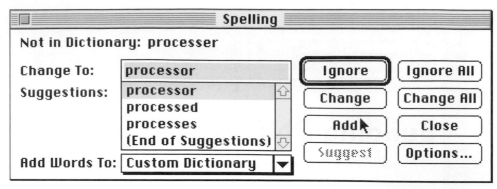

Figure 32.2. Example of Spell Checking. Here, *processor* was incorrectly typed and *processer* identified by the spell checker as a questionable word. The program suggests likely substitutes, which can be accepted by the user. The user also has the options of making no change, typing in another word, or asking the program to correct all occurrences of *processer*. An unrecognized term that is correctly spelled can be added to the program spelling dictionary.

Table 32.3 *TIPS FOR USING A WORD PROCESSOR*

- Do not type returns at the ends of lines.
- Do not overuse multiple fonts and formatting commands within a single document.
- Always proofread your documents; do not rely solely on spell checking to eliminate mistakes.
- Type only one space between sentences.
- Do not line up lists or tables with the space bar; use tabs instead.
- Do not type the letter "l" instead of the number "1."
- Create or buy and use templates when possible.
- Use a word processor to prepare text for desktop publishing.

There are many differences between using a typewriter and a word processing program. For an excellent discussion and explanations of these guidelines, see: *The PC is not a Typewriter*, Robin Williams, Peachpit Press, 1992, or *The Mac is not a Typewriter*, Robin Williams, Peachpit Press, 1990.

record numbers and help organize the results of repeated calculations, Figure 32.4. The resulting matrix of values provides a clear record of both raw and processed data. Traditional spreadsheets were prepared on gridded paper using a calculator. Spreadsheet software provides an electronic ledger, eliminating the need for repetitive use of a hand-held calculator and providing data entry checks, virtually instantaneous calculations, and automated generation of graphs and reports, Table 32.4.

Spreadsheet files (which are also called **worksheets**) provide a blank grid or table of **cells**, *each of which is uniquely designated by a row and column coordinate*, Figure 32.5. The user then enters labels or other text, numerical values, and mathematical formulas into the grid. Once formulas are entered, calculations are automatically performed on specified cells, columns, or rows of numbers, and the results are shown in specified cells. Large spreadsheet programs can handle tens of thousands of rows and columns; more than a billion cells per spreadsheet! In many cases, spreadsheet formats can be saved as templates for recording similar data, Table 32.5.

Table 32.4 *WHEN TO USE A SPREADSHEET*

Use a spreadsheet when you want to organize text and numbers into a grid and perform calculations. Applications include:

- tables
- graphs
- budgets and ordering
- course grading
- any repetitive calculations

You do not need a spreadsheet for:

- single calculations
- data that needs sorting but not manipulation
- simple graphs from previously calculated numbers

This font is Arial 12. It is good for text.

Futura Extra Bold 18 is good for headlines.

This font is Script MT Bold 16. It is good for invitations.

Figure 32.3. Examples of Three Fonts.

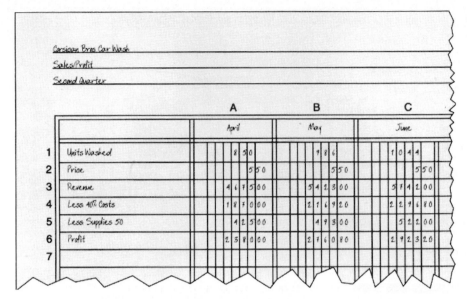

Figure 32.4.   Example of a Paper Spreadsheet. (From *Introduction to Computers & Software*, R.A. Szymanski, D.P. Szymanski, and D.M. Pulschen, Prentice Hall, p. 193, 1996.)

Spreadsheet programs generate a variety of output formats. They can be used to produce both numerical and text tables easily, Figure 32.6. Most can automatically graph numerical data in several formats, which are discussed later in more detail.

One of the most powerful aspects of spreadsheet software is that changes in data entries will produce instant recalculation of results in other parts of the spreadsheet. In paper spreadsheets, correcting a single data entry could result in changing values for many other cells. Computer-

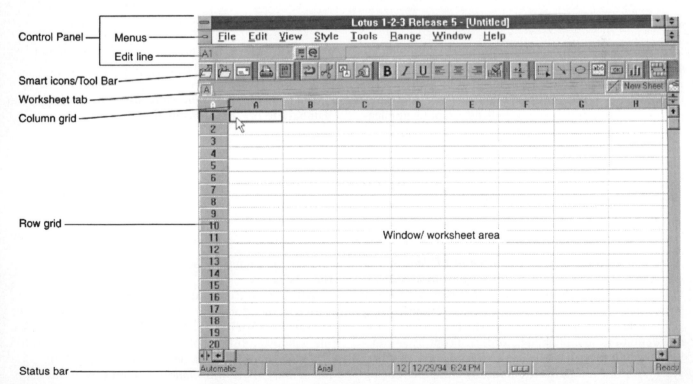

Figure 32.5.   A Typical Spreadsheet Screen. Cell A1 is selected, designating column A and row 1. (Adapted from *Introduction to Computers & Software*, R.A. Szymanski, D.P. Szymanski, and D.M. Pulschen, Prentice Hall, p. 197, 1996.)

Weekly Expense Report

Purpose of Trip	Emp No.	Ext.	Div.	Dept.	Project	Week Ending	Rate/Mi.
Mail Check To:							

| Cities Visited From: | | | | | | |
| To: | | | | | | |

TRAVEL		Date:	7/13	7/14	7/15	7/16	7/17		Totals
	Miles								
	Air Fare	345.00							345.00
	Auto Rental					196.55			196.55
	Tolls & Parking	15.00		45.00					60.00
	Air, Rail, Bus								
	Taxi, Limousine (Incl. Tip)	20.00				20.00			40.00
	Lodging, Hotel, Motel	150.00	150.00	150.00	150.00	150.00			750.00
		$530.00	$150.00	$195.00	$150.00	$366.55			$1,391.55

OTHER									
	Telephone	12.10	15.76	33.15	4.75	5.20			70.96
	Personal Meals	55.00	33.50	45.00	60.00	30.00			223.50
	Banquet Meals								
	Business Meals		180.00						180.00
	Entertainment				35.00	45.00			80.00
	Other Expenses	15.00		43.98					58.98

| **TOTAL** | $612.10 | $379.26 | $317.13 | $249.75 | $446.75 | | | $2,004.99 |

REMARKS	PLACE/LOCATION	Date	Items/Services Purchased/Persons Entertained (Names, Titles, Business Relationships	Business Purpose	Amount

Employee Signature / Date _____   Manager Signature / Date _____

(a)

Figure 32.6.   Sample Output from a Spreadsheet Program. a. A formatted travel expense report. Calculations for subtotals and totals are automated. **b.** Same travel report after a change in one data cell ($45 for parking on 7/15). Note the updated calculation cells. Created with ClarisWorks 5.0.

ized spreadsheets make all the programmed changes automatically. They are especially useful for financial calculations such as budgets, which can be manipulated to determine the results of shifts in funding and priorities.

There are a number of specialized scientific applications for spreadsheets that provide organized and consistent data handling and presentation. Spreadsheets allow you to look for patterns in large amounts of data and are ideal for the types of repetitive statistical calculations that are frequently applied to data.

D. Databases

A **database** *is an organized collection of data that is accessed through database management software.* **Database management systems** *allow the user to sort and report data in a variety of formats.* An example of a printed database is a dictionary, which catalogs words and definitions in alphabetical order. In contrast, a computerized dictionary database provides a dynamic filing system for information, automatically cross-referencing at multiple levels. For example, a computer database could be used to quickly identify words with Greek roots, words with more than three syllables, or definitions related to medicine. Database management software excels at detecting patterns in large quantities of data.

There are many significant advantages of using a database to organize information:

- Ease of access
- Sorting and manipulation capabilities
- Consistency of format
- Prevention of data duplication
- Data security
- Creation of complex reports and other documents

Table 32.6 summarizes some of the uses of a database in the laboratory.

Although different database management software varies tremendously in complexity, all databases have the same type of basic structure, Figure 32.7. A database file contains a full set of related information. Each file is divided into **records**, *which are the basic organizational units of the information.* Each record in turn contains multiple information **fields**, *which provide the individual data items, which can be text, numbers, or*

Weekly Expense Report

Mail Check To:

Purpose of Trip	Emp No.	Ext.	Div.	Dept.	Project	Week Ending	Rate/Mi.

| Cities Visited | From: | | | | | | |
| | To: | | | | | | |

	Date:	7/13	7/14	7/15	7/16	7/17			Totals
Miles									
Air Fare		345.00							345.00
Auto Rental						196.55			196.55
Tolls & Parking		15.00							15.00
Air, Rail, Bus									
Taxi, Limousine (Incl. Tip)		20.00				20.00			40.00
Lodging, Hotel, Motel		150.00	150.00	150.00	150.00	150.00			750.00
		$530.00	$150.00	$150.00	$150.00	$366.55			$1,346.55
Telephone		12.10	15.76	33.15	4.75	5.20			70.96
Personal Meals		55.00	33.50	45.00	60.00	30.00			223.50
Banquet Meals									
Business Meals			180.00						180.00
Entertainment					35.00	45.00			80.00
Other Expenses		15.00		43.98					58.98
TOTAL		$612.10	$379.26	$272.13	$249.75	$446.75			$1,959.99

PLACE/LOCATION	Date	Items/Services Purchased/Persons Entertained (Names, Titles, Business Relationships	Business Purpose	Amount

TRAVEL · OTHER · REMARKS

Employee Signature / Date _____   Manager Signature / Date _____

(b)

Figure 32.6. *(continued)*

other values. One analogy would be a phonebook, which could represent a database file. Each individual entry is a record, which includes several data fields (i.e., name, address, phone number).

Table 32.5 TIPS FOR USING A SPREADSHEET

Computerized spreadsheets can save huge amounts of time performing repetitive calculations. However, the results will only be valid if the spreadsheet is appropriately programmed for the desired functions. The general steps in developing a data spreadsheet are:

1. Create the spreadsheet template, with labels and formulas.
2. Input the data into the appropriate cells.
3. Create the necessary report forms and graphs.
4. Modify data and formulas as necessary.

Some general guidelines:

- Always check a few random calculations by hand to be sure that the spreadsheet is performing the correct manipulations.
- "#####" as a cell value indicates that the correct number is too large for the cell size (enlarge the cell to see the number).
- Sets of add-on templates that address many common tasks are commercially available.
- Spreadsheets can be used to prepare text tables (more easily than using a word processor, in some cases).

Successful use of databases requires some thought about the problem you would like to solve, Table 32.7. For example, does your laboratory need to keep clear documentation on batches of medium used for tissue culture? Who would access these records, what would they need to know, and would they like to receive regular updated reports? Once you understand these needs, you can design and create a database structure, first on paper and then with the database management software.

Table 32.6 WHEN TO USE A DATABASE

Use a database when you want to collect and sort data. Applications include:

- any list you need to sort, such as a bibliography
- quality control records
- batch records for medium, cell lines, or other laboratory materials
- lists of orders, vendors, and supplies
- addresses for labels and form letters
- collections of digital images and text
- personal information manager, with to-do lists and calendar

You do not need a database for:

- sets of numbers which require calculations
- small numbers of data items
- organizing data that does not need sorting

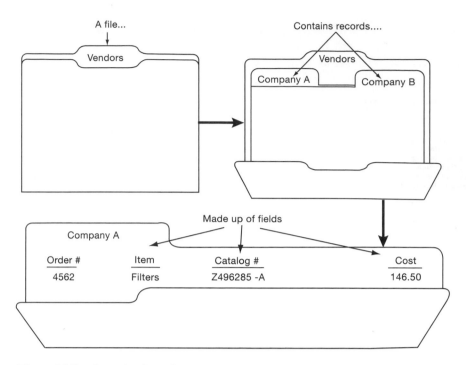

Figure 32.7.   Organization of Data in a Database. The units of organization are files, records, and fields. (Adapted from *Introduction to Computers & Software*, R.A. Szymanski, D.P. Szymanski, and D.M. Pulschen, Prentice Hall, p. 172, 1996.)

Each field will have specified characteristics, such as a name, size, and type of entry, Figure 32.8. Fields should be placed in logical order to minimize the time needed for data entry (e.g., by following the same sequence as the blanks on a data form). Once the database structure is created, then data records can be entered. Because many data errors are the result of faulty entry into the database, the software usually provides methods for

Table 32.7 Tips for Using Database Management Software

Database management software provides a powerful tool for sorting and organizing data. Before creating a database, take time to plan your database design on paper and include fields for any and all data you are likely to need in the immediate future. The structure of a database can be altered later, but it saves time to plan ahead.

- Use available templates for routine data collections, such as address books.
- Take advantage of the ability of database software to import data from other applications, which can save data entry time.
- For easier sorting, set up phone numbers and zip codes as text, not numbers.
- Because most database programs automatically save new records, make an initial backup copy if you need dated versions of the database.
- Consider security for network data.
- If data seems to mysteriously disappear when you open a database, check if filters (see Figure 32.9) have been applied.

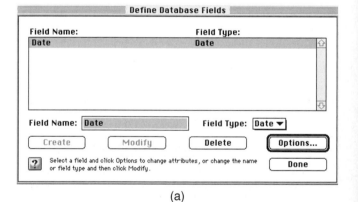

Figure 32.8.   Setting Up a Database. Most database management software leads you through the process of creating fields and specifying their characteristics. This example was created in ClarisWorks 5.0. **a.** Field definition screen. **b.** Selection of field content options. In this case, the date field must be filled with the current date.

detecting errors through data validation procedures. The user can designate the format and range of data which are appropriate in each field, Figure 32.8b.

After the records are entered, the user can exploit the data processing abilities of the software. Records can be sorted and organized at many levels. The user can specify a filter, which shows (or hides) records with designated characteristics. For example, someone checking the media batch records can choose to look at only those batches of a single medium type that were formulated by a specific person. These records can then be sorted by date, Figure 32.9. Database programs can provide

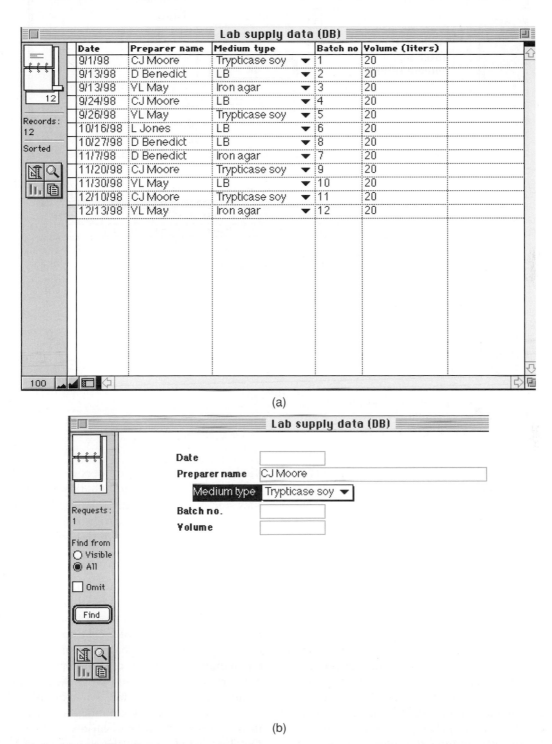

(a)

(b)

Figure 32.9.   Filtering and Sorting a Database. a. The raw data as entered. Field names are indicated across the tops of the columns. **b.** A filtering screen, where the user indicates which subset of records should be displayed. **c.** A sort screen, where the user indicates which fields should be applied to order the selected data. **d.** The final product of the filtering and sorting of **a.**

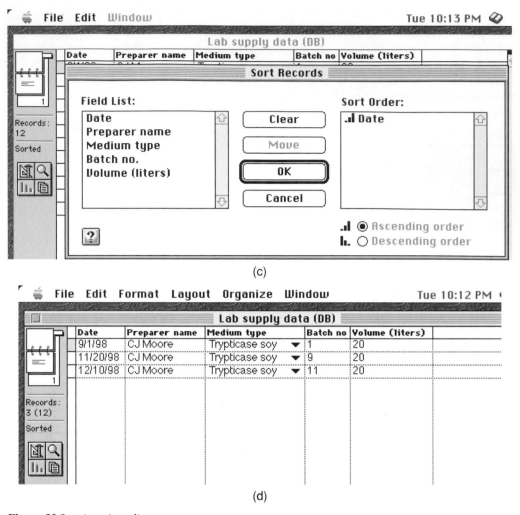

Figure 32.9.   *(continued)*

output in the form of specialized reports and/or screen displays. A database **query** *is the process of extracting a desired subset of data from a database and then presenting it in a specific format.*

An important feature of database management software is the ability to maintain, add, and update records as necessary. It is essential that the information stored in a laboratory database be kept current. The simplest and most common type of database is called a **flatfile database**, *which means that each database is stored in a separate computer file.* Each database file represents a simple table of information that can be internally manipulated, and data are kept current by direct entry into the database.

One approach to updating complex, interrelated data is the use of relational database software. In a **relational database management system**, *various database files are linked by common fields. Updating information in one database automatically updates the other files as well.* Relational databases are useful when the same information, such as names and addresses, is needed in multiple databases. This information need only be entered or mod-

ified in one database, and becomes available to all the related databases. Although this capability can save time and effort, designing relational databases requires some skill in computer programming, and is generally performed by specialists in this field. Relational database management software is also expensive compared with flatfile programs.

Databases have become a major part of business in this country. Anyone who has placed a mail order and then been inundated by mail offers from many companies has become part of a customer name and address database that has probably been sold to those other companies. Many types of commercial databases collect and provide information on a wide range of topics, sometimes for a user fee. For example, bibliographic databases, which catalog publications, currently number in the thousands, with a variety of formats available. Databases that contain the sequences of DNA nucleotides for specific genes are readily available on the Internet. It is increasingly simple to obtain information, both public and private, from database sources.

Figure 32.10.   Examples of Public Domain Clip Art.

E. Graphics

i. INTRODUCTION

Drawing and graphing, including the creation, manipulation, and printing of graphic images, is routine in many laboratories. There are many types of graphics software to make these tasks easier and the results more professional in appearance. **Computer graphics** *are images created or modified through the application of graphics software.* The graphics needed in laboratory work may be original or imported from external sources, such as the Internet or clip-art collections. **Clip art** *refers to simple prepared images that are pasted into documents and used to enhance presentations,* Figure 32.10.

Several major types of graphics programs are available:

*   Paint programs—also called raster graphics or bit-mapped graphics
*   Draw programs—also called vector graphics or object-oriented graphics
*   Illustration programs—usually providing both paint and draw capabilities
*   Presentation graphics programs—including table-, graph-, and chart-making programs
*   Multimedia programs

ii. PAINT PROGRAMS

Paint programs *provide the ability to create and modify complex images using computerized tools which are similar to those used by an artist on paper.* The resulting graphics are "painted" into a file that contains information about the graphic content of each screen pixel. These images are also called **bit-mapped graphics**, *with each computer bit of information specified for the image,* Figure 32.11. Another, less-widely used term for these pictures is **raster graphics**, *indicating that the picture is a collection of dots (or pixels or bits) that together form an image.* Because information about each pixel is stored in the image file, paint programs create very large files. These files are almost always stored in one of a variety of compressed formats, in order to conserve disk space, Table 32.8. **File compression** *means that the data has been abbreviated according to a specified formula, which depends on the file format.* File compression leads to a loss of information about stored images, so a file format must be chosen carefully to retain the image quality needed in the final product.

Although paint programs allow the user to create completely new images, a major application of these bit-mapping programs is image editing. Image editing pro-

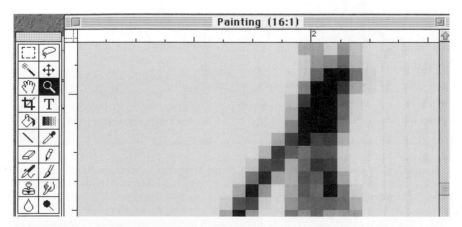

Figure 32.11.   Bit-Mapped Graphics. Notice the pattern of individual pixels.

Table 32.8 *POPULAR GRAPHICS FILE FORMATS*

Graphics files can be saved and stored in many formats. Listed here are several that you are likely to encounter. There are many books devoted to computer graphics that explain the compression characteristics and relative advantages and disadvantages of these and many other file formats.

GIF—Graphics Interchange Format—can be used for vector or bit-mapped graphics; current standard format for drawings on the Internet.

JPEG—Joint Photographic Experts Group—highly compressed bit-mapped images; current standard format for photographs on the Internet.

TIFF—Tagged Image File Format—widely supported and versatile graphics format; current standard for scanned images.

BMP—Bit-Mapped Paint—standard file format for the Paint program that is supplied with the Windows operating system.

PIC or **PICT**—Picture—can be used for both bit-mapped and vector graphics; well-suited for on-screen presentations.

grams can also be purchased separately. **Image editing programs** *provide the capabilities to take previously existing images, such as scanned photographs, and modify them.* This manipulation can range from simple cleanup of film scratches to major changes in photographic content. Using these programs, it is now possible to manufacture "photographs" of nonexistent events.

iii. DRAW PROGRAMS

Draw programs provide a different approach to the creation of graphic images, by using vector graphics. **Vector graphics** *are images based on a set of lines, which can be described and stored by the software as mathematical*

equations. These line images require significantly less disk storage space than bit-mapped images. Another term for this process is **object-oriented graphics**, *images that are created as a set of separate objects, each described separately within the image file*, Figure 32.12. **Draw programs** *create vector graphics*, and are ideal for detailed images containing clean lines with precise dimensions, such as blueprints and maps. It is relatively easy to make large-scale changes in vector graphic images because individual objects within the picture can be moved, deleted, and manipulated.

Another major advantage of drawing programs is that the images created can be easily resized because

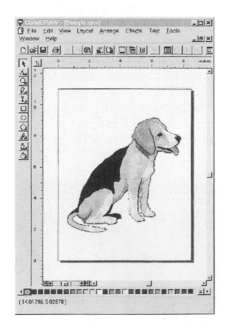

Figure 32.12. Vector Graphics: How A Draw Program Works. In a vector graphics program, the user creates separate geometric objects, and then combines them to form a picture. A dissection of the picture is shown on the right. The dotted rectangle indicates the selection of one of the objects, the dog's eye, by the user. The program is CorelDraw. (From *Introduction to Computers and Information Systems*, 6th ed., Larry Long and Nancy Long, Prentice Hall, 1999.) *(See color insert C-2.)*

they are described by equations that will maintain the appropriate dimensions and clarity of the image. Because bit-mapped graphics specify an image pixel-by-pixel, image quality degrades as the picture is enlarged. The basic differences between the products of these program types are shown in Figure 32.13. The software suite ClarisWorks contains both Draw and Paint programs, which were each used to draw a circle of the same dimensions. The resulting file sizes are 22K for the Draw circle and 319K for the Paint circle. In Figure 32.13b, the difference in the line quality is clear when the circles are each enlarged to 800%. Table 32.9 summarizes the optimal uses of paint and draw programs, and Table 32.10 offers tips for using graphics programs.

iv. PRESENTATION AND MULTIMEDIA AUTHORING PROGRAMS

Presentation graphics programs *provide capabilities to create and incorporate both text and graphic images into "slide" presentations, which can be projected from a computer screen, printed as handouts, or used to make actual slides.* The biggest advantage of using these programs is the ability to set and maintain a constant format throughout a presentation. Title cards, graphs and tables, video, sounds, and text can all be included, Figure 32.14.

Multimedia authoring *is the creation of presentations that use more than one medium to convey information.* Although this is a complex artistic field, there are some

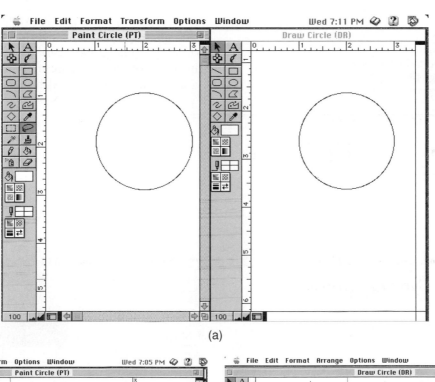

(a)

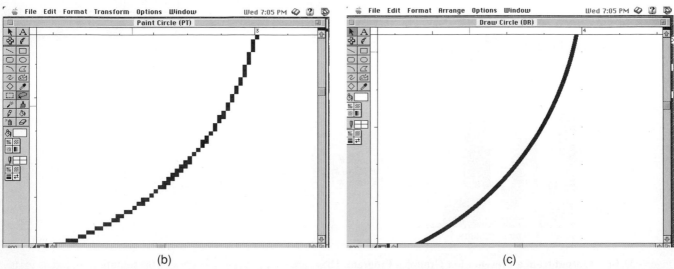

(b)   (c)

Figure 32.13.   Comparison of the Results of Paint and Draw Programs. a. Circles drawn in a paint (left) and a draw (right) program. Note the similarity of the tools shown for each program, with paint providing extra brush effects. In **b.**, each circle is expanded to 800% of the original dimensions.

Table 32.9   *WHEN TO USE PAINT AND DRAW PROGRAMS*

Use a graphics program when you want to create or modify artwork, photographs, or animations. Depending on your desired product, choose either a paint- or draw-type program.

Use a paint program to:

- create complex pictures with flowing lines
- maintain control over subtle differences in color and texture
- take advantage of artistic techniques such as air brushes and filters
- manipulate scanned photographic images
- avoid clean "cartoon" lines

Use a draw program to create:

- simple geometric shapes, such as lines, boxes, and circles
- images with precise dimensions
- charts and graphs
- maps and blueprints
- layouts of multiple objects
- images that will need to be enlarged
- smaller image storage files

Table 32.10   *TIPS FOR USING GRAPHICS PROGRAMS*

Graphics programs provide a wide variety of creation capabilities and storage formats for artwork.

- Decide on the final image quality needed before choosing a program.
- Consider the file size of your final product—storage and transport can present difficulties for larger files.
- Remember that file compression results in permanent loss of graphic information; always keep a copy of the uncompressed image for later recovery.
- Explore the possibilities of clip art, which is available in many file formats.
- Do not forget copyright issues when importing images from the Internet and other sources.

relatively simple programs available (e.g., HyperCard) that allow the user to create presentations in the form of sets of individual "cards," each of which contains information in a specific medium (e.g., text, graphics, animation, or sound). Although this program can be

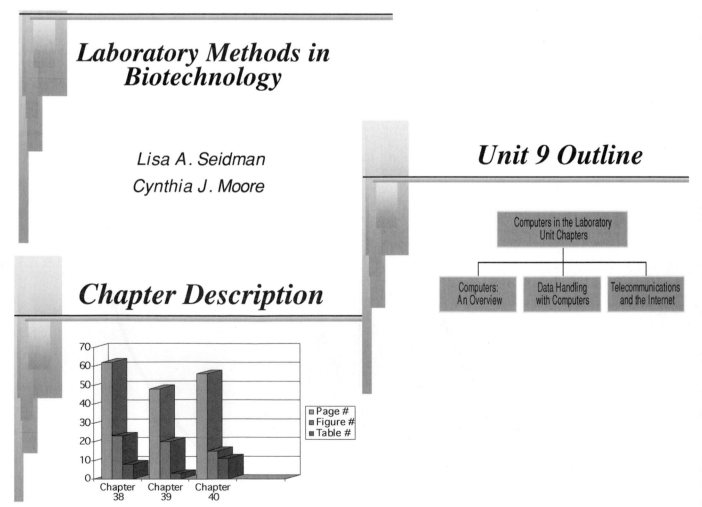

Figure 32.14.   Output from a Presentation Graphics Program. These are examples of the types of presentation materials that can be prepared using Microsoft Power Point.

used without programming experience, especially for text applications, it is powerful enough to have been the creation medium for the popular computer game, *Myst*.

III. SPECIALIZED APPLICATIONS FOR DATA HANDLING

A. Introduction

In many cases, you will have data that require very specific forms of processing and output. There are specialized software packages designed to analyze and compare DNA sequences, for example, or to calculate peak areas for chromatography data. These are constantly being updated and improved, and are beyond the scope of this book. Two examples of specific and widespread needs will now be discussed: graphs and statistics. The final section addresses a common practical dilemma: when to choose a database instead of a spreadsheet to record and manipulate data sets.

B. Graphing

Graphing is a standard procedure in every laboratory. Graphs help the viewer see relationships and patterns among data items. The construction of a graph can actually suggest an interpretation of the data, Figure 32.15. For these reasons, data which do not reveal patterns or are not closely related are usually best presented in tables. There are specialized computer programs for creating complex graphs and charts, although spreadsheet software generally provides an adequate variety of graph types. Most programs will allow the user to compare graphic presentations for effectiveness.

The most common types of scientific graphs are bar and line graphs. Bar graphs are frequently used and are best suited for comparing values of separate items. Line graphs are most effective when used to illustrate trends

among related items. The general types of scientific graphs are shown in Figures 32.16–32.19. Table 32.11 offers some tips for creating graphs.

C. Statistics

Spreadsheet software is adequate for simple statistical tests such as standard deviations and correlation coefficients. More advanced analyses, however, may require one of the statistical analysis software packages, such as SAS® or SPSS®. These programs provide complex curve-fitting equations, customized calculations, and sophisticated data analysis capabilities. The user needs a solid knowledge of statistics in order to apply these programs effectively. Many of these packages, such as SAS® software, provide programming systems for organization of complex data.

D. Databases versus Spreadsheets

It is common to collect a data set and then be uncertain of the optimal computer application for storing, manipulating, and reporting the results. One of the most likely

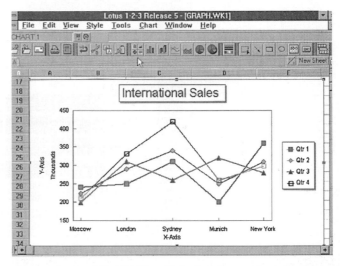

Figure 32.16. Line Graphs Show Continuous Data Items and Trends. (From *Introduction to Computers & Software*, R.A. Szymanski, D.P. Szymanski, and D.M. Pulschen, Prentice Hall, p. 237, 1996.) *(See color insert C-3.)*

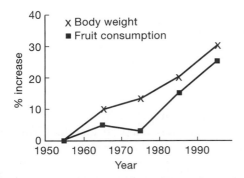

Figure 32.15. Line Graphs Suggest Relationships Between Data Sets. In these hypothetical data sets (which do reflect real trends in American society), the graph implies a relationship between obesity and fruit intake.

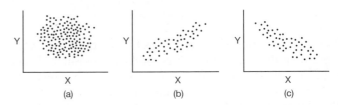

Figure 32.17. Scatter Plots Show the General Distribution of Data Points That Are Not Adequately Defined by a Line. a. No relationship. **b.** Positive relationship. **c.** Negative relationship.

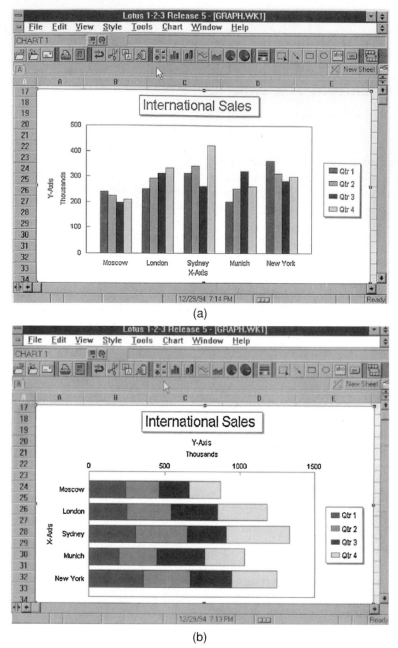

Figure 32.18. Bar Graphs Show Proportions or Comparisons among Small Numbers of Data Items. a. A horizontal grouped bar graph. **b.** A vertical stacked bar graph. **c.** A stacked three-dimensional bar graph. (From *Introduction to Computers & Software*, R.A. Szymanski, D.P. Szymanski, and D.M. Pulschen, Prentice Hall, 1996, p. 235-236 .) **d.** A histogram, which is a type of bar graph used to display the distribution of a variable. *(See color insert C-3.)*

sources of confusion and possible error is choosing between the use of a database and a spreadsheet. There are situations where either can be used successfully, and other situations where the data will need to be transferred to the other type of program in order to achieve the desired results. For example, if you are collecting information about vendors, catalog numbers, dates of purchase, and prices for lab supplies, entering the information into a database will allow you to easily determine which items are purchased from a single vendor and when you last purchased a particular item. You will probably not be able to use the price information for budgeting purposes unless you are using a spreadsheet. Because data transfer procedures are potential sources of errors in data entries and also a lot of unnecessary work, it is better to choose the proper program in the beginning. Table 32.12 provides a summary of the similarities and differences between these two types of program.

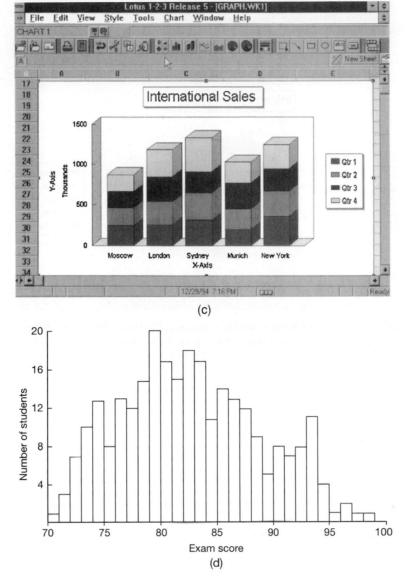

(c)

(d)

Figure 32.18.   *(continued) (See color insert C-3.)*

Table 32.11   TIPS FOR CREATING EASY-TO-READ GRAPHS

All Graphs:

- Be sure each part of the graph is easily identified.
- Label each axis clearly.
- Use appropriate, evenly spaced markings on axes.
- Do not combine an excessive number of colors.
- Do not use graphic tricks unless needed to explain the data.
- Limit the number of lines, bars, or segments in a single graph.

Line Graphs:

- Use line graphs to plot only continuous functions.
- Plot the dependent variable (what was measured) on the Y-axis.
- Do not connect scattered data points with lines.
- Use large, easily distinguished symbols for data points.
- Do not extrapolate beyond the measured data points.

Table 32.12   DATABASE VERSUS SPREADSHEET— SIMPLE GUIDELINES

Both will:

- store and organize data
- perform simple data sorting (alphabetical or numerical)
- allow editing of data sets

Database management software excels at:

- customizing data input screens for ease of use
- creating specialized report formats
- performing complex data sorting, based on multiple fields and/or logical values

Spreadsheets software excels at:

- performing multiple calculations
- making graphs and charts
- revising data and automatically revising calculations
- allowing data manipulation to create "what-if" scenarios

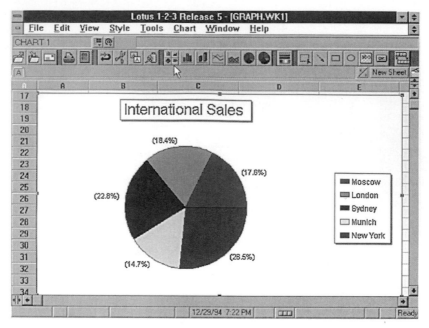

Figure 32.19. Pie Charts Show Data Items as Portions of a Whole. (From *Introduction to Computers & Software*, R.A. Szymanski, D.P. Szymanski, and D.M. Pulschen, Prentice Hall, p. 239, 1996.) *(See color insert C-3.)*

PRACTICE PROBLEMS

1. Which of the following data would best be entered into a database and which should be placed into a spreadsheet?
   a. budget projections for the next fiscal year
   b. vendor addresses
   c. batch records
   d. records of the use of medium batches
   e. time sheets
   f. laboratory maintenance assignments

2. What type of graph would best present the following types of data?
   a. percent of your worktime spent among four different projects
   b. number of mice developing tumors over time after treatment with a carcinogenic chemical
   c. comparison of the growth rate of cells in two types of medium
   d. decrease in the effectiveness of a continuous filtration system with increasing use

DISCUSSION PROBLEMS

1. Think about your current laboratory work. What are two specific applications of the following software that could enhance your productivity?
   a. Word processing
   b. Spreadsheet
   c. Database

2. Design a simple database that would contain the information required on a typical component label (see Figure 5.4, p. 69). Specify the characteristics of the fields where data will be recorded.

3. Why is it inappropriate to write your lab notebook in a word-processing program?

ANSWERS TO PRACTICE PROBLEMS

1. a. spreadsheet
   b. database
   c. database (or spreadsheet if calculations are needed)
   d. spreadsheet
   e. spreadsheet
   f. database

2. a. pie chart
   b. line graph
   c. bar graph or line graph
   d. line graph

CHAPTER 33

Telecommunications and the Internet

I. TELECOMMUNICATIONS

A. Introduction to Computer Networks

Many of the computers that you encounter in laboratories are part of a network. A **network** *is simply two or more computers that are electronically connected for the purpose of sharing programs, files, information, and/or other resources.* A network can be local or global. The telephone system is one example of a giant global computer network. Networks provide the resources for **telecommunications**, *the electronic transport of data in the forms of text, audio, and video over significant distances.* By electronic transmission of data, you can avoid the need to exchange paper and disk copies by mail.

Because of the rapidly changing technologies involved, the terminology for networks can be confusing. Two terms that are frequently used are *intranet* and *internet*. An **intranet** *is an internal, usually private network.* An **internet** *is two or more connected networks.* The **Internet** (with a capital *I*) *is a specific internet connected with a special set of communication protocols* (see p. 717).

In order to be part of a network, computer hardware must be connected to the other computers in the network. This requires a modem (or a network interface card, in many instances), a phone connection or cable, and communications software. To exchange information with other computers, you create communications documents using software and send them electronically through the cable or phone lines. This can be done either through direct connection or through indirect routes. Local telecommunications are frequently han-

dled with direct cables, whereas long distance connections are handled through telephone lines and satellite transmission, Figure 33.1.

B. Local Networks

Local networks provide an economical means for a relatively small group to share resources. These networks can provide a central facility for group-licensed software, storage of large data files, access to expensive peripheral devices, and a central link to outside networks. This eliminates both the expense of buying high-end laser printers for every laboratory or office in the group and the need to store large application programs on multiple computers.

A **LAN** *is a local area network, generally operating within a building, school, or company.* The computers are physically connected by network or fiber optic cables, allowing programs, printers, and other peripherals to be shared within the facilities. LANs are private and limited to a 5-mile radius. To use this type of network, all participating computers must contain a **network interface card (NIC)**, *which is computer hardware that provides an appropriate interface between the computer and the network.* **Ethernet** *is currently the best-known type of LAN.*

A LAN system will usually have one or more **servers**, *which are central computers that act as a common resource for data storage and program access.* The server includes both hardware and software to provide user access to the network. Because local networks are private, they usually require users to **log in**, *by entering their name*

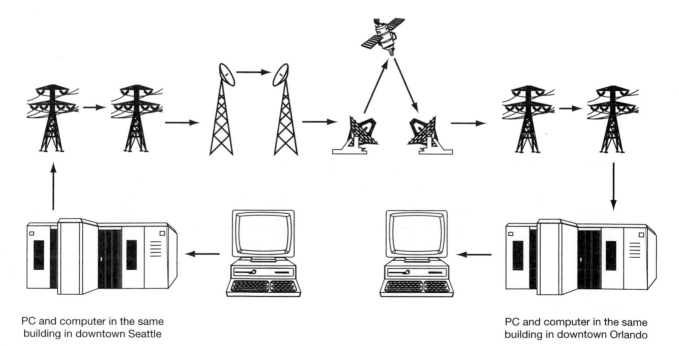

PC and computer in the same
building in downtown Seattle

PC and computer in the same
building in downtown Orlando

Figure 33.1.   Transmission of Data Through a Long Distance Network. (Adapted from *Introduction to Computers and Information Systems*, 5th ed., Larry Long and Nancy Long, Prentice Hall, p. C173, 1997.)

and password into the computer. Once inside the network, users have access to other computers, programs, and equipment that are part of the local network.

C. Global Networks

In order to connect to a larger network, a computer must have access that goes beyond direct cable connections. Most laboratories have direct access to the Internet and both public and private global networks. Through the use of satellite transmissions, networks can communicate throughout the world, as long as there are suitable telephone connections, Figure 33.1. Most private networks require authorization for access (in the form of an assigned password); there are many public networks as well. The structure of these networks is constantly evolving, and so this book will only provide an overview of only two major aspects of computer networking: the Internet and e-mail. The next section will discuss the general nature of the Internet and how it can be used to access information.

II. THE INTERNET

A. What is the Internet?

The computer has become an important part of scientific and other types of communication largely through the technology of the Internet. The **Internet**, *which is short for interconnected networks, consists of tens of thousands of loosely connected computer networks that use a common set of guidelines called IP protocols to send and receive information.* Scientists have used the Internet since the early 1980s to exchange data and documents instantly among universities, government, and other research laboratories. This technology has now become a common part of everyday life. There were an estimated 60 million Internet users in 1997 and there will probably be at least 200 million users in more than 130 countries by the year 2000.

In one sense, the Internet can be compared with a library, which holds vast quantities of information in various forms (e.g., print, video, audio, and combinations of these media). Unlike a standard library, however, the Internet does not have a single physical location, and there is no official group that controls and organizes the Internet. All the specifications needed to create an electronic site on the Internet are publicly available; therefore, anyone with a minimum of training can create their own entry into the Internet "library." In this sense, the Internet is truly democratic. There are pitfalls to this structure, which will be discussed shortly.

In another sense, the Internet provides a massive communication medium. Its resources allow contacts and discussion with scientists around the world, as well

Table 33.1   USES FOR THE INTERNET

The Internet and other network resources have many applications in the laboratory. These include:

- exchanging information with other laboratories
- performing on-line research using databases
- obtaining bibliographic information
- reading on-line journals
- checking vendor catalogs and ordering
- accessing newsgroups and discussions of research topics
- obtaining information about companies and schools
- finding contact information for scientists
- job hunting and hiring

as data exchange and other collaborative processes. It is frequently easier and faster to obtain advice and clarification about procedures and information through network contact than by telephone or regular mail. Some of the many uses of the Internet are listed in Table 33.1.

B. Information Transfer on the Internet

Computers cannot communicate over a network unless they share a common language. **Communication software** *are the programs that provide this common language and allow networks to function.* These programs direct the network interface card or modem to contact other computers and instruct the computer how to send and receive data through the network. This software specifies an appropriate communication **protocol**, *which is a specific common language for information exchange.* By definition, using the Internet requires **IP (Internet Protocol)**, *which is the standard for Internet communication.*

We do not have to understand how protocols work in order to use the Internet. A basic knowledge of the processes involved, however, helps when a problem arises. IP is a system for packaging, addressing, and sending data through the Internet. Every computer has a unique Internet address, consisting of four numbers separated by periods (e.g., 123.231.12.1). These numbers are used within the IP system to address and send information to specific computers. One drawback to IP software is that it can only send small amounts of data in each addressed "package." Because most data transfers are larger than a single package, Internet users also need a program called **TCP (Transmission Control Protocol)**, *which works with IP software to make sure that data packages arrive and are reassembled in the correct order, and that the data are intact.* TCP software acts as a system checker, without which data transfer on the Internet would be unreliable. New computers come equipped with **TCP/IP software**, *which provides the basic means of Internet access.*

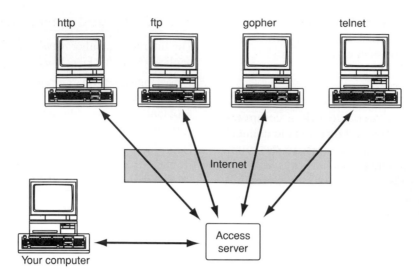

Figure 33.2.   Internet Connection Through an Access System.

C. Access to the Internet

Most laboratory computers have a network connection, or access system, that is essentially invisible to the user, either through a server or in some other configuration, Figure 33.2. Large companies and schools frequently provide high-speed, large-volume connections that allow extensive access to network services. In contrast, home computers usually connect to the Internet by having a modem that establishes telephone contact with another computer that has direct access to the Internet. **Internet Service Providers (ISPs)** *provide telephone or other connections to the Internet for a fee.*

Internet tools or services *provide many means of access and data exchange over the Internet.* The most common of these are described in Table 33.2. Each of these was developed to meet a specific need for information transfer. The World Wide Web has become the most popular part of the Internet because of its unique structure, described later, and because the software for using the Web integrates all the other tools as well, creating a simpler interface for users.

Telnet, one of the oldest internet services, *is a terminal emulation program, which means that it allows your computer to function as a terminal for a remote computer.* In practical terms, this allows you to log in and access distant information. This was frequently used with **File Transfer Protocol (FTP)**, which provides an efficient means to transfer files between computers. Until late 1994, FTP was the most common use of the Internet, and it is still used for downloading files from the Internet. **Downloading** *is copying files or programs from a distant location to your computer.* Downloading can be done at **anonymous FTP sites**, *which are Internet sites designed to allow users without passwords to copy files from archives.* Because FTP is a text-based protocol that is not especially user-friendly, you generally need to determine ahead of time which files you need to download from an anonymous FTP site. Use of FTP

Table 33.2 *THE MAJOR INTERNET TOOLS*

These are the major tools for accessing parts of the Internet through different types of software and communication protocols. All are available through most hookups.

- **Telnet**—Telnet is one of the oldest network services. It allows users to log in at other computers and access their contents. Telnet is still used for text-based programs such as "Pine" (a free e-mail program).
- **FTP (File Transfer Protocol)**—FTP is a standard protocol for moving files between computers. It provides efficient data transfer, but it is not user friendly.
- **Gopher**—The direct predecessor to the World Wide Web, Gopher was developed in the early 1990s to act as an information search and retrieval service.
- **HTTP (Hypertext Transfer Protocol)**—HTTP is the standard protocol for accessing the World Wide Web. The Web is now the most commonly used Internet service.
- **Usenet**—Usenet is another of the oldest services of the Internet, providing a loose organization of more than 22,000 discussion groups.
- **IRC (Internet Relay Chat)**—This is an evolving service that allows real-time conversations over the Internet.
- **E-mail**—Electronic mail is one of the oldest and most popular tools on the Internet.

is becoming less frequent with the advent of more user-friendly downloading systems.

From the beginning, the Internet has served as a forum for public discussions on a wide variety of topics. Many of these are loosely organized under the **Usenet** service, *which provides more than 20,000 discussion groups or newsgroups, each generally focused on a single topic of interest.* These groups are open to the public. **Newsgroups** *allow participants to post and respond to messages in an open forum.* Usenet is older than the Internet itself, and additional newsgroup services, some private, have also developed in recent years. A more recent advance in discussion groups is **IRC**, or **Internet Relay Chat**, *which* *allows interactive, real-time typed "conversations" among the participants.*

Gopher *is a menu-driven information search and retrieval system for the Internet.* It originated at the University of Minnesota (whose mascot is the Golden Gopher), first as a campus system, and then as a global service. It was the first relatively user-friendly means to find information on the Internet. It has largely been supplanted by the World Wide Web.

III. THE WORLD WIDE WEB

A. Hypertext and the Web

By early 1995, the most commonly accessed part of the Internet was the World Wide Web. The **World Wide Web** (WWW or simply the Web) currently *is the predominant means for finding information on the Internet.* Although the Web uses a text-based interface to connect documents, the form of this text is different from previous access methods. The Web uses a data transfer process called **Hypertext Transport Protocol (HTTP),** *which is a relatively sophisticated method for presenting both textual and graphic information to network users.* One of the most valuable aspects of the Web is the way in which information is connected. When you read a book, you move through printed materials in an essentially linear fashion, even when you skip around. Traditional print authors organize material to help the reader understand specific topic areas. **Hypertext** *was created as a nonlinear way to organize information.* It is designed to allow readers to receive and process information according to their own needs.

Hypertext provides an organization of information that is virtually unique to the computer medium. The closest print analogy would be a stack of index cards (one of the first hypertext programs was called HyperCard, for this reason). Each index card holds a piece of information, and the cards can be shuffled into many different arrangements. Because the sequence of cards is not fixed, you can reorganize the cards according to your interests and as many ways as you want. This process is simple to achieve using a computer, and relatively difficult on paper.

By definition hypertext has a flexible form of organization, although web designers arrange hypertext in certain standard formats. Information is provided in **web pages,** *which are the basic units (the index cards) of information storage in hypertext.* Unlike the cards, these web pages can vary in size from a sentence to multiple pages, and each can contain a combination of text, pictures, sounds, or video. Web pages are connected by **links,** *which are highlighted words or phrases that allow the reader to move from page to page with the click of a button,* Figure 33.3. The links act as a continuous menu of ideas. The user controls the flow of information by choosing a link to follow from the current web page. *This process is called* **browsing** *or* **navigating**.

Hypertext authors create an information network, consisting of a variety of associated ideas connected by links, Figure 33.4. The user then chooses the most useful personal pathway through the information, ignoring sections that are already understood and going into depth on the topics of most interest. Another advantage to hypertext on the Internet is that it provides simultaneous access to many users. Each can read or print the information at their own pace.

B. Web Sites

The purpose of hypertext is the subdivision of information into digestible segments, each of which is found on a web page. Each web page is part of a larger **web site,** *which is a collection of web pages that fulfill an author's purpose.* This purpose may be to present a body of information, to explain the policies of a company, or to sell products. Web pages within a site are linked together, usually in some logical order, and then each page can link to other web sites, as shown in Figure 33.3. Each web page can be identified by its address or URL.

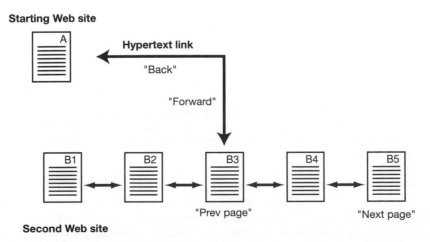

Figure 33.3.   Using Hypertext Links. Choosing a link on one web page leads to a second, potentially distant web page.

(a)

(b)

Figure 33.4.   Hypertext in Action. a. This is the home page for the Prentice Hall Science and Math Division web site. The underlined phrases represent links to additional information on specific topics. For example, if you are interested in writing a textbook for Prentice Hall, you would select the link to *on-line guide*, which would immediately connect you to **b.** (http://www. prenhall.com/divisions/esm/) *(See color insert C-4.)*

A **URL (Uniform Resource Locator)** *is a unique address for each web page on the Internet.* As discussed earlier, every computer attached to the Internet has a numerical IP address. Because these are not easy for users to remember, each computer also has a text name. These are part of a **Domain Name System (DNS)**, *which is a* *way of distributing the responsibility for naming web pages by assigning domains to networks.* Domain names have from three to five parts, separated by periods (e.g., "biosgi.wustl.edu"). Each of the three parts of the name is a separate domain. There is a hierarchy of domains moving from right to left in the address, Figure 33.5. The

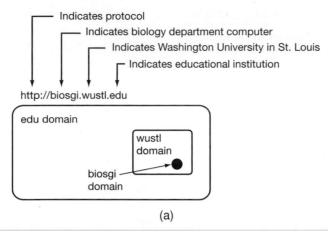

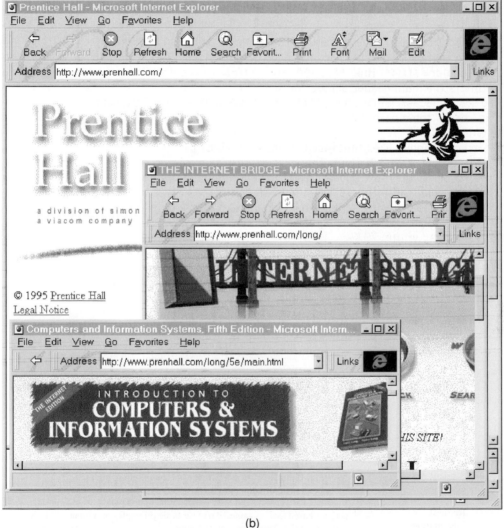

(b)

Figure 33.5.   The Hierarchy Provided by URLs. a. The structure of a URL. **b.** Three levels of
URLs at the Prentice Hall server site are shown. [**b.** from *Computers*, 5th ed., Larry Long and
Nancy Long, Prentice Hall, p. PC47, 1998.]

final domain name is the largest subset, usually referring to a type of organization or a country, Table 33.3.
The next domain to the left is then assigned to a specific
institution or organization, then the next to a specific
section of the institution, continuing until each network
server has a unique designation. The exception to this
scheme is that many domain names on the Web begin
with "www," which refers to the Web itself.

Table 33.3 *WHAT DO THOSE DOMAIN NAMES MEAN?*

There were originally six high-level domains on the Internet. These are still in use and designate the following types of networks:

- .com—Commercial or business sites
- .edu—Educational institution sites (usually universities)
- .gov—Government sites
- .mil—Military sites
- .net—Network sites
- .org—Nonprofit or private organization sites

Other domain names with less standardization are also in use. In addition, there are now hundreds of country domain codes, as more countries are represented on the Internet. The United States is one of the few countries that does not routinely use its national domain code (".us").

Once each computer has a unique name, then the specific web sites and web pages on that computer receive a designation. Many URLs for web pages are quite long, in order to provide a unique address for the many associated pages of a site, Figure 33.5. The main domain names for the site are followed by a backslash (\), more information, usually more backslashes, and more information, frequently ending in ".htm" or ".html." **Hypertext Markup Language (HTML)** *is the programming language used to create and format web pages.* HTML is a relatively simple programming language, which is why anyone with network access can create web pages, Figure 33.6. Designing effective web sites requires significant time and thought, in terms of developing the specific pages and link structures, Figure 33.7.

```
<html><head>
<title>Modern Genetics for All Students</title></head>
<!-- main table - this is the recommended default size-->
<body bgcolor="ffffff" link=#cc0000 vlink=#ff6666> <center><table width=500><tr><td>
<!--this is the main header across the page-->
<img src="images/Mainheader.gif" alt="Modern Genetics for All Students">
<p><p><p><p><table border=0><tr align=top>
<td width=20%>
<h3><a href="Aboutfake.html">About this project</a><p>
<a href="Usingfake.html">Using this curriculum</a><p>
<a href="Curricfake.html">Modern Genetics curriculum</a><p>
<a href="Studentfake.html">Student web activities</a><p>
<a href="Requestfake.html">Request information</a><p>
Return to genetics <br>home page</h3></td><td width=5%>
 </td><td width=75%>
<!--this will be a centered picture at the top of the page--><center>
<img src="images/hpphoto.gif" alt="Happy Students with Fingers in the Goo">
```

(a)

Modern Genetics for All Students

About this project

Using this curriculum

Modern Genetics curriculum

Student web activities

Request information

Return to genetics home page

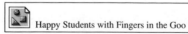
Happy Students with Fingers in the Goo

Modern Genetics for All Students is an on-going high school biology curriculum project. Working together, teachers and scientists have created and field-tested an integrated twelve week unit of activity-based lessons centered on four basic principles of human genetics, molecular biology, and bioethics.

http://biodec.wustl.edu/sciout/modgen
Washington University in St. Louis
Last modified: June 20, 1998
All contents copyright © 1998

(b)

Figure 33.6.   HTML Coding a. and the Resulting Web Page b.

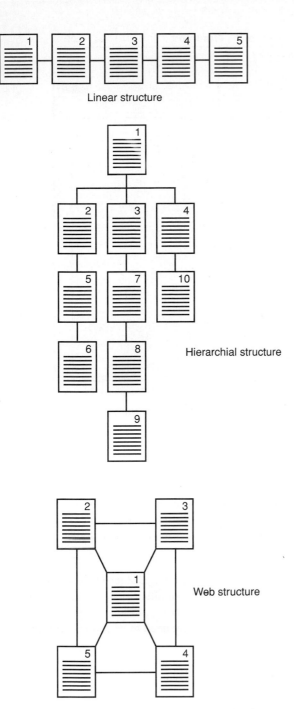

Figure 33.7. Examples of Web Site Structural Design.

C. Navigating the Web

Effective use of the Internet requires an understanding of the process of **navigating**, or **browsing**. *These are synonyms for moving through the Web by selecting URLs or links.* This is accomplished with the use of **browser software**, such as Netscape Navigator or Microsoft Internet Explorer. *These programs provide a user-friendly interface between the Internet and the user.* A major strength of these browsers is their abilities to integrate different Internet access tools such as FTP

and HTTP. In order to access sites on the Internet, the address or URL is typed into the browser, along with the selected tool. For example, all web URLs begin with "http://" followed by the domain names. FTP requests similarly begin "ftp://".

One of the most helpful features of web browsers is that the user only needs to type in a URL once. That URL can then be saved as a **bookmark**, *which enters the URL into a permanent address list in the browser.* Bookmarks are designated by text titles, which can be edited by the user. The next time the site is accessed, the user can simply select the bookmark itself and the browser will navigate to the site. Users can both create bookmark files that store their favorite site addresses and exchange these files with other users. Bookmark files make URLs invisible to the user, Figure 33.8.

One of the best features of the Internet is the fact that its content can be updated virtually instantly. This makes it a great place to get the latest news. It has the drawback that web sites can and do change or disappear as their authors lose interest or switch to a new computer system. When exchanging bookmarks with others, try to be sure that the URLs are still in operation. With this warning, Table 33.4 lists some currently active sites that provide excellent information about the Internet itself.

An important function of the Internet is to provide information. The Internet contains many **search engines**, *which are programs that search topic indexes on the Internet and find web pages that match the search criteria that you provide.* Search engines work by a variety of mechanisms and so each program may give you a different subset of web pages that match your criteria. More information about search engines is provided in Table 33.5.

IV. E-MAIL

A. What is E-Mail?

Electronic or **e-mail** *is a system that allows users to send and receive messages from other computers.* E-mail is unquestionably the most popular application for net-

Table 33.4 USEFUL SITES FOR INTERNET INFORMATION

- **An introduction to the Internet—**
  http://www.stimulate.com/education/internet101.html
- **Internet Beginner's Guide—**
  http://www.vschool.net/outlinks/begin.html
- **Internet update newsletter—**
  http://www.itworks.be/EyeonIT/current.html
- **Glossary of Internet terms—**
  http://www.matisse.net/files/glossary.html
- **List of computer and Internet-related books—**
  http://hoganbooks.com/freebook/webbooks.html

Note: Internet sites move to new addresses with some regularity.

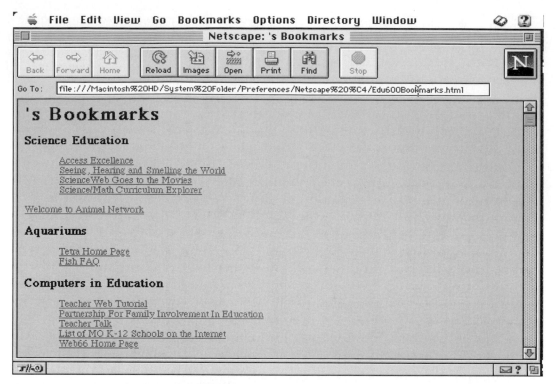

Figure 33.8. Example of a Bookmark File. Bookmarks have been divided into headings. Each underlined term acts as a link to the associated web page.

Table 33.5 EFFECTIVE USE OF SEARCH ENGINES

Current web browsers provide users with access to multiple search engines using the Search command. You can also locate search engines by providing their URL. It is essential to realize that different engines will provide different results for a search. This is due to differences in the way these engines create indexes for web pages. Some search engines only cover web pages by authors who have applied for inclusion; others automatically search entire subsets of the Internet. An excellent web site for learning more about search engines, how they work, and effective search strategies is "Search Engine Facts and Fun," at http://www.searchenginewatch.com/facts/index.html. Some of the more popular search engines are:

Alta Vista	http://altavista.digital.com/
Excite	http://www.excite.com/
HotBot	http://www.hotbot.com/
Infoseek	http://www.infoseek.com/
Lycos	http://www.lycos.com/
WebCrawler	http://www.webcrawler.com/
Yahoo	http://www.yahoo.com/

works. Messages can be sent and received in seconds, with good reliability. This system can be both blessing and curse; recipients can no longer claim that a letter took weeks to arrive. E-mail is an excellent communication medium for many, though not all, purposes, Table 33.6.

E-mail messages are simple to compose. In most e-mail software, you simply fill out a form similar to an interoffice memo, Figure 33.9. If you are familiar with URLs, e-mail addresses will also look familiar. A typical

Table 33.6 WHY E-MAIL?

E-mail can be incredibly useful or incredibly annoying, depending on how it is used. There are both advantages and disadvantages to using e-mail in place of either the telephone or the regular mail. These include:

Advantages over the telephone:

• It is less expensive for long distance communication.
• The message can be simultaneously transmitted to many recipients.
• The recipient does not need to be present to receive the message.
• The recipient is not interrupted by the telephone.
• A message is available in file form to be printed or permanently saved.

Advantages over regular mail:

• It is faster than "snail mail."
• It is possible to almost carry on a "conversation" with the exchange of multiple quick messages.
• It is easier to send e-mail.
• Messages can be relatively informal.
• A message can be sent simultaneously to many recipients.

Disadvantages:

• There is no guarantee that the receiving computer will be able to read your message.
• E-mail can be less private than other types of mail.
• E-mail cannot be retrieved once it is sent.
• It is relatively easy to send junk e-mail to multiple recipients.

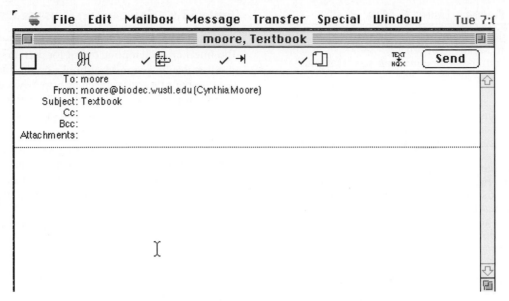

Figure 33.9.   Example of an E-Mail Message Form.

address looks like "moore@biodec.wustl.edu." The information to the right of the "@" sign shows the computer domain address and the text to the left of the "@" sign gives the unique name of the individual.

B. Mailing Lists

One powerful function of e-mail is the ability to simultaneously send the same message to large numbers of individuals. Most e-mail software can create a **listserv**, *which is a mailing list with a unique mailing designation that contains many addresses*. The user can then compose a single message, type in one address for the mailing list, and the message is sent to everyone on the list. This is a popular mechanism for group collaborations. Users can set up their own mailing lists, and they can request to **subscribe** to other lists, *which will add their names to the requested list*. This allows the user to stay current with various groups and topics by subscribing to lists that address specific interests. There are two major drawbacks to participation in mailing lists. One is that you may receive more e-mail than anticipated. If the volume or content is unwelcome, all groups have mechanisms to **unsubscribe**, or *be removed from the list*. The other problem is that most users eventually make the embarrassing mistake of sending an e-mail message intended for a specific individual in the group, but accidentally addressed to the entire group. Someone will usually point this out. You can avoid this by always checking the "TO:" line in the message header.

C. Personal Expression in E-Mail

One of the disadvantages of e-mail (i.e., the lack of direct human contact) has led to the development of various techniques and shortcuts for conveying emotional content over the Internet. **Emoticons (emotion icons)** *are keyboard-generated symbols that indicate your feelings in an easily interpretable way*. Somewhat more likely to be confusing is **cyberspeak**, *which includes both computer jargon and various abbreviations that are frequently used to indicate emotional states*. For example, if someone writes "ROTFLOL" in an e-mail message, you may be mystified unless you know that this is a cyberspeak expression meaning "rolling on the floor laughing out loud." Some common examples are provided in Table 33.7.

E-mail has an informal feel that sometimes encourages users to apply less common sense to their messages than they might in regular letters. Remember that e-mail messages can (and frequently should) be saved. Table 33.8 offers some tips for using e-mail effectively.

Table 33.7   EMOTIONS AND CYPERSPEAK

An entire body of special symbols and abbreviations have developed in cyberspace in order to express emotions to the recipient of a message. Some examples:

Emoticons:

:-)	I'm happy	:-D	I'm laughing
:-(	I'm not happy	;-)	I'm winking
:-|	I'm bored	@->->-	Here's a flower for you

Abbreviations:

IMHO	In my humble opinion
LOL	Laughing out loud
RTFM	Read the manual
TPTB	The powers that be
TTYL	Talk to you later

Table 33.8 TIPS FOR USING E-MAIL

- Check your e-mail regularly.
- Remember that e-mail is not private.
- Do not write anything in an e-mail message that you would not write in a letter.
- DON'T SHOUT! The use of all capital letters is considered rude, at best.
- Check your spelling.
- Be careful when using sarcasm in e-mail; remember that the recipient of your message cannot necessarily guess your intentions.
- Avoid inflammatory remarks (flaming).
- Do not send unrequested or inappropriate e-mail to multiple recipients (this is called **spamming** and you are likely to be flamed for it).
- Be careful when answering a message from a mailing list— your private response could be sent to the entire group.

V. NETIQUETTE

As in all forms of human expression, there are limits to the types of behavior that are appropriate on the Internet. The lack of central governance, issues of free speech, and a general sense of anonymity create a true sense of freedom on the Internet. Although this atmosphere has frequently been compared with the Wild West, there are still standards of basic ethical behavior. **Netiquette (network etiquette)** *is the uncodified but generally accepted set of standards for behavior while online.* As in any situation, there will be individuals who do not follow these suggestions, but it is important to remember that the Internet is not necessarily as anonymous as it seems. **Flaming**, for example, *is the Internet term for making offensive personal comments in a discussion group or on a mailing list.* Although it is not a rare

Table 33.9 ETHICAL AND SENSIBLE USE OF THE INTERNET

The Internet is a tempting outlet for certain types of questionable behavior. Most books on the Internet as well as a number of websites (such as http://www.etiquette.net) discuss proper behavior and ethical standards. Here are some generally accepted guidelines for acceptable use of these resources.

- Remember that you are dealing with human beings.
- Always assume that material you see on the Internet is copyrighted unless otherwise informed; do not copy or disseminate materials obtained from the Web without permission.
- Do not exploit free work-related Internet resources for personal gain.
- Do not misrepresent yourself in discussions.
- Always check the FAQ file before asking questions in a discussion.

occurrence, it is not the type of behavior that will win the respect of others. Some guidelines for sensible use of networking opportunities are presented in Table 33.9.

DISCUSSION PROBLEMS

1. What would be the advantages and disadvantages for a biotechnology company to be connected to a global network?

2. Why might a company choose to provide an internal network (LAN) in addition to an Internet connection?

3. Find four examples of URLs that would be useful in your current laboratory.

INDEX

A

Abscissa, 99, 152
Absolute error, 241, 251
Absolute zero, 326, 327, 337
Absorbance, 374–76, 378–79, 430, 432
  calculation of, 378
  relationship between transmittance and, 378
Absorbance scale accuracy, *See* Photometric accuracy
Absorbance spectrum, 375, 430
  and resolution, 390–91
Absorption, 430, 442, 539, 545, 634
Absorption coefficient, 404
  molar, 405
Absorptivity, 404, 430
Absorptivity constant, 404–5, 430
  calculating from a standard curve, 406
  determining concentration based on, 407–8
  determining from a calibration curve, 407
  molar, 405
Academic research laboratories, 52–53
Acceptance, 180
Acceptance criteria, 46
Accepted reference value, 251
Accidents, 20
Accuracy, 46, 80, 230, 237, 239, 251
  in assays, 415
  photometric, 388, 430, 432
  wavelength, 387–88, 433
Acetic acid, 490, 492, 500, 507, 617, 624
Acetone, 489, 492, 615, 617, 624
Acetone-dry ice bath, 609
Acetonitrile, 624
ACGIH (American Conference of Governmental Industrial
    Hygienists), 622, 634–35
Acid-base error, 344, 365
Acidic solution, 365
Acids, 342, 363, 365, 442, 634
  strong, 342, 367
  weak, 342, 367
Acquired immunodeficiency syndrome (AIDS), 652, 659
Acrylamide, 497, 506, 589, 619, 624
ACS (American Chemical Society), 511, 624
Activated carbon, 442, 514, 517
Activity, 78, 365
  biological, 46
  specific, 590
  star, 447, 499
Acute exposure, 631, 634
Acute toxic agent, 618
Acute toxicity, 634
Adapters, 565, 570
Adsorption, 442, 539, 545, 634
  and protein solutions, 491
Adsorption chromatography, 584, 589
Adulterated, 46
Aerosols, 310, 321, 545, 640, 659
  laboratory procedures generating, 648
  reducing exposure to, 649

Affinity, 579, 589
Affinity chromatography, 589
Agarose, 497, 506, 582–83, 589
α-helix, 442, 486
AIDS (acquired immunodeficiency syndrome), 652, 659
Airborne particle retention, 545
Air displacement design, micropipettor, 308, 309, 310, 321
Airfoils, 625, 634
Air purification respirators, 604, 611
Alcohol error, 344, 365
Aliquot, 442, 525
Alkaline error, 348, 365
Alkalinity, 442
Allergens, 618–19, 634
Allergies, 600, 611, 619, 634
  fungal, 651
Alpha helix, 442, 486
Alternating current (AC), 260, 276
Ambient temperature, 337
American Association for Accreditation of Laboratory Animal Care
    (AAALAC), 656
American Chemical Society (ACS), 511, 624
American Conference of Governmental Industrial Hygienists
    (ACGIH), 622, 634–35
American National Standards Institute (ANSI), 22, 46, 231, 251, 635
American Society for Testing and Materials (ASTM), 231, 252, 511
Amino acid sequencing, 79
Amino group, 442, 484
Ammeter, 276, 360, 365
Ammonium acetate, 506
Ammonium hydroxide, 624
Ammonium persulfate, 506
Ammonium sulfate, 490, 492
Amount, 99, 131, 442, 450
Amperes, 260, 276
Ampicillin, 506
Amplification, 271, 276
Amplifier, 271, 276
Analog, 271, 276, 664
Analog to digital converter (A/D converter), 271–72, 276
Analysis, 589
  trace, 447
  See also Job Safety Analysis (JSA); Product purity and yield
    analysis; Qualitative analysis; Quantitative analysis;
    Uncertainty analysis
Analyte, 251, 379, 430
Analytical balance, 284, 286, 295
Analytical centrifugation, 552
Analytical laboratory documents, 61
Analytical ultracentrifuge, 570
Anaphylactic shock, 619, 635
Angstrom, 99
Animals:
  humane handling of, 656
  transgenic, 9
  use/handling of, 22
  and worker safety, 656–57
Anion, 442

CD-ROMs, 683
Ceiling limit, 635
Cell constant (K), 360–61, 365
Cell culture, 2, 9
    primary, 654
Cell homogenate, 580, 581, 589
Cell homogenization methods, 581
Cell lines, 2, 38–40, 654, 659
Cells, 379, 430
    primary, 654, 661
    spreadsheets, 665, 700
Celsius scale, 326, 327, 337
Center for Biologic Evaluation and Research (CBER), 2, 36–37
Center for Devices and Radiological Health (CDRH), 2, 36–37
Center for Drug Evaluation and Research (CDER), 2, 36–37
Center for Food Safety and Applied Nutrition (CFSAN), 2, 37, 41
Centers for Disease Control and Prevention (CDC), 641–42, 659
    Internet information, 662
Centi-, 99, 112
Centigrade, *See* Celsius scale
Centimeter, 99, 112
    cubic, 99
Central nervous system, 635
Central processing unit (CPU), 665, 674–75
Central tendency measurement, *See* Measurement, central tendency
Centrifugal force, 548–50, 570
Centrifugation, 570
    analytical, 552
    applications of, 551–52
    basic principles of, 548–52
        centrifugal force, 548–50
        rate of sedimentation, 550–51
        sedimentation, 548
    buoyant density gradient, 570
    continuous, 556
    continuous flow, 571
    density, 554–56
        buoyant density, 555–56
        collection of fractions, 556
        formation of gradients, 554–55
        rate zonal, 555
    density gradient, 554, 571
    differential, 553–54
    high speed, 553, 571
    isopycnic density gradient, 571
        preparative, 551–52, 571
    low speed, 553, 571
    maintenance/trouble-shooting, 567
    operation, 562
        modes of, 553–56
    preparative, 551–52
    principles/instrumentation, rate zonal density gradient, 571
    rotors, 4, 20, 548, 557–64, 571
    safety, 567–69
    tubes/bottles/adapters, 565–66, 567
    ultraspeed, 553
Centrifuges, 548, 552, 570
    chemical corrosion, 562–64
    design of, 552–53
    high-speed, 553
    low-speed, 553
    modes of operation, 556
    operating, 562
    ultraspeed, 553
Certified reference materials, 234, 252
CFR (Code of Federal Regulations), 2, 46, 635
CFSAN (Center for Food Safety and Applied Nutrition), 2, 37, 41
Chain of custody, 46, 69
    documentation, 69
    forms, 61
Change-of-state indicators, 333, 337
Chatroom, 665
Chelating agents, 443, 490
Chelators, 443, 499
Chemical Abstracts Service number (CAS number), 635

Chemical container, primary, 629, 636
Chemical corrosion, 562–64, 570
Chemical corrosive, 617
Chemical disinfectants, antimicrobial activity of, 650
Chemical emergency response, 630–31
Chemical exposure, symptoms of, 631
Chemical fume hoods, 625–27, 635
Chemical hazards:
    chemical emergency response, 630–31
    classification of, 614
    corrosives, 617
        safe handling of, 617
    flammables, 615
        safe handling of, 615, 616
    reactives, 615–17
        safe handling of, 616–17
    skin exposure, limiting, 627–29
    storage of, 629, 630
    toxic:
        acute versus chronic, 617–18
        embryotoxins/teratogens, 620–21
        irritants/sensitizers/allergens, 618–19
        mutagens/carcinogens, 619–20
        neurotoxins, 619
        safe handling of, 618
    waste materials, handling of, 629–30
    work area, preparing, 625
    See also Fume hood
Chemical hygiene plan (CHP), 2, 23
Chemical laboratory safety, *See* Chemical safety
Chemical purity, grades of, 476
Chemical resistance chart for plastic, 527–33
Chemical safety:
    emergency response, 630–31
    labeling, 629
    routes of exposure:
        ingestion, 624, 635
        inhalation, 622–23, 662–63
        injection, 624
        skin and eye contact, 623–24
        toxicity measurements, 621–22
    spills:
        chemical spill kits, 632–33
        handling, 632–33
        preventing, 631–32
        procedures, 633
    storage, 629
    See also Chemical hazards; Hazardous chemicals
Chemical spill kit, 632–33, 635
Chip, 268, 276, 665
Chloramphenicol, 506
Chlorates, 617
Chloroform, 499, 624
Chromatin, 443, 498–99
Chromatography, 584–86, 589
    adsorption, 584, 589
    affinity, 589
    chromatographic methods, 79
    classification scheme, 584
    column, 585, 589
    gas liquid (GLC/GC), 584
    gas solid (GSC), 584
    gel permeation (GPC), 584, 585, 587, 589
    high-performance liquid (HPLC), 444, 476, 512, 585, 586, 589
    hydrophobic interaction, 590
    instruments, spectrophotometer/photometers as detectors for, 424–25
    ion exchange, 584, 590
    liquid, 584, 590
    method by molecular characteristics, 584
    partition, 584, 590
    permeation, 584
    size exclusion, 585, 590
    solvents, 40
    thin-layer (TLC), 585, 590